Introductory Algebra

MEDIA UPDATE EDITION

Julie Miller
Daytona State College

Molly O'Neill
Daytona State College

Nancy Hyde
Formerly of Broward College

McGraw Hill

*Connect
Learn
Succeed*™

The McGraw·Hill Companies

Connect
Learn
Succeed™

INTRODUCTORY ALGEBRA: MEDIA UPDATE, SECOND EDITION

Published by McGraw-Hill, a business unit of The McGraw-Hill Companies, Inc., 1221 Avenue of the Americas, New York, NY 10020. Copyright © 2013 by The McGraw-Hill Companies, Inc. All rights reserved. Printed in the United States of America. Previous editions © 2009 and 2007. No part of this publication may be reproduced or distributed in any form or by any means, or stored in a database or retrieval system, without the prior written consent of The McGraw-Hill Companies, Inc., including, but not limited to, in any network or other electronic storage or transmission, or broadcast for distance learning.

Some ancillaries, including electronic and print components, may not be available to customers outside the United States.

This book is printed on acid-free paper.

1 2 3 4 5 6 7 8 9 0 DOW/DOW 1 0 9 8 7 6 5 4 3 2

ISBN 978–0–07–340630–5
MHID 0–07–340630–9

ISBN 978–0–07–754376–1 (Annotated Instructor's Edition)
MHID 0–07–754376–9

Vice President, Editor-in-Chief: *Marty Lange*
Vice President, EDP: *Kimberly Meriwether David*
Senior Director of Development: *Kristine Tibbetts*
Editorial Director: *Stewart K. Mattson*
Executive Editor: *Dawn R. Bercier*
Sponsoring Editor: *Mary Ellen Rahn*
Developmental Editor: *Emily Williams*
Marketing Manager: *Peter A. Vanaria*
Lead Project Manager: *Peggy J. Selle*
Senior Buyer: *Sherry L. Kane*
Senior Media Project Manager: *Sandra M. Schnee*
Senior Designer: *Laurie B. Janssen*
Cover Illustration: *Imagineering Media Services Inc.*
Lead Photo Research Coordinator: *Carrie K. Burger*
Compositor: *Aptara®, Inc.*
Typeface: *10/12 Times Ten Roman*
Printer: *R. R. Donnelley*

All credits appearing on page or at the end of the book are considered to be an extension of the copyright page.

Library of Congress Cataloging-in-Publication Data

Miller, Julie, 1962–
    Introductory algebra : media update / Julie Miller, Molly O'Neill, Nancy Hyde. — 2nd ed.
        p. cm.
    Includes index.
    ISBN 978–0–07–340630–5—ISBN 0–07–340630–9 (hard copy : alk. paper)
    1. Algebra—Textbooks.   I. O'Neill, Molly, 1953–   II. Hyde, Nancy.   III. Title.
QA152.3.M576 2013
512—dc23
                            2011030993

www.mhhe.com

McGraw Hill **connect**®
|MATH

Hosted by **ALEKS Corp.**

Get Better Results with high-quality digital content and an easy-to-use platform!

The Miller/O'Neill/Hyde series now offers a complete digital solution for your course needs! Introducing Connect Math Hosted by ALEKS Corp., McGraw-Hill's premier eLearning system.

With Connect Math Hosted by ALEKS Corp., instructors and students will experience:

► Intuitive and easy-to-use platform

► High-quality digital exercises

► Comprehensive Guided Solutions and Solve It examples that are consistent with the authors' approach, terminology, and methodology

► Integrated, media-rich eBook

► Integrated ALEKS® Assessment to identify strengths and weaknesses

► Flexible gradebook and assignment creation

► Flexible gradebook including reports on time spent working in Connect Math Hosted by ALEKS Corp.

► Administer course consistency with Master Templates

Julie Miller's Lecture Videos and Dynamic Math Animations

Julie Miller's NEW lecture videos and dynamic math animations are also available within Connect Math Hosted by ALEKS Corp. Students can learn and review the material right alongside Julie Miller as she narrates and teaches the learning objectives and brings the math concepts to life!

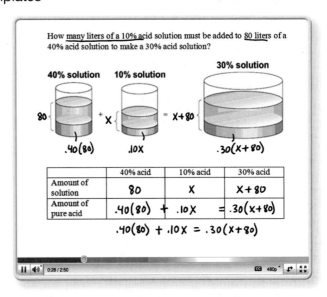

For more information, please contact your McGraw-Hill representative at **www.mhhe.com.**

ALEKS is a registered trademark of ALEKS Corporation.

McGraw Hill ■ CONNECT®
|MATH

Hosted by **ALEKS Corp.**

Connect Math Hosted by ALEKS Corporation is an exciting, new ehomework platform combining the strengths of McGraw-Hill Higher Education and ALEKS Corporation. Connect Math Hosted by ALEKS Corporation is the first platform on the market to combine an artificially-intelligent, diagnostic assessment with an intuitive ehomework platform designed to meet your needs.

Connect Math Hosted by ALEKS Corporation is the culmination of a one-of-a-kind market development process involving full-time and adjunct Math faculty at every step of the process. This process enables us to provide you with a solution that best meets your needs.

Connect Math Hosted by ALEKS Corporation is built by Math educators for Math educators!

1 *Your students want a well-organized homepage where key information is easily viewable.*

Modern Student Homepage

▶ This homepage provides a dashboard for students to immediately view their assignments, grades, and announcements for their course. (Assignments include HW, quizzes, and tests.)

▶ Students can access their assignments through the course Calendar to stay up-to-date and organized for their class.

Modern, intuitive, and simple interface.

2 *You want a way to identify the strengths and weaknesses of your class at the beginning of the term rather than after the first exam.*

Integrated ALEKS® Assessment

▶ This artificially-intelligent (AI), diagnostic assessment identifies precisely what a student knows and is ready to learn next.

▶ Detailed assessment reports provide instructors with specific information about where students are struggling most.

▶ This AI-driven assessment is the only one of its kind in an online homework platform.

Recommended to be used as the first assignment in any course.

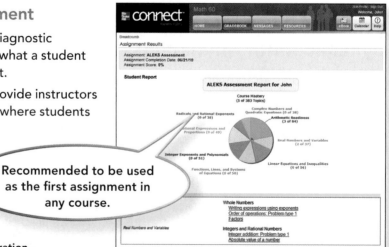

ALEKS is a registered trademark of ALEKS Corporation.

Resources for Online Homework

3 *Your students want an assignment page that is easy to use and includes lots of extra help resources.*

Efficient Assignment Navigation

- ▶ Students have access to immediate feedback and help while working through assignments.
- ▶ Students have direct access to a media-rich eBook for easy referencing.
- ▶ Students can view detailed, step-by-step solutions written by instructors who teach the course, providing a unique solution to each and every exercise.

4 *You want a more intuitive and efficient assignment creation process because of your busy schedule.*

Assignment Creation Process

- ▶ Instructors can select textbook-specific questions organized by chapter, section, and objective.
- ▶ Drag-and-drop functionality makes creating an assignment quick and easy.
- ▶ Instructors can preview their assignments for efficient editing.

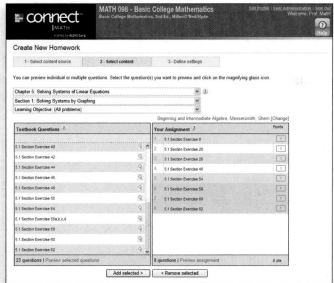

www.connectmath.com

connect
|MATH

Hosted by ALEKS Corp.

⑤ Your students want an interactive eBook with rich functionality integrated into the product.

Integrated Media-Rich eBook

▶ A Web-optimized eBook is seamlessly integrated within ConnectPlus Math Hosted by ALEKS Corp. for ease of use.

▶ Students can access videos, images, and other media in context within each chapter or subject area to enhance their learning experience.

▶ Students can highlight, take notes, or even access shared instructor highlights/notes to learn the course material.

▶ The integrated eBook provides students with a cost-saving alternative to traditional textbooks.

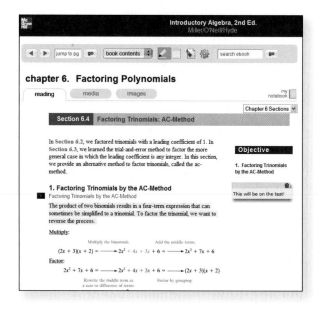

⑥ You want a flexible gradebook that is easy to use.

Flexible Instructor Gradebook

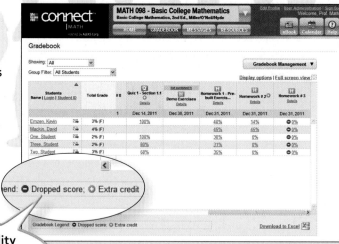

▶ Based on instructor feedback, Connect Math Hosted by ALEKS Corp.'s straightforward design creates an intuitive, visually pleasing grade management environment.

▶ Assignment types are color-coded for easy viewing.

▶ The gradebook allows instructors the flexibility to import and export additional grades.

Instructors have the ability to drop grades as well as assign extra credit.

Built by Math Educators
for Math Educators

7 *You want algorithmic content that was developed by math faculty to ensure the content is pedagogically sound and accurate.*

Digital Content Development Story

As the usage of online homework progresses and evolves, McGraw-Hill understands the need to involve instructors while developing the digital content to ensure that what students see in the online homework system is consistent with what they see in their textbooks. For the Miller, O'Neill, and Hyde developmental math series, we partnered with instructors that have taught from the series to help ensure a seamless transition from print to digital offerings.

The development of McGraw-Hill's Connect Math Hosted by ALEKS Corporation content involved collaboration between McGraw-Hill, our authors, experienced instructors, and ALEKS Corporation, a company known for its high-quality digital content. The result of this process, outlined below, is accurate content created with your students in mind. It is available in a simple-to-use interface with all the functionality tools needed to manage your course.

1. McGraw-Hill selected experienced instructors to work as Digital Contributors.

2. The Digital Contributors selected the textbook exercises to be included in the algorithmic content to ensure appropriate coverage of the textbook content.

3. The Digital Contributors created detailed, stepped-out solutions for use in the Guided Solution and Solve It features, matching the voice of the authors.

4. The Digital Contributors provided detailed instructions for authoring the algorithm specific to each exercise to maintain the original intent and integrity of each unique exercise.

5. Each algorithm was reviewed by the Contributor, went through a detailed quality control process by ALEKS Corporation, and was copyedited prior to being posted live.

Solutions in Connect Math Hosted by ALEKS match the procedure and language of the text.

Previous | 1 2 3 4 5

Question 6 of 7 (1 point)

Factor completely.

$$100u^2 - 80u + 16$$

Step 1:
Factor out the GCF of 4.

Step 2:
$$100u^2 - 80u + 16 = 4(25u^2 - 20u + 4)$$

Step 3:
Can we factor again? Yes.

Step 4:
Notice that the first and last terms in the trinomial are perfect squares.
$$25u^2 - 20u + 4 = (5u)^2 - 20u + (-2)^2$$

The middle term is $2 \cdot 5u \cdot (-2) = -20u$.

Print whole assignment | Print this page only

Tutorial help

Try Another
Solve It
Guided Solution
Show Example
Ask My Instructor
Link to Textbook

Result = Truly Vetted, Consistent Digital Content That Is Built By Math Educators And Supported By ALEKS Corporation.

Lead Digital Contributors

Donna Gerken, *Miami Dade College*
Nicole Lloyd, *Lansing Community College*
Stephen Toner, *Victor Valley College*

Digital Contributors

Jody Harris, *Broward College*
Lizette Hernandez Foley, *Broward College*
Linda Schott, *Ozarks Technical Community College*
Michael Larkin, *Pacific University*
Alina Coronel, *Miami Dade College*

www.connectmath.com

ALEKS 360: A Total Course Solution

With **eBook** Integration

Connect Learn Succeed™

A cost-effective total course solution: fully integrated, interactive eBook combined with ALEKS individualized assessment and learning.

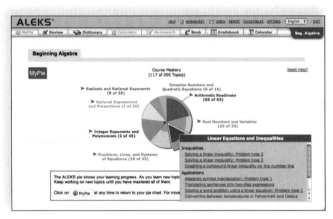

Individualized Learning

- The ALEKS Pie summarizes a student's current knowledge and provides individualized learning on the exact topics the student is **ready to learn**

- Artificial intelligence successfully targets gaps by assessing precisely a student's knowledge and periodically reassessing for long-term retention

- Adaptive, open-response environment avoids multiple-choice and includes problems, explanations, and realistic answer input tools

Interactive eBook

- eBook access provides worked examples, videos, and additional support

- Robust virtual features include highlighting, bookmarking, and note-taking capabilities

- Students can easily access the eBook, multimedia resources, and their notes from within their ALEKS Student Accounts

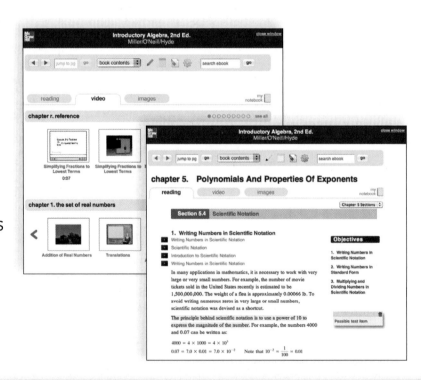

Learn More: www.aleks.com/highered/math/aleks360

ALEKS Course Management Tools

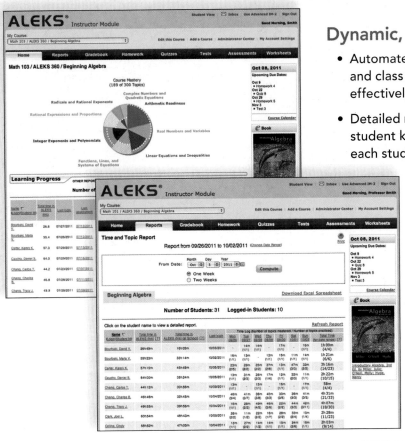

Dynamic, Automated Reporting

- Automated reports dynamically track student and class learning progress so instructors can effectively direct classroom instruction

- Detailed reports identify precisely what each student knows, and more importantly, what each student is ready to learn next

  - Time and Topic Report offers up-to-the-minute daily progress, including time logged, topics attempted, and topics mastered

Course Control and Customization

- Align ALEKS topics with a textbook or course syllabus

- Create and customize course objectives and modules

- Set due dates for course objectives to pace student progress

- Assign automatically-graded homework, quizzes, and tests

- Seamlessly track and adjust student scores with the customizable gradebook

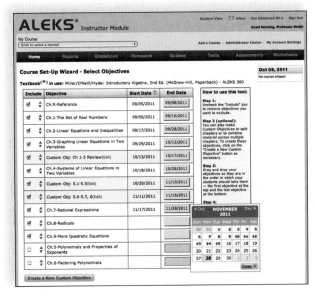

About the Authors

Julie Miller has been on the faculty of the Mathematics Department at Daytona State College for 19 years, where she has taught developmental and upper level courses. Prior to her work at Daytona State College, she worked as a software engineer for General Electric in the area of flight and radar simulation. Julie earned a bachelor of science in applied mathematics from Union College in Schenectady, New York, and a master of science in mathematics from the University of Florida. In addition to this textbook, she has authored several course supplements for college algebra, trigonometry, and precalculus, as well as several short works of fiction and nonfiction for young readers.

"My father is a medical researcher, and I got hooked on math and science when I was young and would visit his laboratory. I can remember using graph paper to plot data points for his experiments and doing simple calculations. He would then tell me what the peaks and features in the graph meant in the context of his experiment. I think that applications and hands-on experience made math come alive for me and I'd like to see math come alive for my students."

—Julie Miller

Molly O'Neill is also from Daytona State College, where she has taught for 21 years in the Mathematics Department. She has taught a variety of courses from developmental mathematics to calculus. Before she came to Florida, Molly taught as an adjunct instructor at the University of Michigan–Dearborn, Eastern Michigan University, Wayne State University, and Oakland Community College. Molly earned a bachelor of science in mathematics and a master of arts and teaching from Western Michigan University in Kalamazoo, Michigan. Besides this textbook, she has authored several course supplements for college algebra, trigonometry, and precalculus and has reviewed texts for developmental mathematics.

"I differ from many of my colleagues in that math was not always easy for me. But in seventh grade I had a teacher who taught me that if I follow the rules of mathematics, even I could solve math problems. Once I understood this, I enjoyed math to the point of choosing it for my career. I now have the greatest job because I get to do math everyday and I have the opportunity to influence my students just as I was influenced. Authoring these texts has given me another avenue to reach even more students."

—Molly O'Neill

Nancy Hyde served as a full-time faculty member of the Mathematics Department at Broward College for 24 years. During this time she taught the full spectrum of courses from developmental math through differential equations. She received a bachelor of science degree in math education from Florida State University and master's degree in math education from Florida Atlantic University. She has conducted workshops and seminars for both students and teachers on the use of technology in the classroom. In addition to this textbook, she has authored a graphing calculator supplement for College Algebra.

"I grew up in Brevard County, Florida, where my father worked at Cape Canaveral. I was always excited by mathematics and physics in relation to the space program. As I studied higher levels of mathematics I became more intrigued by its abstract nature and infinite possibilities. It is enjoyable and rewarding to convey this perspective to students while helping them to understand mathematics."

—Nancy Hyde

Contents

Chapter R Reference 1

R.1 Study Tips 2

R.2 Fractions 6

R.3 Decimals and Percents 20

R.4 Introduction to Geometry 29

Chapter 1 The Set of Real Numbers 45

1.1 Sets of Numbers and the Real Number Line 46

1.2 Exponents, Square Roots, and the Order of Operations 56

1.3 Addition of Real Numbers 66

1.4 Subtraction of Real Numbers 74

Problem Recognition Exercises: Addition and Subtraction of Real Numbers 82

1.5 Multiplication and Division of Real Numbers 82

1.6 Properties of Real Numbers and Simplifying Expressions 93

Group Activity: Evaluating Formulas Using a Calculator 106

Chapter 1 Summary 107

Chapter 1 Review Exercises 111

Chapter 1 Test 113

Chapter 2 Linear Equations and Inequalities 115

2.1 Addition, Subtraction, Multiplication, and Division Properties of Equality 116

2.2 Solving Linear Equations 127

2.3 Linear Equations: Clearing Fractions and Decimals 135

Problem Recognition Exercises: Equations and Expressions 142

2.4 Applications of Linear Equations: Introduction to Problem Solving 143

2.5 Applications Involving Percents 153

2.6 Formulas and Applications of Geometry 159

2.7 Mixture Applications and Uniform Motion 169

2.8 Linear Inequalities 178

Group Activity: Computing Body Mass Index (BMI) 192

Chapter 2 Summary 193

Chapter 2 Review Exercises 199

Chapter 2 Test 202

Chapters 1–2 Cumulative Review Exercises 203

Chapter 3 Graphing Linear Equations in Two Variables 205

3.1 Rectangular Coordinate System 206

3.2 Linear Equations in Two Variables 215

3.3 Slope of a Line and Rate of Change 230

3.4 Slope-Intercept Form of a Line 243

Problem Recognition Exercises: Linear Equations in Two Variables 252

3.5 Point-Slope Formula 253

3.6 Applications of Linear Equations and Modeling 261

3.7 Introduction to Functions 268

Group Activity: Modeling a Linear Equation 279

Chapter 3 Summary 280

Chapter 3 Review Exercises 286

Chapter 3 Test 290

Chapters 1–3 Cumulative Review Exercises 293

Chapter 4 Systems of Linear Equations in Two Variables 295

4.1 Solving Systems of Equations by the Graphing Method 296

4.2 Solving Systems of Equations by the Substitution Method 307

4.3 Solving Systems of Equations by the Addition Method 316

Problem Recognition Exercises: Systems of Equations 325

4.4 Applications of Linear Equations in Two Variables 326

4.5 Linear Inequalities and Systems of Inequalities in Two Variables 335

Group Activity: Creating Linear Models from Data 345

Chapter 4 Summary 347

Chapter 4 Review Exercises 352

Chapter 4 Test 356

Chapters 1–4 Cumulative Review Exercises 357

Chapter 5 Polynomials and Properties of Exponents 359

5.1 Exponents: Multiplying and Dividing Common Bases 360

5.2 More Properties of Exponents 370

5.3 Definitions of b^0 and b^{-n} 375

5.4 Scientific Notation 383

Problem Recognition Exercises: Properties of Exponents 389

5.5 Addition and Subtraction of Polynomials 390

5.6 Multiplication of Polynomials and Special Products 398

5.7 Division of Polynomials 407

Problem Recognition Exercises: Operations on Polynomials 415

Group Activity: The Pythagorean Theorem and a Geometric "Proof" 416

Chapter 5 Summary 417

Chapter 5 Review Exercises 420

Chapter 5 Test 423

Chapters 1–5 Cumulative Review Exercises 424

Chapter 6 Factoring Polynomials 427

6.1 Greatest Common Factor and Factoring by Grouping 428

6.2 Factoring Trinomials of the Form $x^2 + bx + c$ 437

6.3 Factoring Trinomials: Trial-and-Error Method 443

6.4 Factoring Trinomials: AC-Method 452

6.5 Difference of Squares and Perfect Square Trinomials 458

6.6 Sum and Difference of Cubes 464

Problem Recognition Exercises: Factoring Strategy 470

6.7 Solving Equations Using the Zero Product Rule 472

Problem Recognition Exercises: Polynomial Expressions and Polynomial Equations 478

6.8 Applications of Quadratic Equations 479

Group Activity: Building a Factoring Test 487

Chapter 6 Summary 488

Chapter 6 Review Exercises 493

Chapter 6 Test 495

Chapters 1–6 Cumulative Review Exercises 496

Chapter 7 Rational Expressions 497

7.1 Introduction to Rational Expressions 498

7.2 Multiplication and Division of Rational Expressions 508

7.3 Least Common Denominator 514

7.4 Addition and Subtraction of Rational Expressions 520

Problem Recognition Exercises: Operations on Rational Expressions 529

7.5 Complex Fractions 530

7.6 Rational Equations 537

Problem Recognition Exercises: Comparing Rational Equations and Rational Expressions 547

7.7 Applications of Rational Equations and Proportions 548

7.8 Variation 559

Group Activity: Computing Monthly Mortgage Payments 567

Chapter 7 Summary 568

Chapter 7 Review Exercises 574

Chapter 7 Test 577

Chapters 1–7 Cumulative Review Exercises 578

Chapter 8 Radicals 581

8.1 Introduction to Roots and Radicals 582

8.2 Simplifying Radicals 593

8.3 Addition and Subtraction of Radicals 601

8.4 Multiplication of Radicals 608

8.5 Division of Radicals and Rationalization 614

Problem Recognition Exercises: Operations on Radicals 623

8.6 Radical Equations 624

8.7 Rational Exponents 630

Group Activity: Approximating Square Roots 637

Chapter 8 Summary 638

Chapter 8 Review Exercises 642

Chapter 8 Test 645

Chapters 1–8 Cumulative Review Exercises 646

Chapter 9 More Quadratic Equations 649

9.1 The Square Root Property 650

9.2 Completing the Square 655

9.3 Quadratic Formula 661

Problem Recognition Exercises: Solving Quadratic Equations 669

9.4 Graphing Quadratic Functions 670

Group Activity: Maximizing Volume 682

Chapter 9 Summary 683

Chapter 9 Review Exercises 685

Chapter 9 Test 687

Chapters 1–9 Cumulative Review Exercises 688

Student Answer Appendix SA–1

Credits C–1

Index I–1

How Will Miller/O'Neill/Hyde Help Your Students *Get Better Results?*

Better Clarity, Quality, and Accuracy!

Julie Miller, Molly O'Neill, and Nancy Hyde know what students need to be successful in mathematics. Better results come from clarity in their exposition, quality of step-by-step worked examples, and accuracy of exercise sets, but it takes more than just great authors to build a textbook series to help students achieve success in mathematics. Our authors worked with a strong mathematical team of instructors from around the country to ensure clarity, quality, and accuracy.

> "The authors' writing style is clear and concise. The tone is appropriate for our course. Students will find it easy to read and follow."
> —Carla Kulinsky, *Salt Lake Community College*

Better Exercise Sets!

A comprehensive set of exercises are available for every student level. Julie Miller, Molly O'Neill, and Nancy Hyde worked with a national board of advisors from across the country to ensure the series will offer the appropriate depth and breadth of exercises for your students. New to this edition, **Problem Recognition Exercises** were created in direct response to student need and resulted in improved student performance on tests.

Our exercise sets help students progress from skill development to conceptual understanding. Student tested and instructor approved, the Miller/O'Neill/Hyde exercise sets will help your students get better results.

- ▶ **Problem Recognition Exercises**
- ▶ **Skill Practice Exercises**
- ▶ **Study Skills Exercises**
- ▶ **Mixed Exercises**
- ▶ **Expanding Your Skills Exercises**

> "The number, type, various levels of difficulty and quality of practice exercises are excellent in this book. If students go over most of the available resources in this book, they can excel at the next math level. No doubt."
> —Pauline Chow, *Harrisburg Area Community College*

Better Step-By-Step Pedagogy!

The second edition provides enhanced step-by-step learning tools available to help students *get better results*.

- ▶ **Worked Examples** provide an "easy-to-understand" approach, clearly guiding each student through a step-by-step approach to master each practice exercise for better comprehension.
- ▶ **TIPS** offer students extra cautious direction to help improve understanding through hints and further insight.
- ▶ **Avoiding Mistakes** boxes alert students to common errors and provide practical ways to avoid them.

  These learning aids will help students get better results by learning how to work through a problem using a clearly defined step-by-step methodology that has been class-tested and student approved.

> "These are useful reminders to the instructor as well as a resource for students."
> —Patricia Roux, *Delgado Community College*

Formula for Student Success

Step-by-Step Worked Examples

▶ Do you get the feeling that there is a disconnect between your students' classwork and homework?

▶ Do your students have trouble finding worked examples that match the practice exercises?

▶ Would you like your students to see examples in the textbook that match the ones you use in class?

Miller/O'Neill/Hyde's worked examples offer a clear, concise methodology that replicates the mathematical processes used in the authors' classroom lectures!

"The wording is clear, consistent, and concise in the worked out examples. Having seen the text itself, I know the use of pedagogy here gets my top grade!"
—Linda Murphy, *Northern Essex Community College*

PROCEDURE Solving a Proportion

Step 1 Set the cross products equal to each other.
Step 2 Divide both sides of the equation by the number being multiplied by the variable.
Step 3 Check the solution in the original proportion.

Example 4 Solving a Linear Equation

Solve the equation. $7 + 3 = 2(p - 3)$

Solution:

$$7 + 3 = 2(p - 3)$$

$$10 = 2p - 6 \qquad$$ **Step 1:** Simplify both sides of the equation by clearing parentheses and combining *like* terms.

Step 2: The variable terms are already on one side.

$$10 + 6 = 2p - 6 + 6 \qquad$$ **Step 3:** Add 6 to both sides to collect the constant terms on the other side.

$$16 = 2p$$

$$\frac{16}{2} = \frac{2p}{2} \qquad$$ **Step 4:** Divide both sides by 2 to obtain a coefficient of 1 for p.

$$8 = p \qquad$$ **Step 5:** Check:

$$7 + 3 = 2(p - 3)$$

$$10 \stackrel{?}{=} 2(8 - 3)$$

$$10 \stackrel{?}{=} 2(5)$$

The solution is 8. $\qquad 10 \stackrel{?}{=} 10$ ✔ True

Skill Practice

Solve the equation.

4. $12 + 2 = 7(3 - y)$

Good step-by-step examples with easy to follow explanations."
—Lynn Beckett-Lemus, *El Camino College*

"All of the worked examples are good and easy to understand. There are plenty of examples given. Also, it appears that there is at least one example for each different particular type of exercise in each section—very good"
—Mike Kirby, *Tidewater Community College*

Better Learning Tools

Avoiding Mistakes Boxes

Avoiding Mistakes boxes are integrated throughout the textbook to alert students to common errors and how to avoid them.

> "The Avoiding Mistakes is probably one of my favorite "supplements" to the text. This idea is an EXCELLENT tool. I wouldn't change anything with the exception of adding more when appropriate."
> —Mark Billiris, *St. Petersburg Community College*

If the slopes of two lines are known, then we can compare the slopes to determine if the lines are parallel, perpendicular, or neither.

Skill Practice

Determine if lines l_1 and l_2 are parallel, perpendicular, or neither.

9. l_1: $(-2, -3)$ and $(4, -1)$
   l_2: $(0, 2)$ and $(-3, 1)$

Avoiding Mistakes
You can check that two lines are perpendicular by checking that the product of their slopes is -1.
$$4\left(-\frac{1}{4}\right) = -1$$

Example 7 Determining If Lines Are Parallel, Perpendicular, or Neither

Lines l_1 and l_2 pass through the given points. Determine if l_1 and l_2 are parallel, perpendicular, or neither.

$$l_1: \ (2, -7) \text{ and } (4, 1) \qquad l_2: \ (-3, 1) \text{ and } (1, 0)$$

Solution:

Find the slope of each line.

$$l_1: \ \underset{(x_1, y_1)}{(2, -7)} \text{ and } \underset{(x_2, y_2)}{(4, 1)} \qquad\qquad l_2: \ \underset{(x_1, y_1)}{(-3, 1)} \text{ and } \underset{(x_2, y_2)}{(1, 0)}$$

$$m_1 = \frac{1 - (-7)}{4 - 2} \qquad\qquad m_2 = \frac{0 - 1}{1 - (-3)}$$

$$m_1 = \frac{8}{2} \qquad\qquad m_2 = \frac{-1}{4}$$

$$m_1 = 4 \qquad\qquad m_2 = -\frac{1}{4}$$

One slope is the opposite of the reciprocal of the other slope. Therefore, the lines are perpendicular.

TIP Boxes

Teaching tips are usually only revealed in the classroom. Not anymore. Tip boxes offer students helpful hints and extra direction to help improve understanding and further insight.

TIP: The solution to a system of linear equations can be confirmed by graphing. The system from Example 2 is graphed here.

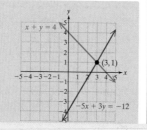

> "The nice thing is that as I am reading through the text, I think to myself, "I would say this," and then the authors will say that very thing in a tip off to the side."
> —Barb Elzey, *Bluegrass Community and Technical College*

Concept Connection Boxes

Concept Connections help students understand the conceptual meaning of the problems they are solving—a vital skill in mathematics.

> "The concept connections seem to ask almost the same questions I ask my students when we are in class. I think they are helpful."
> —Lori Grady, *University of Wisconsin–Whitewater*

2. Dependent and Inconsistent Systems of Linear Equations

When two lines are drawn in a rectangular coordinate system, three geometric relationships are possible:

1. Two lines may intersect at *exactly one point*.

2. Two lines may intersect at *no point*. This occurs if the lines are parallel.

Concept Connections

3. If the linear equations in a system represent a pair of parallel lines, how many solutions does the system have?

4. If the linear equations in a system represent a pair of perpendicular lines, how many solutions does the system have?

New to this Edition

▶ Do your students have trouble with problem solving?

▶ Do you want to help students overcome math anxiety?

▶ Do you want to help your students improve performance on math assessments?

Problem Recognition Exercises!

Problem Recognition Exercises present a collection of problems that look similar to a student upon first glance, but are actually quite different in the manner of their individual solutions. Students sharpen critical thinking skills and better develop their "solution recall" to help them distinguish the method needed to solve an exercise—an essential skill in mathematics. Problem Recognition Exercises, tested in a developmental mathematics classroom, were created in direct response to student need to improve performance in testing where different problem types are mixed.

> "Excellent problems. The PREs review all different exercise possibilities and require students to identify what they see."
> —Patricia Roux, *Delgado Community College*

> "I think this is a great way for students to have to think through the different topics. These problems are significantly different than they have been presented previously, so the student cannot use the visual clues they have used previously."
> —Ron Koehn, *Southwestern Oklahoma State University*

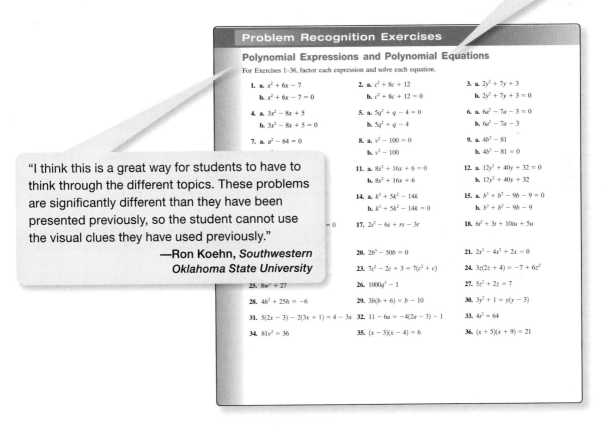

Problem Recognition Exercises

Polynomial Expressions and Polynomial Equations

For Exercises 1–36, factor each expression and solve each equation.

1. a. $x^2 + 6x - 7$
   b. $x^2 + 6x - 7 = 0$

2. a. $c^2 + 8c + 12$
   b. $c^2 + 8c + 12 = 0$

3. a. $2y^2 + 7y + 3$
   b. $2y^2 + 7y + 3 = 0$

4. a. $3x^2 - 8x + 5$
   b. $3x^2 - 8x + 5 = 0$

5. a. $5q^2 + q - 4 = 0$
   b. $5q^2 + q - 4$

6. a. $6a^2 - 7a - 3 = 0$
   b. $6a^2 - 7a - 3$

7. a. $a^2 - 64 = 0$

8. a. $v^2 - 100 = 0$
   b. $v^2 - 100$

9. a. $4b^2 - 81$
   b. $4b^2 - 81 = 0$

11. a. $8x^2 + 16x + 6 = 0$
    b. $8x^2 + 16x + 6$

12. a. $12y^2 + 40y + 32 = 0$
    b. $12y^2 + 40y + 32$

14. a. $k^3 + 5k^2 - 14k$
    b. $k^3 + 5k^2 - 14k = 0$

15. a. $b^3 + b^2 - 9b - 9 = 0$
    b. $b^3 + b^2 - 9b - 9$

17. $2s^2 - 6s + rs - 3r$

18. $6t^2 + 3t + 10tu + 5u$

20. $2b^3 - 50b = 0$

21. $2x^3 - 4x^2 + 2x = 0$

23. $7c^2 - 2c + 3 = 7(c^2 + c)$

24. $3z(2z + 4) = -7 + 6z^2$

25. $8w^3 + 27$

26. $1000q^3 - 1$

27. $5z^2 + 2z = 7$

28. $4h^2 + 25h = -6$

29. $3b(b + 6) = b - 10$

30. $3y^2 + 1 = y(y - 3)$

31. $5(2x - 3) - 2(3x + 1) = 4 - 3x$

32. $11 - 6a = -4(2a - 3) - 1$

33. $4s^2 = 64$

34. $81v^2 = 36$

35. $(x - 3)(x - 4) = 6$

36. $(x + 5)(x + 9) = 21$

New and Improved Applications!

Class-Tested and Student Approved!

New and improved applications have been developed by an advisory team. The Miller/O'Neill/Hyde Board of Advisors Team partnered with our authors to bring you the *best applications* from every region of the country! These applications include real data and topics which are more relevant and interesting to today's student.

Objective 2: Applications Involving Principal and Interest

15. Shanelle invested $10,000, and at the end of 1 yr, she received $805 in interest. She invested part of the money in an account earning 10% simple interest and the remaining money in an account earning 7% simple interest. How much did she invest in each account? **(See Example 2.)**

	10% Account	7% Account	Total
Principal invested			
Interest earned			

16. $12,000 was invested in two accounts, one earning 12% simple interest and the other earning 8% simple interest. If the total interest at the end of 1 yr was $1240, how much was invested in each account?

	12% Account	8% Account	Total
Principal invested			
Interest earned			

17. Troy borrowed a total of $12,000 in two different loans to help pay for his new Chevy Silverado. One loan charges 9% simple interest, and the other charges 6% simple interest. If he is charged $810 in interest after 1 yr, find the amount borrowed at each rate.

18. Blake has a total of $4000 to invest in two accounts. One account earns 2% simple interest, and the other earns 5% simple interest. How much should be invested in both accounts to earn exactly $155 at the end of 1 yr?

NEW Group Activities!

Each chapter concludes with a Group Activity selected by objective to promote classroom discussion and collaboration—helping students not only to solve problems but to explain their solutions for better mathematical mastery. Group Activities are great for instructors and adjuncts—bringing a more interactive approach to teaching mathematics! All required materials, activity time, and suggested group sizes are provided in the directions of the activity. Activities include: Investigating Probability, Tracking Stocks, Using Card Games with Fractions, and more!

Group Activity

Modeling a Linear Equation

Materials: Yardstick or other device for making linear measurements

Estimated Time: 15–20 minutes

Group Size: 3

1. The members of each group should measure the length of their arms (in inches) from elbow to wrist. Record this measurement as x and the person's height (in inches) as y. Write these values as ordered pairs for each member of the group. Then write the ordered pairs on the board.

2. Next, copy the ordered pairs collected from all groups in the class and plot the ordered pairs. (This is called a "scatter diagram.")

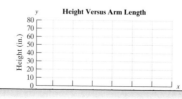

Dynamic Math Animations

The Miller/O'Neill/Hyde author team has developed a series of Flash animations to illustrate difficult concepts where static images and text fall short. The animations leverage the use of on-screen movement and morphing shapes to enhance conceptual learning.

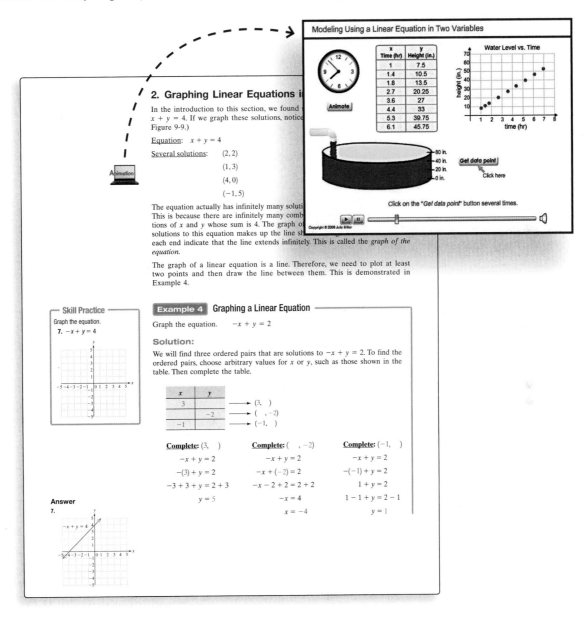

Through their classroom experience, the authors recognize that such media assets are great teaching tools for the classroom and excellent for online learning. The Miller/O'Neill/Hyde animations are interactive and quite diverse in their use. Some provide a virtual laboratory for which an application is simulated and where students can collect data points for analysis and modeling. Others provide interactive question-and-answer sessions to test conceptual learning. For word problem applications, the animations ask students to estimate answers and practice "number sense."

360° Development Process

McGraw-Hill's 360° Development Process is an ongoing, never-ending, market-oriented approach to building accurate and innovative print and digital products. It is dedicated to continual large-scale and incremental improvement driven by multiple customer feedback loops and checkpoints. The process is initiated during the early planning stages of our new products, and is intensified during development and production. Then the process begins again upon publication in anticipation of the next edition.

A key principle in the development of any mathematics text is its ability to adapt to teaching specifications in a universal way. The only way to do so is by contacting those universal voices—and learning from their suggestions. We are confident that our book has the most current content the industry has to offer, thus pushing our desire for accuracy to the highest standard possible. In order to accomplish this, we have moved through an arduous road to production. Extensive and open-minded advice is critical in the production of a superior text.

Here is a brief overview of the initiatives included in the *Basic College Mathematics*, Second Edition, 360° Development Process:

Board of Advisors

A hand-picked group of trusted teachers active in the basic math course served as chief advisors and consultants to the authors and editorial team with regards to manuscript development. The Board of Advisors reviewed parts of the manuscript; served as a sounding board for pedagogical, media, and design concerns; consulted on organizational changes; and attended a focus group to confirm the manuscript's readiness for publication.

Basic College Mathematics

Vernon Bridges, *Durham Technical Community College*

Lynette King, *Gadsden State Community College*

Sharon Morrison, *St. Petersburg College*

Deanna Murphy, *Lane County Community College*

Rod Oberdick, *Delaware Technical and Community College*

Matthew Robinson, *Tallahassee Community College*

Pat Rome, *Delgado Community College–City Park*

Introductory Algebra

Mark Billiris, *St. Petersburg Community College*

Pauline Chow, *Harrisburg Community College*

John Close, *Salt Lake Community College*

Barbara Elzy, *Bluegrass Community College*

Lori Grady, *University of Wisconsin–Whitewater*

Patricia Roux, *Delgado Community College*

Mike Kirby, *Tidewater Community College*

Intermediate Algebra

Susan Dimick, *Spokane Community College*

Sue Duff, *Guilford Technical Community College*

Alicia Giovinazzo, *Miami Dade Community College*

Charlotte Newsome, *Tidewater Community College*

Ena Salter, *Manatee Community College*

Acknowledgments and Reviewers

The development of this textbook series would never have been possible without the creative ideas and feedback offered by many reviewers. We are especially thankful to the following instructors for their careful review of the manuscript.

Manuscript Review Panels

Over 400 teachers and academics from across the country reviewed the various drafts of the manuscript to give feedback on content, design, pedagogy, and organization. This feedback was summarized by the book team and used to guide the direction of the text.

Special "*thank you*" to our Manuscript Class-Testers

Vernon Bridges, *Durham Technical Community College*
Susan Dimick, *Spokane Community College*
Lori Grady, *University of Wisconsin–Whitewater*

Rod Oberdick, *Delaware Technical and Community College*
Matthew Robinson, *Tallahassee Community College*
Pat Rome, *Delgado Community College–City Park*

Reviewers of the Miller/O'Neill/Hyde Developmental Mathematics Series

Darla Aguilar, *Pima Community College–Desert Vista*
Ebrahim Ahmadizadeh, *Northampton Community College*
Sara Alford, *North Central Texas College*
Theresa Allen, *University of Idaho*
Sheila Anderson, *Housatonic Community College*
Lane Andrew, *Arapahoe Community College*
Jan Archibald, *Ventura College*
Yvonne Aucoin, *Tidewater Community College–Norfolk*
Eric Aurand, *Mohave Community College*
Sohrab Bakhtyari, *St. Petersburg College*
Anna Bakman, *Los Angeles Trade Technical*
Andrew Ball, *Durham Technical Community College*
Russell Banks, *Guilford Technical Community College*
Suzanne Battista, *St. Petersburg College*
Kevin Baughn, *Kirtland Community College*
Sarah Baxter, *Gloucester County College*
Lynn Beckett-Lemus, *El Camino College*
Edward Bender, *Century College*
Emilie Berglund, *Utah Valley State College*
Rebecca Berthiaume, *Edison College–Fort Myers*
John Beyers, *Miami Dade College–Hialeah*
Leila Bicksler, *Delgado Community College–West Bank*
Norma Bisulca, *University of Maine–Augusta*
Kaye Black, *Bluegrass Community and Technical College*
Deronn Bowen, *Broward College–Central*
Timmy Bremer, *Broome Community College*
Donald Bridgewater, *Broward College*
Peggy Brock, *TVI Community College*
Kelly Brooks, *Pierce College*
Susan D. Caire, *Delgado Community College–West Bank*
Peter Carlson, *Delta College*

Judy Carter, *North Shore Community College*
Veena Chadha, *University of Wisconsin–Eau Claire*
Zhixiong Chen, *New Jersey City University*
Tyrone Clinton, *Saint Petersburg College–Gibbs*
Eugenia Cox, *Palm Beach Community College*
Julane Crabtree, *Johnson Community College*
Mark Crawford, *Waubonsee Community College*
Natalie Creed, *Gaston College*
Anabel Darini, *Suffolk County Community College–Brentwood*
Antonio David, *Del Mar College*
Ron Davis, *Kennedy-King College–Chicago*
Laurie Delitsky, *Nassau Community College*
Patti D'Emidio, *Montclair State University*
Bob Denton, *Orange Coast College*
Robert Diaz, *Fullerton College*
Eileen Doran, *Palm Beach Community College*
Deborah Doucette, *Erie Community College–North Campus—Williamsville*
Thomas Drucker, *University of Wisconsin–Whitewater*
Michael Dubrowsky, *Wayne Community College*
Barbara Duncan, *Hillsborough Community College–Dale Mabry*
Jeffrey Dyess, *Bishop State Community College*
Elizabeth Eagle, *University of North Carolina–Charlotte*
Sabine Eggleston, *Edison College–Fort Myers*
Lynn Eisenberg, *Rowan-Cabarrus Community College*
Barb Elzey, *Bluegrass Community and Technical College*
Nerissa Felder, *Polk Community College*
Mark Ferguson, *Chemeketa Community College*
Diane Fisher, *Lousiana State University–Eunice*
David French, *Tidewater Community College–Chesapeake*

Reviewers of the Miller/O'Neill/Hyde Developmental Mathematics Series *(continued)*

Dot French, *Community College of Philadelphia*
Deborah Fries, *Wor-Wic Community College*
Robert Frye, *Polk Community College*
Jesse M. Fuson, *Mountain State University*
Patricia Gary, *North Virginia Community College–Manassas*
Calvin Gatson, *Alabama State University*
Donna Gerken, *Miami Dade College–Kendall*
Mehrnaz Ghaffarian, *Tarrant County College South*
Mark Glucksman, *El Camino College*
Judy Godwin, *Collin County Community College*
William Graesser, *Ivy Tech Community College*
Victoria Gray, *Scott Community College*
Edna Greenwood, *Tarrant County College–Northwest*
Kimberly Gregor, *Delaware Technical Community College–Wilmington*
Vanetta Grier-Felix, *Seminole Community College*
Kathy Grigsby, *Moraine Valley Community College*
Joseph Guiciardi, *Community College of Allegheny County–Monroeville*
Susan Haley, *Florence-Darlington Technical College*
Mary Lou Hammond, *Spokane Community College*
Joseph Harris, *Gulf Coast Community College*
Lloyd Harris, *Gulf Coast Community College*
Mary Harris, *Harrisburg Area Community College–Lancaster*
Susan Harrison, *University of Wisconsin–Eau Claire*
Kristen Hathcock, *Barton County Community College*
Marie Hoover, *University of Toledo*
Linda Hoppe, *Jefferson College*
Joe Howe, *St. Charles County Community College*
Juan Jimenez, *Springfield Technical Community College*
Jennifer Johnson, *Delgado Community College*
Yolanda Johnson, *Tarrant County College South*
Shelbra Jones, *Wake Technical Community College*
Joe Jordan, *John Tyler Community College*
Cheryl Kane, *University of Nebraska–Lincoln*
Ismail Karahouni, *Lamar University*
Mike Karahouni, *Lamar University–Beaumont*
Joanne Kawczenski, *Luzerne County Community College*
Eliane Keane, *Miami Dade College–North*
Miriam Keesey, *San Diego State University*
Joe Kemble, *Lamar University–Beaumont*
Patrick Kimani, *Morrisville State College*
Sonny Kirby, *Gadsden State Community College*
Vicky Kirkpatrick, *Lane Community College*
Marcia Kleinz, *Atlantic Cape Community College*
Ron Koehn, *Southwestern Oklahoma State University*
Jeff Koleno, *Lorain County Community College*
Rosa Kontos, *Bergen Community College*
Randa Kress, *Idaho State University*

Gayle Krzemie, *Pikes Peak Community College*
Gayle Kulinsky, Carla, *Salt Lake Community College*
Linda Kuroski, *Erie Community College*
Catherine Laberta, *Erie Community College–North Campus—Williamsville*
Joyce Langguth, *University of Missouri–St. Louis*
Betty Larson, *South Dakota State University*
Katie Lathan, *Tri-County Technical College*
Kathryn Lavelle, *Westchester Community College*
Patricia Lazzarino, *North Virginia Community College–Manassas*
Julie Letellier, *University of Wisconsin–Whitewater*
Mickey Levendusky, *Pima Community College*
Barbara Little, *Central Texas College*
David Liu, *Central Oregon Community College*
Maureen Loiacano, *Montgomery College*
Wanda Long, *St. Charles County Community College*
Kerri Lookabill, *Mountain State University*
Jessica Lowenfield, *Nassau Community College*
Diane Lussier, *Pima Community College*
Mark Marino, *Erie Community College–North Campus—Williamsville*
Dorothy Marshall, *Edison College–Fort Myers*
Diane Masarik, *University of Wisconsin–Whitewater*
Louise Mataox, *Miami Dade College*
Cindy McCallum, *Tarrant County College South*
Joyce McCleod, *Florida Community College–South Campus*
Roger McCoach, *County College of Morris*
Stephen F. McCune, *Austin State University*
Ennis McKenna, Hazel, *Utah Valley State College*
Harry McLaughlin, *Montclair State University*
Valerie Melvin, *Cape Fear Community College*
Richard Moore, *St. Petersburg College–Seminole*
Elizabeth Morrison, *Valencia Community College*
Sharon Morrison, *St. Petersburg College*
Shauna Mullins, *Murray State University*
Linda Murphy, *Northern Essex Community College*
Michael Murphy, *Guilford Technical Community College*
Kathy Nabours, *Riverside Community College*
Roya Namavar, *Rogers State University*
Tony Nelson, *Tulsa Community College*
Melinda Nevels, *Utah Valley State College*
Charlotte Newsom, *Tidewater Community College–Virginia Beach*
Brenda Norman, *Tidewater Community College*
David Norwood, *Alabama State University*
Rhoda Oden, *Gadsden State Community College*
Tammy Payton, *North Idaho College*
Melissa Pedone, *Valencia Community College–Osceola*

Shirley Pereira, *Grossmont College*

Pete Peterson, *John Tyler Community College*

Suzie Pickle, *St. Petersburg College*

Sheila Pisa, *Riverside Community College–Moreno Valley*

Marilyn Platt, *Gaston College*

Richard Ponticelli, *North Shore Community College*

Tammy Potter, *Gadsden State Community College*

Joel Rappaport, *Florida Community College*

Sherry Ray, *Oklahoma City Community College*

Angelia Reynolds, *Gulf Coast Community College*

Suellen Robinson, *North Shore Community College*

Jeri Rogers, *Seminole Community College–Oviedo*

Trisha Roth, *Gloucester County College*

Richard Rupp, *Del Mark College*

Dave Ruszkiewicz, *Milwaukee Area Technical College*

Nancy Sattler, *Terra Community College*

Vicki Schell, *Pensacola Junior College*

Nyeita Schult, *St. Petersburg College*

Wendiann Sethi, *Seton Hall University*

Dustin Sharp, *Pittsburg Community College*

Marvin Shubert, *Hagerstown Community College*

Plamen Simeonov, *University of Houston–Downtown*

Carolyn Smith, *Armstrong Atlantic State University*

Melanie Smith, *Bishop State Community College*

John Squires, *Cleveland State Community College*

Sharon Staver, Judith, *Florida Community College–South Campus*

Sharon Steuer, *Nassau Community College*

Trudy Streilein, *North Virginia Community College–Annandale*

Gretchen Syhre, *Hawkeye Community College*

Katalin Szucs, *Pittsburg Community College*

Mike Tiano, *Suffolk County Community College*

Stephen Toner, *Victor Valley College*

Mary Lou Townsend, *Wor-Wic Community College*

Susan Twigg, *Wor-Wic Community College*

Matthew Utz, *University of Arkansas–Fort Smith*

Joan Van Glabek, *Edison College–Fort Myers*

John Van Kleef, *Guilford Technical Community College*

Diane Veneziale, *Burlington County College–Pemberton*

Andrea Vorwark, *Metropolitan Community College–Maple Woods*

Edward Wagner, *Central Texas College*

David Wainaina, *Coastal Carolina Community College*

James Wang, *University of Alabama*

Richard Watkins, *Tidewater Community College–Virginia Beach*

Sharon Wayne, *Patrick Henry Community College*

Leben Wee, *Montgomery College*

Betty Vix Weinberger, *Delgado Community College–West Bank*

Christine Wetzel-Ulrich, *Northampton Community College*

Jackie Wing, *Angelina College*

Michelle Wolcott, *Pierce College*

Deborah Wolfson, *Suffolk County Community College–Brentwood*

Mary Wolyniak, *Broome Community College*

Rick Woodmansee, *Cosumnes River College*

Susan Working, *Grossmont College*

Karen Wyrick, *Cleveland State Community College*

Alan Yang, *Columbus State Community College*

William Young, Jr, *Century College*

Vasilis Zafiris, *University of Houston*

Vivian Zimmerman, *Prairie State College*

Special thanks go to Jon Weerts for preparing the *Instructor's Solutions Manual* and the *Student's Solutions Manual* and to Rebecca Hubiak for her work ensuring accuracy. Many thanks to Cindy Reed for her work in the video series, and to Ethel Wheland for advising us on the Instructor Notes.

Finally, we are forever grateful to the many people behind the scenes at McGraw-Hill without whom we would still be on page 1. To our developmental editor (and math instructor extraordinaire), Emilie Berglund, thanks for your day-to-day support and understanding of the world of developmental mathematics. To David Millage, our sponsoring editor and overall team captain, thanks for keeping the train on the track. Where did you find enough hours in the day? To Torie Anderson and Sabina Navsariwala, we greatly appreciate your countless hours of support and creative ideas promoting all of our efforts. To our director of development and champion, Kris Tibbetts, thanks for being there in our time of need. To Pat Steele, where would we be without your watchful eye over our manuscript? To our editorial director, Stewart Mattson, we're grateful for your experience and energizing new ideas. Thanks for believing in us. To Jeff Huettman and Amber Bettcher, we give our greatest appreciation for the exciting technology so critical to student success. To Peggy Selle, thanks for keeping watch over the whole team as the project came together. Thank you to our wonderful designer Laurie Janssen—not only did Laurie help develop a better textbook series by delivering a clean, clear design framework for the mathematics content, Laurie also designed the best covers of the Miller/O'Neill/Hyde series to date.

Most importantly, we give special thanks to all the students and instructors who use *Basic College Mathematics* in their classes.

Supplements

For the Instructor

Instructor's Resource Manual

The *Instructor's Resource Manual* (*IRM*), written by the authors, is a printable electronic supplement available through Connect Math Hosted by ALEKS Corp. The *IRM* includes discovery-based classroom activities, worksheets for drill and practice, materials for a student portfolio, and tips for implementing successful cooperative learning. Numerous classroom activities are available for each section of text and can be used as a complement to the lectures or can be assigned for work outside of class. The activities are designed for group or individual work and take about 5–10 minutes each. With increasing demands on faculty schedules, these ready-made lessons offer a convenient means for both full-time and adjunct faculty to promote active learning in the classroom.

Instructor's Test Bank

Among the supplements is a **computerized test bank** utilizing Brownstone Diploma® algorithm-based testing software to create customized exams quickly. This user-friendly program enables instructors to search for questions by topic, format, or difficulty level; to edit existing questions or to add new ones; and to scramble questions and answer keys for multiple versions of a single test. Hundreds of text-specific, open-ended, and multiple-choice questions are included in the question bank. Sample chapter tests are also provided.

Annotated Instructor's Edition

In the *Annotated Instructor's Edition* (*AIE*), **answers to all exercises and tests appear adjacent to each exercise,** in a color used *only* for annotations. The *AIE* also contains **Instructor Notes** that appear in the margin. The notes may assist with lecture preparation. Also found in the *AIE* are icons within the Practice Exercises that serve to guide instructors in their preparation of homework assignments and lessons.

Another significant feature new to this edition is the inclusion of ***Classroom Examples*** for the instructor. In the Annotated Instructor's Edition of the text, we include references to even-numbered exercises at the end of the section for instructors to use as *Classroom Examples*. These exercises mirror the examples in the text. Therefore, if an instructor covers these exercises as classroom examples, then all the major objectives in that section will have been covered. This feature was added because we recognize the growing demands on faculty time, and to assist new faculty, adjunct faculty, and graduate assistants. Furthermore, because these exercises appear in the student edition of the text, students will not waste valuable class time copying down complicated examples from the board.

Instructor's Solutions Manual

The *Instructor's Solutions Manual* provides comprehensive, worked-out solutions to all exercises in the Chapter Openers; the Practice Exercises; the Problem Recognition Exercises; the end-of-chapter Review Exercises; the Chapter Tests; and the Cumulative Review Exercises.

For the Student

NEW Lecture Videos created by Julie Miller

Julie Miller began creating these lecture videos for her own students to use when they were absent and unable to attend one of her lectures. She found them to be so helpful that she decided to create the lecture videos for her entire developmental math book series. In these new videos, Julie walks students through the learning objectives using the same language and procedures outlined in the book. Students are able to learn and review right alongside the author! Students can also access the note files that accompany the videos so that they can take their notes while following the video just like a classroom lecture. These videos as well as the exercise videos are available online through Connect Math Hosted by ALEKS Corp.

   The videos are closed-captioned for the hearing-impaired, subtitled in Spanish, and meet the Americans with Disabilities Act Standards for Accessible Design. Instructors may use them as resources in a learning center, for online courses, and to provide additional help for students who require extra practice.

ALEKS Prep for Developmental Mathematics

ALEKS Prep for Beginning Algebra and Prep for Intermediate Algebra focus on prerequisite and introductory material for Beginning Algebra and Intermediate Algebra. These prep products can be used during the first 3 weeks of a course to prepare students for future success in the course and to increase retention and pass rates. Backed by two decades of National Science Foundation funded research, ALEKS interacts with students much like a human tutor, with the ability to precisely asses a student's preparedness and provide instruction on the topics the student is most likely to learn.

ALEKS Prep Course Products Feature:

- Artificial Intelligence Targets Gaps in Individual Student's Knowledge
- Assessment and Learning Directed Toward Individual Student's Needs
- Open Response Environment with Realistic Input Tools
- Unlimited Online Access—PC and Mac Compatible

Free trial at www.aleks.com/free_trial/instructor

Student's Solutions Manual

The *Student's Solutions Manual* provides comprehensive, worked-out solutions to the odd-numbered exercises in the Practice Exercise sets; the Problem Recognition Exercises, the end-of-chapter Review Exercises, the Chapter Tests, and the Cumulative Review Exercises. Answers to the odd- and even-numbered entries to the Chapter Opener Puzzles are also provided.

Exercise Video Series

The video series is based on exercises from the textbook. Each presenter works through selected problems, following the solution methodology employed in the text. The video series is available online as part of Connect Math Hosted by ALEKS Corp. The videos are closed-captioned for the hearing impaired, are subtitled in Spanish, and meet the Americans with Disabilities Act Standards for Accessible Design.

Reference

R

CHAPTER OUTLINE

R.1 Study Tips 2

R.2 Fractions 6

R.3 Decimals and Percents 20

R.4 Introduction to Geometry 29

Chapter R

This chapter is a reference chapter that provides a review of the basic operations on fractions, decimals, percents, and geometry. This chapter also provides a section of valuable study tips.

As you read through Chapter R, try to become familiar with the features of this textbook. Then match the feature in column B with its description in column A. Then fill in the numbered blanks with the matching letters to complete the riddle.

Column A

1. Allows you to check your work as you do your homework
2. Shows you how to avoid common errors
3. Provides an online tutorial and exercise supplement
4. Outlines key concepts for each section in the chapter
5. Provides exercises that allow you to distinguish between different types of problems
6. Offers helpful hints and insight
7. Offers practice exercises that go along with each example

Column B

a. Tips
c. ALEKS
l. Skill Practice Exercises
m. Problem Recognition Exercises
d. Answers to odd exercises
i. Chapter Summary
e. Avoiding Mistakes

Where do mathematicians shop? At the $\dfrac{d}{1}\dfrac{e}{2}\dfrac{c}{3}\dfrac{i}{4}\dfrac{m}{5}\dfrac{a}{6}\dfrac{l}{7}\dfrac{l}{7}$.

Section R.1 Study Tips

Objectives

1. Before the Course
2. During the Course
3. Preparation for Exams
4. Where to Go for Help

In taking a course in algebra, you are making a commitment to yourself, your instructor, and your classmates. Following some or all of the study tips below can help you be successful in this endeavor. The features of this text that will assist you are printed in blue.

1. Before the Course

- Purchase the necessary materials for the course before the course begins or on the first day.
- Obtain a three-ring binder to keep and organize your notes, homework, tests, and any other materials acquired in the class. We call this type of notebook a portfolio.
- Arrange your schedule so that you have enough time to attend class and to do homework. A common rule is to set aside at least 2 hours for homework for every hour spent in class. That is, if you are taking a 4-credit-hour course, plan on at least 8 hours a week for homework. A 4-credit-hour course will then take *at least* 12 hours each week—about the same as a part-time job. If you experience difficulty in mathematics, plan for more time.
- Communicate with your employer and family members the importance of your success in this course so that they can support you.
- Be sure to find out the type of calculator (if any) that your instructor requires.

2. During the Course

- Read the section in the text *before* the lecture to familiarize yourself with the material and terminology.
- Attend every class, and be on time.
- Take notes in class. Write down all of the examples that the instructor presents. Read the notes after class, and add any comments to make your notes clearer to you. Use a tape recorder to record the lecture if the instructor permits the recording of lectures.
- Ask questions in class.
- Read the section in the text *after* the lecture, and pay special attention to the Tip boxes and Avoiding Mistakes boxes.
- After you read an example, try the accompanying Skill Practice problem in the margin. The skill practice problem mirrors the example and tests your understanding of what you have read.
- Do homework every night. Even if your class does not meet every day, you should still do some work every night to keep the material fresh in your mind.
- Check your homework with the answers that are supplied in the back of this text. Correct the exercises that do not match, and circle or star those that you cannot correct yourself. This way you can easily find them and ask your instructor the next day.
- Write the definition and give an example of each Key Term found at the beginning of the Practice Exercises.
- The Problem Recognition Exercises are located in most chapters. These provide additional practice distinguishing among a variety of problem types. Sometimes the most difficult part of learning mathematics is retaining all that you learn. These exercises are excellent tools for retention of material.

- Form a study group with fellow students in your class, and exchange phone numbers. You will be surprised by how much you can learn by talking about mathematics with other students.
- If you use a calculator in your class, read the Calculator Connections boxes to learn how and when to use your calculator.
- Ask your instructor where you might obtain extra help if necessary.

3. Preparation for Exams

- Look over your homework. Pay special attention to the exercises you have circled or starred to be sure that you have learned that concept.
- Read through the Summary at the end of the chapter. Be sure that you understand each concept and example. If not, go to the section in the text and reread that section.
- Give yourself enough time to take the Chapter Test uninterrupted. Then check the answers. For each problem you answered incorrectly, go to the Review Exercises and do all of the problems that are similar.
- To prepare for the final exam, complete the Cumulative Review Exercises at the end of each chapter, starting with Chapter 2. If you complete the cumulative reviews after finishing each chapter, then you will be preparing for the final exam throughout the course. The Cumulative Review Exercises are another excellent tool for helping you retain material.

4. Where to Go for Help

- At the first sign of trouble, see your instructor. Most instructors have specific office hours set aside to help students. Don't wait until after you have failed an exam to seek assistance.
- Get a tutor. Most colleges and universities have free tutoring available.
- When your instructor and tutor are unavailable, use the Student Solutions Manual for step-by-step solutions to the odd-numbered problems in the exercise sets.
- Work with another student from your class.
- Work on the computer. Many mathematics tutorial programs and websites are available on the Internet, including the website that accompanies this text.

Section R.1 Practice Exercises

Objective 1: Before the Course

1. To motivate yourself to complete a course, it is helpful to have a very clear reason for taking the course. List your goals for taking this course.

2. Budgeting enough time to do homework and to study for a class is one of the most important steps to success in a class. Use a weekly calendar to help you plan your time for your studies this week. Also write other obligations such as the time required for your job, for your family, for sleeping, and for eating. Be realistic when estimating the time for each activity.

 Writing Translating Expression Geometry Scientific Calculator Video NS&E

3. Taking 12 credit-hours is the equivalent of a full-time job. Often students try to work too many hours while taking classes at school.

   a. Write down how many hours you work per week and the number of credit hours you are taking this term.

   number of hours worked per week _____

   number of credit-hours this term _____

   b. The table gives a recommended limit to the number of hours you should work for the number of credit hours you are taking at school. (Keep in mind that other responsibilities in your life such as your family might also make it necessary to limit your hours at work even more.) How do your numbers from part (a) compare to those in the table? Are you working too many hours?

Number of Credit-Hours	Maximum Number of Hours of Work per Week
3	40
6	30
9	20
12	10
15	0

4. It is important to establish a place where you can study—someplace that has few distractions and is readily available. Answer the questions about the space that you have chosen.

   a. Is there enough room to spread out books and paper to do homework?

   b. Is this space available anytime?

   c. Are there any distractions? Can you be interrupted?

   d. Is the furniture appropriate for studying? That is, is there a comfortable chair and good lighting?

5. Organization is an important ingredient to success. A calendar or pocket planner is a valuable resource for keeping track of assignments and test dates. Write the date of the first test in this class.

Objective 2: During the Course

6. Taking notes can help in many ways. Good notes provide examples for reference as you do your homework. Also, taking notes keeps your mind on track during the lecture and helps make you an active listener. Here are some tips to help you take better notes in class.

   a. In your next math class, take notes by drawing a vertical line about $\frac{3}{4}$ of the way across the paper as shown. On the left side, write down what your instructor puts on the board or overhead. On the right side, make your own comments about important words, procedures, or questions that you have.

   b. Revisit your notes as soon as possible after class to fill in the missing parts that you recall from lecture but did not have time to write down.

   c. Be sure that you label each page with the date, chapter, section, and topic. This will make your notes easier to study from when you study for a test.

7. Many instructors use a variety of styles to accommodate all types of learners. From the following list, try to identify the type of learner that best describes you.

Auditory Learner   Do you learn best by listening to your instructor's lectures? Do you tape the lecture so that you can listen to it as many times as you need? Do you talk aloud when doing homework or studying for a test?

Visual Learner   Do you learn best by seeing problems worked out on the board? Do you understand better if there is a picture or illustration accompanying the problem? Do you take notes in class?

Tactile Learner   Do you learn best with hands-on projects? Do you prefer having some sort of physical objects to manipulate? Do you prefer to move around the classroom in a lab situation?

Objective 3: Preparation for Exams

8. When taking a test, go through the test and work all the problems that you know first. Then go back and work on the problems that were more difficult. Give yourself a time limit for each problem (maybe 3 to 5 minutes the first time through the test). Circle the importance of each statement.

	not important	somewhat important	very important
a. Read through the entire test first.	1	2	3
b. If time allows, go back and check each problem.	1	2	3
c. Write out all steps instead of doing the work in your head.	1	2	3

9. One way to lessen test anxiety is to feel prepared for the exam. Check the ways that you think might be helpful in preparing for a test.

_____ Do the chapter test in the text.

_____ Get in a study group with your peers and go over the major topics.

_____ Write your own pretest.

_____ Use the online component that goes with your textbook.

_____ Write an outline of the major topics. Then find an example for each topic.

_____ Make flash cards with a definition, property, or rule on one side and an example on the other.

_____ Re-read your notes that you took in class.

_____ Write down the definitions of all key terms introduced in the chapter.

10. The following list gives symptoms of math anxiety.

- Experiencing loss of sleep and worrying about an upcoming exam.
- Mind becoming blank when answering a question or taking a test.
- Becoming nervous about asking questions in class.
- Experiencing anxiety that interferes with studying.
- Being afraid or embarrassed to let the instructor see your work.
- Becoming physically ill during a test.
- Having sweaty palms or shaking hands when asked a math question.

Have you experienced any of these symptoms? If so, how many and how often?

11. If you think that you have math anxiety, read the following list for some possible solutions. Check the activities that you can realistically try to help you overcome this problem.

_____ Read a book on math anxiety.

_____ Search the Web for helpful tips on handling math anxiety.

_____ See a counselor to discuss your anxiety.

_____ See your instructor to inform him or her about your situation.

_____ Evaluate your time management to see if you are trying to do too much. Then adjust your schedule accordingly.

Writing   Translating Expression   Geometry   Scientific Calculator   Video   NSE

Objective 4: Where to Go for Help

12. Does your college offer free tutoring? If so, write down the room number and the hours of the tutoring center.

13. Does your instructor have office hours? If so, write down your professor's office number and office hours.

14. Is there a supplement to your text? If so, find out its price and where you can get it.

15. Find out how to access the online tutoring available with this text.

Section R.2 Fractions

Objectives

1. Basic Definitions
2. Prime Factorization
3. Simplifying Fractions to Lowest Terms
4. Multiplying Fractions
5. Dividing Fractions
6. Adding and Subtracting Fractions
7. Operations on Mixed Numbers

1. Basic Definitions

The study of algebra involves many of the operations and procedures used in arithmetic. Therefore, we begin this text by reviewing the basic operations of addition, subtraction, multiplication, and division on fractions and mixed numbers.

In day-to-day life, the numbers we use for counting are

the **natural numbers**: 1, 2, 3, 4, . . .

and

the **whole numbers**: 0, 1, 2, 3, . . .

Whole numbers are used to count the number of whole units in a quantity. A fraction is used to express part of a whole unit. If a child gains $2\frac{1}{2}$ lb, the child has gained two whole pounds plus a portion of a pound. To express the additional half pound mathematically, we may use the fraction, $\frac{1}{2}$.

> **DEFINITION A Fraction and Its Parts**
>
> **Fractions** are numbers of the form $\frac{a}{b}$, where $\frac{a}{b} = a \div b$ and b does not equal zero.
>
> In the fraction $\frac{a}{b}$, the **numerator** is a, and the **denominator** is b.

The denominator of a fraction indicates how many equal parts divide the whole. The numerator indicates how many parts are being represented. For instance, suppose Jack wants to plant carrots in $\frac{2}{5}$ of a rectangular garden. He can divide the garden into five equal parts and use two of the parts for carrots (Figure R-1).

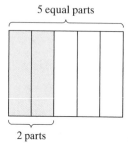

5 equal parts

2 parts

The shaded region represents $\dfrac{2}{5}$ of the garden.

Figure R-1

Instructor Note: Tell the students that the word "improper" does not imply that there is anything wrong with a fraction.

DEFINITION Proper Fractions, Improper Fractions, and Mixed Numbers

1. If the numerator of a fraction is less than the denominator, the fraction is a **proper fraction**. A proper fraction represents a quantity that is less than a whole unit.
2. If the numerator of a fraction is greater than or equal to the denominator, then the fraction is an **improper fraction**. An improper fraction represents a quantity greater than or equal to a whole unit.
3. A **mixed number** is a whole number added to a proper fraction.

Proper Fractions: $\dfrac{3}{5}$ $\dfrac{1}{8}$

Improper Fractions: $\dfrac{7}{5}$ $\dfrac{8}{8}$

Mixed Numbers: $1\frac{1}{5}$ $2\frac{3}{8}$

Concept Connections

1. Write a proper or improper fraction associated with the shaded region.

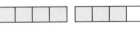

2. Which is greater, $\dfrac{3}{2}$ or $3\dfrac{1}{2}$?

2. Prime Factorization

To perform operations on fractions it is important to understand the concept of a factor. For example, when the numbers 2 and 6 are multiplied, the result (called the **product**) is 12.

$$2 \times 6 = 12$$

factors   product

The numbers 2 and 6 are said to be **factors** of 12. (In this context, we refer only to natural number factors.) The number 12 is said to be factored when it is written as the product of two or more natural numbers. For example, 12 can be factored in several ways:

$$12 = 1 \times 12 \qquad 12 = 2 \times 6 \qquad 12 = 3 \times 4 \qquad 12 = 2 \times 2 \times 3$$

A natural number greater than 1 that has only two factors, 1 and itself, is called a **prime number**. The first several prime numbers are 2, 3, 5, 7, 11, and 13. A natural number greater than 1 that is not prime is called a **composite number**. That is, a composite number has factors other than itself and 1. The first several composite numbers are 4, 6, 8, 9, 10, 12, 14, 15, and 16.

Avoiding Mistakes

The number 1 is neither prime nor composite.

Answers

1. $\dfrac{7}{4}$ **2.** $3\dfrac{1}{2}$

Skill Practice

Write the number as a product of prime factors.

3. 40   **4.** 60

Classroom Example: p. 18, Exercise 34

Example 1 Writing a Natural Number as a Product of Prime Factors

Write each number as a product of prime factors.

a. 12   **b.** 30

Solution:

a. $12 = 2 \times 2 \times 3$ Divide 12 by prime numbers until only prime numbers are obtained.

$$\begin{array}{r} 2\overline{)12} \\ 2\overline{)6} \\ 3 \end{array}$$

Or use a factor tree

b. $30 = 2 \times 3 \times 5$

$$\begin{array}{r} 2\overline{)30} \\ 3\overline{)15} \\ 5 \end{array}$$

3. Simplifying Fractions to Lowest Terms

The process of factoring numbers can be used to reduce or simplify fractions to lowest terms. A fractional portion of a whole can be represented by infinitely many fractions. For example, Figure R-2 shows that $\frac{1}{2}$ is equivalent to $\frac{2}{4}, \frac{3}{6}, \frac{4}{8}$, and so on.

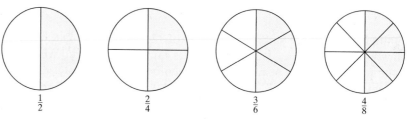

$$\frac{1}{2} \qquad \frac{2}{4} \qquad \frac{3}{6} \qquad \frac{4}{8}$$

Figure R-2

The fraction $\frac{1}{2}$ is said to be in **lowest terms** because the numerator and denominator share no common factor other than 1.

To simplify a fraction to lowest terms, we use the following important principle.

PROPERTY Fundamental Principle of Fractions

Suppose that a number, c, is a common factor in the numerator and denominator of a fraction. Then

$$\frac{a \times c}{b \times c} = \frac{a}{b} \times \frac{c}{c} = \frac{a}{b} \times 1 = \frac{a}{b}$$

Answers

3. $2 \times 2 \times 2 \times 5$
4. $2 \times 2 \times 3 \times 5$

To simplify a fraction, we begin by factoring the numerator and denominator into prime factors. This will help identify the common factors.

| Example 2 | **Simplifying a Fraction to Lowest Terms** |

Simplify $\dfrac{45}{30}$ to lowest terms.

Solution:

$\dfrac{45}{30} = \dfrac{3 \times 3 \times 5}{2 \times 3 \times 5}$ Factor the numerator and denominator.

$= \dfrac{3}{2} \times \dfrac{3}{3} \times \dfrac{5}{5}$ Apply the fundamental principle of fractions.

$= \dfrac{3}{2} \times 1 \times 1$ Any nonzero number divided by itself is 1.

$= \dfrac{3}{2}$ Any number multiplied by 1 is itself.

Skill Practice

5. Simplify to lowest terms.
$\dfrac{20}{50}$

Classroom Example: p. 18, Exercise 42

Instructor Note: Students often confuse writing a fraction in lowest terms with writing a mixed number. The fraction $\frac{3}{2}$ is in lowest terms.

In Example 2, we showed numerous steps to reduce fractions to lowest terms. However, the process is often simplified. Notice that the same result can be obtained by dividing out the greatest common factor from the numerator and denominator. (The **greatest common factor** is the largest factor that is common to both numerator and denominator.)

$\dfrac{45}{30} = \dfrac{3 \times 15}{2 \times 15}$ The greatest common factor of 45 and 30 is 15.

$= \dfrac{3 \times \cancel{15}^{1}}{2 \times \cancel{15}_{1}}$ The symbol ╱ is often used to show that a common factor has been divided out.

$= \dfrac{3}{2}$ Notice that "dividing out" the common factor of 15 has the same effect as dividing the numerator and denominator by 15. This is often done mentally.

$\dfrac{\cancel{45}^{3}}{\cancel{30}_{2}} = \dfrac{3}{2}$ ← 45 divided by 15 equals 3. ← 30 divided by 15 equals 2.

Avoiding Mistakes

In Example 3, the common factor 14 in the numerator and denominator simplifies to 1. It is important to remember to write the factor of 1 in the numerator. The simplified form of the fraction is $\frac{1}{3}$.

| Example 3 | **Simplifying a Fraction to Lowest Terms** |

Simplify $\dfrac{14}{42}$ to lowest terms.

Solution:

$\dfrac{14}{42} = \dfrac{1 \times 14}{3 \times 14}$ The greatest common factor of 14 and 42 is 14.

$= \dfrac{1 \times \cancel{14}^{1}}{3 \times \cancel{14}_{1}}$

$= \dfrac{1}{3}$ $\dfrac{\cancel{14}^{1}}{\cancel{42}_{3}} = \dfrac{1}{3}$ ← 14 divided by 14 equals 1. ← 42 divided by 14 equals 3.

Skill Practice

6. Simplify to lowest terms.
$\dfrac{32}{12}$

Classroom Example: p. 18, Exercise 46

Answers
5. $\dfrac{2}{5}$ 6. $\dfrac{8}{3}$

4. Multiplying Fractions

> **PROCEDURE** Multiplying Fractions
>
> If b is not zero and d is not zero, then
> $$\frac{a}{b} \times \frac{c}{d} = \frac{a \times c}{b \times d}$$
>
> To multiply fractions, multiply the numerators and multiply the denominators.

Skill Practice

Multiply.

7. $\dfrac{2}{7} \times \dfrac{3}{4}$

Classroom Example: p. 18, Exercise 56

Example 4 Multiplying Fractions

Multiply the fractions: $\dfrac{1}{4} \times \dfrac{1}{2}$

Solution:

$$\frac{1}{4} \times \frac{1}{2} = \frac{1 \times 1}{4 \times 2} = \frac{1}{8}$$ Multiply the numerators. Multiply the denominators.

Notice that the product $\frac{1}{4} \times \frac{1}{2}$ represents a quantity that is $\frac{1}{4}$ of $\frac{1}{2}$. Taking $\frac{1}{4}$ of a quantity is equivalent to dividing the quantity by 4. One-half of a pie divided into four pieces leaves pieces that each represent $\frac{1}{8}$ of the pie (Figure R-3).

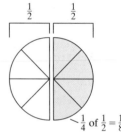

$\frac{1}{4}$ of $\frac{1}{2} = \frac{1}{8}$

Figure R-3

Skill Practice

Multiply.

8. $\dfrac{8}{9} \times \dfrac{3}{4}$ **9.** $\dfrac{4}{5} \times \dfrac{5}{4}$

10. $10 \times \dfrac{1}{10}$

Classroom Example: p. 18, Exercise 60

Example 5 Multiplying Fractions

Multiply the fractions.

a. $\dfrac{7}{10} \times \dfrac{15}{14}$ **b.** $\dfrac{2}{13} \times \dfrac{13}{2}$ **c.** $5 \times \dfrac{1}{5}$

Solution:

a. $\dfrac{7}{10} \times \dfrac{15}{14} = \dfrac{7 \times 15}{10 \times 14}$ Multiply the numerators. Multiply the denominators.

$= \dfrac{\overset{3}{\cancel{105}}}{\underset{4}{\cancel{140}}}$ Divide out the common factor, 35.

$= \dfrac{3}{4}$

> **TIP:** The same result can be obtained by dividing out common factors *before* multiplying.
> $$\frac{\overset{1}{\cancel{7}}}{\underset{2}{\cancel{10}}} \times \frac{\overset{3}{\cancel{15}}}{\underset{2}{\cancel{14}}} = \frac{3}{4}$$

Answers

7. $\dfrac{3}{14}$ **8.** $\dfrac{2}{3}$ **9.** 1   **10.** 1

Instructor Note: You can also show multiplication of fractions using shading: the overlapped portion is the product.

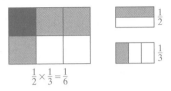

$$\frac{1}{2} \times \frac{1}{3} = \frac{1}{6}$$

b. $\dfrac{2}{13} \times \dfrac{13}{2} = \dfrac{2 \times 13}{13 \times 2} = \dfrac{\overset{1}{2} \times \overset{1}{13}}{\underset{1}{13} \times \underset{1}{2}} = \dfrac{1}{1} = 1$

Multiply $1 \times 1 = 1$.

Multiply $1 \times 1 = 1$.

c. $5 \times \dfrac{1}{5} = \dfrac{5}{1} \times \dfrac{1}{5}$

The whole number 5 can be written as $\frac{5}{1}$.

$= \dfrac{\overset{1}{5} \times 1}{1 \times \underset{1}{5}} = \dfrac{1}{1} = 1$

Multiply and simplify to lowest terms.

5. Dividing Fractions

Before we divide fractions, we need to know how to find the reciprocal of a fraction. Notice from Example 5 that $\frac{2}{13} \times \frac{13}{2} = 1$ and $5 \times \frac{1}{5} = 1$. The numbers $\frac{2}{13}$ and $\frac{13}{2}$ are said to be reciprocals because their product is 1. Likewise the numbers 5 and $\frac{1}{5}$ are reciprocals.

DEFINITION The Reciprocal of a Number

Two nonzero numbers are **reciprocals** of each other if their product is 1. Therefore, the reciprocal of the fraction

$$\frac{a}{b} \text{ is } \frac{b}{a} \qquad \text{because} \qquad \frac{a}{b} \times \frac{b}{a} = 1$$

Number	Reciprocal	Product
$\dfrac{2}{15}$	$\dfrac{15}{2}$	$\dfrac{2}{15} \times \dfrac{15}{2} = 1$
$\dfrac{1}{8}$	$\dfrac{8}{1}$ (or equivalently 8)	$\dfrac{1}{8} \times 8 = 1$
$6 \left(\text{or equivalently } \dfrac{6}{1}\right)$	$\dfrac{1}{6}$	$6 \times \dfrac{1}{6} = 1$

Concept Connections

Give the reciprocal.

11. $\dfrac{9}{4}$ **12.** 7   **13.** $\dfrac{1}{15}$

To understand the concept of dividing fractions, consider a pie that is half-eaten. Suppose the remaining half must be divided among three people, that is, $\frac{1}{2} \div 3$. However, dividing by 3 is equivalent to taking $\frac{1}{3}$ of the remaining $\frac{1}{2}$ of the pie (Figure R-4).

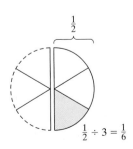

$$\frac{1}{2} \div 3 = \frac{1}{2} \cdot \frac{1}{3} = \frac{1}{6}$$

Figure R-4

Answers

11. $\dfrac{4}{9}$ **12.** $\dfrac{1}{7}$ **13.** 15

This example illustrates that dividing two numbers is equivalent to multiplying the first number by the reciprocal of the second number.

PROCEDURE   Dividing Fractions

Let a, b, c, and d be numbers such that b, c, and d are not zero. Then,

$$\frac{a}{b} \div \frac{c}{d} = \frac{a}{b} \times \frac{d}{c}$$

multiply · reciprocal

To divide fractions, multiply the first fraction by the reciprocal of the second fraction.

Skill Practice

Divide.

14. $\dfrac{12}{25} \div \dfrac{8}{15}$ **15.** $\dfrac{1}{4} \div 2$

Classroom Examples: p. 18,
Exercises 58 and 62

Example 6   Dividing Fractions

Divide the fractions.

   a. $\dfrac{8}{5} \div \dfrac{3}{10}$ **b.** $\dfrac{12}{13} \div 6$

Solution:

a. $\dfrac{8}{5} \div \dfrac{3}{10} = \dfrac{8}{5} \times \dfrac{10}{3}$ Multiply by the reciprocal of $\frac{3}{10}$, which is $\frac{10}{3}$.

$\qquad\qquad = \dfrac{8 \times \overset{2}{\cancel{10}}}{\underset{1}{\cancel{5}} \times 3} = \dfrac{16}{3}$ Multiply and simplify to lowest terms.

b. $\dfrac{12}{13} \div 6 = \dfrac{12}{13} \div \dfrac{6}{1}$ Write the whole number 6 as $\frac{6}{1}$.

$\qquad\qquad = \dfrac{12}{13} \times \dfrac{1}{6}$ Multiply by the reciprocal of $\frac{6}{1}$, which is $\frac{1}{6}$.

$\qquad\qquad = \dfrac{\overset{2}{\cancel{12}} \times 1}{13 \times \underset{1}{\cancel{6}}} = \dfrac{2}{13}$ Multiply and simplify to lowest terms.

---

6. Adding and Subtracting Fractions

PROCEDURE   Adding and Subtracting Fractions

Two fractions can be added or subtracted if they have a common denominator. Let a, b, and c be numbers such that b does not equal zero. Then,

$$\frac{a}{b} + \frac{c}{b} = \frac{a+c}{b} \qquad \text{and} \qquad \frac{a}{b} - \frac{c}{b} = \frac{a-c}{b}$$

To add or subtract fractions with the same denominator, add or subtract the numerators and write the result over the common denominator.

Answers

14. $\dfrac{9}{10}$ **15.** $\dfrac{1}{8}$

| Example 7 | Adding and Subtracting Fractions with the Same Denominator |

Add or subtract as indicated.

a. $\dfrac{1}{12} + \dfrac{7}{12}$ **b.** $\dfrac{13}{5} - \dfrac{3}{5}$

Solution:

a. $\dfrac{1}{12} + \dfrac{7}{12} = \dfrac{1+7}{12}$ Add the numerators.

$= \dfrac{8}{12}$

$= \dfrac{2}{3}$ Simplify to lowest terms.

TIP: The sum $\frac{1}{12} + \frac{7}{12}$ can be visualized as the sum of the pink and blue sections of the figure.

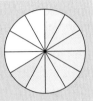

---

b. $\dfrac{13}{5} - \dfrac{3}{5} = \dfrac{13-3}{5}$ Subtract the numerators.

$= \dfrac{10}{5}$ Simplify.

$= 2$ Simplify to lowest terms.

Skill Practice

Add or subtract as indicated.

16. $\dfrac{2}{3} + \dfrac{5}{3}$ **17.** $\dfrac{5}{8} - \dfrac{1}{8}$

Classroom Examples: p. 19, Exercises 76 and 78

In Example 7, we added and subtracted fractions with the same denominators. To add or subtract fractions with different denominators, we must first become familiar with the idea of a least common multiple between two or more numbers. The **least common multiple (LCM)** of two numbers is the smallest whole number that is a multiple of each number. For example, the LCM of 6 and 9 is 18.

multiples of 6:   6, 12, 18, 24, 30, 36, …

multiples of 9:   9, 18, 27, 36, 45, 54, …

Listing the multiples of two or more given numbers can be a cumbersome way to find the LCM. Therefore, we offer the following method to find the LCM of two numbers.

PROCEDURE Finding the LCM of Two Numbers

Step 1 Write each number as a product of prime factors.

Step 2 The LCM is the product of unique prime factors from *both* numbers. Use repeated factors the maximum number of times they appear in *either* factorization.

Answers

16. $\dfrac{7}{3}$ **17.** $\dfrac{1}{2}$

Skill Practice

Find the LCM.

18. 10 and 25

Classroom Example: p. 19, Exercise 80

Example 8 Finding the LCM of Two Numbers

Find the LCM of 9 and 15.

Solution:

	3's	5's
9 =	③ × ③	
15 =	3 ×	⑤

$$LCM = 3 \times 3 \times 5 = 45$$

For the factors of 3 and 5, we circle the greatest number of times each occurs. The LCM is the product.

To add or subtract fractions with *different* denominators, we must first write each fraction as an equivalent fraction with a common denominator. A common denominator may be *any* common multiple of the denominators. However, we will use the least common denominator. The **least common denominator (LCD)** of two or more fractions is the LCM of the denominators of the fractions. The following steps outline the procedure to write a fraction as an equivalent fraction with a common denominator.

PROCEDURE Writing Equivalent Fractions

To write a fraction as an equivalent fraction with a common denominator, multiply the numerator and denominator by the factors from the common denominator that are missing from the denominator of the original fraction.

Note: Multiplying the numerator and denominator by the *same* nonzero quantity will not change the value of the fraction.

Skill Practice

19. Write each of the fractions $\frac{5}{8}$ and $\frac{5}{12}$ as an equivalent fraction with the LCD as its denominator.

20. Subtract. $\frac{5}{8} - \frac{5}{12}$

Classroom Examples: p. 19, Exercises 84 and 88

Example 9 Writing Equivalent Fractions

a. Write each of the fractions $\frac{1}{9}$ and $\frac{1}{15}$ as an equivalent fraction with the LCD as its denominator.

b. Subtract $\frac{1}{9} - \frac{1}{15}$.

Solution:

From Example 8, we know that the LCM for 9 and 15 is 45. Therefore, the LCD of $\frac{1}{9}$ and $\frac{1}{15}$ is 45.

a. $\dfrac{1}{9} = \dfrac{}{45}$ $\dfrac{1 \times 5}{9 \times 5} = \dfrac{5}{45}$ So, $\dfrac{1}{9}$ is equivalent to $\dfrac{5}{45}$.

What number must we multiply 9 by to get 45?   Multiply numerator and denominator by 5.

$\dfrac{1}{15} = \dfrac{}{45}$ $\dfrac{1 \times 3}{15 \times 3} = \dfrac{3}{45}$ So, $\dfrac{1}{15}$ is equivalent to $\dfrac{3}{45}$.

What number must we multiply 15 by to get 45?   Multiply numerator and denominator by 3.

Answers

18. 50   **19.** $\frac{5}{8} = \frac{15}{24}$ and $\frac{5}{12} = \frac{10}{24}$

20. $\frac{5}{24}$

b. $\dfrac{1}{9} - \dfrac{1}{15}$

$= \dfrac{5}{45} - \dfrac{3}{45}$ Write $\frac{1}{9}$ and $\frac{1}{15}$ as equivalent fractions with the same denominator.

$= \dfrac{2}{45}$ Subtract.

Example 10   **Adding and Subtracting Fractions**

Simplify. $\dfrac{5}{12} + \dfrac{3}{4} - \dfrac{1}{2}$

Solution:

$\dfrac{5}{12} + \dfrac{3}{4} - \dfrac{1}{2}$

To find the LCD, we have:

LCD $= 2 \times 2 \times 3 = 12$

	2's	3's
12 =	(2 × 2)	③
4 =	2 × 2	
2 =	2	

$= \dfrac{5}{12} + \dfrac{3 \times 3}{4 \times 3} - \dfrac{1 \times 6}{2 \times 6}$ Write each fraction as an equivalent fraction with the LCD as its denominator.

$= \dfrac{5}{12} + \dfrac{9}{12} - \dfrac{6}{12}$

$= \dfrac{5 + 9 - 6}{12}$ Add and subtract the numerators.

$= \dfrac{\overset{2}{\cancel{8}}}{\underset{3}{\cancel{12}}}$ Simplify to lowest terms.

$= \dfrac{2}{3}$

Skill Practice

21. Add. $\dfrac{2}{3} + \dfrac{1}{2} + \dfrac{5}{6}$

Classroom Example: p. 19, Exercise 96

---

7. Operations on Mixed Numbers

Recall that a mixed number is a whole number added to a fraction. The number $3\frac{1}{2}$ represents the sum of three wholes plus a half, that is, $3\frac{1}{2} = 3 + \frac{1}{2}$. For this reason, any mixed number can be converted to an improper fraction by using addition.

$$3\tfrac{1}{2} = 3 + \dfrac{1}{2} = \dfrac{6}{2} + \dfrac{1}{2} = \dfrac{7}{2}$$

Concept Connections

22. Write as an improper fraction.

$$2\dfrac{1}{4}$$

TIP: A shortcut to writing a mixed number as an improper fraction is to multiply the whole number by the denominator of the fraction. Then add this value to the numerator of the fraction, and write the result over the denominator.

$3\tfrac{1}{2} \longrightarrow$ Multiply the whole number by the denominator: $3 \times 2 = 6$.

Add the numerator: $6 + 1 = 7$.

Write the result over the denominator: $\frac{7}{2}$.

To add, subtract, multiply, or divide mixed numbers, we will first write the mixed number as an improper fraction.

Answers

21. 2   **22.** $\dfrac{9}{4}$

Skill Practice

23. Subtract. $2\frac{3}{4} - 1\frac{1}{3}$

Classroom Example: p. 20, Exercise 112

Example 11 Operations on Mixed Numbers

Subtract. $5\frac{1}{3} - 2\frac{1}{4}$

Solution:

$5\frac{1}{3} - 2\frac{1}{4}$

$= \frac{16}{3} - \frac{9}{4}$ Write the mixed numbers as improper fractions.

$= \frac{16 \times 4}{3 \times 4} - \frac{9 \times 3}{4 \times 3}$ The LCD is 12. Multiply numerators and denominators by the missing factors from the denominators.

$= \frac{64}{12} - \frac{27}{12}$

$= \frac{37}{12}$ or $3\frac{1}{12}$ Subtract the fractions.

TIP: An improper fraction can also be written as a mixed number. Both answers are acceptable. Note that

$$\frac{37}{12} = \frac{36}{12} + \frac{1}{12} = 3 + \frac{1}{12}, \text{ or } 3\frac{1}{12}$$

This can easily be found by dividing.

$$\frac{37}{12} \longrightarrow \begin{array}{r} 3 \\ 12\overline{)37} \\ -36 \\ \hline 1 \end{array} \longrightarrow 3\frac{1}{12}$$

quotient — remainder — divisor

Skill Practice

24. Divide. $5\frac{5}{6} \div 3\frac{2}{3}$

Classroom Example: p. 19, Exercise 108

Avoiding Mistakes

Remember that when dividing (or multiplying) fractions, a common denominator is not necessary.

Example 12 Operations on Mixed Numbers

Divide. $7\frac{1}{2} \div 3$

Solution:

$7\frac{1}{2} \div 3$

$= \frac{15}{2} \div \frac{3}{1}$ Write the mixed number and whole number as fractions.

$= \frac{\overset{5}{15}}{2} \times \frac{1}{\underset{1}{3}}$ Multiply by the reciprocal of $\frac{3}{1}$, which is $\frac{1}{3}$.

$= \frac{5}{2}$ or $2\frac{1}{2}$ The answer may be written as an improper fraction or as a mixed number.

Answers

23. $\frac{17}{12}$ or $1\frac{5}{12}$ **24.** $\frac{35}{22}$ or $1\frac{13}{22}$

Section R.2 Practice Exercises

Objective 1: Basic Definitions

For Exercises 1–8, identify the numerator and denominator of the fraction. Then determine if the fraction is a proper fraction or an improper fraction.

1. $\frac{7}{8}$ Numerator: 7; denominator: 8; proper

2. $\frac{2}{3}$ Numerator: 2; denominator: 3; proper

3. $\frac{9}{5}$ Numerator: 9; denominator: 5; improper

4. $\frac{5}{2}$ Numerator: 5; denominator: 2; improper

5. $\frac{6}{6}$ Numerator: 6; denominator: 6; improper

6. $\frac{4}{4}$ Numerator: 4; denominator: 4; improper

7. $\frac{12}{1}$ Numerator: 12; denominator: 1; improper

8. $\frac{5}{1}$ Numerator: 5; denominator: 1; improper

For Exercises 9–16, write a proper or improper fraction associated with the shaded region of each figure.

9. $\frac{3}{4}$

10. $\frac{4}{5}$

11. $\frac{4}{3}$

12. $\frac{5}{4}$

13. $\frac{1}{6}$

14. $\frac{1}{8}$

15. $\frac{2}{2}$

16. $\frac{4}{4}$

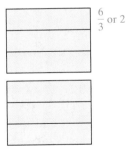

For Exercises 17–20, write both an improper fraction and a mixed number associated with the shaded region of each figure.

17. $\frac{5}{2}$ or $2\frac{1}{2}$

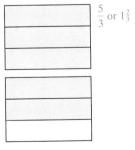

18. $\frac{5}{3}$ or $1\frac{2}{3}$

19. $\frac{6}{2}$ or 3

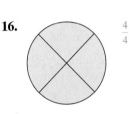

20. $\frac{6}{3}$ or 2

21. Explain the difference between the set of whole numbers and the set of natural numbers. The set of whole numbers includes the number 0 and the set of natural numbers does not.

22. Explain the difference between a proper fraction and an improper fraction. A proper fraction represents a number less than one unit. An improper fraction represents a number greater than or equal to a whole unit.

23. Write a fraction that simplifies to $\frac{1}{2}$. (Answers may vary.) For example: $\frac{2}{4}$

24. Write a fraction that simplifies to $\frac{1}{3}$. (Answers may vary.) For example: $\frac{2}{6}$

Objective 2: Prime Factorization

For Exercises 25–32, identify the number as either a prime number or a composite number.

25. 5 Prime   **26.** 9 Composite   **27.** 4 Composite   **28.** 2 Prime

29. 39 Composite   **30.** 23 Prime   **31.** 53 Prime   **32.** 51 Composite

For Exercises 33–40, write the number as a product of prime factors. **(See Example 1.)**

33. 36 $2 \times 2 \times 3 \times 3$ **34.** 70 $2 \times 5 \times 7$ **35.** 42 $2 \times 3 \times 7$ **36.** 35 5×7

37. 110 $2 \times 5 \times 11$ **38.** 136 $2 \times 2 \times 2 \times 17$ **39.** 135 $3 \times 3 \times 3 \times 5$ **40.** 105 $3 \times 5 \times 7$

Objective 3: Simplifying Fractions to Lowest Terms

For Exercises 41–52, simplify each fraction to lowest terms. **(See Examples 2–3.)**

41. $\dfrac{3}{15}$ $\dfrac{1}{5}$ **42.** $\dfrac{8}{12}$ $\dfrac{2}{3}$ **43.** $\dfrac{6}{16}$ $\dfrac{3}{8}$ **44.** $\dfrac{12}{20}$ $\dfrac{3}{5}$

45. $\dfrac{42}{48}$ $\dfrac{7}{8}$ **46.** $\dfrac{35}{80}$ $\dfrac{7}{16}$ **47.** $\dfrac{48}{64}$ $\dfrac{3}{4}$ **48.** $\dfrac{32}{48}$ $\dfrac{2}{3}$

49. $\dfrac{110}{176}$ $\dfrac{5}{8}$ **50.** $\dfrac{70}{120}$ $\dfrac{7}{12}$ **51.** $\dfrac{150}{200}$ $\dfrac{3}{4}$ **52.** $\dfrac{119}{210}$ $\dfrac{17}{30}$

Objectives 4–5: Multiplying and Dividing Fractions

For Exercises 53–54, determine if the statement is true or false. If it is false, rewrite as a true statement.

53. When multiplying or dividing fractions, it is necessary to have a common denominator.
False: When adding or subtracting fractions, it is necessary to have a common denominator.

54. When dividing two fractions, it is necessary to multiply the first fraction by the reciprocal of the second fraction. True

For Exercises 55–66, multiply or divide as indicated. **(See Examples 4–6.)**

55. $\dfrac{10}{13} \times \dfrac{26}{15}$ $\dfrac{4}{3}$ **56.** $\dfrac{15}{28} \times \dfrac{7}{9}$ $\dfrac{5}{12}$ **57.** $\dfrac{3}{7} \div \dfrac{9}{14}$ $\dfrac{2}{3}$ **58.** $\dfrac{7}{25} \div \dfrac{1}{5}$ $\dfrac{7}{5}$

59. $\dfrac{9}{10} \times 5$ $\dfrac{9}{2}$ **60.** $\dfrac{3}{7} \times 14$ 6 **61.** $\dfrac{12}{5} \div 4$ $\dfrac{3}{5}$ **62.** $\dfrac{20}{6} \div 5$ $\dfrac{2}{3}$

63. $\dfrac{5}{2} \times \dfrac{10}{21} \times \dfrac{7}{5}$ $\dfrac{5}{3}$ **64.** $\dfrac{55}{9} \times \dfrac{18}{32} \times \dfrac{24}{11}$ $\dfrac{15}{2}$ **65.** $\dfrac{9}{100} \div \dfrac{13}{1000}$ $\dfrac{90}{13}$ **66.** $\dfrac{1000}{17} \div \dfrac{10}{3}$ $\dfrac{300}{17}$

67. Stephen's take-home pay is $4200 a month. If his rent is $\frac{1}{4}$ of his pay, how much is his rent?
$1050

68. Gus decides to save $\frac{1}{3}$ of his pay each month. If his monthly pay is $2112, how much will he save each month? $704

69. On a college basketball team, one-third of the team graduated with honors. If the team has 12 members, how many graduated with honors?
4 graduated with honors.

70. Shontell had only enough paper to print out $\frac{3}{5}$ of her book report before school. If the report is 10 pages long, how many pages did she print out?
6 pages

71. Natalie has 4 yd of material with which she can make holiday aprons. If it takes $\frac{1}{2}$ yd of material per apron, how many aprons can she make? 8 aprons

72. There are 4 cups of oatmeal in a box. If each serving is $\frac{1}{3}$ of a cup, how many servings are contained in the box? 12 servings

 73. Gail buys 6 lb of mixed nuts to be divided into decorative jars that will each hold $\frac{3}{4}$ lb of nuts. How many jars will she be able to fill? 8 jars

74. Troy has a $\frac{7}{8}$-in. nail that he must hammer into a board. Each strike of the hammer moves the nail $\frac{1}{16}$ in. into the board. How many strikes of the hammer must he make to drive the nail completely into the board? Troy must make 14 strikes.

Objective 6: Adding and Subtracting Fractions

For Exercises 75–78, add or subtract as indicated. **(See Example 7.)**

75. $\frac{5}{14} + \frac{1}{14}$ $\frac{3}{7}$

76. $\frac{9}{5} + \frac{1}{5}$ 2

77. $\frac{17}{24} - \frac{5}{24}$ $\frac{1}{2}$

78. $\frac{11}{18} - \frac{5}{18}$ $\frac{1}{3}$

For Exercises 79–82, find the least common multiple for each list of numbers. **(See Example 8.)**

79. 6, 15 30

80. 12, 30 60

81. 20, 8, 4 40

82. 24, 40, 30 120

For Exercises 83–98, add or subtract as indicated. **(See Examples 9–10.)**

83. $\frac{1}{8} + \frac{3}{4}$ $\frac{7}{8}$

84. $\frac{3}{16} + \frac{1}{2}$ $\frac{11}{16}$

85. $\frac{3}{8} - \frac{3}{10}$ $\frac{3}{40}$

86. $\frac{12}{35} - \frac{1}{10}$ $\frac{17}{70}$

87. $\frac{7}{26} - \frac{2}{13}$ $\frac{3}{26}$

88. $\frac{11}{24} - \frac{5}{16}$ $\frac{7}{48}$

89. $\frac{7}{18} + \frac{5}{12}$ $\frac{29}{36}$

90. $\frac{3}{16} + \frac{9}{20}$ $\frac{51}{80}$

91. $\frac{3}{4} - \frac{1}{20}$ $\frac{7}{10}$

92. $\frac{1}{6} - \frac{1}{24}$ $\frac{1}{8}$

93. $\frac{5}{12} + \frac{5}{16}$ $\frac{35}{48}$

94. $\frac{3}{25} + \frac{8}{35}$ $\frac{61}{175}$

95. $\frac{1}{6} + \frac{3}{4} - \frac{5}{8}$ $\frac{7}{24}$

96. $\frac{1}{2} + \frac{2}{3} - \frac{5}{12}$ $\frac{3}{4}$

 97. $\frac{4}{7} + \frac{1}{2} + \frac{3}{4}$ $\frac{51}{28}$ or $1\frac{23}{28}$

98. $\frac{9}{10} + \frac{4}{5} + \frac{3}{4}$ $\frac{49}{20}$ or $2\frac{9}{20}$

Objective 7: Operations on Mixed Numbers

For Exercises 99–116, perform the indicated operations. **(See Examples 11–12.)**

99. $4\frac{3}{5} \div \frac{1}{10}$ 46

100. $2\frac{4}{5} \div \frac{7}{11}$ $\frac{22}{5}$ or $4\frac{2}{5}$

101. $3\frac{1}{5} \times \frac{7}{8}$ $\frac{14}{5}$ or $2\frac{4}{5}$

102. $2\frac{1}{2} \times \frac{4}{5}$ 2

103. $3\frac{1}{5} \times 2\frac{7}{8}$ $\frac{46}{5}$ or $9\frac{1}{5}$

104. $2\frac{1}{2} \times 1\frac{4}{5}$ $\frac{9}{2}$ or $4\frac{1}{2}$

105. $1\frac{2}{9} \div 7\frac{1}{3}$ $\frac{1}{6}$

106. $2\frac{2}{5} \div 1\frac{2}{7}$ $\frac{28}{15}$ or $1\frac{13}{15}$

107. $1\frac{2}{9} \div 6$ $\frac{11}{54}$

108. $2\frac{2}{5} \div 2$ $\frac{6}{5}$ or $1\frac{1}{5}$

109. $2\frac{1}{8} + 1\frac{3}{8}$ $\frac{7}{2}$ or $3\frac{1}{2}$

110. $1\frac{3}{14} + 1\frac{1}{14}$ $\frac{16}{7}$ or $2\frac{2}{7}$

 Writing   _Translating Expression_   _Geometry_   Scientific Calculator   _Video_   NS&E

111. $3\frac{1}{2} - 1\frac{7}{8}$ $\frac{13}{8}$ or $1\frac{5}{8}$

112. $5\frac{1}{3} - 2\frac{3}{4}$ $\frac{31}{12}$ or $2\frac{7}{12}$

113. $1\frac{1}{6} + 3\frac{3}{4}$ $\frac{59}{12}$ or $4\frac{11}{12}$

114. $4\frac{1}{2} + 2\frac{2}{3}$ $\frac{43}{6}$ or $7\frac{1}{6}$

115. $1 - \frac{7}{8}$ $\frac{1}{8}$

116. $2 - \frac{3}{7}$ $\frac{11}{7}$ or $1\frac{4}{7}$

117. A board $26\frac{3}{8}$ in. long must be cut into three pieces of equal length. Find the length of each piece. $8\frac{19}{24}$ in.

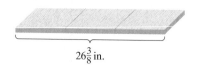

$26\frac{3}{8}$ in.

118. A futon, when set up as a sofa, measures $3\frac{5}{6}$ ft wide. When it is opened to be used as a bed, the width is increased by $1\frac{3}{4}$ ft. What is the total width of this bed? $5\frac{7}{12}$ ft

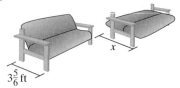

$3\frac{5}{6}$ ft

x

119. A plane trip from Orlando to Detroit takes $2\frac{3}{4}$ hr. If the plane traveled for $1\frac{1}{6}$ hr, how much time remains for the flight? $1\frac{7}{12}$ hr

120. Antonio bought $3\frac{3}{4}$ lb of smoked turkey for sandwiches. If he made 10 sandwiches, how much turkey did he put in each sandwich? $\frac{3}{8}$ lb

121. José ordered two seafood platters for a party. One platter has $1\frac{1}{2}$ lb of shrimp, and the other has $\frac{3}{4}$ lb of shrimp. How many pounds of shrimp does he have altogether? $2\frac{1}{4}$ lb

122. Ayako took a trip to the store $5\frac{1}{2}$ miles away. If she rode the bus for $4\frac{5}{6}$ miles and walked the rest of the way, how far did she have to walk? $\frac{2}{3}$ mile

123. If Tampa, Florida, averages $6\frac{1}{4}$ in. of rain during each summer month, how much total rain would be expected in June, July, August, and September? 25 in.

124. Pete started working out and found that he lost approximately $\frac{3}{4}$ in. off his waistline every month. How much would he lose around his waist in 6 months? $\frac{9}{2}$ in. or $4\frac{1}{2}$ in.

Section R.3   Decimals and Percents

Objectives

1. Introduction to a Place Value System
2. Converting Fractions to Decimals
3. Converting Decimals to Fractions
4. Converting Percents to Decimals and Fractions
5. Converting Decimals and Fractions to Percents
6. Applications of Percents

1. Introduction to a Place Value System

In a **place value** number system, each digit in a numeral has a particular value determined by its location in the numeral (Figure R-5).

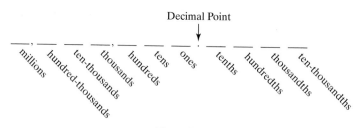

Decimal Point

millions, hundred-thousands, ten-thousands, thousands, hundreds, tens, ones . tenths, hundredths, thousandths, ten-thousandths

Figure R-5

For example, the number 197.215 represents

$$(1 \times 100) + (9 \times 10) + (7 \times 1) + \left(2 \times \frac{1}{10}\right) + \left(1 \times \frac{1}{100}\right) + \left(5 \times \frac{1}{1000}\right)$$

Each of the digits 1, 9, 7, 2, 1, and 5 is multiplied by 100, 10, 1, $\frac{1}{10}$, $\frac{1}{100}$, and $\frac{1}{1000}$, respectively, depending on its location in the numeral 197.215.

 Writing   Translating Expression   Geometry   Scientific Calculator   Video   NSE

By obtaining a common denominator and adding fractions, we have

$$197.215 = 100 + 90 + 7 + \frac{200}{1000} + \frac{10}{1000} + \frac{5}{1000}$$

$$= 197 + \frac{215}{1000} \quad \text{or} \quad 197\frac{215}{1000}$$

Because 197.215 is equal to the mixed number $197\frac{215}{1000}$, we read 197.215 as one hundred ninety-seven *and* two hundred fifteen thousandths. The decimal point is read as the word *and*.

If there are no digits to the right of the decimal point, we usually omit the decimal point. For example, the number 7125. is written simply as 7125 without a decimal point.

2. Converting Fractions to Decimals

In Section R.2, we learned that a fraction represents part of a whole unit. Likewise, the digits to the right of the decimal point represent a fraction of a whole unit. In this section, we will learn how to convert a fraction to a decimal number and vice versa.

> **PROCEDURE** Converting a Fraction to a Decimal
>
> To convert a fraction to a decimal, divide the numerator of the fraction by the denominator of the fraction.

Example 1 Converting Fractions to Decimals

Convert each fraction to a decimal. **a.** $\frac{7}{40}$ **b.** $\frac{2}{3}$

Solution:

a. $\frac{7}{40} = 0.175$

$$\begin{array}{r} 0.175 \\ 40\overline{)7.000} \\ \underline{40} \\ 300 \\ \underline{280} \\ 200 \\ \underline{200} \\ 0 \end{array}$$

The number 0.175 is said to be a **terminating decimal** because there are no nonzero digits to the right of the last digit, 5.

b. $\frac{2}{3} = 0.666\ldots$

$$\begin{array}{r} 0.666\ldots \\ 3\overline{)2.00000} \\ \underline{18} \\ 20 \\ \underline{18} \\ 20 \\ \underline{18} \\ 2\ldots \end{array}$$

The pattern 0.666 . . . continues indefinitely. Therefore, we say that this is a **repeating decimal**.

For a repeating decimal, a horizontal bar is often used to denote the repeating pattern after the decimal point. Hence, $\frac{2}{3} = 0.\overline{6}$.

Concept Connections

1. Give the place value of the underlined digit.
   42.6<u>7</u>5
2. Write the number form of "four hundred eighteen and seventy-two thousandths."

Skill Practice

Convert to a decimal.

3. $\frac{5}{8}$ **4.** $\frac{1}{6}$

Classroom Examples: p. 27, Exercises 14 and 16

Instructor Note: If a fraction in lowest terms has a denominator whose prime factorization includes *only* 2's and/or 5's, it will terminate. If it contains any other factors, it will repeat.

Answers

1. Hundredths   2. 418.072
3. 0.625   4. $0.1\overline{6}$

Sometimes it is useful to round a decimal number to a desired place value. This is demonstrated in Example 2.

Skill Practice

Round to the indicated place value.
5. 0.624 hundredths place
6. 1.6̄2̄ ten-thousandths place

Classroom Example: p. 27,
Exercise 20

Example 2 **Rounding Decimal Numbers**

Round the numbers to the indicated place value.

a. 0.175 to the hundredths place   **b.** 0.5̄4̄ to the thousandths place

Solution:

To round decimal numbers to a given place value, we actually look at the digit to the right of that position. If the digit to the right is 5 or greater, we round up. If the digit to the right is less than 5, we truncate the number beyond the given place value.

hundredths place
↓

a. 0.17$\boxed{5}$ The digit to the right of the hundredths place is 5 or greater.

≈ 0.18 Round up.

b. $0.\overline{54} = 0.545454\ldots$ This is a repeating decimal. We write out several digits.

thousandths place
↓

0.545$\boxed{4}$54 . . .   The digit to the right of the thousandths place is less than 5.

≈ 0.545

3. Converting Decimals to Fractions

For a terminating decimal, use the word name to write the number as a fraction or mixed number.

Skill Practice

Convert to a fraction.
7. 0.107
8. 11.25

Classroom Example: p. 27,
Exercise 30

Example 3 **Converting Terminating Decimals to Fractions**

Convert each decimal to a fraction.

a. 0.0023   **b.** 50.06

Solution:

a. 0.0023 is read as twenty-three ten-thousandths. Thus,

$$0.0023 = \frac{23}{10,000}$$

b. 50.06 is read as fifty and six hundredths. Thus,

$$50.06 = 50\frac{6}{100}$$

$$= 50\frac{3}{50}$$ Simplify the fraction to lowest terms.

$$= \frac{2503}{50}$$ Write the mixed number as a fraction.

Answers
5. 0.62   **6.** 1.6263
7. $\frac{107}{1000}$ **8.** $\frac{45}{4}$

Repeating decimals also can be written as fractions. However, the procedure to convert a repeating decimal to a fraction requires some knowledge of algebra. Table R-1 shows some common repeating decimals and an equivalent fraction for each.

Table R-1

$0.\overline{1} = \dfrac{1}{9}$	$0.\overline{4} = \dfrac{4}{9}$	$0.\overline{7} = \dfrac{7}{9}$
$0.\overline{2} = \dfrac{2}{9}$	$0.\overline{5} = \dfrac{5}{9}$	$0.\overline{8} = \dfrac{8}{9}$
$0.\overline{3} = \dfrac{3}{9} = \dfrac{1}{3}$	$0.\overline{6} = \dfrac{6}{9} = \dfrac{2}{3}$	$0.\overline{9} = \dfrac{9}{9} = 1$

4. Converting Percents to Decimals and Fractions

The concept of percent (%) is widely used in a variety of mathematical applications. The word *percent* means "per 100." Therefore, we can write percents as fractions.

$6\% = \dfrac{6}{100}$ A sales tax of 6% means that 6 cents in tax is charged for every 100 cents spent.

$91\% = \dfrac{91}{100}$ The fact that 91% of the population is right-handed means that 91 people out of 100 are right-handed.

The quantity $91\% = \dfrac{91}{100}$ can be written as $91 \times \dfrac{1}{100}$ or as 91×0.01.

Notice that the % symbol implies "division by 100" or, equivalently, "multiplication by $\frac{1}{100}$." Thus, we have the following rule to convert a percent to a fraction (or to a decimal).

PROCEDURE Converting a Percent to a Decimal or Fraction

Replace the % symbol by $\div 100$ (or equivalently $\times \dfrac{1}{100}$ or $\times 0.01$).

Example 4   Converting Percents to Decimals

Convert the percents to decimals.

a. 78%   **b.** 412%   **c.** 0.045%

Solution:

a. $78\% = 78 \times 0.01 = 0.78$

b. $412\% = 412 \times 0.01 = 4.12$

c. $0.045\% = 0.045 \times 0.01 = 0.00045$

TIP: Multiplying by 0.01 is equivalent to dividing by 100. This has the effect of moving the decimal point two places to the left.

Skill Practice

Convert the percent to a decimal.
9. 29%
10. 3.5%
11. 100%

Classroom Example: p. 27, Exercise 40

Answers
9. 0.29   **10.** 0.035   **11.** 1.00

Classroom Example: p. 28,
Exercise 44

Example 5   **Converting Percents to Fractions**

Convert the percents to fractions.

a. 52%   **b.** $33\frac{1}{3}$%   **c.** 6.5%

Solution:

a. $52\% = 52 \times \dfrac{1}{100}$ Replace the % symbol by $\frac{1}{100}$.

$= \dfrac{52}{100}$ Multiply.

$= \dfrac{13}{25}$ Simplify to lowest terms.

b. $33\frac{1}{3}\% = 33\frac{1}{3} \times \dfrac{1}{100}$ Replace the % symbol by $\frac{1}{100}$.

$= \dfrac{100}{3} \times \dfrac{1}{100}$ Write the mixed number as a fraction $33\frac{1}{3} = \frac{100}{3}$.

$= \dfrac{100}{300}$ Multiply the fractions.

$= \dfrac{1}{3}$ Simplify to lowest terms.

c. $6.5\% = 6.5 \times \dfrac{1}{100}$ Replace the % symbol by $\frac{1}{100}$.

$= \dfrac{65}{10} \times \dfrac{1}{100}$ Write 6.5 as an improper fraction.

$= \dfrac{65}{1000}$ Multiply the fractions.

$= \dfrac{13}{200}$ Simplify to lowest terms.

5. Converting Decimals and Fractions to Percents

To convert a percent to a decimal or fraction, we replace the % symbol by $\div$ 100. To convert a decimal or fraction to a percent, we reverse this process.

PROCEDURE   Converting Decimals and Fractions to Percents

Multiply the fraction or decimal by 100%.

Answers

12. $\dfrac{3}{10}$ **13.** $\dfrac{241}{200}$ **14.** $\dfrac{1}{40}$

Example 6 **Converting Decimals to Percents**

Convert the decimals to percents.

a. 0.92   **b.** 10.80   **c.** 0.005

Solution:

a. $0.92 = 0.92 \times 100\% = 92\%$ Multiply by 100%.

b. $10.80 = 10.80 \times 100\% = 1080\%$ Multiply by 100%.

c. $0.005 = 0.005 \times 100\% = 0.5\%$ Multiply by 100%.

Skill Practice

Convert the decimal to a percent.

15. 0.56

16. 4.36

17. 0.002

Classroom Examples: p. 28, Exercises 52 and 56

Example 7 **Converting Fractions to Percents**

Convert the fractions to percents.

a. $\dfrac{2}{5}$ **b.** $\dfrac{5}{3}$

Solution:

a. $\dfrac{2}{5} = \dfrac{2}{5} \times 100\%$ Multiply by 100%.

$= \dfrac{2}{5} \times \dfrac{100}{1}\%$ Write the whole number as a fraction.

$= \dfrac{2}{\cancel{5}_{1}} \times \dfrac{\overset{20}{\cancel{100}}}{1}\%$ Multiply fractions.

$= \dfrac{40}{1}\%$ or 40% Simplify.

b. $\dfrac{5}{3} = \dfrac{5}{3} \times 100\%$ Multiply by 100%.

$= \dfrac{5}{3} \times \dfrac{100}{1}\%$ Write the whole number as a fraction.

$= \dfrac{500}{3}\%$ Multiply fractions.

$= \dfrac{500}{3}\%$ or $166.\overline{6}\%$ The value $\frac{500}{3}$ can be written in decimal form by dividing 500 by 3.

Skill Practice

Convert the fraction to a percent.

18. $\dfrac{7}{8}$ **19.** $\dfrac{5}{6}$

Classroom Examples: p. 28, Exercises 64 and 72

6. Applications of Percents

Many applications involving percents involve finding a percent of some base number. For example, suppose a textbook is discounted 25%. If the book originally cost $60, find the amount of the discount.

In this example, we must find 25% of $60. In this context, the word *of* means multiply.

$$\underset{\underset{0.25}{\downarrow}}{25\%} \; \underset{\underset{\times}{\downarrow}}{\text{of}} \; \underset{\underset{60}{\downarrow}}{\$60} = 15 \qquad \text{The amount of the discount is \$15.}$$

Answers

15. 56%   **16.** 436%   **17.** 0.2%

18. 87.5%   **19.** $83.\overline{3}\%$ or $83\frac{1}{3}\%$

Note that the *decimal form* of a percent is always used in calculations. Therefore, 25% was converted to 0.25 *before* multiplying by $60.

Skill Practice

20. The sales tax rate for a certain county is 6%. Find the amount of sales tax on a $52.00 fishing pole.

Classroom Example: p. 28, Exercise 76

Example 8 Applying Percentages

Shauna received a raise, so now her new salary is 105% of her old salary. Find Shauna's new salary if her old salary was $36,000 per year.

Solution:

The new salary is 105% of $36,000.

$$1.05 \times 36,000 = 37,800$$

The new salary is $37,800 per year.

In some applications, it is necessary to convert a fractional part of a whole to a percent of the whole.

Skill Practice

21. Eduardo answered 66 questions correctly on a test with 75 questions. What percent of the questions does 66 represent?

Classroom Example: p. 28, Exercise 80

Example 9 Finding a Percentage

Union College in Schenectady, New York, accepts approximately 520 students each year from 3500 applicants. What percent does 520 represent? Round to the nearest tenth of a percent.

Solution:

$$\frac{520}{3500} \approx 0.149$$

Convert the fractional part of the total number of applicants to decimal form. (*Note:* Rounding the decimal form of the quotient to the thousandths place gives us the nearest tenth of a percent.)

$$= 0.149 \times 100\%$$ Convert the decimal to a percent.

$$= 14.9\%$$ Simplify.

Approximately 14.9% of the applicants to Union College are accepted.

Answers
20. $3.12   **21.** 88%

Calculator Connections

Topic: Approximating Repeating Decimals on a Calculator

Calculators can display only a limited number of digits on the calculator screen. Therefore, repeating decimals and terminating decimals with a large number of digits will be truncated or rounded to fit the calculator display. For example, the fraction $\frac{2}{3} = 0.\overline{6}$ may be entered into the calculator as [2] [÷] [3]. The result may appear as 0.6666666667 or as 0.6666666666. The fraction $\frac{2}{11}$ equals the repeating decimal $0.\overline{18}$. However, the calculator converts $\frac{2}{11}$ to the terminating decimal 0.1818181818.

```
2/3
       .6666666667
2/11
       .1818181818
```

Calculator Exercises

Without using a calculator, find a repeating decimal to represent each of the following fractions. Then use a calculator to confirm your answer.

1. $\frac{4}{9}$ $0.\overline{4}$ **2.** $\frac{7}{11}$ $0.\overline{63}$ **3.** $\frac{3}{22}$ $0.1\overline{36}$ **4.** $\frac{5}{13}$ $0.\overline{384615}$

Section R.3 Practice Exercises

Boost *your* GRADE at ALEKS.com!

ALEKS version 3.0

• Practice Problems
• Self-Tests
• NetTutor

• e-Professors
• Videos

For additional exercises, see Classroom Activities R.3A–R.3D in the *Instructor's Resource Manual* at www.mhhe.com/moh.

Objective 1: Introduction to a Place Value System

For Exercises 1–8, write the name of the place value for the underlined digit.

1. 481.24
Tens

2. 1345.42
Tens

3. 2912.032
Hundreds

4. 4208.03
Hundreds

5. 2.381
Tenths

6. 8.249
Tenths

7. 21.413
Hundredths

8. 82.794
Hundredths

9. The first 10 Roman numerals are: I, II, III, IV, V, VI, VII, VIII, IX, X. Is this numbering system a place value system? Explain your answer. No, the symbols I, V, X, and so on each represent certain values but the values are not dependent on the position of the symbol within the number.

10. Write the number in decimal form. $3(100) + 7(10) + 6 + \dfrac{1}{100} + \dfrac{5}{1000}$ 376.015

Objective 2: Converting Fractions to Decimals

For Exercises 11–18, convert each fraction to a terminating decimal or a repeating decimal. **(See Example 1.)**

11. $\dfrac{7}{10}$
0.7

12. $\dfrac{9}{10}$
0.9

13. $\dfrac{9}{25}$
0.36

14. $\dfrac{3}{25}$
0.12

15. $\dfrac{11}{9}$
$1.\overline{2}$

16. $\dfrac{16}{9}$
$1.\overline{7}$

17. $\dfrac{7}{33}$
$0.\overline{21}$

18. $\dfrac{2}{11}$
$0.\overline{18}$

For Exercises 19–26, round the decimal to the given place value. **(See Example 2.)**

19. 214.059; tenths 214.1

20. 1004.165; hundredths 1004.17

21. 39.26849; thousandths 39.268

22. 0.059499; thousandths 0.059

23. 39,918.2; thousands 40,000

24. 599,621.5; thousands 600,000

25. $0.7\overline{2}$; hundredths 0.73

26. $0.3\overline{4}$; thousandths 0.343

Objective 3: Converting Decimals to Fractions

For Exercises 27–38, convert each decimal to a fraction or a mixed number. **(See Example 3.)**

27. 0.45 $\dfrac{9}{20}$

28. 0.65 $\dfrac{13}{20}$

29. 0.181 $\dfrac{181}{1000}$

30. 0.273 $\dfrac{273}{1000}$

31. 2.04 $\dfrac{51}{25}$ or $2\dfrac{1}{25}$

32. 6.02 $\dfrac{301}{50}$ or $6\dfrac{1}{50}$

33. 13.007 $\dfrac{13,007}{1000}$ or $13\dfrac{7}{1000}$

34. 12.003 $\dfrac{12,003}{1000}$ or $12\dfrac{3}{1000}$

35. $0.\overline{5}$ (*Hint*: Refer to Table R-1)
$\dfrac{5}{9}$

36. $0.\overline{8}$
$\dfrac{8}{9}$

37. $1.\overline{1}$
$\dfrac{10}{9}$ or $1\dfrac{1}{9}$

38. $2.\overline{3}$
$\dfrac{7}{3}$ or $2\dfrac{1}{3}$

Objective 4: Converting Percents to Decimals and Fractions

For Exercises 39–48, convert each percent to a decimal and to a fraction. **(See Examples 4–5.)**

39. The sale price is 30% off of the original price. 0.3, $\dfrac{3}{10}$

40. An HMO (health maintenance organization) pays 80% of all doctors' bills. 0.8, $\dfrac{4}{5}$

Writing   Translating Expression   Geometry   Scientific Calculator   Video   NS

41. The building will be 75% complete by spring. $0.75, \frac{3}{4}$

42. Chan plants roses in 25% of his garden. $0.25, \frac{1}{4}$

43. The bank pays $3\frac{3}{4}$% interest on a checking account. $0.0375, \frac{3}{80}$

44. A credit union pays $4\frac{1}{2}$% interest on a savings account. $0.045, \frac{9}{200}$

45. Kansas received 15.7% of its annual rainfall in 1 week. $0.157, \frac{157}{1000}$

46. Social Security withholds 6.2% of an employee's gross pay. $0.062, \frac{31}{500}$

47. The world population in 2008 was 270% of the world population in 1950. $2.7, \frac{27}{10}$

48. The cost of a home is 140% of its cost 10 years ago. $1.40, \frac{7}{5}$

Objective 5: Converting Decimals and Fractions to Percents

49. Explain how to convert a decimal to a percent.
Multiply by 100%.

50. Explain how to convert a percent to a decimal.
Replace the % symbol by ÷ 100, by × $\frac{1}{100}$ or by × 0.01.

For Exercises 51–62, convert the decimal to a percent. (See Example 6.)

51. 0.05 5%   **52.** 0.06 6%   **53.** 0.90 90%   **54.** 0.70 70%

55. 1.2 120%   **56.** 4.8 480%   **57.** 7.5 750%   **58.** 9.3 930%

59. 0.135 13.5% **60.** 0.536 53.6% **61.** 0.003 0.3%   **62.** 0.002 0.2%

For Exercises 63–74, convert the fraction to a percent. (See Example 7.)

63. $\frac{3}{50}$ 6%   **64.** $\frac{23}{50}$ 46%   **65.** $\frac{9}{2}$ 450%   **66.** $\frac{7}{4}$ 175%

67. $\frac{5}{8}$ 62.5%   **68.** $\frac{1}{8}$ 12.5%   **69.** $\frac{5}{16}$ 31.25%   **70.** $\frac{7}{16}$ 43.75%

71. $\frac{5}{6}$ $83.\overline{3}$%   **72.** $\frac{4}{15}$ $26.\overline{6}$%   **73.** $\frac{14}{15}$ $93.\overline{3}$%   **74.** $\frac{5}{18}$ $27.\overline{7}$%

Objective 6: Applications of Percents

75. A suit that costs $140 is discounted by 30%. How much is the discount? (See Example 8.) $42

76. Louise completed 40% of her task that takes a total of 60 hr to finish. How many hours did she complete? 24 hr

77. Tom's federal taxes amount to 27% of his income. If Tom earns $12,500 per quarter, how much will he pay in taxes for that quarter? $3375

78. A tip of $7 is left for a meal that costs $56. What percent of the cost does the tip represent? 12.5%

79. Jamie paid $5.95 in sales tax on a textbook that costs $85. Find the percent of the sales tax. (See Example 9.) 7%

80. Sue saves $37.50 each week out of her paycheck of $625. What percent of her paycheck does her savings represent? 6%

 Writing   Translating Expression   Geometry   Scientific Calculator   Video   NSE

For Exercises 81–84, refer to the graph. The pie graph shows a family budget based on a net income of $2400 per month.

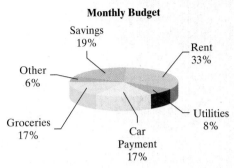

Monthly Budget

81. Determine the amount spent on rent. $792

82. Determine the amount spent on car payments. $408

83. Determine the amount spent on utilities. $192

84. How much more money is spent than saved? $1488

 85. By the end of the year, Felipe will have 75% of his mortgage paid. If the mortgage was originally for $90,000, how much will have been paid at the end of the year? $67,500

 86. A certificate of deposit (CD) earns 4% interest in 1 year. If Mr. Patel has $12,000 invested in the CD, how much interest will he receive at the end of the year? $480

Introduction to Geometry

Section R.4

1. Perimeter

In this section, we present several facts and formulas that may be used throughout the text in applications of geometry. One of the most important uses of geometry involves the measurement of objects of various shapes. We begin with an introduction to perimeter, area, and volume for several common shapes and objects.

Objectives

1. Perimeter
2. Area
3. Volume
4. Angles
5. Triangles

   Perimeter is defined as the distance around a figure. If we were to put up a fence around a field, the perimeter would determine the amount of fencing. For example, in Figure R-6 the distance around the field is 300 ft. For a polygon (a closed figure constructed from line segments), the perimeter is the sum of the lengths of the sides. For a circle, the distance around the outside is called the **circumference**.

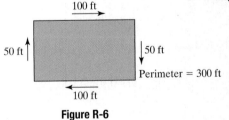

Figure R-6

Rectangle	Square	Triangle	Circle

$$P = 2\ell + 2w \qquad P = 4s \qquad P = a + b + c \qquad \text{Circumference: } C = 2\pi r$$

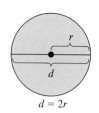

$d = 2r$

For a circle, r represents the length of a radius—the distance from the center to any point on the circle. The length of a diameter, d, of a circle is twice that of a radius. Thus, $d = 2r$. The number π is a constant equal to the circumference of a circle divided by the length of a diameter. That is, $\pi = \frac{C}{d}$. The value of π is often approximated by 3.14 or $\frac{22}{7}$.

Skill Practice

1. Find the perimeter of the square.

7.25 in.

2. Find the circumference. Use 3.14 for π.

2 in.

Classroom Examples: p. 38, Exercises 4 and 10

TIP: If a calculator is used to find the circumference of a circle, use the π key to get a more accurate answer.

Example 1 **Finding Perimeter and Circumference**

Find the perimeter or circumference as indicated. Use 3.14 for π.

a. Perimeter of the rectangle

3.1 ft
5.5 ft

b. Circumference of the circle

6 cm

Solution:

a. $P = 2\ell + 2w$

$\quad = 2(5.5 \text{ ft}) + 2(3.1 \text{ ft})$ Substitute $\ell = 5.5$ ft and $w = 3.1$ ft.

$\quad = 11 \text{ ft} + 6.2 \text{ ft}$

$\quad = 17.2 \text{ ft}$ The perimeter is 17.2 ft.

b. $C = 2\pi r$

$\quad = 2(3.14)(6 \text{ cm})$ Substitute 3.14 for π and $r = 6$ cm.

$\quad = 6.28(6 \text{ cm})$

$\quad = 37.68 \text{ cm}$ The circumference is 37.68 cm.

---

2. Area

The **area** of a geometric figure is the number of square units that can be enclosed within the figure. In applications, we would find the area of a region if we were laying carpet or putting down sod for a lawn. For example, the rectangle shown in Figure R-7 encloses 6 square inches (6 in.²).

The formulas used to compute the area for several common geometric shapes are given here:

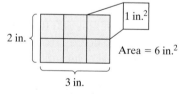

2 in.
3 in.
1 in.²
Area = 6 in.²

Figure R-7

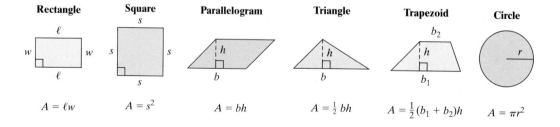

Rectangle	Square	Parallelogram	Triangle	Trapezoid	Circle
$A = \ell w$	$A = s^2$	$A = bh$	$A = \frac{1}{2}bh$	$A = \frac{1}{2}(b_1 + b_2)h$	$A = \pi r^2$

Answers

1. 29 in.   **2.** 12.56 in.

Example 2	Finding Area

Find the area.

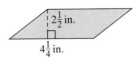

Solution:

$A = bh$ The figure is a parallelogram.

$= (4\frac{1}{4} \text{ in.})(2\frac{1}{2} \text{ in.})$ Substitute $b = 4\frac{1}{4}$ in. and $h = 2\frac{1}{2}$ in.

$= \left(\frac{17}{4} \text{ in.}\right)\left(\frac{5}{2} \text{ in.}\right)$

$= \frac{85}{8} \text{ in.}^2 \text{ or } 10\frac{5}{8} \text{ in.}^2$

TIP: Notice that the units of area are given in square units such as square inches (in.^{2}), square feet (ft^2), square yards (yd^2), square centimeters (cm^2), and so on.

Example 3	Finding Area

Find the area.

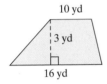

10 yd

3 yd

16 yd

Solution:

$A = \frac{1}{2}(b_1 + b_2)h$ The figure is a trapezoid.

$= \frac{1}{2}(16 \text{ yd} + 10 \text{ yd})(3 \text{ yd})$ Substitute $b_1 = 16$ yd, $b_2 = 10$ yd, and $h = 3$ yd.

$= \frac{1}{2}(26 \text{ yd})(3 \text{ yd})$

$= (13 \text{ yd})(3 \text{ yd})$

$= 39 \text{ yd}^2$ The area is 39 yd^2.

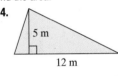
TIP: Notice that several of the formulas presented thus far involve multiple operations. The order in which we perform the arithmetic is called the **order of operations** and is covered in detail in Section 1.2. We will follow these guidelines in the order given below:

1. Perform operations within parentheses first.
2. Evaluate expressions with exponents.
3. Perform multiplication or division in order from left to right.
4. Perform addition or subtraction in order from left to right.

Answers

3. $\frac{9}{16}$ cm^2 **4.** 30 m^2

Skill Practice

5. Find the area of the circular region. Use 3.14 for π.

10 in.

Classroom Example: p. 38, Exercise 22

Example 4   **Finding Area of a Circle**

Find the area of a circular fountain if the radius is 25 ft. Use 3.14 for π.

25 ft

Solution:

$$A = \pi r^2$$

$$\approx (3.14)(25 \text{ ft})^2$$

$$= (3.14)(625 \text{ ft}^2) \qquad \text{Substitute 3.14 for } \pi \text{ and } r = 25 \text{ ft.}$$

$$= 1962.5 \text{ ft}^2 \qquad \text{The area of the fountain is } 1962.5 \text{ ft}^2.$$

3. Volume

The **volume** of a solid is the number of cubic units that can be enclosed within a solid. The solid shown in Figure R-8 contains 18 cubic inches (18 in.³). In applications, volume might refer to the amount of water in a swimming pool.

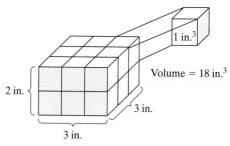

2 in.

3 in.

3 in.

1 in.³

Volume = 18 in.³

Figure R-8

The formulas used to compute the volume of several common solids are given here:

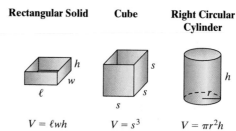

Rectangular Solid	Cube	Right Circular Cylinder
h, w, ℓ	s, s, s	h, r
$V = \ell wh$	$V = s^3$	$V = \pi r^2 h$

TIP: Notice that the volume formulas for the three figures just shown are given by the product of the area of the base and the height of the figure:

$V = \ell wh$	$V = s \cdot s \cdot s$	$V = \pi r^2 h$
↑	↑	↑
Area of Rectangular Base	Area of Square Base	Area of Circular Base

Right Circular Cone   **Sphere**

$V = \frac{1}{3}\pi r^2 h$ $V = \frac{4}{3}\pi r^3$

Answer

5. 314 in.²

Example 5   **Finding Volume**

Find the volume.

Solution:

$V = s^3$ The object is a cube.

$= (1\tfrac{1}{2}\text{ ft})^3$ Substitute $s = 1\tfrac{1}{2}$ ft.

$= \left(\dfrac{3}{2}\text{ ft}\right)^3$

$= \left(\dfrac{3}{2}\text{ ft}\right)\left(\dfrac{3}{2}\text{ ft}\right)\left(\dfrac{3}{2}\text{ ft}\right)$

$= \dfrac{27}{8}\text{ ft}^3$, or $3\tfrac{3}{8}\text{ ft}^3$

TIP: Notice that the units of volume are cubic units such as cubic inches (in.³), cubic feet (ft³), cubic yards (yd³), cubic centimeters (cm³), and so on.

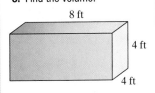

Example 6   **Finding Volume**

Find the volume. Round to the nearest whole unit.

Solution:

$V = \dfrac{1}{3}\pi r^2 h$ The object is a right circular cone.

$\approx \dfrac{1}{3}(3.14)(4\text{ cm})^2(12\text{ cm})$ Substitute 3.14 for π, $r = 4$ cm, and $h = 12$ cm.

$= \dfrac{1}{3}(3.14)(16\text{ cm}^2)(12\text{ cm})$

$= 200.96\text{ cm}^3$

$\approx 201\text{ cm}^3$ Round to the nearest whole unit.

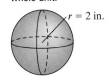

Example 7   **Finding Volume in an Application**

An underground gas tank is in the shape of a right circular cylinder.

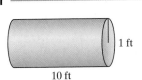

a. Find the volume of the tank. Use 3.14 for π.

b. Find the cost to fill the tank with gasoline if gasoline costs $35/ft³.

Classroom Example: p. 40, Exercise 52

Solution:

a. $V = \pi r^2 h$

$\approx (3.14)(1 \text{ ft})^2(10 \text{ ft})$ Substitute 3.14 for π, $r = 1$ ft, and $h = 10$ ft.

$= (3.14)(1 \text{ ft}^2)(10 \text{ ft})$

$= 31.4 \text{ ft}^3$ The tank holds 31.4 ft³ of gasoline.

b. $\text{Cost} = (35/\text{ft}^3)(31.4 \text{ ft}^3)$

$= \$1099$ It will cost \$1099 to fill the tank.

4. Angles

Applications involving angles and their measure come up often in the study of algebra, trigonometry, calculus, and applied sciences. The most common unit to measure an angle is the degree (°). Several angles and their corresponding degree measure are shown in Figure R-9.

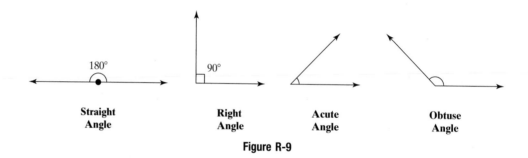

Figure R-9

- An angle that measures 90° is a **right angle** (right angles are often marked with a square or corner symbol, □).
- An angle that measures 180° is called a **straight angle**.
- An angle that measures between 0° and 90° is called an **acute angle**.
- An angle that measures between 90° and 180° is called an **obtuse angle**.
- Two angles with the same measure are **equal angles** (or **congruent angles**).

The measure of an angle will be denoted by the symbol m written in front of the angle. Therefore, the measure of $\angle A$ is denoted $m(\angle A)$.

- Two angles are **complementary** if their sum is 90°.
- Two angles are **supplementary** if their sum is 180°.

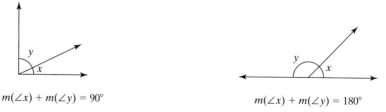

$m(\angle x) + m(\angle y) = 90°$ $m(\angle x) + m(\angle y) = 180°$

Answers

8. 339 cm³   **9.** 2034 cm³

10. For example, 70° and 20°.

11. For example, 80° and 100°.

When two lines intersect, four angles are formed (Figure R-10). In Figure R-10, $\angle a$ and $\angle b$ are a pair of **vertical angles**. Another set of vertical angles is the pair $\angle c$ and $\angle d$. An important property of vertical angles is that the measures of two vertical angles are *equal*. In the figure, $m(\angle a) = m(\angle b)$ and $m(\angle c) = m(\angle d)$.

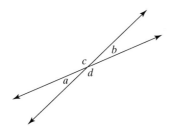

Figure R-10

Parallel lines are lines that lie in the same plane and do not intersect. In Figure R-11, the lines L_1 and L_2 are parallel lines. If a line intersects two parallel lines, the line forms eight angles with the parallel lines.

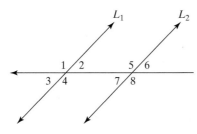

Figure R-11

The measures of angles 1–8 in Figure R-11 have the following special properties.

L_1 and L_2 are Parallel	Name of Angles	Property
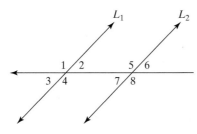	The following pairs of angles are called **alternate interior angles:** $\angle 2$ and $\angle 7$ $\angle 4$ and $\angle 5$	Alternate interior angles are equal in measure. $m(\angle 2) = m(\angle 7)$ $m(\angle 4) = m(\angle 5)$
	The following pairs of angles are called **alternate exterior angles:** $\angle 1$ and $\angle 8$ $\angle 3$ and $\angle 6$	Alternate exterior angles are equal in measure. $m(\angle 1) = m(\angle 8)$ $m(\angle 3) = m(\angle 6)$
	The following pairs of angles are called **corresponding angles:** $\angle 1$ and $\angle 5$ $\angle 2$ and $\angle 6$ $\angle 3$ and $\angle 7$ $\angle 4$ and $\angle 8$	Corresponding angles are equal in measure. $m(\angle 1) = m(\angle 5)$ $m(\angle 2) = m(\angle 6)$ $m(\angle 3) = m(\angle 7)$ $m(\angle 4) = m(\angle 8)$

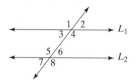
Example 8 **Finding Unknown Angles in a Diagram**

Find the measure of each angle and explain how the angle is related to the given angle of 70°.

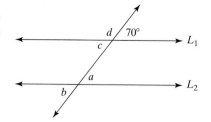

   a. $\angle a$ **b.** $\angle b$

   c. $\angle c$ **d.** $\angle d$

Solution:

a. $m(\angle a) = 70°$ $\angle a$ is a corresponding angle to the given angle of 70°.

b. $m(\angle b) = 70°$ $\angle b$ and the given angle of 70° are alternate exterior angles.

c. $m(\angle c) = 70°$ $\angle c$ and the given angle of 70° are vertical angles.

d. $m(\angle d) = 110°$ $\angle d$ is the supplement of the given angle of 70°.

5. Triangles

Triangles are categorized by the measures of the angles (Figure R-12) and by the number of equal sides or angles (Figure R-13).

- An **acute triangle** is a triangle in which all three angles are acute.
- A **right triangle** is a triangle in which one angle is a right angle.
- An **obtuse triangle** is a triangle in which one angle is obtuse.

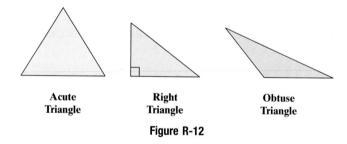

Acute Triangle   Right Triangle   Obtuse Triangle

Figure R-12

- An **equilateral triangle** is a triangle in which all three sides (and all three angles) are equal in measure.
- An **isosceles triangle** is a triangle in which two sides are equal in measure (the angles opposite the equal sides are also equal in measure).
- A **scalene triangle** is a triangle in which no sides (or angles) are equal in measure.

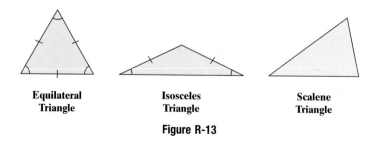

Equilateral Triangle   Isosceles Triangle   Scalene Triangle

Figure R-13

The following important property is true for all triangles.

PROPERTY Sum of the Angles in a Triangle
The sum of the measures of the angles of a triangle is 180°.

Example 9 Finding Unknown Angles in a Diagram

Find the measure of each angle in the figure.

a. $\angle a$

b. $\angle b$

c. $\angle c$

d. $\angle d$

e. $\angle e$

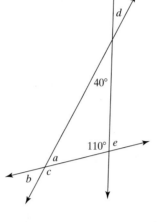

Skill Practice
For Exercises 13–17, refer to the figure. Find the measure of the indicated angle.

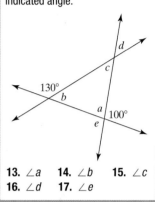

13. $\angle a$ **14.** $\angle b$ **15.** $\angle c$
16. $\angle d$ **17.** $\angle e$

Classroom Example: p. 42, Exercise 98

Solution:

a. $m(\angle a) = 30°$ The sum of the angles in a triangle is 180°.

b. $m(\angle b) = 30°$ $\angle a$ and $\angle b$ are vertical angles and are equal.

c. $m(\angle c) = 150°$ $\angle c$ and $\angle a$ are supplementary angles ($\angle c$ and $\angle b$ are also supplementary).

d. $m(\angle d) = 40°$ $\angle d$ and the given angle of 40° are vertical angles.

e. $m(\angle e) = 70°$ $\angle e$ and the given angle of 110° are supplementary angles.

Answers
13. 80° **14.** 50° **15.** 50°
16. 50° **17.** 100°

Section R.4 Practice Exercises

Objective 1: Perimeter

1. Identify which of the following units could be measures of perimeter. b, e, i

   a. Square inches (in.²)   **b.** Meters (m)   **c.** Cubic feet (ft³)

   d. Cubic meters (m³)   **e.** Miles (mi)   **f.** Square centimeters (cm²)

   g. Square yards (yd²)   **h.** Cubic inches (in.³)   **i.** Kilometers (km)

2. Identify which of the following units could be measures of circumference. b, e, i

   a. Square inches (in.²)   **b.** Meters (m)   **c.** Cubic feet (ft³)

   d. Cubic meters (m³)   **e.** Miles (mi)   **f.** Square centimeters (cm²)

   g. Square yards (yd²)   **h.** Cubic inches (in.³)   **i.** Kilometers (km)

 Writing   Translating Expression   Geometry Scientific Calculator   Video NS E

For Exercises 3–10, find the perimeter or circumference of each figure. Use 3.14 for π. **(See Example 1.)**

3.
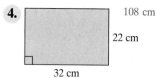
32 m
6 m
10 m

4.
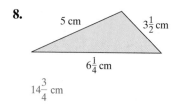
108 cm
22 cm
32 cm

5.
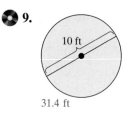
17.2 mi
4.3 mi
4.3 mi

6.
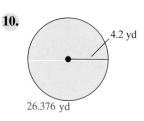
1 ft
0.25 ft
0.25 ft

7.
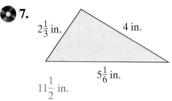
$2\frac{1}{3}$ in.
4 in.
$5\frac{1}{6}$ in.
$11\frac{1}{2}$ in.

8.
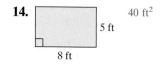
5 cm
$3\frac{1}{2}$ cm
$6\frac{1}{4}$ cm
$14\frac{3}{4}$ cm

9.
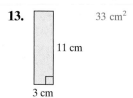
10 ft
31.4 ft

10.
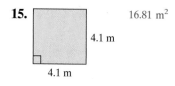
4.2 yd
26.376 yd

Objective 2: Area

11. Identify which of the following units could be measures of area.   a, f, g

    a. Square inches (in.^{2})   **b.** Meters (m)   **c.** Cubic feet (ft^3)

    d. Cubic meters (m^3)   **e.** Miles (mi)   **f.** Square centimeters (cm^2)

    g. Square yards (yd^2)   **h.** Cubic inches (in.^{3})   **i.** Kilometers (km)

12. Would you measure area or perimeter to determine the amount of carpeting needed for a room?   Area

For Exercises 13–26, find the area. Use 3.14 for π. **(See Examples 2–4.)**

13.
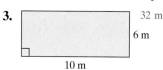
33 cm^2
11 cm
3 cm

14.
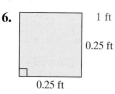
40 ft^2
5 ft
8 ft

15.
16.81 m^2
4.1 m
4.1 m

16.
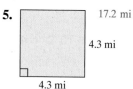
37.21 in.^{2}
6.1 in.
6.1 in.

17.
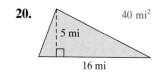
84 in.^{2}
6 in.
14 in.

18.
0.0004 m^2
0.01 m
0.04 m

19.
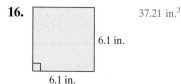
10.12 km^2
2.3 km
8.8 km

20.
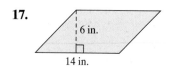
40 mi^2
5 mi
16 mi

21.
13.8474 ft^2
4.2 ft

22.
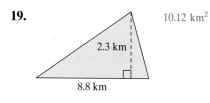
132.665 cm^2
6.5 cm

23.
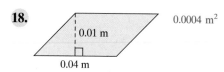
8 in.
66 in.^{2}
6 in.
14 in.

24.
66 in.^{2}
6 in.
14 in.
8 in.

25.
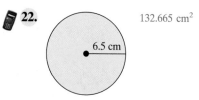
9 ft
31.5 ft^2
7 ft

26.
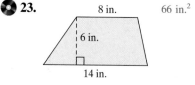
4 km
3 km
6 km^2

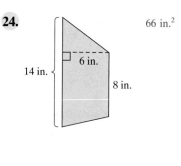
Writing   Translating Expression   Geometry   Scientific Calculator   Video   NS&E

Objective 3: Volume

27. Identify which of the following units could be measures of volume.   c, d, h

    a. Square inches (in.^{2})   **b.** Meters (m)   **c.** Cubic feet (ft^3)

    d. Cubic meters (m^3)   **e.** Miles (mi)   **f.** Square centimeters (cm^2)

    g. Square yards (yd^2)   **h.** Cubic inches (in.^{3})   **i.** Kilometers (km)

28. Would you measure perimeter, area, or volume to determine the amount of water needed to fill a swimming pool?   Volume

For Exercises 29–36, find the volume of each figure. Use 3.14 for π. **(See Examples 5–7.)**

29. 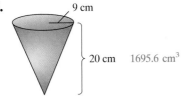 3.5 cm   307.72 cm^3 8 cm

30. 2 ft   75.36 ft^3 6 ft

31. 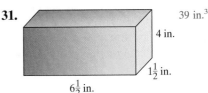 39 in.^{3} 4 in. $1\frac{1}{2}$ in. $6\frac{1}{2}$ in.

32. 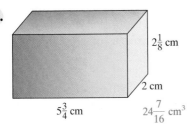 $2\frac{1}{8}$ cm   2 cm $5\frac{3}{4}$ cm $24\frac{7}{16}$ cm^3

33. 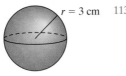 $r = 3$ cm   113.04 cm^3

34. 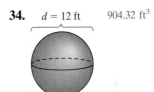 $d = 12$ ft   904.32 ft^3

35. 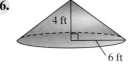 9 cm   20 cm   1695.6 cm^3

36. 150.72 ft^3 4 ft   6 ft

37. A florist sells balloons and needs to know how much helium to order. Each balloon is approximately spherical with a radius of 9 in. How much helium is needed to fill one balloon?
3052.08 in.^{3}

38. Find the volume of a spherical ball whose radius is 3 in. Use 3.14 for π.   113.04 in.^{3}

39. Find the volume of a snow cone in the shape of a right circular cone whose radius is 3 cm and whose height is 12 cm. Use 3.14 for π.
113.04 cm^3

40. A landscaping supply company has a pile of gravel in the shape of a right circular cone whose radius is 10 yd and whose height is 18 yd. Find the volume of the gravel. Use 3.14 for π.
1884 yd^3

Mixed Exercises: Perimeter, Area, and Volume

41. A wall measuring 20 ft by 8 ft can be painted for $40.

    a. What is the price per square foot? $0.25/ft^2

    b. At this rate, how much would it cost to paint the remaining three walls that measure 20 ft by 8 ft, 16 ft by 8 ft, and 16 ft by 8 ft? $104

42. Suppose it costs $336 to carpet a 16 ft by 12 ft room.

    a. What is the price per square foot? $1.75/ft^2

    b. At this rate, how much would it cost to carpet a room that is 20 ft by 32 ft? $1120

43. If you were to purchase fencing for a garden, would you measure the perimeter or area of the garden?   Perimeter

44. If you were to purchase sod (grass) for your front yard, would you measure the perimeter or area of the yard?   Area

45. How much fence is needed to enclose a triangularly shaped garden whose sides measure 12 ft, 22 ft, and 20 ft?   54 ft

46. A regulation soccer field is 100 yd long by 60 yd wide. Find the perimeter of the field.   320 yd

47. a. An American football field is 360 ft long by 160 ft wide. What is the area of the field?   57,600 ft^2

    b. How many pieces of sod, each 1 ft wide and 3 ft long, are needed to sod an entire field? (*Hint:* First find the area of a piece of sod.)   19,200 pieces

48. The Transamerica tower in San Francisco is a pyramid with triangular sides (excluding the "wings"). Each side measures 145 ft wide with a height of 853 ft. What is the area of each side?   61,842.5 ft^2

49. a. Find the area of a circular pizza that is 8 in. in diameter (the radius is 4 in.). Use 3.14 for π.   50.24 in.^{2}

    b. Find the area of a circular pizza that is 12 in. in diameter (the radius is 6 in.).   113.04 in.^{2}

    c. Assume that the 8-in. diameter and 12-in. diameter pizzas are both the same thickness. Which would provide more pizza, two 8-in. pizzas or one 12-in. pizza?   One 12-in. pizza

50. a. Find the area of a rectangular pizza that is 12 in. by 8 in.   96 in.^{2}

    b. Find the area of a circular pizza that has a 16-in. diameter. Use 3.14 for π.   200.96 in.^{2}

    c. Assume that the two pizzas have the same thickness. Which would provide more pizza? Two rectangular pizzas or one circular pizza?   One circular pizza

51. Find the volume of a soup can in the shape of a right circular cylinder if its radius is 3.2 cm and its height is 9 cm. Use 3.14 for π.   289.3824 cm^3

52. Find the volume of a coffee mug whose radius is 2.5 in. and whose height is 6 in. Use 3.14 for π.   117.75 in.^{3}

Objective 4: Angles

For Exercises 53–58, answer True or False. If an answer is false, explain why.

53. The sum of the measures of two right angles equals the measure of a straight angle.   True

54. Two right angles are complementary.   False; they are supplementary.

55. Two right angles are supplementary.   True

56. Two acute angles cannot be supplementary.   True

57. Two obtuse angles cannot be supplementary.   True

58. An obtuse angle and an acute angle can be supplementary.   True

59. If possible, find two acute angles that are supplementary.   Not possible

60. If possible, find two acute angles that are complementary. Answers may vary.   For example: 30°, 60°

61. If possible, find an obtuse angle and an acute angle that are supplementary. Answers may vary.   For example: 100°, 80°

62. If possible, find two obtuse angles that are supplementary.   Not possible

63. Refer to the figure.

    a. State all the pairs of vertical angles.   ∠1 and ∠3, ∠2 and ∠4

    b. State all the pairs of supplementary angles.   ∠1 and ∠2, ∠2 and ∠3, ∠3 and ∠4, ∠1 and ∠4

    c. If the measure of ∠4 is 80°, find the measures of ∠1, ∠2, and ∠3. $m(\angle 1) = 100°, m(\angle 2) = 80°, m(\angle 3) = 100°$

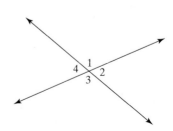

Writing   Translating Expression   Geometry   Scientific Calculator   Video   NS&E

64. Refer to the figure.

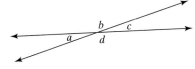

a. State all the pairs of vertical angles.
∠b and ∠d, ∠a and ∠c

b. State all the pairs of supplementary angles.
∠a and ∠b, ∠b and ∠c, ∠c and ∠d, ∠a and ∠d

c. If the measure of ∠a is 25°, find the measures of ∠b, ∠c, and ∠d.
$m(\angle b) = 155°, m(\angle c) = 25°, m(\angle d) = 155°$

For Exercises 65–68, find the complement of each angle.

65. 33° 57° **66.** 87° 3° **67.** 12° 78° **68.** 45° 45°

For Exercises 69–72, find the supplement of each angle.

69. 33° 147° **70.** 87° 93° **71.** 122° 58° **72.** 90° 90°

For Exercises 73–80, refer to the figure. Assume that L_1 and L_2 are parallel lines. **(See Example 8.)**

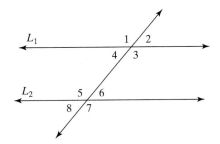

73. $m(\angle 5) = m(\angle \underline{\;7\;})$ Reason: Vertical angles have equal measures.

74. $m(\angle 5) = m(\angle \underline{\;3\;})$ Reason: Alternate interior angles have equal measures.

75. $m(\angle 5) = m(\angle \underline{\;1\;})$ Reason: Corresponding angles have equal measures.

76. $m(\angle 7) = m(\angle \underline{\;3\;})$ Reason: Corresponding angles have equal measures.

77. $m(\angle 7) = m(\angle \underline{\;1\;})$ Reason: Alternate exterior angles have equal measures.

78. $m(\angle 7) = m(\angle \underline{\;5\;})$ Reason: Vertical angles have equal measures.

79. $m(\angle 3) = m(\angle \underline{\;5\;})$ Reason: Alternate interior angles have equal measures.

80. $m(\angle 3) = m(\angle \underline{\;1\;})$ Reason: Vertical angles have equal measures.

81. Find the measure of angles a–g in the figure. Assume that L_1 and L_2 are parallel.
$m(\angle a) = 45°, m(\angle b) = 135°, m(\angle c) = 45°, m(\angle d) = 135°,$
$m(\angle e) = 45°, m(\angle f) = 135°, m(\angle g) = 45°$

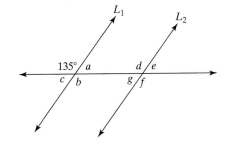

82. Find the measure of angles a–g in the figure. Assume that L_1 and L_2 are parallel.
$m(\angle a) = 65°, m(\angle b) = 115°, m(\angle c) = 115°, m(\angle d) = 65°,$
$m(\angle e) = 115°, m(\angle f) = 115°, m(\angle g) = 65°$

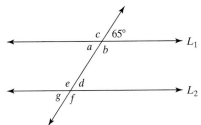

Objective 5: Triangles

For Exercises 83–86, identify the triangle as equilateral, isosceles, or scalene.

83.

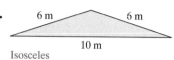

6 in.   10 in.

8 in.

Scalene

84.

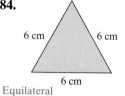

6 cm   6 cm

6 cm

Equilateral

85.

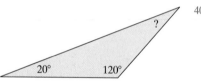

6 m   6 m

10 m

Isosceles

86.

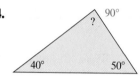

15 ft

7 ft

9 ft

Scalene

87. True or False? If a triangle is equilateral, then it is not scalene.   True

88. True or False? If a triangle is isosceles, then it is also scalene.   False; an isosceles triangle has two sides that are the same length. A scalene triangle has three sides of different length.

89. What angle is its own complement?   45°

90. What angle is its own supplement?   90°

91. Can a triangle be both a right triangle and an obtuse triangle? Explain.   No, a 90° angle plus an angle greater than 90° would make the sum of the angles greater than 180°.

92. Can a triangle be both a right triangle and an isosceles triangle? Explain.
Yes, the sides forming the right angle can be equal.

For Exercises 93–96, find the measure of the missing angles.

93.

40°

?

20°   120°

94.

90°

?

40°   50°

95.

?   37°

53°

96.

63°

27°

?

97. Refer to the figure. Find the measure of angles
a–j. **(See Example 9.)** $m(\angle a) = 80°$, $m(\angle b) = 80°$, $m(\angle c) = 100°$, $m(\angle d) = 100°$, $m(\angle e) = 65°$, $m(\angle f) = 115°$, $m(\angle g) = 115°$, $m(\angle h) = 35°$, $m(\angle i) = 145°$, $m(\angle j) = 145°$

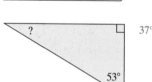

b
c d
a

f   65°
e   g   35°   i
           j   h

98. Refer to the figure. Find the measure of angles *a–j*.
$m(\angle a) = 120°$, $m(\angle b) = 60°$, $m(\angle c) = 60°$, $m(\angle d) = 40°$, $m(\angle e) = 40°$, $m(\angle f) = 140°$, $m(\angle g) = 140°$, $m(\angle h) = 160°$, $m(\angle i) = 20°$, $m(\angle j) = 160°$

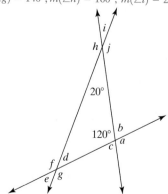

i
h   j

20°

120°   b
c   a

f   d
e   g

99. Refer to the figure. Find the measure of angles *a*–*k*. Assume that L_1 and L_2 are parallel.

$m(\angle a) = 70°, m(\angle b) = 65°, m(\angle c) = 65°, m(\angle d) = 110°, m(\angle e) = 70°,$
$m(\angle f) = 110°, m(\angle g) = 115°, m(\angle h) = 115°, m(\angle i) = 65°,$
$m(\angle j) = 70°, m(\angle k) = 65°$

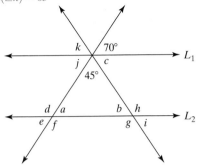

100. Refer to the figure. Find the measure of angles *a*–*k*. Assume that L_1 and L_2 are parallel.

$m(\angle a) = 75°, m(\angle b) = 65°, m(\angle c) = 40°, m(\angle d) = 65°, m(\angle e) = 65°,$
$m(\angle f) = 40°, m(\angle g) = 40°, m(\angle h) = 140°, m(\angle i) = 140°,$
$m(\angle j) = 115°, m(\angle k) = 115°$

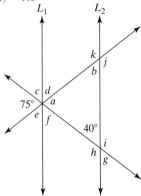

Expanding Your Skills

For Exercises 101–102, find the perimeter.

101.

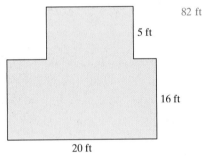

82 ft

102.

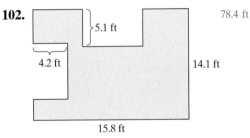

78.4 ft

For Exercises 103–106, find the area of the shaded region. Use 3.14 for π.

103.

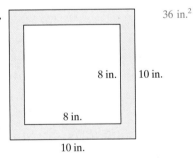

36 in.²

104.

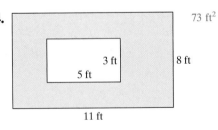

73 ft²

105.

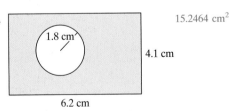

15.2464 cm²

106.

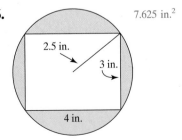

7.625 in.²

The Set of Real Numbers

1

CHAPTER OUTLINE

1.1 Sets of Numbers and the Real Number Line   46

1.2 Exponents, Square Roots, and the Order of Operations   56

1.3 Addition of Real Numbers   66

1.4 Subtraction of Real Numbers   74

   Problem Recognition Exercises: Addition and Subtraction of Real Numbers   82

1.5 Multiplication and Division of Real Numbers   82

1.6 Properties of Real Numbers and Simplifying Expressions   93

   Group Activity: Evaluating Formulas Using a Calculator   106

Chapter 1

In Chapter 1, we present operations on real numbers. The skills that you will learn in this chapter are particularly important as you continue in algebra. Pay careful attention to the order of operations.

Use what you learn about operations on real numbers and the order of operations to complete the puzzle. Fill in each blank box with one of the four basic operations, $+$, $-$, $\bullet$, or $\div$ so that the statement is true both going across and going down. Pay careful attention to the order of operations.

-10	$+$	2	$+$	-3	$=$	-11
$+$		$\bullet$		$\bullet$		$-$
12	$\div$	-2	$\bullet$	3	$=$	-18
$+$		$+$		$+$		$+$
6	$-$	4	$-$	10	$=$	-8
$=$		$=$		$=$		$=$
8	$\bullet$	0	$-$	1	$=$	-1

Objectives

1. The Set of Real Numbers
2. Inequalities
3. Opposite of a Real Number
4. Absolute Value of a Real Number

1. The Set of Real Numbers

The numbers we work with on a day-to-day basis are all part of the set of **real numbers**. The real numbers encompass zero, all positive, and all negative numbers, including those represented by fractions and decimal numbers. The set of real numbers can be represented graphically on a horizontal number line with a point labeled as 0. Positive real numbers are graphed to the right of 0, and negative real numbers are graphed to the left of 0. Zero is neither positive nor negative. Each point on the number line corresponds to exactly one real number. For this reason, this number line is called the *real number line* (Figure 1-1).

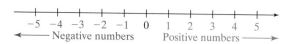

Figure 1-1

Skill Practice

1. Plot the numbers on a real number line.
   $\{-1, \frac{3}{4}, -2.5, \frac{10}{3}\}$

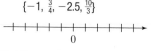

Classroom Example: p. 53, Exercise 4

Example 1 Plotting Points on the Real Number Line

Plot the points on the real number line that represent the following real numbers.

a. -3 **b.** $\dfrac{3}{2}$ **c.** -4.7 **d.** $\dfrac{16}{5}$

Solution:

a. Because -3 is negative, it lies three units to the left of 0.

b. The fraction $\frac{3}{2}$ can be expressed as the mixed number $1\frac{1}{2}$, which lies half-way between 1 and 2 on the number line.

c. The negative number -4.7 lies $\frac{7}{10}$ units to the left of -4 on the number line.

d. The fraction $\frac{16}{5}$ can be expressed as the mixed number $3\frac{1}{5}$, which lies $\frac{1}{5}$ unit to the right of 3 on the number line.

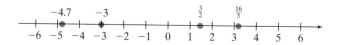

TIP: The natural numbers are used for counting. For this reason, they are sometimes called the "counting numbers."

In mathematics, a well-defined collection of elements is called a **set**. "Well-defined" means the set is described in such a way that it is clear whether an element is in the set. The symbols { } are used to enclose the elements of the set. For example, the set {A, B, C, D, E} represents the set of the first five letters of the alphabet.

Several sets of numbers are used extensively in algebra and are *subsets* (or part) of the set of real numbers.

> **DEFINITION** Natural Numbers, Whole Numbers, and Integers
>
> The set of **natural numbers** is $\{1, 2, 3, \ldots\}$
> The set of **whole numbers** is $\{0, 1, 2, 3, \ldots\}$
> The set of **integers** is $\{\ldots -3, -2, -1, 0, 1, 2, 3, \ldots\}$

Answer

1.
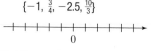

Notice that the set of whole numbers includes the natural numbers. Therefore, every natural number is also a whole number. The set of integers includes the set of whole numbers. Therefore, every whole number is also an integer.

Fractions are also among the numbers we use frequently. A number that can be written as a fraction whose numerator is an integer and whose denominator is a nonzero integer is called a *rational number*.

Instructor Note: Memory Device: Rational = RATIO-nal can be written as a ratio of two integers

> **DEFINITION Rational Numbers**
>
> The set of **rational numbers** is the set of numbers that can be expressed in the form $\frac{p}{q}$, where both p and q are integers and q does not equal 0.

We also say that a rational number $\frac{p}{q}$ is a *ratio* of two integers, p and q, where q is not equal to zero.

Example 2 Identifying Rational Numbers

Show that the following numbers are rational numbers by finding an equivalent ratio of two integers.

  a. $\dfrac{-2}{3}$ **b.** -12 **c.** 0.5 **d.** $0.\overline{6}$

Solution:

  a. The fraction $\frac{-2}{3}$ is a rational number because it can be expressed as the ratio of -2 and 3.

  b. The number -12 is a rational number because it can be expressed as the ratio of -12 and 1, that is, $-12 = \frac{-12}{1}$. In this example, we see that an integer is also a rational number.

  c. The terminating decimal 0.5 is a rational number because it can be expressed as the ratio of 5 and 10. That is, $0.5 = \frac{5}{10}$. In this example, we see that a terminating decimal is also a rational number.

  d. The repeating decimal $0.\overline{6}$ is a rational number because it can be expressed as the ratio of 2 and 3. That is, $0.\overline{6} = \frac{2}{3}$. In this example, we see that a repeating decimal is also a rational number.

Skill Practice

Show that each number is rational by finding an equivalent ratio of two integers.

2. $\dfrac{3}{7}$ **3.** -5

4. 0.3 **5.** $0.\overline{3}$

Classroom Examples: p. 53, Exercises 6 and 8

TIP: Any rational number can be represented by a terminating decimal or by a repeating decimal.

Some real numbers, such as the number π, cannot be represented by the ratio of two integers. These numbers are called irrational numbers and in decimal form are nonterminating, nonrepeating decimals. The value of π, for example, can be approximated as $\pi \approx 3.1415926535897932$. However, the decimal digits continue forever with no repeated pattern. Another example of an irrational number is $\sqrt{3}$ (read as "the positive square root of 3"). The expression $\sqrt{3}$ is a number that when multiplied by itself is 3. There is no rational number that satisfies this condition. Thus, $\sqrt{3}$ is an irrational number.

> **DEFINITION Irrational Numbers**
>
> The set of **irrational numbers** is a subset of the real numbers whose elements cannot be written as a ratio of two integers.
>
> *Note:* An irrational number cannot be written as a terminating decimal or as a repeating decimal.

Answers

2. ratio of 3 and 7
3. ratio of -5 and 1
4. ratio of 3 and 10
5. ratio of 1 and 3

The set of real numbers consists of both the rational and the irrational numbers. The relationship among these important sets of numbers is illustrated in Figure 1-2:

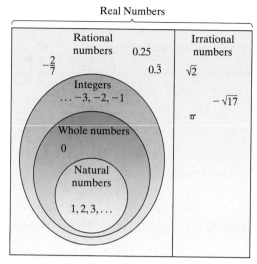

Figure 1-2

Example 3 **Classifying Numbers by Set**

Check the set(s) to which each number belongs. The numbers may belong to more than one set.

	Natural Numbers	Whole Numbers	Integers	Rational Numbers	Irrational Numbers	Real Numbers
5						
$\dfrac{-47}{3}$						
1.48						
$\sqrt{7}$						
0						

Solution:

	Natural Numbers	Whole Numbers	Integers	Rational Numbers	Irrational Numbers	Real Numbers
5	✔	✔	✔	✔ (ratio of 5 and 1)		✔
$\dfrac{-47}{3}$				✔ (ratio of −47 and 3)		✔
1.48				✔ (ratio of 148 and 100)		✔
$\sqrt{7}$					✔	✔
0		✔	✔	✔ (ratio of 0 and 1)		✔

2. Inequalities

The relative size of two real numbers can be compared using the real number line. Suppose a and b represent two real numbers. We say that a is less than b, denoted $a < b$, if a lies to the left of b on the number line.

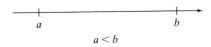

$a < b$

We say that a is greater than b, denoted $a > b$, if a lies to the right of b on the number line.

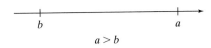

$a > b$

Table 1-1 summarizes the relational operators that compare two real numbers a and b.

Table 1-1

Mathematical Expression	Translation	Example
$a < b$	a is less than b.	$2 < 3$
$a > b$	a is greater than b.	$5 > 1$
$a \leq b$	a is less than or equal to b.	$4 \leq 4$
$a \geq b$	a is greater than or equal to b.	$10 \geq 9$
$a = b$	a is equal to b.	$6 = 6$
$a \neq b$	a is not equal to b.	$7 \neq 0$
$a \approx b$	a is approximately equal to b.	$2.3 \approx 2$

The symbols $<, >, \leq, \geq$, and $\neq$ are called *inequality signs*, and the expressions $a < b, a > b, a \leq b, a \geq b$, and $a \neq b$ are called **inequalities**.

Example 4 **Ordering Real Numbers**

The average temperatures (in degrees Celsius) for selected cities in the United States and Canada in January are shown in Table 1-2.

Table 1-2

City	Temp (°C)
Prince George, British Columbia	-12.5
Corpus Christi, Texas	13.4
Parkersburg, West Virginia	-0.9
San Jose, California	9.7
Juneau, Alaska	-5.7
New Bedford, Massachusetts	-0.2
Durham, North Carolina	4.2

Plot a point on the real number line representing the temperature of each city. Compare the temperatures between the following cities, and fill in the blank with the appropriate inequality sign: $<$ or $>$.

Skill Practice

Fill in the blanks with the appropriate inequality sign: $<$ or $>$.

11. -11 _____ 20

12. -3 _____ -6

13. 0 _____ -9

14. -6.2 _____ -1.8

Classroom Example: p. 54, Exercise 34

Answers

11. $<$ **12.** $>$ **13.** $>$ **14.** $<$

Solution:

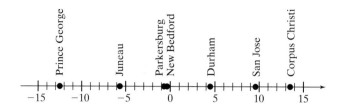

a. Temperature of San Jose $\boxed{<}$ temperature of Corpus Christi

b. Temperature of Juneau $\boxed{>}$ temperature of Prince George

c. Temperature of Parkersburg $\boxed{<}$ temperature of New Bedford

d. Temperature of Parkersburg $\boxed{>}$ temperature of Prince George

3. Opposite of a Real Number

To gain mastery of any algebraic skill, it is necessary to know the meaning of key definitions and key symbols. Two important definitions are the *opposite* of a real number and the *absolute value* of a real number.

> **DEFINITION   The Opposite of a Real Number**
>
> Two numbers that are the same distance from 0 but on opposite sides of 0 on the number line are called **opposites** of each other. Symbolically, we denote the opposite of a real number a as $-a$.

Skill Practice

18. Find the opposite of 224.
19. Find the opposite of -3.4.

Classroom Examples: p. 54, Exercises 36 and 38

Example 5 Finding the Opposite of a Real Number

a. Find the opposite of 5.   **b.** Find the opposite of $-\frac{4}{7}$.

Solution:

a. The opposite of 5 is -5.   **b.** The opposite of $-\frac{4}{7}$ is $\frac{4}{7}$.

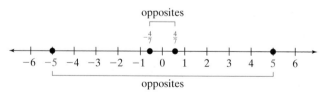

Skill Practice

20. Evaluate $-(2.8)$.

21. Evaluate $-\left(-\dfrac{1}{5}\right)$.

Classroom Example: p. 55, Exercise 46

Example 6 Finding the Opposite of a Real Number

a. Evaluate $-(0.46)$.   **b.** Evaluate $-\left(-\dfrac{11}{3}\right)$.

Solution:

a. $-(0.46) = -0.46$ The expression $-(0.46)$ represents the opposite of 0.46.

b. $-\left(-\dfrac{11}{3}\right) = \dfrac{11}{3}$ The expression $-(-\frac{11}{3})$ represents the opposite of $-\frac{11}{3}$.

Answers

15. $<$ **16.** $>$ **17.** $>$
18. -224 **19.** 3.4   **20.** -2.8
21. $\dfrac{1}{5}$

4. Absolute Value of a Real Number

The concept of absolute value will be used to define the addition of real numbers in Section 1.3.

> **DEFINITION Informal Definition of the Absolute Value of a Real Number**
>
> The **absolute value** of a real number a, denoted $|a|$, is the distance between a and 0 on the number line.
>
> *Note:* The absolute value of any real number is positive or zero.

Instructor Note: Remind students that because absolute value is a measure of *distance* it will always be nonnegative. We don't measure distance with negative numbers.

For example, $|3| = 3$ and $|-3| = 3$.

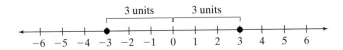

Example 7 **Finding the Absolute Value of a Real Number**

Evaluate the absolute value expressions.

  a. $|-4|$ **b.** $\left|\frac{1}{2}\right|$ **c.** $|-6.2|$ **d.** $|0|$

Solution:

  a. $|-4| = 4$ -4 is 4 units from 0 on the number line.

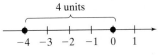

  b. $\left|\frac{1}{2}\right| = \frac{1}{2}$ $\frac{1}{2}$ is $\frac{1}{2}$ unit from 0 on the number line.

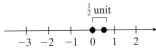

  c. $|-6.2| = 6.2$ -6.2 is 6.2 units from 0 on the number line.

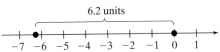

  d. $|0| = 0$ 0 is 0 units from 0 on the number line.

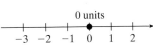

Skill Practice

Evaluate.

  22. $|-99|$ **23.** $\left|\dfrac{7}{8}\right|$

  24. $|-1.4|$ **25.** $|1|$

Classroom Examples: p. 55, Exercises 52 and 56

The absolute value of a number a is its distance from 0 on the number line. The definition of $|a|$ may also be given symbolically depending on whether a is negative or nonnegative.

> **DEFINITION Absolute Value of a Real Number**
>
> Let a be a real number. Then
>
>   **1.** If a is nonnegative (that is, $a \geq 0$), then $|a| = a$.
>
>   **2.** If a is negative (that is, $a < 0$), then $|a| = -a$.

Answers

22. 99   **23.** $\dfrac{7}{8}$

24. 1.4   **25.** 1

This definition states that if a is a nonnegative number, then $|a|$ equals a itself. If a is a negative number, then $|a|$ equals the opposite of a. For example:

$|9| = 9$ Because 9 is positive, then $|9|$ equals the number 9 itself.

$|-7| = 7$ Because -7 is negative, then $|-7|$ equals the opposite of -7, which is 7.

Answers
26. False **27.** True

Example 8 Comparing Absolute Value Expressions

Determine if the statements are true or false.

a. $|3| \le 3$ **b.** $-|5| = |-5|$

Solution:

a. $|3| \le 3$ True. The symbol $\le$ means "less than *or* equal to." Since $|3|$ is equal to 3, then $|3| \le 3$ is a true statement.

b. $-|5| = |-5|$ False. On the left-hand side, $-|5|$ is the opposite of $|5|$. Hence $-|5| = -5$. On the right-hand side, $|-5| = 5$. Therefore, the original statement simplifies to $-5 = 5$, which is false.

Calculator Connections

Topic: Approximating Irrational Numbers on a Calculator

Scientific and graphing calculators approximate irrational numbers by using rational numbers in the form of terminating decimals. For example, consider approximating π and $\sqrt{3}$.

Scientific Calculator:

Enter: $\boxed{\pi}$ or $\boxed{2^{nd}}$ $\boxed{\pi}$ **Result:** $\boxed{3.141592654}$

Enter: 3 $\boxed{\sqrt{x}}$ **Result:** $\boxed{1.732050808}$

Graphing Calculator:

Enter: $\boxed{2^{nd}}$ $\boxed{\pi}$ $\boxed{\text{ENTER}}$

Enter: $\boxed{2^{nd}}$ $\boxed{\sqrt{\ }}$ 3 $\boxed{\text{ENTER}}$

```
π
         3.141592654
√(3)
         1.732050808
```

Note that when writing approximations, we use the symbol, $\approx$.

$$\pi \approx 3.141592654 \quad \text{and} \quad \sqrt{3} \approx 1.732050808$$

Calculator Exercises

Use a calculator to approximate the irrational numbers. Remember to use the appropriate symbol, $\approx$, when expressing answers.

1. $\sqrt{12}$
≈ 3.464101615

2. $\sqrt{99}$
≈ 9.949874371

3. $4 \cdot \pi$
≈ 12.56637061

4. $\sqrt{\pi}$
≈ 1.772453851

Section 1.1   Practice Exercises

Study Skills Exercises

1. In this text, we will provide skills for you to enhance your learning experience. In the first four chapters, each set of Practice Exercises begins with an activity that focuses on one of eight areas: learning about your course, using your text, taking notes, doing homework, taking an exam (test and math anxiety), managing your time, recognizing your learning style, and studying for the final exam.

   Each activity requires only a few minutes and will help you to pass this course and become a better math student. Many of these skills can be carried over to other disciplines and help you to become a model college student. To begin, write down the following information:

   a. Instructor's name

   b. Instructor's office number

   c. Instructor's telephone number

   d. Instructor's e-mail address

   e. Instructor's office hours

   f. Days of the week that the class meets

   g. The room number in which the class meets

   h. Is there a lab requirement for this course? How often must you attend lab and where is it located?

2. Define the key terms:

   a. real numbers

   b. set

   c. natural numbers

   d. whole numbers

   e. integers

   f. rational numbers

   g. irrational numbers

   h. inequality

   i. opposite

   j. absolute value

Objective 1: The Set of Real Numbers

3. Plot the numbers on a real number line: $\{1, -2, -\pi, 0, -\frac{5}{2}, 5.1\}$ **(See Example 1.)**

4. Plot the numbers on a real number line: $\{3, -4, \frac{1}{8}, -1.7, -\frac{4}{3}, 1.75\}$

For Exercises 5–20, describe each number as (a) a terminating decimal, (b) a repeating decimal, or (c) a nonterminating, nonrepeating decimal. Then classify the number as a rational number or as an irrational number. **(See Example 2.)**

5. 0.29 a; rational

6. 3.8 a; rational

7. $\frac{1}{9}$ b; rational

8. $\frac{1}{3}$ b; rational

9. $\frac{1}{8}$ a; rational

10. $\frac{1}{5}$ a; rational

11. 2π c; irrational

12. 3π c; irrational

13. -0.125 a; rational

14. -3.24 a; rational

15. -3 a; rational

16. -6 a; rational

17. $0.\overline{2}$ b; rational

18. $0.\overline{6}$ b; rational

19. $\sqrt{6}$ c; irrational

20. $\sqrt{10}$ c; irrational

21. List three numbers that are real numbers but not rational numbers.
For example: $\pi, -\sqrt{2}, \sqrt{3}$

22. List three numbers that are real numbers but not irrational numbers.
For example: $-\frac{1}{2}, 5, 0.\overline{3}$

23. List three numbers that are integers but not natural numbers.
For example: $-4, -1, 0$

24. List three numbers that are integers but not whole numbers.
For example: $-5, -2, -1$

25. List three numbers that are rational numbers but not integers.
For example: $-\frac{3}{4}, \frac{1}{2}, 0.206$

For Exercises 26–32, let $A = \{-\frac{3}{2}, \sqrt{11}, -4, 0.\overline{6}, 0, \sqrt{7}, 1\}$ **(See Example 3.)**

26. Are all of the numbers in set A real numbers?
Yes

27. List all of the rational numbers in set A. $-\frac{3}{2}, -4, 0.\overline{6}, 0, 1$

28. List all of the whole numbers in set A.
$0, 1$

29. List all of the natural numbers in set A. 1

30. List all of the irrational numbers in set A.
$\sqrt{11}, \sqrt{7}$

31. List all of the integers in set A. $-4, 0, 1$

32. Plot the real numbers from set A on a number line. (*Hint:* $\sqrt{11} \approx 3.3$ and $\sqrt{7} \approx 2.6$)

Objective 2: Inequalities

33. The LPGA Samsung World Championship of women's golf scores for selected players are given in the table. Compare the scores and fill in the blank with the appropriate inequality sign: $<$ or $>$.
(See Example 4.)

  a. Kane's score ___>___ Pak's score.

  b. Sorenstam's score ___>___ Davies' score.

  c. Pak's score ___<___ McCurdy's score.

  d. Kane's score ___>___ Davies' score.

LPGA Golfers	Final Score with Respect to Par
Annika Sorenstam	7
Laura Davies	−4
Lorie Kane	0
Cindy McCurdy	3
Se Ri Pak	−8

34. The elevations of selected cities in the United States are shown in the figure. Compare the elevations and fill in the blank with the appropriate inequality sign: $<$ or $>$. (A negative number indicates that the city is below sea level.)

  a. Elevation of Tucson ___>___ elevation of Cincinnati.

  b. Elevation of New Orleans ___<___ elevation of Chicago.

  c. Elevation of New Orleans ___<___ elevation of Houston.

  d. Elevation of Chicago ___>___ elevation of Cincinnati.

Objective 3: Opposite of a Real Number

For Exercises 35–42, find the opposite of each number. **(See Example 5.)**

35. 18 -18

36. 2 -2

37. -6.1 6.1

38. -2.5 2.5

39. $-\frac{5}{8}$ $\frac{5}{8}$

40. $-\frac{1}{3}$ $\frac{1}{3}$

41. $\frac{7}{3}$ $-\frac{7}{3}$

42. $\frac{1}{9}$ $-\frac{1}{9}$

 Writing   Translating Expression   Geometry   Scientific Calculator   Video   NS&E

The opposite of a is denoted as $-a$. For Exercises 43–50, simplify. **(See Example 6.)**

43. $-(-3)$ 3

44. $-(-5.1)$ 5.1

45. $-\left(\dfrac{7}{3}\right)$ $-\dfrac{7}{3}$

46. $-(-7)$ 7

47. $-(-8)$ 8

48. $-(36)$ -36

49. $-(72.1)$ -72.1

50. $-\left(\dfrac{9}{10}\right)$ $-\dfrac{9}{10}$

Objective 4: Absolute Value of a Real Number

For Exercises 51–62, simplify. **(See Example 7.)**

51. $|-2|$ 2

52. $|-7|$ 7

53. $|-1.5|$ 1.5

54. $|-3.7|$ 3.7

55. $-|-1.5|$ -1.5

56. $-|-3.7|$ -3.7

57. $\left|\dfrac{3}{2}\right|$ $\dfrac{3}{2}$

58. $\left|\dfrac{7}{4}\right|$ $\dfrac{7}{4}$

59. $-|10|$ -10

60. $-|20|$ -20

61. $-\left|-\dfrac{1}{2}\right|$ $-\dfrac{1}{2}$

62. $-\left|-\dfrac{11}{3}\right|$ $-\dfrac{11}{3}$

For Exercises 63–64, answer true or false. If a statement is false, explain why.

63. If n is positive, then $|n|$ is negative.
False, $|n|$ is never negative.

64. If m is negative, then $|m|$ is negative.
False, $|m|$ is never negative.

For Exercises 65–88, determine if the statements are true or false. Use the real number line to justify the answer. **(See Example 8.)**

65. $5 > 2$ True

66. $8 < 10$ True

67. $6 < 6$ False

68. $19 > 19$ False

69. $-7 \geq -7$ True

70. $-1 \leq -1$ True

71. $\dfrac{3}{2} \leq \dfrac{1}{6}$ False

72. $-\dfrac{1}{4} \geq -\dfrac{7}{8}$ True

73. $-5 > -2$ False

74. $6 < -10$ False

75. $8 \neq 8$ False

76. $10 \neq 10$ False

77. $|-2| \geq |-1|$ True

78. $|3| \leq |-1|$ False

79. $\left|-\dfrac{1}{9}\right| = \left|\dfrac{1}{9}\right|$ True

80. $\left|-\dfrac{1}{3}\right| = \left|\dfrac{1}{3}\right|$ True

81. $|7| \neq |-7|$ False

82. $|-13| \neq |13|$ False

83. $-1 < |-1|$ True

84. $-6 < |-6|$ True

85. $|-8| \geq |8|$ True

86. $|-11| \geq |11|$ True

87. $|-2| \leq |2|$ True

88. $|-21| \leq |21|$ True

Expanding Your Skills

89. For what numbers, a, is $-a$ positive?
For all $a < 0$

90. For what numbers, a, is $|a| = a$?
For all $a \geq 0$

Writing   Translating Expression   Geometry   Scientific Calculator   Video   NS&E

Exponents, Square Roots, and the Order of Operations

Objectives

1. Variables and Expressions
2. Evaluating Algebraic Expressions
3. Exponential Expressions
4. Square Roots
5. Order of Operations
6. Translations

Concept Connections

1. Why is π not a variable?

1. Variables and Expressions

A **variable** is a symbol or letter such as x, y, and z, used to represent an unknown number. **Constants** are values that do not vary such as the numbers 3, -1.5, $\frac{2}{7}$, and π. An algebraic **expression** is a collection of variables and constants under algebraic operations. For example, $\frac{3}{x}$, $y + 7$, and $t - 1.4$ are algebraic expressions.

The symbols used to show the four basic operations of addition, subtraction, multiplication, and division are summarized in Table 1-3.

Table 1-3

Operation	Symbols	Translation
Addition	$a + b$	**sum** of a and b a plus b b added to a b more than a a increased by b the total of a and b
Subtraction	$a - b$	**difference** of a and b a minus b b subtracted from a a decreased by b b less than a a less b
Multiplication	$a \times b, a \cdot b, a(b), (a)b, (a)(b), ab$ (*Note:* We rarely use the notation $a \times b$ because the symbol, $\times$, might be confused with the variable, x.)	**product** of a and b a times b a multiplied by b
Division	$a \div b, \dfrac{a}{b}, a/b, b\overline{)a}$	**quotient** of a and b a divided by b b divided into a ratio of a and b a over b a per b

2. Evaluating Algebraic Expressions

The value of an algebraic expression depends on the values of the variables within the expression.

Skill Practice

Evaluate the algebraic expressions when $x = 5$ and $y = 2$.

2. $20 - y$ **3.** xy

Classroom Examples: p. 63, Exercises 10 and 14

Answers

1. The symbol π is not a variable because its value does not vary. It is always equal to the irrational number 3.14159. . . .

2. 18   **3.** 10

Example 1 **Evaluating an Algebraic Expression**

Evaluate the algebraic expression when $p = 4$ and $q = \frac{3}{4}$.

a. $100 - p$ **b.** pq

Solution:

a. $100 - p$

$100 - (\quad)$ When substituting a number for a variable, use parentheses.

$= 100 - (4)$ Substitute $p = 4$ in the parentheses.

$= 96$ Subtract.

b. pq

$= (\ )(\ )$	When substituting a number for a variable, use parentheses.
$= (4)\left(\dfrac{3}{4}\right)$	Substitute $p = 4$ and $q = \frac{3}{4}$.
$= \dfrac{\overset{1}{4}}{1} \cdot \dfrac{3}{\underset{1}{4}}$	Write the whole number as a fraction.
$= \dfrac{3}{1}$	Multiply fractions.
$= 3$	Simplify.

3. Exponential Expressions

In algebra, repeated multiplication can be expressed using exponents. The expression, $4 \cdot 4 \cdot 4$ can be written as

exponent

4^3

base

In the expression 4^3, 4 is the base, and 3 is the exponent, or power. The exponent indicates how many factors of the base to multiply.

DEFINITION Definition of b^n

Let b represent any real number and n represent a positive integer. Then,

$$b^n = \underbrace{b \cdot b \cdot b \cdot b \ldots \cdot b}_{n \text{ factors of } b}$$

b^n is read as "b to the nth power."

b is called the **base**, and n is called the **exponent**, or **power**.

b^2 is read as "b squared," and b^3 is read as "b cubed."

The exponent, n, is the number of times the base, b, is used as a factor.

> **TIP:** A number or variable with no exponent shown implies that there is an exponent of 1. That is, $b = b^1$.

Example 2   **Evaluating Exponential Expressions**

Translate the expression into words and then evaluate the expression.

a. 2^5 **b.** 5^2 **c.** $\left(\dfrac{3}{4}\right)^3$ **d.** 1^6

Solution:

a. The expression 2^5 is read as "two to the fifth power."
$2^5 = (2)(2)(2)(2)(2) = 32$

b. The expression 5^2 is read as "five to the second power" or "five, squared."
$5^2 = (5)(5) = 25$

Skill Practice

Evaluate.

4. 4^3 **5.** 2^4

6. $\left(\dfrac{2}{3}\right)^2$ **7.** $(1)^7$

Classroom Examples: p. 64, Exercises 32 and 34

Answers

4. 64   **5.** 16   **6.** $\dfrac{4}{9}$ **7.** 1

c. The expression $\left(\frac{3}{4}\right)^3$ is read as "three-fourths to the third power" or "three-fourths, cubed."

$$\left(\frac{3}{4}\right)^3 = \left(\frac{3}{4}\right)\left(\frac{3}{4}\right)\left(\frac{3}{4}\right) = \frac{27}{64}$$

d. The expression 1^6 is read as "one to the sixth power."
$$1^6 = (1)(1)(1)(1)(1)(1) = 1$$

4. Square Roots

The inverse operation to squaring a number is to find its **square roots**. For example, finding a square root of 9 is equivalent to asking "what number(s) when squared equals 9?" The symbol, $\sqrt{}$, (called a *radical sign*) is used to find the *principal* square root of a number. By definition, the principal square root of a number is nonnegative. Therefore, $\sqrt{9}$ is the nonnegative number that when squared equals 9. Hence $\sqrt{9} = 3$ because 3 is nonnegative and $(3)^2 = 9$.

Skill Practice

Evaluate.
  8. $\sqrt{81}$ **9.** $\sqrt{100}$
  10. $\sqrt{1}$ **11.** $\sqrt{\dfrac{9}{25}}$

Classroom Examples: p. 64, Exercises 42 and 50

Example 3 **Evaluating Square Roots**

Evaluate the square roots.

  a. $\sqrt{64}$ **b.** $\sqrt{121}$ **c.** $\sqrt{0}$ **d.** $\sqrt{\dfrac{4}{9}}$

Solution:

  a. $\sqrt{64} = 8$ Because $(8)^2 = 64$

  b. $\sqrt{121} = 11$ Because $(11)^2 = 121$

  c. $\sqrt{0} = 0$ Because $(0)^2 = 0$

  d. $\sqrt{\dfrac{4}{9}} = \dfrac{2}{3}$ Because $\dfrac{2}{3} \cdot \dfrac{2}{3} = \dfrac{4}{9}$

A perfect square is a number whose square root is a rational number. If a number is not a perfect square, its square root is an irrational number that can be approximated on a calculator.

TIP: To simplify square roots, it is advisable to become familiar with the following perfect squares and square roots.

$0^2 = 0 \rightarrow \sqrt{0} = 0$	$7^2 = 49 \rightarrow \sqrt{49} = 7$
$1^2 = 1 \rightarrow \sqrt{1} = 1$	$8^2 = 64 \rightarrow \sqrt{64} = 8$
$2^2 = 4 \rightarrow \sqrt{4} = 2$	$9^2 = 81 \rightarrow \sqrt{81} = 9$
$3^2 = 9 \rightarrow \sqrt{9} = 3$	$10^2 = 100 \rightarrow \sqrt{100} = 10$
$4^2 = 16 \rightarrow \sqrt{16} = 4$	$11^2 = 121 \rightarrow \sqrt{121} = 11$
$5^2 = 25 \rightarrow \sqrt{25} = 5$	$12^2 = 144 \rightarrow \sqrt{144} = 12$
$6^2 = 36 \rightarrow \sqrt{36} = 6$	$13^2 = 169 \rightarrow \sqrt{169} = 13$

Answers
  8. 9   **9.** 10
  10. 1   **11.** $\dfrac{3}{5}$

5. Order of Operations

When algebraic expressions contain numerous operations, it is important to evaluate the operations in the proper order. Parentheses ( ), brackets [ ], and braces { } are used for grouping numbers and algebraic expressions. It is important to recognize that operations must be done within parentheses and other grouping symbols first. Other grouping symbols include absolute value bars, radical signs, and fraction bars.

> **PROCEDURE** Order of Operations
>
> **Step 1** Simplify expressions within parentheses and other grouping symbols first. These include absolute value bars, fraction bars, and radicals. If imbedded parentheses are present, start with the innermost parentheses.
> **Step 2** Evaluate expressions involving exponents, radicals, and absolute values.
> **Step 3** Perform multiplication or division in the order that they occur from left to right.
> **Step 4** Perform addition or subtraction in the order that they occur from left to right.

Instructor Note: Radical signs act as grouping symbols. $\sqrt{30+6} = \sqrt{(30+6)}$. Perform the operations inside the radical first, then apply the square root.

Example 4 **Applying the Order of Operations**

Simplify the expressions.

a. $17 - 3 \cdot 2 + 2^2$ **b.** $\dfrac{1}{2}\left(\dfrac{5}{6} - \dfrac{3}{4}\right)$

Skill Practice

Simplify the expressions.
12. $14 - 3 \cdot 2 + 3^2$
13. $\dfrac{13}{4} - \dfrac{1}{4}(10-2)$

Classroom Examples: p. 64, Exercises 56 and 62

Solution:

a. $17 - 3 \cdot 2 + 2^2$

$\quad = 17 - 3 \cdot 2 + 4$ Simplify exponents.

$\quad = 17 - 6 + 4$ Multiply before adding or subtracting.

$\quad = 11 + 4$ Add or subtract from left to right.

$\quad = 15$

b. $\dfrac{1}{2}\left(\dfrac{5}{6} - \dfrac{3}{4}\right)$ Subtract fractions within the parentheses.

$\quad = \dfrac{1}{2}\left(\dfrac{10}{12} - \dfrac{9}{12}\right)$ The least common denominator is 12.

$\quad = \dfrac{1}{2}\left(\dfrac{1}{12}\right)$

$\quad = \dfrac{1}{24}$ Multiply fractions.

Answers

12. 17   **13.** $\dfrac{5}{4}$

Avoiding Mistakes

In Example 5(a), division is performed before multiplication because it occurs first as we read from left to right.

Example 5 Applying the Order of Operations

Simplify the expressions.

a. $25 - 12 \div 3 \cdot 4$

b. $6.2 - |-2.1| + \sqrt{15 - 6}$

c. $28 - 2[(6 - 3)^2 + 4]$

Solution:

a. $25 - 12 \div 3 \cdot 4$ — Multiply or divide in order from left to right.

$= 25 - 4 \cdot 4$ — Notice that the operation $12 \div 3$ is performed first (not $3 \cdot 4$).

$= 25 - 16$ — Multiply $4 \cdot 4$ before subtracting.

$= 9$ — Subtract.

b. $6.2 - |-2.1| + \sqrt{15 - 6}$

$= 6.2 - |-2.1| + \sqrt{9}$ — Simplify within the square root.

$= 6.2 - (2.1) + 3$ — Simplify the absolute value and square root.

$= 4.1 + 3$ — Add or subtract from left to right.

$= 7.1$ — Add.

c. $28 - 2[(6 - 3)^2 + 4]$

$= 28 - 2[(3)^2 + 4]$ — Simplify within the inner parentheses first.

$= 28 - 2[(9) + 4]$ — Simplify exponents.

$= 28 - 2[13]$ — Add within the square brackets.

$= 28 - 26$ — Multiply before subtracting.

$= 2$ — Subtract.

6. Translations

Example 6 Translating from English Form to Algebraic Form

Translate each English phrase to an algebraic expression.

a. The quotient of x and 5

b. The difference of p and the square root of q

c. Seven less than n

d. Seven less n

e. Eight more than the absolute value of w

f. x subtracted from 18

Solution:

a. $\dfrac{x}{5}$ or $x \div 5$ The quotient of x and 5

b. $p - \sqrt{q}$ The difference of p and the square root of q

c. $n - 7$ Seven less than n

d. $7 - n$ Seven less n

e. $|w| + 8$ Eight more than the absolute value of w

f. $18 - x$ x subtracted from 18

> **Avoiding Mistakes**
>
> Recall that "a less than b" is translated as $b - a$. Therefore, the statement "seven less than n" must be translated as $n - 7$, not $7 - n$.

Example 7   **Translating from English Form to Algebraic Form**

Translate each English phrase into an algebraic expression. Then evaluate the expression for $a = 6$, $b = 4$, and $c = 20$.

a. The product of a and the square root of b

b. Twice the sum of b and c

c. The difference of twice a and b

Solution:

a. The product of a and the square root of b

$a\sqrt{b}$

$= (\ \ )\sqrt{(\ \ )}$ Use parentheses to substitute a number for a variable.

$= (6)\sqrt{(4)}$ Substitute $a = 6$ and $b = 4$.

$= 6 \cdot 2$ Simplify the radical first.

$= 12$ Multiply.

b. Twice the sum of b and c

$2(b + c)$ To compute "twice the sum of b and c," it is necessary to take the sum first and then multiply by 2. To ensure the proper order, the sum of b and c must be enclosed in parentheses. The proper translation is $2(b + c)$.

$= 2((\ \ ) + (\ \ ))$ Use parentheses to substitute a number for a variable.

$= 2((4) + (20))$ Substitute $b = 4$ and $c = 20$.

$= 2(24)$ Simplify within the parentheses first.

$= 48$ Multiply.

c. The difference of twice a and b

$2a - b$

$= 2(\ \ ) - (\ \ )$ Use parentheses to substitute a number for a variable.

$= 2(6) - (4)$ Substitute $a = 6$ and $b = 4$.

$= 12 - 4$ Multiply first.

$= 8$ Subtract.

Skill Practice

Translate each English phrase to an algebraic expression. Then evaluate the expression for $x = 3$, $y = 9$, $z = 10$.

23. The quotient of the square root of y and x.

24. One-half the sum of x and y.

25. The difference of z and twice x.

Answers

23. $\dfrac{\sqrt{y}}{x}$; 1   **24.** $\dfrac{1}{2}(x + y)$; 6

25. $z - 2x$; 4

Calculator Connections

Topic: Evaluating Exponential Expressions on a Calculator

On a calculator, we enter exponents greater than the second power by using the key labeled y^x or $\wedge$. For example, evaluate 2^4 and 10^6:

Scientific Calculator:

Enter: 2 y^x 4 =   **Result:** | 16 |

Enter: 10 y^x 6 =   **Result:** | 1000000 |

Graphing Calculator:

```
2^4
              16
10^6
         1000000
```

Topic: Applying the Order of Operations on a Calculator

Most calculators also have the capability to enter several operations at once. However, it is important to note that fraction bars and radicals require user-defined parentheses to ensure that the proper order of operations is followed. For example, evaluate the following expressions on a calculator:

a. $130 - 2(5 - 1)^3$ **b.** $\dfrac{18 - 2}{11 - 9}$ **c.** $\sqrt{25 - 9}$

Scientific Calculator:

Enter: 130 − 2 × (5 − 1) y^x 3 =   **Result:** | 2 |

Enter: (18 − 2) ÷ (11 − 9) =   **Result:** | 8 |

Enter: (25 − 9) $\sqrt{\ }$ **Result:** | 4 |

Graphing Calculator:

```
130-2*(5-1)^3
              2
(18-2)/(11-9)
              8
√(25-9)
              4
```

Calculator Exercises

Simplify the expression without the use of a calculator. Then enter the expression into the calculator to verify your answer.

1. $\dfrac{4 + 6}{8 - 3}$ 2   **2.** $110 - 5(2 + 1) - 4$ 91   **3.** $100 - 2(5 - 3)^3$ 84

4. $3 + (4 - 1)^2$ 12   **5.** $(12 - 6 + 1)^2$ 49   **6.** $3 \cdot 8 - \sqrt{32 + 2^2}$ 18

7. $\sqrt{18 - 2}$ 4   **8.** $(4 \cdot 3 - 3 \cdot 3)^3$ 27   **9.** $\dfrac{20 - 3^2}{26 - 2^2}$ 0.5

Section 1.2 Practice Exercises

Boost *your* GRADE at ALEKS.com!

ALEKS®
version 3.0

- Practice Problems
- Self-Tests
- NetTutor

- e-Professors
- Videos

For additional exercises, see Classroom Activities 1.2A–1.2C in the *Instructor's Resource Manual* at www.mhhe.com/moh.

Study Skills Exercises

1. Sometimes you may run into a problem with homework or you find that you are having trouble keeping up with the pace of the class. A tutor can be a good resource.

   a. Does your college offer tutoring?

   b. Is it free?

   c. Where would you go to sign up for a tutor?

2. Define the key terms:

   a. variable   **b.** constant   **c.** expression   **d.** base

   e. exponent   **f.** square root   **g.** order of operations

Review Exercises

3. Which of the following are rational numbers. $\left\{-4, 5.\overline{6}, \sqrt{29}, 0, \pi, 4.02, \dfrac{7}{9}\right\}$ $-4, 5.\overline{6}, 0, 4.02, \dfrac{7}{9}$

4. Evaluate. $|-56|$ 56

5. Evaluate. $|9.2|$ 9.2

6. Evaluate. $-|-14|$ -14

7. Find the opposite of 19. -19

8. Find the opposite of -34.2.   34.2

Objective 2: Evaluating Algebraic Expressions

For Exercises 9–14, evaluate the expressions for the given substitutions. **(See Example 1.)**

9. $y - 3$ when $y = 18$ 15

10. $3q$ when $q = 5$ 15

11. $\dfrac{15}{t}$ when $t = 5$ 3

12. $8 + w$ when $w = 12$ 20

13. $5 + 6d$ when $d = \dfrac{2}{3}$ 9

14. $\dfrac{6}{5}h - 1$ when $h = 10$ 11

Objective 3: Exponential Expressions

For Exercises 15–20, write each of the products using exponents.

15. $\dfrac{1}{6} \cdot \dfrac{1}{6} \cdot \dfrac{1}{6} \cdot \dfrac{1}{6}$ $\left(\dfrac{1}{6}\right)^4$

16. $10 \cdot 10 \cdot 10 \cdot 10 \cdot 10 \cdot 10$ 10^6

17. $a \cdot a \cdot a \cdot b \cdot b$ $a^3 b^2$

18. $7 \cdot x \cdot x \cdot y \cdot y$ $7x^2 y^2$

19. $5c \cdot 5c \cdot 5c \cdot 5c \cdot 5c$ $(5c)^5$

20. $3 \cdot w \cdot z \cdot z \cdot z \cdot z$ $3wz^4$

21. **a.** For the expression $5x^3$, what is the base for the exponent 3? x

   b. Does 5 have an exponent? If so, what is it?   Yes, 1

22. **a.** For the expression $2y^4$, what is the base for the exponent 4? y

   b. Does 2 have an exponent? If so, what is it?   Yes, 1

Writing   *Translating Expression*   *Geometry*   *Scientific Calculator*   *Video*   NS&E

For Exercises 23–30, write each expression in expanded form using the definition of an exponent.

23. x^3 $x \cdot x \cdot x$ **24.** y^4 $y \cdot y \cdot y \cdot y$ **25.** $(2b)^3$ $2b \cdot 2b \cdot 2b$ **26.** $(8c)^2$ $8c \cdot 8c$

27. $10y^5$ $10 \cdot y \cdot y \cdot y \cdot y \cdot y$ **28.** x^2y^3 $x \cdot x \cdot y \cdot y \cdot y$ **29.** $2wz^2$ $2 \cdot w \cdot z \cdot z$ **30.** $3a^3b$ $3 \cdot a \cdot a \cdot a \cdot b$

For Exercises 31–38, simplify the expressions. **(See Example 2.)**

31. 6^2 36 **32.** 5^3 125 **33.** $\left(\dfrac{1}{7}\right)^2$ $\dfrac{1}{49}$ **34.** $\left(\dfrac{1}{2}\right)^5$ $\dfrac{1}{32}$

35. $(0.2)^3$ 0.008 **36.** $(0.8)^2$ 0.64 **37.** 2^6 64 **38.** 13^2 169

Objective 4: Square Roots

For Exercises 39–50, simplify the square roots. **(See Example 3.)**

39. $\sqrt{81}$ 9 **40.** $\sqrt{64}$ 8 **41.** $\sqrt{4}$ 2 **42.** $\sqrt{9}$ 3

43. $\sqrt{144}$ 12 **44.** $\sqrt{49}$ 7 **45.** $\sqrt{16}$ 4 **46.** $\sqrt{36}$ 6

47. $\sqrt{\dfrac{1}{9}}$ $\dfrac{1}{3}$ **48.** $\sqrt{\dfrac{1}{64}}$ $\dfrac{1}{8}$ **49.** $\sqrt{\dfrac{25}{81}}$ $\dfrac{5}{9}$ **50.** $\sqrt{\dfrac{49}{100}}$ $\dfrac{7}{10}$

Objective 5: Order of Operations

For Exercises 51–80, use the order of operations to simplify the expressions. **(See Examples 4–5.)**

51. $8 + 2 \cdot 6$ 20 **52.** $7 + 3 \cdot 4$ 19 **53.** $(8 + 2) \cdot 6$ 60 **54.** $(7 + 3) \cdot 4$ 40

55. $4 + 2 \div 2 \cdot 3 + 1$ 8 **56.** $5 + 12 \div 2 \cdot 6 - 1$ 40 **57.** $81 - 4 \cdot 3 + 3^2$ 78 **58.** $100 - 25 \cdot 2 - 5^2$ 25

59. $\dfrac{1}{4} \cdot \dfrac{2}{3} - \dfrac{1}{6}$ 0 **60.** $\dfrac{3}{4} \cdot \dfrac{2}{3} + \dfrac{2}{3}$ $\dfrac{7}{6}$ **61.** $\left(\dfrac{11}{6} - \dfrac{3}{8}\right) \cdot \dfrac{4}{5}$ $\dfrac{7}{6}$ **62.** $\left(\dfrac{9}{8} - \dfrac{1}{3}\right) \cdot \dfrac{3}{4}$ $\dfrac{19}{32}$

63. $3[5 + 2(8 - 3)]$ 45 **64.** $2[4 + 3(6 - 4)]$ 20 **65.** $10 + |-6|$ 16 **66.** $18 + |-3|$ 21

67. $21 - |8 - 2|$ 15 **68.** $12 - |6 - 1|$ 7 **69.** $2^2 + \sqrt{9} \cdot 5$ 19 **70.** $3^2 + \sqrt{16} \cdot 2$ 17

71. $\sqrt{9 + 16} - 2$ 3 **72.** $\sqrt{36 + 13} - 5$ 2 **73.** $[4^2 \cdot (6 - 4) \div 8] + [7 \cdot (8 - 3)]$ 39

74. $(18 \div \sqrt{4}) \cdot \{[(9^2 - 1) \div 2] - 15\}$ 225 **75.** $48 - 13 \cdot 3 + [(50 - 7 \cdot 5) + 2]$ 26

76. $80 \div 16 \cdot 2 + (6^2 - |-2|)$ 44 **77.** $\dfrac{7 + 3(8 - 2)}{(7 + 3)(8 - 2)}$ $\dfrac{5}{12}$ **78.** $\dfrac{16 - 8 \div 4}{4 + 8 \div 4 - 2}$ $\dfrac{7}{2}$

79. $\dfrac{15 - 5(3 \cdot 2 - 4)}{10 - 2(4 \cdot 5 - 16)}$ $\dfrac{5}{2}$ **80.** $\dfrac{5(7 - 3) + 8(6 - 4)}{4[7 + 3(2 \cdot 9 - 8)]}$ $\dfrac{9}{37}$

81. The area of a rectangle is given by $A = \ell w$, where ℓ is the length of the rectangle and w is the width. Find the area for the rectangle shown. 57,600 ft²

160 ft

360 ft

82. The perimeter of a rectangle is given by $P = 2\ell + 2w$. Find the perimeter for the rectangle shown. 1040 ft

83. The area of a trapezoid is given by $A = \frac{1}{2}(b_1 + b_2)h$, where b_1 and b_2 are the lengths of the two parallel sides and h is the height. A window is in the shape of a trapezoid. Find the area of the trapezoid with dimensions shown in the figure. 21 ft^2

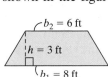

 84. The volume of a rectangular solid is given by $V = \ell w h$, where ℓ is the length of the box, w is the width, and h is the height. Find the volume of the box shown in the figure. 1000 yd^3

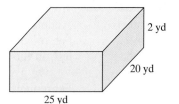

Objective 6: Translations

For Exercises 85–96, translate each English phrase into an algebraic expression. **(See Example 6.)**

85. The product of 3 and x $3x$

86. The sum of b and 6 $b + 6$

87. The quotient of x and 7 $\frac{x}{7}$ or $x \div 7$

88. Four divided by k $\frac{4}{k}$ or $4 \div k$

89. The difference of 2 and a $2 - a$

90. Three subtracted from t $t - 3$

91. x more than twice y $2y + x$

92. Nine decreased by the product of 3 and p $9 - 3p$

93. Four times the sum of x and 12 $4(x + 12)$

94. Twice the difference of x and 3 $2(x - 3)$

95. Q less than 3 $3 - Q$

96. Fourteen less than t $t - 14$

For Exercises 97–104, translate the English phrase into an algebraic expression. Then evaluate the expression for $x = 4$, $y = 2$, and $z = 10$. **(See Example 7.)**

97. Two times y cubed $2y^3$; 16

98. Three times z squared $3z^2$; 300

99. The absolute value of the difference of z and 8 $|z - 8|$; 2

100. The absolute value of the difference of x and 3 $|x - 3|$; 1

101. The product of 5 and the square root of x $5\sqrt{x}$; 10

102. The square root of the difference of z and 1 $\sqrt{z - 1}$; 3

103. The value x subtracted from the product of y and z $yz - x$; 16

104. The difference of z and the product of x and y $z - xy$; 2

Expanding Your Skills

For Exercises 105–108, use the order of operations to simplify the expressions.

105. $\dfrac{\sqrt{\frac{1}{9}} + \frac{2}{3}}{\sqrt{\frac{4}{25}} + \frac{3}{5}}$ 1

106. $\dfrac{5 - \sqrt{9}}{\sqrt{\frac{4}{9}} + \frac{1}{3}}$ 2

107. $\dfrac{|-2|}{|-10| - |2|}$ $\frac{1}{4}$

108. $\dfrac{|-4|^2}{2^2 + \sqrt{144}}$ 1

109. Some students use the following common memorization device (mnemonic) to help them remember the order of operations: the acronym PEMDAS or **P**lease **E**xcuse **M**y **D**ear **A**unt **S**ally to remember **P**arentheses, **E**xponents, **M**ultiplication, **D**ivision, **A**ddition, and **S**ubtraction. The problem with this mnemonic is that it suggests that multiplication is done before division and similarly, it suggests that addition is performed before subtraction. Explain why following this acronym may give incorrect answers for the expressions:

a. $36 \div 4 \cdot 3$
$36 \div 4 \cdot 3$ Division must be performed before multiplication.
$= 9 \cdot 3$
$= 27$

b. $36 - 4 + 3$
$36 - 4 + 3$ Subtraction must be performed before addition.
$= 32 + 3$
$= 35$

110. If you use the acronym **P**lease **E**xcuse **M**y **D**ear **A**unt **S**ally to remember the order of operations, what must you keep in mind about the last four operations?   Multiplication or division is performed in order from left to right. Addition or subtraction is performed in order from left to right.

111. Explain why the acronym **P**lease **E**xcuse **D**r. **M**ichael **S**mith's **A**unt could also be used as a memory device for the order of operations.   This is acceptable, provided division and multiplication are performed in order from left to right, and subtraction and addition are performed in order from left to right.

 Writing   Translating Expression   Geometry   Scientific Calculator   Video   NS/E

Section 1.3 Addition of Real Numbers

Objectives

1. Addition of Real Numbers and the Number Line
2. Addition of Real Numbers
3. Translations
4. Applications Involving Addition of Real Numbers

1. Addition of Real Numbers and the Number Line

Adding real numbers can be visualized on the number line. To add a positive number, move to the right on the number line. To add a negative number, move to the left on the number line. The following example may help to illustrate the process.

On a winter day in Detroit, suppose the temperature starts out at 5 degrees Fahrenheit ($5°F$) at noon, and then drops $12°$ two hours later when a cold front passes through. The resulting temperature can be represented by the expression $5° + (-12°)$. On the number line, start at 5 and count 12 units to the left (Figure 1-3). The resulting temperature at 2:00 P.M. is $-7°F$.

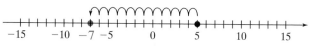

Figure 1-3

Skill Practice

Use the number line to add the numbers.

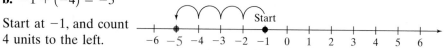

1. $-2 + 4$
2. $-2 + (-3)$
3. $5 + (-6)$

Classroom Examples: p. 71, Exercises 10 and 12

TIP: Note that we move to the left when we add a negative number, and we move to the right when we add a positive number.

Example 1 Using the Number Line to Add Real Numbers

Use the number line to add the numbers.

a. $-5 + 2$ **b.** $-1 + (-4)$ **c.** $4 + (-7)$

Solution:

a. $-5 + 2 = -3$

Start at -5, and count 2 units to the right.

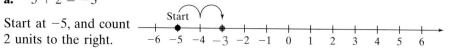

b. $-1 + (-4) = -5$

Start at -1, and count 4 units to the left.

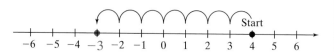

c. $4 + (-7) = -3$

Start at 4, and count 7 units to the left.

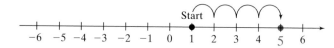

2. Addition of Real Numbers

When adding large numbers or numbers that involve fractions or decimals, counting units on the number line can be cumbersome. Study the following example to determine a pattern for adding two numbers with the *same* sign.

$1 + 4 = 5$

$-1 + (-4) = -5$

Answers

1. 2   **2.** -5 **3.** -1

PROCEDURE Adding Numbers with the *Same* Sign

To add two numbers with the *same* sign, add their absolute values and apply the common sign.

Study the following example to determine a pattern for adding two numbers with *different* signs.

$1 + (-4) = -3$

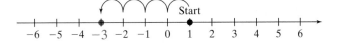

$-1 + 4 = 3$

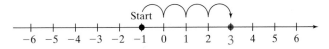

Instructor Note: Memory Device:
Same Sign ⇒ Sum
Different Sign ⇒ Difference

PROCEDURE Adding Numbers with *Different* Signs

To add two numbers with *different* signs, subtract the smaller absolute value from the larger absolute value. Then apply the sign of the number having the larger absolute value.

Example 2 **Adding Real Numbers with the Same Sign**

Add.

a. $-12 + (-14)$ **b.** $-8.8 + (-3.7)$ **c.** $-\dfrac{4}{3} + \left(-\dfrac{6}{7}\right)$

Solution:

a.
$$-12 + (-14)$$

$$= -(12 + 14)$$

common sign is negative

$$= -26$$

First find the absolute value of the addends.
$|-12| = 12$ and $|-14| = 14$.

Add their absolute values and apply the common sign (in this case, the common sign is negative).

The sum is -26.

b.
$$-8.8 + (-3.7)$$

$$= -(8.8 + 3.7)$$

common sign is negative

$$= -12.5$$

First find the absolute value of the addends.
$|-8.8| = 8.8$ and $|-3.7| = 3.7$.

Add their absolute values and apply the common sign (in this case, the common sign is negative).

The sum is -12.5.

Skill Practice

Add the numbers.
 4. $-5 + (-25)$
 5. $-14.8 + (-9.7)$
 6. $-\dfrac{1}{2} + \left(-\dfrac{5}{8}\right)$

Classroom Examples: pp. 71–72, Exercises 22 and 52

Answers

4. -30 **5.** -24.5 **6.** $-\dfrac{9}{8}$

c. $-\dfrac{4}{3} + \left(-\dfrac{6}{7}\right)$ The least common denominator (LCD) is 21.

$= -\dfrac{4 \cdot 7}{3 \cdot 7} + \left(-\dfrac{6 \cdot 3}{7 \cdot 3}\right)$ Write each fraction with the LCD.

$= -\dfrac{28}{21} + \left(-\dfrac{18}{21}\right)$ Find the absolute value of the addends.

$$\left|-\dfrac{28}{21}\right| = \dfrac{28}{21} \text{ and } \left|-\dfrac{18}{21}\right| = \dfrac{18}{21}.$$

$= -\left(\dfrac{28}{21} + \dfrac{18}{21}\right)$ Add their absolute values and apply the common sign (in this case, the common sign is negative).

common sign is negative

$= -\dfrac{46}{21}$ The sum is $-\dfrac{46}{21}$.

Example 3   **Adding Real Numbers with Different Signs**

Add.   **a.** $12 + (-17)$ **b.** $-8 + 8$

Solution:

a. $12 + (-17)$ First find the absolute value of the addends. $|12| = 12$ and $|-17| = 17$.

The absolute value of -17 is greater than the absolute value of 12. Therefore, the sum is negative.

$= -(17 - 12)$ Next, subtract the smaller absolute value from the larger absolute value.

Apply the sign of the number with the larger absolute value.

$= -5$

b. $-8 + 8$ First find the absolute value of the addends. $|-8| = 8$ and $|8| = 8$.

$= (8 - 8)$ The absolute values are equal. Therefore, their difference is 0. The number zero is neither positive nor negative.

$= 0$

Example 4   **Adding Real Numbers with Different Signs**

Add.   **a.** $-10.6 + 20.4$ **b.** $\dfrac{2}{15} + \left(-\dfrac{4}{5}\right)$

Answers

7. 1   **8.** 0   **9.** 9.2   **10.** $-\dfrac{1}{2}$

Solution:

a. $-10.6 + 20.4$ First find the absolute value of the addends.
$|-10.6| = 10.6$ and $|20.4| = 20.4$.

The absolute value of 20.4 is greater than the absolute value of -10.6. Therefore, the sum is positive.

$= +(20.4 - 10.6)$ Next, subtract the smaller absolute value from the larger absolute value.

Apply the sign of the number with the larger absolute value.

$= 9.8$

b. $\dfrac{2}{15} + \left(-\dfrac{4}{5}\right)$ The least common denominator is 15.

$= \dfrac{2}{15} + \left(-\dfrac{4 \cdot 3}{5 \cdot 3}\right)$ Write each fraction with the LCD.

$= \dfrac{2}{15} + \left(-\dfrac{12}{15}\right)$ Find the absolute value of the addends.

$$\left|\dfrac{2}{15}\right| = \dfrac{2}{15} \text{ and } \left|-\dfrac{12}{15}\right| = \dfrac{12}{15}.$$

The absolute value of $-\dfrac{12}{15}$ is greater than the absolute value of $\dfrac{2}{15}$. Therefore, the sum is negative.

$= -\left(\dfrac{12}{15} - \dfrac{2}{15}\right)$ Next, subtract the smaller absolute value from the larger absolute value.

Apply the sign of the number with the larger absolute value.

$= -\dfrac{10}{15}$ Subtract.

$= -\dfrac{2}{3}$ Simplify by reducing to lowest terms. $-\dfrac{\overset{2}{\cancel{10}}}{\underset{3}{\cancel{15}}} = -\dfrac{2}{3}$

3. Translations

Example 5   **Translating Expressions Involving the Addition of Real Numbers**

Translate each English phrase into an algebraic expression. Then simplify the result.

a. The sum of $-12, -8, 9$, and -1

b. Negative three-tenths added to $-\dfrac{7}{8}$

c. The sum of -12 and its opposite

Solution:

a. The sum of $-12, -8, 9$, and -1

$\underbrace{-12 + (-8)} + 9 + (-1)$

$= \underbrace{-20 + 9} + (-1)$ Add from left to right.

$= \underbrace{-11 + (-1)}$

$= -12$

b. Negative three-tenths added to $-\frac{7}{8}$

$$-\frac{7}{8} + \left(-\frac{3}{10}\right)$$

$$= -\frac{35}{40} + \left(-\frac{12}{40}\right) \qquad \text{The common denominator is 40.}$$

$$= -\frac{47}{40} \qquad \text{The numbers have the same signs. Add their absolute values and keep the common sign. } -\left(\frac{35}{40} + \frac{12}{40}\right).$$

c. The sum of -12 and its opposite

$$-12 + (12)$$

$$= 0 \qquad \text{Add.}$$

TIP: The sum of any number and its opposite is 0.

4. Applications Involving Addition of Real Numbers

Example 6 Adding Real Numbers in Applications

a. A running back on a football team gains 4 yd. On the next play, the quarterback is sacked and loses 13 yd. Write a mathematical expression to describe this situation and then simplify the result.

b. A student has $120 in her checking account. After depositing her paycheck of $215, she writes a check for $255 to cover her portion of the rent and another check for $294 to cover her car payment. Write a mathematical expression to describe this situation and then simplify the result.

Solution:

a. $4 + (-13)$ The loss of 13 yd can be interpreted as adding -13 yd.

     $= -9$ The football team has a net loss of 9 yd.

b. $\underbrace{120 + 215} + (-255) + (-294)$ Writing a check is equivalent to adding a negative amount to the bank account.

    $= \underbrace{335 + (-255)} + (-294)$ Use the order of operations. Add from left to right.

    $= \quad 80 + (-294)$

    $= \quad\quad -214$ The student has overdrawn her account by $214.

Skill Practice

14. GE stock was priced at $32.00 per share at the beginning of the month. After the first week, the price went up $2.15 per share. At the end of the second week it went down $3.28 per share. Write a mathematical expression to describe the price of the stock and find the price of the stock at the end of the 2-week period.

Classroom Example: p. 73, Exercise 94

Answer

14. $32.00 + 2.15 + (-3.28)$; $30.87 per share

Section 1.3 Practice Exercises

Study Skills Exercise

1. It is very important to attend class every day. Math is cumulative in nature, and you must master the material learned in the previous class to understand today's lesson. Because this is so important, many instructors have an attendance policy that may affect your final grade. Write down the attendance policy for your class.

Review Exercises

Plot the points in set A on a number line. Then for Exercises 2–7 place the appropriate inequality ($<$, $>$) between the numbers.

$$A = \left\{ -2, \frac{3}{4}, -\frac{5}{2}, 3, \frac{9}{2}, 1.6, 0 \right\}$$

2. $-2 \;\square\; 0$ $<$

3. $\dfrac{9}{2} \;\square\; \dfrac{3}{4}$ $>$

4. $-2 \;\square\; -\dfrac{5}{2}$ $>$

5. $0 \;\square\; -\dfrac{5}{2}$ $>$

6. $\dfrac{3}{4} \;\square\; 1.6$ $<$

7. $\dfrac{3}{4} \;\square\; -\dfrac{5}{2}$ $>$

8. Evaluate the expressions.

    a. $-(-8)$ 8   **b.** $-|-8|$ -8

Objective 1: Addition of Real Numbers and the Number Line

For Exercises 9–16, add the numbers using the number line. **(See Example 1.)**

9. $-2 + (-4)$ -6

10. $-3 + (-5)$ -8

11. $-7 + 10$ 3

12. $-2 + 9$ 7

13. $6 + (-3)$ 3

14. $8 + (-2)$ 6

15. $2 + (-5)$ -3

16. $7 + (-3)$ 4

Objective 2: Addition of Real Numbers

For Exercises 17–70, add. **(See Examples 2–4.)**

17. $-19 + 2$ -17

18. $-25 + 18$ -7

19. $-4 + 11$ 7

20. $-3 + 9$ 6

21. $-16 + (-3)$ -19

22. $-12 + (-23)$ -35

23. $-2 + (-21)$ -23

24. $-13 + (-1)$ -14

25. $0 + (-5)$ -5

26. $0 + (-4)$ -4

27. $-3 + 0$ -3

28. $-8 + 0$ -8

29. $-16 + 16$ 0

30. $11 + (-11)$ 0

31. $41 + (-41)$ 0

32. $-15 + 15$ 0

33. $4 + (-9)$ -5

34. $6 + (-9)$ -3

35. $7 + (-2) + (-8)$ -3

36. $2 + (-3) + (-6)$ -7

37. $-17 + (-3) + 20$ 0

38. $-9 + (-6) + 15$ 0

39. $-3 + (-8) + (-12)$ -23

40. $-8 + (-2) + (-13)$ -23

41. $-42 + (-3) + 45 + (-6)$ -6

42. $36 + (-3) + (-8) + (-25)$ 0

43. $-5 + (-3) + (-7) + 4 + 8$ -3

44. $-13 + (-1) + 5 + 2 + (-20)$ -27

45. $23.81 + (-2.51)$ 21.3   **46.** $-9.23 + 10.53$ 1.3   **47.** $-\dfrac{2}{7} + \dfrac{1}{14}$ $-\dfrac{3}{14}$ **48.** $-\dfrac{1}{8} + \dfrac{5}{16}$ $\dfrac{3}{16}$

49. $\dfrac{2}{3} + \left(-\dfrac{5}{6}\right)$ $-\dfrac{1}{6}$ **50.** $\dfrac{1}{2} + \left(-\dfrac{3}{4}\right)$ $-\dfrac{1}{4}$ **51.** $-\dfrac{7}{8} + \left(-\dfrac{1}{16}\right)$ $-\dfrac{15}{16}$ **52.** $-\dfrac{1}{9} + \left(-\dfrac{4}{3}\right)$ $-\dfrac{13}{9}$

53. $-\dfrac{1}{4} + \dfrac{3}{10}$ $\dfrac{1}{20}$ **54.** $-\dfrac{7}{6} + \dfrac{7}{8}$ $-\dfrac{7}{24}$ **55.** $-2.1 + \left(-\dfrac{3}{10}\right)$ -2.4 or $-\dfrac{12}{5}$ **56.** $-8.3 + \left(-\dfrac{9}{10}\right)$ -9.2 or $-\dfrac{46}{5}$

57. $\dfrac{3}{4} + (-0.5)$ $\dfrac{1}{4}$ or 0.25   **58.** $-\dfrac{3}{2} + 0.45$ $-\dfrac{21}{20}$ or -1.05 **59.** $8.23 + (-8.23)$ 0   **60.** $-7.5 + 7.5$ 0

61. $-\dfrac{7}{8} + 0$ $-\dfrac{7}{8}$ **62.** $0 + \left(-\dfrac{21}{22}\right)$ $-\dfrac{21}{22}$ **63.** $-\dfrac{3}{2} + \left(-\dfrac{1}{3}\right) + \dfrac{5}{6}$ -1 **64.** $-\dfrac{7}{8} + \dfrac{7}{6} + \dfrac{7}{12}$ $\dfrac{7}{8}$

65. $-\dfrac{2}{3} + \left(-\dfrac{1}{9}\right) + 2$ $\dfrac{11}{9}$ **66.** $-\dfrac{1}{4} + \left(-\dfrac{3}{2}\right) + 2$ $\dfrac{1}{4}$ **67.** $-47.36 + 24.28$ -23.08 **68.** $-0.015 + (0.0026)$ -0.0124

69. $-0.000617 + (-0.0015)$ -0.002117 **70.** $-5315.26 + (-314.89)$ -5630.15

71. State the rule for adding two numbers with different signs. To add two numbers with different signs, subtract the smaller absolute value from the larger absolute value and apply the sign of the number with the larger absolute value.

72. State the rule for adding two numbers with the same signs.
To add two numbers with the same sign, add their absolute values and apply the common sign.

For Exercises 73–80, evaluate the expression for $x = -3$, $y = -2$, and $z = 16$.

73. $x + y + \sqrt{z}$ -1 **74.** $2z + x + y$ 27   **75.** $y + 3\sqrt{z}$ 10   **76.** $-\sqrt{z} + y$ -6

77. $|x| + |y|$ 5   **78.** $z + x + |y|$ 15   **79.** $-x + y$ 1   **80.** $x + (-y) + z$ 15

Objective 3: Translations

For Exercises 81–90, translate the English phrase into an algebraic expression. Then simplify the result.
(See Example 5.)

81. The sum of -6 and -10 $-6 + (-10)$; -16

82. The sum of -3 and 5 $-3 + 5$; 2

83. Negative three increased by 8 $-3 + 8$; 5

84. Twenty-one increased by 4 $21 + 4$; 25

85. Seventeen more than -21 $-21 + 17$; -4

86. Twenty-four more than -7 $-7 + 24$; 17

87. Three times the sum of -14 and 20 $3(-14 + 20)$; 18

88. Two times the sum of 6 and -10 $2(6 + (-10))$; -8

89. Five more than the sum of -7 and -2 $(-7 + (-2)) + 5$; -4

90. Negative six more than the sum of 4 and -1 $(4 + (-1)) + (-6)$; -3

 Writing   Translating Expression   Geometry   Scientific Calculator   Video   NS E

Objective 4: Applications Involving Addition of Real Numbers

91. The temperature in Minneapolis, Minnesota, began at −5°F (5° below zero) at 6:00 A.M. By noon the temperature had risen 13°, and by the end of the day, the temperature had dropped 11° from its noontime high. Write an expression using addition that describes the change in temperatures during the day. Then evaluate the expression to give the temperature at the end of the day.
−5 + 13 + (−11); −3°F

92. The temperature in Toronto, Ontario, Canada, began at 4°F. A cold front went through at noon, and the temperature dropped 9°. By 4:00 P.M. the temperature had risen 2° from its noontime low. Write an expression using addition that describes the changes in temperature during the day. Then evaluate the expression to give the temperature at the end of the day. 4 + (−9) + 2; −3°F

93. During a football game, the Nebraska Cornhuskers lost 2 yd, gained 6 yd, and then lost 5 yd. Write an expression using addition that describes the team's total loss or gain and evaluate the expression. **(See Example 6.)**
−2 + 6 + (−5); −1 yd or 1-yd loss

94. During a football game, the University of Oklahoma's team gained 3 yd, lost 5 yd, and then gained 14 yd. Write an expression using addition that describes the team's total loss or gain and evaluate the expression.
3 + (−5) + 14; 12-yd gain

95. Yoshima has $52.23 in her checking account. She writes a check for groceries for $52.95. **(See Example 6.)**

    a. Write an addition problem that expresses Yoshima's transaction.
52.23 + (−52.95)

    b. Is Yoshima's account overdrawn? Yes

96. Mohammad has $40.02 in his checking account. He writes a check for a pair of shoes for $40.96.

    a. Write an addition problem that expresses Mohammad's transaction. 40.02 + (−40.96)

    b. Is Mohammad's account overdrawn? Yes

97. In the game show *Jeopardy*, a contestant responds to six questions with the following outcomes:

$$\$100, \$200, -\$500, \$300, \$100, -\$200$$

    a. Write an expression using addition to describe the contestant's scoring activity.
100 + 200 + (−500) + 300 + 100 + (−200)

    b. Evaluate the expression from part (a) to determine the contestant's final outcome. $0

98. A company that has been in business for 5 yr has the following profit and loss record.

    a. Write an expression using addition to describe the company's profit/loss activity.
−50,000 + (−32,000) + (−5000) + 13,000 + 26,000

    b. Evaluate the expression from part (a) to determine the company's net profit or loss.
−$48,000

Year	Profit/Loss ($)
1	−50,000
2	−32,000
3	−5000
4	13,000
5	26,000

 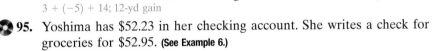

Section 1.4 | Subtraction of Real Numbers

Objectives

1. **Subtraction of Real Numbers**
2. **Translations**
3. **Applications Involving Subtraction**
4. **Applying the Order of Operations**

1. Subtraction of Real Numbers

In Section 1.3, we learned the rules for adding real numbers. Subtraction of real numbers is defined in terms of the addition process. For example, consider the following subtraction problem and the corresponding addition problem:

$$6 - 4 = 2 \quad \Leftrightarrow \quad 6 + (-4) = 2$$

In each case, we start at 6 on the number line and move to the left 4 units. That is, adding the opposite of 4 produces the same result as subtracting 4. This is true in general. To subtract two real numbers, add the opposite of the second number to the first number.

> **PROCEDURE** Subtracting Real Numbers
>
> If a and b are real numbers, then $a - b = a + (-b)$.

$$\left. \begin{array}{l} 10 - 4 = 10 + (-4) = 6 \\ -10 - 4 = -10 + (-4) = -14 \end{array} \right\} \quad \text{Subtracting 4 is the same as adding } -4.$$

$$\left. \begin{array}{l} 10 - (-4) = 10 + (4) = 14 \\ -10 - (-4) = -10 + (4) = -6 \end{array} \right\} \quad \text{Subtracting } -4 \text{ is the same as adding 4.}$$

Skill Practice

Subtract.
1. $1 - (-3)$
2. $-2 - 2$
3. $-6 - (-11)$
4. $8 - 15$

Classroom Examples: p. 80, Exercises 16, 18, and 20

Example 1 Subtracting Integers

Subtract the numbers.

a. $4 - (-9)$ **b.** $-6 - 9$ **c.** $-11 - (-5)$ **d.** $7 - 10$

Solution:

a. $4 - (-9)$

$= 4 + (9) = 13$

     Take the opposite of -9.

Change subtraction to addition.

b. $-6 - 9$

$= -6 + (-9) = -15$

     Take the opposite of 9.

Change subtraction to addition.

c. $-11 - (-5)$

$= -11 + (5) = -6$

     Take the opposite of -5.

Change subtraction to addition.

d. $7 - 10$

$= 7 + (-10) = -3$

     Take the opposite of 10.

Change subtraction to addition.

Answers

1. 4   **2.** -4 **3.** 5   **4.** -7

Example 2 Subtracting Real Numbers

a. $\dfrac{3}{20} - \left(-\dfrac{4}{15}\right)$ **b.** $-2.3 - 6.04$

Solution:

a. $\dfrac{3}{20} - \left(-\dfrac{4}{15}\right)$ The least common denominator is 60.

$\dfrac{9}{60} - \left(-\dfrac{16}{60}\right)$ Write equivalent fractions with the LCD.

$\dfrac{9}{60} + \left(\dfrac{16}{60}\right)$ Rewrite subtraction in terms of addition.

$\dfrac{25}{60}$ Add.

$\dfrac{5}{12}$ Reduce to lowest terms.

b. $-2.3 - 6.04$

$-2.3 + (-6.04)$ Rewrite subtraction in terms of addition.

-8.34 Add.

Classroom Example: p. 80, Exercise 40

2. Translations

Example 3 Translating Expressions Involving Subtraction

Write an algebraic expression for each English phrase and then simplify the result.

a. The difference of -7 and -5

b. 12.4 subtracted from -4.7

c. -24 decreased by the sum of -10 and 13

d. Seven-fourths less than one-third

Solution:

a. The difference of -7 and -5

$-7 - (-5)$

$= -7 + (5)$ Rewrite subtraction in terms of addition.

$= -2$ Simplify.

b. 12.4 subtracted from -4.7

$-4.7 - 12.4$

$= -4.7 + (-12.4)$ Rewrite subtraction in terms of addition.

$= -17.1$ Simplify.

Skill Practice

Subtract.

5. $\dfrac{1}{6} - \left(-\dfrac{7}{12}\right)$

6. $-7.5 - 1.5$

Skill Practice

Write an algebraic expression for each phrase and then simplify.

7. 8 less than -10

8. -7.2 subtracted from -8.2

9. 10 more than the difference of -2 and 3

10. Two-fifths decreased by four-thirds

Classroom Examples: p. 80, Exercises 64 and 70

TIP: Recall that "b subtracted from a" is translated as $a - b$. In Example 3(b), -4.7 is written first and then 12.4.

Answers

5. $\dfrac{3}{4}$ **6.** -9

7. $-10 - 8$; -18

8. $-8.2 - (-7.2)$; -1

9. $(-2 - 3) + 10$; 5

10. $\dfrac{2}{5} - \dfrac{4}{3}$; $-\dfrac{14}{15}$

TIP: Parentheses must be used around the sum of -10 and 13 so that -24 is decreased by the entire quantity $(-10 + 13)$.

c. -24 decreased by the sum of -10 and 13

$-24 - (-10 + 13)$

$\quad = -24 - (3)$ Simplify inside parentheses.

$\quad = -24 + (-3)$ Rewrite subtraction in terms of addition.

$\quad = -27$ Simplify.

d. Seven-fourths less than one-third

$$\frac{1}{3} - \frac{7}{4}$$

$$= \frac{1}{3} + \left(-\frac{7}{4}\right) \qquad \text{Rewrite subtraction in terms of addition.}$$

$$= \frac{4}{12} + \left(-\frac{21}{12}\right) \qquad \text{The common denominator is 12.}$$

$$= -\frac{17}{12} \quad \text{or} \quad -1\frac{5}{12}$$

3. Applications Involving Subtraction

Skill Practice

11. During Harold's first round on *Jeopardy*, he got the $100, $200, and $400 questions correct but he got the $300 and $500 questions incorrect. Determine Harold's score for this round.

Classroom Example: p. 80, Exercise 72

Example 4   **Using Subtraction of Real Numbers in an Application**

During one of his turns on *Jeopardy*, Harold selected the category "Show Tunes." He got the $200, $600, and $1000 questions correct, but he got the $400 and $800 questions incorrect. Write an expression that determines Harold's score. Then simplify the expression to find his total winnings for that category.

Solution:

$200 + 600 + 1000 - 400 - 800$

$\quad = 200 + 600 + 1000 + (-400) + (-800)$ Add the positive numbers.
                                                      Add the negative numbers.

$\quad = 1800 + (-1200)$

$\quad = 600$ Harold won $600.

Answer

11. -100, Harold lost $100.

Example 5 **Using Subtraction of Real Numbers in an Application**

The highest recorded temperature in North America was 134°F, recorded on July 10, 1913, in Death Valley, California. The lowest temperature of −81°F was recorded on February 3, 1947, in Snag, Yukon, Canada.

Find the difference between the highest and lowest recorded temperatures in North America.

Solution:

$134 - (-81)$

$= 134 + (81)$ Rewrite subtraction in terms of addition.

$= 215$ Add.

The difference between the highest and lowest temperatures is 215°F.

Skill Practice

12. The record high temperature for the state of Montana occurred in 1937 and was 117°F. The record low occurred in 1954 and was −70°F. Find the difference between the highest and lowest temperatures.

Classroom Example: p. 81, Exercise 74

4. Applying the Order of Operations

Example 6 **Applying the Order of Operations**

Simplify the expressions.

a. $-6 + \{10 - [7 - (-4)]\}$ **b.** $5 - \sqrt{35 - (-14)} - 2$

Skill Practice

Simplify the expressions.
13. $-11 - \{8 - [2 - (-3)]\}$
14. $(12 - 5)^2 + \sqrt{4 - (-21)}$

Classroom Examples: p. 81, Exercises 86 and 92

Solution:

a. $-6 + \{10 - [7 - (-4)]\}$ Work inside the inner brackets first.

$= -6 + \{10 - [7 + (4)]\}$ Rewrite subtraction in terms of addition.

$= -6 + \{10 - (11)\}$ Simplify the expression inside braces.

$= -6 + \{10 + (-11)\}$ Rewrite subtraction in terms of addition.

$= -6 + (-1)$

$= -7$ Add.

b. $5 - \sqrt{35 - (-14)} - 2$ Work inside the radical first.

$= 5 - \sqrt{35 + (14)} - 2$ Rewrite subtraction in terms of addition.

$= 5 - \sqrt{49} - 2$

$= 5 - 7 - 2$ Simplify the radical.

$= 5 + (-7) + (-2)$ Rewrite subtraction in terms of addition.

$= -2 + (-2)$ Add from left to right.

$= -4$

Answers
12. 187°F   **13.** −14   **14.** 54

Skill Practice

Simplify the expressions.

15. $\left(-1 + \dfrac{1}{4}\right) - \left(\dfrac{3}{4} - \dfrac{1}{2}\right)$

16. $4 - 2\,|6 + (-8)| + (4)^2$

Classroom Examples: p. 81,
Exercises 90 and 94

Example 7   Applying the Order of Operations

Simplify the expressions.

a. $\left(-\dfrac{5}{8} - \dfrac{2}{3}\right) - \left(\dfrac{1}{8} + 2\right)$ **b.** $-6 - |7 - 11| + (-3 + 7)^2$

Solution:

a. $\left(-\dfrac{5}{8} - \dfrac{2}{3}\right) - \left(\dfrac{1}{8} + 2\right)$ Work inside the parentheses first.

$= \left[-\dfrac{5}{8} + \left(-\dfrac{2}{3}\right)\right] - \left(\dfrac{1}{8} + 2\right)$ Rewrite subtraction in terms of addition.

$= \left[-\dfrac{15}{24} + \left(-\dfrac{16}{24}\right)\right] - \left(\dfrac{1}{8} + \dfrac{16}{8}\right)$ Get a common denominator in each parentheses.

$= \left(-\dfrac{31}{24}\right) - \left(\dfrac{17}{8}\right)$ Add fractions in each parentheses.

$= \left(-\dfrac{31}{24}\right) + \left(-\dfrac{17}{8}\right)$ Rewrite subtraction in terms of addition.

$= -\dfrac{31}{24} + \left(-\dfrac{51}{24}\right)$ Get a common denominator.

$= -\dfrac{82}{24}$ Add.

$= -\dfrac{41}{12}$ Reduce to lowest terms.

b. $-6 - |7 - 11| + (-3 + 7)^2$ Simplify within absolute value bars and parentheses first.

$= -6 - |7 + (-11)| + (-3 + 7)^2$ Rewrite subtraction in terms of addition.

$= -6 - |-4| + (4)^2$

$= -6 - (4) + 16$ Simplify absolute value and exponent.

$= -6 + (-4) + 16$ Rewrite subtraction in terms of addition.

$= -10 + 16$ Add from left to right.

$= 6$

Answers
15. -1 **16.** 16

Calculator Connections

Topic: Operations with Signed Numbers on a Calculator

Most calculators can add, subtract, multiply, and divide signed numbers. It is important to note, however, that the key used for the negative sign is different from the key used for subtraction. On a scientific calculator, the `+/-` key or `+c-` key is used to enter a negative number or to change the sign of an existing number. On a graphing calculator, the `(-)` key is used. These keys should not be confused with the `-` key which is used for subtraction. For example, try simplifying the following expressions.

a. $-7 + (-4) - 6$ **b.** $-3.1 - (-0.5) + 1.1$

Scientific Calculator:

Enter: 7 [+◦-] [+] [(] 4 [+◦-] [)] [−] 6 [=]   **Result:** | −17 |

Enter: 3.1 [+◦-] [−] [(] 0.5 [+◦-] [)] [+] 1.1 [=]   **Result:** | −1.5 |

Graphing Calculator:

```
-7+(-4)-6
             -17
-3.1-(-0.5)+1.1
             -1.5
```

Calculator Exercises

Simplify the expression without the use of a calculator. Then use the calculator to verify your answer.

1. $-8 + (-5)$ −13

2. $4 + (-5) + (-1)$ −2

3. $627 - (-84)$ 711

4. $-0.06 - 0.12$
   −0.18

5. $-3.2 - (-14.5)$ 11.3

6. $-472 + (-518)$ −990

7. $-12 - 9 + 4$ −17

8. $209 - 108 + (-63)$
   38

Section 1.4   Practice Exercises

Boost _your_ GRADE at ALEKS.com!

ALEKS® version 3.0

- Practice Problems
- Self-Tests
- NetTutor

- e-Professors
- Videos

For additional exercises, see Classroom Activities 1.4A–1.4C in the _Instructor's Resource Manual_ at www.mhhe.com/moh.

Study Skills Exercise

1. Some instructors allow the use of calculators. What is your instructor's policy regarding calculators in class, on the homework, and on tests?

Helpful Hint: If you are not permitted to use a calculator on tests, it is a good idea to do your homework in the same way, without a calculator.

Review Exercises

 For Exercises 2–5, translate each English phrase into an algebraic expression.

2. The square root of 6 $\sqrt{6}$

3. The square of x x^2

4. Negative seven increased by 10 $-7 + 10$

5. Two more than $-b$ $-b + 2$

For Exercises 6–8, simplify the expression.

6. $4^2 - 6 \div 2$ 13

7. $1 + 36 \div 9 \cdot 2$ 9

8. $14 - |10 - 6|$ 10

Objective 1: Subtraction of Real Numbers

For Exercises 9–14, fill in the blank to make each statement correct.

9. $5 - 3 = 5 +$ ____−3____

10. $8 - 7 = 8 +$ ____−7____

11. $-2 - 12 = -2 +$ ____−12____

12. $-4 - 9 = -4 +$ ____−9____

13. $7 - (-4) = 7 +$ ____4____

14. $13 - (-4) = 13 +$ ____4____

Writing   Translating Expression   Geometry   Scientific Calculator   Video   NS&E

For Exercises 15–60, subtract. **(See Examples 1–2.)**

15. $3 - 5$ $_{-2}$
16. $9 - 12$ $_{-3}$
17. $3 - (-5)$ $_8$
18. $9 - (-12)$ $_{21}$

19. $-3 - 5$ $_{-8}$
20. $-9 - 12$ $_{-21}$
21. $-3 - (-5)$ $_2$
22. $-9 - (-5)$ $_{-4}$

23. $23 - 17$ $_6$
24. $14 - 2$ $_{12}$
25. $23 - (-17)$ $_{40}$
26. $14 - (-2)$ $_{16}$

27. $-23 - 17$ $_{-40}$
28. $-14 - 2$ $_{-16}$
29. $-23 - (-23)$ $_0$
30. $-14 - (-14)$ $_0$

31. $-6 - 14$ $_{-20}$
32. $-9 - 12$ $_{-21}$
33. $-7 - 17$ $_{-24}$
34. $-8 - 21$ $_{-29}$

35. $13 - (-12)$ $_{25}$
36. $20 - (-5)$ $_{25}$
37. $-14 - (-9)$ $_{-5}$
38. $-21 - (-17)$ $_{-4}$

39. $-\dfrac{6}{5} - \dfrac{3}{10}$ $_{-\frac{3}{2}}$
40. $-\dfrac{2}{9} - \dfrac{5}{3}$ $_{-\frac{17}{9}}$
41. $\dfrac{3}{8} - \left(-\dfrac{4}{3}\right)$ $_{\frac{41}{24}}$
42. $\dfrac{7}{10} - \left(-\dfrac{5}{6}\right)$ $_{\frac{23}{15}}$

43. $\dfrac{1}{2} - \dfrac{1}{10}$ $_{\frac{2}{5}}$
44. $\dfrac{2}{7} - \dfrac{3}{14}$ $_{\frac{1}{14}}$
45. $-\dfrac{11}{12} - \left(-\dfrac{1}{4}\right)$ $_{-\frac{2}{3}}$
46. $-\dfrac{7}{8} - \left(-\dfrac{1}{6}\right)$ $_{-\frac{17}{24}}$

47. $6.8 - (-2.4)$ $_{9.2}$
48. $7.2 - (-1.9)$ $_{9.1}$
49. $3.1 - 8.82$ $_{-5.72}$
50. $1.8 - 9.59$ $_{-7.79}$

51. $-4 - 3 - 2 - 1$ $_{-10}$
52. $-10 - 9 - 8 - 7$ $_{-34}$
53. $6 - 8 - 2 - 10$ $_{-14}$
54. $20 - 50 - 10 - 5$ $_{-45}$

55. $-36.75 - 14.25$ $_{-51}$
56. $-84.21 - 112.16$ $_{-196.37}$

57. $-112.846 + (-13.03) - 47.312$ $_{-173.188}$
58. $-96.473 + (-36.02) - 16.617$ $_{-149.11}$

59. $0.085 - (-3.14) + 0.018$ $_{3.243}$
60. $0.00061 - (-0.00057) + 0.0014$ $_{0.00258}$

Objective 2: Translations

For Exercises 61–70, translate each English phrase into an algebraic expression. Then evaluate the expression. **(See Example 3.)**

61. Six minus -7 $_{6 - (-7);\ 13}$
62. Eighteen minus -1 $_{18 - (-1);\ 19}$

63. Eighteen subtracted from 3 $_{3 - 18;\ -15}$
64. Twenty-one subtracted from 8 $_{8 - 21;\ -13}$

65. The difference of -5 and -11 $_{-5 - (-11);\ 6}$
66. The difference of -2 and -18 $_{-2 - (-18);\ 16}$

67. Negative thirteen subtracted from -1
$_{-1 - (-13);\ 12}$
68. Negative thirty-one subtracted from -19
$_{-19 - (-31);\ 12}$

69. Twenty less than -32 $_{-32 - 20;\ -52}$
70. Seven less than -3 $_{-3 - 7;\ -10}$

Objective 3: Applications Involving Subtraction

71. On the game, *Jeopardy*, Jasper selected the category "The Last." He got the first four questions correct (worth $200, $400, $600, and $800) but then missed the last question (worth $1000). Write an expression that determines Jasper's score. Then simplify the expression to find his total winnings for that category. **(See Example 4.)** $_{200 + 400 + 600 + 800 - 1000;\ \$1000}$

72. On Courtney's turn in *Jeopardy*, she chose the category "Birds of a Feather." She already had $1200 when she selected a Double Jeopardy question. She wagered $500 but guessed incorrectly (therefore she lost $500). On her next turn, she got the $800 question correct. Write an expression that determines Courtney's score. Then simplify the expression to find her total winnings for that category. $_{1200 - 500 + 800;\ \$1500}$

73. In Ohio, the highest temperature ever recorded was 113°F and the lowest was −39°F. Find the difference between the highest and lowest temperatures. (*Source: Information Please Almanac*) **(See Example 5.)**   152°F

74. In Mississippi, the highest temperature ever recorded was 115°F and the lowest was −19°F. Find the difference between the highest and lowest temperatures. (*Source: Information Please Almanac*)   134°F

75. The highest mountain in the world is Mt. Everest, located in the Himalayas. Its height is 8848 meters (m) (29,028 ft). The lowest recorded depth in the ocean is located in the Marianas Trench in the Pacific Ocean. Its "height" relative to sea level is −11,033 m (−36,198 ft). Determine the difference in elevation, in meters, between the highest mountain in the world and the deepest ocean trench. (*Source: Information Please Almanac*)   19,881 m

76. The lowest point in North America is located in Death Valley, California, at an elevation of −282 ft (−86 m). The highest point in North America is Mt. McKinley, Alaska, at an elevation of 20,320 ft (6194 m). Find the difference in elevation, in feet, between the highest and lowest points in North America. (*Source: Information Please Almanac*)   20,602 ft

Objective 4: Applying the Order of Operations

For Exercises 77–96, perform the indicated operations. **(See Examples 6–7.)**

77. $6 + 8 - (-2) - 4 + 1$ 13

78. $-3 - (-4) + 1 - 2 - 5$ −5

79. $-1 - 7 + (-3) - 8 + 10$ −9

80. $13 - 7 + 4 - 3 - (-1)$ 8

81. $2 - (-8) + 7 + 3 - 15$ 5

82. $8 - (-13) + 1 - 9$ 13

83. $-6 + (-1) + (-8) + (-10)$ −25

84. $-8 + (-3) + (-5) + (-2)$ −18

85. $-4 - \{11 - [4 - (-9)]\}$ −2

86. $15 - \{25 + 2[3 - (-1)]\}$ −18

87. $-\dfrac{13}{10} + \dfrac{8}{15} - \left(-\dfrac{2}{5}\right)$ $-\dfrac{11}{30}$

88. $\dfrac{11}{14} - \left(-\dfrac{9}{7}\right) - \dfrac{3}{2}$ $\dfrac{4}{7}$

89. $\left(\dfrac{2}{3} - \dfrac{5}{9}\right) - \left(\dfrac{4}{3} - (-2)\right)$ $-\dfrac{29}{9}$

90. $\left(-\dfrac{9}{8} - \dfrac{1}{4}\right) - \left(-\dfrac{5}{6} + \dfrac{1}{8}\right)$ $-\dfrac{2}{3}$

91. $\sqrt{29 + (-4)} - 7$ −2

92. $8 - \sqrt{98 + (-3) + 5}$ −2

93. $|10 + (-3)| - |-12 + (-6)|$ −11

94. $|6 - 8| + |12 - 5|$ 9

95. $\dfrac{3 - 4 + 5}{4 + (-2)}$ 2

96. $\dfrac{12 - 14 + 6}{6 + (-2)}$ 1

For Exercises 97–104, evaluate the expressions for $a = -2$, $b = -6$, and $c = -1$.

97. $(a + b) - c$ −7

98. $(a - b) + c$ 3

99. $a - (b + c)$ 5

100. $a + (b - c)$ −7

101. $(a - b) - c$ 5

102. $(a + b) + c$ −9

103. $a - (b - c)$ 3

104. $a + (b + c)$ −9

 Writing   Translating Expression   Geometry   Scientific Calculator   Video   NS E

Problem Recognition Exercises

Addition and Subtraction of Real Numbers

1. State the rule for adding two negative numbers.
Add their absolute values and apply a negative sign.

2. State the rule for adding a negative number to a positive number. Subtract the smaller absolute value from the larger absolute value. Apply the sign of the number with the larger absolute value.

For Exercises 3–40, perform the indicated operations.

3. $65 - 24$ 41

4. $42 - 29$ 13

5. $13 - (-18)$ 31

6. $22 - (-24)$ 46

7. $4.8 - 6.1$ -1.3

8. $3.5 - 7.1$ -3.6

9. $4 + (-20)$ -16

10. $5 + (-12)$ -7

11. $\dfrac{1}{3} - \dfrac{5}{12}$ $-\dfrac{1}{12}$

12. $\dfrac{3}{8} - \dfrac{1}{12}$ $\dfrac{7}{24}$

13. $-32 - 4$ -36

14. $-51 - 8$ -59

15. $-6 + (-6)$ -12

16. $-25 + (-25)$ -50

17. $-4 - \left(-\dfrac{5}{6}\right)$ $-\dfrac{19}{6}$

18. $-2 - \left(-\dfrac{2}{5}\right)$ $-\dfrac{8}{5}$

19. $-60 + 55$ -5

20. $-55 + 23$ -32

21. $-18 - (-18)$ 0

22. $-3 - (-3)$ 0

23. $-3.5 - 4.2$ -7.7

24. $-6.6 - 3.9$ -10.5

25. $-90 + (-24)$ -114

26. $-35 + (-21)$ -56

27. $-14 + (-2) - 16$ -32

28. $-25 + (-6) - 15$ -46

29. $-4.2 + 1.2 + 3.0$ 0

30. $-4.6 + 8.6 + (-4.0)$ 0

31. $-10 - 8 - 6 - 4 - 2$ -30

32. $-100 - 90 - 80 - 70 - 60$ -400

33. $-8 - [4 - (-3)]^2$ -57

34. $10 - (-3 + 9)^2$ -26

35. $18 - |9^2 + (-100)|$ -1

36. $14 - |5^2 + (-5)|$ -6

37. $6 - [1 - (2 - 3)]$ 4

38. $9 - [3 - (-2 - 5)]$ -1

39. $\left[-\dfrac{2}{3} - \left(-\dfrac{5}{6}\right)\right]^2$ $\dfrac{1}{36}$

40. $\left[-\dfrac{3}{4} - \left(-\dfrac{7}{8}\right)\right]^2$ $\dfrac{1}{64}$

Section 1.5   Multiplication and Division of Real Numbers

Objectives

1. **Multiplication of Real Numbers**
2. **Exponential Expressions**
3. **Division of Real Numbers**
4. **Applying the Order of Operations**

1. Multiplication of Real Numbers

Multiplication of real numbers can be interpreted as repeated addition. For example:

$$3(4) = 4 + 4 + 4 = 12 \qquad \text{Add 3 groups of 4.}$$

$$3(-4) = -4 + (-4) + (-4) = -12 \qquad \text{Add 3 groups of } -4.$$

These results suggest that the product of a positive number and a negative number is *negative.* Consider the following pattern of products.

$$
\begin{aligned}
4 \times \;\;\; 3 &= \;\;\;12 \\
4 \times \;\;\; 2 &= \;\;\;\;\,8 \\
4 \times \;\;\; 1 &= \;\;\;\;\,4 \\
4 \times \;\;\; 0 &= \;\;\;\;\,0 \\
4 \times -1 &= \;\;-4 \\
4 \times -2 &= \;\;-8 \\
4 \times -3 &= -12
\end{aligned}
$$

The pattern decreases by 4 with each row.

Thus, the product of a positive number and a negative number must be *negative* for the pattern to continue.

Now suppose we have a product of two negative numbers. To determine the sign, consider the following pattern of products.

$$-4 \times 3 = -12$$
$$-4 \times 2 = -8$$
$$-4 \times 1 = -4$$
$$-4 \times 0 = 0$$
$$-4 \times -1 = 4$$
$$-4 \times -2 = 8$$
$$-4 \times -3 = 12$$

The pattern increases by 4 with each row.

Thus, the product of two negative numbers must be *positive* for the pattern to continue.

From the first four rows, we see that the product increases by 4 for each row. For the pattern to continue, it follows that the product of two negative numbers must be *positive*.

We now summarize the rules for multiplying real numbers.

PROCEDURE Multiplying Real Numbers

- The product of two real numbers with the *same* sign is positive.
  Examples: $(5)(6) = 30$
  $(-4)(-10) = 40$
- The product of two real numbers with *different* signs is negative.
  Examples: $(-2)(5) = -10$
  $(4)(-9) = -36$
- The product of any real number and zero is zero.
  Examples: $(8)(0) = 0$
  $(0)(-6) = 0$

Example 1 Multiplying Real Numbers

Multiply the real numbers.

a. $-8(-4)$ **b.** $-2.5(-1.7)$ **c.** $-7(10)$

d. $\frac{1}{2}(-8)$ **e.** $0(-8.3)$ **f.** $-\frac{2}{7}\left(-\frac{7}{2}\right)$

Solution:

a. $-8(-4) = 32$

b. $-2.5(-1.7) = 4.25$

Same signs. Product is positive.

c. $-7(10) = -70$

d. $\frac{1}{2}(-8) = -4$

Different signs. Product is negative.

e. $0(-8.3)$
$= 0$ The product of any real number and zero is zero.

f. $-\frac{2}{7}\left(-\frac{7}{2}\right)$

$= \frac{14}{14}$ *Same signs.* Product is positive.

$= 1$ Reduce to lowest terms.

Skill Practice

Multiply.
1. $-9(-3)$ **2.** $-1.5(-1.5)$
3. $-6(4)$ **4.** $\frac{1}{3}(-15)$
5. $0(-4.1)$ **6.** $-\frac{5}{9}\left(-\frac{9}{5}\right)$

Classroom Examples: p. 90, Exercises 8 and 18

Answers
1. 27   **2.** 2.25   **3.** −24
4. −5   **5.** 0   **6.** 1

Observe the pattern for repeated multiplications.

$$(-1)(-1) \qquad (-1)(-1)(-1) \qquad (-1)(-1)(-1)(-1) \qquad (-1)(-1)(-1)(-1)(-1)$$
$$= 1 \qquad\quad = (1)(-1) \qquad\quad = (1)(-1)(-1) \qquad\quad = (1)(-1)(-1)(-1)$$
$$\qquad\quad = -1 \qquad\qquad = (-1)(-1) \qquad\qquad = (-1)(-1)(-1)$$
$$\qquad\qquad\qquad\qquad = 1 \qquad\qquad\quad = (1)(-1)$$
$$\qquad\qquad\qquad\qquad\qquad\qquad = -1$$

The pattern demonstrated in these examples indicates that

- The product of an even number of negative factors is positive.
- The product of an odd number of negative factors is negative.

Concept Connections

7. Without actually computing, determine if the product shown is positive or negative. Explain.

$(-526)(420)(-105)$

2. Exponential Expressions

Recall that for any real number b and any positive integer, n:

$$b^n = \underbrace{b \cdot b \cdot b \cdot b \ldots \cdot b}_{n \text{ factors of } b}$$

Be particularly careful when evaluating exponential expressions involving negative numbers. An exponential expression with a negative base is written with parentheses around the base, such as $(-2)^4$.

To evaluate $(-2)^4$, the base -2 is multiplied four times:

$$(-2)^4 = (-2)(-2)(-2)(-2) = 16$$

If parentheses are *not* used, the expression -2^4 has a different meaning:

- The expression -2^4 has a base of 2 (not -2) and can be interpreted as $-1 \cdot 2^4$.

$$-2^4 = -1(2)(2)(2)(2) = -16$$

- The expression -2^4 can also be interpreted as the opposite of 2^4.

$$-2^4 = -(2 \cdot 2 \cdot 2 \cdot 2) = -16$$

Skill Practice

Simplify.

8. $(-7)^2$ **9.** -7^2

10. $\left(-\dfrac{2}{3}\right)^3$ **11.** -0.2^3

Classroom Examples: p. 90, Exercises 20, 22, 24, and 26

Example 2   Evaluating Exponential Expressions

Simplify.

a. $(-5)^2$ **b.** -5^2 **c.** $\left(-\dfrac{1}{2}\right)^3$ **d.** -0.4^3

Solution:

a. $(-5)^2 = (-5)(-5) = 25$ Multiply two factors of -5.

b. $-5^2 = -1(5)(5) = -25$ Multiply -1 by two factors of 5.

Avoiding Mistakes

The negative sign is not part of the base unless it is in parentheses with the base. Thus, in the expression -5^2, the exponent applies only to 5 and not to the negative sign.

Answers

7. The product is positive because there is an even number of negative factors.

8. 49   **9.** -49

10. $-\dfrac{8}{27}$ **11.** -0.008

c. $\left(-\dfrac{1}{2}\right)^3 = \left(-\dfrac{1}{2}\right)\left(-\dfrac{1}{2}\right)\left(-\dfrac{1}{2}\right) = -\dfrac{1}{8}$ Multiply three factors of $-\dfrac{1}{2}$.

d. $-0.4^3 = -1(0.4)(0.4)(0.4) = -0.064$ Multiply -1 by three factors of 0.4.

3. Division of Real Numbers

Two numbers are *reciprocals* if their product is 1. For example, $-\frac{2}{7}$ and $-\frac{7}{2}$ are reciprocals because $-\frac{2}{7}\left(-\frac{7}{2}\right) = 1$. Symbolically, if a is a nonzero real number, then the reciprocal of a is $\frac{1}{a}$ because $a \cdot \frac{1}{a} = 1$. This definition also implies that a number and its reciprocal have the same sign.

> **DEFINITION The Reciprocal of a Real Number**
>
> Let a be a nonzero real number. Then, the **reciprocal** of a is $\frac{1}{a}$.

Recall that to subtract two real numbers, we add the opposite of the second number to the first number. In a similar way, division of real numbers is defined in terms of multiplication. To divide two real numbers, we multiply the first number by the reciprocal of the second number.

> **DEFINITION Division of Real Numbers**
>
> Let a and b be real numbers such that $b \neq 0$. Then, $a \div b = a \cdot \frac{1}{b}$.

Consider the quotient $10 \div 5$. The reciprocal of 5 is $\frac{1}{5}$, so we have

$$10 \div 5 = 2 \qquad \text{or equivalently,} \qquad 10 \cdot \dfrac{1}{5} = 2$$

(multiply — reciprocal)

Because division of real numbers can be expressed in terms of multiplication, then the sign rules that apply to multiplication also apply to division.

> **PROCEDURE Dividing Real Numbers**
>
> - The quotient of two real numbers with the *same* sign is positive.
>
>   Examples: $24 \div 4 = 6$
> $-36 \div -9 = 4$
>
> - The quotient of two real numbers with *different* signs is negative.
>
>   Examples: $100 \div (-5) = -20$
> $-12 \div 4 = -3$

TIP: If the numerator and denominator of a fraction are both negative, then the quotient is positive. Therefore, $\dfrac{-9}{-5}$ can be simplified to $\dfrac{9}{5}$.

Example 3 Dividing Real Numbers

Divide the real numbers.

a. $200 \div (-10)$ **b.** $\dfrac{-48}{16}$ **c.** $\dfrac{-6.25}{-1.25}$ **d.** $\dfrac{-9}{-5}$

Solution:

a. $200 \div (-10) = -20$ *Different signs.* Quotient is negative.

b. $\dfrac{-48}{16} = -3$ *Different signs.* Quotient is negative.

c. $\dfrac{-6.25}{-1.25} = 5$ *Same signs.* Quotient is positive.

d. $\dfrac{-9}{-5} = \dfrac{9}{5}$ *Same signs.* Quotient is positive.

Because 5 does not divide into 9 evenly the answer can be left as a fraction.

Example 4 Dividing Real Numbers

Divide the real numbers.

a. $15 \div -25$ **b.** $-\dfrac{3}{14} \div \dfrac{9}{7}$

Solution:

a. $15 \div -25$ *Different signs.* Quotient is negative.

$= \dfrac{15}{-25}$

$= -\dfrac{3}{5}$

TIP: If the numerator and denominator of a fraction have opposite signs, then the quotient will be negative. Therefore, a fraction has the same value whether the negative sign is written in the numerator, in the denominator, or in front of the fraction.

$$\frac{-3}{5} = \frac{3}{-5} = -\frac{3}{5}$$

b. $-\dfrac{3}{14} \div \dfrac{9}{7}$ *Different signs.* Quotient is negative.

$= -\dfrac{3}{14} \cdot \dfrac{7}{9}$ Multiply by the reciprocal of $\frac{9}{7}$ which is $\frac{7}{9}$.

$= -\dfrac{\overset{1}{3}}{\underset{2}{14}} \cdot \dfrac{\overset{1}{7}}{\underset{3}{9}}$ Divide out common factors.

$= -\dfrac{1}{6}$ Multiply the fractions.

Multiplication can be used to check any division problem. If $\frac{a}{b} = c$, then $bc = a$ (provided that $b \neq 0$). For example,

$$\frac{8}{-4} = -2 \; \rightarrow \; \underline{\text{Check:}} \quad (-4)(-2) = 8 \; ✔$$

This relationship between multiplication and division can be used to investigate division problems involving the number zero.

1. The quotient of 0 and any nonzero number is 0. For example:

$$\frac{0}{6} = 0 \qquad \text{because } 6 \cdot 0 = 0 \; ✔$$

2. The quotient of any nonzero number and 0 is undefined. For example,

$$\frac{6}{0} = ?$$

Finding the quotient $\frac{6}{0}$ is equivalent to asking, "What number times zero will equal 6?" That is, $(0)(?) = 6$. No real number satisfies this condition. Therefore, we say that division by zero is undefined.

3. The quotient of 0 and 0 cannot be determined. Evaluating an expression of the form $\frac{0}{0} = ?$ is equivalent to asking, "What number times zero will equal 0?" That is, $(0)(?) = 0$. Any real number will satisfy this requirement; however, expressions involving $\frac{0}{0}$ are usually discussed in advanced mathematics courses.

Concept Connections

Simplify.

18. $\dfrac{0}{2}$ **19.** $\dfrac{4}{0}$

20. $-25 \div 0$ **21.** $0 \div (-8.2)$

Instructor Note: Memory Device: Zero under the line, division undefined.

PROPERTY   Division Involving Zero

Let a represent a nonzero real number. Then,

1. $\dfrac{0}{a} = 0$ **2.** $\dfrac{a}{0}$ is undefined

---

4.   Applying the Order of Operations

Example 5   **Applying the Order of Operations**

Simplify. $-8 + 8 \div (-2) \div (-6)$

Solution:

$$-8 + 8 \div (-2) \div (-6)$$

$$= -8 + (-4) \div (-6) \qquad \text{Perform division before addition.}$$

$$= -8 + \frac{4}{6} \qquad\qquad \text{The quotient of } -4 \text{ and } -6 \text{ is positive } \tfrac{4}{6} \text{ or } \tfrac{2}{3}.$$

$$= -\frac{8}{1} + \frac{2}{3} \qquad\qquad \text{Write } -8 \text{ as a fraction.}$$

$$= -\frac{24}{3} + \frac{2}{3} \qquad\qquad \text{Get a common denominator.}$$

$$= -\frac{22}{3} \qquad\qquad\quad \text{Add.}$$

Skill Practice

Simplify.

22. $-36 + 36 \div (-4) \div (-3)$

Classroom Example: p. 91, Exercise 120

Answers

18. 0   **19.** Undefined
20. Undefined   **21.** 0   **22.** -33

— Skill Practice —

Simplify.

23. $\dfrac{100 - 3[-1 + (2 - 6)^2]}{|20 - 25|}$

Classroom Example: p. 92, Exercise 134

Example 6 **Applying the Order of Operations**

Simplify. $\dfrac{24 - 2[-3 + (5 - 8)]^2}{2|-12 + 3|}$

Solution:

$\dfrac{24 - 2[-3 + (5 - 8)]^2}{2|-12 + 3|}$ Simplify numerator and denominator separately.

$= \dfrac{24 - 2[-3 + (-3)]^2}{2|-9|}$ Simplify within the inner parentheses and absolute value.

$= \dfrac{24 - 2[-6]^2}{2(9)}$ Simplify within brackets, [ ]. Simplify the absolute value.

$= \dfrac{24 - 2(36)}{2(9)}$ Simplify exponents.

$= \dfrac{24 - 72}{18}$ Perform multiplication before subtraction.

$= \dfrac{-48}{18}$ or $-\dfrac{8}{3}$ Reduce to lowest terms.

— Skill Practice —

Given $a = -7$, evaluate the expressions.

24. a^2 **25.** $-a^2$

Classroom Example: p. 92, Exercise 136

Example 7 **Evaluating an Algebraic Expression**

Given $y = -6$, evaluate the expressions.

a. y^2 **b.** $-y^2$

Solution:

a. y^2

$= (\ \ )^2$ When substituting a number for a variable, use parentheses.

$= (-6)^2$ Substitute $y = -6$.

$= 36$ Square -6, that is, $(-6)(-6) = 36$.

b. $-y^2$

$= -(\ \ )^2$ When substituting a number for a variable, use parentheses.

$= -(-6)^2$ Substitute $y = -6$.

$= -(36)$ Square -6.

$= -36$ Multiply by -1.

Answers

23. 11   **24.** 49   **25.** −49

Calculator Connections

Topic: Evaluating Exponential Expressions with Positive and Negative Bases

Be particularly careful when raising a negative number to an even power on a calculator. For example, the expressions $(-4)^2$ and -4^2 have different values. That is, $(-4)^2 = 16$ and $-4^2 = -16$. Verify these expressions on a calculator.

Scientific Calculator:

To evaluate $(-4)^2$

Enter: (4 +○-) x^2 **Result:** | 16 |

To evaluate -4^2 on a scientific calculator, it is important to square 4 first and then take its opposite.

Enter: 4 x^2 +○-   **Result:** | −16 |

Graphing Calculator:

```
(-4)²
          16
-4²
         -16
```

The graphing calculator allows for several methods of denoting the multiplication of two real numbers. For example, consider the product of −8 and 4.

```
-8*4
          -32
-8(4)
          -32
(-8)(4)
          -32
```

Calculator Exercises

Simplify the expression without the use of a calculator. Then use the calculator to verify your answer.

1. $-6(5)$ −30   **2.** $\dfrac{-5.2}{2.6}$ −2   **3.** $(-5)(-5)(-5)(-5)$ 625   **4.** $(-5)^4$ 625   **5.** -5^4 −625

6. -2.4^2 −5.76   **7.** $(-2.4)^2$ 5.76   **8.** $(-1)(-1)(-1)$ −1   **9.** $\dfrac{-8.4}{-2.1}$ 4   **10.** $90 \div (-5)(2)$ −36

Section 1.5   Practice Exercises

Boost *your* GRADE at ALEKS.com!

ALEKS version 3.0

- Practice Problems
- Self-Tests
- NetTutor
- e-Professors
- Videos

For additional exercises, see Classroom Activities 1.5A–1.5D in the *Instructor's Resource Manual* at www.mhhe.com/moh.

Study Skills Exercises

1. Look through the text, and write down a page number that contains:

   a. An Avoiding Mistakes box _____

   b. A Tip box _____

   c. A key term (shown in bold) _____

2. Define the key term **reciprocal of a real number**.

Review Exercises

For Exercises 3–6, determine if the expression is true or false.

3. $6 + (-2) > -5 + 6$ True

4. $|-6| + |-14| \leq |-3| + |-17|$ True

5. $\sqrt{36} - |-6| > 0$ False

6. $\sqrt{9} + |-3| \leq 0$ False

Objective 1: Multiplication of Real Numbers

For Exercises 7–18, multiply the real numbers. **(See Example 1.)**

7. $8(-7)$ -56

8. $(-3) \cdot 4$ -12

9. $(-6) \cdot 7$ -42

10. $9(-5)$ -45

11. $(-11)(-13)$ 143

12. $(-5)(-26)$ 130

13. $(-30)(-8)$ 240

14. $(-16)(-8)$ 128

15. $(-2.2)(5.8)$ -12.76

16. $(9.1)(-4.5)$ -40.95

17. $\left(-\dfrac{2}{3}\right)\left(-\dfrac{9}{8}\right)$ $\dfrac{3}{4}$

18. $\left(-\dfrac{5}{4}\right)\left(-\dfrac{12}{25}\right)$ $\dfrac{3}{5}$

Objective 2: Exponential Expressions

For Exercises 19–26, simplify the exponential expression. **(See Example 2.)**

19. $(-6)^2$ 36

20. $(-10)^2$ 100

21. -6^2 -36

22. -10^2 -100

23. $\left(-\dfrac{3}{5}\right)^3$ $-\dfrac{27}{125}$

24. $\left(-\dfrac{5}{2}\right)^3$ $-\dfrac{125}{8}$

25. $(-0.2)^4$ 0.0016

26. $(-0.1)^4$ 0.0001

Objective 3: Division of Real Numbers

For Exercises 27–34, divide the real numbers. **(See Examples 3–4.)**

27. $\dfrac{54}{-9}$ -6

28. $\dfrac{-27}{3}$ -9

29. $\dfrac{-100}{-10}$ 10

30. $\dfrac{-120}{-40}$ 3

31. $\dfrac{-14}{-7}$ 2

32. $\dfrac{-21}{-3}$ 7

33. $\dfrac{13}{-65}$ $-\dfrac{1}{5}$

34. $\dfrac{7}{-77}$ $-\dfrac{1}{11}$

For Exercises 35–42, show how multiplication can be used to check the division problems.

35. $\dfrac{14}{-2} = -7$
$(-2)(-7) = 14$

36. $\dfrac{-18}{-6} = 3$
$(-6)(3) = -18$

37. $\dfrac{0}{-5} = 0$
$-5 \cdot 0 = 0$

38. $\dfrac{0}{-4} = 0$
$-4 \cdot 0 = 0$

39. $\dfrac{6}{0}$ is undefined
No number multiplied by zero equals 6.

40. $\dfrac{-4}{0}$ is undefined
No number multiplied by zero equals -4.

41. $-24 \div (-6) = 4$
$(-6)(4) = -24$

42. $-18 \div 2 = -9$
$(2)(-9) = -18$

Mixed Exercises

For Exercises 43–114, multiply or divide as indicated.

43. $2 \cdot 3$ 6

44. $8 \cdot 6$ 48

45. $2(-3)$ -6

46. $8(-6)$ -48

47. $(-2)3$ -6

48. $(-8)6$ -48

49. $(-2)(-3)$ 6

50. $(-8)(-6)$ 48

51. $24 \div 3$ 8

52. $52 \div 2$ 26

53. $24 \div (-3)$ -8

54. $52 \div (-2)$ -26

55. $(-24) \div 3$ -8

56. $(-52) \div 2$ -26

57. $(-24) \div (-3)$ 8

58. $(-52) \div (-2)$ 26

 Writing   Translating Expression   Geometry   Scientific Calculator   Video   NSE

59. $-6 \cdot 0$ 0

60. $-8 \cdot 0$ 0

61. $-18 \div 0$ Undefined

62. $-42 \div 0$ Undefined

63. $0\left(-\dfrac{2}{5}\right)$ 0

64. $0\left(-\dfrac{1}{8}\right)$ 0

65. $0 \div \left(-\dfrac{1}{10}\right)$ 0

66. $0 \div \left(\dfrac{4}{9}\right)$ 0

67. $\dfrac{-9}{6}$ $-\dfrac{3}{2}$

68. $\dfrac{-15}{10}$ $-\dfrac{3}{2}$

69. $\dfrac{-30}{-100}$ $\dfrac{3}{10}$

70. $\dfrac{-250}{-1000}$ $\dfrac{1}{4}$

71. $\dfrac{26}{-13}$ -2

72. $\dfrac{52}{-4}$ -13

73. $1.72(-4.6)$ -7.912

74. $361.3(-14.9)$ -5383.37

75. $-0.02(-4.6)$ 0.092

76. $-0.06(-2.15)$ 0.129

77. $\dfrac{14.4}{-2.4}$ -6

78. $\dfrac{50.4}{-6.3}$ -8

79. $\dfrac{-5.25}{-2.5}$ 2.1

80. $\dfrac{-8.5}{-27.2}$ 0.3125

81. $(-3)^2$ 9

82. $(-7)^2$ 49

83. -3^2 -9

84. -7^2 -49

85. $\left(-\dfrac{4}{3}\right)^3$ $-\dfrac{64}{27}$

86. $\left(-\dfrac{1}{5}\right)^3$ $-\dfrac{1}{125}$

87. $(-0.2)^3$ -0.008

88. $(-0.1)^6$ 0.000001

89. -0.2^4 -0.0016

90. -0.1^4 -0.0001

91. $87 \div (-3)$ -29

92. $96 \div (-6)$ -16

93. $-4(-12)$ 48

94. $(-5)(-11)$ 55

95. $2.8(-5.1)$ -14.28

96. $(7.21)(-0.3)$ -2.163

97. $(-6.8) \div (-0.02)$ 340

98. $(-12.3) \div (-0.03)$ 410

99. $\left(-\dfrac{2}{15}\right)\left(\dfrac{25}{3}\right)$ $-\dfrac{10}{9}$

100. $\left(-\dfrac{5}{16}\right)\left(\dfrac{4}{9}\right)$ $-\dfrac{5}{36}$

101. $\left(-\dfrac{7}{8}\right) \div \left(-\dfrac{9}{16}\right)$ $\dfrac{14}{9}$

102. $\left(-\dfrac{22}{23}\right) \div \left(-\dfrac{11}{3}\right)$ $\dfrac{6}{23}$

103. $(-2)(-5)(-3)$
-30

104. $(-6)(-1)(-10)$
-60

105. $(-8)(-4)(-1)(-3)$
96

106. $(-6)(-3)(-1)(-5)$
90

107. $100 \div (-10) \div (-5)$
2

108. $150 \div (-15) \div (-2)$
5

109. $-12 \div (-6) \div (-2)$
-1

110. $-36 \div (-2) \div 6$
3

111. $\dfrac{2}{5} \cdot \dfrac{1}{3} \cdot \left(-\dfrac{10}{11}\right)$
$-\dfrac{4}{33}$

112. $\left(-\dfrac{9}{8}\right) \cdot \left(-\dfrac{2}{3}\right) \cdot \left(1\dfrac{5}{12}\right)$
$\dfrac{17}{16}$ or $1\dfrac{1}{16}$

113. $\left(1\dfrac{1}{3}\right) \div 3 \div \left(-\dfrac{7}{9}\right)$
$-\dfrac{4}{7}$

114. $-\dfrac{7}{8} \div \left(3\dfrac{1}{4}\right) \div (-2)$
$\dfrac{7}{52}$

Objective 4: Applying the Order of Operations

For Exercises 115–134, perform the indicated operations. **(See Examples 5–6.)**

115. $12 \div (-2)(4)$ -24

116. $(-6) \cdot 7 \div (-2)$ 21

117. $\left(-\dfrac{12}{5}\right) \div (-6) \cdot \left(-\dfrac{1}{8}\right)$ $-\dfrac{1}{20}$

118. $10 \cdot \dfrac{1}{3} \div \dfrac{25}{6}$ $\dfrac{4}{5}$

119. $8 - 2^3 \cdot 5 + 3 - (-6)$ -23

120. $-14 \div (-7) - 8 \cdot 2 + 3^3$ 13

121. $-(2 - 8)^2 \div (-6) \cdot 2$ 12

122. $-(3 - 5)^2 \cdot 6 \div (-4)$ 6

123. $\dfrac{6(-4) - 2(5 - 8)}{-6 - 3 - 5}$ $\dfrac{9}{7}$

124. $\dfrac{3(-4) - 5(9 - 11)}{-9 - 2 - 3}$ $\dfrac{1}{7}$

125. $\dfrac{-4 + 5}{(-2) \cdot 5 + 10}$ Undefined

126. $\dfrac{-3 + 10}{2(-4) + 8}$ Undefined

Writing   Translating Expression   Geometry   Scientific Calculator   Video   NS&E

127. $-4 - 3[2 - (-5 + 3)] - 8 \cdot 2^2$ **128.** $-6 - 5[-4 - (6 - 12)] + (-5)^2$ **129.** $-|-1| - |5|$ -6
  -48 9

130. $-|-10| - |6|$ -16 **131.** $\dfrac{|2 - 9| - |5 - 7|}{10 - 15}$ -1 **132.** $\dfrac{|-2 + 6| - |3 - 5|}{13 - 11}$ 1

133. $\dfrac{6 - 3[2 - (6 - 8)]^2}{-2|2 - 5|}$ 7 **134.** $\dfrac{12 - 4[-6 - (5 - 8)]^2}{4|6 - 10|}$ $-\dfrac{3}{2}$

For Exercises 135–140, evaluate the expression for $x = -2$, $y = -4$, and $z = 6$. **(See Example 7.)**

135. $x^2 - 2y$ 12 **136.** $3y^2 - z$ 42 **137.** $4(2x - z)$ -40

138. $6(3x + y)$ -60 **139.** $\dfrac{3x + 2y}{y}$ $\dfrac{7}{2}$  **140.** $\dfrac{2z - y}{x}$ -8

141. Is the expression $\dfrac{10}{5x}$ equal to 10/5x? Explain.
No. The first expression is equivalent to $10 \div (5x)$. The
second is $10 \div 5 \cdot x$.

142. Is the expression 10/(5x) equal to $\dfrac{10}{5x}$? Explain.
Yes, the parentheses indicate that the divisor is the quantity $5x$.

For Exercises 143–150, translate the English phrase into an algebraic expression. Then evaluate the expression.

143. The product of -3.75 and 0.3
$-3.75(0.3); -1.125$

144. The product of -0.4 and -1.258
$(-0.4)(-1.258); 0.5032$

145. The quotient of $\frac{16}{5}$ and $\left(-\frac{8}{9}\right)$ $\dfrac{16}{5} \div \left(-\dfrac{8}{9}\right); -\dfrac{18}{5}$

146. The quotient of $\left(-\frac{3}{14}\right)$ and $\frac{1}{7}$ $-\dfrac{3}{14} \div \dfrac{1}{7}; -\dfrac{3}{2}$

147. The number -0.4 plus the quantity
6 times -0.42 $-0.4 + 6(-0.42); -2.92$

148. The number 0.5 plus the quantity -2 times 0.125
$0.5 + (-2)(0.125); 0.25$

149. The number $-\frac{1}{4}$ minus the quantity
6 times $-\frac{1}{3}$ $-\dfrac{1}{4} - 6\left(-\dfrac{1}{3}\right); \dfrac{7}{4}$

150. Negative five minus the quantity $\left(-\frac{5}{6}\right)$ times $\frac{3}{8}$
$-5 - \left(-\dfrac{5}{6}\right)\dfrac{3}{8}; -\dfrac{75}{16}$

151. For 3 weeks, Jim pays $2 a week for lottery tickets. Jim has one winning ticket for $3. Write an expression that describes his net gain or loss. How much money has Jim won or lost? $3(-2) + 3 = -3$; loss of $3

152. Stephanie pays $2 a week for 6 weeks for lottery tickets. Stephanie has one winning ticket for $5. Write an expression that describes her net gain or loss. How much money has Stephanie won or lost?
$-2(6) + 5 = -7$; loss of $7

153. Evaluate the expressions in parts (a) and (b).

  a. $-4 - 3 - 2 - 1$ -10

  b. $-4(-3)(-2)(-1)$ 24

  c. Explain the difference between the operations in parts (a) and (b).   In part (a), we subtract; in part (b), we multiply.

154. Evaluate the expressions in parts (a) and (b).

  a. $-10 - 9 - 8 - 7$ -34

  b. $-10(-9)(-8)(-7)$ 5040

  c. Explain the difference between the operations in parts (a) and (b).
In part (a), we subtract; in part (b), we multiply.

Properties of Real Numbers and Simplifying Expressions

1. Commutative Properties of Real Numbers

When getting dressed in the morning, it makes no difference whether you put on your left shoe first and then your right shoe, or vice versa. This example illustrates a process in which the order does not affect the outcome. Such a process or operation is said to be *commutative*.

In algebra, the operations of addition and multiplication are commutative because the order in which we add or multiply two real numbers does not affect the result. For example:

$$10 + 5 = 5 + 10 \quad \text{and} \quad 10 \cdot 5 = 5 \cdot 10$$

Objectives

1. Commutative Properties of Real Numbers
2. Associative Properties of Real Numbers
3. Identity and Inverse Properties of Real Numbers
4. Distributive Property of Multiplication over Addition
5. Simplifying Algebraic Expressions

> **PROPERTY Commutative Properties of Real Numbers**
>
> If a and b are real numbers, then
>
> **1.** $a + b = b + a$ **commutative property of addition**
>
> **2.** $ab = ba$ **commutative property of multiplication**

It is important to note that although the operations of addition and multiplication are commutative, subtraction and division are *not* commutative. For example:

$$\underbrace{10 - 5}_{5} \neq \underbrace{5 - 10}_{-5} \quad \text{and} \quad \underbrace{10 \div 5}_{2} \neq \underbrace{5 \div 10}_{\frac{1}{2}}$$

Example 1 **Applying the Commutative Property of Addition**

Use the commutative property of addition to rewrite each expression.

a. $-3 + (-7)$ **b.** $3x^3 + 5x^4$

Solution:

a. $-3 + (-7) = -7 + (-3)$

b. $3x^3 + 5x^4 = 5x^4 + 3x^3$

Skill Practice

Use the commutative property of addition to rewrite each expression.

1. $-5 + 9$
2. $7y + x$

Classroom Example: p. 103, Exercise 16

Recall that subtraction is not a commutative operation. However, if we rewrite $a - b$, as $a + (-b)$, we can apply the commutative property of addition. This is demonstrated in Example 2.

Answers
1. $9 + (-5)$ 2. $x + 7y$

Example 2 **Applying the Commutative Property of Addition**

Rewrite the expression in terms of addition. Then apply the commutative property of addition.

a. $5a - 3b$ **b.** $z^2 - \dfrac{1}{4}$

Solution:

a. $5a - 3b$

$= 5a + (-3b)$ Rewrite subtraction as addition of $-3b$.

$= -3b + 5a$ Apply the commutative property of addition.

b. $z^2 - \dfrac{1}{4}$

$= z^2 + \left(-\dfrac{1}{4}\right)$ Rewrite subtraction as addition of $-\frac{1}{4}$.

$= -\dfrac{1}{4} + z^2$ Apply the commutative property of addition.

Example 3 **Applying the Commutative Property of Multiplication**

Use the commutative property of multiplication to rewrite each expression.

a. $12(-6)$ **b.** $x \cdot 4$

Solution:

a. $12(-6) = -6(12)$

b. $x \cdot 4 = 4 \cdot x$ (or simply $4x$)

2. Associative Properties of Real Numbers

The associative property of real numbers states that the manner in which three or more real numbers are grouped under addition or multiplication will not affect the outcome. For example:

$$(5 + 10) + 2 = 5 + (10 + 2) \quad \text{and} \quad (5 \cdot 10)2 = 5(10 \cdot 2)$$
$$15 + 2 = 5 + 12 \qquad\qquad (50)2 = 5(20)$$
$$17 = 17 \qquad\qquad 100 = 100$$

PROPERTY **Associative Properties of Real Numbers**

If a, b, and c represent real numbers, then

1. $(a + b) + c = a + (b + c)$ **associative property of addition**

2. $(ab)c = a(bc)$ **associative property of multiplication**

Example 4 **Applying the Associative Property**

Use the associative property of multiplication to rewrite each expression. Then simplify the expression if possible.

a. $(y + 5) + 6$ **b.** $4(5z)$ **c.** $-\dfrac{3}{2}\left(-\dfrac{2}{3}w\right)$

Solution:

a. $(y + 5) + 6$

$= y + (5 + 6)$ Apply the associative property of addition.

$= y + 11$ Simplify.

b. $4(5z)$

$= (4 \cdot 5)z$ Apply the associative property of multiplication.

$= 20z$ Simplify.

c. $-\dfrac{3}{2}\left(-\dfrac{2}{3}w\right)$

$= \left[-\dfrac{3}{2}\left(-\dfrac{2}{3}\right)\right]w$ Apply the associative property of multiplication.

$= 1w$ Simplify.

$= w$

Note: In most cases, a detailed application of the associative property will not be shown. Instead, the process will be written in one step, such as

$$(y + 5) + 6 = y + 11, \quad 4(5z) = 20z \quad \text{and} \quad -\dfrac{3}{2}\left(-\dfrac{2}{3}w\right) = w$$

Skill Practice

Use the associative property of addition or multiplication to rewrite each expression. Simplify if possible.

7. $(x + 4) + 3$

8. $-2(4x)$

9. $\dfrac{5}{4}\left(\dfrac{4}{5}t\right)$

Classroom Examples: p. 103, Exercises 28 and 30

3. Identity and Inverse Properties of Real Numbers

The number 0 has a special role under the operation of addition. Zero added to any real number does not change the number. Therefore, the number 0 is said to be the *additive identity* (also called the *identity element of addition*). For example:

$$-4 + 0 = -4 \qquad 0 + 5.7 = 5.7 \qquad 0 + \dfrac{3}{4} = \dfrac{3}{4}$$

The number 1 has a special role under the operation of multiplication. Any real number multiplied by 1 does not change the number. Therefore, the number 1 is said to be the *multiplicative identity* (also called the *identity element of multiplication*). For example:

$$(-8)1 = -8 \qquad 1(-2.85) = -2.85 \qquad 1\left(\dfrac{1}{5}\right) = \dfrac{1}{5}$$

PROPERTY **Identity Properties of Real Numbers**

If a is a real number, then

1. $a + 0 = 0 + a = a$ **identity property of addition**

2. $a \cdot 1 = 1 \cdot a = a$ **identity property of multiplication**

Answers

7. $x + (4 + 3); x + 7$

8. $(-2 \cdot 4)x; -8x$

9. $\left(\dfrac{5}{4} \cdot \dfrac{4}{5}\right)t; t$

The sum of a number and its opposite equals 0. For example, $-12 + 12 = 0$. For any real number, a, the opposite of a (also called the *additive inverse* of a) is $-a$ and $a + (-a) = -a + a = 0$. The inverse property of addition states that the sum of any number and its additive inverse is the identity element of addition, 0. For example:

Number	Additive Inverse (Opposite)	Sum
9	-9	$9 + (-9) = 0$
-21.6	21.6	$-21.6 + 21.6 = 0$
$\dfrac{2}{7}$	$-\dfrac{2}{7}$	$\dfrac{2}{7} + \left(-\dfrac{2}{7}\right) = 0$

If b is a nonzero real number, then the reciprocal of b (also called the *multiplicative inverse* of b) is $\frac{1}{b}$. The inverse property of multiplication states that the product of b and its multiplicative inverse is the identity element of multiplication, 1. Symbolically, we have $b \cdot \frac{1}{b} = \frac{1}{b} \cdot b = 1$. For example:

Number	Multiplicative Inverse (Reciprocal)	Product
7	$\dfrac{1}{7}$	$7 \cdot \dfrac{1}{7} = 1$
3.14	$\dfrac{1}{3.14}$	$3.14\left(\dfrac{1}{3.14}\right) = 1$
$-\dfrac{3}{5}$	$\dfrac{5}{3}$	$-\dfrac{3}{5}\left(-\dfrac{5}{3}\right) = 1$

Concept Connections

Fill in the blanks. State whether the property used is an inverse property or an identity property.

10. $5 + \square = 0$

11. $-8 \cdot \square = -8$

12. $\dfrac{1}{2} \cdot \square = 1$

13. $\dfrac{a}{b} + \square = \dfrac{a}{b}$

PROPERTY Inverse Properties of Real Numbers

If a is a real number and b is a nonzero real number, then

1. $a + (-a) = -a + a = 0$ **inverse property of addition**

2. $b \cdot \dfrac{1}{b} = \dfrac{1}{b} \cdot b = 1$ **inverse property of multiplication**

4. Distributive Property of Multiplication over Addition

The operations of addition and multiplication are related by an important property called the **distributive property of multiplication over addition**. Consider the expression $6(2 + 3)$. The order of operations indicates that the sum $2 + 3$ is evaluated first, and then the result is multiplied by 6:

$$6(2 + 3)$$
$$= 6(5)$$
$$= 30$$

Answers

10. -5; inverse   11. 1; identity
12. 2; inverse   13. 0; identity

Notice that the same result is obtained if the factor of 6 is multiplied by each of the numbers 2 and 3, and then their products are added:

6(2 + 3)   The factor of 6 is distributed to the numbers 2 and 3.

$= 6(2) + 6(3)$

$=\ \ 12 + 18$

$=\ \ \ \ \ 30$

The distributive property of multiplication over addition states that this is true in general.

> **TIP:** The mathematical definition of the distributive property is consistent with the everyday meaning of the word *distribute*. To distribute means to "spread out from one to many." In the mathematical context, the factor a is distributed to both b and c in the parentheses.

PROPERTY Distributive Property of Multiplication over Addition

If a, b, and c are real numbers, then

$$a(b + c) = ab + ac \qquad \text{and} \qquad (b + c)a = ab + ac$$

Example 5   Applying the Distributive Property

Apply the distributive property: $2(a + 6b + 7)$

Solution:

$2(a + 6b + 7)$

$= 2(a + 6b + 7)$

$= 2(a) + 2(6b) + 2(7)$ Apply the distributive property.

$= 2a + 12b + 14$ Simplify.

Skill Practice

14. Apply the distributive property.

$7(x + 4y + z)$

Classroom Example: p. 103, Exercise 44

> **TIP:** Notice that the parentheses are removed after the distributive property is applied. Sometimes this is referred to as *clearing parentheses.*

Because the difference of two expressions $a - b$ can be written in terms of addition as $a + (-b)$, the distributive property can be applied when the operation of subtraction is present within the parentheses. For example:

$5(y - 7)$

$= 5[y + (-7)]$ Rewrite subtraction as addition of -7.

$= 5[y + (-7)]$ Apply the distributive property.

$= 5(y) + 5(-7)$

$= 5y + (-35), \text{ or } 5y - 35$ Simplify.

Answers

14. $7x + 28y + 7z$

── **Skill Practice** ──

Use the distributive property to rewrite each expression.

15. $-(12x + 8y - 3z)$

16. $-6(-3a + 7b)$

Classroom Example: p. 103, Exercise 56

TIP: Notice that a negative factor preceding the parentheses changes the signs of all the terms to which it is multiplied.

$$-1(-3a + 2b + 5c)$$
$$= +3a - 2b - 5c$$

Example 6   **Applying the Distributive Property**

Use the distributive property to rewrite each expression.

a. $-(-3a + 2b + 5c)$ **b.** $-6(2 - 4x)$

Solution:

a. $-(-3a + 2b + 5c)$

$\quad = -1(-3a + 2b + 5c)$ The negative sign preceding the parentheses can be interpreted as taking the opposite of the quantity that follows or as $-1(-3a + 2b + 5c)$

$\quad = -1(-3a + 2b + 5c)$

$\quad = -1(-3a) + (-1)(2b) + (-1)(5c)$ Apply the distributive property.

$\quad = 3a + (-2b) + (-5c)$ Simplify.

$\quad = 3a - 2b - 5c$

b. $-6(2 - 4x)$

$\quad = -6[2 + (-4x)]$ Change subtraction to addition of $-4x$.

$\quad = -6[2 + (-4x)]$ Apply the distributive property. Notice that multiplying by -6 changes the signs of all terms to which it is applied.

$\quad = -6(2) + (-6)(-4x)$

$\quad = -12 + 24x$ Simplify.

Note: In most cases, the distributive property will be applied without as much detail as shown in Examples 5 and 6. Instead, the distributive property will be applied in one step.

$$2(a + 6b + 7)$$
$$1 \text{ step} = 2a + 12b + 14$$

$$-(3a + 2b + 5c)$$
$$1 \text{ step} = -3a - 2b - 5c$$

$$-6(2 - 4x)$$
$$1 \text{ step} = -12 + 24x$$

5. Simplifying Algebraic Expressions

An algebraic expression is the sum of one or more terms. A term is a constant or the product or quotient of constants and variables. For example, the expression

$$-7x^2 + xy - 100 \quad \text{or} \quad -7x^2 + xy + (-100)$$

consists of the terms $-7x^2$, xy, and -100.

The terms $-7x^2$ and xy are **variable terms** and the term -100 is called a **constant term**. It is important to distinguish between a term and the factors within a term. For example, the quantity xy is one term, and the values x and y are factors within the term. The constant factor in a term is called the *numerical coefficient* (or simply **coefficient**) of the term. In the terms $-7x^2$, xy, and -100, the coefficients are -7, 1, and -100, respectively.

Answers

15. $-12x - 8y + 3z$

16. $18a - 42b$

Terms are *like* terms if they each have the same variables and the corresponding variables are raised to the same powers. For example:

Like **Terms**	**Unlike Terms**	
$-3b$ and $5b$	$-5c$ and $7d$	(different variables)
$9p^2q^3$ and p^2q^3	$4p^2q^3$ and $8p^3q^2$	(different powers)
$5w$ and $2w$	$5w$ and 2	(different variables)

Example 7 Identifying Terms, Factors, Coefficients and *Like* Terms

a. List the terms of the expression $5x^2 - 3x + 2$.

b. Identify the coefficient of the term $6yz^3$.

c. Which of the pairs are *like* terms: $8b, 3b^2$ or $4c^2d, -6c^2d$?

Solution:

a. The terms of the expression $5x^2 - 3x + 2$ are $5x^2$, $-3x$, and 2.

b. The coefficient of $6yz^3$ is 6.

c. $4c^2d$ and $-6c^2d$ are *like* terms.

Classroom Example: p. 104, Exercise 74

Skill Practice

17. List the terms in the expression.

$4xy - 9x^2 + 15$

18. Identify the coefficients of each term in the expression.

$2a - b + c - 80$

19. Which of the pairs are *like* terms?

$5x^3, 5x$ or $-7x^2, 11x^2$

Two terms can be added or subtracted only if they are *like* terms. To add or subtract *like* terms, we use the distributive property as shown in Example 8.

Example 8 Using the Distributive Property to Add and Subtract *Like* Terms

Add or subtract as indicated.

a. $7x + 2x$ **b.** $-2p + 3p - p$

Solution:

a. $7x + 2x$

$= (7 + 2)x$ Apply the distributive property.

$= 9x$ Simplify.

b. $-2p + 3p - p$

$= -2p + 3p - 1p$ Note that $-p$ equals $-1p$.

$= (-2 + 3 - 1)p$ Apply the distributive property.

$= (0)p$ Simplify.

$= 0$

Skill Practice

Simplify by adding *like* terms.

20. $8x + 3x$

21. $-6a + 4a + a$

Classroom Example: p. 104, Exercise 84

Although the distributive property is used to add and subtract *like* terms, it is tedious to write each step. Observe that adding or subtracting *like* terms is a matter of adding or subtracting the coefficients and leaving the variable factors unchanged. This can be shown in one step, a shortcut that we will use throughout the text. For example:

$$7x + 2x = 9x \qquad -2p + 3p - 1p = 0p = 0 \qquad -3a - 6a = -9a$$

Answers

17. $4xy, -9x^2, 15$
18. $2, -1, 1, -80$
19. $-7x^2$ and $11x^2$ are *like* terms.
20. $11x$ **21.** $-a$

Example 9   Using the Distributive Property to Add and Subtract *Like* Terms

Simplify by combining *like* terms.

a. $3yz + 5 - 2yz + 9$ **b.** $1.2w^3 + 5.7w^3$

Solution:

a. $3yz + 5 - 2yz + 9$

$= 3yz - 2yz + 5 + 9$ Arrange *like* terms together. Notice that constants such as 5 and 9 are *like* terms.

$= 1yz + 14$ Combine *like* terms.

$= yz + 14$

b. $1.2w^3 + 5.7w^3$

$= 6.9w^3$ Combine *like* terms.

Examples 10 and 11 illustrate how the distributive property is used to clear parentheses.

Example 10   Clearing Parentheses and Combining *Like* Terms

Simplify by clearing parentheses and combining *like* terms. $5 - 2(3x + 7)$

Solution:

$5 - 2(3x + 7)$ The order of operations indicates that we must perform multiplication before subtraction.

It is important to understand that a factor of -2 (not 2) will be multiplied to all terms within the parentheses. To see why this is so, we may rewrite the subtraction in terms of addition.

$= 5 + (-2)(3x + 7)$ Change subtraction to addition.

$= 5 + (-2)(3x + 7)$ A factor of -2 is to be distributed to terms in the parentheses.

$= 5 + (-2)(3x) + (-2)(7)$ Apply the distributive property.

$= 5 + (-6x) + (-14)$ Simplify.

$= 5 + (-14) + (-6x)$ Arrange *like* terms together.

$= -9 + (-6x)$ Combine *like* terms.

$= -9 - 6x$ Simplify by changing addition of the opposite to subtraction.

Answers

22. $9pq - 15$
23. $13.4x^2$
24. $-10x + 44$

| **Example 11** Clearing Parentheses and Combining *Like* Terms | **Skill Practice** |

Simplify by clearing parentheses and combining *like* terms.

a. $10(5y + 2) - 6(y - 1)$ **b.** $\frac{1}{4}(4k + 2) - \frac{1}{2}(6k + 1)$

c. $-(4s - 6t) - (3t + 5s) - 2s$

Clear the parentheses and combine *like* terms.

25. $5(2y + 3) - 2(3y + 1)$

26. $\frac{1}{2}(8x + 4) + \frac{1}{3}(3x - 9)$

27. $-4(x + 2y) - (2x - y) - 5x$

Classroom Examples: p. 105, Exercises 100 and 108

Solution:

a. $10(5y + 2) - 6(y - 1)$

$= 50y + 20 - 6y + 6$ Apply the distributive property. Notice that a factor of -6 is distributed through the second parentheses and changes the signs.

$= 50y - 6y + 20 + 6$ Arrange *like* terms together.

$= 44y + 26$ Combine *like* terms.

b. $\frac{1}{4}(4k + 2) - \frac{1}{2}(6k + 1)$

$= \frac{4}{4}k + \frac{2}{4} - \frac{6}{2}k - \frac{1}{2}$ Apply the distributive property. Notice that a factor of $-\frac{1}{2}$ is distributed through the second parentheses and changes the signs.

$= k + \frac{1}{2} - 3k - \frac{1}{2}$ Simplify fractions.

$= k - 3k + \frac{1}{2} - \frac{1}{2}$ Arrange *like* terms together.

$= -2k + 0$ Combine *like* terms.

$= -2k$

c. $-(4s - 6t) - (3t + 5s) - 2s$

$= -1(4s - 6t) - 1(3t + 5s) - 2s$ Notice that a factor of -1 is distributed through each parentheses.

$= -4s + 6t - 3t - 5s - 2s$ Apply the distributive property.

$= -4s - 5s - 2s + 6t - 3t$ Arrange *like* terms together.

$= -11s + 3t$ Combine *like* terms.

Answers

25. $4y + 13$ **26.** $5x - 1$

27. $-11x - 7y$

┌─ **Skill Practice** ─┐
Clear the parentheses and
combine *like* terms.
28. $6 - 5[-2y - 4(2y - 5)]$

Classroom Example: p. 105,
Exercise 116

┌─ **Avoiding Mistakes** ─┐
First clear the innermost parentheses and combine *like* terms within
the brackets. Then use the distributive property to clear the brackets.

Answer
28. $50y - 94$

┌───┐
│ **Example 12** **Clearing Parentheses and Combining *Like* Terms** │
│   │
│ Simplify by clearing parentheses and combining *like* terms.   │
│   │
│ $$-7a - 4[3a - 2(a + 6)] - 4$$ │
│   │
│ **Solution:**   │
│   │
│ $-7a - 4[3a - 2(a + 6)] - 4$ │
│   │
│ $= -7a - 4[3a - 2a - 12] - 4$ Apply the distributive property to clear │
│   the innermost parentheses.   │
│   │
│ $= -7a - 4[a - 12] - 4$ Simplify within brackets by combining *like* │
│   terms.   │
│   │
│ $= -7a - 4a + 48 - 4$ Apply the distributive property to clear   │
│   the brackets.   │
│   │
│ $= -11a + 44$ Combine *like* terms   │
└───┘

Section 1.6 Practice Exercises

Study Skills Exercises

1. Write down the page number(s) for the Chapter Summary for this chapter. Describe one way in which you can use the Summary found at the end of each chapter.

2. Define the key term:

   a. commutative properties   **b. associative properties**   **c. identity properties**

   d. inverse properties   **e. distributive property of multiplication over addition**

   f. variable term   **g. constant term**   **h. coefficient**   **i. *like* terms**

Review Exercises

For Exercises 3–14, perform the indicated operations.

3. $(-6) + 14$ 8

4. $(-2) + 9$ 7

5. $-13 - (-5)$ -8

6. $-1 - (-19)$ 18

7. $18 \div (-4)$ $-\frac{9}{2}$ or -4.5

8. $-27 \div 5$ $-\frac{27}{5}$ or -5.4

9. $-3 \cdot 0$ 0

10. $0(-15)$ 0

11. $\frac{1}{2} + \frac{3}{8}$ $\frac{7}{8}$

12. $\frac{25}{21} - \frac{6}{7}$ $\frac{1}{3}$

13. $\left(-\frac{3}{5}\right)\left(\frac{4}{27}\right)$ $-\frac{4}{45}$

14. $\left(-\frac{11}{12}\right) \div \left(-\frac{5}{4}\right)$ $\frac{11}{15}$

 Writing   Translating Expression   Geometry Scientific Calculator   Video NS&E

Objective 1: Commutative Properties of Real Numbers

For Exercises 15–22, rewrite each expression using the commutative property of addition or the commutative property of multiplication. **(See Examples 1 and 3.)**

15. $5 + (-8)$ $-8 + 5$ **16.** $7 + (-2)$ $-2 + 7$ **17.** $8 + x$ $x + 8$ **18.** $p + 11$ $11 + p$

19. $5(4)$ $4(5)$ **20.** $10(8)$ $8(10)$ **21.** $x(-12)$ $-12x$ **22.** $y(-23)$ $-23y$

For Exercises 23–26, rewrite each expression using addition. Then apply the commutative property of addition. **(See Example 2.)**

23. $x - 3$
$x + (-3); -3 + x$

24. $y - 7$
$y + (-7); -7 + y$

25. $4p - 9$
$4p + (-9); -9 + 4p$

26. $3m - 12$
$3m + (-12); -12 + 3m$

Objective 2: Associative Properties of Real Numbers

For Exercises 27–34, use the associative property of addition or multiplication to rewrite each expression. Then simplify the expression if possible. **(See Example 4.)**

27. $(x + 4) + 9$
$x + (4 + 9); x + 13$

28. $-3 + (5 + z)$
$(-3 + 5) + z; 2 + z$

29. $-5(3x)$
$(-5 \cdot 3)x; -15x$

30. $-12(4z)$
$(-12 \cdot 4)z; -48z$

31. $\dfrac{6}{11}\left(\dfrac{11}{6}x\right)$
$\left(\dfrac{6}{11} \cdot \dfrac{11}{6}\right)x; x$

32. $\dfrac{3}{5}\left(\dfrac{5}{3}x\right)$
$\left(\dfrac{3}{5} \cdot \dfrac{5}{3}\right)x; x$

33. $-4\left(-\dfrac{1}{4}t\right)$
$\left(-4 \cdot -\dfrac{1}{4}\right)t; t$

34. $-5\left(-\dfrac{1}{5}w\right)$
$\left(-5 \cdot -\dfrac{1}{5}\right)w; w$

Objective 3: Identity and Inverse Properties of Real Numbers

35. What is another name for multiplicative inverse?
Reciprocal

36. What is another name for additive inverse?
Opposite

37. What is the additive identity?   0

38. What is the multiplicative identity?   1

Objective 4: Distributive Property of Multiplication over Addition

For Exercises 39–58, use the distributive property to clear parentheses. **(See Examples 5–6.)**

39. $6(5x + 1)$
$30x + 6$

40. $2(x + 7)$
$2x + 14$

41. $-2(a + 8)$
$-2a - 16$

42. $-3(2z + 9)$
$-6z - 27$

43. $3(5c - d)$
$15c - 3d$

44. $4(w - 13z)$
$4w - 52z$

45. $-7(y - 2)$
$-7y + 14$

46. $-2(4x - 1)$
$-8x + 2$

47. $-\dfrac{2}{3}(x - 6)$ $-\dfrac{2}{3}x + 4$

48. $-\dfrac{1}{4}(2b - 8)$ $-\dfrac{1}{2}b + 2$

49. $\dfrac{1}{3}(m - 3)$ $\dfrac{1}{3}m - 1$

50. $\dfrac{2}{5}(n - 5)$ $\dfrac{2}{5}n - 2$

51. $-(2p + 10)$
$-2p - 10$

52. $-(7q + 1)$
$-7q - 1$

53. $-2(-3w - 5z + 8)$
$6w + 10z - 16$

54. $-4(-7a - b - 3)$
$28a + 4b + 12$

55. $4(x + 2y - z)$
$4x + 8y - 4z$

56. $-6(2a - b + c)$
$-12a + 6b - 6c$

57. $-(-6w + x - 3y)$
$6w - x + 3y$

58. $-(-p - 5q - 10r)$
$p + 5q + 10r$

Mixed Exercises

For Exercises 59–62, use the associative property or distributive property to clear parentheses.

59. $2(3 + x)$ $6 + 2x$ **60.** $5(4 + y)$ $20 + 5y$ **61.** $4(6z)$ $24z$ **62.** $8(2p)$ $16p$

For Exercises 63–71, match the statement with the property that describes it.

63. $6 \cdot \dfrac{1}{6} = 1$ b

64. $7(4 \cdot 9) = (7 \cdot 4)9$ f

65. $2(3 + k) = 6 + 2k$ i

66. $3 \cdot 7 = 7 \cdot 3$ c

67. $5 + (-5) = 0$ g

68. $18 \cdot 1 = 18$ e

69. $(3 + 7) + 19 = 3 + (7 + 19)$ d

70. $23 + 6 = 6 + 23$ a

71. $3 + 0 = 3$ h

a. Commutative property of addition

b. Inverse property of multiplication

c. Commutative property of multiplication

d. Associative property of addition

e. Identity property of multiplication

f. Associative property of multiplication

g. Inverse property of addition

h. Identity property of addition

i. Distributive property of multiplication over addition

Objective 5: Simplifying Algebraic Expressions

For Exercises 72–75, for each expression list the terms and their coefficients. **(See Example 7.)**

72. $3xy - 6x^2 + y - 17$

Term	Coefficient
$3xy$	3
$-6x^2$	-6
y	1
-17	-17

73. $2x - y + 18xy + 5$

Term	Coefficient
$2x$	2
$-y$	-1
$18xy$	18
5	5

74. $x^4 - 10xy + 12 - y$

Term	Coefficient
x^4	1
$-10xy$	-10
12	12
$-y$	-1

75. $-x + 8y - 9x^2y - 3$

Term	Coefficient
$-x$	-1
$8y$	8
$-9x^2y$	-9
-3	-3

76. Explain why $12x$ and $12x^2$ are not *like* terms.
The exponents on x are different.

77. Explain why $3x$ and $3xy$ are not *like* terms.
The variable factors are different.

78. Explain why $7z$ and $\sqrt{13}z$ are *like* terms.
The variables are the same and raised to the same power.

79. Explain why πx and $8x$ are *like* terms.
The variables are the same and raised to the same power.

80. Write three different *like* terms.
For example: $5y, -2y, y$

81. Write three terms that are not *like*.
For example: $5y, -2x, 6$

For Exercises 82–90, simplify by combining *like* terms. **(See Examples 8–9.)**

82. $5k - 10k - 12k + 16 + 7$
$-17k + 23$

83. $-4p - 2p + 8p - 15 + 3$
$2p - 12$

84. $9x - 7x^2 + 12x + 14x^2$
$7x^2 + 21x$

85. $2y^2 - 8y + y - 5y^2 - 3y^2$
$-6y^2 - 7y$

86. $4ab^2 + 2a^2b - 6ab^2 + 5 + 3a^2b - 2$
$-2ab^2 + 5a^2b + 3$

 Writing    Translating Expression   Geometry   Scientific Calculator   Video   NS E

87. $8x^3y - 5xy + 3 - 7 + 6xy - x^3y$
$7x^3y + xy - 4$

88. $\frac{1}{4}a + b - \frac{3}{4}a - 5b$
$-\frac{1}{2}a - 4b$

89. $\frac{2}{5} + 2t - \frac{3}{5} + t - \frac{6}{5}$
$3t - \frac{7}{5}$

90. $2.8z - 8.1z + 6 - 15.2$
$-5.3z - 9.2$

For Exercises 91–118, simplify by clearing parentheses and combining *like* terms. **(See Examples 10–12.)**

91. $-3(2x - 4) + 10$
$-6x + 22$

92. $-2(4a + 3) - 14$
$-8a - 20$

93. $4(w + 3) - 12$ $4w$

94. $5(2r + 6) - 30$ $10r$

95. $5 - 3(x - 4)$
$-3x + 17$

96. $4 - 2(3x + 8)$
$-6x - 12$

97. $-3(2t + 4) + 8(2t - 4)$
$10t - 44$

98. $-5(5y + 9) + 3(3y + 6)$
$-16y - 27$

99. $2(w - 5) - (2w + 8)$
-18

100. $6(x + 3) - (6x - 5)$ 23

101. $-\frac{1}{3}(6t + 9) + 10$
$-2t + 7$

102. $-\frac{3}{4}(8 + 4q) + 7$
$-3q + 1$

103. $10(5.1a - 3.1) + 4$
$51a - 27$

104. $100(-3.14p - 1.05) + 212$
$-314p + 107$

105. $-4m + 2(m - 3) + 2m$ -6

106. $-3b + 4(b + 2) - 8b$
$-7b + 8$

107. $\frac{1}{2}(10q - 2) + \frac{1}{3}(2 - 3q)$
$4q - \frac{1}{3}$

108. $\frac{1}{5}(15 - 4p) - \frac{1}{10}(10p + 5)$
$-\frac{9}{5}p + \frac{5}{2}$

109. $7n - 2(n - 3) - 6 + n$ $6n$

110. $8k - 4(k - 1) + 7 - k$
$3k + 11$

111. $6(x + 3) - 12 - 4(x - 3)$
$2x + 18$

112. $5(y - 4) + 3 - 6(y - 7)$
$-y + 25$

113. $6.1(5.3z - 4.1) - 5.8$
$32.33z - 30.81$

114. $-3.6(1.7q - 4.2) + 14.6$
$-6.12q + 29.72$

115. $6 + 2[-8 - 3(2x + 4)] + 10x$
$-2x - 34$

116. $-3 + 5[-3 - 4(y + 2)] - 8y$
$-28y - 58$

117. $1 - 3[2(z + 1) - 5(z - 2)]$
$9z - 35$

118. $1 - 6[3(2t + 2) - 8(t + 2)]$
$12t + 61$

Expanding Your Skills

For Exercises 119–126, determine if the expressions are equivalent. If two expressions are not equivalent, state why.

119. $3a + b, b + 3a$
Equivalent

120. $4y + 1, 1 + 4y$
Equivalent

121. $2c + 7, 9c$
Not equivalent. The terms are not *like* terms and cannot be combined.

122. $5z + 4, 9z$
Not equivalent. The terms are not *like* terms and cannot be combined.

123. $5x - 3, 3 - 5x$
Not equivalent; subtraction is not commutative.

124. $6d - 7, 7 - 6d$
Not equivalent; subtraction is not commutative.

125. $5x - 3, -3 + 5x$
Equivalent

126. $8 - 2x, -2x + 8$
Equivalent

127. Which grouping of terms is easier to compute, $(14\frac{2}{7} + 2\frac{1}{3}) + \frac{2}{3}$ or $14\frac{2}{7} + (2\frac{1}{3} + \frac{2}{3})$?
$14\frac{2}{7} + (2\frac{1}{3} + \frac{2}{3})$ is easier.

128. Which grouping of terms is easier to compute, $(5\frac{1}{8} + 18\frac{2}{5}) + 1\frac{3}{5}$ or $5\frac{1}{8} + (18\frac{2}{5} + 1\frac{3}{5})$?
$5\frac{1}{8} + (18\frac{2}{5} + 1\frac{3}{5})$ is easier.

129. As a small child in school, the great mathematician Karl Friedrich Gauss (1777–1855) was said to have found the sum of the integers from 1 to 100 mentally:

$$1 + 2 + 3 + 4 + \cdots + 99 + 100$$

Rather than adding the numbers sequentially, he added the numbers in pairs:

$$(1 + 99) + (2 + 98) + (3 + 97) + \cdots + 100$$

a. Use this technique to add the integers from 1 to 10. 55

$$1 + 2 + 3 + 4 + 5 + 6 + 7 + 8 + 9 + 10$$

b. Use this technique to add the integers from 1 to 20. 210

Group Activity

Evaluating Formulas Using a Calculator

Materials: A calculator

Estimated Time: 15 minutes

Group Size: 2

In this chapter, we learned one of the most important concepts in mathematics—the order of operations. The proper order of operations is required whenever we evaluate any mathematical expression. The following formulas are taken from applications from science, math, statistics, and business. These are just some samples of what you may encounter as you work your way through college.

For Exercises 1–8, substitute the given values into the formula. Then use a calculator and the proper order of operations to simplify the result. Round to three decimal places if necessary.

1. $F = \dfrac{9}{5}C + 32$ (biology) $C = 35$ $F = 95$

2. $V = \dfrac{nRT}{P}$ (chemistry) $n = 1.00, R = 0.0821, T = 273.15, P = 1.0$ $V = 22.426$

3. $R = k\left(\dfrac{L}{r^2}\right)$ (electronics) $k = 0.05, L = 200, r = 0.5$ $R = 40$

4. $m = \dfrac{y_2 - y_1}{x_2 - x_1}$ (mathematics) $x_1 = -8.3, x_2 = 3.3, y_1 = 4.6, y_2 = -9.2$ $m = -1.190$

5. $z = \dfrac{\bar{x} - \mu}{\dfrac{\sigma}{\sqrt{n}}}$ (statistics) $\bar{x} = 69, \mu = 55, \sigma = 20, n = 25$ $z = 3.5$

6. $S = R\left[\dfrac{(1 + i)^n - 1}{i}\right]$ (finance) $R = 200, i = 0.08, n = 30$ $S = 22{,}656.642$

7. $x = \dfrac{-b + \sqrt{b^2 - 4ac}}{2a}$ (mathematics) $a = 2, b = -7, c = -15$ $x = 5$

8. $h = \dfrac{1}{2}gt^2 + v_0 t + h_0$ (physics) $g = -32, t = 2.4, v_0 = 192, h_0 = 288$ $h = 656.64$

Chapter 1 Summary

Section 1.1 — Sets of Numbers and the Real Number Line

Key Concepts

Natural numbers: $\{1, 2, 3, \ldots\}$

Whole numbers: $\{0, 1, 2, 3, \ldots\}$

Integers: $\{\ldots -3, -2, -1, 0, 1, 2, 3, \ldots\}$

Rational numbers: The set of numbers that can be expressed in the form $\frac{p}{q}$, where p and q are integers and q does not equal 0. In decimal form, rational numbers are terminating or repeating decimals.

Irrational numbers: A subset of the real numbers whose elements cannot be written as a ratio of two integers. In decimal form, irrational numbers are nonterminating, nonrepeating decimals.

Real numbers: The set of both the rational numbers and the irrational numbers.

$a < b$	"a is less than b."
$a > b$	"a is greater than b."
$a \leq b$	"a is less than or equal to b."
$a \geq b$	"a is greater than or equal to b."

Two numbers that are the same distance from zero but on opposite sides of zero on the number line are called **opposites**. The opposite of a is denoted $-a$.

The **absolute value** of a real number, a, denoted $|a|$, is the distance between a and 0 on the number line.

If $a \geq 0$, $|a| = a$

If $a < 0$, $|a| = -a$

Examples

Example 1

$-5, 0,$ and 4 are integers.

$-\dfrac{5}{2}, -0.5,$ and $0.\overline{3}$ are rational numbers.

$\sqrt{7}, -\sqrt{2},$ and π are irrational numbers.

Example 2

All real numbers can be located on the real number line.

Example 3

$5 < 7$	"5 is less than 7."
$-2 > -10$	"-2 is greater than -10."
$y \leq 3.4$	"y is less than or equal to 3.4."
$x \geq \frac{1}{2}$	"x is greater than or equal to $\frac{1}{2}$."

Example 4

5 and -5 are opposites.

Example 5

$|7| = 7$

$|-7| = 7$

| Section 1.2 | Exponents, Square Roots, and the Order of Operations |

Key Concepts

A **variable** is a symbol or letter used to represent an unknown number.

A **constant** is a value that is not variable.

An algebraic **expression** is a collection of variables and constants under algebraic operations.

$$b^n = \underbrace{b \cdot b \cdot b \cdot b \ldots \cdot b}_{n \text{ factors of } b}$$

b is the **base**,
n is the **exponent**

$\sqrt{x}$ is the positive **square root** of x.

The Order of Operations

1. Simplify expressions within parentheses and other grouping symbols first.
2. Evaluate expressions involving exponents, radicals, and absolute values.
3. Perform multiplication or division in the order that they occur from left to right.
4. Perform addition or subtraction in the order that they occur from left to right.

Examples

Example 1

Variables:	x, y, z, a, b
Constants:	$2, -3, \pi$
Expressions:	$2x + 5, 3a + b^2$

Example 2

$5^3 = 5 \cdot 5 \cdot 5 = 125$

Example 3

$\sqrt{49} = 7$

Example 4

$10 + 5(3 - 1)^2 - \sqrt{5 - 1}$

$= 10 + 5(2)^2 - \sqrt{4}$ Work within grouping symbols.

$= 10 + 5(4) - 2$ Simplify exponents and radicals.

$= 10 + 20 - 2$ Perform multiplication.

$= 30 - 2$ Add and subtract, left to right

$= 28$

| Section 1.3 | Addition of Real Numbers |

Key Concepts

Addition of Two Real Numbers

Same Signs.   Add the absolute values of the numbers and apply the common sign to the sum.

Different Signs.   Subtract the smaller absolute value from the larger absolute value. Then apply the sign of the number having the larger absolute value.

Examples

Example 1

$-3 + (-4) = -7$

$-1.3 + (-9.1) = -10.4$

Example 2

$-5 + 7 = 2$

$\dfrac{2}{3} + \left(-\dfrac{7}{3}\right) = -\dfrac{5}{3}$

Section 1.4　Subtraction of Real Numbers

Key Concepts

Subtraction of Two Real Numbers

Add the opposite of the second number to the first number. That is,

$$a - b = a + (-b)$$

Examples

Example 1

$7 - (-5) = 7 + (5) = 12$

$-3 - 5 = -3 + (-5) = -8$

$-11 - (-2) = -11 + (2) = -9$

Section 1.5　Multiplication and Division of Real Numbers

Key Concepts

Multiplication and Division of Two Real Numbers

Same Signs.
Product is positive.
Quotient is positive.

Different Signs.
Product is negative.
Quotient is negative.

The **reciprocal** of a number a is $\frac{1}{a}$.

Multiplication and Division Involving Zero

The product of any real number and 0 is 0.

The quotient of 0 and any nonzero real number is 0.

The quotient of any nonzero real number and 0 is undefined.

Examples

Example 1

$(-5)(-2) = 10$　　　$\dfrac{-20}{-4} = 5$

Example 2

$(-3)(7) = -21$　　　$\dfrac{-4}{8} = -\dfrac{1}{2}$

Example 3

The reciprocal of -6 is $-\frac{1}{6}$.

Example 4

$4 \cdot 0 = 0$

$0 \div 4 = 0$

$4 \div 0$ is undefined.

| Section 1.6 | **Properties of Real Numbers and Simplifying Expressions** |

Key Concepts

Properties of Real Numbers

Commutative Properties.

$a + b = b + a$

$ab = ba$

Associative Properties.

$(a + b) + c = a + (b + c)$

$(ab)c = a(bc)$

Identity Properties.

$0 + a = a$

$1 \cdot a = a$

Inverse Properties.

$a + (-a) = 0$

$b \cdot \dfrac{1}{b} = 1 \text{ for } b \neq 0$

Distributive Property of Multiplication over Addition.

$a(b + c) = ab + ac$

A **term** is a constant or the product or quotient of constants and variables. The **coefficient** of a term is the numerical factor of the term.

Like terms have the same variables, and the corresponding variables have the same powers.

Terms can be added or subtracted if they are *like* terms. Sometimes it is necessary to clear parentheses before adding or subtracting *like* terms.

Examples

Example 1

$(-5) + (-7) = (-7) + (-5)$

$3 \cdot 8 = 8 \cdot 3$

Example 2

$(2 + 3) + 10 = 2 + (3 + 10)$

$(2 \cdot 4) \cdot 5 = 2 \cdot (4 \cdot 5)$

Example 3

$0 + (-5) = -5$

$1(-8) = -8$

Example 4

$1.5 + (-1.5) = 0$

$6 \cdot \dfrac{1}{6} = 1$

Example 5

$$-2(x - 3y) = (-2)x + (-2)(-3y)$$
$$= -2x + 6y$$

Example 6

$-2x$ is a term with coefficient -2.
yz^2 is a term with coefficient 1.

$3x$ and $-5x$ are *like* terms.
$4a^2b$ and $4ab$ are not *like* terms.

Example 7

$$-2w - 4(w - 2) + 3$$
$$= -2w - 4w + 8 + 3 \qquad \text{Clear parentheses.}$$
$$= -6w + 11 \qquad \text{Combine } like \text{ terms.}$$

Chapter 1   Review Exercises

Section 1.1

1. Given the set $\{7, \frac{1}{3}, -4, 0, -\sqrt{3}, -0.\overline{2}, \pi, 1\}$,

    a. List the natural numbers. $7, 1$

    b. List the integers. $7, -4, 0, 1$

    c. List the whole numbers. $7, 0, 1$

    d. List the rational numbers. $7, \frac{1}{3}, -4, 0, -0.\overline{2}, 1$

    e. List the irrational numbers. $-\sqrt{3}, \pi$

    f. List the real numbers. $7, \frac{1}{3}, -4, 0, -\sqrt{3}, -0.\overline{2}, \pi, 1$

For Exercises 2–5, determine the absolute value.

2. $\left|\dfrac{1}{2}\right|$ $\frac{1}{2}$ **3.** $|-6|$ 6 **4.** $|-\sqrt{7}|$ $\sqrt{7}$ **5.** $|0|$ 0

For Exercises 6–13, identify whether the inequality is true or false.

6. $-6 > -1$ **7.** $0 < -5$ **8.** $-10 \leq 0$
   False   False   True

9. $5 \neq -5$ **10.** $7 \geq 7$ **11.** $7 \geq -7$
   True   True   True

12. $0 \leq -3$ False   **13.** $-\dfrac{2}{3} \leq -\dfrac{2}{3}$ True

Section 1.2

↔ For Exercises 14–19, translate the English phrases into algebraic expressions.

14. The product of x and $\dfrac{2}{3}$ $x \cdot \frac{2}{3}$ or $\frac{2}{3}x$

15. The quotient of 7 and y $\frac{7}{y}$ or $7 \div y$

16. The sum of 2 and $3b$ $2 + 3b$

17. The difference of a and 5 $a - 5$

18. Two more than $5k$ $5k + 2$

19. Seven less than $13z$ $13z - 7$

For Exercises 20–23, evaluate the expressions for $x = 8$, $y = 4$, and $z = 1$.

20. $x - 2y$ 0 **21.** $x^2 - y$ 60

22. $\sqrt{x + z}$ 3 **23.** $\sqrt{x + 2y}$ 4

For Exercises 24–29, simplify the expressions.

24. 6^3 216 **25.** 15^2 225 **26.** $\sqrt{36}$ 6

27. $\dfrac{1}{\sqrt{100}}$ $\frac{1}{10}$ **28.** $\left(\dfrac{1}{4}\right)^2$ $\frac{1}{16}$ **29.** $\left(\dfrac{3}{2}\right)^3$ $\frac{27}{8}$

For Exercises 30–33, perform the indicated operations.

30. $15 - 7 \cdot 2 + 12$ **31.** $|-11| + |5| - (7 - 2)$
   13 11

32. $4^2 - (5 - 2)^2$ **33.** $22 - 3(8 \div 4)^2$
   7 10

Section 1.3

For Exercises 34–46, add.

34. $-6 + 8$ 2 **35.** $14 + (-10)$ 4

36. $21 + (-6)$ 15 **37.** $-12 + (-5)$ -17

38. $\dfrac{2}{7} + \left(-\dfrac{1}{9}\right)$ $\frac{11}{63}$ **39.** $\left(-\dfrac{8}{11}\right) + \left(\dfrac{1}{2}\right)$ $-\frac{5}{22}$

40. $\left(-\dfrac{1}{10}\right) + \left(-\dfrac{5}{6}\right)$ $-\frac{14}{15}$ **41.** $\left(-\dfrac{5}{2}\right) + \left(-\dfrac{1}{5}\right)$ $-\frac{27}{10}$

42. $-8.17 + 6.02$ -2.15 **43.** $2.9 + (-7.18)$ -4.28

44. $13 + (-2) + (-8)$ 3 **45.** $-5 + (-7) + 20$ 8

46. $2 + 5 + (-8) + (-7) + 0 + 13 + (-1)$ 4

47. Under what conditions will the expression $a + b$ be negative?   When a and b are both negative or when a and b have different signs and the number with the larger absolute value is negative.

48. The high temperatures (in degrees Celsius) for the province of Alberta, Canada, during a week in January were $-8, -1, -4, -3, -4, 0,$ and 7. What was the average high temperature for that week? Round to the nearest tenth of a degree. $-1.9°C$

Section 1.4

For Exercises 49–61, subtract.

49. $13 - 25$ -12 **50.** $31 - (-2)$ 33

51. $-8 - (-7)$ -1 **52.** $-2 - 15$ -17

53. $\left(-\dfrac{7}{9}\right) - \dfrac{5}{6}$ $-\frac{29}{18}$ **54.** $\dfrac{1}{3} - \dfrac{9}{8}$ $-\frac{19}{24}$

55. $7 - 8.2$ -1.2

56. $-1.05 - 3.2$ -4.25

57. $-16.1 - (-5.9)$ -10.2

58. $7.09 - (-5)$ 12.09

59. $\dfrac{11}{2} - \left(-\dfrac{1}{6}\right) - \dfrac{7}{3}$ $\dfrac{10}{3}$

60. $-\dfrac{4}{5} - \dfrac{7}{10} - \left(-\dfrac{13}{20}\right)$ $-\dfrac{17}{20}$

61. $6 - 14 - (-1) - 10 - (-21) - 5$ -1

62. Under what conditions will the expression $a - b$ be negative? If $a < b$

→ For Exercises 63–67, write an algebraic expression and simplify.

63. -18 subtracted from -7
   $-7 - (-18)$; 11

64. The difference of -6 and 41
   $-6 - 41$; -47

65. Seven decreased by 13
   $7 - 13$; -6

66. Five subtracted from the difference of 20 and -7
   $(20 - (-7)) - 5$; 22

67. The sum of 6 and -12, decreased by 21
   $(6 + (-12)) - 21$; -27

68. In Nevada, the highest temperature ever recorded was 125°F and the lowest was -50°F. Find the difference between the highest and lowest temperatures. (*Source: Information Please Almanac*)
   175°F

Section 1.5

For Exercises 69–86, multiply or divide as indicated.

69. $10(-17)$ -170

70. $(-7)13$ -91

71. $(-52) \div 26$ -2

72. $(-48) \div (-16)$ 3

73. $\dfrac{7}{4} \div \left(-\dfrac{21}{2}\right)$ $-\dfrac{1}{6}$

74. $\dfrac{2}{3}\left(-\dfrac{12}{11}\right)$ $-\dfrac{8}{11}$

75. $-\dfrac{21}{5} \cdot 0$ 0

76. $\dfrac{3}{4} \div 0$ Undefined

77. $0 \div (-14)$ 0

78. $(-0.45)(-5)$ 2.25

79. $\dfrac{-21}{14}$ $-\dfrac{3}{2}$

80. $\dfrac{-13}{-52}$ $\dfrac{1}{4}$

81. $(5)(-2)(3)$ -30

82. $(-6)(-5)(15)$ 450

83. $\left(-\dfrac{1}{2}\right)\left(\dfrac{7}{8}\right)\left(-\dfrac{4}{7}\right)$ $\dfrac{1}{4}$

84. $\left(\dfrac{12}{13}\right)\left(-\dfrac{1}{6}\right)\left(\dfrac{13}{14}\right)$ $-\dfrac{1}{7}$

85. $40 \div 4 \div (-5)$ -2

86. $\dfrac{10}{11} \div \dfrac{7}{11} \div \dfrac{5}{9}$ $\dfrac{18}{7}$

For Exercises 87–92, perform the indicated operations.

87. $9 - 4[-2(4 - 8) - 5(3 - 1)]$ 17

88. $\dfrac{8(-3) - 6}{-7 - (-2)}$ 6

89. $\dfrac{2}{3} - \left(\dfrac{3}{8} + \dfrac{5}{6}\right) \div \dfrac{5}{3}$ $-\dfrac{7}{120}$

90. $5.4 - (0.3)^2 \div 0.09$ 4.4

91. $\dfrac{5 \frac{1}{120}[3 - (-4)^2]}{36 \div (-2)(3)}$ $-\dfrac{1}{3}$

92. $|-8 + 5| - \sqrt{5^2 - 3^2}$ -1

For Exercises 93–96, evaluate the expression with the given substitution.

93. $3(x + 2) \div y$ for $x = 4$ and $y = -9$ -2

94. $a^2 - bc$ for $a = -6$, $b = 5$, and $c = 2$ 26

95. $w + xy - \sqrt{z}$
    for $w = 12$, $x = 6$, $y = -5$, and $z = 25$
    -23

96. $(u - v)^2 + (u^2 - v^2)$ for $u = 5$ and $v = -3$ 80

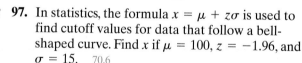 **97.** In statistics, the formula $x = \mu + z\sigma$ is used to find cutoff values for data that follow a bell-shaped curve. Find x if $\mu = 100$, $z = -1.96$, and $\sigma = 15$. 70.6

For Exercises 98–104, answer true or false. If a statement is false, explain why.

98. If n is positive, then $-n$ is negative. True

99. If m is negative, then m^4 is negative. False, any nonzero real number raised to an even power is positive.

100. If m is negative, then m^3 is negative. True

101. If $m > 0$ and $n > 0$, then $mn > 0$. True

102. If $p < 0$ and $q < 0$, then $pq < 0$. False, the product of two negative numbers is positive.

103. A number and its reciprocal have the same signs. True

104. A nonzero number and its opposite have different signs. True

Section 1.6

For Exercises 105–112, answers may vary.

105. Give an example of the commutative property of addition.
For example: $2 + 3 = 3 + 2$

106. Give an example of the associative property of addition.
For example: $(2 + 3) + 4 = 2 + (3 + 4)$

107. Give an example of the inverse property of addition.
For example: $5 + (-5) = 0$

108. Give an example of the identity property of addition.
For example: $7 + 0 = 7$

109. Give an example of the commutative property of multiplication.
For example: $5 \cdot 2 = 2 \cdot 5$

110. Give an example of the associative property of multiplication.
For example: $(8 \cdot 2)10 = 8(2 \cdot 10)$

111. Give an example of the inverse property of multiplication.
For example: $3 \cdot \dfrac{1}{3} = 1$

112. Give an example of the identity property of multiplication.
For example: $8 \cdot 1 = 8$

113. Explain why $5x - 2y$ is the same as $-2y + 5x$.
$5x - 2y = 5x + (-2y)$, then use the commutative property of addition.

114. Explain why $3a - 9y$ is the same as $-9y + 3a$.
$3a - 9y = 3a + (-9y)$, then use the commutative property of addition.

115. List the terms of the expression:
$3y + 10x - 12 + xy$
$3y, 10x, -12, xy$

116. Identify the coefficients for the terms listed in Exercise 115.
$3, 10, -12, 1$

For Exercises 117–118, simplify by combining *like* terms.

117. $3a + 3b - 4b + 5a - 10$
$8a - b - 10$

118. $-6p + 2q + 9 - 13q - p + 7$
$-7p - 11q + 16$

For Exercises 119–120, use the distributive property to clear the parentheses.

119. $-2(4z + 9)$
$-8z - 18$

120. $5(4w - 8y + 1)$
$20w - 40y + 5$

For Exercises 121–126, simplify the expression.

121. $2p - (p + 5) + 3$ $p - 2$

122. $6(h + 3) - 7h - 4$ $-h + 14$

123. $\dfrac{1}{2}(-6q) + q - 4\left(3q + \dfrac{1}{4}\right)$ $-14q - 1$

124. $0.3b + 12(0.2 - 0.5b)$ $-5.7b + 2.4$

125. $-4[2(x + 1) - (3x + 8)]$ $4x + 24$

126. $5[(7y - 3) + 3(y + 8)]$ $50y + 105$

Chapter 1   Test

1. Is $0.\overline{315}$ a rational number or an irrational number? Explain your reasoning.
Rational, all repeating decimals are rational numbers.

2. Plot the points on a number line: $|3|, 0, -2, 0.5,$ $\left|-\frac{3}{2}\right|, \sqrt{16}.$

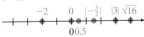

3. Use the number line in Exercise 2 to identify whether the statements are true or false.

a. $|3| < -2$ False

b. $0 \le \left|-\dfrac{3}{2}\right|$ True

c. $-2 < 0.5$ True

d. $|3| \ge \left|-\dfrac{3}{2}\right|$ True

4. Use the definition of exponents to expand the expressions:

a. $(4x)^3$ $(4x)(4x)(4x)$ **b.** $4x^3$ $4 \cdot x \cdot x \cdot x$

5. **a.** Translate the expression into an English phrase: $2(a - b)$. (Answers may vary.)
Twice the difference of a and b
   b. Translate the expression into an English phrase: $2a - b$. (Answers may vary.)
b subtracted from twice a

6. Translate the phrase into an algebraic expression: "The quotient of the square root of c and the square of d."
$\dfrac{\sqrt{c}}{d^2}$ or $\sqrt{c} \div d^2$

For Exercises 7–23, perform the indicated operations.

7. $18 + (-12)$ ⠀6

8. $-10 + (-9)$ ⠀-19

9. $-15 - (-3)$ ⠀-12

10. $21 - (-7)$ ⠀28

11. $-\dfrac{1}{8} + \left(-\dfrac{3}{4}\right)$ ⠀$-\dfrac{7}{8}$

12. $-10.06 - (-14.72)$ ⠀4.66

13. $-14 + (-2) - 16$ ⠀-32

14. $-84 \div 7$ ⠀-12

15. $38 \div 0$ ⠀Undefined

16. $7(-4)$ ⠀-28

17. $-22 \cdot 0$ ⠀0

18. $(-16)(-2)(-1)(-3)$ ⠀96

19. $\dfrac{2}{5} \div \left(-\dfrac{7}{10}\right) \cdot \left(-\dfrac{7}{6}\right)$ ⠀$\dfrac{2}{3}$

20. $(8 - 10) \cdot \dfrac{3}{2} + (-5)$ ⠀-8

21. $8 - [(2 - 4) - (8 - 9)]$ ⠀9

22. $\dfrac{\sqrt{5^2 - 4^2}}{|-12 + 3|}$ ⠀$\dfrac{1}{3}$

23. $\dfrac{|4 - 10|}{2 - 3(5 - 1)}$ ⠀$-\dfrac{3}{5}$

24. Identify the property that justifies each statement.

   a. $6(-8) = (-8)6$
   Commutative property of multiplication

   b. $5 + 0 = 5$
   Identity property of addition

   c. $(2 + 3) + 4 = 2 + (3 + 4)$
   Associative property of addition

   d. $\dfrac{1}{7} \cdot 7 = 1$
   Inverse property of multiplication

   e. $8[7(-3)] = (8 \cdot 7)(-3)$
   Associative property of multiplication

For Exercises 25–29, simplify the expression.

25. $-5x - 4y + 3 - 7x + 6y - 7$ ⠀$-12x + 2y - 4$

26. $-3(4m + 8p - 7)$ ⠀$-12m - 24p + 21$

27. $3k - 20 + (-9k) + 12$ ⠀$-6k - 8$

28. $4(p - 5) - (8p + 3)$ ⠀$-4p - 23$

29. $\dfrac{1}{2}(12p - 4) + \dfrac{1}{3}(2 - 6p)$ ⠀$4p - \dfrac{4}{3}$

For Exercises 30–33, evaluate the expression given the values $x = 4$ and $y = -3$ and $z = -7$.

30. $y^2 - x$ ⠀5

31. $3x - 2y$ ⠀18

32. $y(x - 2)$ ⠀-6

33. $-y^2 - 4x + z$ ⠀-32

For Exercises 34–35, translate the English statement to an algebraic expression. Then simplify the expression.

34. Subtract -4 from 12 ⠀$12 - (-4); 16$

35. Find the difference of 6 and 8 ⠀$6 - 8; -2$

Writing ⠀ Translating Expression ⠀ Geometry ⠀ Scientific Calculator ⠀ Video ⠀

Linear Equations and Inequalities

2

CHAPTER OUTLINE

2.1 Addition, Subtraction, Multiplication, and Division Properties of Equality 116

2.2 Solving Linear Equations 127

2.3 Linear Equations: Clearing Fractions and Decimals 135

Problem Recognition Exercises: Equations and Expressions 142

2.4 Applications of Linear Equations: Introduction to Problem Solving 143

2.5 Applications Involving Percents 153

2.6 Formulas and Applications of Geometry 159

2.7 Mixture Applications and Uniform Motion 169

2.8 Linear Inequalities 178

Group Activity: Computing Body Mass Index (BMI) 192

Chapter 2

In this chapter, we learn how to solve linear equations and inequalities in one variable. This important category of equations and inequalities is used in a variety of applications.

Use your skills at solving equations and simplifying expressions to complete this puzzle.

Across

2. Solve. $\dfrac{z}{-6} = -81$

4. Simplify. $15x - 5(x - 49) - 10x + 2000$

5. Smallest positive integer that can be used to clear decimals.
   $0.06x + 1.8 = 2 - 2.251$

Down

1. Solve. $2a - 3 = 45$

3. $-10^2 \cdot (-2)^3 \cdot 11$

4. Solve. $4(2x - 1) = 5x + 56$

5. Smallest positive integer that can be used to clear fractions.
   $\dfrac{1}{2}y - \dfrac{3}{4} = \dfrac{5}{3}y + 1$

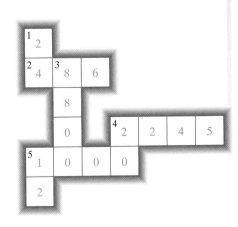

| Section 2.1 | **Addition, Subtraction, Multiplication, and Division Properties of Equality** |

Objectives

1. **Definition of a Linear Equation in One Variable**
2. **Addition and Subtraction Properties of Equality**
3. **Multiplication and Division Properties of Equality**
4. **Translations**

1. Definition of a Linear Equation in One Variable

An *equation* is a statement that indicates that two quantities are equal. The following are equations.

$$x = 5 \qquad y + 2 = 12 \qquad -4z = 28$$

All equations have an equal sign. Furthermore, notice that the equal sign separates the equation into two parts, the left-hand side and the right-hand side. A **solution to an equation** is a value of the variable that makes the equation a true statement. Substituting a solution into an equation for the variable makes the right-hand side equal to the left-hand side.

Equation	Solution	Check	
$x = 5$	5	$x = 5$ $\downarrow$ $5 = 5$ ✔	Substitute 5 for x. Right-hand side equals left-hand side.
$y + 2 = 12$	10	$y + 2 = 12$ $\downarrow$ $10 + 2 = 12$ ✔	Substitute 10 for y. Right-hand side equals left-hand side.
$-4z = 28$	-7	$-4z = 28$ $\downarrow$ $-4(-7) = 28$ ✔	Substitute -7 for z. Right-hand side equals left-hand side.

Avoiding Mistakes

Be sure to notice the difference between solving an equation versus simplifying an expression. For example, $2x + 1 = 7$ is an equation, whose solution is 3, while $2x + 1 + 7$ is an expression that simplifies to $2x + 8$.

Skill Practice

Determine if the number given is a solution to the equation.

1. $4x - 1 = 7; \quad 3$
2. $-2y + 5 = 9; \quad -2$

Classroom Example: p. 125, Exercise 14

| Example 1 | **Determining Whether a Number Is a Solution to an Equation** |

Determine whether the given number is a solution to the equation.

a. $4x + 7 = 5; \quad -\frac{1}{2}$ **b.** $-6w + 14 = 4; \quad 3$

Solution:

a. $4x + 7 = 5$

     $4(-\frac{1}{2}) + 7 \stackrel{?}{=} 5$ Substitute $-\frac{1}{2}$ for x.

       $-2 + 7 \stackrel{?}{=} 5$ Simplify.

           $5 \stackrel{?}{=} 5$ ✔   Right-hand side equals the left-hand side.
 Thus, $-\frac{1}{2}$ *is a solution* to the equation $4x + 7 = 5$.

b. $-6w + 14 = 4$

     $-6(3) + 14 \stackrel{?}{=} 4$ Substitute 3 for w.

     $-18 + 14 \stackrel{?}{=} 4$ Simplify.

           $-4 \neq 4$ Right-hand side does not equal left-hand side.
 Thus, 3 is *not a solution* to the equation $-6w + 14 = 4$.

Answers

1. No   2. Yes

In the study of algebra, you will encounter a variety of equations. In this chapter, we will focus on a specific type of equation called a linear equation in one variable.

DEFINITION   Linear Equation in One Variable

Let a and b be real numbers such that $a \neq 0$. A **linear equation in one variable** is an equation that can be written in the form

$$ax + b = 0$$

Notice that a linear equation in one variable has only one variable. Furthermore, because the variable has an implied exponent of 1, a linear equation is sometimes called a *first-degree equation*.

<u>Linear Equation in One Variable</u>	<u>*Not* a Linear Equation in One Variable</u>
$2x + 3 = 0$	$4x^2 + 8 = 0$ (exponent for x is not 1)
$\frac{1}{5}a + \frac{2}{7} = 0$	$\frac{3}{4}a + \frac{5}{8}b = 0$ (more than one variable)

2. Addition and Subtraction Properties of Equality

If two equations have the same solution then the equations are equivalent. For example, the following equations are equivalent because the solution for each equation is 6.

<u>Equivalent Equations</u>	<u>Check the Solution 6</u>
$2x - 5 = 7$	$2(6) - 5 \overset{?}{=} 7 \Rightarrow 12 - 5 \overset{?}{=} 7 ✔$
$2x = 12$	$2(6) \overset{?}{=} 12 \Rightarrow \quad 12 \overset{?}{=} 12 ✔$
$x = 6$	$6 \overset{?}{=} 6 \Rightarrow \quad 6 \overset{?}{=} 6 ✔$

To solve a linear equation, $ax + b = 0$, the goal is to find *all* values of x that make the equation true. One general strategy for solving an equation is to rewrite it as an equivalent but simpler equation. This process is repeated until the equation can be written in the form $x =$ number. The addition and subtraction properties of equality help us do this.

PROPERTY   Addition and Subtraction Properties of Equality

Let a, b, and c represent algebraic expressions.

1. **Addition property of equality:**   If $a = b$,
      then $a + c = b + c$

2. ***Subtraction property of equality:**   If $a = b$,
      then $a - c = b - c$

*The subtraction property of equality follows directly from the addition property, because subtraction is defined in terms of addition.

If $a + (-c) = b + (-c)$

then, $a - c = b - c$

Answers

3. Yes **4.** No **5.** Yes **6.** No

The addition and subtraction properties of equality indicate that adding or subtracting the same quantity on each side of an equation results in an equivalent equation. This is true because if two quantities are increased or decreased by the same amount, then the resulting quantities will also be equal (Figure 2-1).

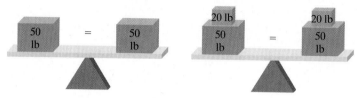

Figure 2-1

Skill Practice

Solve the equations.

7. $v - 7 = 2$

8. $x + 4 = 4$

Classroom Examples: p. 125, Exercises 16 and 18

Example 2   **Applying the Addition and Subtraction Properties of Equality**

Solve the equations.

a. $p - 4 = 11$ **b.** $w + 5 = -2$

Solution:

In each equation, the goal is to isolate the variable on one side of the equation. To accomplish this, we use the fact that the sum of a number and its opposite is zero and the difference of a number and itself is zero.

a. $p - 4 = 11$

$p - 4 + 4 = 11 + 4$ To isolate p, add 4 to both sides ($-4 + 4 = 0$).

$p + 0 = 15$ Simplify.

$p = 15$ Check by substituting $p = 15$ into the original equation.

Check: $p - 4 = 11$

$15 - 4 \stackrel{?}{=} 11$

The solution is 15. $11 \stackrel{?}{=} 11$ ✔ True

b. $w + 5 = -2$

$w + 5 - 5 = -2 - 5$ To isolate w, subtract 5 from both sides. $(5 - 5 = 0)$.

$w + 0 = -7$ Simplify.

$w = -7$ Check by substituting $w = -7$ into the original equation.

Check: $w + 5 = -2$

$-7 + 5 \stackrel{?}{=} -2$

The solution is -7. $-2 \stackrel{?}{=} -2$ ✔ True

Answers

7. 9   **8.** 0

In some situations, rather than writing the solution to an equation, we may write the solution *set* to an equation. The **solution set** is the set of all solutions to an equation. For instance, in Example 2(b), the solution is -7, but the *solution set* would be written as $\{-7\}$. Check to determine how your instructor prefers that you write the answer.

Example 3 Applying the Addition and Subtraction Properties of Equality

Solve the equations.

a. $\dfrac{9}{4} = q - \dfrac{3}{4}$ **b.** $-1.2 + z = 4.6$

Skill Practice

Solve the equations.

9. $\dfrac{1}{4} = a - \dfrac{2}{3}$

10. $-8.1 + w = 11.5$

Classroom Example: p. 125, Exercise 30

Solution:

a.
$$\dfrac{9}{4} = q - \dfrac{3}{4}$$

$$\dfrac{9}{4} + \dfrac{3}{4} = q - \dfrac{3}{4} + \dfrac{3}{4} \qquad \text{To isolate } q, \text{ add } \tfrac{3}{4} \text{ to both sides } \left(-\tfrac{3}{4} + \tfrac{3}{4} = 0\right).$$

$$\dfrac{12}{4} = q + 0 \qquad \text{Simplify.}$$

TIP: The variable may be isolated on either side of the equation.

$$3 = q \quad \text{or equivalently,} \quad q = 3$$

The solution is 3.

Check: $\dfrac{9}{4} = q - \dfrac{3}{4}$

$\dfrac{9}{4} \overset{?}{=} 3 - \dfrac{3}{4}$ Substitute $q = 3$.

$\dfrac{9}{4} \overset{?}{=} \dfrac{12}{4} - \dfrac{3}{4}$ Common denominator

$\dfrac{9}{4} \overset{?}{=} \dfrac{9}{4}$ ✔ True

b.
$$-1.2 + z = 4.6$$

$$-1.2 + 1.2 + z = 4.6 + 1.2 \qquad \text{To isolate } z, \text{ add } 1.2 \text{ to both sides.}$$

$$0 + z = 5.8$$

$$z = 5.8$$

Check: $-1.2 + z = 4.6$

$-1.2 + 5.8 \overset{?}{=} 4.6$ Substitute $z = 5.8$.

$4.6 \overset{?}{=} 4.6$ ✔ True

The solution is 5.8.

3. Multiplication and Division Properties of Equality

Adding or subtracting the same quantity to both sides of an equation results in an equivalent equation. In a similar way, multiplying or dividing both sides of an equation by the same nonzero quantity also results in an equivalent equation. This is stated formally as the multiplication and division properties of equality.

Answers

9. $\dfrac{11}{12}$ **10.** 19.6

Concept Connections

11. The division property of equality states that if

$$a = b, \text{ then } \frac{a}{c} = \frac{b}{c}$$

provided that $c \neq 0$. Why is the condition $c \neq 0$ necessary?

PROPERTY   Multiplication and Division Properties of Equality

Let a, b, and c represent algebraic expressions.

1. **Multiplication property of equality:**   If $a = b$,
   then $ac = bc$

2. ***Division property of equality:***   If $a = b$
   then $\dfrac{a}{c} = \dfrac{b}{c}$ (provided $c \neq 0$)

*The division property of equality follows directly from the multiplication property because division is defined as multiplication by the reciprocal.

If $a \cdot \dfrac{1}{c} = b \cdot \dfrac{1}{c}$ $(c \neq 0)$

then, $\dfrac{a}{c} = \dfrac{b}{c}$

To understand the multiplication property of equality, suppose we start with a true equation such as $10 = 10$. If both sides of the equation are multiplied by a constant such as 3, the result is also a true statement (Figure 2-2).

$$10 = 10$$
$$3 \cdot 10 = 3 \cdot 10$$
$$30 = 30$$

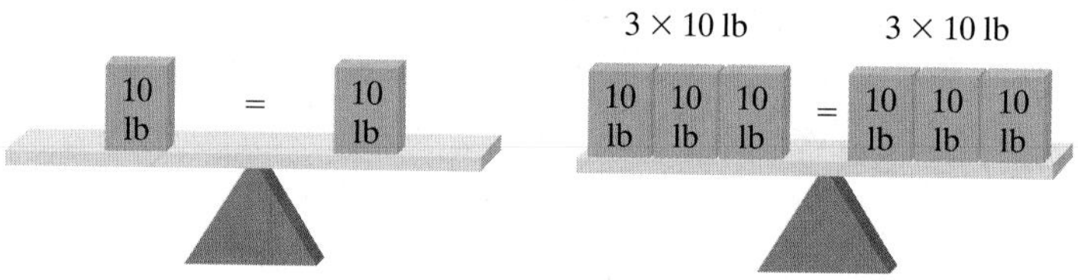

Figure 2-2

Similarly, if both sides of the equation are divided by a nonzero real number such as 2, the result is also a true statement (Figure 2-3).

$$10 = 10$$
$$\frac{10}{2} = \frac{10}{2}$$
$$5 = 5$$

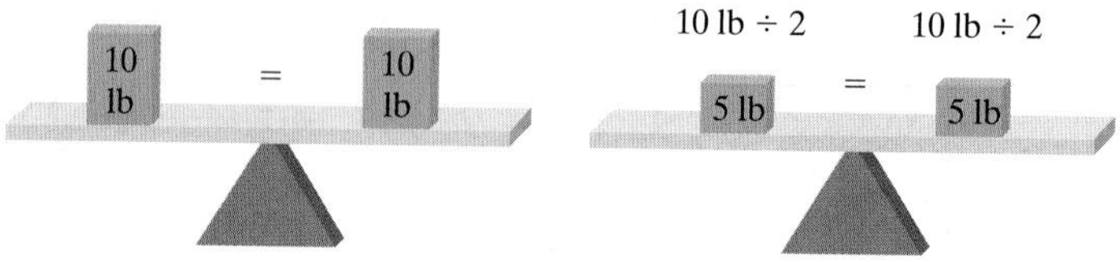

Figure 2-3

TIP: The product of a number and its reciprocal is always 1. For example:

$$\frac{1}{5}(5) = 1$$

$$-\frac{7}{2}\left(-\frac{2}{7}\right) = 1$$

To solve an equation in the variable x, the goal is to write the equation in the form $x = $ number. In particular, notice that we desire the coefficient of x to be 1. That is, we want to write the equation as $1x = $ number. Therefore, to solve an equation such as $5x = 15$, we can multiply both sides of the equation by the reciprocal of the x-term coefficient. In this case, multiply both sides by the reciprocal of 5, which is $\frac{1}{5}$.

$$5x = 15$$

$$\frac{1}{5}(5x) = \frac{1}{5}(15) \qquad \text{Multiply by } \frac{1}{5}.$$

$$1x = 3 \qquad \text{The coefficient of the } x\text{-term is now 1.}$$

$$x = 3$$

Answer

11. $c \neq 0$ because division by zero is undefined.

The division property of equality can also be used to solve the equation $5x = 15$ by dividing both sides by the coefficient of the x-term. In this case, divide both sides by 5 to make the coefficient of x equal to 1.

$$5x = 15$$

$$\frac{5x}{5} = \frac{15}{5} \qquad \text{Divide by 5.}$$

$$1x = 3 \qquad \text{The coefficient on the } x\text{-term is now 1.}$$

$$x = 3$$

> **TIP:** The quotient of a nonzero real number and itself is always 1. For example:
>
> $$\frac{5}{5} = 1$$
>
> $$\frac{-3.5}{-3.5} = 1$$

Example 4 Applying the Multiplication and Division Properties of Equality

Solve the equations using the multiplication or division property of equality.

a. $12x = 60$ **b.** $48 = -8w$ **c.** $-x = 8$

Skill Practice

Solve the equations.
12. $4x = -20$
13. $100 = -4p$
14. $-y = -11$

Classroom Examples: p. 125, Exercises 38 and 50

Solution:

a. $12x = 60$

$$\frac{12x}{12} = \frac{60}{12} \qquad$$ To obtain a coefficient of 1 for the x-term, divide both sides by 12.

$$1x = 5 \qquad \text{Simplify.}$$

$$x = 5 \qquad \underline{\text{Check:}} \quad 12x = 60$$

$$12(5) \stackrel{?}{=} 60$$

The solution is 5. $\qquad\qquad 60 \stackrel{?}{=} 60 ✔ \quad \text{True}$

b. $48 = -8w$

$$\frac{48}{-8} = \frac{-8w}{-8} \qquad$$ To obtain a coefficient of 1 for the w-term, divide both sides by -8.

$$-6 = 1w \qquad \text{Simplify.}$$

$$-6 = w \qquad \underline{\text{Check:}} \ 48w = -8w$$

$$48 \stackrel{?}{=} -8(-6)$$

The solution is -6. $\qquad\qquad 48 \stackrel{?}{=} 48 ✔ \quad \text{True}$

c. $-x = 8 \qquad$ Note that $-x$ is equivalent to $-1 \cdot x$.

$$-1x = 8$$

$$\frac{-1x}{-1} = \frac{8}{-1} \qquad$$ To obtain a coefficient of 1 for the x-term, divide by -1.

$$x = -8 \qquad \underline{\text{Check:}} \ -x = 8$$

$$-(-8) \stackrel{?}{=} 8$$

The solution is -8. $\qquad\qquad 8 \stackrel{?}{=} 8 ✔ \quad \text{True}$

> **TIP:** In Example 4(c), we could also have *multiplied* both sides by -1 to create a coefficient of 1 on the x-term.
>
> $$-x = 8$$
>
> $$(-1)(-x) = (-1)8$$
>
> $$x = -8$$

Answers
12. -5 **13.** -25 **14.** 11

Example 5 **Applying the Multiplication and Division Properties of Equality**

Solve the equations by using the multiplication or division property of equality.

a. $-\dfrac{2}{9}q = \dfrac{1}{3}$ **b.** $-3.43 = -0.7z$ **c.** $\dfrac{d}{6} = -4$

Solution:

a.
$$-\frac{2}{9}q = \frac{1}{3}$$

$$\left(-\frac{9}{2}\right)\left(-\frac{2}{9}q\right) = \frac{1}{3}\left(-\frac{9}{2}\right)$$ To obtain a coefficient of 1 for the q-term, multiply by the reciprocal of $-\frac{2}{9}$, which is $-\frac{9}{2}$.

$$1q = -\frac{3}{2}$$ Simplify. The product of a number and its reciprocal is 1.

$$q = -\frac{3}{2}$$ Check: $-\dfrac{2}{9}q = \dfrac{1}{3}$

$$-\frac{2}{9}\left(-\frac{3}{2}\right) \stackrel{?}{=} \frac{1}{3}$$

The solution is $-\dfrac{3}{2}$. $\dfrac{1}{3} \stackrel{?}{=} \dfrac{1}{3}$ ✔ True

TIP: When applying the multiplication or division property of equality to obtain a coefficient of 1 for the variable term, we will generally use the following convention:

- If the coefficient of the variable term is expressed as a fraction, we will usually multiply both sides by its reciprocal, as in Example 5(a).
- If the coefficient of the variable term is an integer or decimal, we will divide both sides by the coefficient itself, as in Example 5(b).

b. $-3.43 = -0.7z$

$$\frac{-3.43}{-0.7} = \frac{-0.7z}{-0.7}$$ To obtain a coefficient of 1 for the z-term, divide by -0.7.

$$4.9 = 1z$$ Simplify.

$$4.9 = z$$

$$z = 4.9$$ Check: $-3.43 = -0.7z$

$$-3.43 \stackrel{?}{=} -0.7(4.9)$$

The solution is 4.9. $-3.43 \stackrel{?}{=} -3.43$ ✔ True

c. $\dfrac{d}{6} = -4$

$$\frac{1}{6}d = -4$$ $\dfrac{d}{6}$ is equivalent to $\dfrac{1}{6}d$.

$$\frac{6}{1} \cdot \frac{1}{6}d = -4 \cdot \frac{6}{1}$$ To obtain a coefficient of 1 for the d-term, multiply by the reciprocal of $\frac{1}{6}$, which is $\frac{6}{1}$.

$1d = -24$ Simplify.

$d = -24$ Check: $\dfrac{d}{6} = -4$

$\dfrac{-24}{6} \overset{?}{=} -4$

The solution is -24. $-4 \overset{?}{=} -4$ ✔ True

It is important to distinguish between cases where the addition or subtraction properties of equality should be used to isolate a variable versus those in which the multiplication or division property of equality should be used. Remember the goal is to isolate the variable term and obtain a coefficient of 1. Compare the equations:

$$5 + x = 20 \quad \text{and} \quad 5x = 20$$

In the first equation, the relationship between 5 and x is addition. Therefore, we want to reverse the process by subtracting 5 from both sides. In the second equation, the relationship between 5 and x is multiplication. To isolate x, we reverse the process by dividing by 5 or equivalently, multiplying by the reciprocal, $\frac{1}{5}$.

$$5 + x = 20 \quad \text{and} \quad 5x = 20$$

$$5 - 5 + x = 20 - 5 \qquad \dfrac{5x}{5} = \dfrac{20}{5}$$

$$x = 15 \qquad\qquad x = 4$$

4. Translations

| Example 6 | Translating to a Linear Equation |

Write an algebraic equation to represent each English sentence. Then solve the equation.

a. The quotient of a number and 4 is 6.

b. The product of a number and 4 is 6.

c. Negative twelve is equal to the sum of -5 and a number.

d. The value 1.4 subtracted from a number is 5.7.

Solution:

For each case we will let x represent the unknown number.

a. The quotient of a number and 4 is 6.

$$\dfrac{x}{4} = 6$$

$$4 \cdot \dfrac{x}{4} = 4 \cdot 6 \qquad \text{Multiply both sides by 4.}$$

$$\dfrac{4}{1} \cdot \dfrac{x}{4} = 4 \cdot 6$$

$$x = 24 \qquad \text{Check: } \dfrac{24}{4} \overset{?}{=} 6 \text{ ✔ True}$$

Answers

18. $\dfrac{x}{-2} = 8$; The number is -16.

19. $-3x = -24$; The number is 8.

20. $y + 6 = -20$; The number is -26.

21. $13 = x - 5$; The number is 18.

b. The product of a number and 4 is 6.

$$4x = 6$$

$$\frac{4x}{4} = \frac{6}{4} \qquad \text{Divide both sides by 4.}$$

$$x = \frac{3}{2} \qquad \underline{\text{Check:}} \quad 4\left(\frac{3}{2}\right) \stackrel{?}{=} 6 \; ✔ \quad \text{True}$$

c. Negative twelve is equal to the sum of -5 and a number.

$$-12 = -5 + x$$

$$-12 + 5 = -5 + 5 + x \qquad \text{Add 5 to both sides.}$$

$$-7 = x \qquad \underline{\text{Check:}} \quad -12 \stackrel{?}{=} -5 + (-7) \; ✔ \quad \text{True}$$

d. The value 1.4 subtracted from a number is 5.7.

$$x - 1.4 = 5.7$$

$$x - 1.4 + 1.4 = 5.7 + 1.4 \qquad \text{Add 1.4 to both sides.}$$

$$x = 7.1 \qquad \underline{\text{Check:}} \quad 7.1 - 1.4 \stackrel{?}{=} 5.7 \; ✔ \quad \text{True}$$

Section 2.1   Practice Exercises

Study Skills Exercises

1. After getting a test back, it is a good idea to correct the test so that you do not make the same errors again. One recommended approach is to use a clean sheet of paper, and divide the paper down the middle vertically as shown. For each problem that you missed on the test, rework the problem correctly on the left-hand side of the paper. Then give a written explanation on the right-hand side of the paper. To reinforce the correct procedure, do four more problems of that type.

Take the time this week to make corrections from your last test.

Perform the correct math here.	Explain the process here.
$\downarrow$	$\downarrow$
$2 + 4(5)$ $= 2 + 20$ $= 22$	Do multiplication before addition.

2. Define the key terms:

   a. linear equation in one variable   **b.** solution to an equation

   c. addition property of equality   **d.** subtraction property of equality

   e. multiplication property of equality   **f.** division property of equality   **g.** solution set

Objective 1: Definition of a Linear Equation in One Variable

For Exercises 3–6, identify the following as either an expression or an equation.

3. $x - 4 + 5x$ **4.** $8x + 2 = 7$ **5.** $9 = 2x - 4$ **6.** $3x^2 + x = -3$
    Expression   Equation   Equation   Equation

7. Explain how to determine if a number is a solution to an equation.
    Substitute the value into the equation and determine if the right-hand side is equal to the left-hand side.

8. Explain why the equations $6x = 12$ and $x = 2$ are *equivalent equations*.
    They are equivalent equations because their solutions are the same. The solution is 2 in both cases.

 Writing   ←→ Translating Expression   Geometry   Scientific Calculator   Video   NS E

For Exercises 9–14, determine whether the given number is a solution to the equation. **(See Example 1.)**

9. $x - 1 = 5$; 4   No

10. $x - 2 = 1$; -1 No

11. $5x = -10$; -2 Yes

12. $3x = 21$;   7   Yes

13. $3x + 9 = 3$; -2 Yes

14. $2x - 1 = -3$; -1 Yes

Objective 2: Addition and Subtraction Properties of Equality

For Exercises 15–34, solve the equations using the addition or subtraction property of equality. Be sure to check your answers. **(See Examples 2–3.)**

15. $x + 6 = 5$ -1

16. $x - 2 = 10$ 12

17. $q - 14 = 6$ 20

18. $w + 3 = -5$ -8

19. $2 + m = -15$ -17

20. $-6 + n = 10$ 16

21. $-23 = y - 7$ -16

22. $-9 = -21 + b$ 12

23. $4 + c = 4$ 0

24. $-13 + b = -13$ 0

25. $4.1 = 2.8 + a$ 1.3

26. $5.1 = -2.5 + y$ 7.6

27. $5 = z - \dfrac{1}{2}$ $\dfrac{11}{2}$ or $5\dfrac{1}{2}$

28. $-7 = p + \dfrac{2}{3}$ $-\dfrac{23}{3}$ or $-7\dfrac{2}{3}$

29. $x + \dfrac{5}{2} = \dfrac{1}{2}$ -2

30. $x - \dfrac{2}{3} = \dfrac{7}{3}$ 3

31. $-6.02 + c = -8.15$ -2.13

32. $p + 0.035 = -1.12$ -1.155

33. $3.245 + t = -0.0225$ -3.2675

34. $-1.004 + k = 3.0589$ 4.0629

Objective 3: Multiplication and Division Properties of Equality

For Exercises 35–54, solve the equations using the multiplication or division property of equality. Be sure to check your answers. **(See Examples 4–5.)**

35. $6x = 54$ 9

36. $2w = 8$ 4

37. $12 = -3p$ -4

38. $6 = -2q$ -3

39. $-5y = 0$ 0

40. $-3k = 0$ 0

41. $-\dfrac{y}{5} = 3$ -15

42. $-\dfrac{z}{7} = 1$ -7

43. $\dfrac{4}{5} = -t$ $-\dfrac{4}{5}$

44. $-\dfrac{3}{7} = -h$ $\dfrac{3}{7}$

45. $\dfrac{2}{5}a = -4$ -10

46. $\dfrac{3}{8}b = -9$ -24

47. $-\dfrac{1}{5}b = -\dfrac{4}{5}$ 4

48. $-\dfrac{3}{10}w = \dfrac{2}{5}$ $-\dfrac{4}{3}$

49. $-41 = -x$ 41

50. $32 = -y$ -32

51. $3.81 = -0.03p$ -127

52. $2.75 = -0.5q$ -5.5

53. $5.82y = -15.132$ -2.6

54. $-32.3x = -0.4522$ 0.014

Objective 4: Translations

For Exercises 55–64, write an algebraic equation to represent each English sentence. (Let x represent the unknown number.) Then solve the equation. **(See Example 6.)**

55. The sum of negative eight and a number is forty-two. $-8 + x = 42$; The number is 50.

56. The sum of thirty-one and a number is thirteen. $31 + x = 13$; The number is -18.

57. The difference of a number and negative six is eighteen. $x - (-6) = 18$; The number is 12.

58. The sum of negative twelve and a number is negative fifteen. $-12 + x = -15$; The number is -3.

59. The product of a number and seven is the same as negative sixty-three. $x \cdot 7 = -63$ or $7x = -63$; The number is -9.

60. The product of negative three and a number is the same as twenty-four. $-3x = 24$; The number is -8.

✏️ Writing   ↔ Translating Expression   📐 Geometry   🖩 Scientific Calculator   💿 Video   NS E

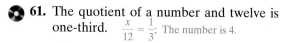**61.** The quotient of a number and twelve is one-third. $\frac{x}{12} = \frac{1}{3}$; The number is 4.

62. Eighteen is equal to the quotient of a number and two. $18 = \frac{x}{2}$; The number is 36.

63. The sum of a number and $\frac{5}{8}$ is $\frac{13}{8}$. $x + \frac{5}{8} = \frac{13}{8}$; The number is 1.

64. The difference of a number and $\frac{2}{3}$ is $\frac{1}{3}$. $x - \frac{2}{3} = \frac{1}{3}$; The number is 1.

Mixed Exercises

For Exercises 65–92, solve the equation using the appropriate property of equality.

65. $a - 9 = 1$ 10

66. $b - 2 = -4$ −2

67. $-9x = 1$ $-\frac{1}{9}$

68. $-2k = -4$ 2

69. $-\frac{2}{3}h = 8$ −12

70. $\frac{3}{4}p = 15$ 20

71. $\frac{2}{3} + t = 8$ $\frac{22}{3}$

72. $\frac{3}{4} + y = 15$ $\frac{57}{4}$

73. $\frac{r}{3} = -12$ −36

74. $\frac{d}{-4} = 5$ −20

75. $k + 16 = 32$ 16

76. $-18 = -9 + t$ −9

77. $16k = 32$ 2

78. $-18 = -9t$ 2

79. $7 = -4q$ $-\frac{7}{4}$

80. $-3s = 10$ $-\frac{10}{3}$

81. $-4 + q = 7$ 11

82. $s - 3 = 10$ 13

83. $-\frac{1}{3}d = 12$ −36

84. $-\frac{2}{5}m = 10$ −25

85. $4 = \frac{1}{2} + z$ $\frac{7}{2}$

86. $3 = \frac{1}{4} + p$ $\frac{11}{4}$

87. $1.2y = 4.8$ 4

88. $4.3w = 8.6$ 2

89. $4.8 = 1.2 + y$ 3.6

90. $8.6 = w - 4.3$ 12.9

91. $0.0034 = y - 0.405$ 0.4084

92. $-0.98 = m + 1.0034$ −1.9834

For Exercises 93–100, determine if the equation is a linear equation in one variable. Answer yes or no.

93. $4p + 5 = 0$ Yes

94. $3x - 5y = 0$ No

95. $4 + 2a^2 = 5$ No

96. $-8t = 7$ Yes

97. $x - 4 = 9$ Yes

98. $2x^3 + y = 0$ No

99. $19b = -3$ Yes

100. $13 + x = 19$ Yes

Expanding Your Skills

For Exercises 101–106, construct an equation with the given solution. Answers will vary.

101. $y = 6$ For example: $y + 9 = 15$

102. $x = 2$ For example: $1 + x = 3$

103. $p = -4$ For example: $2p = -8$

104. $t = -10$ For example: $-4t = 40$

105. $a = 0$ For example: $5a + 5 = 5$

106. $k = 1$ For example: $6k + 1 = 7$

For Exercises 107–110, simplify by collecting the *like* terms. Then solve the equation.

107. $5x - 4x + 7 = 8 - 2$ −1

108. $2 + 3 = 2y + 1 - y$ 4

109. $6p - 3p = 15 + 6$ 7

110. $12 - 20 = 2t + 2t$ −2

Solving Linear Equations

1. Solving Linear Equations Involving Multiple Steps

In Section 2.1, we studied a one-step process to solve linear equations by using the addition, subtraction, multiplication, and division properties of equality. In Example 1, we solve the equation $-2w - 7 = 11$. Solving this equation will require multiple steps. To understand the proper steps, always remember the ultimate goal—to isolate the variable. Therefore, we will first isolate the *term* containing the variable before dividing both sides by -2.

Objectives

1. **Solving Linear Equations Involving Multiple Steps**
2. **Procedure for Solving a Linear Equation in One Variable**
3. **Conditional Equations, Identities, and Contradictions**

Example 1 — Solving a Linear Equation

Solve the equation. $-2w - 7 = 11$

Solution:

$$-2w - 7 = 11$$

$$-2w - 7 + 7 = 11 + 7 \quad \text{Add 7 to both sides of the equation. This isolates the } w\text{-term.}$$

$$-2w = 18$$

$$\frac{-2w}{-2} = \frac{18}{-2} \quad \text{Next, apply the division property of equality to obtain a coefficient of 1 for } w. \text{ Divide by } -2 \text{ on both sides.}$$

$$w = -9$$

Check:

$$-2w - 7 = 11$$

$$-2(-9) - 7 \overset{?}{=} 11 \quad \text{Substitute } w = -9 \text{ in the original equation.}$$

$$18 - 7 \overset{?}{=} 11$$

$$11 \overset{?}{=} 11 \checkmark \quad \text{True. The solution is } -9.$$

Skill Practice

Solve the equation.

1. $-5y - 5 = 10$

Classroom Example: p. 133, Exercise 14

Example 2 — Solving a Linear Equation

Solve the equation. $2 = \frac{1}{5}x + 3$

Solution:

$$2 = \frac{1}{5}x + 3$$

$$2 - 3 = \frac{1}{5}x + 3 - 3 \quad \text{Subtract 3 from both sides. This isolates the } x\text{-term.}$$

$$-1 = \frac{1}{5}x \quad \text{Simplify.}$$

$$5(-1) = 5 \cdot \left(\frac{1}{5}x\right) \quad \text{Next, apply the multiplication property of equality to obtain a coefficient of 1 for } x.$$

$$-5 = 1x$$

$$-5 = x \quad \text{Simplify. The answer checks in the original equation.}$$

The solution is -5.

Skill Practice

Solve the equation.

2. $2 = \frac{1}{2}a - 7$

Classroom Example: p. 133, Exercise 22

Answers

1. -3 **2.** 18

In Example 3, the variable x appears on both sides of the equation. In this case, apply the addition or subtraction property of equality to collect the variable terms on one side of the equation and the constant terms on the other side. Then use the multiplication or division property of equality to get a coefficient equal to 1.

Skill Practice

Solve the equation.

3. $10x - 3 = 4x - 2$

Classroom Example: p. 133, Exercise 28

Example 3   **Solving a Linear Equation**

Solve the equation. $6x - 4 = 2x - 8$

Solution:

$$6x - 4 = 2x - 8$$

$$6x - 2x - 4 = 2x - 2x - 8 \qquad \text{Subtract } 2x \text{ from both sides leaving } 0x \text{ on the right-hand side.}$$

$$4x - 4 = 0x - 8 \qquad \text{Simplify.}$$

$$4x - 4 = -8 \qquad \text{The } x\text{-terms have now been combined on one side of the equation.}$$

$$4x - 4 + 4 = -8 + 4 \qquad \text{Add 4 to both sides of the equation. This combines the constant terms on the } \textit{other}$$

$$4x = -4 \qquad \text{side of the equation.}$$

$$\frac{4x}{4} = \frac{-4}{4} \qquad \text{To obtain a coefficient of 1 for } x, \text{ divide both sides of the equation by 4.}$$

$$x = -1 \qquad \underline{\text{Check}}:$$

$$6x - 4 = 2x - 8$$

$$6(-1) - 4 \stackrel{?}{=} 2(-1) - 8$$

$$-6 - 4 \stackrel{?}{=} -2 - 8$$

The solution is -1. $\qquad\qquad -10 \stackrel{?}{=} -10 \ ✔ \ \text{True}$

TIP: It is important to note that the variable may be isolated on either side of the equation. We will solve the equation from Example 3 again, this time isolating the variable on the right-hand side.

$$6x - 4 = 2x - 8$$

$$6x - 6x - 4 = 2x - 6x - 8 \qquad \text{Subtract } 6x \text{ on both sides.}$$

$$0x - 4 = -4x - 8$$

$$-4 = -4x - 8$$

$$-4 + 8 = -4x - 8 + 8 \qquad \text{Add 8 to both sides.}$$

$$4 = -4x$$

$$\frac{4}{-4} = \frac{-4x}{-4} \qquad \text{Divide both sides by } -4.$$

$$-1 = x \quad \text{or equivalently } x = -1$$

Answer

3. $\dfrac{1}{6}$

2. Procedure for Solving a Linear Equation in One Variable

In some cases, it is necessary to simplify both sides of a linear equation before applying the properties of equality. Therefore, we offer the following steps to solve a linear equation in one variable.

PROCEDURE Solving a Linear Equation in One Variable

Step 1 Simplify both sides of the equation.
- Clear parentheses
- Combine *like* terms

Step 2 Use the addition or subtraction property of equality to collect the variable terms on one side of the equation.

Step 3 Use the addition or subtraction property of equality to collect the constant terms on the other side of the equation.

Step 4 Use the multiplication or division property of equality to make the coefficient of the variable term equal to 1.

Step 5 Check your answer.

Example 4 Solving a Linear Equation

Solve the equation. $7 + 3 = 2(p - 3)$

Solution:

$$7 + 3 = 2(p - 3)$$

$$10 = 2p - 6$$

Step 1: Simplify both sides of the equation by clearing parentheses and combining *like* terms.

Step 2: The variable terms are already on one side.

$$10 + 6 = 2p - 6 + 6$$

Step 3: Add 6 to both sides to collect the constant terms on the other side.

$$16 = 2p$$

$$\frac{16}{2} = \frac{2p}{2}$$

Step 4: Divide both sides by 2 to obtain a coefficient of 1 for p.

$$8 = p$$

Step 5: Check:

$$7 + 3 = 2(p - 3)$$

$$10 \overset{?}{=} 2(8 - 3)$$

$$10 \overset{?}{=} 2(5)$$

The solution is 8.

$$10 \overset{?}{=} 10 \ \checkmark \ \text{True}$$

Skill Practice

Solve the equation.

4. $12 + 2 = 7(3 - y)$

Classroom Example: p. 134, Exercise 44

Answer

4. 1

Skill Practice

Solve the equation.

5. $1.5t + 2.3 = 3.5t - 1.9$

Classroom Example: p. 134, Exercise 46

Example 5 Solving a Linear Equation

Solve the equation. $2.2y - 8.3 = 6.2y + 12.1$

Solution:

$$2.2y - 8.3 = 6.2y + 12.1$$

Step 1: The right- and left-hand sides are already simplified.

$$2.2y - 2.2y - 8.3 = 6.2y - 2.2y + 12.1$$
$$-8.3 = 4.0y + 12.1$$

Step 2: Subtract $2.2y$ from both sides to collect the variable terms on one side of the equation.

$$-8.3 - 12.1 = 4.0y + 12.1 - 12.1$$
$$-20.4 = 4.0y$$

Step 3: Subtract 12.1 from both sides to collect the constant terms on the other side.

$$\frac{-20.4}{4.0} = \frac{4.0y}{4.0}$$
$$-5.1 = y$$

Step 4: To obtain a coefficient of 1 for the y-term, divide both sides of the equation by 4.0.

$$y = -5.1$$

Step 5: Check:

$$2.2y - 8.3 = 6.2y + 12.1$$
$$2.2(-5.1) - 8.3 \stackrel{?}{=} 6.2(-5.1) + 12.1$$
$$-11.22 - 8.3 \stackrel{?}{=} -31.62 + 12.1$$
$$-19.52 \stackrel{?}{=} -19.52 \; ✔ \;\text{True}$$

The solution is -5.1.

Skill Practice

Solve the equation.

6. $4(2y - 1) + y = 6y + 3 - y$

Classroom Example: p. 134, Exercise 50

Example 6 Solving a Linear Equation

Solve the equation. $2 + 7x - 5 = 6(x + 3) + 2x$

Solution:

$$2 + 7x - 5 = 6(x + 3) + 2x$$
$$-3 + 7x = 6x + 18 + 2x$$

Step 1: Add *like* terms on the left. Clear parentheses on the right.

$$-3 + 7x = 8x + 18$$

Combine *like* terms.

$$-3 + 7x - 7x = 8x - 7x + 18$$

Step 2: Subtract $7x$ from both sides.

$$-3 = x + 18$$

Simplify.

$$-3 - 18 = x + 18 - 18$$

Step 3: Subtract 18 from both sides.

$$-21 = x$$

$$x = -21$$

Step 4: Because the coefficient of the x term is already 1, there is no need to apply the multiplication or division property of equality.

The solution is -21.

Step 5: The check is left to the reader.

Answers

5. 2.1 **6.** $\dfrac{7}{4}$

Example 7 Solving a Linear Equation

Solve the equation. $9 - (z - 3) + 4z = 4z - 5(z + 2) - 6$

Solution:

$$9 - (z - 3) + 4z = 4z - 5(z + 2) - 6$$

$9 - z + 3 + 4z = 4z - 5z - 10 - 6$	**Step 1:** Clear parentheses.
$12 + 3z = -z - 16$	Combine *like* terms.
$12 + 3z + z = -z + z - 16$	**Step 2:** Add z to both sides.
$12 + 4z = -16$	
$12 - 12 + 4z = -16 - 12$	**Step 3:** Subtract 12 from both sides.
$4z = -28$	
$\dfrac{4z}{4} = \dfrac{-28}{4}$	**Step 4:** Divide both sides by 4.
$z = -7$	**Step 5:** The check is left for the reader.

The solution is -7.

Skill Practice

Solve the equation.
7. $10 - (x + 5) + 3x$
$= 6x - 5(x - 1) - 3$

Classroom Example: p. 134, Exercise 54

3. Conditional Equations, Identities, and Contradictions

The solutions to a linear equation are the values of x that make the equation a true statement. A linear equation has one unique solution. Some types of equations, however, have no solution while others have infinitely many solutions.

I. Conditional Equations

An equation that is true for some values of the variable but false for other values is called a **conditional equation**. The equation $x + 4 = 6$, for example, is true on the condition that $x = 2$. For other values of x, the statement $x + 4 = 6$ is false.

II. Contradictions

Some equations have no solution, such as $x + 1 = x + 2$. There is no value of x, that when increased by 1 will equal the same value increased by 2. If we tried to solve the equation by subtracting x from both sides, we get the contradiction $1 = 2$. This indicates that the equation has no solution. An equation that has no solution is called a **contradiction**.

$$x + 1 = x + 2$$
$$x - x + 1 = x - x + 2$$
$$1 = 2 \quad \text{(contradiction)} \qquad \text{No solution.}$$

Concept Connections

8. Write a conditional equation. Answers may vary.
9. Write an equation that is a contradiction. Answers may vary.
10. Write an equation that is an identity. Answers may vary.

III. Identities

An equation that has all real numbers as its solution set is called an **identity**. For example, consider the equation, $x + 4 = x + 4$. Because the left- and right-hand

Answers

7. -3 **8.** For example: $x = 5$
9. For example: $x + 3 = x + 8$
10. For example: $x - 9 = x - 9$

sides are equal, any real number substituted for x will result in equal quantities on both sides. If we subtract x from both sides of the equation, we get the identity $4 = 4$. In such a case, the solution is the set of all real numbers.

$$x + 4 = x + 4$$

$$x - x + 4 = x - x + 4$$

$$4 = 4 \quad \text{(identity)} \qquad \text{The solution is all real numbers.}$$

Skill Practice

Solve the equation. Identify the equation as a conditional equation, a contradiction, or an identity.

11. $4(2t + 1) - 1 = 8t + 3$

12. $3x - 5 = 4x + 1 - x$

13. $6(v - 2) = 2v - 4$

Classroom Examples: p. 134, Exercises 60, 62, and 64

Example 8 **Identifying Conditional Equations, Contradictions, and Identities**

Solve the equation. Identify each equation as a conditional equation, a contradiction, or an identity.

a. $4k - 5 = 2(2k - 3) + 1$ **b.** $2(b - 4) = 2b - 7$ **c.** $3x + 7 = 2x - 5$

Solution:

a.
$$4k - 5 = 2(2k - 3) + 1$$
$$4k - 5 = 4k - 6 + 1 \qquad \text{Clear parentheses.}$$
$$4k - 5 = 4k - 5 \qquad \text{Combine } like \text{ terms.}$$
$$4k - 4k - 5 = 4k - 4k - 5 \qquad \text{Subtract } 4k \text{ from both sides.}$$
$$-5 = -5 \quad \text{(Identity)}$$

This is an identity. The solution is all real numbers.

b.
$$2(b - 4) = 2b - 7$$
$$2b - 8 = 2b - 7 \qquad \text{Clear parentheses.}$$
$$2b - 2b - 8 = 2b - 2b - 7 \qquad \text{Subtract } 2b \text{ from both sides.}$$
$$-8 = -7 \quad \text{(Contradiction)}$$

This is a contradiction. There is no solution.

c.
$$3x + 7 = 2x - 5$$
$$3x - 2x + 7 = 2x - 2x - 5 \qquad \text{Subtract } 2x \text{ from both sides.}$$
$$x + 7 = -5 \qquad \text{Simplify.}$$
$$x + 7 - 7 = -5 - 7 \qquad \text{Subtract } 7 \text{ from both sides.}$$
$$x = -12 \quad \text{(Conditional equation)}$$

This is a conditional equation. The solution is -12. (The equation is true only on the condition that $x = -12$.)

Answers

11. All real numbers; the equation is an identity.

12. No solution; the equation is a contradiction.

13. 2; the equation is a conditional equation.

Section 2.2 Practice Exercises

Study Skills Exercises

1. Several strategies are given here about taking notes. Which would you do first to help make the most of note-taking? Put them in order of importance to you by labeling them with the numbers 1–6.

_____ Read your notes after class and complete any abbreviations or incomplete sentences.

_____ Highlight important terms and definitions.

_____ Review your notes from the previous class.

_____ Bring pencils (more than one) and paper to class.

_____ Sit in class where you can clearly read the board and hear your instructor.

_____ Keep your focus on the instructor looking for phrases such as, "The most important point is . . ." and "Here is where the problem usually occurs."

2. Define the key terms:

   a. conditional equation   **b. contradiction**   **c. identity**

Review Exercises

For Exercises 3–6, simplify the expressions by clearing parentheses and combining *like* terms.

3. $5z + 2 - 7z - 3z$ $-5z + 2$

4. $10 - 4w + 7w - 2 + w$ $4w + 8$

5. $-(-7p + 9) + (3p - 1)$ $10p - 10$

6. $8y - (2y + 3) - 19$ $6y - 22$

7. Explain the difference between simplifying an expression and solving an equation.
   To simplify an expression, clear parentheses and combine *like* terms. To solve an equation, use the addition, subtraction, multiplication, and division properties of equality to isolate the variable.

For Exercises 8–12, solve the equations using the addition, subtraction, multiplication, or division property of equality.

8. $5w = -30$ -6

9. $-7y = 21$ -3

10. $x + 8 = -15$ -23

11. $z - 23 = -28$ -5

12. $-\dfrac{9}{8} = -\dfrac{3}{4}k$ $\dfrac{3}{2}$

Objective 1: Solving Linear Equations Involving Multiple Steps

For Exercises 13–36, solve the equations using the steps outlined in the text. **(See Examples 1–3.)**

13. $6z + 1 = 13$ 2

14. $5x + 2 = -13$ -3

15. $3y - 4 = 14$ 6

16. $-7w - 5 = -19$ 2

17. $-2p + 8 = 3$ $\dfrac{5}{2}$

18. $2b - \dfrac{1}{4} = 5$ $\dfrac{21}{8}$

19. $0.2x + 3.1 = -5.3$ -42

20. $-1.8 + 2.4a = -6.6$ -2

21. $\dfrac{5}{8} = \dfrac{1}{4} - \dfrac{1}{2}p$ $-\dfrac{3}{4}$

22. $\dfrac{6}{7} = \dfrac{1}{7} + \dfrac{5}{3}r$ $\dfrac{3}{7}$

23. $7w - 6w + 1 = 10 - 4$ 5

24. $5v - 3 - 4v = 13$ 16

25. $11h - 8 - 9h = -16$ -4

26. $6u - 5 - 8u = -7$ 1

27. $3a + 7 = 2a - 19$ -26

28. $6b - 20 = 14 + 5b$ 34

29. $-4r - 28 = -58 - r$ 10

30. $-6x - 7 = -3 - 8x$ 2

31. $-2z - 8 = -z$ -8

32. $-7t + 4 = -6t$ 4

33. $\dfrac{5}{6}x + \dfrac{2}{3} = -\dfrac{1}{6}x - \dfrac{5}{3}$ $-\dfrac{7}{3}$

34. $\dfrac{3}{7}x - \dfrac{1}{4} = -\dfrac{4}{7}x - \dfrac{5}{4}$ -1

35. $3y - 2 = 5y - 2$ 0

36. $4 + 10t = -8t + 4$ 0

Objective 2: Procedure for Solving a Linear Equation in One Variable

For Exercises 37–58, solve the equations using the steps outlined in the text. **(See Examples 4–7.)**

37. $4q + 14 = 2$ -3

38. $6 = 7m - 1$ 1

39. $-9 = 4n - 1$ -2

40. $-\dfrac{1}{2} - 4x = 8$ $-\dfrac{17}{8}$

41. $3(2p - 4) = 15$ $\dfrac{9}{2}$

42. $4(t + 15) = 20$ -10

43. $6(3x + 2) - 10 = -4$ $-\dfrac{1}{3}$

44. $4(2k + 1) - 1 = 5$ $\dfrac{1}{4}$

45. $3.4x - 2.5 = 2.8x + 3.5$ 10

46. $5.8w + 1.1 = 6.3w + 5.6$ -9

47. $17(s + 3) = 4(s - 10) + 13$ -6

48. $5(4 + p) = 3(3p - 1) - 9$ 8

49. $6(3t - 4) + 10 = 5(t - 2) - (3t + 4)$ 0

50. $-5y + 2(2y + 1) = 2(5y - 1) - 7$ 1

51. $5 - 3(x + 2) = 5$ -2

52. $1 - 6(2 - h) = 7$ 3

53. $3(2z - 6) - 4(3z + 1) = 5 - 2(z + 1)$ $-\dfrac{25}{4}$

54. $-2(4a + 3) - 5(2 - a) = 3(2a + 3) - 7$ -2

55. $-2[(4p + 1) - (3p - 1)] = 5(3 - p) - 9$ $\dfrac{10}{3}$

56. $5 - (6k + 1) = 2[(5k - 3) - (k - 2)]$ $\dfrac{3}{7}$

57. $3(-0.9n + 0.5) = -3.5n + 1.3$ -0.25

58. $7(0.4m - 0.1) = 5.2m + 0.86$ -0.65

Objective 3: Conditional Equations, Identities, and Contradictions

For Exercises 59–64, identify the equation as a conditional equation, a contradiction, or an identity. Then describe the solution. **(See Example 8.)**

59. $2(k - 7) = 2k - 13$
Contradiction; no solution

60. $5h + 4 = 5(h + 1) - 1$
Identity; all real numbers

61. $7x + 3 = 6(x - 2)$
Conditional equation; -15

62. $3y - 1 = 1 + 3y$
Contradiction; no solution

63. $3 - 5.2p = -5.2p + 3$
Identity; all real numbers

64. $2(q + 3) = 4q + q - 9$
Conditional equation; 5

65. A conditional linear equation has (choose one): One solution, no solution, or infinitely many solutions.   One solution

66. An equation that is a contradiction has (choose one): One solution, no solution, or infinitely many solutions.   No solution

67. An equation that is an identity has (choose one): One solution, no solution, or infinitely many solutions.   Infinitely many solutions

68. If the only solution to a linear equation is 5, then is the equation a conditional equation, an identity, or a contradiction?   Conditional equation

Mixed Exercises

For Exercises 69–92, find the solution, if possible.

69. $4p - 6 = 8 + 2p$ 7

70. $\dfrac{1}{2}t - 2 = 3$ 10

71. $2k - 9 = -8$ $\dfrac{1}{2}$

72. $3(y - 2) + 5 = 5$ 2

73. $7(w - 2) = -14 - 3w$ 0

74. $0.24 = 0.4m$ 0.6

75. $2(x + 2) - 3 = 2x + 1$
All real numbers

76. $n + \dfrac{1}{4} = -\dfrac{1}{2}$
$-\dfrac{3}{4}$

77. $0.5b = -23$
-46

78. $3(2r + 1) = 6(r + 2) - 6$
No solution

79. $8 - 2q = 4$
2

80. $\dfrac{x}{7} - 3 = 1$
28

81. $2 - 4(y - 5) = -4$
$\dfrac{13}{2}$

82. $4 - 3(4p - 1) = -8$
$\dfrac{5}{4}$

83. $0.4(a + 20) = 6$
-5

84. $2.2r - 12 = 3.4$
7

85. $10(2n + 1) - 6 = 20(n - 1) + 12$
No solution

86. $\dfrac{2}{5}y + 5 = -3$ -20

87. $c + 0.123 = 2.328$
2.205

88. $4(2z + 3) = 8(z - 3) + 36$
All real numbers

89. $\dfrac{4}{5}t - 1 = \dfrac{1}{5}t + 5$
10

90. $6g - 8 = 4 - 3g$
$\dfrac{4}{3}$

91. $8 - (3q + 4) = 6 - q$
-1

92. $6w - (8 + 2w) = 2(w - 4)$
0

Expanding Your Skills

93. Suppose -5 is a solution to the equation $x + a = 10$. Find the value of a.
$a = 15$

94. Suppose 6 is a solution to the equation $x + a = -12$. Find the value of a.
$a = -18$

95. Suppose 3 is a solution to the equation $ax = 12$. Find the value of a.
$a = 4$

96. Suppose 11 is a solution to the equation $ax = 49.5$. Find the value of a.
$a = 4.5$

97. Write an equation that is an identity.
Answers may vary.
For example: $5x + 2 = 2 + 5x$

98. Write an equation that is a contradiction.
Answers may vary.
For example: $4x - 3 = 4x + 1$

Linear Equations: Clearing Fractions and Decimals

Section 2.3

1. Solving Linear Equations with Fractions

Linear equations that contain fractions can be solved in different ways. The first procedure, illustrated here, uses the method outlined in Section 2.2.

Objectives

1. **Solving Linear Equations with Fractions**
2. **Solving Linear Equations with Decimals**

$$\frac{5}{6}x - \frac{3}{4} = \frac{1}{3}$$

$$\frac{5}{6}x - \frac{3}{4} + \frac{3}{4} = \frac{1}{3} + \frac{3}{4}$$ To isolate the variable term, add $\dfrac{3}{4}$ to both sides.

$$\frac{5}{6}x = \frac{4}{12} + \frac{9}{12}$$ Find the common denominator on the right-hand side.

$$\frac{5}{6}x = \frac{13}{12}$$ Simplify.

$$\frac{6}{5}\left(\frac{5}{6}x\right) = \frac{\overset{1}{\cancel{6}}}{5}\left(\frac{13}{\underset{2}{\cancel{12}}}\right)$$ Multiply by the reciprocal of $\dfrac{5}{6}$, which is $\dfrac{6}{5}$.

$$x = \frac{13}{10}$$ The solution is $\dfrac{13}{10}$.

Sometimes it is simpler to solve an equation with fractions by eliminating the fractions first using a process called **clearing fractions**. To clear fractions in the equation $\frac{5}{6}x - \frac{3}{4} = \frac{1}{3}$, we can multiply both sides of the equation by the least common denominator (LCD) of all terms in the equation. In this case, the LCD of $\frac{5}{6}x$, $-\frac{3}{4}$, and $\frac{1}{3}$ is 12. Because each denominator in the equation is a factor of 12, we can simplify common factors to leave integer coefficients for each term.

Skill Practice

Solve the equation by clearing fractions.

1. $\frac{2}{5}y + \frac{1}{2} = -\frac{7}{10}$

Classroom Example: p. 141, Exercise 18

TIP: Recall that the multiplication property of equality indicates that multiplying both sides of an equation by a nonzero constant results in an equivalent equation.

Example 1   **Solving a Linear Equation by Clearing Fractions**

Solve the equation by clearing fractions first. $\frac{5}{6}x - \frac{3}{4} = \frac{1}{3}$

Solution:

$$\frac{5}{6}x - \frac{3}{4} = \frac{1}{3}$$

$$12\left(\frac{5}{6}x - \frac{3}{4}\right) = 12\left(\frac{1}{3}\right) \quad \text{Multiply both sides of the equation by the LCD, 12.}$$

$$\frac{\overset{2}{\cancel{12}}}{1}\left(\frac{5}{\cancel{6}}x\right) - \frac{\overset{3}{\cancel{12}}}{1}\left(\frac{3}{\cancel{4}}\right) = \frac{\overset{4}{\cancel{12}}}{1}\left(\frac{1}{\cancel{3}}\right) \quad \text{Apply the distributive property (recall that } 12 = \frac{12}{1}\text{).}$$

$$2(5x) - 3(3) = 4(1) \quad \text{Simplify common factors to clear the fractions.}$$

$$10x - 9 = 4$$

$$10x - 9 + 9 = 4 + 9 \quad \text{Add 9 to both sides.}$$

$$10x = 13$$

$$\frac{10x}{10} = \frac{13}{10} \quad \text{Divide both sides by 10.}$$

$$x = \frac{13}{10} \quad \text{The solution is } \frac{13}{10}.$$

TIP: The fractions in this equation can be eliminated by multiplying both sides of the equation by *any* common multiple of the denominators. For example, try multiplying both sides of the equation by 24:

$$24\left(\frac{5}{6}x - \frac{3}{4}\right) = 24\left(\frac{1}{3}\right)$$

$$\frac{\overset{4}{24}}{1}\left(\frac{5}{6}x\right) - \frac{\overset{6}{24}}{1}\left(\frac{3}{4}\right) = \frac{\overset{8}{24}}{1}\left(\frac{1}{3}\right)$$

$$20x - 18 = 8$$

$$20x = 26$$

$$\frac{20x}{20} = \frac{26}{20}$$

$$x = \frac{13}{10}$$

Answer

1. -3

In this section, we combine the process for clearing fractions and decimals with the general strategies for solving linear equations. To solve a linear equation, it is important to follow the steps listed below.

PROCEDURE Solving a Linear Equation in One Variable

Step 1 Simplify both sides of the equation.
- Clear parentheses
- Consider clearing fractions and decimals (if any are present) by multiplying both sides of the equation by a common denominator of all terms.
- Combine *like* terms

Step 2 Use the addition or subtraction property of equality to collect the variable terms on one side of the equation.

Step 3 Use the addition or subtraction property of equality to collect the constant terms on the other side of the equation.

Step 4 Use the multiplication or division property of equality to make the coefficient of the variable term equal to 1.

Step 5 Check your answer.

Example 2 Solving a Linear Equation with Fractions

Solve the equation. $\dfrac{1}{6}x - \dfrac{2}{3} = \dfrac{1}{5}x - 1$

Solution:

$$\frac{1}{6}x - \frac{2}{3} = \frac{1}{5}x - 1$$

The LCD of $\frac{1}{6}x$, $-\frac{2}{3}$, and $\frac{1}{5}x$ is 30.

$$30\left(\frac{1}{6}x - \frac{2}{3}\right) = 30\left(\frac{1}{5}x - 1\right)$$

Multiply by the LCD, 30.

$$\frac{\overset{5}{\cancel{30}}}{1} \cdot \frac{1}{\cancel{6}}x - \frac{\overset{10}{\cancel{30}}}{1} \cdot \frac{2}{\cancel{3}} = \frac{\overset{6}{\cancel{30}}}{1} \cdot \frac{1}{\cancel{5}}x - 30(1)$$

Apply the distributive property (recall $30 = \frac{30}{1}$).

$$5x - 20 = 6x - 30$$

Clear fractions.

$$5x - 6x - 20 = 6x - 6x - 30$$

Subtract $6x$ from both sides.

$$-x - 20 = -30$$

$$-x - 20 + 20 = -30 + 20$$

Add 20 to both sides.

$$-x = -10$$

$$\frac{-x}{-1} = \frac{-10}{-1}$$

Divide both sides by -1.

$$x = 10$$

The solution is 10.

— Skill Practice —

Solve the equation.

2. $\dfrac{2}{5}x - \dfrac{1}{2} = \dfrac{7}{4} + \dfrac{3}{10}x$

Classroom Example: p. 141, Exercise 22

Answer

2. $\dfrac{45}{2}$

Skill Practice

Solve the equation.

3. $\frac{1}{5}(z + 1) + \frac{1}{4}(z + 3) = 2$

Classroom Example: p. 141, Exercise 26

Example 3 Solving a Linear Equation with Fractions

Solve the equation. $\frac{1}{3}(x + 7) - \frac{1}{2}(x + 1) = 4$

Solution:

$$\frac{1}{3}(x + 7) - \frac{1}{2}(x + 1) = 4$$

$$\frac{1}{3}x + \frac{7}{3} - \frac{1}{2}x - \frac{1}{2} = 4 \qquad \text{Clear parentheses.}$$

$$6\left(\frac{1}{3}x + \frac{7}{3} - \frac{1}{2}x - \frac{1}{2}\right) = 6(4) \qquad \text{The LCD of } \frac{1}{3}x, \frac{7}{3}, -\frac{1}{2}x, \text{ and } -\frac{1}{2} \text{ is 6.}$$

$$\frac{\overset{2}{\cancel{6}}}{1} \cdot \frac{1}{\cancel{3}}x + \frac{\overset{2}{\cancel{6}}}{1} \cdot \frac{7}{\cancel{3}} + \frac{\overset{3}{\cancel{6}}}{1}\left(-\frac{1}{\cancel{2}}x\right) + \frac{\overset{3}{\cancel{6}}}{1}\left(-\frac{1}{\cancel{2}}\right) = 6(4) \qquad \text{Apply the distributive property.}$$

$$2x + 14 - 3x - 3 = 24$$

$$-x + 11 = 24 \qquad \text{Combine } like \text{ terms.}$$

$$-x + 11 - 11 = 24 - 11 \qquad \text{Subtract 11.}$$

$$-x = 13$$

$$\frac{-x}{-1} = \frac{13}{-1} \qquad \text{Divide by } -1.$$

$$x = -13 \qquad \text{The check is left to the reader.}$$

> **TIP:** In Example 3 both parentheses and fractions are present within the equation. In such a case, we recommend that you clear parentheses first. Then clear the fractions.

Skill Practice

Solve the equation.

4. $\frac{x + 1}{4} + \frac{x + 2}{6} = 1$

Classroom Example: p. 141, Exercise 32

Example 4 Solving a Linear Equation with Fractions

Solve. $\frac{x - 2}{5} - \frac{x - 4}{2} = 2$

Solution:

$$\frac{x - 2}{5} - \frac{x - 4}{2} = \frac{2}{1} \qquad \text{The LCD of } \frac{x-2}{5}, \frac{x-4}{2}, \text{ and } \frac{2}{1} \text{ is 10.}$$

$$10\left(\frac{x - 2}{5} - \frac{x - 4}{2}\right) = 10\left(\frac{2}{1}\right) \qquad \text{Multiply both sides by 10.}$$

$$\frac{\overset{2}{\cancel{10}}}{1} \cdot \left(\frac{x - 2}{\cancel{5}}\right) - \frac{\overset{5}{\cancel{10}}}{1} \cdot \left(\frac{x - 4}{\cancel{2}}\right) = \frac{10}{1} \cdot \left(\frac{2}{1}\right) \qquad \text{Apply the distributive property.}$$

$$2(x - 2) - 5(x - 4) = 20 \qquad \text{Clear fractions.}$$

$$2x - 4 - 5x + 20 = 20 \qquad \text{Apply the distributive property.}$$

$$-3x + 16 = 20 \qquad \text{Simplify both sides of the equation.}$$

$$-3x + 16 - 16 = 20 - 16 \qquad \text{Subtract 16 from both sides.}$$

$$-3x = 4$$

$$\frac{-3x}{-3} = \frac{4}{-3} \qquad \text{Divide both sides by } -3.$$

$$x = -\frac{4}{3} \qquad \text{The check is left to the reader.}$$

> **Avoiding Mistakes**
>
> In Example 4, several of the fractions in the equation have two terms in the numerator. It is important to enclose these fractions in parentheses when clearing fractions. In this way, we will remember to use the distributive property to multiply the factors shown in blue with both terms from the numerator of the fractions.

Answers

3. $\frac{7}{3}$ **4.** 1

2. Solving Linear Equations with Decimals

The same procedure used to clear fractions in an equation can be used to **clear decimals**. For example, consider the equation

$$2.5x + 3 = 1.7x - 6.6$$

Recall that any terminating decimal can be written as a fraction. Therefore, the equation can be interpreted as

$$\frac{25}{10}x + 3 = \frac{17}{10}x - \frac{66}{10}$$

A convenient common denominator of all terms is 10. Therefore, we can multiply the original equation by 10 to clear decimals. The result is

$$25x + 30 = 17x - 66$$

Multiplying by the appropriate power of 10 moves the decimal points so that all coefficients become integers.

Example 5 **Solving a Linear Equation Containing Decimals**

Solve the equation by clearing decimals. $2.5x + 3 = 1.7x - 6.6$

Solution:

$2.5x + 3 = 1.7x - 6.6$	
$10(2.5x + 3) = 10(1.7x - 6.6)$	Multiply both sides of the equation by 10.
$25x + 30 = 17x - 66$	Apply the distributive property.
$25x - 17x + 30 = 17x - 17x - 66$	Subtract $17x$ from both sides.
$8x + 30 = -66$	
$8x + 30 - 30 = -66 - 30$	Subtract 30 from both sides.
$8x = -96$	
$\dfrac{8x}{8} = \dfrac{-96}{8}$	Divide both sides by 8.
$x = -12$	

Skill Practice

Solve.
5. $1.2w + 3.5 = 2.1 + w$

Classroom Example: p. 141, Exercise 40

TIP: Notice that multiplying a decimal number by 10 has the effect of moving the decimal point one place to the right. Similarly, multiplying by 100 moves the decimal point two places to the right, and so on.

Answer

5. -7

Skill Practice

Solve.

6. $0.25(x + 2) - 0.15(x + 3) = 4$

Classroom Example: p. 141,
Exercise 46

TIP: The terms with the most digits following the decimal point are $-0.45x$ and -4.05. Each of these is written to the hundredths place. Therefore, we multiply both sides by 100.

Answer
6. 39.5

Example 6 **Solving a Linear Equation Containing Decimals**

Solve the equation by clearing decimals. $0.2(x + 4) - 0.45(x + 9) = 12$

Solution:

$$0.2(x + 4) - 0.45(x + 9) = 12$$

$0.2x + 0.8 - 0.45x - 4.05 = 12$	Clear parentheses first.
$100(0.2x + 0.8 - 0.45x - 4.05) = 100(12)$	Multiply both sides by 100.
$20x + 80 - 45x - 405 = 1200$	Apply the distributive property.
$-25x - 325 = 1200$	Simplify both sides.
$-25x - 325 + 325 = 1200 + 325$	Add 325 to both sides.
$-25x = 1525$	
$\dfrac{-25x}{-25} = \dfrac{1525}{-25}$	Divide both sides by -25.
$x = -61$	The check is left to the reader.

Section 2.3 Practice Exercises

Boost *your* GRADE at ALEKS.com!

ALEKS version 3.0

- Practice Problems
- Self-Tests
- NetTutor
- e-Professors
- Videos

For additional exercises, see Classroom Activities 2.3A–2.3B in the *Instructor's Resource Manual* at www.mhhe.com/moh.

Study Skills Exercises

1. Instructors vary in what they emphasize on tests. For example, test material may come from the textbook, notes, handouts, or homework. What does your instructor emphasize?

2. Define the key terms:

   a. clearing fractions   **b. clearing decimals**

Review Exercises

For Exercises 3–6, solve the equation.

3. $5(x + 2) - 3 = 4x + 5$ –2

4. $-2(2x - 4x) = 6 + 18$ 6

5. $3(2y + 3) - 4(-y + 1) = 7y - 10$ –5

6. $-(3w + 4) + 5(w - 2) - 3(6w - 8) = 10$ 0

7. Solve the equation and describe the solution set. $7x + 2 = 7(x - 12)$ No solution

8. Solve the equation and describe the solution set. $2(3x - 6) = 3(2x - 4)$ All real numbers

Objective 1: Solving Linear Equations with Fractions

For Exercises 9–14, determine which of the values could be used to clear fractions or decimals in the given equation.

9. $\dfrac{2}{3}x - \dfrac{1}{6} = \dfrac{x}{9}$

   Values: 6, 9, 12, 18, 24, 36   18, 36

10. $\dfrac{1}{4}x - \dfrac{2}{7} = \dfrac{1}{2}x + 2$

   Values: 4, 7, 14, 21, 28, 42   28

11. $0.02x + 0.5 = 0.35x + 1.2$
   Values: 10; 100; 1000; 10,000
   100; 1000; 10,000

 Writing   Translating Expression   Geometry   Scientific Calculator   Video   NS E

12. $0.003 - 0.002x = 0.1x$

Values: 10; 100; 1000; 10,000

1000; 10,000

13. $\frac{1}{6}x + \frac{7}{10} = x$

Values: 3, 6, 10, 30, 60

30, 60

14. $2x - \frac{5}{2} = \frac{x}{3} - \frac{1}{4}$

Values: 2, 3, 4, 6, 12, 24

12, 24

For Exercises 15–36, solve the equation. **(See Examples 1–4.)**

15. $\frac{1}{2}x + 3 = 5$ 4

16. $\frac{1}{3}y - 4 = 9$ 39

17. $\frac{1}{6}y + 2 = \frac{5}{12}$ $-\frac{19}{2}$

18. $\frac{2}{15}z + 3 = \frac{7}{5}$ -12

19. $\frac{1}{3}q + \frac{3}{5} = \frac{1}{15}q - \frac{2}{5}$ $-\frac{15}{4}$

20. $\frac{3}{7}x - 5 = \frac{24}{7}x + 7$ -4

21. $\frac{12}{5}w + 7 = 31 - \frac{3}{5}w$ 8

22. $-\frac{1}{9}p - \frac{5}{18} = -\frac{1}{6}p + \frac{1}{3}$ 11

23. $\frac{1}{4}(3m - 4) - \frac{1}{5} = \frac{1}{4}m + \frac{3}{10}$ 3

24. $\frac{1}{25}(20 - t) = \frac{4}{25}t - \frac{3}{5}$ 7

25. $\frac{1}{6}(5s + 3) = \frac{1}{2}(s + 11)$ 15

26. $\frac{1}{12}(4n - 3) = \frac{1}{4}(2n + 1)$ -3

27. $\frac{2}{3}x + 4 = \frac{2}{3}x - 6$ No solution

28. $-\frac{1}{9}a + \frac{2}{9} = \frac{1}{3} - \frac{1}{9}a$ No solution

29. $\frac{1}{6}(2c - 1) = \frac{1}{3}c - \frac{1}{6}$ All real numbers

30. $\frac{3}{2}b - 1 = \frac{1}{8}(12b - 8)$ All real numbers

31. $\frac{2x + 1}{3} + \frac{x - 1}{3} = 5$ 5

32. $\frac{4y - 2}{5} - \frac{y + 4}{5} = -3$ -3

33. $\frac{3w - 2}{6} = 1 - \frac{w - 1}{3}$ 2

34. $\frac{z - 7}{4} = \frac{6z - 1}{8} - 2$ $\frac{3}{4}$

35. $\frac{x + 3}{3} - \frac{x - 1}{2} = 4$ -15

36. $\frac{5y - 1}{2} - \frac{y + 4}{5} = 1$ 1

Objective 2: Solving Linear Equations with Decimals

For Exercises 37–54, solve the equation. **(See Examples 5–6.)**

37. $9.2y - 4.3 = 50.9$ 6

38. $-6.3x + 1.5 = -4.8$ 1

39. $21.1w + 4.6 = 10.9w + 35.2$ 3

40. $0.05z + 0.2 = 0.15z - 10.5$ 107

41. $0.2p - 1.4 = 0.2(p - 7)$ All real numbers

42. $0.5(3q + 87) = 1.5q + 43.5$ All real numbers

43. $0.20x + 53.60 = x$ 67

44. $z + 0.06z = 3816$ 3600

45. $0.15(90) + 0.05p = 0.10(90 + p)$ 90

46. $0.25(60) + 0.10x = 0.15(60 + x)$ 120

47. $0.40(y + 10) - 0.60(y + 2) = 2$ 4

48. $0.75(x - 2) + 0.25(x + 4) = 0.5$ 1

49. $0.4x + 0.2 = -3.6 - 0.6x$ -3.8

50. $0.12x + 3 - 0.8x = 0.22x - 0.6$ 4

51. $0.06(x - 0.5) = 0.06x + 0.01$ No solution

52. $0.125x = 0.025(5x + 1)$ No solution

53. $-3.5x + 1.3 = -0.3(9x - 5)$ -0.25

54. $x + 4 = 2(0.4x + 1.3)$ -7

Mixed Exercises

For Exercises 55–64, solve the equation.

55. $0.2x - 1.8 = -3$ -6

56. $9.8h + 2 = 3.8h + 20$ 3

57. $\frac{1}{4}(x + 4) = \frac{1}{5}(2x + 3)$ $\frac{8}{3}$ or $2\frac{2}{3}$

58. $\frac{2}{3}(y - 1) = \frac{3}{4}(3y - 2)$ $\frac{10}{19}$

59. $0.3(x + 6) - 0.7(x + 2) = 4$ -9

60. $0.05(2t - 1) - 0.03(4t - 1) = 0.2$ -11

 Writing   ←→ Translating Expression   Geometry   Scientific Calculator    Video   NS🎧E

61. $\dfrac{2k + 5}{4} = 2 - \dfrac{k + 2}{3}$ $\dfrac{1}{10}$

62. $\dfrac{3d - 4}{6} + 1 = \dfrac{d + 1}{8}$ $-\dfrac{5}{9}$

63. $\dfrac{1}{8}v + \dfrac{2}{3} = \dfrac{1}{6}v + \dfrac{3}{4}$ -2

64. $\dfrac{2}{5}z - \dfrac{1}{4} = \dfrac{3}{10}z + \dfrac{1}{2}$ $\dfrac{15}{2}$

Expanding Your Skills

For Exercises 65–68, solve the equation.

65. $\dfrac{1}{2}a + 0.4 = -0.7 - \dfrac{3}{5}a$ -1

66. $\dfrac{3}{4}c - 0.11 = 0.23(c - 5)$ -2

67. $0.8 + \dfrac{7}{10}b = \dfrac{3}{2}b - 0.8$ 2

68. $0.78 - \dfrac{1}{25}h = \dfrac{3}{5}h - 0.5$ 2

Problem Recognition Exercises

Equations and Expressions

For Exercises 1–30, identify each exercise as an expression or an equation. Then simplify the expression or solve the equation.

1. $2b + 23 - 6b - 5$
Expression; $-4b + 18$

2. $10p - 9 + 2p - 3 + 8p - 18$
Expression; $20p - 30$

3. $\dfrac{y}{4} = -2$ Equation; -8

4. $-\dfrac{x}{2} = 7$ Equation; -14

5. $3(4h - 2) - (5h - 8) = 8 - (2h + 3)$ Equation; $\dfrac{1}{3}$

6. $7y - 3(2y + 5) = 7 - (10 - 10y)$
Equation; $-\dfrac{4}{3}$

7. $3(8z - 1) + 10 - 6(5 + 3z)$
Expression; $6z - 23$

8. $-5(1 - x) - 3(2x + 3) + 5$
Expression; $-x - 9$

9. $6c + 3(c + 1) = 10$
Equation; $\dfrac{7}{9}$

10. $-9 + 5(2y + 3) = -7$
Equation; $-\dfrac{13}{10}$

11. $0.5(2a - 3) - 0.1 = 0.4(6 + 2a)$
Equation; 20

12. $0.07(2v - 4) = 0.1(v - 4)$
Equation; -3

13. $-\dfrac{5}{9}w + \dfrac{11}{12} = \dfrac{23}{36}$
Equation; $\dfrac{1}{2}$

14. $\dfrac{3}{8}t - \dfrac{5}{8} = \dfrac{1}{2}t + \dfrac{1}{8}$
Equation; -6

15. $\dfrac{3}{4}x + \dfrac{1}{2} - \dfrac{1}{8}x + \dfrac{5}{4}$
Expression; $\dfrac{5}{8}x + \dfrac{7}{4}$

16. $\dfrac{7}{3}(6 - 12t) + \dfrac{1}{2}(4t + 8)$
Expression; $-26t + 18$

17. $2z - 7 = 2(z - 13)$
Equation; no solution

18. $-6x + 2(x + 1) = -2(2x + 3)$
Equation; no solution

19. $\dfrac{2x - 1}{4} + \dfrac{3x + 2}{6} = 2$
Equation; $\dfrac{23}{12}$

20. $\dfrac{w - 4}{6} - \dfrac{3w - 1}{2} = -1$
Equation; $\dfrac{5}{8}$

21. $4b - 8 - b = -3b + 2(3b - 4)$
Equation; all real numbers

22. $-k - 41 - 2 - k = -2(20 + k) - 3$
Equation; all real numbers

23. $\dfrac{4}{3}(6y - 3) = 0$
Equation; $\dfrac{1}{2}$

24. $\dfrac{1}{2}(2c - 4) + 3 = \dfrac{1}{3}(6c + 3)$
Equation; 0

25. $3(x + 6) - 7(x + 2) - 4(1 - x)$
Expression; 0

26. $-10(2k + 1) - 4(4 - 5k) + 25$
Expression; -1

27. $3 - 2[4a - 5(a + 1)]$
Expression; $2a + 13$

28. $-9 - 4[3 - 2(q + 3)]$
Expression; $8q + 3$

29. $4 + 2[8 - (6 + x)] = -2(x - 1) - 4 + x$
Equation; 10

30. $-1 - 5[2 + 3(w - 2)] = 5(w + 4)$ Equation; $-\dfrac{1}{20}$

Writing   Translating Expression   Geometry   Scientific Calculator   Video   NS E

Applications of Linear Equations: Introduction to Problem Solving

1. Problem-Solving Strategies

Linear equations can be used to solve many real-world applications. However, with "word problems," students often do not know where to start. To help organize the problem-solving process, we offer the following guidelines:

Objectives

1. **Problem-Solving Strategies**
2. **Translations Involving Linear Equations**
3. **Consecutive Integer Problems**
4. **Applications of Linear Equations**

Problem-Solving Flowchart for Word Problems

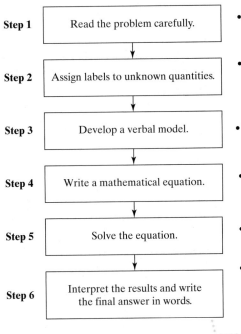

Step 1 Read the problem carefully.
- Familiarize yourself with the problem. Identify the unknown, and if possible, estimate the answer.

Step 2 Assign labels to unknown quantities.
- Identify the unknown quantity or quantities. Let x or another variable represent one of the unknowns. Draw a picture and write down relevant formulas.

Step 3 Develop a verbal model.
- Write an equation in *words*.

Step 4 Write a mathematical equation.
- Replace the verbal model with a mathematical equation using x or another variable.

Step 5 Solve the equation.
- Solve for the variable using the steps for solving linear equations.

Step 6 Interpret the results and write the final answer in words.
- Once you have obtained a numerical value for the variable, recall what it represents in the context of the problem. Can this value be used to determine other unknowns in the problem? Write an answer to the word problem in *words*.

Instructor Note: Remind students that x is a convenient variable, but any letter can be used. For example, for an unknown width use w.

Avoiding Mistakes

Once you have reached a solution to a word problem, verify that it is reasonable in the context of the problem.

2. Translations Involving Linear Equations

We have already practiced translating an English sentence to a mathematical equation. Recall from Section 1.2 that several key words translate to the algebraic operations of addition, subtraction, multiplication, and division.

Example 1 Translating to a Linear Equation

The sum of a number and negative eleven is negative fifteen. Find the number.

Solution:

	Step 1: Read the problem.
Let x represent the unknown number.	**Step 2:** Label the unknown.

$$\overset{\text{the sum of}}{\underset{\downarrow}{(\text{a number})}} + (-11) \overset{\text{is}}{\underset{\downarrow}{=}} (-15)$$

Step 3: Develop a verbal model.

$$x + (-11) = -15 \qquad \textbf{Step 4: Write an equation.}$$

$$x + (-11) + 11 = -15 + 11 \qquad \textbf{Step 5: Solve the equation.}$$

$$x = -4$$

The number is -4.   **Step 6:** Write the final answer in words.

Skill Practice

1. The sum of a number and negative seven is 12. Find the number.

Classroom Example: p. 150, Exercise 10

Answer

1. The number is 19.

Avoiding Mistakes

It is important to remember that subtraction is not a commutative operation. Therefore, the order in which two real numbers are subtracted affects the outcome. The expression "forty less than five times a number" must be translated as: $5x - 40$ (not $40 - 5x$). Similarly, "fifty-two less than the number" must be translated as: $x - 52$ (not $52 - x$).

Example 2 **Translating to a Linear Equation**

Forty less than five times a number is fifty-two less than the number. Find the number.

Solution:

Let x represent the unknown number.

$$\left(\begin{array}{c} 5 \text{ times} \\ \text{a number} \end{array}\right) \overset{\text{less}}{-} (40) \overset{\text{is}}{=} \left(\begin{array}{c} \text{the} \\ \text{number} \end{array}\right) \overset{\text{less}}{-} (52)$$

$$5x - 40 = x - 52$$

$$5x - 40 = x - 52$$

$$5x - x - 40 = x - x - 52$$

$$4x - 40 = -52$$

$$4x - 40 + 40 = -52 + 40$$

$$4x = -12$$

$$\frac{4x}{4} = \frac{-12}{4}$$

$$x = -3$$

The number is -3.

Step 1: Read the problem.

Step 2: Label the unknown.

Step 3: Develop a verbal model.

Step 4: Write an equation.

Step 5: Solve the equation.

Step 6: Write the final answer in words.

Example 3 **Translating to a Linear Equation**

Twice the sum of a number and six is two more than three times the number. Find the number.

Solution:

Let x represent the unknown number.

$$\underset{\text{twice}}{2} \ \underset{\text{the sum}}{(x + 6)} \ \underset{\text{is}}{=} \ \underset{\text{2 more than}}{3x + 2}$$

$$\underset{\begin{array}{c}\text{three times}\\\text{a number}\end{array}}{}$$

Step 1: Read the problem.

Step 2: Label the unknown.

Step 3: Develop a verbal model.

Step 4: Write an equation.

Answers

2. The number is -8.

3. The number is -10.

$$2(x + 6) = 3x + 2$$

Step 5: Solve the equation.

$$2x + 12 = 3x + 2$$

$$2x - 2x + 12 = 3x - 2x + 2$$

$$12 = x + 2$$

$$12 - 2 = x + 2 - 2$$

$$10 = x$$

The number is 10.

Step 6: Write the final answer in words.

> **Avoiding Mistakes**
>
> It is important to enclose "the sum of a number and six" within parentheses so that the entire quantity is multiplied by 2. Forgetting the parentheses would imply that only the x-term is multiplied by 2.
>
> Correct: $2(x + 6)$

3. Consecutive Integer Problems

The word *consecutive* means "following one after the other in order without gaps." The numbers 6, 7, and 8 are examples of three **consecutive integers**. The numbers $-4, -2, 0$, and 2 are examples of **consecutive even integers**. The numbers 23, 25, and 27 are examples of **consecutive odd integers**.

Notice that any two consecutive integers differ by 1. Therefore, if x represents an integer, then $(x + 1)$ represents the next larger consecutive integer (Figure 2-4).

Instructor Note: Explain to students that we could let x be the larger of two consecutive integers. Then $x - 1$ is the next smaller integer. However, using x and $x + 1$ generally causes less confusion for students.

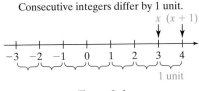

Figure 2-4

Any two consecutive even integers differ by 2. Therefore, if x represents an even integer, then $(x + 2)$ represents the next consecutive larger even integer (Figure 2-5).

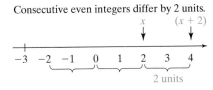

Figure 2-5

Likewise, any two consecutive odd integers differ by 2. If x represents an odd integer, then $(x + 2)$ is the next larger odd integer (Figure 2-6).

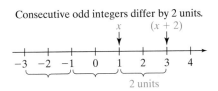

Figure 2-6

Classroom Example: p. 150, Exercise 22

Skill Practice

4. The sum of two consecutive even integers is 66. Find the integers.

Example 4 Solving an Application Involving Consecutive Integers

The sum of two consecutive odd integers is -188. Find the integers.

Solution:

In this example we have two unknown integers. We can let x represent either of the unknowns.

| | **Step 1:** | Read the problem. |

Suppose x represents the first odd integer. **Step 2:** Label the variables.

Then $(x + 2)$ represents the second odd integer.

$$\left(\begin{matrix}\text{First} \\ \text{integer}\end{matrix}\right) + \left(\begin{matrix}\text{second} \\ \text{integer}\end{matrix}\right) = (\text{total})$$

Step 3: Write an equation in words.

$$x \quad + \quad (x + 2) \quad = -188$$

Step 4: Write a mathematical equation.

$$x + (x + 2) = -188$$

$$2x + 2 = -188$$

Step 5: Solve for x.

$$2x + 2 - 2 = -188 - 2$$

$$2x = -190$$

$$\frac{2x}{2} = \frac{-190}{2}$$

$$x = -95$$

The first integer is $x = -95$.

Step 6: Interpret the results and write the answer in words.

The second integer is $x + 2 = -95 + 2 = -93$.

The two integers are -95 and -93.

TIP: With word problems, it is advisable to check that the answer is reasonable.

The numbers -95 and -93 are consecutive odd integers. Furthermore, their sum is -188 as desired.

Skill Practice

5. Five times the smallest of three consecutive integers is 17 less than twice the sum of the integers. Find the integers.

Classroom Example: p. 151, Exercise 30

Example 5 Solving an Application Involving Consecutive Integers

Ten times the smallest of three consecutive integers is twenty-two more than three times the sum of the integers. Find the integers.

Solution:

| | **Step 1:** | Read the problem. |

Let x represent the first integer. **Step 2:** Label the variables.

$x + 1$ represents the second consecutive integer.

$x + 2$ represents the third consecutive integer.

Answers

4. The integers are 32 and 34.
5. The integers are 11, 12, and 13.

$$\begin{pmatrix} 10 \text{ times} \\ \text{the first} \\ \text{integer} \end{pmatrix} = \begin{pmatrix} 3 \text{ times} \\ \text{the sum of} \\ \text{the integers} \end{pmatrix} + 22$$

Step 3: Write an equation in words.

10 times the first integer | is | 3 times | 22 more than

$$10x = 3\underbrace{[(x) + (x + 1) + (x + 2)]}_{\text{the sum of the integers}} + 22$$

Step 4: Write a mathematical equation.

$$10x = 3(x + x + 1 + x + 2) + 22$$

Step 5: Solve the equation.

$$10x = 3(3x + 3) + 22$$

Clear parentheses.

$$10x = 9x + 9 + 22$$

Combine *like* terms.

$$10x = 9x + 31$$

$$10x - 9x = 9x - 9x + 31$$

Isolate the x-terms on one side.

$$x = 31$$

The first integer is $x = 31$.

The second integer is $x + 1 = 31 + 1 = 32$.

The third integer is $x + 2 = 31 + 2 = 33$.

Step 6: Interpret the results and write the answer in words.

The three integers are 31, 32, and 33.

4. Applications of Linear Equations

Example 6 Using a Linear Equation in an Application

A carpenter cuts a 6-ft board in two pieces. One piece must be three times as long as the other. Find the length of each piece.

Solution:

In this problem, one piece must be three times as long as the other. Thus, if x represents the length of one piece, then $3x$ can represent the length of the other.

Step 1: Read the problem completely.

x represents the length of the smaller piece. $3x$ represents the length of the longer piece.

Step 2: Label the unknowns. Draw a figure.

3x   x

$$\left(\begin{array}{c}\text{Length of}\\\text{one piece}\end{array}\right) + \left(\begin{array}{c}\text{length of}\\\text{other piece}\end{array}\right) = \left(\begin{array}{c}\text{total length}\\\text{of the board}\end{array}\right)$$

Step 3: Set up a verbal equation.

$$x \quad + \quad 3x \quad = \quad 6$$

Step 4: Write an equation.

$$4x = 6$$

Step 5: Solve the equation.

$$\frac{4x}{4} = \frac{6}{4}$$

$$x = 1.5$$

The smaller piece is $x = 1.5$ ft.

Step 6: Interpret the results.

The longer piece is $3x$ or $3(1.5 \text{ ft}) = 4.5$ ft.

> **TIP:** The variable can represent either unknown. In Example 6, if we let x represent the length of the longer piece of pipe, then $\frac{1}{3}x$ would represent the length of the smaller piece. The equation would become $x + \frac{1}{3}x = 6$. Try solving this equation and interpreting the result.

Skill Practice

7. There are 40 students in an algebra class. There are 4 more women than men. How many women and how many men are in the class?

Classroom Example: p. 151, Exercise 38

Example 7   **Using a Linear Equation in an Application**

The hit movies *Spider-Man 3* and *Pirates of the Caribbean: At World's End* together brought in $265.8 million during their opening weekends. *Spider-Man 3* earned $78.3 million less than twice what *Pirates of the Caribbean* earned. How much revenue did each movie bring in during its opening weekend?

Solution:

In this example, we have two unknowns. The variable, x, can represent *either* quantity. However, the revenue from *Spider-Man 3* is given in terms of the revenue for *Pirates of the Caribbean: At World's End.*

Step 1: Read the problem.

Let x represent the revenue for *Pirates of the Caribbean.*

Step 2: Label the variables.

Then $2x - 78.3$ represents the revenue for *Spider-Man 3.*

Answer

7. There are 22 women and 18 men.

$$\left(\begin{array}{c}\text{Revenue from}\\ \textit{Pirates}\end{array}\right) + \left(\begin{array}{c}\text{revenue from}\\ \textit{Spider-Man}\end{array}\right) = \left(\begin{array}{c}\text{total}\\ \text{revenue}\end{array}\right)$$

Step 3: Write an equation in words.

$$x \quad + \quad (2x - 78.3) \quad = \quad 265.8$$

Step 4: Write an equation.

$$3x - 78.3 = 265.8$$

Step 5: Solve the equation.

$$3x - 78.3 + 78.3 = 265.8 + 78.3$$
$$3x = 344.1$$
$$\frac{3x}{3} = \frac{344.1}{3}$$
$$x = 114.7$$

- Revenue from *Pirates of the Caribbean:* $x = 114.7$
- Revenue from *Spider-Man 3:* $2x - 78.3 = 2(114.7) - 78.3 = 151.1$

The revenue from *Pirates of the Caribbean: At World's End* was $114.7 million for its opening weekend. The revenue for *Spider-Man 3* was $151.1 million.

Section 2.4 Practice Exercises

Boost *your* GRADE at ALEKS.com!

- Practice Problems
- Self-Tests
- NetTutor
- e-Professors
- Videos

For additional exercises, see Classroom Activities 2.4A–2.4C in the *Instructor's Resource Manual* at www.mhhe.com/moh.

Study Skills Exercises

1. After doing a section of homework, check the odd-numbered answers in the back of the text. Choose a method to identify the exercises that gave you trouble (i.e., circle the number or put a star by the number). List some reasons why it is important to label these problems.

2. Define the key terms:

   a. consecutive integers   **b. consecutive even integers**   **c. consecutive odd integers**

Objective 2: Translations Involving Linear Equations

For Exercises 3–8, write an expression representing the unknown quantity.

3. In a math class, the number of students who received an "A" in the class was 5 more than the number of students who received a "B." If x represents the number of "B" students, write an expression for the number of "A" students. $x + 5$

4. At a recent motorcycle rally, the number of men exceeded the number of women by 216. If x represents the number of women, write an expression for the number of men. $x + 216$

5. Anna is three times as old as Jake. If x represents Jake's age, write an expression for Anna's age. $3x$

6. Rebecca downloaded twice as many songs as Nigel. If x represents the number of songs downloaded by Nigel, write an expression for the number downloaded by Rebecca. $2x$

7. Sidney made $20 more than three times Casey's weekly salary. If x represents Casey's weekly salary, write an expression for Sidney's weekly salary. $3x + 20$

8. David scored 26 points less than twice the number of points Rich scored in a video game. If x represents the number of points scored by Rich, write an expression representing the number of points scored by David. $2x - 26$

For Exercises 9–18, use the problem-solving flowchart on page 143. (**See Examples 1–3.**)

9. Six less than a number is –10. Find the number. The number is -4.

10. Fifteen less than a number is 41. Find the number. The number is 56.

11. Twice the sum of a number and seven is eight. Find the number. The number is -3.

12. Twice the sum of a number and negative two is sixteen. Find the number. The number is 10.

13. A number added to five is the same as twice the number. Find the number. The number is 5.

14. Three times a number is the same as the difference of twice the number and seven. Find the number. The number is -7.

15. The sum of six times a number and ten is equal to the difference of the number and fifteen. Find the number. The number is -5.

16. The difference of fourteen and three times a number is the same as the sum of the number and negative ten. Find the number. The number is 6.

17. If the difference of a number and four is tripled, the result is six more than the number. Find the number. The number is 9.

18. Twice the sum of a number and eleven is twenty-two less than three times the number. Find the number. The number is 44.

Objective 3: Consecutive Integer Problems

19. a. If x represents the smallest of three consecutive integers, write an expression to represent each of the next two consecutive integers. $x + 1, x + 2$

   b. If x represents the largest of three consecutive integers, write an expression to represent each of the previous two consecutive integers. $x - 1, x - 2$

20. a. If x represents the smallest of three consecutive odd integers, write an expression to represent each of the next two consecutive odd integers. $x + 2, x + 4$

   b. If x represents the largest of three consecutive odd integers, write an expression to represent each of the previous two consecutive odd integers. $x - 2, x - 4$

For Exercises 21–30, use the problem-solving flowchart from page 143. (**See Examples 4–5.**)

21. The sum of two consecutive integers is -67. Find the integers. The integers are -34 and -33.

22. The sum of two consecutive odd integers is 52. Find the integers. The integers are 25 and 27.

23. The sum of two consecutive odd integers is 28. Find the integers. The integers are 13 and 15.

24. The sum of three consecutive even integers is 66. Find the integers. The integers are 20, 22, and 24.

25. The perimeter of a pentagon (a five-sided polygon) is 80 in. The five sides are represented by consecutive integers. Find the measures of the sides. The sides are 14 in., 15 in., 16 in., 17 in., and 18 in.

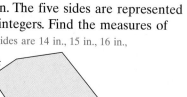

26. The perimeter of a triangle is 96 in. The lengths of the sides are represented by consecutive integers. Find the measures of the sides. The sides are 31 in., 32 in., and 33 in.

27. The sum of three consecutive even integers is 48 more than twice the smallest of the three integers. Find the integers.   The integers are 42, 44, and 46.

28. The sum of three consecutive odd integers is 89 more than twice the largest integer. Find the integers.   The integers are 91, 93, and 95.

29. Eight times the sum of three consecutive odd integers is 210 more than 10 times the middle integer. Find the integers.   The integers are 13, 15, and 17.

30. Five times the sum of three consecutive even integers is 140 more than ten times the smallest. Find the integers.   The integers are 22, 24, and 26.

Objective 4: Applications of Linear Equations

For Exercises 31–42, use the problem-solving flowchart (page 143) to solve the problems.

31. A board is 86 cm in length and must be cut so that one piece is 20 cm longer than the other piece. Find the length of each piece. (**See Example 6.**)   The lengths of the pieces are 33 cm and 53 cm.

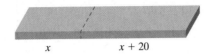

x $x + 20$

32. A rope is 54 in. in length and must be cut into two pieces. If one piece must be twice as long as the other, find the length of each piece.   The lengths of the pieces are 18 in. and 36 in.

33. Karen's age is 12 years more than Clarann's age. The sum of their ages is 58. Find their ages.   Karen's age is 35, and Clarann's age is 23.

34. Maria's age is 15 years less than Orlando's age. The sum of their ages is 29. Find their ages.   Maria's age is 7, and Orlando's age is 22.

35. For a recent year, 31 more Democrats than Republicans were in the U.S. House of Representatives. If the total number of representatives in the House from these two parties was 433, find the number of representatives from each party.   There were 201 Republicans and 232 Democrats.

36. For a recent year, the number of men in the U.S. Senate totaled 4 more than 5 times the number of women. Find the number of men and the number of women in the Senate given that the Senate has 100 members.   There were 84 men and 16 women.

37. Approximately 5.816 million people watch *The Oprah Winfrey Show*. This is 1.118 million more than watch *The Dr. Phil Show*. How many watch *The Dr. Phil Show*? (*Source: Neilson Media Research*) (**See Example 7.**)   4.698 million watch *The Dr. Phil Show*.

38. Two of the largest Internet retailers are e-Bay and Amazon.com. Recently, the estimated U.S. sales of e-Bay were $0.1 billion less than twice the sales of Amazon.com. Given the total sales of $5.6 billion, determine the sales of e-Bay and Amazon.com.   e-Bay's sales were $3.7 billion and Amazon.com's sales were $1.9 billion.

39. The longest river in Africa is the Nile. It is 2455 km longer than the Congo River, also in Africa. The sum of the lengths of these rivers is 11,195 km. What is the length of each river?   The Congo River is 4370 km long, and the Nile River is 6825 km.

40. The average depth of the Gulf of Mexico is three times the depth of the Red Sea. The difference between the average depths is 1078 m. What is the average depth of the Gulf of Mexico and the average depth of the Red Sea?   The average depth of the Red Sea is 539 m and that of the Gulf of Mexico is 1617 m.

41. Asia and Africa are the two largest continents in the world. The land area of Asia is approximately 14,514,000 km² larger than the land area of Africa. Together their total area is 74,644,000 km². Find the land area of Asia and the land area of Africa.   The area of Africa is 30,065,000 km². The area of Asia is 44,579,000 km².

42. Mt. Everest, the highest mountain in the world, is 2654 m higher than Mt. McKinley, the highest mountain in the United States. If the sum of their heights is 15,042 m, find the height of each mountain.   Mt. McKinley is 6194 m high. Mt. Everest is 8848 m high.

 Writing    Translating Expression   Geometry   Scientific Calculator   Video   NS&E

Mixed Exercises

43. A group of hikers walked from Hawk Mt. Shelter to Blood Mt. Shelter along the Appalachian Trail, a total distance of 20.5 mi. It took 2 days for the walk. The second day the hikers walked 4.1 mi less than they did on the first day. How far did they walk each day?   They walked 12.3 mi on the first day and 8.2 mi on the second.

44. $120 is to be divided among three restaurant servers. Angie made $10 more than Marie. Gwen, who went home sick, made $25 less than Marie. How much money should each server get?   Marie made $45, Angie made $55, and Gwen made $20.

45. A 4-ft piece of PVC pipe is cut into three pieces. The longest piece is 12 in. longer than twice the smallest piece. The middle piece is 6 in. less than the longest piece. How long is each piece?   The pieces are 6 in., 18 in., and 24 in.

46. A 6-ft piece of copper wire must be cut into three pieces. The middle piece is 16 in. longer than the shortest piece. The longest piece is twice as long as the middle piece. How long is each piece?   The pieces are 6 in., 22 in., and 44 in.

47. Three consecutive integers are such that three times the largest exceeds the sum of the two smaller integers by 47. Find the integers.   42, 43, and 44

48. Four times the smallest of three consecutive odd integers is 236 more than the sum of the other two integers. Find the integers.   121, 123, and 125

49. In a recent year, the estimated earnings for Jennifer Lopez was $2.5 million more than half of the earnings for the band U2. If the total earnings were $106 million, what were the earnings for Jennifer Lopez and U2? (*Source: Forbes*)   Jennifer Lopez made $37 million, and U2 made $69 million.

50. Two of the longest-running TV series are *Gunsmoke* and *The Simpsons. Gunsmoke* ran 97 fewer episodes than twice the number of *The Simpsons*. If the total number of episodes is 998, how many of each show was produced?   There are 633 episodes of *Gunsmoke* and 365 episodes of *The Simpsons.*

51. Five times the difference of a number and three is four less than four times the number. Find the number.   The number is 11.

52. Three times the difference of a number and seven is one less than twice the number. Find the number.   The number is 20.

53. The sum of the page numbers on two facing pages in a book is 941. What are the page numbers?
The page numbers are 470 and 471.

54. Three raffle tickets are represented by three consecutive integers. If the sum of the three integers is 2,666,031, find the numbers.   The ticket numbers are 888,676; 888,677; and 888,678.

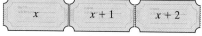

55. If three is added to five times a number, the result is forty-three more than the number. Find the number.   The number is 10.

56. If seven is added to three times a number, the result is thirty-one more than the number.   The number is 12.

57. The deepest point in the Pacific Ocean is 676 m more than twice the deepest point in the Arctic Ocean. If the deepest point in the Pacific is 10,920 m, how many meters is the deepest point in the Arctic Ocean?   The deepest point in the Arctic Ocean is 5122 m.

58. The area of Greenland is 201,900 km² less than three times the area of New Guinea. What is the area of New Guinea if the area of Greenland is 2,175,600 km²?   The area of New Guinea is 792,500 km².

59. The sum of twice a number and $\frac{3}{4}$ is the same as the difference of four times the number and $\frac{1}{8}$. Find the number.   The number is $\frac{7}{16}$.

60. The difference of a number and $-\frac{11}{12}$ is the same as the difference of three times the number and $\frac{1}{6}$. Find the number.   The number is $\frac{13}{24}$.

61. The product of a number and 3.86 is equal to 7.15 more than the number. Find the number.   The number is 2.5.

62. The product of a number and 4.6 is 33.12 less than the number. Find the number.   The number is -9.2.

Applications Involving Percents

1. Solving Basic Percent Equations

Recall from Section R.3 that the word *percent* means "per hundred." For example:

Percent	**Interpretation**
63% of homes have a computer	63 out of 100 homes have a computer.
5% sales tax	5¢ in tax is charged for every 100¢ in merchandise.
15% commission	$15 is earned in commission for every $100 sold.

Objectives

1. Solving Basic Percent Equations
2. Applications Involving Simple Interest
3. Applications Involving Discount and Markup

Percents come up in a variety of applications in day-to-day life. Many such applications follow the basic percent equation:

$$\text{Amount} = (\text{percent})(\text{base}) \quad \text{Basic percent equation}$$

In Example 1, we apply the basic percent equation to compute **sales tax**.

Example 1   **Computing Sales Tax**

A new digital camera costs $429.95.

a. Compute the sales tax if the tax rate is 4%.

b. Determine the total cost, including tax.

Skill Practice

1. Find the amount of tax on a portable CD player that sells for $89. Assume the tax rate is 6%.
2. Find the total cost including tax.

Classroom Example: p. 157, Exercise 18

Solution:

	Step 1:	Read the problem.
a. Let x represent the amount of tax.	**Step 2:**	Label the variable.
Apply the percent equation to compute sales tax.	**Step 3:**	Write a verbal equation.

$$\underset{\downarrow}{\text{Amount}} = \underset{\downarrow}{(\text{percent})} \cdot \underset{\downarrow}{(\text{base})}$$

Sales tax = (tax rate)(price of merchandise)

$x = (0.04)(\$429.95)$	**Step 4:**	Write a mathematical equation.
$x = \$17.198$	**Step 5:**	Solve the equation.
$x = \$17.20$		Round to the nearest cent.
The tax on the merchandise is $17.20.	**Step 6:**	Interpret the results.

Avoiding Mistakes

Be sure to use the decimal form of a percentage within an equation.

$$4\% = 0.04$$

b. The total cost is found by:

total cost = cost of merchandise + amount of tax

Therefore the total cost is $429.95 + $17.20 = $447.15.

Answers

1. The amount of tax is $5.34.
2. The total cost is $94.34.

In Example 2, we solve a problem in which the percent is unknown.

Example 2 Finding an Unknown Percent

A group of 240 college men were asked what intramural sport they most enjoyed playing. The results are in the graph. What percent of the men surveyed preferred tennis?

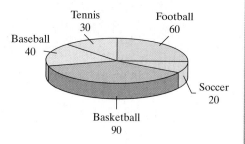

Solution:

	Step 1: Read the problem.
Let x represent the unknown percent.	**Step 2:** Label the variable.
The problem can be rephrased as:	
30 is what percent of 240?	**Step 3:** Write an equation in words.
$30 = x \cdot 240$	**Step 4:** Write a mathematical equation.
$30 = 240x$	**Step 5:** Solve the equation.
$\dfrac{30}{240} = \dfrac{240x}{240}$	Divide both sides by 240.
$0.125 = x$	
$0.125 \times 100\% = 12.5\%$	**Step 6:** Interpret the results. Change the value of x to a percent by multiplying by 100%.

In this survey, 12.5% of men prefer tennis.

Example 3 Solving a Percent Equation with an Unknown Base

Andrea spends 20% of her monthly paycheck on rent each month. If her rent payment is $750, what is her monthly paycheck?

Solution:

	Step 1: Read the problem.
Let x represent the amount of Andrea's monthly paycheck.	**Step 2:** Label the variables.
The problem can be rephrased as:	
$750 is 20% of what number?	**Step 3:** Write an equation in words.
$750 = 0.20 \cdot x$	**Step 4:** Write a mathematical equation.

$$750 = 0.20x$$

Step 5: Solve the equation.

$$\frac{750}{0.20} = \frac{0.20x}{0.20}$$

Divide both sides by 0.20.

$$3750 = x$$

Andrea's monthly paycheck is $3750.   **Step 6:** Interpret the results.

2. Applications Involving Simple Interest

One important application of percents is in computing simple interest on a loan or on an investment.

Simple interest is interest that is earned on principal (the original amount of money invested in an account). The following formula is used to compute simple interest:

$$\begin{pmatrix} \text{Simple} \\ \text{interest} \end{pmatrix} = \begin{pmatrix} \text{principal} \\ \text{invested} \end{pmatrix} \begin{pmatrix} \text{annual} \\ \text{interest rate} \end{pmatrix} \begin{pmatrix} \text{time} \\ \text{in years} \end{pmatrix}$$

This formula is often written symbolically as $I = Prt$.

For example, to find the simple interest earned on $2000 invested at 7.5% interest for 3 years, we have

$$I = Prt$$
$$\text{Interest} = (\$2000)(0.075)(3)$$
$$= \$450$$

Example 4   **Applying Simple Interest**

Jorge wants to save money for his daughter's college education. If Jorge needs to have $4340 at the end of 4 years, how much money would he need to invest at a 6% simple interest rate?

Solution:

Step 1: Read the problem.

Let P represent the original amount invested.   **Step 2:** Label the variables.

$$\begin{pmatrix} \text{Original} \\ \text{principal} \end{pmatrix} + (\text{interest}) = (\text{total})$$

Step 3: Write an equation in words.

$$(P) \quad + \quad (Prt) \quad = (\text{total})$$

$$P \quad + \quad P(0.06)(4) = 4340$$

Step 4: Write a mathematical equation.

$$P + 0.24P = 4340$$

$$1.24P = 4340$$

Step 5: Solve the equation.

$$\frac{1.24P}{1.24} = \frac{4340}{1.24}$$

$$P = 3500$$

The original investment should be $3500.   **Step 6:** Interpret the results and write the answer in words.

3. Applications Involving Discount and Markup

Applications involving percent increase and percent decrease are abundant in many real-world settings. Sales tax for example is essentially a markup by a state or local government. It is important to understand that percent increase or decrease is always computed on the original amount given.

In Example 5, we illustrate an example of percent decrease in an application where merchandise is discounted.

Skill Practice

6. An iPod is on sale for $151.20. This is after a 20% discount. What was the original cost of the iPod?

Classroom Example: p. 158, Exercise 36

Example 5 Applying Percents to a Discount Problem

After a 38% discount, a used treadmill costs $868 on e-Bay. What was the original cost of the treadmill?

Solution:

	Step 1:	Read the problem.
Let x be the original cost of the treadmill.	**Step 2:**	Label the variables.

$$\begin{pmatrix} \text{Original} \\ \text{cost} \end{pmatrix} - (\text{discount}) = \begin{pmatrix} \text{sale} \\ \text{price} \end{pmatrix}$$ **Step 3:** Write an equation in words.

$$x \qquad - 0.38(x) \quad = 868$$ **Step 4:** Write a mathematical equation. The discount is a percent of the *original* amount.

$$x - 0.38x = 868$$ **Step 5:** Solve the equation.

$$0.62x = 868$$ Combine *like* terms.

$$\frac{0.62x}{0.62} = \frac{868}{0.62}$$ Divide by 0.62.

$$x = 1400$$

The original cost of the treadmill was $1400. **Step 6:** Interpret the result.

Answer

6. The iPod originally cost $189.

Section 2.5 Practice Exercises

Boost *your* GRADE at ALEKS.com!

ALEKS version 3.0

• Practice Problems
• Self-Tests
• NetTutor

• e-Professors
• Videos

For additional exercises, see Classroom Activities 2.5A–2.5D in the *Instructor's Resource Manual* at www.mhhe.com/moh.

Study Skills Exercise

1. Define the key terms:

   a. sales tax   **b. simple interest**

Review Exercises

2. List the six steps to solve an application. See page 143.

For Exercises 3–4, use the steps for problem solving to solve these applications.

3. Find two consecutive integers such that three times the larger is the same as 45 more than the smaller.
The numbers are 21 and 22.

4. The height of the Great Pyramid of Giza is 17 m more than twice the height of the pyramid found in Saqqara. If the difference in their heights is 77 m, find the height of each pyramid.
The pyramid at Saqqara is 60 m high, and the Great Pyramid is 137 m high.

Objective 1: Solving Basic Percent Equations

For Exercises 5–16, find the missing values.

5. 45 is what percent of 360? 12.5%

6. 338 is what percent of 520? 65%

7. 544 is what percent of 640? 85%

8. 576 is what percent of 800? 72%

9. What is 0.5% of 150? 0.75

10. What is 9.5% of 616? 58.52

11. What is 142% of 740? 1050.8

12. What is 156% of 280? 436.8

13. 177 is 20% of what number? 885

14. 126 is 15% of what number? 840

15. 275 is 12.5% of what number? 2200

16. 594 is 45% of what number? 1320

17. A Craftsman drill is on sale for $99.99. If the sales tax rate is 7%, how much will Molly have to pay for the drill?
(See Example 1.) Molly will have to pay $106.99.

18. Patrick purchased four new tires that were regularly priced at $94.99 each, but are on sale for $20 off per tire. If the sales tax rate is 6%, how much will be charged to Patrick's VISA card?
Patrick will have to pay $317.96.

For Exercises 19–20, use the graph showing the distribution for leading forms of cancer in men. (*Source:* Centers for Disease Control)

19. If there are 700,000 cases of cancer in men in the United States, approximately how many are prostate cancer?
Approximately 231,000 cases

20. Approximately how many cases of lung cancer would be expected in 700,000 cancer cases among men in the United States?
Approximately 91,000 cases

21. There were 14,000 cases of cancer of the pancreas diagnosed out of 700,000 cancer cases. What percent is this? **(See Example 2.)** 2%

Percent of Cancer Cases by Type (Men)

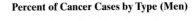

All other sites 33%
Prostate 33%
Melanoma 4%
Bladder 6%
Colon 11%
Lung 13%

22. There were 21,000 cases of leukemia diagnosed out of 700,000 cancer cases. What percent is this? 3%

 Writing   Translating Expression   Geometry Scientific Calculator   Video NS E

23. Javon is in a 28% tax bracket for his federal income tax. If the amount of money that he paid for federal income tax was $23,520, what was his taxable income? **(See Example 3.)** Javon's taxable income was $84,000.

24. In a recent survey of college-educated adults, 155 indicated that they regularly work more than 50 hr a week. If this represents 31% of those surveyed, how many people were in the survey? There were 500 people in the survey.

Objective 2: Applications Involving Simple Interest

For Exercises 25–32, solve these equations involving simple interest.

25. How much interest will Pam earn in 4 years if she invests $3000 in an account that pays 3.5% simple interest? Pam will earn $420.

26. How much interest will Roxanne have to pay if she borrows $2000 for 2 yr at a simple interest rate of 4%? Roxanne will have to pay $160.

27. Bob borrowed some money for 1 yr at 5% simple interest. If he had to pay back a total of $1260, how much did he originally borrow? **(See Example 4.)** Bob borrowed $1200.

28. Mike borrowed some money for 2 yr at 6% simple interest. If he had to pay back a total of $3640, how much did he originally borrow? Mike borrowed $3250.

29. If $1500 grows to $1950 after 5 yr, find the simple interest rate. The rate is 6%.

30. If $9000 grows to $10,440 in 2 yr, find the simple interest rate. The rate is 8%.

31. Perry is planning a vacation to Europe in 2 yr. How much should he invest in a certificate of deposit that pays 3% simple interest to get the $3500 that he needs for the trip? Round to the nearest dollar. Perry needs to invest $3302.

32. Sherica invested in a mutual fund and at the end of 20 yr she has $14,300 in her account. If the mutual fund returned an average yield of 8%, how much did she originally invest? Sherica invested $5500.

Objective 3: Applications Involving Discount and Markup

33. A Pioneer car CD/MP3 player costs $170. Circuit City has it on sale for 12% off with free installation.

a. What is the discount on the CD/MP3 player?
$20.40
b. What is the sale price? $149.60

34. A laptop computer, originally selling for $899.00, is on sale for 10% off.

a. What is the discount on the laptop? $89.90

b. What is the sale price? $809.10

35. A Sony digital camera is on sale for $400.00. This price is 15% off the original price. What was the original price? Round to the nearest cent. **(See Example 5.)** The original price was $470.59.

36. The *Star Wars: Episode III* DVD is on sale for $18. If this represents an 18% discount rate, what was the original price of the DVD? The original price was $21.95.

37. The original price of an Audio Jukebox was $250. It is on sale for $220. What percent discount does this represent? The discount rate is 12%.

38. During the holiday season, the Xbox 360 sold for $425.00 in stores. This product was in such demand that it sold for $800 online. What percent markup does this represent? (Round to the nearest whole percent.) The markup rate is 88%.

39. In one area, the cable company marked up the monthly cost by 6%. The new cost is $63.60 per month. What was the cost before the increase?

The original cost was $60.

40. A doctor ordered a dosage of medicine for a patient. After 2 days, she increased the dosage by 20% and the new dosage came to 18 cc. What was the original dosage? The original dosage was 15 cc.

 Writing Translating Expression Geometry Scientific Calculator Video NS&E

Mixed Exercises

41. Sun Lei bought a laptop computer for $1800. The total cost, including tax, came to $1890. What is the tax rate?   The tax rate is 5%.

42. Jamie purchased a compact disk and paid $18.26. If the disk price is $16.99, what is the sales tax rate (round to the nearest tenth of a percent)?   The tax rate is 7.5%.

43. For a recent year, admission to Walt Disney World cost $74.37, including taxes of 11%. What was the original ticket price before taxes?   The ticket price was $67.

44. A hotel room rented for 5 nights costs $706.25 including 13% in taxes. Find the original price of the room (before tax) for the 5 nights. Then find the price per night.   The 5 nights (without tax) costs $625. The price per night is $125.

45. Deon purchased a house and sold it for a 24% profit. If he sold the house for $260,400, what was the original purchase price?   The original price was $210,000.

46. To meet the rising cost of energy, the yearly membership at a YMCA had to be increased by 12.5% from the past year. The yearly membership fee is currently $450. What was the cost of membership last year?   The cost was $400.

47. Alina earns $1600 per month plus a 12% commission on pharmaceutical sales. If she sold $25,000 in pharmaceuticals one month, what was her salary that month?   Alina made $4600 that month.

48. Dan sold a beachfront home for $650,000. If his commission rate is 4%, what did he earn in commission?   Dan earned $26,000.

49. Diane sells women's sportswear at a department store. She earns a regular salary and, as a bonus, she receives a commission of 4% on all sales over $200. If Diane earned an extra $25.80 last week in commission, how much merchandise did she sell over $200?   Diane sold $645 over $200 worth of merchandise.

50. For selling software, Tom received a bonus commission based on sales over $500. If he received $180 in commission for selling a total of $2300 worth of software, what is his commission rate?   Tom's commission rate is 10%.

Formulas and Applications of Geometry

Section 2.6

1. Formulas and Literal Equations

Objectives

1. **Formulas and Literal Equations**
2. **Geometry Applications**

Literal equations are equations that contain several variables. A formula is a literal equation with a specific application. For example, the perimeter of a triangle (distance around the triangle) can be found by the formula $P = a + b + c$, where a, b, and c are the lengths of the sides (Figure 2-7).

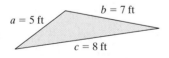

$a = 5$ ft $b = 7$ ft $c = 8$ ft

$$P = a + b + c$$
$$= 5\text{ ft} + 7\text{ ft} + 8\text{ ft}$$
$$= 20\text{ ft}$$

Figure 2-7

In this section, we will learn how to rewrite formulas to solve for a different variable within the formula. Suppose, for example, that the perimeter of a triangle is known and two of the sides are known (say, sides a and b). Then the third side, c, can be found by subtracting the lengths of the known sides from the perimeter (Figure 2-8).

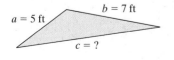

$a = 5$ ft $b = 7$ ft $c = ?$

If the perimeter is 20 ft, then
$$c = P - a - b$$
$$= 20\text{ ft} - 5\text{ ft} - 7\text{ ft}$$
$$= 8\text{ ft}$$

Figure 2-8

Writing   Translating Expression   Geometry   Scientific Calculator   Video   NS E

To solve a formula for a different variable, we use the same properties of equality outlined in the earlier sections of this chapter. For example, consider the two equations $2x + 3 = 11$ and $wx + y = z$. Suppose we want to solve for x in each case:

$2x + 3 = 11$			$wx + y = z$	
$2x + 3 - 3 = 11 - 3$	Subtract 3.		$wx + y - y = z - y$	Subtract y.
$2x = 8$			$wx = z - y$	
$\dfrac{2x}{2} = \dfrac{8}{2}$	Divide by 2.		$\dfrac{wx}{w} = \dfrac{z - y}{w}$	Divide by w.
$x = 4$			$x = \dfrac{z - y}{w}$	

The equation on the left has only one variable and we are able to simplify the equation to find a numerical value for x. The equation on the right has multiple variables. Because we do not know the values of w, y, and z, we are not able to simplify further. The value of x is left as a formula in terms of w, y, and z.

Classroom Examples: p. 165, Exercises 16 and 24

Example 1   Solving for an Indicated Variable

Solve for the indicated variables.

a. $d = rt$ for t **b.** $5x + 2y = 12$ for y

Solution:

a. $d = rt$ for t The goal is to isolate the variable t.

$\dfrac{d}{r} = \dfrac{rt}{r}$ Because the relationship between r and t is multiplication, we reverse the process by dividing both sides by r.

$\dfrac{d}{r} = t$, or equivalently $t = \dfrac{d}{r}$

b. $5x + 2y = 12$ for y The goal is to solve for y.

$5x - 5x + 2y = 12 - 5x$ Subtract $5x$ from both sides to isolate the y-term.

$2y = -5x + 12$ $-5x + 12$ is the same as $12 - 5x$.

$\dfrac{2y}{2} = \dfrac{-5x + 12}{2}$ Divide both sides by 2 to isolate y.

$y = \dfrac{-5x + 12}{2}$

TIP: In the expression $\dfrac{-5x + 12}{2}$ do not try to divide the 2 into the 12. The divisor of 2 is dividing the entire quantity, $-5x + 12$ (not just the 12).

We may, however, apply the divisor to each term individually in the numerator. That is, $\dfrac{-5x + 12}{2}$ can be written in several different forms. Each is correct.

$$y = \dfrac{-5x + 12}{2} \quad \text{or} \quad y = \dfrac{-5x}{2} + \dfrac{12}{2} \Rightarrow y = -\dfrac{5x}{2} + 6$$

Example 2 Solving Formulas for an Indicated Variable

The formula $C = \frac{5}{9}(F - 32)$ is used to find the temperature, C, in degrees Celsius for a given temperature expressed in degrees Fahrenheit, F. Solve the formula $C = \frac{5}{9}(F - 32)$ for F.

Solution:

$$C = \frac{5}{9}(F - 32)$$

$$C = \frac{5}{9}F - \frac{5}{9} \cdot 32 \qquad \text{Clear parentheses.}$$

$$C = \frac{5}{9}F - \frac{160}{9} \qquad \text{Multiply: } \frac{5}{9} \cdot \frac{32}{1} = \frac{160}{9}.$$

$$9(C) = 9\left(\frac{5}{9}F - \frac{160}{9}\right) \qquad \text{Multiply by the LCD to clear fractions.}$$

$$9C = \frac{9}{1} \cdot \frac{5}{9}F - \frac{9}{1} \cdot \frac{160}{9} \qquad \text{Apply the distributive property.}$$

$$9C = 5F - 160 \qquad \text{Simplify.}$$

$$9C + 160 = 5F - 160 + 160 \qquad \text{Add } 160 \text{ to both sides.}$$

$$9C + 160 = 5F$$

$$\frac{9C + 160}{5} = \frac{5F}{5} \qquad \text{Divide both sides by } 5.$$

$$\frac{9C + 160}{5} = F$$

The answer may be written in several forms:

$$F = \frac{9C + 160}{5} \quad \text{or} \quad F = \frac{9C}{5} + \frac{160}{5} \quad \Rightarrow \quad F = \frac{9}{5}C + 32$$

Skill Practice

3. Solve for the indicated variable.

$$y = \frac{1}{3}(x - 7) \text{ for } x.$$

Classroom Example: p. 165, Exercise 32

2. Geometry Applications

In Section R.4, we presented numerous facts and formulas related to geometry. Sometimes these are needed to solve applications in geometry.

Example 3 Solving a Geometry Application Involving Perimeter

The length of a rectangular lot is 1 m less than twice the width. If the perimeter is 190 m, find the length and width.

Solution:

Step 1: Read the problem.

Let x represent the width of the rectangle. **Step 2:** Label the variables.

Then $2x - 1$ represents the length.

x

$2x - 1$

Skill Practice

4. The length of a rectangle is 10′ less than twice the width. If the perimeter is 178′ find the length and width.

Classroom Example: p. 166, Exercise 46

Answers

3. $x = 3y + 7$
4. The length is 56′, and the width is 33′.

$$P = 2l + 2w \qquad \textbf{Step 3:} \quad \text{Perimeter formula}$$

$$190 = 2(2x - 1) + 2(x) \qquad \textbf{Step 4:} \quad \text{Write an equation in terms of } x.$$

$$190 = 4x - 2 + 2x \qquad\qquad\quad \textbf{Step 5:} \quad \text{Solve for } x.$$

$$190 = 6x - 2$$

$$192 = 6x$$

$$\frac{192}{6} = \frac{6x}{6}$$

$$32 = x$$

The width is $x = 32$.

The length is $2x - 1 = 2(32) - 1 = 63$.   **Step 6:**   Interpret the results and write the answer in words.

The width of the rectangular lot is 32 m and the length is 63 m.

Recall some facts about angles.

- Two angles are complementary if the sum of their measures is 90°.
- Two angles are supplementary if the sum of their measures is 180°.
- The sum of the measures of the angles within a triangle is 180°.

Skill Practice

5. Two complementary angles are constructed so that one measures 1° less than six times the other. Find the measures of the angles.

Classroom Example: p. 166, Exercise 52

Example 4   **Solving a Geometry Application Involving Complementary Angles**

Two complementary angles are drawn such that one angle is 4° more than seven times the other angle. Find the measure of each angle.

Solution:

                                                   Step 1:   Read the problem.

Let x represent the measure of one angle.   **Step 2:**   Label the variables.

Then $7x + 4$ represents the measure of the other angle.

The angles are complementary, so their sum must be 90°.

$$\begin{pmatrix} \text{Measure of} \\ \text{first angle} \end{pmatrix} + \begin{pmatrix} \text{measure of} \\ \text{second angle} \end{pmatrix} = 90° \qquad \textbf{Step 3:} \quad \text{Create a verbal equation.}$$

$$x \qquad + \qquad 7x + 4 \qquad = 90 \qquad \textbf{Step 4:} \quad \text{Write a mathematical equation.}$$

$$8x + 4 = 90 \qquad \textbf{Step 5:} \quad \text{Solve for } x.$$

$$8x = 86$$

$$\frac{8x}{8} = \frac{86}{8}$$

$$x = 10.75$$

Answer

5. 13° and 77°

One angle is $x = 10.75$.

Step 6: Interpret the results and write the answer in words.

The other angle is $7x + 4 = 7(10.75) + 4 = 79.25$.

The angles are $10.75°$ and $79.25°$.

Example 5 Solving a Geometry Application

One angle in a triangle is twice as large as the smallest angle. The third angle is $10°$ more than seven times the smallest angle. Find the measure of each angle.

Solution:

Step 1: Read the problem.

Let x represent the measure of the smallest angle.

Step 2: Label the variables.

Then $2x$ and $7x + 10$ represent the measures of the other two angles.

The sum of the angles must be $180°$.

$$\begin{pmatrix}\text{Measure of}\\\text{first angle}\end{pmatrix} + \begin{pmatrix}\text{measure of}\\\text{second angle}\end{pmatrix} + \begin{pmatrix}\text{measure of}\\\text{third angle}\end{pmatrix} = 180°$$

Step 3: Create a verbal equation.

$$x \quad + \quad 2x \quad + \quad (7x + 10) = 180$$

Step 4: Write a mathematical equation.

$$x + 2x + 7x + 10 = 180$$

Step 5: Solve for x.

$$10x + 10 = 180$$
$$10x = 170$$
$$x = 17$$

Step 6: Interpret the results and write the answer in words.

The smallest angle is $x = 17$.

The other angles are $2x = 2(17) = 34$

$$7x + 10 = 7(17) + 10 = 129$$

The angles are $17°$, $34°$, and $129°$.

Answer

6. $25°$, $50°$, and $105°$

Example 6   Solving a Geometry Application Involving
                Circumference

The distance around a circular garden is 188.4 ft.
Find the radius to the nearest tenth of a foot
(Figure 2-9). Use 3.14 for π.

$C = 188.4$ ft

Figure 2-9

Solution:

$$C = 2\pi r \qquad \text{Use the formula for the circumference of a circle.}$$

$$188.4 = 2\pi r \qquad \text{Substitute 188.4 for } C.$$

$$\frac{188.4}{2\pi} = \frac{2\pi r}{2\pi} \qquad \text{Divide both sides by } 2\pi.$$

$$\frac{188.4}{2\pi} = r$$

$$r \approx \frac{188.4}{2(3.14)}$$

$$= 30.0$$

The radius is approximately 30.0 ft.

Answer

7. 2.0 cm

Calculator Connections

Topic: Using the π Key on a Calculator

In Example 6 we could have obtained a more accurate result if we had used the π key on the calculator.

  Note that parentheses are required to divide 188.4 by the quantity 2π. This guarantees that the calculator follows the implied order of operations. Without parentheses, the calculator would divide 188.4 by 2 and then multiply the result by π.

Scientific Calculator

Enter: 188.4 [÷] [(] [2] [×] [π] [)] [=]   **Result:** | 29.98479128 | correct

Enter: 188.4 [÷] [2] [×] [π] [=]   **Result:** | 295.938028 | incorrect

Graphing Calculator

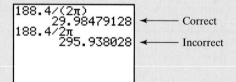

```
188.4/(2π)
         29.98479128   ←—— Correct
188.4/2π
         295.938028    ←—— Incorrect
```

Calculator Exercises

Approximate the expressions with a calculator. Round to three decimal places if necessary.

1. $\dfrac{880}{2\pi}$ 140.056   **2.** $\dfrac{1600}{\pi(4)^2}$ 31.831   **3.** $\dfrac{20}{5\pi}$ 1.273   **4.** $\dfrac{10}{7\pi}$ 0.455

Section 2.6	Practice Exercises

Boost *your* GRADE at ALEKS.com!

ALEKS version 3.0

- Practice Problems
- Self-Tests
- NetTutor
- e-Professors
- Videos

For additional exercises, see Classroom Activities 2.6A–2.6C in the *Instructor's Resource Manual* at www.mhhe.com/moh.

Study Skills Exercises

1. A good technique for studying for a test is to choose four problems from each section of the chapter and write the problems along with the directions on a 3×5 card. On the back of the card, put the page number where you found that problem. Then shuffle the cards and test yourself on the procedure to solve each problem. If you find one that you do not know how to solve, look at the page number and do several of that type. Write four problems you would choose for this section.

2. Define the key term: **literal equation**

Review Exercises

For Exercises 3–8, solve the equation.

3. $3(2y + 3) - 4(-y + 1) = 7y - 10$ -5

4. $-(3w + 4) + 5(w - 2) - 3(6w - 8) = 10$ 0

5. $\frac{1}{2}(x - 3) + \frac{3}{4} = 3x - \frac{3}{4}$ 0

6. $\frac{5}{6}x + \frac{1}{2} = \frac{1}{4}(x - 4)$ $-\frac{18}{7}$

7. $0.5(y + 2) - 0.3 = 0.4y + 0.5$ -2

8. $0.25(500 - x) + 0.15x = 75$ 500

Objective 1: Formulas and Literal Equations

For Exercises 9–40, solve for the indicated variable. **(See Examples 1–2.)**

9. $P = a + b + c$ for a
$a = P - b - c$

10. $P = a + b + c$ for b
$b = P - a - c$

11. $x = y - z$ for y
$y = x + z$

12. $c + d = e$ for d
$d = e - c$

13. $p = 250 + q$ for q
$q = p - 250$

14. $y = 35 + x$ for x
$x = y - 35$

15. $A = bh$ for b $b = \frac{A}{h}$

16. $d = rt$ for r $r = \frac{d}{t}$

17. $PV = nrt$ for t $t = \frac{PV}{nr}$

18. $P_1 V_1 = P_2 V_2$ for V_1 $V_1 = \frac{P_2 V_2}{P_1}$

19. $x - y = 5$ for x $x = 5 + y$

20. $x + y = -2$ for y $y = -2 - x$

21. $3x + y = -19$ for y
$y = -3x - 19$

22. $x - 6y = -10$ for x
$x = 6y - 10$

23. $2x + 3y = 6$ for y
$y = \frac{-2x + 6}{3}$ or $y = -\frac{2}{3}x + 2$

24. $5x + 2y = 10$ for y
$y = \frac{-5x + 10}{2}$ or $y = -\frac{5}{2}x + 5$

25. $-2x - y = 9$ for x
$x = \frac{y + 9}{-2}$ or $x = -\frac{1}{2}y - \frac{9}{2}$

26. $3x - y = -13$ for x
$x = \frac{y - 13}{3}$ or $x = \frac{1}{3}y - \frac{13}{3}$

27. $4x - 3y = 12$ for y
$y = \frac{-4x + 12}{-3}$ or $y = \frac{4}{3}x - 4$

28. $6x - 3y = 4$ for y
$y = \frac{-6x + 4}{-3}$ or $y = 2x - \frac{4}{3}$

29. $ax + by = c$ for y
$y = \frac{-ax + c}{b}$ or $y = -\frac{a}{b}x + \frac{c}{b}$

30. $ax + by = c$ for x
$x = \frac{-by + c}{a}$ or $x = -\frac{b}{a}y + \frac{c}{a}$

31. $A = P(1 + rt)$ for t
$t = \frac{A - P}{Pr}$ or $t = \frac{A}{Pr} - \frac{1}{r}$

32. $P = 2(L + w)$ for L
$L = \frac{P - 2w}{2}$ or $L = \frac{P}{2} - w$

33. $a = 2(b + c)$ for c
$c = \frac{a - 2b}{2}$ or $c = \frac{a}{2} - b$

34. $3(x + y) = z$ for x
$x = \frac{z - 3y}{3}$ or $x = \frac{z}{3} - y$

35. $Q = \frac{x + y}{2}$ for y
$y = 2Q - x$

36. $Q = \frac{a - b}{2}$ for a
$a = 2Q + b$

37. $M = \frac{a}{S}$ for a
$a = MS$

38. $A = \frac{1}{3}(a + b + c)$ for c
$c = 3A - a - b$

39. $P = I^2 R$ for R
$R = \frac{P}{I^2}$

40. $F = \frac{GMm}{d^2}$ for m
$m = \frac{Fd^2}{GM}$

Objective 2: Geometry Applications

For Exercises 41–62, use the problem-solving flowchart (page 143) from Section 2.4.

41. The perimeter of a rectangular garden is 24 ft. The length is 2 ft more than the width. Find the length and the width of the garden. **(See Example 3.)** The length is 7 ft, and the width is 5 ft.

42. In a small rectangular wallet photo, the width is 7 cm less than the length. If the border (perimeter) of the photo is 34 cm, find the length and width. The length is 12 cm, and the width is 5 cm.

43. The length of a rectangular parking area is four times the width. The perimeter is 300 yd. Find the length and width of the parking area. The length is 120 yd and the width is 30 yd.

44. The width of Jason's workbench is $\frac{1}{2}$ the length. The perimeter is 240 in. Find the length and the width of the workbench. The length is 80 in. and the width is 40 in.

45. A builder buys a rectangular lot of land such that the length is 5 m less than two times the width. If the perimeter is 590 m, find the length and the width. The length is 195 m, and the width is 100 m.

46. The perimeter of a rectangular pool is 140 yd. If the length is 20 yd less than twice the width, find the length and the width. The length is 40 yd, and the width is 30 yd.

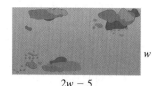

$2w - 5$

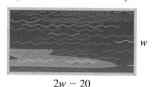

$2w - 20$

47. A triangular parking lot has two sides that are the same length and the third side is 5 m longer. If the perimeter is 71 m, find the lengths of the sides. The sides are 22 m, 22 m, and 27 m.

48. The perimeter of a triangle is 16 ft. One side is 3 ft longer than the shortest side. The third side is 1 ft longer than the shortest side. Find the lengths of all the sides. The sides are 4 ft, 5 ft, and 7 ft.

49. Sometimes memory devices are helpful for remembering mathematical facts. Recall that the sum of two complementary angles is 90°. That is, two complementary angles when added together form a right angle or "corner." The words *Complementary* and *Corner* both start with the letter "*C*." Derive your own memory device for remembering that the sum of two supplementary angles is 180°.   "Adjacent supplementary angles form a straight angle." The words *Supplementary* and *Straight* both begin with the same letter.

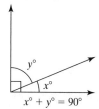

$x° + y° = 90°$

Complementary angles form a "Corner"

$x° + y° = 180°$

Supplementary angles . . .

50. What do you know about the measures of two vertical angles?   The measures are the same.

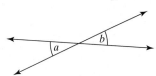

51. Two angles are complementary. One angle is 20° less than the other angle. Find the measures of the angles. **(See Example 4.)** The angles are 55° and 35°.

52. Two angles are complementary. One angle is 4° less than three times the other angle. Find the measures of the angles. The angles are 23.5° and 66.5°.

53. Two angles are supplementary. One angle is three times as large as the other angle. Find the measures of the angles. The angles are 45° and 135°.

54. Two angles are supplementary. One angle is 6° more than four times the other. Find the measures of the two angles. The angles are 34.8° and 145.2°.

55. Find the measures of the vertical angles labeled in the figure by first solving for x.

$x = 20$; the vertical angles measure 37°.

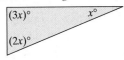

56. Find the measures of the vertical angles labeled in the figure by first solving for y.

$y = 40$; the vertical angles measure 146°.

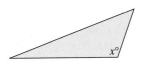

57. The largest angle in a triangle is three times the smallest angle. The middle angle is two times the smallest angle. Given that the sum of the angles in a triangle is 180°, find the measure of each angle. **(See Example 5.)**

The measures of the angles are 30°, 60°, and 90°.

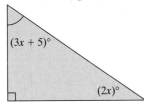

58. The smallest angle in a triangle measures 90° less than the largest angle. The middle angle measures 60° less than the largest angle. Find the measure of each angle.

The measures of the angles are 20°, 50°, and 110°.

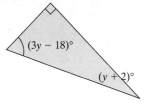

59. The smallest angle in a triangle is half the largest angle. The middle angle measures 30° less than the largest angle. Find the measure of each angle.

The measures of the angles are 42°, 54°, and 84°.

60. The largest angle of a triangle is three times the middle angle. The smallest angle measures 10° less than the middle angle. Find the measure of each angle.

The measures of the angles are 38°, 28°, and 114°.

61. Find the value of x and the measure of each angle labeled in the figure.

$x = 17$; the measures of the angles are 34° and 56°.

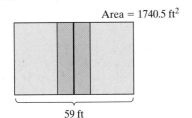

62. Find the value of y and the measure of each angle labeled in the figure.

$y = 26.5$; the measures of the angles are 28.5° and 61.5°.

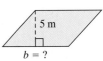

63. a. A rectangle has length l and width w. Write a formula for the area.

$A = lw$

b. Solve the formula for the width, w. $w = \dfrac{A}{l}$

c. The area of a rectangular volleyball court is 1740.5 ft^2 and the length is 59 ft. Find the width.

The width is 29.5 ft.

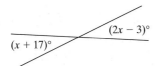

Area = 1740.5 ft^2

59 ft

64. a. A parallelogram has height h and base b. Write a formula for the area.

$A = bh$

b. Solve the formula for the base, b. $b = \dfrac{A}{h}$

c. Find the base of the parallelogram pictured if the area is 40 m^2.

The base is 8 m.

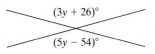

5 m

$b = ?$

65. a. A rectangle has length *l* and width *w*. Write a formula for the perimeter.
$P = 2l + 2w$

b. Solve the formula for the length, *l*. $l = \dfrac{P - 2w}{2}$

c. The perimeter of the soccer field at Giants Stadium is 338 m. If the width is 66 m, find the length. The length is 103 m.

Perimeter = 338 m

66 m

66. a. A triangle has height *h* and base *b*. Write a formula for the area. $A = \dfrac{1}{2}bh$

b. Solve the formula for the height, *h*. $h = \dfrac{2A}{b}$

c. Find the height of the triangle pictured if the area is 12 km². The height is 4 km.

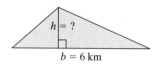

$h = ?$

$b = 6$ km

 67. a. A circle has a radius of *r*. Write a formula for the circumference. **(See Example 6.)**
$C = 2\pi r$

b. Solve the formula for the radius, *r*. $r = \dfrac{C}{2\pi}$

c. The circumference of the circular Buckingham Fountain in Chicago is approximately 880 ft. Find the radius. Round to the nearest foot.

The radius is approximately 140 ft.

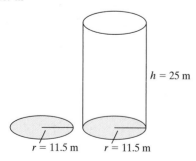

68. a. The length of each side of a square is *s*. Write a formula for the perimeter of the square. $P = 4s$

b. Solve the formula for the length of a side, *s*. $s = \dfrac{P}{4}$

c. The Pyramid of Khufu (known as the Great Pyramid) at Giza has a square base. If the distance around the bottom is 921.6 m, find the length of the sides at the bottom of the pyramid. The length of each side at the bottom of the pyramid is 230.4 m.

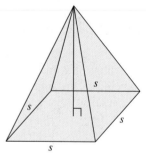

s

s

s

s

Expanding Your Skills

For Exercises 69–70, find the indicated area or volume. Be sure to include the proper units and round each answer to two decimal places if necessary.

 69. a. Find the area of a circle with radius 11.5 m.
415.48 m²

b. Find the volume of a right circular cylinder with radius 11.5 m and height 25 m.
10,386.89 m³

$h = 25$ m

$r = 11.5$ m $r = 11.5$ m

70. a. Find the area of a parallelogram with base 30 in. and height 12 in.
360 in.²

b. Find the area of a triangle with base 30 in. and height 12 in.
180 in.²

c. Compare the areas found in parts (a) and (b).
The area of the triangle is one-half the area of the parallelogram.

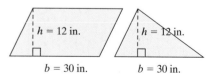

$h = 12$ in. $h = 12$ in.

$b = 30$ in. $b = 30$ in.

 Writing   Translating Expression   Geometry   Scientific Calculator   Video NS&E

Mixture Applications and Uniform Motion

1. Applications Involving Cost

In Examples 1 and 2, we will look at different kinds of mixture problems. The first example "mixes" two types of movie tickets, adult tickets that sell for $8 and children's tickets that sell for $6. Furthermore, there were 300 tickets sold for a total revenue of $2040. Before attempting the problem, we should try to gain some familiarity. Let's try a few combinations to see how many of each type of ticket might have been sold.

Suppose 100 adult tickets were sold and 200 children's tickets were sold (a total of 300 tickets).

- 100 adult tickets at $8 each gives $100(\$8) = \$800$
- 200 children's tickets at $6 each gives $200(\$6) = \$1200$

Total revenue: $2000 (not enough)

Suppose 150 adult tickets were sold and 150 children's tickets were sold (a total of 300 tickets).

- 150 adult tickets at $8 each gives $150(\$8) = \$1200$
- 150 children's tickets at $6 each gives $150(\$6) = \$900$

Total revenue: $2100 (too much)

As you can see, the trial-and-error process can be tedious and time-consuming. Therefore we will use algebra to determine the correct combination of each type of ticket.

Suppose we let x represent the number of adult tickets, then the number of children's tickets is the *total minus x*. That is

$$\begin{pmatrix} \text{Number of} \\ \text{children's tickets} \end{pmatrix} = \begin{pmatrix} \text{total number} \\ \text{of tickets} \end{pmatrix} - \begin{pmatrix} \text{number of} \\ \text{adult tickets, } x \end{pmatrix}$$

Number of children's tickets = $300 - x$.

Notice that the number of tickets sold times the price per ticket gives the revenue.

- x adult tickets at $8 each gives a revenue of: $x(\$8)$ or simply $8x$.
- $300 - x$ children's tickets at $6 each gives: $(300 - x)(\$6)$ or $6(300 - x)$

This will help us set up an equation in Example 1.

Objectives

1. Applications Involving Cost
2. Applications Involving Mixtures
3. Applications Involving Uniform Motion

Concept Connections

1. Suppose 120 sodas were sold at a concession stand. Let x represent the number of diet sodas sold. Write an expression for the number of nondiet sodas sold.

| Example 1 | Solving a Mixture Problem Involving Ticket Sales |

At one showing of *Shrek 3*, there were 300 tickets sold. Adult tickets cost $8 and tickets for children cost $6. If the total revenue from ticket sales was $2040, determine the number of each type of ticket sold.

Solution:

Let x represent the number of adult tickets sold.

$300 - x$ is the number of children's tickets.

Step 1: Read the problem.

Step 2: Label the variables.

	$8 Tickets	**$6 Tickets**	**Total**
Number of tickets	x	$300 - x$	300
Revenue	$8x$	$6(300 - x)$	2040

Skill Practice

2. At a Performing Arts Center, seats in the orchestra section cost $18 and seats in the balcony cost $12. If there were 120 seats sold for one performance, for a total revenue of $1920, how many of each type of seat were sold?

Classroom Example: p. 174, Exercise 14

Answers

1. $120 - x$
2. There were 80 seats in the orchestra section, and there were 40 in the balcony.

$$\left(\begin{array}{c}\text{Revenue from}\\\text{adult tickets}\end{array}\right) + \left(\begin{array}{c}\text{revenue from}\\\text{children's tickets}\end{array}\right) = \left(\begin{array}{c}\text{total}\\\text{revenue}\end{array}\right)$$

Step 3: Write an equation in words.

$$8x \quad + \quad 6(300 - x) \quad = \quad 2040$$

Step 4: Write a mathematical equation.

$$8x + 6(300 - x) = 2040$$

Step 5: Solve the equation.

$$8x + 1800 - 6x = 2040$$
$$2x + 1800 = 2040$$
$$2x = 240$$
$$x = 120$$

Step 6: Interpret the results.

There were 120 adult tickets sold.
The number of children's tickets is $300 - x$ which is 180.

> **Avoiding Mistakes**
>
> Check that the answer is reasonable. 120 adult tickets and 180 children's tickets makes 300 total tickets.
>   Furthermore, 120 adult tickets at $8 each amounts to $960, and 180 children's tickets at $6 amounts to $1080. The total revenue is $2040 as expected.

2. Applications Involving Mixtures

| **Example 2** | **Solving a Mixture Application** |

How many liters (L) of a 60% antifreeze solution must be added to 8 L of a 10% antifreeze solution to produce a 20% antifreeze solution?

Solution:

The information can be organized in a table. Notice that an algebraic equation is derived from the second row of the table. This relates the number of liters of pure antifreeze in each container.

Step 1: Read the problem.

Step 2: Label the variables.

	60% Antifreeze	**10% Antifreeze**	**Final Mixture: 20% Antifreeze**
Number of liters of solution	x	8	$(8 + x)$
Number of liters of pure antifreeze	$0.60x$	$0.10(8)$	$0.20(8 + x)$

The amount of pure antifreeze in the final solution equals the sum of the amounts of antifreeze in the first two solutions.

$$\left(\begin{array}{c}\text{Pure antifreeze}\\\text{from solution 1}\end{array}\right) + \left(\begin{array}{c}\text{pure antifreeze}\\\text{from solution 2}\end{array}\right) = \left(\begin{array}{c}\text{pure antifreeze}\\\text{in the final solution}\end{array}\right)$$

Step 3: Write an equation in words.

$$0.60x \quad + \quad 0.10(8) \quad = \quad 0.20(8 + x)$$
$$0.60x + 0.10(8) = 0.20(8 + x)$$

Step 4: Write a mathematical equation.

> **Skill Practice**
>
> 3. How many gallons of a 5% bleach solution must be added to 10 gallons (gal) of a 20% bleach solution to produce a solution that is 15% bleach?

Classroom Example: p. 175, Exercise 24

Answer

3. 5 gal is needed.

$$0.6x + 0.8 = 1.6 + 0.2x$$ **Step 5:** Solve the equation.

$$0.6x - 0.2x + 0.8 = 1.6 + 0.2x - 0.2x$$ Subtract $0.2x$.

$$0.4x + 0.8 = 1.6$$

$$0.4x + 0.8 - 0.8 = 1.6 - 0.8$$ Subtract 0.8.

$$0.4x = 0.8$$

$$\frac{0.4x}{0.4} = \frac{0.8}{0.4}$$ Divide by 0.4.

$$x = 2$$ **Step 6:** Interpret the result.

Therefore, 2 L of 60% antifreeze solution is necessary to make a final solution that is 20% antifreeze.

3. Applications Involving Uniform Motion

The formula (distance) = (rate)(time) or simply, $d = rt$, relates the distance traveled to the rate of travel and the time of travel.

For example, if a car travels at 60 mph for 3 hr, then

$$d = (60 \text{ mph})(3 \text{ hours})$$

$$= 180 \text{ miles}$$

If a car travels at 60 mph for x hr, then

$$d = (60 \text{ mph})(x \text{ hours})$$

$$= 60x \text{ miles}$$

Example 3   Solving an Application Involving Distance, Rate, and Time

One bicyclist rides 4 mph faster than another bicyclist. The faster rider takes 3 hr to complete a race, while the slower rider takes 4 hr. Find the speed for each rider.

Solution:

Step 1: Read the problem.

The problem is asking us to find the speed of each rider.

Let x represent the speed of the slower rider. Then $(x + 4)$ is the speed of the faster rider.

Step 2: Label the variables and organize the information given in the problem. A distance-rate-time chart may be helpful.

	Distance	**Rate**	**Time**
Faster rider	$3(x + 4)$	$x + 4$	3
Slower rider	$4(x)$	x	4

To complete the first column, we can use the relationship, $d = rt$.

─ **Skill Practice** ─

4. An express train travels 25 mph faster than a cargo train. It takes the express train 6 hr to travel a route, and it takes 9 hr for the cargo train to travel the same route. Find the speed of each train.

Classroom Example: p. 176, Exercise 32

Answer

4. The express train travels 75 mph, and the cargo train travels 50 mph.

Because the riders are riding in the same race, their distances are equal.

$$\begin{pmatrix} \text{Distance} \\ \text{by faster rider} \end{pmatrix} = \begin{pmatrix} \text{distance} \\ \text{by slower rider} \end{pmatrix}$$ **Step 3:** Set up a verbal model.

$$3(x + 4) = 4(x)$$ **Step 4:** Write a mathematical equation.

$$3x + 12 = 4x$$ **Step 5:** Solve the equation.

$$12 = x$$ Subtract $3x$ from both sides.

The variable x represents the slower rider's rate. The quantity $x + 4$ is the faster rider's rate. Thus, if $x = 12$, then $x + 4 = 16$.

The slower rider travels 12 mph and the faster rider travels 16 mph.

> **Avoiding Mistakes**
>
> Check that the answer is reasonable. If the slower rider rides at 12 mph for 4 hr, he travels 48 mi. If the faster rider rides at 16 mph for 3 hr, he also travels 48 mi as expected.

> **Skill Practice**
>
> **5.** A Piper Cub airplane has an average air speed that is 10 mph faster than a Cessna 150 airplane. If the combined distance traveled by these two small planes is 690 miles after 3 hr, what is the average speed of each plane?

Classroom Example: p. 176, Exercise 38

Example 4 **Solving an Application Involving Distance, Rate, and Time**

Two families that live 270 mi apart plan to meet for an afternoon picnic at a park that is located between their two homes. Both families leave at 9.00 A.M., but one family averages 12 mph faster than the other family. If the families meet at the designated spot $2\frac{1}{2}$ hr later, determine

a. The average rate of speed for each family.

b. The distance each family traveled to the picnic.

Solution:

For simplicity, we will call the two families, Family A and Family B. Let Family A be the family that travels at the slower rate (Figure 2-10).

Step 1: Read the problem and draw a sketch.

270 miles

Family A   Family B

Figure 2-10

Let x represent the rate of Family A.

Then $(x + 12)$ is the rate of Family B.

Step 2: Label the variables.

	Distance	**Rate**	**Time**
Family A	2.5x	x	2.5
Family B	2.5(x + 12)	x + 12	2.5

To complete the first column, we can use the relationship $d = rt$.

Answer

5. The Cessna's speed is 110 mph, and the Piper Cub's speed is 120 mph.

To set up an equation, recall that the total distance between the two families is given as 270 miles.

$$\begin{pmatrix} \text{Distance} \\ \text{traveled by} \\ \text{Family A} \end{pmatrix} + \begin{pmatrix} \text{distance} \\ \text{traveled by} \\ \text{Family B} \end{pmatrix} = \begin{pmatrix} \text{total} \\ \text{distance} \end{pmatrix}$$

Step 3: Create a verbal equation.

$$2.5x \quad + \quad 2.5(x + 12) \quad = \quad 270$$

Step 4: Write a mathematical equation.

$$2.5x + 2.5(x + 12) = 270$$
$$2.5x + 2.5x + 30 = 270$$
$$5.0x + 30 = 270$$
$$5x = 240$$
$$x = 48$$

Step 5: Solve for x.

a. Family A traveled 48 (mph).

Family B traveled $x + 12 = 48 + 12 = 60$ (mph).

Step 6: Interpret the results and write the answer in words.

b. To compute the distance each family traveled, use $d = rt$.

Family A traveled: (48 mph)(2.5 hr) = 120 mi

Family B traveled: (60 mph)(2.5 hr) = 150 mi

Section 2.7 Practice Exercises

Study Skills

1. The following is a list of steps to help you solve word problems. Check those that you follow on a regular basis when solving a word problem. Place an asterisk next to the steps that you need to improve.

_____ Read through the entire problem before writing anything down.

_____ Write down exactly what you are being asked to find.

_____ Write down what is known and assign variables to what is unknown.

_____ Draw a figure or diagram if it will help you understand the problem.

_____ Highlight key words like total, sum, difference, etc.

_____ Translate the word problem to a mathematical problem.

_____ After solving, check that your answer makes sense.

Review Exercises

For Exercises 2–3, solve for the indicated variable.

2. $ax - by = c$ for x $x = \dfrac{c + by}{a}$ **3.** $cd = r$ for c $c = \dfrac{r}{d}$

Writing   Translating Expression   Geometry   Scientific Calculator   Video   NS&E

For Exercises 4–6, solve the equation.

4. $-2d + 11 = 4 - d$ 7

5. $3(2y + 5) - 8(y - 1) = 3y + 3$ 4

6. $0.02x + 0.04(10 - x) = 1.26$ −43

Objective 1: Applications Involving Cost

For Exercises 7–12, write an algebraic expression as indicated.

7. Two numbers total 200. Let t represent one of the numbers. Write an algebraic expression for the other number. $200 - t$

8. The total of two numbers is 43. Let s represent one of the numbers. Write an algebraic expression for the other number. $43 - s$

9. Olivia needs to bring 100 cookies to her friend's party. She has already baked x cookies. Write an algebraic expression for the number of cookies Olivia still needs to bake. $100 - x$

10. Rachel needs a mixture of 55 pounds (lb) of nuts consisting of peanuts and cashews. Let p represent the number of pounds of peanuts in the mixture. Write an algebraic expression for the number of pounds of cashews that she needs to add. $55 - p$

11. Max has a total of $3000 in two bank accounts. Let y represent the amount in one account. Write an algebraic expression for the amount in the other account. $3000 - y$

12. Roberto has a total of $7500 in two savings accounts. Let z represent the amount in one account. Write an algebraic expression for the amount in the other account. $7500 - z$

13. A church had an ice cream social and sold tickets for $3 and $2. When the social was over, 81 tickets had been sold totaling $215. How many of each type of ticket did the church sell? **(See Example 1.)**   53 tickets were sold at $3 and 28 tickets were sold at $2.

	$3 Tickets	$2 Tickets	Total
Number of tickets			
Cost of tickets			

14. Anna is a teacher at an elementary school. She purchased 72 tickets to take the first-grade children and some parents on a field trip to the zoo. She purchased children's tickets for $10 each and adults' tickets for $18 each. She spent a total of $856. How many of each ticket did she buy?   Anna purchased 17 adult tickets and 55 children's tickets.

	Adults	Children	Total
Number of tickets			
Cost of tickets			

15. Josh downloaded 25 tunes from an online site for his MP3 player. Some songs cost $0.90 each while others were $1.50 each. He spent a total of $27.30. How many of each type of song did he download?   Josh downloaded 17 songs for $0.90 and 8 songs for $1.50

16. During the past year, Mrs. Singh purchased 32 books at a wholesale club store. She purchased softcover books for $4.50 each and hardcover books for $13.50 each. The total cost of the books was $243. How many of each type of book did she purchase?   Mrs. Singh purchased 21 softcover books and 11 hardcover books.

17. Angelina purchased 45 bottled drinks for her graduation party. She purchased soda for $1.60 each and bottles of flavored water for $2.00 each. The total cost of the drinks was $80. How many of each type did she buy?   Angelina purchased 25 bottles of soda and 20 bottles of flavored water.

18. Mr. Garvey purchased 58 food items at a fast food restaurant for his Little League team. He purchased hamburgers for $2.50 each and french fries for $1.50 each. He spent a total of $127. How many hamburgers and how many french fries did he purchase?   Mr. Garvey purchased 40 hamburgers and 18 french fries.

Objective 2: Applications Involving Mixtures

19. A container holds 7 ounces (oz) of liquid. Let x represent the number of ounces of liquid in another container. Write an expression for the total amount of liquid. $x + 7$

20. A bucket contains 2.5 L of a bleach solution. Let n represent the number of liters of bleach solution in a second bucket. Write an expression for the total amount of bleach solution. $n + 2.5$

21. If Miguel invests $2000 in a certificate of deposit and d dollars in a stock, write an expression for the total amount he invested. $d + 2000$

22. James has $5000 in one savings account. Let y represent the amount he has in another savings account. Write an expression for the total amount of money in both accounts. $y + 5000$

23. How many ounces of a 50% antifreeze solution must be mixed with 10 oz of an 80% antifreeze solution to produce a 60% antifreeze solution? **(See Example 2.)** 20 oz of 50% antifreeze solution

	50% Antifreeze	80% Antifreeze	Final Mixture: 60% Antifreeze
Number of ounces of solution			
Number of ounces of pure antifreeze			

24. How many liters of a 10% alcohol solution must be mixed with 12 L of a 5% alcohol solution to produce an 8% alcohol solution? 18 L of 10% alcohol solution

	10% Alcohol	5% Alcohol	Final Mixture: 8% Alcohol
Number of liters of solution			
Number of liters of pure alcohol			

25. A pharmacist needs to mix a 1% saline (salt) solution with 24 milliliters (mL) of a 16% saline solution to obtain a 9% saline solution. How many milliliters of the 1% solution must she use? The pharmacist needs to use 21 mL of the 1% saline solution.

26. A landscaper needs to mix a 75% pesticide solution with 30 gal of a 25% pesticide solution to obtain a 60% pesticide solution. How many gallons of the 75% solution must he use? The landscaper needs to use 70 gal of the 75% pesticide solution.

27. To clean a concrete driveway, a contractor needs a solution that is 30% acid. How many ounces of a 50% acid solution must be mixed with 15 ounces of a 21% solution to obtain a 30% acid solution? The contractor needs to mix 6.75 oz of 50% acid solution.

28. A veterinarian needs a mixture that contains 12% of a certain medication to treat an injured bird. How many milliliters of a 16% solution should be mixed with 6 mL of a 7% solution to obtain a solution that is 12% medication? The veterinarian needs to use 7.5 mL of the 16% solution.

Objective 3: Applications Involving Uniform Motion

29. a. If a car travels 60 mph for 5 hr, find the distance traveled. 300 mi

b. If a car travels at x mi per hour for 5 hr, write an expression that represents the distance traveled. $5x$

c. If a car travels at $x + 12$ mph for 5 hr, write an expression that represents the distance traveled. $5(x + 12)$ or $5x + 60$

30. a. If a plane travels 550 mph for 2.5 hr, find the distance traveled. 1375 mi

b. If a plane travels at x mi per hour for 2.5 hr, write an expression that represents the distance traveled. $2.5x$

c. If a plane travels at $x - 100$ mph for 2.5 hr, write an expression that represents the distance traveled. $2.5(x - 100)$ or $2.5x - 250$

 Writing Translating Expression Geometry Scientific Calculator Video NS&E

31. A woman can walk 2 mph faster down a trail to Cochita Lake than she can on the return trip uphill. It takes her 2 hr to get to the lake and 4 hr to return. What is her speed walking down to the lake? **(See Example 3.)** She walks 4 mph to the lake.

	Distance	Rate	Time
Downhill to the lake			
Uphill from the lake			

32. A car travels 20 mph slower in a bad rain storm than in sunny weather. The car travels the same distance in 2 hr in sunny weather as it does in 3 hr in rainy weather. Find the speed of the car in sunny weather. The car travels 60 mph in sunny weather.

	Distance	Rate	Time
Rain storm			
Sunny weather			

33. Bryan hiked up to the top of City Creek in 3 hr and then returned down the canyon to the trailhead in another 2 hr. His speed downhill was 1 mph faster than his speed uphill. How far up the canyon did he hike? Bryan hiked 6 mi up the canyon.

34. Kevin hiked up Lamb's Canyon in 2 hr and then ran back down in 1 hr. His speed running downhill was 2.5 mph greater than his speed hiking uphill. How far up the canyon did he hike? Kevin hiked 5 mi up the canyon.

35. Hazel and Emilie fly from Atlanta to San Diego. The flight from Atlanta to San Diego is against the wind and takes 4 hr. The return flight with the wind takes 3.5 hr. If the wind speed is 40 mph, find the speed of the plane in still air. The plane travels 600 mph in still air.

36. A boat on the Potomac River travels the same distance downstream in $\frac{2}{3}$ hr as it does going upstream in 1 hr. If the speed of the current is 3 mph, find the speed of the boat in still water. The speed of the boat is 15 mph in still water.

37. Two cars are 200 mi apart and traveling toward each other on the same road. They meet in 2 hr. One car is traveling 4 mph faster than the other. What is the speed of each car? **(See Example 4.)** The slower car travels 48 mph and the faster car travels 52 mph.

38. Two cars are 238 mi apart and traveling toward each other along the same road. They meet in 2 hr. One car is traveling 5 mph slower than the other. What is the speed of each car? The cars are traveling 62 mph and 57 mph.

39. After Hurricane Katrina, a rescue vehicle leaves a station at noon and heads for New Orleans. An hour later a second vehicle traveling 10 mph faster leaves the same station. By 4:00 P.M., the first vehicle reaches its destination, and the second is still 10 mi away. How fast is each vehicle? The speeds of the vehicles are 40 mph and 50 mph.

40. A truck leaves a truck stop at 9:00 A.M. and travels toward Sturgis, Wyoming. At 10:00 A.M., a motorcycle leaves the same truck stop and travels the same route. The motorcycle travels 15 mph faster than the truck. By noon, the truck has traveled 20 mi further than the motorcycle. How fast is each vehicle? The speed of the motorcycle is 65 mph and the speed of the truck is 50 mph.

41. Two boats traveling the same direction leave a harbor at noon. After 2 hr, they are 40 mi apart. If one boat travels twice as fast as the other, find the rate of each boat. The rates of the boats are 20 mph and 40 mph.

42. Two canoes travel down a river, starting at 9:00 A.M. One canoe travels twice as fast as the other. After 3.5 hr, the canoes are 5.25 mi apart. Find the speed of each canoe. The canoes travel 1.5 mph and 3 mph.

Mixed Exercises

43. A certain granola mixture is 10% peanuts.

  a. If a container has 20 lb of granola, how many pounds of peanuts are there? 2 lb

  b. If a container has x pounds of granola, write an expression that represents the number of pounds of peanuts in the granola. $0.10x$

  c. If a container has $x + 3$ lb of granola, write an expression that represents the number of pounds of peanuts.
  $0.10(x + 3) = 0.10x + 0.30$

44. A certain blend of coffee sells for $9.00 per pound.

  a. If a container has 20 lb of coffee, how much will it cost. $180

  b. If a container has x lb of coffee, write an expression that represents the cost. $9.00x$

  c. If a container has $40 - x$ lb of this coffee, write an expression that represents the cost.
  $9(40 - x) = 360 - 9x$

✏️ Writing   ↔ Translating Expression   ◿ Geometry   🖩 Scientific Calculator   💿 Video   NS&E

45. The Coffee Company mixes coffee worth $12 per pound with coffee worth $8 per pound to produce 50 lb of coffee worth $8.80 per pound. How many pounds of the $12 coffee and how many pounds of the $8 coffee must be used? 10 lb of coffee sold at $12 per pound and 40 lb of coffee sold at $8 per pound.

	$12 Coffee	$8 Coffee	Total
Number of pounds			
Value of coffee			

46. The Nut House sells pecans worth $4 per pound and cashews worth $6 per pound. How many pounds of pecans and how many pounds of cashews must be mixed to form 16 lb of a nut mixture worth $4.50 per pound? 12 lb of pecans and 4 lb of cashews

	$4 Pecans	$6 Cashews	Total
Number of pounds			
Value of nuts			

47. A boat in distress, 21 nautical miles from a marina, travels toward the marina at 3 knots (nautical miles per hour). A coast guard cruiser leaves the marina and travels toward the boat at 25 knots. How long will it take for the boats to reach each other? The boats will meet in $\frac{3}{4}$ hr (45 min).

48. An air traffic controller observes a plane heading from New York to San Francisco traveling at 450 mph. At the same time, another plane leaves San Francisco and travels 500 mph to New York. If the distance between the airports is 2850 mi, how long will it take for the planes to pass each other? The planes will pass in 3 hr.

49. Surfer Sam purchased a total of 21 items at the surf shop. He bought wax for $3.00 per package and sunscreen for $8.00 per bottle. He spent a total amount of $88.00. How many of each item did he purchase? Sam purchased 16 packages of wax and 5 bottles of sunscreen.

50. Tonya Toast loves jam. She purchased 30 jars of gourmet jam for $178.50. She bought raspberry jam for $6.25 per jar and strawberry jam for $5.50 per jar. How many jars of each did she purchase? Tonya purchased 18 jars of raspberry and 12 jars of strawberry.

51. How many quarts of 85% chlorine solution must be mixed with 5 quarts of 25% chlorine solution to obtain a 45% chlorine solution? 2.5 quarts of 85% chlorine solution

52. How many liters of a 58% sugar solution must be added to 14 L of a 40% sugar solution to obtain a 50% sugar solution? 17.5 L of 58% sugar solution

Expanding Your Skills

53. How much pure water must be mixed with 12 L of a 40% alcohol solution to obtain a 15% alcohol solution? (*Hint:* Pure water is 0% alcohol.) 20 L of water

54. How much pure water must be mixed with 10 oz of a 60% alcohol solution to obtain a 25% alcohol solution? 14 oz of water

55. Amtrak Acela Express is a high speed train that runs in the United States between Washington, D.C., and Boston. In Japan, a bullet train along the Sanyo line operates at an average speed of 60 km/hr faster than the Amtrak Acela Express. It takes the Japanese bullet train 2.7 hr to travel the same distance as the Acela Express can travel in 3.375 hr. Find the speed of each train. The Japanese bullet train travels 300 km/hr and the Acela Express travels 240 km/hr.

56. Amtrak Acela Express is a high-speed train along the northeast corridor between Washington, D.C., and Boston. Since its debut, it cuts the travel time from 4 hr 10 min to 3 hr 20 min. On average, if the Acela Express is 30 mph faster than the old train, find the speed of the Acela Express. (*Hint:* 4 hr 10 min = $4\frac{1}{6}$ hr.) 150 mph

Section 2.8   Linear Inequalities

Objectives

1. Graphing Linear Inequalities
2. Set-Builder Notation and Interval Notation
3. Addition and Subtraction Properties of Inequality
4. Multiplication and Division Properties of Inequality
5. Solving Inequalities of the Form $a < x < b$
6. Applications of Linear Inequalities

1. Graphing Linear Inequalities

Consider the following two statements.

$$2x + 7 = 11 \quad \text{and} \quad 2x + 7 < 11$$

The first statement is an equation (it has an $=$ sign). The second statement is an inequality (it has an inequality symbol, $<$). In this section, we will learn how to solve linear *inequalities*, such as $2x + 7 < 11$.

> **DEFINITION   A Linear Inequality in One Variable**
>
> A **linear inequality in one variable**, x, is defined as any relationship of the form:
>
> $$ax + b < 0, \ ax + b \leq 0, \ ax + b > 0, \ \text{or} \ ax + b \geq 0, \text{where } a \neq 0.$$

The following inequalities are linear equalities in one variable.

$$2x - 3 < 0 \qquad -4z - 3 > 0 \qquad a \leq 4 \qquad 5.2y \geq 10.4$$

The number line is a useful tool to visualize the solution set of an equation or inequality. For example, the solution set to the equation $x = 2$ is {2} and may be graphed as a single point on the number line.

$$x = 2 \qquad \begin{array}{c} \text{number line from } -6 \text{ to } 6 \text{ with point at } 2 \end{array}$$

The solution set to an inequality is the set of real numbers that make the inequality a true statement. For example, the solution set to the inequality $x \geq 2$ is all real numbers 2 or greater. Because the solution set has an infinite number of values, we cannot list all of the individual solutions. However, we can graph the solution set on the number line.

$$x \geq 2 \qquad \begin{array}{c} \text{number line from } -6 \text{ to } 6 \text{ with bracket at } 2 \end{array}$$

Instructor Note: Some students may be acquainted with the use of open and closed circles, ○ or ●, used as endpoints on their graphs.

The square bracket symbol, [, is used on the graph to indicate that the point $x = 2$ is included in the solution set. By convention, square brackets, either [or], are used to *include* a point on a number line. Parentheses, (or), are used to *exclude* a point on a number line.

The solution set of the inequality $x > 2$ includes the real numbers greater than 2 but not including 2. Therefore, a (symbol is used on the graph to indicate that $x = 2$ is not included.

$$x > 2 \qquad \begin{array}{c} \text{number line from } -6 \text{ to } 6 \text{ with parenthesis at } 2 \end{array}$$

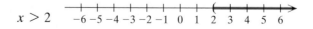

In Example 1, we demonstrate how to graph linear inequalities. To graph an inequality means that we graph its *solution set*. That is, we graph all of the values on the number line that make the inequality true.

Example 1 Graphing Linear Inequalities

Graph the solution sets.

a. $x > -1$ **b.** $c \le \dfrac{7}{3}$ **c.** $3 > y$

Solution:

a. $x > -1$

The solution set is the set of all real numbers strictly greater than -1. Therefore, we graph the region on the number line to the right of -1. Because $x = -1$ is not included in the solution set, we use the (symbol at $x = -1$.

b. $c \le \dfrac{7}{3}$ is equivalent to $c \le 2\frac{1}{3}$.

The solution set is the set of all real numbers less than or equal to $2\frac{1}{3}$. Therefore, graph the region on the number line to the left of and including $2\frac{1}{3}$. Use the symbol] to indicate that $c = 2\frac{1}{3}$ is included in the solution set.

c. $3 > y$ This inequality reads "3 is greater than y." This is equivalent to saying, "y is less than 3." The inequality $3 > y$ can also be written as $y < 3$.

$y < 3$

The solution set is the set of real numbers less than 3. Therefore, graph the region on the number line to the left of 3. Use the symbol) to denote that the endpoint, 3, is not included in the solution.

TIP: Some textbooks use a closed circle or an open circle (● or ○) rather than a bracket or parenthesis to denote inclusion or exclusion of a value on the real number line. For example, the solution sets for the inequalities $x > -1$ and $c \le \frac{7}{3}$ are graphed here.

$x > -1$

$c \le \dfrac{7}{3}$

A statement that involves more than one inequality is called a **compound inequality**. One type of compound inequality is used to indicate that one number is between two others. For example, the inequality $-2 < x < 5$ means that $-2 < x$ and $x < 5$. In words, this is easiest to understand if we read the variable first: x is greater than -2 and x is less than 5. The numbers satisfied by these two conditions are those between -2 and 5.

Skill Practice

Graph the solution sets.
1. $y < 0$
2. $x \ge -\dfrac{5}{4}$
3. $5 \ge a$

Classroom Examples: p. 188, Exercises 8 and 10

Instructor Note: Students sometimes confuse the meaning of $x > 1$ to think it means the set $\{2, 3, 4 \ldots\}$. They don't think about the real number line. They don't see that 1.00000001 is a solution.

Answers
1.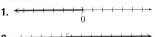
2.
3.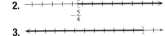

Skill Practice

Graph the solution set.

4. $0 \le y \le 8.5$

Classroom Example: p. 188, Exercise 6

Concept Connections

5. Translate the phrase below to set-builder notation.

The set of all b such that b is greater than or equal to -20.

6. Translate the phrase below to interval notation.

The set of all x such that x is less than -30.

Example 2 Graphing a Compound Inequality

Graph the solution set of the inequality: $-4.1 < y \le -1.7$

Solution:

$-4.1 < y \le -1.7$ means that

$-4.1 < y$ and $y \le -1.7$

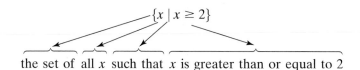

Shade the region of the number line greater than -4.1 and less than or equal to -1.7.

---

2. Set-Builder Notation and Interval Notation

Graphing the solution set to an inequality is one way to define the set. Two other methods are to use **set-builder notation** or **interval notation**.

Set-Builder Notation

The solution to the inequality $x \ge 2$ can be expressed in set-builder notation as follows:

$$\{x \mid x \ge 2\}$$

the set of all x such that x is greater than or equal to 2

Interval Notation

To understand interval notation, first think of a number line extending infinitely far to the right and infinitely far to the left. Sometimes we use the infinity symbol, ∞, or negative infinity symbol, $-\infty$, to label the far right and far left ends of the number line (Figure 2-11).

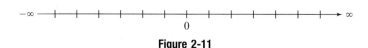

Figure 2-11

To express the solution set of an inequality in interval notation, sketch the graph first. Then use the endpoints to define the interval.

Inequality	Graph	Interval Notation
$x \ge 2$		$[2, \infty)$

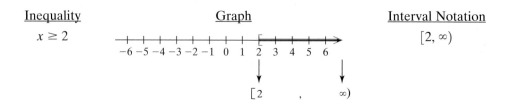

The graph of the solution set $x \ge 2$ begins at 2 and extends infinitely far to the right. The corresponding interval notation begins at 2 and extends to ∞. Notice that a square bracket $[$ is used at 2 for both the graph and the interval notation. A parenthesis is always used at ∞ and for $-\infty$, because there is no endpoint.

Answers

4.

5. $\{b \mid b \ge -20\}$

6. $(-\infty, -30)$

PROCEDURE Using Interval Notation

- The endpoints used in interval notation are always written from left to right. That is, the smaller number is written first, followed by a comma, followed by the larger number.
- A parenthesis, (or), indicates that an endpoint is excluded from the set.
- A square bracket, [or], indicates that an endpoint is included in the set.
- Parentheses, (and), are always used with $-\infty$ and ∞, respectively.

In Table 2-1, we present examples of eight different scenarios for interval notation and the corresponding graph.

Table 2-1

Interval Notation	Graph	Interval Notation	Graph
(a, ∞)	a	$[a, \infty)$	a
$(-\infty, a)$	a	$(-\infty, a]$	a
(a, b)	$a \quad b$	$[a, b]$	$a \quad b$
$(a, b]$	$a \quad b$	$[a, b)$	$a \quad b$

Example 3 Using Set-Builder Notation and Interval Notation

Complete the chart.

Set-Builder Notation	Graph	Interval Notation
	$-6\,-5\,-4\,-3\,-2\,-1\ \ 0\ \ 1\ \ 2\ \ 3\ \ 4\ \ 5\ \ 6$	
		$[-\frac{1}{2}, \infty)$
$\{y \mid -2 \le y < 4\}$		

Solution:

Set-Builder Notation	Graph	Interval Notation
$\{x \mid x < -3\}$	$-6\,-5\,-4\,-3\,-2\,-1\ \ 0\ \ 1\ \ 2\ \ 3\ \ 4\ \ 5\ \ 6$	$(-\infty, -3)$
$\{x \mid x \ge -\frac{1}{2}\}$	$-6\,-5\,-4\,-3\,-2\,-1\ \ 0\ \ 1\ \ 2\ \ 3\ \ 4\ \ 5\ \ 6$ $\quad -\frac{1}{2}$	$[-\frac{1}{2}, \infty)$
$\{y \mid -2 \le y < 4\}$	$-6\,-5\,-4\,-3\,-2\,-1\ \ 0\ \ 1\ \ 2\ \ 3\ \ 4\ \ 5\ \ 6$	$[-2, 4)$

3. Addition and Subtraction Properties of Inequality

The process to solve a linear inequality is very similar to the method used to solve linear equations. Recall that adding or subtracting the same quantity to both sides of an equation results in an equivalent equation. The addition and subtraction properties of inequality state that the same is true for an inequality.

PROPERTY Addition and Subtraction Properties of Inequality

Let a, b, and c represent real numbers.

1. *Addition Property of Inequality: If $a < b$,
                                             then $a + c < b + c$

2. *Subtraction Property of Inequality: If $a < b$,
                                             then $a - c < b - c$

*These properties may also be stated for $a \le b$, $a > b$, and $a \ge b$.

To illustrate the addition and subtraction properties of inequality, consider the inequality $5 > 3$. If we add or subtract a real number such as 4 to both sides, the left-hand side will still be greater than the right-hand side. (See Figure 2-12.)

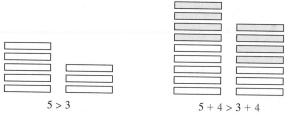

$5 > 3$ $5 + 4 > 3 + 4$

Figure 2-12

Classroom Example: p. 189, Exercise 40

Instructor Note: Students should be encouraged to check the solution set by replacing several values less than 1 into the original problem.

Instructor Note: Ask students whether 1 is a solution to the inequality.

Answer

10.

$(-\infty, -6)$

Example 4   Solving a Linear Inequality

Solve the inequality and graph the solution set. Express the solution set in set-builder notation and in interval notation.

$$-2p + 5 < -3p + 6$$

Solution:

$$-2p + 5 < -3p + 6$$

$-2p + 3p + 5 < -3p + 3p + 6$ Addition property of inequality (add $3p$ to both sides).

$p + 5 < 6$ Simplify.

$p + 5 - 5 < 6 - 5$ Subtraction property of inequality.

$p < 1$

Graph:

$\begin{array}{c}\xleftarrow{\hspace{2cm}}\\ -6\ -5\ -4\ -3\ -2\ -1\ \ 0\ \ 1\ \ 2\ \ 3\ \ 4\ \ 5\ \ 6\end{array}$

Set-builder notation: $\{p \mid p < 1\}$

Interval notation: $(-\infty, 1)$

TIP: The solution to an inequality gives a set of values that make the original inequality true. Therefore, you can test your final answer by using *test points*. That is, pick a value in the proposed solution set and verify that it makes the original inequality true. Furthermore, any test point picked outside the solution set should make the original inequality false. For example,

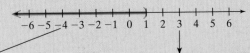

Pick $p = -4$ as an arbitrary test point within the proposed solution set.

$$-2p + 5 < -3p + 6$$
$$-2(-4) + 5 \overset{?}{<} -3(-4) + 6$$
$$8 + 5 \overset{?}{<} 12 + 6$$
$$13 < 18 \; ✔ \quad \text{True}$$

Pick $p = 3$ as an arbitrary test point outside the proposed solution set.

$$-2p + 5 < -3p + 6$$
$$-2(3) + 5 \overset{?}{<} -3(3) + 6$$
$$-6 + 5 \overset{?}{<} -9 + 6$$
$$-1 \overset{?}{<} -3 \quad \text{False}$$

4. Multiplication and Division Properties of Inequality

Multiplying both sides of an equation by the same quantity results in an equivalent equation. However, the same is not always true for an inequality. If you multiply or divide an inequality by a negative quantity, the direction of the inequality symbol must be reversed.

For example, consider multiplying or dividing the inequality, $4 < 5$ by -1.

$$\text{Multiply/Divide} \quad 4 < 5$$
$$\text{by } -1 \quad -4 > -5$$

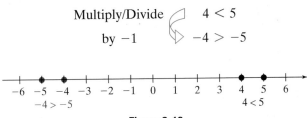

Figure 2-13

The number 4 lies to the left of 5 on the number line. However, -4 lies to the right of -5 (Figure 2-13). Changing the sign of two numbers changes their relative position on the number line. This is stated formally in the multiplication and division properties of inequality.

Instructor Note: Show that this works for other situations. Start with two negative numbers or one positive and one negative. Show that multiplying or dividing by -1 changes the direction of the inequality sign.

PROPERTY **Multiplication and Division Properties of Inequality**

Let a, b, and c represent real numbers.

*If c is positive and $a < b$, then $ac < bc$ and $\dfrac{a}{c} < \dfrac{b}{c}$

*If c is negative and $a < b$, then $ac > bc$ and $\dfrac{a}{c} > \dfrac{b}{c}$

The second statement indicates that if both sides of an inequality are multiplied or divided by a negative quantity, the inequality sign must be reversed.

*These properties may also be stated for $a \le b$, $a > b$, and $a \ge b$.

Classroom Example: p. 189,
Exercise 66

Skill Practice

Solve. Graph the solution set and express the solution in interval notation.

11. $-5p + 2 > 22$

Example 5   Solving a Linear Inequality

Solve the inequality $-5x - 3 \le 12$. Graph the solution set and write the answer in interval notation.

Solution:

$$-5x - 3 \le 12$$

$$-5x - 3 + 3 \le 12 + 3 \qquad \text{Add 3 to both sides.}$$

$$-5x \le 15$$

$$\frac{-5x}{-5} \ge \frac{15}{-5} \qquad \begin{array}{l}\text{Divide by } -5. \text{ Reverse the direction of the}\\ \text{inequality sign.}\end{array}$$

$$x \ge -3$$

Interval notation: $[-3, \infty)$

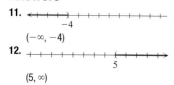

TIP: The inequality $-5x - 3 \le 12$ could have been solved by isolating x on the right-hand side of the inequality. This would create a positive coefficient on the variable term and eliminate the need to divide by a negative number.

$$-5x - 3 \le 12$$

$$-3 \le 5x + 12$$

$$-15 \le 5x \qquad \text{Notice that the coefficient of } x \text{ is positive.}$$

$$\frac{-15}{5} \le \frac{5x}{5} \qquad \begin{array}{l}\text{Do not reverse the inequality sign because we are}\\ \text{dividing by a positive number.}\end{array}$$

$$-3 \le x, \text{ or equivalently, } x \ge -3$$

Skill Practice

Solve. Graph the solution set and express the solution in interval notation.

12. $1.1x - 0.8 > 0.1x + 4.2$

Classroom Example: p. 190,
Exercise 82

Example 6   Solving a Linear Inequality

Solve the inequality. Graph the solution set and write the answer in interval notation.

$$1.4x + 4.5 < 0.2x - 0.3$$

Solution:

$$1.4x + 4.5 < 0.2x - 0.3$$

$$1.4x - 0.2x + 4.5 < 0.2x - 0.2x - 0.3 \qquad \text{Subtract } 0.2x \text{ from both sides.}$$

$$1.2x + 4.5 < -0.3 \qquad \text{Simplify.}$$

$$1.2x + 4.5 - 4.5 < -0.3 - 4.5 \qquad \text{Subtract 4.5 from both sides.}$$

$$1.2x < -4.8 \qquad \text{Simplify.}$$

$$\frac{1.2x}{1.2} < \frac{-4.8}{1.2} \qquad \begin{array}{l}\text{Divide by 1.2. The direction of the}\\ \text{inequality sign is } not \text{ reversed because}\\ \text{we divided by a positive number.}\end{array}$$

$$x < -4$$

Interval notation: $(-\infty, -4)$

Answers

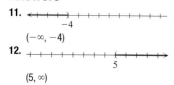

11.

$(-\infty, -4)$

12.

$(5, \infty)$

Example 7 Solving Linear Inequalities

Solve the inequality $-\frac{1}{4}k + \frac{1}{6} \le 2 + \frac{2}{3}k$. Graph the solution set and write the answer in interval notation.

Solution:

$$-\frac{1}{4}k + \frac{1}{6} \le 2 + \frac{2}{3}k$$

$$12\left(-\frac{1}{4}k + \frac{1}{6}\right) \le 12\left(2 + \frac{2}{3}k\right)$$

Multiply both sides by 12 to clear fractions. (Because we multiplied by a positive number, the inequality sign is not reversed.)

$$\frac{12}{1}\left(-\frac{1}{4}k\right) + \frac{12}{1}\left(\frac{1}{6}\right) \le 12(2) + \frac{12}{1}\left(\frac{2}{3}k\right)$$

Apply the distributive property.

$$-3k + 2 \le 24 + 8k$$

Simplify.

$$-3k - 8k + 2 \le 24 + 8k - 8k$$

Subtract $8k$ from both sides.

$$-11k + 2 \le 24$$

$$-11k + 2 - 2 \le 24 - 2$$

Subtract 2 from both sides.

$$-11k \le 22$$

$$\frac{-11k}{-11} \ge \frac{22}{-11}$$

Divide both sides by -11.
Reverse the inequality sign.

$$k \ge -2$$

Graph:

$$\xleftarrow{\hspace{0.3em}}\begin{array}{c}\;\;\;\;\;\;\;\;\,[\\ +\!+\!+\!+\!+\!+\!+\!+\!+\!+\!+\\ -4\;-3\;-2\;-1\;\;0\;\;1\;\;2\;\;3\;\;4\end{array}\xrightarrow{\hspace{0.3em}}$$

Interval notation: $[-2, \infty)$

Skill Practice

Solve. Graph the solution set and express the solution in interval notation.

13. $\frac{1}{5}t + 7 \le \frac{1}{2}t - 2$

Classroom Example: p. 190, Exercise 78

5. Solving Inequalities of the Form $a < x < b$

To solve a compound inequality of the form $a < x < b$ we can work with the inequality as a three-part inequality and isolate the variable, x, as demonstrated in Example 8.

Example 8 Solving a Compound Inequality
of the Form $a < x < b$

Solve the inequality: $-3 \le 2x + 1 < 7$. Graph the solution and write the answer in interval notation.

Solution:

To solve the compound inequality $-3 \le 2x + 1 < 7$ isolate the variable x in the middle. The operations performed on the middle portion of the inequality must also be performed on the left-hand side and right-hand side.

Skill Practice

Solve. Graph the solution set and express the solution in interval notation.

14. $-3 \le -5 + 2y < 11$

Classroom Example: p. 189, Exercise 58

Answers

13.

$[30, \infty)$

14.

$[1, 8)$

$$-3 \le 2x + 1 < 7$$

$$-3 - 1 \le 2x + 1 - 1 < 7 - 1 \qquad \text{Subtract 1 from all three parts of the inequality.}$$

$$-4 \le 2x < 6 \qquad \text{Simplify.}$$

$$\frac{-4}{2} \le \frac{2x}{2} < \frac{6}{2} \qquad \text{Divide by 2 in all three parts of the inequality.}$$

$$-2 \le x < 3$$

Graph:

Interval notation: $[-2, 3)$

6. Applications of Linear Inequalities

Table 2-2 provides several commonly used translations to express inequalities.

Table 2-2

English Phrase	Mathematical Inequality
a is less than b	$a < b$
a is greater than b a exceeds b	$a > b$
a is less than or equal to b a is at most b a is no more than b	$a \le b$
a is greater than or equal to b a is at least b a is no less than b	$a \ge b$

Skill Practice

Translate the English phrase into a mathematical inequality.

15. Bill needs a score of at least 92 on the final exam. Let x represent Bill's score.
16. Fewer than 19 cars are in the parking lot. Let c represent the number of cars.
17. The heights, h, of women who wear petite size clothing are typically between 58 in. and 63 in., inclusive.

Classroom Examples: p. 190, Exercises 96 and 98

Example 9   **Translating Expressions Involving Inequalities**

Translate the English phrases into mathematical inequalities.

a. Claude's annual salary, s, is no more than $40,000.

b. A citizen must be at least 18 years old to vote. (Let a represent a citizen's age.)

c. An amusement park ride has a height requirement between 48 in. and 70 in. (Let h represent height in inches.)

Solution:

a. $s \le 40,000$ Claude's annual salary, s, is no more than $40,000.

b. $a \ge 18$ A citizen must be at least 18 years old to vote.

c. $48 < h < 70$ An amusement park ride has a height requirement between 48 in. and 70 in.

Answers
15. $x \ge 92$
16. $c < 19$
17. $58 \le h \le 63$

Linear inequalities are found in a variety of applications. Example 10 can help you determine the minimum grade you need on an exam to get an A in your math course.

Example 10 **Solving an Application with Linear Inequalities**

To earn an A in a math class, Alsha must average at least 90 on all of her tests. Suppose Alsha has scored 79, 86, 93, 90, and 95 on her first five math tests. Determine the minimum score she needs on her sixth test to get an A in the class.

Solution:

Let x represent the score on the sixth exam.   Label the variable.

$$\left(\begin{array}{c}\text{Average of} \\ \text{all tests}\end{array}\right) \geq 90$$ Create a verbal model.

$$\frac{79 + 86 + 93 + 90 + 95 + x}{6} \geq 90$$ The average score is found by taking the sum of the test scores and dividing by the number of scores.

$$\frac{443 + x}{6} \geq 90$$ Simplify.

$$6\left(\frac{443 + x}{6}\right) \geq (90)6$$ Multiply both sides by 6 to clear fractions.

$$443 + x \geq 540$$ Solve the inequality.

$$x \geq 540 - 443$$ Subtract 443 from both sides.

$$x \geq 97$$ Interpret the results.

Alsha must score at least 97 on her sixth exam to receive an A in the course.

Section 2.8   Practice Exercises

Study Skills Exercises

1. Find the page numbers for the Chapter Review Exercises, the Chapter Test, and the Cumulative Review Exercises for this chapter.

   Chapter Review Exercises _____   Chapter Test _____

   Cumulative Review Exercises _____

   Compare these features and state the advantages of each.

2. Define the key terms:

   a. linear inequality in one variable   **b. compound inequality**

   c. set-builder notation   **d. interval notation**

Review Problems

3. Solve the equation. $3(x + 2) - (2x - 7) = -(5x - 1) - 2(x + 6)$ −3

4. Solve the equation. $6 - 8(x + 3) + 5x = 5x - (2x - 5) + 13$ −6

 Writing   Translating Expression    Geometry   Scientific Calculator   Video   NS&E

Objectives 1–2: Graphing Linear Inequalities; Set-Builder Notation and Interval Notation

For Exercises 5–10, graph each inequality and write the solution set in interval notation. **(See Examples 1–3.)**

Set-Builder Notation	Graph	Interval Notation
5. $\{x \mid x \geq 6\}$		$[6, \infty)$
6. $\left\{x \mid \dfrac{1}{2} < x \leq 4\right\}$		$\left(\dfrac{1}{2}, 4\right]$
7. $\{x \mid x \leq 2.1\}$		$(-\infty, 2.1]$
8. $\left\{x \mid x > \dfrac{7}{3}\right\}$		$\left(\dfrac{7}{3}, \infty\right)$
9. $\{x \mid -2 < x \leq 7\}$		$(-2, 7]$
10. $\{x \mid x < -5\}$		$(-\infty, -5)$

For Exercises 11–16, write each set in set-builder notation and in interval notation. **(See Examples 1–3.)**

Set-Builder Notation	Graph	Interval Notation
11. $\left\{x \mid x > \dfrac{3}{4}\right\}$		$\left(\dfrac{3}{4}, \infty\right)$
12. $\{x \mid x \leq -0.3\}$		$(-\infty, -0.3]$
13. $\{x \mid -1 < x < 8\}$		$(-1, 8)$
14. $\{x \mid x \geq 0\}$		$[0, \infty)$
15. $\{x \mid x \leq -14\}$		$(-\infty, -14]$
16. $\{x \mid 0 < x \leq 9\}$		$(0, 9]$

For Exercises 17–22, graph each set and write the set in set-builder notation. **(See Examples 1–3.)**

Set-Builder Notation	Graph	Interval Notation
17. $\{x \mid x \geq 18\}$		$[18, \infty)$
18. $\{x \mid -10 \leq x \leq -2\}$		$[-10, -2]$
19. $\{x \mid x < -0.6\}$		$(-\infty, -0.6)$
20. $\left\{x \mid x < \dfrac{5}{3}\right\}$		$\left(-\infty, \dfrac{5}{3}\right)$
21. $\{x \mid -3.5 \leq x < 7.1\}$		$[-3.5, 7.1)$
22. $\{x \mid x \geq -10\}$		$[-10, \infty)$

Objectives 3–4: Properties of Inequality

For Exercises 23–30, solve the equation in part (a). For part (b), solve the inequality and graph the solution set. **(See Examples 4–7.)**

23. a. $x + 3 = 6$ $x = 3$ **24. a.** $y - 6 = 12$ $y = 18$ **25. a.** $p - 4 = 9$ $p = 13$ **26. a.** $k + 8 = 10$ $k = 2$

b. $x + 3 > 6$ $x > 3$ **b.** $y - 6 \geq 12$ $y \geq 18$ **b.** $p - 4 \leq 9$ $p \leq 13$ **b.** $k + 8 < 10$ $k < 2$

27. a. $4c = -12$ $c = -3$ **28. a.** $5d = -35$ $d = -7$ **29. a.** $-10z = 15$ $z = -\dfrac{3}{2}$ **30. a.** $-2w = 14$ $w = -7$

    b. $4c < -12$ $c < -3$ **b.** $5d > -35$ $d > -7$ **b.** $-10z \le 15$ $z \ge -\dfrac{3}{2}$ **b.** $-2w < 14$ $w > -7$

Objective 5: Solving Inequalities of the Form $a < x < b$

For Exercises 31–36, graph the solution. **(See Example 8.)**

31. $-1 < y \le 4$

32. $2.5 \le t < 5.7$

33. $0 < x + 3 < 8$ $-3 < x < 5$

34. $-2 \le x - 4 \le 3$ $2 \le x \le 7$

35. $8 \le 4x \le 24$ $2 \le x \le 6$

36. $-9 < 3x < 12$ $-3 < x < 4$

Mixed Exercises

For Exercises 37–84, solve the inequality. Graph the solution set and write the set in interval notation.

37. $x + 5 \le 6$ $(-\infty, 1]$

38. $y - 7 < 6$ $(-\infty, 13)$

39. $3q - 7 > 2q + 3$ $(10, \infty)$

40. $5r + 4 \ge 4r - 1$ $[-5, \infty)$

41. $4 < 1 + z$ $(3, \infty)$

42. $3 > z - 6$ $(-\infty, 9)$

43. $2 \ge a - 6$ $(-\infty, 8]$

44. $7 \le b + 12$ $[-5, \infty)$

45. $3c > 6$ $(2, \infty)$

46. $4d \le 12$ $(-\infty, 3]$

47. $-3c > 6$ $(-\infty, -2)$

48. $-4d \le 12$ $[-3, \infty)$

49. $-h \le -14$ $[14, \infty)$

50. $-q > -7$ $(-\infty, 7)$

51. $12 \ge -\dfrac{x}{2}$ $[-24, \infty)$

52. $6 < -\dfrac{m}{3}$ $(-\infty, -18)$

53. $-2 \le p + 1 < 4$ $[-3, 3)$

54. $0 < k + 7 < 6$ $(-7, -1)$

55. $-3 < 6h - 3 < 12$ $\left(0, \dfrac{5}{2}\right)$

56. $-6 \le 4a - 2 \le 12$ $\left[-1, \dfrac{7}{2}\right]$

57. $5 < \dfrac{1}{2}x < 6$ $(10, 12)$

58. $-6 \le 3x \le 12$ $[-2, 4]$

59. $-5 \le 4x - 1 < 15$ $[-1, 4)$

60. $-2 < \dfrac{1}{3}x - 2 \le 2$ $(0, 12]$

61. $54 \le 0.6z$ $[90, \infty)$

62. $28 < -0.7w$ $(-\infty, -40)$

63. $-\dfrac{2}{3}y < 6$ $(-9, \infty)$

64. $\dfrac{3}{4}x \le -12$ $(-\infty, -16]$

65. $-2x - 4 \le 11$ $\left[-\dfrac{15}{2}, \infty\right)$

66. $-3x + 1 > 0$ $\left(-\infty, \dfrac{1}{3}\right)$

67. $-12 > 7x + 9$ $(-\infty, -3)$

68. $8 < 2x - 10$ $(9, \infty)$

69. $-7b - 3 \le 2b$ $\left[-\dfrac{1}{3}, \infty\right)$

70. $3t \geq 7t - 35$ $\left(-\infty, \frac{35}{4}\right]$

71. $4n + 2 < 6n + 8$ $(-3, \infty)$

72. $2w - 1 \leq 5w + 8$ $[-3, \infty)$

73. $8 - 6(x - 3) > -4x + 12$ $(-\infty, 7)$ **74.** $3 - 4(h - 2) > -5h + 6$ $(-5, \infty)$ **75.** $3(x + 1) - 2 \leq \frac{1}{2}(4x - 8)$ $(-\infty, -5]$

76. $8 - (2x - 5) \geq \frac{1}{3}(9x - 6)$ $(-\infty, 3]$ **77.** $\frac{7}{6}p + \frac{4}{3} \geq \frac{11}{6}p - \frac{7}{6}$ $\left(-\infty, \frac{15}{4}\right]$ **78.** $\frac{1}{3}w - \frac{1}{2} \leq \frac{5}{6}w + \frac{1}{2}$ $[-2, \infty)$

79. $\frac{y - 6}{3} > y + 4$ $(-\infty, -9)$ **80.** $\frac{5t + 7}{2} < t - 4$ $(-\infty, -5)$ **81.** $-1.2a - 0.4 < -0.4a + 2$ $(-3, \infty)$

82. $-0.4c + 1.2 > -2c - 0.4$ $(-1, \infty)$ **83.** $-2x + 5 \geq -x + 5$ $(-\infty, 0]$ **84.** $4x - 6 < 5x - 6$ $(0, \infty)$

For Exercises 85–88, determine whether the given number is a solution to the inequality.

85. $-2x + 5 < 4;$ $x = -2$ No

86. $-3y - 7 > 5;$ $y = 6$ No

87. $4(p + 7) - 1 > 2 + p;$ $p = 1$ Yes

88. $3 - k < 2(-1 + k);$ $k = 4$ Yes

Objective 6: Applications of Linear Inequalities

89. Let x represent a student's average in a math class. The grading scale is given here.

A	$93 \leq x \leq 100$
B+	$89 \leq x < 93$
B	$84 \leq x < 89$
C+	$80 \leq x < 84$
C	$75 \leq x < 80$
F	$0 \leq x < 75$

a. Write the range of scores corresponding to each letter grade in interval notation.
A $[93, 100]$; B+ $[89, 93)$; B $[84, 89)$; C+ $[80, 84)$; C $[75, 80)$; F $[0, 75)$
b. If Stephan's average is 84.01, what grade will he receive?
B
c. If Estella's average is 79.89, what grade will she receive?
C

90. Let x represent a student's average in a science class. The grading scale is given here.

A	$90 \leq x \leq 100$
B+	$86 \leq x < 90$
B	$80 \leq x < 86$
C+	$76 \leq x < 80$
C	$70 \leq x < 76$
D+	$66 \leq x < 70$
D	$60 \leq x < 66$
F	$0 \leq x < 60$

a. Write the range of scores corresponding to each letter grade in interval notation.
A $[90, 100]$; B+ $[86, 90)$; B $[80, 86)$; C+ $[76, 80)$; C $[70, 76)$; D+ $[66, 70)$; D $[60, 66)$; F $[0, 60)$
b. If Jacque's average is 89.99, what is her grade?
B+
c. If Marc's average is 66.01, what is his grade?
D+

For Exercises 91–100, translate the English phrase into a mathematical inequality. **(See Example 9.)**

91. The length of a fish, L, was at least 10 in.
$L \geq 10$

92. Tasha's average test score, t, exceeded 90.
$t > 90$

93. The wind speed, w, exceeded 75 mph. $w > 75$

94. The height of a cave, h, was no more than 2 ft.
$h \leq 2$

95. The temperature of the water in Blue Spring, t, is no more than 72°F. $t \leq 72$

96. The temperature on the tennis court, t, was no less than 100°F. $t \geq 100$

97. The length of the hike, L, was no less than 8 km. $L \geq 8$

98. The depth, d, of a certain pool was at most 10 ft. $d \leq 10$

 Writing   Translating Expression   Geometry   Scientific Calculator   Video   NSE

99. The snowfall, h, in Monroe County is between 2 inches and 5 inches. $2 < h < 5$

100. The cost, c, of carpeting a room is between $300 and $400. $300 < c < 400$

101. The average summer rainfall for Miami, Florida, for June, July, and August is 7.4 in. per month. If Miami receives 5.9 in. of rain in June and 6.1 in. in July, how much rain is required in August to exceed the 3-month summer average? **(See Example 10.)** More than 10.2 in. of rain is needed.

102. The average winter snowfall for Burlington, Vermont, for December, January, and February is 18.7 in. per month. If Burlington receives 22 in. of snow in December and 24 in. in January, how much snow is required in February to exceed the 3-month winter average? More than 10.1 in. of snow is needed.

103. An artist paints wooden birdhouses. She buys the birdhouses for $9 each. However, for large orders, the price per birdhouse is discounted by a percentage off the original price. Let x represent the number of birdhouses ordered. The corresponding discount is given in the table.

Size of Order	Discount
$x \le 49$	0%
$50 \le x \le 99$	5%
$100 \le x \le 199$	10%
$x \ge 200$	20%

a. If the artist places an order for 190 birdhouses, compute the total cost. $1539

b. Which costs more: 190 birdhouses or 200 birdhouses? Explain your answer. 200 birdhouses cost $1440. It is cheaper to purchase 200 birdhouses because the discount is greater.

104. A wholesaler sells T-shirts to a surf shop at $8 per shirt. However, for large orders, the price per shirt is discounted by a percentage off the original price. Let x represent the number of shirts ordered. The corresponding discount is given in the table.

Number of Shirts Ordered	Discount
$x \le 24$	0%
$25 \le x \le 49$	2%
$50 \le x \le 99$	4%
$100 \le x \le 149$	6%
$x \ge 150$	8%

a. If the surf shop orders 50 shirts, compute the total cost. $384

b. Which costs more: 148 shirts or 150 shirts? Explain your answer. It costs $1112.96 for 148 shirts and $1104.00 for 150 shirts. 150 shirts cost less than 148 shirts because the discount is greater.

105. Maggie sells lemonade at an art show. She has a fixed cost of $75 to cover the registration fee for the art show. In addition, her cost to produce each lemonade is $0.17. If x represents the number of lemonades, then the total cost to produce x lemonades is given by:

$$\text{Cost} = 75 + 0.17x$$

If Maggie sells each lemonade for $2, then her revenue (the amount she brings in) for selling x lemonades is given by:

$$\text{Revenue} = 2.00x$$

a. Write an inequality that expresses the number of lemonades, x, that Maggie must sell to make a profit. Profit is realized when the revenue is greater than the cost (Revenue > Cost). $2.00x > 75 + 0.17x$

b. Solve the inequality in part (a). $x \ge 41$; profit occurs when 41 or more lemonades are sold.

106. Two rental car companies rent subcompact cars at a discount. Company A rents for $14.95 per day plus 22 cents per mile. Company B rents for $18.95 a day plus 18 cents per mile. Let x represent the number of miles driven in one day.

The cost to rent a subcompact car for one day from Company A is: $\text{Cost}_A = 14.95 + 0.22x$

The cost to rent a subcompact car for one day from Company B is: $\text{Cost}_B = 18.95 + 0.18x$

a. Write an inequality that expresses the number of miles, x, for which the daily cost to rent from Company A is less than the daily cost to rent from Company B. $14.95 + 0.22x < 18.95 + 0.18x$

b. Solve the inequality in part (a). $x < 100$; Company A costs less than Company B if the mileage is less than 100 miles.

 Writing   ◄► Translating Expression   Geometry   Scientific Calculator   Video   NS E

Expanding Your Skills

For Exercises 107–112, solve the inequality. Graph the solution set and write the set in interval notation.

107. $3(x + 2) - (2x - 7) \leq (5x - 1) - 2(x + 6)$ $[13, \infty)$

108. $6 - 8(y + 3) + 5y > 5y - (2y - 5) + 13$ $(-\infty, -6)$

109. $-2 - \dfrac{w}{4} \leq \dfrac{1 + w}{3}$ $[-4, \infty)$

110. $\dfrac{z - 3}{4} - 1 > \dfrac{z}{2}$ $(-\infty, -7)$

111. $-0.703 < 0.122p - 2.472$ $(14.5, \infty)$

112. $3.88 - 1.335t \geq 5.66$ $(-\infty, -1.3]$

Group Activity

Computing Body Mass Index (BMI)

Materials: Calculator

Estimated Time: 10 minutes

Group Size: 2

Body mass index is a statistical measure of an individual's weight in relation to the person's height. It is computed by

$$\text{BMI} = \frac{703W}{h^2}$$ where W is a person's weight in *pounds*.
h is the person's height in *inches*.

The NIH categorizes body mass indices as follows:

1. Compute the body mass index for a person 5′4″ tall weighing 160 lb. Is this person's weight considered ideal?

27.5, No, the person is considered overweight.

Body Mass Index (BMI)	Weight Status
$18.5 \leq \text{BMI} \leq 24.9$	considered ideal
$25.0 \leq \text{BMI} \leq 29.9$	considered overweight
$\text{BMI} \geq 30.0$	considered obese

2. At the time that basketball legend Michael Jordan played for the Chicago Bulls, he was 210 lb and stood 6′6″ tall. What was Michael Jordan's body mass index?

24.3

3. For a fixed height, body mass index is a function of a person's weight only. For example, for a person 72 in. tall (6 ft), solve the following inequality to determine the person's ideal weight range.

136.4 lb $\leq W \leq$ 183.6 lb

$$18.5 \leq \frac{703W}{(72)^2} \leq 24.9$$

4. At the time that professional bodybuilder, Jay Cutler, won the Mr. Olympia contest he was 260 lb and stood 5′10″ tall.

a. What was Jay Cutler's body mass index?   37.3

b. As a body builder, Jay Cutler has an extraordinarily small percentage of body fat. Yet, according to the chart, would he be considered overweight or obese? Why do you think that the formula is not an accurate measurement of Mr. Cutler's weight status? Jay Cutler's BMI was 37.3 which placed him in the "obese" category. However, BMI is meant to be used as a simple means of classifying physically inactive individuals with an average body composition. The BMI formula does not account for other factors affecting a person's weight such as fitness level, muscle mass, bone structure, and gender.

 Writing    Translating Expression   Geometry   Scientific Calculator   Video   NS E

Chapter 2 Summary

Addition, Subtraction, Multiplication, and Division Properties of Equality

Key Concepts

An equation is an algebraic statement that indicates two expressions are equal. A **solution to an equation** is a value of the variable that makes the equation a true statement. The set of all solutions to an equation is the solution set of the equation.

A **linear equation in one variable** can be written in the form $ax + b = 0$, where $a \neq 0$.

Addition Property of Equality:

If $a = b$, then $a + c = b + c$

Subtraction Property of Equality:

If $a = b$, then $a - c = b - c$

Multiplication Property of Equality:

If $a = b$, then $ac = bc$

Division Property of Equality:

If $a = b$, then $\dfrac{a}{c} = \dfrac{b}{c}$ $(c \neq 0)$

Examples

Example 1

$2x + 1 = 9$ is an equation with solution 4.

Check: $\quad 2(4) + 1 \overset{?}{=} 9$

$$8 + 1 \overset{?}{=} 9$$

$$9 \overset{?}{=} 9 \quad \checkmark \quad \text{True}$$

Example 2

$$x - 5 = 12$$
$$x - 5 + 5 = 12 + 5$$
$$x = 17 \qquad \text{The solution is 17.}$$

Example 3

$$z + 1.44 = 2.33$$
$$z + 1.44 - 1.44 = 2.33 - 1.44$$
$$z = 0.89 \qquad \text{The solution is 0.89.}$$

Example 4

$$\frac{3}{4}x = 12$$
$$\frac{4}{3} \cdot \frac{3}{4}x = 12 \cdot \frac{4}{3}$$
$$x = 16 \qquad \text{The solution is 16.}$$

Example 5

$$16 = 8y$$
$$\frac{16}{8} = \frac{8y}{8}$$
$$2 = y \qquad \text{The solution is 2.}$$

Section 2.2 Solving Linear Equations

Key Concepts

Steps for Solving a Linear Equation in One Variable:

1. Simplify both sides of the equation.
   - Clear parentheses
   - Combine *like* terms
2. Use the addition or subtraction property of equality to collect the variable terms on one side of the equation.
3. Use the addition or subtraction property of equality to collect the constant terms on the other side of the equation.
4. Use the multiplication or division property of equality to make the coefficient of the variable term equal to 1.
5. Check your answer.

A **conditional equation** is true for some values of the variable but is false for other values.

An equation that has all real numbers as its solution set is an **identity**.

An equation that has no solution is a **contradiction**.

Examples

Example 1

$$5y + 7 = 3(y - 1) + 2$$

$5y + 7 = 3y - 3 + 2$ Clear parentheses.

$5y + 7 = 3y - 1$ Combine *like* terms.

$2y + 7 = -1$ Isolate the variable term.

$2y = -8$ Isolate the constant term.

$y = -4$ Divide both sides by 2.

Check:

$$5(-4) + 7 \stackrel{?}{=} 3[(-4) - 1] + 2$$

$$-20 + 7 \stackrel{?}{=} 3(-5) + 2$$

$$-13 \stackrel{?}{=} -15 + 2$$

The solution is -4. $-13 \stackrel{?}{=} -13$ ✔ True

Example 2

$x + 5 = 7$ is a conditional equation because it is true only on the condition that $x = 2$.

Example 3

$$x + 4 = 2(x + 2) - x$$

$$x + 4 = 2x + 4 - x$$

$$x + 4 = x + 4$$

$4 = 4$ is an identity

The solution is all real numbers.

Example 4

$$y - 5 = 2(y + 3) - y$$

$$y - 5 = 2y + 6 - y$$

$$y - 5 = y + 6$$

$-5 = 6$ is a contradiction

There is no solution.

Section 2.3 Linear Equations: Clearing Fractions and Decimals

Key Concepts

Steps for Solving a Linear Equation in One Variable:

1. Simplify both sides of the equation.
   - Clear parentheses
   - Consider clearing fractions or decimals (if any are present) by multiplying both sides of the equation by a common denominator of all terms
   - Combine *like* terms
2. Use the addition or subtraction property of equality to collect the variable terms on one side of the equation.
3. Use the addition or subtraction property of equality to collect the constant terms on the other side of the equation.
4. Use the multiplication or division property of equality to make the coefficient of the variable term equal to 1.
5. Check your answer.

Examples

Example 1

$$\frac{1}{2}x - 2 - \frac{3}{4}x = \frac{7}{4}$$

$$\frac{4}{1}\left(\frac{1}{2}x - 2 - \frac{3}{4}x\right) = \frac{4}{1}\left(\frac{7}{4}\right) \quad \text{Multiply by the LCD.}$$

$$2x - 8 - 3x = 7 \quad \text{Apply distributive property.}$$

$$-x - 8 = 7 \quad \text{Combine } \textit{like} \text{ terms.}$$

$$-x = 15 \quad \text{Add 8 to both sides.}$$

$$x = -15 \quad \text{Divide by } -1.$$

The solution -15 checks in the original equation.

Example 2

$$-1.2x - 5.1 = 16.5$$

$$10(-1.2x - 5.1) = 10(16.5) \quad \text{Multiply both sides by 10.}$$

$$-12x - 51 = 165$$

$$-12x = 216$$

$$\frac{-12x}{-12} = \frac{216}{-12}$$

$$x = -18$$

The solution is -18.

Section 2.4 Applications of Linear Equations: Introduction to Problem Solving

Key Concepts

Problem-Solving Steps for Word Problems:

1. Read the problem carefully.
2. Assign labels to unknown quantities.
3. Develop a verbal model.
4. Write a mathematical equation.
5. Solve the equation.
6. Interpret the results and write the answer in words.

Examples

Example 1

The perimeter of a triangle is 54 m. The lengths of the sides are represented by three consecutive even integers. Find the lengths of the three sides.

1. Read the problem.

2. Let x represent one side, $x + 2$ represent the second side, and $x + 4$ represent the third side.

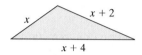

3. (First side) + (second side) + (third side) = perimeter

4. $x + (x + 2) + (x + 4) = 54$

5. $3x + 6 = 54$
   $3x = 48$
   $x = 16$

6. $x = 16$ represents the length of the shortest side. The lengths of the other sides are given by $x + 2 = 18$ and $x + 4 = 20$.

   The lengths of the three sides are 16 m, 18 m, and 20 m.

Section 2.5 Applications Involving Percents

Key Concepts

The following formula will help you solve basic percent problems.

$$\text{Amount} = (\text{percent})(\text{base})$$

One common use of percents is in computing **sales tax**.

Another use of percent is in computing **simple interest** using the formula:

$$\begin{pmatrix} \text{Simple} \\ \text{interest} \end{pmatrix} = (\text{principal}) \begin{pmatrix} \text{annual} \\ \text{interest} \\ \text{rate} \end{pmatrix} \begin{pmatrix} \text{time in} \\ \text{years} \end{pmatrix}$$

or $I = Prt$.

Examples

Example 1

A dinette set costs $1260.00 after a 5% sales tax is included. What was the price before tax?

$$\begin{pmatrix} \text{Price} \\ \text{before tax} \end{pmatrix} + (\text{tax}) = \begin{pmatrix} \text{total} \\ \text{price} \end{pmatrix}$$

$$x \quad\quad + 0.05x = 1260$$
$$1.05x = 1260$$
$$x = 1200$$

The dinette set cost $1200 before tax.

Example 2

John Li invests $5400 at 2.5% simple interest. How much interest does he earn after 5 years?

$$I = Prt$$
$$I = (\$5400)(0.025)(5)$$
$$I = \$675$$

Section 2.6 Formulas and Applications of Geometry

Key Concepts

A **literal equation** is an equation that has more than one variable. Often such an equation can be manipulated to solve for different variables.

Formulas from Section R.4 can be used in applications involving geometry.

Examples

Example 1

$$P = 2a + b, \text{ solve for } a.$$
$$P - b = 2a + b - b$$
$$P - b = 2a$$
$$\frac{P - b}{2} = \frac{2a}{2}$$
$$\frac{P - b}{2} = a \quad \text{or} \quad a = \frac{P - b}{2}$$

Example 2

Find the length of a side of a square whose perimeter is 28 ft.

Use the formula $P = 4s$. Substitute 28 for P and solve:

$$P = 4s$$
$$28 = 4s$$
$$7 = s$$

The length of a side of the square is 7 ft.

Section 2.7   Mixture Applications and Uniform Motion

Key Concepts

Example 1 illustrates a mixture problem.

Example 1

How much 80% disinfectant solution should be mixed with 8 L of a 30% disinfectant solution to make a 40% solution?

	80% Solution	30% Solution	40% Solution
Amount of Solution	x	8	$x + 8$
Amount of Pure Disinfectant	$0.80x$	$0.30(8)$	$0.40(x + 8)$

$0.80x + 0.30(8) = 0.40(x + 8)$

$\quad 0.80x + 2.4 = 0.40x + 3.2$

$\quad 0.40x + 2.4 = 3.2 \qquad$ Subtract $0.40x$.

$\qquad\quad 0.40x = 0.80 \qquad$ Subtract 2.4.

$\qquad\qquad\;\; x = 2 \qquad\quad$ Divide by 0.40.

2 L of 80% solution is needed.

Examples

Example 2 illustrates a uniform motion problem.

Example 2

Jack and Diane participate in a bicycle race. Jack rides the first half of the race in 1.5 hr. Diane rides the second half at a rate 5 mph slower than Jack and completes her portion in 2 hr. How fast does each person ride?

	Distance	Rate	Time
Jack	$1.5x$	x	1.5
Diane	$2(x - 5)$	$x - 5$	2

$$\left(\begin{array}{c}\text{Distance}\\\text{Jack rides}\end{array}\right) = \left(\begin{array}{c}\text{Distance}\\\text{Diane rides}\end{array}\right)$$

$\quad 1.5x \;\; = \;\; 2(x - 5)$

$\qquad 1.5x = 2x - 10$

$\;\; -0.5x = -10 \qquad$ Subtract $2x$.

$\qquad\quad x = 20 \qquad$ Divide by -0.5.

Jack's speed is x. Jack rides 20 mph. Diane's speed is $x - 5$, which is 15 mph.

Section 2.8   Linear Inequalities

Key Concepts

A **linear inequality in one variable**, x, is any relationship in the form: $ax + b < 0$, $ax + b > 0$, $ax + b \leq 0$, or $ax + b \geq 0$, where $a \neq 0$.

The solution set to an inequality can be expressed as a graph or in **set-builder notation** or in **interval notation**.

When graphing an inequality or when writing interval notation, a parenthesis, (or), is used to denote that an endpoint is *not included* in a solution set. A square bracket, [or], is used to show that an endpoint *is included* in a solution set. Parenthesis (or) are always used with $-\infty$ and ∞, respectively.

The inequality $a < x < b$ is used to show that x is greater than a and less than b. That is, x is *between* a and b.

Multiplying or dividing an inequality by a negative quantity requires the direction of the inequality sign to be reversed.

Examples

Example 1

$\qquad -2x + 6 \geq 14$

$-2x + 6 - 6 \geq 14 - 6 \qquad$ Subtract 6.

$\qquad\quad -2x \geq 8 \qquad\qquad$ Simplify.

$\qquad\quad \dfrac{-2x}{-2} \leq \dfrac{8}{-2} \qquad\quad$ Divide by -2. Reverse the inequality sign.

$\qquad\qquad x \leq -4$

Set-builder notation: $\{x \,|\, x \leq -4\}$

Graph: ←———————|————→
$\qquad\qquad\qquad\;\; -4$

Interval notation: $(-\infty, -4]$

Chapter 2 Review Exercises

Section 2.1

1. Label the following as either an expression or an equation:

   a. $3x + y = 10$
   Equation
   b. $9x + 10y - 2xy$
   Expression
   c. $4(x + 3) = 12$
   Equation
   d. $-5x = 7$
   Equation

2. Explain how to determine whether an equation is linear in one variable.   A linear equation can be written in the form $ax + b = 0, \quad a \neq 0$.

3. Identify which equations are linear.

   a. $4x^2 + 8 = -10$
   Nonlinear
   b. $x + 18 = 72$
   Linear
   c. $-3 + 2y^2 = 0$
   Nonlinear
   d. $-4p - 5 = 6p$
   Linear

4. For the equation, $4y + 9 = -3$, determine if the given numbers are solutions.

   a. $y = 3$ No
   b. $y = 0$ No
   c. $y = -3$ Yes
   d. $y = -2$ No

For Exercises 5–12, solve the equation using the addition property, subtraction property, multiplication property, or division property of equality.

5. $a + 6 = -2$ -8
6. $6 = z - 9$ 15

7. $-\dfrac{3}{4} + k = \dfrac{9}{2}$ $\dfrac{21}{4}$
8. $0.1r = 7$ 70

9. $-5x = 21$ $-\dfrac{21}{5}$
10. $\dfrac{t}{3} = -20$ -60

11. $-\dfrac{2}{5}k = \dfrac{4}{7}$ $-\dfrac{10}{7}$
12. $-m = -27$ 27

13. The quotient of a number and negative six is equal to negative ten. Find the number.
    The number is 60.

14. The difference of a number and $-\frac{1}{8}$ is $\frac{5}{12}$. Find the number.   The number is $\dfrac{7}{24}$.

15. Four subtracted from a number is negative twelve. Find the number.
    The number is -8.

16. Six subtracted from a number is negative eight. Find the number.
    The number is -2.

Section 2.2

For Exercises 17–28, solve the equation.

17. $4d + 2 = 6$ 1
18. $5c - 6 = -9$ $-\dfrac{3}{5}$

19. $-7c = -3c - 9$ $\dfrac{9}{4}$
20. $-28 = 5w + 2$ -6

21. $\dfrac{b}{3} + 1 = 0$ -3
22. $\dfrac{2}{3}h - 5 = 7$ 18

23. $-3p + 7 = 5p + 1$ $\dfrac{3}{4}$
24. $4t - 6 = -12t + 16$ $\dfrac{11}{8}$

25. $4a - 9 = 3(a - 3)$ 0
26. $3(2c + 5) = -2(c - 8)$ $\dfrac{1}{8}$

27. $7b + 3(b - 1) + 16 = 2(b + 8)$ $\dfrac{3}{8}$

28. $2 + (17 - x) + 2(x - 1) = 4(x + 2) - 8$ $\dfrac{17}{3}$

29. Explain the difference between an equation that is a contradiction and an equation that is an identity.   A contradiction has no solution and an identity is true for all real numbers.

30. Label each equation as a conditional equation, a contradiction, or an identity.

    a. $x + 3 = 3 + x$
    Identity
    b. $3x - 19 = 2x + 1$
    Conditional equation
    c. $5x + 6 = 5x - 28$
    Contradiction
    d. $2x - 8 = 2(x - 4)$
    Identity
    e. $-8x - 9 = -8(x - 9)$
    Contradiction

Section 2.3

For Exercises 31–48, solve the equation.

31. $\dfrac{x}{8} - \dfrac{1}{4} = \dfrac{1}{2}$ 6
32. $\dfrac{y}{15} - \dfrac{2}{3} = \dfrac{4}{5}$ 22

33. $\dfrac{x + 5}{2} - \dfrac{2x + 10}{9} = 5$ 13

34. $\dfrac{x - 6}{3} - \dfrac{2x + 8}{2} = 12$ -27

35. $\dfrac{1}{10}p - 3 = \dfrac{2}{5}p$ -10
36. $\dfrac{1}{4}y - \dfrac{3}{4} = \dfrac{1}{2}y + 1$ -7

37. $-\dfrac{1}{4}(2 - 3t) = \dfrac{3}{4}$ $\dfrac{5}{3}$
38. $\dfrac{2}{7}(w + 4) = \dfrac{1}{2}$ $-\dfrac{9}{4}$

39. $17.3 - 2.7q = 10.55$ 2.5

40. $4.9z + 4.6 = 3.2z - 2.2$ -4

41. $5.74a + 9.28 = 2.24a - 5.42$ -4.2

42. $62.84t - 123.66 = 4(2.36 + 2.4t)$ 2.5

43. $0.05x + 0.10(24 - x) = 0.75(24)$ -312

44. $0.20(x + 4) + 0.65x = 0.20(854)$ 200

45. $100 - (t - 6) = -(t - 1)$ No solution

46. $3 - (x + 4) + 5 = 3x + 10 - 4x$ No solution

47. $5t - (2t + 14) = 3t - 14$ All real numbers

48. $9 - 6(2z + 1) = -3(4z - 1)$ All real numbers

Section 2.4

49. Twelve added to the sum of a number and two is forty-four. Find the number.   The number is 30.

50. Twenty added to the sum of a number and six is thirty-seven. Find the number.   The number is 11.

51. Three times a number is the same as the difference of twice the number and seven. Find the number.   The number is −7.

52. Eight less than five times a number is forty-eight less than the number. Find the number.
The number is −10.

53. Three times the largest of three consecutive even integers is 76 more than the sum of the other two integers. Find the integers.
The integers are 66, 68, and 70.

54. Ten times the smallest of three consecutive integers is 213 more than the sum of the other two integers. Find the integers.
The integers are 27, 28, and 29.

55. The perimeter of a triangle is 78 in. The lengths of the sides are represented by three consecutive integers. Find the lengths of the sides of the triangle.   The sides are 25 in., 26 in., and 27 in.

56. The perimeter of a pentagon (a five-sided polygon) is 190 cm. The five sides are represented by consecutive integers. Find the lengths of the sides.   The sides are 36 cm, 37 cm, 38 cm, 39 cm, and 40 cm.

57. The minimum salary for a major league baseball player in 1985 was $60,000. This was twice the minimum salary in 1980. What was the minimum salary in 1980?   The minimum salary was $30,000 in 1980.

58. The state of Indiana has approximately 2.1 million more people than Kentucky. Together their population totals 10.3 million. Approximately how many people are in each state?   Indiana has 6.2 million people and Kentucky has 4.1 million.

Section 2.5

For Exercises 59–68, solve the problems involving percents.

59. What is 35% of 68?   23.8

60. What is 4% of 720?   28.8

61. 53.5 is what percent of 428?   12.5%

62. 68.4 is what percent of 72?   95%

63. 24 is 15% of what number?   160

64. 8.75 is 0.5% of what number?   1750

65. A novel is discounted 30%. The sale price is $20.65. What was the original price?   The novel originally cost $29.50.

66. A couple spent a total of $50.40 for dinner. This included a 20% tip and 6% sales tax on the price of the meal. What was the price of the dinner before tax and tip?   The dinner was $40 before tax and tip.

67. Anna Tsao invested $3000 in an account paying 8% simple interest.

    a. How much interest will she earn in $3\frac{1}{2}$ years?
    $840

    b. What will her balance be at that time?
    $3840

68. Eduardo invested money in an account earning 4% simple interest. At the end of 5 years, he had a total of $14,400. How much money did he originally invest? $12,000

Section 2.6

For Exercises 69–76, solve for the indicated variable.

69. $C = K - 273$ for K $K = C + 273$

70. $K = C + 273$ for C $C = K - 273$

71. $P = 4s$ for s $s = \dfrac{P}{4}$

72. $P = 3s$ for s $s = \dfrac{P}{3}$

73. $y = mx + b$ for x $x = \dfrac{y - b}{m}$

74. $a + bx = c$ for x $x = \dfrac{c - a}{b}$

75. $2x + 5y = -2$ for y $y = \dfrac{-2x - 2}{5}$

76. $4(a + b) = Q$ for b $b = \dfrac{Q - 4a}{4}$ or $b = \dfrac{Q}{4} - a$

For Exercises 77–83, use the appropriate geometry formula to solve the problem.

77. The volume of a cone is given by the formula $V = \frac{1}{3}\pi r^2 h$.

    a. Solve the formula for h. $h = \frac{3V}{\pi r^2}$

    b. Find the height of a right circular cone whose volume is 47.8 in.³ and whose radius is 3 in. Round to the nearest tenth of an inch.
The height is 5.1 in.

78. Find the height of a parallelogram whose area is 42 m² and whose base is 6 m.   The height is 7 m.

79. The smallest angle of a triangle is 2° more than $\frac{1}{4}$ of the largest angle. The middle angle is 2° less than the largest angle. Find the measure of each angle.   The angles are 22°, 78°, and 80°.

80. One angle is 6° less than twice a second angle. If the two angles are complementary, what are their measures?   The angles are 32° and 58°.

81. A rectangular window has width 1 ft less than its length. The perimeter is 18 ft. Find the length and the width of the window.
The length is 5 ft, and the width is 4 ft.

82. Find the measure of the vertical angles by first solving for x. $x = 20$. The angle measure is 65°.

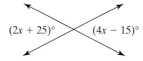

$(2x + 25)°$ $(4x - 15)°$

83. Find the measure of angle y.
The measure of angle y is 53°.

$37°$

y

Section 2.7

84. In stormy conditions, a delivery truck can travel a route in 14 hr. In good weather, the same trip can be made in 10.5 hr because the truck travels 15 km/hr faster. Find the speed of the truck in stormy weather and the speed in good weather.
The truck travels 45 km/hr in bad weather and 60 km/hr in good weather.

85. Winston and Gus ride their bicycles in a relay. Each rider rides the same distance. Winston rides 3 mph faster than Gus and finishes the course in 2.5 hr. Gus finishes in 3 hr. How fast does each person ride?   Gus rides 15 mph and Winston rides 18 mph.

86. Two cars leave a rest stop on Interstate I-10 at the same time. One heads east and the other heads west. One car travels 55 mph and the other 62 mph. How long will it take for them to be 327.6 mi apart?   The cars will be 327.6 mi apart after 2.8 hr (2 hr and 48 min).

87. Two hikers begin at the same time at opposite ends of a 9-mi trail and walk toward each other. One hiker walks 2.5 mph and the other walks 1.5 mph. How long will it be before they meet?
They meet in 2.25 hr (2 hr and 15 min).

88. How much ground beef with 24% fat should be mixed with 8 lb of ground sirloin that is 6% fat to make a mixture that is 9.6% fat?   2 lb of 24% fat content beef is needed.

89. A soldering compound with 40% lead (the rest is tin) must be combined with 80 lb of solder that is 75% lead to make a compound that is 68% lead? How much solder with 40% lead should be used?   20 lb of the 40% solder should be used.

Section 2.8

For Exercises 90–92, graph the inequalities and write the set in interval notation.

90. $\{x \mid x > -2\}$ $(-2, \infty)$ 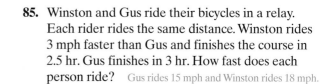

91. $\left\{x \mid x \leq \frac{1}{2}\right\}$ $\left(-\infty, \frac{1}{2}\right]$

92. $\{x \mid -1 < x \leq 4\}$ $(-1, 4]$

93. A landscaper buys potted geraniums from a nursery at a price of $5 per plant. However, for large orders, the price per plant is discounted by a percentage off the original price. Let x represent the number of potted plants ordered. The corresponding discount is given in the following table.

Number of Plants	Discount
$x \leq 99$	0%
$100 \leq x \leq 199$	2%
$200 \leq x \leq 299$	4%
$x \geq 300$	6%

a. Find the cost to purchase 130 plants. $637

b. Which costs more, 300 plants or 295 plants? Explain your answer.   300 plants cost $1410, and 295 plants cost $1416. 300 plants cost less.

For Exercises 94–103, solve the inequality. Graph the solution set and express the answer in interval notation.

94. $c + 6 < 23$ $(-\infty, 17)$

95. $3w - 4 > -5$ $\left(-\dfrac{1}{3}, \infty\right)$

96. $-2x - 7 \geq 5$ $(-\infty, -6]$

97. $5(y + 2) \leq -4$ $\left(-\infty, -\dfrac{14}{5}\right]$

98. $-\dfrac{3}{7}a \leq -21$ $[49, \infty)$

99. $1.3 > 0.4t - 12.5$ $(-\infty, 34.5)$

100. $4k + 23 < 7k - 31$ $(18, \infty)$

101. $\dfrac{6}{5}h - \dfrac{1}{5} \leq \dfrac{3}{10} + h$ $\left(-\infty, \dfrac{5}{2}\right]$

102. $-6 < 2b \leq 14$ $(-3, 7]$

103. $-2 \leq z + 4 \leq 9$ $[-6, 5]$

104. The summer average rainfall for Bermuda for June, July, and August is 5.3 in. per month. If Bermuda receives 6.3 in. of rain in June and 7.1 in. in July, how much rain is required in August to exceed the 3-month summer average?   More than 2.5 in. is required.

105. Reggie sells hot dogs at a ballpark. He has a fixed cost of $33 to use the concession stand at the park. In addition, the cost for each hot dog is $0.40. If x represents the number of hot dogs sold, then the total cost is given by

$$\text{Cost} = 33 + 0.40x$$

If Reggie sells each hot dog for $1.50, then his revenue (the amount he brings in) for selling x hot dogs is given by

$$\text{Revenue} = 1.50x$$

a. Write an inequality that expresses the number of hot dogs, x, that Reggie must sell to make a profit. Profit is realized when the revenue is greater than the cost (revenue > cost). $1.50x > 33 + 0.4x$

b. Solve the inequality in part (a). $x > 30$; a profit is realized if more than 30 hot dogs are sold.

Chapter 2   Test

1. Which of the equations have $x = -3$ as a solution?

a. $4x + 1 = 10$

b. $6(x - 1) = x - 21$

c. $5x - 2 = 2x + 1$

d. $\dfrac{1}{3}x + 1 = 0$ b, d

2. a. Simplify: $3x - 1 + 2x + 8$ $5x + 7$

  b. Solve: $3x - 1 = 2x + 8$ 9

For Exercises 3–13, solve the equation.

3. $t + 3 = -13$ -16

4. $8 = p - 4$ 12

5. $\dfrac{t}{8} = -\dfrac{2}{9}$ $-\dfrac{16}{9}$

6. $-3x + 5 = -2$ $\dfrac{7}{3}$

7. $2(p - 4) = p + 7$ 15

8. $2 + d = 2 - 3(d - 5) - 2$ $\dfrac{13}{4}$

9. $\dfrac{3}{7} + \dfrac{2}{5}x = -\dfrac{1}{5}x + 1\dfrac{20}{21}$

10. $3h + 1 = 3(h + 1)$ No solution

11. $\dfrac{3x + 1}{2} - \dfrac{4x - 3}{3} = 1$ -3

12. $0.5c - 1.9 = 2.8 + 0.6c$ -47

13. $-5(x + 2) + 8x = -2 + 3x - 8$ All real numbers

14. Solve the equation for y: $3x + y = -4$ $y = -3x - 4$

15. Solve $C = 2\pi r$ for r. $r = \dfrac{C}{2\pi}$

16. 13% of what is 11.7?   90

17. One number is four plus one-half of another. The sum of the numbers is 31. Find the numbers.   The numbers are 18 and 13.

 Writing   Translating Expression   Geometry   Scientific Calculator   Video   NS&E

18. The perimeter of a pentagon (a five-sided polygon) is 315 in. The five sides are represented by consecutive integers. Find the measures of the sides. The sides are 61 in., 62 in., 63 in., 64 in., and 65 in.

19. The total bill for a pair of basketball shoes (including sales tax) is $87.74. If the tax rate is 7%, find the cost of the shoes before tax. The cost was $82.00.

20. A couple purchased two hockey tickets and two basketball tickets for $153.92. A hockey ticket cost $4.32 more than a basketball ticket. What were the prices of the individual tickets? Each basketball ticket was $36.32, and each hockey ticket was $40.64.

21. Clarita borrowed money at a 6% simple interest rate. If she paid back a total of $8000 at the end of 10 yr, how much did she originally borrow? Clarita originally borrowed $5000.

22. The length of a soccer field for international matches is 40 m less than twice its width. If the perimeter is 370 m, what are the dimensions of the field? The field is 110 m long and 75 m wide.

23. Given the triangle, find the measures of each angle by first solving for y. $y = 30$; The measures of the angles are 30°, 39°, and 111°.

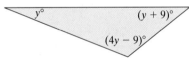

24. Paula mixes macadamia nuts that cost $9.00 per pound with 50 lb of peanuts that cost $5.00 per pound. How many pounds of macadamia nuts should she mix to make a nut mixture that costs $6.50 per pound? Paula needs 30 lb of macadamia nuts.

25. Two families leave their homes at the same time to meet for lunch. The families live 210 mi apart, and one family drives 5 mph slower than the other. If it takes them 2 hr to meet at a point between their homes, how fast does each family travel? One family travels 55 mph and the other travels 50 mph.

26. Two angles are complementary. One angle is 26° more than the other angle. What are the measures of the angles? The measures of the angles are 32° and 58°.

27. Graph the inequalities and write the sets in interval notation.

    a. $\{x \mid x < 0\}$ $(-\infty, 0)$

    b. $\{x \mid -2 \leq x < 5\}$ $[-2, 5)$

For Exercises 28–31, solve the inequality. Graph the solution and write the solution set in interval notation.

28. $5x + 14 > -2x$ $(-2, \infty)$

29. $2(3 - x) \geq 14$ $(-\infty, -4]$

30. $3(2y - 4) + 1 > 2(2y - 3) - 8$ $\left(-\dfrac{3}{2}, \infty\right)$

31. $-13 \leq 3p + 2 \leq 5$ $[-5, 1]$

32. The average winter snowfall for Syracuse, New York, for December, January, and February is 27.5 in. per month. If Syracuse receives 24 in. of snow in December and 32 in. in January, how much snow is required in February to exceed the 3-month average? More than 26.5 in. is required.

Chapters 1–2   Cumulative Review Exercises

For Exercises 1–5, perform the indicated operations.

1. $\left| -\dfrac{1}{5} + \dfrac{7}{10} \right|$ $\dfrac{1}{2}$

2. $5 - 2[3 - (4 - 7)]$ -7

3. $-\dfrac{2}{3} + \left(\dfrac{1}{2}\right)^2$ $-\dfrac{5}{12}$

4. $-3^2 + (-5)^2$ 16

5. $\sqrt{5 - (-20)} - 3^2$ 4

For Exercises 6–7, translate the mathematical expressions and simplify the results.

6. The square root of the difference of five squared and nine $\sqrt{5^2 - 9}$; 4

7. The sum of -14 and 12 $-14 + 12$; -2

8. List the terms of the expression: $-7x^2y + 4xy - 6$ $-7x^2y, 4xy, -6$

9. Simplify: $-4[2x - 3(x + 4)] + 5(x - 7)$ $9x + 13$

For Exercises 10–15, solve the equations.

10. $8t - 8 = 24$ 4

11. $-2.5x - 5.2 = 12.8$ -7.2

12. $-5(p - 3) + 2p = 3(5 - p)$ All real numbers

13. $\dfrac{x + 3}{5} - \dfrac{x + 2}{2} = 2$ -8

14. $\dfrac{2}{9}x - \dfrac{1}{3} = x + \dfrac{1}{9}$ $-\dfrac{4}{7}$

15. $-0.6w = 48$ -80

16. The sum of two consecutive odd integers is 156. Find the integers.   The numbers are 77 and 79.

17. The total bill for a man's three-piece suit (including sales tax) is $374.50. If the tax rate is 7%, find the cost of the suit before tax.
The cost before tax was $350.00.

18. The area of a triangle is 41 cm². Find the height of the triangle if the base is 12 cm.

The height is $\frac{41}{6}$ cm or $6\frac{5}{6}$ cm.

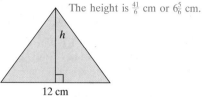

For Exercises 19–20, solve the inequality. Graph the solution set on a number line and express the solution in interval notation.

19. $-3x - 3(x + 1) < 9$ $(-2, \infty)$

20. $-6 \le 2x - 4 \le 14$ $[-1, 9]$

Graphing Linear Equations in Two Variables

3

CHAPTER OUTLINE

3.1 Rectangular Coordinate System 206

3.2 Linear Equations in Two Variables 215

3.3 Slope of a Line and Rate of Change 230

3.4 Slope-Intercept Form of a Line 243

Problem Recognition Exercises: Linear Equations in Two Variables 252

3.5 Point-Slope Formula 253

3.6 Applications of Linear Equations and Modeling 261

3.7 Introduction to Functions 268

Group Activity: Modeling a Linear Equation 279

Chapter 3

In this chapter, we study graphing and focus on the graphs of lines. As you work through this chapter, make note of the key terms. Then try to find them in this word search puzzle.

origin
axis
line
parallel
intercept

slope
point
quadrant
ordered pair

Section 3.1   Rectangular Coordinate System

Objectives

1. **Interpreting Graphs**
2. **Plotting Points in a Rectangular Coordinate System**
3. **Applications of Plotting and Identifying Points**

1. Interpreting Graphs

Mathematics is a powerful tool used by scientists and has directly contributed to the highly technical world in which we live. Applications of mathematics have led to advances in the sciences, business, computer technology, and medicine.

One fundamental application of mathematics is the graphical representation of numerical information (or **data**). For example, Table 3-1 represents the number of clients admitted to a drug and alcohol rehabilitation program over a 12-month period.

Table 3-1

	Month	**Number of Clients**
Jan.	1	55
Feb.	2	62
March	3	64
April	4	60
May	5	70
June	6	73
July	7	77
Aug.	8	80
Sept.	9	80
Oct.	10	74
Nov.	11	85
Dec.	12	90

In table form, the information is difficult to picture and interpret. It appears that on a monthly basis, the number of clients fluctuates. However, when the data are represented in a graph, an upward trend is clear (Figure 3-1).

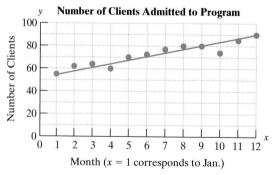

Month ($x = 1$ corresponds to Jan.)

Figure 3-1

From the increase in clients shown in this graph, management for the rehabilitation center might make plans for the future. If the trend continues, management might consider expanding its facilities and increasing its staff to accommodate the expected increase in clients.

Example 1 Interpreting a Graph

Refer to Figure 3-1 and Table 3-1.

   a. For which month was the number of clients the greatest?

   b. How many clients were served in the first month (January)?

   c. Which month corresponds to 60 clients served?

   d. Between which two months did the number of clients decrease?

   e. Between which two months did the number of clients remain the same?

Solution:

   a. Month 12 (December) corresponds to the highest point on the graph, which represents the most clients.

   b. In month 1 (January), there were 55 clients served.

   c. Month 4 (April).

   d. The number of clients decreased between months 3 and 4 and between months 9 and 10.

   e. The number of clients remained the same between months 8 and 9.

Classroom Example: p. 210, Exercise 4

2. Plotting Points in a Rectangular Coordinate System

In Example 1, two variables are represented, time and the number of clients. To picture two variables, we use a graph with two number lines drawn at right angles to each other (Figure 3-2). This forms a **rectangular coordinate system**. The horizontal line is called the **x-axis**, and the vertical line is called the **y-axis**. The point where the lines intersect is called the **origin**. On the x-axis, the numbers to the right of the origin are positive and the numbers to the left are negative. On the y-axis, the numbers above the origin are positive and the numbers below are negative. The x- and y-axes divide the graphing area into four regions called **quadrants**.

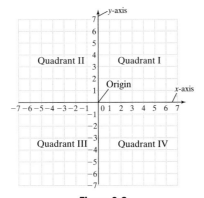

Figure 3-2

Points graphed in a rectangular coordinate system are defined by two numbers as an **ordered pair**, (x, y). The first number (called the **x-coordinate**, or the abscissa) is the horizontal position from the origin. The second number (called the **y-coordinate**, or the ordinate) is the vertical position from the origin. Example 2 shows how points are plotted in a rectangular coordinate system.

Answers
1. Month 10 (October)
2. 30
3. Month 2 (February)
4. Months 1 and 6 (January and June)

5. Plot the points.

$A(3, 4)$ $B(-2, 2)$

$C(4, 0)$ $D\left(\dfrac{5}{2}, -\dfrac{1}{2}\right)$

$E(-5, -2)$

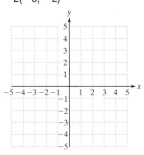

Classroom Example: p. 211, Exercise 10

| **Example 2** | **Plotting Points in a Rectangular Coordinate System** |

Plot the points.

a. $(4, 5)$ **b.** $(-4, -5)$ **c.** $(-1, 3)$ **d.** $(3, -1)$

e. $\left(\dfrac{1}{2}, -\dfrac{7}{3}\right)$ **f.** $(-2, 0)$ **g.** $(0, 0)$ **h.** $(\pi, 1.1)$

Solution:

See Figure 3-3.

a. The ordered pair $(4, 5)$ indicates that $x = 4$ and $y = 5$. Beginning at the origin, move 4 units in the positive x-direction (4 units to the right), and from there move 5 units in the positive y-direction (5 units up). Then plot the point. The point is in Quadrant I.

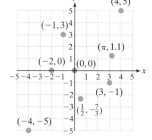

Figure 3-3

b. The ordered pair $(-4, -5)$ indicates that $x = -4$ and $y = -5$. Move 4 units in the negative x-direction (4 units to the left), and from there move 5 units in the negative y-direction (5 units down). Then plot the point. The point is in Quadrant III.

c. The ordered pair $(-1, 3)$ indicates that $x = -1$ and $y = 3$. Move 1 unit to the left and 3 units up. The point is in Quadrant II.

d. The ordered pair $(3, -1)$ indicates that $x = 3$ and $y = -1$. Move 3 units to the right and 1 unit down. The point is in Quadrant IV.

TIP: Notice that changing the order of the x- and y-coordinates changes the location of the point. The point $(-1, 3)$ for example is in Quadrant II, whereas $(3, -1)$ is in Quadrant IV (Figure 3-3). This is why points are represented by *ordered* pairs. The order of the coordinates is important.

e. The improper fraction $-\dfrac{7}{3}$ can be written as the mixed number $-2\dfrac{1}{3}$. Therefore, to plot the point $(\dfrac{1}{2}, -\dfrac{7}{3})$ move to the right $\dfrac{1}{2}$ unit, and down $2\dfrac{1}{3}$ units. The point is in Quadrant IV.

Avoiding Mistakes

Points that lie on either of the axes do not lie in any quadrant.

f. The point $(-2, 0)$ indicates $y = 0$. Therefore, the point is on the x-axis.

g. The point $(0, 0)$ is at the origin.

h. The irrational number, π, can be approximated as 3.14. Thus, the point $(\pi, 1.1)$ is located approximately 3.14 units to the right and 1.1 units up. The point is in Quadrant I.

Answer

5.

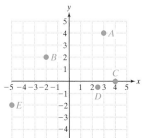

3. Applications of Plotting and Identifying Points

The effective use of graphs for mathematical models requires skill in identifying points and interpreting graphs.

Example 3   Determining Points from a Graph

A map of a national park is drawn so that the origin is placed at the ranger station (Figure 3-4). Four fire observation towers are located at points A, B, C, and D. Estimate the coordinates of the fire towers relative to the ranger station (all distances are in miles).

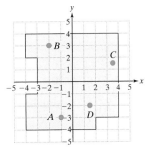

Figure 3-4

Solution:

Point A: $(-1, -3)$

Point B: $(-2, 3)$

Point C: $(3\frac{1}{2}, 1\frac{1}{2})$ or $(\frac{7}{2}, \frac{3}{2})$ or $(3.5, 1.5)$

Point D: $(1\frac{1}{2}, -2)$ or $(\frac{3}{2}, -2)$ or $(1.5, -2)$

Skill Practice

6. Towers are located at points A, B, C, and D. Estimate the coordinates of the towers.

Classroom Example: p. 212, Exercise 26

Example 4   Plotting Points in an Application

The daily low temperatures (in degrees Fahrenheit) for one week in January for Sudbury, Ontario, Canada, are given in Table 3-2.

a. Write an ordered pair for each row in the table using the day number as the x-coordinate and the temperature as the y-coordinate.

b. Plot the ordered pairs from part (a) on a rectangular coordinate system.

Table 3-2

Day Number, x	Temperature (°F), y
1	-3
2	-5
3	1
4	6
5	5
6	0
7	-4

Solution:

a. Each ordered pair represents the day number and the corresponding low temperature for that day.

$(1, -3)$ $(2, -5)$ $(3, 1)$ $(4, 6)$ $(5, 5)$ $(6, 0)$ $(7, -4)$

b.

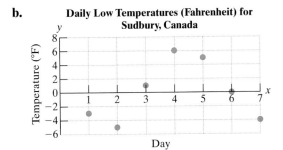

TIP: The graph in Example 4(b) shows only Quadrants I and IV because all x-coordinates are positive.

Skill Practice

7. The table shows the number of homes sold in one town for a 6-month period. Plot the ordered pairs.

Month, x	Number Sold, y
1	20
2	25
3	28
4	40
5	45
6	30

Classroom Example: p. 213, Exercise 28

Answers

6. $A(5, 4\frac{1}{2})$
   $B(0, 3)$
   $C(-4, -2)$
   $D(2, -4)$

7.

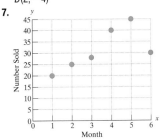

Section 3.1   Practice Exercises

Study Skills Exercises

1. Before you proceed further in Chapter 3, make your test corrections for the Chapter 2 test. See Exercise 1 of Section 2.1 for instructions.

2. Define the key terms:

   a. data   **b. ordered pair**   **c. origin**   **d. quadrant**   **e. rectangular coordinate system**

   f. *x*-axis   **g. *y*-axis**   **h. *x*-coordinate**   **i. *y*-coordinate**

Objective 1: Interpreting Graphs

For Exercises 3–6, refer to the graphs to answer the questions.
(See Example 1.)

3. The number of patients served by a certain hospice care center for the first 12 months after it opened is shown in the graph.

   a. For which month was the number of patients greatest?
   Month 10

   b. How many patients did the center serve in the first month?
   30

   c. Between which months did the number of patients decrease?
   Between months 3 and 5 and between months 10 and 12

   d. Between which two months did the number of patients remain the same?   Months 8 and 9

   e. Which month corresponds to 40 patients served?   Month 3

   f. Approximately how many patients were served during the 10th month?   80

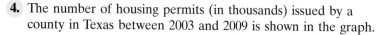

Number of Patients Served by Month

4. The number of housing permits (in thousands) issued by a county in Texas between 2003 and 2009 is shown in the graph.

   a. For which year was the number of permits greatest?   2009

   b. How many permits did the county issue in 2003?
   6 thousand or 6000

   c. Between which years did the number of permits decrease?
   Between 2003 and 2004

   d. Between which two years did the number of permits remain the same?   Between 2007 and 2008

   e. Which year corresponds to 7000 permits issued?   2006

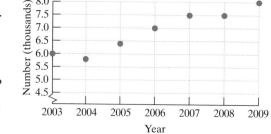

Number of Housing Permits

5. The price per share of a stock (in dollars) over a period of 5 days is shown in the graph.

   a. Interpret the meaning of the ordered pair (1, 89.25).
   On day 1 the price per share was $89.25.

   b. What was the increase in price between day 3 and day 4?
   $1.75

   c. What was the decrease in price between day 4 and day 5?
   −$2.75

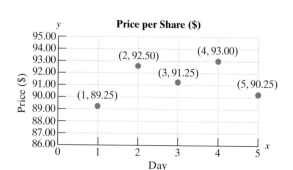

Price per Share ($)

6. The price per share of a stock (in dollars) over a period of 5 days is shown in the graph.

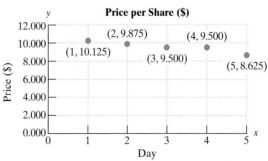

Price per Share ($)

a. Interpret the meaning of the ordered pair (1, 10.125).
   On day 1 the price per share was $10.125.

b. What was the decrease between day 4 and day 5?
   −$0.875

Objective 2: Plotting Points in a Rectangular Coordinate System

7. Plot the points on a rectangular coordinate system.
   (See Example 2.)

   a. (2, 6)   b. (6, 2)   c. (−7, 3)

   d. (−7, −3)   e. (0, −3)   f. (−3, 0)

   g. (6, −4)   h. (0, 5)

8. Plot the points on a rectangular coordinate system.

   a. (4, 5)   b. (−4, 5)   c. (−6, 0)

   d. (6, 0)   e. (4, −5)   f. (−4, −5)

   g. (0, −2)   h. (0, 0)

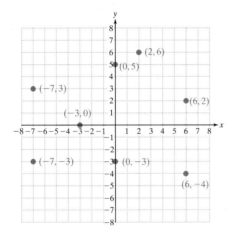

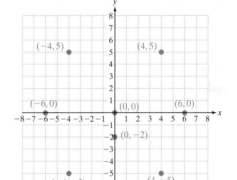

9. Plot the points on a rectangular coordinate system.

   a. (−1, 5)   b. (0, 4)   c. $\left(-2, -\frac{3}{2}\right)$

   d. (2, −0.75) e. (4, 2)   f. (−6, 0)

10. Plot the points on a rectangular coordinate system.

   a. (7, 0)   b. (−3, −1)   c. $\left(6\frac{3}{5}, 1\right)$

   d. (0, 1.5)   e. $\left(\frac{1}{4}, -4\right)$ f. $\left(-\frac{1}{4}, 4\right)$

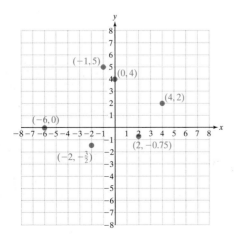

Writing   Translating Expression   Geometry   Scientific Calculator   Video   NS&E

For Exercises 11–18, identify the quadrant in which the given point is found.

11. $(13, -2)$ IV
12. $(25, 16)$ I
13. $(-8, 14)$ II
14. $(-82, -71)$ III

15. $(-5, -19)$ III
16. $(-31, 6)$ II
17. $\left(\dfrac{5}{2}, \dfrac{7}{4}\right)$ I
18. $(9, -40)$ IV

19. Explain why the point $(0, -5)$ is *not* located in Quadrant IV.
$(0, -5)$ lies on the *y*-axis.

20. Explain why the point $(-1, 0)$ is *not* located in Quadrant II.
$(-1, 0)$ lies on the *x*-axis.

21. Where is the point $(\frac{7}{8}, 0)$ located?
$(\frac{7}{8}, 0)$ is located on the *x*-axis.

22. Where is the point $(0, \frac{6}{5})$ located?
$(0, \frac{6}{5})$ is located on the *y*-axis.

Objective 3: Applications of Plotting and Identifying Points

For Exercises 23–24, refer to the graph. **(See Example 3.)**

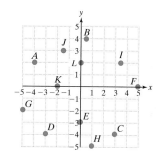

23. Estimate the coordinates of the points *A*, *B*, *C*, *D*, *E*, and *F*.
$A(-4, 2), B(\frac{1}{2}, 4), C(3, -4), D(-3, -4), E(0, -3), F(5, 0)$

24. Estimate the coordinates of the points *G*, *H*, *I*, *J*, *K*, and *L*.
$G(-5, -2), H(1, -5), I(3\frac{1}{2}, 2), J(-1\frac{1}{2}, 3), K(-2, 0), L(0, 2)$

25. A map of a park is laid out with the visitor center located at the origin. Five visitors are in the park located at points *A*, *B*, *C*, *D*, and *E*. All distances are in meters.

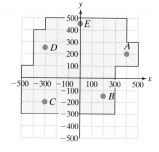

  a. Estimate the coordinates of each hiker. **(See Example 3.)**
  $A(400, 200), B(200, -150), C(-300, -200), D(-300, 250), E(0, 450)$
  b. How far apart are visitors *C* and *D*?   450 m

26. A townhouse has a sprinkler system in the backyard. With the water source at the origin, the sprinkler heads are located at points *A*, *B*, *C*, *D*, and *E*. All distances are in feet.

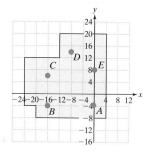

  a. Estimate the coordinates of each sprinkler head.
  $A(0, -4), B(-16, -4), C(-16, 6), D(-8, 14), E(0, 8)$
  b. How far is the distance from sprinkler head B to C?   10 ft

27. A movie theater has kept records of popcorn sales versus movie attendance.

  a. Use the table shown on page 213 to write the corresponding ordered pairs using the movie attendance as the *x*-variable and sales of popcorn as the *y*-variable. Interpret the meaning of the first ordered pair.
  (See Example 4.)
  $(250, 225), (175, 193), (315, 330), (220, 209), (450, 570), (400, 480), (190, 185)$; the ordered pair $(250, 225)$ means that 250 people produce \$225 in popcorn sales.

b. Plot the data points on a rectangular coordinate system.

Movie Attendance (Number of People)	Sales of Popcorn ($)
250	225
175	193
315	330
220	209
450	570
400	480
190	185

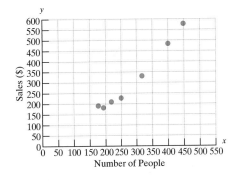

28. The age and systolic blood pressure (in millimeters of mercury, mm Hg) for eight different women are given in the table.

a. Write the corresponding ordered pairs using the woman's age as the *x*-variable and the systolic blood pressure as the *y*-variable. Interpret the meaning of the first ordered pair.   (57, 149), (41, 120), (71, 158), (36, 115), (64, 151), (25, 110), (40, 118), (77, 165); the ordered pair (57, 149) means that a 57-year-old woman has a systolic blood pressure of 149 mm Hg.

b. Plot the data points on a rectangular coordinate system.

Age (Years)	Systolic Blood Pressure (mm Hg)
57	149
41	120
71	158
36	115
64	151
25	110
40	118
77	165

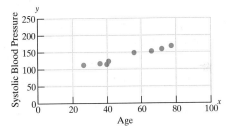

29. The income level defining the poverty line for an individual is given for selected years between 1980 and 2005. Let *x* represent the number of years since 1980. Let *y* represent the income defining the poverty level. (*Source: U.S. Department of the Census*)

(0, 4300)   (5, 5600)   (10, 6800)

(15, 7900)   (20, 9000)   (25, 10,500)

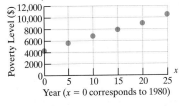

a. Interpret the meaning of the ordered pair (10, 6800).
   (10, 6800) means that in 1990, the poverty level was $6800.
b. Plot the points on a rectangular coordinate system.

30. The following ordered pairs give the population of the U.S. colonies from 1700 to 1770. Let *x* represent the year, where *x* = 0 corresponds to 1700, *x* = 10 corresponds to 1710, and so on. Let *y* represent the population of the colonies.
(*Source: Information Please Almanac*)

(0, 251000)   (10, 332000)   (20, 466000)

(30, 629000)   (40, 906000)   (50, 1171000)

(60, 1594000)   (70, 2148000)

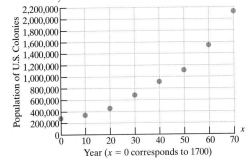

a. Interpret the meaning of the ordered pair (10, 332000).
   (10, 332000); In 1710 the population of the U.S. colonies was 332,000.
b. Plot the points on a rectangular coordinate system.

 Writing   Translating Expression   Geometry   Scientific Calculator   Video   NS E

31. The following table shows the average temperature in degrees Celsius for Montreal, Quebec, Canada, by month.

Month, x		Temperature (°C), y
Jan.	1	−10.2
Feb.	2	−9.0
March	3	−2.5
April	4	5.7
May	5	13.0
June	6	18.3
July	7	20.9
Aug.	8	19.6
Sept.	9	14.8
Oct.	10	8.7
Nov.	11	2.0
Dec.	12	−6.9

a. Write the corresponding ordered pairs, letting $x = 1$ correspond to the month of January. (1, −10.2), (2, −9.0), (3, −2.5), (4, 5.7), (5, 13.0), (6, 18.3), (7, 20.9), (8, 19.6), (9, 14.8), (10, 8.7), (11, 2.0), (12, −6.9).

b. Plot the ordered pairs on a rectangular coordinate system.

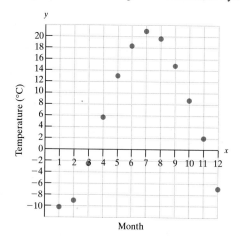

32. The table shows the average temperature in degrees Fahrenheit for Fairbanks, Alaska, by month.

Month, x		Temperature (°F), y
Jan.	1	−12.8
Feb.	2	−4.0
March	3	8.4
April	4	30.2
May	5	48.2
June	6	59.4
July	7	61.5
Aug.	8	56.7
Sept.	9	45.0
Oct.	10	25.0
Nov.	11	6.1
Dec.	12	−10.1

a. Write the corresponding ordered pairs, letting $x = 1$ correspond to the month of January. (1, −12.8), (2, −4.0), (3, 8.4), (4, 30.2), (5, 48.2), (6, 59.4), (7, 61.5), (8, 56.7), (9, 45.0), (10, 25.0), (11, 6.1), (12, −10.1).

b. Plot the ordered pairs on a rectangular coordinate system.

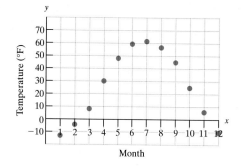

Expanding Your Skills

33. The data in the table give the percent of males and females who have completed 4 or more years of college education for selected years. Let x represent the number of years since 1960. Let y represent the percent of men and the percent of women that completed 4 or more years of college.

Year	x	Percent, y Men	Percent, y Women
1960	0	9.7	5.8
1970	10	13.5	8.1
1980	20	20.1	12.8
1990	30	24.4	18.4
2000	40	27.8	23.6
2005	45	28.9	26.5

a. Plot the data points for men and for women on the same graph.

b. Is the percentage of men with 4 or more years of college increasing or decreasing?   Increasing

c. Is the percentage of women with 4 or more years of college increasing or decreasing?   Increasing

34. Use the data and graph from Exercise 33 to answer the questions.

a. In which year was the difference in percentages between men and women with 4 or more years of college the greatest?   1980

b. In which year was the difference in percentages between men and women the least?   2005

c. If the trend continues beyond the data in the graph, does it seem possible that in the future, the percentage of women with 4 or more years of college will be greater than or equal to the percentage of men?   Yes

Percent of Males/Females with 4 or More Years of College, United States

---

Linear Equations in Two Variables

Section 3.2

1. Definition of a Linear Equation in Two Variables

Recall that an equation in the form $ax + b = 0$, where $a \neq 0$, is called a linear equation in one variable. A solution to such an equation is a value of x that makes the equation a true statement. For example, $3x + 6 = 0$ has a solution of -2.

In this section, we will look at linear equations in *two* variables.

Objectives

1. **Definition of a Linear Equation in Two Variables**
2. **Graphing Linear Equations in Two Variables by Plotting Points**
3. ***x*- and *y*-Intercepts**
4. **Horizontal and Vertical Lines**

DEFINITION   Linear Equation in Two Variables

Let A, B, and C be real numbers such that A and B are not both zero. Then, an equation that can be written in the form:

$$Ax + By = C$$

is called a **linear equation in two variables**.

The equation $x + y = 4$ is a linear equation in two variables. A solution to such an equation is an ordered pair (x, y) that makes the equation a true statement. Several solutions to the equation $x + y = 4$ are listed here:

Solution:	Check:
(x, y)	$x + y = 4$
$(2, 2)$	$(2) + (2) = 4$ ✔
$(1, 3)$	$(1) + (3) = 4$ ✔
$(4, 0)$	$(4) + (0) = 4$ ✔
$(-1, 5)$	$(-1) + (5) = 4$ ✔

 Writing   Translating Expression   Geometry   Scientific Calculator   Video   NS&E

By graphing these ordered pairs, we see that the solution points line up (Figure 3-5).

Notice that there are infinitely many solutions to the equation $x + y = 4$ so they cannot all be listed. Therefore, to visualize all solutions to the equation $x + y = 4$, we draw the line through the points in the graph. Every point on the line represents an ordered pair solution to the equation $x + y = 4$, and the line represents the set of *all* solutions to the equation.

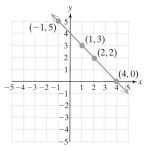

Figure 3-5

Example 1 **Determining Solutions to a Linear Equation**

For the linear equation, $4x - 5y = 8$, determine whether the given ordered pair is a solution.

a. $(2, 0)$ **b.** $(3, 1)$ **c.** $\left(1, -\dfrac{4}{5}\right)$

Solution:

a. $4x - 5y = 8$

$4(2) - 5(0) \overset{?}{=} 8$ Substitute $x = 2$ and $y = 0$.

$8 - 0 \overset{?}{=} 8$ ✔ True   The ordered pair $(2, 0)$ is a solution.

b. $4x - 5y = 8$

$4(3) - 5(1) \overset{?}{=} 8$ Substitute $x = 3$ and $y = 1$.

$12 - 5 \neq 8$ The ordered pair $(3, 1)$ is *not* a solution.

c. $4x - 5y = 8$

$4(1) - 5\left(-\dfrac{4}{5}\right) \overset{?}{=} 8$ Substitute $x = 1$ and $y = -\dfrac{4}{5}$.

$4 + 4 \overset{?}{=} 8$ ✔ True   The ordered pair $\left(1, -\dfrac{4}{5}\right)$ is a solution.

---

2. Graphing Linear Equations in Two Variables by Plotting Points

In this section, we will graph linear equations in two variables.

> **DEFINITION** **The Graph of an Equation in Two Variables**
>
> The graph of an equation in two variables is the graph of all ordered pair solutions to the equation.

The word *linear* means "relating to or resembling a line." It is not surprising then that the solution set for any linear equation in two variables forms a line in a rectangular coordinate system. Because two points determine a line, to graph a linear equation it is sufficient to find two solution points and draw the line between them. We will find three solution points and use the third point as a check point. This process is demonstrated in Example 2.

| Example 2 | Graphing a Linear Equation |

Graph the equation $x - 2y = 8$.

Solution:

We will find three ordered pairs that are solutions to $x - 2y = 8$. To find the ordered pairs, choose arbitrary values of x or y, such as those shown in the table. Then complete the table to find the corresponding ordered pairs.

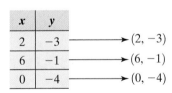

x	y
2	
	−1
0	

→ (2,)
→ (, −1)
→ (0,)

TIP: Usually we try to choose arbitrary values that will be convenient to graph.

From the first row, substitute $x = 2$:

$x - 2y = 8$
$(2) - 2y = 8$
$-2y = 8 - 2$
$-2y = 6$
$y = -3$

From the second row, substitute $y = -1$:

$x - 2y = 8$
$x - 2(-1) = 8$
$x + 2 = 8$
$x = 8 - 2$
$x = 6$

From the third row, substitute $x = 0$:

$x - 2y = 8$
$(0) - 2y = 8$
$-2y = 8$
$y = -4$

The completed table is shown below with the corresponding ordered pairs.

x	y
2	−3
6	−1
0	−4

→ (2, −3)
→ (6, −1)
→ (0, −4)

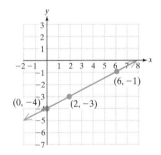

To graph the equation, plot the three solutions and draw the line through the points (Figure 3-6).

Figure 3-6

TIP: Only two points are needed to graph a line. However, in Example 2, we found a third ordered pair, (0, −4). Notice that this point "lines up" with the other two points. If the three points do not line up, then we know that a mistake was made in solving for at least one of the ordered pairs.

In Example 2, the original values for x and y given in the table were picked arbitrarily by the authors. It is important to note, however, that once you pick an arbitrary value for x, the corresponding y-value is determined by the equation. Similarly, once you pick an arbitrary value for y, the x-value is determined by the equation.

Skill Practice

4. Graph the equation.
   $2x + y = 6$

Classroom Example: p. 225, Exercise 18

Answer

4.

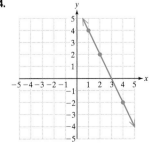

Classroom Example: p. 225,
Exercise 22

Skill Practice

5. Graph the equation.
$$2x + 3y = 12$$

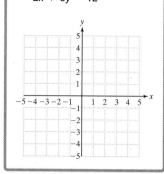

Answer

5.

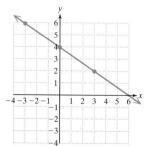

Example 3 **Graphing a Linear Equation**

Graph the equation $4x + 3y = 15$.

Solution:

We will find three ordered pairs that are solutions to the equation $4x + 3y = 15$. In the table, we have selected arbitrary values for x and y and must complete the ordered pairs. Notice that in this case, we are choosing zero for x and zero for y to illustrate that the resulting equation is often easy to solve.

x	y	
0		→ (0,)
	0	→ (, 0)
3		→ (3,)

From the first row, substitute $x = 0$:

$$4x + 3y = 15$$
$$4(0) + 3y = 15$$
$$3y = 15$$
$$y = 5$$

From the second row, substitute $y = 0$:

$$4x + 3y = 15$$
$$4x + 3(0) = 15$$
$$4x = 15$$
$$x = \frac{15}{4} \text{ or } 3\frac{3}{4}$$

From the third row, substitute $x = 3$:

$$4x + 3y = 15$$
$$4(3) + 3y = 15$$
$$12 + 3y = 15$$
$$3y = 3$$
$$y = 1$$

The completed table is shown with the corresponding ordered pairs.

x	y	
0	5	→ (0, 5)
$3\frac{3}{4}$	0	→ ($3\frac{3}{4}$, 0)
3	1	→ (3, 1)

To graph the equation, plot the three solutions and draw the line through the points (Figure 3-7).

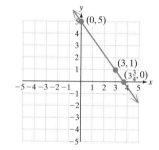

Figure 3-7

Example 4 **Graphing a Linear Equation in Two Variables**

Graph the equation $y = -\dfrac{1}{3}x + 1$.

Solution:

Because the y-variable is isolated in the equation, it is easy to substitute a value for x and simplify the right-hand side to find y. Since any number for x can be picked, choose numbers that are multiples of 3. These will simplify easily when multiplied by $-\frac{1}{3}$.

x	y
3	
0	
−3	

$$y = -\frac{1}{3}x + 1$$

Let $x = 3$:   Let $x = 0$:   Let $x = -3$:

$y = -\frac{1}{3}(3) + 1$ $y = -\frac{1}{3}(0) + 1$ $y = -\frac{1}{3}(-3) + 1$

$y = -1 + 1$ $y = 0 + 1$ $y = 1 + 1$

$y = 0$ $y = 1$ $y = 2$

x	y
3	0
0	1
−3	2

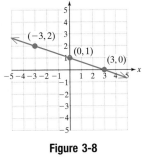

The line through the three ordered pairs $(3, 0)$, $(0, 1)$, and $(-3, 2)$ is shown in Figure 3-8. The line represents the set of all solutions to the equation $y = -\frac{1}{3}x + 1$.

Figure 3-8

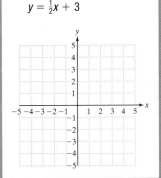
3. *x*- and *y*-Intercepts

The x- and y-intercepts are the points where the graph intersects the x- and y-axes, respectively. From Example 4, we see that the x-intercept is at the point $(3, 0)$ and the y-intercept is at the point $(0, 1)$. See Figure 3-8. Notice that a y-intercept is a point on the y-axis and must have an x-coordinate of 0. Likewise, an x-intercept is a point on the x-axis and must have a y-coordinate of 0.

DEFINITION   *x*- and *y*-Intercepts

An **x-intercept** of a graph is a point $(a, 0)$ where the graph intersects the x-axis.

A **y-intercept** of a graph is a point $(0, b)$ where the graph intersects the y-axis.

In some applications, an x-intercept is defined as the x-coordinate of a point of intersection that a graph makes with the x-axis. For example, if an x-intercept is at the point $(3, 0)$, it is sometimes stated simply as 3 (the y-coordinate is assumed to be 0). Similarly, a y-intercept is sometimes defined as the y-coordinate of a point of intersection that a graph makes with the y-axis. For example, if a y-intercept is at the point $(0, 7)$, it may be stated simply as 7 (the x-coordinate is assumed to be 0).

Although any two points may be used to graph a line, in some cases it is convenient to use the x- and y-intercepts of the line. To find the x- and y-intercepts of any two-variable equation in x and y, follow these steps:

Answer

6.

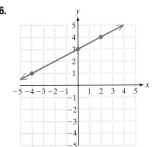

PROCEDURE   Finding *x*- and *y*-Intercepts

Step 1   Find the x-intercept(s) by substituting $y = 0$ into the equation and solving for x.

Step 2   Find the y-intercept(s) by substituting $x = 0$ into the equation and solving for y.

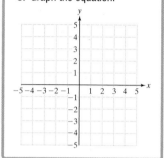
Example 5 Finding the x- and y-Intercepts of a Line

Given the equation $-3x + 2y = 8$,

a. Find the x-intercept.

b. Find the y-intercept.

c. Graph the equation.

Solution:

a. To find the x-intercept, substitute $y = 0$.

$$-3x + 2y = 8$$
$$-3x + 2(0) = 8$$
$$-3x = 8$$
$$\frac{-3x}{-3} = \frac{8}{-3}$$
$$x = -\frac{8}{3}$$

The x-intercept is $\left(-\frac{8}{3}, 0\right)$.

b. To find the y-intercept, substitute $x = 0$.

$$-3x + 2y = 8$$
$$-3(0) + 2y = 8$$
$$2y = 8$$
$$y = 4$$

The y-intercept is $(0, 4)$.

c. The line through the ordered pairs $\left(-\frac{8}{3}, 0\right)$ and $(0, 4)$ is shown in Figure 3-9. Note that the point $\left(-\frac{8}{3}, 0\right)$ can be written as $\left(-2\frac{2}{3}, 0\right)$.
  The line represents the set of all solutions to the equation $-3x + 2y = 8$.

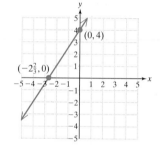

Figure 3-9

Example 6 Finding the x- and y-Intercepts of a Line

Given the equation $4x + 5y = 0$,

a. Find the x-intercept.

b. Find the y-intercept.

c. Graph the equation.

Solution:

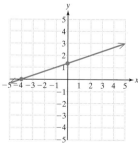
a. To find the x-intercept, substitute $y = 0$.

$$4x + 5y = 0$$
$$4x + 5(0) = 0$$
$$4x = 0$$
$$x = 0$$

The x-intercept is $(0, 0)$.

b. To find the y-intercept, substitute $x = 0$.

$$4x + 5y = 0$$
$$4(0) + 5y = 0$$
$$5y = 0$$
$$y = 0$$

The y-intercept is $(0, 0)$.

c. Because the *x*-intercept and the *y*-intercept are the same point (the origin), one or more additional points are needed to graph the line. In the table, we have arbitrarily selected additional values for *x* and *y* to find two more points on the line.

x	*y*
−5	
	2

Let $x = -5$:

$$4x + 5y = 0$$
$$4(-5) + 5y = 0$$
$$-20 + 5y = 0$$
$$5y = 20$$
$$y = 4$$

Let $y = 2$:

$$4x + 5y = 0$$
$$4x + 5(2) = 0$$
$$4x + 10 = 0$$
$$4x = -10$$
$$x = -\frac{10}{4}$$
$$x = -\frac{5}{2}$$

The line through the ordered pairs $(0, 0)$, $(-5, 4)$, and $\left(-\frac{5}{2}, 2\right)$ is shown in Figure 3-10. Note that the point $\left(-\frac{5}{2}, 2\right)$ can be written as $\left(-2\frac{1}{2}, 2\right)$.

The line represents the set of all solutions to the equation $4x + 5y = 0$.

x	*y*
−5	4
$-\frac{5}{2}$	2
0	0

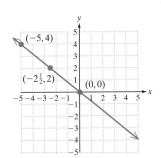

Figure 3-10

4. Horizontal and Vertical Lines

Recall that a linear equation can be written in the form of $Ax + By = C$, where *A* and *B* are not both zero. However, if *A* or *B* is 0, then the line is either parallel to the *x*-axis (horizontal) or parallel to the *y*-axis (vertical), respectively.

> **DEFINITION Equations of Vertical and Horizontal Lines**
>
> **1.** A **vertical line** can be represented by an equation of the form, $x = k$, where *k* is a constant.
>
> **2.** A **horizontal line** can be represented by an equation of the form, $y = k$, where *k* is a constant.

Skill Practice

Given the equation

$2x - 3y = 0$,

10. Find the *x*-intercept.

11. Find the *y*-intercept.

12. Graph the equation. (*Hint:* You may need to find an additional point.)

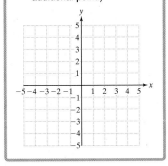

Classroom Example: p. 228, Exercise 56

Instructor Note: Have students go back to the definition of a linear equation in two variables at the beginning of Section 3.2. Ask how equations like $x = 3$ and $y = -2$ fit the requirements of a linear equation in 2 variables?

Answers

10. (0, 0)   **11.** (0, 0)

12.

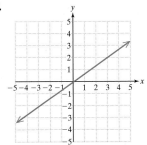

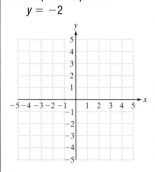
Example 7   **Graphing a Horizontal Line**

Graph the equation $y = 3$.

Solution:

Because this equation is in the form $y = k$, the line is horizontal and must cross the y-axis at $y = 3$ (Figure 3-11).

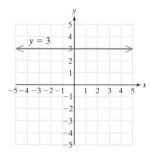

Figure 3-11

Alternative Solution:

Create a table of values for the equation $y = 3$. The choice for the y-coordinate must be 3, but x can be any real number.

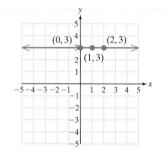

x	y
0	3
1	3
2	3

x can be any number. y must be 3.

TIP: Notice that a horizontal line has a y-intercept, but does not have an x-intercept (unless the horizontal line is the x-axis itself).

Answers

13.

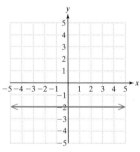

14.

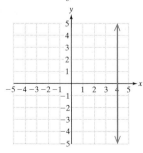

Example 8   **Graphing a Vertical Line**

Graph the equation $4x = -8$.

Solution:

Because the equation does not have a y-variable, we can solve the equation for x.

$4x = -8$ is equivalent to $x = -2$

This equation is in the form $x = k$, indicating that the line is vertical and must cross the x-axis at $x = -2$ (Figure 3-12).

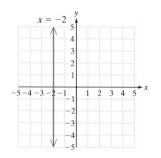

Figure 3-12

Alternative Solution:

Create a table of values for the equation $x = -2$. The choice for the x-coordinate must be -2, but y can be any real number.

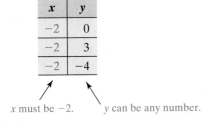

x	y
-2	0
-2	3
-2	-4

x must be -2. y can be any number.

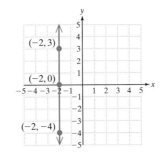

TIP: Notice that a vertical line has an x-intercept but does not have a y-intercept (unless the vertical line is the y-axis itself).

Calculator Connections

Topic: Graphing Linear Equations on an Appropriate Viewing Window

A viewing window of a graphing calculator shows a portion of a rectangular coordinate system. The standard viewing window for many calculators shows the x-axis between -10 and 10 and the y-axis between -10 and 10 (Figure 3-13). Furthermore, the scale defined by the "tick" marks on both the x- and y-axes is usually set to 1.

The "Standard Viewing Window"

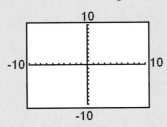

Figure 3-13

To graph an equation in x and y on a graphing calculator, the equation must be written with the y-variable isolated. For example, to graph the equation $x + 3y = 3$, we solve for y by applying the steps for solving a literal equation. The result, $y = -\frac{1}{3}x + 1$, can now be entered into a graphing calculator. To enter the equation $y = -\frac{1}{3}x + 1$ use parentheses around the fraction $\frac{1}{3}$. The *Graph* option displays the graph of the line.

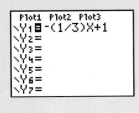

Sometimes the standard viewing window does not provide an adequate display for the graph of an equation. For example, the graph of $y = -x + 15$ is visible only in a small portion of the upper right corner of the standard viewing window.

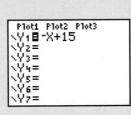

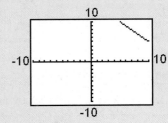

To see where this line crosses the x- and y-axes, we can change the viewing window to accommodate larger values of x and y. Most calculators have a *Range* feature or *Window* feature that allows the user to change the minimum and maximum x- and y-values.

To get a better picture of the equation $y = -x + 15$, change the minimum x-value to -10 and the maximum x-value to 20. Similarly, use a minimum y-value of -10 and a maximum y-value of 20.

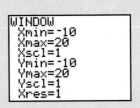

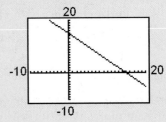

Calculator Exercises

For Exercises 1–8, graph the equations on the standard viewing window.

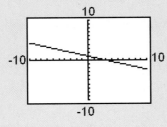

1. $y = -2x + 5$ **2.** $y = 3x - 1$

3. $y = \dfrac{1}{2}x - \dfrac{7}{2}$ **4.** $y = -\dfrac{3}{4}x + \dfrac{5}{3}$

5. $4x - 7y = 21$ **6.** $2x + 3y = 12$

7. $-3x - 4y = 6$ **8.** $-5x + 4y = 10$

For Exercises 9–12, graph the equations on the given viewing window.

9. $y = 3x + 15$

Window: $-10 \le x \le 10$
                     $-5 \le y \le 20$

10. $y = -2x - 25$

Window: $-30 \le x \le 30$
                     $-30 \le y \le 30$

Xscl = 3 (sets the *x*-axis tick marks to increments of 3)

Yscl = 3 (sets the *y*-axis tick marks to increments of 3)

11. $y = -0.2x + 0.04$

Window: $-0.1 \le x \le 0.3$
                     $-0.1 \le y \le 0.1$

Xscl = 0.01 (sets the *x*-axis tick marks to increments of 0.01)

Yscl = 0.01 (sets the *y*-axis tick marks to increments of 0.01)

12. $y = 0.3x - 0.5$

Window: $-1 \le x \le 3$
                     $-1 \le y \le 1$

Xscl = 0.1 (sets the *x*-axis tick marks to increments of 0.1)

Yscl = 0.1 (sets the *y*-axis tick marks to increments of 0.1)

Section 3.2   Practice Exercises

- Practice Problems
- Self-Tests
- NetTutor

- e-Professors
- Videos

For additional exercises, see Classroom Activities 3.2A–3.2C in the *Instructor's Resource Manual* at www.mhhe.com/moh.

Study Skills Exercises

1. Check your progress by answering these questions.

Yes _____ No _____   Did you have sufficient time to study for the test on Chapter 2? If not, what could you have done to create more time for studying?

Yes _____ No _____   Did you work all of the assigned homework problems in Chapter 2?

Yes _____ No _____   If you encountered difficulty, did you see your instructor or tutor for help?

Yes _____ No _____   Have you taken advantage of the textbook supplements such as the *Student Solutions Manual*?

2. Define the key terms:

a. horizontal line   **b. linear equation in two variables**   **c. vertical line**

d. *x*-intercept   **e. *y*-intercept**

Review Exercises

For Exercises 3–8, refer to the figure to give the coordinates of the labeled points, and state the quadrant or axis where the point is located.

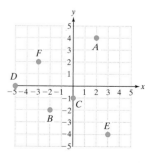

3. *A*   (2, 4); quadrant I   **4.** *B*   (−2, −2); quadrant III

5. *C*   (0, −1); *y*-axis   **6.** *D*   (−5, 0); *x*-axis

7. *E*   (3, −4); quadrant IV   **8.** *F*   (−3, 2); quadrant II

Objective 1: Definition of a Linear Equation in Two Variables

For Exercises 9–17, determine if the given ordered pair is a solution to the equation. **(See Example 1.)**

9. $x - y = 6$; $(8, 2)$ Yes

10. $y = 3x - 2$; $(1, 1)$ Yes

11. $y = -\dfrac{1}{3}x + 3$; $(-3, 4)$ Yes

12. $y = -\dfrac{5}{2}x + 5$; $(-2, 0)$ No

13. $4x + 5y = 20$; $(-5, -4)$ No

14. $y = 7$; $(0, 7)$ Yes

15. $y = -2$; $(-2, 6)$ No

16. $x = 1$; $(0, 1)$ No

17. $x = -5$; $(-5, 6)$ Yes

Objective 2: Graphing Linear Equations in Two Variables by Plotting Points

For Exercises 18–31, complete each table, and graph the corresponding ordered pairs. Draw the line defined by the points to represent all solutions to the equation. **(See Examples 2–4.)**

18. $x + y = 3$

x	y
2	1
0	3
-1	4
3	0

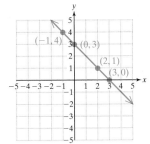

19. $x + y = -2$

x	y
1	-3
-2	0
-3	1
-4	2

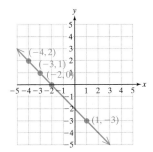

20. $y = 5x + 1$

x	y
1	6
0	1
-1	-4

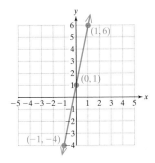

21. $y = -3x - 3$

x	y
-2	3
-1	0
-4	9

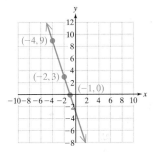

22. $2x - 3y = 6$

x	y
0	-2
3	0
2	-2/3

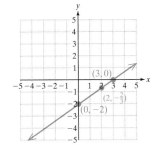

23. $4x + 2y = 8$

x	y
0	4
2	0
3	-2

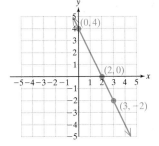

24. $y = \dfrac{2}{7}x - 5$

x	y
7	-3
-7	-7
0	-5

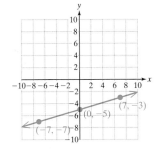

25. $y = -\dfrac{3}{5}x - 2$

x	y
0	-2
5	-5
10	-8

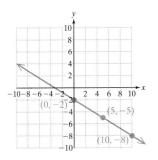

Writing   Translating Expression   Geometry   Scientific Calculator   Video   NS&E

26. $y = 3$

x	y
2	3
0	3
−1	3

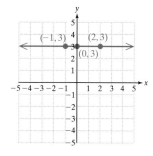

27. $y = -2$

x	y
0	−2
−3	−2
5	−2

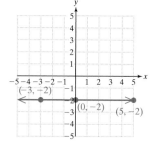

28. $x = -4$

x	y
−4	1
−4	−2
−4	4

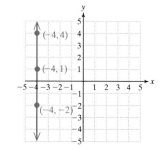

29. $x = \frac{3}{2}$

x	y
3/2	−1
3/2	2
3/2	−3

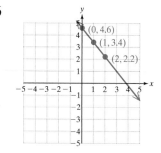

30. $y = -3.4x + 5.8$

x	y
0	5.8
1	2.4
2	−1

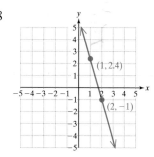

31. $y = -1.2x + 4.6$

x	y
0	4.6
1	3.4
2	2.2

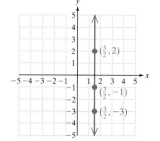

For Exercises 32–43, graph the lines by making a table of at least three ordered pairs and plotting the points.

32. $x = y + 2$

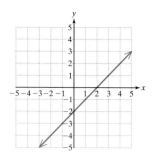

33. $x - y = 4$

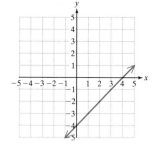

34. $-3x + y = -6$

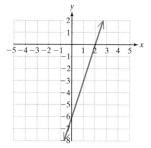

35. $2x - 5y = 10$

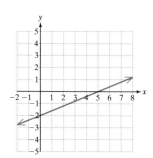

36. $y = 4x$

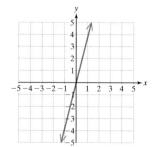

37. $y = -2x$

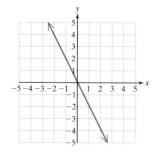

Writing   Translating Expression   Geometry   Scientific Calculator   Video   NS&E

38. $y = -\dfrac{1}{2}x + 3$

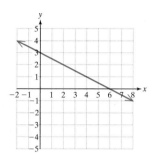

39. $y = \dfrac{1}{4}x - 2$

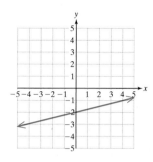

40. $x + y = 0$

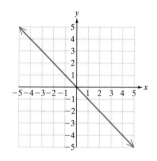

41. $-x + y = 0$

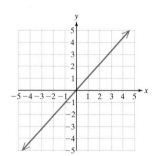

42. $50x - 40y = 200$

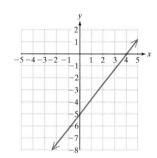

43. $-30x - 20y = 60$

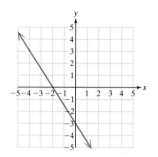

Objective 3: *x*- and *y*-Intercepts

44. The *x*-intercept is on which axis?   *x*-axis

45. The *y*-intercept is on which axis?   *y*-axis

For Exercises 46–49, estimate the coordinates of the *x*- and *y*-intercepts.

46.

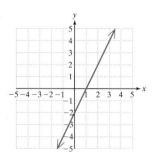

x-intercept: $(1, 0)$;
y-intercept: $(0, -2)$

47.

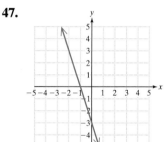

x-intercept: $(-1, 0)$;
y-intercept: $(0, -3)$

48.

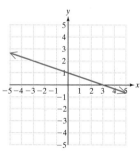

x-intercept: $(3, 0)$;
y-intercept: $(0, 1)$

49.

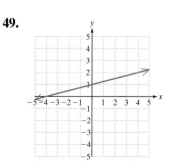

x-intercept: $(-4, 0)$;
y-intercept: $(0, 1)$

Writing   Translating Expression   Geometry   Scientific Calculator   Video   NS&E

For Exercises 50–61, find the *x*- and *y*-intercepts (if they exist), and graph the line. **(See Examples 5–6.)**

50. $5x + 2y = 5$ *x*-intercept: $(1, 0)$; *y*-intercept: $\left(0, \frac{5}{2}\right)$

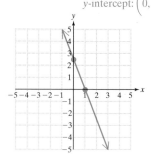

51. $4x - 3y = -9$ *x*-intercept: $\left(-\frac{9}{4}, 0\right)$; *y*-intercept: $(0, 3)$

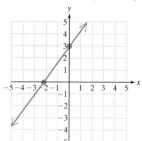

52. $y = \frac{2}{3}x - 1$ *x*-intercept: $\left(\frac{3}{2}, 0\right)$; *y*-intercept: $(0, -1)$

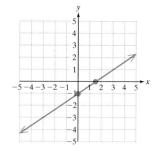

53. $y = -\frac{3}{4}x + 2$ *x*-intercept: $\left(\frac{8}{3}, 0\right)$; *y*-intercept: $(0, 2)$

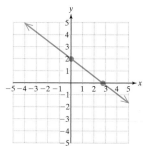

54. $x - 3 = y$ *x*-intercept: $(3, 0)$; *y*-intercept: $(0, -3)$

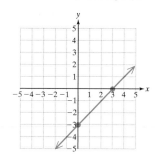

55. $2x + 8 = y$ *x*-intercept: $(-4, 0)$; *y*-intercept: $(0, 8)$

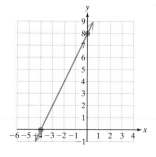

56. $-3x + y = 0$ *x*-intercept: $(0, 0)$; *y*-intercept: $(0, 0)$

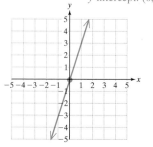

57. $2x - 2y = 0$ *x*-intercept: $(0, 0)$; *y*-intercept: $(0, 0)$

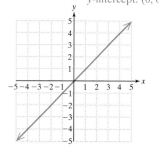

58. $25y = 10x + 100$ *x*-intercept: $(-10, 0)$; *y*-intercept: $(0, 4)$

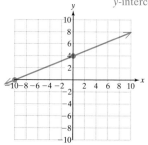

59. $20x = -40y + 200$
x-intercept: $(10, 0)$; *y*-intercept: $(0, 5)$

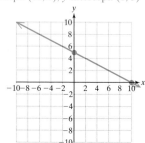

60. $x = 2y$
x-intercept: $(0, 0)$; *y*-intercept: $(0, 0)$

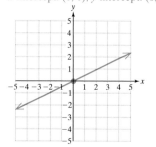

61. $x = -5y$
x-intercept: $(0, 0)$; *y*-intercept: $(0, 0)$

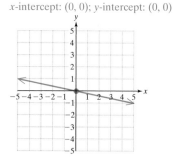

Objective 4: Horizontal and Vertical Lines

For Exercises 62–65, answer true or false. If the statement is false, rewrite it to be true.

62. The line $x = 3$ is horizontal.   False, $x = 3$ is vertical

63. The line $y = -4$ is horizontal.   True

64. A line parallel to the y-axis is vertical.   True

65. A line perpendicular to the x-axis is vertical.   True

For Exercises 66–74,

a. Identify the equation as representing a horizontal or vertical line.

b. Graph the line.

c. Identify the x- and y-intercepts if they exist. **(See Examples 7–8.)**

66. $x = 3$ a. Vertical c. x-intercept: $(3, 0)$; no y-intercept

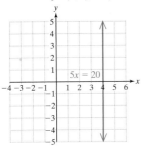

67. $y = -1$ a. Horizontal c. no x-intercept; y-intercept: $(0, -1)$

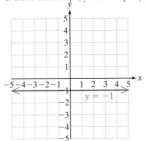

68. $-2y = 8$ a. Horizontal c. no x-intercept; y-intercept: $(0, -4)$

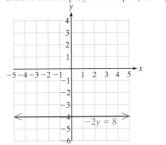

69. $5x = 20$ a. Vertical c. x-intercept: $(4, 0)$; no y-intercept

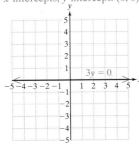

70. $x + 3 = 7$ a. Vertical c. x-intercept: $(4, 0)$; no y-intercept

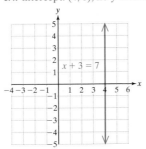

71. $y - 8 = -13$ a. Horizontal c. no x-intercept; y-intercept $(0, -5)$

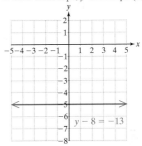

72. $3y = 0$ a. Horizontal c. All points on the x-axis are x-intercepts; y-intercept: $(0, 0)$

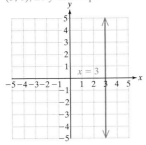

73. $5x = 0$ a. Vertical c. All points on the y-axis are y-intercepts; x-intercept: $(0, 0)$

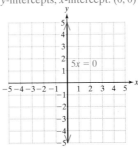

74. $2x + 7 = 10$ a. Vertical c. x-intercept: $(\frac{3}{2}, 0)$; no y-intercept

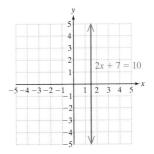

75. Explain why not every line has both an x- and a y-intercept.   A horizontal line may not have an x-intercept. A vertical line may not have a y-intercept.

76. Which of the lines has an x-intercept?

  a. $2x - 3y = 6$ **b.** $x = 5$ **c.** $2y = 8$ **d.** $-x + y = 0$ a, b, d

77. Which of the lines has a y-intercept?

  a. $y = 2$ **b.** $x + y = 0$ **c.** $2x - 10 = 2$ **d.** $x + 4y = 8$ a, b, d

 Writing   Translating Expression   Geometry   Scientific Calculator   Video   NS&E

Expanding Your Skills

78. The store "CDs R US" sells all compact disks for $13.99. The following equation represents the revenue, y, (in dollars) generated by selling x CDs.

$$y = 13.99x \quad (x \geq 0)$$

a. Find y when $x = 13$. $\quad y = 181.87$

b. Find x when $y = 279.80$. $\quad x = 20$

c. Write the ordered pairs from parts (a) and (b), and interpret their meaning in the context of the problem. $(13, 181.87)$ Selling 13 compact disks yields $181.87 in revenue. $(20, 279.80)$ Selling 20 compact disks yields $279.80 in revenue.

d. Graph the ordered pairs and the line defined by the points.

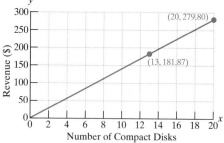

79. The value of a car depreciates once it is driven off of the dealer's lot. For a Hyundai Accent, the value of the car is given by the equation $y = -1531x + 11{,}599 \, (x \geq 0)$ where y is the value of the car in dollars x years after its purchase. (*Source: Kelly Blue Book*)

a. Find y when $x = 1$. $\quad y = 10{,}068$

b. Find x when $y = 7006$. $\quad x = 3$

c. Write the ordered pairs from parts (a) and (b), and interpret their meaning in the context of the problem. $(1, 10068)$ One year after purchase the value of the car is $10,068. $(3, 7006)$ Three years after purchase the value of the car is $7006.

Section 3.3 | Slope of a Line and Rate of Change

Objectives

1. **Introduction to Slope**
2. **Slope Formula**
3. **Parallel and Perpendicular Lines**
4. **Applications of Slope: Rate of Change**

1. Introduction to Slope

The x- and y-intercepts represent the points where a line crosses the x- and y-axes. Another important feature of a line is its slope. Geometrically, the slope of a line measures the "steepness" of the line. For example, two ski runs are depicted by the lines in Figure 3-14.

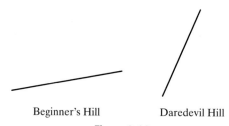

Beginner's Hill   Daredevil Hill

Figure 3-14

By visual inspection, Daredevil Hill is "steeper" than Beginner's Hill. To measure the slope of a line quantitatively, consider two points on the line. The **slope** of the line is the ratio of the vertical change (change in y) between the two points and the horizontal change (change in x). As a memory device, we might think of the slope of a line as "rise over run." See Figure 3-15.

$$\text{Slope} = \frac{\text{change in } y}{\text{change in } x} = \frac{\text{rise}}{\text{run}}$$

Change in x (run)

Change in y (rise)

Figure 3-15

To move from point A to point B on Beginner's Hill, rise 2 ft and move to the right 6 ft (Figure 3-16).

To move from point A to point B on Daredevil Hill, rise 12 ft and move to the right 6 ft (Figure 3-17).

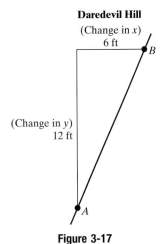

Daredevil Hill

Figure 3-17

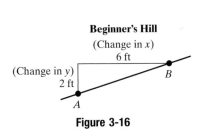

Figure 3-16

$$\text{Slope} = \frac{\text{change in } y}{\text{change in } x} = \frac{2 \text{ ft}}{6 \text{ ft}} = \frac{1}{3}$$

$$\text{Slope} = \frac{\text{change in } y}{\text{change in } x} = \frac{12 \text{ ft}}{6 \text{ ft}} = \frac{2}{1} = 2$$

The slope of Daredevil Hill is greater than the slope of Beginner's Hill, confirming the observation that Daredevil Hill is steeper. On Daredevil Hill there is a 12-ft change in elevation for every 6 ft of horizontal distance (a 2:1 ratio). On Beginner's Hill there is only a 2-ft change in elevation for every 6 ft of horizontal distance (a 1:3 ratio).

Example 1 **Finding Slope in an Application**

Determine the slope of the ramp up the stairs.

Solution:

$$\text{Slope} = \frac{\text{change in } y}{\text{change in } x} = \frac{8 \text{ ft}}{16 \text{ ft}}$$

$\frac{8}{16} = \frac{1}{2}$ Write the ratio for the slope and simplify.

The slope is $\frac{1}{2}$.

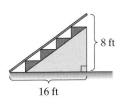

Skill Practice

1. Determine the slope of the aircraft's takeoff path.

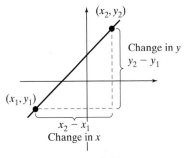

Classroom Example: p. 238, Exercise 12

2. Slope Formula

The slope of a line may be found using any two points on the line—call these points (x_1, y_1) and (x_2, y_2). The change in y between the points can be found by taking the difference of the y-values: $y_2 - y_1$. The change in x can be found by taking the difference of the x-values in the same order: $x_2 - x_1$ (Figure 3-18).

The slope of a line is often symbolized by the letter m and is given by the following formula.

Figure 3-18

Answer

1. $\frac{500}{6000} = \frac{1}{12}$

> **FORMULA** Slope Formula
>
> The **slope** of a line passing through the distinct points (x_1, y_1) and (x_2, y_2) is
>
> $$m = \frac{y_2 - y_1}{x_2 - x_1} \quad \text{provided } x_2 - x_1 \neq 0$$

— Skill Practice —

Find the slope of the line through the given points.

2. $(-5, 2)$ and $(1, 3)$

Classroom Example: p. 239, Exercise 38

Example 2 Finding the Slope of a Line Given Two Points

Find the slope of the line through the points $(-1, 3)$ and $(-4, -2)$.

Solution:

To use the slope formula, first label the coordinates of each point and then substitute the coordinates into the slope formula.

$$\underset{(x_1, y_1)}{(-1, 3)} \quad \text{and} \quad \underset{(x_2, y_2)}{(-4, -2)} \qquad \text{Label the points.}$$

$$m = \frac{y_2 - y_1}{x_2 - x_1} = \frac{(-2) - (3)}{(-4) - (-1)} \qquad \text{Apply the slope formula.}$$

$$= \frac{-5}{-3}$$

$$= \frac{5}{3} \qquad \text{Simplify to lowest terms.}$$

The slope of the line can be verified from the graph (Figure 3-19).

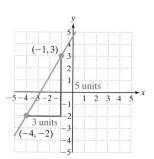

Figure 3-19

> **TIP:** The slope formula is not dependent on which point is labeled (x_1, y_1) and which point is labeled (x_2, y_2). In Example 2, reversing the order in which the points are labeled results in the same slope.
>
> $$\underset{(x_2, y_2)}{(-1, 3)} \quad \text{and} \quad \underset{(x_1, y_1)}{(-4, -2)} \qquad \text{Label the points.}$$
>
> $$m = \frac{(3) - (-2)}{(-1) - (-4)} = \frac{5}{3} \qquad \text{Apply the slope formula.}$$

Answer

2. $\dfrac{1}{6}$

When you apply the slope formula, you will see that the slope of a line may be positive, negative, zero, or undefined.

- Lines that increase, or rise, from left to right have a positive slope.
- Lines that decrease, or fall, from left to right have a negative slope.
- Horizontal lines have a slope of zero.
- Vertical lines have an undefined slope.

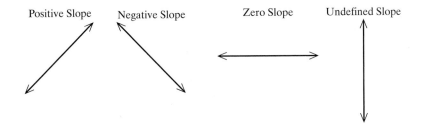

Positive Slope   Negative Slope   Zero Slope   Undefined Slope

Example 3 **Finding the Slope of a Line Given Two Points**

Find the slope of the line passing through the points $(-5, 0)$ and $(2, -3)$.

Solution:

$$\underset{(x_1, y_1)}{(-5, 0)} \quad \text{and} \quad \underset{(x_2, y_2)}{(2, -3)} \quad \text{Label the points.}$$

$$m = \frac{y_2 - y_1}{x_2 - x_1} = \frac{(-3) - (0)}{(2) - (-5)} \quad \text{Apply the slope formula.}$$

$$= \frac{-3}{7} \quad \text{or} \quad -\frac{3}{7} \quad \text{Simplify.}$$

By graphing the points $(-5, 0)$ and $(2, -3)$, we can verify that the slope is $-\frac{3}{7}$ (Figure 3-20). Notice that the line slopes downward from left to right.

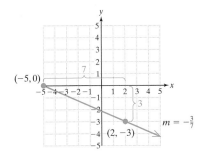

Figure 3-20

Concept Connections

3. Label each line as having a slope that is positive, negative, zero, or undefined.

a.

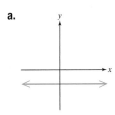

b.

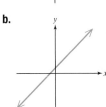

c.

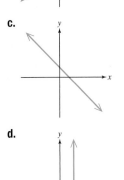

d.

Skill Practice

Find the slope of the line through the given points.

4. $(0, -8)$ and $(-2, -2)$

Classroom Example: p. 239, Exercise 40

Answers

3. a. Zero   **b.** Positive
   c. Negative   **d.** Undefined
4. -3

Example 4 Determining the Slope of a Vertical Line

Find the slope of the line passing through the points $(2, -1)$ and $(2, 4)$.

Solution:

$$(2, -1) \quad \text{and} \quad (2, 4)$$
$$(x_1, y_1) \qquad\qquad (x_2, y_2) \qquad \text{Label the points.}$$

$$m = \frac{y_2 - y_1}{x_2 - x_1} = \frac{(4) - (-1)}{(2) - (2)} \qquad \text{Apply the slope formula.}$$

$$m = \frac{5}{0} \quad \text{Undefined}$$

Because the slope, m, is undefined, we expect the points to form a vertical line as shown in Figure 3-21.

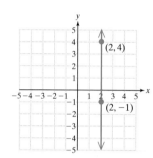

Figure 3-21

Example 5 Determine the Slope of a Horizontal Line

Find the slope of the line passing through the points $(3, -2)$ and $(-4, -2)$.

Solution:

$$(3, -2) \quad \text{and} \quad (-4, -2)$$
$$(x_1, y_1) \qquad\qquad (x_2, y_2) \qquad \text{Label the points.}$$

$$m = \frac{y_2 - y_1}{x_2 - x_1} = \frac{(-2) - (-2)}{(-4) - (3)} \qquad \text{Apply the slope formula.}$$

$$m = \frac{-2 + 2}{-4 - 3} = \frac{0}{-7} = 0$$

Because the slope is 0, we expect the points to form a horizontal line, as shown in Figure 3-22.

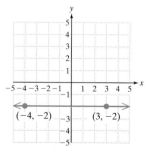

Figure 3-22

Answers

5. Undefined   **6.** 0

3. Parallel and Perpendicular Lines

Lines in the same plane that do not intersect are called **parallel lines**. Parallel lines have the same slope and different *y*-intercepts (Figure 3-23).

Lines that intersect at a right angle are **perpendicular lines**. If two lines are perpendicular then the slope of one line is the opposite of the reciprocal of the slope of the other line (provided neither line is vertical) (Figure 3-24).

Instructor Note: If you use calculators for graphing lines, try putting two lines with slopes that are similar ($y = x$ and $y = \frac{19}{20}x + 1$). They will appear parallel, but clearly we can show they are not by inspecting the equations. Sometimes students don't value what we do by hand until they see it does not always appear correctly on the calculator.

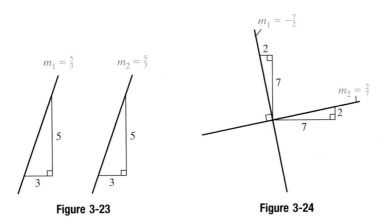

Figure 3-23   **Figure 3-24**

PROPERTY Slopes of Parallel Lines

If m_1 and m_2 represent the slopes of two parallel (nonvertical) lines, then

$$m_1 = m_2.$$

See Figure 3-23.

PROPERTY Slopes of Perpendicular Lines

If $m_1 \neq 0$ and $m_2 \neq 0$ represent the slopes of two perpendicular lines, then

$$m_1 = -\frac{1}{m_2} \text{ or equivalently, } m_1 m_2 = -1. \text{ See Figure 3-24.}$$

Instructor Note: Ask students how we know that all vertical lines are parallel and all horizontal lines are parallel.

Example 6 **Determining the Slope of Parallel and Perpendicular Lines**

Suppose a given line has a slope of $-\frac{1}{4}$.

a. Find the slope of a line parallel to the given line.

b. Find the slope of a line perpendicular to the given line.

Solution:

a. Parallel lines must have the same slope. The slope of a line parallel to the given line is: $m = -\dfrac{1}{4}$.

b. Perpendicular lines must have opposite and reciprocal slopes. The slope of a line perpendicular to the given line is: $m = +\dfrac{4}{1}$ or simply, $m = 4$.

Skill Practice

A given line has a slope of $\frac{5}{3}$.

7. Find the slope of a line parallel to the given line.
8. Find the slope of a line perpendicular to the given line.

Classroom Example: p. 240, Exercise 54

Answers

7. $\dfrac{5}{3}$ **8.** $-\dfrac{3}{5}$

If the slopes of two lines are known, then we can compare the slopes to determine if the lines are parallel, perpendicular, or neither.

Example 7 **Determining If Lines Are Parallel, Perpendicular, or Neither**

Lines l_1 and l_2 pass through the given points. Determine if l_1 and l_2 are parallel, perpendicular, or neither.

$$l_1: \ (2, -7) \text{ and } (4, 1) \qquad l_2: \ (-3, 1) \text{ and } (1, 0)$$

Solution:

Find the slope of each line.

l_1: $(2, -7)$ and $(4, 1)$
$(x_1, y_1)(x_2, y_2)$

$$m_1 = \frac{1 - (-7)}{4 - 2}$$

$$m_1 = \frac{8}{2}$$

$$m_1 = 4$$

l_2: $(-3, 1)$ and $(1, 0)$
$(x_1, y_1)(x_2, y_2)$

$$m_2 = \frac{0 - 1}{1 - (-3)}$$

$$m_2 = \frac{-1}{4}$$

$$m_2 = -\frac{1}{4}$$

One slope is the opposite of the reciprocal of the other slope. Therefore, the lines are perpendicular.

4. Applications of Slope: Rate of Change

In many applications, the interpretation of slope refers to the *rate of change* of the y-variable to the x-variable.

Example 8 **Interpreting Slope in an Application**

The annual median income for males in the United States for selected years is shown in Figure 3-25. The trend is approximately linear. Find the slope of the line and interpret the meaning of the slope in the context of the problem.

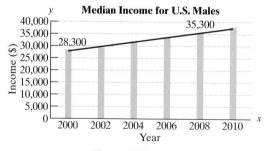

Figure 3-25

Source: U.S. Department of the Census

Solution:

To determine the slope we need to know two points on the line. From the graph, the median income for males in the year 2000 was approximately $28,300. This gives us the ordered pair (2000, 28,300). In the year 2008, the income was $35,300. This gives the ordered pair (2008, 35,300).

(2000, 28,300)   and   (2008, 35,300)
(x_1, y_1) (x_2, y_2) Label the points.

$m = \dfrac{y_2 - y_1}{x_2 - x_1} = \dfrac{35,300 - 28,300}{2008 - 2000}$ Apply the slope formula.

$= \dfrac{7000}{8}$

$= 875$ Simplify.

The slope is 875. This tells us the rate of change of the y-variable (income) to the x-variable (years). This means that men's median income in the United States increased at a rate of $875 per year during this time period.

Section 3.3 Practice Exercises

Boost *your* GRADE at ALEKS.com!

ALEKS version 3.0

• Practice Problems
• Self-Tests
• NetTutor

• e-Professors
• Videos

For additional exercises, see Classroom Activities 3.3A–3.3C in the *Instructor's Resource Manual* at www.mhhe.com/moh.

Study Skills Exercises

1. Make up a practice test for yourself. Use examples or exercises from the text. Be sure to cover each concept that was presented.

2. Define the key terms: **a. parallel lines**   **b. perpendicular lines**   **c. slope**

Review Exercises

For Exercises 3–8, find the x- and y-intercepts (if they exist). Then graph the line.

3. $x - 3y = 6$ x-intercept: (6, 0); y-intercept: (0, −2)

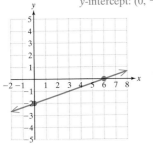

4. $x - 5 = 2$ x-intercept: (7, 0); y-intercept: none

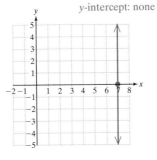

5. $y = \dfrac{2}{3}x$ x-intercept: (0, 0); y-intercept: (0, 0)

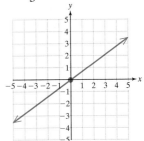

6. $2y - 3 = 0$ x-intercept: none; y-intercept: $(0, \frac{3}{2})$

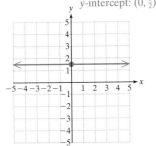

7. $4x + y = 8$ x-intercept: (2, 0); y-intercept: (0, 8)

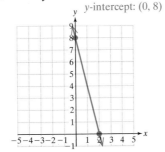

8. $2x = 4y$ x-intercept: (0, 0); y-intercept: (0, 0)

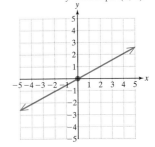

 Writing   Translating Expression   Geometry   Scientific Calculator   Video   NS&E

Objective 1: Introduction to Slope

9. Determine the pitch (slope) of the roof.
(See Example 1.)

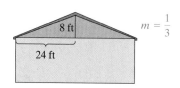

$m = \dfrac{1}{3}$

10. Determine the slope of the stairs.

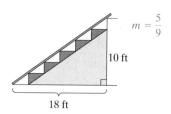

$m = \dfrac{5}{9}$

11. Calculate the slope of the handrail.

$m = \dfrac{6}{11}$

12. Determine the slope of the treadmill.

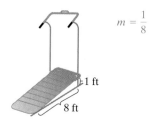

$m = \dfrac{1}{8}$

Objective 2: Slope Formula

For Exercises 13–16, fill in the blank with the appropriate term: zero, negative, positive, or undefined.

13. The slope of a line parallel to the *y*-axis is <u>Undefined</u>.

14. The slope of a horizontal line is <u>Zero</u>.

15. The slope of a line that rises from left to right is <u>Positive</u>.

16. The slope of a line that falls from left to right is <u>Negative</u>.

For Exercises 17–25, determine if the slope is positive, negative, zero, or undefined.

17. Negative

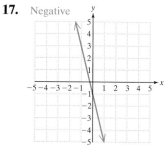

18. Undefined

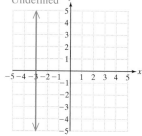

19. Zero

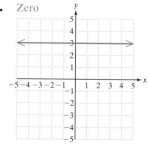

20. Negative

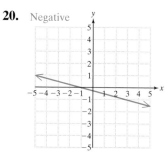

21. Undefined

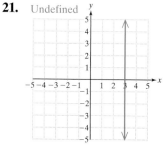

22. Zero

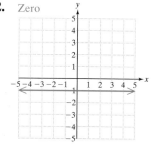

23. Positive

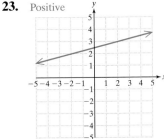

24. Positive

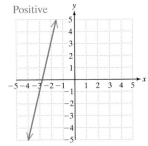

25. Negative

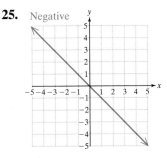

For Exercises 26–34, determine the slope by using the slope formula and any two points on the line. Check your answer by drawing a right triangle and labeling the "rise" and "run."

26. $m = 2$

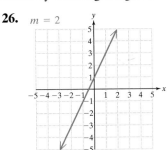

27. $m = \dfrac{1}{2}$

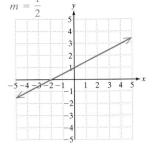

28. $m = -\dfrac{1}{3}$

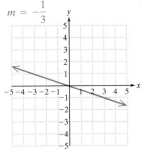

29. $m = -3$

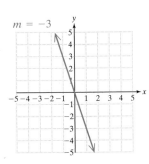

30. $m = 0$

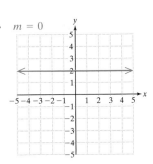

31. $m = 0$

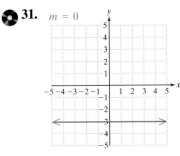

32. The slope is undefined.

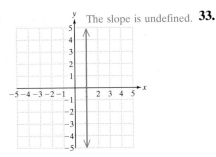

33. The slope is undefined.

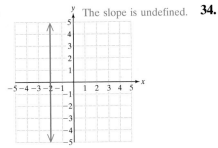

34. $m = -1$

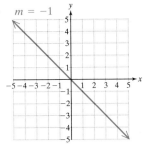

For Exercises 35–52, find the slope of the line that passes through the two points. **(See Examples 2–5.)**

35. $(2, 4)$ and $(-1, 3)$ $\dfrac{1}{3}$

36. $(0, 4)$ and $(3, 0)$ $-\dfrac{4}{3}$

37. $(-2, 3)$ and $(-1, 0)$ -3

38. $(-3, -4)$ and $(1, -5)$ $-\dfrac{1}{4}$

39. $(1, 5)$ and $(-4, 2)$ $\dfrac{3}{5}$

40. $(-6, -1)$ and $(-2, -3)$ $-\dfrac{1}{2}$

41. $(5, 3)$ and $(-2, 3)$ Zero

42. $(0, -1)$ and $(-4, -1)$ Zero

43. $(2, -7)$ and $(2, 5)$ Undefined

44. $(-4, 3)$ and $(-4, -4)$ Undefined

45. $\left(\dfrac{1}{2}, \dfrac{3}{5}\right)$ and $\left(\dfrac{1}{4}, -\dfrac{4}{5}\right)$ $\dfrac{28}{5}$

46. $\left(-\dfrac{2}{7}, \dfrac{1}{3}\right)$ and $\left(\dfrac{8}{7}, -\dfrac{5}{6}\right)$ $-\dfrac{49}{60}$

47. $(3, -1)$ and $(-5, 6)$ $-\dfrac{7}{8}$

48. $(-6, 5)$ and $(-10, 4)$ $\dfrac{1}{4}$

49. $(6.8, -3.4)$ and $(-3.2, 1.1)$
-0.45 or $-\dfrac{9}{20}$

50. $(-3.15, 8.25)$ and $(6.85, -4.25)$
-1.25 or $-\dfrac{5}{4}$

51. $(1994, 3.5)$ and $(2000, 2.6)$
-0.15 or $-\dfrac{3}{20}$

52. $(1988, 4.65)$ and $(1998, 9.25)$
0.46 or $\dfrac{23}{50}$

Objective 3: Parallel and Perpendicular Lines

For Exercises 53–60, the slope of a line is given. **(See Example 6.)**

a. Determine the slope of a line parallel to the given line.

b. Determine the slope of a line perpendicular to the given line.

53. $m = -2$ a. -2 b. $\dfrac{1}{2}$

54. $m = \dfrac{2}{3}$ a. $\dfrac{2}{3}$ b. $-\dfrac{3}{2}$

55. $m = 0$ a. 0 b. undefined

56. The slope is undefined.   a. undefined b. 0

57. $m = \dfrac{4}{5}$ a. $\dfrac{4}{5}$ b. $-\dfrac{5}{4}$

58. $m = -4$ a. -4 b. $\dfrac{1}{4}$

59. The slope is undefined.
a. undefined b. 0

60. $m = 0$ a. 0 b. undefined

For Exercises 61–66, let m_1 and m_2 represent the slopes of two lines. Determine if the lines are parallel, perpendicular, or neither. **(See Example 6.)**

61. $m_1 = -2, m_2 = \dfrac{1}{2}$
Perpendicular

62. $m_1 = \dfrac{2}{3}, m_2 = \dfrac{3}{2}$
Neither

63. $m_1 = 1, m_2 = \dfrac{4}{4}$
Parallel

64. $m_1 = \dfrac{3}{4}, m_2 = -\dfrac{8}{6}$
Perpendicular

65. $m_1 = \dfrac{2}{7}, m_2 = -\dfrac{2}{7}$
Neither

66. $m_1 = 5, m_2 = 5$
Parallel

For Exercises 67–72, find the slopes of the lines l_1 and l_2 defined by the two given points. Then determine whether l_1 and l_2 are parallel, perpendicular, or neither. **(See Example 7.)**

67. l_1: $(2, 4)$ and $(-1, -2)$
l_2: $(1, 7)$ and $(0, 5)$
l_1: $m = 2$, l_2: $m = 2$; parallel

68. l_1: $(0, 0)$ and $(-2, 4)$
l_2: $(1, -5)$ and $(-1, -1)$
l_1: $m = -2$, l_2: $m = -2$; parallel

69. l_1: $(1, 9)$ and $(0, 4)$
l_2: $(5, 2)$ and $(10, 1)$
l_1: $m = 5$, l_2: $m = -\dfrac{1}{5}$; perpendicular

70. l_1: $(3, -4)$ and $(-1, -8)$
l_2: $(5, -5)$ and $(-2, 2)$
l_1: $m = 1$, l_2: $m = -1$; perpendicular

71. l_1: $(4, 4)$ and $(0, 3)$
l_2: $(1, 7)$ and $(-1, -1)$
l_1: $m = \dfrac{1}{4}$, l_2: $m = 4$; neither

72. l_1: $(3, 5)$ and $(-2, -5)$
l_2: $(2, 0)$ and $(-4, -3)$
l_1: $m = 2$, l_2: $m = \dfrac{1}{2}$; neither

Objective 4: Applications of Slope: Rate of Change

73. In 1980, there were 304 thousand male inmates in federal and state prisons. By 2005, the number increased to 1479 thousand. Let x represent the year, and let y represent the number of prisoners (in thousands). **(See Example 8.)**

a. Using the ordered pairs $(1980, 304)$ and $(2005, 1479)$, find the slope of the line. $m = 47$

b. Interpret the slope in the context of this problem.
The number of male prisoners increased at a rate of 47 thousand per year during this time period.

Number of Male State and Federal Prisoners (in thousands) 1980–2005

(Source: U.S. Bureau of Justice Statistics)

74. In the year 1980, there were 12 thousand female inmates in federal and state prisons. By 2005, the number increased to 102 thousand.

Let *x* represent the year, and let *y* represent the number of prisoners (in thousands).

a. Using the ordered pairs (1980, 12) and (2005, 102), find the slope of the line.   *m* = 3.6

b. Interpret the slope in the context of this problem.
The number of female prisoners increased at a rate of 3.6 thousand per year during this time period.

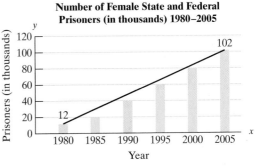

Number of Female State and Federal Prisoners (in thousands) 1980–2005

(*Source:* U.S. Bureau of Justice Statistics)

75. In 1985, the U.S. Postal Service charged $0.22 for first class letters and cards up to 1 oz. By 2007, the price had increased to $0.41. In the graph, *x* represents the year, and *y* represents the cost for 1 oz of first class postage.

a. Determine the slope of the line. Round to three decimal places.   *m* ≈ 0.009

b. Interpret the meaning of the slope in the context of this problem.

The slope *m* = 0.009 means that postage increased by about $0.009 per year (or equivalently 0.9¢ per year) during this time.

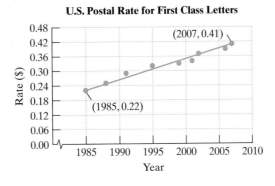

U.S. Postal Rate for First Class Letters

76. The distance, *d* (in miles), between a lightning strike and an observer is given by the equation $d = 0.2t$, where *t* is the time (in seconds) between seeing lightning and hearing thunder.

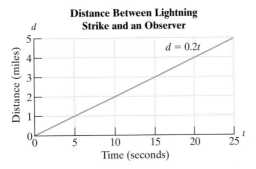

Distance Between Lightning Strike and an Observer

a. If an observer counts 5 sec between seeing lightning and hearing thunder, how far away was the lightning strike?   1 mi

b. If an observer counts 10 sec between seeing lightning and hearing thunder, how far away was the lightning strike?   2 mi

c. If an observer counts 15 sec between seeing lightning and hearing thunder, how far away was the lightning strike?   3 mi

d. What is the slope of the line? Interpret the meaning of the slope in the context of this problem.   *m* = 0.2; The distance between a lightning strike and an observer increases by 0.2 mi for every additional second between seeing lightning and hearing thunder.

Writing   Translating Expression   Geometry   Scientific Calculator   Video   NS E

Mixed Exercises

For Exercises 77–80, determine the slope of the line passing through points A and B.

77. Point A is located 3 units up and 4 units to the right of point B. $m = \dfrac{3}{4}$

78. Point A is located 2 units up and 5 units to the left of point B. $m = -\dfrac{2}{5}$

79. Point A is located 5 units to the right of point B. $m = 0$

80. Point A is located 3 units down from point B. The slope is undefined.

81. Graph the line through the point $(1, -2)$ having slope $\frac{2}{3}$. Then give two other points on the line.

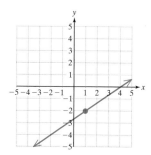

For example: $(4, 0)$ and $(-2, -4)$

82. Graph the line through the point $(-2, -3)$ having slope $\frac{3}{4}$. Then give two other points on the line.

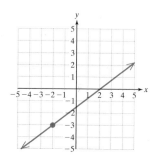

For example: $(2, 0)$ and $(-6, -6)$

83. Graph the line through the point $(2, 2)$ having slope -3. Then give two other points on the line.

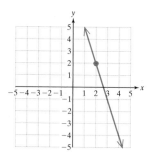

For example: $(3, -1)$ and $(1, 5)$

84. Graph the line through the point $(-1, 3)$ having slope -2. Then give two other points on the line.

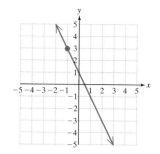

For example: $(0, 1)$ and $(-2, 5)$

For Exercises 85–90, draw a line as indicated. Answers may vary.

85. Draw a line with a positive slope and a positive y-intercept.

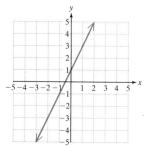

86. Draw a line with a positive slope and a negative y-intercept.

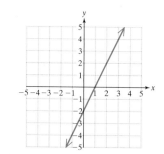

87. Draw a line with a negative slope and a negative y-intercept.

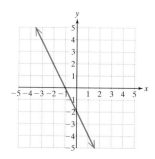

88. Draw a line with a negative slope and positive y-intercept.

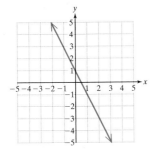

89. Draw a line with a zero slope and a positive y-intercept.

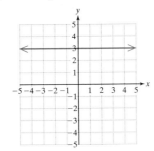

90. Draw a line with undefined slope and a negative x-intercept.

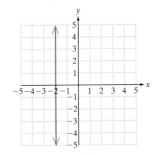

Expanding Your Skills

91. Determine the slope between the points $(a + b, 4m - n)$ and $(a - b, m + 2n)$. $\frac{3m - 3n}{2b}$ or $\frac{-3m + 3n}{-2b}$

92. Determine the slope between the points $(3c - d, s + t)$ and $(c - 2d, s - t)$. $\frac{2t}{2c + d}$ or $\frac{-2t}{-2c - d}$

93. Determine the x-intercept of the line $ax + by = c$. $\left(\frac{c}{a}, 0\right)$

94. Determine the y-intercept of the line $ax + by = c$. $\left(0, \frac{c}{b}\right)$

95. Find another point on the line that contains the point $(2, -1)$ and has a slope of $\frac{2}{5}$. For example: $(7, 1)$

96. Find another point on the line that contains the point $(-3, 4)$ and has a slope of $\frac{1}{4}$. For example: $(1, 5)$

Slope-Intercept Form of a Line

Section 3.4

1. Slope-Intercept Form of a Line

In Section 3.2, we learned that the solutions to an equation of the form $Ax + By = C$ (where A and B are not both zero) represent a line in a rectangular coordinate system. An equation of a line written in this way is said to be in **standard form**. In this section, we will learn a new form, called **slope-intercept form**, which is useful in determining the slope and y-intercept of a line.

Let $(0, b)$ represent the y-intercept of a line. Let (x, y) represent any other point on the line. Then the slope of the line can be found as follows:

Let $(0, b)$ represent (x_1, y_1), and let (x, y) represent (x_2, y_2). Apply the slope formula.

$$m = \frac{(y_2 - y_1)}{(x_2 - x_1)} \rightarrow m = \frac{y - b}{x - 0}$$ Apply the slope formula.

$$m = \frac{y - b}{x}$$ Simplify.

$$mx = \left(\frac{y - b}{x}\right)x$$ Multiply by x to clear fractions.

$$mx = y - b$$

$$mx + b = y - b + b$$ To isolate y, add b to both sides.

$$mx + b = y \quad \text{or} \quad y = mx + b$$ The equation is in slope-intercept form.

Objectives

1. **Slope-Intercept Form of a Line**
2. **Graphing a Line from Its Slope and y-Intercept**
3. **Determining Whether Two Lines Are Parallel, Perpendicular, or Neither**
4. **Writing an Equation of a Line Using Slope-Intercept Form**

> **DEFINITION** Slope-Intercept Form of a Line
> $y = mx + b$ is the slope-intercept form of a line.
> m is the slope and the point $(0, b)$ is the y-intercept.

Skill Practice

Identify the slope and the y-intercept.
1. $y = 4x + 6$
2. $y = 3.5x - 4.2$
3. $y = -7$

Classroom Examples: p. 248, Exercises 12 and 16

Example 1 Identifying the Slope and y-Intercept of a Line

For each equation, identify the slope and y-intercept.

a. $y = 3x - 1$ **b.** $y = -2.7x + 5$ **c.** $y = 4x$

Solution:

Each equation is written in slope-intercept form, $y = mx + b$. The slope is the coefficient of x, and the y-intercept is determined by the constant term.

a. $y = 3x - 1$ The slope is 3. The y-intercept is $(0, -1)$.

b. $y = -2.7x + 5$ The slope is -2.7. The y-intercept is $(0, 5)$.

c. $y = 4x$ can be written as $y = 4x + 0$. The slope is 4.
The y-intercept is $(0, 0)$.

Given the equation of a line, we can write the equation in slope-intercept form by solving the equation for the y-variable. This is demonstrated in the next example.

Skill Practice

Given the equation of the line $2x - 6y = -3$.
4. Write the equation in slope-intercept form.
5. Identify the slope and the y-intercept.

Classroom Example: p. 248, Exercise 22

Example 2 Identifying the Slope and y-Intercept of a Line

Given the equation of the line $-5x - 2y = 6$,

a. Write the equation in slope-intercept form.

b. Identify the slope and y-intercept.

Solution:

a. Write the equation in slope-intercept form, $y = mx + b$, by solving for y.

$$-5x - 2y = 6$$

$$-2y = 5x + 6 \qquad \text{Add } 5x \text{ to both sides.}$$

$$\frac{-2y}{-2} = \frac{5x + 6}{-2} \qquad \text{Divide both sides by } -2.$$

$$y = \frac{5x}{-2} + \frac{6}{-2} \qquad \text{Divide each term by } -2 \text{ and simplify.}$$

$$y = -\frac{5}{2}x - 3 \qquad \text{Slope-intercept form.}$$

b. The slope is $-\frac{5}{2}$, and the y-intercept is $(0, -3)$.

Answers

1. slope: 4; y-intercept: $(0, 6)$
2. slope: 3.5; y-intercept: $(0, -4.2)$
3. slope: 0; y-intercept: $(0, -7)$
4. $y = \frac{1}{3}x + \frac{1}{2}$
5. slope is $\frac{1}{3}$; y-intercept is $\left(0, \frac{1}{2}\right)$

2. Graphing a Line from Its Slope and y-Intercept

Slope-intercept form is a useful tool to graph a line. The y-intercept is a known point on the line. The slope indicates the direction of the line and can be used to find a second point. Using slope-intercept form to graph a line is demonstrated in the next example.

Example 3 Graphing a Line Using the Slope and *y*-Intercept

Graph the equation of the line $y = -\frac{5}{2}x - 3$ by using the slope and *y*-intercept.

Solution:

First plot the *y*-intercept, $(0, -3)$.

The slope, $m = -\frac{5}{2}$, can be written as

$$m = \frac{-5}{2} \quad \longleftarrow \text{The change in } y \text{ is } -5.$$
$$\phantom{m = \frac{-5}{2}} \quad \longleftarrow \text{The change in } x \text{ is } 2.$$

To find a second point on the line, start at the *y*-intercept and move down 5 units and to the right 2 units. Then draw the line through the two points (Figure 3-26).

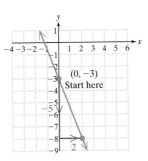

Figure 3-26

Similarly, the slope can be written as

$$m = \frac{5}{-2} \quad \longleftarrow \text{The change in } y \text{ is } 5.$$
$$\phantom{m = \frac{5}{-2}} \quad \longleftarrow \text{The change in } x \text{ is } -2.$$

To find a second point, start at the *y*-intercept and move up 5 units and to the left 2 units. Then draw the line through the two points (Figure 3-27).

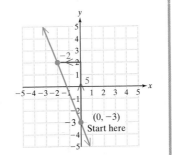

Figure 3-27

Skill Practice

6. Graph the equation by using the slope and the *y*-intercept.
$$y = 2x - 3$$

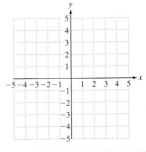

Classroom Example: p. 249, Exercise 42

Skill Practice

7. Graph the equation by using the slope and the *y*-intercept.
$$y = -\frac{1}{4}x$$

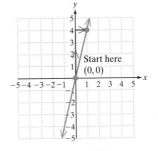

Classroom Example: p. 250, Exercise 50

Example 4 Graphing a Line from Its Slope and *y*-Intercept

Graph the equation of the line $y = 4x$ by using the slope and *y*-intercept.

Solution:

The line can be written as $y = 4x + 0$. Therefore, we can plot the *y*-intercept at $(0, 0)$. The slope $m = 4$ can be written as

$$m = \frac{4}{1}$$

with The change in *y* is 4.

The change in *x* is 1.

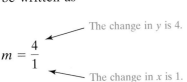

Figure 3-28

To find a second point on the line, start at the *y*-intercept and move up 4 units and to the right 1 unit. Then draw the line through the two points (Figure 3-28).

Answers

6.

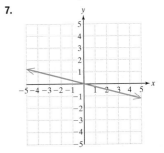

7.

3. Determining Whether Two Lines Are Parallel, Perpendicular, or Neither

The slope-intercept form provides a means to find the slope of a line by inspection. Recall that if the slopes of two lines are known, then we can compare the slopes to determine if the lines are parallel, perpendicular, or neither parallel nor perpendicular. (Two distinct nonvertical lines are parallel if their slopes are equal. Two lines are perpendicular if the slope of one line is the opposite of the reciprocal of the slope of the other line.)

Skill Practice

For each pair of lines determine if they are parallel, perpendicular, or neither.

8. $y = 3x - 5$
$y = -3x - 15$

9. $x - 5y = 10$
$5x - 1 = -y$

10. $y = \dfrac{5}{6}x - \dfrac{1}{2}$
$y = \dfrac{5}{6}x + \dfrac{1}{2}$

11. $y = -5$
$x = 6$

Classroom Examples: p. 250, Exercises 56, 62, and 64

Example 5   **Determining If Two Lines Are Parallel, Perpendicular, or Neither**

For each pair of lines, determine if they are parallel, perpendicular, or neither.

a. l_1: $y = 3x - 5$ **b.** l_1: $x - 3y = -9$
  l_2: $y = 3x + 1$ l_2: $3x = -y + 4$

c. l_1: $y = \frac{3}{2}x + 2$ **d.** l_1: $x = 2$
  l_2: $y = \frac{2}{3}x + 1$ l_2: $2y = 8$

Solution:

a. l_1: $y = 3x - 5$ The slope of l_1 is 3.
  l_2: $y = 3x + 1$ The slope of l_2 is 3.

Because the slopes are the same, the lines are parallel.

b. First write the equation of each line in slope-intercept form.

l_1: $x - 3y = -9$ l_2: $3x = -y + 4$

$\qquad -3y = -x - 9$ $3x + y = 4$

$\qquad \dfrac{-3y}{-3} = \dfrac{-x}{-3} - \dfrac{9}{-3}$ $y = -3x + 4$

$\qquad\qquad y = \dfrac{1}{3}x + 3$

l_1: $y = \frac{1}{3}x + 3$ The slope of l_1 is $\frac{1}{3}$.

l_2: $y = -3x + 4$ The slope of l_2 is -3.

The slope of $\frac{1}{3}$ is the opposite of the reciprocal of -3. Therefore, the lines are perpendicular.

c. l_1: $y = \frac{3}{2}x + 2$ The slope of l_1 is $\frac{3}{2}$.

l_2: $y = \frac{2}{3}x + 1$ The slope of l_2 is $\frac{2}{3}$.

The slopes are not the same. Therefore, the lines are not parallel. The values of the slopes are reciprocals, but they are not opposite in sign. Therefore, the lines are not perpendicular. The lines are neither parallel nor perpendicular.

d. The equation $x = 2$ represents a vertical line because the equation is in the form $x = k$.

The equation $2y = 8$ can be simplified to $y = 4$, which represents a horizontal line.

In this example, we do not need to analyze the slopes because vertical lines and horizontal lines are perpendicular.

4. Writing an Equation of a Line Using Slope-Intercept Form

Answers

8. Neither   **9.** Perpendicular
10. Parallel   **11.** Perpendicular

The slope-intercept form of a line can be used to write an equation of a line when the slope is known and the y-intercept is known.

Example 6 Writing an Equation of a Line Using Slope-Intercept Form

Write an equation of the line whose slope is $\frac{2}{3}$ and whose y-intercept is $(0, 8)$.

Solution:

The slope is given as $m = \frac{2}{3}$, and the y-intercept $(0, b)$ is given as $(0, 8)$. Substitute the values $m = \frac{2}{3}$ and $b = 8$ into the slope-intercept form of a line.

$$y = mx + b$$
$$y = \frac{2}{3}x + 8$$

Skill Practice

12. Write an equation of the line whose slope is -4 and y-intercept is $(0, -10)$.

Classroom Example: p. 251, Exercise 70

Example 7 Writing an Equation of a Line Using Slope-Intercept Form

Write an equation of the line that passes through the origin and whose slope is -2.

Solution:

The slope is given as $m = -2$. Furthermore, the line passes through the origin. Therefore, the y-intercept is $(0, 0)$ and the corresponding value of b is 0. The slope-intercept form of the line becomes:

$$y = mx + b$$
$$y = -2x + 0 \text{ or equivalently } y = -2x.$$

Skill Practice

13. Write an equation of the line that passes through the origin and has slope $\frac{6}{5}$.

Classroom Example: p. 251, Exercise 76

Answers

12. $y = -4x - 10$

13. $y = \frac{6}{5}x$

Calculator Connections

Topic: Using the *ZSquare* Option in Zoom

In Example 5(b) we found that the equations $y = \frac{1}{3}x + 3$ and $y = -3x + 4$ represent perpendicular lines. We can verify our results by graphing the lines on a graphing calculator.

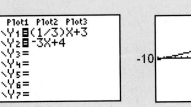

Notice that the lines do not appear perpendicular in the calculator display. That is, they do not appear to form a right angle at the point of intersection. Because many calculators have a rectangular screen, the standard viewing window is elongated in the horizontal direction. To eliminate this distortion, try using a *ZSquare* option, which is located under the Zoom menu. This feature will set the viewing window so that equal distances on the display denote an equal number of units on the graph.

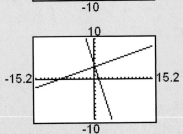

Calculator Exercises

For each pair of lines, determine if the lines are parallel, perpendicular, or neither. Then use a square viewing window to graph the lines on a graphing calculator to verify your results.

1. $x + y = 1$
   $x - y = -3$ Perpendicular

2. $3x + y = -2$
   $6x + 2y = 6$ Parallel

3. $2x - y = 4$
   $3x + 2y = 4$ Neither

4. Graph the lines defined by $y = x + 1$ and $y = 0.99x + 3$. Are these lines parallel? Explain.
   The lines may appear parallel; however, they are not parallel because the slopes are different.

5. Graph the lines defined by $y = -2x - 1$ and $y = -2x - 0.99$. Are these lines the same? Explain.
   The lines may appear to coincide on a graph; however, they are not the same line because the y-intercepts are different.

6. Graph the line defined by $y = 0.001x + 3$. Is this line horizontal? Explain.
   The line may appear to be horizontal, but it is not. The slope is 0.001 rather than 0.

Section 3.4 Practice Exercises

Study Skills Exercises

1. When taking a test, go through the test and do all the problems that you know first. Then go back and work on the problems that were more difficult. Give yourself a time limit for how much time you spend on each problem (maybe 3 to 5 min the first time through). Circle the importance of each statement.

	not important	somewhat important	very important
a. Read through the entire test first.	1	2	3
b. If time allows, go back and check each problem.	1	2	3
c. Write out all steps instead of doing the work in your head.	1	2	3

2. Define the key terms:

   a. **slope-intercept form of a line**   b. **standard form of a line**

Review Exercises

For Exercises 3–10, determine the *x*- and *y*-intercepts, if they exist.

3. $x - 5y = 10$
   x-intercept: $(10, 0)$;
   y-intercept: $(0, -2)$

4. $3x + y = -12$
   x-intercept: $(-4, 0)$;
   y-intercept: $(0, -12)$

5. $3y = -9$
   x-intercept: none;
   y-intercept: $(0, -3)$

6. $2 + y = 5$
   x-intercept: none;
   y-intercept: $(0, 3)$

7. $-4x = 6y$
   x-intercept: $(0, 0)$;
   y-intercept: $(0, 0)$

8. $-x + 3 = 8$
   x-intercept: $(-5, 0)$;
   y-intercept: none

9. $5x = 20$
   x-intercept: $(4, 0)$;
   y-intercept: none

10. $y = \frac{1}{2}x$
    x-intercept: $(0, 0)$;
    y-intercept: $(0, 0)$

Objective 1: Slope-Intercept Form of a Line

For Exercises 11–30, identify the slope and *y*-intercept, if they exist. **(See Examples 1–2.)**

11. $y = -2x + 3$
    $m = -2$; *y*-intercept: $(0, 3)$

12. $y = \frac{2}{3}x + 5$
    $m = \frac{2}{3}$; *y*-intercept: $(0, 5)$

13. $y = x - 2$
    $m = 1$; *y*-intercept: $(0, -2)$

14. $y = -x + 6$
    $m = -1$; *y*-intercept: $(0, 6)$

15. $y = -x$
    $m = -1$; *y*-intercept: $(0, 0)$

16. $y = -5x$
    $m = -5$; *y*-intercept: $(0, 0)$

17. $y = \frac{3}{4}x - 1$
    $m = \frac{3}{4}$; *y*-intercept: $(0, -1)$

18. $y = x - \frac{5}{3}$
    $m = 1$; *y*-intercept: $\left(0, -\frac{5}{3}\right)$

19. $2x - 5y = 4$
    $m = \frac{2}{5}$; *y*-intercept: $\left(0, -\frac{4}{5}\right)$

20. $3x + 2y = 9$
    $m = -\frac{3}{2}$; *y*-intercept: $\left(0, \frac{9}{2}\right)$

21. $3x - y = 5$
    $m = 3$; *y*-intercept: $(0, -5)$

22. $7x - 3y = -6$
    $m = \frac{7}{3}$; *y*-intercept: $(0, 2)$

23. $x + y = 6$
    $m = -1$; *y*-intercept: $(0, 6)$

24. $x - y = 1$
    $m = 1$; *y*-intercept: $(0, -1)$

25. $x + 6 = 8$
    Undefined slope; no *y*-intercept

26. $-4 + x = 1$
    Undefined slope; no *y*-intercept

27. $-8y = 2$
    $m = 0$; *y*-intercept: $\left(0, -\frac{1}{4}\right)$

28. $1 - y = 9$
    $m = 0$; *y*-intercept: $(0, -8)$

29. $3y - 2x = 0$
    $m = \frac{2}{3}$; *y*-intercept: $(0, 0)$

30. $5x = 6y$
    $m = \frac{5}{6}$; *y*-intercept: $(0, 0)$

Objective 2: Graphing a Line from Its Slope and *y*-Intercept

For Exercises 31–34, graph the line using the slope and *y*-intercept. **(See Examples 3–4.)**

31. Graph the line through the point $(0, 2)$, having a slope of -4.

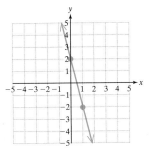

32. Graph the line through the point $(0, -1)$, having a slope of -3.

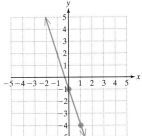

33. Graph the line through the point $(0, -5)$, having a slope of $\frac{3}{2}$.

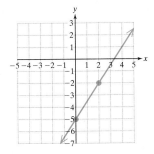

34. Graph the line through the point $(0, 3)$, having a slope of $-\frac{1}{4}$.

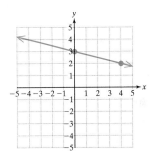

For Exercises 35–40, match the equation with the graph (a–f) by identifying if the slope is positive or negative and if the *y*-intercept is positive, negative, or zero.

35. $y = 2x + 3$ b

36. $y = -3x - 2$ d

37. $y = -\frac{1}{3}x + 3$ e

38. $y = \frac{1}{2}x - 2$ a

39. $y = x$ c

40. $y = -2x$ f

a.

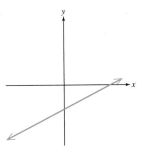

b.

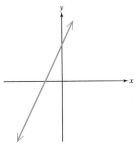

c.

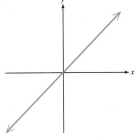

d.

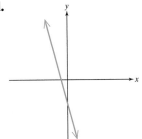

e.

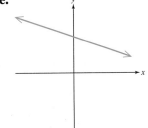

f.

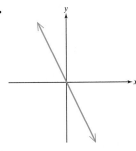

For Exercises 41–52, write each equation in slope-intercept form (if possible) and graph the line. **(See Examples 3–4.)**

41. $x - 2y = 6$
$y = \dfrac{1}{2}x - 3$
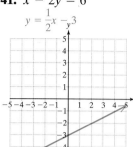

42. $5x - 2y = 2$
$y = \dfrac{5}{2}x - 1$
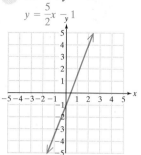

43. $2x + y = 9$
$y = -2x + 9$
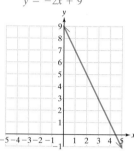

44. $-6x + y = 8$
$y = 6x + 8$
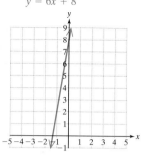

45. $2x = -4y + 6$
$y = -\dfrac{1}{2}x + \dfrac{3}{2}$
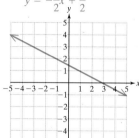

46. $3x = y - 7$
$y = 3x + 7$
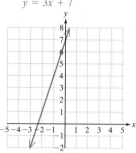

47. $x + y = 0$
$y = -x$
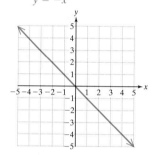

48. $x - y = 0$
$y = x$
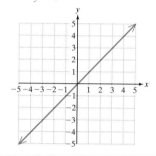

49. $5y = 4x$
$y = \dfrac{4}{5}x$
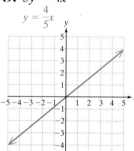

50. $-2x = 5y$
$y = -\dfrac{2}{5}x$
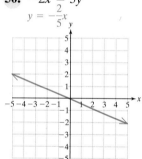

51. $3y + 2 = 0$
$y = -\dfrac{2}{3}$
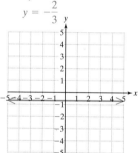

52. $1 + 5y = 6$
$y = 1$
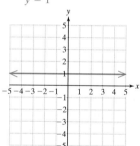

Objective 3: Determining Whether Two Lines Are Parallel, Perpendicular, or Neither

For Exercises 53–68, determine if the equations represent parallel lines, perpendicular lines, or neither. **(See Example 5.)**

53. l_1: $y = -2x - 3$
l_2: $y = \dfrac{1}{2}x + 4$
Perpendicular

54. l_1: $y = \dfrac{4}{3}x - 2$
l_2: $y = -\dfrac{3}{4}x + 6$
Perpendicular

55. l_1: $y = \dfrac{4}{5}x - \dfrac{1}{2}$
l_2: $y = \dfrac{5}{4}x - \dfrac{2}{3}$
Neither

56. l_1: $y = \dfrac{1}{5}x + 1$
l_2: $y = 5x - 3$
Neither

57. l_1: $y = -9x + 6$
l_2: $y = -9x - 1$
Parallel

58. l_1: $y = 4x - 1$
l_2: $y = 4x + \dfrac{1}{2}$
Parallel

59. l_1: $x = 3$
l_2: $y = \dfrac{7}{4}$
Perpendicular

60. l_1: $y = \dfrac{2}{3}$
l_2: $x = 6$
Perpendicular

61. l_1: $2x = 4$
l_2: $6 = x$
Parallel

62. l_1: $2y = 7$
l_2: $y = 4$
Parallel

63. l_1: $2x + 3y = 6$
l_2: $3x - 2y = 12$
Perpendicular

64. l_1: $4x + 5y = 20$
l_2: $5x - 4y = 60$
Perpendicular

65. l_1: $4x + 2y = 6$ 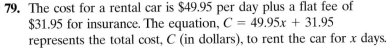 **66.** l_1: $3x + y = 5$ **67.** l_1: $y = \dfrac{1}{5}x - 3$ **68.** l_1: $y = \dfrac{1}{3}x + 2$

   l_2: $4x + 8y = 16$ l_2: $x + 3y = 18$ l_2: $2x - 10y = 20$ l_2: $-x + 3y = 12$
   Neither   Neither   Parallel   Parallel

Objective 4: Writing an Equation of a Line Using Slope-Intercept Form

For Exercises 69–78, write an equation of the line given the following information. Write the answer in slope-intercept form if possible. **(See Examples 6–7.)**

69. The slope is $-\frac{1}{3}$, and the y-intercept is $(0, 2)$.
   $y = -\frac{1}{3}x + 2$

70. The slope is $\frac{2}{3}$, and the y-intercept is $(0, -1)$.
   $y = \frac{2}{3}x - 1$

71. The slope is 10, and the y-intercept is $(0, -19)$.
   $y = 10x - 19$

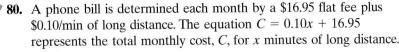 **72.** The slope is -14, and the y-intercept is $(0, 2)$.
   $y = -14x + 2$

73. The slope is 0, and the y-intercept is -11.
   $y = -11$

74. The slope is 0, and the y-intercept is $\dfrac{6}{7}$.
   $y = \frac{6}{7}$

75. The slope is 5, and the line passes through the origin.
   $y = 5x$

76. The slope is -3, and the line passes through the origin.
   $y = -3x$

77. The slope is 6, and the line passes through the point $(0, -2)$.
   $y = 6x - 2$

78. The slope is -4, and the line passes through the point $(0, -3)$.
   $y = -4x - 3$

Expanding Your Skills

79. The cost for a rental car is $49.95 per day plus a flat fee of $31.95 for insurance. The equation, $C = 49.95x + 31.95$ represents the total cost, C (in dollars), to rent the car for x days.

   a. Identify the slope. Interpret the meaning of the slope in the context of this problem.
   $m = 49.95$. The cost increases $49.95 per day.

   b. Identify the C-intercept. Interpret the meaning of the C-intercept in the context of this problem.
   $(0, 31.95)$. The base fee for renting a car is $31.95.

   c. Use the equation to determine how much it would cost to rent the car for 1 week. $381.60

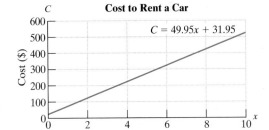
Cost to Rent a Car
$C = 49.95x + 31.95$

80. A phone bill is determined each month by a $16.95 flat fee plus $0.10/min of long distance. The equation $C = 0.10x + 16.95$ represents the total monthly cost, C, for x minutes of long distance.

   a. Identify the slope. Interpret the meaning of the slope in the context of this problem.
   $m = 0.10$. The cost increases by $0.10/min of long distance.

   b. Identify the C-intercept. Interpret the meaning of the C-intercept in the context of this problem.
   $(0, 16.95)$. The monthly bill for 0 min is $16.95.

   c. Use the equation to determine the total cost of 234 min of long distance.
   The cost is $40.35

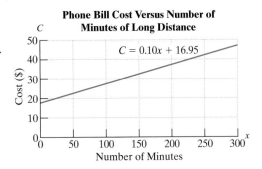
Phone Bill Cost Versus Number of Minutes of Long Distance
$C = 0.10x + 16.95$

81. A linear equation is written in standard form if it can be written as $Ax + By = C$, where A and B are not both zero. Write the equation $Ax + By = C$ in slope-intercept form to show that the slope is given by the ratio, $-\frac{A}{B}$. ($B \neq 0$.)
   $y = -\frac{A}{B}x + \frac{C}{B}$; the slope is $-\frac{A}{B}$.

For Exercises 82–85, use the result of Exercise 81 to find the slope of the line.

82. $2x + 5y = 8$
   $m = -\dfrac{2}{5}$

83. $6x + 7y = -9$
   $m = -\dfrac{6}{7}$

84. $4x - 3y = -5$
   $m = \dfrac{4}{3}$

85. $11x - 8y = 4$
   $m = \dfrac{11}{8}$

Problem Recognition Exercises

Linear Equations in Two Variables

For Exercises 1–20, choose the equation(s) from the column on the right whose graph satisfies the condition described. Give all possible answers.

1. Line whose slope is positive.   a, c, d

2. Line whose slope is negative.   b, f, h

3. Line that passes through the origin.   a

4. Line that contains the point (3, –2).   f

5. Line whose y-intercept is (0, 4).   b, f

6. Line whose y-intercept is (0, –5).   c

7. Line whose slope is $\dfrac{1}{2}$.   c, d

8. Line whose slope is –2.   f

9. Line whose slope is 0.   e

10. Line whose slope is undefined.   g

11. Line that is parallel to the line with equation $y = -\dfrac{2}{3}x + 4$.   b

12. Line perpendicular to the line with equation $y = 2x + 9$.   h

13. Line that is vertical.   g

14. Line that is horizontal.   e

15. Line whose x-intercept is (10, 0).   c

16. Line whose x-intercept is (6, 0).   b, h

17. Line that is parallel to the x-axis.   e

18. Line that is perpendicular to the y-axis.   e

19. Line with a negative slope and positive y-intercept.   b, f, h

20. Line with a positive slope and negative y-intercept.   c, d

a. $y = 5x$

b. $2x + 3y = 12$

c. $y = \dfrac{1}{2}x - 5$

d. $3x - 6y = 10$

e. $2y = -8$

f. $y = -2x + 4$

g. $3x = 1$

h. $x + 2y = 6$

Point-Slope Formula

1. Writing an Equation of a Line Using the Point-Slope Formula

Objectives

1. Writing an Equation of a Line Using the Point-Slope Formula
2. Writing an Equation of a Line Given Two Points
3. Writing an Equation of a Line Parallel or Perpendicular to Another Line
4. Different Forms of Linear Equations: A Summary

In Section 3.4, the slope-intercept form of a line was used as a tool to construct an equation of a line. Another useful tool to determine an equation of a line is the point-slope formula. The point-slope formula can be derived from the slope formula as follows:

Suppose a line passes through a given point (x_1, y_1) and has slope m. If (x, y) is any other point on the line, then:

$$m = \frac{y - y_1}{x - x_1} \qquad \text{Slope formula}$$

$$m(x - x_1) = \frac{y - y_1}{x - x_1}(x - x_1) \qquad \text{Clear fractions.}$$

$$m(x - x_1) = y - y_1$$

$$y - y_1 = m(x - x_1) \qquad \text{Point-slope formula}$$

> **FORMULA** Point-Slope Formula
>
> The **point-slope formula** is given by
>
> $$y - y_1 = m(x - x_1)$$
>
> where m is the slope of the line and (x_1, y_1) is a known point on the line.

Example 1 demonstrates how to use the point-slope formula to find an equation of a line when a point on the line and slope are given.

Example 1 Writing an Equation of a Line Using the Point-Slope Formula

Use the point-slope formula to write an equation of the line having a slope of 3 and passing through the point $(-2, -4)$. Write the answer in slope-intercept form.

Solution:

The slope of the line is given: $m = 3$.

A point on the line is given: $(x_1, y_1) = (-2, -4)$.

Skill Practice

1. Use the point-slope formula to write an equation of the line having a slope of -4 and passing through $(-1, 5)$. Write the answer in slope-intercept form.

Classroom Example: p. 258, Exercise 12

Answer

1. $y = -4x + 1$

Instructor Note: Show students an alternative method.

$y = mx + b$ replace m, x, y
$-4 = 3(-2) + b$
$-4 = -6 + b$
$2 = b$

$y = mx + b$ replace m, b
$y = 3x + 2$

The point-slope formula:

$$y - y_1 = m(x - x_1)$$

$y - (-4) = 3[x - (-2)]$ Substitute $m = 3$, $x_1 = -2$, and $y_1 = -4$.

$y + 4 = 3(x + 2)$ Simplify. Because the final answer is required in slope-intercept form, simplify the equation and solve for y.

$y + 4 = 3x + 6$ Apply the distributive property.

$y = 3x + 2$ Slope-intercept form

The equation $y = 3x + 2$ from Example 1 is graphed in Figure 3-29. Notice that the line does indeed pass through the point $(-2, -4)$.

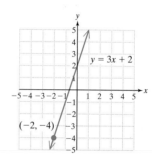

Figure 3-29

2. Writing an Equation of a Line Given Two Points

Example 2 is similar to Example 1; however, the slope must first be found from two given points.

Classroom Example: p. 259, Exercise 26

TIP: The point-slope formula can be applied using either given point for (x_1, y_1). In Example 2, using the point $(4, -1)$ for (x_1, y_1) produces the same result.

$y - y_1 = m(x - x_1)$
$y - (-1) = -1(x - 4)$
$y + 1 = -x + 4$
$y = -x + 3$

Example 2   Writing an Equation of a Line Given Two Points

Use the point-slope formula to find an equation of the line passing through the points $(-2, 5)$ and $(4, -1)$. Write the final answer in slope-intercept form.

Solution:

Given two points on a line, the slope can be found with the slope formula.

$\underset{(x_1, y_1)}{(-2, 5)}$ and $\underset{(x_2, y_2)}{(4, -1)}$ Label the points.

$$m = \frac{y_2 - y_1}{x_2 - x_1} = \frac{(-1) - (5)}{(4) - (-2)} = \frac{-6}{6} = -1$$

To apply the point-slope formula, use the slope, $m = -1$ and either given point. We will choose the point $(-2, 5)$ as (x_1, y_1).

$y - y_1 = m(x - x_1)$

$y - 5 = -1[x - (-2)]$ Substitute $m = -1$, $x_1 = -2$, and $y_1 = 5$.

$y - 5 = -1(x + 2)$ Simplify.

$y - 5 = -x - 2$

$y = -x + 3$

Answer

2. $y = 2x - 3$

The solution to Example 2 can be checked by graphing the line $y = -x + 3$ using the slope and y-intercept. Notice that the line passes through the points $(-2, 5)$ and $(4, -1)$ as expected. See Figure 3-30.

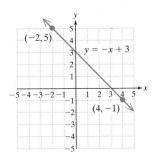

Figure 3-30

3. Writing an Equation of a Line Parallel or Perpendicular to Another Line

Example 3 Writing an Equation of a Line Parallel to Another Line

Use the point-slope formula to find an equation of the line passing through the point $(-1, 0)$ and parallel to the line $y = -4x + 3$. Write the final answer in slope-intercept form.

Solution:

Figure 3-31 shows the line $y = -4x + 3$ (pictured in black) and a line parallel to it (pictured in blue) that passes through the point $(-1, 0)$. The equation of the given line, $y = -4x + 3$, is written in slope-intercept form, and its slope is easily identified as -4. The line parallel to the given line must also have a slope of -4.

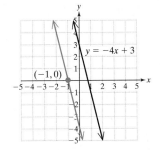

Figure 3-31

Apply the point-slope formula using $m = -4$ and the point $(x_1, y_1) = (-1, 0)$.

$$y - y_1 = m(x - x_1)$$
$$y - 0 = -4[x - (-1)]$$
$$y = -4(x + 1)$$
$$y = -4x - 4$$

TIP: When writing an equation of a line, slope-intercept form or standard form is usually preferred. For instance, the solution to Example 3 can be written as follows.

Slope-intercept form:
$y = -4x - 4$

Standard form:
$4x + y = -4$

Example 4 Writing an Equation of a Line Perpendicular to Another Line

Use the point-slope formula to find an equation of the line passing through the point $(-3, 1)$ and perpendicular to the line $3x + y = -2$. Write the final answer in slope-intercept form.

Solution:

The given line can be written in slope-intercept form as $y = -3x - 2$. The slope of this line is -3. Therefore, the slope of a line perpendicular to the given line is $\frac{1}{3}$.

Apply the point-slope formula with $m = \frac{1}{3}$, and $(x_1, y_1) = (-3, 1)$.

$y - y_1 = m(x - x_1)$	Point-slope formula
$y - (1) = \frac{1}{3}[x - (-3)]$	Substitute $m = \frac{1}{3}$, $x_1 = -3$, and $y_1 = 1$.
$y - 1 = \frac{1}{3}(x + 3)$	To write the final answer in slope-intercept form, simplify the equation and solve for y.
$y - 1 = \frac{1}{3}x + 1$	Apply the distributive property.
$y = \frac{1}{3}x + 2$	Add 1 to both sides.

A sketch of the perpendicular lines $y = \frac{1}{3}x + 2$ and $y = -3x - 2$ is shown in Figure 3-32. Notice that the line $y = \frac{1}{3}x + 2$ passes through the point $(-3, 1)$.

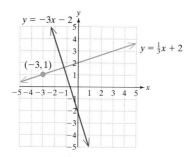

Figure 3-32

4. Different Forms of Linear Equations: A Summary

A linear equation can be written in several different forms, as summarized in Table 3-3.

Table 3-3

Form	Example	Comments
Standard Form $Ax + By = C$	$4x + 2y = 8$	A and B must not both be zero.
Horizontal Line $y = k$ (k is constant)	$y = 4$	The slope is zero, and the y-intercept is $(0, k)$.
Vertical Line $x = k$ (k is constant)	$x = -1$	The slope is undefined, and the x-intercept is $(k, 0)$.
Slope-Intercept Form $y = mx + b$ the slope is m y-intercept is $(0, b)$	$y = -3x + 7$ Slope $= -3$ y-intercept is $(0, 7)$	Solving a linear equation for y results in slope-intercept form. The coefficient of the x-term is the slope, and the constant defines the location of the y-intercept.
Point-Slope Formula $y - y_1 = m(x - x_1)$	$m = -3$ $(x_1, y_1) = (4, 2)$ $y - 2 = -3(x - 4)$	This formula is typically used to build an equation of a line when a point on the line is known and the slope of the line is known.

Concept Connections

5. Explain why the point-slope formula cannot be used to write an equation for a vertical line.

6. Explain how to write an equation for a vertical line.

Answers

5. The point-slope formula requires a value for the slope and a vertical line has an undefined slope.

6. The form $x = k$ must be used. Then the appropriate value is substituted for k.

Although standard form and slope-intercept form can be used to express an equation of a line, often the slope-intercept form is used to give a *unique* representation of the line. For example, the following linear equations are all written in standard form, yet they each define the same line.

$$2x + 5y = 10$$

$$-4x - 10y = -20$$

$$6x + 15y = 30$$

$$\frac{2}{5}x + y = 2$$

The line can be written uniquely in slope-intercept form as: $y = -\frac{2}{5}x + 2$.

Although it is important to understand and apply slope-intercept form and the point-slope formula, they are not necessarily applicable to all problems, particularly when dealing with a horizontal or vertical line.

Example 5 **Writing an Equation of a Line**

Find an equation of the line passing through the point $(2, -4)$ and parallel to the *x*-axis.

Solution:

Because the line is parallel to the *x*-axis, the line must be horizontal. Recall that all horizontal lines can be written in the form $y = k$, where k is a constant. A quick sketch can help find the value of the constant. See Figure 3-33.

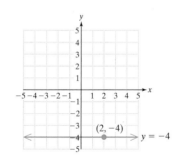

Figure 3-33

Because the line must pass through a point whose *y*-coordinate is -4, then the equation of the line must be $y = -4$.

Section 3.5 Practice Exercises

Boost *your* GRADE at ALEKS.com!

ALEKS version 3.0

- Practice Problems
- Self-Tests
- NetTutor

- e-Professors
- Videos

For additional exercises, see Classroom Activity 3.5A in the *Instructor's Resource Manual* at www.mhhe.com/moh.

Study Skills Exercises

1. Prepare a one-page summary sheet with the most important information that you need for the test. On the day of the test, look at this sheet several times to refresh your memory instead of trying to memorize new information.

2. Define the key term: **point-slope formula**

 Writing Translating Expression Geometry Scientific Calculator Video NS E

Review Exercises

For Exercises 3–6, graph the equations.

3. $2x - 3y = -3$

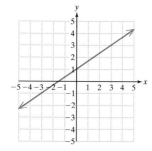

4. $y = -2x$

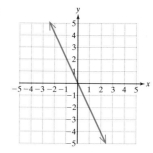

5. $3 - y = 9$

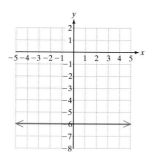

6. $y = \dfrac{4}{5}x$

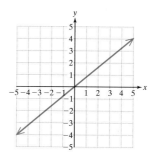

For Exercises 7–10, find the slope of the line that passes through the given points.

7. $(1, -3)$ and $(2, 6)$ 9

8. $(2, -4)$ and $(-2, 4)$ −2

9. $(-2, 5)$ and $(5, 5)$ 0

10. $(6.1, 2.5)$ and $(6.1, -1.5)$ Undefined

Objective 1: Writing an Equation of a Line Using the Point-Slope Formula

For Exercises 11–22, use the point-slope formula (if possible) to write an equation of the line given the following information. **(See Example 1.)**

11. The slope is 3, and the line passes through the point $(-2, 1)$.
$y = 3x + 7$ or $3x - y = -7$

12. The slope is -2, and the line passes through the point $(1, -5)$.
$y = -2x - 3$ or $2x + y = -3$

13. The slope is -4, and the line passes through the point $(-3, -2)$.
$y = -4x - 14$ or $4x + y = -14$

14. The slope is 5, and the line passes through the point $(-1, -3)$.
$y = 5x + 2$ or $5x - y = -2$

15. The slope is $-\frac{1}{2}$, and the line passes through $(-1, 0)$. $y = -\dfrac{1}{2}x - \dfrac{1}{2}$ or $x + 2y = -1$

16. The slope is $-\frac{3}{4}$, and the line passes through $(2, 0)$. $y = -\dfrac{3}{4}x + \dfrac{3}{2}$ or $3x + 4y = 6$

17. The slope is $\frac{1}{4}$, and the line passes through the point $(-8, 6)$. $y = \dfrac{1}{4}x + 8$ or $x - 4y = -32$

18. The slope is $\frac{2}{5}$, and the line passes through the point $(-5, 4)$. $y = \dfrac{2}{5}x + 6$ or $2x - 5y = -30$

19. The slope is 4.5, and the line passes through the point $(5.2, -2.2)$.
$y = 4.5x - 25.6$ or $45x - 10y = 256$

20. The slope is -3.6, and the line passes through the point $(10.0, 8.2)$.
$y = -3.6x + 44.2$ or $36x + 10y = 442$

21. The slope is 0, and the line passes through the point $(3, -2)$.
$y = -2$

22. The slope is 0, and the line passes through the point $(0, 5)$.
$y = 5$

Objective 2: Writing an Equation of a Line Given Two Points

For Exercises 23–28, use the point-slope formula to write an equation of the line given the following information. **(See Example 2.)**

23. The line passes through the points $(-2, -6)$ and $(1, 0)$. $y = 2x - 2$ or $2x - y = 2$

24. The line passes through the points $(-2, 5)$ and $(0, 1)$. $y = -2x + 1$ or $2x + y = 1$

25. The line passes through the points $(0, -4)$ and $(-1, -3)$. $y = -x - 4$ or $x + y = -4$

26. The line passes through the points $(1, -3)$ and $(-7, 2)$. $y = -\dfrac{5}{8}x - \dfrac{19}{8}$ or $5x + 8y = -19$

27. The line passes through the points $(2.2, -3.3)$ and $(12.2, -5.3)$.
$y = -0.2x - 2.86$ or $20x + 100y = -286$

28. The line passes through the points $(4.7, -2.2)$ and $(-0.3, 6.8)$.
$y = -1.8x + 6.26$ or $180x + 100y = 626$

For Exercises 29–34, find an equation of the line through the given points. Write the final answer in slope-intercept form.

29. 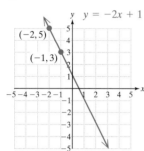 $y = -2x + 1$

30. 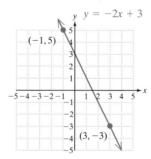 $y = -2x + 3$

31. 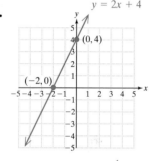 $y = 2x + 4$

32. 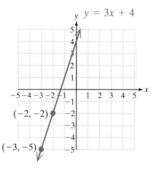 $y = 3x + 4$

33. 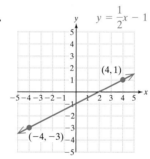 $y = \dfrac{1}{2}x - 1$

34. 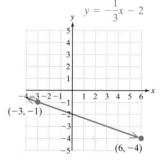 $y = -\dfrac{1}{3}x - 2$

Objective 3: Writing an Equation of a Line Parallel or Perpendicular to Another Line

For Exercises 35–44, use the point-slope formula to write an equation of the line given the following information. **(See Examples 3–4.)**

35. The line passes through the point $(-3, 1)$ and is parallel to the line $y = 4x + 3$. $y = 4x + 13$ or $4x - y = -13$

36. The line passes through the point $(4, -1)$ and is parallel to the line $y = 3x + 1$. $y = 3x - 13$ or $3x - y = 13$

37. The line passes through the point $(4, 0)$ and is parallel to the line $3x + 2y = 8$. $y = -\dfrac{3}{2}x + 6$ or $3x + 2y = 12$

38. The line passes through the point $(2, 0)$ and is parallel to the line $5x + 3y = 6$. $y = -\dfrac{5}{3}x + \dfrac{10}{3}$ or $5x + 3y = 10$

39. The line passes through the point $(-5, 2)$ and is perpendicular to the line $y = \frac{1}{2}x + 3$.
$y = -2x - 8$ or $2x + y = -8$

40. The line passes through the point $(-2, -2)$ and is perpendicular to the line $y = \frac{1}{3}x - 5$.
$y = -3x - 8$ or $3x + y = -8$

41. The line passes through the point $(0, -6)$ and is perpendicular to the line $-5x + y = 4$. $y = -\dfrac{1}{5}x - 6$
or $x + 5y = -30$

42. The line passes through the point $(0, -8)$ and is perpendicular to the line $2x - y = 5$.
$y = -\frac{1}{2}x - 8$ or $x + 2y = -16$

43. The line passes through the point $(4, 4)$ and is parallel to the line $3x - y = 6$.
$y = 3x - 8$ or $3x - y = 8$

44. The line passes through the point $(-1, -7)$ and is parallel to the line $5x + y = -5$.
$y = -5x - 12$ or $5x + y = -12$

Objective 4: Different Forms of Linear Equations: A Summary

For Exercises 45–50, match the form or formula on the left with its name on the right.

45. $x = k$ iv

46. $y = mx + b$ v

47. $m = \dfrac{y_2 - y_1}{x_2 - x_1}$ vi

48. $y - y_1 = m(x - x_1)$ ii

49. $y = k$ iii

50. $Ax + By = C$ i

i.   Standard form

ii. Point-slope formula

iii. Horizontal line

iv. Vertical line

v.   Slope-intercept form

vi. Slope formula

For Exercises 51–60, find an equation for the line given the following information. **(See Example 5.)**

51. The line passes through the point $(3, 1)$ and is parallel to the line $y = -4$. See the figure.
$y = 1$

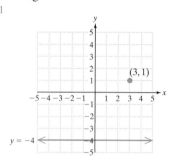

52. The line passes through the point $(-1, 1)$ and is parallel to the line $y = 2$. See the figure.
$y = 1$

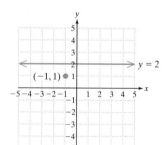

53. The line passes through the point $(2, 6)$ and is perpendicular to the line $y = 1$. (*Hint:* Sketch the line first.)
$x = 2$

54. The line passes through the point $(0, 3)$ and is perpendicular to the line $y = -5$. (*Hint:* Sketch the line first.)
$x = 0$

55. The line passes through the point $(2, 2)$ and is perpendicular to the line $x = 0$. $y = 2$

56. The line passes through the point $(5, -2)$ and is perpendicular to the line $x = 0$. $y = -2$

57. The slope is undefined, and the line passes through the point $(-6, -3)$. $x = -6$

58. The slope is undefined, and the line passes through the point $(2, -1)$. $x = 2$

59. The line passes through the points $(-4, 0)$ and $(-4, 3)$. $x = -4$

60. The line passes through the points $(1, 3)$ and $(1, -4)$. $x = 1$

Applications of Linear Equations and Modeling

1. Interpreting a Linear Equation in Two Variables

Linear equations can often be used to describe (or model) the relationship between two variables in a real-world event.

Example 1   Interpreting a Linear Equation

The number of tigers in India decreased from 1900 to 2005. This decrease can be approximated by the equation $y = -350x + 42{,}000$. The variable y represents the number of tigers left in India, and x represents the number of years since 1900.

a. Use the equation to predict the number of tigers in 1960.

b. Use the equation to predict the number of tigers in 2010.

c. Determine the slope of the line. Interpret the meaning of the slope in terms of the number of tigers and the year.

d. Determine the x-intercept. Interpret the meaning of the x-intercept in terms of the number of tigers.

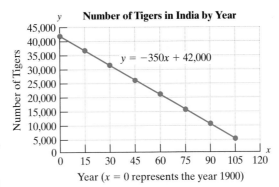

Objectives

1. Interpreting a Linear Equation in Two Variables
2. Writing a Linear Equation Using Observed Data Points
3. Writing a Linear Equation Given a Fixed Value and a Rate of Change

Skill Practice

The cost y (in dollars) for a local move by a small moving company is given by $y = 60x + 100$, where x is the number of hours required for the move.

1. How much would be charged for a move that required 3 hr?
2. How much would be charged for a move that required 8 hr?
3. What is the slope of the line and what does it mean in the context of this problem?
4. Determine the y-intercept and interpret its meaning in the context of this problem.

Classroom Example: p. 264, Exercise 10

Solution:

a. The year 1960 is 60 yr since 1900. Substitute $x = 60$ into the equation.

$y = -350x + 42{,}000$

$y = -350(60) + 42{,}000$

$\quad = 21{,}000$ There were approximately 21,000 tigers in India in 1960.

b. The year 2010 is 110 yr since 1900. Substitute $x = 110$.

$y = -350(110) + 42{,}000$

$\quad = 3500$ There will be approximately 3500 tigers in India in 2010.

c. The slope is -350.   The slope means that the tiger population is decreasing by 350 tigers per year.

d. To find the x-intercept, substitute $y = 0$.

$y = -350x + 42{,}000$

$0 = -350x + 42{,}000$ Substitute 0 for y.

$-42{,}000 = -350x$

$120 = x$

The x-intercept is $(120, 0)$. This means that 120 yr after the year 1900, the tiger population would be expected to reach zero. That is, in the year 2020, there will be no tigers left in India if this linear trend continues.

Answers

1. $280   2. $580
3. 60; This means that for each additional hour of service, the cost of the move goes up by $60.
4. (0, 100); The $100 charge is a fixed fee in addition to the hourly rate.

2. Writing a Linear Equation Using Observed Data Points

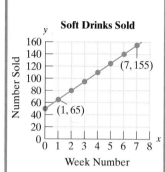
Example 2 **Writing a Linear Equation from Observed Data Points**

The monthly sales of hybrid cars sold in the United States are given for a recent year. The sales for the first 8 months of the year are shown in Figure 3-34. The value $x = 0$ represents January, $x = 1$ represents February, and so on.

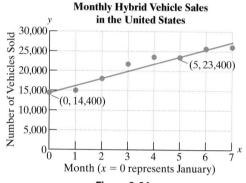

Figure 3-34

a. Use the data points from Figure 3-34 to find a linear equation that represents the monthly sales of hybrid cars in the United States. Let *x* represent the month number and let *y* represent the number of vehicles sold.

b. Use the linear equation in part (a) to estimate the number of hybrid vehicles sold in month 7 (August).

Solution:

a. The ordered pairs $(0, 14{,}400)$ and $(5, 23{,}400)$ are given in the graph. Use these points to find the slope.

$$\underset{(x_1, y_1)}{(0, 14{,}400)} \quad \text{and} \quad \underset{(x_2, y_2)}{(5, 23{,}400)} \qquad \text{Label the points.}$$

$$m = \frac{y_2 - y_1}{x_2 - x_1} = \frac{23{,}400 - 14{,}400}{5 - 0}$$

$$= \frac{9000}{5}$$

$$= 1800 \qquad \text{The slope is 1800. This indicates that sales increased by approximately 1800 per month during this time period.}$$

With $m = 1800$, and the *y*-intercept given as $(0, 14{,}400)$, we have the following linear equation in slope-intercept form.

$$y = 1800x + 14{,}400$$

b. To approximate the sales in month number 7, substitute $x = 7$ into the equation from part (a).

$$y = 1800(7) + 14{,}400 \qquad \text{Substitute } x = 7.$$

$$= 27{,}000$$

The monthly sales for August (month 7) would be 27,000 vehicles.

3. Writing a Linear Equation Given a Fixed Value and a Rate of Change

Another way to look at the equation $y = mx + b$ is to identify the term mx as the variable term and the term b as the constant term. The value of the term mx will change with the value of x (this is why the slope, m, is called a *rate of change*). However, the term b will remain constant regardless of the value of x. With these ideas in mind, we can write a linear equation if the rate of change and the constant are known.

Example 3   Finding a Linear Equation

A stack of posters to advertise a school play costs $19.95 plus $1.50 per poster at the printer.

a. Write a linear equation to compute the cost, c, of buying x posters.

b. Use the equation to compute the cost of 125 posters.

Solution:

a. The constant cost is $19.95. The variable cost is $1.50 per poster. If m is replaced with 1.50 and b is replaced with 19.95, the equation is

$$c = 1.50x + 19.95 \qquad \text{where } c \text{ is the cost (in dollars) of buying } x \text{ posters.}$$

b. Because x represents the number of posters, substitute $x = 125$.

$$c = 1.50(125) + 19.95$$
$$= 187.5 + 19.95$$
$$= 207.45$$

The total cost of buying 125 posters is $207.45.

Calculator Connections

Topic: Using the Evaluate Feature on a Graphing Calculator

In Example 3, the equation $c = 1.50x + 19.95$ was used to represent the cost, c, to buy x posters. To graph this equation on a graphing calculator, first replace the variable c by y.

$$y = 1.50x + 19.95$$

We enter the equation into the calculator and set the viewing window.

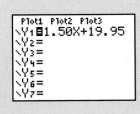

To evaluate the equation for a user-defined value of x, use the *Value* feature in the CALC menu.

In this case, we entered $x = 125$, and the calculator returned $y = 207.45$.

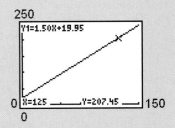

Calculator Exercises

Use a graphing calculator to graph the lines on an appropriate viewing window. Evaluate the equation at the given values of x.

1. $y = -4.6x + 27.1$ at $x = 3$ 13.3

2. $y = -3.6x - 42.3$ at $x = 0$ -42.3

3. $y = 40x + 105$ at $x = 6$ 345

4. $y = 20x - 65$ at $x = 8$ 95

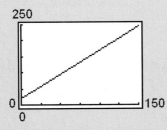

Section 3.6   Practice Exercises

Study Skills Exercise

1. On test day, take a look at any formulas or important points that you had to memorize before you enter the classroom. Then when you sit down to take your test, write these formulas on the test or on scrap paper. This is called a memory dump. Write down the formulas from Chapter 3.

Review Exercises

2. Determine the slope of the line defined by $2x - 8y = 15$. $m = \dfrac{1}{4}$

For Exercises 3–8, find the x- and y-intercepts of the lines, if possible.

3. $5x + 6y = 30$ x-intercept: $(6, 0)$;
   y-intercept: $(0, 5)$

4. $3x + 4y = 1$ x-intercept: $(\frac{1}{3}, 0)$;
   y-intercept: $(0, \frac{1}{4})$

5. $y = -2x - 4$ x-intercept: $(-2, 0)$;
   y-intercept: $(0, -4)$

6. $y = 5x$
   x-intercept: $(0, 0)$;
   y-intercept: $(0, 0)$

7. $y = -9$
   x-intercept: none;
   y-intercept: $(0, -9)$

8. $x = 2$ x-intercept: $(2, 0)$;
   y-intercept: none

Objective 1: Interpreting a Linear Equation in Two Variables

9. The minimum hourly wage, y (in dollars per hour), in the United States can be approximated by the equation $y = 0.14x + 1.60$. In this equation, x represents the number of years since 1970 ($x = 0$ represents 1970, $x = 5$ represents 1975, and so on). **(See Example 1.)**

   a. Use the equation to approximate the minimum wage in the year 1980. $3.00

   b. Use the equation to predict the minimum wage in 2010.
   $7.20

   c. Determine the y-intercept. Interpret the meaning of the y-intercept in the context of this problem.   The y-intercept is $(0, 1.6)$. This indicates that the minimum wage was $1.60 per hour in the year 1970.

   d. Determine the slope. Interpret the meaning of the slope in the context of this problem.
   The slope is 0.14. This indicates that the minimum wage has risen approximately $0.14 per year during this period.

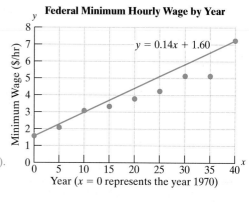

Federal Minimum Hourly Wage by Year

$y = 0.14x + 1.60$

Minimum Wage ($/hr)

Year ($x = 0$ represents the year 1970)

10. The graph depicts the rise in the number of jail inmates in the United States since 1995. Two linear equations are given: one to describe the number of female inmates and one to describe the number of male inmates by year.

   Let y represent the number of inmates (in thousands). Let x represent the number of years since 1995.

   a. What is the slope of the line representing the number of female inmates? Interpret the meaning of the slope in the context of this problem. $m = 4$. The number of female inmates has increased by 4 thousand per year since 1995.

   b. What is the slope of the line representing the number of male inmates? Interpret the meaning of the slope in the context of this problem. $m = 19.2$. The number of male inmates has increased by 19.2 thousand per year since 1995.

   c. Which group, males or females, has the larger slope? What does this imply about the rise in the number of male and female prisoners?   Males. The number of male inmates is increasing at a faster rate than the number of female inmates.

   d. Assuming this trend continues, use the equation to predict the number of female inmates in 2015.
   131 thousand or 131,000

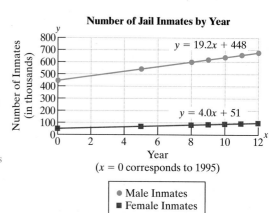

Number of Jail Inmates by Year

$y = 19.2x + 448$

Number of Inmates (in thousands)

$y = 4.0x + 51$

Year ($x = 0$ corresponds to 1995)

● Male Inmates
■ Female Inmates

(*Source:* U.S. Bureau of Justice Statistics)

Writing   Translating Expression   Geometry   Scientific Calculator   Video   NS&E

11. The average daily temperature in January for cities along the eastern seaboard of the United States and Canada generally decreases for cities farther north. A city's latitude in the northern hemisphere is a measure of how far north it is on the globe.

The average temperature, y (measured in degrees Fahrenheit), can be described by the equation

$y = -2.333x + 124.0$ where x is the latitude of the city.

City	x Latitude (°N)	y Average Daily Temperature (°F)
Jacksonville, FL	30.3	52.4
Miami, FL	25.8	67.2
Atlanta, GA	33.8	41.0
Baltimore, MD	39.3	31.8
Boston, MA	42.3	28.6
Atlantic City, NJ	39.4	30.9
New York, NY	40.7	31.5
Portland, ME	43.7	20.8
Charlotte, NC	35.2	39.3
Norfolk, VA	36.9	39.1

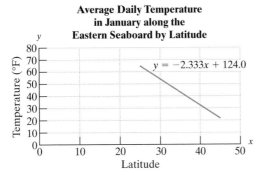

(*Source:* U.S. National Oceanic and Atmospheric Administration)

a. Use the equation to predict the average daily temperature in January for Philadelphia, Pennsylvania, whose latitude is 40.0°N. Round to one decimal place. 30.7°

b. Use the equation to predict the average daily temperature in January for Edmundston, New Brunswick, Canada, whose latitude is 47.4°N. Round to one decimal place. 13.4°

c. What is the slope of the line? Interpret the meaning of the slope in terms of latitude and temperature.
$m = -2.333$. The average temperature in January decreases at a rate of 2.333° per 1° of latitude.

d. From the equation, determine the value of the x-intercept. Round to one decimal place. Interpret the meaning of the x-intercept in terms of latitude and temperature.
(53.2, 0). At 53.2° latitude, the average temperature in January is 0°.

12. The graph shows the number of points scored by Shaquille O'Neal and by Allen Iverson according to the number of minutes played for several games. Two linear equations are given: one to describe the number of points scored by O'Neal and one to describe the number of points scored by Iverson. In both equations, y represents the number of points scored and x represents the number of minutes played.

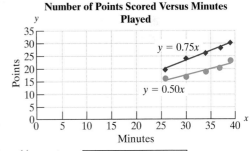

a. What is the slope of the line representing the number of points scored by O'Neal? Interpret the meaning of the slope in the context of this problem. 0.75 or equivalently $\frac{3}{4}$. O'Neal would expect to score 0.75 point for each minute played. This is equivalent to scoring 3 points for 4 min played.

b. What is the slope of the line representing the number of points scored by Iverson? Interpret the meaning of the slope in the context of this problem. 0.50 or equivalently $\frac{1}{2}$. Iverson would expect to score 0.5 point for each minute played. This is equivalent to scoring 1 point for 2 min played.

c. According to these linear equations, approximately how many points would each player expect to score if he played for 36 min? Round to the nearest point. O'Neal: 27 points; Iverson: 18 points

13. The electric bill charge for a certain utility company is $0.095 per kilowatt-hour. The total cost, y, depends on the number of kilowatt-hours, x, according to the equation $y = 0.095x$, $x \geq 0$.

a. Determine the cost of using 1000 kilowatt-hours. $95

b. Determine the cost of using 2000 kilowatt-hours. $190

c. Determine the y-intercept. Interpret the meaning of the y-intercept in the context of this problem.
(0, 0). For 0 kilowatt-hours used, the cost is $0.

d. Determine the slope. Interpret the meaning of the slope in the context of this problem.
$m = 0.095$. The cost increases by $0.095 for each kilowatt-hour used.

14. For a recent year, children's admission to the Minnesota State Fair was $8. Ride tickets were $0.75 each. The equation $y = 0.75x + 8$ represented the cost, y, in dollars to be admitted to the fair and to purchase x ride tickets.

a. Determine the slope of the line represented by $y = 0.75x + 8$. Interpret the meaning of the slope in the context of this problem.
$m = 0.75$; The slope means that the cost increases at a rate of 75¢ per ride.

b. Determine the y-intercept. Interpret its meaning in the context of this problem.
(0, 8); The cost was $8 if 0 rides were purchased.

c. Use the equation to determine how much money a child needed for admission and to ride 10 rides.
$15.50

Objective 2: Writing a Linear Equation Using Observed Data Points

15. The average length of stay for community hospitals decreased in the United States from 1980 to 2005. Let x represent the number of years since 1980. Let y represent the average length of a hospital stay in days. **(See Example 2.)**

a. Find a linear equation that relates the average length of hospital stays versus the year. $y = -0.08x + 7.6$

b. Use the linear equation found in part (a) to predict the average length of stay in community hospitals in the year 2010. Round to the nearest day. 5 days

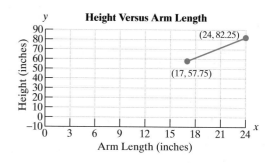

Average Length of Hospital Stay by Year

16. The figure depicts a relationship between a person's height, y (in inches), and the length of the person's arm, x (measured in inches from shoulder to wrist).

a. Use the points $(17, 57.75)$ and $(24, 82.25)$ to find a linear equation relating height to arm length.
$y = 3.5x - 1.75$

b. What is the slope of the line? Interpret the slope in the context of this problem. $m = 3.5$. For each additional inch in length of a person's arm, the person's height increases by 3.5 in.

c. Use the equation from part (a) to estimate the height of a person whose arm length is 21.5 in.
73.5 in. or 6 ft $1\frac{1}{2}$ in.

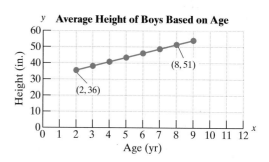

Height Versus Arm Length

17. The graph shows the average height for boys based on age. Let x represent a boy's age, and let y represent his height (in inches).

a. Find a linear equation that represents the height of a boy versus his age. $y = 2.5x + 31$

b. Use the linear equation found in part (a) to predict the average height of a 5-year-old boy. 43.5 in.

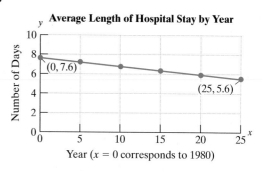

Average Height of Boys Based on Age

(*Source:* National Parenting Council)

18. Wind energy is one type of renewable energy that does not produce dangerous greenhouse gases as a by-product. The graph shows the consumption of wind energy in the United States for selected years. The variable y represents the amount of wind energy in trillions of Btu, and the variable x represents the number of years since 2000.

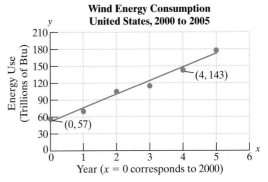

Wind Energy Consumption
United States, 2000 to 2005

(*Source:* United States Department of Energy)

a. Use the points $(0, 57)$ and $(4, 143)$ to determine the slope of the line. $m = 21.5$
The slope means that the consumption of wind energy in the

b. Interpret the slope in the context of this problem?
United States increased by 21.5 trillion Btu per year.

c. Use the points $(0, 57)$ and $(4, 143)$ to find a linear equation relating the consumption of wind energy, y, to the number of years, x, since 2000. $y = 21.5x + 57$

d. If this linear trend continues beyond the observed data values, use the equation in part (c) to predict the consumption of wind energy in the year 2010. 272 trillion Btu

Objective 3: Writing a Linear Equation Given a Fixed Value and a Rate of Change

19. The cost to rent a car, y, for 1 day is $20 plus $0.25 per mile. **(See Example 3.)**

a. Write a linear equation to compute the cost, y, of driving a car x miles for 1 day.
$y = 0.25x + 20$

b. Use the equation to compute the cost of driving 258 miles in the rental car for 1 day.
$84.50

20. A phone bill is determined each month by a $18.95 flat fee plus $0.08 per minute of long distance.

a. Write a linear equation to compute the monthly cost of a phone bill, y, if x minutes of long distance are used.
$y = 0.08x + 18.95$

b. Use the equation to compute the phone bill for a month in which 1 hr and 27 min of long distance was used.
$25.91

21. The cost to rent a 10 ft by 10 ft storage space is $90 per month plus a nonrefundable deposit of $105.

a. Write a linear equation to compute the cost, y, of renting a 10 ft by 10 ft space for x months.
$y = 90x + 105$

b. What is the cost of renting such a storage space for 1 year (12 months)?
$1185.00

22. An air-conditioning and heating company has a fixed monthly cost of $5000. Furthermore, each service call costs the company $25.

a. Write a linear equation to compute the total cost, y, for 1 month if x service calls are made.
$y = 25x + 5000$

b. Use the equation to compute the cost for 1 month if 150 service calls are made.
$8750.00

23. A bakery that specializes in bread rents a booth at a flea market. The daily cost to rent the booth is $100. Each loaf of bread costs the bakery $0.80 to produce.

a. Write a linear equation to compute the total cost, y, for 1 day if x loaves of bread are produced.
$y = 0.8x + 100$

b. Use the equation to compute the cost for 1 day if 200 loaves of bread are produced.
$260.00

24. A beverage company rents a booth at an art show to sell lemonade. The daily cost to rent a booth is $35. Each lemonade costs $0.50 to produce.

a. Write a linear equation to compute the total cost, y, for 1 day if x lemonades are produced.
$y = 0.5x + 35$

b. Use the equation to compute the cost for 1 day if 350 lemonades are produced.
$210.00

 Writing   Translating Expression   Geometry Scientific Calculator   Video NS/E

Section 3.7 Introduction to Functions

Objectives

1. **Definition of a Relation**
2. **Definition of a Function**
3. **Vertical Line Test**
4. **Function Notation**
5. **Domain and Range of a Function**
6. **Applications of Functions**

1. Definition of a Relation

The number of points scored by LeBron James during the first six games of a recent basketball season is shown in Table 3-4.

Table 3-4

Game, x	Number of Points, y		Ordered Pair
1	26	$\longrightarrow$	$(1, 26)$
2	35	$\longrightarrow$	$(2, 35)$
3	16	$\longrightarrow$	$(3, 16)$
4	34	$\longrightarrow$	$(4, 34)$
5	19	$\longrightarrow$	$(5, 19)$
6	38	$\longrightarrow$	$(6, 38)$

Each ordered pair from Table 3-4 shows a correspondence, or relationship, between the game number and the number of points scored by LeBron James. The set of ordered pairs: $\{(1, 26), (2, 35), (3, 16), (4, 34), (5, 19), (6, 38)\}$ defines a relation between the game number and the number of points scored.

DEFINITION Relation in x and y

Any set of ordered pairs, (x, y), is called a **relation** in x and y. Furthermore:
- The set of first components in the ordered pairs is called the **domain** of the relation.
- The set of second components in the ordered pairs is called the **range** of the relation.

Skill Practice

1. Find the domain and range of the relation. $\{(0, 1), (4, 5), (-6, 8), (4, 13), (-8, 8)\}$

Classroom Example: p. 276, Exercise 6

Instructor Note: Tell students that repeated values in the domain or range are not listed more than once.

Example 1 Finding the Domain and Range of a Relation

Find the domain and range of the relation linking the game number to the number of points scored by James in the first six games of the season:

$$\{(1, 26), (2, 35), (3, 16), (4, 34), (5, 19), (6, 38)\}$$

Solution:

Domain: $\{1, 2, 3, 4, 5, 6\}$ (Set of first coordinates)

Range: $\{26, 35, 16, 34, 19, 38\}$ (Set of second coordinates)

The domain consists of the game numbers for the first six games of the season. The range represents the corresponding number of points.

Answer

1. Domain: $\{0, 4, -6, -8\}$; Range: $\{1, 5, 8, 13\}$

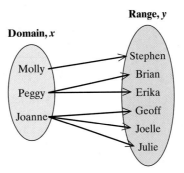

Example 2 Finding the Domain and Range of a Relation

The three women represented in Figure 3-35 each have children. Molly has one child, Peggy has two children, and Joanne has three children.

a. If the set of mothers is given as the domain and the set of children is the range, write a set of ordered pairs defining the relation given in Figure 3-35.

b. Write the domain and range of the relation.

Figure 3-35

Solution:

a. {(Molly, Stephen), (Peggy, Brian), (Peggy, Erika), (Joanne, Geoff), (Joanne, Joelle), (Joanne, Julie)}

b. Domain: {Molly, Peggy, Joanne}
   Range: {Stephen, Brian, Erika, Geoff, Joelle, Julie}

Skill Practice

Given the relation represented by the figure:

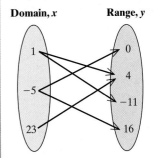

2. Write the relation as a set of ordered pairs.
3. Write the domain and range of the relation.

Classroom Example: p. 276, Exercise 12

2. Definition of a Function

In mathematics, a special type of relation, called a function, is used extensively.

> **DEFINITION Function**
>
> Given a relation in x and y, we say "y is a **function** of x" if for each element x in the domain, there is exactly one value of y in the range.
>
> *Note:* This means that no two ordered pairs may have the same first coordinate and different second coordinates.

To understand the difference between a relation that is a function and one that is not a function, consider Example 3.

Example 3 Determining Whether a Relation Is a Function

Determine whether the following relations are functions:

a. {(2, −3), (4, 1), (3, −1), (2, 4)}   **b.** {(−3, 1), (0, 2), (4, −3), (1, 5), (−2, 1)}

Solution:

a. This relation is defined by the set of ordered pairs.

same x-values

{(2, −3), (4, 1), (3, −1), (2, 4)}

different y-values

When $x = 2$, there are two possibilities for y: $y = −3$ and $y = 4$.

This relation is *not* a function because for $x = 2$, there is more than one corresponding element in the range.

b. This relation is defined by the set of ordered pairs: {(−3, 1), (0, 2), (4, −3), (1, 5), (−2, 1)}. Notice that no two ordered pairs have the same value of x but different values of y. Therefore, this relation *is* a function.

Skill Practice

Determine whether the following relations are functions. If the relation is not a function, state why.

4. {(0, −7), (4, 9), (−2, −7), $\left(\dfrac{1}{3}, \dfrac{1}{2}\right)$, (4, 10)}

5. {(−8, −3), (4, −3), (−12, 7), (−1, −1)}

Classroom Example: p. 276, Exercise 16

Answers

2. {(1, 4), (1, −11), (−5, 0), (−5, 16), (23, 4)}
3. Domain: {1, −5, 23}; Range: {0, 4, −11, 16}
4. Not a function because the domain element, 4, has two different y-values: (4, 9) and (4, 10).
5. Function

In Example 2, the relation linking the set of mothers with their respective children is *not* a function. The domain elements, "Peggy" and "Joanne," each have more than one child. Because these *x*-values in the domain have more than one corresponding *y*-value in the range, the relation is not a function.

3. Vertical Line Test

A relation that is not a function has at least one domain element, *x*, paired with more than one range element, *y*. For example, the ordered pairs (2, 1) and (2, 4) do not make a function. On a graph, these two points are aligned vertically in the *xy*-plane, and a vertical line drawn through one point also intersects the other point (Figure 3-36). Thus, if a vertical line drawn through a graph of a relation intersects the graph in more than one point, the relation cannot be a function. This idea is stated formally as the **vertical line test**.

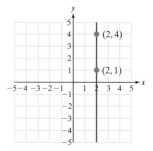

Figure 3-36

> **PROCEDURE** Using the Vertical Line Test
>
> Consider a relation defined by a set of points (*x*, *y*) on a rectangular coordinate system. Then the graph defines *y* as a function of *x* if no vertical line intersects the graph in more than one point.

The vertical line test also implies that if any vertical line drawn through the graph of a relation intersects the relation in more than one point, then the relation does *not* define *y* as a function of *x*.

The vertical line test can be demonstrated by graphing the ordered pairs from the relations in Example 3 (Figure 3-37 and Figure 3-38).

$$\{(2, -3), (4, 1), (3, -1), (2, 4)\} \qquad \{(-3, 1), (0, 2), (4, -3), (1, 5), (-2, 1)\}$$

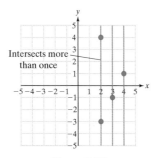

Figure 3-37

Not a Function

A vertical line intersects in more than one point.

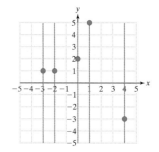

Figure 3-38

Function

No vertical line intersects more than once.

The relations in Examples 1, 2, and 3 consist of a finite number of ordered pairs. A relation may, however, consist of an *infinite* number of points defined by an equation or by a graph. For example, the equation $y = x + 1$ defines infinitely many ordered pairs whose *y*-coordinate is one more than its *x*-coordinate. These ordered pairs cannot all be listed but can be depicted in a graph.

The vertical line test is especially helpful in determining whether a relation is a function based on its graph.

| Example 4 | **Using the Vertical Line Test** |

Use the vertical line test to determine whether the following relations are functions.

a.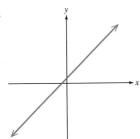

b.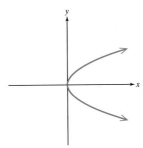

Skill Practice

Use the vertical line test to determine if the following relations are functions.

6.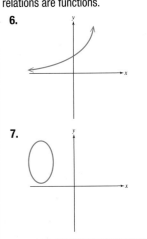

7.

Classroom Examples: pp. 276–277, Exercises 20 and 22

Solution:

a.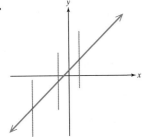

b.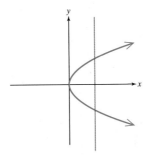

| **Function** | **Not a Function** |
| No vertical line intersects more than once. | A vertical line intersects in more than one point. |

4. Function Notation

A function is defined as a relation with the added restriction that each value of the domain corresponds to only one value in the range. In mathematics, functions are often given by rules or equations to define the relationship between two or more variables. For example, the equation, $y = x + 1$ defines the set of ordered pairs such that the y-value is one more than the x-value.

When a function is defined by an equation, we often use **function notation**. For example, the equation $y = x + 1$ may be written in function notation as

$$f(x) = x + 1$$

where f is the name of the function, x is an input value from the domain of the function, and $f(x)$ is the function value (or y-value) corresponding to x.

The notation $f(x)$ is read as "f of x" or "the value of the function, f, at x."

A function may be evaluated at different values of x by substituting values of x from the domain into the function. For example, for the function defined by $f(x) = x + 1$ we can evaluate f at $x = 3$ by using substitution.

$f(x) = x + 1$

$f(3) = (3) + 1$

$f(3) = 4$ This is read as "f of 3 equals 4."

Thus, when $x = 3$, the corresponding function value is 4. This can also be interpreted as an ordered pair: $(3, 4)$

The names of functions are often given by either lowercase letters or uppercase letters such as f, g, h, p, k, M, and so on.

Avoiding Mistakes

The notation $f(x)$ is read as "f of x" and does *not* imply multiplication.

Answers

6. Function   **7.** Not a function

Example 5   **Evaluating a Function**

Given the function defined by $h(x) = x^2 - 2$, find the function values.

a. $h(0)$ **b.** $h(1)$ **c.** $h(2)$ **d.** $h(-1)$ **e.** $h(-2)$

Solution:

a. $h(x) = x^2 - 2$

$h(0) = (0)^2 - 2$ Substitute $x = 0$ into the function.

$= 0 - 2$

$= -2$ $h(0) = -2$ means that when $x = 0$, $y = -2$, yielding the ordered pair $(0, -2)$.

b. $h(x) = x^2 - 2$

$h(1) = (1)^2 - 2$ Substitute $x = 1$ into the function.

$= 1 - 2$

$= -1$ $h(1) = -1$ means that when $x = 1$, $y = -1$, yielding the ordered pair $(1, -1)$.

c. $h(x) = x^2 - 2$

$h(2) = (2)^2 - 2$ Substitute $x = 2$ into the function.

$= 4 - 2$

$= 2$ $h(2) = 2$ means that when $x = 2$, $y = 2$, yielding the ordered pair $(2, 2)$.

d. $h(x) = x^2 - 2$

$h(-1) = (-1)^2 - 2$ Substitute $x = -1$ into the function.

$= 1 - 2$

$= -1$ $h(-1) = -1$ means that when $x = -1$, $y = -1$, yielding the ordered pair $(-1, -1)$.

e. $h(x) = x^2 - 2$

$h(-2) = (-2)^2 - 2$ Substitute $x = -2$ into the function.

$= 4 - 2$

$= 2$ $h(-2) = 2$ means that when $x = -2$, $y = 2$, yielding the ordered pair $(-2, 2)$.

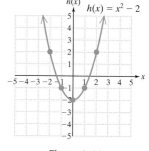

Figure 3-39

The rule $h(x) = x^2 - 2$ is equivalent to the equation $y = x^2 - 2$. The function values $h(0), h(1), h(2), h(-1)$, and $h(-2)$ correspond to the y-values in the ordered pairs $(0, -2), (1, -1), (2, 2), (-1, -1)$, and $(-2, 2)$, respectively. These points can be used to sketch a graph of the function (Figure 3-39).

Answers

8. -4 **9.** 0 **10.** 24

11. -6 **12.** 6

5. Domain and Range of a Function

A function is a relation, and it is often necessary to determine its domain and range. Consider a function defined by the equation $y = f(x)$. The domain of f is the set of all x-values within the ordered pairs that make up the function. The range is the set of y-values.

Example 6 Finding the Domain and Range of a Function

Find the domain and range of the functions based on the graph of the function. Express the answers in interval notation.

a.
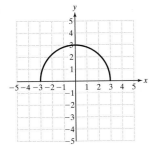

b.
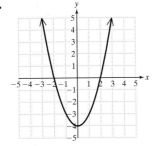

Solution:

a.
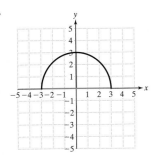

The horizontal "span" of the graph is determined by the x-values of the points. This is the domain. In this graph, the x-values in the domain are bounded between -3 and 3. (Shown in blue.)

Domain: $[-3, 3]$

The vertical "span" of the graph is determined by the y-values of the points. This is the range.

The y-values in the range are bounded between 0 and 3. (Shown in red.)

Range: $[0, 3]$

b.
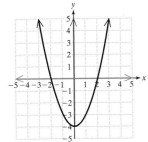

The function extends infinitely far to the left and right. The domain is shown in blue.

Domain: $(-\infty, \infty)$

The y-values extend infinitely far in the positive direction, but are bounded below at $y = -4$. (Shown in red.)

Range: $[-4, \infty)$

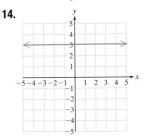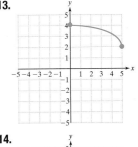
Answers

13. Domain: $[0, 5]$
Range: $[2, 4]$
14. Domain: $(-\infty, \infty)$
Range: $\{3\}$

6. Applications of Functions

Example 7 Using a Function in an Application

The score a student receives on an exam is a function of the number of hours the student spends studying. The function defined by

$$P(x) = \frac{100x^2}{40 + x^2} \quad (x \geq 0)$$

indicates that a student will achieve a score of $P\%$ after studying for x hours.

a. Evaluate $P(0)$, $P(10)$, and $P(20)$.

b. Interpret the function values from part (a) in the context of this problem.

Solution:

a. $P(x) = \dfrac{100x^2}{40 + x^2}$

$$P(0) = \frac{100(0)^2}{40 + (0)^2} \qquad P(10) = \frac{100(10)^2}{40 + (10)^2} \qquad P(20) = \frac{100(20)^2}{40 + (20)^2}$$

$$P(0) = \frac{0}{40} \qquad P(10) = \frac{10{,}000}{140} \qquad P(20) = \frac{40{,}000}{440}$$

$$P(0) = 0 \qquad P(10) = \frac{500}{7} \approx 71.4 \qquad P(20) = \frac{1000}{11} \approx 90.9$$

b. $P(0) = 0$ means that for 0 hr spent studying, the student will receive 0% on the exam.

$P(10) \approx 71.4$ means that for 10 hr spent studying, the student will receive approximately 71.4% on the exam.

$P(20) \approx 90.9$ means that for 20 hr spent studying, the student will receive approximately 90.9% on the exam.

The graph of $P(x) = \dfrac{100x^2}{40 + x^2}$ is shown in Figure 3-40.

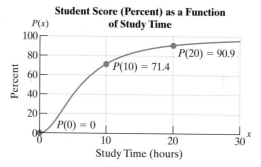

Student Score (Percent) as a Function of Study Time

Figure 3-40

Calculator Connections

Topic: Graphing Functions

A graphing calculator can be used to graph a function. We replace $f(x)$ by y and enter the defining expression into the calculator. For example,

$$f(x) = \frac{1}{4}x^3 - x^2 - x + 4 \quad \text{becomes} \quad y = \frac{1}{4}x^3 - x^2 - x + 4.$$

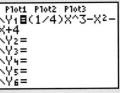

Calculator Exercises

Use a graphing calculator to graph the following functions.

1. $f(x) = x^2 - 5x + 2$ **2.** $g(x) = -x^2 + 4x + 5$

3. $m(x) = \frac{1}{3}x^3 + x^2 - 3x - 1$ **4.** $n(x) = x^3 - 9x$

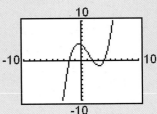

Section 3.7   Practice Exercises

Boost *your* GRADE at ALEKS.com!

ALEKS version 3.0

- Practice Problems
- Self-Tests
- NetTutor
- e-Professors
- Videos

For additional exercises, see Classroom Activities 3.7A–3.7D in the *Instructor's Resource Manual* at www.mhhe.com/moh.

Study Skills Exercises

1. List three ways that your friends or family can help support you in your academic pursuits.

2. Define the key terms:

    a. domain   **b. function**   **c. function notation**

    d. range   **e. relation**   **f. vertical line test**

Review Exercises

3. Given the equation $y = 3x - 5$:

    a. Find the x- and y-intercepts.

    b. Find the slope of the line.

    c. Graph the line.

x-intercept: $\left(\frac{5}{3}, 0\right)$; y-intercept: $(0, -5)$

3

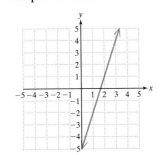

4. Given the equation $2x - 5y = 12$:

    a. Find the x- and y-intercepts.

    b. Find the slope of the line.

    c. Graph the line.

x-intercept: $(6, 0)$; y-intercept: $\left(0, -\frac{12}{5}\right)$

$\frac{2}{5}$

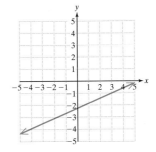

 Writing   Translating Expression   Geometry   Scientific Calculator   Video   NS&E

Objective 1: Definition of a Relation

For Exercises 5–14, determine the domain and range of the relation. **(See Examples 1–2.)**

5. $\{(4, 2), (3, 7), (4, 1), (0, 6)\}$
Domain: {4, 3, 0}; range: {2, 7, 1, 6}

6. $\{(-3, -1), (-2, 6), (1, 3), (1, -2)\}$
Domain: {−3, −2, 1}; range: {−1, 6, 3, −2}

7. $\{(\frac{1}{2}, 3), (0, 3), (1, 3)\}$
Domain: {$\frac{1}{2}$, 0, 1}; range: {3}

8. $\{(9, 6), (4, 6), (-\frac{1}{3}, 6)\}$
Domain: {9, 4, $-\frac{1}{3}$}; range: {6}

9. $\{(0, 0), (5, 0), (-8, 2), (8, 5)\}$
Domain: {0, 5, −8, 8}; range: {0, 2, 5}

10. $\{(\frac{1}{2}, -\frac{1}{2}), (-4, 0), (0, -\frac{1}{2}), (\frac{1}{2}, 0)\}$
Domain: {$\frac{1}{2}$, −4, 0}; range: {$-\frac{1}{2}$, 0}

11.

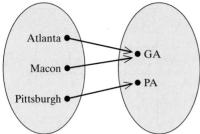

Domain: {Atlanta, Macon, Pittsburgh};
range: {GA, PA}

12.

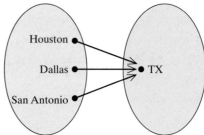

Domain: {Houston, Dallas, San Antonio};
range: {TX}

13.

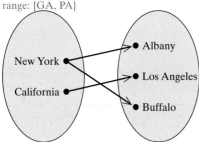

Domain: {New York, California};
range: {Albany, Los Angeles, Buffalo}

14.

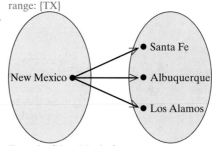

Domain: {New Mexico};
range: {Santa Fe, Albuquerque, Los Alamos}

Objective 2: Definition of a Function

15. How can you tell if a set of ordered pairs represents a function?
The relation is a function if each element in the domain has exactly one corresponding element in the range.

16. Refer back to Exercises 6, 8, 10, 12, and 14. Identify which relations are functions.
The relations in Exercises 8 and 12 are functions.

17. Refer back to Exercises 5, 7, 9, 11, and 13. Identify which relations are functions. **(See Example 3.)**
The relations in Exercises 7, 9, and 11 are functions.

Objective 3: Vertical Line Test

18. How can you tell from the graph of a relation if the relation is a function?
The graph represents a function if no vertical line intersects the graph more than once.

For Exercises 19–27, determine if the relation defines y as a function of x. **(See Example 4.)**

19. Yes

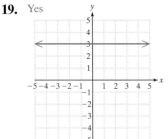

20. Yes

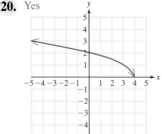

21. No

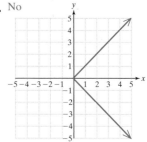

Writing   Translating Expression   Geometry   Scientific Calculator   Video   NS&E

22. No
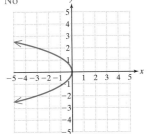

23. No
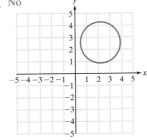

24. No
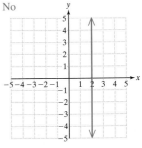

25. Yes
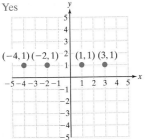

26. No
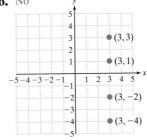

27. Yes
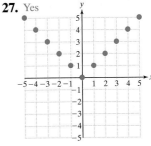

Objective 4: Function Notation

28. Explain how you would evaluate $f(x) = 3x^2$ at $x = -1$.

Substitute -1 for x and simplify the result.

For Exercises 29–36, evaluate the given functions. **(See Example 5.)**

29. Let $f(x) = 2x - 5$.

Find: $f(0)$ –5
     $f(2)$ –1
     $f(-3)$ –11

30. Let $g(x) = x^2 + 1$.

Find: $g(0)$ 1
     $g(-1)$ 2
     $g(3)$ 10

31. Let $h(x) = \dfrac{1}{x + 4}$.

Find: $h(1)$ $\frac{1}{5}$
     $h(0)$ $\frac{1}{4}$
     $h(-2)$ $\frac{1}{2}$

32. Let $p(x) = \sqrt{x + 4}$.

Find: $p(0)$ 2
     $p(-4)$ 0
     $p(5)$ 3

33. Let $m(x) = |5x - 7|$.

Find: $m(0)$ 7
     $m(1)$ 2
     $m(2)$ 3

34. Let $w(x) = |2x - 3|$.

Find: $w(0)$ 3
     $w(1)$ 1
     $w(2)$ 1

35. Let $n(x) = \sqrt{x - 2}$.

Find: $n(2)$ 0
     $n(3)$ 1
     $n(6)$ 2

36. Let $t(x) = \dfrac{1}{x - 3}$.

Find: $t(1)$ $-\frac{1}{2}$
     $t(-1)$ $-\frac{1}{4}$
     $t(2)$ -1

Objective 5: Domain and Range of a Function

For Exercises 37–40, match the domain and range given with a possible graph. **(See Example 6.)**

37. Domain: $(-\infty, \infty)$

Range: $[1, \infty)$ b

38. Domain: $[-4, 4]$

Range: $[-2, 2]$ a

39. Domain: $[-2, \infty)$

Range: $(-\infty, \infty)$ c

40. Domain: $(-\infty, \infty)$

Range: $(-\infty, \infty)$ d

a.
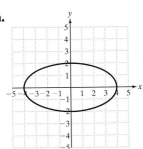

b.
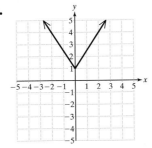

c.
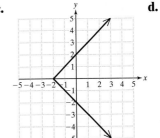

d.
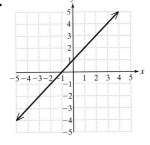

 Writing   Translating Expression   Geometry   Scientific Calculator   Video   NS&E

For Exercises 41–44, translate the expressions into English phrases.

41. $f(6) = 2$
The function value at $x = 6$ is 2.

42. $f(-2) = -14$
The function value at $x = -2$ is -14.

43. $g\left(\dfrac{1}{2}\right) = \dfrac{1}{4}$
The function value at $x = \frac{1}{2}$ is $\frac{1}{4}$.

44. $h(k) = k^2$
The function value at $x = k$ is k^2.

45. Consider a function defined by $y = f(x)$. The function value $f(2) = 7$ corresponds to what ordered pair?
$(2, 7)$

46. Consider a function defined by $y = f(x)$. The function value $f(-3) = -4$ corresponds to what ordered pair?
$(-3, -4)$

47. Consider a function defined by $y = f(x)$. The function value $f(0) = 8$ corresponds to what ordered pair?
$(0, 8)$

48. Consider a function defined by $y = f(x)$. The function value $f(4) = 0$ corresponds to what ordered pair?
$(4, 0)$

Objective 6: Applications of Functions

49. In the absence of air resistance, the speed, s (in feet per second: ft/sec), of an object in free fall is a function of the number of seconds, t, after it was dropped: **(See Example 7.)**

$$s(t) = 32t$$

a. Find $s(1)$, and interpret the meaning of this function value in terms of speed and time.
$s(1) = 32$. The speed of an object 1 sec after being dropped is 32 ft/sec.

b. Find $s(2)$, and interpret the meaning in terms of speed and time.
$s(2) = 64$. The speed of an object 2 sec after being dropped is 64 ft/sec.

c. Find $s(10)$, and interpret the meaning in terms of speed and time.
$s(10) = 320$. The speed of an object 10 sec after being dropped is 320 ft/sec.

d. A ball dropped from the top of the Sears Tower in Chicago falls for approximately 9.2 sec. How fast was the ball going the instant before it hit the ground? 294.4 ft/sec

50. The number of people diagnosed with skin cancer, $N(x)$, can be approximated by $N(x) = 45{,}625(1 + 0.029x)$. For this function, x represents the number of years since 2003. (*Source:* Centers for Disease Control)

a. Evaluate $N(0)$ and interpret its meaning in the context of this problem.
$N(0) = 45{,}625$; This means that in the year 2003 (when $x = 0$), the number of people diagnosed with skin cancer was approximately 45,625.

b. Evaluate $N(7)$ and interpret its meaning in the context of this problem. Round to the nearest whole number. $N(7) = 54{,}887$; This means that in the year 2010 (when $x = 7$), the number of people diagnosed with skin cancer will be approximately 54,887.

51. A punter kicks a football straight up with an initial velocity of 64 ft/sec. The height of the ball, h (in feet), is a function of the number of seconds, t, after the ball is kicked:

$$h(t) = -16t^2 + 64t + 3$$

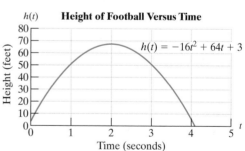

a. Find $h(0)$, and interpret the meaning of the function value in terms of time and height.
$h(0) = 3$. The initial height of the ball is 3 ft.

b. Find $h(1)$, and interpret the meaning in terms of time and height.
$h(1) = 51$. The height of the ball 1 sec after being kicked is 51 ft.

c. Find $h(2)$, and interpret the meaning in terms of time and height.
$h(2) = 67$. The height of the ball 2 sec after being kicked is 67 ft.

d. Find $h(4)$, and interpret the meaning in terms of time and height.
$h(4) = 3$. The height of the ball 4 sec after being kicked is 3 ft.

52. For people 16 years old and older, the maximum recommended heart rate, M (in beats per minute: beats/min), is a function of a person's age, x (in years).

$$M(x) = 220 - x \text{ for } x \geq 16$$

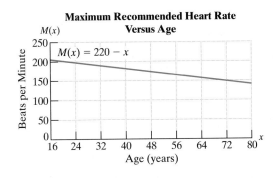

Maximum Recommended Heart Rate Versus Age

a. Find $M(16)$, and interpret the meaning in terms of maximum recommended heart rate and age.
   $M(16) = 204$. A 16-year-old adult's maximum recommended heart rate is 204 beats/min.
b. Find $M(30)$, and interpret the meaning in terms of maximum recommended heart rate and age.
   $M(30) = 190$. A 30-year-old adult's maximum recommended heart rate is 190 beats/min.
c. Find $M(60)$, and interpret the meaning in terms of maximum recommended heart rate and age.
   $M(60) = 160$. A 60-year-old adult's maximum recommended heart rate is 160 beats/min.
d. Find your own maximum recommended heart rate.   Answers will vary.

Group Activity

Modeling a Linear Equation

Materials: Yardstick or other device for making linear measurements

Estimated Time: 15–20 minutes

Group Size: 3

1. The members of each group should measure the length of their arms (in inches) from elbow to wrist. Record this measurement as x and the person's height (in inches) as y. Write these values as ordered pairs for each member of the group. Then write the ordered pairs on the board.

2. Next, copy the ordered pairs collected from all groups in the class and plot the ordered pairs. (This is called a "scatter diagram.")   Answers will vary throughout this exercise.

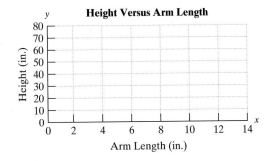

Height Versus Arm Length

3. Select two ordered pairs that seem to follow the upward trend of the data. Using these data points, determine the slope of the line.

Slope: _____

4. Using the data points and slope from question 3, find an equation of the line through the two points. Write the equation in slope-intercept form, $y = mx + b$.

Equation: _____

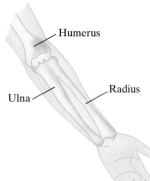

5. Using the equation from question 4, estimate the height of a person whose arm length from elbow to wrist is 8.5 in.

6. Suppose a crime scene investigator uncovers a partial skeleton and identifies a bone as a human ulna (the ulna is one of two bones in the forearm and extends from elbow to wrist). If the length of the bone is 12 in., estimate the height of the person before death. Would you expect this person to be male or female?

Chapter 3   Summary

Section 3.1   Rectangular Coordinate System

Key Concepts

Graphical representation of numerical **data** is often helpful to study problems in real-world applications.

A **rectangular coordinate system** is made up of a horizontal line called the **x-axis** and a vertical line called the **y-axis**. The point where the lines meet is the **origin**. The four regions of the plane are called **quadrants**.

The point (x, y) is an **ordered pair**. The first element in the ordered pair is the point's horizontal position from the origin. The second element in the ordered pair is the point's vertical position from the origin.

Example

Example 1

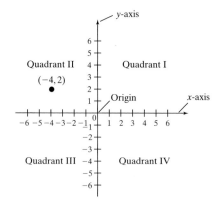

Section 3.2 Linear Equations in Two Variables

Key Concepts

An equation written in the form $Ax + By = C$ (where A and B are not both zero) is a **linear equation in two variables**.

A solution to a linear equation in x and y is an ordered pair (x, y) that makes the equation a true statement. The graph of the set of all solutions of a linear equation in two variables is a line in a rectangular coordinate system.

A linear equation can be graphed by finding at least two solutions and graphing the line through the points.

Examples

Example 1

Graph the equation $2x + y = 2$.

Select arbitrary values of x or y such as those shown in the table. Then complete the table to find the corresponding ordered pairs.

x	y	
0	2	$\longrightarrow (0, 2)$
−1	4	$\longrightarrow (-1, 4)$
1	0	$\longrightarrow (1, 0)$

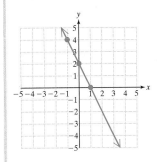

An ***x*-intercept** of a graph is a point $(a, 0)$ where the graph intersects the x-axis.

A ***y*-intercept** of a graph is a point $(0, b)$ where the graph intersects the y-axis.

Example 2

For the line $2x + y = 2$, the x-intercept is $(1, 0)$ and the y-intercept is $(0, 2)$.

A **vertical line** can be represented by an equation of the form $x = k$.

A **horizontal line** can be represented by an equation of the form $y = k$.

Example 3

$x = 3$ represents a vertical line

$y = 3$ represents a horizontal line

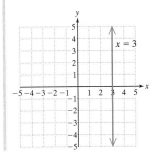

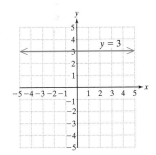

Section 3.3 Slope of a Line and Rate of Change

Key Concepts

The **slope**, m, of a line between two points (x_1, y_1) and (x_2, y_2) is given by

$$m = \frac{y_2 - y_1}{x_2 - x_1} \quad \text{or} \quad \frac{\text{change in } y}{\text{change in } x}$$

The slope of a line may be positive, negative, zero, or undefined.

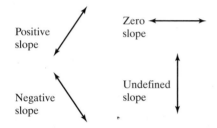

Positive slope

Negative slope

Zero slope

Undefined slope

If m_1 and m_2 represent the slopes of two **parallel lines** (nonvertical), then $m_1 = m_2$.

If $m_1 \neq 0$ and $m_2 \neq 0$ represent the slopes of two nonvertical **perpendicular lines**, then

$$m_1 = -\frac{1}{m_2} \quad \text{or equivalently, } m_1 m_2 = -1.$$

Examples

Example 1

Find the slope of the line between $(1, -5)$ and $(-3, 7)$.

$$m = \frac{7 - (-5)}{-3 - 1} = \frac{12}{-4} = -3$$

Example 2

The slope of the line $y = -2$ is 0 because the line is horizontal.

Example 3

The slope of the line $x = 4$ is undefined because the line is vertical.

Example 4

The slopes of two distinct lines are given. Determine whether the lines are parallel, perpendicular, or neither.

a. $m_1 = -7$ and $m_2 = -7$ Parallel

b. $m_1 = -\frac{1}{5}$ and $m_2 = 5$ Perpendicular

c. $m_1 = -\frac{3}{2}$ and $m_2 = -\frac{2}{3}$ Neither

Section 3.4 Slope-Intercept Form of a Line

Key Concepts

The **slope-intercept form** of a line is

$$y = mx + b$$

where m is the slope of the line and $(0, b)$ is the y-intercept.

Slope-intercept form is used to identify the slope and y-intercept of a line when the equation is given.

Slope-intercept form can also be used to graph a line.

Examples

Example 1

Find the slope and y-intercept.

$$7x - 2y = 4$$
$$-2y = -7x + 4 \qquad \text{Solve for } y.$$
$$\frac{-2y}{-2} = \frac{-7x}{-2} + \frac{4}{-2}$$
$$y = \frac{7}{2}x - 2$$

The slope is $\frac{7}{2}$. The y-intercept is $(0, -2)$.

Example 2

Graph the line.

$$y = \frac{7}{2}x - 2$$

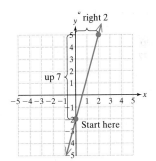

Section 3.5 Point-Slope Formula

Key Concepts

The **point-slope formula** is used primarily to construct an equation of a line given a point and the slope.

Equations of Lines—A Summary:

Standard form: $Ax + By = C$
Horizontal line: $y = k$
Vertical line: $x = k$
Slope-intercept form: $y = mx + b$
Point-slope formula: $y - y_1 = m(x - x_1)$

Examples

Example 1

Find an equation of the line passing through the point $(6, -4)$ and having a slope of $-\frac{1}{2}$.

Label the given information:
$m = -\frac{1}{2}$ and $(x_1, y_1) = (6, -4)$

$$y - y_1 = m(x - x_1)$$
$$y - (-4) = -\frac{1}{2}(x - 6)$$
$$y + 4 = -\frac{1}{2}x + 3$$
$$y = -\frac{1}{2}x - 1$$

Section 3.6   Applications of Linear Equations and Modeling

Key Concepts

Linear equations can often be used to describe or model the relationship between variables in a real-world event. In such applications, the slope may be interpreted as a rate of change.

Examples

Example 1

The number of drug-related arrests for a small city has been growing approximately linearly since 1980.

Let y represent the number of drug arrests, and let x represent the number of years after 1980.

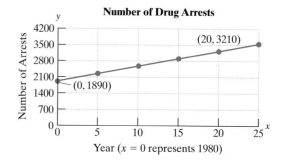

Number of Drug Arrests

Year ($x = 0$ represents 1980)

a. Use the ordered pairs (0, 1890) and (20, 3210) to find an equation of the line shown in the graph.

$$m = \frac{y_2 - y_1}{x_2 - x_1} = \frac{3210 - 1890}{20 - 0}$$

$$= \frac{1320}{20} = 66$$

The slope is 66, indicating that the number of drug arrests is increasing at a rate of 66 per year. $m = 66$, and the y-intercept is (0, 1890). Hence:

$$y = mx + b \implies y = 66x + 1890$$

b. Use the equation in part (a) to predict the number of drug-related arrests in the year 2010. (The year 2010 is 30 years after 1980. Hence, $x = 30$.)

$$y = 66(30) + 1890$$

$$y = 3870$$

The number of drug arrests is predicted to be 3870 by the year 2010.

Section 3.7   Introduction to Functions

Key Concepts

Any set of ordered pairs, (x, y), is called a **relation** in x and y.

The **domain** of a relation is the set of first components in the ordered pairs in the relation. The **range** of a relation is the set of second components in the ordered pairs.

Given a relation in x and y, we say "y is a **function** of x" if for each element x in the domain, there is exactly one value y in the range.

Vertical Line Test for Functions

Consider any relation defined by a set of points (x, y) on a rectangular coordinate system. Then the graph defines y as a function of x if no vertical line intersects the graph in more than one point.

Function Notation

$f(x)$ is the value of the function, f, at x.

Examples

Example 1

Find the domain and range of the relation.

$\{(0, 0), (1, 1), (2, 4), (3, 9), (-1, 1), (-2, 4), (-3, 9)\}$

Domain: $\{0, 1, 2, 3, -1, -2, -3\}$

Range: $\{0, 1, 4, 9\}$

Example 2

Function: $\{(1, 3), (2, 5), (6, 3)\}$

Not a function: $\{(1, 3), (2, 5), (1, -2)\}$

different y-values for the same x-value

Example 3

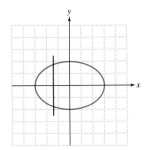 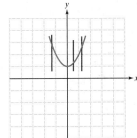

Not a Function
Vertical line intersects more than once.

Function
No vertical line intersects more than once.

Example 4

Given $f(x) = -3x^2 + 5x$, find $f(-2)$.

$f(-2) = -3(-2)^2 + 5(-2)$

$= -12 - 10$

$= -22$

Chapter 3 Review Exercises

Section 3.1

1. Graph the points on a rectangular coordinate system.

a. $\left(\dfrac{1}{2}, 5\right)$ **b.** $(-1, 4)$ **c.** $(2, -1)$

d. $(0, 3)$ **e.** $(0, 0)$ **f.** $\left(-\dfrac{8}{5}, 0\right)$

g. $(-2, -5)$ **h.** $(3, 1)$

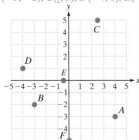

2. Estimate the coordinates of the points A, B, C, D, E, and F.

$A(4, -3)$; $B(-3, -2)$; $C(\frac{5}{2}, 5)$; $D(-4, 1)$; $E(-\frac{1}{2}, 0)$; $F(0, -5)$

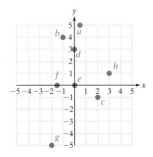

For Exercises 3–8, determine the quadrant in which the given point is found.

3. $(-2, -10)$ III   **4.** $(-4, 6)$ II

5. $(3, -5)$ IV   **6.** $\left(\dfrac{1}{2}, \dfrac{7}{5}\right)$ I

7. $(\pi, -2.7)$ IV   **8.** $(-1.2, -6.8)$ III

9. On which axis is the point $(2, 0)$ found?   x-axis

10. On which axis is the point $(0, -3)$ found?   y-axis

11. The price per share of a stock (in dollars) over a period of 5 days is shown in the graph.

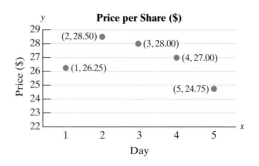

a. Interpret the meaning of the ordered pair $(1, 26.25)$.   On day 1, the price was $26.25.

b. On which day was the price the highest?   Day 2

c. What was the increase in price between day 1 and day 2? $2.25

12. The number of space shuttle launches for selected years is given by the ordered pairs. Let x represent the number of years since 1995. Let y represent the number of launches.

$(1, 7)$ $(2, 8)$ $(3, 5)$ $(4, 3)$

$(5, 5)$ $(6, 6)$ $(7, 5)$ $(8, 1)$

a. Interpret the meaning of the ordered pair $(8, 1)$.   In 2003 (8 years after 1995), there was only one space shuttle launch (this was the year that the Columbia and its crew were lost).

b. Plot the points on a rectangular coordinate system.

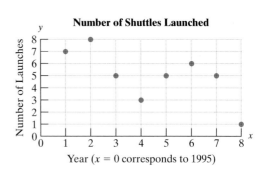

Section 3.2

For Exercises 13–16, determine if the given ordered pair is a solution to the equation.

13. $5x - 3y = 12$; $(0, 4)$ No

14. $2x - 4y = -6$; $(3, 0)$ No

15. $y = \dfrac{1}{3}x - 2$; $(9, 1)$ Yes

16. $y = -\dfrac{2}{5}x + 1$; $(-10, 5)$ Yes

For Exercises 17–20, complete the table and graph the corresponding ordered pairs. Graph the line through the points to represent all solutions to the equation.

17. $3x - y = 5$

x	y
2	1
3	4
1	-2

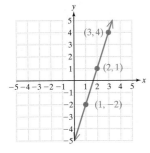

18. $\dfrac{1}{2}x + 3y = 6$

x	y
0	2
-2	7/3
-6	3

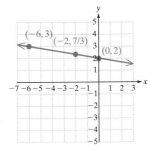

19. $y = \dfrac{2}{3}x - 1$

x	y
0	-1
3	1
-6	-5

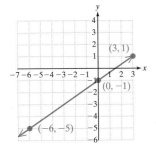

20. $y = -2x - 3$

x	y
0	-3
-3	3
1	-5

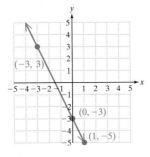

For Exercises 21–24, graph the equation.

21. $x + 2y = 4$

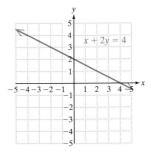

22. $x - y = 5$

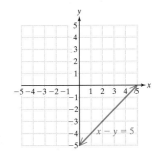

23. $y = 3x$

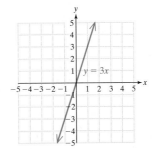

24. $y = \dfrac{1}{4}x$

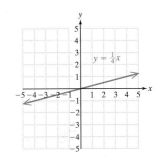

For Exercises 25–28, identify the line as horizontal or vertical. Then graph the equation.

25. $3x - 2 = 10$ Vertical

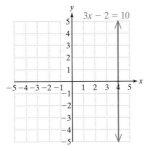

26. $2x + 1 = -2$ Vertical

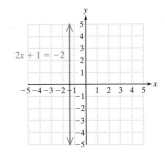

27. $6y + 1 = 13$
Horizontal

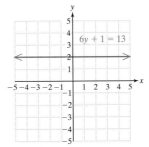

28. $5y - 1 = 14$
Horizontal

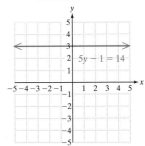

For Exercises 29–36, find the *x*- and *y*-intercepts if they exist.

29. $-4x + 8y = 12$ x-intercept: $(-3, 0)$; y-intercept: $\left(0, \frac{3}{2}\right)$

30. $2x + y = 6$ x-intercept: $(3, 0)$; y-intercept: $(0, 6)$

31. $y = 8x$ x-intercept: $(0, 0)$; y-intercept: $(0, 0)$

32. $5x - y = 0$ x-intercept: $(0, 0)$; y-intercept: $(0, 0)$

33. $6y = -24$ x-intercept: none; y-intercept: $(0, -4)$

34. $2y - 3 = 1$ x-intercept: none; y-intercept: $(0, 2)$

35. $2x + 5 = 0$ x-intercept: $\left(-\frac{5}{2}, 0\right)$; y-intercept: none

36. $-3x + 1 = 0$ x-intercept: $\left(\frac{1}{3}, 0\right)$; y-intercept: none

Section 3.3

37. What is the slope of the ladder leaning up against the wall? $m = \dfrac{12}{5}$

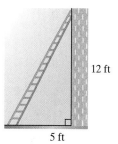

12 ft

5 ft

38. Point *A* is located 4 units down and 2 units to the right of point *B*. What is the slope of the line through points *A* and *B*? -2

39. Determine the slope of the line that passes through the points $(7, -9)$ and $(-5, -1)$. $-\dfrac{2}{3}$

40. Determine the slope of the line that has *x*- and *y*-intercepts of $(-1, 0)$ and $(0, 8)$. 8

41. Determine the slope of the line that passes through the points $(3, 0)$ and $(3, -7)$. Undefined

42. Determine the slope of the horizontal line given by $y = -1$. 0

43. A given line has a slope of -5.

   a. What is the slope of a line parallel to the given line? -5

   b. What is the slope of a line perpendicular to the given line? $\dfrac{1}{5}$

44. A given line has a slope of 0.

   a. What is the slope of a line parallel to the given line? 0

   b. What is the slope of a line perpendicular to the given line? Undefined

For Exercises 45–48, find the slopes of the lines l_1 and l_2 from the two given points. Then determine whether l_1 and l_2 are parallel, perpendicular, or neither.

45. l_1: $(3, 7)$ and $(0, 5)$ $m_1 = \dfrac{2}{3}$

   l_2: $(6, 3)$ and $(-3, -3)$ $m_2 = \dfrac{2}{3}$; parallel

46. l_1: $(-2, 1)$ and $(-1, 9)$ $m_1 = 8$

   l_2: $(0, -6)$ and $(2, 10)$ $m_2 = 8$; parallel

47. l_1: $\left(0, \frac{5}{6}\right)$ and $(2, 0)$ $m_1 = -\dfrac{5}{12}$

   l_2: $\left(0, \frac{6}{5}\right)$ and $\left(-\frac{1}{2}, 0\right)$ $m_2 = \dfrac{12}{5}$; perpendicular

48. l_1: $(1, 1)$ and $(1, -8)$ $m_1 =$ undefined

   l_2: $(4, -5)$ and $(7, -5)$ $m_2 = 0$; perpendicular

Section 3.4

For Exercises 49–54, write each equation in slope-intercept form. Identify the slope and the *y*-intercept.

49. $5x - 2y = 10$ $y = \dfrac{5}{2}x - 5$; $m = \dfrac{5}{2}$; y-intercept: $(0, -5)$

50. $3x + 4y = 12$ $y = -\dfrac{3}{4}x + 3$; $m = -\dfrac{3}{4}$; y-intercept: $(0, 3)$

51. $x - 3y = 0$ $y = \dfrac{1}{3}x$; $m = \dfrac{1}{3}$; y-intercept: $(0, 0)$

52. $5y - 8 = 4$ $y = \dfrac{12}{5}$; $m = 0$; y-intercept: $\left(0, \frac{12}{5}\right)$

53. $2y = -5$ $y = -\dfrac{5}{2}$; $m = 0$; y-intercept: $\left(0, -\frac{5}{2}\right)$

54. $y - x = 0$ $y = x$; $m = 1$; y-intercept: $(0, 0)$

For Exercises 55–59, determine whether the equations represent parallel lines, perpendicular lines, or neither.

55. l_1: $y = \dfrac{3}{5}x + 3$

   l_2: $y = \dfrac{5}{3}x + 1$

   Neither

56. l_1: $2x - 5y = 10$

   l_2: $5x + 2y = 20$

   Perpendicular

57. l_1: $3x + 2y = 6$

   l_2: $-6x - 4y = 4$

   Parallel

58. l_1: $y = \dfrac{1}{4}x - 3$

   l_2: $-x + 4y = 8$

   Parallel

59. l_1: $2x = 4$

   l_2: $y = 6$ Perpendicular

60. Write an equation of the line whose slope is $-\frac{4}{3}$ and whose *y*-intercept is $(0, -1)$.

$y = -\frac{4}{3}x - 1$ or $4x + 3y = -3$

61. Write an equation of the line that passes through the origin and has a slope of 5.

$y = 5x$ or $5x - y = 0$

Section 3.5

62. Write a linear equation in two variables in slope-intercept form. (Answers may vary.)

For example: $y = 3x + 2$

63. Write a linear equation in two variables in standard form. (Answers may vary.)

For example: $5x + 2y = -4$

64. Write the slope formula to find the slope of the line between the points (x_1, y_1) and (x_2, y_2).

$m = \dfrac{y_2 - y_1}{x_2 - x_1}$

✏ Writing   ↔ Translating Expression   ◣ Geometry   🖩 Scientific Calculator   ◉ Video   NS🌐E

65. Write the point-slope formula. $y - y_1 = m(x - x_1)$

66. Write an equation of a vertical line (answers may vary).   For example: $x = 6$

67. Write an equation of a horizontal line (answers may vary).   For example: $y = -5$

For Exercises 68–73, write an equation of a line given the following information.

68. The slope is -6, and the line passes through the point $(-1, 8)$. $y = -6x + 2$ or $6x + y = 2$

69. The slope is $\frac{2}{3}$, and the line passes through the point $(5, 5)$. $y = \frac{2}{3}x + \frac{5}{3}$ or $2x - 3y = -5$

70. The line passes through the points $(0, -4)$ and $(8, -2)$. $y = \frac{1}{4}x - 4$ or $x - 4y = 16$

71. The line passes through the points $(2, -5)$ and $(8, -5)$. $y = -5$

72. The line passes through the point $(5, 12)$ and is perpendicular to the line $y = -\frac{5}{6}x - 3$. $y = \frac{6}{5}x + 6$ or $6x - 5y = -30$

73. The line passes through the point $(-6, 7)$ and is parallel to the line $4x - y = 0$. $y = 4x + 31$ or $4x - y = -31$

Section 3.6

74. The graph shows the average height for girls based on age (*Source:* National Parenting Council). Let x represent a girl's age, and let y represent her height (in inches).

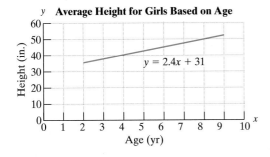

a. Use the equation to estimate the average height of a 7-year-old girl.   47.8 in.

b. What is the slope of the line? Interpret the meaning of the slope in the context of the problem.   The slope is 2.4 and indicates that the average height for girls increases at a rate of 2.4 in. per year.

75. The number of drug prescriptions has increased between 1995 and 2007 (see graph). Let x represent the number of years since 1995. Let y represent the number of prescriptions (in millions).

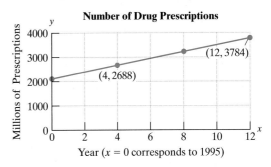

a. Using the ordered pairs $(4, 2688)$ and $(12, 3784)$ find the slope of the line. $m = 137$

b. Interpret the meaning of the slope in the context of this problem.   The number of prescriptions increased by 137 million per year during this time period.

c. Find a linear equation that represents the number of prescriptions, y, versus the year, x. $y = 137x + 2140$

d. Predict the number of prescriptions for the year 2010.   4195 million

76. A water purification company charges $20 per month and a $55 installation fee.

a. Write a linear equation to compute the total cost, y, of renting this system for x months. $y = 20x + 55$

b. Use the equation from part (a) to determine the total cost to rent the system for 9 months. $235

77. A small cleaning company has a fixed monthly cost of $700 and a variable cost of $8 per service call.

a. Write a linear equation to compute the total cost, y, of making x service calls in one month. $y = 8x + 700$

b. Use the equation from part (a) to determine the total cost of making 80 service calls. $1340

Section 3.7

For Exercises 78–83, state the domain and range of each relation. Then determine whether the relation is a function.

78. $\{(6, 3), (10, 3), (-1, 3), (0, 3)\}$
Domain: $\{6, 10, -1, 0\}$; range: $\{3\}$; function

79. $\{(2, 0), (2, 1), (2, -5), (2, 2)\}$
Domain: $\{2\}$; range: $\{0, 1, -5, 2\}$; not a function

80. Domain: $[-4, 5]$; range: $[-3, 3]$; not a function

81. Domain: $(-\infty, \infty)$; range: $[-2, \infty)$; function

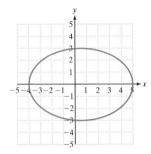

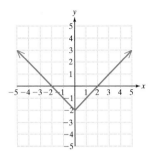

82. $\{(4, 23), (3, -2), (-6, 5), (4, 6)\}$
Domain: $\{4, 3, -6\}$; range: $\{23, -2, 5, 6\}$; not a function

83. $\{(3, 0), (-4, \frac{1}{2}), (0, 3), (2, -12)\}$
Domain: $\{3, -4, 0, 2\}$; range: $\{0, \frac{1}{2}, 3, -12\}$; function

84. Given the function defined by $f(x) = x^3$ find:

   a. $f(0)$ 0 **b.** $f(2)$ 8 **c.** $f(-3)$ -27

   d. $f(-1)$ -1 **e.** $f(4)$ 64

85. Given the function defined by $g(x) = \dfrac{x}{5 - x}$. Find:

   a. $g(0)$ 0 **b.** $g(4)$ 4 **c.** $g(-1)$ $-\dfrac{1}{6}$

   d. $g(3)$ $\dfrac{3}{2}$ **e.** $g(-5)$ $-\dfrac{1}{2}$

86. The landing distance that a certain plane will travel on a runway is determined by the initial landing speed at the instant the plane touches down. The following function relates landing distance, $D(x)$, to initial landing speed, x, where $x \geq 15$.

$$D(x) = \frac{1}{10}x^2 - 3x + 22$$ where D is in feet and x is in feet per second.

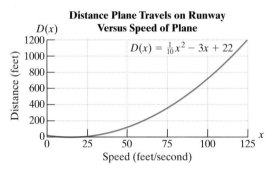

Distance Plane Travels on Runway Versus Speed of Plane

a. Find $D(90)$, and interpret the meaning of the function value in terms of landing speed and length of the runway. $D(90) = 562$. A plane traveling 90 ft/sec when it touches down will require 562 ft of runway.

b. Find $D(110)$, and interpret the meaning in terms of landing speed and length of the runway. $D(110) = 902$. A plane traveling 110 ft/sec when it touches down will require 902 ft of runway.

Chapter 3   Test

1. In which quadrant is the given point found?

   a. $\left(-\dfrac{7}{2}, 4\right)$ **b.** $(4.6, -2)$ **c.** $(-37, -45)$

       II   IV   III

2. What is the y-coordinate for a point on the x-axis? 0

3. What is the x-coordinate for a point on the y-axis? 0

4. The following table depicts a boy's height versus his age. Let x represent the boy's age and y represent his height.

Age (years), x	Height (inches), y
5	46
7	50
9	55
11	60

  a. Write the data as ordered pairs and interpret the meaning of the first ordered pair.

$(5, 46)$ At age 5 the boy's height was 46 in. $(7, 50)$ $(9, 55)$ $(11, 60)$

  b. Graph the ordered pairs on a rectangular coordinate system.

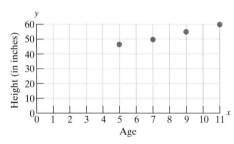

  c. From the graph and table estimate the boy's height at age 10. 57.5 in.

  d. The data appear to follow an upward trend up to the boy's teenage years. Do you think this trend will continue? Would it be reasonable to use these data to predict the boy's height at age 25? No, his height will maximize in his teen years.

5. Determine whether the ordered pair is a solution to the equation $2x - y = 6$.

a. $(0, 6)$ No

b. $(4, 2)$ Yes

c. $(3, 0)$ Yes

d. $\left(\frac{9}{2}, 3\right)$ Yes

6. Given the equation $y = \frac{1}{4}x - 2$, complete the table. Plot the ordered pairs and graph the line through the points to represent the set of all solutions to the equation.

x	y
0	-2
4	-1
6	$-\frac{1}{2}$

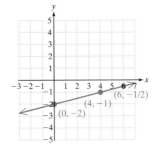

7. If x represents an adult's age, then the person's maximum recommended heart rate, y, during exercise is approximated by the equation

$$y = 220 - x \quad (x \geq 18)$$

a. Use the equation to find the maximum recommended heart rate for a person who is 18 years old.   202 beats per minute

b. Use the equation to complete the following ordered pairs: $(20, \ )$, $(30, \ )$, $(40, \ )$, $(50, \ )$, $(60, \ )$.   (20, 200)   (30, 190)
(40, 180)   (50, 170)   (60, 160)

For Exercises 8–9, determine whether the equation represents a horizontal or vertical line. Then graph the line.

8. $-6y = 18$ Horizontal

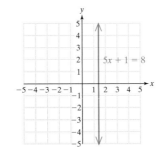

9. $5x + 1 = 8$ Vertical

For Exercises 10–13, determine the x- and y-intercepts if they exist.

10. $-4x + 3y = 6$
x-intercept: $\left(-\frac{3}{2}, 0\right)$; y-intercept: $(0, 2)$

11. $2y = 6x$
x-intercept: $(0, 0)$;
y-intercept: $(0, 0)$

12. $x = 4$
x-intercept: $(4, 0)$;
y-intercept: none

13. $y - 3 = 0$
x-intercept: none;
y-intercept: $(0, 3)$

14. What is the slope of the hill? $\frac{2}{5}$

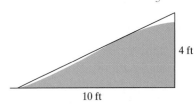

15. a. Find the slope of the line that passes through the points $(-2, 0)$ and $(-5, -1)$. $\frac{1}{3}$

b. Find the slope of the line $4x - 3y = 9$. $\frac{4}{3}$

16. a. What is the slope of a line parallel to the line $x + 4y = -16$? $-\frac{1}{4}$

b. What is the slope of a line perpendicular to the line $x + 4y = -16$?   4

17. a. What is the slope of the line $x = 5$?   Undefined

b. What is the slope of the line $y = -3$?   0

For Exercises 18–19, graph the equations.

18. $y = 8x + 2$

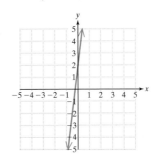

19. $2x + 9y = 0$

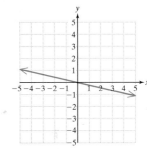

20. Determine whether the equations represent parallel lines, perpendicular lines, or neither.

l_1: $2y = 3x - 3$ l_2: $4x = -6y + 1$
Perpendicular

21. Write an equation of the line that has y-intercept $\left(0, \frac{1}{2}\right)$ and slope $\frac{1}{4}$. $y = \frac{1}{4}x + \frac{1}{2}$ or $x - 4y = -2$

22. Write an equation of the line that passes through the points $(2, 8)$, and $(4, 1)$.
$y = -\frac{7}{2}x + 15$ or $7x + 2y = 30$

23. Write an equation of the line that passes through the point $(2, -6)$ and is parallel to the x-axis. $y = -6$

24. Write an equation of the line that passes through the point $(3, 0)$ and is parallel to the line $2x + 6y = -5$. $y = -\frac{1}{3}x + 1$ or $x + 3y = 3$

25. Write an equation of the line that passes through the point $(-3, -1)$ and is perpendicular to the line $x + 3y = 9$. $y = 3x + 8$ or $3x - y = -8$

26. Hurricane Floyd dumped rain at an average rate of $\frac{3}{4}$ in./hr on Southport, North Carolina. Further inland, in Lumberton, North Carolina, the storm dropped $\frac{1}{2}$ in. of rain per hour. The following graph depicts the total amount of rainfall (in inches) versus the time (in hours) for both locations in North Carolina.

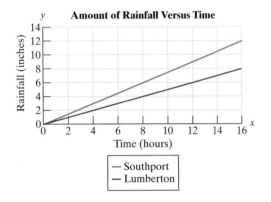

a. What is the slope of the line representing the rainfall for Southport? $\frac{3}{4}$

b. What is the slope of the line representing the rainfall for Lumberton? $\frac{1}{2}$

27. To attend a state fair, the cost is $10 per person to cover exhibits and musical entertainment. There is an additional cost of $1.50 per ride.

a. Write an equation that gives the total cost, y, of visiting the state fair and going on x rides. $y = 1.5x + 10$

b. Use the equation from part (a) to determine the cost of going to the state fair and going on 10 rides. $25

28. The number of medical doctors for selected years is shown in the graph. Let x represent the number of years since 1980, and let y represent the number of medical doctors (in thousands) in the United States.

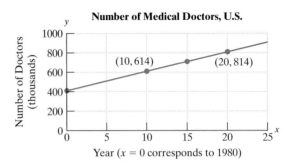

a. Find the slope of the line shown in the graph. Interpret the meaning of the slope in the context of this problem. $m = 20$; The slope indicates that there is an increase of 20 thousand medical doctors per year.

b. Find an equation of the line. $y = 20x + 414$

c. Use the equation from part (b) to predict the number of medical doctors in the United States for the year 2010.   1014 thousand or, equivalently, 1,014,000

29. Given the relation $\{(0, -1), (2, 3), (-15, -8), (4, 4), (9, -1)\}$:

a. State the domain. $\{0, 2, -15, 4, 9\}$

b. State the range. $\{-1, 3, -8, 4\}$

c. Determine whether the relation is a function.   Function

30. Given the function defined by $f(x) = x^2 - x$, find

a. $f(3)$ 6   **b.** $f(-3)$ 12

Chapters 1–3 Cumulative Review Exercises

1. Identify the numbers as rational or irrational.

   a. -3 **b.** $\dfrac{5}{4}$ **c.** $\sqrt{10}$ **d.** 0

    Rational   Rational   Irrational   Rational

2. Write the opposite and the absolute value for each number.

   a. $-\dfrac{2}{3}$ $\dfrac{2}{3}; \dfrac{2}{3}$ **b.** 5.3 $-5.3; 5.3$

3. Simplify the expression using the order of operations: $32 \div 2 \cdot 4 + 5$ 69

4. Add: $3 + (-8) + 2 + (-10)$ -13

5. Subtract: $16 - 5 - (-7)$ 18

For Exercises 6–7, translate the English phrase into an algebraic expression. Then evaluate the expression.

6. The quotient of $\dfrac{3}{4}$ and $-\dfrac{7}{8}$. $\dfrac{3}{4} \div -\dfrac{7}{8}; -\dfrac{6}{7}$

7. The product of -2.1 and -6. $(-2.1)(-6); 12.6$

8. Name the property that is illustrated by the following statement. $6 + (8 + 2) = (6 + 8) + 2$

   The associative property of addition

For Exercises 9–12, solve the equation.

9. $6x - 10 = 14$ **10.** $3(m + 2) - 3 = 2m + 8$

    4   5

11. $\dfrac{2}{3}y - \dfrac{1}{6} = y + \dfrac{4}{3}$ **12.** $1.7z + 2 = -2(0.3z + 1.3)$

    $-\dfrac{9}{2}$ -2

13. The area of Texas is $267{,}277$ mi^2. If this is 712 mi^2 less than 29 times the area of Maine, find the area of Maine. 9241 mi^2

14. For the formula $3a + b = c$, solve for a. $a = \dfrac{c - b}{3}$

15. Graph the equation $-6x + 2y = 0$.

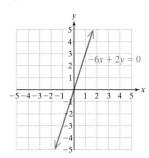

16. Find the x- and y-intercepts of $-2x + 4y = 4$.
   x-intercept: $(-2, 0)$; y-intercept: $(0, 1)$

17. Write the equation in slope-intercept form. Then identify the slope and the y-intercept.
   $3x + 2y = -12$
   $y = -\dfrac{3}{2}x - 6$;   slope: $-\dfrac{3}{2}$; y-intercept: $(0, -6)$

18. Explain why the line $2x + 3 = 5$ has only one intercept. $2x + 3 = 5$ can be written as $x = 1$, which represents a vertical line. A vertical line of the form $x = k \ (k \neq 0)$ has an x-intercept of $(k, 0)$ and no y-intercept.

19. Find an equation of a line passing through $(2, -5)$ with slope -3. $y = -3x + 1$ or $3x + y = 1$

20. Find an equation of the line passing through $(0, 6)$ and $(-3, 4)$. $y = \dfrac{2}{3}x + 6$ or $2x - 3y = -18$

Systems of Linear Equations in Two Variables

4

CHAPTER OUTLINE

4.1 Solving Systems of Equations by the Graphing Method 296

4.2 Solving Systems of Equations by the Substitution Method 307

4.3 Solving Systems of Equations by the Addition Method 316

  Problem Recognition Exercises: Systems of Equations 325

4.4 Applications of Linear Equations in Two Variables 326

4.5 Linear Inequalities and Systems of Inequalities in Two Variables 335

  Group Activity: Creating Linear Models from Data 345

Chapter 4

This chapter is devoted to solving systems of linear equations and inequalities. We will present a graphical method for solving systems and two algebraic methods. Applications of systems of equations in two variables involve two variables that are subject to two conditions.

The puzzle shown is a unique crossword puzzle. It consists of four words. First you will have to determine the four words. Then you will have to place them in the puzzle so that they fit into the boxes.

- Three words in the puzzle are the three methods shown in this chapter for solving systems of equations in two variables.

- The fourth word is the number of solutions for an inconsistent system of equations.

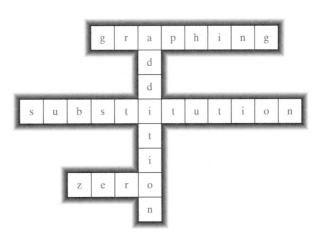

Solving Systems of Equations by the Graphing Method

Objectives

1. Determining Solutions to a System of Linear Equations
2. Dependent and Inconsistent Systems of Linear Equations
3. Solving Systems of Linear Equations by Graphing

1. Determining Solutions to a System of Linear Equations

Recall from Section 3.2 that a linear equation in two variables has an infinite number of solutions. The set of all solutions to a linear equation forms a line in a rectangular coordinate system. Two or more linear equations form a **system of linear equations**. For example, here are three systems of equations:

$$x - 3y = -5 \qquad y = \tfrac{1}{4}x - \tfrac{3}{4} \qquad 5a + b = 4$$
$$2x + 4y = 10 \qquad -2x + 8y = -6 \qquad -10a - 2b = 8$$

A **solution to a system of linear equations** is an ordered pair that is a solution to *each* individual linear equation.

--- **Skill Practice** ---

Determine whether the ordered pair is a solution to the system.
$5x - 2y = 24$
$2x + y = 6$
1. $(6, 3)$
2. $(4, -2)$

Classroom Examples: p. 302, Exercises 4 and 6

Avoiding Mistakes

It is important to test an ordered pair in *both* equations to determine if the ordered pair is a solution.

Example 1 Determining Solutions to a System of Linear Equations

Determine whether the ordered pairs are solutions to the system.

$$x + y = 4$$
$$-2x + y = -5$$

a. $(3, 1)$ **b.** $(0, 4)$

Solution:

a. Substitute the ordered pair $(3, 1)$ into both equations:

$$x + y = 4 \longrightarrow (3) + (1) \overset{?}{=} 4 \; \checkmark \qquad \text{True}$$
$$-2x + y = -5 \longrightarrow -2(3) + (1) \overset{?}{=} -5 \; \checkmark \qquad \text{True}$$

Because the ordered pair $(3, 1)$ is a solution to each equation, it is a solution to the *system* of equations.

b. Substitute the ordered pair $(0, 4)$ into both equations.

$$x + y = 4 \longrightarrow (0) + (4) \overset{?}{=} 4 \; \checkmark \qquad \text{True}$$
$$-2x + y = -5 \longrightarrow -2(0) + (4) \overset{?}{=} -5 \qquad \text{False}$$

Because the ordered pair $(0, 4)$ is not a solution to the second equation, it is *not* a solution to the system of equations.

A solution to a system of two linear equations may be interpreted graphically as a point of intersection between the two lines. Using slope-intercept form to graph the lines from Example 1, we have

$$x + y = 4 \longrightarrow y = -x + 4$$
$$-2x + y = -5 \longrightarrow y = 2x - 5$$

Notice that the lines intersect at $(3, 1)$ (Figure 4-1).

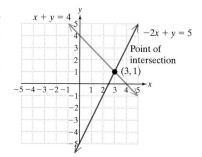

Figure 4-1

2. Dependent and Inconsistent Systems of Linear Equations

When two lines are drawn in a rectangular coordinate system, three geometric relationships are possible:

1. Two lines may intersect at *exactly one point.*

2. Two lines may intersect at *no point.* This occurs if the lines are parallel.

3. Two lines may intersect at *infinitely many points* along the line. This occurs if the equations represent the same line (the lines coincide).

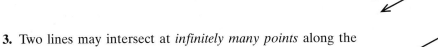

If a system of linear equations has one or more solutions, the system is said to be **consistent**. If a linear equation has no solution, it is said to be **inconsistent**.

If two equations represent the same line, then all points along the line are solutions to the system of equations. In such a case, the system is characterized as a **dependent system**. An **independent system** is one in which the two equations represent different lines.

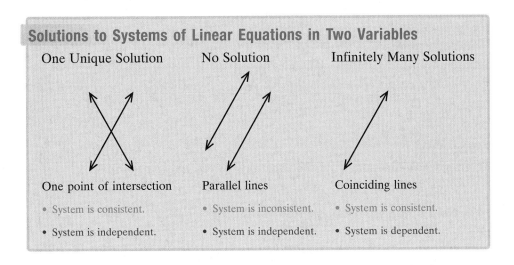

Solutions to Systems of Linear Equations in Two Variables

One Unique Solution	No Solution	Infinitely Many Solutions
One point of intersection	Parallel lines	Coinciding lines
• System is consistent.	• System is inconsistent.	• System is consistent.
• System is independent.	• System is independent.	• System is dependent.

3. Solving Systems of Linear Equations by Graphing

One way to find a solution to a system of equations is to graph the equations and find the point (or points) of intersection. This is called the *graphing method* to solve a system of equations.

Answers

3. Zero **4.** One

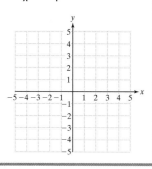
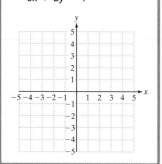
Answers

5.

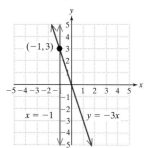

6.

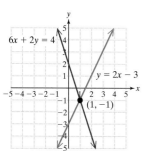

Example 2 Solving a System of Linear Equations by Graphing

Solve the system by the graphing method. $y = 2x$

$$y = 2$$

Solution:

The equation $y = 2x$ is written in slope-intercept form as $y = 2x + 0$. The line passes through the origin, with a slope of 2.

The line $y = 2$ is a horizontal line and has a slope of 0.

Because the lines have different slopes, the lines must be different and nonparallel. From this, we know that the lines must intersect at exactly one point. Graph the lines to find the point of intersection (Figure 4-2).

The point (1, 2) appears to be the point of intersection. This can be confirmed by substituting $x = 1$ and $y = 2$ into both original equations.

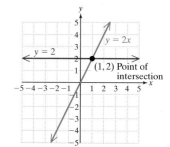

Figure 4-2

$$y = 2x \qquad (2) \overset{?}{=} 2(1) ✔ \quad \text{True}$$
$$y = 2 \qquad (2) \overset{?}{=} 2 ✔ \quad \text{True}$$

The solution is (1, 2).

Example 3 Solving a System of Linear Equations by Graphing

Solve the system by the graphing method.

$$x - 2y = -2$$
$$-3x + 2y = 6$$

Solution:

To graph each equation, write the equation in slope-intercept form: $y = mx + b$.

Equation 1

$x - 2y = -2$

$-2y = -x - 2$

$\dfrac{-2y}{-2} = \dfrac{-x}{-2} - \dfrac{2}{-2}$

$y = \dfrac{1}{2}x + 1$

Equation 2

$-3x + 2y = 6$

$2y = 3x + 6$

$\dfrac{2y}{2} = \dfrac{3x}{2} + \dfrac{6}{2}$

$y = \dfrac{3}{2}x + 3$

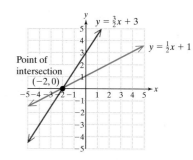

Figure 4-3

From their slope-intercept forms, we see that the lines have different slopes, indicating that the lines are different and nonparallel. Therefore, the lines must intersect at exactly one point. Graph the lines to find that point (Figure 4-3).

The point $(-2, 0)$ appears to be the point of intersection. This can be confirmed by substituting $x = -2$ and $y = 0$ into both equations.

$$x - 2y = -2 \longrightarrow (-2) - 2(0) \stackrel{?}{=} -2 \; ✔ \quad \text{True}$$

$$-3x + 2y = 6 \longrightarrow -3(-2) + 2(0) \stackrel{?}{=} 6 \; ✔ \quad \text{True}$$

The solution is $(-2, 0)$.

TIP: In Examples 2 and 3, the lines could also have been graphed by using the x- and y-intercepts or by using a table of points. However, the advantage of writing the equations in slope-intercept form is that we can compare the slopes and y-intercepts of each line.

1. If the slopes differ, the lines are different and nonparallel and must cross in exactly one point.

2. If the slopes are the same and the y-intercepts are different, the lines are parallel and will not intersect.

3. If the slopes are the same and the y-intercepts are the same, the two equations represent the same line.

Example 4 **Graphing an Inconsistent System**

Solve the system by graphing.

$$-x + 3y = -6$$

$$6y = 2x + 6$$

Solution:

To graph the lines, write each equation in slope-intercept form.

Equation 1	**Equation 2**
$-x + 3y = -6$	$6y = 2x + 6$
$3y = x - 6$	
$\dfrac{3y}{3} = \dfrac{x}{3} - \dfrac{6}{3}$	$\dfrac{6y}{6} = \dfrac{2x}{6} + \dfrac{6}{6}$
$y = \dfrac{1}{3}x - 2$	$y = \dfrac{1}{3}x + 1$

Because the lines have the same slope but different y-intercepts, they are parallel (Figure 4-4). Two parallel lines do not intersect, which implies that the system has no solution. The system is inconsistent.

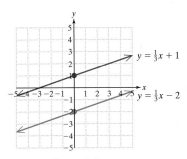

Figure 4-4

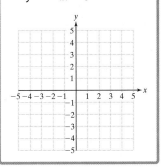
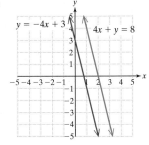

Skill Practice

Solve the system by graphing.

8. $x - 3y = 4$

$y = \frac{1}{3}x - \frac{4}{3}$

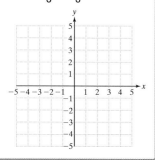

Classroom Example: p. 304,
Exercise 36

Answer

8.

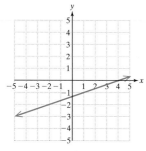

$\left\{ (x, y) \mid y = \frac{1}{3}x - \frac{4}{3} \right\}$

The system is dependent.

Example 5 **Graphing a Dependent System**

Solve the system by graphing.

$$x + 4y = 8$$

$$y = -\frac{1}{4}x + 2$$

Solution:

Write the first equation in slope-intercept form. The second equation is already in slope-intercept form.

Equation 1	**Equation 2**
$x + 4y = 8$	$y = -\frac{1}{4}x + 2$
$4y = -x + 8$	
$\dfrac{4y}{4} = \dfrac{-x}{4} + \dfrac{8}{4}$	
$y = -\dfrac{1}{4}x + 2$	

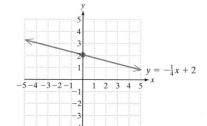

Figure 4-5

Notice that the slope-intercept forms of the two lines are identical. Therefore, the equations represent the same line (Figure 4-5). The system is dependent, and the solution to the system of equations is the set of all points on the line.

Because the ordered pairs in the solution set cannot all be listed, we can write the solution in set-builder notation: $\{(x, y) \mid y = -\frac{1}{4}x + 2\}$. This can be read as "the set of all ordered pairs" (x, y) such that the ordered pairs satisfy the equation $y = -\frac{1}{4}x + 2$.

In summary:

- There are infinitely many solutions to the system of equations.
- The solution set is $\{(x, y) \mid y = -\frac{1}{4}x + 2\}$.
- The system is dependent.

Calculator Connections

Topic: Graphing Systems of Linear Equations in Two Variables

The solution to a system of equations can be found by using either a *Trace* feature or an *Intersect* feature on a graphing calculator to find the point of intersection between two graphs.

For example, consider the system:

$$-2x + y = 6$$

$$5x + y = -1$$

First graph the equations together on the same viewing window. Recall that to enter the equations into the calculator, the equations must be written with the *y*-variable isolated.

Isolate *y*.

$$-2x + y = 6 \longrightarrow y = 2x + 6$$

$$5x + y = -1 \longrightarrow y = -5x - 1$$

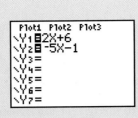

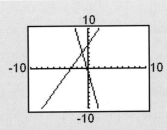

By inspection of the graph, it appears that the solution is $(-1, 4)$. The *Trace* option on the calculator may come close to $(-1, 4)$ but may not show the exact solution (Figure 4-6). However, an *Intersect* feature on a graphing calculator may provide the exact solution (Figure 4-7). See your user's manual for further details.

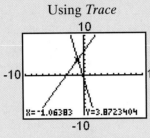

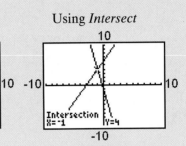

Using *Trace*

Using *Intersect*

Figure 4-6

Figure 4-7

Calculator Exercises

Use a graphing calculator to graph each linear equation on the same viewing window. Use a *Trace* or *Intersect* feature to find the point(s) of intersection.

1. $y = 2x - 3$ (2, 1)
$y = -4x + 9$

2. $y = -\dfrac{1}{2}x + 2$ (6, −1)
$y = \dfrac{1}{3}x - 3$

3. $x + y = 4$ (Example 1)
$-2x + y = -5$ (3, 1)

4. $x - 2y = -2$ (Example 3)
$-3x + 2y = 6$
(−2, 0)

5. $-x + 3y = -6$ (Example 4)
$6y = 2x + 6$
No solution

6. $x + 4y = 8$ (Example 5)
$y = -\dfrac{1}{4}x + 2$ Dependent system

Section 4.1 Practice Exercises

Boost *your* GRADE at ALEKS.com!

ALEKS® version 3.0

- Practice Problems
- Self-Tests
- NetTutor

- e-Professors
- Videos

For additional exercises, see Classroom Activity 4.1A, in the *Instructor's Resource Manual* at www.mhhe.com/moh.

Study Skills Exercises

1. Figure out your grade at this point. Are you earning the grade that you want? If not, maybe organizing a study group would help.

In a study group, check the activities that you might try to help you learn and understand the material.

_____ Quiz each other by asking each other questions.

_____ Practice teaching each other.

_____ Share and compare class notes.

_____ Support and encourage each other.

_____ Work together on exercises and sample problems.

 Writing Translating Expression Geometry Scientific Calculator Video NS̊E

2. Define the key terms:

    a. system of linear equations   **b. solution to a system of linear equations**   **c. consistent system**

    d. inconsistent system   **e. dependent system**   **f. independent system**

Objective 1: Determining Solutions to a System of Linear Equations

For Exercises 3–10, determine if the given point is a solution to the system. **(See Example 1.)**

3. $3x - y = 7$ $(2, -1)$
$x - 2y = 4$ Yes

4. $x - y = 3$ $(4, 1)$
$x + y = 5$ Yes

5. $4y = -3x + 12$ $(0, 4)$
$y = \dfrac{2}{3}x - 4$ No

6. $y = -\dfrac{1}{3}x + 2$ $(9, -1)$
$x = 2y + 6$ No

7. $3x - 6y = 9$ $\left(4, \dfrac{1}{2}\right)$
$x - 2y = 3$ Yes

8. $x - y = 4$ $(6, 2)$
$3x - 3y = 12$ Yes

9. $\dfrac{1}{3}x = \dfrac{2}{5}y - \dfrac{4}{5}$ $(0, 2)$
$\dfrac{3}{4}x + \dfrac{1}{2}y = 2$ No

10. $\dfrac{1}{4}x + \dfrac{1}{2}y = \dfrac{3}{2}$ $(4, 1)$
$y = \dfrac{3}{2}x - 6$ No

Objective 2: Dependent and Inconsistent Systems of Linear Equations

For Exercises 11–14, match the graph of the system of equations with the appropriate description of the solution.

11. b
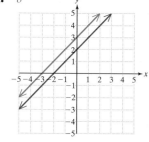

a. The solution is $(1, 3)$.

b. No solution.

c. There are infinitely many solutions.

d. The solution is $(0, 0)$.

12. c
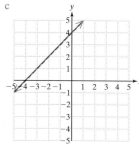

13. d
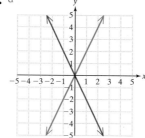

14. a
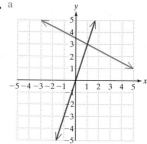

15. Graph each system of equations. **(See Examples 2–5.)**

a. $y = 2x - 3$
$y = 2x + 5$

b. $y = 2x + 1$
$y = 4x - 5$

c. $y = 3x - 5$
$y = 3x - 5$

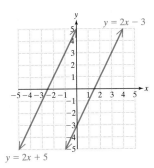

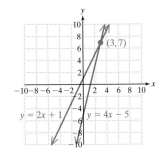

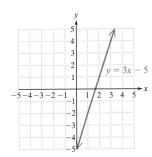

For Exercises 16–26, determine which system of equations (a, b, or c) makes the statement true. (*Hint:* Refer to the graphs from Exercise 15.)

a. $y = 2x - 3$
$y = 2x + 5$

b. $y = 2x + 1$
$y = 4x - 5$

c. $y = 3x - 5$
$y = 3x - 5$

16. The lines are parallel.   a

17. The lines coincide.   c

18. The lines intersect at exactly one point.   b

19. The system is inconsistent.   a

20. The system is dependent.   c

21. The lines have the same slope but different y-intercepts.   a

22. The lines have the same slope and same y-intercept.   c

23. The lines have different slopes.   b

24. The system has exactly one solution.   b

25. The system has infinitely many solutions.   c

26. The system has no solution.   a

Objective 3: Solving Systems of Linear Equations by Graphing

For Exercises 27–52, solve the systems by graphing. If a system does not have a unique solution, identify the system as inconsistent or dependent. **(See Examples 2–5.)**

27. $y = -x + 4$
$y = x - 2$

28. $y = 3x + 2$
$y = 2x$

29. $2x + y = 0$
$3x + y = 1$

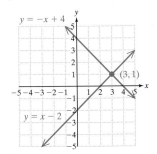

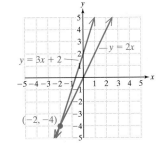

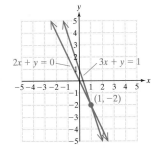

30. $x + y = -1$

$2x - y = -5$

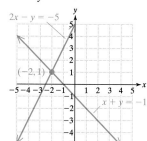

31. $2x + y = 6$

$x = 1$

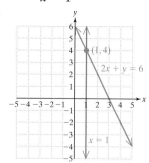

32. $4x + 3y = 9$

$x = 3$

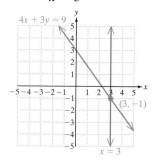

33. $-6x - 3y = 0$ No solution; inconsistent system

$4x + 2y = 4$

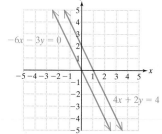

34. $2x - 6y = 12$ No solution; inconsistent system

$-3x + 9y = 12$

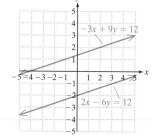

35. $-2x + y = 3$ Infinitely many solutions $\{(x, y) | -2x + y = 3\}$; dependent system

$6x - 3y = -9$

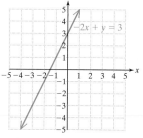

36. $x + 3y = 0$ Infinitely many solutions $\{(x, y) | x + 3y = 0\}$; dependent system

$-2x - 6y = 0$

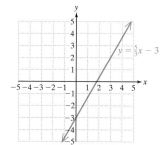

37. $y = 6$

$2x + 3y = 12$

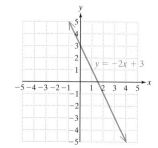

38. $y = -2$

$x - 2y = 10$

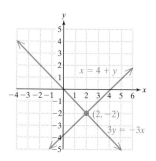

39. $-5x + 3y = -9$ Infinitely many solutions $\{(x, y) | y = \frac{5}{3}x - 3\}$; dependent system

$y = \frac{5}{3}x - 3$

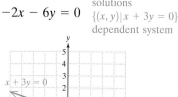

40. $4x + 2y = 6$ Infinitely many solutions $\{(x, y) | y = -2x + 3\}$; dependent system

$y = -2x + 3$

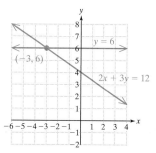

41. $x = 4 + y$

$3y = -3x$

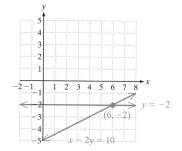

 Writing  Translating Expression   Geometry   Scientific Calculator   Video NS&E

42. $3y = 4x$

$x - y = -1$

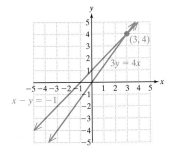

43. $-x + y = 3$ No solution;
inconsistent system

$4y = 4x + 6$

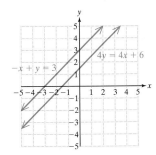

44. $x - y = 4$ No solution;
inconsistent system

$3y = 3x + 6$

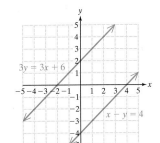

45. $x = 4$

$2y = 4$

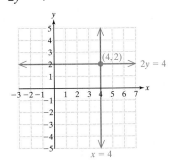

46. $-3x = 6$

$y = 2$

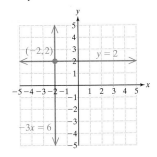

47. $4x + 4y = 8$ No solution;
inconsistent system

$5x + 5y = 5$

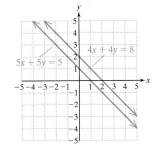

48. $2x + 3y = 8$ No solution;
inconsistent system

$-4x - 6y = 6$

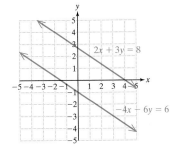

49. $2x + y = 4$

$4x - 2y = -4$

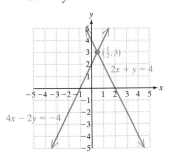

50. $6x + 6y = 3$

$2x - y = 4$

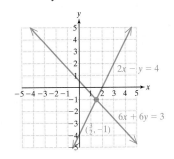

51. $y = 0.5x + 2$ Infinitely many
solutions

$-x + 2y = 4$ $\{(x, y) | y = 0.5x + 2\}$;
dependent system

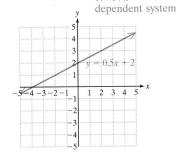

52. $3x - 4y = 6$ Infinitely many solutions
$\{(x, y) | 3x - 4y = 6\}$;

$-6x + 8y = -12$ dependent system

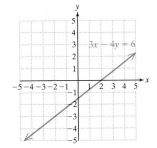

Writing   Translating Expression   Geometry   Scientific Calculator   Video   NS E

53. Two tennis instructors have two different fee schedules. Owen charges $25 per lesson plus a one-time court fee of $20 at the tennis club. Joan charges $30 per lesson but does not require a court fee. The total cost, y, depends on the number of lessons, x, according to the equations

Owen: $y = 25x + 20$

Joan: $y = 30x$

From the graph, determine the number of lessons for which the total cost is the same for both instructors.

4 lessons will cost $120 for each instructor.

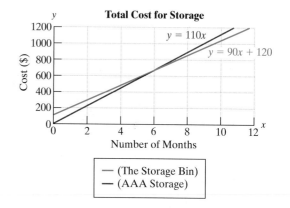

Cost of Tennis Instruction
$y = 30x$
$y = 25x + 20$
— Owen
— Joan
Cost ($)
Number of Lessons

54. The cost to rent a 10 ft by 10 ft storage space is different for two different storage companies. The Storage Bin charges $90 per month plus a nonrefundable deposit of $120. AAA Storage charges $110 per month with no deposit. The total cost, y, to rent a 10 ft by 10 ft space depends on the number of months, x, according to the equations

The Storage Bin: $y = 90x + 120$

AAA Storage: $y = 110x$

From the graph, determine the number of months required for which the cost to rent space is equal for both companies.   For 6 months, the cost is $660 for each company.

Total Cost for Storage
$y = 110x$
$y = 90x + 120$
Cost ($)
Number of Months
— (The Storage Bin)
— (AAA Storage)

For the systems graphed in Exercises 55–56, explain why the ordered pair cannot be a solution to the system of equations.

55. $(-3, 1)$ The point of intersection is below the x-axis and cannot have a positive y-coordinate.

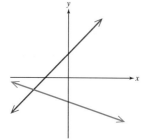

56. $(-1, -4)$ The point of intersection is above the x-axis and cannot have a negative y-coordinate.

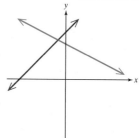

Expanding Your Skills

57. Write a system of linear equations whose solution is $(2, 1)$.   For example: $4x + y = 9$
$-2x - y = -5$

58. Write a system of linear equations whose solution is $(1, 4)$.   For example: $x - 3y = -11$
$3x + y = 7$

59. One equation in a system of linear equations is $x + y = 4$. Write a second equation such that the system will have no solution. (Answers may vary.)
For example: $2x + 2y = 1$

60. One equation in a system of linear equations is $x - y = 3$. Write a second equation such that the system will have infinitely many solutions. (Answers may vary.)
For example: $3x - 3y = 9$

 Writing   Translating Expression   Geometry   Scientific Calculator   Video   NS&E

Solving Systems of Equations by the Substitution Method

1. Solving Systems of Linear Equations by the Substitution Method

In Section 4.1, we used the graphing method to find the solution set to a system of equations. However, sometimes it is difficult to determine the solution using this method because of limitations in the accuracy of the graph. This is particularly true when the coordinates of a solution are not integer values or when the solution is a point not sufficiently close to the origin. Identifying the coordinates of the point $(\frac{3}{17}, -\frac{23}{9})$ or $(-251, 8349)$, for example, might be difficult from a graph.

In this section and the next, we will cover two algebraic methods to solve a system of equations that do not require graphing. The first method, called the *substitution method*, is demonstrated in Examples 1–5.

Objectives

1. Solving Systems of Linear Equations by the Substitution Method
2. Solutions to Systems of Linear Equations: A Summary
3. Applications of the Substitution Method

Example 1 Solving a System of Linear Equations Using the Substitution Method

Solve by using the substitution method.

$$x = 2y - 3$$
$$-4x + 3y = 2$$

Solution:

The variable x has been isolated in the first equation. The quantity $2y - 3$ is equal to x and therefore can be substituted for x in the second equation. This leaves the second equation in terms of y only.

First equation: $x = \underbrace{2y - 3}$

Second equation: $-4x + 3y = 2$

$-4(2y - 3) + 3y = 2$ This equation now contains only one variable.

$-8y + 12 + 3y = 2$ Solve the resulting equation.

$-5y + 12 = 2$

$-5y = -10$

$y = 2$

To find x, substitute $y = 2$ back into the first equation.

$$x = 2y - 3$$
$$x = 2(2) - 3$$
$$x = 1$$

Check the ordered pair $(1, 2)$ in both original equations.

$x = 2y - 3 \longrightarrow 1 \overset{?}{=} 2(2) - 3 ✔$ True

$-4x + 3y = 2 \longrightarrow -4(1) + 3(2) \overset{?}{=} 2 ✔$ True

The solution is $(1, 2)$ because it checks in both original equations.

Skill Practice

Solve the system by the substitution method.
1. $2x + 3y = -2$
    $y = x + 1$

Classroom Example: p. 314, Exercise 14

Instructor Note: Remind students that they can also graph a system of equations to determine if an answer is reasonable.

Answer
1. $(-1, 0)$

In Example 1, we eliminated the x-variable from the second equation by substituting an equivalent expression for x. The resulting equation was relatively simple to solve because it had only one variable. This is the premise of the substitution method.

The substitution method can be summarized as follows.

> **PROCEDURE** **Solving a System of Equations by the Substitution Method**
>
> **Step 1** Isolate one of the variables from one equation.
> **Step 2** Substitute the quantity found in step 1 into the other equation.
> **Step 3** Solve the resulting equation.
> **Step 4** Substitute the value found in step 3 back into the equation in step 1 to find the value of the remaining variable.
> **Step 5** Check the solution in both original equations and write the answer as an ordered pair.

— **Skill Practice** —

Solve the system by the substitution method.

2. $x + y = 3$
$-2x + 3y = 9$

Classroom Example: p. 314, Exercise 22

Example 2 **Solving a System of Linear Equations Using the Substitution Method**

Solve the system using the substitution method.

$$x + y = 4$$
$$-5x + 3y = -12$$

Solution:

The x- or y-variable in the first equation is easy to isolate because the coefficients are both 1. While either variable can be isolated, we arbitrarily choose to solve for the x-variable.

$x + y = 4 \longrightarrow x = \underbrace{4 - y}$ **Step 1:** Solve the first equation for x.

$-5(4 - y) + 3y = -12$ **Step 2:** Substitute $4 - y$ for x in the other equation.

$-20 + 5y + 3y = -12$ **Step 3:** Solve for y.

$-20 + 8y = -12$

$8y = 8$

$y = 1$

$x = 4 - y$ **Step 4:** Substitute $y = 1$ into the equation $x = 4 - y$.

$x = 4 - 1$

$x = 3$

Step 5: Check the ordered pair $(3, 1)$ in both original equations.

$$x + y = 4 \qquad (3) + (1) \stackrel{?}{=} 4 \checkmark \qquad \text{True}$$

$$-5x + 3y = -12 \qquad -5(3) + 3(1) \stackrel{?}{=} -12 \checkmark \qquad \text{True}$$

The solution is $(3, 1)$ because it checks in both original equations.

Answer

2. $(0, 3)$

TIP: The solution to a system of linear equations can be confirmed by graphing. The system from Example 2 is graphed here.

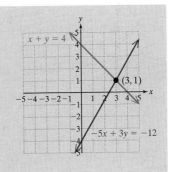

Example 3   **Solving a System of Linear Equations Using the Substitution Method**

Solve the system by using the substitution method.

$$3x + 5y = 17$$
$$2x - y = -6$$

Solution:

The y-variable in the second equation is the easiest variable to isolate because its coefficient is -1.

$$3x + 5y = 17$$
$$2x - y = -6 \longrightarrow -y = -2x - 6$$

$$y = \underbrace{2x + 6}$$

Step 1: Solve the second equation for y.

$$3x + 5(2x + 6) = 17$$

Step 2: Substitute the quantity $2x + 6$ for y in the other equation.

$$3x + 10x + 30 = 17$$ **Step 3:** Solve for x.

$$13x + 30 = 17$$

$$13x = 17 - 30$$

$$13x = -13$$

$$x = -1$$

$$y = 2x + 6$$ **Step 4:** Substitute $x = -1$ into the equation $y = 2x + 6$.

$$y = 2(-1) + 6$$

$$y = -2 + 6$$

$$y = 4$$

Step 5: The ordered pair $(-1, 4)$ can be checked in the original equations to verify the answer.

$$3x + 5y = 17 \longrightarrow 3(-1) + 5(4) \overset{?}{=} 17 \longrightarrow -3 + 20 \overset{?}{=} 17 \checkmark \text{ True}$$

$$2x - y = -6 \longrightarrow 2(-1) - (4) \overset{?}{=} -6 \longrightarrow -2 - 4 \overset{?}{=} -6 \checkmark \text{ True}$$

The solution is $(-1, 4)$.

Skill Practice

Solve the system by the substitution method.

3. $x + 4y = 11$
     $2x - 5y = -4$

Classroom Example: p. 314, Exercise 24

Avoiding Mistakes

Do not substitute $y = 2x + 6$ into the same equation from which it came. This mistake will result in an identity:

$$2x - y = -6$$
$$2x - (2x + 6) = -6$$
$$2x - 2x - 6 = -6$$
$$-6 = -6$$

Answer

3. $(3, 2)$

Recall from Section 4.1 that a system of linear equations may represent two parallel lines. In such a case, there is no solution to the system.

Classroom Example: p. 314, Exercise 26

Example 4   Solving an Inconsistent System Using Substitution

Solve the system by using the substitution method.

$$2x + 3y = 6$$
$$y = -\tfrac{2}{3}x + 4$$

Solution:

$$2x + 3y = 6$$

$y = -\tfrac{2}{3}x + 4$	**Step 1:** The variable y is already isolated in the second equation.
$2x + 3(-\tfrac{2}{3}x + 4) = 6$	**Step 2:** Substitute $y = -\tfrac{2}{3}x + 4$ from the second equation into the first equation.
$2x - 2x + 12 = 6$	**Step 3:** Solve the resulting equation.
$12 = 6$ (contradiction)	

The equation results in a contradiction. There are no values of x and y that will make 12 equal to 6. Therefore, there is no solution, and the system is inconsistent.

Instructor Note: Remind students that in Chapter 2 when we solved equations in one variable, a contradiction had no solution. For example, the equation $x + 3 = x + 2$ has no solution.

TIP: The answer to Example 4 can be verified by writing each equation in slope-intercept form and graphing the lines.

Equation 1

$2x + 3y = 6$

$3y = -2x + 6$

$\dfrac{3y}{3} = \dfrac{-2x}{3} + \dfrac{6}{3}$

$y = -\dfrac{2}{3}x + 2$

Equation 2

$y = -\dfrac{2}{3}x + 4$

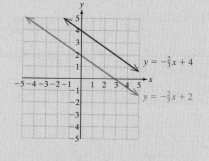

The equations indicate that the lines have the same slope but different y-intercepts. Therefore, the lines must be parallel. There is no point of intersection, indicating that the system has no solution.

Recall that a system of two linear equations may represent the same line. In such a case, the solution is the set of all points on the line.

Answer

4. No solution.

Example 5 Solving a Dependent System Using Substitution

Solve the system by using the substitution method.

$$\frac{1}{2}x - \frac{1}{4}y = 1$$

$$6x - 3y = 12$$

Solution:

$\frac{1}{2}x - \frac{1}{4}y = 1$ To make the first equation easier to work with, we have
                              the option of clearing fractions.
$6x - 3y = 12$

$\frac{1}{2}x - \frac{1}{4}y = 1 \xrightarrow{\text{Multiply by 4.}} 4\left(\frac{1}{2}x\right) - 4\left(\frac{1}{4}y\right) = 4(1) \longrightarrow 2x - y = 4$

Now the system becomes:

$2x - y = 4$ The y-variable in the first equation is the easiest to
                    isolate because its coefficient is -1.
$6x - 3y = 12$

$2x - y = 4 \xrightarrow{\text{Solve for } y.} -y = -2x + 4 \rightarrow y = 2x - 4$ **Step 1:** Isolate one of the
$6x - 3y = 12$ variables.

$6x - 3(2x - 4) = 12$ **Step 2:** Substitute $y = 2x - 4$ from the
                            first equation into the second
                            equation.

$6x - 6x + 12 = 12$ **Step 3:** Solve the resulting equation.
$\qquad 12 = 12$ (identity)

Because the equation produces an identity, all values of x make this equation true. Thus, x can be any real number. Substituting any real number, x, into the equation $y = 2x - 4$ produces an ordered pair on the line $y = 2x - 4$. Hence, the solution set to the system of equations is the set of all ordered pairs on the line $y = 2x - 4$. This can be written as $\{(x, y)\,|\,y = 2x - 4\}$. The system is dependent.

Skill Practice

Solve the system by using the substitution method.

5. $2x + \frac{1}{3}y = -\frac{1}{3}$

$\quad 12x + 2y = -2$

Classroom Example: p. 315, Exercise 40

TIP: The solution to Example 5 can be verified by writing each equation in slope-intercept form and graphing the lines.

Equation 1

Clear $\frac{1}{2}x - \frac{1}{4}y = 1$
fractions $2x - y = 4$

$-y = -2x + 4$

$y = 2x - 4$

Equation 2

$6x - 3y = 12$

$-3y = -6x + 12$

$\dfrac{-3y}{-3} = \dfrac{-6x}{-3} + \dfrac{12}{-3}$

$y = 2x - 4$

Notice that the slope-intercept forms for both equations are identical. The equations represent the same line, indicating that the system is dependent. Each point on the line is a solution to the system of equations.

Answer

5. Infinitely many solutions;
$\quad \{(x, y)\,|\,12x + 2y = -2\};$
$\quad$ dependent system

2. Solutions to Systems of Linear Equations: A Summary

The following summary reviews the three different geometric relationships between two lines and the solutions to the corresponding systems of equations.

> **PROCEDURE** Interpreting Solutions to a System of Two Linear Equations
>
> - The lines may intersect at one point (yielding one unique solution).
> - The lines may be parallel and have no point of intersection (yielding no solution). This is detected algebraically when a contradiction (false statement) is obtained (for example, $0 = -3$ and $12 = 6$).
> - The lines may be the same and intersect at all points on the line (yielding an infinite number of solutions). This is detected algebraically when an identity is obtained (for example, $0 = 0$ and $12 = 12$).

3. Applications of the Substitution Method

In Chapter 2, we solved word problems using one linear equation and one variable. In this chapter, we investigate application problems with two unknowns. In such a case, we can use two variables to represent the unknown quantities. However, if two variables are used, we must write a system of *two* distinct equations.

Classroom Example: p. 315, Exercise 52

Skill Practice

6. One number is 16 more than another. Their sum is 92. Use a system of equations to find the numbers.

Example 6 Applying the Substitution Method

One number is 3 more than 4 times another. Their sum is 133. Find the numbers.

Solution:

We can use two variables to represent the two unknown numbers.

Let x represent one number.
Let y represent the other number. Label the variables.

We must now write two equations. Each of the first two sentences gives a relationship between x and y:

One number is 3 more than 4 times another. $\longrightarrow x = 4y + 3$ (first equation)

Their sum is 133. $\longrightarrow x + y = 133$ (second equation)

	Step 1: Notice that x is already isolated in the first equation.
$(4y + 3) + y = 133$	**Step 2:** Substitute $x = 4y + 3$ into the second equation, $x + y = 133$.
$5y + 3 = 133$	**Step 3:** Solve the resulting equation.
$5y = 130$	
$y = 26$	

Answer

6. One number is 38, and the other number is 54.

$x = 4y + 3$

$x = 4(26) + 3$ **Step 4:** To solve for x, substitute $y = 26$ into the equation
$x = 4y + 3$.

$x = 104 + 3$

$x = 107$

One number is 26, and the other is 107.

TIP: Check that the numbers 26 and 107 meet the conditions of Example 6.
- 4 times 26 is 104. Three more than 104 is 107. ✔
- The sum of the numbers should be 133: $26 + 107 = 133$. ✔

Example 7 **Using the Substitution Method in a Geometry Application**

Two angles are supplementary. The measure of one angle is 15° more than twice the measure of the other angle. Find the measures of the two angles.

Solution:

Let x represent the measure of one angle.
Let y represent the measure of the other angle.

The sum of the measures of supplementary angles is 180° ⟶ $x + y = 180$

The measure of one angle is 15° more than
twice the other angle ⟶ $x = 2y + 15$

$x + y = 180$

$x = 2y + 15$ **Step 1:** The x-variable in the second equation is already isolated.

$(2y + 15) + y = 180$ **Step 2:** Substitute $2y + 15$ into the first equation for x.

$2y + 15 + y = 180$ **Step 3:** Solve the resulting equation.

$3y + 15 = 180$

$3y = 165$

$y = 55$

$x = 2y + 15$ **Step 4:** Substitute $y = 55$ into the equation
$x = 2y + 15$.

$x = 2(55) + 15$

$x = 110 + 15$

$x = 125$

One angle is 55°, and the other is 125°.

TIP: Check that the angles 55° and 125° meet the conditions of Example 7.

- Because $55° + 125° = 180°$, the angles are supplementary. ✔
- The angle 125° is 15° more than twice 55°: $125° = 2(55°) + 15°$. ✔

Answer

7. The measures of the angles are 23° and 67°.

Section 4.2 Practice Exercises

Review Exercises

For Exercises 1–6, write each pair of lines in slope-intercept form. Then identify whether the lines intersect in exactly one point or if the lines are parallel or coinciding.

1. $2x - y = 4$ $y = 2x - 4$
$-2y = -4x + 8$
$y = 2x - 4$; coinciding lines

2. $x - 2y = 5$ $y = \frac{1}{2}x - \frac{5}{2}$
$3x = 6y + 15$
$y = \frac{1}{2}x - \frac{5}{2}$; coinciding lines

3. $2x + 3y = 6$ $y = -\frac{2}{3}x + 2$
$x - y = 5$
$y = x - 5$; intersecting lines

4. $x - y = -1$ $y = x + 1$
$x + 2y = 4$ $y = -\frac{1}{2}x + 2$;
intersecting lines

5. $2x = \frac{1}{2}y + 2$ $y = 4x - 4$
$4x - y = 13$ $y = 4x - 13$; parallel lines

6. $4y = 3x$ $y = \frac{3}{4}x$
$3x - 4y = 15$ $y = \frac{3}{4}x - \frac{15}{4}$; parallel lines

Objective 1: Solving Systems of Linear Equations by the Substitution Method

For Exercises 7–10, solve each system using the substitution method. (See Example 1.)

7. $3x + 2y = -3$
$y = 2x - 12$
$(3, -6)$

8. $4x - 3y = -19$
$y = -2x + 13$
$(2, 9)$

9. $x = -4y + 16$
$3x + 5y = 20$
$(0, 4)$

10. $x = -y + 3$
$-2x + y = 6$
$(-1, 4)$

11. Given the system: $4x - 2y = -6$
$3x + y = 8$

  a. Which variable from which equation is easiest to isolate and why? y in the second equation is easiest to isolate because its coefficient is 1.
  b. Solve the system using the substitution method.
  $(1, 5)$

12. Given the system: $x - 5y = 2$
$11x + 13y = 22$

  a. Which variable from which equation is easiest to isolate and why? x in the first equation is easiest to isolate because its coefficient is 1.
  b. Solve the system using the substitution method.
  $(2, 0)$

For Exercises 13–48, solve each system using the substitution method. (See Examples 1–5.)

13. $x = 3y - 1$ $(5, 2)$
$2x - 4y = 2$

14. $2y = x + 9$ $(-1, 4)$
$y = -3x + 1$

15. $-2x + 5y = 5$ $(10, 5)$
$x = 4y - 10$

16. $y = -2x + 27$ $(11, 5)$
$3x - 7y = -2$

17. $4x - y = -1$ $\left(\frac{1}{2}, 3\right)$
$2x + 4y = 13$

18. $5x - 3y = -2$ $\left(\frac{1}{5}, 1\right)$
$10x - y = 1$

19. $4x - 3y = 11$ $(5, 3)$
$x = 5$

20. $y = -3x - 9$ $(-7, 12)$
$y = 12$

21. $4x = 8y + 4$ $(1, 0)$
$5x - 3y = 5$

22. $3y = 6x - 6$ $(2, 2)$
$-3x + y = -4$

23. $x - 3y = -11$ $(1, 4)$
$6x - y = 2$

24. $-2x - y = 9$ $(-6, 3)$
$x + 7y = 15$

25. $3x + 2y = -1$
$\frac{3}{2}x + y = 4$
No solution; inconsistent system

26. $5x - 2y = 6$
$-\frac{5}{2}x + y = 5$
No solution; inconsistent system

27. $10x - 30y = -10$
$2x - 6y = -2$
Infinitely many solutions; $\{(x, y)\,|\,2x - 6y = -2\}$; dependent system

28. $3x + 6y = 6$
$-6x - 12y = -12$
Infinitely many solutions; $\{(x, y)\,|\,3x + 6y = 6\}$; dependent system

29. $2x + y = 3$
$y = -7$
$(5, -7)$

30. $-3x = 2y + 23$
$x = -1$
$(-1, -10)$

31. $x + 2y = -2$ $\left(-5, \frac{3}{2}\right)$
$4x = -2y - 17$

32. $x + y = 1$ $\left(-\frac{1}{3}, \frac{4}{3}\right)$
$2x - y = -2$

33. $y = -\frac{1}{2}x - 4$ $(2, -5)$
$y = 4x - 13$

34. $y = \frac{2}{3}x - 3$ $(3, -1)$
$y = 6x - 19$

35. $y = 6$ $(-4, 6)$
$y - 4 = -2x - 6$

36. $x = 9$ $(9, -1)$
$x - 3 = 6y + 12$

37. $3x + 2y = 4$

$2x - 3y = -6$

$(0, 2)$

38. $4x + 3y = 4$

$-2x + 5y = -2$

$(1, 0)$

39. $y = 0.25x + 1$
Infinitely many solutions;
$-x + 4y = 4$
$\{(x, y) | y = 0.25x + 1\}$;
dependent system

40. $y = 0.75x - 3$
Infinitely many solutions;
$-3x + 4y = -12$
$\{(x, y) | y = 0.75x - 3\}$;
dependent system

41. $11x + 6y = 17$

$5x - 4y = 1$

$(1, 1)$

42. $3x - 8y = 7$

$10x - 5y = 45$

$(5, 1)$

43. $x + 2y = 4$

$4y = -2x - 8$

No solution; inconsistent system

44. $-y = x - 6$

$2x + 2y = 4$

No solution; inconsistent system

45. $2x = 3 - y$

$x + y = 4$ $(-1, 5)$

46. $2x = 4 + 2y$

$3x + y = 10$ $(3, 1)$

47. $\dfrac{x}{3} + \dfrac{y}{2} = -4$

$x - 3y = 6$ $(-6, -4)$

48. $x - 2y = -5$

$\dfrac{2x}{3} + \dfrac{y}{3} = 0$ $(-1, 2)$

Objective 3: Applications of the Substitution Method

For Exercises 49–58, set up a system of linear equations and solve for the indicated quantities. **(See Examples 6–7.)**

49. Two numbers have a sum of 106. One number is 10 less than the other. Find the numbers.
The numbers are 48 and 58.

 **50.** Two positive numbers have a difference of 8. The larger number is 2 less than 3 times the smaller number. Find the numbers.
The numbers are 13 and 5.

51. The difference between two positive numbers is 26. The larger number is three times the smaller. Find the numbers.
The numbers are 13 and 39.

52. The sum of two numbers is 956. One number is 94 less than 6 times the other. Find the numbers.
The numbers are 150 and 806.

 53. Two angles are supplementary. One angle is 15° more than 10 times the other angle. Find the measure of each angle.
The angles are 165° and 15°.

 54. Two angles are complementary. One angle is 1° less than 6 times the other angle. Find the measure of each angle. The angles are 13° and 77°.

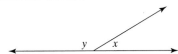

 55. Two angles are complementary. One angle is 10° more than 3 times the other angle. Find the measure of each angle.
The angles are 70° and 20°.

56. Two angles are supplementary. One angle is 5° less than twice the other angle. Find the measure of each angle.
The angles are $118\frac{1}{3}°$ and $61\frac{2}{3}°$.

57. In a right triangle, one of the acute angles is 6° less than the other acute angle. Find the measure of each acute angle. The angles are 42° and 48°.

58. In a right triangle, one of the acute angles is 9° less than twice the other acute angle. Find the measure of each acute angle. The angles are 57° and 33°.

Expanding Your Skills

59. The following system of equations is dependent and has infinitely many solutions. Find three ordered pairs that are solutions to the system of equations.

$$y = 2x + 3$$

$$-4x + 2y = 6$$

For example: $(0, 3), (1, 5), (-1, 1)$

60. The following system of equations is dependent and has infinitely many solutions. Find three ordered pairs that are solutions to the system of equations.

$$y = -x + 1$$

$$2x + 2y = 2$$

For example: $(0, 1), (1, 0), (-1, 2)$

✏️ Writing ↔ Translating Expression ✏️ Geometry 🖩 Scientific Calculator 💿 Video NS E

Section 4.3 | Solving Systems of Equations by the Addition Method

Objectives

1. Solving a System of Linear Equations by the Addition Method
2. Summary of Methods for Solving Linear Equations in Two Variables

1. Solving a System of Linear Equations by the Addition Method

Thus far in Chapter 4 we have used the graphing method and the substitution method to solve a system of linear equations in two variables. In this section, we present another algebraic method to solve a system of linear equations, called the *addition method* (sometimes called the *elimination method*). The purpose of the addition method is to eliminate one variable.

Skill Practice

Solve the system using the addition method.

1. $x + y = 13$
$2x - y = 2$

Classroom Example: p. 322, Exercise 10

Example 1   **Using the Addition Method**

Solve the system using the addition method.

$$x + y = -2$$
$$x - y = -6$$

Solution:

Notice that the coefficients of the y-variables are opposites:

Coefficient is 1.

$$x + 1y = -2$$
$$x - 1y = -6$$

Coefficient is -1.

Because the coefficients of the y-variables are opposites, we can add the two equations to eliminate the y-variable.

$$x + y = -2$$
$$\underline{x - y = -6}$$
$$2x \quad\quad = -8 \quad\leftarrow \text{After adding the equations, we have one equation and one variable.}$$

$$2x = -8 \quad\quad \text{Solve the resulting equation.}$$
$$x = -4$$

To find the value of y, substitute $x = -4$ into *either* of the original equations.

$$x + y = -2 \quad\quad \text{First equation}$$
$$(-4) + y = -2$$
$$y = -2 + 4$$
$$y = 2$$

The solution is $(-4, 2)$.

TIP: Notice that the value $x = -4$ could have been substituted into the second equation to obtain the same value for y.

$$x - y = -6$$
$$(-4) - y = -6$$
$$-y = -6 + 4$$
$$-y = -2$$
$$y = 2$$

Check:

$$x + y = -2 \longrightarrow (-4) + (2) \overset{?}{=} -2 \longrightarrow -2 \overset{?}{=} -2 \ ✔ \ \text{True}$$
$$x - y = -6 \longrightarrow (-4) - (2) \overset{?}{=} -6 \longrightarrow -6 \overset{?}{=} -6 \ ✔ \ \text{True}$$

Answer

1. $(5, 8)$

It is important to note that the addition method works on the premise that the two equations have *opposite* values for the coefficients of one of the variables. Sometimes it is necessary to manipulate the original equations to create two coefficients that are opposites. This is accomplished by multiplying one or both equations by an appropriate constant. The process is outlined as follows.

PROCEDURE Solving a System of Equations by the Addition Method

Step 1 Write both equations in standard form: $Ax + By = C$.
Step 2 Clear fractions or decimals (optional).
Step 3 Multiply one or both equations by nonzero constants to create opposite coefficients for one of the variables.
Step 4 Add the equations from step 3 to eliminate one variable.
Step 5 Solve for the remaining variable.
Step 6 Substitute the known value from step 5 into one of the original equations to solve for the other variable.
Step 7 Check the solution in both equations.

Instructor Note: Point out that in step 3 it does not matter whether we create opposites for the *x*-variables or the *y*-variables.

Example 2 **Solving a System of Linear Equations Using the Addition Method**

Solve the system using the addition method.

$$3x + 5y = 17$$
$$2x - y = -6$$

Solution:

$3x + 5y = 17$ **Step 1:** Both equations are already written in standard form.

$2x - y = -6$ **Step 2:** There are no fractions or decimals.

Notice that neither the coefficients of x nor the coefficients of y are opposites. However, multiplying the second equation by 5 creates the term $-5y$ in the second equation. This is the opposite of the term $+5y$ in the first equation.

$$\begin{array}{ll} 3x + 5y = 17 & \quad\quad 3x + 5y = 17 \\ 2x - y = -6 \xrightarrow[\text{Multiply by 5.}]{} & \quad\underline{10x - 5y = -30} \\ & \quad\quad 13x = -13 \end{array}$$

Step 3: Multiply the second equation by 5.

Step 4: Add the equations.

$$13x = -13$$ **Step 5:** Solve the equation.
$$x = -1$$

$3x + 5y = 17$ First equation

$3(-1) + 5y = 17$

$-3 + 5y = 17$

$5y = 20$

$y = 4$

Step 6: Substitute $x = -1$ into one of the original equations.

The solution is $(-1, 4)$.

Step 7: Check the solution in both original equations.

Check:

$$3x + 5y = 17 \longrightarrow 3(-1) + 5(4) \stackrel{?}{=} 17 \longrightarrow -3 + 20 \stackrel{?}{=} 17 \checkmark \text{ True}$$
$$2x - y = -6 \longrightarrow 2(-1) - (4) \stackrel{?}{=} -6 \longrightarrow -2 - 4 \stackrel{?}{=} -6 \checkmark \text{ True}$$

Skill Practice

Solve the system using the addition method.
2. $4x + 3y = 3$
$x - 2y = 9$

Classroom Example: p. 322, Exercise 14

Instructor Note: Remind students to multiply the chosen constant on *both* sides of the equation.

Answer
2. $(3, -3)$

In Example 3, the system of equations uses the variables a and b instead of x and y. In such a case, we will write the solution as an ordered pair with the variables written in alphabetical order, such as (a, b).

Classroom Example: p. 323,
Exercise 18

Skill Practice

Solve the system using the addition method.

3. $8n = 4 - 5m$
$7m + 6n = -10$

Example 3 Solve a System of Linear Equations Using the Addition Method

Solve the system using the addition method.

$$5b = 7a + 8$$
$$-4a - 2b = -10$$

Solution:

Step 1: Write the equations in standard form.

The first equation becomes: $5b = 7a + 8 \longrightarrow -7a + 5b = 8$

The system becomes: $-7a + 5b = 8$
$-4a - 2b = -10$

Step 2: There are no fractions or decimals.

Step 3: We need to obtain opposite coefficients on either the a or b term.

Notice that neither the coefficients of a nor the coefficients of b are opposites. However, it is possible to change the coefficients of b to 10 and -10 (this is because the LCM of 5 and 2 is 10). This is accomplished by multiplying the first equation by 2 and the second equation by 5.

$$-7a + 5b = 8 \xrightarrow{\text{Multiply by 2.}} -14a + 10b = 16$$
$$-4a - 2b = -10 \xrightarrow{\text{Multiply by 5.}} \underline{-20a - 10b = -50}$$
$$-34a \qquad = -34$$

Step 4: Add the equations.

$$-34a = -34$$

Step 5: Solve the resulting equation.

$$\frac{-34a}{-34} = \frac{-34}{-34}$$

$$a = 1$$

$5b = 7a + 8$ First equation

Step 6: Substitute $a = 1$ into one of the original equations.

$5b = 7(1) + 8$

$5b = 15$

$b = 3$

The solution is $(1, 3)$.

Step 7: Check the solution in the original equations.

Check:

$$5b = 7a + 8 \longrightarrow 5(3) \stackrel{?}{=} 7(1) + 8 \longrightarrow 15 \stackrel{?}{=} 7 + 8 \; ✔ \; \text{True}$$
$$-4a - 2b = -10 \longrightarrow -4(1) - 2(3) \stackrel{?}{=} -10 \longrightarrow -4 - 6 \stackrel{?}{=} -10 \; ✔ \; \text{True}$$

Answer

3. $(-4, 3)$

Example 4 **Solving a System of Linear Equations Using the Addition Method**

Solve the system using the addition method.

$$34x - 22y = 4$$

$$17x - 88y = -19$$

Skill Practice

Solve the system using the addition method.

4. $15x - 16y = 1$
$ 45x + 4y = 16$

Classroom Example: p. 322, Exercise 22

Solution:

The equations are already in standard form. There are no fractions or decimals to clear.

$$34x - 22y = 4 \longrightarrow 34x - 22y = 4$$

$$17x - 88y = -19 \xrightarrow{\text{Multiply by } -2.} \underline{-34x + 176y = 38}$$
$$154y = 42$$

Solve for y.　$154y = 42$

$$\frac{154y}{154} = \frac{42}{154}$$

Simplify.　$y = \dfrac{3}{11}$

To find the value of x, we normally substitute y into one of the original equations and solve for x. In this example, we will show an alternative method for finding x. By repeating the addition method, this time eliminating y, we can solve for x. This approach allows us to avoid substitution of the fractional value for y.

$$34x - 22y = 4 \xrightarrow{\text{Multiply by } -4.} -136x + 88y = -16$$

$$17x - 88y = -19 \longrightarrow \underline{17x - 88y = -19}$$
$$-119x = -35$$

Solve for x.　$-119x = -35$

$$\frac{-119x}{-119} = \frac{-35}{-119}$$

Simplify.　$x = \dfrac{5}{17}$

The solution is $\left(\frac{5}{17}, \frac{3}{11}\right)$. These values can be checked in the original equations:

$$34x - 22y = 4 \qquad\qquad 17x - 88y = -19$$

$$34\left(\frac{5}{17}\right) - 22\left(\frac{3}{11}\right) \stackrel{?}{=} 4 \qquad\qquad 17\left(\frac{5}{17}\right) - 88\left(\frac{3}{11}\right) \stackrel{?}{=} -19$$

$$10 - 6 \stackrel{?}{=} 4 ✔ \text{ True} \qquad\qquad 5 - 24 \stackrel{?}{=} -19 ✔ \text{ True}$$

Answer

4. $\left(\dfrac{1}{3}, \dfrac{1}{4}\right)$

Example 5 **Solving a System of Linear Equations**

Solve the system using the addition method.

$$2x - 5y = 10$$
$$\frac{1}{2}x - \frac{5}{4}y = 1$$

Solution:

$2x - 5y = 10$

$\frac{1}{2}x - \frac{5}{4}y = 1$ **Step 1:** The equations are in standard form.

Step 2: Multiply both sides of the second equation by 4 to clear fractions.

$$\frac{1}{2}x - \frac{5}{4}y = 1 \longrightarrow 4\left(\frac{1}{2}x - \frac{5}{4}y\right) = 4(1) \longrightarrow 2x - 5y = 4$$

Now the system becomes $2x - 5y = 10$
                           $2x - 5y = 4$

To make either the x-coefficients or y-coefficients opposites, multiply either equation by -1.

$$2x - 5y = 10 \xrightarrow{\text{Multiply by } -1.} -2x + 5y = -10$$ **Step 3:** Create opposite coefficients.
$$2x - 5y = 4 \longrightarrow \underline{2x - 5y = 4}$$
$$0 = -6$$ **Step 4:** Add the equations.

Because the result is a contradiction, there is no solution, and the system of equations is inconsistent. Writing each line in slope-intercept form verifies that the lines are parallel (Figure 4-8).

$$2x - 5y = 10 \xrightarrow{\text{slope-intercept form}} y = \frac{2}{5}x - 2$$

$$\frac{1}{2} - \frac{5}{4}y = 1 \xrightarrow{\text{slope-intercept form}} y = \frac{2}{5}x - \frac{4}{5}$$

There is no solution.

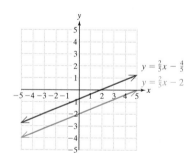

Figure 4-8

| Example 6 | **Solving a System of Linear Equations** |

Solve the system by the addition method.

$$3x - y = 4$$
$$2y = 6x - 8$$

Classroom Example: p. 323, Exercise 34

Solution:

$3x - y = 4 \longrightarrow 3x - y = 4$ **Step 1:** Write the equations in standard form.

$2y = 6x - 8 \longrightarrow -6x + 2y = -8$ **Step 2:** There are no fractions or decimals.

Notice that the equations differ exactly by a factor of -2, which indicates that these two equations represent the same line. Multiply the first equation by 2 to create opposite coefficients for the variables.

$$\begin{array}{l} 3x - y = 4 \xrightarrow{\text{Multiply by 2.}} 6x - 2y = 8 \\ -6x + 2y = -8 \underline{-6x + 2y = -8} \\ 0 = 0 \end{array}$$

Step 3: Create opposite coefficients.

Step 4: Add the equations.

Because the resulting equation is an identity, the original equations represent the same line. This can be confirmed by writing each equation in slope-intercept form.

$$3x - y = 4 \longrightarrow -y = -3x + 4 \longrightarrow y = 3x - 4$$
$$-6x + 2y = -8 \longrightarrow 2y = 6x - 8 \longrightarrow y = 3x - 4$$

The solution is the set of all points on the line, or equivalently, $\{(x, y) | y = 3x - 4\}$.

2. Summary of Methods for Solving Linear Equations in Two Variables

If no method of solving a system of linear equations is specified, you may use the method of your choice. However, we recommend the following guidelines:

1. If one of the equations is written with a variable isolated, the substitution method is a good choice. For example:

$$2x + 5y = 2 \qquad \text{or} \qquad y = \frac{1}{3}x - 2$$
$$x = y - 6 \qquad\qquad\qquad x - 6y = 9$$

2. If both equations are written in standard form, $Ax + By = C$, where none of the variables has coefficients of 1 or -1, then the addition method is a good choice.

$$4x + 5y = 12$$
$$5x + 3y = 15$$

3. If both equations are written in standard form, $Ax + By = C$, and at least one variable has a coefficient of 1 or -1, then either the substitution method or the addition method is a good choice.

Section 4.3 Practice Exercises

Boost *your* GRADE at ALEKS.com!

- Practice Problems
- Self-Tests
- NetTutor
- e-Professors
- Videos

For additional exercises, see Classroom Activities 4.3A–4.3B in the *Instructor's Resource Manual* at www.mhhe.com/moh.

Study Skills Exercise

1. Now that you have learned three methods of solving a system of linear equations with two variables, choose a system and solve it all three ways. There are two advantages to this. One is to check your answer (you should get the same answer using all three methods). The second advantage is to show you which method is the easiest for you to use.

   Solve the system by using the graphing method, the substitution method, and the addition method.

   $$2x + y = -7$$
   $$x - 10 = 4y$$

Review Exercises

For Exercises 2–5, check whether the given ordered pair is a solution to the system.

2. $x + y = 8$ (5, 3) Yes
   $y = x - 2$

3. $x = y + 1$ (3, 2) No
   $-x + 2y = 0$

4. $3x + 2y = 14$ (5, −2) No
   $5x - 2y = 29$

5. $x = 2y - 11$ (−3, 4) Yes
   $-x + 5y = 23$

Objective 1: Solving a System of Linear Equations by the Addition Method

For Exercises 6–7, answer as true or false.

6. Given the system $5x - 4y = 1$
                      $7x - 2y = 5$

   a. To eliminate the *y*-variable using the addition method, multiply the second equation by 2. False, multiply by −2.

   b. To eliminate the *x*-variable, multiply the first equation by 7 and the second equation by −5. True

7. Given the system $3x + 5y = -1$
                      $9x - 8y = -26$

   a. To eliminate the *x*-variable using the addition method, multiply the first equation by −3. True

   b. To eliminate the *y*-variable, multiply the first equation by 8 and the second equation by −5. False, multiply the second equation by 5.

8. Given the system $3x - 4y = 2$
                      $17x + y = 35$

   a. Which variable, *x* or *y*, is easier to eliminate using the addition method? *y* would be easier.

   b. Solve the system using the addition method.
      (2, 1)

9. Given the system $-2x + 5y = -15$
                      $6x - 7y = 21$

   a. Which variable, *x* or *y*, is easier to eliminate using the addition method? *x* would be easier.

   b. Solve the system using the addition method.
      (0, −3)

For Exercises 10–25, solve the systems using the addition method. **(See Examples 1–4.)**

10. $x + 2y = 8$ (2, 3)
     $5x - 2y = 4$

11. $2x - 3y = 11$ (4, −1)
     $-4x + 3y = -19$

12. $a + b = 3$ (5, −2)
     $3a + b = 13$

13. $-2u + 6v = 10$ (4, 3)
     $-2u + v = -5$

14. $-3x + y = 1$ (0, 1)
     $-6x - 2y = -2$

15. $5m - 2n = 4$ (2, 3)
    $3m + n = 9$

16. $3x - 5y = 13$ $(1, -2)$

$x - 2y = 5$

17. $7a + 2b = -1$ $(1, -4)$

$3a - 4b = 19$

18. $6c - 2d = -2$ $(1, 4)$

$5c = -3d + 17$

19. $2s + 3t = -1$ $(1, -1)$

$5s = 2t + 7$

20. $6y - 4z = -2$ $(3, 5)$

$4y + 6z = 42$

21. $4k - 2r = -4$ $(-4, -6)$

$3k - 5r = 18$

22. $2x + 3y = 6$ $\left(\dfrac{21}{5}, -\dfrac{4}{5}\right)$

$x - y = 5$

23. $6x + 6y = 8$ $\left(\dfrac{7}{9}, \dfrac{5}{9}\right)$

$9x - 18y = -3$

24. $2x - 5y = 4$ $\left(\dfrac{8}{9}, -\dfrac{4}{9}\right)$

$3x - 3y = 4$

25. $6x - 5y = 7$ $\left(\dfrac{7}{16}, -\dfrac{7}{8}\right)$

$4x - 6y = 7$

26. In solving a system of equations, suppose you get the statement $0 = 5$. How many solutions will the system have? What can you say about the graphs of these equations?
The system will have no solution. The lines are parallel.

27. In solving a system of equations, suppose you get the statement $0 = 0$. How many solutions will the system have? What can you say about the graphs of these equations?
There are infinitely many solutions. The lines coincide.

28. In solving a system of equations, suppose you get the statement $3 = 3$. How many solutions will the system have? What can you say about the graphs of these equations?
There are infinitely many solutions. The lines coincide.

29. In solving a system of equations, suppose you get the statement $2 = -5$. How many solutions will the system have? What can you say about the graphs of these equations?
The system will have no solution. The lines are parallel.

30. Suppose in solving a system of linear equations, you get the statement $x = 0$. How many solutions will the system have? What can you say about the graphs of these equations?
The system will have one solution. The lines intersect at a point whose x-coordinate is 0.

31. Suppose in solving a system of linear equations, you get the statement $y = 0$. How many solutions will the system have? What can you say about the graphs of these equations?
The system will have one solution. The lines intersect at a point whose y-coordinate is 0.

For Exercises 32–43, solve the system of equations using the addition method. (See Examples 5–6.)

32. $-2x + y = -5$ No solution; inconsistent system

$8x - 4y = 12$

33. $x - 3y = 2$ No solution; inconsistent system

$-5x + 15y = 10$

34. $x + 2y = 2$ Infinitely many solutions; $\{(x, y)|x + 2y = 2\}$; dependent system

$-3x - 6y = -6$

35. $4x - 3y = 6$ Infinitely many solutions; $\{(x, y)|4x - 3y = 6\}$; dependent system

$-12x + 9y = -18$

36. $3a + 2b = 11$ $(1, 4)$

$7a - 3b = -5$

37. $4y + 5z = -2$

$5y - 3z = 16$

$(2, -2)$

38. $3x - 5y = 7$

$5x - 2y = -1$

$(-1, -2)$

39. $4s + 3t = 9$ $(0, 3)$

$3s + 4t = 12$

40. $2x + 2 = -3y + 9$

$3x - 10 = -4y$

$(2, 1)$

41. $-3x + 6 + 7y = 5$ $(5, 2)$

$5y = 2x$

42. $4x - 5y = 0$

$8(x - 1) = 10y$

No solution; inconsistent system

43. $y = 2x + 1$

$-3(2x - y) = 0$

No solution; inconsistent system

Objective 2: Summary of Methods for Solving Linear Equations in Two Variables

For Exercises 44–63, solve the system of equations by either the addition method or the substitution method.

44. $5x - 2y = 4$ $(2, 3)$

$y = -3x + 9$

45. $-x = 8y + 5$ $(-5, 0)$

$4x - 3y = -20$

46. $0.1x + 0.1y = 0.6$ $(3.5, 2.5)$

$0.1x - 0.1y = 0.1$

47. $0.1x + 0.1y = 0.2$ $(2.5, -0.5)$

$0.1x - 0.1y = 0.3$

48. $3x = 5y - 9$ $\left(\dfrac{1}{3}, 2\right)$

$2y = 3x + 3$

49. $10x - 5 = 3y$ $\left(\dfrac{1}{2}, 0\right)$

$4x + 5y = 2$

50. $y = -5x - 5$ $(-2, 5)$

$6x - 3 = -3y$

51. $4x + 5y = -2$ $(-3, 2)$

$3x = -2y - 5$

52. $x = -\dfrac{1}{2}$ $\left(-\dfrac{1}{2}, 1\right)$

$6x - 5y = -8$

 Writing Translating Expression Geometry Scientific Calculator Video NS

53. $4x - 2y = 1$ $\left(\frac{7}{4}, 3\right)$

$y = 3$

54. $0.02x + 0.04y = 0.12$ $(0, 3)$

$0.03x - 0.05y = -0.15$

55. $-0.04x + 0.03y = 0.03$ $(0, 1)$

$-0.06x - 0.02y = -0.02$

56. $8x - 16y = 24$

$2x - 4y = 0$

No solution; inconsistent system

57. $y = -\frac{1}{2}x - 5$

$2x + 4y = -8$

No solution; inconsistent system

58. $\frac{m}{2} + \frac{n}{5} = \frac{13}{10}$

$3m - 3n = m - 10$

$(1, 4)$

59. $\frac{a}{4} - \frac{3b}{2} = \frac{15}{2}$ $(0, -5)$

$a + 2b = -10$

60. $2m - 6n = m + 4$ $(4, 0)$

$3m + 8 = 5m - n$

61. $m - 3n = 10$ $(4, -2)$

$3m + 12n = -12$

62. $9a - 2b = 8$

$18a + 6 = 4b + 22$

Infinitely many solutions;
$\{(a, b) | 9a - 2b = 8\}$; dependent system

63. $a = 5 + 2b$

$3a - 6b = 15$

Infinitely many solutions;
$\{(a, b) | a = 5 + 2b\}$; dependent system

For Exercises 64–67, set up a system of linear equations, and solve for the indicated quantities.

64. The sum of two positive numbers is 26. Their difference is 14. Find the numbers.
The numbers are 20 and 6.

65. The difference of two positive numbers is 2. The sum of the numbers is 36. Find the numbers.
The numbers are 17 and 19.

66. Eight times the smaller of two numbers plus 2 times the larger number is 44. Three times the smaller number minus 2 times the larger number is zero. Find the numbers.
The numbers are 4 and 6.

67. Six times the smaller of two numbers minus the larger number is -9. Ten times the smaller number plus five times the larger number is 5. Find the numbers.
The numbers are -1 and 3.

For Exercises 68–70, solve the system by using each of the three methods: (a) the graphing method, (b) the substitution method, and (c) the addition method.

68. $2x + y = 1$

$-4x - 2y = -2$

Infinitely many solutions;
$\{(x, y) | 2x + y = 1\}$

69. $3x + y = 6$

$-2x + 2y = 4$

$(1, 3)$

70. $2x - 2y = 6$

$5y = 5x + 5$

No solution

Expanding Your Skills

71. Explain why a system of linear equations cannot have exactly two solutions. One line within the system of equations would have to "bend" for the system to have exactly two points of intersection. This is not possible.

72. The solution to the system of linear equations is $(1, 2)$. Find A and B.

$Ax + 3y = 8$ $A = 2, B = -4$

$x + By = -7$

73. The solution to the system of linear equations is $(-3, 4)$. Find A and B.

$4x + Ay = -32$ $A = -5, B = 2$

$Bx + 6y = 18$

Problem Recognition Exercises

Systems of Equations

For Exercises 1–6 determine the number of solutions to the system without solving the system. Explain your answers.

1. $y = -4x + 2$

$y = -4x + 2$
Infinitely many solutions. The equations represent the same line.

2. $y = -4x + 6$

$y = -4x + 1$
No solution. The equations represent parallel lines.

3. $y = 4x - 3$

$y = -4x + 5$
One solution. The equations represent intersecting lines.

4. $y = 7$

$2x + 3y = 1$
One solution. The equations represent intersecting lines.

5. $2x + 3y = 1$

$2x + 3y = 8$
No solution. The equations represent parallel lines.

6. $8x - 2y = 6$

$12x - 3y = 9$
Infinitely many solutions. The equations represent the same line.

For Exercises 7–26, solve the system using the method of your choice.

7. $x = -2y + 5$ $(5, 0)$

$2x - 4y = 10$

8. $y = -3x - 4$ $(1, -7)$

$2x - y = 9$

9. $3x - 2y = 22$ $(4, -5)$

$5x + 2y = 10$

10. $-4x + 2y = -2$ $(2, 3)$

$4x - 5y = -7$

11. $\dfrac{1}{3}x + \dfrac{1}{2}y = \dfrac{2}{3}$ $(2, 0)$

$-\dfrac{2}{3}x + y = -\dfrac{4}{3}$

12. $\dfrac{1}{4}x + \dfrac{2}{5}y = 6$ $(8, 10)$

$\dfrac{1}{2}x - \dfrac{1}{10}y = 3$

13. $2c + 7d = -1$ $\left(2, -\dfrac{5}{7}\right)$

$c = 2$

14. $-3w + 5z = -6$ $\left(-\dfrac{14}{3}, -4\right)$

$z = -4$

15. $y = 0.4x - 0.3$

$-4x + 10y = 20$
No solution; inconsistent system

16. $x = -0.5y + 0.1$

$-10x - 5y = 2$
No solution; inconsistent system

17. $3a + 7b = -3$ $(-1, 0)$

$-11a + 3b = 11$

18. $2v - 5w = 10$ $(5, 0)$

$9v + 7w = 45$

19. $y = 2x - 14$

$4x - 2y = 28$
Infinitely many solutions; $\{(x, y)\,|\,y = 2x - 14\}$; dependent system

20. $x = 5y - 9$

$-2x + 10y = 18$
Infinitely many solutions; $\{(x, y)\,|\,x = 5y - 9\}$; dependent system

21. $x + y = 3200$ $(2200, 1000)$

$0.06x + 0.04y = 172$

22. $x + y = 4500$

$0.07x + 0.05y = 291$
$(3300, 1200)$

23. $3x + y - 7 = x - 4$ $(5, -7)$

$3x - 4y + 4 = -6y + 5$

24. $7y - 8y - 3 = -3x + 4$

$10x - 5y - 12 = 13$ $(2, -1)$

25. $3x - 6y = -1$ $\left(\dfrac{2}{3}, \dfrac{1}{2}\right)$

$9x + 4y = 8$

26. $8x - 2y = 5$ $\left(\dfrac{1}{4}, -\dfrac{3}{2}\right)$

$12x + 4y = -3$

Section 4.4 Applications of Linear Equations in Two Variables

Objectives

1. Applications Involving Cost
2. Applications Involving Principal and Interest
3. Applications Involving Mixtures
4. Applications Involving Distance, Rate, and Time

1. Applications Involving Cost

In Sections 2.4–2.7, we solved several applied problems by setting up a linear equation in one variable. When solving an application that involves two unknowns, sometimes it is convenient to use a system of linear equations in two variables.

Skill Practice

1. Lynn went to a fast-food restaurant and spent $9.00. She purchased 4 hamburgers and 5 orders of fries. The next day, Ricardo went to the same restaurant and purchased 10 hamburgers and 7 orders of fries. He spent $18.10. Use a system of equations to determine the cost of a burger and the cost of an order of fries.

Classroom Example: p. 332, Exercise 12

Example 1 Using a System of Linear Equations Involving Cost

At a movie theater a couple buys one large popcorn and two drinks for $5.75. A group of teenagers buys two large popcorns and five drinks for $13.00. Find the cost of one large popcorn and the cost of one drink.

Solution:

In this application we have two unknowns, which we can represent by x and y.

Let x represent the cost of one large popcorn.
Let y represent the cost of one drink.

We must now write two equations. Each of the first two sentences in the problem gives a relationship between x and y:

$$\left(\begin{array}{c}\text{Cost of 1}\\\text{large popcorn}\end{array}\right) + \left(\begin{array}{c}\text{cost of 2}\\\text{drinks}\end{array}\right) = \left(\begin{array}{c}\text{total}\\\text{cost}\end{array}\right) \longrightarrow x + 2y = 5.75$$

$$\left(\begin{array}{c}\text{Cost of 2}\\\text{large popcorns}\end{array}\right) + \left(\begin{array}{c}\text{cost of 5}\\\text{drinks}\end{array}\right) = \left(\begin{array}{c}\text{total}\\\text{cost}\end{array}\right) \longrightarrow 2x + 5y = 13.00$$

To solve this system, we may either use the substitution method or the addition method. We will use the substitution method by solving for x in the first equation.

$$x + 2y = 5.75 \longrightarrow x = -2y + 5.75 \qquad \text{Isolate } x \text{ in the first equation.}$$

$$2x + 5y = 13.00$$

$$2(-2y + 5.75) + 5y = 13.00 \qquad \begin{array}{l}\text{Substitute } x = -2y + 5.75\\\text{into the other equation.}\end{array}$$

$$-4y + 11.50 + 5y = 13.00$$

$$y + 11.50 = 13.00 \qquad \text{Solve for } y.$$

$$y = 1.50$$

$$x = -2y + 5.75$$

$$x = -2(1.50) + 5.75 \qquad \begin{array}{l}\text{Substitute } y = 1.50 \text{ into the}\\\text{equation}\\x = -2y + 5.75.\end{array}$$

$$x = -3.00 + 5.75$$

$$x = 2.75$$

The cost of one large popcorn is $2.75 and the cost of one drink is $1.50.

Check by verifying that the solutions meet the specified conditions.

1 popcorn + 2 drinks = 1($2.75) + 2($1.50) = $5.75 ✔ True
2 popcorns + 5 drinks = 2($2.75) + 5($1.50) = $13.00 ✔ True

Answer

1. The cost of a burger is $1.25 and the cost of an order of fries is $0.80.

2. Applications Involving Principal and Interest

In Section 2.4, we learned that simple interest is interest computed on the principal amount of money invested (or borrowed). Simple interest, I, is found by using the formula

$$I = Prt \quad \text{where } P \text{ is the principal,}$$
$$r \text{ is the annual interest rate, and}$$
$$t \text{ is the time in years.}$$

If the amount of time is taken to be 1 year, we have: $I = Pr(1)$ or simply $I = Pr$.

In Example 2, we apply the concept of simple interest to two accounts to produce a desired amount of interest after 1 yr.

Example 2 **Using a System of Linear Equations Involving Investments**

Joanne has a total of $6000 to deposit in two accounts. One account earns 3.5% simple interest and the other earns 2.5% simple interest. If the total amount of interest at the end of 1 yr is $195, find the amount she deposited in each account.

Solution:

Let x represent the principal deposited in the 2.5% account.
Let y represent the principal deposited in the 3.5% account.

	2.5% Account	3.5% Account	Total
Principal	x	y	6000
Interest $(I = Pr)$	0.025x	0.035y	195

Each row of the table yields an equation in x and y:

$$\begin{pmatrix} \text{Principal} \\ \text{invested} \\ \text{at } 2.5\% \end{pmatrix} + \begin{pmatrix} \text{principal} \\ \text{invested} \\ \text{at } 3.5\% \end{pmatrix} = \begin{pmatrix} \text{total} \\ \text{principal} \end{pmatrix} \longrightarrow x + y = 6000$$

$$\begin{pmatrix} \text{Interest} \\ \text{earned} \\ \text{at } 2.5\% \end{pmatrix} + \begin{pmatrix} \text{interest} \\ \text{earned} \\ \text{at } 3.5\% \end{pmatrix} = \begin{pmatrix} \text{total} \\ \text{interest} \end{pmatrix} \longrightarrow 0.025x + 0.035y = 195$$

We will choose the addition method to solve the system of equations. First multiply the second equation by 1000 to clear decimals.

Multiply by -25.

$$x + y = 6000 \longrightarrow \quad x + y = 6000 \quad \longrightarrow -25x - 25y = -150{,}000$$
$$0.025x + 0.035y = 195 \longrightarrow 25x + 35y = 195{,}000 \longrightarrow \underline{\quad 25x + 35y = \quad 195{,}000}$$

Multiply by 1000.

$$10y = \quad 45{,}000$$

$10y = 45{,}000$ After eliminating the x-variable, solve for y.

$$\frac{10y}{10} = \frac{45{,}000}{10}$$

$y = 4500$ The amount invested in the 3.5% account is $4500.

Classroom Example: p. 332, Exercise 16

Skill Practice

2. Addie has a total of $8000 in two accounts. One pays 5% interest, and the other pays 6.5% interest. At the end of one year, she earned $475 interest. Use a system of equations to determine the amount invested in each account.

Answer

2. $3000 is invested at 5%, and $5000 is invested at 6.5%.

$x + y = 6000$ Substitute $y = 4500$ into the equation $x + y = 6000$.

$x + 4500 = 6000$

$x = 1500$ The amount invested in the 2.5% account is $1500.

Joanne deposited $1500 in the 2.5% account and $4500 in the 3.5% account.

To check, verify that the conditions of the problem have been met.

1. The sum of $1500 and $4500 is $6000 as desired. ✔ True

2. The interest earned on $1500 at 2.5% is: $0.025(\$1500) = \$37.50$
   The interest earned on $4500 at 3.5% is: $0.035(\$4500) = \$157.50$

   Total interest: $195.00 ✔ True

3. Applications Involving Mixtures

Example 3 **Using a System of Linear Equations in a Mixture Application**

Classroom Example: p. 333, Exercise 22

A 10% alcohol solution is mixed with a 40% alcohol solution to produce 30 L of a 20% alcohol solution. Find the number of liters of 10% solution and the number of liters of 40% solution required for this mixture.

Solution:

Each solution contains a percentage of alcohol plus some other mixing agent such as water. Before we set up a system of equations to model this situation, it is helpful to have background understanding of the problem. In Figure 4-9, the liquid depicted in blue is pure alcohol and the liquid shown in gray is the mixing agent (such as water). Together these liquids form a solution. (Realistically the mixture may not separate as shown, but this image may be helpful for your understanding.)

Let x represent the number of liters of 10% solution.
Let y represent the number of liters of 40% solution.

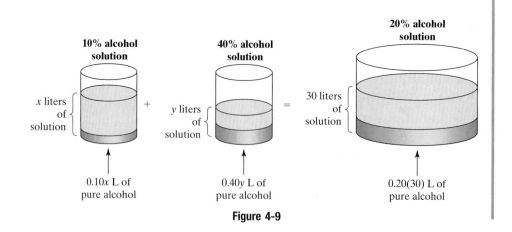

Figure 4-9

Skill Practice

3. How many ounces of 20% and 35% acid solution should be mixed together to obtain 15 oz of 30% acid solution.

Answer

3. 10 oz of the 35% solution, and 5 oz of the 20% solution.

The information given in the statement of the problem can be organized in a chart.

	10% Alcohol	40% Alcohol	20% Alcohol
Number of liters of solution	x	y	30
Number of liters of pure alcohol	$0.10x$	$0.40y$	$0.20(30) = 6$

From the first row, we have

$$\left(\begin{array}{c} \text{Amount of} \\ \text{10\% solution} \end{array}\right) + \left(\begin{array}{c} \text{amount of} \\ \text{40\% solution} \end{array}\right) = \left(\begin{array}{c} \text{total amount} \\ \text{of 20\% solution} \end{array}\right) \rightarrow x + y = 30$$

From the second row, we have

$$\left(\begin{array}{c} \text{Amount of} \\ \text{alcohol in} \\ \text{10\% solution} \end{array}\right) + \left(\begin{array}{c} \text{amount of} \\ \text{alcohol in} \\ \text{40\% solution} \end{array}\right) = \left(\begin{array}{c} \text{total amount of} \\ \text{alcohol in} \\ \text{20\% solution} \end{array}\right) \rightarrow 0.10x + 0.40y = 6$$

We will solve the system with the addition method by first clearing decimals.

$$
\begin{array}{llll}
 & & & \text{Multiply by } -1. \\
x + y = 30 & \longrightarrow & x + y = 30 & \longrightarrow & -x - y = -30 \\
0.10x + 0.40y = 6 & \longrightarrow & x + 4y = 60 & \longrightarrow & \underline{x + 4y = 60} \\
 & \text{Multiply by 10.} & & & 3y = 30
\end{array}
$$

$3y = 30$ After eliminating the x-variable, solve for y.

$y = 10$ 10 L of 40% solution is needed.

$x + y = 30$ Substitute $y = 10$ into either of the original equations.

$x + (10) = 30$

$x = 20$ 20 L of 10% solution is needed.

10 L of 40% solution must be mixed with 20 L of 10% solution.

4. Applications Involving Distance, Rate, and Time

The following formula relates the distance traveled to the rate and time of travel.

$$d = rt \qquad \text{distance} = \text{rate} \cdot \text{time}$$

For example, if a car travels at 60 mph for 3 hr, then

$$d = (60 \text{ mph})(3 \text{ hr})$$
$$= 180 \text{ mi}$$

If a car travels at 60 mph for x hr, then

$$d = (60 \text{ mph})(x \text{ hr})$$
$$= 60x \text{ mi}$$

The relationship $d = rt$ is used in Example 4.

Classroom Example: p. 333, Exercise 28

Skill Practice

4. Dan and Cheryl paddled their canoe 40 mi in 5 hr with the current and 16 mi in 8 hr against the current. Find the speed of the current and the speed of the canoe in still water.

Example 4 **Using a System of Linear Equations in a Distance, Rate, and Time Application**

A plane travels with a tail wind from Kansas City, Missouri, to Denver, Colorado, a distance of 600 mi in 2 hr. The return trip against a head wind takes 3 hr. Find the speed of the plane in still air, and find the speed of the wind.

Solution:

Let p represent the speed of the plane in still air.
Let w represent the speed of the wind.

Notice that when the plane travels with the wind, the net speed is $p + w$. When the plane travels against the wind, the net speed is $p - w$.

The information given in the problem can be organized in a chart.

	Distance	Rate	Time
With a tail wind	600	$p + w$	2
Against a head wind	600	$p - w$	3

To set up two equations in p and w, recall that $d = rt$.

From the first row, we have

$$\begin{pmatrix} \text{Distance} \\ \text{with the wind} \end{pmatrix} = \begin{pmatrix} \text{rate with} \\ \text{the wind} \end{pmatrix}\begin{pmatrix} \text{time traveled} \\ \text{with the wind} \end{pmatrix} \longrightarrow 600 = (p + w) \cdot 2$$

From the second row, we have

$$\begin{pmatrix} \text{Distance} \\ \text{against the wind} \end{pmatrix} = \begin{pmatrix} \text{rate against} \\ \text{the wind} \end{pmatrix}\begin{pmatrix} \text{time traveled} \\ \text{against the wind} \end{pmatrix} \longrightarrow 600 = (p - w) \cdot 3$$

Using the distributive property to clear parentheses produces the following system:

$$2p + 2w = 600$$
$$3p - 3w = 600$$

The coefficients of the w-variable can be changed to 6 and -6 by multiplying the first equation by 3 and the second equation by 2.

$$2p + 2w = 600 \xrightarrow{\text{Multiply by 3.}} 6p + 6w = 1800$$
$$3p - 3w = 600 \xrightarrow[\text{Multiply by 2.}]{} \underline{6p - 6w = 1200}$$
$$12p \qquad = 3000$$

$$12p = 3000$$

$$\frac{12p}{12} = \frac{3000}{12}$$

$$p = 250 \qquad \text{The speed of the plane in still air is 250 mph.}$$

Answer

4. The speed of the canoe in still water is 5 mph. The speed of the current is 3 mph.

TIP: To create opposite coefficients on the *w*-variables, we could have divided the first equation by 2 and divided the second equation by 3:

$$2p + 2w = 600 \xrightarrow{\text{Divide by 2.}} p + w = 300$$
$$3p - 3w = 600 \xrightarrow{\text{Divide by 3.}} p - w = 200$$
$$2p \qquad\ = 500$$
$$p = 250$$

$$2p + 2w = 600 \qquad \text{Substitute } p = 250 \text{ into the first equation.}$$
$$2(250) + 2w = 600$$
$$500 + 2w = 600$$
$$2w = 100$$
$$w = 50 \qquad \text{The speed of the wind is 50 mph.}$$

The speed of the plane in still air is 250 mph. The speed of the wind is 50 mph.

Section 4.4 Practice Exercises

Review Exercises

For Exercises 1–4, solve each system of equations by three different methods:

   a. Graphing method   **b.** Substitution method   **c.** Addition method

1. $-2x + y = 6$
   $2x + y = 2$ $(-1, 4)$

2. $x - y = 2$
   $x + y = 6$ $(4, 2)$

3. $y = -2x + 6$
   $4x - 2y = 8$ $\left(\frac{5}{2}, 1\right)$

4. $2x = y + 4$
   $4x = 2y + 8$ Infinitely many solutions; $\{(x, y)\,|\,2x = y + 4\}$; dependent system

For Exercises 5–8, set up a system of linear equations in two variables to solve for the unknown quantities.

5. One number is eight more than twice another. Their sum is 20. Find the numbers.
The numbers are 4 and 16.

6. The difference of two positive numbers is 264. The larger number is three times the smaller number. Find the numbers.
The numbers are 132 and 396.

7. Two angles are complementary. The measure of one angle is 10° less than nine times the measure of the other. Find the measure of each angle.
The angles are 80° and 10°.

8. Two angles are supplementary. The measure of one angle is 9° more than twice the measure of the other angle. Find the measure of each angle.
The angles are 123° and 57°.

Objective 1: Applications Involving Cost

9. Kent bought three DVDs and two CDs for $62.50. Demond bought one DVD and four CDs for $72.50. Find the cost of one DVD and the cost of one CD. **(See Example 1.)**

DVDs are $10.50 each, and CDs are $15.50 each.

10. Tanya bought three adult tickets and one child's ticket to a movie for $23.00. Li bought two adult tickets and five children's tickets for $30.50. Find the cost of one adult ticket and the cost of one children's ticket.   Adult tickets cost $6.50 each, and children's tickets cost $3.50 each.

11. Linda bought 100 shares of a technology stock and 200 shares of a mutual fund for $3800. Her sister, Sandy, bought 300 shares of technology stock and 50 shares of a mutual fund for $5350. Find the cost per share of the technology stock, and the cost per share of the mutual fund.

Technology stock costs $16 per share, and the mutual fund costs $11 per share.

12. Two video games and three DVDs can be rented for $19.15. Four video games and one DVD can be rented for $17.35. Find the cost to rent one video game and the cost to rent one DVD.   It costs $3.29 to rent a video game and $4.19 to rent a DVD.

13. Patricia buys a combination of 42¢ stamps and 59¢ stamps at the Post Office. If she spends exactly $22.70 on 50 stamps, how many of each type did she buy?

Patricia bought forty 42¢ stamps and ten 59¢ stamps.

14. Bennett purchased some beef and some chicken for a family barbeque. The beef cost $6.00 per pound and the chicken cost $4.50 per pound. He bought a total of 18 lb of meat and spent $96. How much of each type of meat did he purchase?

He bought 10 lb of beef and 8 lb of chicken.

Objective 2: Applications Involving Principal and Interest

15. Shanelle invested $10,000, and at the end of 1 yr, she received $805 in interest. She invested part of the money in an account earning 10% simple interest and the remaining money in an account earning 7% simple interest. How much did she invest in each account? **(See Example 2.)**   Shanelle invested $3500 in the 10% account and $6500 in the 7% account.

	10% Account	7% Account	Total
Principal invested			
Interest earned			

16. $12,000 was invested in two accounts, one earning 12% simple interest and the other earning 8% simple interest. If the total interest at the end of 1 yr was $1240, how much was invested in each account? $7000 was invested in the 12% account and $5000 was invested in the 8% account.

	12% Account	8% Account	Total
Principal invested			
Interest earned			

17. Troy borrowed a total of $12,000 in two different loans to help pay for his new Chevy Silverado. One loan charges 9% simple interest, and the other charges 6% simple interest. If he is charged $810 in interest after 1 yr, find the amount borrowed at each rate.

$9000 was borrowed at 6%, and $3000 was borrowed at 9%.

18. Blake has a total of $4000 to invest in two accounts. One account earns 2% simple interest, and the other earns 5% simple interest. How much should be invested in both accounts to earn exactly $155 at the end of 1 yr?

$1500 must be invested in the 2% account and $2500 invested in the 5% account.

19. Suppose a rich uncle dies and leaves you an inheritance of $30,000. You decide to invest part of the money in a relatively safe bond fund that returns 8%. You invest the rest of the money in a riskier stock fund that you hope will return 12% at the end of 1 yr. If you need $3120 at the end of 1 yr to make a down payment on a car, how much should you invest at each rate?

Invest $12,000 in the bond fund and $18,000 in the stock fund.

20. As part of his retirement strategy, John plans to invest $200,000 in two different funds. He projects that the moderately high risk investments should return, over time, about 9% per year, while the low risk investments should return about 4% per year. If he wants a supplemental income of $12,000 a year, how should he divide his investments?

He should invest $80,000 at 9% and $120,000 at 4%.

Objective 3: Applications Involving Mixtures

21. How much 50% disinfectant solution must be mixed with a 40% disinfectant solution to produce 25 gal of a 46% disinfectant solution? **(See Example 3.)** 15 gal of the 50% mixture should be mixed with 10 gal of the 40% mixture.

	50% Mixture	40% Mixture	46% Mixture
Amount of solution			
Amount of disinfectant			

22. How many gallons of 20% antifreeze solution and a 10% antifreeze solution must be mixed to obtain 40 gal of a 16% antifreeze solution? 24 gal of the 20% mixture should be mixed with 16 gal of the 10% mixture.

	20% Mixture	10% Mixture	16% Mixture
Amount of solution			
Amount of antifreeze			

23 How much 45% disinfectant solution must be mixed with a 30% disinfectant solution to produce 20 gal of a 39% disinfectant solution? 12 gal of the 45% disinfectant solution should be mixed with 8 gal of the 30% disinfectant solution.

24. How many gallons of a 25% antifreeze solution and a 15% antifreeze solution must be mixed to obtain 15 gal of a 23% antifreeze solution? 12 gal of the 25% antifreeze solution should be mixed with 3 gal of the 15% antifreeze solution.

25. A nurse needs 50 mL of a 16% salt solution for a patient. She can only find a 13% salt solution and an 18% salt solution in the supply room. How many milliliters of the 13% solution should be mixed with the 18% solution to produce the desired amount of the 16% solution? She should mix 20 mL of the 13% solution with 30 mL of the 18% solution.

26. Meadowsilver Dairy keeps two kinds of milk on hand, skim milk that has 0.3% butterfat and whole milk that contains 3.3% butterfat. How many gallons of each type of milk does the company need to produce 300 gallons of 1% milk for the P&A grocery store? It needs 230 gal of skim milk and 70 gal of whole milk.

Objective 4: Applications Involving Distance, Rate, and Time

27. It takes a boat 2 hr to go 16 mi downstream with the current and 4 hr to return against the current. Find the speed of the boat in still water and the speed of the current. **(See Example 4.)** The speed of the boat in still water is 6 mph, and the speed of the current is 2 mph.

	Distance	Rate	Time
Downstream			
Return			

28. A boat takes 1.5 hr to go 12 mi upstream against the current. It can go 24 mi downstream with the current in the same amount of time. Find the speed of the current and the speed of the boat in still water. The speed of the boat is 12 mph, and the speed of the current is 4 mph.

	Distance	Rate	Time
Upstream			
Downstream			

29. A plane can fly 960 mi with the wind in 3 hr. It takes the same amount of time to fly 840 mi against the wind. What is the speed of the plane in still air and the speed of the wind? The speed of the plane in still air is 300 mph, and the wind is 20 mph.

30. A plane flies 720 mi with the wind in 3 hr. The return trip takes 4 hr. What is the speed of the wind and the speed of the plane in still air? The speed of the plane in still air is 210 mph, and the wind is 30 mph.

31. Tony Markins flew from JFK Airport to London. It took him 6 hr to fly with the wind, and 8 hr on the return flight against the wind. If the distance is approximately 3600 mi, determine the speed of the plane in still air and the speed of the wind. The speed of the plane in still air is 525 mph and the speed of the wind is 75 mph.

32. A riverboat cruise upstream on the Mississippi River from New Orleans, Louisiana, to Natchez, Mississippi, takes 10 hr and covers 140 mi. The return trip downstream with the current takes only 7 hr. Find the speed of the riverboat in still water and the speed of the current. The speed of the boat in still water is 17 mph and the speed of the current is 3 mph.

Mixed Exercises

33. Debi has $2.80 in a collection of dimes and nickels. The number of nickels is five more than the number of dimes. Find the number of each type of coin.
There are 17 dimes and 22 nickels.

34. A child is collecting state quarters and new $1 coins. If she has a total of 25 coins, and the number of quarters is nine more than the number of dollar coins, how many of each type of coin does she have?
She has eight $1 coins and 17 quarters.

35. In the 1961–1962 NBA basketball season, Wilt Chamberlain of the Philadelphia Warriors made 2432 baskets. Some of the baskets were free throws (worth 1 point each) and some were field goals (worth 2 points each). The number of field goals was 762 more than the number of free throws.

  a. How many field goals did he make and how many free throws did he make?
  835 free throws and 1597 field goals
  b. What was the total number of points scored?
  4029 points
  c. If Wilt Chamberlain played 80 games during this season, what was the average number of points per game?
  Approximately 50 points per game

36. In the 1971–1972 NBA basketball season, Kareem Abdul-Jabbar of the Milwaukee Bucks made 1663 baskets. Some of the baskets were free throws (worth 1 point each) and some were field goals (worth 2 points each). The number of field goals he scored was 151 more than twice the number of free throws.

  a. How many field goals did he make and how many free throws did he make?
  504 free throws and 1159 field goals
  b. What was the total number of points scored?
  2822 points
  c. If Kareem Abdul-Jabbar played 81 games during this season, what was the average number of points per game?
  Approximately 35 points per game

37. A small plane can fly 350 mi with a tailwind in $1\frac{3}{4}$ hr. In the same amount of time, the same plane can travel only 210 mi with a headwind. What is the speed of the plane in still air and the speed of the wind?
The speed of the plane in still air is 160 mph, and the wind is 40 mph.

38. A plane takes 2 hr to travel 1000 mi with the wind. It can travel only 880 mi against the wind in the same amount of time. Find the speed of the wind and the speed of the plane in still air.
The speed of the plane in still air is 470 mph, and the wind is 30 mph.

39. At the holidays, Erica likes to sell a candy/nut mixture to her neighbors. She wants to combine candy that costs $1.80 per pound with nuts that cost $1.20 per pound. If Erica needs 20 lb of mixture that will sell for $1.56 per pound, how many pounds of candy and how many pounds of nuts should she use?   12 lb of candy should be mixed with 8 lb of nuts.

40. Mary Lee's natural food store sells a combination of teas. The most popular is a mixture of a tea that sells for $3.00 per pound with one that sells for $4.00 per pound. If she needs 40 lb of tea that will sell for $3.65 per pound, how many pounds of each tea should she use?   14 lb of $3 per pound tea should be mixed with 26 lb of $4 per pound tea.

41. A total of $60,000 is invested in two accounts, one that earns 5.5% simple interest, and one that earns 6.5% simple interest. If the total interest at the end of 1 yr is $3750, find the amount invested in each account.
$15,000 was invested in the 5.5% account, and $45,000 was invested in the 6.5% account.

42. Jacques borrows a total of $15,000. Part of the money is borrowed from a bank that charges 12% simple interest per year. Jacques borrows the remaining part of the money from his sister and promises to pay her 7% simple interest per year. If Jacques' total interest for the year is $1475, find the amount he borrowed from each source.
Jacques borrowed $8500 at 12%, and $6500 at 7%.

43. In the 1994 Super Bowl, the Dallas Cowboys scored four more points than twice the number of points scored by the Buffalo Bills. If the total number of points scored by both teams was 43, find the number of points scored by each team.

Dallas scored 30 points, and Buffalo scored 13 points.

44. In the 1973 Super Bowl, the Miami Dolphins scored twice as many points as the Washington Redskins. If the total number of points scored by both teams was 21, find the number of points scored by each team.

Miami scored 14 points, and Washington scored 7 points.

Expanding Your Skills

45. In a survey conducted among 500 college students, 340 said that the campus lacked adequate lighting. If $\frac{4}{5}$ of the women and $\frac{1}{2}$ of the men said that they thought the campus lacked adequate lighting, how many men and how many women were in the survey? There were 300 women and 200 men in the survey.

46. During a 1-hr television program, there were 22 commercials. Some commercials were 15 sec and some were 30 sec long. Find the number of 15-sec commercials and the number of 30-sec commercials if the total playing time for commercials was 9.5 min.

There are six 15-sec commercials and sixteen 30-sec commercials.

Linear Inequalities and Systems of Inequalities in Two Variables

Section 4.5

1. Graphing Linear Inequalities in Two Variables

A **linear inequality in two variables** x and y is an inequality that can be written in one of the following forms: $ax + by < c$, $ax + by > c$, $ax + by \leq c$, or $ax + by \geq c$.

A solution to a linear inequality in two variables is an ordered pair that makes the inequality true. For example, solutions to the inequality $x + y < 3$ are ordered pairs (x, y) such that the sum of the x- and y-coordinates is less than 3. Several such examples are $(0, 0), (-2, -2), (3, -2)$, and $(-4, 1)$. There are actually infinitely many solutions to this inequality, and therefore it is convenient to express the solution set as a graph. The shaded area in Figure 4-10 represents all solutions (x, y), whose coordinates total less than 3.

Objectives

1. Graphing Linear Inequalities in Two Variables
2. Graphing Systems of Linear Inequalities in Two Variables

Concept Connections

Determine which of the ordered pairs are solutions to the inequality $y < 4x - 2$.
1. $(3, -1)$ **2.** $(0, 0)$

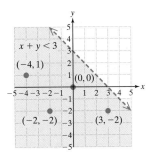

Figure 4-10

To graph a linear inequality in two variables we will use a process called the **test point method**. To use the test point method, first graph the related equation. In this case, the related equation represents a line in the xy-plane. Then choose a test point *not* on the line to determine which side of the line to shade. This process is demonstrated in Example 1.

Answers
1. Solution
2. Not a solution

Classroom Example: p. 342, Exercise 12

Example 1 Graphing a Linear Inequality in Two Variables

Graph the solution set. $2x + y \leq 3$

Solution:

$2x + y \leq 3 \longrightarrow 2x + y = 3$ **Step 1:** Set up the related equation.

Step 2: Graph the related equation.

Graph the line by either setting up a table of points, or by using the slope-intercept form (Figure 4-11).

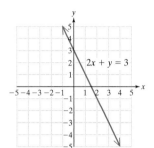

Figure 4-11

Table:

x	y
1	1
0	3
$\frac{3}{2}$	0

Slope-intercept form:

$2x + y = 3$

$y = -2x + 3$

Step 3: The solution to $2x + y \leq 3$ includes points for which $2x + y$ is less than *or equal to* 3. Because equality is included, points on the line $2x + y = 3$ are included. A solid line shows that the points on the line are included.

Now we must determine which side of the line to shade. To do so, we choose an arbitrary test point *not* on the line. The point $(0, 0)$ is a convenient choice.

<u>Test point:</u> $(0, 0)$

$2x + y \leq 3$

$2(0) + (0) \overset{?}{\leq} 3$

$0 \overset{?}{\leq} 3$ ✔ True

The test point $(0, 0)$ is true in the original inequality. This means that the region from which the test point was taken is part of the solution set. Therefore, shade below the line (Figure 4-12).

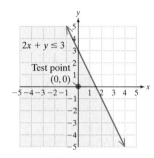

Figure 4-12

TIP: If a point above the line is selected as a test point, notice that it will *not* make the original inequality true. For example, test the point $(2, 2)$.

$2x + y \leq 3$

$2(2) + (2) \overset{?}{\leq} 3$

$6 \overset{?}{\leq} 3$ False

A false result tells us to shade the *other* side of the line.

Answer

3.

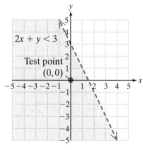

Now suppose the inequality from Example 1 had the strict inequality symbol, $<$. That is, consider the inequality $2x + y < 3$. The boundary line $2x + y = 3$ is *not* included in the solution set, because the expression $2x + y$ must be *strictly less than* 3 (not equal to 3). To show that the boundary line is not included in the solution set, we draw a dashed line (Figure 4-13).

Figure 4-13

The test point method to graph linear inequalities in two variables is summarized as follows:

Instructors Note: Tell the students that one test point is sufficient, but you can choose two—one below the line and one above. The second point can serve as a check.

> **PROCEDURE Test Point Method: Summary**
>
> **Step 1** Set up the related equation.
> **Step 2** Graph the related equation from step 1. The equation will be a boundary line in the xy-plane.
> - If the original inequality is a strict inequality, $<$ or $>$, then the line is *not* part of the solution set. Graph the line as a *dashed line.*
> - If the original inequality is not strict, $\leq$ or $\geq$, then the line *is* part of the solution set. Graph the line as a *solid line.*
> **Step 3** Choose a point not on the line and substitute its coordinates into the original inequality.
> - If the test point makes the inequality true, then the region it represents is part of the solution set. Shade that region.
> - If the test point makes the inequality false, then the other region is part of the solution set and should be shaded.

Example 2 **Graphing a Linear Inequality in Two Variables**

Graph the solution set. $4x - 2y > 6$

Solution:

$4x - 2y > 6 \longrightarrow 4x - 2y = 6$ **Step 1:** Set up the related equation.

Figure 4-14

Step 2: Graph the equation. Draw a dashed line because the inequality is strict, $>$ (Figure 4-14).

Table:

x	y
0	-3
$\frac{3}{2}$	0
2	1

Slope-intercept form:

$$4x - 2y = 6$$
$$-2y = -4x + 6$$
$$y = 2x - 3$$

Step 3: Choose a test point. Again $(0, 0)$ is a good choice because, when substituted into the original inequality, the arithmetic will be minimal.

$$4x - 2y > 6$$
$$\overset{?}{4(0) - 2(0) > 6}$$
$$\overset{?}{0 > 6} \quad \text{False}$$

The test point from above the line does not check in the original inequality. Therefore, shade below the line (Figure 4-15).

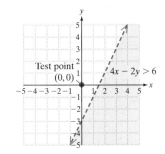

Figure 4-15

— **Skill Practice** —

Graph the solution set.
4. $6x - 2y < -6$

Classroom Example: p. 342, Exercise 18

Answer

4.

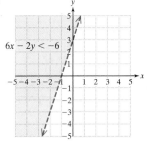

TIP: An inequality can also be graphed by first solving the inequality for y. Then,

- Shade *below* the line if the inequality is of the form $y < mx + b$ or $y \leq mx + b$.
- Shade *above* the line if the inequality is of the form $y > mx + b$ or $y \geq mx + b$.

From Example 2, we have

$$4x - 2y > 6$$
$$-2y > -4x + 6$$
$$\frac{-2y}{-2} < \frac{-4x}{-2} + \frac{6}{-2} \qquad \text{Reverse the inequality sign.}$$
$$y < 2x - 3 \qquad \text{Shade below the line.}$$

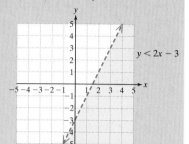

Classroom Example: p. 343, Exercise 26

Skill Practice

Graph the solution set.

5. $-2y < x$

Example 3 Graphing a Linear Inequality in Two Variables

Graph the solution set. $2y \geq 5x$

Solution:

$2y \geq 5x \longrightarrow 2y = 5x$ **Step 1:** Set up the related equation.

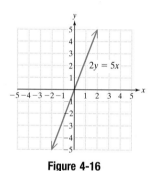

Figure 4-16

Step 2: Graph the equation. Draw a solid line because the symbol $\geq$ is used (Figure 4-16).

Table:

x	y
0	0
1	$\frac{5}{2}$
-1	$-\frac{5}{2}$

Slope-intercept form:

$$2y = 5x$$
$$y = \frac{5}{2}x$$

Step 3: The point $(0, 0)$ cannot be used as a test point because it is on the boundary line. Choose a different point such as $(1, 1)$.

$$2y \geq 5x$$
$$2(1) \overset{?}{\geq} 5(1)$$
$$2 \overset{?}{\geq} 5 \quad \text{False}$$

The test point from below the line does not check in the original inequality. Therefore, shade above the line (Figure 4-17).

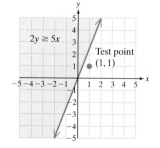

Figure 4-17

Answer

5.

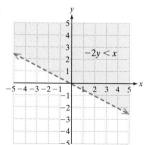

 **Example 4** Graphing a Linear Inequality in Two Variables

Graph the solution set. $2x > -4$

Solution:

$2x > -4 \longrightarrow 2x = -4$ **Step 1:** Set up the related equation.

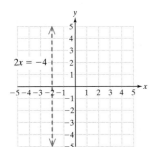

Step 2: Graph the equation. The equation represents a vertical line.

$$2x = -4$$

$$x = -2$$

Draw a dashed vertical line (Figure 4-18).

Figure 4-18

Step 3: Choose a test point such as $(0, 0)$.

$$2x > -4$$
$$2(0) \overset{?}{>} -4$$
$$0 \overset{?}{>} -4 \; ✔ \quad \text{True}$$

The test point from the right of the line checks in the original inequality. Therefore, shade to the right of the line (Figure 4-19).

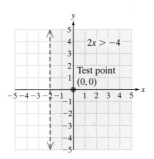

Figure 4-19

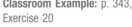 **Skill Practice**

Graph the solution set.

6. $4x \geq 12$

Classroom Example: p. 343, Exercise 20

2. Graphing Systems of Linear Inequalities in Two Variables

In Sections 4.1–4.4, we studied systems of linear equations in two variables. Graphically, a solution to such a system is a point of intersection between two lines. In this section, we will study systems of linear *inequalities* in two variables. Graphically, the solution set to such a system is the intersection (or "overlap") of the shaded regions of each individual inequality.

Answer

6.

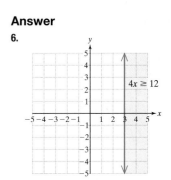

Skill Practice

Graph the solution set.

7. $x - 3y \geq 3$
   $y > -2x + 4$

Classroom Example: p. 344,
Exercise 42

Example 5   Graphing a System of Linear Inequalities

Graph the solution set. $y > \dfrac{1}{2}x - 2$

$$x + y \leq 1$$

Solution:

Sketch each inequality.

$y > \dfrac{1}{2}x - 2 \xrightarrow{\text{Related equation}} y = \dfrac{1}{2}x - 2$ $\qquad$ $x + y \leq 1 \xrightarrow{\text{Related equation}} x + y = 1$

The line $y = \dfrac{1}{2}x - 2$ is drawn in red in Figure 4-20. Substituting the test point $(0, 0)$ into the inequality results in a true statement. Therefore, we shade above the line.

The line $x + y = 1$ is drawn in blue in Figure 4-21. Substituting the test point $(0, 0)$ into the inequality results in a true statement. Therefore, we shade below the line.

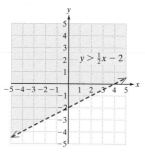

Figure 4-20

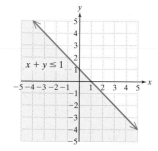

Figure 4-21

Next, we draw these regions on the same graph. The intersection ("overlap") is shown in purple (Figure 4-22).

In Figure 4-23, we show the solution to the system of inequalities. Notice that the portions of the lines not bounding the solution are dashed.

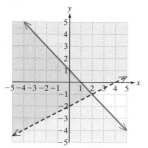

Figure 4-22

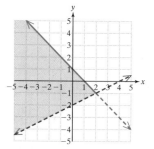

Figure 4-23

Answer

7.

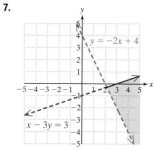

Section 4.5 Practice Exercises

Study Skills Exercise

1. Define the key terms:

   a. linear inequality in two variables   **b. test point method**

Review Exercises

For Exercises 2–4, graph the equations.

2. $x = -3$

3. $y = \frac{3}{5}x + 2$

4. $y = -\frac{4}{3}x$

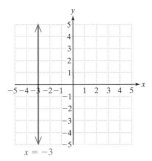

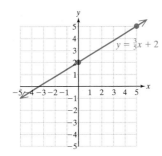

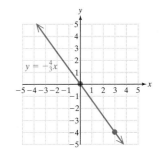

Objective 1: Graphing Linear Inequalities in Two Variables

5. When is a solid line used in the graph of a linear inequality in two variables?
   When the inequality symbol is ≤ or ≥

6. When is a dashed line used in the graph of a linear inequality in two variables?
   When the inequality symbol is < or >

7. What does the shaded region represent in the graph of a linear inequality in two variables?
   All of the points in the shaded region are solutions to the inequality.

8. When graphing a linear inequality in two variables, how do you determine which side of the boundary line to shade?   Choose a test point not on the line. Substitute the coordinates of the test point into the inequality. If this results in a true statement, shade the region represented by the test point. If the test point makes a false statement, shade the other region.

9. Which is the graph of $-2x - y \leq 2$?   a

   a.
   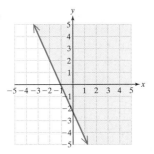

   b.
   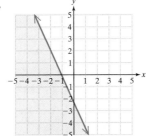

   c.
   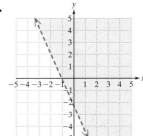

10. Which is the graph of $-3x + y > -1$?   c

a.
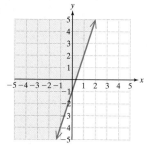

b.
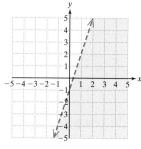

c.
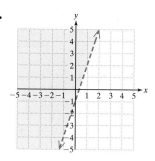

For Exercises 11–16, graph the solution set. Then write three ordered pairs that are solutions to the inequality. **(See Examples 1–4.)**

11. $y \geq -x + 5$

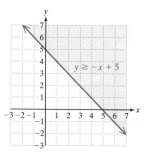

For example: $(0, 5)$ $(2, 7)$ $(-1, 8)$

12. $y \leq 2x - 1$

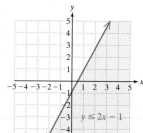

For example: $(0, -1)$ $(2, -2)$ $(-1, -5)$

13. $y < 4x$

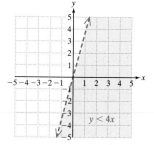

For example: $(1, -1)$ $(3, 0)$ $(-2, -9)$

14. $y > -5x$

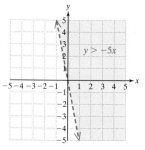

For example: $(2, 1)$ $(0, 4)$ $(-1, 8)$

15. $3x + 7y \leq 14$

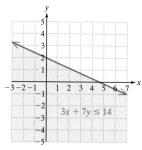

For example: $(0, 0)$ $(0, 2)$ $(-1, -3)$

16. $5x - 6y \geq 18$

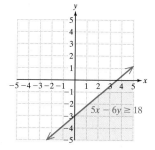

For example: $(0, -3)$ $(4, -4)$ $(-1, -6)$

For Exercises 17–34, graph the solution set. **(See Examples 1–4.)**

17. $x - y > 6$

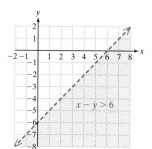

18. $x + y < 5$

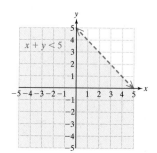

19. $x \geq -1$

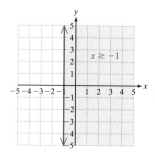

20. $x \le 6$

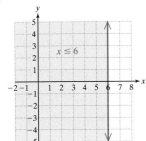

21. $y < 3$

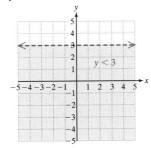

22. $y > -3$

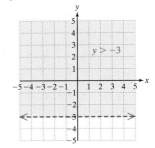

23. $y \le -\dfrac{3}{4}x + 2$

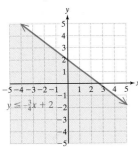

24. $y \ge \dfrac{2}{3}x + 1$

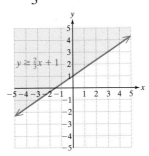

25. $y - 2x > 0$

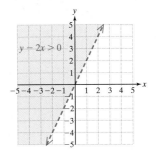

26. $y + 3x < 0$

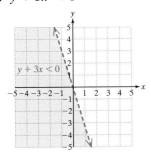

27. $x \le 0$

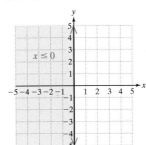

28. $y \le 0$

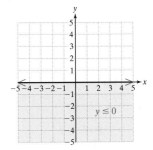

29. $y \ge 0$

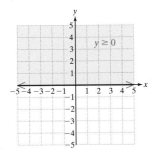

30. $x \ge 0$

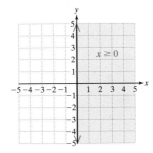

31. $-x \le \dfrac{1}{2}y - 2$

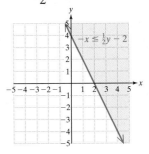

32. $-3 + 2x \le -y$

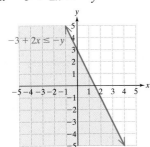

33. $2x > 3y$

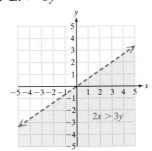

34. $-4x > 5y$

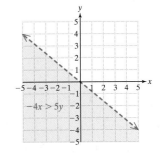

 Writing   Translating Expression   Geometry   Scientific Calculator   Video   NS&E

35. a. Describe the graph of the inequality $x + y > 4$. Find three solutions to the inequality (answers will vary).
   The set of ordered pairs above the line $x + y = 4$, for example, $(6, 3)(-2, 8)(0, 5)$

   b. Describe the graph of the equation $x + y = 4$. Find three solutions to the equation (answers will vary).
   The set of ordered pairs on the line $x + y = 4$, for example, $(0, 4)(4, 0)(2, 2)$

   c. Describe the graph of the inequality $x + y < 4$. Find three solutions to the inequality (answers will vary).
   The set of ordered pairs below the line $x + y = 4$, for example, $(0, 0)(-2, 1)(3, 0)$

36. a. Describe the graph of the inequality $x + y < 3$. Find three solutions to the inequality (answers will vary).
   The set of ordered pairs below the line $x + y = 3$, for example, $(0, 2)(-1, -1)(3, -2)$

   b. Describe the graph of the equation $x + y = 3$. Find three solutions to the equation (answers will vary).
   The set of ordered pairs on the line $x + y = 3$, for example, $(0, 3)(3, 0)(1, 2)$

   c. Describe the graph of the inequality $x + y > 3$. Find three solutions to the inequality (answers will vary).
   The set of ordered pairs above the line $x + y = 3$, for example, $(4, 0)(-1, 6)(2, 2)$

Objective 2: Graphing Systems of Linear Inequalities in Two Variables

For Exercises 37–54, graph the solution set. **(See Example 5.)**

37. $2x + y < 3$

   $y \geq x + 3$

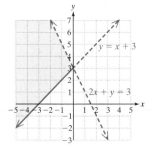

38. $x + y < 3$

   $y - x \geq 0$

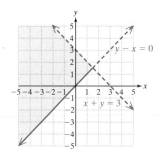

39. $x + y \geq -3$

   $x - 2y \geq 6$

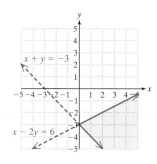

40. $y \geq -3x + 4$

   $x + y \leq 4$

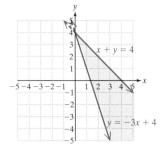

41. $2x + 3y < 6$

   $3x + y > -5$

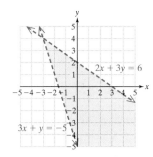

42. $-2x - y < 5$

   $x + 2y \geq 2$

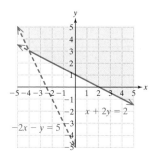

43. $y > 2x$

   $y > -4x$

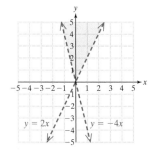

44. $2y \geq 6x$

   $y \leq x$

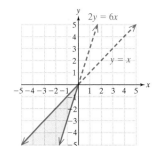

45. $y < \frac{1}{2}x - 1$

   $5x + y \leq -12$

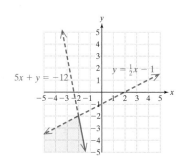

46. $y \geq \dfrac{1}{3}x + 2$

$4x + y < -2$

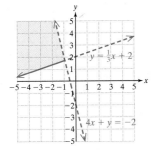

47. $y < 4$

$4x + 3y \geq 12$

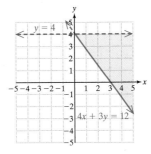

48. $x \geq -3$

$2x + 4y < 4$

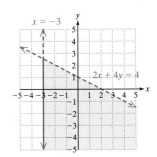

49. $x > -4$

$y \leq 3$

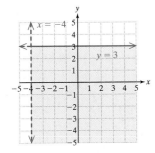

50. $x \leq 3$

$y > 1$

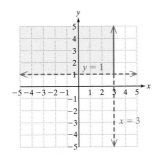

51. $2x \geq 5$

$6 > 3y$

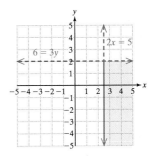

52. $4y \geq 6$

$8 > 2x$

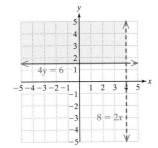

53. $x \geq -4$

$x \leq 1$

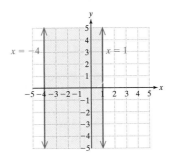

54. $y \geq -2$

$y \leq 3$

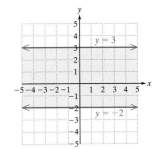

Group Activity

Creating Linear Models from Data

Materials: Two pieces of rope for each group. The ropes should be of different thicknesses. The piece of thicker rope should be between 4 and 5 ft long. The thinner piece of rope should be 8 to 12 in. shorter than the thicker rope. You will also need a yardstick or other device for making linear measurements.

Estimated Time: 30–35 minutes

Group Size: 4 (2 pairs)

Writing   Translating Expression   Geometry   Scientific Calculator   Video   NS E

1. Each group of 4 should divide into two pairs, and each pair will be given a piece of rope. Each pair will measure the initial length of rope. Then students will tie a series of knots in the rope and measure the new length after each knot is tied. (*Hint:* Try to tie the knots with an equal amount of force each time. Also, as the ropes are straightened for measurement, try to use the same amount of tension in the rope.) The results should be recorded in the table.

Thick Rope		Thin Rope		
Number of Knots, *x*	Length (in.), *y*	Number of Knots, *x*	Length (in.), *y*	
0		0		
1		1		
2		2		
3		3		
4		4		

Answers will vary throughout this exercise.

2. Graph each set of data points. Use a different color pen or pencil for each set of points. Does it appear that each set of data follows a linear trend? Draw a line through each set of points.

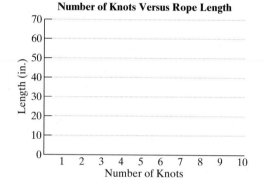

Number of Knots Versus Rope Length

3. Each time a knot is tied, the rope decreases in length. Using the results from question 1, compute the average amount of length lost per knot tied.

For the thick rope, the length decreases by _____ inches per knot tied.

For the thin rope, the length decreases by _____ inches per knot tied.

4. For each set of data points, find an equation of the line through the points. Write the equation in slope-intercept form, $y = mx + b$.

[*Hint:* The slope of the line will be negative and will be represented by the amount of length lost per knot (see question 3). The value of b will be the original length of the rope.]

Equation for the thick rope: _____

Equation for the thin rope: _____

5. Next, you will try to predict the number of knots that you need to tie in each rope so that the ropes will be equal in length. To do this, solve the system of equations in question 4.

Solution to the system of equations: (_____, _____)

number of knots, *x*   length, *y*

Interpret the meaning of the ordered pair in terms of the number of knots tied and the length of the ropes.

6. Check your answer from question 5 by actually tying the required number of knots in each rope. After doing this, are the ropes the same length? What is the length of each rope? Does this match the length predicted from question 5?

Chapter 4   Summary

Section 4.1   Solving Systems of Equations by the Graphing Method

Key Concepts

A **system of two linear equations** can be solved by graphing.

A **solution to a system of linear equations** is an ordered pair that satisfies each equation in the system. Graphically, this represents a point of intersection of the lines.

There may be one solution, infinitely many solutions, or no solution.

One solution   Infinitely many solutions   No solution
Consistent   Consistent   Inconsistent
Independent   Dependent   Independent

A system of equations is **consistent** if there is at least one solution. A system is **inconsistent** if there is no solution.

A linear system in x and y is **dependent** if two equations represent the same line. The solution set is the set of all points on the line.

If two linear equations represent different lines, then the system of equations is **independent**.

Examples

Example 1

Solve by graphing.

$$x + y = 3$$
$$2x - y = 0$$

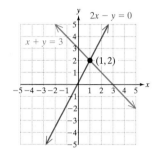

The solution is $(1, 2)$.

Example 2

Solve by graphing.

$$3x - 2y = 2$$
$$-6x + 4y = 4$$

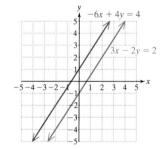

There is no solution. The system is inconsistent.

Example 3

Solve by graphing.

$$x + 2y = 2$$
$$-3x - 6y = -6$$

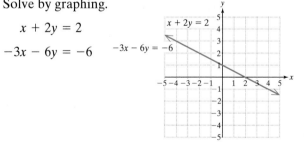

The system is dependent, and the solution set consists of all points on the line, given by
$$\{(x, y) \mid x + 2y = 2\}.$$

Section 4.2 Solving Systems of Equations by the Substitution Method

Key Concepts

Steps to Solve a System of Equations by the Substitution Method:

1. Isolate one of the variables from one equation.
2. Substitute the quantity found in step 1 into the other equation.
3. Solve the resulting equation.
4. Substitute the value found in step 3 back into the equation in step 1 to find the remaining variable.
5. Check the solution in both original equations and write the answer as an ordered pair.

An inconsistent system has no solution and is detected algebraically by a contradiction (such as $0 = 3$).

If two linear equations represent the same line, the system is dependent. This is detected algebraically by an identity (such as $0 = 0$).

Examples

Example 1

Solve by the substitution method.

$$x + 4y = -11$$
$$3x - 2y = -5$$

Isolate x in the first equation: $x = -4y - 11$
Substitute into the second equation.

$$3(-4y - 11) - 2y = -5 \qquad \text{Solve the equation.}$$
$$-12y - 33 - 2y = -5$$
$$-14y = 28$$
$$y = -2$$

$$\begin{aligned} & \text{Substitute} \\ x = -4y - 11 \qquad & y = -2. \\ x = -4(-2) - 11 \qquad & \text{Solve for } x. \\ x = -3 \end{aligned}$$

The solution is $(-3, -2)$ and checks in both original equations.

Example 2

Solve by the substitution method.

$$3x + y = 4$$
$$-6x - 2y = 2$$

Isolate y in the first equation: $y = -3x + 4$.
Substitute into the second equation.

$$-6x - 2(-3x + 4) = 2$$
$$-6x + 6x - 8 = 2$$
$$-8 = 2 \qquad \text{Contradiction}$$

The system is inconsistent and has no solution.

Example 3

Solve by the substitution method.

$$y = x + 2 \qquad y \text{ is already isolated.}$$
$$x - y = -2$$

$$x - (x + 2) = -2 \qquad \text{Substitute } y = x + 2 \text{ into the}$$
$$x - x - 2 = -2 \qquad \text{second equation.}$$
$$-2 = -2 \qquad \text{Identity}$$

The system is dependent. The solution set is all points on the line $y = x + 2$ or $\{(x, y) \mid y = x + 2\}$.

Section 4.3 Solving Systems of Equations by the Addition Method

Key Concepts

Solving a System of Linear Equations

by the Addition Method:

1. Write both equations in standard form: $Ax + By = C$.
2. Clear fractions or decimals (optional).
3. Multiply one or both equations by a nonzero constant to create opposite coefficients for one of the variables.
4. Add the equations to eliminate one variable.
5. Solve for the remaining variable.
6. Substitute the known value into one of the original equations to solve for the other variable.
7. Check the solution in both equations.

Examples

Example 1

Solve by using the addition method.

$$5x = -4y - 7 \qquad \text{Write the first equation in}$$
$$6x - 3y = 15 \qquad \text{standard form.}$$

Multiply by 3.
$$5x + 4y = -7 \xrightarrow{\quad\quad} 15x + 12y = -21$$
$$6x - 3y = 15 \xrightarrow[\text{Multiply by 4.}]{\quad\quad} \underline{24x - 12y = 60}$$
$$\qquad\qquad\qquad\qquad 39x \qquad\quad = 39$$
$$\qquad\qquad\qquad\qquad\qquad x = 1$$

$$5x = -4y - 7$$
$$5(1) = -4y - 7$$
$$5 = -4y - 7$$
$$12 = -4y$$
$$-3 = y \qquad \text{The solution is } (1, -3) \text{ and checks in both original equations.}$$

Section 4.4   Applications of Linear Equations in Two Variables

Examples

Example 1

A riverboat travels 36 mi with the current in 2 hr. The return trip takes 3 hr against the current. Find the speed of the current and the speed of the boat in still water.

Let x represent the speed of the boat in still water.
Let y represent the speed of the current.

	Distance	Rate	Time
Against current	36	$x - y$	3
With current	36	$x + y$	2

Distance = (rate)(time)

$$36 = (x - y) \cdot 3 \longrightarrow 36 = 3x - 3y$$
$$36 = (x + y) \cdot 2 \longrightarrow 36 = 2x + 2y$$

$$36 = 3x - 3y \xrightarrow{\text{Multiply by 2.}} 72 = 6x - 6y$$
$$36 = 2x + 2y \xrightarrow[\text{Multiply by 3.}]{} 108 = 6x + 6y$$
$$\overline{}$$
$$180 = 12x$$

$$15 = x$$

$$36 = 2(15) + 2y$$
$$36 = 30 + 2y$$
$$6 = 2y$$
$$3 = y$$

The speed of the boat in still water is 15 mph, and the speed of the current is 3 mph.

Example 2

Diane borrows a total of $15,000. Part of the money is borrowed from a lender that charges 8% simple interest. She borrows the rest of the money from her mother and will pay back the money at 5% interest. If the total interest after one year is $900, how much did she borrow from each source?

	8%	5%	Total
Principal	x	y	15,000
Interest	0.08x	0.05y	900

$$x + y = 15,000$$
$$0.08x + 0.05y = 900$$

Substitute $x = 15,000 - y$ into the second equation.

$$0.08(15,000 - y) + 0.05y = 900$$
$$1200 - 0.08y + 0.05y = 900$$
$$1200 - 0.03y = 900$$
$$-0.03y = -300$$
$$y = 10,000$$

$$x = 15,000 - 10,000$$
$$= 5,000$$

The amount borrowed at 8% is $5,000.
The amount borrowed from her mother is $10,000.

Section 4.5   Linear Inequalities and Systems of Inequalities in Two Variables

Key Concepts

A **linear inequality in two variables** can be written in one of the forms: $ax + by < c$, $ax + by > c$, $ax + by \leq c$, or $ax + by \geq c$.

Steps for Using the Test Point Method to Solve a Linear Inequality in Two Variables:

1. Set up the related *equation*.
2. Graph the related equation. This will be a line in the xy-plane.
   - If the original inequality is a strict inequality, $<$ or $>$, then the line is *not* part of the solution set. Therefore, graph the boundary as a dashed line.

   - If the original inequality is not strict, $\leq$ or $\geq$, then the line *is* part of the solution set. Therefore, graph the boundary as a solid line.

3. Choose a point not on the line and substitute its coordinates into the original inequality.
   - If the test point makes the inequality true, then the region it represents is part of the solution set. Shade that region.
   - If the test point makes the inequality false, then the other region is part of the solution set and should be shaded.

Example

Example 1

Graph the inequality. $2x - y < 4$

1. The related equation is $2x - y = 4$.
2. Graph the equation $2x - y = 4$ (dashed line).

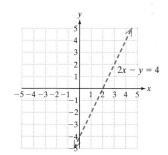

3. Choose an arbitrary test point not on the line such as $(0, 0)$.

$$2x - y < 4$$
$$2(0) - (0) \overset{?}{<} 4$$
$$0 \overset{?}{<} 4 \; ✔ \quad \text{True}$$

Shade the region represented by the test point (in this case, above the line).

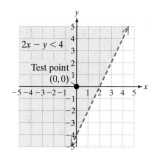

Chapter 4   Review Exercises

Section 4.1

For Exercises 1–4, determine if the ordered pair is a solution to the system.

1. $x - 4y = -4$ $(4, 2)$

    $x + 2y = 8$ Yes

2. $x - 6y = 6$ $(12, 1)$

    $-x + y = 4$ No

3. $3x + y = 9$ $(1, 3)$

    $y = 3$ No

4. $2x - y = 8$ $(2, -4)$

    $x = 2$ Yes

For Exercises 5–10, identify whether the system represents intersecting lines, parallel lines, or coinciding lines by comparing their slopes and y-intercepts.

5. $y = -\dfrac{1}{2}x + 4$

    $y = x - 1$
Intersecting lines (the lines have different slopes)

6. $y = -3x + 4$

    $y = 3x + 4$
Intersecting lines (the lines have different slopes)

7. $y = -\dfrac{4}{7}x + 3$

    $y = -\dfrac{4}{7}x - 5$
Parallel lines (the lines have the same slope but different y-intercepts)

8. $y = 5x - 3$

    $y = \dfrac{1}{5}x - 3$
Intersecting lines (the lines have different slopes)

9. $y = 9x - 2$

    $9x - y = 2$
Coinciding lines (the lines have the same slope and same y-intercept)

10. $x = -5$

    $y = 2$
Intersecting lines (the lines have different slopes)

For Exercises 11–18, solve the systems by graphing. If a system does not have a unique solution, identify the system as inconsistent or dependent.

11. $y = -\dfrac{2}{3}x - 2$ $(0, -2)$

    $-x + 3y = -6$

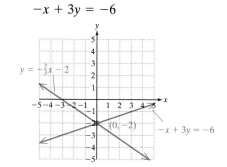

12. $y = -2x - 1$ $(-2, 3)$

    $2x + 3y = 5$

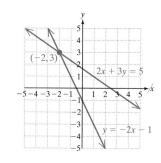

13. $4x = -2y + 10$

    $2x + y = 5$

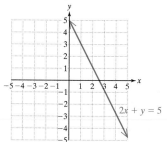

Infinitely many solutions $\{(x, y) \mid 2x + y = 5\}$; dependent system

14. $10y = 2x - 10$

    $-x + 5y = -5$

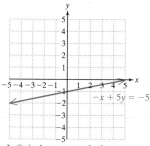

Infinitely many solutions $\{(x, y) \mid -x + 5y = -5\}$; dependent system

15. $6x - 3y = 9$

    $y = -1$ $(1, -1)$

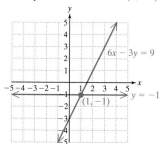

16. $5x + y = -11$

    $x = -1$ $(-1, -6)$

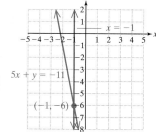

17. $x - 7y = 14$

    $-2x + 14y = 14$

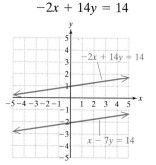

No solution; inconsistent system

18. $y = -5x + 6$

    $10x + 2y = 6$

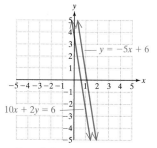

No solution; inconsistent system

19. A rental car company rents a compact car for $20 a day plus $0.25 per mile. A midsize car rents for $30 a day plus $0.20 per mile.

The cost, y_c, to rent a compact car for one day is given by the equation:

$y_c = 20 + 0.25x$ where x is the number of miles driven

The cost, y_m, to rent a midsize car for one day is given by the equation:

$y_m = 30 + 0.20x$ where x is the number of miles driven

Find the number of miles at which the cost to rent either car for a day would be the same. Confirm your answer with the graph.   200 mi

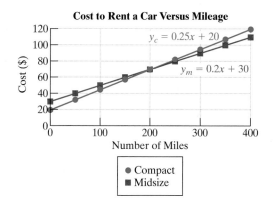

Cost to Rent a Car Versus Mileage

● Compact
■ Midsize

Section 4.2

For Exercises 20–23, solve the systems using the substitution method.

20. $6x + y = 2$ $\left(\frac{2}{3}, -2\right)$

$y = 3x - 4$

21. $2x + 3y = -5$ $(-4, 1)$

$x = y - 5$

22. $2x + 6y = 10$

$x = -3y + 6$

No solution; inconsistent system

23. $4x + 2y = 4$

$y = -2x + 2$

Infinitely many solutions; $\{(x, y)|y = -2x + 2\}$; dependent system

24. Given the system:

$x + 2y = 11$

$5x + 4y = 40$

a. Which variable from which equation is easiest to isolate and why? x in the first equation is easiest to isolate because its coefficient is 1.

b. Solve the system using the substitution method. $\left(6, \frac{5}{2}\right)$

25. Given the system:

$4x - 3y = 9$

$2x + y = 12$

a. Which variable from which equation is easiest to isolate and why? y in the second equation is easiest to isolate because its coefficient is 1.

b. Solve the system using the substitution method. $\left(\frac{9}{2}, 3\right)$

For Exercises 26–29, solve the systems using the substitution method.

26. $3x - 2y = 23$ $(5, -4)$

$x + 5y = -15$

27. $x + 5y = 20$ $(0, 4)$

$3x + 2y = 8$

28. $x - 3y = 9$

$5x - 15y = 45$

Infinitely many solutions; $\{(x, y)|x - 3y = 9\}$; dependent system

29. $-3x + y = 15$

$6x - 2y = 12$

No solution; inconsistent system

30. The difference of two positive numbers is 42. The larger number is 2 more than 6 times the smaller number. Find the numbers.   The numbers are 50 and 8.

31. In a right triangle, one of the acute angles is 6° less than the other acute angle. Find the measure of each acute angle.   The angles are 42° and 48°.

32. Two angles are supplementary. One angle measures 14° less than two times the other angle. Find the measure of each angle.   The angles are $115\frac{1}{3}°$ and $64\frac{2}{3}°$.

Section 4.3

33. Explain the process for solving a system of two equations using the addition method.
See page 317.

34. Given the system:

$3x - 5y = 1$

$2x - y = -4$

a. Which variable, x or y, is easier to eliminate using the addition method? (Answers may vary.)

b. Solve the system using the addition method. $(-3, -2)$

35. Given the system:

$9x - 2y = 14$

$4x + 3y = 14$

a. Which variable, x or y, is easier to eliminate using the addition method? (Answers may vary.)

b. Solve the system using the addition method. $(2, 2)$

For Exercises 36–43, solve the systems using the addition method.

36. $2x + 3y = 1$ $(2, -1)$

$x - 2y = 4$

37. $x + 3y = 0$ $(-6, 2)$

$-3x - 10y = -2$

38. $8x + 8 = -6y + 6$

$10x = 9y - 8$ $\left(-\frac{1}{2}, \frac{1}{3}\right)$

39. $12x = 5y + 5$ $\left(\frac{1}{4}, -\frac{2}{5}\right)$

$5y = -1 - 4x$

40. $-4x - 6y = -2$

$6x + 9y = 3$

Infinitely many solutions;
$\{(x, y) \mid -4x - 6y = -2\}$; dependent system

41. $-8x - 4y = 16$

$10x + 5y = 5$

No solution; inconsistent system

42. $\frac{1}{2}x - \frac{3}{4}y = -\frac{1}{2}$

$\frac{1}{3}x + y = -\frac{10}{3}$

$(-4, -2)$

43. $0.5x - 0.2y = 0.5$

$0.4x + 0.7y = 0.4$

$(1, 0)$

44. Given the system:

$$4x + 9y = -7$$

$$y = 2x - 13$$

a. Which method would you choose to solve the system, the substitution method or the addition method? Explain your choice. (Answers may vary.)

b. Solve the system. $(5, -3)$

45. Given the system:

$$5x - 8y = -2$$

$$3x - y = -5$$

a. Which method would you choose to solve the system, the substitution method or the addition method? Explain your choice. (Answers may vary.)

b. Solve the system. $(-2, -1)$

Section 4.4

46. Miami Metrozoo charges $11.50 for adult admission and $6.75 for children under 12. The total bill before tax for a school group of 60 people is $443. How many adults and how many children were admitted?

There were 8 adult tickets and 52 children's tickets sold.

47. As part of his retirement strategy Winston plans to invest $600,000 in two different funds. He projects that the high-risk investments should return, over time, about 12% per year, while the low-risk investments should return about 4% per year. If he wants a supplemental income of $30,000 a year, how should he divide his investments?

He should invest $75,000 at 12% and $525,000 at 4%.

48. Suppose that whole milk with 4% fat is mixed with 1% low fat milk to make a 2% reduced fat milk. How much of the whole milk should be mixed with the low fat milk to make 60 gal of 2% reduced fat milk?   20 gal of whole milk should be mixed with 40 gal of low fat milk.

49. A boat travels 80 mi downstream with the current in 4 hr and 80 mi upstream against the current in 5 hr. Find the speed of the current and the speed of the boat in still water.
The speed of the boat is 18 mph, and that of the current is 2 mph.

50. A plane travels 870 miles against a headwind in 3 hr. Traveling with a tailwind, the plane travels 700 mi in 2 hr. Find the speed of the plane in still air and the speed of the wind.
The plane's speed in still air is 320 mph. The wind speed is 30 mph.

51. At Conseco Fieldhouse, home of the Indiana Pacers, the total cost of a soft drink and a hot dog is $8.00. The price of the hot dog is $1.00 more than the cost of the soft drink. Find the cost of a soft drink and the cost of a hot dog.
A hot dog costs $4.50 and a drink costs $3.50

52. In a recent election 5700 votes were cast and 3675 voters voted for the winning candidate. If $\frac{5}{8}$ of the women and $\frac{2}{3}$ of the men voted for the winning candidate, how many men and how many women voted?
3000 women and 2700 men

53. Ray played two rounds of golf at Pebble Beach for a total score of 154. If his score in the second round is 10 more than his score in the first round, find the scores for each round.
The score was 72 on the first round and 82 on the second round.

Section 4.5

For Exercises 54–57, graph the inequalities. Then write three ordered pairs that are in the solution set (answers may vary).

54. $y < 3x - 1$

55. $y > -2x + 6$

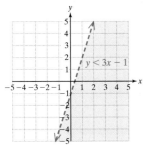

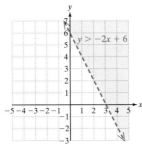

For example: $(1, -1)(0, -4)(2, 0)$

For example: $(5, 5)(4, 0)(0, 7)$

56. $-2x - 3y \geq 8$

57. $4x - 2y \leq 10$

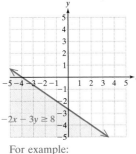

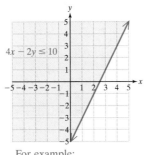

For example:
$(-4, 0)(-2, -2)(1, -4)$

For example:
$(0, 0)(0, -5)(-1, 1)$

For Exercises 58–63, graph the solution set.

58. $x - 5y \geq 0$

59. $7x - y \leq 0$

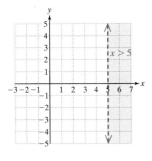

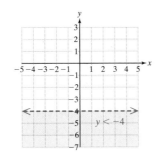

60. $x > 5$

61. $y < -4$

62. $y \geq 0$

63. $x \geq 0$

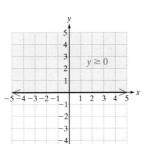

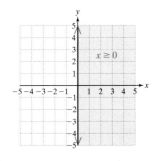

For Exercises 64–67, graph the solution set.

64. $2x - y \geq 8$
$x + y \leq 3$

65. $y \leq x - 1$
$x + 2y \geq 4$

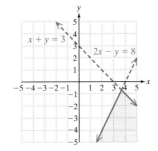

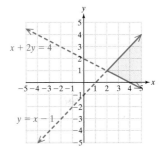

66. $y \leq 2x$
$-2x - y > -3$

67. $y \leq 4$
$2x - y < 1$

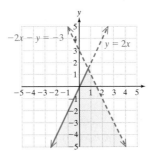

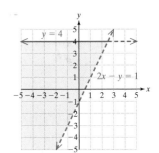

Chapter 4 Test

1. Write each line in slope-intercept form. Then determine if the lines represent intersecting lines, parallel lines, or coinciding lines.

$$5x + 2y = -6 \quad y = -\frac{5}{2}x - 3$$

$$-\frac{5}{2}x - y = -3 \quad y = -\frac{5}{2}x + 3$$

Parallel lines

For Exercises 2–3 solve the system by graphing.

2. $y = 2x - 4$

   $-2x + 3y = 0$

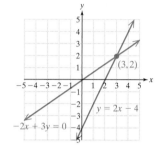

3. $2x + 4y = 6$

   $2y - 3 = -x$

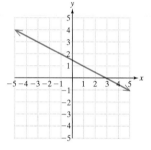

Infinitely many solutions;
$\{(x, y) \mid 2x + 4y = 6\}$; dependent system

4. Solve the system using the substitution method.

$$x = 5y - 2$$

$$2x + y = -4 \quad (-2, 0)$$

5. In the 2005 WNBA (basketball) season, the league's leading scorer was Sheryl Swoopes from the Houston Comets. Swoopes scored 17 points more than the second leading scorer, Lauren Jackson from the Seattle Storm. Together they scored a total of 1211 points. How many points did each player score?
   Swoopes scored 614 points and Jackson scored 597.

6. Solve the system using the addition method.

$$3x - 6y = 8$$
$$2x + 3y = 3 \quad \left(2, -\frac{1}{3}\right)$$

7. How many milliliters of a 50% acid solution and how many milliliters of a 20% acid solution must be mixed to produce 36 mL of a 30% acid solution?   12 mL of the 50% acid solution should be mixed with 24 mL of the 20% solution.

8. **a.** How many solutions does a system of two linear equations have if the equations represent parallel lines?   No solution

   b. How many solutions does a system of two linear equations have if the equations represent coinciding lines?   Infinitely many solutions

   c. How many solutions does a system of two linear equations have if the equations represent intersecting lines?   One solution

For Exercises 9–14, solve the systems using any method.

9. $\frac{1}{3}x + y = \frac{7}{3}$

   $x = \frac{3}{2}y - 11$

   $(-5, 4)$

10. $2(x - 6) = y$

    $2x - \frac{1}{2}y = x + 5$

    No solution

11. $3x - 4y = 29 \quad (3, -5)$

    $2x + 5y = -19$

12. $2x = 6y - 14 \quad (-1, 2)$

    $2y = 3 - x$

13. $-0.25x - 0.05y = 0.2$ Infinitely many solutions;

    $10x + 2y = -8$ $\{(x, y) \mid 10x + 2y = -8\}$

14. $3x + 3y = -2y - 7 \quad (1, -2)$

    $-3y = 10 - 4x$

15. At Best Buy, Latrell buys four CDs and two DVDs for $54 from the sale rack. Kendra buys two CDs and three DVDs from the same rack for $49. What is the price per CD and the price per DVD?
    CDs cost $8 each and DVDs cost $11 each.

16. The cost to ride the trolley one-way in San Diego is $2.25. Kelly and Hazel had to buy eight tickets for their group.

    a. What was the total amount of money required? $18 was required.

    b. Kelly and Hazel had only quarters and $1 bills. They also determined that they used twice as many quarters as $1 bills. How many quarters and how many $1 bills did they use?
    They used 24 quarters and 12 $1 bills.

17. Suppose a total of $5000 is borrowed from two different loans. One loan charges 10% simple interest, and the other charges 8% simple interest. How much was borrowed at each rate if $424 in interest is charged at the end of 1 yr?
    $1200 was borrowed at 10%, and $3800 was borrowed at 8%.

18. During the first 13 yr of his football career, Jerry Rice scored a total of 166 touchdowns. One touchdown was scored on a kickoff return, and the remaining 165 were scored rushing or receiving. The number of receiving touchdowns he scored was 5 more than 15 times the number of rushing touchdowns he scored. How many receiving touchdowns and how many rushing touchdowns did he score?   He scored 155 receiving touchdowns and 10 touchdowns rushing.

19. A plane travels 910 mi in 2 hr against the wind and 1090 mi in 2 hr with the same wind. Find the speed of the plane in still air and the speed of the wind.   The plane travels 500 mph in still air, and the wind speed is 45 mph.

20. The number of calories in a piece of cake is 20 less than 3 times the number of calories in a scoop of ice cream. Together, the cake and ice cream have 460 calories. How many calories are in each?
The cake has 340 calories, and the ice cream has 120 calories.

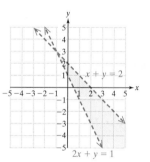

21. How much 10% acid solution should be mixed with a 25% acid solution to create 100 mL of a 16% acid solution?
60 mL of 10% solution and 40 mL of 25% solution.

22. Graph the solution set. $5x - y \geq -6$.

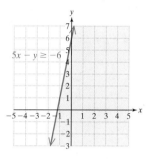

23. Graph the solution set.

$$2x + y > 1$$
$$x + y < 2$$

Chapters 1–4   Cumulative Review Exercises

1. Simplify.

$$\frac{|2 - 5| + 10 \div 2 + 3}{\sqrt{10^2 - 8^2}} \quad \frac{11}{6}$$

2. Solve for x. $\frac{1}{3}x - \frac{3}{4} = \frac{1}{2}(x + 2)$ $-\frac{21}{2}$

3. Solve for a. $-4(a + 3) + 2 = -5(a + 1) + a$
No solution

4. Solve for y. $3x - 2y = 6$ $y = \frac{3}{2}x - 3$

5. Solve for z. Graph the solution set on a number line and write the solution in interval notation:

$$-2(3z + 1) \leq 5(z - 3) + 10$$

$\left[\frac{3}{11}, \infty\right)$ ⊢────────→
                        $\frac{3}{11}$

6. The largest angle in a triangle is 110°. Of the remaining two angles, one is 4° less than the other angle. Find the measure of the three angles.
The angles are 37°, 33°, and 110°.

7. Two hikers start at opposite ends of an 18-mi trail and walk toward each other. One hiker walks predominately down hill and averages 2 mph faster than the other hiker. Find the average rate of each hiker if they meet in 3 hr.
The rates of the hikers are 2 mph and 4 mph.

8. Jesse Ventura became the 38th governor of Minnesota by receiving 37% of the votes. If approximately 2,060,000 votes were cast, how many did Mr. Ventura get?
Jesse Ventura received approximately 762,200 votes.

9. The YMCA wants to raise $2500 for its summer program for disadvantaged children. If the YMCA has already raised $900, what percent of its goal has been achieved? 36% of the goal has been achieved.

10. Two angles are complementary. One angle measures 17° more than the other angle. Find the measure of each angle. The angles are 36.5° and 53.5°.

11. Solve for x. $z = \dfrac{x - m}{5}$ $x = 5z + m$

12. Solve for y. $2x - 3y = 6$ $y = \dfrac{2}{3}x - 2$

13. The slope of a given line is $-\frac{2}{3}$.

   a. What is the slope of a line parallel to the given line? $-\dfrac{2}{3}$

   b. What is the slope of a line perpendicular to the given line? $\dfrac{3}{2}$

14. Find an equation of the line passing through the point $(2, -3)$ and having a slope of -3. Write the final answer in slope-intercept form. $y = -3x + 3$

15. Sketch the following equations on the same graph.

   a. $2x + 5y = 10$

   b. $2y = 4$

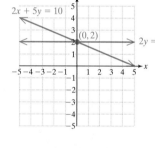

   c. Find the point of intersection and check the solution in each equation.
   $(0, 2)$

16. Solve the system of equations by using the substitution method.

$$2x + 5y = 10 \quad (0, 2)$$
$$2y = 4$$

17. a. Graph the line $2x + y = 3$.

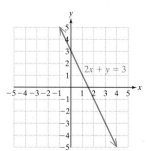

b. Graph the inequality $2x + y < 3$.

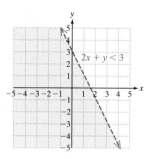

c. Explain the difference between the graphs in parts (a) and (b). Part (a) represents the solutions to an equation. Part (b) represents the solutions to a strict inequality.

18. How many gallons of a 15% antifreeze solution should be mixed with a 60% antifreeze solution to produce 60 gal of a 45% antifreeze solution? 20 gal of the 15% solution should be mixed with 40 gal of the 60% solution.

19. Use a system of linear equations to solve for x and y. x is 27°; y is 63°

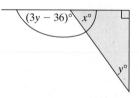

20. In 1920, the average speed for the winner of the Indianapolis 500 car race was 88.6 mph. By 1990, the average speed of the winner was 186.0 mph.

   a. Find the slope of the line shown in the figure. Round to one decimal place. 1.4

   b. Interpret the meaning of the slope in the context of this problem. Between 1920 and 1990, the winning speed in the Indianapolis 500 increased on average by 1.4 mph per year.

Average Speed for the Winning Car in the Indianapolis 500

Polynomials and Properties of Exponents

5

CHAPTER OUTLINE

5.1 Exponents: Multiplying and Dividing Common Bases 360

5.2 More Properties of Exponents 370

5.3 Definitions of b^0 and b^{-n} 375

5.4 Scientific Notation 383

Problem Recognition Exercises: Properties of Exponents 389

5.5 Addition and Subtraction of Polynomials 390

5.6 Multiplication of Polynomials and Special Products 398

5.7 Division of Polynomials 407

Problem Recognition Exercises: Operations on Polynomials 415

Group Activity: The Pythagorean Theorem and a Geometric "Proof" 416

Chapter 5

In Chapter 5, we begin by covering the properties and applications of expressions involving exponents, including scientific notation. Then we introduce the concept of a polynomial and learn how to add, subtract, multiply, and divide polynomials.

The following is a Sudoku puzzle. Use the clues given to fill in the boxes labeled A–E. Then fill in the remaining part of the grid so that every row, every column, and every 2 × 3 box contains the digits 1 through 6.

A. The number of terms in a trinomial

B. The constant term in the polynomial: $4x^2 - 3x + 5$

C. The value of the missing exponent: $x \cdot x^5 \cdot x^{-4} = x^?$

D. The degree of the polynomial: $-2y^6 - 3y^4 + 5y^2 + 9$

E. Simplified form of $(-2)^2 \cdot 3^0$

1	5	6	2	4	A 3
2	4	3	5	1	6
4	3	1	6	B 5	2
5	6	C 2	4	3	1
3	2	4	1	6	5
D 6	1	5	3	2	E 4

359

Objectives

1. Review of Exponential Notation
2. Evaluating Expressions with Exponents
3. Multiplying and Dividing Common Bases
4. Simplifying Expressions with Exponents
5. Applications of Exponents

1. Review of Exponential Notation

Recall that an **exponent** is used to show repeated multiplication of the **base**.

> **DEFINITION** Definition of b^n
>
> Let b represent any real number and n represent a positive integer. Then,
> $$b^n = \underbrace{b \cdot b \cdot b \cdot b \cdot \ldots b}_{n \text{ factors of } b}$$

Skill Practice

For each expression, identify the base and exponent.

1. 8^3
2. $\left(-\dfrac{1}{4}\right)^2$
3. 0.2^4

Classroom Example: p. 367, Exercise 2

Example 1 Evaluating Expressions with Exponents

For each expression, identify the exponent and base. Then evaluate the expression.

a. 6^2 **b.** $\left(-\dfrac{1}{2}\right)^3$ **c.** 0.8^4

Solution:

Expression	Base	Exponent	Result
a. 6^2	6	2	$(6)(6) = 36$
b. $\left(-\dfrac{1}{2}\right)^3$	$-\dfrac{1}{2}$	3	$\left(-\dfrac{1}{2}\right)\left(-\dfrac{1}{2}\right)\left(-\dfrac{1}{2}\right) = -\dfrac{1}{8}$
c. 0.8^4	0.8	4	$(0.8)(0.8)(0.8)(0.8) = 0.4096$

Note that if no exponent is explicitly written for an expression, then the expression has an implied exponent of 1. For example, $x = x^1$.

Consider an expression such as $3y^6$. The factor 3 has an exponent of 1, and the factor y has an exponent of 6. That is, the expression $3y^6$ is interpreted as 3^1y^6.

2. Evaluating Expressions with Exponents

Recall from Section 1.2 that particular care must be taken when evaluating exponential expressions involving negative numbers. An exponential expression with a negative base is written with parentheses around the base, such as $(-3)^2$.

To evaluate $(-3)^2$, we have: $\quad (-3)^2 = (-3)(-3) = 9$

If no parentheses are present, the expression -3^2, is the *opposite* of 3^2, or equivalently, $-1 \cdot 3^2$.

$$-3^2 = -1(3^2) = -1(3)(3) = -9$$

Answers

1. Base 8; exponent 3
2. Base $-\dfrac{1}{4}$; exponent 2
3. Base 0.2; exponent 4

Example 2 Evaluating Expressions with Exponents

Evaluate each expression.

a. -5^4 **b.** $(-5)^4$ **c.** $(-0.2)^3$ **d.** -0.2^3

Solution:

a. -5^4

$= -1 \cdot 5^4$ 5 is the base with exponent 4.

$= -1 \cdot 5 \cdot 5 \cdot 5 \cdot 5$ Multiply -1 times four factors of 5.

$= -625$

b. $(-5)^4$

$= (-5)(-5)(-5)(-5)$ Parentheses indicate that -5 is the base with exponent 4.

$= 625$ Multiply four factors of -5.

c. $(-0.2)^3$ Parentheses indicate that -0.2 is the base with exponent 3.

$= (-0.2)(-0.2)(-0.2)$ Multiply three factors of -0.2.

$= -0.008$

d. -0.2^3

$= -1 \cdot 0.2^3$ 0.2 is the base with exponent 3.

$= -1 \cdot 0.2 \cdot 0.2 \cdot 0.2$ Multiply -1 times three factors of 0.2.

$= -0.008$

Skill Practice

Evaluate.

4. -2^4 **5.** $(-2)^4$
6. $(-0.1)^3$ **7.** -0.1^3

Classroom Examples: p. 367,
Exercises 24 and 26

Concept Connections

8. Will the value of a negative base with an odd exponent be negative or positive?
9. Will the value of a negative base with an even exponent be negative or positive?

Example 3 Evaluating Expressions with Exponents

Evaluate each expression for $a = 2$ and $b = -3$.

a. $5a^2$ **b.** $(5a)^2$ **c.** $5ab^2$ **d.** $(b + a)^2$

Solution:

a. $5a^2$

$= 5(\ )^2$ Use parentheses to substitute a number for a variable.

$= 5(2)^2$ Substitute $a = 2$.

$= 5(4)$ Simplify.

$= 20$

b. $(5a)^2$

$= [5(\ )]^2$ Use parentheses to substitute a number for a variable. The original parentheses are replaced with brackets.

$= [5(2)]^2$ Substitute $a = 2$.

$= (10)^2$ Simplify inside the parentheses first.

$= 100$

Skill Practice

Evaluate each expression for
$x = 2$ and $y = -5$.
10. $6x^2$ **11.** $(6x)^2$
12. $2xy^2$ **13.** $(y - x)^2$

Classroom Examples: p. 368,
Exercises 48 and 52

Answers

4. -16 **5.** 16   **6.** -0.001
7. -0.001 **8.** Negative
9. Positive   **10.** 24   **11.** 144
12. 100   **13.** 49

Avoiding Mistakes

In the expression $5ab^2$, the exponent, 2, applies only to the variable b. The constant 5 and the variable a both have an implied exponent of 1.

c. $5ab^2$

$= 5(2)(-3)^2$ Substitute $a = 2, b = -3$.

$= 5(2)(9)$ Simplify.

$= 90$ Multiply.

d. $(b + a)^2$

$= [(-3) + (2)]^2$ Substitute $b = -3$ and $a = 2$.

$= (-1)^2$ Simplify within the parentheses first.

$= 1$

Avoiding Mistakes

Be sure to follow the order of operations. In Example 3(d), it would be incorrect to square the terms within the parentheses before adding.

3. Multiplying and Dividing Common Bases

In this section, we investigate the effect of multiplying or dividing two quantities with the same base. For example, consider the expressions: $x^5 x^2$ and $\frac{x^5}{x^2}$. Simplifying each expression, we have:

$$x^5 x^2 = (x \cdot x \cdot x \cdot x \cdot x)(x \cdot x) = \overbrace{x \cdot x \cdot x \cdot x \cdot x \cdot x \cdot x}^{7 \text{ factors of } x} = x^7$$

$$\frac{x^5}{x^2} = \frac{x \cdot x \cdot x \cdot \cancel{x} \cdot \cancel{x}}{\cancel{x} \cdot \cancel{x}} = \frac{x \cdot x \cdot x}{1} = x^3$$

These examples suggest that to multiply two quantities with the same base, we add the exponents. To divide two quantities with the same base, we subtract the exponent in the denominator from the exponent in the numerator. These rules are stated formally in the following two properties.

PROPERTY   **Multiplication of Like Bases**

Assume that b is a real number and that m and n represent positive integers. Then,

$$b^m b^n = b^{m+n}$$

PROPERTY   **Division of Like Bases**

Assume that $b \neq 0$ is a real number and that m and n represent positive integers such that $m > n$. Then,

$$\frac{b^m}{b^n} = b^{m-n}$$

| Example 4 | **Simplifying Expressions with Exponents** |

Simplify the expressions.

a. $w^3 w^4$ **b.** $2^3 \cdot 2^4$

Solution:

a. $w^3 w^4$ $(w \cdot w \cdot w)(w \cdot w \cdot w \cdot w)$

$= w^{3+4}$ To multiply like bases, add the exponents.

$= w^7$

b. $2^3 \cdot 2^4$ $(2 \cdot 2 \cdot 2)(2 \cdot 2 \cdot 2 \cdot 2)$

$= 2^{3+4}$ To multiply like bases, add the exponents (the base is unchanged).

$= 2^7$ or 128

Avoiding Mistakes

When we multiply like bases, we add the exponents. The base does not change. In Example 4(b), we have $2^3 \cdot 2^4 = 2^7$.

| Example 5 | **Simplifying Expressions with Exponents** |

Simplify the expressions.

a. $\dfrac{t^6}{t^4}$ **b.** $\dfrac{5^6}{5^4}$

Solution:

a. $\dfrac{t^6}{t^4}$ $\dfrac{t \cdot t \cdot t \cdot t \cdot t \cdot t}{t \cdot t \cdot t \cdot t}$

$= t^{6-4}$ To divide like bases subtract the exponents.

$= t^2$

b. $\dfrac{5^6}{5^4}$ $\dfrac{5 \cdot 5 \cdot 5 \cdot 5 \cdot 5 \cdot 5}{5 \cdot 5 \cdot 5 \cdot 5}$

$= 5^{6-4}$ To divide like bases subtract the exponents (the base is unchanged).

$= 5^2$ or 25

Answers
14. q^{12} **15.** 8^{12}
16. y^7 **17.** 3^7

Example 6   Simplifying Expressions with Exponents

Simplify the expressions.

a. $\dfrac{z^4 z^5}{z^3}$ **b.** $\dfrac{10^7}{10^2 \cdot 10}$

Solution:

a. $\dfrac{z^4 z^5}{z^3}$

$= \dfrac{z^{4+5}}{z^3}$ Add the exponents in the numerator (the base is unchanged).

$= \dfrac{z^9}{z^3}$

$= z^{9-3}$ Subtract the exponents.

$= z^6$

b. $\dfrac{10^7}{10^2 \cdot 10}$

$= \dfrac{10^7}{10^2 \cdot 10^1}$ Note that 10 is equivalent to 10^1.

$= \dfrac{10^7}{10^{2+1}}$ Add the exponents in the denominator (the base is unchanged).

$= \dfrac{10^7}{10^3}$

$= 10^{7-3}$ Subtract the exponents.

$= 10^4$ or 10,000   Simplify.

4. Simplifying Expressions with Exponents

Example 7   Simplifying Expressions with Exponents

Use the commutative and associative properties of real numbers and the properties of exponents to simplify the expressions.

a. $(3p^2 q^4)(2pq^5)$ **b.** $\dfrac{16w^9 z^3}{4w^8 z}$

Solution:

a. $(3p^2 q^4)(2pq^5)$

$= (3 \cdot 2)(p^2 p)(q^4 q^5)$ Apply the associative and commutative properties of multiplication to group coefficients and like bases.

$= (3 \cdot 2)p^{2+1} q^{4+5}$ Add the exponents when multiplying like bases.

$= 6p^3 q^9$ Simplify.

Instructor Note: Remind students to divide coefficients but subtract exponents.

b. $\dfrac{16w^9z^3}{4w^8z}$

$$= \left(\dfrac{16}{4}\right)\left(\dfrac{w^9}{w^8}\right)\left(\dfrac{z^3}{z}\right) \qquad \text{Group coefficients and like bases.}$$

$$= 4w^{9-8}z^{3-1} \qquad\qquad \text{Subtract the exponents when dividing like bases.}$$

$$= 4wz^2 \qquad\qquad\qquad \text{Simplify.}$$

5. Applications of Exponents

Simple interest on an investment or loan is computed by the formula $I = Prt$, where P is the amount of principal, r is the annual interest rate, and t is the time in years. Simple interest is based only on the original principal. However, in most day-to-day applications, the interest computed on money invested or borrowed is compound interest. **Compound interest** is computed on the original principal and on the interest already accrued.

Suppose \$1000 is invested at 8% interest for 3 yr. Compare the total amount in the account if the money earns simple interest versus if the interest is compounded annually.

Simple Interest

The simple interest earned is given by $I = Prt$

$$= (1000)(0.08)(3)$$

$$= \$240$$

The total amount in the account after 3 yr is \$1240.

Compound Interest (Annual)

To compute interest compounded annually over a period of 3 yr, compute the interest earned in the first year. Then add the interest earned in the first year to the principal. This value then becomes the principal on which to base the interest earned in the second year. We repeat this process, finding the interest for the second and third years based on the principal and interest earned in the preceding years. This process is outlined using a table.

Year	Interest Earned $I = Prt$	Total Amount in the Account
First year	$I = (\$1000)(0.08)(1) = \$80$	$\$1000 + \$80 = \$1080$
Second year	$I = (\$1080)(0.08)(1) = \$86.40$	$\$1080 + \$86.40 = \$1166.40$
Third year	$I = (\$1166.40)(0.08)(1) \approx \$93.31$	$\$1166.40 + 93.31 = \mathbf{\$1259.71}$

The total amount in the account found by compounding interest annually is \$1259.71. The difference in the account balance for interest compounded annually versus for simple interest is $\$1259.71 - \$1240 = \$19.71$.

The total amount, A, in an account earning compound annual interest may be computed quickly using the following formula:

$A = P(1 + r)^t$ where P is the amount of principal, r is the annual interest rate (expressed in decimal form), and t is the number of years.

Concept Connections

22. Identify the values for the variables P, r, and t if $19,000 is invested for 10 years at 7% interest compounded annually.

For example, for $1000 invested at 8% interest compounded annually for 3 yr, we have $P = 1000$, $r = 0.08$, and $t = 3$.

$$A = P(1 + r)^t$$
$$A = 1000(1 + 0.08)^3$$
$$= 1000(1.08)^3$$
$$= 1000(1.259712)$$
$$= 1259.712$$

Rounding to the nearest cent, we have $A = \$1259.71$, as expected.

Skill Practice

23. Find the amount in an account after 3 yr if the initial investment is $4000 invested at 5% interest compounded annually.

Classroom Example: p. 369, Exercise 114

Example 8 — Using Exponents in an Application

Find the amount in an account after 8 yr if the initial investment is $7000, invested at 2.25% interest compounded annually.

Solution:

Identify the values for each variable.

$P = 7000$

$r = 0.0225$ Note that the decimal form of a percent is used for calculations.

$t = 8$

$A = P(1 + r)^t$

$\qquad = 7000(1 + 0.0225)^8$ Substitute.

$\qquad = 7000(1.0225)^8$ Simplify inside the parentheses.

$\qquad = 7000(1.194831142)$ Approximate $(1.0225)^8$.

$\qquad = 8363.82$ Multiply (round to the nearest cent).

The amount in the account after 8 yr is $8363.82.

Answers

22. $P = 19000$; $r = 0.07$; $t = 10$
23. $4630.50

Calculator Connections

In Example 8, it was necessary to evaluate the expression $(1.0225)^8$. Recall that the $\boxed{\land}$ or $\boxed{y^x}$ key may be used to enter expressions with exponents.

Scientific Calculator

Enter: 1.0225 $\boxed{y^x}$ 8 $\boxed{=}$ **Result:** $\boxed{\qquad 1.194831142 \qquad}$

Graphing Calculator

```
1.0225^8
        1.194831142
```

Calculator Exercises

Use a calculator to evaluate the expressions.

1. $(1.06)^5$ **2.** $(1.02)^{40}$ **3.** $5000(1.06)^5$

4. $2000(1.02)^{40}$ **5.** $3000(1 + 0.06)^2$ **6.** $1000(1 + 0.05)^3$

Section 5.1 Practice Exercises

For this exercise set, assume all variables represent nonzero real numbers.

Study Skills Exercise

1. Define the key terms:

    a. exponent   **b. base**   **c. simple interest**   **d. compound interest**

Objective 1: Review of Exponential Notation

For Exercises 2–9, identify the base and the exponent. **(See Example 1.)**

2. c^3 Base: c; exponent: 3

3. x^4 Base: x; exponent: 4

4. 5^2 Base: 5; exponent: 2

5. 3^5 Base: 3; exponent: 5

6. $(-4)^8$ Base: -4; exponent: 8

7. $(-1)^4$ Base: -1; exponent: 4

8. x Base: x; exponent: 1

9. q Base: q; exponent: 1

10. What base corresponds to the exponent 5 in the expression $x^3 y^5 z^2$? y

11. What base corresponds to the exponent 2 in the expression $w^3 v^2$? v

12. What is the exponent for the factor of 2 in the expression $2x^3$?   1

13. What is the exponent for the factor of p in the expression pq^7?   1

For Exercises 14–22, write the expression using exponents.

14. $(4n)(4n)(4n)$ $(4n)^3$

15. $(-6b)(-6b)$ $(-6b)^2$

16. $4 \cdot n \cdot n \cdot n$ $4n^3$

17. $-6 \cdot b \cdot b$ $-6b^2$

18. $(x-5)(x-5)(x-5)$ $(x-5)^3$

19. $(y+2)(y+2)(y+2)(y+2)$ $(y+2)^4$

20. $\dfrac{4}{x \cdot x \cdot x \cdot x \cdot x}$ $\dfrac{4}{x^5}$

21. $\dfrac{-2}{t \cdot t \cdot t}$ $\dfrac{-2}{t^3}$

22. $\dfrac{5 \cdot x \cdot x \cdot x}{(y-7)(y-7)}$ $\dfrac{5x^3}{(y-7)^2}$

Objective 2: Evaluating Expressions with Exponents

For Exercises 23–30, evaluate the two expressions and compare the answers. Do the expressions have the same value? **(See Example 2.)**

23. -5^2 and $(-5)^2$
No; $-5^2 = -25$ and $(-5)^2 = 25$

24. -3^4 and $(-3)^4$
No; $-3^4 = -81$ and $(-3)^4 = 81$

25. -2^5 and $(-2)^5$
Yes; $-2^5 = -32$ and $(-2)^5 = -32$

26. -5^3 and $(-5)^3$
Yes; $-5^3 = -125$ and $(-5)^3 = -125$

27. $\left(\dfrac{1}{2}\right)^3$ and $\dfrac{1}{2^3}$
Yes; $\left(\dfrac{1}{2}\right)^3 = \dfrac{1}{8}$ and $\dfrac{1}{2^3} = \dfrac{1}{8}$

28. $\left(\dfrac{1}{5}\right)^2$ and $\dfrac{1}{5^2}$
Yes; $\left(\dfrac{1}{5}\right)^2 = \dfrac{1}{25}$ and $\dfrac{1}{5^2} = \dfrac{1}{25}$

29. $\left(\dfrac{3}{10}\right)^2$ and $(0.3)^2$
Yes; $\left(\dfrac{3}{10}\right)^2 = \dfrac{9}{100}$ and $(0.3)^2 = 0.09$

30. $\left(\dfrac{7}{10}\right)^3$ and $(0.7)^3$
Yes; $\left(\dfrac{7}{10}\right)^3 = \dfrac{343}{1000}$ and $(0.7)^3 = 0.343$

For Exercises 31–38, evaluate the expressions.

31. 16^1 16

32. 20^1 20

33. $(-1)^{21}$ -1

34. $(-1)^{30}$ 1

35. $\left(-\dfrac{1}{3}\right)^2$ $\dfrac{1}{9}$

36. $\left(-\dfrac{1}{4}\right)^3$ $-\dfrac{1}{64}$

37. $-\left(\dfrac{2}{5}\right)^2$ $-\dfrac{4}{25}$

38. $-\left(\dfrac{3}{5}\right)^2$ $-\dfrac{9}{25}$

For Exercises 39–46, simplify using the order of operations.

39. $3 \cdot 2^4$ 48
40. $2 \cdot 0^5$ 0
41. $-4(-1)^7$ 4
42. $-3(-1)^4$ -3

43. $6^2 - 3^3$ 9
44. $4^3 + 2^3$ 72
45. $2 \cdot 3^2 + 4 \cdot 2^3$ 50
46. $6^2 - 3 \cdot 1^3$ 33

For Exercises 47–58, evaluate each expression for $a = -4$ and $b = 5$. **(See Example 3.)**

47. $-4b^2$ -100
48. $3a^2$ 48
49. $(-4b)^2$ 400
50. $(3a)^2$ 144

51. $(a + b)^2$ 1
52. $(a - b)^2$ 81
53. $a^2 + 2ab + b^2$ 1
54. $a^2 - 2ab + b^2$ 81

55. $-10ab^2$ 1000
56. $-6a^3b$ 1920
57. $-10a^2b$ -800
58. $-a^2b$ -80

Objective 3: Multiplying and Dividing Common Bases

59. Expand the following expressions first. Then simplify using exponents.

a. $x^4 \cdot x^3$
$(x \cdot x \cdot x \cdot x)(x \cdot x \cdot x) = x^7$
b. $5^4 \cdot 5^3$
$(5 \cdot 5 \cdot 5 \cdot 5)(5 \cdot 5 \cdot 5) = 5^7$

60. Expand the following expressions first. Then simplify using exponents.

a. $y^2 \cdot y^4$
$(y \cdot y)(y \cdot y \cdot y \cdot y) = y^6$
b. $3^2 \cdot 3^4$
$(3 \cdot 3)(3 \cdot 3 \cdot 3 \cdot 3) = 3^6$

For Exercises 61–72, simplify the expressions. Write the answers in exponent form. **(See Example 4.)**

61. z^5z^3 z^8
62. w^4w^7 w^{11}
63. $a \cdot a^8$ a^9
64. p^4p p^5

65. $4^5 \cdot 4^9$ 4^{14}
66. $6^7 \cdot 6^5$ 6^{12}
67. $\left(\frac{2}{3}\right)^3\left(\frac{2}{3}\right)$ $\left(\frac{2}{3}\right)^4$
68. $\left(\frac{1}{x}\right)\left(\frac{1}{x}\right)^2$ $\left(\frac{1}{x}\right)^3$

69. $c^5c^2c^7$ c^{14}
70. $b^7b^2b^8$ b^{17}
71. $x \cdot x^4 \cdot x^{10} \cdot x^3$ x^{18}
72. $z^7 \cdot z^{11} \cdot z^{60} \cdot z$ z^{79}

73. Expand the expressions. Then simplify.

a. $\dfrac{p^8}{p^3}$
$\dfrac{p \cdot p \cdot p \cdot p \cdot p \cdot p \cdot p \cdot p}{p \cdot p \cdot p} = p^5$
b. $\dfrac{8^8}{8^3}$
$\dfrac{8 \cdot 8 \cdot 8 \cdot 8 \cdot 8 \cdot 8 \cdot 8 \cdot 8}{8 \cdot 8 \cdot 8} = 8^5$

74. Expand the expressions. Then simplify.

a. $\dfrac{w^5}{w^2}$
$\dfrac{w \cdot w \cdot w \cdot w \cdot w}{w \cdot w} = w^3$
b. $\dfrac{4^5}{4^2}$
$\dfrac{4 \cdot 4 \cdot 4 \cdot 4 \cdot 4}{4 \cdot 4} = 4^3$

For Exercises 75–90, simplify the expressions. Write the answers in exponent form. **(See Examples 5–6.)**

75. $\dfrac{x^8}{x^6}$ x^2
76. $\dfrac{z^5}{z^4}$ z
77. $\dfrac{a^{10}}{a}$ a^9
78. $\dfrac{b^{12}}{b}$ b^{11}

79. $\dfrac{7^{13}}{7^6}$ 7^7
80. $\dfrac{2^6}{2^4}$ 2^2
81. $\dfrac{5^8}{5}$ 5^7
82. $\dfrac{3^5}{3}$ 3^4

83. $\dfrac{y^{13}}{y^{12}}$ y
84. $\dfrac{w^7}{w^6}$ w
85. $\dfrac{h^3h^8}{h^7}$ h^4
86. $\dfrac{n^5n^4}{n^2}$ n^7

87. $\dfrac{7^2 \cdot 7^6}{7}$ 7^7
88. $\dfrac{5^3 \cdot 5^8}{5}$ 5^{10}
89. $\dfrac{10^{20}}{10^3 \cdot 10^8}$ 10^9
90. $\dfrac{3^{15}}{3^2 \cdot 3^{10}}$ 3^3

Objective 4: Simplifying Expressions with Exponents (Mixed Exercises)

For Exercises 91–112, use the commutative and associative properties of real numbers and the properties of exponents to simplify the expressions. **(See Example 7.)**

91. $(2x^3)(3x^4)$ $6x^7$
92. $(10y)(2y^3)$ $20y^4$
93. $(5a^2b)(8a^3b^4)$ $40a^5b^5$
94. $(10xy^3)(3x^4y)$ $30x^5y^4$

95. $(r^6s^4)(13r^2s)$
$13r^8s^5$
96. $(6p^2q^8)(7p^5q^3)$
$42p^7q^{11}$
97. $s^3 \cdot t^5 \cdot t \cdot t^{10} \cdot s^6$
s^9t^{16}
98. $c \cdot c^4 \cdot d^2 \cdot c^3 \cdot d^3$
c^8d^5

99. $(-2v^2)(3v)(5v^5)$
$-30v^8$

100. $(10q^5)(-3q^8)(q)$
$-30q^{14}$

101. $\left(\dfrac{2}{3}m^{13}n^8\right)(24m^7n^2)$
$16m^{20}n^{10}$

102. $\left(\dfrac{1}{4}c^6d^6\right)(28c^2d^7)$
$7c^8d^{13}$

103. $\dfrac{14c^4d^5}{7c^3d}$ $2cd^4$

104. $\dfrac{36h^5k^2}{9h^3k}$ $4h^2k$

105. $\dfrac{z^3z^{11}}{z^4z^6}$ z^4

106. $\dfrac{w^{12}w^2}{w^4w^5}$ w^5

107. $\dfrac{25h^3jk^5}{12h^2k}$ $\dfrac{25hjk^4}{12}$

108. $\dfrac{15m^5np^{12}}{4mp^9}$ $\dfrac{15m^4np^3}{4}$

109. $(-4p^6q^8r^4)(2pqr^2)$
$-8p^7q^9r^6$

110. $(-5a^4bc)(-10a^2b)$
$50a^6b^2c$

111. $\dfrac{-12s^2tu^3}{4su^2}$ $-3stu$

112. $\dfrac{15w^5x^{10}y^3}{-15w^4x}$ $-wx^9y^3$

Objective 5: Applications of Exponents

 Use the formula $A = P(1 + r)^t$ for Exercises 113–116. **(See Example 8.)**

113. Find the amount in an account after 2 yr if the initial investment is \$5000, invested at 7% interest compounded annually.   \$5724.50

114. Find the amount in an account after 5 yr if the initial investment is \$2000, invested at 4% interest compounded annually.   \$2433.31

115. Find the amount in an account after 3 yr if the initial investment is \$4000, invested at 6% interest compounded annually.   \$4764.06

116. Find the amount in an account after 4 yr if the initial investment is \$10,000, invested at 5% interest compounded annually.   \$12,155.06

For Exercises 117–120, use the geometry formulas found in Section R.4.

117. Find the area of the pizza shown in the figure. Round to the nearest square inch.   201 in.^{2}

16 in.

118. Find the volume of the sphere shown in the figure. Round to the nearest cubic centimeter.
113 cm^3

$r = 3$ cm

119. Find the volume of a spherical balloon that is 8 in. in diameter. Round to the nearest cubic inch.
268 in.^{3}

120. Find the area of a circular pool 50 ft in diameter. Round to the nearest square foot.
1963 ft^2

Expanding Your Skills

For Exercises 121–128, simplify the expressions using the addition or subtraction rules of exponents. Assume that a, b, m, and n represent positive integers.

121. x^nx^{n+1} x^{2n+1}

122. y^ay^{2a} y^{3a}

123. $p^{3m+5}p^{-m-2}$ p^{2m+3}

124. $q^{4b-3}q^{-4b+4}$ q

125. $\dfrac{z^{b+1}}{z^b}$ z

126. $\dfrac{w^{5n+3}}{w^{2n}}$ w^{3n+3}

127. $\dfrac{r^{3a+3}}{r^{3a}}$ r^3

128. $\dfrac{t^{3+2m}}{t^{2m}}$ t^3

Section 5.2 More Properties of Exponents

Objectives

1. Power Rule for Exponents
2. The Properties
   $(ab)^m = a^m b^m$ and
   $\left(\dfrac{a}{b}\right)^m = \dfrac{a^m}{b^m}$

1. Power Rule for Exponents

The expression $(x^2)^3$ indicates that the quantity x^2 is cubed.

$$(x^2)^3 = (x^2)(x^2)(x^2) = (x \cdot x)(x \cdot x)(x \cdot x) = x^6$$

From this example, it appears that to raise a base to successive powers, we multiply the exponents and leave the base unchanged. This is stated formally as the power rule for exponents.

> **PROPERTY Power Rule for Exponents**
>
> Assume that b is a real number and that m and n represent positive integers. Then,
>
> $$(b^m)^n = b^{m \cdot n}$$

Skill Practice

Simplify the expressions.

1. $(y^3)^5$
2. $(2^8)^{10}$
3. $(q^5 q^4)^3$

Classroom Examples: p. 373, Exercises 12 and 22

Example 1 Simplifying Expressions with Exponents

Simplify the expressions.

a. $(s^4)^2$ **b.** $(3^4)^2$ **c.** $(x^2 x^5)^4$

Solution:

a. $(s^4)^2$

$= s^{4 \cdot 2}$ Multiply exponents (the base is unchanged).

$= s^8$

b. $(3^4)^2$

$= 3^{4 \cdot 2}$ Multiply exponents (the base is unchanged).

$= 3^8$ or 6561

c. $(x^2 x^5)^4$

$= (x^7)^4$ Simplify inside the parentheses by adding exponents.

$= x^{7 \cdot 4}$ Multiply exponents (the base is unchanged).

$= x^{28}$

2. The Properties $(ab)^m = a^m b^m$ and $\left(\frac{a}{b}\right)^m = \frac{a^m}{b^m}$

Consider the following expressions and their simplified forms:

$$(xy)^3 = (xy)(xy)(xy) = (x \cdot x \cdot x)(y \cdot y \cdot y) = x^3 y^3$$

$$\left(\frac{x}{y}\right)^3 = \left(\frac{x}{y}\right)\left(\frac{x}{y}\right)\left(\frac{x}{y}\right) = \left(\frac{x \cdot x \cdot x}{y \cdot y \cdot y}\right) = \frac{x^3}{y^3}$$

The expressions were simplified using the commutative and associative properties of multiplication. The simplified forms for each expression could have been reached in one step by applying the exponent to each factor inside the parentheses.

Answers

1. y^{15} **2.** 2^{80} **3.** q^{27}

PROPERTY Power of a Product and Power of a Quotient

Assume that a and b are real numbers. Let m represent a positive integer. Then,

$$(ab)^m = a^m b^m$$

$$\left(\frac{a}{b}\right)^m = \frac{a^m}{b^m}, \quad b \neq 0$$

Avoiding Mistakes

The power rule of exponents can be applied to a product of bases but in general cannot be applied to a sum or difference of bases.

$$(ab)^n = a^n b^n$$

but $\quad (a + b)^n \neq a^n + b^n$

Applying these properties of exponents, we have

$$(xy)^3 = x^3 y^3 \quad \text{and} \quad \left(\frac{x}{y}\right)^3 = \frac{x^3}{y^3}$$

Example 2 Simplifying Expressions with Exponents

Simplify the expressions.

a. $(-2xyz)^4$ 　　**b.** $(5x^2y^7)^3$ 　　**c.** $\left(\frac{2}{5}\right)^3$ 　　**d.** $\left(\frac{1}{3xy^4}\right)^2$

Skill Practice

Simplify the expressions.

4. $(3abc)^5$ 　　**5.** $(-2t^2w^4)^3$

6. $\left(\frac{3}{4}\right)^3$ 　　**7.** $\left(\frac{2x^3}{y^5}\right)^2$

Classroom Examples: p. 374, Exercises 28 and 32

Solution:

a. $(-2xyz)^4$

$= (-2)^4 x^4 y^4 z^4$ 　　Raise each factor within parentheses to the fourth power.

$= 16x^4 y^4 z^4$

b. $(5x^2 y^7)^3$

$= 5^3 (x^2)^3 (y^7)^3$ 　　Raise each factor within parentheses to the third power.

$= 125 x^6 y^{21}$ 　　Multiply exponents and simplify.

Instructor Note: Remind students that they can write exponents of 1 where appropriate to avoid confusion. For example $(5x^2y^7)^3 = (5^1x^2y^7)^3$.

c. $\left(\frac{2}{5}\right)^3$

$= \frac{(2)^3}{(5)^3}$ 　　Apply the exponent to each factor in parentheses.

$= \frac{8}{125}$ 　　Simplify.

d. $\left(\frac{1}{3xy^4}\right)^2$

$= \frac{1^2}{3^2 x^2 (y^4)^2}$ 　　Square each factor within parentheses.

$= \frac{1}{9x^2 y^8}$ 　　Multiply exponents and simplify.

Answers

4. $3^5 a^5 b^5 c^5$ or $243 a^5 b^5 c^5$

5. $-8t^6 w^{12}$ 　**6.** $\frac{27}{64}$ 　**7.** $\frac{4x^6}{y^{10}}$

The properties of exponents can be used along with the properties of real numbers to simplify complicated expressions.

Example 3   Simplifying Expressions with Exponents

Simplify the expression. $\dfrac{(x^2)^6(x^3)}{(x^7)^2}$

Solution:

$\dfrac{(x^2)^6(x^3)}{(x^7)^2}$ Clear parentheses by applying the power rule.

$= \dfrac{x^{2\cdot6}x^3}{x^{7\cdot2}}$ Multiply exponents.

$= \dfrac{x^{12}x^3}{x^{14}}$

$= \dfrac{x^{12+3}}{x^{14}}$ Add exponents in the numerator.

$= \dfrac{x^{15}}{x^{14}}$

$= x^{15-14}$ Subtract exponents.

$= x$ Simplify.

Example 4   Simplifying Expressions with Exponents

Simplify the expression. $(3cd^2)(2cd^3)^3$

Solution:

$(3cd^2)(2cd^3)^3$ Clear parentheses by applying the power rule.

$= 3cd^2 \cdot 2^3 c^3 d^9$ Raise each factor in the second parentheses to the third power.

$= 3 \cdot 2^3 cc^3 d^2 d^9$ Group like factors.

$= 3 \cdot 8c^{1+3} d^{2+9}$ Add exponents on like bases.

$= 24c^4 d^{11}$ Simplify.

Answers
8. k^{10} **9.** $64x^{16} y^{17}$

Example 5 **Simplifying Expressions with Exponents**

Simplify the expression. $\left(\dfrac{x^7yz^4}{8xz^3}\right)^2$

Solution:

$\left(\dfrac{x^7yz^4}{8xz^3}\right)^2$

$=\left(\dfrac{x^{7-1}yz^{4-3}}{8}\right)^2$ First simplify inside the parentheses by subtracting exponents on like bases.

$=\left(\dfrac{x^6yz}{8}\right)^2$

$=\dfrac{(x^6)^2y^2z^2}{8^2}$ Apply the power rule of exponents.

$=\dfrac{x^{12}y^2z^2}{64}$

Skill Practice

Simplify the expression.

10. $\left(\dfrac{w^2xy^4}{6xy^3}\right)^2$

Classroom Example: p. 374, Exercise 70

Answer

10. $\dfrac{w^4y^2}{36}$

Section 5.2 Practice Exercises

Boost *your* GRADE at ALEKS.com!

ALEKS version 3.0

- Practice Problems
- Self-Tests
- NetTutor
- e-Professors
- Videos

For additional exercises, see Classroom Activity 5.2A in the *Instructor's Resource Manual* at www.mhhe.com/moh.

For this exercise set assume all variables represent nonzero real numbers.

Review Exercises

For Exercises 1–8, simplify.

1. $4^2 \cdot 4^7$ 4^9

2. $5^8 \cdot 5^3 \cdot 5$ 5^{12}

3. $a^{13} \cdot a \cdot a^6$ a^{20}

4. $y^{14}y^3$ y^{17}

5. $\dfrac{d^{13}d}{d^5}$ d^9

6. $\dfrac{3^8 \cdot 3}{3^2}$ 3^7

7. $\dfrac{7^{11}}{7^5}$ 7^6

8. $\dfrac{z^4}{z^3}$ z

 **9.** Explain when to add exponents versus when to subtract exponents.   When multiplying expressions with the same base, add the exponents. When dividing expressions with the same base, subtract the exponents.

 **10.** Explain when to add exponents versus when to multiply exponents.   When multiplying expressions with the same base, add the exponents. When raising an expression with an exponent to a power, multiply the exponents.

Objective 1: Power Rule for Exponents

For Exercises 11–22, simplify and write answers in exponent form. **(See Example 1.)**

11. $(5^3)^4$ 5^{12}

12. $(2^8)^7$ 2^{56}

13. $(12^3)^2$ 12^6

14. $(6^4)^4$ 6^{16}

15. $(y^7)^2$ y^{14}

16. $(z^6)^4$ z^{24}

17. $(w^5)^5$ w^{25}

18. $(t^3)^6$ t^{18}

19. $(a^2a^4)^6$ a^{36}

20. $(z \cdot z^3)^2$ z^8

21. $(y^3y^4)^2$ y^{14}

22. $(w^5w)^4$ w^{24}

 Writing   Translating Expression   Geometry   Scientific Calculator   Video   NS E

23. Evaluate the two expressions and compare the answers: $(2^2)^3$ and $(2^3)^2$. They are both equal to 2^6.

24. Evaluate the two expressions and compare the answers: $(4^4)^2$ and $(4^2)^4$. They are both equal to 4^8.

25. Evaluate the two expressions and compare the answers. Which expression is greater? Why?

$$2^{(2^4)} \quad \text{and} \quad (2^2)^4$$

$2^{(2^4)} = 2^{16}$ $(2^2)^4 = 2^8$; the expression $2^{(2^4)}$ is greater than $(2^2)^4$.

26. Evaluate the two expressions and compare the answers. Which expression is greater? Why?

$$3^{(2^4)} \quad \text{and} \quad (3^2)^4$$

$3^{(2^4)} = 3^{16}$ $(3^2)^4 = 3^8$; the expression $3^{(2^4)}$ is greater than $(3^2)^4$.

Objective 2: The Properties $(ab)^m = a^m b^m$ and $\left(\frac{a}{b}\right)^m = \frac{a^m}{b^m}$

For Exercises 27–42, use the appropriate property to clear the parentheses. **(See Example 2.)**

27. $(5w)^2$ $25w^2$

28. $(4y)^3$ $64y^3$

29. $(srt)^4$ $s^4 r^4 t^4$

30. $(wxy)^6$ $w^6 x^6 y^6$

31. $\left(\frac{2}{r}\right)^4$ $\frac{16}{r^4}$

32. $\left(\frac{1}{t}\right)^8$ $\frac{1}{t^8}$

33. $\left(\frac{x}{y}\right)^5$ $\frac{x^5}{y^5}$

34. $\left(\frac{w}{z}\right)^7$ $\frac{w^7}{z^7}$

35. $(-3a)^4$ $81a^4$

36. $(2x)^5$ $32x^5$

37. $(-3abc)^3$ $-27a^3 b^3 c^3$

38. $(-5xyz)^2$ $25x^2 y^2 z^2$

39. $\left(-\frac{4}{x}\right)^3$ $-\frac{64}{x^3}$

40. $\left(-\frac{1}{w}\right)^4$ $\frac{1}{w^4}$

41. $\left(-\frac{a}{b}\right)^2$ $\frac{a^2}{b^2}$

42. $\left(-\frac{r}{s}\right)^3$ $-\frac{r^3}{s^3}$

Mixed Exercises

For Exercises 43–74, simplify the expressions. **(See Examples 3–5.)**

43. $(6u^2 v^4)^3$ $6^3 u^6 v^{12}$ or $216\, u^6 v^{12}$

44. $(3a^5 b^2)^6$ $3^6 a^{30} b^{12}$ or $729 a^{30} b^{12}$

45. $5(x^2 y)^4$ $5x^8 y^4$

46. $18(u^3 v^4)^2$ $18u^6 v^8$

47. $(-h^4)^7$ $-h^{28}$

48. $(-k^6)^3$ $-k^{18}$

49. $(-m^2)^6$ m^{12}

50. $(-n^3)^8$ n^{24}

51. $\left(\frac{4}{rs^4}\right)^5$ $\frac{4^5}{r^5 s^{20}}$ or $\frac{1024}{r^5 s^{20}}$

52. $\left(\frac{2}{h^7 k}\right)^3$ $\frac{8}{h^{21} k^3}$

53. $\left(\frac{3p}{q^3}\right)^5$ $\frac{3^5 p^5}{q^{15}}$ or $\frac{243 p^5}{q^{15}}$

54. $\left(\frac{5x^2}{y^3}\right)^4$ $\frac{5^4 x^8}{y^{12}}$ or $\frac{625 x^8}{y^{12}}$

55. $\frac{y^8 (y^3)^4}{(y^2)^3}$ y^{14}

56. $\frac{(w^3)^2 (w^4)^5}{(w^4)^2}$ w^{18}

57. $(x^2)^5 (x^3)^7$ x^{31}

58. $(y^3)^4 (y^2)^5$ y^{22}

59. $(2a^2 b)^3 (5a^4 b^3)^2$ $200a^{14} b^9$

60. $(4c^3 d^5)^2 (3cd^3)^2$ $144c^8 d^{16}$

61. $(-2p^2 q^4)^4$ $16p^8 q^{16}$

62. $(-7x^4 y^5)^2$ $49x^8 y^{10}$

63. $(-m^7 n^3)^5$ $-m^{35} n^{15}$

64. $(-a^3 b^6)^7$ $-a^{21} b^{42}$

65. $\frac{(5a^3 b)^4 (a^2 b)^4}{(5ab)^2}$ $25a^{18} b^6$

66. $\frac{(6s^3)^2 (s^4 t^5)^2}{(3s^4 t^2)^2}$ $4s^6 t^6$

67. $\left(\frac{2c^3 d^4}{3c^2 d}\right)^2$ $\frac{4c^2 d^6}{9}$

68. $\left(\frac{x^3 y^5 z}{5xy^2}\right)^2$ $\frac{x^4 y^6 z^2}{25}$

69. $(2c^3 d^2)^5 \left(\frac{c^6 d^8}{4c^2 d}\right)^3$ $\frac{c^{27} d^{31}}{2}$

70. $\left(\frac{s^5 t^6}{2s^2 t}\right)^2 (10s^3 t^3)^2$ $25s^{12} t^{16}$

71. $\left(\frac{-3a^3 b}{c^2}\right)^3$ $-\frac{27a^9 b^3}{c^6}$

72. $\left(\frac{-4x^2}{y^4 z}\right)^3$ $-\frac{64x^6}{y^{12} z^3}$

73. $\frac{(-8b^6)^2 (b^3)^5}{4b}$ $16b^{26}$

74. $\frac{(-6a^2)^2 (a^3)^4}{9a}$ $4a^{15}$

Expanding Your Skills

For Exercises 75–82, simplify the expressions using the addition or subtraction properties of exponents. Assume that a, b, m, and n represent positive integers.

75. $(x^m)^2$ x^{2m}

76. $(y^3)^n$ y^{3n}

77. $(5a^{2n})^3$ $125a^{6n}$

78. $(3b^4)^m$ $3^m b^{4m}$

79. $\left(\frac{m^2}{n^3}\right)^b$ $\frac{m^{2b}}{n^{3b}}$

80. $\left(\frac{x^5}{y^3}\right)^m$ $\frac{x^{5m}}{y^{3m}}$

81. $\left(\frac{3a^3}{5b^4}\right)^n$ $\frac{3^n a^{3n}}{5^n b^{4n}}$

82. $\left(\frac{4m^6}{3n^2}\right)^b$ $\frac{4^b m^{6b}}{3^b n^{2b}}$

Writing ⟷ *Translating Expression* ◥ *Geometry* 🖩 *Scientific Calculator* 💿 *Video* NS&E

Definitions of b^0 and b^{-n}

In Sections 5.1 and 5.2, we learned several rules that enable us to manipulate expressions containing *positive* integer exponents. In this section, we present definitions that can be used to simplify expressions with negative exponents or with an exponent of zero.

Objectives

1. **Definition of b^0**
2. **Definition of b^{-n}**
3. **Properties of Integer Exponents: A Summary**

1. Definition of b^0

To begin, consider the following pattern.

$3^3 = 27$ Divide by 3.
$3^2 = 9$ Divide by 3.
$3^1 = 3$ Divide by 3.
$3^0 = 1$

As the exponents decrease by 1, the resulting expressions are divided by 3.

For the pattern to continue, we define $3^0 = 1$.

This pattern suggests that we should define an expression with a zero exponent as follows.

> **DEFINITION** Definition of b^0
>
> Let b be a nonzero real number. Then, $b^0 = 1$.

Avoiding Mistakes

$b^0 = 1$ provided that b is not zero. Therefore, the expression 0^0 cannot be simplified by this rule.

Example 1   Simplifying Expressions with a Zero Exponent

Simplify (assume $z \neq 0$).

a. 4^0 **b.** $(-4)^0$ **c.** -4^0

d. z^0 **e.** $-4z^0$ **f.** $(4z)^0$

Solution:

a. $4^0 = 1$ By definition

b. $(-4)^0 = 1$ By definition

c. $-4^0 = -1 \cdot 4^0 = -1 \cdot 1 = -1$ The exponent 0 applies only to 4.

d. $z^0 = 1$ By definition

e. $-4z^0 = -4 \cdot z^0 = -4 \cdot 1 = -4$ The exponent 0 applies only to z.

f. $(4z)^0 = 1$ The parentheses indicate that the exponent, 0, applies to both factors 4 and z.

Skill Practice

Evaluate the expressions. Assume all variables represent nonzero real numbers.

  1. 7^0 **2.** $(-7)^0$
  3. -5^0 **4.** y^0
  5. $-2x^0$ **6.** $(2x)^0$

Classroom Examples: p. 381, Exercises 14, 18, and 24

The definition of b^0 is consistent with the other properties of exponents learned thus far. For example, we know that $1 = \frac{5^3}{5^3}$. If we subtract exponents, the result is 5^0.

Subtract exponents.

$$1 = \frac{5^3}{5^3} = 5^{3-3} = 5^0.$$ Therefore, 5^0 must be defined as 1.

Answers

1. 1   **2.** 1   **3.** -1
4. 1   **5.** -2 **6.** 1

2. Definition of b^{-n}

To understand the concept of a *negative* exponent, consider the following pattern.

$3^3 = 27$ — Divide by 3.
$3^2 = 9$ — Divide by 3.
$3^1 = 3$ — Divide by 3. As the exponents decrease by 1, the resulting expressions are divided by 3.
$3^0 = 1$ — Divide by 3.
Divide by 3.

$3^{-1} = \dfrac{1}{3}$ ⟵ For the pattern to continue, we define $3^{-1} = \dfrac{1}{3^1} = \dfrac{1}{3}$.

$3^{-2} = \dfrac{1}{9}$ ⟵ For the pattern to continue, we define $3^{-2} = \dfrac{1}{3^2} = \dfrac{1}{9}$.

$3^{-3} = \dfrac{1}{27}$ ⟵ For the pattern to continue, we define $3^{-3} = \dfrac{1}{3^3} = \dfrac{1}{27}$.

This pattern suggests that $3^{-n} = \frac{1}{3^n}$ for all integers, n. In general, we have the following definition involving negative exponents.

> **DEFINITION** Definition of b^{-n}
>
> Let n be an integer and b be a nonzero real number. Then,
>
> $$b^{-n} = \left(\dfrac{1}{b}\right)^n \quad \text{or} \quad \dfrac{1}{b^n}$$

The definition of b^{-n} implies that to evaluate b^{-n}, take the reciprocal of the base and change the sign of the exponent.

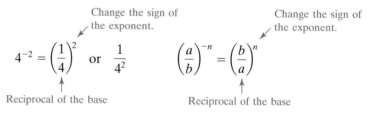

$$4^{-2} = \left(\dfrac{1}{4}\right)^2 \quad \text{or} \quad \dfrac{1}{4^2} \qquad \left(\dfrac{a}{b}\right)^{-n} = \left(\dfrac{b}{a}\right)^n$$

Change the sign of the exponent.   Change the sign of the exponent.
Reciprocal of the base   Reciprocal of the base

Classroom Example: p. 382, Exercise 32

Skill Practice

Simplify.

7. p^{-4} **8.** 3^{-3}

9. $(-5)^{-2}$

Avoiding Mistakes

A negative exponent does *not* affect the sign of the base.

Answers

7. $\dfrac{1}{p^4}$ **8.** $\dfrac{1}{3^3}$ or $\dfrac{1}{27}$

9. $\dfrac{1}{(-5)^2}$ or $\dfrac{1}{25}$

Example 2 Simplifying Expressions with Negative Exponents

Simplify. Assume that $c \neq 0$.   **a.** c^{-3} **b.** 5^{-1} **c.** $(-3)^{-4}$

Solution:

a. $c^{-3} = \dfrac{1}{c^3}$ By definition

b. $5^{-1} = \dfrac{1}{5^1}$ By definition

      $= \dfrac{1}{5}$ Simplify.

c. $(-3)^{-4} = \dfrac{1}{(-3)^4}$ The base is -3 and must be enclosed in parentheses.

      $= \dfrac{1}{81}$ Simplify. Note that $(-3)^4 = (-3)(-3)(-3)(-3) = 81$.

Example 3 Simplifying Expressions with Negative Exponents

Simplify. Assume that $y \neq 0$. **a.** $\left(\dfrac{1}{4}\right)^{-2}$ **b.** $\left(-\dfrac{3}{5}\right)^{-3}$ **c.** $\dfrac{1}{y^{-5}}$

Solution:

a. $\left(\dfrac{1}{4}\right)^{-2} = 4^2$ Take the reciprocal of the base, and change the sign of the exponent.

$= 16$ Simplify.

b. $\left(-\dfrac{3}{5}\right)^{-3} = \left(-\dfrac{5}{3}\right)^{3}$ Take the reciprocal of the base, and change the sign of the exponent.

$= -\dfrac{125}{27}$ Simplify.

c. $\dfrac{1}{y^{-5}} = \left(\dfrac{1}{y}\right)^{-5}$ Apply the power of a quotient rule from Section 5.2.

$= (y)^5$ Take the reciprocal of the base, and change the sign of the exponent.

$= y^5$

Skill Practice

Simplify. Assume $w \neq 0$.

10. $\left(\dfrac{1}{3}\right)^{-1}$

11. $\left(-\dfrac{2}{5}\right)^{-2}$

12. $\dfrac{1}{w^{-7}}$

Classroom Examples: p. 382, Exercises 30 and 46

Example 4 Simplifying Expressions with Negative Exponents

Simplify. Assume that $x \neq 0$. **a.** $(5x)^{-3}$ **b.** $5x^{-3}$ **c.** $-5x^{-3}$

Solution:

a. $(5x)^{-3} = \left(\dfrac{1}{5x}\right)^{3}$ Take the reciprocal of the base, and change the sign of the exponent.

$= \dfrac{(1)^3}{(5x)^3}$ Apply the exponent of 3 to each factor within parentheses.

$= \dfrac{1}{125x^3}$ Simplify.

b. $5x^{-3} = 5 \cdot x^{-3}$ Note that the exponent, -3, applies only to x.

$= 5 \cdot \dfrac{1}{x^3}$ Rewrite x^{-3} as $\dfrac{1}{x^3}$.

$= \dfrac{5}{x^3}$ Multiply.

c. $-5x^{-3} = -5 \cdot x^{-3}$ Note that the exponent, -3, applies only to x, and that -5 is a coefficient.

$= -5 \cdot \dfrac{1}{x^3}$ Rewrite x^{-3} as $\dfrac{1}{x^3}$.

$= -\dfrac{5}{x^3}$ Multiply.

Skill Practice

Simplify. Assume $w \neq 0$.

13. $(2w)^{-4}$

14. $2w^{-4}$

15. $-2w^{-4}$

Classroom Examples: p. 382, Exercises 36, 38, and 42

Answers

10. 3   **11.** $\dfrac{25}{4}$ **12.** w^7

13. $\dfrac{1}{16w^4}$ **14.** $\dfrac{2}{w^4}$ **15.** $-\dfrac{2}{w^4}$

It is important to note that the definition of b^{-n} is consistent with the other properties of exponents learned thus far. For example, consider the expression

$$\frac{x^4}{x^7} = \frac{\cancel{x} \cdot \cancel{x} \cdot \cancel{x} \cdot \cancel{x}}{\cancel{x} \cdot \cancel{x} \cdot \cancel{x} \cdot \cancel{x} \cdot x \cdot x \cdot x} = \frac{1}{x^3}$$

Subtract exponents.

Hence, $x^{-3} = \dfrac{1}{x^3}$

By subtracting exponents, we have $\quad \dfrac{x^4}{x^7} = x^{4-7} = x^{-3}$

3. Properties of Integer Exponents: A Summary

The definitions of b^0 and b^{-n} allow us to extend the properties of exponents learned in Sections 5.1 and 5.2 to include integer exponents. These are summarized in Table 5-1.

Table 5-1

Properties of Integer Exponents		
Assume that a and b are real numbers ($b \neq 0$) and that m and n represent integers.		
Property	**Example**	**Details/Notes**
Multiplication of Like Bases $b^m b^n = b^{m+n}$	$b^2 b^4 = b^{2+4} = b^6$	$b^2 b^4 = (b \cdot b)(b \cdot b \cdot b \cdot b) = b^6$
Division of Like Bases $\dfrac{b^m}{b^n} = b^{m-n}$	$\dfrac{b^5}{b^2} = b^{5-2} = b^3$	$\dfrac{b^5}{b^2} = \dfrac{\cancel{b} \cdot \cancel{b} \cdot b \cdot b \cdot b}{\cancel{b} \cdot \cancel{b}} = b^3$
The Power Rule $(b^m)^n = b^{m \cdot n}$	$(b^4)^2 = b^{4 \cdot 2} = b^8$	$(b^4)^2 = (b \cdot b \cdot b \cdot b)(b \cdot b \cdot b \cdot b) = b^8$
Power of a Product $(ab)^m = a^m b^m$	$(ab)^3 = a^3 b^3$	$(ab)^3 = (ab)(ab)(ab)$ $= (a \cdot a \cdot a)(b \cdot b \cdot b) = a^3 b^3$
Power of a Quotient $\left(\dfrac{a}{b}\right)^m = \dfrac{a^m}{b^m}$	$\left(\dfrac{a}{b}\right)^3 = \dfrac{a^3}{b^3}$	$\left(\dfrac{a}{b}\right)^3 = \left(\dfrac{a}{b}\right)\left(\dfrac{a}{b}\right)\left(\dfrac{a}{b}\right) = \dfrac{a \cdot a \cdot a}{b \cdot b \cdot b} = \dfrac{a^3}{b^3}$
Definitions		
Assume that b is a real number ($b \neq 0$) and that n represents an integer.		
Definition	**Example**	**Details/Notes**
$b^0 = 1$	$(4)^0 = 1$	Any nonzero quantity raised to the zero power equals 1.
$b^{-n} = \left(\dfrac{1}{b}\right)^n = \dfrac{1}{b^n}$	$b^{-5} = \left(\dfrac{1}{b}\right)^5 = \dfrac{1}{b^5}$	To simplify a negative exponent, take the reciprocal of the base and make the exponent positive.

Example 5	**Simplifying Expressions with Exponents**

Simplify the expressions. Write the answers with positive exponents only. Assume all variables are nonzero.

a. $\dfrac{a^3b^{-2}}{c^{-5}}$ **b.** $\dfrac{x^2x^{-7}}{x^3}$ **c.** $\dfrac{z^2}{w^{-4}w^4z^{-8}}$

Solution:

a. $\dfrac{a^3b^{-2}}{c^{-5}}$

$= \dfrac{a^3}{1} \cdot \dfrac{b^{-2}}{1} \cdot \dfrac{1}{c^{-5}}$

$= \dfrac{a^3}{1} \cdot \dfrac{1}{b^2} \cdot \dfrac{c^5}{1}$ Simplify negative exponents.

$= \dfrac{a^3c^5}{b^2}$ Multiply.

b. $\dfrac{x^2x^{-7}}{x^3}$

$= \dfrac{x^{2+(-7)}}{x^3}$ Add the exponents in the numerator.

$= \dfrac{x^{-5}}{x^3}$ Simplify.

$= x^{-5-3}$ Subtract the exponents.

$= x^{-8}$

$= \dfrac{1}{x^8}$ Simplify the negative exponent.

c. $\dfrac{z^2}{w^{-4}w^4z^{-8}}$

$= \dfrac{z^2}{w^{-4+4}z^{-8}}$ Add the exponents in the denominator.

$= \dfrac{z^2}{w^0z^{-8}}$

$= \dfrac{z^2}{(1)z^{-8}}$ Recall that $w^0 = 1$.

$= z^{2-(-8)}$ Subtract the exponents.

$= z^{10}$ Simplify.

Classroom Examples: pp. 382–383,
Exercises 82 and 92

Skill Practice

Simplify the expressions. Assume
all variables are nonzero.

19. $(-5x^{-2}y^3)^{-2}$

20. $\left(\dfrac{3x^{-3}y^{-2}}{4xy^{-3}}\right)^{-2}$

Example 6 **Simplifying Expressions with Exponents**

Simplify the expressions. Write the answers with positive exponents only. Assume
that all variables are nonzero.

a. $(-4ab^{-2})^{-3}$ **b.** $\left(\dfrac{2p^{-4}q^3}{5p^2q}\right)^{-2}$

Solution:

a. $(-4ab^{-2})^{-3}$

$\quad = (-4)^{-3}a^{-3}(b^{-2})^{-3}$ Apply the power rule of exponents.

$\quad = (-4)^{-3}a^{-3}b^6$

$\quad = \dfrac{1}{(-4)^3}\cdot\dfrac{1}{a^3}\cdot b^6$ Simplify the negative exponents.

$\quad = \dfrac{1}{-64}\cdot\dfrac{1}{a^3}\cdot b^6$ Simplify.

$\quad = -\dfrac{b^6}{64a^3}$ Multiply fractions.

b. $\left(\dfrac{2p^{-4}q^3}{5p^2q}\right)^{-2}$ First simplify within the parentheses.

$\quad = \left(\dfrac{2p^{-4-2}q^{3-1}}{5}\right)^{-2}$ Divide like bases by subtracting exponents.

$\quad = \left(\dfrac{2p^{-6}q^2}{5}\right)^{-2}$ Simplify.

$\quad = \dfrac{(2p^{-6}q^2)^{-2}}{(5)^{-2}}$ Apply the power rule of a quotient.

$\quad = \dfrac{2^{-2}(p^{-6})^{-2}(q^2)^{-2}}{5^{-2}}$ Apply the power rule of a product.

$\quad = \dfrac{2^{-2}p^{12}q^{-4}}{5^{-2}}$ Simplify.

$\quad = \dfrac{5^2p^{12}}{2^2q^4}$ Simplify the negative exponents.

$\quad = \dfrac{25p^{12}}{4q^4}$ Simplify.

Answers

19. $\dfrac{x^4}{25y^6}$ **20.** $\dfrac{16x^8}{9y^2}$

Example 7   Simplifying an Expression with Exponents

Simplify the expression $2^{-1} + 3^{-1} + 5^0$. Write the answer with positive exponents only.

Solution:

$2^{-1} + 3^{-1} + 5^0$

$= \dfrac{1}{2} + \dfrac{1}{3} + 1$ Simplify negative exponents. Simplify $5^0 = 1$.

$= \dfrac{3}{6} + \dfrac{2}{6} + \dfrac{6}{6}$ The least common denominator is 6.

$= \dfrac{11}{6}$ Simplify.

Skill Practice

Simplify the expressions.

21. $2^{-1} + 4^{-2} + 3^0$

Classroom Example: p. 383, Exercise 96

Answer

21. $\dfrac{25}{16}$

Section 5.3   Practice Exercises

Boost your GRADE at ALEKS.com!

ALEKS version 3.0

- Practice Problems
- Self-Tests
- NetTutor
- e-Professors
- Videos

For additional exercises, see Classroom Activities 5.3A–5.3B in the *Instructor's Resource Manual* at www.mhhe.com/moh.

For this set of exercises, assume all variables represent nonzero real numbers.

Study Skills Exercise

1. To help you remember the properties of exponents, write them on 3×5 cards. On each card, write a property on one side and an example using that property on the other side. Keep these cards with you, and when you have a spare moment (such as waiting at the doctor's office), pull out these cards and go over the properties.

Review Exercises

For Exercises 2–9, simplify the expressions.

2. $b^3 b^8$ b^{11} **3.** $c^7 c^2$ c^9 **4.** $\dfrac{x^6}{x^2}$ x^4 **5.** $\dfrac{y^9}{y^8}$ y

6. $\dfrac{9^4 \cdot 9^8}{9}$ 9^{11} **7.** $\dfrac{3^{14}}{3^3 \cdot 3^5}$ 3^6 or 729   **8.** $(6ab^3 c^2)^5$ $6^5 a^5 b^{15} c^{10}$ or $7776 a^5 b^{15} c^{10}$ **9.** $(7w^7 z^2)^4$ $7^4 w^{28} z^8$ or $2401 w^{28} z^8$

Objective 1: Definition of b^0

10. Simplify.

    a. 8^0 1   **b.** $\dfrac{8^4}{8^4}$ 1

11. Simplify.

    a. d^0 1   **b.** $\dfrac{d^3}{d^3}$ 1

12. Simplify.

    a. m^0 1   **b.** $\dfrac{m^5}{m^5}$ 1

For Exercises 13–24, simplify the expression. **(See Example 1.)**

13. p^0 1   **14.** k^0 1   **15.** 5^0 1   **16.** 2^0 1

17. -4^0 -1 **18.** -1^0 -1 **19.** $(-6)^0$ 1   **20.** $(-2)^0$ 1

21. $(8x)^0$ 1   **22.** $(-3y^3)^0$ 1   **23.** $-7x^0$ -7 **24.** $6y^0$ 6

✏️ Writing   ↔️ Translating Expression   📐 Geometry   🖩 Scientific Calculator   💿 Video   NS&E

Objective 2: Definition of b^{-n}

25. Simplify and write the answers with positive exponents.

a. t^{-5} $\dfrac{1}{t^5}$ **b.** $\dfrac{t^3}{t^8}$ $\dfrac{1}{t^5}$

26. Simplify and write the answers with positive exponents.

a. 4^{-3} $\dfrac{1}{4^3}$ **b.** $\dfrac{4^2}{4^5}$ $\dfrac{1}{4^3}$

For Exercises 27–46, simplify. **(See Examples 2–4.)**

27. $\left(\dfrac{2}{7}\right)^{-3}$ $\dfrac{343}{8}$

28. $\left(\dfrac{5}{4}\right)^{-1}$ $\dfrac{4}{5}$

29. $\left(-\dfrac{1}{5}\right)^{-2}$ 25

30. $\left(-\dfrac{1}{3}\right)^{-3}$ -27

31. a^{-3} $\dfrac{1}{a^3}$

32. c^{-5} $\dfrac{1}{c^5}$

33. 12^{-1} $\dfrac{1}{12}$

34. 4^{-2} $\dfrac{1}{16}$

35. $(4b)^{-2}$ $\dfrac{1}{16b^2}$

36. $(3z)^{-1}$ $\dfrac{1}{3z}$

37. $6x^{-2}$ $\dfrac{6}{x^2}$

38. $7y^{-1}$ $\dfrac{7}{y}$

39. $(-8)^{-2}$ $\dfrac{1}{64}$

40. -8^{-2} $-\dfrac{1}{64}$

41. $-3y^{-4}$ $\dfrac{-3}{y^4}$

42. $-6a^{-2}$ $\dfrac{-6}{a^2}$

43. $(-t)^{-3}$ $-\dfrac{1}{t^3}$

44. $(-r)^{-5}$ $-\dfrac{1}{r^5}$

45. $\dfrac{1}{a^{-5}}$ a^5

46. $\dfrac{1}{b^{-6}}$ b^6

Objective 3: Properties of Integer Exponents: A Summary

47. Explain what is wrong with the following logic. $\dfrac{x^4}{x^{-6}} = x^{4-6} = x^{-2}$ $\dfrac{x^4}{x^{-6}} = x^{4-(-6)} = x^{10}$

48. Explain what is wrong with the following logic. $\dfrac{y^5}{y^{-3}} = y^{5-3} = y^2$ $\dfrac{y^5}{y^{-3}} = y^{5-(-3)} = y^8$

49. Explain what is wrong with the following logic. $2a^{-3} = \dfrac{1}{2a^3}$ $2a^{-3} = 2 \cdot \dfrac{1}{a^3} = \dfrac{2}{a^3}$

50. Explain what is wrong with the following logic. $5b^{-2} = \dfrac{1}{5b^2}$ $5b^{-2} = 5 \cdot \dfrac{1}{b^2} = \dfrac{5}{b^2}$

Mixed Exercises

For Exercises 51–94, simplify the expression. Write the answer with positive exponents only. **(See Examples 5–6.)**

51. $x^{-8}x^4$ $\dfrac{1}{x^4}$

52. s^5s^{-6} $\dfrac{1}{s}$

53. $a^{-8}a^8$ 1

54. q^3q^{-3} 1

55. $y^{17}y^{-13}$ y^4

56. $b^{20}b^{-14}$ b^6

57. $(m^{-6}n^9)^3$ $\dfrac{n^{27}}{m^{18}}$

58. $(c^4d^{-5})^{-2}$ $\dfrac{d^{10}}{c^8}$

59. $(-3j^{-5}k^6)^4$ $\dfrac{81k^{24}}{j^{20}}$

60. $(6xy^{-11})^{-3}$ $\dfrac{y^{33}}{6^3x^3}$ or $\dfrac{y^{33}}{216x^3}$

61. $\dfrac{p^3}{p^9}$ $\dfrac{1}{p^6}$

62. $\dfrac{q^2}{q^{10}}$ $\dfrac{1}{q^8}$

63. $\dfrac{r^{-5}}{r^{-2}}$ $\dfrac{1}{r^3}$

64. $\dfrac{u^{-2}}{u^{-6}}$ u^4

65. $\dfrac{a^2}{a^{-6}}$ a^8

66. $\dfrac{p^3}{p^{-5}}$ p^8

67. $\dfrac{y^{-2}}{y^6}$ $\dfrac{1}{y^8}$

68. $\dfrac{s^{-4}}{s^3}$ $\dfrac{1}{s^7}$

69. $\dfrac{7^3}{7^2 \cdot 7^8}$ $\dfrac{1}{7^7}$

70. $\dfrac{3^4 \cdot 3}{3^7}$ $\dfrac{1}{9}$

71. $\dfrac{a^2a}{a^3}$ 1

72. $\dfrac{t^5}{t^2t^3}$ 1

73. $\dfrac{a^{-1}b^2}{a^3b^8}$ $\dfrac{1}{a^4b^6}$

74. $\dfrac{k^{-4}h^{-1}}{k^6h}$ $\dfrac{1}{k^{10}h^2}$

75. $\dfrac{w^{-8}(w^2)^{-5}}{w^3}$ $\dfrac{1}{w^{21}}$

76. $\dfrac{p^2p^{-7}}{(p^2)^3}$ $\dfrac{1}{p^{11}}$

77. $\dfrac{3^{-2}}{3}$ $\dfrac{1}{27}$

78. $\dfrac{5^{-1}}{5}$ $\dfrac{1}{25}$

79. $\left(\dfrac{p^{-1}q^5}{p^{-6}}\right)^0$ 1

80. $\left(\dfrac{ab^{-4}}{a^{-5}}\right)^0$ 1

81. $(8x^3y^0)^{-2}$ $\dfrac{1}{64x^6}$

82. $(3u^2v^0)^{-3}$ $\dfrac{1}{27u^6}$

83. $(-8y^{-12})(2y^{16}z^{-2})$
$\dfrac{-16y^4}{z^2}$

84. $(5p^{-2}q^5)(-2p^{-4}q^{-1})$
$\dfrac{-10q^4}{p^6}$

85. $\dfrac{-18a^{10}b^6}{108a^{-2}b^6}$ $-\dfrac{a^{12}}{6}$

86. $\dfrac{-35x^{-4}y^{-3}}{-21x^2y^{-3}}$ $\dfrac{5}{3x^6}$

87. $\dfrac{(-4c^{12}d^7)^2}{(5c^{-3}d^{10})^{-1}}$ $80c^{21}d^{24}$

88. $\dfrac{(s^3t^{-2})^4}{(3s^{-4}t^6)^{-2}}$ $9s^4t^4$

89. $\left(\dfrac{2}{p^6p^3}\right)^{-3}$ $\dfrac{p^{27}}{8}$

90. $\left(\dfrac{5x}{x^7}\right)^{-2}$ $\dfrac{x^{12}}{25}$

91. $\left(\dfrac{5cd^{-3}}{10d^5}\right)^{-2}$ $\dfrac{4d^{16}}{c^2}$

92. $\left(\dfrac{4m^{10}n^4}{2m^{12}n^{-2}}\right)^{-1}$ $\dfrac{m^2}{2n^6}$

93. $(2xy^3)\left(\dfrac{9xy}{4x^3y^2}\right)$ $\dfrac{9y^2}{2x}$

94. $(-3a^3)\left(\dfrac{ab}{27a^4b^2}\right)$ $-\dfrac{1}{9b}$

For Exercises 95–102, simplify the expression. **(See Example 7.)**

95. $5^{-1} + 2^{-2}$ $\dfrac{9}{20}$

96. $4^{-2} + 8^{-1}$ $\dfrac{3}{16}$

97. $10^0 - 10^{-1}$ $\dfrac{9}{10}$

98. $3^0 - 3^{-2}$ $\dfrac{8}{9}$

99. $2^{-2} + 1^{-2}$ $\dfrac{5}{4}$

100. $4^{-1} + 8^{-1}$ $\dfrac{3}{8}$

101. $4 \cdot 5^0 - 2 \cdot 3^{-1}$ $\dfrac{10}{3}$

102. $2 \cdot 4^0 - 3 \cdot 4^{-1}$ $\dfrac{5}{4}$

Scientific Notation

1. Writing Numbers in Scientific Notation

Objectives

1. Writing Numbers in Scientific Notation
2. Writing Numbers in Standard Form
3. Multiplying and Dividing Numbers in Scientific Notation

In many applications in mathematics, it is necessary to work with very large or very small numbers. For example, the number of movie tickets sold in the United States recently is estimated to be 1,500,000,000. The weight of a flea is approximately 0.00066 lb. To avoid writing numerous zeros in very large or small numbers, scientific notation was devised as a shortcut.

The principle behind scientific notation is to use a power of 10 to express the magnitude of the number. For example, the numbers 4000 and 0.07 can be written as:

$$4000 = 4 \times 1000 = 4 \times 10^3$$

$$0.07 = 7.0 \times 0.01 = 7.0 \times 10^{-2} \quad \text{Note that } 10^{-2} = \frac{1}{100} = 0.01$$

> **DEFINITION Scientific Notation**
>
> A positive number expressed in the form: $a \times 10^n$, where $1 \le a < 10$ and n is an integer is said to be written in **scientific notation**.

To write a positive number in scientific notation, we apply the following guidelines:

1. Move the decimal point so that its new location is to the right of the first nonzero digit. The number should now be greater than or equal to 1 but less than 10. Count the number of places that the decimal point is moved.

2. If the original number is *large* (greater than or equal to 10), use the number of places the decimal point was moved as a *positive* power of 10.

$$450,000 \quad = 4.5 \times 100,000 = 4.5 \times 10^5$$

5 places

3. If the original number is *small* (between 0 and 1), use the number of places the decimal point was moved as a *negative* power of 10.

$$0.0002 \quad = 2.0 \times 0.0001 = 2.0 \times 10^{-4}$$

4 places

✎ Writing   ⇄ Translating Expression   ◹ Geometry   🖩 Scientific Calculator   ● Video   NS&E

4. If the original number is greater than or equal to 1 but less than 10, use 0 as the power of 10.

$$7.592 = 7.592 \times 10^0$$ *Note:* A number between 1 and 10 is seldom written in scientific notation.

5. If the original number is negative, then $-10 < a \le -1$.

$$-450,000 = -4.5 \times 100,000 = -4.5 \times 10^5$$

5 places

Example 1 **Writing Numbers in Scientific Notation**

Write the numbers in scientific notation.

a. 53,000   **b.** 0.00053

Solution:

a. $53,000. = 5.3 \times 10^4$ To write 53,000 in scientific notation, the decimal point must be moved four places to the left. Because 53,000 is larger than 10, a *positive* power of 10 is used.

b. $0.00053 = 5.3 \times 10^{-4}$ To write 0.00053 in scientific notation, the decimal point must be moved four places to the right. Because 0.00053 is less than 1, a *negative* power of 10 is used.

Example 2 **Writing Numbers in Scientific Notation**

Write the numerical values in scientific notation.

a. The number of movie tickets sold in the United States recently is estimated to be 1,500,000,000.

b. The weight of a flea is approximately 0.00066 lb.

c. The temperature on a January day in Fargo dropped to $-43°F$.

d. A bench is 8.2 ft long.

Solution:

a. $1,500,000,000 = 1.5 \times 10^9$ **b.** $0.00066 \text{ lb} = 6.6 \times 10^{-4} \text{ lb}$

c. $-43°F = -4.3 \times 10^1 \text{ °F}$ **d.** $8.2 \text{ ft} = 8.2 \times 10^0 \text{ ft}$

---

2. Writing Numbers in Standard Form

Example 3 **Writing Numbers in Standard Form**

Write the numerical values in standard form.

a. The mass of a proton is approximately 1.67×10^{-24} g.

b. The "nearby" star Vega is approximately 1.552×10^{14} miles from Earth.

Solution:

a. 1.67×10^{-24} g $= 0.000\ 000\ 000\ 000\ 000\ 000\ 000\ 001\ 67$ g

Because the power of 10 is negative, the value of 1.67×10^{-24} is a decimal number between 0 and 1. Move the decimal point 24 places to the *left*.

b. 1.552×10^{14} miles $= 155,200,000,000,000$ miles

Because the power of 10 is a positive integer, the value of 1.552×10^{14} is a large number greater than 10. Move the decimal point 14 places to the *right*.

Skill Practice

Write the numerical values in standard form.

5. The probability of winning the California Super Lotto Jackpot is 5.5×10^{-8}.

6. The Sun's mass is 2×10^{30} kilograms.

Classroom Examples: p. 388, Exercises 40 and 50

3. Multiplying and Dividing Numbers in Scientific Notation

To multiply or divide two numbers in scientific notation, use the commutative and associative properties of multiplication to group the powers of 10. For example:

$$400 \times 2000 = (4 \times 10^2)(2 \times 10^3) = (4 \cdot 2) \times (10^2 \cdot 10^3) = 8 \times 10^5$$

$$\frac{0.00054}{150} = \frac{5.4 \times 10^{-4}}{1.5 \times 10^2} = \left(\frac{5.4}{1.5}\right) \times \left(\frac{10^{-4}}{10^2}\right) = 3.6 \times 10^{-6}$$

Example 4 **Multiplying and Dividing Numbers in Scientific Notation**

Multiply or divide as indicated.

a. $(8.7 \times 10^4)(2.5 \times 10^{-12})$ **b.** $\dfrac{4.25 \times 10^{13}}{8.5 \times 10^{-2}}$

Skill Practice

Multiply or divide as indicated

7. $(7 \times 10^5)(5 \times 10^3)$

8. $\dfrac{1 \times 10^{-2}}{4 \times 10^{-7}}$

Classroom Examples: p. 388, Exercises 54 and 66

Solution:

a. $(8.7 \times 10^4)(2.5 \times 10^{-12})$

$= (8.7 \cdot 2.5) \times (10^4 \cdot 10^{-12})$ Commutative and associative properties of multiplication

$= 21.75 \times 10^{-8}$ The number 21.75 is not in proper scientific notation because 21.75 is not between 1 and 10.

$= (2.175 \times 10^1) \times 10^{-8}$ Rewrite 21.75 as 2.175×10^1.

$= 2.175 \times (10^1 \times 10^{-8})$ Associative property of multiplication

$= 2.175 \times 10^{-7}$ Simplify.

b. $\dfrac{4.25 \times 10^{13}}{8.5 \times 10^{-2}}$

$= \left(\dfrac{4.25}{8.5}\right) \times \left(\dfrac{10^{13}}{10^{-2}}\right)$ Commutative and associative properties

$= 0.5 \times 10^{15}$ The number 0.5×10^{15} is not in proper scientific notation because 0.5 is not between 1 and 10.

$= (5.0 \times 10^{-1}) \times 10^{15}$ Rewrite 0.5 as 5.0×10^{-1}.

$= 5.0 \times (10^{-1} \times 10^{15})$ Associative property of multiplication

$= 5.0 \times 10^{14}$ Simplify.

Answers

5. 0.000 000 055
6. 2,000,000,000,000,000,000,000,000,000,000
7. 3.5×10^9
8. 2.5×10^4

Calculator Connections

Topic: Using Scientific Notation

Both scientific and graphing calculators can perform calculations involving numbers written in scientific notation. Most calculators use an **EE** key or an **EXP** key to enter the power of 10.

Scientific Calculator

Enter: 2.7 **EE** 5 **=**   or   2.7 **EXP** 5 **=**   **Result:** | 270000 |

Enter: 7.1 **EE** 3 **+○-** **=**   or   7.1 **EXP** 3 **+○-** **=**   **Result:** | 0.0071 |

Graphing Calculator

```
2.7E5
              270000
7.1E-3
               .0071
```

We recommend that you use parentheses to enclose each number written in scientific notation when performing calculations. Try using your calculator to perform the calculations from Example 4.

  a. $(8.7 \times 10^4)(2.5 \times 10^{-12})$ **b.** $\dfrac{4.25 \times 10^{13}}{8.5 \times 10^{-2}}$

Scientific Calculator

Enter: **(** 8.7 **EE** 4 **)**   **×**   **(** 2.5 **EE** 12 **+○-** **)**   **=**   **Result:** | 0.000000218 |

Enter: **(** 4.25 **EE** 13 **)**   **÷**   **(** 8.5 **EE** 2 **+○-** **)**   **=**   **Result:** | 5E14 |

Notice that the answer to part (b) is shown on the calculator in scientific notation. The calculator does not have enough room to display 14 zeros. Also notice that the calculator rounds the answer to part (a). The exact answer is 2.175×10^{-7} or 0.0000002175.

Graphing Calculator

```
(8.7E4)*(2.5E-12
)
           2.175E-7
(4.25E13)/(8.5E-
2)
              5E14
```

Avoiding Mistakes

A display of 5E14 on a calculator does not mean 5^{14}. It is scientific notation and means 5×10^{14}.

Calculator Exercises

Use a calculator to perform the indicated operations:

1. $(5.2 \times 10^6)(4.6 \times 10^{-3})$

2. $(2.19 \times 10^{-8})(7.84 \times 10^{-4})$

3. $\dfrac{4.76 \times 10^{-5}}{2.38 \times 10^9}$

4. $\dfrac{8.5 \times 10^4}{4.0 \times 10^{-1}}$

5. $\dfrac{(9.6 \times 10^7)(4.0 \times 10^{-3})}{2.0 \times 10^{-2}}$

6. $\dfrac{(5.0 \times 10^{-12})(6.4 \times 10^{-5})}{(1.6 \times 10^{-8})(4.0 \times 10^2)}$

Section 5.4 Practice Exercises

Boost *your* **GRADE at ALEKS.com!**

ALEKS version 3.0

• Practice Problems
• Self-Tests
• NetTutor

• e-Professors
• Videos

For additional exercises, see Classroom Activities 5.4A–5.4B in the *Instructor's Resource Manual* at www.mhhe.com/moh.

Study Skills Exercise

1. Define the key term: **scientific notation**

Review Exercises

For Exercises 2–13, simplify the expression. Assume all variables represent nonzero real numbers.

2. a^3a^{-4} $\dfrac{1}{a}$

3. b^5b^8 b^{13}

4. $10^3 \cdot 10^{-4}$ $\dfrac{1}{10}$

5. $10^5 \cdot 10^8$ 10^{13}

6. $\dfrac{x^3}{x^6}$ $\dfrac{1}{x^3}$

7. $\dfrac{y^2}{y^7}$ $\dfrac{1}{y^5}$

8. $(c^4d^2)^3$ $c^{12}d^6$

9. $(x^5y^{-3})^4$ $\dfrac{x^{20}}{y^{12}}$

10. $\dfrac{z^9z^4}{z^3}$ z^{10}

11. $\dfrac{w^{-2}w^5}{w^{-1}}$ w^4

12. $\dfrac{10^9 \cdot 10^4}{10^3}$ 10^{10}

13. $\dfrac{10^{-2} \cdot 10^5}{10^{-1}}$ 10^4

Objective 1: Writing Numbers in Scientific Notation

14. Explain how scientific notation might be valuable in studying astronomy. Answers may vary.
   For example: Scientific notation would be helpful in writing the distance between Earth and the stars or planets.

15. Explain how you would write the number 0.000 000 000 23 in scientific notation.   Move the decimal point between the 2 and 3 and multiply by 10^{-10}; 2.3×10^{-10}.

16. Explain how you would write the number 23,000,000,000,000 in scientific notation.   Move the decimal point between the 2 and 3 and multiply by 10^{13}; 2.3×10^{13}.

For Exercises 17–28, write the number in scientific notation. **(See Example 1.)**

17. 50,000 5×10^4

18. 900,000 9×10^5

19. 208,000 2.08×10^5

20. 420,000,000 4.2×10^8

21. 6,010,000 6.01×10^6

22. 75,000 7.5×10^4

23. 0.000008 8×10^{-6}

24. 0.003 3×10^{-3}

25. 0.000125 1.25×10^{-4}

26. 0.00000025 2.5×10^{-7}

27. 0.006708 6.708×10^{-3}

28. 0.02004 2.004×10^{-2}

For Exercises 29–34, write the numbers in scientific notation. **(See Example 2.)**

29. The mass of a proton is approximately 0.000 000 000 000 000 000 000 0017 g. 1.7×10^{-24} g

30. The total combined salaries of the president, vice president, senators, and representatives of the United States federal government is approximately \$85,000,000. $\$8.5 \times 10^7$

31. The Bill Gates Foundation has over \$27,000,000,000 from which it makes contributions to global charities. $\$2.7 \times 10^{10}$

32. One gram is equivalent to 0.0035 oz. 3.5×10^{-3} oz

33. In the world's largest tanker disaster, *Amoco Cadiz* spilled 68,000,000 gal of oil off Portsall, France, causing widespread environmental damage over 100 miles of Brittany coast.
   6.8×10^7 gal; 1.0×10^2 miles

34. The human heart pumps about 1400 L of blood per day. That means that it pumps approximately 10,000,000 L per year. 1.4×10^3 L; 1.0×10^7 L

Objective 2: Writing Numbers in Standard Form

35. Explain how you would write the number 3.1×10^{-9} in standard form. Move the decimal point nine places to the left; 0.000 000 0031.

36. Explain how you would write the number 3.1×10^9 in standard form. Move the decimal point nine places to the right; 3,100,000,000.

For Exercises 37–52, write the numbers in standard form. **(See Example 3.)**

37. 5×10^{-5} 0.00005

38. 2×10^{-7} 0.0000002

39. 2.8×10^3 2800

40. 9.1×10^6 9,100,000

41. 6.03×10^{-4} 0.000603

42. 7.01×10^{-3} 0.00701

43. 2.4×10^6 2,400,000

44. 3.1×10^4 31,000

45. 1.9×10^{-2} 0.019

46. 2.8×10^{-6} 0.0000028

47. 7.032×10^3 7032

48. 8.205×10^2 820.5

49. One picogram (pg) is equal to 1×10^{-12} g. 0.000 000 000 001 g

50. A nanometer (nm) is approximately 3.94×10^{-8} in. 0.000 000 0394 in.

51. A normal diet contains between 1.6×10^3 Cal and 2.8×10^3 Cal per day. 1600 calories and 2800 calories

52. The total land area of Texas is approximately 2.62×10^5 square miles. 262,000 square miles

Objective 3: Multiplying and Dividing Numbers in Scientific Notation

For Exercises 53–72, multiply or divide as indicated. Write the answers in scientific notation. **(See Example 4.)**

53. $(2.5 \times 10^6)(2.0 \times 10^{-2})$ 5.0×10^4

54. $(2.0 \times 10^{-7})(3.0 \times 10^{13})$ 6.0×10^6

55. $(1.2 \times 10^4)(3 \times 10^7)$ 3.6×10^{11}

56. $(3.2 \times 10^{-3})(2.5 \times 10^8)$ 8×10^5

57. $\dfrac{7.7 \times 10^6}{3.5 \times 10^2}$ 2.2×10^4

58. $\dfrac{9.5 \times 10^{11}}{1.9 \times 10^3}$ 5×10^8

59. $\dfrac{9.0 \times 10^{-6}}{4.0 \times 10^7}$ 2.25×10^{-13}

60. $\dfrac{7.0 \times 10^{-2}}{5.0 \times 10^9}$ 1.4×10^{-11}

61. $(8.0 \times 10^{10})(4.0 \times 10^3)$ 3.2×10^{14}

62. $(6.0 \times 10^{-4})(3.0 \times 10^{-2})$ 1.8×10^{-5}

63. $(3.2 \times 10^{-4})(7.6 \times 10^{-7})$ 2.432×10^{-10}

64. $(5.9 \times 10^{12})(3.6 \times 10^9)$ 2.124×10^{22}

65. $\dfrac{2.1 \times 10^{11}}{7.0 \times 10^{-3}}$ 3.0×10^{13}

66. $\dfrac{1.6 \times 10^{14}}{8.0 \times 10^{-5}}$ 2.0×10^{18}

67. $\dfrac{5.7 \times 10^{-2}}{9.5 \times 10^{-8}}$ 6.0×10^5

68. $\dfrac{2.72 \times 10^{-6}}{6.8 \times 10^{-4}}$ 4.0×10^{-3}

69. $6{,}000{,}000{,}000 \times 0.0000000023$ 1.38×10^1

70. $0.000055 \times 40{,}000$ 2.2×10^0

71. $\dfrac{0.0000000003}{6000}$ 5.0×10^{-14}

72. $\dfrac{420{,}000}{0.0000021}$ 2.0×10^{11}

Mixed Exercises

73. If a piece of paper is 3.0×10^{-3} in. thick, how thick is a stack of 1.25×10^3 pieces of paper? 3.75 in.

74. A box of staples contains 5.0×10^3 staples and weighs 15 oz. How much does one staple weigh? Write your answer in scientific notation. 3.0×10^{-3} oz

 Writing   Translating Expression   Geometry  Scientific Calculator Video NS&E

75. Bill Gates owned approximately 1,100,000,000 shares of Microsoft stock. If the stock price was $27 per share, how much was Bill Gates' stock worth? 2.97×10^{10}

76. A state lottery had a jackpot of 5.2×10^7. This week the winner was a group of office employees that included 13 people. How much would each person receive?
4.0×10^6 or $4,000,000$

77. Dinosaurs became extinct about 65 million years ago.

   a. Write the number 65 million in scientific notation. 6.5×10^7

   b. How many days is 65 million years?
   2.3725×10^{10} days

   c. How many hours is 65 million years?
   5.694×10^{11} hr

   d. How many seconds is 65 million years?
   2.04984×10^{15} sec

78. The Earth is 111,600,000 km from the Sun.

   a. Write the number 111,600,000 in scientific notation. 1.116×10^8

   b. If there are 1000 m in a kilometer, how many meters is the Earth from the Sun?
   1.116×10^{11} m

   c. If there are 100 cm in a meter, how many centimeters is the Earth from the Sun?
   1.116×10^{13} cm

Problem Recognition Exercises

Properties of Exponents

Simplify completely. Assume that all variables represent nonzero real numbers.

1. $t^3 t^5$ t^8

2. $2^3 2^5$ 2^8 or 256

3. $\dfrac{y^7}{y^2}$ y^5

4. $\dfrac{p^9}{p^3}$ p^6

5. $(r^2 s^4)^2$ $r^4 s^8$

6. $(ab^3 c^2)^3$ $a^3 b^9 c^6$

7. $\dfrac{w^4}{w^{-2}}$ w^6

8. $\dfrac{m^{-14}}{m^2}$ $\dfrac{1}{m^{16}}$

9. $\dfrac{y^{-7} x^4}{z^{-3}}$ $\dfrac{x^4 z^3}{y^7}$

10. $\dfrac{a^3 b^{-6}}{c^{-8}}$ $\dfrac{a^3 c^8}{b^6}$

11. $(2.5 \times 10^{-3})(5.0 \times 10^5)$
1.25×10^3

12. $(3.1 \times 10^6)(4.0 \times 10^{-2})$
1.24×10^5

13. $\dfrac{4.8 \times 10^7}{6.0 \times 10^{-2}}$ 8.0×10^8

14. $\dfrac{5.4 \times 10^{-2}}{9.0 \times 10^6}$ 6.0×10^{-9}

15. $\dfrac{1}{p^{-6} p^{-8} p^{-1}}$ p^{15}

16. $p^6 p^8 p$ p^{15}

17. $\dfrac{v^9}{v^{11}}$ $\dfrac{1}{v^2}$

18. $(c^5 d^4)^{10}$ $c^{50} d^{40}$

19. $\left(\dfrac{1}{2}\right)^{-1} + \left(\dfrac{1}{3}\right)^0$ 3

20. $\left(\dfrac{1}{4}\right)^0 - \left(\dfrac{1}{5}\right)^{-1}$ -4

21. $(2^5 b^{-3})^{-3}$ $\dfrac{b^9}{2^{15}}$

22. $(3^{-2} y^3)^{-2}$ $\dfrac{81}{y^6}$

23. $\left(\dfrac{3x}{2y}\right)^{-4}$ $\dfrac{16y^4}{81x^4}$

24. $\left(\dfrac{6c}{5d^3}\right)^{-2}$ $\dfrac{25d^6}{36c^2}$

25. $(3ab^2)(a^2 b)^3$ $3a^7 b^5$

26. $(4x^2 y^3)^3 (xy^2)$ $64x^7 y^{11}$

27. $\left(\dfrac{xy^2}{x^3 y}\right)^4$ $\dfrac{y^4}{x^8}$

28. $\left(\dfrac{a^3 b}{a^5 b^3}\right)^5$ $\dfrac{1}{a^{10} b^{10}}$

29. $\dfrac{(t^{-2})^3}{t^{-4}}$ $\dfrac{1}{t^2}$

30. $\dfrac{(p^3)^{-4}}{p^{-5}}$ $\dfrac{1}{p^7}$

31. $\left(\dfrac{2w^2 x^3}{3y^0}\right)^3$ $\dfrac{8w^6 x^9}{27}$

32. $\left(\dfrac{5a^0 b^4}{4c^3}\right)^2$ $\dfrac{25b^8}{16c^6}$

33. $\dfrac{q^3 r^{-2}}{s^{-1} t^5}$ $\dfrac{q^3 s}{r^2 t^5}$

34. $\dfrac{n^{-3} m^2}{p^{-3} q^{-1}}$ $\dfrac{m^2 p^3 q}{n^3}$

35. $\dfrac{(y^{-3})^2 (y^5)}{(y^{-3})^{-4}}$ $\dfrac{1}{y^{13}}$

36. $\dfrac{(w^2)^{-4} (w^{-2})}{(w^5)^{-4}}$ w^{10}

37. $\left(\dfrac{-2a^2 b^{-3}}{a^{-4} b^{-5}}\right)^{-3}$ $-\dfrac{1}{8a^{18} b^6}$

38. $\left(\dfrac{-3x^{-4} y^3}{2x^5 y^{-2}}\right)^{-2}$ $\dfrac{4x^{18}}{9y^{10}}$

39. $(5h^{-2} k^0)^3 (5k^{-2})^{-4}$ $\dfrac{k^8}{5h^6}$

40. $(6m^3 n^{-5})^{-4} (6m^0 n^{-2})^5$ $\dfrac{6n^{10}}{m^{12}}$

 Writing   Translating Expression   Geometry   Scientific Calculator   Video   NS E

Section 5.5	Addition and Subtraction of Polynomials

Objectives

1. Introduction to Polynomials
2. Addition of Polynomials
3. Subtraction of Polynomials
4. Polynomials and Applications to Geometry

1. Introduction to Polynomials

One commonly used algebraic expression is called a polynomial. A **polynomial** in one variable, x, is defined as a single term or a sum of terms of the form ax^n, where a is a real number and the exponent, n, is a nonnegative integer. For each term, a is called the **coefficient**, and n is called the **degree of the term**. For example:

Term (Expressed in the Form ax^n)	Coefficient	Degree
$-12z^7$	-12	7
$x^3 \rightarrow$ rewrite as $1x^3$	1	3
$10w \rightarrow$ rewrite as $10w^1$	10	1
$7 \rightarrow$ rewrite as $7x^0$	7	0

If a polynomial has exactly one term, it is categorized as a **monomial**. A two-term polynomial is called a **binomial**, and a three-term polynomial is called a **trinomial**. Usually the terms of a polynomial are written in descending order according to degree. The term with highest degree is called the **leading term**, and its coefficient is called the **leading coefficient**. The **degree of a polynomial** is the greatest degree of all of its terms. Thus, when written in descending order, the leading term determines the degree of the polynomial.

Instructor Note: Ask students why a constant has degree 0.

	Expression	Descending Order	Leading Coefficient	Degree of Polynomial
Monomials	$-3x^4$	$-3x^4$	-3	4
	17	17	17	0
Binomials	$4y^3 - 6y^5$	$-6y^5 + 4y^3$	-6	5
	$\dfrac{1}{2} - \dfrac{1}{4}c$	$-\dfrac{1}{4}c + \dfrac{1}{2}$	$-\dfrac{1}{4}$	1
Trinomials	$4p - 3p^3 + 8p^6$	$8p^6 - 3p^3 + 4p$	8	6
	$7a^4 - 1.2a^8 + 3a^3$	$-1.2a^8 + 7a^4 + 3a^3$	-1.2	8

Skill Practice

a. Write the polynomial in descending order; **b.** State the degree of the polynomial; and **c.** State the coefficient of the leading term.

1. $5x^3 - x + 8x^4 + 3x^2$

Classroom Example: p. 396, Exercise 14

| Example 1 | Identifying the Parts of a Polynomial |

Given: $4.5a - 2.7a^{10} + 1.6 - 3.7a^5$

a. List the terms of the polynomial, and state the coefficient and degree of each term.

b. Write the polynomial in descending order.

c. State the degree of the polynomial and the leading coefficient.

Answers

1. a. $8x^4 + 5x^3 + 3x^2 - x$
   b. 4   **c.** 8

Solution:

a. term: $4.5a$ coefficient: 4.5 degree: 1

   term: $-2.7a^{10}$ coefficient: -2.7 degree: 10

   term: 1.6 coefficient: 1.6 degree: 0

   term: $-3.7a^{5}$ coefficient: -3.7 degree: 5

b. $-2.7a^{10} - 3.7a^{5} + 4.5a + 1.6$

c. The degree of the polynomial is 10 and the leading coefficient is -2.7.

Polynomials may have more than one variable. In such a case, the degree of a term is the sum of the exponents of the variables contained in the term. For example, the term, $32x^{2}y^{5}z$, has degree 8 because the exponents applied to x, y, and z are 2, 5, and 1, respectively. The following polynomial has a degree of 11 because the highest degree of its terms is 11.

$$32x^{2}y^{5}z \quad - \quad 2x^{3}y \quad + \quad 2x^{2}yz^{8} \quad + \quad 7$$

$$\begin{array}{cccc} \uparrow & \uparrow & \uparrow & \uparrow \\ \text{degree} & \text{degree} & \text{degree} & \text{degree} \\ 8 & 4 & 11 & 0 \end{array}$$

2. Addition of Polynomials

Recall that two terms are *like* terms if they each have the same variables, and the corresponding variables are raised to the same powers.

Same exponents   Same exponents

Like Terms: $3x^{2}, -7x^{2}$ $-5yz^{3}, yz^{3}$

Same variables   Same variables

Different exponents

Unlike Terms: $9z^{2}, 12z^{6}$ $\dfrac{1}{3}w^{6}, \dfrac{2}{5}p^{6}$ $4y, 7$

Different variables   Different variables

Recall that the distributive property is used to add or subtract *like* terms. For example,

$3x^{2} + 9x^{2} - 2x^{2}$

$= (3 + 9 - 2)x^{2}$ Apply the distributive property.

$= (10)x^{2}$ Simplify.

$= 10x^{2}$

Example 2 Adding Polynomials

Add the polynomials.

a. $3x^2y + 5x^2y$ **b.** $(-3c^3 + 5c^2 - 7c) + (11c^3 + 6c^2 + 3)$

Solution:

a. $3x^2y + 5x^2y$

$= (3 + 5)x^2y$ Apply the distributive property.

$= (8)x^2y$

$= 8x^2y$ Simplify.

b. $(-3c^3 + 5c^2 - 7c) + (11c^3 + 6c^2 + 3)$

$= -3c^3 + 11c^3 + 5c^2 + 6c^2 - 7c + 3$ Clear parentheses, and group
                                             like terms.

$= 8c^3 + 11c^2 - 7c + 3$ Combine *like* terms.

TIP: It is the distributive property that enables us to add *like* terms. We shorten the process by adding the coefficients of *like* terms.

Instructor Note: Remind students that when adding terms, exponents do *not* change. $2x^3 + 3x^3 \neq 5x^6$

TIP: Polynomials can also be added by combining *like* terms in columns. The sum of the polynomials from Example 2(b) is shown here.

$$\begin{array}{rl} -3c^3 + 5c^2 - 7c + 0 \\ + \underline{11c^3 + 6c^2 + 0c + 3} \\ 8c^3 + 11c^2 - 7c + 3 \end{array}$$

Place holders such as 0 and 0c may be used to help line up *like* terms.

Example 3 Adding Polynomials

Add the polynomials. $(4w^2 - 2x) + (3w^2 - 4x^2 + 6x)$

Solution:

$(4w^2 - 2x) + (3w^2 - 4x^2 + 6x)$

$= 4w^2 + 3w^2 - 4x^2 - 2x + 6x$ Clear parentheses and group
                                     like terms.

$= 7w^2 - 4x^2 + 4x$

TIP: To add these polynomials vertically, align *like* terms in the same column.

$$\begin{array}{rl} 4w^2 \qquad - 2x \\ + \underline{3w^2 - 4x^2 + 6x} \\ 7w^2 - 4x^2 + 4x \end{array}$$

Answers

2. $15a^2b^3$
3. $12q^2 + 4q - 5$
4. $2x^2 - 4xy - 4y^2$

3. Subtraction of Polynomials

Subtraction of two polynomials requires us to find the opposite of the polynomial being subtracted. To find the opposite of a polynomial, take the opposite of each term. This is equivalent to multiplying the polynomial by -1.

Example 4   **Finding the Opposite of a Polynomial**

Find the opposite of the polynomials.

a. $5x$ **b.** $3a - 4b - c$ **c.** $5.5y^4 - 2.4y^3 + 1.1y$

Solution:

Expression	Opposite	Simplified Form
a. $5x$	$-(5x)$	$-5x$
b. $3a - 4b - c$	$-(3a - 4b - c)$	$-3a + 4b + c$
c. $5.5y^4 - 2.4y^3 + 1.1y$	$-(5.5y^4 - 2.4y^3 + 1.1y)$	$-5.5y^4 + 2.4y^3 - 1.1y$

Skill Practice

Find the opposite of the polynomials.

5. $x - 3$
6. $3y^2 - 2xy + 6x + 2$
7. $-2.1w^3 + 4.9w^2 - 1.9w$

Classroom Example: p. 396, Exercise 46

TIP: Notice that the sign of each term is changed when finding the opposite of a polynomial.

Subtraction of two polynomials is similar to subtracting real numbers. Add the opposite of the second polynomial to the first polynomial.

DEFINITION   **Subtraction of Polynomials**

If A and B are polynomials, then $A - B = A + (-B)$.

Example 5   **Subtracting Polynomials**

Subtract the polynomials. $(-4p^4 + 5p^2 - 3) - (11p^2 + 4p - 6)$

Solution:

$(-4p^4 + 5p^2 - 3) - (11p^2 + 4p - 6)$

$= (-4p^4 + 5p^2 - 3) + (-11p^2 - 4p + 6)$ Add the opposite of the second polynomial.

$= -4p^4 + 5p^2 - 11p^2 - 4p - 3 + 6$ Group *like* terms.

$= -4p^4 - 6p^2 - 4p + 3$ Combine *like* terms.

Skill Practice

Subtract the polynomials.
8. $(x^2 + 3x - 2)$
   $- (4x^2 + 6x + 1)$

Classroom Example: p. 397, Exercise 54

TIP: Two polynomials can also be subtracted in columns by adding the opposite of the second polynomial to the first polynomial. Place holders (shown in red) may be used to help line up *like* terms.

$$
\begin{array}{r}
-4p^4 + 0p^3 + 5p^2 + 0p - 3 \\
-(0p^4 + 0p^3 + 11p^2 + 4p - 6)
\end{array}
\quad \xrightarrow{\text{Add the opposite}} \quad
\begin{array}{r}
-4p^4 + 0p^3 + 5p^2 + 0p - 3 \\
+ \ -0p^4 - 0p^3 - 11p^2 - 4p + 6 \\
\hline
-4p^4 \qquad\quad - 6p^2 - 4p + 3
\end{array}
$$

The difference of the polynomials is $-4p^4 - 6p^2 - 4p + 3$.

Answers

5. $-x + 3$
6. $-3y^2 + 2xy - 6x - 2$
7. $2.1w^3 - 4.9w^2 + 1.9w$
8. $-3x^2 - 3x - 3$

Skill Practice

Subtract the polynomials.

9. $(-3y^2 + xy + 2x^2)$
$- (-2y^2 - 3xy - 8x^2)$

Classroom Example: p. 397,
Exercise 60

Example 6 **Subtracting Polynomials**

Subtract the polynomials. $(a^2 - 2ab + 7b^2) - (-8a^2 - 6ab + 2b^2)$

Solution:

$(a^2 - 2ab + 7b^2) - (-8a^2 - 6ab + 2b^2)$

$= (a^2 - 2ab + 7b^2) + (8a^2 + 6ab - 2b^2)$ Add the opposite of the second polynomial.

$= a^2 + 8a^2 - 2ab + 6ab + 7b^2 - 2b^2$ Group *like* terms.

$= 9a^2 + 4ab + 5b^2$ Combine *like* terms.

Concept Connections

Determine if the polynomial is equivalent to $-(a - b)$.

10. $a + b$ 11. $-a + b$
12. $b - a$ 13. $-a - b$

TIP: To subtract these polynomials vertically, align *like* terms in the same column. Then add the opposite of the second polynomial to the first.

$$a^2 - 2ab + 7b^2 \qquad \xrightarrow{\text{Add the opposite}} \qquad a^2 - 2ab + 7b^2$$
$$- (-8a^2 - 6ab + 2b^2) \qquad \qquad + 8a^2 + 6ab - 2b^2$$
$$\overline{\qquad\qquad\qquad\qquad\qquad\qquad 9a^2 + 4ab + 5b^2}$$

The difference is $9a^2 + 4ab + 5b^2$.

Subtracting polynomials can also be accomplished by clearing parentheses and combining *like* terms.

$(a^2 - 2ab + 7b^2) - (-8a^2 - 6ab + 2b^2)$

$= a^2 - 2ab + 7b^2 + 8a^2 + 6ab - 2b^2$ Clear parentheses.

$= 9a^2 + 4ab + 5b^2$ Combine *like* terms.

Skill Practice

14. Subtract $\dfrac{3}{4}x^2 + \dfrac{2}{5}$ from $x^2 + 3x$.

Classroom Example: p. 397,
Exercise 72

Example 7 **Subtracting Polynomials**

Subtract $\frac{1}{3}t^4 + \frac{1}{2}t^2$ from $t^2 - 4$, and simplify the result.

Solution:

To subtract a from b, we write $b - a$. Thus, to subtract $\overbrace{\frac{1}{3}t^4 + \frac{1}{2}t^2}^{a}$ from $\overbrace{t^2 - 4}^{b}$, we have

Avoiding Mistakes

Example 7 involves subtracting two *expressions*. This is not an equation. Therefore, we cannot clear fractions.

$\overset{b}{(t^2 - 4)} - \overset{a}{\left(\dfrac{1}{3}t^4 + \dfrac{1}{2}t^2\right)}$

$= t^2 - 4 - \dfrac{1}{3}t^4 - \dfrac{1}{2}t^2$ Apply the distributive property.

$= -\dfrac{1}{3}t^4 + t^2 - \dfrac{1}{2}t^2 - 4$ Group *like* terms in descending order.

$= -\dfrac{1}{3}t^4 + \dfrac{2}{2}t^2 - \dfrac{1}{2}t^2 - 4$ The t^2-terms are the only *like* terms.

Get a common denominator for the t^2-terms.

$= -\dfrac{1}{3}t^4 + \dfrac{1}{2}t^2 - 4$ Add *like* terms.

Answers

9. $-y^2 + 4xy + 10x^2$
10. No   11. Yes
12. Yes   13. No
14. $\dfrac{1}{4}x^2 + 3x - \dfrac{2}{5}$

4. Polynomials and Applications to Geometry

Example 8 Subtracting Polynomials in Geometry

If the perimeter of the triangle in Figure 5-1 can be represented by the polynomial $2x^2 + 5x + 6$, find a polynomial that represents the length of the missing side.

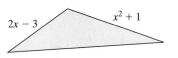

Figure 5-1

Solution:

The missing side of the triangle can be found by subtracting the sum of the two known sides from the perimeter.

$$\begin{pmatrix} \text{Length} \\ \text{of missing} \\ \text{side} \end{pmatrix} = (\text{perimeter}) - \begin{bmatrix} \text{sum of the} \\ \text{two known sides} \end{bmatrix}$$

$$\begin{pmatrix} \text{Length} \\ \text{of missing} \\ \text{side} \end{pmatrix} = (2x^2 + 5x + 6) - [(2x - 3) + (x^2 + 1)]$$

$= 2x^2 + 5x + 6 - [2x - 3 + x^2 + 1]$ Clear inner parentheses.

$= 2x^2 + 5x + 6 - (x^2 + 2x - 2)$ Combine *like* terms within [ ].

$= 2x^2 + 5x + 6 - x^2 - 2x + 2$ Apply the distributive property.

$= 2x^2 - x^2 + 5x - 2x + 6 + 2$ Group *like* terms.

$= x^2 + 3x + 8$ Combine *like* terms.

The polynomial $x^2 + 3x + 8$ represents the length of the missing side.

Classroom Example: p. 397, Exercise 76

Skill Practice

15. If the perimeter of the triangle is represented by the polynomial $6x - 9$, find the polynomial that represents the missing side.

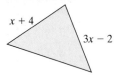

Answer

15. $2x - 11$

Section 5.5 Practice Exercises

Boost *your* GRADE at ALEKS.com!

ALEKS version 3.0

- Practice Problems
- Self-Tests
- NetTutor

- e-Professors
- Videos

For additional exercises, see Classroom Activities 5.5A–5.5C in the *Instructor's Resource Manual* at www.mhhe.com/moh.

Study Skills Exercise

1. Define the key terms:

   a. polynomial   **b.** coefficient   **c.** degree of term

   d. monomial   **e.** binomial   **f.** trinomial

   g. leading term   **h.** leading coefficient   **i.** degree of polynomial

Review Exercises

For Exercises 2–7, simplify the expression.

2. $\dfrac{p^3 \cdot 4p}{p^2}$ $4p^2$

3. $(3x)^2(5x^{-4})$ $\dfrac{45}{x^2}$

4. $(6y^{-3})(2y^9)$ $12y^6$

5. $\dfrac{8t^{-6}}{4t^{-2}}$ $\dfrac{2}{t^4}$

6. $\dfrac{8^3 \cdot 8^{-4}}{8^{-2} \cdot 8^6}$ $\dfrac{1}{8^5}$

7. $\dfrac{3^4 \cdot 3^{-8}}{3^{12} \cdot 3^{-4}}$ $\dfrac{1}{3^{12}}$

 Writing   Translating Expression   Geometry   Scientific Calculator   Video   NSE

8. Explain the difference between 3.0×10^7 and 3^7. 3.0×10^7 is in scientific notation in which 10 is raised to the seventh power. 3^7 is not in scientific notation and 3 is being raised to the seventh power.

9. Explain the difference between 4.0×10^{-2} and 4^{-2}. 4.0×10^{-2} is in scientific notation in which 10 is raised to the -2 power. 4^{-2} is not in scientific notation and 4 is being raised to the -2 power.

Objective 1: Introduction to Polynomials

10. Write the polynomial in descending order. $10 - 8a - a^3 + 2a^2 + a^5$ $\quad a^5 - a^3 + 2a^2 - 8a + 10$

11. Write the polynomial in descending order:

$$6 + 7x^2 - 7x^4 + 9x \quad -7x^4 + 7x^2 + 9x + 6$$

12. Write the polynomial in descending order:

$$\frac{1}{2}y + y^2 - 12y^4 + y^3 - 6 \quad -12y^4 + y^3 + y^2 + \frac{1}{2}y - 6$$

For Exercises 13–24, categorize the expression as a monomial, a binomial, or a trinomial. Then identify the coefficient and degree of the leading term. **(See Example 1.)**

13. $10a^2 + 5a$
Binomial; 10; 2

14. $7z + 13z^2 - 15$
Trinomial; 13; 2

15. $6x^2$
Monomial; 6; 2

16. 9
Monomial; 9; 0

17. $2t - t^4$
Binomial; -1; 4

18. $7x + 2$
Binomial; 7; 1

19. $12y^4 - 3y + 1$
Trinomial; 12; 4

20. $5bc^2$
Monomial; 5; 3

21. 23
Monomial; 23; 0

22. $4 - 2c$
Binomial; -2; 1

23. $-32xyz$
Monomial; -32; 3

24. $w^4 - w^2$
Binomial; 1; 4

Objective 2: Addition of Polynomials

25. Explain why the terms $3x$ and $3x^2$ are not *like* terms. The exponents on the x-factors are different.

26. Explain why the terms $4w^3$ and $4z^3$ are not *like* terms. The variables are different.

For Exercises 27–42, add the polynomials. **(See Examples 2–3.)**

27. $23x^2y + 12x^2y$ $\quad 35x^2y$

28. $-5ab^3 + 17ab^3$ $\quad 12ab^3$

29. $(6y + 3x) + (4y - 3x)$ $\quad 10y$

30. $(2z - 5h) + (-3z + h)$ $\quad -z - 4h$

31. $3b^5d^2 + (5b^5d^2 - 9d)$ $\quad 8b^5d^2 - 9d$

32. $4c^2d^3 + (3cd - 10c^2d^3)$ $\quad -6c^2d^3 + 3cd$

33. $(7y^2 + 2y - 9) + (-3y^2 - y)$ $\quad 4y^2 + y - 9$

34. $(-3w^2 + 4w - 6) + (5w^2 + 2)$ $\quad 2w^2 + 4w - 4$

35. $\quad 6a + 2b - 5c$ $\quad 4a - 8c$
$+\ \underline{-2a - 2b - 3c}$

36. $\quad -13x + 5y + 10z$ $\quad -16x + 2y + 12z$
$+\ \underline{-3x - 3y + \ 2z}$

37. $\left(\frac{2}{5}a + \frac{1}{4}b - \frac{5}{6}\right) + \left(\frac{3}{5}a - \frac{3}{4}b - \frac{7}{6}\right)$ $\quad a - \frac{1}{2}b - 2$

38. $\left(\frac{5}{9}x + \frac{1}{10}y\right) + \left(-\frac{4}{9}x + \frac{3}{10}y\right)$ $\quad \frac{1}{9}x + \frac{2}{5}y$

39. $\left(z - \frac{8}{3}\right) + \left(\frac{4}{3}z^2 - z + 1\right)$ $\quad \frac{4}{3}z^2 - \frac{5}{3}$

40. $\left(-\frac{7}{5}r + 1\right) + \left(-\frac{3}{5}r^2 + \frac{7}{5}r + 1\right)$ $\quad -\frac{3}{5}r^2 + 2$

41. $\quad 7.9t^3 \qquad\quad + 2.6t - 1.1$
$+\ \underline{\quad\ -3.4t^2 + 3.4t - 3.1}$
$7.9t^3 - 3.4t^2 + 6t - 4.2$

42. $\quad 0.34y^2 \qquad\quad + 1.23$
$+\ \underline{\quad\quad\ 3.42y - 7.56}$
$0.34y^2 + 3.42y - 6.33$

Objective 3: Subtraction of Polynomials

For Exercises 43–48, find the opposite of each polynomial. **(See Example 4.)**

43. $4h - 5$
$-4h + 5$

44. $5k - 12$
$-5k + 12$

45. $-2m^2 + 3m - 15$
$2m^2 - 3m + 15$

46. $-n^2 - 6n + 9$
$n^2 + 6n - 9$

47. $3v^3 + 5v^2 + 10v + 22$
$-3v^3 - 5v^2 - 10v - 22$

48. $7u^4 + 3v^2 + 17$
$-7u^4 - 3v^2 - 17$

 ✎ Writing ←→ Translating Expression   Geometry Scientific Calculator Video  NS&E

For Exercises 49–68, subtract the polynomials. **(See Examples 5–6.)**

49. $4a^3b^2 - 12a^3b^2$ $-8a^3b^2$

50. $5yz^4 - 14yz^4$ $-9yz^4$

51. $-32x^3 - 21x^3$ $-53x^3$

52. $-23c^5 - 12c^5$ $-35c^5$

53. $(7a - 7) - (12a - 4)$ $-5a - 3$

54. $(4x + 3v) - (-3x + v)$ $7x + 2v$

55. $(4k + 3) - (-12k - 6)$ $16k + 9$

56. $(3h - 15) - (8h + 13)$ $-5h - 28$

57. $25s - (23s + 14)$ $2s - 14$

58. $3x^2 - (-x^2 - 12)$ $4x^2 + 12$

59. $(5t^2 - 3t - 2) - (2t^2 + t + 1)$ $3t^2 - 4t - 3$

60. $(k^2 + 2k + 1) - (3k^2 - 6k + 2)$ $-2k^2 + 8k - 1$

61. $\quad 10r - 6s + 2t$
$- (12r - 3s - \ t)$
$\overline{-2r - 3s + 3t}$

62. $\quad a - 14b + 7c$
$- (-3a - \ 8b + 2c)$
$\overline{4a - 6b + 5c}$

63. $\left(\dfrac{7}{8}x + \dfrac{2}{3}y - \dfrac{3}{10}\right) - \left(\dfrac{1}{8}x + \dfrac{1}{3}y\right)$ $\frac{3}{4}x + \frac{1}{3}y - \frac{3}{10}$

64. $\left(r - \dfrac{1}{12}s\right) - \left(\dfrac{1}{2}r - \dfrac{5}{12}s - \dfrac{4}{11}\right)$ $\frac{1}{2}r + \frac{1}{3}s + \frac{4}{11}$

65. $\left(\dfrac{2}{3}h^2 - \dfrac{1}{5}h - \dfrac{3}{4}\right) - \left(\dfrac{4}{3}h^2 - \dfrac{4}{5}h + \dfrac{7}{4}\right)$ $-\frac{2}{3}h^2 + \frac{3}{5}h - \frac{5}{2}$

66. $\left(\dfrac{3}{8}p^3 - \dfrac{5}{7}p^2 - \dfrac{2}{5}\right) - \left(\dfrac{5}{8}p^3 - \dfrac{2}{7}p^2 + \dfrac{7}{5}\right)$
$-\frac{1}{4}p^3 - \frac{3}{7}p^2 - \frac{9}{5}$

67. $\quad 4.5x^4 - 3.1x^2 \qquad\quad - 6.7$
$- (2.1x^4 \qquad\quad + 4.4x \qquad\ )$
$\overline{2.4x^4 - 3.1x^2 - 4.4x - 6.7}$

68. $\quad 1.3c^3 \qquad\qquad + 4.8$
$- (\qquad 4.3c^2 - 2c - 2.2)$
$\overline{1.3c^3 - 4.3c^2 + 2c + 7}$

69. Find the difference of $(4b^3 + 6b - 7)$ and $(-12b^2 + 11b + 5)$. $4b^3 + 12b^2 - 5b - 12$

70. Find the difference of $(-5y^2 + 3y - 21)$ and $(-4y^2 - 5y + 23)$. $-y^2 + 8y - 44$

71. Subtract $(3x^3 - 5x + 10)$ from $(-2x^2 + 6x - 21)$. **(See Example 7.)** $-3x^3 - 2x^2 + 11x - 31$

72. Subtract $(7a^5 - 2a^3 - 5a)$ from $(3a^5 - 9a^2 + 3a - 8)$. $-4a^5 + 2a^3 - 9a^2 + 8a - 8$

Objective 4: Polynomials and Applications to Geometry

73. Find a polynomial that represents the perimeter of the figure. $4y^3 + 2y^2 + 2$

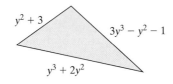

74. Find a polynomial that represents the perimeter of the figure. $9t^3 + t^2$

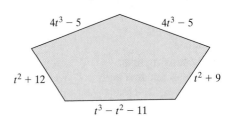

75. If the perimeter of the figure can be represented by the polynomial $5a^2 - 2a + 1$, find a polynomial that represents the length of the missing side. **(See Example 8.)** $3a^2 - 3a + 5$

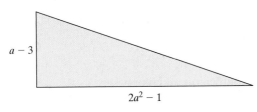

76. If the perimeter of the figure can be represented by the polynomial $6w^3 - 2w - 3$, find a polynomial that represents the length of the missing side. $4w^3 - 6w^2 - 2w$

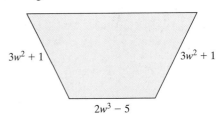

Mixed Exercises

For Exercises 77–92, perform the indicated operation.

77. $(2ab^2 + 9a^2b) + (7ab^2 - 3ab + 7a^2b)$ $\quad 9ab^2 - 3ab + 16a^2b$

78. $(8x^2y - 3xy - 6xy^2) + (3x^2y - 12xy)$
$11x^2y - 15xy - 6xy^2$

79.
$$\begin{array}{r} 4z^5 \quad\quad + z^3 - 3z + 13 \\ -(\quad - z^4 - 8z^3 \quad\quad + 15) \\ \hline \end{array}$$
$4z^5 + z^4 + 9z^3 - 3z - 2$

80.
$$\begin{array}{r} -15t^4 \quad\quad - 23t^2 + 16t \\ -(\quad 21t^3 + 18t^2 + \quad t) \\ \hline \end{array}$$
$-15t^4 - 21t^3 - 41t^2 + 15t$

81. $(9x^4 + 2x^3 - x + 5) + (9x^3 - 3x^2 + 8x + 3) - (7x^4 - x + 12)$
$2x^4 + 11x^3 - 3x^2 + 8x - 4$

82. $(-6y^3 - 9y^2 + 23) - (7y^2 + 2y - 11) + (3y^3 - 25)$
$-3y^3 - 16y^2 - 2y + 9$

83. $(5w^2 - 3w + 2) + (-4w + 6) - (7w^2 - 10)$
$-2w^2 - 7w + 18$

84. $(10u^3 - 5u^2 + 4) - (2u^3 + 5u^2 + u) - (u^3 - 3u + 9)$
$7u^3 - 10u^2 + 2u - 5$

85. $(7p^2q - 3pq^2) - (8p^2q + pq) + (4pq - pq^2)$
$-p^2q - 4pq^2 + 3pq$

86. $(12c^2d - 2cd + 8cd^2) - (-c^2d + 4cd) - (5cd - 2cd^2)$
$13c^2d - 11cd + 10cd^2$

87. $(5x - 2x^3) + (2x^3 - 5x)$ $\quad 0$

88. $(p^2 - 4p + 2) - (2 + p^2 - 4p)$ $\quad 0$

89.
$$\begin{array}{r} 2a^2b - 4ab + \quad ab^2 \\ -(2a^2b + \quad ab - 5ab^2) \\ \hline \end{array}$$
$-5ab + 6ab^2$

90.
$$\begin{array}{r} -3xy + \quad 7xy^2 + 5x^2y \\ + -8xy - 11xy^2 + 3x^2y \\ \hline \end{array}$$
$-11xy - 4xy^2 + 8x^2y$

91. $[(3y^2 - 5y) - (2y^2 + y - 1)] + (10y^2 - 4y - 5)$
$11y^2 - 10y - 4$

92. $(12c^3 - 5c^2 - 2c) + [(7c^3 - 2c^2 + c) - (4c^3 + 4c)]$
$15c^3 - 7c^2 - 5c$

Expanding Your Skills

93. Write a binomial of degree 3. (Answers may vary.)
For example, $x^3 + 6$

94. Write a trinomial of degree 6. (Answers may vary.)
For example, $5x^6 + x - 4$

95. Write a monomial of degree 5. (Answers may vary.)
For example, $8x^5$

96. Write a monomial of degree 1. (Answers may vary.)
For example, $3x$

97. Write a trinomial with the leading coefficient -6. (Answers may vary.)
For example, $-6x^2 + 2x + 5$

98. Write a binomial with the leading coefficient 13. (Answers may vary.) For example, $13x + 7$

Section 5.6 Multiplication of Polynomials and Special Products

Objectives

1. Multiplication of Polynomials
2. Special Case Products: Difference of Squares and Perfect Square Trinomials
3. Applications to Geometry

1. Multiplication of Polynomials

The properties of exponents covered in Sections 5.1–5.3 can be used to simplify many algebraic expressions including the multiplication of monomials. To multiply monomials, first use the associative and commutative properties of multiplication to group coefficients and like bases. Then simplify the result by using the properties of exponents.

Example 1 Multiplying Monomials

Multiply.

a. $(3x^4)(4x^2)$ **b.** $(-4c^5d)(2c^2d^3e)$ **c.** $\left(\dfrac{1}{3}a^4b^3\right)\left(\dfrac{3}{4}b^7\right)$

Solution:

a. $(3x^4)(4x^2)$

$= (3 \cdot 4)(x^4 x^2)$ Group coefficients and like bases.

$= 12x^6$ Multiply the coefficients and add the exponents on x.

b. $(-4c^5d)(2c^2d^3e)$

$= (-4 \cdot 2)(c^5c^2)(dd^3)(e)$ Group coefficients and like bases.

$= -8c^7d^4e$ Simplify.

c. $\left(\frac{1}{3}a^4b^3\right)\left(\frac{3}{4}b^7\right)$

$= \left(\frac{1}{3} \cdot \frac{3}{4}\right)(a^4)(b^3b^7)$ Group coefficients and like bases.

$= \frac{1}{4}a^4b^{10}$ Simplify.

The distributive property is used to multiply polynomials: $a(b + c) = ab + ac$.

Example 2 **Multiplying a Polynomial by a Monomial**

Multiply.

a. $2t(4t - 3)$ **b.** $-3a^2\left(-4a^2 + 2a - \frac{1}{3}\right)$

Solution:

a. $2t(4t - 3)$ Multiply each term of the binomial by $2t$.

$= (2t)(4t) + 2t(-3)$ Apply the distributive property.

$= 8t^2 - 6t$ Simplify each term.

b. $-3a^2\left(-4a^2 + 2a - \frac{1}{3}\right)$ Multiply each term of the trinomial by $-3a^2$.

$= (-3a^2)(-4a^2) + (-3a^2)(2a) + (-3a^2)\left(-\frac{1}{3}\right)$ Apply the distributive property.

$= 12a^4 - 6a^3 + a^2$ Simplify each term.

Thus far, we have illustrated polynomial multiplication involving monomials. Next, the distributive property will be used to multiply polynomials with more than one term.

$(x + 3)(x + 5) = x(x + 5) + 3(x + 5)$ Apply the distributive property.

$= x(x + 5) + 3(x + 5)$ Apply the distributive property again.

$= (x)(x) + (x)(5) + (3)(x) + (3)(5)$

$= x^2 + 5x + 3x + 15$

$= x^2 + 8x + 15$ Combine *like* terms.

Note: Using the distributive property results in multiplying each term of the first polynomial by each term of the second polynomial.

$$(x + 3)(x + 5) = (x)(x) + (x)(5) + (3)(x) + (3)(5)$$
$$= x^2 + 5x + 3x + 15$$
$$= x^2 + 8x + 15$$

Example 3 **Multiplying a Polynomial by a Polynomial**

Multiply the polynomials. $(c - 7)(c + 2)$

Solution:

$(c - 7)(c + 2)$ Multiply each term in the first polynomial by each term in the second. That is, apply the distributive property.

$$= (c)(c) + (c)(2) + (-7)(c) + (-7)(2)$$
$$= c^2 + 2c - 7c - 14 \qquad \text{Simplify.}$$
$$= c^2 - 5c - 14 \qquad \text{Combine } like \text{ terms.}$$

> **TIP:** Notice that the product of two *binomials* equals the sum of the products of the **First** terms, the **Outer** terms, the **Inner** terms, and the **Last** terms. The acronym **FOIL** (First Outer Inner Last) can be used as a memory device to multiply two binomials.

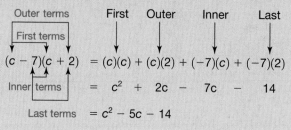

Example 4 **Multiplying a Polynomial by a Polynomial**

Multiply the polynomials. $(10x + 3y)(2x - 4y)$

Solution:

$(10x + 3y)(2x - 4y)$ Multiply each term in the first polynomial by each term in the second. That is, apply the distributive property.

$$= (10x)(2x) + (10x)(-4y) + (3y)(2x) + (3y)(-4y)$$
$$= 20x^2 - 40xy + 6xy - 12y^2 \qquad \text{Simplify each term.}$$
$$= 20x^2 - 34xy - 12y^2 \qquad \text{Combine } like \text{ terms.}$$

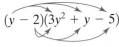 **Multiplying a Polynomial by a Polynomial**

Multiply the polynomials. $(y - 2)(3y^2 + y - 5)$

Solution:

$(y - 2)(3y^2 + y - 5)$ Multiply each term in the first polynomial by each term in the second.

$= (y)(3y^2) + (y)(y) + (y)(-5) + (-2)(3y^2) + (-2)(y) + (-2)(-5)$

$= 3y^3 + y^2 - 5y - 6y^2 - 2y + 10$ Simplify each term.

$= 3y^3 - 5y^2 - 7y + 10$ Combine *like* terms.

> **TIP:** Multiplication of polynomials can be performed vertically by a process similar to column multiplication of real numbers. For example,
>
> $$\begin{array}{r} 235 \\ \times\ 21 \\ \hline 235 \\ 4700 \\ \hline 4935 \end{array}$$
>
> $$\begin{array}{r} 3y^2\ +\ y\ -\ 5 \\ \times\qquad\quad y\ -\ 2 \\ \hline -6y^2\ -\ 2y\ +\ 10 \\ 3y^3\ +\ y^2\ -\ 5y\ +\ \ 0 \\ \hline 3y^3\ -\ 5y^2\ -\ 7y\ +\ 10 \end{array}$$

Note: When multiplying by the column method, it is important to *align like* terms vertically before adding terms.

Skill Practice

Multiply.

8. $(2y + 4)(3y^2 - 5y + 2)$

Classroom Example: p. 404, Exercise 44

Avoiding Mistakes

It is important to note that the acronym FOIL does not apply to Example 5 because the product does not involve two binomials.

2. Special Case Products: Difference of Squares and Perfect Square Trinomials

In some cases the product of two binomials takes on a special pattern.

I. The first special case occurs when multiplying the sum and difference of the same two terms. For example:

$(2x + 3)(2x - 3)$

$= 4x^2 - 6x + 6x - 9$

$= 4x^2 - 9$

Notice that the middle terms are opposites. This leaves only the difference between the square of the first term and the square of the second term. For this reason, the product is called a *difference of squares.*

Note: The binomials $2x + 3$ and $2x - 3$ are called **conjugates**. In one expression, $2x$ and 3 are added, and in the other, $2x$ and 3 are subtracted.

II. The second special case involves the square of a binomial. For example:

$(3x + 7)^2$

$= (3x + 7)(3x + 7)$

$= 9x^2 + 21x + 21x + 49$

$= 9x^2 + 42x + 49$

$= (3x)^2 + 2(3x)(7) + (7)^2$

When squaring a binomial, the product will be a trinomial called a *perfect square trinomial.* The first and third terms are formed by squaring each term of the binomial. The middle term equals twice the product of the terms in the binomial.

Note: The expression $(3x - 7)^2$ also expands to a perfect square trinomial, but the middle term will be negative:

$(3x - 7)(3x - 7) = 9x^2 - 21x - 21x + 49 = 9x^2 - 42x + 49$

Answer

8. $6y^3 + 2y^2 - 16y + 8$

> **FORMULA** Special Case Product Formulas
>
> 1. $(a + b)(a - b) = a^2 - b^2$ The product is called a **difference of squares**.
>
> 2. $\begin{aligned}(a + b)^2 &= a^2 + 2ab + b^2 \\ (a - b)^2 &= a^2 - 2ab + b^2\end{aligned}$ The product is called a **perfect square trinomial**.

Instructor Note: Have students substitute some numbers to confirm that $(a + b)^2 \neq a^2 + b^2$.

You should become familiar with these special case products because they will be used again in the next chapter to factor polynomials.

Skill Practice

Multiply the conjugates.

9. $(a + 7)(a - 7)$

10. $\left(\dfrac{4}{5}x - 10\right)\left(\dfrac{4}{5}x + 10\right)$

Classroom Examples: p. 405, Exercises 54 and 58

Example 6 Finding Special Products

Multiply the conjugates.

a. $(x - 9)(x + 9)$ **b.** $\left(\dfrac{1}{2}p - 6\right)\left(\dfrac{1}{2}p + 6\right)$

Solution:

a. $(x - 9)(x + 9)$ Apply the formula: $(a + b)(a - b) = a^2 - b^2$.

$\overset{a^2 - b^2}{}$

$= (x)^2 - (9)^2$ Substitute $a = x$ and $b = 9$.

$= x^2 - 81$

> **TIP:** The product of two conjugates can be checked by applying the distributive property:
>
> $(x - 9)(x + 9)$
>
> $= x^2 + 9x - 9x - 81$
>
> $= x^2 - 81$

b. $\left(\dfrac{1}{2}p - 6\right)\left(\dfrac{1}{2}p + 6\right)$ Apply the formula: $(a + b)(a - b) = a^2 - b^2$.

$\overset{a^2 - b^2}{}$

$= \left(\dfrac{1}{2}p\right)^2 - (6)^2$ Substitute $a = \dfrac{1}{2}p$ and $b = 6$.

$= \dfrac{1}{4}p^2 - 36$ Simplify each term.

Skill Practice

Square the binomials.

11. $(2x + 3)^2$

12. $(3c^2 - 4)^2$

Classroom Examples: p. 405, Exercises 70 and 72

Answers

9. $a^2 - 49$
10. $\dfrac{16}{25}x^2 - 100$
11. $4x^2 + 12x + 9$
12. $9c^4 - 24c^2 + 16$

Example 7 Finding Special Products

Square the binomials.

a. $(3w - 4)^2$ **b.** $(5x^2 + 2)^2$

Solution:

a. $(3w - 4)^2$ Apply the formula: $(a - b)^2 = a^2 - 2ab + b^2$.

$\overset{a^2 - 2ab + b^2}{}$

$= (3w)^2 - 2(3w)(4) + (4)^2$ Substitute $a = 3w$, $b = 4$.

$= 9w^2 - 24w + 16$ Simplify each term.

TIP: The square of a binomial can be checked by explicitly writing the product of the two binomials and applying the distributive property:

$$(3w - 4)^2 = (3w - 4)(3w - 4) = 9w^2 - 12w - 12w + 16$$
$$= 9w^2 - 24w + 16$$

b. $(5x^2 + 2)^2$ Apply the formula:
$(a + b)^2 = a^2 + 2ab + b^2$.

$a^2 + 2ab + b^2$

$= (5x^2)^2 + 2(5x^2)(2) + (2)^2$ Substitute $a = 5x^2$, $b = 2$.

$= 25x^4 + 20x^2 + 4$ Simplify each term.

Avoiding Mistakes

The property for squaring two factors is different than the property for squaring two terms: $(ab)^2 = a^2b^2$ but $(a + b)^2 = a^2 + 2ab + b^2$

3. Applications to Geometry

Example 8 **Using Special Case Products in an Application of Geometry**

Find a polynomial that represents the volume of the cube (Figure 5-2).

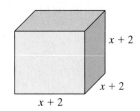

Figure 5-2

Solution:

Volume = (length)(width)(height)

$V = (x + 2)(x + 2)(x + 2)$ or $V = (x + 2)^3$

To expand $(x + 2)(x + 2)(x + 2)$, multiply the first two factors. Then multiply the result by the last factor.

$V = \underbrace{(x + 2)(x + 2)}(x + 2)$

$= (x^2 + 4x + 4)(x + 2)$ ◄——

TIP: $(x + 2)(x + 2) = (x + 2)^2$ and results in a perfect square trinomial.
$(x + 2)^2 = (x)^2 + 2(x)(2) + (2)^2$
$= x^2 + 4x + 4$

$= (x^2)(x) + (x^2)(2) + (4x)(x) + (4x)(2) + (4)(x) + (4)(2)$ Apply the distributive property.

$= x^3 + 2x^2 + 4x^2 + 8x + 4x + 8$ Group *like* terms.

$= x^3 + 6x^2 + 12x + 8$ Combine *like* terms.

The volume of the cube can be represented by

$$V = (x + 2)^3 = x^3 + 6x^2 + 12x + 8.$$

Skill Practice

13. Find the polynomial that represents the volume of the cube.

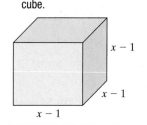

Classroom Example: p. 406, Exercise 84

Answer

13. The volume of the cube can be represented by
$x^3 - 3x^2 + 3x - 1$.

Section 5.6 Practice Exercises

Study Skills Exercise

1. Define the key terms:

    a. conjugates   **b. difference of squares**   **c. perfect square trinomial**

Review Exercises

For Exercises 2–9, simplify the expressions (if possible).

2. $4x + 5x$ $9x$ **3.** $2y^2 - 4y^2$ $-2y^2$ **4.** $(4x)(5x)$ $20x^2$ **5.** $(2y^2)(-4y^2)$ $-8y^4$

6. $-5a^3b - 2a^3b$ $-7a^3b$ **7.** $7uvw^2 + uvw^2$ $8uvw^2$ **8.** $(-5a^3b)(-2a^3b)$ $10a^6b^2$ **9.** $(7uvw^2)(uvw^2)$ $7u^2v^2w^4$

Objective 1: Multiplication of Polynomials

For Exercises 10–18, multiply the expressions. **(See Example 1.)**

10. $8(4x)$ $32x$ **11.** $-2(6y)$ $-12y$ **12.** $-10(5z)$ $-50z$

13. $7(3p)$ $21p$ **14.** $(x^{10})(4x^3)$ $4x^{13}$ **15.** $(a^{13}b^4)(12ab^4)$ $12a^{14}b^8$

16. $(4m^3n^7)(-3m^6n)$ $-12m^9n^8$ **17.** $(2c^7d)(-c^3d^{11})$ $-2c^{10}d^{12}$ **18.** $(-5u^2v)(-8u^3v^2)$ $40u^5v^3$

For Exercises 19–52, multiply the polynomials. **(See Examples 2–5.)**

19. $8pq(2pq - 3p + 5q)$
$16p^2q^2 - 24p^2q + 40pq^2$

20. $5ab(2ab + 6a - 3b)$
$10a^2b^2 + 30a^2b - 15ab^2$

21. $(k^2 - 13k - 6)(-4k)$
$-4k^3 + 52k^2 + 24k$

22. $(h^2 + 5h - 12)(-2h)$
$-2h^3 - 10h^2 + 24h$

23. $-15pq(3p^2 + p^3q^2 - 2q)$
$-45p^3q - 15p^4q^3 + 30pq^2$

24. $-4u^2v(2u - 5uv^3 + v)$
$-8u^3v + 20u^3v^4 - 4u^2v^2$

25. $(y - 10)(y + 9)$
$y^2 - y - 90$

26. $(x + 5)(x - 6)$
$x^2 - x - 30$

27. $(m - 12)(m - 2)$
$m^2 - 14m + 24$

28. $(n - 7)(n - 2)$
$n^2 - 9n + 14$

29. $(3p - 2)(4p + 1)$
$12p^2 - 5p - 2$

30. $(7q + 11)(q - 5)$
$7q^2 - 24q - 55$

31. $(-4w + 8)(-3w + 2)$
$12w^2 - 32w + 16$

32. $(-6z + 10)(-2z + 4)$
$12z^2 - 44z + 40$

33. $(p - 3w)(p - 11w)$
$p^2 - 14pw + 33w^2$

34. $(y - 7x)(y - 10x)$
$y^2 - 17xy + 70x^2$

35. $(6x - 1)(2x + 5)$
$12x^2 + 28x - 5$

36. $(3x + 7)(x - 8)$
$3x^2 - 17x - 56$

37. $(4a - 9)(2a - 1)$
$8a^2 - 22a + 9$

38. $(3b + 5)(b - 5)$
$3b^2 - 10b - 25$

39. $(3t - 7)(3t + 1)$
$9t^2 - 18t - 7$

40. $(5w - 2)(2w - 5)$
$10w^2 - 29w + 10$

41. $(3m + 4n)(m + 8n)$
$3m^2 + 28mn + 32n^2$

42. $(7y + z)(3y + 5z)$
$21y^2 + 38yz + 5z^2$

43. $(5s + 3)(s^2 + s - 2)$
$5s^3 + 8s^2 - 7s - 6$

44. $(t - 4)(2t^2 - t + 6)$
$2t^3 - 9t^2 + 10t - 24$

45. $(3w - 2)(9w^2 + 6w + 4)$
$27w^3 - 8$

46. $(z + 5)(z^2 - 5z + 25)$
$z^3 + 125$

47. $(p^2 + p - 5)(p^2 + 4p - 1)$
$p^4 + 5p^3 - 2p^2 - 21p + 5$

48. $(-x^2 - 2x + 4)(x^2 + 2x - 6)$
$-x^4 - 4x^3 + 6x^2 + 20x - 24$

49. $\begin{array}{r} 3a^2 - 4a + 9 \\ \times \quad 2a - 5 \\ \hline 6a^3 - 23a^2 + 38a - 45 \end{array}$

50. $\begin{array}{r} 7x^2 - 3x - 4 \\ \times \quad 5x + 1 \\ \hline 35x^3 - 8x^2 - 23x - 4 \end{array}$

51. $\begin{array}{r} 4x^2 - 12xy + 9y^2 \\ \times \quad 2x - 3y \\ \hline 8x^3 - 36x^2y + 54xy^2 - 27y^3 \end{array}$

52. $\begin{array}{r} 25a^2 + 10ab + b^2 \\ \times \quad 5a + b \\ \hline 125a^3 + 75a^2b + 15ab^2 + b^3 \end{array}$

Objective 2: Special Case Products: Difference of Squares and Perfect Square Trinomials

For Exercises 53–64, multiply the conjugates. **(See Example 6.)**

53. $(3a - 4b)(3a + 4b)$
$9a^2 - 16b^2$

54. $(5y + 7x)(5y - 7x)$
$25y^2 - 49x^2$

55. $(9k + 6)(9k - 6)$
$81k^2 - 36$

56. $(2h - 5)(2h + 5)$
$4h^2 - 25$

57. $\left(\dfrac{1}{2} - t\right)\left(\dfrac{1}{2} + t\right)$
$\dfrac{1}{4} - t^2$

58. $\left(r + \dfrac{1}{4}\right)\left(r - \dfrac{1}{4}\right)$
$r^2 - \dfrac{1}{16}$

59. $(u^3 + 5v)(u^3 - 5v)$
$u^6 - 25v^2$

60. $(8w^2 - x)(8w^2 + x)$
$64w^4 - x^2$

61. $(2 - 3a)(2 + 3a)$
$4 - 9a^2$

62. $(1 - 4x^2)(1 + 4x^2)$
$1 - 16x^4$

63. $\left(\dfrac{2}{3} - p\right)\left(\dfrac{2}{3} + p\right)$
$\dfrac{4}{9} - p^2$

64. $\left(\dfrac{1}{8} - q\right)\left(\dfrac{1}{8} + q\right)$
$\dfrac{1}{64} - q^2$

For Exercises 65–76, square the binomials. **(See Example 7.)**

65. $(a + 5)^2$
$a^2 + 10a + 25$

66. $(a - 3)^2$
$a^2 - 6a + 9$

67. $(x - y)^2$
$x^2 - 2xy + y^2$

68. $(x + y)^2$
$x^2 + 2xy + y^2$

69. $(2c + 5)^2$
$4c^2 + 20c + 25$

70. $(5d - 9)^2$
$25d^2 - 90d + 81$

71. $(3t^2 - 4s)^2$
$9t^4 - 24st^2 + 16s^2$

72. $(u^2 + 4v)^2$
$u^4 + 8u^2v + 16v^2$

73. $(7 - t)^2$
$t^2 - 14t + 49$

74. $(4 + w)^2$
$w^2 + 8w + 16$

75. $(3 + 4q)^2$
$16q^2 + 24q + 9$

76. $(2 - 3b)^2$
$9b^2 - 12b + 4$

77. a. Evaluate $(2 + 4)^2$ by working within the parentheses first.   36

   b. Evaluate $2^2 + 4^2$.   20

   c. Compare the answers to parts (a) and (b) and make a conjecture about $(a + b)^2$ and $a^2 + b^2$.
   $(a + b)^2 \neq a^2 + b^2$ in general.

78. a. Evaluate $(6 - 5)^2$ by working within the parentheses first.   1

   b. Evaluate $6^2 - 5^2$.   11

   c. Compare the answers to parts (a) and (b) and make a conjecture about $(a - b)^2$ and $a^2 - b^2$.
   $(a - b)^2 \neq a^2 - b^2$ in general.

79. a. Simplify $(3x + y)^2$. $9x^2 + 6xy + y^2$

   b. Simplify $(3xy)^2$. $9x^2y^2$

Objective 3: Applications to Geometry

80. Find a polynomial expression that represents the area of the rectangle shown in the figure.
$4x^2 - 25$

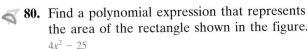

81. Find a polynomial expression that represents the area of the rectangle shown in the figure.
$36 - y^2$

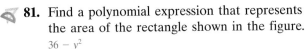

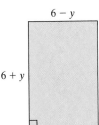

82. Find a polynomial expression that represents the area of the square shown in the figure.

$16p^2 + 40p + 25$

4p + 5

83. Find a polynomial expression that represents the area of the square shown in the figure.

$49q^2 - 42q + 9$

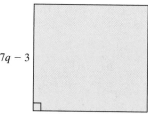

7q − 3

84. Find a polynomial that represents the volume of the cube shown in the figure.

(Recall: $V = s^3$) $27p^3 - 135p^2 + 225p - 125$

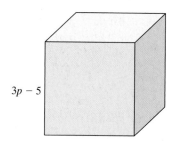

3p − 5

85. Find a polynomial that represents the volume of the rectangular solid shown in the figure.

(See Example 8.) (Recall: $V = lwh$) $r^3 - 15r^2 + 63r - 49$

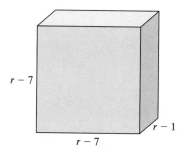

r − 7

r − 1

r − 7

86. Find a polynomial that represents the area of the triangle shown in the figure.

(Recall: $A = \frac{1}{2}bh$) $15a^5 - 6a^2$

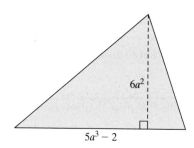

6a²

5a³ − 2

87. Find a polynomial that represents the area of the triangle shown in the figure. $3t^3 - 12t$

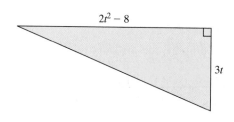

2t² − 8

3t

Mixed Exercises

For Exercises 88–117, multiply the expressions.

88. $(7x + y)(7x - y)$ $49x^2 - y^2$

89. $(9w - 4z)(9w + 4z)$ $81w^2 - 16z^2$

90. $(5s + 3t)^2$ $25s^2 + 30st + 9t^2$

91. $(5s - 3t)^2$ $25s^2 - 30st + 9t^2$

92. $(7x - 3y)(3x - 8y)$
$21x^2 - 65xy + 24y^2$

93. $(5a - 4b)(2a - b)$
$10a^2 - 13ab + 4b^2$

94. $\left(\frac{2}{3}t + 2\right)(3t + 4)$ $2t^2 + \frac{26}{3}t + 8$

95. $\left(\frac{1}{5}s + 6\right)(5s - 3)$ $s^2 + \frac{147}{5}s - 18$

96. $(5z + 3)(z^2 + 4z - 1)$
$5z^3 + 23z^2 + 7z - 3$

97. $(2k - 5)(2k^2 + 3k + 5)$
$4k^3 - 4k^2 - 5k - 25$

98. $(3a - 2)(5a + 1 + 2a^2)$
$6a^3 + 11a^2 - 7a - 2$

99. $(u + 4)(2 - 3u + u^2)$
$u^3 + u^2 - 10u + 8$

100. $(y^2 + 2y + 4)(y - 5)$
$y^3 - 3y^2 - 6y - 20$

101. $(w^2 - w + 6)(w + 2)$
$w^3 + w^2 + 4w + 12$

102. $\left(\dfrac{1}{3}m - n\right)^2$ $\dfrac{1}{9}m^2 - \dfrac{2}{3}mn + n^2$

103. $\left(\dfrac{2}{5}p - q\right)^2$ $\dfrac{4}{25}p^2 - \dfrac{4}{5}pq + q^2$

104. $6w^2(7w - 14)$ $42w^3 - 84w^2$

105. $4v^3(v + 12)$ $4v^4 + 48v^3$

106. $(4y - 8.1)(4y + 8.1)$
$16y^2 - 65.61$

107. $(2h + 2.7)(2h - 2.7)$
$4h^2 - 7.29$

108. $(3c^2 + 4)(7c^2 - 8)$
$21c^4 + 4c^2 - 32$

109. $(5k^3 - 9)(k^3 - 2)$
$5k^6 - 19k^3 + 18$

110. $(3.1x + 4.5)^2$
$9.61x^2 + 27.9x + 20.25$

111. $(2.5y + 1.1)^2$
$6.25y^2 + 5.5y + 1.21$

112. $(k - 4)^3$
$k^3 - 12k^2 + 48k - 64$

113. $(h + 3)^3$
$h^3 + 9h^2 + 27h + 27$

114. $(5x + 3)^3$
$125x^3 + 225x^2 + 135x + 27$

115. $(2a - 4)^3$
$8a^3 - 48a^2 + 96a - 64$

116. $(y^2 + 2y + 1)(2y^2 - y + 3)$
$2y^4 + 3y^3 + 3y^2 + 5y + 3$

117. $(2w^2 - w - 5)(3w^2 + 2w + 1)$
$6w^4 + w^3 - 15w^2 - 11w - 5$

Expanding Your Skills

For Exercises 118–121, multiply the expressions containing more than two factors.

118. $2a(3a - 4)(a + 5)$ $6a^3 + 22a^2 - 40a$

119. $5x(x + 2)(6x - 1)$ $30x^3 + 55x^2 - 10x$

120. $(x - 3)(2x + 1)(x - 4)$ $2x^3 - 13x^2 + 17x + 12$

121. $(y - 2)(2y - 3)(y + 3)$ $2y^3 - y^2 - 15y + 18$

122. What binomial when multiplied by $(3x + 5)$ will produce a product of $6x^2 - 11x - 35$? [*Hint:* Let the quantity $(a + b)$ represent the unknown binomial.] Then find a and b such that $(3x + 5)(a + b) = 6x^2 - 11x - 35$. $2x - 7$

123. What binomial when multiplied by $(2x - 4)$ will produce a product of $2x^2 + 8x - 24$? $x + 6$

For Exercises 124–126, determine what values of k would create a perfect square trinomial.

124. $x^2 + kx + 25$ $k = 10$ or -10

125. $w^2 + kw + 9$ $k = 6$ or -6

126. $a^2 + ka + 16$ $k = 8$ or -8

Division of Polynomials

Section 5.7

Division of polynomials will be presented in this section as two separate cases: The first case illustrates division by a monomial divisor. The second case illustrates long division by a polynomial with two or more terms.

Objectives

1. **Division by a Monomial**
2. **Long Division**

1. Division by a Monomial

To divide a polynomial by a monomial, divide each individual term in the polynomial by the divisor and simplify the result.

PROCEDURE Dividing a Polynomial by a Monomial

If a, b, and c are polynomials such that $c \neq 0$, then

$$\frac{a + b}{c} = \frac{a}{c} + \frac{b}{c} \quad \text{Similarly,} \quad \frac{a - b}{c} = \frac{a}{c} - \frac{b}{c}$$

- **Skill Practice** -

Divide the polynomials.

1. $(36a^4 - 48a^3 + 12a^2) \div (6a^3)$

2. $\dfrac{-15x^3y^4 + 25x^2y^3 - 5xy^2}{-5xy^2}$

Classroom Examples: p. 413, Exercises 26 and 28

Example 1 Dividing a Polynomial by a Monomial

Divide the polynomials.

a. $\dfrac{5a^3 - 10a^2 + 20a}{5a}$

b. $(12y^2z^3 - 15yz^2 + 6y^2z) \div (-6y^2z)$

Solution:

a. $\dfrac{5a^3 - 10a^2 + 20a}{5a}$

$= \dfrac{5a^3}{5a} - \dfrac{10a^2}{5a} + \dfrac{20a}{5a}$ Divide each term in the numerator by $5a$.

$= a^2 - 2a + 4$ Simplify each term using the properties of exponents.

b. $(12y^2z^3 - 15yz^2 + 6y^2z) \div (-6y^2z)$

$= \dfrac{12y^2z^3 - 15yz^2 + 6y^2z}{-6y^2z}$

$= \dfrac{12y^2z^3}{-6y^2z} - \dfrac{15yz^2}{-6y^2z} + \dfrac{6y^2z}{-6y^2z}$ Divide each term by $-6y^2z$.

$= -2z^2 + \dfrac{5z}{2y} - 1$ Simplify each term.

Instructor Note: In Example 1(b), mention to students that the third term in the quotient is -1 not 0.

2. Long Division

If the divisor has two or more terms, a *long division* process similar to the division of real numbers is used. Take a minute to review the long division process for real numbers by dividing 2273 by 5.

$$
\begin{array}{r}
454 \leftarrow \text{Quotient} \\
5\overline{)2273} \\
-20 \\
\hline
27 \\
-25 \\
\hline
23 \\
-20 \\
\hline
3 \leftarrow \text{Remainder}
\end{array}
$$

Therefore, $2273 \div 5 = 454\frac{3}{5}$

A similar procedure is used for long division of polynomials as shown in Example 2.

Answers

1. $6a - 8 + \dfrac{2}{a}$

2. $3x^2y^2 - 5xy + 1$

| Example 2 | **Using Long Division to Divide Polynomials**

Divide the polynomials using long division: $(2x^2 - x + 3) \div (x - 3)$

Skill Practice

Divide the polynomials using long division.

3. $(3x^2 + 2x - 5) \div (x + 2)$

Solution:

$x - 3 \overline{)2x^2 - x + 3}$ Divide the leading term in the dividend by the leading term in the divisor.

$$\frac{2x^2}{x} = 2x$$

This is the first term in the quotient.

$\begin{array}{r} 2x \\ x - 3 \overline{)2x^2 - x + 3} \\ -(2x^2 - 6x) \end{array}$ Multiply $2x$ by the divisor $2x(x - 3) = 2x^2 - 6x$ and subtract the result.

$\begin{array}{r} 2x \\ x - 3 \overline{)2x^2 - x + 3} \\ -2x^2 + 6x \\ \hline 5x \end{array}$ Subtract the quantity $2x^2 - 6x$. To do this, add the opposite.

$\begin{array}{r} 2x + 5 \\ x - 3 \overline{)2x^2 - x + 3} \\ -2x^2 + 6x \downarrow \\ \hline 5x + 3 \end{array}$ Bring down the next column, and repeat the process.
Divide the leading term by x: $(5x)/x = 5$. Place 5 in the quotient.

$\begin{array}{r} 2x + 5 \\ x - 3 \overline{)2x^2 - x + 3} \\ -2x^2 + 6x \\ \hline 5x + 3 \\ -(5x - 15) \end{array}$ Multiply the divisor by 5: $5(x - 3) = 5x - 15$ and subtract the result.

$\begin{array}{r} 2x + 5 \\ x - 3 \overline{)2x^2 - x + 3} \\ -2x^2 + 6x \\ \hline 5x + 3 \\ -5x + 15 \\ \hline 18 \end{array}$ Subtract the quantity $5x - 15$ by adding the opposite.
The remainder is 18.

Summary:

The quotient is $2x + 5$
The remainder is 18
The divisor is $x - 3$
The dividend is $2x^2 - x + 3$

The solution to a long division problem is usually written in the form:

$$\text{quotient} + \frac{\text{remainder}}{\text{divisor}}$$

Hence

$$(2x^2 - x + 3) \div (x - 3) = 2x + 5 + \frac{18}{x - 3}$$

The division of polynomials can be checked in the same fashion as the division of real numbers. To check Example 2, we have:

Classroom Example: p. 413, Exercise 34

Answer

3. $3x - 4 + \dfrac{3}{x + 2}$

$$\text{Dividend} = (\text{divisor})(\text{quotient}) + \text{remainder}$$

$$2x^2 - x + 3 \overset{?}{=} (x - 3)(2x + 5) + (18)$$

$$\overset{?}{=} 2x^2 + 5x - 6x - 15 + (18)$$

$$= 2x^2 - x + 3 ✔$$

Skill Practice

Divide the polynomials using long division.

4. $\dfrac{9x^3 + 11x + 10}{3x + 2}$

Classroom Example: p. 414, Exercise 50

Example 3 Using Long Division to Divide Polynomials

Divide the polynomials using long division: $(2w^3 + 8w^2 - 16) \div (2w + 4)$

Solution:

First note that the dividend has a missing power of w and can be written as $2w^3 + 8w^2 + 0w - 16$. The term $0w$ is a place holder for the missing term. It is helpful to use the place holder to keep the powers of w lined up.

$$
\begin{array}{r}
w^2 \phantom{{}+ 0w - 16} \\
2w + 4 \overline{)\, 2w^3 + 8w^2 + 0w - 16 } \\
-(2w^3 + 4w^2) \phantom{{} + 0w - 16}
\end{array}
$$

Divide $2w^3 \div 2w = w^2$. This is the first term of the quotient.
Then multiply $w^2(2w + 4) = 2w^3 + 4w^2$.

$$
\begin{array}{r}
w^2 \phantom{{}+ 0w - 16} \\
2w + 4 \overline{)\, 2w^3 + 8w^2 + 0w - 16 } \\
-2w^3 - 4w^2 \phantom{{} + 0w - 16} \\
\hline
4w^2 + 0w \phantom{{}- 16}
\end{array}
$$

Subtract by adding the opposite.

Bring down the next column, and repeat the process.

$$
\begin{array}{r}
w^2 + 2w \phantom{{}- 16} \\
2w + 4 \overline{)\, 2w^3 + 8w^2 + 0w - 16 } \\
-2w^3 - 4w^2 \phantom{{} + 0w - 16} \\
\hline
4w^2 + 0w \phantom{{}- 16} \\
-(4w^2 + 8w) \phantom{{}- 16}
\end{array}
$$

Divide $4w^2$ by the leading term in the divisor. $4w^2 \div 2w = 2w$. Place $2w$ in the quotient.
Multiply $2w(2w + 4) = 4w^2 + 8w$.

$$
\begin{array}{r}
w^2 + 2w \phantom{{}- 16} \\
2w + 4 \overline{)\, 2w^3 + 8w^2 + 0w - 16 } \\
-2w^3 - 4w^2 \phantom{{} + 0w - 16} \\
\hline
4w^2 + 0w \phantom{{}- 16} \\
-4w^2 - 8w \phantom{{}- 16} \\
\hline
-8w - 16
\end{array}
$$

Subtract by adding the opposite.

Bring down the next column and repeat.

$$
\begin{array}{r}
w^2 + 2w - 4 \\
2w + 4 \overline{)\, 2w^3 + 8w^2 + 0w - 16 } \\
-2w^3 - 4w^2 \phantom{{} + 0w - 16} \\
\hline
4w^2 + 0w \phantom{{}- 16} \\
-4w^2 - 8w \phantom{{}- 16} \\
\hline
-8w - 16 \\
-(-8w - 16)
\end{array}
$$

Divide $-8w$ by the leading term in the divisor. $-8w \div 2w = -4$. Place -4 in the quotient.
Multiply $-4(2w + 4) = -8w - 16$.

$$
\begin{array}{r}
w^2 + 2w - 4 \\
2w + 4 \overline{)\, 2w^3 + 8w^2 + 0w - 16 } \\
-2w^3 - 4w^2 \phantom{{} + 0w - 16} \\
\hline
4w^2 + 0w \phantom{{}- 16} \\
-4w^2 - 8w \phantom{{}- 16} \\
\hline
-8w - 16 \\
8w + 16 \\
\hline
0
\end{array}
$$

Subtract by adding the opposite.

The remainder is 0.

The quotient is $w^2 + 2w - 4$, and the remainder is 0.

Answer

4. $3x^2 - 2x + 5$

In Example 3, the remainder is zero. Therefore, we say that $2w + 4$ divides *evenly* into $2w^3 + 8w^2 - 16$. For this reason, the divisor and quotient are factors of $2w^3 + 8w^2 - 16$. To check, we have

$$\text{Dividend} = (\text{divisor})(\text{quotient}) + \text{remainder}$$

$$2w^3 + 8w^2 - 16 \stackrel{?}{=} (2w + 4)(w^2 + 2w - 4) + 0$$

$$\stackrel{?}{=} 2w^3 + 4w^2 - 8w + 4w^2 + 8w - 16$$

$$= 2w^3 + 8w^2 - 16 \checkmark$$

Example 4 **Using Long Division to Divide Polynomials**

Divide the polynomials using long division.

$$\frac{2y + y^4 - 5}{1 + y^2}$$

Solution:

First note that both the dividend and divisor should be written in descending order:

$$\frac{y^4 + 2y - 5}{y^2 + 1}$$

Also note that the dividend and the divisor have missing powers of y. Leave place holders.

$$y^2 + 0y + 1\overline{\smash{)}y^4 + 0y^3 + 0y^2 + 2y - 5}$$

$$\begin{array}{r} y^2 \\ y^2 + 0y + 1\overline{\smash{)}y^4 + 0y^3 + 0y^2 + 2y - 5} \\ -(y^4 + 0y^3 + y^2) \end{array}$$
Divide $y^4 \div y^2 = y^2$. This is the first term of the quotient.

Multiply $y^2(y^2 + 0y + 1) = y^4 + 0y^3 + y^2$.

$$\begin{array}{r} y^2 \\ y^2 + 0y + 1\overline{\smash{)}y^4 + 0y^3 + 0y^2 + 2y - 5} \\ \underline{-y^4 - 0y^3 - y^2} \\ -y^2 + 2y - 5 \end{array}$$
Subtract by adding the opposite.

Bring down the next columns.

$$\begin{array}{r} y^2 \qquad -1 \\ y^2 + 0y + 1\overline{\smash{)}y^4 + 0y^3 + 0y^2 + 2y - 5} \\ \underline{-y^4 - 0y^3 - y^2} \\ -y^2 + 2y - 5 \\ -(-y^2 - 0y - 1) \end{array}$$
Divide $-y^2 \div y^2 = -1$.

Multiply $-1(y^2 + 0y + 1) = -y^2 - 0y - 1$.

$$\begin{array}{r} y^2 \qquad -1 \\ y^2 + 0y + 1\overline{\smash{)}y^4 + 0y^3 + 0y^2 + 2y - 5} \\ \underline{-y^4 - 0y^3 - y^2} \\ -y^2 + 2y - 5 \\ \underline{y^2 + 0y + 1} \\ 2y - 4 \end{array}$$
Subtract by adding the opposite.

Remainder

Therefore, $\dfrac{y^4 + 2y - 5}{y^2 + 1} = y^2 - 1 + \dfrac{2y - 4}{y^2 + 1}$

Classroom Example: p. 414, Exercise 56

Skill Practice

Divide the polynomials using long division.

5. $(4 - x^2 + x^3) \div (2 + x^2)$

Answer

5. $x - 1 + \dfrac{-2x + 6}{x^2 + 2}$

Skill Practice

Divide the polynomials using the appropriate method of division.

6. $\dfrac{6x^3 - x^2 + 3x - 5}{2x + 3}$

7. $\dfrac{9w^3 - 18w^2 + 6w + 12}{3w}$

Classroom Examples: p. 414, Exercises 64 and 68

Answers

6. $3x^2 - 5x + 9 + \dfrac{-32}{2x + 3}$

7. $3w^2 - 6w + 2 + \dfrac{4}{w}$

Example 5 **Determining Whether Long Division Is Necessary**

Determine whether long division is necessary for each division of polynomials.

a. $\dfrac{2p^5 - 8p^4 + 4p - 16}{p^2 - 2p + 1}$

b. $\dfrac{2p^5 - 8p^4 + 4p - 16}{2p^2}$

c. $(3z^3 - 5z^2 + 10) \div (15z^3)$

d. $(3z^3 - 5z^2 + 10) \div (3z + 1)$

Solution:

a. $\dfrac{2p^5 - 8p^4 + 4p - 16}{p^2 - 2p + 1}$ The divisor has three terms. Use long division.

b. $\dfrac{2p^5 - 8p^4 + 4p - 16}{2p^2}$ The divisor has one term. No long division.

c. $(3z^3 - 5z^2 + 10) \div (15z^3)$ The divisor has one term. No long division.

d. $(3z^3 - 5z^2 + 10) \div (3z + 1)$ The divisor has two terms. Use long division.

> **TIP:** Recall that
> - Long division is used when the divisor has *two or more terms*.
> - If the divisor has *one term*, then divide each term in the dividend by the monomial divisor.

Section 5.7 Practice Exercises

Boost *your* GRADE at ALEKS.com!

ALEKS version 3.0

- Practice Problems
- Self-Tests
- NetTutor
- e-Professors
- Videos

For additional exercises, see Classroom Activities 5.7A–5.7B in the *Instructor's Resource Manual* at www.mhhe.com/moh.

Review Exercises

For Exercises 1–10, perform the indicated operations.

1. $(6z^5 - 2z^3 + z - 6) - (10z^4 + 2z^3 + z^2 + z)$
$6z^5 - 10z^4 - 4z^3 - z^2 - 6$

2. $(7a^2 + a - 6) + (2a^2 + 5a + 11)$ $9a^2 + 6a + 5$

3. $(10x + y)(x - 3y)$ $10x^2 - 29xy - 3y^2$

4. $8b^2(2b^2 - 5b + 12)$ $16b^4 - 40b^3 + 96b^2$

5. $(10x + y) + (x - 3y)$ $11x - 2y$

6. $(2w^3 + 5)^2$ $4w^6 + 20w^3 + 25$

7. $\left(\dfrac{4}{3}y^2 - \dfrac{1}{2}y + \dfrac{3}{8}\right) - \left(\dfrac{1}{3}y^2 + \dfrac{1}{4}y - \dfrac{1}{8}\right)$ $y^2 - \dfrac{3}{4}y + \dfrac{1}{2}$

8. $\left(\dfrac{7}{8}w - 1\right)\left(\dfrac{7}{8}w + 1\right)$ $\dfrac{49}{64}w^2 - 1$

9. $(a + 3)(a^2 - 3a + 9)$ $a^3 + 27$

10. $(2x + 1)(5x - 3)$ $10x^2 - x - 3$

Objective 1: Division by a Monomial

11. There are two methods for dividing polynomials. Explain when long division is used.

Use long division when the divisor is a polynomial with two or more terms.

12. Explain how to check a polynomial division problem.

Multiply the quotient by the divisor and add the remainder. The result should equal the dividend.

 Writing   Translating Expression   Geometry Scientific Calculator   Video   NS&E

13. a. Divide $\dfrac{15t^3 + 18t^2}{3t}$ $5t^2 + 6t$

    b. Check by multiplying the quotient by the divisor.

14. a. Divide $(-9y^4 + 6y^2 - y) \div (3y)$ $-3y^3 + 2y - \dfrac{1}{3}$

    b. Check by multiplying the quotient by the divisor.

For Exercises 15–30, divide the polynomials. **(See Example 1.)**

15. $(6a^2 + 4a - 14) \div (2)$ $3a^2 + 2a - 7$

16. $\dfrac{4b^2 + 16b - 12}{4}$ $b^2 + 4b - 3$

17. $\dfrac{-5x^2 - 20x + 5}{-5}$ $x^2 + 4x - 1$

18. $\dfrac{-3y^3 + 12y - 6}{-3}$ $y^3 - 4y + 2$

19. $\dfrac{3p^3 - p^2}{p}$ $3p^2 - p$

20. $(7q^4 + 5q^2) \div q$ $7q^3 + 5q$

21. $(4m^2 + 8m) \div 4m^2$ $1 + \dfrac{2}{m}$

22. $\dfrac{n^2 - 8}{n}$ $n - \dfrac{8}{n}$

23. $\dfrac{14y^4 - 7y^3 + 21y^2}{-7y^2}$ $-2y^2 + y - 3$

24. $(25a^5 - 5a^4 + 15a^3 - 5a) \div (-5a)$ $-5a^4 + a^3 - 3a^2 + 1$

25. $(4x^3 - 24x^2 - x + 8) \div (4x)$ $x^2 - 6x - \dfrac{1}{4} + \dfrac{2}{x}$

26. $\dfrac{20w^3 + 15w^2 - w + 5}{10w}$ $2w^2 + \dfrac{3}{2}w - \dfrac{1}{10} + \dfrac{1}{2w}$

27. $\dfrac{-a^3b^2 + a^2b^2 - ab^3}{-a^2b^2}$ $a - 1 + \dfrac{b}{a}$

28. $(3x^4y^3 - x^2y^2 - xy^3) \div (-x^2y^2)$ $-3x^2y + 1 + \dfrac{y}{x}$

29. $(6t^4 - 2t^3 + 3t^2 - t + 4) \div (2t^3)$
$3t - 1 + \dfrac{3}{2t} - \dfrac{1}{2t^2} + \dfrac{2}{t^3}$

30. $\dfrac{2y^3 - 2y^2 + 3y - 9}{2y^2}$ $y - 1 + \dfrac{3}{2y} - \dfrac{9}{2y^2}$

Objective 2: Long Division

31. a. Divide $(z^2 + 7z + 11) \div (z + 5)$ $z + 2 + \dfrac{1}{z + 5}$

    b. Check by multiplying the quotient by the divisor and adding the remainder.

32. a. Divide $\dfrac{2w^2 - 7w + 3}{w - 4}$ $2w + 1 + \dfrac{7}{w - 4}$

    b. Check by multiplying the quotient by the divisor and adding the remainder.

For Exercises 33–56, divide the polynomials. **(See Examples 2–4.)**

33. $\dfrac{t^2 + 4t + 5}{t + 1}$ $t + 3 + \dfrac{2}{t + 1}$

34. $(3x^2 + 8x + 5) \div (x + 2)$ $3x + 2 + \dfrac{3}{x + 2}$

35. $(7b^2 - 3b - 4) \div (b - 1)$ $7b + 4$

36. $\dfrac{w^2 - w - 2}{w - 2}$ $w + 1$

37. $\dfrac{5k^2 - 29k - 6}{5k + 1}$ $k - 6$

38. $(4y^2 + 25y - 21) \div (4y - 3)$ $y + 7$

39. $(4p^3 + 12p^2 + p - 12) \div (2p + 3)$ $2p^2 + 3p - 4$

40. $\dfrac{12a^3 - 2a^2 - 17a - 5}{3a + 1}$ $4a^2 - 2a - 5$

 Writing   Translating Expression   Geometry   Scientific Calculator   Video   NS E

41. $\dfrac{-k - 6 + k^2}{1 + k}$ $k - 2 + \dfrac{-4}{k + 1}$

42. $(1 + h^2 + 3h) \div (2 + h)$ $h + 1 + \dfrac{-1}{h + 2}$

43. $(4x^3 - 8x^2 + 15x - 16) \div (2x - 3)$
$2x^2 - x + 6 + \dfrac{2}{2x - 3}$

44. $\dfrac{3b^3 + b^2 + 17b - 49}{3b - 5}$ $b^2 + 2b + 9 + \dfrac{-4}{3b - 5}$

45. $\dfrac{3y^3 + 5y^2 + y + 1}{3y - 1}$ $y^2 + 2y + 1 + \dfrac{2}{3y - 1}$

46. $\dfrac{4t^3 + 4t^2 - 9t + 3}{2t + 3}$ $2t^2 - t - 3 + \dfrac{12}{2t + 3}$

47. $\dfrac{9 + a^2}{a + 3}$ $a - 3 + \dfrac{18}{a + 3}$

48. $(3 + m^2) \div (m + 3)$ $m - 3 + \dfrac{12}{m + 3}$

49. $(4x^3 - 3x - 26) \div (x - 2)$ $4x^2 + 8x + 13$

50. $(4y^3 + y + 1) \div (2y + 1)$ $2y^2 - y + 1$

51. $(w^4 + 5w^3 - 5w^2 - 15w + 7) \div (w^2 - 3)$
$w^2 + 5w - 2 + \dfrac{1}{w^2 - 3}$

52. $\dfrac{p^4 - p^3 - 4p^2 - 2p - 15}{p^2 + 2}$ $p^2 - p - 6 + \dfrac{-3}{p^2 + 2}$

53. $\dfrac{2n^4 + 5n^3 - 11n^2 - 20n + 12}{2n^2 + 3n - 2}$
$n^2 + n - 6$

54. $(6y^4 - 5y^3 - 8y^2 + 16y - 8) \div (2y^2 - 3y + 2)$
$3y^2 + 2y - 4$

55. $(5x^3 - 4x - 9) \div (5x^2 + 5x + 1)$
$x - 1 + \dfrac{-8}{5x^2 + 5x + 1}$

56. $\dfrac{3a^3 - 5a + 16}{3a^2 - 6a + 7}$ $a + 2 + \dfrac{2}{3a^2 - 6a + 7}$

57. Show that $(x^3 - 8) \div (x - 2)$ is *not* $(x^2 + 4)$.
Multiply $(x - 2)(x^2 + 4) = x^3 - 2x^2 + 4x - 8$, which does not equal $x^3 - 8$.

58. Explain why $(y^3 + 27) \div (y + 3)$ is *not* $(y^2 + 9)$.
To check, multiply $(y + 3)(y^2 + 9) = y^3 + 3y^2 + 9y + 27$, which does not equal $y^3 + 27$.

Mixed Exercises

For Exercises 59–70, determine which method to use to divide the polynomials: monomial division or long division. Then use that method to divide the polynomials. **(See Example 5.)**

59. $\dfrac{9a^3 + 12a^2}{3a}$ Monomial division; $3a^2 + 4a$

60. $\dfrac{3y^2 + 17y - 12}{y + 6}$ Long division; $3y - 1 + \dfrac{-6}{y + 6}$

61. $(p^3 + p^2 - 4p - 4) \div (p^2 - p - 2)$
Long division; $p + 2$

62. $(q^3 + 1) \div (q + 1)$
Long division; $q^2 - q + 1$

63. $\dfrac{t^4 + t^2 - 16}{t + 2}$
Long division; $t^3 - 2t^2 + 5t - 10 + \dfrac{4}{t + 2}$

64. $\dfrac{-8m^5 - 4m^3 + 4m^2}{-2m^2}$
Monomial division; $4m^3 + 2m - 2$

65. $(w^4 + w^2 - 5) \div (w^2 - 2)$
Long division; $w^2 + 3 + \dfrac{1}{w^2 - 2}$

66. $(2k^2 + 9k + 7) \div (k + 1)$
Long division; $2k + 7$

67. $\dfrac{n^3 - 64}{n - 4}$
Long division; $n^2 + 4n + 16$

68. $\dfrac{15s^2 + 34s + 28}{5s + 3}$
Long division; $3s + 5 + \dfrac{13}{5s + 3}$

69. $(9r^3 - 12r^2 + 9) \div (-3r^2)$
Monomial division; $-3r + 4 - \dfrac{3}{r^2}$

70. $(6x^4 - 16x^3 + 15x^2 - 5x + 10) \div (3x + 1)$
Long division; $2x^3 - 6x^2 + 7x - 4 + \dfrac{14}{3x + 1}$

Expanding Your Skills

For Exercises 71–78, divide the polynomials and note any patterns.

71. $(x^2 - 1) \div (x - 1)$
$x + 1$

72. $(x^3 - 1) \div (x - 1)$
$x^2 + x + 1$

73. $(x^4 - 1) \div (x - 1)$
$x^3 + x^2 + x + 1$

74. $(x^5 - 1) \div (x - 1)$
$x^4 + x^3 + x^2 + x + 1$

75. $x^2 \div (x - 1)$ $x + 1 + \dfrac{1}{x - 1}$

76. $x^3 \div (x - 1)$ $x^2 + x + 1 + \dfrac{1}{x - 1}$

77. $x^4 \div (x - 1)$
$x^3 + x^2 + x + 1 + \dfrac{1}{x - 1}$

78. $x^5 \div (x - 1)$
$x^4 + x^3 + x^2 + x + 1 + \dfrac{1}{x - 1}$

✏️ Writing ↔ Translating Expression 📐 Geometry 🖩 Scientific Calculator 💿 Video NS E

Problem Recognition Exercises

Operations on Polynomials

Perform the indicated operations and simplify.

1. $(2x - 4)(x^2 - 2x + 3)$
$2x^3 - 8x^2 + 14x - 12$

2. $(3y^2 + 8)(-y^2 - 4)$
$-3y^4 - 20y^2 - 32$

3. $(2x - 4) + (x^2 - 2x + 3)$
$x^2 - 1$

4. $(3y^2 + 8) - (-y^2 - 4)$
$4y^2 + 12$

5. $(6y - 7)^2$
$36y^2 - 84y + 49$

6. $(3z + 2)^2$
$9z^2 + 12z + 4$

7. $(6y - 7)(6y + 7)$
$36y^2 - 49$

8. $(3z + 2)(3z - 2)$
$9z^2 - 4$

9. $(4x + y)^2$
$16x^2 + 8xy + y^2$

10. $(2a + b)^2$
$4a^2 + 4ab + b^2$

11. $(4xy)^2$
$16x^2y^2$

12. $(2ab)^2$
$4a^2b^2$

13. $(-2x^4 - 6x^3 + 8x^2) \div (2x^2)$ $-x^2 - 3x + 4$

14. $(-15m^3 + 12m^2 - 3m) \div (-3m)$ $5m^2 - 4m + 1$

15. $(m^3 - 4m^2 - 6) - (3m^2 + 7m) + (-m^3 - 9m + 6)$
$-7m^2 - 16m$

16. $(n^4 + 2n^2 - 3n) + (4n^2 + 2n - 1) - (4n^5 + 6n - 3)$
$-4n^5 + n^4 + 6n^2 - 7n + 2$

17. $(8x^3 + 2x + 6) \div (x - 2)$ $8x^2 + 16x + 34 + \dfrac{74}{x - 2}$

18. $(-4x^3 + 2x^2 - 5) \div (x - 3)$ $-4x^2 - 10x - 30 + \dfrac{-95}{x - 3}$

19. $(2x - y)(3x^2 + 4xy - y^2)$ $6x^3 + 5x^2y - 6xy^2 + y^3$

20. $(3a + b)(2a^2 - ab + 2b^2)$ $6a^3 - a^2b + 5ab^2 + 2b^3$

21. $(x + y^2)(x^2 - xy^2 + y^4)$ $x^3 + y^6$

22. $(m^2 + 1)(m^4 - m^2 + 1)$ $m^6 + 1$

23. $(a^2 + 2b) - (a^2 - 2b)$
$4b$

24. $(y^3 - 6z) - (y^3 + 6z)$
$-12z$

25. $(a^2 + 2b)(a^2 - 2b)$
$a^4 - 4b^2$

26. $(y^3 - 6z)(y^3 + 6z)$
$y^6 - 36z^2$

27. $(8u + 3v)^2$
$64u^2 + 48uv + 9v^2$

28. $(2p - t)^2$
$4p^2 - 4pt + t^2$

29. $\dfrac{8p^2 + 4p - 6}{2p - 1}$
$4p + 4 + \dfrac{-2}{2p - 1}$

30. $\dfrac{4v^2 - 8v + 8}{2v + 3}$
$2v - 7 + \dfrac{29}{2v + 3}$

31. $\dfrac{12x^3y^7}{3xy^5}$ $4x^2y^2$

32. $\dfrac{-18p^2q^4}{2pq^3}$ $-9pq$

33. $(2a - 9)(5a - 6)$
$10a^2 - 57a + 54$

34. $(7a + 1)(4a - 3)$
$28a^2 - 17a - 3$

35. $\left(\dfrac{3}{7}x - \dfrac{1}{2}\right)\left(\dfrac{3}{7}x + \dfrac{1}{2}\right)$
$\dfrac{9}{49}x^2 - \dfrac{1}{4}$

36. $\left(\dfrac{2}{5}y + \dfrac{4}{3}\right)\left(\dfrac{2}{5}y - \dfrac{4}{3}\right)$
$\dfrac{4}{25}y^2 - \dfrac{16}{9}$

37. $\left(\dfrac{1}{9}x^3 + \dfrac{2}{3}x^2 + \dfrac{1}{6}x - 3\right) - \left(\dfrac{4}{3}x^3 + \dfrac{1}{9}x^2 + \dfrac{2}{3}x + 1\right)$
$-\dfrac{11}{9}x^3 + \dfrac{5}{9}x^2 - \dfrac{1}{2}x - 4$

38. $\left(\dfrac{1}{10}y^2 - \dfrac{3}{5}y - \dfrac{1}{15}\right) - \left(\dfrac{7}{5}y^2 + \dfrac{3}{10}y - \dfrac{1}{3}\right)$
$-\dfrac{13}{10}y^2 - \dfrac{9}{10}y + \dfrac{4}{15}$

39. $(0.05x^2 - 0.16x - 0.75) + (1.25x^2 - 0.14x + 0.25)$
$1.3x^2 - 0.3x - 0.5$

40. $(1.6w^3 + 2.8w + 6.1) + (3.4w^3 - 4.1w^2 - 7.3)$
$5w^3 - 4.1w^2 + 2.8w - 1.2$

Group Activity

The Pythagorean Theorem and a Geometric "Proof"

Estimated Time: 10–15 minutes

Group Size: 2

Right triangles occur in many applications of mathematics. By definition, a right triangle is a triangle that contains a 90° angle. The two shorter sides in a right triangle are referred to as the "legs," and the longest side is called the "hypotenuse." In the triangle, the legs are labeled as a and b, and the hypotenuse is labeled as c.

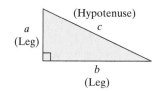

Right triangles have an important property that the sum of the squares of the two legs of a right triangle equals the square of the hypotenuse. This fact is referred to as the Pythagorean theorem. In symbols, the Pythagorean theorem is stated as:

$$a^2 + b^2 = c^2$$

1. The following triangles are right triangles. Verify that $a^2 + b^2 = c^2$. (The units may be left off when performing these calculations.)

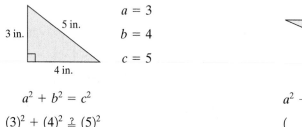

$a = 3$
$b = 4$
$c = 5$

$a =$ _____
$b =$ _____
$c =$ _____

$a^2 + b^2 = c^2$

$(3)^2 + (4)^2 \stackrel{?}{=} (5)^2$

$9 + 16 = 25$ ✔

$a^2 + b^2 = c^2$

$(\underline{\quad})^2 + (\underline{\quad})^2 \stackrel{?}{=} (\underline{\quad})^2$

$\underline{\quad} + \underline{\quad} = \underline{\quad}$ ✔

2. The following geometric "proof" of the Pythagorean theorem uses addition, subtraction, and multiplication of polynomials. Consider the square figure. The length of each side of the large outer square is $(a + b)$. Therefore, the area of the large outer square is $(a + b)^2$.

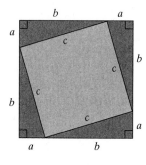

The area of the large outer square can also be found by adding the area of the inner square (pictured in light gray) plus the area of the four right triangles (pictured in dark gray).

Area of inner square: c^2 Area of the four right triangles: $4 \cdot (\frac{1}{2} a\ b)$

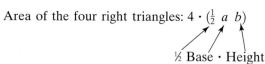

½ Base · Height

3. Now equate the two expressions representing the area of the large outer square:

$$\left(\begin{array}{c}\text{Area of outer}\\\text{square}\end{array}\right) = \left(\begin{array}{c}\text{area of inner}\\\text{square}\end{array}\right) + \left(\begin{array}{c}\text{4 times the area}\\\text{of the right triangles}\end{array}\right)$$

$(a + b)^2 = c^2 + 4 \cdot (\frac{1}{2}ab)$;

$a^2 + 2ab + b^2 = c^2 + 2ab$;

$a^2 + b^2 = c^2$

$(a + b)^2 \quad = \quad c^2 \quad + \quad 4 \cdot (\frac{1}{2}ab)$

$\underline{\qquad} = \underline{\qquad} + \underline{\qquad}$ ⟵ Clear parentheses on both sides of the equation.

$\underline{\qquad} = \underline{\qquad}$ ⟵ Subtract $2ab$ from both sides.

Chapter 5   Summary

Section 5.1   Exponents: Multiplying and Dividing Common Bases

Key Concepts

Definition

$$b^n = \underbrace{b \cdot b \cdot b \cdot b \cdot \ldots b}_{n \text{ factors of } b} \qquad \begin{array}{l} b \text{ is the base,} \\ n \text{ is the exponent} \end{array}$$

Multiplying Like Bases

$b^m b^n = b^{m+n}$ (m, n positive integers)

Dividing Like Bases

$\dfrac{b^m}{b^n} = b^{m-n}$ ($b \neq 0$, m, n, positive integers)

Examples

Example 1

$3^4 = 3 \cdot 3 \cdot 3 \cdot 3 = 81$ 3 is the base
4 is the exponent

Example 2

Compare: $(-5)^2$ versus -5^2

versus $\begin{cases} (-5)^2 = (-5)(-5) = 25 \\ \\ -5^2 = -1(5^2) = -1(5)(5) = -25 \end{cases}$

Example 3

Simplify: $x^3 \cdot x^4 \cdot x^2 \cdot x = x^{10}$

Example 4

Simplify: $\dfrac{c^4 d^{10}}{cd^5} = c^{4-1}d^{10-5} = c^3 d^5$

Section 5.2   More Properties of Exponents

Key Concepts

Power Rule for Exponents

$(b^m)^n = b^{mn}$ ($b \neq 0$, m, n positive integers)

Power of a Product and Power of a Quotient

Assume m and n are positive integers and a and b are real numbers where $b \neq 0$.
Then,

$(ab)^m = a^m b^m$ and $\left(\dfrac{a}{b}\right)^m = \dfrac{a^m}{b^m}$

Examples

Example 1

Simplify: $(x^4)^5 = x^{20}$

Example 2

Simplify: $(4uv^2)^3 = 4^3 u^3 (v^2)^3 = 64u^3 v^6$

Example 3

Simplify: $\left(\dfrac{p^5 q^3}{5pq^2}\right)^2 = \left(\dfrac{p^{5-1}q^{3-2}}{5}\right)^2 = \left(\dfrac{p^4 q}{5}\right)^2$

$$= \dfrac{p^8 q^2}{25}$$

Section 5.3   Definitions of b^0 and b^{-n}

Key Concepts

Definitions

If b is a nonzero real number and n is an integer, then:

1. $b^0 = 1$

2. $b^{-n} = \left(\dfrac{1}{b}\right)^n = \dfrac{1}{b^n}$

Examples

Example 1

Simplify: $4^0 = 1$

Example 2

Simplify: $y^{-7} = \dfrac{1}{y^7}$

Example 3

Simplify: $\dfrac{8a^0 b^{-2}}{c^{-5}d}$

$$= \dfrac{8(1)c^5}{b^2 d} = \dfrac{8c^5}{b^2 d}$$

Section 5.4   Scientific Notation

Key Concepts

A positive number written in **scientific notation** is expressed in the form:

$a \times 10^n$ where $1 \le a < 10$ and n is an integer.

Examples

Example 1

Write the numbers in scientific notation:

$35{,}000 = 3.5 \times 10^4$

$0.000\,000\,548 = 5.48 \times 10^{-7}$

Example 2

Multiply: $(3.5 \times 10^4)(2.0 \times 10^{-6})$

$$= 7.0 \times 10^{-2}$$

Example 3

Divide: $\dfrac{8.4 \times 10^{-9}}{2.1 \times 10^3} = 4.0 \times 10^{-9-3} = 4.0 \times 10^{-12}$

| Section 5.5 | Addition and Subtraction of Polynomials |

Key Concepts

A **polynomial** in one variable is a finite sum of terms of the form ax^n, where a is a real number and the exponent, n, is a nonnegative integer. For each term, a is called the **coefficient** of the term and n is the **degree of the term**. The term with highest degree is the **leading term**, and its coefficient is called the **leading coefficient**. The **degree of the polynomial** is the largest degree of all its terms.

To add or subtract polynomials, add or subtract *like* terms.

Examples

Example 1

Given: $4x^5 - 8x^3 + 9x - 5$

Coefficients of each term: $4, -8, 9, -5$

Degree of each term: $\quad 5, 3, 1, 0$

Leading term: $\qquad\quad 4x^5$

Leading coefficient: $\qquad 4$

Degree of polynomial: $\quad 5$

Example 2

Perform the indicated operations:

$$(2x^4 - 5x^3 + 1) - (x^4 + 3) + (x^3 - 4x - 7)$$
$$= 2x^4 - 5x^3 + 1 - x^4 - 3 + x^3 - 4x - 7$$
$$= 2x^4 - x^4 - 5x^3 + x^3 - 4x + 1 - 3 - 7$$
$$= x^4 - 4x^3 - 4x - 9$$

| Section 5.6 | Multiplication of Polynomials and Special Products |

Key Concepts

Multiplying Monomials

Use the commutative and associative properties of multiplication to group coefficients and like bases.

Multiplying Polynomials

Multiply each term in the first polynomial by each term in the second polynomial.

Product of Conjugates

Results in a **difference of squares**

$$(a + b)(a - b) = a^2 - b^2$$

Square of a Binomial

Results in a **perfect square trinomial**

$$(a + b)^2 = a^2 + 2ab + b^2$$
$$(a - b)^2 = a^2 - 2ab + b^2$$

Examples

Example 1

Multiply: $(5a^2b)(-2ab^3)$

$$= (5 \cdot -2)(a^2a)(bb^3)$$
$$= -10a^3b^4$$

Example 2

Multiply: $(x - 2)(3x^2 - 4x + 11)$

$$= 3x^3 - 4x^2 + 11x - 6x^2 + 8x - 22$$
$$= 3x^3 - 10x^2 + 19x - 22$$

Example 3

Multiply: $(3w - 4v)(3w + 4v)$

$$= (3w)^2 - (4v)^2$$
$$= 9w^2 - 16v^2$$

Example 4

Multiply: $(5c - 8d)^2$

$$= (5c)^2 - 2(5c)(8d) + (8d)^2$$
$$= 25c^2 - 80cd + 64d^2$$

Section 5.7 Division of Polynomials

Key Concepts

Division of Polynomials

1. Division by a monomial, use the properties:

$$\frac{a+b}{c} = \frac{a}{c} + \frac{b}{c} \quad \text{and} \quad \frac{a-b}{c} = \frac{a}{c} - \frac{b}{c}$$

2. If the divisor has more than one term, use long division.

Examples

Example 1

Divide: $\dfrac{-3x^2 - 6x + 9}{-3x}$

$$= \frac{-3x^2}{-3x} - \frac{6x}{-3x} + \frac{9}{-3x}$$

$$= x + 2 - \frac{3}{x}$$

Example 2

Divide: $(3x^2 - 5x + 1) \div (x + 2)$

$$
\begin{array}{r}
3x - 11 \\
x+2\overline{)3x^2 - 5x + 1} \\
-(3x^2 + 6x) \\
\hline
-11x + 1 \\
-(-11x - 22) \\
\hline
23
\end{array}
$$

$$3x - 11 + \frac{23}{x+2}$$

Chapter 5   Review Exercises

Section 5.1

For Exercises 1–4, identify the base and the exponent.

1. 5^3
Base: 5; exponent: 3

2. x^4
Base: x; exponent: 4

3. $(-2)^0$
Base: -2; exponent: 0

4. y
Base: y; exponent: 1

5. Evaluate the expressions.

    a. 6^2 36   **b.** $(-6)^2$ 36   **c.** -6^2 −36

6. Evaluate the expressions.

    a. 4^3 64   **b.** $(-4)^3$ −64   **c.** -4^3 −64

For Exercises 7–18, simplify and write the answers in exponent form. Assume that all variables represent nonzero real numbers.

7. $5^3 \cdot 5^{10}$ 5^{13}

8. $a^7 a^4$ a^{11}

9. $x \cdot x^6 \cdot x^2$ x^9

10. $6^3 \cdot 6 \cdot 6^5$ 6^9

11. $\dfrac{10^7}{10^4}$ 10^3

12. $\dfrac{y^{14}}{y^8}$ y^6

13. $\dfrac{b^9}{b}$ b^8

14. $\dfrac{7^8}{7}$ 7^7

15. $\dfrac{k^2 k^3}{k^4}$ k

16. $\dfrac{8^4 \cdot 8^7}{8^{11}}$ 1

17. $\dfrac{2^8 \cdot 2^{10}}{2^3 \cdot 2^7}$ 2^8

18. $\dfrac{q^3 q^{12}}{qq^8}$ q^6

19. Explain why $2^2 \cdot 4^4$ does *not* equal 8^6.
Exponents are added only when multiplying factors with the same base. In such a case, the base does not change.

20. Explain why $\frac{10^5}{5^2}$ does *not* equal 2^3.
Exponents are subtracted only when dividing factors with the same base. In such a case, the base does not change.

For Exercises 21–22, use the formula

$$A = P(1 + r)^t$$

21. Find the amount in an account after 3 yr if the initial investment is $6000, invested at 6% interest compounded annually. $7146.10

22. Find the amount in an account after 2 yr if the initial investment is $20,000, invested at 5% interest compounded annually. $22,050

Section 5.2

For Exercises 23–40, simplify the expressions. Write the answers in exponent form. Assume all variables represent nonzero real numbers.

23. $(7^3)^4$ 7^{12}

24. $(c^2)^6$ c^{12}

25. $(p^4 p^2)^3$ p^{18}

26. $(9^5 \cdot 9^2)^4$ 9^{28}

27. $\left(\dfrac{a}{b}\right)^2$ $\dfrac{a^2}{b^2}$

28. $\left(\dfrac{1}{3}\right)^4$ $\dfrac{1}{3^4}$

29. $\left(\dfrac{5}{c^2 d^5}\right)^2$ $\dfrac{5^2}{c^4 d^{10}}$

30. $\left(-\dfrac{m^2}{4n^6}\right)^5$ $-\dfrac{m^{10}}{4^5 n^{30}}$

31. $(2ab^2)^4$ $2^4 a^4 b^8$

32. $(-x^7 y)^2$ $x^{14} y^2$

33. $\left(\dfrac{-3x^3}{5y^2 z}\right)^3$ $-\dfrac{3^3 x^9}{5^3 y^6 z^3}$

34. $\left(\dfrac{r^3}{s^2 t^6}\right)^5$ $\dfrac{r^{15}}{s^{10} t^{30}}$

35. $\dfrac{a^4(a^2)^8}{(a^3)^3}$ a^{11}

36. $\dfrac{(8^3)^4 \cdot 8^{10}}{(8^4)^5}$ 8^2

37. $\dfrac{(4h^2 k)^2(h^3 k)^4}{(2hk^3)^2}$ $4h^{14}$

38. $\dfrac{(p^3 q)^3(2p^2 q^4)^4}{(8p)(pq^3)^2}$ $2p^{14} q^{13}$

39. $\left(\dfrac{2x^4 y^3}{4xy^2}\right)^2$ $\dfrac{x^6 y^2}{4}$

40. $\left(\dfrac{a^4 b^6}{ab^4}\right)^3$ $a^9 b^6$

Section 5.3

For Exercises 41–62, simplify the expressions. Assume all variables represent nonzero real numbers.

41. 8^0 1

42. $(-b)^0$ 1

43. $-x^0$ -1

44. 1^0 1

45. $2y^0$ 2

46. $(2y)^0$ 1

47. z^{-5} $\dfrac{1}{z^5}$

48. 10^{-4} $\dfrac{1}{10^4}$

49. $(6a)^{-2}$ $\dfrac{1}{36a^2}$

50. $6a^{-2}$ $\dfrac{6}{a^2}$

51. $4^0 + 4^{-2}$ $\dfrac{17}{16}$

52. $9^{-1} + 9^0$ $\dfrac{10}{9}$

53. $t^{-6} t^{-2}$ $\dfrac{1}{t^8}$

54. $r^8 r^{-9}$ $\dfrac{1}{r}$

55. $\dfrac{12x^{-2} y^3}{6x^4 y^{-4}}$ $\dfrac{2y^7}{x^6}$

56. $\dfrac{8ab^{-3} c^0}{10a^{-5} b^{-4} c^{-1}}$ $\dfrac{4a^6 bc}{5}$

57. $(-2m^2 n^{-4})^{-4}$ $\dfrac{n^{16}}{16m^8}$

58. $(3u^{-5} v^2)^{-3}$ $\dfrac{u^{15}}{27v^6}$

59. $\dfrac{(k^{-6})^{-2}(k^3)}{5k^{-6} k^0}$ $\dfrac{k^{21}}{5}$

60. $\dfrac{(3h)^{-2}(h^{-5})^{-3}}{h^{-4} h^8}$ $\dfrac{h^9}{9}$

61. $2 \cdot 3^{-1} - 6^{-1}$ $\dfrac{1}{2}$

62. $2^{-1} - 2^{-2} + 2^0$ $\dfrac{5}{4}$

Section 5.4

63. Write the numbers in scientific notation.

   a. In a recent year there were 97,400,000 packages of M&Ms sold in the United States. 9.74×10^7

   b. The thickness of a piece of paper is 0.0042 in. 4.2×10^{-3} in.

64. Write the numbers in standard form.

   a. A pH of 10 means the hydrogen ion concentration is 1×10^{-10} units. 0.000 000 0001

   b. A fund-raising event for neurospinal research raised 2.56×10^5. $256,000

For Exercises 65–68, perform the indicated operations. Write the answers in scientific notation.

65. $(4.1 \times 10^{-6})(2.3 \times 10^{11})$
9.43×10^5

66. $\dfrac{9.3 \times 10^3}{6.0 \times 10^{-7}}$ 1.55×10^{10}

67. $\dfrac{2000}{0.000008}$ 2.5×10^8

68. $(0.000078)(21,000,000)$ 1.638×10^3

69. Use your calculator to evaluate 5^{20}. Why is scientific notation necessary on your calculator to express the answer? $\approx 9.5367 \times 10^{13}$. This number is too big to fit on most calculator displays.

70. Use your calculator to evaluate $(0.4)^{30}$. Why is scientific notation necessary on your calculator to express the answer? $\approx 1.1529 \times 10^{-12}$. This number is too small to fit on most calculator displays.

71. The average distance between the Earth and Sun is 9.3×10^7 mi.

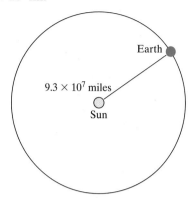

9.3 × 10⁷ miles

Earth

Sun

  a. If the Earth's orbit is approximated by a circle, find the total distance the Earth travels around the Sun in one orbit. (*Hint:* The circumference of a circle is given by $C = 2\pi r$.) Express the answer in scientific notation. $\approx 5.84 \times 10^8$ mi

  b. If the Earth makes one complete trip around the Sun in 1 yr (365 days = 8.76×10^3 hr), find the average speed that the Earth travels around the Sun in miles per hour. Express the answer in scientific notation. $\approx 6.67 \times 10^4$ mph

 72. The average distance between the planet Mercury and the Sun is 3.6×10^7 mi.

  a. If Mercury's orbit is approximated by a circle, find the total distance Mercury travels around the Sun in one orbit. (*Hint:* The circumference of a circle is given by $C = 2\pi r$.) Express the answer in scientific notation. $\approx 2.26 \times 10^8$ mi

  b. If Mercury makes one complete trip around the Sun in 88 days (2.112×10^3 hr), find the average speed that Mercury travels around the Sun in miles per hour. Express the answer in scientific notation. $\approx 1.07 \times 10^5$ mph

Section 5.5

73. For the polynomial $7x^4 - x + 6$

  a. Classify as a monomial, a binomial, or a trinomial.   Trinomial

  b. Identify the degree of the polynomial.   4

  c. Identify the leading coefficient.   7

74. For the polynomial $2y^3 - 5y^7$

  a. Classify as a monomial, a binomial, or a trinomial.   Binomial

  b. Identify the degree of the polynomial.   7

  c. Identify the leading coefficient. -5

For Exercises 75–80, add or subtract as indicated.

75. $(4x + 2) + (3x - 5)$ $7x - 3$

76. $(7y^2 - 11y - 6) - (8y^2 + 3y - 4)$ $-y^2 - 14y - 2$

77. $(9a^2 - 6) - (-5a^2 + 2a)$ $14a^2 - 2a - 6$

78. $\left(5x^3 - \dfrac{1}{4}x^2 + \dfrac{5}{8}x + 2\right) + \left(\dfrac{5}{2}x^3 + \dfrac{1}{2}x^2 - \dfrac{1}{8}x\right)$
$\dfrac{15}{2}x^3 + \dfrac{1}{4}x^2 + \dfrac{1}{2}x + 2$

79. $\begin{array}{r} 8w^4 \qquad\quad - 6w + 3 \\ + \ 2w^4 + 2w^3 - \ \ w + 1 \end{array}$ $10w^4 + 2w^3 - 7w + 4$

80. $\begin{array}{r} -0.02b^5 + b^4 \qquad\quad - 0.7b + 0.3 \\ + \ \ 0.03b^5 \qquad\quad - 0.1b^3 + \quad b + 0.03 \end{array}$
$0.01b^5 + b^4 - 0.1b^3 + 0.3b + 0.33$

81. Subtract $(9x^2 + 4x + 6)$ from $(7x^2 - 5x)$. $-2x^2 - 9x - 6$

82. Find the difference of $(x^2 - 5x - 3)$ and $(6x^2 + 4x + 9)$. $-5x^2 - 9x - 12$

83. Write a trinomial of degree 2 with a leading coefficient of -5. (Answers may vary.) For example, $-5x^2 + 2x - 4$

84. Write a binomial of degree 5 with leading coefficient 6. (Answers may vary.) For example, $6x^5 + 8$

85. Find a polynomial that represents the perimeter of the given rectangle. $6w + 6$

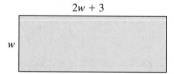

2w + 3

w

Section 5.6

For Exercises 86–103, multiply the expressions.

86. $(25x^4y^3)(-3x^2y)$
$-75x^6y^4$

87. $(9a^6)(2a^2b^4)$
$18a^8b^4$

88. $5c(3c^3 - 7c + 5)$
$15c^4 - 35c^2 + 25c$

89. $(x^2 + 5x - 3)(-2x)$
$-2x^3 - 10x^2 + 6x$

90. $(5k - 4)(k + 1)$
$5k^2 + k - 4$

91. $(4t - 1)(5t + 2)$
$20t^2 + 3t - 2$

92. $(q + 8)(6q - 1)$
$6q^2 + 47q - 8$

93. $(2a - 6)(a + 5)$
$2a^2 + 4a - 30$

94. $\left(7a + \dfrac{1}{2}\right)^2$
$49a^2 + 7a + \dfrac{1}{4}$

95. $(b - 4)^2$
$b^2 - 8b + 16$

96. $(4p^2 + 6p + 9)(2p - 3)$ $8p^3 - 27$

97. $(2w - 1)(-w^2 - 3w - 4)$ $-2w^3 - 5w^2 - 5w + 4$

98. $\begin{array}{r} 2x^2 - 3x + 4 \\ \times \qquad\quad 2x - 1 \end{array}$ $4x^3 - 8x^2 + 11x - 4$

99.
$$\begin{array}{r} 4a^2 + a - 5 \\ \times\ \underline{\quad 3a + 2\quad} \end{array}$$ $12a^3 + 11a^2 - 13a - 10$

100. $(b - 4)(b + 4)$ $b^2 - 16$

101. $\left(\dfrac{1}{3}r^4 - s^2\right)\left(\dfrac{1}{3}r^4 + s^2\right)$ $\dfrac{1}{9}r^8 - s^4$

102. $(-7z^2 + 6)^2$ $49z^4 - 84z^2 + 36$

103. $(2h + 3)(h^4 - h^3 + h^2 - h + 1)$
$2h^5 + h^4 - h^3 + h^2 - h + 3$

104. Find a polynomial that represents the area of the given rectangle $2x^2 + 3x - 20$

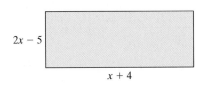

$2x - 5$

$x + 4$

Section 5.7

For Exercises 105–117, divide the polynomials.

105. $\dfrac{20y^3 - 10y^2}{5y}$ $4y^2 - 2y$

106. $(18a^3b^2 - 9a^2b - 27ab^2) \div 9ab$ $2a^2b - a - 3b$

107. $(12x^4 - 8x^3 + 4x^2) \div (-4x^2)$ $-3x^2 + 2x - 1$

108. $\dfrac{10z^7w^4 - 15z^3w^2 - 20zw}{-20z^2w}$ $-\dfrac{z^5w^3}{2} + \dfrac{3zw}{4} + \dfrac{1}{z}$

109. $\dfrac{x^2 + 7x + 10}{x + 5}$ $x + 2$

110. $(2t^2 + t - 10) \div (t - 2)$ $2t + 5$

111. $(2p^2 + p - 16) \div (2p + 7)$ $p - 3 + \dfrac{5}{2p + 7}$

112. $\dfrac{5a^2 + 27a - 22}{5a - 3}$ $a + 6 + \dfrac{-4}{5a - 3}$

113. $\dfrac{b^3 - 125}{b - 5}$ $b^2 + 5b + 25$

114. $(z^3 + 4z^2 + 5z + 20) \div (5 + z^2)$ $z + 4$

115. $(y^4 - 4y^3 + 5y^2 - 3y + 2) \div (y^2 + 3)$ $y^2 - 4y + 2 + \dfrac{9y - 4}{y^2 + 3}$

116. $(3t^4 - 8t^3 + t^2 - 4t - 5) \div (3t^2 + t + 1)$ $t^2 - 3t + 1 + \dfrac{-2t - 6}{3t^2 + t + 1}$

117. $\dfrac{2w^4 + w^3 + 4w - 3}{2w^2 - w + 3}$ $w^2 + w - 1$

Chapter 5 Test

Assume all variables represent nonzero real numbers.

1. Expand the expression using the definition of exponents, then simplify: $\dfrac{3^4 \cdot 3^3}{3^6}$

$\dfrac{(3 \cdot 3 \cdot 3 \cdot 3) \cdot (3 \cdot 3 \cdot 3)}{3 \cdot 3 \cdot 3 \cdot 3 \cdot 3 \cdot 3} = 3$

For Exercises 2–11, simplify the expression. Write the answer with positive exponents only.

2. $9^5 \cdot 9$ 9^6

3. $\dfrac{q^{10}}{q^2}$ q^8

4. $(3a^2b)^3$ $27a^6b^3$

5. $\left(\dfrac{2x}{y^3}\right)^4$ $\dfrac{16x^4}{y^{12}}$

6. $(-7)^0$ 1

7. c^{-3} $\dfrac{1}{c^3}$

8. $\dfrac{14^3 \cdot 14^9}{14^{10} \cdot 14}$ 14

9. $\dfrac{(s^2t)^3(7s^4t)^4}{(7s^2t^3)^2}$ $49s^{18}t$

10. $(2a^0b^{-6})^2$ $\dfrac{4}{b^{12}}$

11. $\left(\dfrac{6a^{-5}b}{8ab^{-2}}\right)^{-2}$ $\dfrac{16a^{12}}{9b^6}$

12. a. Write the number in scientific notation: $43{,}000{,}000{,}000$ 4.3×10^{10}

b. Write the number in standard form: 5.6×10^{-6} $0.000\ 0056$

13. The average amount of water flowing over Niagara Falls is 1.68×10^5 m^3/min.

a. How many cubic meters of water flow over the falls in one day? 2.4192×10^8 m^3

b. How many cubic meters of water flow over the falls in one year? 8.83008×10^{10} m^3

 Writing   Translating Expression   Geometry Scientific Calculator   Video   NS&E

14. Write the polynomial in descending order:
$4x + 5x^3 - 7x^2 + 11$ $5x^3 - 7x^2 + 4x + 11$

    a. Identify the degree of the polynomial.   3

    b. Identify the leading coefficient of the polynomial.   5

15. Perform the indicated operations.

$(7w^2 - 11w - 6) + (8w^2 + 3w + 4) - (-9w^2 - 5w + 2)$ $24w^2 - 3w - 4$

16. Subtract $(3x^2 - 5x^3 + 2x)$ from $(10x^3 - 4x^2 + 1)$.
$15x^3 - 7x^2 - 2x + 1$

For Exercises 17–23, multiply the polynomials.

17. $-2x^3(5x^2 + x - 15)$
$-10x^5 - 2x^4 + 30x^3$

18. $(4a - 3)(2a - 1)$
$8a^2 - 10a + 3$

19. $(4y - 5)(y^2 - 5y + 3)$
$4y^3 - 25y^2 + 37y - 15$

20. $(2 + 3b)(2 - 3b)$
$4 - 9b^2$

21. $(5z - 6)^2$
$25z^2 - 60z + 36$

22. $(5x + 3)(3x - 2)$
$15x^2 - x - 6$

23. $(y^2 - 5y + 2)(y - 6)$ $y^3 - 11y^2 + 32y - 12$

 24. Find the perimeter and the area of the rectangle shown in the figure.   Perimeter: $12x - 2$; area: $5x^2 - 13x - 6$

$5x + 2$

$x - 3$

For Exercises 25–28, divide:

25. $(-12x^8 + x^6 - 8x^3) \div (4x^2)$ $-3x^6 + \dfrac{x^4}{4} - 2x$

26. $\dfrac{2y^2 - 13y + 21}{y - 3}$ $2y - 7$

27. $(-5w^2 + 2w^3 - 2w + 5) \div (2w + 3)$
$w^2 - 4w + 5 + \dfrac{-10}{2w + 3}$

28. $\dfrac{3x^4 + x^3 + 4x - 33}{x^2 + 4}$ $3x^2 + x - 12 + \dfrac{15}{x^2 + 4}$

Chapters 1–5   Cumulative Review Exercises

For Exercises 1–2, simplify completely.

1. $-5 - \dfrac{1}{2}[4 - 3(-7)]$ $-\dfrac{35}{2}$ **2.** $|-3^2 + 5|$ 4

3. Translate the phrase into a mathematical expression and simplify:

The difference of the square of five and the square root of four. $5^2 - \sqrt{4}$; 23

4. Solve for x: $\dfrac{1}{2}(x - 6) + \dfrac{2}{3} = \dfrac{1}{4}x$ $\dfrac{28}{3}$

5. Solve for y: $-2y - 3 = -5(y - 1) + 3y$
No solution

6. For a point in a rectangular coordinate system, in which quadrant are both the x- and y-coordinates negative?   Quadrant III

7. For a point in a rectangular coordinate system, on which axis is the x-coordinate zero and the y-coordinate nonzero? y-axis

8. In a triangle, one angle measures $23°$ more than the smallest angle. The third angle measures $10°$ more than the sum of the other two angles. Find the measure of each angle.
The measures are $31°$, $54°$, $95°$.

9. A snow storm lasts for 9 hr and dumps snow at a rate of $1\frac{1}{2}$ in./hr. If there was already 6 in. of snow on the ground before the storm, the snow depth is given by the equation:

$y = \dfrac{3}{2}x + 6$ where y is the snow depth in inches and $x \geq 0$ is the time in hours.

    a. Find the snow depth after 4 hr.   12 in.

    b. Find the snow depth at the end of the storm.   19.5 in.

    c. How long had it snowed when the total depth of snow was $14\frac{1}{4}$ in.?   5.5 hr

10. Solve the system of equations. $(-3, 4)$

$$5x + 3y = -3$$
$$3x + 2y = -1$$

11. Solve the inequality. Graph the solution set on the real number line and express the solution in interval notation. $2 - 3(2x + 4) \le -2x - (x - 5)$

$[-5, \infty)$

$$\xleftarrow{\hspace{3cm}}\overset{\displaystyle[}{\underset{-5}{\rule{4cm}{0.4pt}}}\xrightarrow{\hspace{3cm}}$$

For Exercises 12–15, perform the indicated operations.

12. $(2x^2 + 3x - 7) - (-3x^2 + 12x + 8)$ $\quad 5x^2 - 9x - 15$

13. $(2y + 3z)(-y - 5z)$ $\quad -2y^2 - 13yz - 15z^2$

14. $(4t - 3)^2$ $\quad 16t^2 - 24t + 9$ **15.** $\left(\dfrac{2}{5}a + \dfrac{1}{3}\right)\left(\dfrac{2}{5}a - \dfrac{1}{3}\right)$

$\quad \dfrac{4}{25}a^2 - \dfrac{1}{9}$

For Exercises 16–17, divide the polynomials.

16. $(12a^4b^3 - 6a^2b^2 + 3ab) \div (-3ab)$

$\quad -4a^3b^2 + 2ab - 1$

17. $\dfrac{4m^3 - 5m + 2}{m - 2}$ $\quad 4m^2 + 8m + 11 + \dfrac{24}{m - 2}$

For Exercises 18–19, use the properties of exponents to simplify the expressions. Write the answers with positive exponents only. Assume all variables represent nonzero real numbers.

18. $\left(\dfrac{2c^2d^4}{8cd^6}\right)^2$ $\quad \dfrac{c^2}{16d^4}$ **19.** $\dfrac{10a^{-2}b^{-3}}{5a^0b^{-6}}$ $\quad \dfrac{2b^3}{a^2}$

20. Perform the indicated operations, and write the final answer in scientific notation. $\quad 4.1 \times 10^3$

$$\dfrac{8.2 \times 10^{-2}}{2.0 \times 10^{-5}}$$

Factoring Polynomials

6

CHAPTER OUTLINE

6.1 Greatest Common Factor and Factoring by Grouping   428

6.2 Factoring Trinomials of the Form $x^2 + bx + c$ 437

6.3 Factoring Trinomials: Trial-and-Error Method   443

6.4 Factoring Trinomials: AC-Method   452

6.5 Difference of Squares and Perfect Square Trinomials   458

6.6 Sum and Difference of Cubes   464

Problem Recognition Exercises: Factoring Strategy   470

6.7 Solving Equations Using the Zero Product Rule   472

Problem Recognition Exercises: Polynomial Expressions and Polynomial Equations   478

6.8 Applications of Quadratic Equations   479

Group Activity: Building a Factoring Test   487

Chapter 6

Chapter 6 is devoted to a mathematical operation called factoring. The applications of factoring are far-reaching, and in this chapter, we use factoring as a tool to solve a type of equation called a quadratic equation.

The following statements are steps from the Factoring Strategy given in this chapter. Fill in the blanks in the sentences to complete the crossword puzzle.

Across

2. If a polynomial has four terms, try factoring by _____.
4. Begin by factoring out the _____ common factor.
5. When finished, check to make sure you have factored _____.

Down

1. If a polynomial has two terms, check if it is the difference of _____.
3. If a polynomial has three terms, check if it is a _____ square trinomial.

Section 6.1 Greatest Common Factor and Factoring by Grouping

Objectives

1. **Identifying the Greatest Common Factor**
2. **Factoring out the Greatest Common Factor**
3. **Factoring out a Negative Factor**
4. **Factoring out a Binomial Factor**
5. **Factoring by Grouping**

1. Identifying the Greatest Common Factor

Chapter 6 is devoted to a mathematical operation called **factoring**. To factor an integer means to write the integer as a product of two or more integers. To factor a polynomial means to express the polynomial as a product of two or more polynomials.

In the product $2 \cdot 5 = 10$, for example, 2 and 5 are factors of 10.

In the product $(3x + 4)(2x - 1) = 6x^2 + 5x - 4$, the quantities $(3x + 4)$ and $(2x - 1)$ are factors of $6x^2 + 5x - 4$.

We begin our study of factoring by factoring integers. The number 20, for example, can be factored as $1 \cdot 20$ or $2 \cdot 10$ or $4 \cdot 5$ or $2 \cdot 2 \cdot 5$. The product $2 \cdot 2 \cdot 5$ (or equivalently $2^2 \cdot 5$) consists only of prime numbers and is called the **prime factorization**.

The **greatest common factor** (denoted **GCF**) of two or more integers is the greatest factor common to each integer. To find the greatest common factor of two or more integers, it is often helpful to express the numbers as a product of prime factors as shown in the next example.

Skill Practice

Find the GCF.

1. 12 and 20
2. 45, 75, and 30

Classroom Example: p. 435, Exercise 6

Example 1 Identifying the GCF

Find the greatest common factor.

a. 24 and 36   **b.** 105, 40, and 60

Solution:

First find the prime factorization of each number. Then find the product of common factors.

a.
$\begin{array}{ll} 2|24 & 2|36 \\ 2|12 & 2|18 \\ 2|6 & 3|9 \\ 3 & 3 \end{array}$

Factors of $24 = 2 \cdot 2 \cdot 2 \cdot 3$ ← Common factors are circled.
Factors of $36 = 2 \cdot 2 \cdot 3 \cdot 3$

The numbers 24 and 36 share two factors of 2 and one factor of 3. Therefore, the greatest common factor is $2 \cdot 2 \cdot 3 = 12$.

b.
$\begin{array}{lll} 5|105 & 5|40 & 5|60 \\ 3|21 & 2|8 & 3|12 \\ 7 & 2|4 & 2|4 \\ & 2 & 2 \end{array}$

Factors of $105 = 3 \cdot 7 \cdot 5$
Factors of $40 = 2 \cdot 2 \cdot 2 \cdot 5$
Factors of $60 = 2 \cdot 2 \cdot 3 \cdot 5$

The greatest common factor is 5.

Answers

1. 4   **2.** 15

In Example 2, we find the greatest common factor of two or more variable terms.

Classroom Example: p. 435, Exercise 10

Example 2 **Identifying the Greatest Common Factor**

Find the GCF among each group of terms.

a. $7x^3, 14x^2, 21x^4$ **b.** $15a^4b, 25a^3b^2$ **c.** $8c^2d^7e, 6c^3d^4$

Solution:

List the factors of each term.

a. $7x^3 = \boxed{7 \cdot x \cdot x} \cdot x$

    $14x^2 = 2 \cdot \boxed{7 \cdot x \cdot x}$ The GCF is $7x^2$.

    $21x^4 = 3 \cdot \boxed{7 \cdot x \cdot x} \cdot x \cdot x$

b. $15a^4b = 3 \cdot \boxed{5 \cdot a \cdot a \cdot a} \cdot a \cdot \boxed{b}$

    $25a^3b^2 = 5 \cdot \boxed{5 \cdot a \cdot a \cdot a} \cdot b \cdot \boxed{b}$ The GCF is $5a^3b$.

TIP: Notice that the expressions $15a^4b$ and $25a^3b^2$ share factors of 5, a, and b. The GCF is the product of the common factors, where each factor is raised to the lowest power to which it occurs in all the original expressions.

$\left. \begin{array}{l} 15a^4b = 3 \cdot 5a^4b \\ 25a^3b^2 = 5^2a^3b^2 \end{array} \right\}$ $\left. \begin{array}{l} \text{Lowest power of 5 is 1:} \quad 5^1 \\ \text{Lowest power of } a \text{ is 3:} \quad a^3 \\ \text{Lowest power of } b \text{ is 1:} \quad b^1 \end{array} \right\}$ The GCF is $5a^3b$.

---

c. $\left. \begin{array}{l} 8c^2d^7e = 2^3c^2d^7e \\ 6c^3d^4 = 2 \cdot 3c^3d^4 \end{array} \right\}$ The common factors are 2, c, and d.

$\left. \begin{array}{l} \text{The lowest power of 2 is 1:} \quad 2^1 \\ \text{The lowest power of } c \text{ is 2:} \quad c^2 \\ \text{The lowest power of } d \text{ is 4:} \quad d^4 \end{array} \right\}$ The GCF is $2c^2d^4$.

Sometimes polynomials share a common binomial factor, as shown in Example 3.

Classroom Example: p. 435, Exercise 14

Example 3 **Finding the Greatest Common Binomial Factor**

Find the greatest common factor between the terms: $3x(a + b)$ and $2y(a + b)$

Solution:

$3x(a + b) \left. \vphantom{\begin{array}{l} a \\ b \end{array}} \right\}$ The only common factor is the binomial $(a + b)$.

$2y(a + b) \left. \vphantom{\begin{array}{l} a \\ b \end{array}} \right\}$ The GCF is $(a + b)$.

Skill Practice

Find the GCF.

  3. $10z^3, 15z^5, 40z$

  4. $6w^3y^5, 21w^4y^2$

  5. $9m^2np^8, 15n^4p^5$

Skill Practice

Find the GCF.

  6. $a(x + 2)$ and $b(x + 2)$

Answers

3. $5z$ **4.** $3w^3y^2$

5. $3np^5$ **6.** $(x + 2)$

2. Factoring out the Greatest Common Factor

The process of factoring a polynomial is the reverse process of multiplying polynomials. Both operations use the distributive property: $ab + ac = a(b + c)$.

Multiply

$$5y(y^2 + 3y + 1) = 5y(y^2) + 5y(3y) + 5y(1)$$
$$= 5y^3 + 15y^2 + 5y$$

Factor

$$5y^3 + 15y^2 + 5y = 5y(y^2) + 5y(3y) + 5y(1)$$
$$= 5y(y^2 + 3y + 1)$$

PROCEDURE **Factoring out the Greatest Common Factor**

Step 1 Identify the GCF of all terms of the polynomial.
Step 2 Write each term as the product of the GCF and another factor.
Step 3 Use the distributive property to remove the GCF.

Note: To check the factorization, multiply the polynomials to remove parentheses.

Skill Practice

Factor out the GCF.
7. $6w + 18$
8. $21m^3 - 7m$

Classroom Examples: p. 435, Exercises 18 and 22

Instructor Note: Tell students that a problem is factored only if the GCF has been removed. The expression $2(2x - 10)$ is not factored completely.

Avoiding Mistakes

In Example 4(b), the GCF, $3w$, is equal to one of the terms of the polynomial. In such a case, you must leave a 1 in place of that term after the GCF is factored out.

Answers
7. $6(w + 3)$ **8.** $7m(3m^2 - 1)$

Example 4 **Factoring out the Greatest Common Factor**

Factor out the GCF.

a. $4x - 20$ **b.** $6w^2 + 3w$

Solution:

a. $4x - 20$ The GCF is 4.

$= 4(x) - 4(5)$ Write each term as the product of the GCF and another factor.

$= 4(x - 5)$ Use the distributive property to factor out the GCF.

TIP: Any factoring problem can be checked by multiplying the factors:

Check: $4(x - 5) = 4x - 20$ ✔

---

b. $6w^2 + 3w$ The GCF is $3w$.

$= 3w(2w) + 3w(1)$ Write each term as the product of $3w$ and another factor.

$= 3w(2w + 1)$ Use the distributive property to factor out the GCF.

Check: $3w(2w + 1) = 6w^2 + 3w$ ✔

Example 5 Factoring out the Greatest Common Factor

Factor out the GCF.

   a. $15y^3 + 12y^4$ **b.** $9a^4b - 18a^5b + 27a^6b$

Solution:

   a. $15y^3 + 12y^4$ The GCF is $3y^3$.

      $= 3y^3(5) + 3y^3(4y)$ Write each term as the product of $3y^3$ and another factor.

      $= 3y^3(5 + 4y)$ Use the distributive property to factor out the GCF.

        Check: $3y^3(5 + 4y) = 15y^3 + 12y^4$ ✔

   b. $9a^4b - 18a^5b + 27a^6b$ The GCF is $9a^4b$.

      $= 9a^4b(1) - 9a^4b(2a) + 9a^4b(3a^2)$ Write each term as the product of $9a^4b$ and another factor.

      $= 9a^4b(1 - 2a + 3a^2)$ Use the distributive property to factor out the GCF.

        Check: $9a^4b(1 - 2a + 3a^2)$
             $= 9a^4b - 18a^5b + 27a^6b$ ✔

Skill Practice

Factor out the GCF.
  9. $9y^2 - 6y^5$
  10. $50s^3t - 40st^2 + 10st$

Classroom Examples: p. 435, Exercises 24 and 34

The greatest common factor of the polynomial $2x + 5y$ is 1. If we factor out the GCF, we have $1(2x + 5y)$. A polynomial whose only factors are itself and 1 is called a **prime polynomial**. A prime polynomial cannot be factored further.

3. Factoring out a Negative Factor

Usually it is advantageous to factor out the *opposite* of the GCF when the leading coefficient of the polynomial is negative. This is demonstrated in the next example. Notice that this *changes the signs* of the remaining terms inside the parentheses.

Instructor Note: Remind students that first we put the polynomial in descending order, then determine if the leading coefficient is negative.

Example 6 Factoring out a Negative Factor

Factor out -3 from the polynomial $-3x^2 + 6x - 33$.

Solution:

   $-3x^2 + 6x - 33$ The GCF is 3. However, in this case, we will factor out the *opposite* of the GCF, -3.

   $= -3(x^2) + (-3)(-2x) + (-3)(11)$ Write each term as the product of -3 and another factor.

   $= -3[x^2 + (-2x) + 11]$ Factor out -3.

   $= -3(x^2 - 2x + 11)$ Simplify. Notice that each sign within the trinomial has changed.

   Check: $-3(x^2 - 2x + 11) = -3x^2 + 6x - 33$ ✔

Skill Practice

  11. Factor out -2 from the polynomial.
     $-2x^2 - 10x + 16$

Classroom Example: p. 436, Exercise 42

Answers
  9. $3y^2(3 - 2y^3)$
 10. $10st(5s^2 - 4t + 1)$
 11. $-2(x^2 + 5x - 8)$

Example 7 **Factoring out a Negative Factor**

Factor out the quantity $-4pq$ from the polynomial $-12p^3q - 8p^2q^2 + 4pq^3$.

Solution:

$-12p^3q - 8p^2q^2 + 4pq^3$ The GCF is $4pq$. However, in this case, we will
factor out the *opposite* of the GCF, $-4pq$.

$= -4pq(3p^2) + (-4pq)(2pq) + (-4pq)(-q^2)$ Write each term as the
product of $-4pq$ and
another factor.

$= -4pq[3p^2 + 2pq + (-q^2)]$ Factor out $-4pq$. Notice that each sign
within the trinomial has changed.

$= -4pq(3p^2 + 2pq - q^2)$ To verify that this is the correct
factorization and that the signs are
correct, multiply the factors.

Check: $-4pq(3p^2 + 2pq - q^2) = -12p^3q - 8p^2q^2 + 4pq^3$ ✔

4. Factoring out a Binomial Factor

The distributive property can also be used to factor out a common factor that consists of more than one term as shown in Example 8.

Example 8 **Factoring out a Binomial Factor**

Factor out the GCF: $2w(x + 3) - 5(x + 3)$

Solution:

$2w(x + 3) - 5(x + 3)$ The greatest common factor is the
quantity $(x + 3)$.

$= (x + 3)(2w) - (x + 3)(5)$ Write each term as the product of $(x + 3)$
and another factor.

$= (x + 3)(2w - 5)$ Use the distributive property to factor
out the GCF.

5. Factoring by Grouping

When two binomials are multiplied, the product before simplifying contains four terms. For example:

$$(x + 4)(3a + 2b) = (x + 4)(3a) + (x + 4)(2b)$$

$$= (x + 4)(3a) + (x + 4)(2b)$$

$$= 3ax + 12a + 2bx + 8b$$

In Example 9, we learn how to reverse this process. That is, given a four-term polynomial, we will factor it as a product of two binomials. The process is called *factoring by grouping*.

PROCEDURE Factoring by Grouping

To factor a four-term polynomial by grouping:

Step 1 Identify and factor out the GCF from all four terms.

Step 2 Factor out the GCF from the first pair of terms. Factor out the GCF from the second pair of terms. (Sometimes it is necessary to factor out the opposite of the GCF.)

Step 3 If the two terms share a common binomial factor, factor out the binomial factor.

Example 9 Factoring by Grouping

Factor by grouping: $3ax + 12a + 2bx + 8b$

Solution:

$3ax + 12a + 2bx + 8b$ **Step 1:** Identify and factor out the GCF from all four terms. In this case, the GCF is 1.

$= 3ax + 12a \mid + 2bx + 8b$ Group the first pair of terms and the second pair of terms.

$= 3a(x + 4) + 2b(x + 4)$ **Step 2:** Factor out the GCF from each pair of terms. *Note:* The two terms now share a common binomial factor of $(x + 4)$.

$= (x + 4)(3a + 2b)$ **Step 3:** Factor out the common binomial factor.

Check: $(x + 4)(3a + 2b) = 3ax + 2bx + 12a + 8b$ ✔

Note: Step 2 results in two terms with a common binomial factor. If the two binomials are different, step 3 cannot be performed. In such a case, the original polynomial may not be factorable by grouping, or different pairs of terms may need to be grouped and inspected.

Skill Practice

Factor by grouping.

14. $5x + 10y + ax + 2ay$

Classroom Example: p. 436, Exercise 58

TIP: One frequently asked question when factoring is whether the order can be switched between the factors. The answer is yes. Because multiplication is commutative, the order in which the factors are written does not matter.

$$(x + 4)(3a + 2b) = (3a + 2b)(x + 4)$$

Answer

14. $(x + 2y)(5 + a)$

Skill Practice

Factor by grouping.

15. $tu - tv - 2u + 2v$

Classroom Example: p. 436, Exercise 60

Avoiding Mistakes

In step 2, the expression $a(x + y) - b(x + y)$ is not yet factored completely because it is a *difference*, not a product. To factor the expression, you must carry it one step further.

$$a(x + y) - b(x + y)$$
$$= (x + y)(a - b)$$

The factored form must be represented as a product.

Example 10   **Factoring by Grouping**

Factor by grouping. $ax + ay - bx - by$

Solution:

$$ax + ay - bx - by$$ **Step 1:** Identify and factor out the GCF from all four terms. In this case, the GCF is 1.

$$= ax + ay \mid - bx - by$$ Group the first pair of terms and the second pair of terms.

$$= a(x + y) - b(x + y)$$ **Step 2:** Factor out a from the first pair of terms.

Factor out $-b$ from the second pair of terms. (This causes sign changes within the second parentheses. The terms in parentheses now match.)

$$= (x + y)(a - b)$$ **Step 3:** Factor out the common binomial factor.

Check: $(x + y)(a - b) = x(a) + x(-b) + y(a) + y(-b)$
$$= ax - bx + ay - by ✔$$

Skill Practice

Factor by grouping.

16. $3ab^2 + 6b^2 - 12ab - 24b$

Classroom Example: p. 436, Exercise 74

Example 11   **Factoring by Grouping**

Factor by grouping. $16w^4 - 40w^3 - 12w^2 + 30w$

Solution:

$$16w^4 - 40w^3 - 12w^2 + 30w$$ **Step 1:** Identify and factor out the GCF from all four terms. In this case, the GCF is $2w$.

$$= 2w[8w^3 - 20w^2 - 6w + 15]$$

$$= 2w[8w^3 - 20w^2 \mid - 6w + 15]$$ Group the first pair of terms and the second pair of terms.

$$= 2w[4w^2(2w - 5) - 3(2w - 5)]$$ **Step 2:** Factor out $4w^2$ from the first pair of terms.

Factor out -3 from the second pair of terms. (This causes sign changes within the second parentheses. The terms in parentheses now match.)

$$= 2w[(2w - 5)(4w^2 - 3)]$$ **Step 3:** Factor out the common binomial factor.

$$= 2w(2w - 5)(4w^2 - 3)$$

Answers

15. $(u - v)(t - 2)$
16. $3b(a + 2)(b - 4)$

Section 6.1 Practice Exercises

Study Skills Exercises

1. The final exam is just around the corner. Your old tests and quizzes provide good material to study for the final exam. Use your old tests to make a list of the chapters on which you need to concentrate. Ask your professor for help if there are still concepts that you do not understand.

2. Define the key terms:

   a. factoring   **b. greatest common factor (GCF)**

   c. prime factorization   **d. prime polynomial**

Objective 1: Identifying the Greatest Common Factor

For Exercises 3–14, identify the greatest common factor. **(See Examples 1–3.)**

3. $28, 63$ 7

4. $24, 40$ 8

5. $42, 30, 60$ 6

6. $20, 52, 32$ 4

7. $3xy, 7y$ y

8. $10mn, 11n$ n

9. $12w^3z, 16w^2z$ $4w^2z$

10. $20cd, 15c^3d$ $5cd$

11. $8x^3y^4z^2, 12xy^5z^4, 6x^2y^8z^3$ $2xy^4z^2$

12. $15r^2s^2t^5, 5r^3s^4t^3, 30r^4s^3t^2$ $5r^2s^2t^2$

13. $7(x - y), 9(x - y)$ $(x - y)$

14. $(2a - b), 3(2a - b)$ $(2a - b)$

Objective 2: Factoring out the Greatest Common Factor

15. a. Use the distributive property to multiply $3(x - 2y)$. $3x - 6y$

    b. Use the distributive property to factor $3x - 6y$. $3(x - 2y)$

16. a. Use the distributive property to multiply $a^2(5a + b)$. $5a^3 + a^2b$

    b. Use the distributive property to factor $5a^3 + a^2b$. $a^2(5a + b)$

For Exercises 17–36, factor out the GCF. **(See Examples 4–5.)**

17. $4p + 12$ $4(p + 3)$

18. $3q - 15$ $3(q - 5)$

19. $5c^2 - 10c + 15$
    $5(c^2 - 2c + 3)$

20. $16d^3 + 24d^2 + 32d$
    $8d(2d^2 + 3d + 4)$

21. $x^5 + x^3$ $x^3(x^2 + 1)$

22. $y^2 - y^3$ $y^2(1 - y)$

23. $t^4 - 4t + 8t^2$
    $t(t^3 - 4 + 8t)$

24. $7r^3 - r^5 + r^4$
    $r^3(7 - r^2 + r)$

25. $2ab + 4a^3b$
    $2ab(1 + 2a^2)$

26. $5u^3v^2 - 5uv$
    $5uv(u^2v - 1)$

27. $38x^2y - 19x^2y^4$
    $19x^2y(2 - y^3)$

28. $100a^5b^3 + 16a^2b$
    $4a^2b(25a^3b^2 + 4)$

29. $6x^3y^5 - 18xy^9z$
    $6xy^5(x^2 - 3y^4z)$

30. $15mp^7q^4 + 12m^4q^3$
    $3mq^3(5p^7q + 4m^3)$

31. $5 + 7y^3$ The expression is
prime because it is not factorable.

32. $w^3 - 5u^3v^2$ The expression is
prime because it is not factorable.

33. $42p^3q^2 + 14pq^2 - 7p^4q^4$ $7pq^2(6p^2 + 2 - p^3q^2)$

34. $8m^2n^3 - 24m^2n^2 + 4m^3n$ $4m^2n(2n^2 - 6n + m)$

35. $t^5 + 2rt^3 - 3t^4 + 4r^2t^2$ $t^2(t^3 + 2rt - 3t^2 + 4r^2)$

36. $u^2v + 5u^3v^2 - 2u^2 + 8uv$ $u(uv + 5u^2v^2 - 2u + 8v)$

Objective 3: Factoring out a Negative Factor

37. For the polynomial $-2x^3 - 4x^2 + 8x$

    a. Factor out $-2x$.
       $-2x(x^2 + 2x - 4)$

    b. Factor out $2x$.
       $2x(-x^2 - 2x + 4)$

38. For the polynomial $-9y^5 + 3y^3 - 12y$

    a. Factor out $-3y$.
       $-3y(3y^4 - y^2 + 4)$

    b. Factor out $3y$.
       $3y(-3y^4 + y^2 - 4)$

39. Factor out -1 from the polynomial
$-8t^2 - 9t - 2$. $-1(8t^2 + 9t + 2)$

40. Factor out -1 from the polynomial
$-6x^3 - 2x - 5$. $-1(6x^3 + 2x + 5)$

For Exercises 41–46, factor out the opposite of the greatest common factor. **(See Examples 6–7.)**

41. $-15p^3 - 30p^2$ $-15p^2(p + 2)$

42. $-24m^3 - 12m^4$ $-12m^3(2 + m)$

43. $-q^4 + 2q^2 - 9q$ $-q(q^3 - 2q + 9)$

44. $-r^3 + 9r^2 - 5r$ $-r(r^2 - 9r + 5)$

45. $-7x - 6y - 2z$ $-1(7x + 6y + 2z)$

46. $-4a + 5b - c$ $-1(4a - 5b + c)$

Objective 4: Factoring out a Binomial Factor

For Exercises 47–52, factor out the GCF. **(See Example 8.)**

47. $13(a + 6) - 4b(a + 6)$
  $(a + 6)(13 - 4b)$

48. $7(x^2 + 1) - y(x^2 + 1)$
  $(x^2 + 1)(7 - y)$

49. $8v(w^2 - 2) + (w^2 - 2)$
  $(w^2 - 2)(8v + 1)$

50. $t(r + 2) + (r + 2)$
  $(r + 2)(t + 1)$

51. $21x(x + 3) + 7x^2(x + 3)$
  $7x(x + 3)^2$

52. $5y^3(y - 2) - 15y(y - 2)$
  $5y(y - 2)(y^2 - 3)$

Objective 5: Factoring by Grouping

For Exercises 53–72, factor by grouping. **(See Examples 9–10.)**

53. $8a^2 - 4ab + 6ac - 3bc$
  $(2a - b)(4a + 3c)$

54. $4x^3 + 3x^2y + 4xy^2 + 3y^3$
  $(4x + 3y)(x^2 + y^2)$

55. $3q + 3p + qr + pr$
  $(q + p)(3 + r)$

56. $xy - xz + 7y - 7z$
  $(y - z)(x + 7)$

57. $6x^2 + 3x + 4x + 2$
  $(2x + 1)(3x + 2)$

58. $4y^2 + 8y + 7y + 14$
  $(y + 2)(4y + 7)$

59. $2t^2 + 6t - 5t - 15$
  $(t + 3)(2t - 5)$

60. $2p^2 - p - 6p + 3$
  $(2p - 1)(p - 3)$

61. $6y^2 - 2y - 9y + 3$
  $(3y - 1)(2y - 3)$

62. $5a^2 + 30a - 2a - 12$
  $(a + 6)(5a - 2)$

63. $b^4 + b^3 - 4b - 4$
  $(b + 1)(b^3 - 4)$

64. $8w^5 + 12w^2 - 10w^3 - 15$
  $(2w^3 + 3)(4w^2 - 5)$

65. $3j^2k + 15k + j^2 + 5$
  $(j^2 + 5)(3k + 1)$

66. $2ab^2 - 6ac + b^2 - 3c$
  $(b^2 - 3c)(2a + 1)$

67. $14w^6x^6 + 7w^6 - 2x^6 - 1$
  $(2x^6 + 1)(7w^6 - 1)$

68. $18p^4q - 9p^5 - 2q + p$
  $(2q - p)(9p^4 - 1)$

69. $ay + bx + by + ax$
  (*Hint:* Rearrange the terms.)
  $(y + x)(a + b)$

70. $2c + 3ay + ac + 6y$
  $(c + 3y)(2 + a)$

71. $vw^2 - 3 + w - 3wv$
  $(vw + 1)(w - 3)$

72. $2x^2 + 6m + 12 + x^2m$
  $(m + 2)(6 + x^2)$

Mixed Exercises

For Exercises 73–78, factor out the GCF first. Then factor by grouping. **(See Example 11.)**

73. $15x^4 + 15x^2y^2 + 10x^3y + 10xy^3$
  $5x(x^2 + y^2)(3x + 2y)$

74. $2a^3b - 4a^2b + 32ab - 64b$
  $2b(a - 2)(a^2 + 16)$

75. $4abx - 4b^2x - 4ab + 4b^2$
  $4b(a - b)(x - 1)$

76. $p^2q - pq^2 - rp^2q + rpq^2$
  $pq(p - q)(1 - r)$

77. $6st^2 - 18st - 6t^4 + 18t^3$
  $6t(t - 3)(s - t^2)$

78. $15j^3 - 10j^2k - 15j^2k^2 + 10jk^3$
  $5j(3j - 2k)(j - k^2)$

79. The formula $P = 2l + 2w$ represents the perimeter, P, of a rectangle given the length, l, and the width, w. Factor out the GCF, and write an equivalent formula in factored form.
  $P = 2(l + w)$

80. The formula $P = 2a + 2b$ represents the perimeter, P, of a parallelogram given the base, b, and an adjacent side, a. Factor out the GCF, and write an equivalent formula in factored form.
  $P = 2(a + b)$

Writing   Translating Expression   Geometry   Scientific Calculator   Video   NS&E

81. The formula $S = 2\pi r^2 + 2\pi rh$ represents the surface area, S, of a cylinder with radius, r, and height, h. Factor out the GCF, and write an equivalent formula in factored form.

$S = 2\pi r(r + h)$

82. The formula $A = P + Prt$ represents the total amount of money, A, in an account that earns simple interest at a rate, r, for t years. Factor out the GCF, and write an equivalent formula in factored form. $A = P(1 + rt)$

Expanding Your Skills

83. Factor out $\frac{1}{7}$ from $\frac{1}{7}x^2 + \frac{3}{7}x - \frac{5}{7}$. $\frac{1}{7}(x^2 + 3x - 5)$

84. Factor out $\frac{1}{5}$ from $\frac{6}{5}y^2 - \frac{4}{5}y + \frac{1}{5}$. $\frac{1}{5}(6y^2 - 4y + 1)$

85. Factor out $\frac{1}{4}$ from $\frac{5}{4}w^2 + \frac{3}{4}w + \frac{9}{4}$. $\frac{1}{4}(5w^2 + 3w + 9)$

86. Factor out $\frac{1}{6}$ from $\frac{1}{6}p^2 - \frac{3}{6}p + \frac{5}{6}$. $\frac{1}{6}(p^2 - 3p + 5)$

87. Write a polynomial that has a GCF of $3x$. (Answers may vary.)   For example, $6x^2 + 9x$

88. Write a polynomial that has a GCF of $7y$. (Answers may vary.)   For example, $14y - 21y^3 + 7y^2$

89. Write a polynomial that has a GCF of $4p^2q$. (Answers may vary.)   For example, $16p^4q^2 + 8p^3q - 4p^2q$

90. Write a polynomial that has a GCF of $2ab^2$. (Answers may vary.)   For example, $18a^2b^3 - 2ab^2$

Factoring Trinomials of the Form $x^2 + bx + c$ Section 6.2

1. Factoring Trinomials with a Leading Coefficient of 1

Objective

1. Factoring Trinomials with a Leading Coefficient of 1

In Section 5.6, we learned how to multiply two binomials. We also saw that such a product often results in a trinomial. For example,

$$\underset{\substack{\text{Product of}\\\text{first terms}}}{} \qquad \underset{\substack{\text{Product of}\\\text{last terms}}}{}$$

$$(x + 3)(x + 7) = x^2 + \underbrace{7x + 3x}_{\substack{\text{Sum of products of inner}\\\text{terms and outer terms}}} + 21 = x^2 + 10x + 21$$

In this section, we want to reverse the process. That is, given a trinomial, we want to *factor* it as a product of two binomials. In particular, we begin our study with the case in which a trinomial has a leading coefficient of 1.

Consider the quadratic trinomial $x^2 + bx + c$. To produce a leading term of x^2, we can construct binomials of the form $(x +\ )(x +\ )$. The remaining terms can be obtained from two integers, p and q, whose product is c and whose sum is b.

$$x^2 + bx + c = (x + \overbrace{p)(x + q}^{\text{Factors of } c}) = x^2 + qx + px + pq$$
$$= x^2 + \underbrace{(q + p)}_{\text{Sum} = b}x + \underbrace{pq}_{\text{Product} = c}$$

This process is demonstrated in Example 1.

Example 1 Factoring a Trinomial of the Form $x^2 + bx + c$

Factor: $x^2 + 4x - 45$

Solution:

$x^2 + 4x - 45 = (x + \square)(x + \square)$ The product of the first terms in the binomials must equal the leading term of the trinomial $x \cdot x = x^2$.

We must fill in the blanks with two integers whose product is -45 and whose sum is 4. The factors must have opposite signs to produce a negative product. The possible factorizations of -45 are:

Product = -45	Sum
$-1 \cdot 45$	44
$-3 \cdot 15$	12
$-5 \cdot 9$	4
$-9 \cdot 5$	-4
$-15 \cdot 3$	-12
$-45 \cdot 1$	-44

$x^2 + 4x - 45 = (x + \square)(x + \square)$

$\qquad\qquad = (x + (-5))(x + 9)$ Fill in the blanks with -5 and 9,

$\qquad\qquad = (x - 5)(x + 9)$ Factored form

Check:
$(x - 5)(x + 9) = x^2 + 9x - 5x - 45$
$\qquad\qquad\qquad\quad = x^2 + 4x - 45 ✔$

One frequently asked question is whether the order of factors can be reversed. The answer is yes because multiplication of polynomials is a commutative operation. Therefore, in Example 1, we can express the factorization as $(x - 5)(x + 9)$ or as $(x + 9)(x - 5)$.

Example 2 Factoring a Trinomial of the Form $x^2 + bx + c$

Factor: $w^2 - 15w + 50$

Solution:

$w^2 - 15w + 50 = (w + \square)(w + \square)$ The product $w \cdot w = w^2$.

Find two integers whose product is 50 and whose sum is -15. To form a positive product, the factors must be either both positive or both negative. The sum must be negative, so we will choose negative factors of 50.

Answers

1. $(x - 7)(x + 2)$
2. $(z - 4)(z - 12)$

$$\underline{\text{Product} = 50} \qquad \underline{\text{Sum}}$$
$$(-1)(-50) \qquad -51$$
$$(-2)(-25) \qquad -27$$
$$(-5)(-10) \qquad -15$$

$$w^2 - 15w + 50 = (w + \square)(w + \square)$$
$$= (w + (-5))(w + (-10))$$
$$= (w - 5)(w - 10) \qquad \text{Factored form}$$

$$\underline{\text{Check:}}$$

$$(w - 5)(w - 10) = w^2 - 10w - 5w + 50$$
$$= w^2 - 15w + 50 \checkmark$$

Practice will help you become proficient in factoring polynomials. As you do your homework, keep these important guidelines in mind:

- To factor a trinomial, write the trinomial in descending order such as $x^2 + bx + c$.
- For all factoring problems, always factor out the GCF from all terms first.

Furthermore, we offer the following rules for determining the signs within the binomial factors.

PROCEDURE Sign Rules for Factoring Trinomials

Given the trinomial $x^2 + bx + c$, the signs within the binomial factors are determined as follows:

Case 1 If c is *positive*, then the signs in the binomials must be the same (either both positive or both negative). The correct choice is determined by the middle term. If the middle term is positive, then both signs must be positive. If the middle term is negative, then both signs must be negative.

c is positive. $\qquad\qquad$ c is positive.

$$x^2 + 6x + 8 \qquad\qquad x^2 - 6x + 8$$
$$(x + 2)(x + 4) \qquad\qquad (x - 2)(x - 4)$$

Same signs $\qquad\qquad$ Same signs

Case 2 If c is *negative*, then the signs in the binomials must be different.

c is negative. $\qquad\qquad$ c is negative.

$$x^2 + 2x - 35 \qquad\qquad x^2 - 2x - 35$$
$$(x + 7)(x - 5) \qquad\qquad (x - 7)(x + 5)$$

Different signs $\qquad\qquad$ Different signs

Skill Practice

Factor.

3. $-5w + w^2 - 6$

4. $30y^3 + 2y^4 + 112y^2$

Classroom Examples: p. 442, Exercises 32 and 40

Example 3 Factoring Trinomials

Factor. **a.** $-8p - 48 + p^2$ **b.** $-40t - 30t^2 + 10t^3$

Solution:

a. $-8p - 48 + p^2$

$= p^2 - 8p - 48$ Write in descending order.

$= (p \quad \Box)(p \quad \Box)$ Find two integers whose product is -48 and whose sum is -8. The numbers are -12 and 4.

$= (p - 12)(p + 4)$ Factored form

b. $-40t - 30t^2 + 10t^3$

$= 10t^3 - 30t^2 - 40t$ Write in descending order.

$= 10t(t^2 - 3t - 4)$ Factor out the GCF.

$= 10t(t \quad \Box)(t \quad \Box)$ Find two integers whose product is -4 and whose sum is -3. The numbers are -4 and 1.

$= 10t(t - 4)(t + 1)$ Factored form

Skill Practice

Factor.

5. $-x^2 + x + 12$

6. $-3a^2 + 15ab - 12b^2$

Classroom Examples: p. 442, Exercises 44 and 48

Avoiding Mistakes

Recall that factoring out -1 from a polynomial changes the signs of all terms within parentheses.

Instructor Note: Show students how a sign difference between two expressions can affect the factors used.
1. $x^2 - 10x + 24$, $x^2 - 10x - 24$
2. $x^2 - 5x + 6$, $x^2 - 5x - 6$
3. $x^2 - 13x + 30$, $x^2 - 13x - 30$

Example 4 Factoring Trinomials

Factor. **a.** $-a^2 + 6a - 8$ **b.** $-2c^2 - 22cd - 60d^2$

Solution:

a. $-a^2 + 6a - 8$ It is generally easier to factor a trinomial with a *positive* leading coefficient. Therefore, we will factor out -1 from all terms.

$= -1(a^2 - 6a + 8)$

$= -1(a \quad \Box)(a \quad \Box)$ Find two integers whose product is 8 and whose sum is -6. The numbers are -4 and -2.

$= -1(a - 4)(a - 2)$

b. $-2c^2 - 22cd - 60d^2$

$= -2(c^2 + 11cd + 30d^2)$ Factor out -2.

$= -2(c \quad \Box d)(c \quad \Box d)$ Notice that the second pair of terms has a factor of d. This will produce a product of d^2.

$= -2(c + 5d)(c + 6d)$ Find two integers whose product is 30 and whose sum is 11. The numbers are 5 and 6.

To factor a trinomial of the form $x^2 + bx + c$, we must find two integers whose product is c and whose sum is b. If no such integers exist, then the trinomial is not factorable and is called a **prime polynomial**.

Answers

3. $(w - 6)(w + 1)$

4. $2y^2(y + 8)(y + 7)$

5. $-(x - 4)(x + 3)$

6. $-3(a - b)(a - 4b)$

Example 5 Factoring Trinomials

Factor: $x^2 - 13x + 8$

Solution:

$x^2 - 13x + 8$ The trinomial is in descending order. The GCF is 1.

$= (x \quad \Box)(x \quad \Box)$ Find two integers whose product is 8 and whose sum is -13. No such integers exist.

The trinomial $x^2 - 13x + 8$ is prime.

Skill Practice

Factor.

7. $x^2 - 7x + 28$

Classroom Example: p. 441, Exercise 18

Answer

7. Prime

Section 6.2 Practice Exercises

Boost *your* GRADE at ALEKS.com!

ALEKS® version 3.0

- Practice Problems
- Self-Tests
- NetTutor
- e-Professors
- Videos

For additional exercises, see Classroom Activity 6.2A in the *Instructor's Resource Manual* at www.mhhe.com/moh.

Study Skills Exercises

1. Sometimes the problems on a test do not appear in the same order as the concepts appear in the text. In order to better prepare for a test, try to practice with problems taken from the book but placed in random order. Choose 30 problems from various chapters, randomize the order, and use the problems to review for the test. Repeat the process several times for additional practice.

2. Define the key term **prime polynomial**.

Review Exercises

For Exercises 3–6, factor completely.

3. $4x^3y^7 - 12x^4y^5 + 8xy^8$ $4xy^5(x^2y^2 - 3x^3 + 2y^3)$

4. $9a^6b^3 - 27a^3b^6 - 3a^2b^2$ $3a^2b^2(3a^4b - 9ab^4 - 1)$

5. $ax + 2bx - 5a - 10b$ $(a + 2b)(x - 5)$

6. $m^2 - mx - 3pm + 3px$ $(m - x)(m - 3p)$

Objective 1: Factoring Trinomials with a Leading Coefficient of 1

For Exercises 7–20, factor completely. **(See Examples 1, 2, and 5.)**

7. $x^2 + 10x + 16$ $(x + 8)(x + 2)$

8. $y^2 + 18y + 80$ $(y + 10)(y + 8)$

9. $z^2 - 11z + 18$ $(z - 9)(z - 2)$

10. $w^2 - 7w + 12$ $(w - 3)(w - 4)$

11. $z^2 - 3z - 18$ $(z - 6)(z + 3)$

12. $w^2 + 4w - 12$ $(w + 6)(w - 2)$

13. $p^2 - 3p - 40$ $(p - 8)(p + 5)$

14. $a^2 - 10a + 9$ $(a - 9)(a - 1)$

15. $t^2 + 6t - 40$ $(t + 10)(t - 4)$

16. $m^2 - 12m + 11$ $(m - 11)(m - 1)$

17. $x^2 - 3x + 20$ Prime

18. $y^2 + 6y + 18$ Prime

19. $n^2 + 8n + 16$ $(n + 4)^2$

20. $v^2 + 10v + 25$ $(v + 5)^2$

For Exercises 21–24, assume that b and c represent positive integers.

21. When factoring a polynomial of the form $x^2 + bx + c$, pick an appropriate combination of signs.   a

   a. ($+$)($+$)   **b.** ($-$)($-$)   **c.** ($+$)($-$)

22. When factoring a polynomial of the form $x^2 + bx - c$, pick an appropriate combination of signs.   c

   a. ($+$)($+$)   **b.** ($-$)($-$)   **c.** ($+$)($-$)

23. When factoring a polynomial of the form $x^2 - bx - c$, pick an appropriate combination of signs.   c

   a. ($+$)($+$)   **b.** ($-$)($-$)   **c.** ($+$)($-$)

24. When factoring a polynomial of the form $x^2 - bx + c$, pick an appropriate combination of signs.   b

   a. ($+$)($+$)   **b.** ($-$)($-$)   **c.** ($+$)($-$)

25. Which is the correct factorization of $y^2 - y - 12$? Explain. $(y - 4)(y + 3)$ or $(y + 3)(y - 4)$
   They are both correct because multiplication of polynomials is a commutative operation.

26. Which is the correct factorization of $x^2 + 14x + 13$? Explain. $(x + 13)(x + 1)$ or $(x + 1)(x + 13)$
   They are both correct because multiplication of polynomials is a commutative operation.

27. Which is the correct factorization of $w^2 + 2w + 1$? Explain. $(w + 1)(w + 1)$ or $(w + 1)^2$
   The expressions are equal and both are correct.

28. Which is the correct factorization of $z^2 - 4z + 4$? Explain. $(z - 2)(z - 2)$ or $(z - 2)^2$
   The expressions are equal and both are correct.

29. In what order should a trinomial be written before attempting to factor it?   Descending order

30. Referring to page 439, write two important guidelines to follow when factoring trinomials.
   To factor a trinomial, write the trinomial in descending order such as $x^2 + bx + c$. For all factoring problems, always factor out the GCF from all terms.

For Exercises 31–48, factor completely. Be sure to factor out the GCF when necessary. **(See Examples 3–4.)**

31. $-13x + x^2 - 30$
   $(x - 15)(x + 2)$

32. $12y - 160 + y^2$
   $(y + 20)(y - 8)$

33. $-18w + 65 + w^2$
   $(w - 13)(w - 5)$

34. $17t + t^2 + 72$
   $(t + 8)(t + 9)$

35. $22t + t^2 + 72$
   $(t + 18)(t + 4)$

36. $10q - 1200 + q^2$
   $(q - 30)(q + 40)$

37. $3x^2 - 30x - 72$
   $3(x - 12)(x + 2)$

38. $2z^2 + 4z - 198$
   $2(z + 11)(z - 9)$

39. $8p^3 - 40p^2 + 32p$
   $8p(p - 1)(p - 4)$

40. $5w^4 - 35w^3 + 50w^2$
   $5w^2(w - 2)(w - 5)$

41. $y^4z^2 - 12y^3z^2 + 36y^2z^2$
   $y^2z^2(y - 6)(y - 6)$ or $y^2z^2(y - 6)^2$

42. $t^4u^2 + 6t^3u^2 + 9t^2u^2$
   $t^2u^2(t + 3)(t + 3)$ or $t^2u^2(t + 3)^2$

43. $-x^2 + 10x - 24$
   $-(x - 4)(x - 6)$

44. $-y^2 - 12y - 35$
   $-(y + 5)(y + 7)$

45. $-m^2 + m + 6$
   $-(m + 2)(m - 3)$

46. $-n^2 + 5n + 6$
   $-(n - 6)(n + 1)$

47. $-4 - 2c^2 - 6c$
   $-2(c + 2)(c + 1)$

48. $-40d - 30 - 10d^2$
   $-10(d + 3)(d + 1)$

Mixed Exercises

For Exercises 49–66, factor completely.

49. $x^3y^3 - 19x^2y^3 + 60xy^3$
   $xy^3(x - 4)(x - 15)$

50. $y^2z^5 + 17yz^5 + 60z^5$
   $z^5(y + 5)(y + 12)$

51. $12p^2 - 96p + 84$
   $12(p - 7)(p - 1)$

52. $5w^2 - 40w - 45$
   $5(w - 9)(w + 1)$

53. $-2m^2 + 22m - 20$
   $-2(m - 10)(m - 1)$

54. $-3x^2 - 36x - 81$
   $-3(x + 9)(x + 3)$

55. $c^2 + 6cd + 5d^2$ $(c + 5d)(c + d)$

56. $x^2 + 8xy + 12y^2$ $(x + 6y)(x + 2y)$

57. $a^2 - 9ab + 14b^2$ $(a - 2b)(a - 7b)$

58. $m^2 - 15mn + 44n^2$
   $(m - 4n)(m - 11n)$

59. $a^2 + 4a + 18$ Prime

60. $b^2 - 6a + 15$ Prime

61. $2q + q^2 - 63$ $(q - 7)(q + 9)$

62. $-32 - 4t + t^2$ $(t - 8)(t + 4)$

63. $x^2 + 20x + 100$ $(x + 10)^2$

64. $z^2 - 24z + 144$ $(z - 12)^2$

65. $t^2 + 18t - 40$ $(t + 20)(t - 2)$

66. $d^2 + 2d - 99$ $(d + 11)(d - 9)$

Writing   Translating Expression   Geometry   Scientific Calculator   Video

67. A student factored a trinomial as $(2x - 4)(x - 3)$. The instructor did not give full credit. Why? The student forgot to factor out the GCF before factoring the trinomial further. The polynomial is not factored completely, because $(2x - 4)$ has a common factor of 2.

68. A student factored a trinomial as $(y + 2)(5y - 15)$. The instructor did not give full credit. Why? The student forgot to factor out the GCF before factoring the trinomial further. The polynomial is not factored completely, because $(5y - 15)$ has a common factor of 5.

69. What polynomial factors as $(x - 4)(x + 13)$? $x^2 + 9x - 52$

70. What polynomial factors as $(q - 7)(q + 10)$? $q^2 + 3q - 70$

Expanding Your Skills

For Exercises 71–74, factor completely.

71. $x^4 + 10x^2 + 9$
$(x^2 + 1)(x^2 + 9)$

72. $y^4 + 4y^2 - 21$
$(y^2 - 3)(y^2 + 7)$

73. $w^4 + 2w^2 - 15$
$(w^2 + 5)(w^2 - 3)$

74. $p^4 - 13p^2 + 40$
$(p^2 - 8)(p^2 - 5)$

75. Find all integers, b, that make the trinomial $x^2 + bx + 6$ factorable. $7, 5, -7, -5$

76. Find all integers, b, that make the trinomial $x^2 + bx + 10$ factorable. $11, 7, -11, -7$

77. Find a value of c that makes the trinomial $x^2 + 6x + c$ factorable.   For example: $c = -16$

78. Find a value of c that makes the trinomial $x^2 + 8x + c$ factorable.   For example, $c = 7$

Factoring Trinomials: Trial-and-Error Method

Section 6.3

In Section 6.2, we learned how to factor trinomials of the form $x^2 + bx + c$. These trinomials have a leading coefficient of 1. In this section and the next, we will consider the more general case in which the leading coefficient may be *any* integer. That is, we will factor quadratic trinomials of the form $ax^2 + bx + c$ (where $a \neq 0$). The method presented in this section is called the trial-and-error method.

Objective

1. Factoring Trinomials by the Trial-and-Error Method

1. Factoring Trinomials by the Trial-and-Error Method

To understand the basis of factoring trinomials of the form $ax^2 + bx + c$, first consider the multiplication of two binomials:

$$\overset{\text{Product of } 2 \cdot 1}{\downarrow} \qquad \overset{\text{Product of } 3 \cdot 2}{\downarrow}$$
$$(2x + 3)(1x + 2) = 2x^2 + \underset{\substack{\text{Sum of products of inner} \\ \text{terms and outer terms}}}{\underline{\mathbf{4x + 3x}}} + 6 = 2x^2 + 7x + 6$$

To factor the trinomial, $2x^2 + 7x + 6$, this operation is reversed.

$$2x^2 + 7x + 6 = \overset{\text{Factors of 2}}{(\square x \quad \square)} \underset{\text{Factors of 6}}{(\square x \quad \square)}$$

We need to fill in the blanks so that the product of the first terms in the binomials is $2x^2$ and the product of the last terms in the binomials is 6. Furthermore, the factors of $2x^2$ and 6 must be chosen so that the sum of the products of the inner terms and outer terms equals $7x$.

To produce the product $2x^2$, we might try the factors $2x$ and x within the binomials:

$$(2x \quad \square)(x \quad \square)$$

To produce a product of 6, the remaining terms in the binomials must either both be positive or both be negative. To produce a positive middle term, we will try positive factors of 6 in the remaining blanks until the correct product is found. The possibilities are $1 \cdot 6, 2 \cdot 3, 3 \cdot 2,$ and $6 \cdot 1$.

$$(2x + 1)(x + 6) = 2x^2 + 12x + 1x + 6 = 2x^2 + 13x + 6 \qquad \text{Wrong middle term}$$

$$(2x + 2)(x + 3) = 2x^2 + 6x + 2x + 6 = 2x^2 + 8x + 6 \qquad \text{Wrong middle term}$$

$$(2x + 3)(x + 2) = 2x^2 + 4x + 3x + 6 = 2x^2 + 7x + 6 \qquad \text{Correct!}$$

$$(2x + 6)(x + 1) = 2x^2 + 2x + 6x + 6 = 2x^2 + 8x + 6 \qquad \text{Wrong middle term}$$

The correct factorization of $2x^2 + 7x + 6$ is $(2x + 3)(x + 2).$ ✔

As this example shows, we factor a trinomial of the form $ax^2 + bx + c$ by shuffling the factors of a and c within the binomials until the correct product is obtained. However, sometimes it is not necessary to test all the possible combinations of factors. In the previous example, the GCF of the original trinomial is 1. Therefore, any binomial factor whose terms share a common factor *greater than 1* does not need to be considered. In this case, the possibilities $(2x + 2)(x + 3)$ and $(2x + 6)(x + 1)$ cannot work.

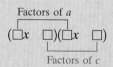

$(2x + 2)(x + 3)$ $(2x + 6)(x + 1)$

Common factor of 2   Common factor of 2

The steps to factor a trinomial by the trial-and-error method are outlined in the following box.

PROCEDURE Trial-and-Error Method to Factor $ax^2 + bx + c$

Step 1 Factor out the GCF.

Step 2 List all pairs of positive factors of a and pairs of positive factors of c. Consider the reverse order for one of the lists of factors.

Step 3 Construct two binomials of the form:

Factors of a

$$(\Box x \quad \Box)(\Box x \quad \Box)$$

Factors of c

Step 4 Test each combination of factors and signs until the correct product is found.

Step 5 If no combination of factors produces the correct product, the trinomial cannot be factored further and is a **prime polynomial**.

Before we begin Example 1, keep these two important guidelines in mind:

- For any factoring problem you encounter, always factor out the GCF from all terms first.
- To factor a trinomial, write the trinomial in the form $ax^2 + bx + c$.

Example 1 **Factoring a Trinomial by the Trial-and-Error Method**

Factor the trinomial by the trial-and-error method: $10x^2 - 9x - 1$

Skill Practice

Factor using the trial-and-error method.

1. $3b^2 + 8b + 4$

Solution:

$10x^2 - 9x - 1$ **Step 1:** Factor out the GCF from all terms. In this case, the GCF is 1.

The trinomial is written in the form $ax^2 + bx + c$.

Classroom Example: p. 450, Exercise 16

To factor $10x^2 - 9x - 1$, two binomials must be constructed in the form:

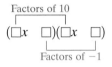

Step 2: To produce the product $10x^2$, we might try $5x$ and $2x$, or $10x$ and $1x$. To produce a product of -1, we will try the factors $(1)(-1)$ and $(-1)(1)$.

Step 3: Construct all possible binomial factors using different combinations of the factors of $10x^2$ and -1.

$(5x + 1)(2x - 1) = 10x^2 - 5x + 2x - 1 = 10x^2 - 3x - 1$ Wrong middle term

$(5x - 1)(2x + 1) = 10x^2 + 5x - 2x - 1 = 10x^2 + 3x - 1$ Wrong middle term

Because the numbers 1 and -1 did not produce the correct trinomial when coupled with $5x$ and $2x$, try using $10x$ and $1x$.

$(10x - 1)(1x + 1) = 10x^2 + 10x - 1x - 1 = 10x^2 + 9x - 1$ Wrong middle term

$(10x + 1)(1x - 1) = 10x^2 - 10x + 1x - 1 = 10x^2 - 9x - 1$ Correct!

Therefore, $10x^2 - 9x - 1 = (10x + 1)(x - 1)$.

In Example 1, the factors of -1 must have opposite signs to produce a negative product. Therefore, one binomial factor is a sum and one is a difference. Determining the correct signs is an important aspect of factoring trinomials. We suggest the following guidelines:

Answer

1. $(3b + 2)(b + 2)$

> **PROCEDURE Sign Rules for the Trial-and-Error Method**
>
> Given the trinomial $ax^2 + bx + c, (a > 0)$, the signs can be determined as follows:
>
> - If c is *positive*, then the signs in the binomials must be the same (either both positive or both negative). The correct choice is determined by the middle term. If the middle term is positive, then both signs must be positive. If the middle term is negative, then both signs must be negative.
>
> $$\overset{c \text{ is positive}}{20x^2 + 43x + 21} \qquad \overset{c \text{ is positive}}{20x^2 - 43x + 21}$$
>
> $$\underset{\text{Same signs}}{(4x + 3)(5x + 7)} \qquad \underset{\text{Same signs}}{(4x - 3)(5x - 7)}$$
>
> - If c is *negative*, then the signs in the binomial must be different. The middle term in the trinomial determines which factor gets the positive sign and which gets the negative sign.
>
> $$\overset{c \text{ is negative}}{x^2 + 3x - 28} \qquad \overset{c \text{ is negative}}{x^2 - 3x - 28}$$
>
> $$\underset{\text{Different signs}}{(x + 7)(x - 4)} \qquad \underset{\text{Different signs}}{(x - 7)(x + 4)}$$

Instructor Note: Give the students this memory device: Look at the sign on the 3rd term. If it is a sum, the signs will be the same in the two binomials. If it is a difference, the signs in the two binomials will be different: sum-same; difference-different.

Skill Practice

Factor.

2. $-25w + 6w^2 + 4$

Classroom Example: p. 451, Exercise 22

Example 2   Factoring a Trinomial

Factor the trinomial: $13y - 6 + 8y^2$

Solution:

$13y - 6 + 8y^2$

$= 8y^2 + 13y - 6$ Write the polynomial in descending order.

$(\square y \; \square)(\square y \; \square)$ **Step 1:** The GCF is 1.

Factors of 8	Factors of 6
$1 \cdot 8$	$1 \cdot 6$
$2 \cdot 4$	$2 \cdot 3$
	$3 \cdot 2$
	$6 \cdot 1$

Step 2: List the positive factors of 8 and positive factors of 6. Consider the reverse order in one list of factors.

$\left. \begin{array}{l} 3 \cdot 2 \\ 6 \cdot 1 \end{array} \right\}$ (reverse order)

Step 3: Construct all possible binomial factors using different combinations of the factors of 8 and 6.

$\left. \begin{array}{l} (2y \; 1)(4y \; 6) \\ (2y \; 2)(4y \; 3) \\ (2y \; 3)(4y \; 2) \\ (2y \; 6)(4y \; 1) \\ (1y \; 1)(8y \; 6) \\ (1y \; 3)(8y \; 2) \end{array} \right\}$ Without regard to signs, these factorizations cannot work because the terms in the binomials share a common factor greater than 1.

Test the remaining factorizations. Keep in mind that to produce a product of -6, the signs within the parentheses must be opposite (one positive and one negative). Also, the sum of the products of the inner terms and outer terms must be combined to form $13y$.

Answer

2. $(6w - 1)(w - 4)$

$(1y \quad 6)(8y \quad 1)$ *Incorrect.*   Wrong middle term.
Regardless of the signs, the product of inner terms, $48y$, and the product of outer terms, $1y$, cannot be combined to form the middle term $13y$.

$(1y \quad 2)(8y \quad 3)$ *Correct.*   The terms $16y$ and $3y$ can be combined to form the middle term $13y$, provided the signs are applied correctly. We require $+16y$ and $-3y$.

The correct factorization of $8y^2 + 13y - 6$ is $(y + 2)(8y - 3)$.

Remember that the first step in any factoring problem is to remove the GCF. By removing the GCF, the remaining terms of the trinomial will be simpler and may have smaller coefficients.

| **Example 3** | Factoring a Trinomial by the Trial-and-Error Method |

Factor the trinomial by the trial-and-error method: $40x^3 - 104x^2 + 10x$

Solution:

$40x^3 - 104x^2 + 10x$

$= 2x(20x^2 - 52x + 5)$ **Step 1:** The GCF is $2x$.

$= 2x(\square x \quad \square)(\square x \quad \square)$ **Step 2:** List the factors of 20 and factors of 5. Consider the reverse order in one list of factors.

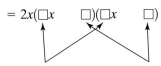

Factors of 20	Factors of 5
$1 \cdot 20$	$1 \cdot 5$
$2 \cdot 10$	$5 \cdot 1$
$4 \cdot 5$	

Step 3: Construct all possible binomial factors using different combinations of the factors of 20 and factors of 5. The signs in the parentheses must both be negative.

$\left. \begin{array}{l} = 2x(1x - 1)(20x - 5) \\ = 2x(2x - 1)(10x - 5) \\ = 2x(4x - 1)(5x - 5) \end{array} \right\}$ *Incorrect.*   Once the GCF has been removed from the original polynomial, the binomial factors cannot contain a GCF greater than 1.

$= 2x(1x - 5)(20x - 1)$ *Incorrect.*   Wrong middle term.
$2x(x - 5)(20x - 1)$
$= 2x(20x^2 - 1x - 100x + 5)$
$= 2x(20x^2 - 101x + 5)$

$= 2x(4x - 5)(5x - 1)$ *Incorrect.*   Wrong middle term.
$2x(4x - 5)(5x - 1)$
$= 2x(20x^2 - 4x - 25x + 5)$
$= 2x(20x^2 - 29x + 5)$

$= 2x(2x - 5)(10x - 1)$ *Correct.* $2x(2x - 5)(10x - 1)$
$= 2x(20x^2 - 2x - 50x + 5)$
$= 2x(20x^2 - 52x + 5)$
$= 40x^3 - 104x^2 + 10x$

The correct factorization is $2x(2x - 5)(10x - 1)$.

Skill Practice

Factor.

3. $8t^3 + 38t^2 + 24t$

Classroom Example: p. 451,
Exercise 32

TIP: Notice that when the GCF, $2x$, is removed from the original trinomial, the new trinomial has smaller coefficients. This makes the factoring process simpler. It is easier to list the factors of 20 and 5 than the factors of 40 and 10.

Answer

3. $2t(4t + 3)(t + 4)$

Often it is easier to factor a trinomial when the leading coefficient is positive. If the leading coefficient is negative, consider factoring out the opposite of the GCF.

Skill Practice

Factor.

4. $-4x^2 + 26xy - 40y^2$

Classroom Example: p. 451, Exercise 36

Example 4   Factoring a Trinomial by the Trial-and-Error Method

Factor: $-45x^2 - 3xy + 18y^2$

Solution:

$-45x^2 - 3xy + 18y^2$

$\quad = -3(15x^2 + xy - 6y^2)$ **Step 1:** Factor out -3 to make the leading coefficient positive.

$\quad = -3(\square x \ \ \square y)(\square x \ \ \square y)$ **Step 2:** List the factors of 15 and 6.

Factors of 15	Factors of 6
$1 \cdot 15$	$1 \cdot 6$
$3 \cdot 5$	$2 \cdot 3$
	$3 \cdot 2$
	$6 \cdot 1$

Step 3: We will construct all binomial combinations, without regard to signs first.

$\left.\begin{array}{l} -3(x \ \ y)(15x \ \ 6y) \\ -3(x \ \ 2y)(15x \ \ 3y) \\ -3(3x \ \ 3y)(5x \ \ 2y) \\ -3(3x \ \ 6y)(5x \ \ y) \end{array}\right\}$ *Incorrect.* The binomials contain a common factor.

Test the remaining factorizations. The signs within parentheses must be opposite to produce a product of $-6y^2$. Also, the sum of the products of the inner terms and outer terms must be combined to form $1xy$.

$-3(x \ \ 3y)(15x \ \ 2y)$ *Incorrect.*   Regardless of signs, $45xy$ and $2xy$ cannot be combined to equal xy.

$-3(x \ \ 6y)(15x \ \ y)$ *Incorrect.*   Regardless of signs, $90xy$ and xy cannot be combined to equal xy.

$-3(3x \ \ y)(5x \ \ 6y)$ *Incorrect.*   Regardless of signs, $5xy$ and $18xy$ cannot be combined to equal xy.

$-3(3x \ \ 2y)(5x \ \ 3y)$ *Correct.*   The terms $10xy$ and $9xy$ can be combined to form xy provided that the signs are applied correctly. We require $10xy$ and $-9xy$.

Avoiding Mistakes

Do not forget to write the GCF in the final answer.

$-3(3x + 2y)(5x - 3y)$ Factored form

Recall that a prime polynomial is a polynomial whose only factors are itself and 1. Not every trinomial is factorable by the methods presented in this text.

Answer

4. $-2(2x - 5y)(x - 4y)$

Example 5 Factoring a Trinomial by the Trial-and-Error Method

Factor the trinomial by the trial-and-error method: $2p^2 - 8p + 3$

Solution:

$2p^2 - 8p + 3$ **Step 1:** The GCF is 1.

$= (1p \ \square)(2p \ \square)$ **Step 2:** List the factors of 2 and the factors of 3.

Factors of 2	Factors of 3
$1 \cdot 2$	$1 \cdot 3$
	$3 \cdot 1$

Step 3: Construct all possible binomial factors using different combinations of the factors of 2 and 3. Because the third term in the trinomial is positive, both signs in the binomial must be the same. Because the middle term coefficient is negative, both signs will be negative.

$(p - 1)(2p - 3) = 2p^2 - 3p - 2p + 3$

$\qquad\qquad = 2p^2 - 5p + 3$ *Incorrect.* Wrong middle term.

$(p - 3)(2p - 1) = 2p^2 - p - 6p + 3$

$\qquad\qquad = 2p^2 - 7p + 3$ *Incorrect.* Wrong middle term.

None of the combinations of factors results in the correct product. Therefore, the polynomial $2p^2 - 8p + 3$ is prime and cannot be factored further.

Skill Practice

Factor.

5. $3a^2 + a + 4$

Classroom Example: p. 451, Exercise 54

In Example 6, we use the trial-and-error method to factor a higher degree trinomial into two binomial factors.

Example 6 Factoring a Higher Degree Trinomial

Factor the trinomial: $3x^4 + 8x^2 + 5$

Solution:

$3x^4 + 8x^2 + 5$ **Step 1:** The GCF is 1.

$= (\square x^2 + \square)(\square x^2 + \square)$ **Step 2:** To produce the product $3x^4$, we must use $3x^2$, and $1x^2$. To produce a product of 5, we will try the factors $(1)(5)$ and $(5)(1)$.

Step 3: Construct all possible binomial factors using the combinations of factors of $3x^4$ and 5.

$(3x^2 + 1)(x^2 + 5) = 3x^4 + 15x^2 + 1x^2 + 5 = 3x^4 + 16x^2 + 5$ Wrong middle term

$(3x^2 + 5)(x^2 + 1) = 3x^4 + 3x^2 + 5x^2 + 5 = 3x^4 + 8x^2 + 5$ Correct!

Therefore, $3x^4 + 8x^2 + 5 = (3x^2 + 5)(x^2 + 1)$

Skill Practice

Factor.

6. $2y^4 - y^2 - 15$

Classroom Example: p. 451, Exercise 40

Answers

5. Prime **6.** $(y^2 - 3)(2y^2 + 5)$

Section 6.3 Practice Exercises

Study Skills Exercises

1. In addition to studying the material for a test, here are some other activities that people use when preparing for a test. Circle the importance of each statement.

	not important	somewhat important	very important
a. Get a good night's sleep the night before the test.	1	2	3
b. Eat a good meal before the test.	1	2	3
c. Wear comfortable clothes on the day of the test.	1	2	3
d. Arrive early to class on the day of the test.	1	2	3

2. Define the key term **prime polynomial**.

Review Exercises

For Exercises 3–8, factor completely.

3. $21a^2b^2 + 12ab^2 - 15a^2b$
   $3ab(7ab + 4b - 5a)$

4. $5uv^2 - 10u^2v + 25u^2v^2$
   $5uv(v - 2u + 5uv)$

5. $mn - m - 2n + 2$
   $(n - 1)(m - 2)$

6. $5x - 10 - xy + 2y$
   $(x - 2)(5 - y)$

7. $6a^2 - 30a - 84$
   $6(a - 7)(a + 2)$

8. $10b^2 + 20b - 240$
   $10(b + 6)(b - 4)$

Objective 1: Factoring Trinomials by the Trial-and-Error Method

For Exercises 9–12, assume a, b, and c represent positive integers.

9. When factoring a polynomial of the form $ax^2 + bx + c$, pick an appropriate combination of signs. a

   a. (+)(+)
   b. (−)(−)
   c. (+)(−)

10. When factoring a polynomial of the form $ax^2 - bx - c$, pick an appropriate combination of signs. c

    a. (+)(+)
    b. (−)(−)
    c. (+)(−)

11. When factoring a polynomial of the form $ax^2 - bx + c$, pick an appropriate combination of signs. b

    a. (+)(+)
    b. (−)(−)
    c. (+)(−)

12. When factoring a polynomial of the form $ax^2 + bx - c$, pick an appropriate combination of signs. c

    a. (+)(+)
    b. (−)(−)
    c. (+)(−)

For Exercises 13–30, factor completely by using the trial-and-error method. **(See Examples 1, 2, and 5.)**

 13. $2y^2 - 3y - 2$
    $(2y + 1)(y - 2)$

14. $2w^2 + 5w - 3$
    $(2w - 1)(w + 3)$

15. $3n^2 + 13n + 4$
    $(3n + 1)(n + 4)$

16. $2a^2 + 7a + 6$
    $(2a + 3)(a + 2)$

17. $5x^2 - 14x - 3$
    $(5x + 1)(x - 3)$

18. $7y^2 + 9y - 10$
    $(7y - 5)(y + 2)$

19. $12c^2 - 5c - 2$
$(4c + 1)(3c - 2)$

20. $6z^2 + z - 12$
$(3z - 4)(2z + 3)$

21. $-12 + 10w^2 + 37w$
$(10w - 3)(w + 4)$

22. $-10 + 10p^2 + 21p$
$(2p + 5)(5p - 2)$

23. $-5q - 6 + 6q^2$
$(3q + 2)(2q - 3)$

24. $17a - 2 + 3a^2$
Prime

25. $6b - 23 + 4b^2$
Prime

26. $8 + 7x^2 - 18x$
$(7x - 4)(x - 2)$

27. $-8 + 25m^2 - 10m$
$(5m + 2)(5m - 4)$

28. $8q^2 + 31q - 4$
$(8q - 1)(q + 4)$

29. $6y^2 + 19xy - 20x^2$
$(6y - 5x)(y + 4x)$

30. $12y^2 - 73yz + 6z^2$
$(12y - z)(y - 6z)$

For Exercises 31–38, factor completely. Be sure to factor out the GCF first. **(See Examples 3–4.)**

31. $2m^2 - 12m - 80$
$2(m + 4)(m - 10)$

32. $3c^2 - 33c + 72$
$3(c - 8)(c - 3)$

33. $2y^5 + 13y^4 + 6y^3$
$y^3(2y + 1)(y + 6)$

34. $3u^8 - 13u^7 + 4u^6$
$u^6(3u - 1)(u - 4)$

35. $-a^2 - 15a + 34$
$-(a + 17)(a - 2)$

36. $-x^2 - 7x - 10$
$-(x + 2)(x + 5)$

37. $80m^2 - 100mp - 30p^2$
$10(4m + p)(2m - 3p)$

38. $60w^2 + 550wz - 500z^2$
$10(w + 10z)(6w - 5z)$

For Exercises 39–44, factor the higher degree polynomial. **(See Example 6.)**

39. $x^4 + 10x^2 + 9$
$(x^2 + 1)(x^2 + 9)$

40. $y^4 + 4y^2 - 21$
$(y^2 - 3)(y^2 + 7)$

41. $w^4 + 2w^2 - 15$
$(w^2 + 5)(w^2 - 3)$

42. $p^4 - 13p^2 + 40$
$(p^2 - 8)(p^2 - 5)$

43. $2x^4 - 7x^2 - 15$
$(2x^2 + 3)(x^2 - 5)$

44. $5y^4 + 11y^2 + 2$
$(5y^2 + 1)(y^2 + 2)$

Mixed Exercises

For Exercises 45–92, factor the trinomial completely.

45. $20z - 18 - 2z^2$
$-2(z - 9)(z - 1)$

46. $25t - 5t^2 - 30$
$-5(t - 2)(t - 3)$

47. $42 - 13q + q^2$
$(q - 7)(q - 6)$

48. $-5w - 24 + w^2$
$(w - 8)(w + 3)$

49. $6t^2 + 7t - 3$
$(2t + 3)(3t - 1)$

50. $4p^2 - 9p + 2$
$(4p - 1)(p - 2)$

51. $4m^2 - 20m + 25$
$(2m - 5)^2$

52. $16r^2 + 24r + 9$
$(4r + 3)^2$

53. $5c^2 - c + 2$
Prime

54. $7s^2 + 2s + 9$
Prime

55. $6x^2 - 19xy + 10y^2$
$(2x - 5y)(3x - 2y)$

56. $15p^2 + pq - 2q^2$
$(3p - q)(5p + 2q)$

57. $12m^2 + 11mn - 5n^2$
$(4m + 5n)(3m - n)$

58. $4a^2 + 5ab - 6b^2$
$(4a - 3b)(a + 2b)$

59. $30r^2 + 5r - 10$
$5(3r + 2)(2r - 1)$

60. $36x^2 - 18x - 4$
$2(6x + 1)(3x - 2)$

61. $4s^2 - 8st + t^2$
Prime

62. $6u^2 - 10uv + 5v^2$
Prime

63. $10t^2 - 23t - 5$
$(2t - 5)(5t + 1)$

64. $16n^2 + 14n + 3$
$(8n + 3)(2n + 1)$

65. $14w^2 + 13w - 12$
$(7w - 4)(2w + 3)$

66. $12x^2 - 16x + 5$
$(6x - 5)(2x - 1)$

67. $x^2 + 7x - 18$
$(x + 9)(x - 2)$

68. $y^2 - 6y - 40$
$(y + 4)(y - 10)$

69. $a^2 - 10a - 24$
$(a - 12)(a + 2)$

70. $b^2 + 6b - 7$
$(b + 7)(b - 1)$

71. $r^2 + 5r - 24$
$(r + 8)(r - 3)$

72. $t^2 + 20t + 100$
$(t + 10)^2$

73. $x^2 + 9xy + 20y^2$
$(x + 5y)(x + 4y)$

74. $p^2 - 13pq + 36q^2$
$(p - 9q)(p - 4q)$

75. $v^2 + 2v + 15$
Prime

76. $x^2 - x - 1$
Prime

77. $a^2 + 21ab + 20b^2$
$(a + 20b)(a + b)$

78. $x^2 - 17xy - 18y^2$
$(x - 18y)(x + y)$

79. $t^2 - 10t + 21$
$(t - 7)(t - 3)$

80. $z^2 - 15z + 36$
$(z - 12)(z - 3)$

81. $5d^3 + 3d^2 - 10d$
$d(5d^2 + 3d - 10)$

82. $3y^3 - y^2 + 12y$
$y(3y^2 - y + 12)$

83. $4b^3 - 4b^2 - 80b$
$4b(b - 5)(b + 4)$

84. $2w^2 + 20w + 42$
$2(w + 7)(w + 3)$

85. $x^2y^2 - 13xy^2 + 30y^2$
$y^2(x - 3)(x - 10)$

86. $p^2q^2 - 14pq^2 + 33q^2$
$q^2(p - 3)(p - 11)$

87. $-12u^3 - 22u^2 + 20u$
$-2u(2u + 5)(3u - 2)$

88. $-18z^4 + 15z^3 + 12z^2$
$-3z^2(3z - 4)(2z + 1)$

89. $8x^4 + 14x^2 + 3$
$(2x^2 + 3)(4x^2 + 1)$

90. $6y^4 - 5y^2 - 4$
$(3y^2 - 4)(2y^2 + 1)$

91. $10z^4 + 9z^2 - 9$
$(5z^2 - 3)(2z^2 + 3)$

92. $6p^4 + 17p^2 + 10$
$(6p^2 + 5)(p^2 + 2)$

 Writing    Translating Expression   Geometry   Scientific Calculator   Video   NS E

Expanding Your Skills

For Exercises 93–96, each pair of trinomials looks similar but differs by one sign. Factor each trinomial, and see how their factored forms differ.

93. a. $x^2 - 10x - 24$ $(x - 12)(x + 2)$

    b. $x^2 - 10x + 24$ $(x - 6)(x - 4)$

95. a. $x^2 - 5x - 6$ $(x - 6)(x + 1)$

    b. $x^2 - 5x + 6$ $(x - 2)(x - 3)$

94. a. $x^2 - 13x - 30$ $(x - 15)(x + 2)$

    b. $x^2 - 13x + 30$ $(x - 10)(x - 3)$

96. a. $x^2 - 10x + 9$ $(x - 9)(x - 1)$

    b. $x^2 + 10x + 9$ $(x + 9)(x + 1)$

Section 6.4   Factoring Trinomials: AC-Method

Objective

1. Factoring Trinomials by the AC-Method

In Section 6.2, we factored trinomials with a leading coefficient of 1. In Section 6.3, we learned the trial-and-error method to factor the more general case in which the leading coefficient is any integer. In this section, we provide an alternative method to factor trinomials, called the ac-method.

1. Factoring Trinomials by the AC-Method

The product of two binomials results in a four-term expression that can sometimes be simplified to a trinomial. To factor the trinomial, we want to reverse the process.

Multiply:

           Multiply the binomials.   Add the middle terms.

$$(2x + 3)(x + 2) = \longrightarrow 2x^2 + 4x + 3x + 6 = \longrightarrow 2x^2 + 7x + 6$$

Factor:

$$2x^2 + 7x + 6 = \longrightarrow 2x^2 + 4x + 3x + 6 = \longrightarrow (2x + 3)(x + 2)$$

          Rewrite the middle term as   Factor by grouping.
          a sum or difference of terms.

To factor a quadratic trinomial, $ax^2 + bx + c$, by the ac-method, we rewrite the middle term, bx, as a sum or difference of terms. The goal is to produce a four-term polynomial that can be factored by grouping. The process is outlined as follows.

Instructor Note: We use the phrase *prime polynomial* to mean that a polynomial is not factorable over the integers.

> **PROCEDURE**   AC-Method: Factoring $ax^2 + bx + c$ $(a \neq 0)$
>
> **Step 1**   Factor out the GCF from all terms.
>
> **Step 2**   Multiply the coefficients of the first and last terms (ac).
>
> **Step 3**   Find two integers whose product is ac and whose sum is b. (If no pair of integers can be found, then the trinomial cannot be factored further and is a **prime polynomial**.)
>
> **Step 4**   Rewrite the middle term, bx, as the sum of two terms whose coefficients are the integers found in step 3.
>
> **Step 5**   Factor the polynomial by grouping.

The ac-method for factoring trinomials is illustrated in Example 1. However, before we begin, keep these two important guidelines in mind:

- For any factoring problem you encounter, always factor out the GCF from all terms first.
- To factor a trinomial, write the trinomial in the form $ax^2 + bx + c$.

Example 1 **Factoring a Trinomial by the AC-Method**

Factor the trinomial by the ac-method: $2x^2 + 7x + 6$

Solution:

$2x^2 + 7x + 6$	**Step 1:** Factor out the GCF from all terms. In this case, the GCF is 1.
$2x^2 + 7x + 6$	**Step 2:** The trinomial is written in the form $ax^2 + bx + c$.
$a = 2, b = 7, c = 6$	Find the product $ac = (2)(6) = 12$.

$$\underline{12} \qquad \underline{12}$$
$$1 \cdot 12 \qquad (-1)(-12)$$
$$2 \cdot 6 \qquad (-2)(-6)$$
$$3 \cdot 4 \qquad (-3)(-4)$$

Step 3: List all factors of ac and search for the pair whose sum equals the value of b. That is, list the factors of 12 and find the pair whose sum equals 7.

The numbers 3 and 4 satisfy both conditions: $3 \cdot 4 = 12$ and $3 + 4 = 7$.

$2x^2 + 7x + 6$

$= 2x^2 + 3x + 4x + 6$ **Step 4:** Write the middle term of the trinomial as the sum of two terms whose coefficients are the selected pair of numbers: 3 and 4.

$= 2x^2 + 3x \mid + 4x + 6$ **Step 5:** Factor by grouping.

$= x(2x + 3) + 2(2x + 3)$

$= (2x + 3)(x + 2)$

$\underline{\text{Check:}} (2x + 3)(x + 2) = 2x^2 + 4x + 3x + 6$

$\qquad\qquad\qquad\qquad = 2x^2 + 7x + 6 \ ✔$

Skill Practice

Factor by the ac-method.

1. $2x^2 + 5x + 3$

Classroom Example: p. 457, Exercise 14

TIP: One frequently asked question is whether the order matters when we rewrite the middle term of the trinomial as two terms (step 3). The answer is no. From the previous example, the two middle terms in step 3 could have been reversed to obtain the same result:

$$2x^2 + 7x + 6$$
$$= 2x^2 + 4x + 3x + 6$$
$$= 2x(x + 2) + 3(x + 2)$$
$$= (x + 2)(2x + 3)$$

This example also points out that the order in which two factors are written does not matter. The expression $(x + 2)(2x + 3)$ is equivalent to $(2x + 3)(x + 2)$ because multiplication is a commutative operation.

Answer

1. $(x + 1)(2x + 3)$

Skill Practice

Factor by the ac-method.

2. $13w + 6w^2 + 6$

Classroom Example: p. 457, Exercise 26

Avoiding Mistakes

Before factoring a trinomial, be sure to write the trinomial in descending order. That is, write it in the form $ax^2 + bx + c$.

Example 2 **Factoring Trinomials by the AC-Method**

Factor the trinomial by the ac-method: $-2x + 8x^2 - 3$

Solution:

$-2x + 8x^2 - 3$ First rewrite the polynomial in the form $ax^2 + bx + c$.

$= 8x^2 - 2x - 3$ **Step 1:** The GCF is 1.

$a = 8, b = -2, c = -3$ **Step 2:** Find the product $ac = (8)(-3) = -24$.

-24	-24
$-1 \cdot 24$	$-24 \cdot 1$
$-2 \cdot 12$	$-12 \cdot 2$
$-3 \cdot 8$	$-8 \cdot 3$
$-4 \cdot 6$	$-6 \cdot 4$

Step 3: List all the factors of -24 and find the pair of factors whose sum equals -2.

The numbers -6 and 4 satisfy both conditions: $(-6)(4) = -24$ and $-6 + 4 = -2$.

$= 8x^2 - 2x - 3$ **Step 4:** Write the middle term of the trinomial as two terms whose coefficients are the selected pair of numbers, -6 and 4.

$= 8x^2 - 6x + 4x - 3$

$= 8x^2 - 6x \mid + 4x - 3$ **Step 5:** Factor by grouping.

$= 2x(4x - 3) + 1(4x - 3)$

$= (4x - 3)(2x + 1)$

Check: $(4x - 3)(2x + 1) = 8x^2 + 4x - 6x - 3$

$= 8x^2 - 2x - 3$ ✔

Skill Practice

Factor by the ac-method.

3. $9y^3 - 30y^2 + 24y$

Classroom Example: p. 457, Exercise 36

Example 3 **Factoring a Trinomial by the AC-Method**

Factor the trinomial by the ac-method: $10x^3 - 85x^2 + 105x$

Solution:

$10x^3 - 85x^2 + 105x$ **Step 1:** Factor out the GCF of $5x$.

$= 5x(2x^2 - 17x + 21)$ The trinomial is in the form $ax^2 + bx + c$.

$a = 2, b = -17, c = 21$ **Step 2:** Find the product $ac = (2)(21) = 42$.

42	42
$1 \cdot 42$	$(-1)(-42)$
$2 \cdot 21$	$(-2)(-21)$
$3 \cdot 14$	$(-3)(-14)$
$6 \cdot 7$	$(-6)(-7)$

Step 3: List all the factors of 42 and find the pair whose sum equals -17.

The numbers -3 and -14 satisfy both conditions: $(-3)(-14) = 42$ and $-3 + (-14) = -17$.

Answers

2. $(2w + 3)(3w + 2)$
3. $3y(3y - 4)(y - 2)$

$= 5x(2x^2 - 17x + 21)$ **Step 4:** Write the middle term of the trinomial as two terms whose coefficients are the selected pair of numbers, -3 and -14.

$= 5x(2x^2 - 3x - 14x + 21)$

$= 5x(2x^2 - 3x \mid - 14x + 21)$ **Step 5:** Factor by grouping.

$= 5x[x(2x - 3) - 7(2x - 3)]$

$= 5x(2x - 3)(x - 7)$

Avoiding Mistakes

Be sure to bring down the GCF in each successive step as you factor.

TIP: Notice when the GCF is removed from the original trinomial, the new trinomial has smaller coefficients. This makes the factoring process simpler because the product ac is smaller. It is much easier to list the factors of 42 than the factors of 1050.

Original trinomial	With the GCF factored out
$10x^3 - 85x^2 + 105x$	$5x(2x^2 - 17x + 21)$
$ac = (10)(105) = 1050$	$ac = (2)(21) = 42$

In most cases, it is easier to factor a trinomial with a positive leading coefficient.

Example 4 **Factoring a Trinomial by the AC-Method**

Factor: $-18x^2 + 21xy + 15y^2$

Solution:

$-18x^2 + 21xy + 15y^2$ **Step 1:** Factor out the GCF.

$= -3(6x^2 - 7xy - 5y^2)$ Factor out -3 to make the leading term positive.

Step 2: The product $ac = (6)(-5) = -30$.

Step 3: The numbers -10 and 3 have a product of -30 and a sum of -7.

$= -3[6x^2 - 10xy + 3xy - 5y^2]$ **Step 4:** Rewrite the middle term, $-7xy$ as $-10xy + 3xy$.

$= -3[6x^2 - 10xy \mid + 3xy - 5y^2]$ **Step 5:** Factor by grouping.

$= -3[2x(3x - 5y) + y(3x - 5y)]$

$= -3(3x - 5y)(2x + y)$ Factored form

Skill Practice

Factor.

4. $-8x^2 - 8xy + 30y^2$

Classroom Example: p. 457, Exercise 34

Recall that a prime polynomial is a polynomial whose only factors are itself and 1. It also should be noted that not every trinomial is factorable by the methods presented in this text.

Answer

4. $-2(2x - 3y)(2x + 5y)$

Skill Practice

Factor.

5. $4x^2 + 5x + 2$

Classroom Example: p. 457,
Exercise 24

Example 5 **Factoring a Trinomial by the AC-Method**

Factor the trinomial by the ac-method: $2p^2 - 8p + 3$

Solution:

$2p^2 - 8p + 3$ **Step 1:** The GCF is 1.

Step 2: The product $ac = 6$.

6	6
$1 \cdot 6$	$(-1)(-6)$
$2 \cdot 3$	$(-2)(-3)$

Step 3: List the factors of 6. Notice that no pair of factors has a sum of -8. Therefore, the trinomial cannot be factored.

The trinomial $2p^2 - 8p + 3$ is a prime polynomial.

In Example 6, we use the ac-method to factor a higher degree trinomial.

Skill Practice

Factor.

6. $3y^4 + 2y^2 - 8$

Classroom Example: p. 457,
Exercise 40

Example 6 **Factoring a Higher Degree Trinomial**

Factor the trinomial. $2x^4 + 5x^2 + 2$

Solution:

$2x^4 + 5x^2 + 2$ **Step 1:** The GCF is 1.

$a = 2, b = 5, c = 2$ **Step 2:** Find the product $ac = (2)(2) = 4$.

Step 3: The numbers 1 and 4 have a product of 4 and a sum of 5.

$2x^4 + x^2 + 4x^2 + 2$ **Step 4:** Rewrite the middle term, $5x^2$, as $x^2 + 4x^2$.

$2x^4 + x^2 + 4x^2 + 2$ **Step 5:** Factor by grouping.

$x^2(2x^2 + 1) + 2(2x^2 + 1)$

$(2x^2 + 1)(x^2 + 2)$ Factored form.

Answers

5. Prime

6. $(3y^2 - 4)(y^2 + 2)$

Section 6.4 Practice Exercises

Boost *your* GRADE at
ALEKS.com!

ALEKS version 3.0

• Practice Problems
• Self-Tests
• NetTutor

• e-Professors
• Videos

For additional exercises, see Classroom Activities
6.4A–6.4B in the *Instructor's Resource Manual*
at www.mhhe.com/moh.

Study Skills Exercise

1. Define the key term **prime polynomial**.

Review Exercises

For Exercises 2–4, factor completely.

2. $5x(x - 2) - 2(x - 2)$
$(x - 2)(5x - 2)$

3. $8(y + 5) + 9y(y + 5)$
$(y + 5)(8 + 9y)$

4. $6ab + 24b - 12a - 48$
$6(a + 4)(b - 2)$

Objective 1: Factoring Trinomials by the AC-Method

For Exercises 5–12, find the pair of integers whose product and sum are given.

5. Product: 12   Sum: 13   12, 1

6. Product: 12   Sum: 7   3, 4

 Writing   Translating Expression   Geometry   Scientific Calculator   Video   NS&E

7. Product: 8   Sum: −9   −8, −1

8. Product: −4   Sum: −3   −4, 1

9. Product: −20   Sum: 1   5, −4

10. Product: −6   Sum: −1   −3, 2

11. Product: −18   Sum: 7   9, −2

⬤ **12.** Product: −72   Sum: −6   −12, 6

For Exercises 13–42, factor the trinomials using the ac-method. **(See Examples 1–6.)**

13. $3x^2 + 13x + 4$ $(x + 4)(3x + 1)$

14. $2y^2 + 7y + 6$ $(2y + 3)(y + 2)$

15. $4w^2 − 9w + 2$ $(w − 2)(4w − 1)$

16. $2p^2 − 3p − 2$ $(p − 2)(2p + 1)$

17. $2m^2 + 5m − 3$ $(m + 3)(2m − 1)$

18. $6n^2 + 7n − 3$ $(2n + 3)(3n − 1)$

⬤ **19.** $8k^2 − 6k − 9$ $(4k + 3)(2k − 3)$

20. $9h^2 − 12h + 4$ $(3h − 2)^2$

21. $4k^2 − 20k + 25$ $(2k − 5)^2$

22. $16h^2 + 24h + 9$ $(4h + 3)^2$

23. $5x^2 + x + 7$ Prime

24. $4y^2 − y + 2$ Prime

25. $10 + 9z^2 − 21z$
$(3z − 5)(3z − 2)$

26. $13x + 4x^2 − 12$
$(4x − 3)(x + 4)$

27. $50y + 24 + 14y^2$
$2(7y + 4)(y + 3)$

28. $−24 + 10w + 4w^2$
$2(2w − 3)(w + 4)$

29. $12y^2 + 8yz − 15z^2$
$(6y − 5z)(2y + 3z)$

30. $20a^2 + 3ab − 9b^2$
$(5a − 3b)(4a + 3b)$

31. $−15w^2 + 22w + 5$
$−(3w − 5)(5w + 1)$

32. $−16z^2 + 34z + 15$
$−(8z + 3)(2z − 5)$

33. $−12x^2 + 20xy − 8y^2$
$−4(x − y)(3x − 2y)$

34. $−6p^2 − 21pq − 9q^2$
$−3(2p + q)(p + 3q)$

35. $18y^3 + 60y^2 + 42y$
$6y(y + 1)(3y + 7)$

36. $8t^3 − 4t^2 − 40t$
$4t(2t − 5)(t + 2)$

37. $a^4 + 5a^2 + 6$
$(a^2 + 2)(a^2 + 3)$

38. $y^4 − 2y^2 − 35$
$(y^2 + 5)(y^2 − 7)$

39. $6x^4 − x^2 − 15$
$(3x^2 − 5)(2x^2 + 3)$

40. $8t^4 + 2t^2 − 3$
$(4t^2 + 3)(2t^2 − 1)$

41. $8p^4 + 37p^2 − 15$
$(8p^2 − 3)(p^2 + 5)$

42. $2a^4 + 11a^2 + 14$
$(2a^2 + 7)(a^2 + 2)$

Mixed Exercises

For Exercises 43–75, factor completely.

43. $20p^2 − 19p + 3$
$(5p − 1)(4p − 3)$

44. $4p^2 + 5pq − 6q^2$
$(p + 2q)(4p − 3q)$

45. $6u^2 − 19uv + 10v^2$
$(3u − 2v)(2u − 5v)$

46. $15m^2 + mn − 2n^2$
$(5m + 2n)(3m − n)$

47. $12a^2 + 11ab − 5b^2$
$(4a + 5b)(3a − b)$

48. $3r^2 − rs − 14s^2$
$(r + 2s)(3r − 7s)$

⬤ **49.** $3h^2 + 19hk − 14k^2$
$(h + 7k)(3h − 2k)$

50. $2x^2 − 13xy + y^2$ Prime

51. $3p^2 + 20pq − q^2$ Prime

52. $3 − 14z + 16z^2$
$(2z − 1)(8z − 3)$

53. $10w + 1 + 16w^2$
$(8w + 1)(2w + 1)$

54. $b^2 + 16 − 8b$
$(b − 4)^2$

55. $1 + q^2 − 2q$
$(q − 1)^2$

⬤ **56.** $25x − 5x^2 − 30$
$−5(x − 2)(x − 3)$

57. $20a − 18 − 2a^2$
$−2(a − 1)(a − 9)$

58. $−6 − t + t^2$
$(t − 3)(t + 2)$

59. $−6 + m + m^2$
$(m + 3)(m − 2)$

60. $72x^2 + 18x − 2$
$2(12x − 1)(3x + 1)$

61. $20y^2 − 78y − 8$
$2(10y + 1)(y − 4)$

62. $p^3 − 6p^2 − 27p$
$p(p + 3)(p − 9)$

63. $w^5 − 11w^4 + 28w^3$
$w^3(w − 4)(w − 7)$

64. $3x^3 + 10x^2 + 7x$
$x(3x + 7)(x + 1)$

⬤ **65.** $4r^3 + 3r^2 − 10r$
$r(4r − 5)(r + 2)$

66. $2p^3 − 38p^2 + 120p$
$2p(p − 15)(p − 4)$

67. $4q^3 − 4q^2 − 80q$
$4q(q − 5)(q + 4)$

68. $x^2y^2 + 14x^2y + 33x^2$
$x^2(y + 3)(y + 11)$

69. $a^2b^2 + 13ab^2 + 30b^2$
$b^2(a + 10)(a + 3)$

70. $−k^2 − 7k − 10$
$−1(k + 2)(k + 5)$

71. $−m^2 − 15m + 34$
$−1(m − 2)(m + 17)$

72. $−3n^2 − 3n + 90$
$−3(n + 6)(n − 5)$

73. $−2h^2 + 28h − 90$
$−2(h − 9)(h − 5)$

74. $x^4 − 7x^2 + 10$
$(x^2 − 2)(x^2 − 5)$

75. $m^4 + 10m^2 + 21$
$(m^2 + 3)(m^2 + 7)$

✎ **76.** Is the expression $(2x + 4)(x − 7)$ factored completely? Explain why or why not.
No. $(2x + 4)$ contains a common factor of 2.

✎ **77.** Is the expression $(3x + 1)(5x − 10)$ factored completely? Explain why or why not.
No. $(5x − 10)$ contains a common factor of 5.

 Writing   ⬅➡ Translating Expression   Geometry   Scientific Calculator   ⬤ Video

Section 6.5 Difference of Squares and Perfect Square Trinomials

Objectives

1. Factoring a Difference of Squares
2. Factoring Perfect Square Trinomials

1. Factoring a Difference of Squares

Up to this point, we have learned several methods of factoring, including:

- Factoring out the greatest common factor from a polynomial
- Factoring a four-term polynomial by grouping
- Factoring trinomials by the ac-method or by the trial-and-error method

In this section, we begin by factoring a special binomial called a difference of squares. Recall from Section 5.6 that the product of two conjugates results in a **difference of squares**:

$$(a + b)(a - b) = a^2 - b^2$$

Therefore, to factor a difference of squares, the process is reversed. Identify a and b and construct the conjugate factors.

> **FORMULA Factored Form of a Difference of Squares**
>
> $$a^2 - b^2 = (a + b)(a - b)$$

We know that the numbers, 1, 4, 9, 16, 25, and so on are perfect squares. It is also important to recognize that a variable expression is a perfect square if its exponent is a multiple of 2. For example:

Perfect Squares

$$x^2 = (x)^2$$
$$x^4 = (x^2)^2$$
$$x^6 = (x^3)^2$$
$$x^8 = (x^4)^2$$
$$x^{10} = (x^5)^2$$

Skill Practice

Factor completely.

1. $a^2 - 64$
2. $25q^2 - 49w^2$
3. $98m^3n - 50mn$

Classroom Examples: p. 463,
Exercises 16 and 18

Example 1 Factoring Differences of Squares

Factor the binomials.

a. $y^2 - 25$ **b.** $49s^2 - 4t^4$ **c.** $18w^2z - 2z$

Solution:

a. $y^2 - 25$ The binomial is a difference of squares.

$\quad = (y)^2 - (5)^2$ Write in the form: $a^2 - b^2$, where $a = y, b = 5$.

$\quad = (y + 5)(y - 5)$ Factor as $(a + b)(a - b)$.

b. $49s^2 - 4t^4$ The binomial is a difference of squares.

$\quad = (7s)^2 - (2t^2)^2$ Write in the form $a^2 - b^2$, where $a = 7s$ and $b = 2t^2$.

$\quad = (7s + 2t^2)(7s - 2t^2)$ Factor as $(a + b)(a - b)$.

Answers

1. $(a + 8)(a - 8)$
2. $(5q + 7w)(5q - 7w)$
3. $2mn(7m + 5)(7m - 5)$

c. $18w^2z - 2z$ The GCF is $2z$.

$= 2z(9w^2 - 1)$ $(9w^2 - 1)$ is a difference of squares.

$= 2z[(3w)^2 - (1)^2]$ Write in the form: $a^2 - b^2$, where $a = 3w, b = 1$.

$= 2z(3w + 1)(3w - 1)$ Factor as $(a + b)(a - b)$.

The difference of squares $a^2 - b^2$ factors as $(a - b)(a + b)$. However, the *sum* of squares is not factorable.

> **PROPERTY**   **Sum of Squares**
>
> Suppose a and b have no common factors. Then the **sum of squares** $a^2 + b^2$ is *not* factorable over the real numbers.
>
> That is, $a^2 + b^2$ is prime over the real numbers.

To see why $a^2 + b^2$ is not factorable, consider the product of binomials:

$$(a + b)(a - b) = a^2 - b^2 \qquad \text{Wrong sign}$$
$$(a + b)(a + b) = a^2 + 2ab + b^2 \qquad \text{Wrong middle term}$$
$$(a - b)(a - b) = a^2 - 2ab + b^2 \qquad \text{Wrong middle term}$$

After exhausting all possibilities, we see that if a and b share no common factors, then the sum of squares $a^2 + b^2$ is a prime polynomial.

Example 2   **Factoring Binomials**

Factor the binomials, if possible.   **a.** $p^2 - 9$ **b.** $p^2 + 9$

Solution:

a. $p^2 - 9$ Difference of squares

    $= (p - 3)(p + 3)$ Factor as $a^2 - b^2 = (a - b)(a + b)$.

b. $p^2 + 9$ Sum of squares

    Prime (cannot be factored)

Skill Practice

Factor the binomials, if possible.

4. $t^2 - 144$

5. $t^2 + 144$

Classroom Example: p. 463, Exercise 24

Some factoring problems require several steps. Always be sure to factor completely.

Example 3   **Factoring the Difference of Squares**

Factor completely. $w^4 - 81$

Solution:

$w^4 - 81$ The GCF is 1. $w^4 - 81$ is a difference of squares.

$= (w^2)^2 - (9)^2$ Write in the form: $a^2 - b^2$, where $a = w^2, b = 9$.

$= (w^2 + 9)(w^2 - 9)$ Factor as $(a + b)(a - b)$.

$= (w^2 + 9)(w + 3)(w - 3)$ Note that $w^2 - 9$ can be factored further as a difference of squares. (The binomial $w^2 + 9$ is a sum of squares and cannot be factored further.)

Skill Practice

Factor completely.

6. $y^4 - 1$

Classroom Example: p. 463, Exercise 36

Answers

4. $(t - 12)(t + 12)$

5. Prime

6. $(y + 1)(y - 1)(y^2 + 1)$

Example 4 Factoring a Polynomial

Factor completely. $y^3 - 5y^2 - 4y + 20$

Solution:

$y^3 - 5y^2 - 4y + 20$ The GCF is 1. The polynomial has four terms. Factor by grouping.

$= y^3 - 5y^2 \mid - 4y + 20$

$= y^2(y - 5) - 4(y - 5)$

$= (y - 5)(y^2 - 4)$ The expression $y^2 - 4$ is a difference of squares and can be factored further as $(y - 2)(y + 2)$.

$= (y - 5)(y - 2)(y + 2)$

Check: $(y - 5)(y - 2)(y + 2) = (y - 5)(y^2 - 2y + 2y - 4)$

$= (y - 5)(y^2 - 4)$

$= (y^3 - 4y - 5y^2 + 20)$

$= y^3 - 5y^2 - 4y + 20$ ✔

2. Factoring Perfect Square Trinomials

Recall from Section 5.6 that the square of a binomial always results in a **perfect square trinomial**.

$$(a + b)^2 = (a + b)(a + b) \xrightarrow{\text{Multiply.}} = a^2 + 2ab + b^2$$
$$(a - b)^2 = (a - b)(a - b) \xrightarrow{\text{Multiply.}} = a^2 - 2ab + b^2$$

For example, $(3x + 5)^2 = (3x)^2 + 2(3x)(5) + (5)^2$

$$= 9x^2 + 30x + 25 \text{ (perfect square trinomial)}$$

We now want to reverse this process by factoring a perfect square trinomial. The trial-and-error method or the ac-method can always be used; however, if we recognize the pattern for a perfect square trinomial, we can use one of the following formulas to reach a quick solution.

> **FORMULA** Factored Form of a Perfect Square Trinomial
> $$a^2 + 2ab + b^2 = (a + b)^2$$
> $$a^2 - 2ab + b^2 = (a - b)^2$$

For example, $9x^2 + 30x + 25$ is a perfect square trinomial with $a = 3x$ and $b = 5$. Therefore, it factors as

$$9x^2 + 30x + 25 = (3x)^2 + 2(3x)(5) + (5)^2 = (3x + 5)^2$$
$$a^2 \quad + 2 \; (a) \; (b) + (b)^2 = (a \; + \; b)^2$$

To apply the formula to factor a perfect square trinomial, we must first be sure that the trinomial is indeed a perfect square trinomial.

PROCEDURE Checking for a Perfect Square Trinomial

Step 1 Determine whether the first and third terms are both perfect squares and have positive coefficients.

Step 2 If this is the case, identify a and b, and determine if the middle term equals $2ab$ or $-2ab$.

Example 5 Factoring Perfect Square Trinomials

Factor the trinomials completely.

a. $x^2 + 14x + 49$ **b.** $25y^2 - 20y + 4$

Solution:

a. $x^2 + 14x + 49$ The GCF is 1.

- The first and third terms are positive.
- The first term is a perfect square: $x^2 = (x)^2$.
- The third term is a perfect square: $49 = (7)^2$.

Perfect squares

$x^2 + 14x + 49$

- The middle term is twice the product of x and 7: $14x = 2(x)(7)$

$= (x)^2 + 2(x)(7) + (7)^2$ The trinomial is in the form $a^2 + 2ab + b^2$, where $a = x$ and $b = 7$.

$= (x + 7)^2$ Factor as $(a + b)^2$.

b. $25y^2 - 20y + 4$ The GCF is 1.

Perfect squares

- The first and third terms are positive.
- The first term is a perfect square: $25y^2 = (5y)^2$.
- The third term is a perfect square: $4 = (2)^2$.

$25y^2 - 20y + 4$

$= (5y)^2 - 2(5y)(2) + (2)^2$ • In the middle: $20y = 2(5y)(2)$

$= (5y - 2)^2$ Factor as $(a - b)^2$.

Skill Practice

Factor completely.

8. $x^2 - 6x + 9$

9. $81w^2 + 72w + 16$

Classroom Examples: p. 464, Exercises 52 and 54

TIP: The sign of the middle term in a perfect square trinomial determines the sign within the binomial of the factored form.

$a^2 + 2ab + b^2 = (a + b)^2$

$a^2 - 2ab + b^2 = (a - b)^2$

Answers

8. $(x - 3)^2$

9. $(9w + 4)^2$

Classroom Examples: p. 464, Exercises 60 and 62

Skill Practice

Factor completely.

10. $5z^3 + 20z^2w + 20zw^2$

11. $40x^2 + 130x + 90$

Example 6 **Factoring Perfect Square Trinomials**

Factor the trinomials completely.

a. $18c^3 - 48c^2d + 32cd^2$ **b.** $5w^2 + 50w + 45$

Solution:

a. $18c^3 - 48c^2d + 32cd^2$

$= 2c(9c^2 - 24cd + 16d^2)$ The GCF is $2c$.

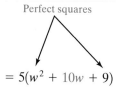

Perfect squares

$= 2c(9c^2 - 24cd + 16d^2)$

- The first and third terms are positive.
- The first term is a perfect square: $9c^2 = (3c)^2$.
- The third term is a perfect square: $16d^2 = (4d)^2$.
- In the middle: $24cd = 2(3c)(4d)$

$= 2c[(3c)^2 - 2(3c)(4d) + (4d)^2]$

$= 2c(3c - 4d)^2$ Factor as $(a - b)^2$.

b. $5w^2 + 50w + 45$

$= 5(w^2 + 10w + 9)$ The GCF is 5.

Perfect squares

$= 5(w^2 + 10w + 9)$

The first and third terms are perfect squares.

$$w^2 = (w)^2 \quad \text{and} \quad 9 = (3)^2$$

However, the middle term is not 2 times the product of w and 3. Therefore, this is not a perfect square trinomial.

$$10w \neq 2(w)(3)$$

$= 5(w + 9)(w + 1)$ To factor, use the trial-and-error method.

TIP: To help you identify a perfect square trinomial, we recommend that you familiarize yourself with the first several perfect squares.

$(1)^2 = 1$	$(6)^2 = 36$	$(11)^2 = 121$
$(2)^2 = 4$	$(7)^2 = 49$	$(12)^2 = 144$
$(3)^2 = 9$	$(8)^2 = 64$	$(13)^2 = 169$
$(4)^2 = 16$	$(9)^2 = 81$	$(14)^2 = 196$
$(5)^2 = 25$	$(10)^2 = 100$	$(15)^2 = 225$

If you do not recognize that a trinomial is a perfect square trinomial, you may still use the trial-and-error method or ac-method to factor it.

Answers

10. $5z(z + 2w)^2$

11. $10(4x + 9)(x + 1)$

Section 6.5 Practice Exercises

Boost *your* GRADE at ALEKS.com!

ALEKS version 3.0

- Practice Problems
- Self-Tests
- NetTutor
- e-Professors
- Videos

For additional exercises, see Classroom Activities 6.5A–6.5B in the *Instructor's Resource Manual* at www.mhhe.com/moh.

Study Skills Exercise

1. Define the key terms:

a. difference of squares   **b. sum of squares**   **c. perfect square trinomial**

Review Exercises

For Exercises 2–10, factor completely.

2. $3x^2 + x - 10$
$(3x - 5)(x + 2)$

3. $6x^2 - 17x + 5$
$(3x - 1)(2x - 5)$

4. $6a^2b + 3a^3b$
$3a^2b(2 + a)$

5. $15x^2y^5 - 10xy^6$
$5xy^5(3x - 2y)$

6. $5p^2q + 20p^2 - 3pq - 12p$
$p(5p - 3)(q + 4)$

7. $ax + ab - 6x - 6b$
$(x + b)(a - 6)$

8. $-6x + 5 + x^2$
$(x - 1)(x - 5)$

9. $6y - 40 + y^2$
$(y + 10)(y - 4)$

10. $a^2 + 7a + 1$
Prime

Objective 1: Factoring a Difference of Squares

11. What binomial factors as $(x - 5)(x + 5)$? $x^2 - 25$

12. What binomial factors as $(n - 3)(n + 3)$? $n^2 - 9$

13. What binomial factors as $(2p - 3q)(2p + 3q)$?
$4p^2 - 9q^2$

14. What binomial factors as $(7x - 4y)(7x + 4y)$?
$49x^2 - 16y^2$

For Exercises 15–38, factor the binomials completely. **(See Examples 1–3.)**

15. $x^2 - 36$ $(x - 6)(x + 6)$

16. $r^2 - 81$ $(r - 9)(r + 9)$

17. $3w^2 - 300$
$3(w + 10)(w - 10)$

18. $t^3 - 49t$
$t(t + 7)(t - 7)$

19. $4a^2 - 121b^2$
$(2a - 11b)(2a + 11b)$

20. $9x^2 - y^2$
$(3x - y)(3x + y)$

21. $49m^2 - 16n^2$
$(7m - 4n)(7m + 4n)$

22. $100a^2 - 49b^2$
$(10a - 7b)(10a + 7b)$

23. $9q^2 + 16$ Prime

24. $36 + s^2$ Prime

25. $y^2 - 4z^2$
$(y + 2z)(y - 2z)$

26. $b^2 - 144c^2$
$(b + 12c)(b - 12c)$

27. $a^2 - b^4$
$(a - b^2)(a + b^2)$

28. $y^4 - x^2$
$(y^2 - x)(y^2 + x)$

29. $25p^2q^2 - 1$
$(5pq - 1)(5pq + 1)$

30. $81s^2t^2 - 1$
$(9st - 1)(9st + 1)$

31. $c^6 - 25$
$(c^3 - 5)(c^3 + 5)$

32. $z^6 - 4$
$(z^3 - 2)(z^3 + 2)$

33. $50 - 32t^2$
$2(5 - 4t)(5 + 4t)$

34. $63 - 7h^2$
$7(3 - h)(3 + h)$

35. $z^4 - 16$
$(z + 2)(z - 2)(z^2 + 4)$

36. $y^4 - 625$
$(y + 5)(y - 5)(y^2 + 25)$

37. $a^4 - \dfrac{1}{81}$
$\left(a + \dfrac{1}{3}\right)\left(a - \dfrac{1}{3}\right)\left(a^2 + \dfrac{1}{9}\right)$

38. $\dfrac{1}{16} - z^4$
$\left(\dfrac{1}{2} + z\right)\left(\dfrac{1}{2} - z\right)\left(\dfrac{1}{4} + z^2\right)$

For Exercises 39–46, factor the polynomials completely. **(See Example 4.)**

39. $x^3 + 5x^2 - 9x - 45$
$(x + 3)(x - 3)(x + 5)$

40. $b^3 + 6b^2 - 4b - 24$
$(b + 2)(b - 2)(b + 6)$

41. $c^3 - c^2 - 25c + 25$
$(c + 5)(c - 5)(c - 1)$

42. $t^3 + 2t^2 - 16t - 32$
$(t + 4)(t - 4)(t + 2)$

43. $2x^2 - 18 + x^2y - 9y$
$(2 + y)(x + 3)(x - 3)$

44. $5a^2 - 5 + a^2b - b$
$(5 + b)(a + 1)(a - 1)$

45. $x^2y^2 - 9x^2 - 4y^2 + 36$
$(x + 2)(x - 2)(y + 3)(y - 3)$

46. $w^2z^2 - w^2 - 25z^2 + 25$
$(w + 5)(w - 5)(z + 1)(z - 1)$

Objective 2: Factoring Perfect Square Trinomials

47. Multiply. $(3x + 5)^2$
$9x^2 + 30x + 25$

48. Multiply. $(2y - 7)^2$
$4y^2 - 28y + 49$

49. a. Which trinomial is a perfect square trinomial? $x^2 + 4x + 4$ or $x^2 + 5x + 4$
a. $x^2 + 4x + 4$ is a perfect square trinomial.
b. Factor the trinomials from part (a).
b. $x^2 + 4x + 4 = (x + 2)^2$;
$x^2 + 5x + 4 = (x + 1)(x + 4)$

50. a. Which trinomial is a perfect square trinomial?
$x^2 + 13x + 36$ or $x^2 + 12x + 36$
a. $x^2 + 12x + 36$ is a perfect square trinomial.
b. Factor the trinomials from part (a).
b. $x^2 + 13x + 36 = (x + 9)(x + 4)$;
$x^2 + 12x + 36 = (x + 6)^2$

For Exercises 51–68, factor completely. (*Hint:* Look for the pattern of a perfect square trinomial.) **(See Examples 5–6.)**

51. $x^2 + 18x + 81$
$(x + 9)^2$

52. $y^2 - 8y + 16$
$(y - 4)^2$

53. $25z^2 - 20z + 4$
$(5z - 2)^2$

54. $36p^2 + 60p + 25$
$(6p + 5)^2$

55. $49a^2 + 42ab + 9b^2$
$(7a + 3b)^2$

56. $25m^2 - 30mn + 9n^2$
$(5m - 3n)^2$

57. $-2y + y^2 + 1$
$(y - 1)^2$

58. $4 + w^2 - 4w$
$(w - 2)^2$

59. $80z^2 + 120zw + 45w^2$
$5(4z + 3w)^2$

60. $36p^2 - 24pq + 4q^2$
$4(3p - q)^2$

61. $9y^2 + 78y + 25$
$(3y + 25)(3y + 1)$

62. $4y^2 + 20y + 9$
$(2y + 9)(2y + 1)$

63. $2a^2 - 20a + 50$
$2(a - 5)^2$

64. $3t^2 + 18t + 27$
$3(t + 3)^2$

65. $4x^2 + x + 9$
Prime

66. $c^2 - 4c + 16$
Prime

67. $4x^2 + 4xy + y^2$
$(2x + y)^2$

68. $100y^2 + 20yz + z^2$
$(10y + z)^2$

Expanding Your Skills

For Exercises 69–76, factor the difference of squares.

69. $(y - 3)^2 - 9$
$y(y - 6)$

70. $(x - 2)^2 - 4$
$x(x - 4)$

71. $(2p + 1)^2 - 36$
$(2p - 5)(2p + 7)$

72. $(4q + 3)^2 - 25$
$8(2q - 1)(q + 2)$

73. $16 - (t + 2)^2$
$(-t + 2)(t + 6)$ or
$-(t - 2)(t + 6)$

74. $81 - (a + 5)^2$
$(-a + 4)(a + 14)$ or
$-(a - 4)(a + 14)$

75. $100 - (2b - 5)^2$
$(-2b + 15)(2b + 5)$ or
$-(2b - 15)(2b + 5)$

76. $49 - (3k - 7)^2$
$3k(-3k + 14)$ or
$-3k(3k - 14)$

77. a. Write a polynomial that represents the area of the shaded region in the figure.
$a^2 - b^2$
b. Factor the expression from part (a).
$(a - b)(a + b)$

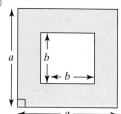

78. a. Write a polynomial that represents the area of the shaded region in the figure.
$g^2 - h^2$
b. Factor the expression from part (a).
$(g - h)(g + h)$

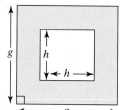

Section 6.6 Sum and Difference of Cubes

Objectives

1. Factoring a Sum or Difference of Cubes
2. Factoring Binomials: A Summary

1. Factoring a Sum or Difference of Cubes

A binomial $a^2 - b^2$ is a difference of squares and can be factored as $(a - b)(a + b)$. Furthermore, if a and b share no common factors, then a sum of squares $a^2 + b^2$ is not factorable over the real numbers. In this section, we will learn that both a difference of cubes, $a^3 - b^3$, and a sum of cubes, $a^3 + b^3$, are factorable.

> **FORMULA Factored Form of a Sum or Difference of Cubes**
>
> **Sum of Cubes:** $a^3 + b^3 = (a + b)(a^2 - ab + b^2)$
>
> **Difference of Cubes:** $a^3 - b^3 = (a - b)(a^2 + ab + b^2)$

Multiplication can be used to confirm the formulas for factoring a sum or difference of cubes:

$$(a + b)(a^2 - ab + b^2) = a^3 - a^2b + ab^2 + a^2b - ab^2 + b^3 = a^3 + b^3 ✔$$

$$(a - b)(a^2 + ab + b^2) = a^3 + a^2b + ab^2 - a^2b - ab^2 - b^3 = a^3 - b^3 ✔$$

To help you remember the formulas for factoring a sum or difference of cubes, keep the following guidelines in mind:

- The factored form is the product of a binomial and a trinomial.
- The first and third terms in the trinomial are the squares of the terms within the binomial factor.
- Without regard to signs, the middle term in the trinomial is the product of terms in the binomial factor.

Square the first term of the binomial.   Product of terms in the binomial

$$x^3 + 8 = (x)^3 + (2)^3 = (x + 2)[(x)^2 - (x)(2) + (2)^2]$$

Square the last term of the binomial.

- The sign within the binomial factor is the same as the sign of the original binomial.
- The first and third terms in the trinomial are always positive.
- The sign of the middle term in the trinomial is opposite the sign within the binomial.

Same sign   Positive

$$x^3 + 8 = (x)^3 + (2)^3 = (x + 2)[(x)^2 - (x)(2) + (2)^2]$$

Opposite signs

To help you recognize a sum or difference of cubes, we recommend that you familiarize yourself with the first several perfect cubes:

Perfect Cubes	**Perfect Cubes**
$1 = (1)^3$	$216 = (6)^3$
$8 = (2)^3$	$343 = (7)^3$
$27 = (3)^3$	$512 = (8)^3$
$64 = (4)^3$	$729 = (9)^3$
$125 = (5)^3$	$1000 = (10)^3$

It is also helpful to recognize that a variable expression is a perfect cube if its exponent is a multiple of 3. For example:

Perfect Cubes

$$x^3 = (x)^3$$
$$x^6 = (x^2)^3$$
$$x^9 = (x^3)^3$$
$$x^{12} = (x^4)^3$$

Example 1 **Factoring a Sum of Cubes**

Factor: $w^3 + 64$

Solution:

$w^3 + 64$ — w^3 and 64 are perfect cubes.

$= (w)^3 + (4)^3$ — Write as $a^3 + b^3$, where $a = w, b = 4$.

$a^3 + b^3 = (a + b)(a^2 - ab + b^2)$ — Apply the formula for a sum of cubes.

$(w)^3 + (4)^3 = (w + 4)[(w)^2 - (w)(4) + (4)^2]$

$= (w + 4)(w^2 - 4w + 16)$ — Simplify.

Skill Practice

Factor.

1. $p^3 + 125$

Classroom Example: p. 469, Exercise 16

Answer

1. $(p + 5)(p^2 - 5p + 25)$

Example 2 **Factoring a Difference of Cubes**

Factor: $27p^3 - 1000q^3$

Solution:

$27p^3 - 1000q^3$ $27p^3$ and $1000q^3$ are perfect cubes.

$(3p)^3 - (10q)^3$ Write as $a^3 - b^3$, where $a = 3p, b = 10q$.

$$a^3 - b^3 = (a - b)(a^2 + ab + b^2)$$ Apply the formula for a difference of cubes.

$(3p)^3 - (10q)^3 = (3p - 10q)[(3p)^2 + (3p)(10q) + (10q)^2]$

$= (3p - 10q)(9p^2 + 30pq + 100q^2)$ Simplify.

2. Factoring Binomials: A Summary

After removing the GCF, the next step in any factoring problem is to recognize what type of pattern it follows. Exponents that are divisible by 2 are perfect squares and those divisible by 3 are perfect cubes. The formulas for factoring binomials are summarized in the following box:

FORMULA **Factored Forms of Binomials**

Difference of Squares: $a^2 - b^2 = (a + b)(a - b)$
Difference of Cubes: $a^3 - b^3 = (a - b)(a^2 + ab + b^2)$
Sum of Cubes: $a^3 + b^3 = (a + b)(a^2 - ab + b^2)$

Example 3 **Factoring Binomials**

Factor completely.

a. $27y^3 + 1$ **b.** $m^2 - \dfrac{1}{4}$ **c.** $z^6 - 8w^3$

Solution:

a. $27y^3 + 1$ Sum of cubes: $27y^3 = (3y)^3$ and $1 = (1)^3$.

$= (3y)^3 + (1)^3$ Write as $a^3 + b^3$, where $a = 3y$ and $b = 1$.

$= (3y + 1)((3y)^2 - (3y)(1) + (1)^2)$ Apply the formula $a^3 + b^3 = (a + b)(a^2 - ab + b^2)$.

$= (3y + 1)(9y^2 - 3y + 1)$ Simplify.

b. $m^2 - \dfrac{1}{4}$ Difference of squares

$= (m)^2 - \left(\dfrac{1}{2}\right)^2$ Write as $a^2 - b^2$, where $a = m$ and $b = \frac{1}{2}$.

$= \left(m + \dfrac{1}{2}\right)\left(m - \dfrac{1}{2}\right)$ Apply the formula $a^2 - b^2 = (a + b)(a - b)$.

c. $z^6 - 8w^3$ Difference of cubes: $z^6 = (z^2)^3$ and $8w^3 = (2w)^3$

$= (z^2)^3 - (2w)^3$ Write as $a^3 - b^3$, where $a = z^2$ and $b = 2w$.

$= (z^2 - 2w)[(z^2)^2 + (z^2)(2w) + (2w)^2]$ Apply the formula $a^3 - b^3 = (a - b)(a^2 + ab + b^2)$.

$= (z^2 - 2w)(z^4 + 2z^2w + 4w^2)$ Simplify.

Each of the factorizations in Example 3 can be checked by multiplying.

Some factoring problems require more than one method of factoring. In general, when factoring a polynomial, be sure to factor completely.

Example 4   **Factoring Polynomials**

Factor completely. $3y^4 - 48$

Solution:

$3y^4 - 48$

$= 3(y^4 - 16)$ Factor out the GCF. The binomial is a difference of squares.

$= 3[(y^2)^2 - (4)^2]$ Write as $a^2 - b^2$, where $a = y^2$ and $b = 4$.

$= 3(y^2 + 4)(y^2 - 4)$ Apply the formula $a^2 - b^2 = (a + b)(a - b)$.

$y^2 + 4$ is a sum of squares and cannot be factored.

$y^2 - 4$ is a difference of squares and can be factored further.

$= 3(y^2 + 4)(y + 2)(y - 2)$

Skill Practice

Factor completely.

6. $2x^4 - 2$

Classroom Example: p. 470, Exercise 58

Example 5   **Factoring Polynomials**

Factor completely. $4x^3 + 4x^2 - 25x - 25$

Solution:

$4x^3 + 4x^2 - 25x - 25$ The GCF is 1.

$= 4x^3 + 4x^2 - 25x - 25$ The polynomial has four terms. Factor by grouping.

$= 4x^2(x + 1) - 25(x + 1)$

$= (x + 1)(4x^2 - 25)$ $4x^2 - 25$ is a difference of squares.

$= (x + 1)(2x + 5)(2x - 5)$

Skill Practice

Factor completely.

7. $x^3 + 6x^2 - 4x - 24$

Classroom Example: p. 470, Exercise 64

Answers

6. $2(x^2 + 1)(x - 1)(x + 1)$
7. $(x + 6)(x + 2)(x - 2)$

Example 6 **Factoring Binomials**

Factor the binomial $x^6 - y^6$ as

a. A difference of cubes   **b.** A difference of squares

Solution:

a. $x^6 - y^6$

Difference of cubes

$= (x^2)^3 - (y^2)^3$ Write as $a^3 - b^3$, where $a = x^2$ and $b = y^2$.

$= (x^2 - y^2)[(x^2)^2 + (x^2)(y^2) + (y^2)^2]$ Apply the formula $a^3 - b^3 = (a - b)(a^2 + ab + b^2)$.

$= (x^2 - y^2)(x^4 + x^2y^2 + y^4)$ Factor $x^2 - y^2$ as a difference of squares.

$= (x + y)(x - y)(x^4 + x^2y^2 + y^4)$

b. $x^6 - y^6$

Difference of squares

$= (x^3)^2 - (y^3)^2$ Write as $a^2 - b^2$, where $a = x^3$ and $b = y^3$.

$= (x^3 + y^3)(x^3 - y^3)$ Apply the formula $a^2 - b^2 = (a + b)(a - b)$.

Sum of cubes   Difference of cubes

Factor $x^3 + y^3$ as a sum of cubes.

Factor $x^3 - y^3$ as a difference of cubes.

$= (x + y)(x^2 - xy + y^2)(x - y)(x^2 + xy + y^2)$

Notice that the expressions x^6 and y^6 are both perfect squares and perfect cubes because both exponents are multiples of 2 and of 3. Consequently, $x^6 - y^6$ can be factored initially as either the difference of squares or as the difference of cubes. In such a case, it is recommended that you factor the expression as a difference of squares first because it factors more completely into polynomials of lower degree.

$$x^6 - y^6 = (x + y)(x^2 - xy + y^2)(x - y)(x^2 + xy + y^2)$$

Section 6.6 Practice Exercises

Study Skills Exercise

1. Define the key terms.

    a. sum of cubes   **b. difference of cubes**

Review Exercises

For Exercises 2–10, factor completely.

2. $600 - 6x^2$
$6(10 - x)(10 + x)$

3. $20 - 5t^2$
$5(2 - t)(2 + t)$

4. $ax + bx + 5a + 5b$
$(a + b)(x + 5)$

5. $2t + 2u + st + su$
$(t + u)(2 + s)$

6. $5y^2 + 13y - 6$
$(5y - 2)(y + 3)$

7. $3v^2 + 5v - 12$
$(3v - 4)(v + 3)$

8. $40a^3b^3 - 16a^2b^2 + 24a^3b$
$8a^2b(5ab^2 - 2b + 3a)$

9. $-c^2 - 10c - 25$
$-(c + 5)^2$

10. $-z^2 + 6z - 9$
$-(z - 3)^2$

Objective 1: Factoring a Sum or Difference of Cubes

11. Identify the expressions that are perfect cubes:
$$\{x^3, 8, 9, y^6, a^4, b^2, 3p^3, 27q^3, w^{12}, r^3s^6\}$$
$x^3, 8, y^6, 27q^3, w^{12}, r^3s^6$

12. Identify the expressions that are perfect cubes:
$$\{z^9, -81, 30, 8, 6x^3, y^{15}, 27a^3, b^2, p^3q^2, -1\}$$
$z^9, 8, y^{15}, 27a^3, -1$

13. From memory, write the formula to factor a sum
of cubes:
$$a^3 + b^3 = \underline{(a + b)(a^2 - ab + b^2)}$$

14. From memory, write the formula to factor a
difference of cubes:
$$a^3 - b^3 = \underline{(a - b)(a^2 + ab + b^2)}$$

For Exercises 15–30, factor the sums or differences of cubes. **(See Examples 1–2.)**

15. $y^3 - 8$
$(y - 2)(y^2 + 2y + 4)$

16. $x^3 + 27$
$(x + 3)(x^2 - 3x + 9)$

17. $1 - p^3$
$(1 - p)(1 + p + p^2)$

18. $q^3 + 1$
$(q + 1)(q^2 - q + 1)$

19. $w^3 + 64$
$(w + 4)(w^2 - 4w + 16)$

20. $8 - t^3$
$(2 - t)(4 + 2t + t^2)$

21. $x^3 - 1000$
$(x - 10)(x^2 + 10x + 100)$

22. $8y^3 - 27$
$(2y - 3)(4y^2 + 6y + 9)$

23. $64t^3 + 1$
$(4t + 1)(16t^2 - 4t + 1)$

24. $125r^3 + 1$
$(5r + 1)(25r^2 - 5r + 1)$

25. $1000a^3 + 27$
$(10a + 3)(100a^2 - 30a + 9)$

26. $216b^3 - 125$
$(6b - 5)(36b^2 + 30b + 25)$

27. $n^3 - \dfrac{1}{8}$
$\left(n - \dfrac{1}{2}\right)\left(n^2 + \dfrac{1}{2}n + \dfrac{1}{4}\right)$

28. $\dfrac{8}{27} + m^3$
$\left(\dfrac{2}{3} + m\right)\left(\dfrac{4}{9} - \dfrac{2}{3}m + m^2\right)$

29. $125m^3 + 8$
$(5m + 2)(25m^2 - 10m + 4)$

30. $27p^3 - 64$
$(3p - 4)(9p^2 + 12p + 16)$

Objective 2: Factoring Binomials: A Summary

For Exercises 31–66, factor completely. **(See Examples 3–6.)**

31. $x^4 - 4$
$(x^2 - 2)(x^2 + 2)$

32. $b^4 - 25$
$(b^2 - 5)(b^2 + 5)$

33. $a^2 + 9$
Prime

34. $w^2 + 36$
Prime

35. $t^3 + 64$
$(t + 4)(t^2 - 4t + 16)$

36. $u^3 + 27$
$(u + 3)(u^2 - 3u + 9)$

37. $g^3 - 4$
Prime

38. $h^3 - 25$
Prime

39. $4b^3 + 108$
$4(b + 3)(b^2 - 3b + 9)$

40. $3c^3 - 24$
$3(c - 2)(c^2 + 2c + 4)$

41. $5p^2 - 125$
$5(p - 5)(p + 5)$

42. $2q^4 - 8$
$2(q^2 - 2)(q^2 + 2)$

43. $\dfrac{1}{64} - 8h^3$
$(\frac{1}{4} - 2h)(\frac{1}{16} + \frac{1}{2}h + 4h^2)$

44. $\dfrac{1}{125} + k^6$
$(\frac{1}{5} + k^2)(\frac{1}{25} - \frac{1}{5}k^2 + k^4)$

45. $x^4 - 16$
$(x - 2)(x + 2)(x^2 + 4)$

46. $p^4 - 81$
$(p - 3)(p + 3)(p^2 + 9)$

47. $q^6 - 64$
$(q - 2)(q^2 + 2q + 4)$
$(q + 2)(q^2 - 2q + 4)$

48. $a^6 - 1$
$(a - 1)(a^2 + a + 1)$
$(a + 1)(a^2 - a + 1)$

49. $\dfrac{4x^2}{9} - w^2$
$\left(\dfrac{2x}{3} - w\right)\left(\dfrac{2x}{3} + w\right)$

50. $\dfrac{16y^2}{25} - x^2$
$\left(\dfrac{4y}{5} - x\right)\left(\dfrac{4y}{5} + x\right)$

51. $x^9 + 64y^3$
$(x^3 + 4y)(x^6 - 4x^3y + 16y^2)$

52. $125w^3 - z^9$
$(5w - z^3)(25w^2 + 5wz^3 + z^6)$

53. $2x^3 + 3x^2 - 2x - 3$
$(2x + 3)(x - 1)(x + 1)$

54. $3x^3 + x^2 - 12x - 4$
$(3x + 1)(x - 2)(x + 2)$

55. $16x^4 - y^4$
$(2x - y)(2x + y)(4x^2 + y^2)$

56. $1 - t^4$
$(1 - t)(1 + t)(1 + t^2)$

57. $81y^4 - 16$
$(3y - 2)(3y + 2)(9y^2 + 4)$

58. $u^5 - 256u$
$u(u - 4)(u + 4)(u^2 + 16)$

59. $a^3 + b^6$
$(a + b^2)(a^2 - ab^2 + b^4)$

60. $u^6 - v^3$
$(u^2 - v)(u^4 + u^2v + v^2)$

61. $x^4 - y^4$
$(x^2 + y^2)(x - y)(x + y)$

62. $a^4 - b^4$
$(a^2 + b^2)(a - b)(a + b)$

63. $k^3 + 4k^2 - 9k - 36$
$(k + 4)(k - 3)(k + 3)$

64. $w^3 - 2w^2 - 4w + 8$
$(w - 2)^2(w + 2)$

65. $2t^3 - 10t^2 - 2t + 10$
$2(t - 5)(t - 1)(t + 1)$

66. $9a^3 + 27a^2 - 4a - 12$
$(a + 3)(3a - 2)(3a + 2)$

Expanding Your Skills

For Exercises 67–70, factor completely.

67. $\dfrac{64}{125}p^3 - \dfrac{1}{8}q^3$
$\left(\dfrac{4}{5}p - \dfrac{1}{2}q\right)\left(\dfrac{16}{25}p^2 + \dfrac{2}{5}pq + \dfrac{1}{4}q^2\right)$

68. $\dfrac{1}{1000}r^3 + \dfrac{8}{27}s^3$
$\left(\dfrac{1}{10}r + \dfrac{2}{3}s\right)\left(\dfrac{1}{100}r^2 - \dfrac{1}{15}rs + \dfrac{4}{9}s^2\right)$

69. $a^{12} + b^{12}$
$(a^4 + b^4)(a^8 - a^4b^4 + b^8)$

70. $a^9 - b^9$
$(a - b)(a^2 + ab + b^2)(a^6 + a^3b^3 + b^6)$

Use Exercises 71–72 to investigate the relationship between division and factoring.

71. a. Use long division to divide $x^3 - 8$ by $(x - 2)$. The quotient is $x^2 + 2x + 4$.

    b. Factor $x^3 - 8$. $(x - 2)(x^2 + 2x + 4)$

72. a. Use long division to divide $y^3 + 27$ by $(y + 3)$. The quotient is $y^2 - 3y + 9$.

    b. Factor $y^3 + 27$. $(y + 3)(y^2 - 3y + 9)$

73. What trinomial multiplied by $(x - 4)$ gives a difference of cubes? $x^2 + 4x + 16$

74. What trinomial multiplied by $(p + 5)$ gives a sum of cubes? $p^2 - 5p + 25$

75. Write a binomial that when multiplied by $(4x^2 - 2x + 1)$ produces a sum of cubes. $2x + 1$

76. Write a binomial that when multiplied by $(9y^2 + 15y + 25)$ produces a difference of cubes. $3y - 5$

Problem Recognition Exercises

Factoring Strategy

PROCEDURE Factoring Strategy

Step 1 Factor out the GCF (Section 6.1).

Step 2 Identify whether the polynomial has two terms, three terms, or more than three terms.

Step 3 If the polynomial has more than three terms, try factoring by grouping (Section 6.1).

Step 4 If the polynomial has three terms, check first for a perfect square trinomial. Otherwise, factor the trinomial with the trial-and-error method or the ac-method (Sections 6.3 or 6.4).

Step 5 If the polynomial has two terms, determine if it fits the pattern for
- A difference of squares: $a^2 - b^2 = (a - b)(a + b)$ (Section 6.5)
- A sum of squares: $a^2 + b^2$ prime
- A difference of cubes: $a^3 - b^3 = (a - b)(a^2 + ab + b^2)$ (Section 6.6)
- A sum of cubes: $a^3 + b^3 = (a + b)(a^2 - ab + b^2)$ (Section 6.6)

Step 6 Be sure to factor the polynomial completely.

Step 7 Check by multiplying.

 Writing   Translating Expression   Geometry   Scientific Calculator   Video   NS&E

1. What is meant by a prime factor?   A prime factor cannot be factored further.

2. What is the first step in factoring any polynomial?   Factor out the GCF.

3. When factoring a binomial, what patterns can you look for?   Look for a difference of squares: $a^2 - b^2$, a difference of cubes: $a^3 - b^3$, or a sum of cubes: $a^3 + b^3$.

4. What technique should be considered when factoring a four-term polynomial?   Grouping

For Exercises 5–73,

a. Factor out the GCF from each polynomial. Then identify the category in which the polynomial best fits. Choose from

- difference of squares
- sum of squares
- difference of cubes
- sum of cubes
- trinomial (perfect square trinomial)
- trinomial (nonperfect square trinomial)
- four terms-grouping
- none of these

Instructor Note: The following exercises involve a sum or difference of cubes: 10, 12, 13, 18, 23, 28, 29.

b. Factor the polynomial completely.

5. $2a^2 - 162$ a. Difference of squares
b. $2(a - 9)(a + 9)$

6. $y^2 + 4y + 3$ a. Nonperfect square trinomial
b. $(y + 3)(y + 1)$

7. $6w^2 - 6w$ a. None of these
b. $6w(w - 1)$

8. $16z^4 - 81$
a. Difference of squares
b. $(2z + 3)(2z - 3)(4z^2 + 9)$

9. $3t^2 + 13t + 4$
a. Nonperfect square trinomial
b. $(3t + 1)(t + 4)$

10. $5r^3 + 5$ a. Sum of cubes
b. $5(r + 1)(r^2 - r + 1)$

11. $3ac + ad - 3bc - bd$
a. Four terms-grouping
b. $(3c + d)(a - b)$

12. $x^3 - 125$ a. Difference of cubes
b. $(x - 5)(x^2 + 5x + 25)$

13. $y^3 + 8$ a. Sum of cubes
b. $(y + 2)(y^2 - 2y + 4)$

14. $7p^2 - 29p + 4$
a. Nonperfect square trinomial
b. $(7p - 1)(p - 4)$

15. $3q^2 - 9q - 12$
a. Nonperfect square trinomial
b. $3(q - 4)(q + 1)$

16. $-2x^2 + 8x - 8$
a. Perfect square trinomial
b. $-2(x - 2)^2$

17. $18a^2 + 12a$
a. None of these
b. $6a(3a + 2)$

18. $54 - 2y^3$
a. Difference of cubes
b. $2(3 - y)(9 + 3y + y^2)$

19. $4t^2 - 100$ a. Difference of squares
b. $4(t - 5)(t + 5)$

20. $4t^2 - 31t - 8$
a. Nonperfect square trinomial
b. $(4t + 1)(t - 8)$

21. $10c^2 + 10c + 10$
a. Nonperfect square trinomial
b. $10(c^2 + c + 1)$

22. $2xw - 10x + 3yw - 15y$
a. Four terms-grouping
b. $(w - 5)(2x + 3y)$

23. $x^3 + 0.001$
a. Sum of cubes
b. $(x + 0.1)(x^2 - 0.1x + 0.01)$

24. $4q^2 - 9$ a. Difference of squares
b. $(2q - 3)(2q + 3)$

25. $64 + 16k + k^2$
a. Perfect square trinomial
b. $(8 + k)^2$

26. $s^2t + 5t + 6s^2 + 30$
a. Four terms-grouping
b. $(t + 6)(s^2 + 5)$

27. $2x^2 + 2x - xy - y$
a. Four terms-grouping
b. $(x + 1)(2x - y)$

28. $w^3 + y^3$ a. Sum of cubes
b. $(w + y)(w^2 - wy + y^2)$

29. $a^3 - c^3$
a. Difference of cubes
b. $(a - c)(a^2 + ac + c^2)$

30. $3y^2 + y + 1$
a. Nonperfect square trinomial
b. Prime

31. $c^2 + 8c + 9$
a. Nonperfect square trinomial
b. Prime

32. $a^2 + 2a + 1$
a. Perfect square trinomial
b. $(a + 1)^2$

33. $b^2 + 10b + 25$
a. Perfect square trinomial
b. $(b + 5)^2$

34. $-t^2 - 4t + 32$
a. Nonperfect square trinomial
b. $-1(t + 8)(t - 4)$

35. $-p^3 - 5p^2 - 4p$
a. Nonperfect square trinomial
b. $-p(p + 4)(p + 1)$

36. $x^2y^2 - 49$ a. Difference of squares
b. $(xy - 7)(xy + 7)$

37. $6x^2 - 21x - 45$
a. Nonperfect square trinomial
b. $3(2x + 3)(x - 5)$

38. $20y^2 - 14y + 2$
a. Nonperfect square trinomial
b. $2(5y - 1)(2y - 1)$

39. $5a^2bc^3 - 7abc^2$
a. None of these
b. $abc^2(5ac - 7)$

40. $8a^2 - 50$
a. Difference of squares
b. $2(2a - 5)(2a + 5)$

41. $t^2 + 2t - 63$
a. Nonperfect square trinomial
b. $(t + 9)(t - 7)$

42. $b^2 + 2b - 80$
a. Nonperfect square trinomial
b. $(b + 10)(b - 8)$

43. $ab + ay - b^2 - by$
a. Four terms-grouping
b. $(b + y)(a - b)$

44. $6x^3y^4 + 3x^2y^5$ a. None of these
b. $3x^2y^4(2x + y)$

45. $14u^2 - 11uv + 2v^2$
a. Nonperfect square trinomial
b. $(7u - 2v)(2u - v)$

46. $9p^2 - 36pq + 4q^2$
a. Nonperfect square trinomial
b. Prime

47. $4q^2 - 8q - 6$
a. Nonperfect square trinomial
b. $2(2q^2 - 4q - 3)$

48. $9w^2 + 3w - 15$
a. Nonperfect square trinomial
b. $3(3w^2 + w - 5)$

49. $9m^2 + 16n^2$ a. Sum of squares
b. Prime

50. $5b^2 - 30b + 45$
a. Perfect square trinomial
b. $5(b - 3)^2$

51. $6r^2 + 11r + 3$
a. Nonperfect square trinomial
b. $(3r + 1)(2r + 3)$

52. $4s^2 + 4s - 15$
a. Nonperfect square trinomial
b. $(2s - 3)(2s + 5)$

53. $16a^4 - 1$
a. Difference of squares
b. $(2a - 1)(2a + 1)(4a^2 + 1)$

54. $p^3 + p^2c - 9p - 9c$
a. Four terms-grouping
b. $(p + c)(p - 3)(p + 3)$

55. $81u^2 - 90uv + 25v^2$
a. Perfect square trinomial
b. $(9u - 5v)^2$

56. $4x^2 + 16$
a. Sum of squares
b. $4(x^2 + 4)$

57. $x^2 - 5x - 6$
a. Nonperfect square trinomial
b. $(x - 6)(x + 1)$

58. $q^2 + q - 7$
a. Nonperfect square trinomial
b. Prime

59. $2ax - 6ay + 4bx - 12by$
a. Four terms-grouping
b. $2(x - 3y)(a + 2b)$

60. $8m^3 - 10m^2 - 3m$
a. Nonperfect square trinomial
b. $m(4m + 1)(2m - 3)$

61. $21x^4y + 41x^3y + 10x^2y$
a. Nonperfect square trinomial
b. $x^2y(3x + 5)(7x + 2)$

62. $2m^4 - 128$
a. Difference of squares
b. $2(m^2 - 8)(m^2 + 8)$

 63. $8uv - 6u + 12v - 9$
a. Four terms-grouping
b. $(4v - 3)(2u + 3)$

64. $4t^2 - 20t + st - 5s$
a. Four terms-grouping
b. $(t - 5)(4t + s)$

65. $12x^2 - 12x + 3$
a. Perfect square trinomial
b. $3(2x - 1)^2$

66. $p^2 + 2pq + q^2$
a. Perfect square trinomial
b. $(p + q)^2$

67. $6n^3 + 5n^2 - 4n$
a. Nonperfect square trinomial
b. $n(2n - 1)(3n + 4)$

68. $4k^3 + 4k^2 - 3k$
a. Nonperfect square trinomial
b. $k(2k - 1)(2k + 3)$

69. $64 - y^2$
a. Difference of squares
b. $(8 - y)(8 + y)$

70. $36b - b^3$ a. Difference of squares
b. $b(6 - b)(6 + b)$

71. $b^2 - 4b + 10$
a. Nonperfect square trinomial
b. Prime

72. $y^2 + 6y + 8$
a. Nonperfect square trinomial
b. $(y + 4)(y + 2)$

73. $c^4 - 12c^2 + 20$
a. Nonperfect square trinomial
b. $(c^2 - 10)(c^2 - 2)$

Section 6.7 Solving Equations Using the Zero Product Rule

Objectives

1. Definition of a Quadratic Equation
2. Zero Product Rule
3. Solving Equations by Factoring

1. Definition of a Quadratic Equation

In Section 2.1, we solved linear equations in one variable. These are equations of the form $ax + b = 0 \, (a \neq 0)$. A linear equation in one variable is sometimes called a first-degree polynomial equation because the highest degree of all its terms is 1. A second-degree polynomial equation in one variable is called a quadratic equation.

> **DEFINITION   A Quadratic Equation in One Variable**
>
> If a, b, and c are real numbers such that $a \neq 0$, then a **quadratic equation** is an equation that can be written in the form
>
> $$ax^2 + bx + c = 0$$

The following equations are quadratic because they can each be written in the form $ax^2 + bx + c = 0, \quad (a \neq 0)$.

$$-4x^2 + 4x = 1 \qquad x(x - 2) = 3 \qquad (x - 4)(x + 4) = 9$$
$$-4x^2 + 4x - 1 = 0 \qquad x^2 - 2x = 3 \qquad x^2 - 16 = 9$$
$$x^2 - 2x - 3 = 0 \qquad x^2 - 25 = 0$$
$$x^2 + 0x - 25 = 0$$

2. Zero Product Rule

One method for solving a quadratic equation is to factor and apply the zero product rule. The **zero product rule** states that if the product of two factors is zero, then one or both of its factors is zero.

> **PROPERTY   Zero Product Rule**
>
> If $ab = 0$, then $a = 0$ or $b = 0$.

Example 1 Using the Zero Product Rule

Solve the equation by using the zero product rule. $(x - 4)(x + 3) = 0$

Solution:

$(x - 4)(x + 3) = 0$ Apply the zero product rule.

$x - 4 = 0$ or $x + 3 = 0$ Set each factor equal to zero.

$x = 4$ or $x = -3$ Solve each equation for x.

<u>Check:</u> $x = 4$ <u>Check:</u> $x = -3$

$(4 - 4)(4 + 3) \stackrel{?}{=} 0$ $(-3 - 4)(-3 + 3) \stackrel{?}{=} 0$

        $(0)(7) \stackrel{?}{=} 0$ ✔ $(-7)(0) \stackrel{?}{=} 0$ ✔

The solutions are 4 and -3.

Skill Practice

1. Solve.

$(x + 1)(x - 8) = 0$

Classroom Example: p. 477, Exercise 14

Example 2 Using the Zero Product Rule

Solve the equation by using the zero product rule. $(2x + 1)(x - 5) = 0$

Solution:

$(2x + 1)(x - 5) = 0$ Apply the zero product rule.

$2x + 1 = 0$ or $x - 5 = 0$ Set each factor equal to zero.

$2x = -1$ or $x = 5$ Solve each equation for x.

$x = -\dfrac{1}{2}$ or $x = 5$

The solutions $-\dfrac{1}{2}$ and 5 check in the original equation.

Skill Practice

2. Solve.

$(4x - 5)(x + 6) = 0$

Classroom Example: p. 477, Exercise 20

Example 3 Using the Zero Product Rule

Solve the equation using the zero product rule. $x(3x - 7) = 0$

Solution:

$x(3x - 7) = 0$ Apply the zero product rule.

$x = 0$ or $3x - 7 = 0$ Set each factor equal to zero.

$x = 0$ or $3x = 7$ Solve each equation for x.

$x = 0$ or $x = \dfrac{7}{3}$

The solutions 0 and $\dfrac{7}{3}$ check in the original equation.

Skill Practice

3. Solve.

$x(4x + 9) = 0$

Classroom Example: p. 477, Exercise 22

Answers

1. $-1, 8$ **2.** $\dfrac{5}{4}, -6$ **3.** $0, -\dfrac{9}{4}$

3. Solving Equations by Factoring

Quadratic equations, like linear equations, arise in many applications in mathematics, science, and business. The following steps summarize the factoring method for solving a quadratic equation.

> **PROCEDURE** Solving a Quadratic Equation by Factoring
>
> **Step 1** Write the equation in the form: $ax^2 + bx + c = 0$.
> **Step 2** Factor the quadratic expression completely.
> **Step 3** Apply the zero product rule. That is, set each factor equal to zero, and solve the resulting equations.
>
> *Note:* The solution(s) found in step 3 may be checked by substitution in the original equation.

─ **Skill Practice** ─

Solve the equation.
4. $2y^2 + 19y = -24$

Classroom Example: p. 477,
Exercise 30

Example 4 Solving a Quadratic Equation

Solve the quadratic equation. $2x^2 - 9x = 5$

Solution:

$$2x^2 - 9x = 5$$

$$2x^2 - 9x - 5 = 0 \qquad \text{Write the equation in the form } ax^2 + bx + c = 0.$$

$$(2x + 1)(x - 5) = 0 \qquad \text{Factor the polynomial completely.}$$

$$2x + 1 = 0 \quad \text{or} \quad x - 5 = 0 \qquad \text{Set each factor equal to zero.}$$

$$2x = -1 \quad \text{or} \quad x = 5 \qquad \text{Solve each equation.}$$

$$x = -\frac{1}{2} \quad \text{or} \quad x = 5$$

$$\underline{\text{Check: } x = -\frac{1}{2}} \qquad\qquad \underline{\text{Check: } x = 5}$$

$$2x^2 - 9x = 5 \qquad\qquad 2x^2 - 9x = 5$$

$$2\left(-\frac{1}{2}\right)^2 - 9\left(-\frac{1}{2}\right) \stackrel{?}{=} 5 \qquad 2(5)^2 - 9(5) \stackrel{?}{=} 5$$

$$2\left(\frac{1}{4}\right) + \frac{9}{2} \stackrel{?}{=} 5 \qquad\qquad 2(25) - 45 \stackrel{?}{=} 5$$

$$\frac{1}{2} + \frac{9}{2} \stackrel{?}{=} 5 \qquad\qquad 50 - 45 \stackrel{?}{=} 5 \checkmark$$

$$\frac{10}{2} \stackrel{?}{=} 5 \checkmark$$

The solutions are $-\frac{1}{2}$ and 5.

Answer

4. $-8, -\frac{3}{2}$

Example 5 Solving a Quadratic Equation

Solve the quadratic equation. $4x^2 + 24x = 0$

Solution:

$$4x^2 + 24x = 0$$ The equation is already in the form $ax^2 + bx + c = 0$. (Note that $c = 0$.)

$$4x(x + 6) = 0$$ Factor completely.

$$4x = 0 \quad \text{or} \quad x + 6 = 0$$ Set each factor equal to zero.

$$x = 0 \quad \text{or} \quad x = -6$$

The solutions are 0 and -6. They both check in the original equation.

Skill Practice

Solve the equation.

5. $5s^2 = 45$

Classroom Example: p. 477, Exercise 46

Example 6 Solving a Quadratic Equation

Solve the quadratic equation. $5x(5x + 2) = 10x + 9$

Solution:

$$5x(5x + 2) = 10x + 9$$

$$25x^2 + 10x = 10x + 9$$ Clear parentheses.

$$25x^2 + 10x - 10x - 9 = 0$$ Set the equation equal to zero.

$$25x^2 - 9 = 0$$ The equation is in the form $ax^2 + bx + c = 0$. (Note that $b = 0$.)

$$(5x - 3)(5x + 3) = 0$$ Factor completely.

$$5x - 3 = 0 \quad \text{or} \quad 5x + 3 = 0$$ Set each factor equal to zero.

$$5x = 3 \quad \text{or} \quad 5x = -3$$ Solve each equation.

$$\frac{5x}{5} = \frac{3}{5} \quad \text{or} \quad \frac{5x}{5} = \frac{-3}{5}$$

$$x = \frac{3}{5} \quad \text{or} \quad x = -\frac{3}{5}$$

The solutions are $\frac{3}{5}$ and $-\frac{3}{5}$. They both check in the original equation.

Skill Practice

Solve the equation.

6. $4z(z + 3) = 4z + 5$

Classroom Example: p. 477, Exercise 54

The zero product rule can be used to solve higher degree polynomial equations provided the equations can be set to zero and written in factored form.

Answers

5. $3, -3$ **6.** $-\frac{5}{2}, \frac{1}{2}$

Classroom Example: p. 477, Exercise 38

Instructor Note: Show students why $x^2 + 7x + 12 = 15$ cannot be solved as: $x^2 + 7x + 12 = 15$
$(x + 4)(x + 3) = 15$
$x + 4 = 15$ or $x + 3 = 15$

Example 7   **Solving Higher Degree Polynomial Equations**

Solve the equation. $-6(y + 3)(y - 5)(2y + 7) = 0$

Solution:

$-6(y + 3)(y - 5)(2y + 7) = 0$ The equation is already in factored form and equal to zero.

Set each factor equal to zero.

Solve each equation for y.

$-6 \neq 0$ or $y + 3 = 0$ or $y - 5 = 0$ or $2y + 7 = 0$

No solution, $y = -3$ or $y = 5$ or $y = -\dfrac{7}{2}$

Notice that when the constant factor is set equal to zero, the result is a contradiction, $-6 = 0$. The constant factor does not produce a solution to the equation. Therefore, the only solutions are $-3, 5$, and $-\frac{7}{2}$. Each solution can be checked in the original equation.

Classroom Example: p. 478, Exercise 68

Example 8   **Solving a Higher Degree Polynomial Equation**

Solve the equation. $w^3 + 5w^2 - 9w - 45 = 0$

Solution:

$w^3 + 5w^2 - 9w - 45 = 0$ This is a higher degree polynomial equation.

$w^3 + 5w^2 \mid - 9w - 45 = 0$ The equation is already set equal to zero. Now factor.

$w^2(w + 5) - 9(w + 5) = 0$ Because there are four terms, try factoring
$(w + 5)(w^2 - 9) = 0$ by grouping.

$(w + 5)(w - 3)(w + 3) = 0$ $w^2 - 9$ is a difference of squares and can be factored further.

$w + 5 = 0$ or $w - 3 = 0$ or $w + 3 = 0$ Set each factor equal to zero.

$w = -5$ or $w = 3$ or $w = -3$ Solve each equation.

The solutions are $-5, -3$ and 3. Each solution checks in the original equation.

Answers

7. $4, -7, \dfrac{9}{2}$ **8.** $-2, -3, 2$

Section 6.7   Practice Exercises

Boost your GRADE at ALEKS.com!

• Practice Problems
• Self-Tests
• NetTutor

• e-Professors
• Videos

For additional exercises, see Classroom Activities 6.7A–6.7B in the *Instructor's Resource Manual* at www.mhhe.com/moh.

Study Skills Exercise

1. Define the key terms:

   a. quadratic equation   **b. zero product rule**

  Writing    Translating Expression   Geometry   Scientific Calculator   Video   NS&E

Review Exercises

For Exercises 2–7, factor completely.

2. $6a - 8 - 3ab + 4b$
$(3a - 4)(2 - b)$

3. $4b^2 - 44b + 120$
$4(b - 5)(b - 6)$

4. $8u^2v^2 - 4uv$
$4uv(2uv - 1)$

5. $3x^2 + 10x - 8$
$(3x - 2)(x + 4)$

6. $3h^2 - 75$
$3(h - 5)(h + 5)$

7. $4x^2 + 16y^2$
$4(x^2 + 4y^2)$

Objective 1: Definition of a Quadratic Equation

For Exercises 8–13, identify the equations as linear, quadratic, or neither.

8. $4 - 5x = 0$
Linear

9. $5x^3 + 2 = 0$
Neither

10. $3x - 6x^2 = 0$
Quadratic

11. $1 - x + 2x^2 = 0$
Quadratic

12. $7x^4 + 8 = 0$
Neither

13. $3x + 2 = 0$
Linear

Objective 2: Zero Product Rule

For Exercises 14–22, solve the equations using the zero product rule. **(See Examples 1–2.)**

14. $(x - 5)(x + 1) = 0$ $\quad 5, -1$

15. $(x + 3)(x - 1) = 0$ $\quad -3, 1$

16. $(3x - 2)(3x + 2) = 0$ $\quad \frac{2}{3}, -\frac{2}{3}$

17. $(2x - 7)(2x + 7) = 0$ $\quad \frac{7}{2}, -\frac{7}{2}$

18. $2(x - 7)(x - 7) = 0$ $\quad 7$

19. $3(x + 5)(x + 5) = 0$ $\quad -5$

20. $(3x - 2)(2x - 3) = 0$ $\quad \frac{2}{3}, \frac{3}{2}$

21. $x(5x - 1) = 0$ $\quad 0, \frac{1}{5}$

22. $x(3x + 8) = 0$ $\quad 0, -\frac{8}{3}$

23. For a quadratic equation of the form $ax^2 + bx + c = 0$, what must be done before applying the zero product rule? The polynomial must be factored completely.

24. What are the requirements needed to use the zero product rule to solve a quadratic equation or higher degree polynomial equation? The equation must have one side equal to zero and the other side factored completely.

Objective 3: Solving Equations by Factoring

For Exercises 25–72, solve the equations. **(See Examples 4–8.)**

25. $p^2 - 2p - 15 = 0$ $\quad 5, -3$

26. $y^2 - 7y - 8 = 0$ $\quad 8, -1$

27. $z^2 + 10z - 24 = 0$ $\quad -12, 2$

28. $w^2 - 10w + 16 = 0$ $\quad 8, 2$

29. $2q^2 - 7q = 4$ $\quad 4, -\frac{1}{2}$

30. $4x^2 - 11x = 3$ $\quad -\frac{1}{4}, 3$

31. $0 = 9x^2 - 4$ $\quad \frac{2}{3}, -\frac{2}{3}$

32. $4a^2 - 49 = 0$ $\quad \frac{7}{2}, -\frac{7}{2}$

33. $2k^2 - 28k + 96 = 0$ $\quad 6, 8$

34. $0 = 2t^2 + 20t + 50$ $\quad -5$

35. $0 = 2m^3 - 5m^2 - 12m$ $\quad 0, -\frac{3}{2}, 4$

36. $3n^3 + 4n^2 + n = 0$ $\quad 0, -\frac{1}{3}, -1$

37. $5(3p + 1)(p - 3)(p + 6) = 0$ $\quad -\frac{1}{3}, 3, -6$

38. $4(2x - 1)(x - 10)(x + 7) = 0$ $\quad \frac{1}{2}, 10, -7$

39. $x(x - 4)(2x + 3) = 0$ $\quad 0, 4, -\frac{3}{2}$

40. $x(3x + 1)(x + 1) = 0$ $\quad 0, -\frac{1}{3}, -1$

41. $-5x(2x + 9)(x - 11) = 0$ $\quad 0, -\frac{9}{2}, 11$

42. $-3x(x + 7)(3x - 5) = 0$ $\quad 0, -7, \frac{5}{3}$

43. $x^3 - 16x = 0$ $\quad 0, 4, -4$

44. $t^3 - 36t = 0$ $\quad 0, 6, -6$

45. $3x^2 + 18x = 0$ $\quad -6, 0$

46. $2y^2 - 20y = 0$ $\quad 0, 10$

47. $16m^2 = 9$ $\quad \frac{3}{4}, -\frac{3}{4}$

48. $9n^2 = 1$ $\quad \frac{1}{3}, -\frac{1}{3}$

49. $2y^3 + 14y^2 = -20y$ $\quad 0, -5, -2$

50. $3d^3 - 6d^2 = 24d$ $\quad 0, -2, 4$

51. $5t - 2(t - 7) = 0$ $\quad -\frac{14}{3}$

52. $8h = 5(h - 9) + 6$ $\quad -13$

53. $2c(c - 8) = -30$ $\quad 5, 3$

54. $3q(q - 3) = 12$ $\quad -1, 4$

55. $b^3 = -4b^2 - 4b$ $\quad 0, -2$

56. $x^3 + 36x = 12x^2$ $\quad 0, 6$

57. $3(a^2 + 2a) = 2a^2 - 9$ $\quad -3$

58. $9(k - 1) = -4k^2$ $\frac{3}{4}, -3$

59. $2n(n + 2) = 6$ $-3, 1$

60. $3p(p - 1) = 18$ $3, -2$

61. $x(2x + 5) - 1 = 2x^2 + 3x + 2$ $\frac{3}{2}$

62. $3z(z - 2) - z = 3z^2 + 4$ $-\frac{4}{7}$

63. $27q^2 = 9q$ $0, \frac{1}{3}$

64. $21w^2 = 14w$ $0, \frac{2}{3}$

65. $3(c^2 - 2c) = 0$ $0, 2$

66. $2(4d^2 + d) = 0$ $0, -\frac{1}{4}$

67. $y^3 - 3y^2 - 4y + 12 = 0$ $3, -2, 2$

68. $t^3 + 2t^2 - 16t - 32 = 0$ $-2, 4, -4$

69. $(x - 1)(x + 2) = 18$ $-5, 4$

70. $(w + 5)(w - 3) = 20$ $-7, 5$

71. $(p + 2)(p + 3) = 1 - p$ $-5, -1$

72. $(k - 6)(k - 1) = -k - 2$ $4, 2$

Problem Recognition Exercises

Polynomial Expressions and Polynomial Equations

For Exercises 1–36, factor each expression and solve each equation.

1. a. $x^2 + 6x - 7$ $(x + 7)(x - 1)$
   b. $x^2 + 6x - 7 = 0$ $-7, 1$

2. a. $c^2 + 8c + 12$ $(c + 6)(c + 2)$
   b. $c^2 + 8c + 12 = 0$ $-6, -2$

3. a. $2y^2 + 7y + 3$ $(2y + 1)(y + 3)$
   b. $2y^2 + 7y + 3 = 0$ $-\frac{1}{2}, -3$

4. a. $3x^2 - 8x + 5$ $(3x - 5)(x - 1)$
   b. $3x^2 - 8x + 5 = 0$ $\frac{5}{3}, 1$

5. a. $5q^2 + q - 4 = 0$ $\frac{4}{5}, -1$
   b. $5q^2 + q - 4$ $(5q - 4)(q + 1)$

6. a. $6a^2 - 7a - 3 = 0$ $-\frac{1}{3}, \frac{3}{2}$
   b. $6a^2 - 7a - 3$ $(3a + 1)(2a - 3)$

7. a. $a^2 - 64 = 0$ $-8, 8$
   b. $a^2 - 64$ $(a + 8)(a - 8)$

8. a. $v^2 - 100 = 0$ $-10, 10$
   b. $v^2 - 100$ $(v + 10)(v - 10)$

9. a. $4b^2 - 81$ $(2b + 9)(2b - 9)$
   b. $4b^2 - 81 = 0$ $-\frac{9}{2}, \frac{9}{2}$

10. a. $36t^2 - 49$ $(6t + 7)(6t - 7)$
   b. $36t^2 - 49 = 0$ $-\frac{7}{6}, \frac{7}{6}$

11. a. $8x^2 + 16x + 6 = 0$ $-\frac{3}{2}, -\frac{1}{2}$
   b. $8x^2 + 16x + 6$ $2(2x + 3)(2x + 1)$

12. a. $12y^2 + 40y + 32 = 0$ $-\frac{4}{3}, -2$
   b. $12y^2 + 40y + 32$ $4(3y + 4)(y + 2)$

13. a. $x^3 - 8x^2 - 20x$ $x(x - 10)(x + 2)$
   b. $x^3 - 8x^2 - 20x = 0$ $0, 10, -2$

14. a. $k^3 + 5k^2 - 14k$ $k(k + 7)(k - 2)$
   b. $k^3 + 5k^2 - 14k = 0$ $0, -7, 2$

15. a. $b^3 + b^2 - 9b - 9 = 0$ $-1, 3, -3$
   b. $b^3 + b^2 - 9b - 9$ $(b + 1)(b - 3)(b + 3)$

16. a. $x^3 - 8x^2 - 4x + 32 = 0$ $8, -2, 2$
   b. $x^3 - 8x^2 - 4x + 32$ $(x - 8)(x + 2)(x - 2)$

17. $2s^2 - 6s + rs - 3r$ $(s - 3)(2s + r)$

18. $6t^2 + 3t + 10tu + 5u$ $(2t + 1)(3t + 5u)$

19. $8x^3 - 2x = 0$ $-\frac{1}{2}, 0, \frac{1}{2}$

20. $2b^3 - 50b = 0$ $-5, 0, 5$

21. $2x^3 - 4x^2 + 2x = 0$ $0, 1$

22. $3t^3 + 18t^2 + 27t = 0$ $-3, 0$

23. $7c^2 - 2c + 3 = 7(c^2 + c)$ $\frac{1}{3}$

24. $3z(2z + 4) = -7 + 6z^2$ $-\frac{7}{12}$

25. $8w^3 + 27$ $(2w + 3)(4w^2 - 6w + 9)$

26. $1000q^3 - 1$ $(10q - 1)(100q^2 + 10q + 1)$

27. $5z^2 + 2z = 7$ $-\frac{7}{5}, 1$

28. $4h^2 + 25h = -6$ $-6, -\frac{1}{4}$

29. $3b(b + 6) = b - 10$ $-\frac{2}{3}, -5$

30. $3y^2 + 1 = y(y - 3)$ $-1, -\frac{1}{2}$

31. $5(2x - 3) - 2(3x + 1) = 4 - 3x$ 3

32. $11 - 6a = -4(2a - 3) - 1$ 0

33. $4s^2 = 64$ $-4, 4$

34. $81v^2 = 36$ $-\frac{2}{3}, \frac{2}{3}$

35. $(x - 3)(x - 4) = 6$ $1, 6$

36. $(x + 5)(x + 9) = 21$ $-2, -12$

Applications of Quadratic Equations

1. Applications of Quadratic Equations

Example 1 Translating to a Quadratic Equation

The product of two consecutive integers is 48 more than the larger integer. Find the integers.

Solution:

Let x represent the first (smaller) integer.

Then $x + 1$ represents the second (larger) integer.   Label the variables.

(First integer)(second integer) = (second integer) + 48   Verbal model

$$x(x + 1) = (x + 1) + 48 \qquad \text{Algebraic equation}$$

$$x^2 + x = x + 49 \qquad \text{Simplify.}$$

$$x^2 + x - x - 49 = 0 \qquad \text{Set one side of the equation equal to zero.}$$

$$x^2 - 49 = 0$$

$$(x - 7)(x + 7) = 0 \qquad \text{Factor.}$$

$$x - 7 = 0 \quad \text{or} \quad x + 7 = 0 \qquad \text{Set each factor equal to zero.}$$

$$x = 7 \quad \text{or} \quad x = -7 \qquad \text{Solve for } x.$$

Recall that x represents the smaller integer. Therefore, there are two possibilities for the pairs of consecutive integers.

If $x = 7$, then the larger integer is $x + 1$ or $7 + 1 = 8$.

If $x = -7$, then the larger integer is $x + 1$ or $-7 + 1 = -6$.

The integers are 7 and 8, or -7 and -6.

Objectives

1. Applications of Quadratic Equations
2. Pythagorean Theorem

Skill Practice

1. The product of two consecutive odd integers is 9 more than 10 times the smaller integer. Find the pair of integers.

Classroom Example: p. 484, Exercise 16

Example 2 Using a Quadratic Equation in a Geometry Application

A rectangular sign has an area of 40 ft^2. If the width is 3 feet shorter than the length, what are the dimensions of the sign?

Solution:

Let x represent the length of the sign.   Label the variables.
Then $x - 3$ represents the width (Figure 6-1).

The problem gives information about the length of the sides and about the area. Therefore, we can form a relationship by using the formula for the area of a rectangle.

$x - 3$

x

Figure 6-1

Skill Practice

2. The length of a rectangle is 5 ft more than the width. The area is 36 ft^2. Find the length and width.

Classroom Example: p. 484, Exercise 18

Answers

1. The integers are 9 and 11 or -1 and 1.
2. The width is 4 ft, and the length is 9 ft.

$A = l \cdot w$ Area equals length times width.

$40 = x(x - 3)$ Set up an algebraic equation.

$40 = x^2 - 3x$ Clear parentheses.

$0 = x^2 - 3x - 40$ Write the equation in the form,
                                              $ax^2 + bx + c = 0$.

$0 = (x - 8)(x + 5)$ Factor.

$0 = x - 8$ or $0 = x + 5$ Set each factor equal to zero.

$8 = x$ or $-5 \cancel{=} x$ Because x represents the length of a
                                              rectangle, reject the negative solution.

The variable x represents the length of the sign. The length is 8 ft.

The expression $x - 3$ represents the width. The width is 8 ft − 3 ft, or 5 ft.

Classroom Example: p. 485,
Exercise 26

Skill Practice

3. An object is launched into the
   air from the ground and its
   height is given by
   $h = -16t^2 + 144t$, where
   h is the height in feet after t
   seconds. Find the time required for the object to hit the
   ground.

Example 3   **Using a Quadratic Equation in an Application**

A stone is dropped off a 64-ft cliff and falls into the ocean below. The height of
the stone above sea level is given by the equation

$h = -16t^2 + 64$ where h is the stone's height in feet, and t is the time
                                      in seconds.

Find the time required for the stone to hit the water.

Solution:

When the stone hits the water, its height is zero. Therefore, substitute $h = 0$ into
the equation.

$h = -16t^2 + 64$ The equation is quadratic.

$0 = -16t^2 + 64$ Substitute $h = 0$.

$0 = -16(t^2 - 4)$ Factor out the GCF.

$0 = -16(t - 2)(t + 2)$ Factor as a difference of squares.

$-16 \cancel{=} 0$ or $t - 2 = 0$ or $t + 2 = 0$ Set each factor to
                                                                          zero.

No solution, $t = 2$ or $t \cancel{=} -2$ Solve for t.

The negative value of t is rejected because the stone cannot fall for a negative
time. Therefore, the stone hits the water after 2 seconds.

Answer

3. 9 seconds

In Example 3, we can analyze the path of the stone as it falls from the cliff. Use the equation $h = -16t^2 + 64$ to compute the height values at various times between 0 and 2 seconds (Table 6-1). The ordered pairs can be graphed where t is used in place of x and h is used in place of y. The graph of the height of the stone versus time is shown in Figure 6-2.

Table 6-1

Time, t (sec)	Height, h (ft)
0.0	64
0.5	60
1.0	48
1.5	28
2.0	0

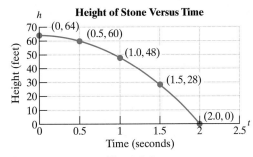

Figure 6-2

2. Pythagorean Theorem

Recall that a right triangle is a triangle that contains a 90° angle. Furthermore, the sum of the squares of the two legs (the shorter sides) of a right triangle equals the square of the hypotenuse (the longest side). This important fact is known as the Pythagorean theorem. The Pythagorean theorem is an enduring landmark of mathematical history from which many mathematical ideas have been built. Although the theorem is named after Pythagoras (sixth century B.C.E.), a Greek mathematician and philosopher, it is thought that the ancient Babylonians were familiar with the principle more than a thousand years earlier.

For the right triangle shown in Figure 6-3, the **Pythagorean theorem** is stated as:

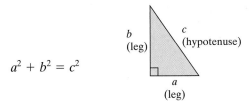

Figure 6-3

In this formula, a and b are the legs of the right triangle and c is the hypotenuse. Notice that the hypotenuse is the longest side of the right triangle and is opposite the 90° angle.

The triangle shown below is a right triangle. Notice that the lengths of the sides satisfy the Pythagorean theorem.

$$a^2 + b^2 = c^2 \qquad \text{Apply the Pythagorean theorem.}$$

$$(4)^2 + (3)^2 = (5)^2 \qquad \text{Substitute } a = 4, b = 3, \text{ and } c = 5.$$

$$16 + 9 = 25$$

$$25 = 25 \checkmark$$

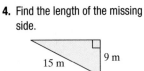
Example 4 Applying the Pythagorean Theorem

Find the length of the missing side of the right triangle.

Solution:

Label the triangle.

$a^2 + b^2 = c^2$ Apply the Pythagorean theorem.

$a^2 + 6^2 = 10^2$ Substitute $b = 6$ and $c = 10$.

$a^2 + 36 = 100$ Simplify.

The equation is quadratic. Set one side of the equation equal to zero.

$a^2 + 36 - 100 = 100 - 100$ Subtract 100 from both sides.

$a^2 - 64 = 0$

$(a + 8)(a - 8) = 0$ Factor.

$a + 8 = 0$ or $a - 8 = 0$ Set each factor equal to zero.

$a \neq -8$ or $a = 8$ Because x represents the length of a side of a triangle, reject the negative solution.

The third side is 8 ft.

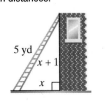
Example 5 Using a Quadratic Equation in an Application

A 13-ft board is used as a ramp to unload furniture off a loading platform. If the distance between the top of the board and the ground is 7 ft less than the distance between the bottom of the board and the base of the platform, find both distances.

Solution:

Let x represent the distance between the bottom of the board and the base of the platform. Then $x - 7$ represents the distance between the top of the board and the ground (Figure 6-4).

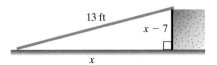

Figure 6-4

$$a^2 + b^2 = c^2 \qquad \text{Pythagorean theorem}$$

$$x^2 + (x - 7)^2 = (13)^2$$

$$x^2 + (x)^2 - 2(x)(7) + (7)^2 = 169$$

$$x^2 + x^2 - 14x + 49 = 169$$

$$2x^2 - 14x + 49 = 169 \qquad \text{Combine \textit{like} terms.}$$

$$2x^2 - 14x + 49 - 169 = 169 - 169 \qquad \text{Set the equation equal to zero.}$$

$$2x^2 - 14x - 120 = 0 \qquad \text{Write the equation in the form } ax^2 + bx + c = 0.$$

$$2(x^2 - 7x - 60) = 0 \qquad \text{Factor.}$$

$$2(x - 12)(x + 5) = 0$$

$$2 \neq 0 \quad \text{or} \quad x - 12 = 0 \quad \text{or} \quad x + 5 = 0 \qquad \text{Set each factor equal to zero.}$$

$$x = 12 \quad \text{or} \quad x \neq -5 \qquad \text{Solve both equations for } x.$$

> **Avoiding Mistakes**
>
> Recall that the square of a binomial results in a perfect square trinomial.
>
> $$(a - b)^2 = a^2 - 2ab + b^2$$
>
> Don't forget the middle term.

Recall that x represents the distance between the bottom of the board and the base of the platform. We reject the negative value of x because a distance cannot be negative. Therefore, the distance between the bottom of the board and the base of the platform is 12 ft. The distance between the top of the board and the ground is $x - 7 = 5$ ft.

Section 6.8 Practice Exercises

Boost _your_ GRADE at ALEKS.com!

ALEKS® version 3.0

- Practice Problems
- Self-Tests
- NetTutor
- e-Professors
- Videos

For additional exercises, see Classroom Activities 6.8A–6.8B in the _Instructor's Resource Manual_ at www.mhhe.com/moh.

Study Skills

1. Define the key term **Pythagorean theorem**.

Review Exercises

For Exercises 2–7, solve the quadratic equations.

2. $(6x + 1)(x + 4) = 0$ $-\frac{1}{6}, -4$

3. $9x(3x + 2) = 0$ $0, -\frac{2}{3}$

4. $4x^2 - 1 = 0$ $-\frac{1}{2}, \frac{1}{2}$

5. $x^2 - 5x = 6$ $6, -1$

6. $x(x - 20) = -100$ 10

7. $6x^2 - 7x - 10 = 0$ $-\frac{5}{6}, 2$

8. Explain what is wrong with the following problem:
$$(x - 3)(x + 2) = 5$$
$$x - 3 = 5 \quad \text{or} \quad x + 2 = 5.$$

A factored expression must be set equal to zero to use the zero product rule.

 Writing   Translating Expression   Geometry   Scientific Calculator   Video   NS E

Objective 1: Applications of Quadratic Equations

9. If eleven is added to the square of a number, the result is sixty. Find all such numbers.
The numbers are 7 and −7.

10. If a number is added to two times its square, the result is thirty-six. Find all such numbers.
The numbers are $-\frac{9}{2}$ and 4.

11. If twelve is added to six times a number, the result is twenty-eight less than the square of the number. Find all such numbers.
The numbers are 10 and −4.

12. The square of a number is equal to twenty more than the number. Find all such numbers.
The numbers are 5 and −4.

13. The product of two consecutive odd integers is sixty-three. Find all such integers. **(See Example 1.)**
The numbers are −9 and −7, or 7 and 9.

14. The product of two consecutive even integers is forty-eight. Find all such integers.
The numbers are 6 and 8, or −8 and −6.

15. The sum of the squares of two consecutive integers is one more than ten times the larger number. Find all such integers.
The numbers are 5 and 6, or −1 and 0.

16. The sum of the squares of two consecutive integers is nine less than ten times the sum of the integers. Find all such integers.   The numbers are 0 and 1, or 9 and 10.

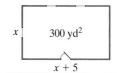 **17.** The length of a rectangular room is 5 yd more than the width. If 300 yd^2 of carpeting cover the room, what are the dimensions of the room? **(See Example 2.)**   The room is 15 yd by 20 yd.

 **18.** The width of a rectangular painting is 2 in. less than the length. The area is 120 in.^{2} Find the length and width.   The painting has length 12 in. and width 10 in.

x 300 yd^2

$x + 5$

x

$x - 2$

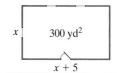 **19.** The width of a rectangular slab of concrete is 3 m less than the length. The area is 28 m^2.

  a. What are the dimensions of the rectangle?
  The slab is 7 m by 4 m.
  b. What is the perimeter of the rectangle?
  22 m

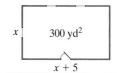 **20.** The width of a rectangular picture is 7 in. less than the length. The area of the picture is 78 in.^{2}

  a. What are the dimensions of the picture?
  The picture is 13 in. by 6 in.
  b. What is the perimeter of the picture?
  38 in.

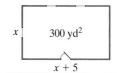 **21.** The base of a triangle is 3 ft more than the height.   The base is 7 ft and the height is 4 ft.

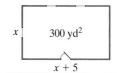 **22.** The height of a triangle is 15 cm more than the base.   The base is 10 cm and the height is 25 cm.

23. In a physics experiment, a ball is dropped off a 144-ft platform. The height of the ball above the ground is given by the equation

$h = -16t^2 + 144$ where h is the ball's height in feet, and t is the time in seconds after the ball is dropped ($t \geq 0$).

Find the time required for the ball to hit the ground. (*Hint:* Let $h = 0$.) **(See Example 3.)**   3 sec

24. A stone is dropped off a 256-ft cliff. The height of the stone above the ground is given by the equation

$h = -16t^2 + 256$ where h is the stone's height in feet, and t is the time in seconds after the stone is dropped ($t \geq 0$).

Find the time required for the stone to hit the ground.   4 sec

25. An object is shot straight up into the air from ground level with an initial speed of 24 ft/sec. The height of the object (in feet) is given by the equation

$$h = -16t^2 + 24t \qquad \text{where } t \text{ is the time in seconds after launch } (t \geq 0).$$

Find the time(s) when the object is at ground level.   0 sec and 1.5 sec

26. A rocket is launched straight up into the air from the ground with initial speed of 64 ft/sec. The height of the rocket (in feet) is given by the equation

$$h = -16t^2 + 64t \qquad \text{where } t \text{ is the time in seconds after launch } (t \geq 0).$$

Find the time(s) when the rocket is at ground level.   0 sec and 4 sec

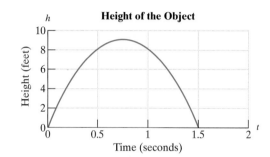

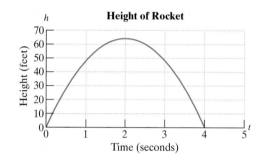

Objective 2: Pythagorean Theorem

27. Sketch a right triangle and label the sides with the words *leg* and *hypotenuse*.

28. State the Pythagorean theorem.

Given a right triangle with legs a and b and hypotenuse c, then $a^2 + b^2 = c^2$.

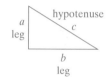

For Exercises 29–32, find the length of the missing side of the right triangle. **(See Example 4.)**

29.

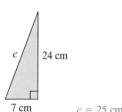

$c = 25$ cm

30.

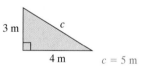

$c = 5$ m

31.

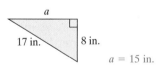

$a = 15$ in.

32.

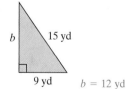

$b = 12$ yd

33. Find the length of the supporting brace.

The brace is 20 in. long.

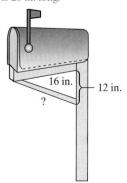

34. Find the height of the airplane above the ground.

The height is 9 km.

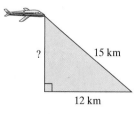

Writing   Translating Expression   Geometry   Scientific Calculator   Video   NS&E

35. Darcy holds the end of a kite string 3 ft (1 yd) off the ground and wants to estimate the height of the kite. Her friend Jenna is 24 yd away from her, standing directly under the kite as shown in the figure. If Darcy has 30 yd of string out, find the height of the kite (ignore the sag in the string).   The kite is 19 yd high.

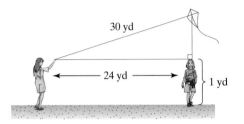

37. A 17-ft ladder rests against the side of a house. The distance between the top of the ladder and the ground is 7 ft more than the distance between the base of the ladder and the bottom of the house. Find both distances. **(See Example 5.)**
The bottom of the ladder is 8 ft from the house. The distance from the top of the ladder to the ground is 15 ft.

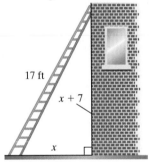

39. One leg of a right triangle is 4 m less than the hypotenuse. The other leg is 2 m less than the hypotenuse. Find the length of the hypotenuse.   The hypotenuse is 10 m.

36. Two cars leave the same point at the same time, one traveling north and the other traveling east. After an hour, one car has traveled 48 mi and the other has traveled 64 mi. How many miles apart were they at that time?
They were 80 mi apart.

38. Two boats leave a marina. One travels east, and the other travels south. After 30 min, the second boat has traveled 1 mi farther than the first boat and the distance between the boats is 5 mi. Find the distance each boat traveled.
The first boat traveled 3 mi; the second boat traveled 4 mi.

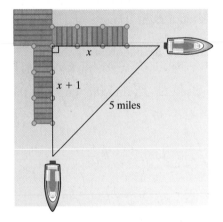

40. The longer leg of a right triangle is 1 cm less than twice the shorter leg. The hypotenuse is 1 cm greater than twice the shorter leg. Find the length of the shorter leg.   The shorter leg is 8 cm.

Group Activity

Building a Factoring Test

Estimated Time: 15–20 minutes

Group Size: 3

Answers will vary throughout this exercise.

In this activity, each group will make a test for this chapter. Then the groups will trade papers and take the test.

For questions 1–8, write a polynomial that has the given conditions.

1. A trinomial with a GCF not equal to 1. The GCF should include a constant and at least one variable.

1. _____

2. A four-term polynomial that is factorable by grouping.

2. _____

3. A factorable trinomial with a leading coefficient of 1. (The trinomial should factor as a product of two binomials.)

3. _____

4. A factorable trinomial with a leading coefficient not equal to 1. (The trinomial should factor as a product of two binomials.)

4. _____

5. A trinomial that requires the GCF to be removed. The resulting trinomial should factor as a product of two binomials.

5. _____

6. A difference of squares.

6. _____

7. A difference of cubes.

7. _____

8. A sum of cubes.

8. _____

9. Write a quadratic *equation* that has solutions $x = 4$ and $x = -7$.

9. _____

10. Write a quadratic *equation* that has solutions $x = 0$ and $x = -\dfrac{2}{3}$.

10. _____

Chapter 6   Summary

Section 6.1   Greatest Common Factor and Factoring by Grouping

Key Concepts

The **greatest common factor** (GCF) is the greatest factor common to all terms of a polynomial. To factor out the GCF from a polynomial, use the distributive property.

A four-term polynomial may be factorable by grouping.

Steps to Factoring by Grouping

1. Identify and factor out the GCF from all four terms.
2. Factor out the GCF from the first pair of terms. Factor out the GCF or its opposite from the second pair of terms.
3. If the two terms share a common binomial factor, factor out the binomial factor.

Examples

Example 1

$3x(a + b) - 5(a + b)$ Greatest common factor is $(a + b)$.

$= (a + b)(3x - 5)$

Example 2

$60xa - 30xb - 80ya + 40yb$

$= 10[6xa - 3xb - 8ya + 4yb]$ Factor out GCF.

$= 10[3x(2a - b) - 4y(2a - b)]$ Factor by grouping.

$= 10(2a - b)(3x - 4y)$

Section 6.2   Factoring Trinomials of the Form $x^2 + bx + c$

Key Concepts

Factoring a Trinomial with a Leading Coefficient of 1

A trinomial of the form $x^2 + bx + c$ factors as

$$x^2 + bx + c = (x \;\; \square)(x \;\; \square)$$

where the remaining terms are given by two integers whose product is c and whose sum is b.

Examples

Example 1

Factor: $x^2 - 14x + 45$

$x^2 - 14x + 45$ The integers -5 and -9 have a product of 45 and a sum of -14.

$= (x \;\; \square)(x \;\; \square)$

$= (x - 5)(x - 9)$

Section 6.3   Factoring Trinomials: Trial-and-Error Method

Key Concepts

Trial-and-Error Method for Factoring Trinomials in the Form $ax^2 + bx + c$ (where $a \neq 0$)

1. Factor out the GCF from all terms.
2. List the pairs of factors of a and the pairs of factors of c. Consider the reverse order in one of the lists.
3. Construct two binomials of the form

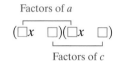

Factors of a

$(\square x \quad \square)(\square x \quad \square)$

Factors of c

4. Test each combination of factors and signs until the product forms the correct trinomial.
5. If no combination of factors produces the correct product, then the trinomial is prime.

Examples

Example 1

$10y^2 + 35y - 20$

$= 5(2y^2 + 7y - 4)$

The pairs of factors of 2 are: $2 \cdot 1$
The pairs of factors of -4 are:

$$-1(4) \qquad 1(-4)$$
$$-2(2) \qquad 2(-2)$$
$$-4(1) \qquad 4(-1)$$

$(2y - 2)(y + 2) = 2y^2 + 2y - 4$	No
$(2y - 4)(y + 1) = 2y^2 - 2y - 4$	No
$(2y + 1)(y - 4) = 2y^2 - 7y - 4$	No
$(2y + 2)(y - 2) = 2y^2 - 2y - 4$	No
$(2y + 4)(y - 1) = 2y^2 + 2y - 4$	No
$(2y - 1)(y + 4) = 2y^2 + 7y - 4$	Yes

$10y^2 + 35y - 20 = 5(2y - 1)(y + 4)$

Section 6.4   Factoring Trinomials: AC-Method

Key Concepts

AC-Method for Factoring Trinomials of the Form $ax^2 + bx + c$ (where $a \neq 0$)

1. Factor out the GCF from all terms.
2. Find the product ac.
3. Find two integers whose product is ac and whose sum is b. (If no pair of integers can be found, then the trinomial is prime.)
4. Rewrite the middle term (bx) as the sum of two terms whose coefficients are the numbers found in step 3.
5. Factor the polynomial by grouping.

Examples

Example 1

$10y^2 + 35y - 20$

$= 5(2y^2 + 7y - 4)$ First factor out GCF.

Identify the product $ac = (2)(-4) = -8$.

Find two integers whose product is -8 and whose sum is 7. The numbers are 8 and -1.

$5[2y^2 + 8y - 1y - 4]$

$= 5[2y(y + 4) - 1(y + 4)]$

$= 5(y + 4)(2y - 1)$

| **Section 6.5** | **Difference of Squares and Perfect Square Trinomials** |

Key Concepts

Factoring a Difference of Squares

$a^2 - b^2 = (a - b)(a + b)$

Factoring a Perfect Square Trinomial

The factored form of a **perfect square trinomial** is the square of a binomial:

$a^2 + 2ab + b^2 = (a + b)^2$

$a^2 - 2ab + b^2 = (a - b)^2$

Examples

Example 1

$25z^2 - 4y^2$

$\quad = (5z - 2y)(5z + 2y)$

Example 2

Factor: $25y^2 + 10y + 1$

$\quad = (5y)^2 + 2(5y)(1) + (1)^2$

$\quad = (5y + 1)^2$

| **Section 6.6** | **Sum and Difference of Cubes** |

Key Concepts

Factoring a Sum or Difference of Cubes

$a^3 - b^3 = (a - b)(a^2 + ab + b^2)$

$a^3 + b^3 = (a + b)(a^2 - ab + b^2)$

Examples

Example 1

$m^3 - 64$

$\quad = (m)^3 - (4)^3$

$\quad = (m - 4)(m^2 + 4m + 16)$

Example 2

$x^6 + 8y^3$

$\quad = (x^2)^3 + (2y)^3$

$\quad = (x^2 + 2y)(x^4 - 2x^2y + 4y^2)$

Section 6.7   Solving Equations Using the Zero Product Rule

Key Concepts

An equation of the form $ax^2 + bx + c = 0$, where $a \neq 0$, is a **quadratic equation**.

The zero product rule states that if $ab = 0$, then $a = 0$ or $b = 0$. The zero product rule can be used to solve a quadratic equation or a higher degree polynomial equation that is factored and set to zero.

Examples

Example 1

The equation $2x^2 - 17x + 30 = 0$ is a quadratic equation.

Example 2

$3w(w - 4)(2w + 1) = 0$

$3w = 0$ or $w - 4 = 0$ or $2w + 1 = 0$

$w = 0$ or $w = 4$ or $w = -\dfrac{1}{2}$

Example 3

$4x^2 = 34x - 60$

$4x^2 - 34x + 60 = 0$

$2(2x^2 - 17x + 30) = 0$

$2(2x - 5)(x - 6) = 0$

$2 \neq 0$ or $2x - 5 = 0$ or $x - 6 = 0$

$x = \dfrac{5}{2}$ or $x = 6$

Section 6.8 Applications of Quadratic Equations

Key Concepts

Use the zero product rule to solve applications.

Examples

Example 1

Find two consecutive integers such that the sum of their squares is 61.

Let x represent one integer.
Let $x + 1$ represent the next consecutive integer.

$$x^2 + (x + 1)^2 = 61$$

$$x^2 + x^2 + 2x + 1 = 61$$

$$2x^2 + 2x - 60 = 0$$

$$2(x^2 + x - 30) = 0$$

$$2(x - 5)(x + 6) = 0$$

$$x = 5 \quad \text{or} \quad x = -6$$

If $x = 5$, then the next consecutive integer is 6.
If $x = -6$, then the next consecutive integer is -5.
The integers are 5 and 6, or -6 and -5.

Some applications involve the Pythagorean theorem.

$$a^2 + b^2 = c^2$$

Example 2

Find the length of the missing side.

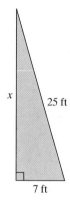

$$x^2 + (7)^2 = (25)^2$$

$$x^2 + 49 = 625$$

$$x^2 - 576 = 0$$

$$(x - 24)(x + 24) = 0$$

$$x = 24 \quad \text{or} \quad x = -24$$

The length of the side is 24 ft.

Chapter 6 Review Exercises

Section 6.1

For Exercises 1–4, identify the greatest common factor for each group of terms.

1. $15a^2b^4, 30a^3b, 9a^5b^3$
$3a^2b$

2. $3(x + 5), x(x + 5)$
$x + 5$

3. $2c^3(3c - 5), 4c(3c - 5)$
$2c(3c - 5)$

4. $-2wyz, -4xyz$
$-2yz$ or $2yz$

For Exercises 5–10, factor out the greatest common factor.

5. $6x^2 + 2x^4 - 8x$
$2x(3x + x^3 - 4)$

6. $11w^3y^3 - 44w^2y^5$
$11w^2y^3(w - 4y^2)$

7. $-t^2 + 5t$
$t(-t + 5)$ or $-t(t - 5)$

8. $-6u^2 - u$
$u(-6u - 1)$ or $-u(6u + 1)$

9. $3b(b + 2) - 7(b + 2)$ $(b + 2)(3b - 7)$

10. $2(5x + 9) + 8x(5x + 9)$ $2(5x + 9)(1 + 4x)$

For Exercises 11–14, factor by grouping.

11. $7w^2 + 14w + wb + 2b$ $(w + 2)(7w + b)$

12. $b^2 - 2b + yb - 2y$ $(b - 2)(b + y)$

13. $60y^2 - 45y - 12y + 9$ $3(4y - 3)(5y - 1)$

14. $6a - 3a^2 - 2ab + a^2b$ $a(2 - a)(3 - b)$

Section 6.2

For Exercises 15–24, factor completely.

15. $x^2 - 10x + 21$
$(x - 3)(x - 7)$

16. $y^2 - 19y + 88$
$(y - 8)(y - 11)$

17. $-6z + z^2 - 72$
$(z - 12)(z + 6)$

18. $-39 + q^2 - 10q$
$(q - 13)(q + 3)$

19. $3p^2w + 36pw + 60w$
$3w(p + 10)(p + 2)$

20. $2m^4 + 26m^3 + 80m^2$
$2m^2(m + 8)(m + 5)$

21. $-t^2 + 10t - 16$
$-(t - 8)(t - 2)$

22. $-w^2 - w + 20$
$-(w - 4)(w + 5)$

23. $a^2 + 12ab + 11b^2$
$(a + b)(a + 11b)$

24. $c^2 - 3cd - 18d^2$
$(c - 6d)(c + 3d)$

Section 6.3

For Exercises 25–28, let a, b, and c represent positive integers.

25. When factoring a polynomial of the form $ax^2 - bx - c$, should the signs of the binomials be both positive, both negative, or different?
Different

26. When factoring a polynomial of the form should the signs of the binomials be both positive, both negative, or different?
Both negative

27. When factoring a polynomial of the form $ax^2 + bx + c$, should the signs of the binomials be both positive, both negative, or different?
Both positive

28. When factoring a polynomial of the form $ax^2 + bx - c$, should the signs of the binomials be both positive, both negative, or different?
Different

For Exercises 29–40, factor the trinomial using the trial-and-error method.

29. $2y^2 - 5y - 12$
$(2y + 3)(y - 4)$

30. $4w^2 - 5w - 6$
$(4w + 3)(w - 2)$

31. $10z^2 + 29z + 10$
$(2z + 5)(5z + 2)$

32. $8z^2 + 6z - 9$
$(4z - 3)(2z + 3)$

33. $2p^2 - 5p + 1$
Prime

34. $5r^2 - 3r + 7$
Prime

35. $10w^2 - 60w - 270$
$10(w - 9)(w + 3)$

36. $3y^2 - 18y - 48$
$3(y - 8)(y + 2)$

37. $9c^2 - 30cd + 25d^2$
$(3c - 5d)^2$

38. $x^2 + 12x + 36$
$(x + 6)^2$

39. $v^4 - 2v^2 - 3$
$(v^2 + 1)(v^2 - 3)$

40. $x^4 + 7x^2 + 10$
$(x^2 + 5)(x^2 + 2)$

41. In Exercises 29–40, which trinomials are perfect square trinomials?
The trinomials in Exercises 37 and 38.

Section 6.4

For Exercises 42–43, find a pair of integers whose product and sum are given.

42. Product: -5 sum: 4 $5, -1$

43. Product: 15 sum: -8 $-3, -5$

For Exercises 44–57, factor the trinomial using the ac-method.

44. $3c^2 - 5c - 2$
$(c - 2)(3c + 1)$

45. $4y^2 + 13y + 3$
$(y + 3)(4y + 1)$

46. $t^2 + 13t + 12$
$(t + 12)(t + 1)$

47. $4x^3 + 17x^2 - 15x$
$x(x + 5)(4x - 3)$

48. $w^3 + 4w^2 - 5w$
$w(w + 5)(w - 1)$

49. $p^2 - 8pq + 15q^2$
$(p - 3q)(p - 5q)$

50. $40v^2 + 22v - 6$
$2(4v + 3)(5v - 1)$

51. $40s^2 + 30s - 100$
$10(4s - 5)(s + 2)$

 Writing  Translating Expression   Geometry Scientific Calculator Video   NS E

52. $a^3b - 10a^2b^2 + 24ab^3$
$ab(a - 6b)(a - 4b)$

53. $2z^6 + 8z^5 - 42z^4$
$2z^4(z + 7)(z - 3)$

54. $3m + 9m^2 - 2$
$(3m - 1)(3m + 2)$

55. $10 + 6p^2 + 19p$
$(3p + 2)(2p + 5)$

56. $49x^2 + 140x + 100$
$(7x + 10)^2$

57. $9w^2 - 6wz + z^2$
$(3w - z)^2$

58. In Exercises 42–57, which trinomials are perfect square trinomials?
The trinomials in Exercises 56 and 57.

Section 6.5

For Exercises 59–60, write the formula to factor each binomial, if possible.

59. $a^2 - b^2$
$(a - b)(a + b)$

60. $a^2 + b^2$
Prime

For Exercises 61–76, factor completely.

61. $a^2 - 49$
$(a - 7)(a + 7)$

62. $d^2 - 64$
$(d - 8)(d + 8)$

63. $100 - 81t^2$
$(10 - 9t)(10 + 9t)$

64. $4 - 25k^2$
$(2 - 5k)(2 + 5k)$

65. $x^2 + 16$
Prime

66. $y^2 + 121$
Prime

67. $y^2 + 12y + 36$
$(y + 6)^2$

68. $t^2 + 16t + 64$
$(t + 8)^2$

69. $9a^2 - 12a + 4$
$(3a - 2)^2$

70. $25x^2 - 40x + 16$
$(5x - 4)^2$

71. $-3v^2 - 12v - 12$
$-3(v + 2)^2$

72. $-2x^2 + 20x - 50$
$-2(x - 5)^2$

73. $2c^4 - 18$
$2(c^2 - 3)(c^2 + 3)$

74. $72x^2 - 2y^2$
$2(6x - y)(6x + y)$

75. $p^3 + 3p^2 - 16p - 48$
$(p + 3)(p - 4)(p + 4)$

76. $4k - 8 - k^3 + 2k^2$
$(k - 2)(2 - k)(2 + k)$
or $-1(k - 2)^2(2 + k)$

Section 6.6

For Exercises 77–78, write the formula to factor each binomial, if possible.

77. $a^3 + b^3$
$(a + b)(a^2 - ab + b^2)$

78. $a^3 - b^3$
$(a - b)(a^2 + ab + b^2)$

For Exercises 79–92, factor completely using the factoring strategy found on page 470.

79. $64 + a^3$
$(4 + a)(16 - 4a + a^2)$

80. $125 - b^3$
$(5 - b)(25 + 5b + b^2)$

81. $p^6 + 8$
$(p^2 + 2)(p^4 - 2p^2 + 4)$

82. $q^6 - \dfrac{1}{27}$
$\left(q^2 - \dfrac{1}{3}\right)\left(q^4 + \dfrac{1}{3}q^2 + \dfrac{1}{9}\right)$

83. $6x^3 - 48$
$6(x - 2)(x^2 + 2x + 4)$

84. $7y^3 + 7$
$7(y + 1)(y^2 - y + 1)$

85. $x^3 - 36x$
$x(x - 6)(x + 6)$

86. $q^4 - 64q$
$q(q - 4)(q^2 + 4q + 16)$

87. $8h^2 + 20$
$4(2h^2 + 5)$

88. $m^2 - 8m$
$m(m - 8)$

89. $x^3 + 4x^2 - x - 4$
$(x + 4)(x + 1)(x - 1)$

90. $5p^4q - 20q^3$
$5q(p^2 - 2q)(p^2 + 2q)$

91. $8n + n^4$
$n(2 + n)(4 - 2n + n^2)$

92. $14m^3 - 14$
$14(m - 1)(m^2 + m + 1)$

Section 6.7

93. For which of the following equations can the zero product rule be applied directly? Explain.

$$(x - 3)(2x + 1) = 0 \quad \text{or} \quad (x - 3)(2x + 1) = 6$$

$(x - 3)(2x + 1) = 0$ can be solved directly by the zero product rule because it is a product of factors set equal to zero.

For Exercises 94–109, solve the equation using the zero product rule.

94. $(4x - 1)(3x + 2) = 0$ $\dfrac{1}{4}, -\dfrac{2}{3}$

95. $(a - 9)(2a - 1) = 0$ $9, \dfrac{1}{2}$

96. $3w(w + 3)(5w + 2) = 0$ $0, -3, -\dfrac{2}{5}$

97. $6u(u - 7)(4u - 9) = 0$ $0, 7, \dfrac{9}{4}$

98. $7k^2 - 9k - 10 = 0$ $-\dfrac{5}{7}, 2$

99. $4h^2 - 23h - 6 = 0$ $-\dfrac{1}{4}, 6$

100. $q^2 - 144 = 0$
$12, -12$

101. $r^2 = 25$
$5, -5$

102. $5v^2 - v = 0$
$0, \dfrac{1}{5}$

103. $x(x - 6) = -8$
$4, 2$

104. $36t^2 + 60t = -25$
$-\dfrac{5}{6}$

105. $9s^2 + 12s = -4$
$-\dfrac{2}{3}$

106. $3(y^2 + 4) = 20y$
$\dfrac{2}{3}, 6$

107. $2(p^2 - 66) = -13p$
$\dfrac{11}{2}, -12$

108. $2y^3 - 18y^2 = -28y$
$0, 7, 2$

109. $x^3 - 4x = 0$
$0, 2, -2$

Section 6.8

110. The base of a parallelogram is 1 ft longer than twice the height. If the area is 78 ft^2, what are the base and height of the parallelogram?
The height is 6 ft, and the base is 13 ft.

111. A ball is tossed into the air from ground level with initial speed of 16 ft/sec. The height of the ball is given by the equation.

$$h = -16t^2 + 16t \quad (t \geq 0)$$

where h is the ball's height in feet, and t is the time in seconds

Find the time(s) when the ball is at ground level.
The ball is at ground level at 0 and 1 sec.

112. Find the length of the ramp.
The ramp is 13 ft long.

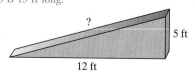

?   5 ft   12 ft

Writing   Translating Expression   Geometry   Scientific Calculator   Video   NS&E

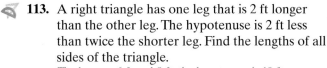

113. A right triangle has one leg that is 2 ft longer than the other leg. The hypotenuse is 2 ft less than twice the shorter leg. Find the lengths of all sides of the triangle.

The legs are 6 ft and 8 ft; the hypotenuse is 10 ft.

114. If the square of a number is subtracted from 60, the result is −4. Find all such numbers.

The numbers are −8 and 8.

115. The product of two consecutive integers is 44 more than 14 times their sum.

The numbers are 29 and 30, or −2 and −1.

116. The base of a triangle is 1 m longer than twice the height. If the area of the triangle is 18 m^2, find the base and height.

The height is 4 m, and the base is 9 m.

Chapter 6   Test

1. Factor out the GCF: $15x^4 - 3x + 6x^3$
$3x(5x^3 - 1 + 2x^2)$

2. Factor by grouping: $7a - 35 - a^2 + 5a$
$(a - 5)(7 - a)$

3. Factor the trinomial: $6w^2 - 43w + 7$
$(6w - 1)(w - 7)$

4. Factor the difference of squares: $169 - p^2$
$(13 - p)(13 + p)$

5. Factor the perfect square trinomial:
$q^2 - 16q + 64$ $(q - 8)^2$

6. Factor the sum of cubes: $8 + t^3$
$(2 + t)(4 - 2t + t^2)$

For Exercises 7–26, factor completely.

7. $a^2 + 12a + 32$
$(a + 4)(a + 8)$

8. $x^2 + x - 42$
$(x + 7)(x - 6)$

9. $2y^2 - 17y + 8$
$(2y - 1)(y - 8)$

10. $6z^2 + 19z + 8$
$(2z + 1)(3z + 8)$

11. $9t^2 - 100$
$(3t - 10)(3t + 10)$

12. $v^2 - 81$
$(v + 9)(v - 9)$

13. $3a^2 + 27ab + 54b^2$
$3(a + 6b)(a + 3b)$

14. $c^4 - 1$
$(c - 1)(c + 1)(c^2 + 1)$

15. $xy - 7x + 3y - 21$
$(y - 7)(x + 3)$

16. $49 + p^2$ Prime

17. $-10u^2 + 30u - 20$
$-10(u - 2)(u - 1)$

18. $12t^2 - 75$
$3(2t - 5)(2t + 5)$

19. $5y^2 - 50y + 125$
$5(y - 5)^2$

20. $21q^2 + 14q$
$7q(3q + 2)$

21. $2x^3 + x^2 - 8x - 4$
$(2x + 1)(x - 2)(x + 2)$

22. $y^3 - 125$
$(y - 5)(y^2 + 5y + 25)$

23. $m^2n^2 - 81$
$(mn - 9)(mn + 9)$

24. $16a^2 - 64b^2$
$16(a - 2b)(a + 2b)$

25. $64x^3 - 27y^6$
$(4x - 3y^2)(16x^2 + 12xy^2 + 9y^4)$

26. $3x^2y - 6xy - 24y$
$3y(x - 4)(x + 2)$

For Exercises 27–31, solve the equation.

27. $(2x - 3)(x + 5) = 0$ $\frac{3}{2}, -5$

28. $x^2 - 7x = 0$ $0, 7$

29. $x^2 - 6x = 16$ $8, -2$

30. $x(5x + 4) = 1$ $\frac{1}{5}, -1$

31. $y^3 + 10y^2 - 9y - 90 = 0$ $3, -3, -10$

32. A tennis court has an area of 312 yd^2. If the length is 2 yd more than twice the width, find the dimensions of the court.

The tennis court is 12 yd by 26 yd.

33. The product of two consecutive odd integers is 35. Find the integers.

The two integers are 5 and 7, or −5 and −7.

34. The height of a triangle is 5 in. less than the length of the base. The area is 42 in^2. Find the length of the base and the height of the triangle.

The base is 12 in., and the height is 7 in.

35. The hypotenuse of a right triangle is 2 ft less than three times the shorter leg. The longer leg is 3 ft less than three times the shorter leg. Find the length of the shorter leg.

The shorter leg is 5 ft.

Chapters 1–6   Cumulative Review Exercises

1. Simplify. $\dfrac{|4 - 25 \div (-5) \cdot 2|}{\sqrt{8^2 + 6^2}}$ $\dfrac{7}{5}$

2. Solve. $5 - 2(t + 4) = 3t + 12$ -3

3. Solve for y. $3x - 2y = 8$ $y = \dfrac{3}{2}x - 4$

4. A child's piggy bank has $3.80 in quarters, dimes, and nickels. The number of nickels is two more than the number of quarters. The number of dimes is three less than the number of quarters. Find the number of each type of coin in the bank.
There are 10 quarters, 12 nickels, and 7 dimes.

5. Solve the inequality. Graph the solution on a number line and write the solution set in interval notation. $[-4, \infty)$

$$-\dfrac{5}{12}x \le \dfrac{5}{3}$$

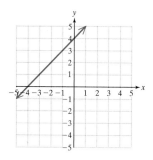

6. Given the equation $y = x + 4$

   a. Is the equation linear?   Yes

   b. Identify the slope.   1

   c. Identify the y-intercept. $(0, 4)$

   d. Identify the x-intercept. $(-4, 0)$

   e. Graph the line.

7. Consider the equation, $x = 5$.

   a. Does the equation present a horizontal or vertical line?   Vertical line

   b. Determine the slope of the line, if it exists.
    Undefined

   c. Identify the x-intercept, if it exists. $(5, 0)$

   d. Identify the y-intercept, if it exists.
    Does not exist

8. Find an equation of the line passing through the point $(-3, 5)$ and having a slope of 3. Write the final answer in slope-intercept form. $y = 3x + 14$

9. Solve the system. $\begin{aligned} 2x - 3y &= 4 \\ 5x - 6y &= 13 \end{aligned}$ $(5, 2)$

For Exercises 10–12, perform the indicated operations.

10. $2\left(\dfrac{1}{3}y^3 - \dfrac{3}{2}y^2 - 7\right) - \left(\dfrac{2}{3}y^3 + \dfrac{1}{2}y^2 + 5y\right)$
  $-\dfrac{7}{2}y^2 - 5y - 14$

11. $(4p^2 - 5p - 1)(2p - 3)$ $8p^3 - 22p^2 + 13p + 3$

12. $(2w - 7)^2$ $4w^2 - 28w + 49$

13. Divide using long division:
$(r^4 + 2r^3 - 5r + 1) \div (r - 3)$
  $r^3 + 5r^2 + 15r + 40 + \dfrac{121}{r - 3}$

14. Simplify. $\dfrac{c^{12}c^{-5}}{c^3}$ c^4

15. Divide. Write the final answer in scientific notation: $\dfrac{8.0 \times 10^{-3}}{5.0 \times 10^{-6}}$ 1.6×10^3

For Exercises 16–19, factor completely.

16. $w^4 - 16$ $(w - 2)(w + 2)(w^2 + 4)$

17. $2ax + 10bx - 3ya - 15yb$ $(a + 5b)(2x - 3y)$

18. $4x^2 - 8x - 5$ $(2x - 5)(2x + 1)$

19. $y^3 - 27$ $(y - 3)(y^2 + 3y + 9)$

20. Solve. $4x(2x - 1)(x + 5) = 0$ $0, \dfrac{1}{2}, -5$

Rational Expressions

CHAPTER OUTLINE

7.1 Introduction to Rational Expressions 498

7.2 Multiplication and Division of Rational Expressions 508

7.3 Least Common Denominator 514

7.4 Addition and Subtraction of Rational Expressions 520

Problem Recognition Exercises: Operations on Rational Expressions 529

7.5 Complex Fractions 530

7.6 Rational Equations 537

Problem Recognition Exercises: Comparing Rational Equations and Rational Expressions 547

7.7 Applications of Rational Equations and Proportions 548

7.8 Variation 559

Group Activity: Computing Monthly Mortgage Payments 567

Chapter 7

In this chapter, we define a rational expression as a ratio of two polynomials. Then we perform operations on rational expressions and solve rational equations.

You may find it helpful to review operations with fractions before you begin this chapter. Try the matching puzzle, then record the letter in the spaces below to complete the sentence

1. Add. $\dfrac{7}{4} + \dfrac{5}{3} + \dfrac{1}{2}$

2. Multiply. $\dfrac{5}{3} \cdot \dfrac{11}{4}$

a. $\dfrac{7}{8}x + \dfrac{2}{3} = \dfrac{1}{2}$

t. $\dfrac{45}{12}$

3. Fraction that is not in lowest terms.

r. $\dfrac{47}{12}$

p. $\dfrac{34}{3} \cdot \dfrac{1}{4}$

4. In this problem, a common denominator is not required.

n. $\dfrac{7}{8} + \dfrac{1}{3} - \dfrac{1}{2}$

o. $\dfrac{55}{12}$

5. Fractions can be eliminated from this problem by multiplying by the LCD.

The enemy of a good student is $\underset{4}{\text{p}}\,\underset{1}{\text{r}}\,\underset{2}{\text{o}}\,\underset{}{\text{c}}\,\underset{1}{\text{r}}\,\underset{5}{\text{a}}\,\underset{}{\text{s}}\,\underset{3}{\text{t}}\,\underset{}{\text{i}}\,\underset{}{\text{n}}\,\underset{5}{\text{a}}\,\underset{3}{\text{t}}\,\underset{}{\text{i}}\,\underset{2}{\text{o}}\,\underset{}{\text{n}}$.

Section 7.1 Introduction to Rational Expressions

Objectives

1. Definition of a Rational Expression
2. Evaluating Rational Expressions
3. Domain of a Rational Expression
4. Simplifying Rational Expressions to Lowest Terms
5. Simplifying a Ratio of −1

1. Definition of a Rational Expression

In Section 1.1, we defined a rational number as the ratio of two integers, $\frac{p}{q}$, where $q \neq 0$.

Examples of rational numbers: $\quad \frac{2}{3}, -\frac{1}{5}, 9$

In a similar way, we define a **rational expression** as the ratio of two polynomials, $\frac{p}{q}$, where $q \neq 0$.

Examples of rational expressions: $\quad \dfrac{3x - 6}{x^2 - 4}, \quad \dfrac{3}{4}, \quad \dfrac{6r^5 + 2r}{7}$

2. Evaluating Rational Expressions

Skill Practice

Evaluate the expression for the given values of x.

$$\frac{x - 3}{x + 5}$$

1. $x = 2$
2. $x = 0$
3. $x = 3$
4. $x = -5$

Classroom Examples: p. 505, Exercises 6 and 10

Instructor Note: Memory Device: Zero **un**der the line is **un**defined.

Example 1 Evaluating Rational Expressions

Evaluate the rational expression (if possible) for the given values of x: $\quad \dfrac{12}{x - 3}$

a. $x = 0$ **b.** $x = 1$ **c.** $x = -3$ **d.** $x = 3$

Solution:

Substitute the given value for the variable. Then use the order of operations to simplify.

a. $\dfrac{12}{x - 3}$

$\dfrac{12}{(0) - 3}$ Substitute $x = 0$.

$= \dfrac{12}{-3}$

$= -4$

b. $\dfrac{12}{x - 3}$

$\dfrac{12}{(1) - 3}$ Substitute $x = 1$.

$= \dfrac{12}{-2}$

$= -6$

c. $\dfrac{12}{x - 3}$

$\dfrac{12}{(-3) - 3}$ Substitute $x = -3$.

$= \dfrac{12}{-6}$

$= -2$

d. $\dfrac{12}{x - 3}$

$\dfrac{12}{(3) - 3}$ Substitute $x = 3$.

$= \dfrac{12}{0}$ Undefined.

Recall that division by zero is undefined.

Answers

1. $-\dfrac{1}{7}$ 2. $-\dfrac{3}{5}$
3. 0 4. Undefined

3. Domain of a Rational Expression

In Section 3.7, we presented the concept of the domain of a function.

> **DEFINITION** Domain of a Function
>
> The **domain** of a function is the set of all real numbers that when substituted into the function produces a real number.

For a rational expression, the domain will consist of the real numbers that when substituted into the expression do not make the denominator equal to zero. Therefore, in Example 1(d), because $\dfrac{12}{x-3}$ is undefined for $x = 3$, the domain is all real numbers except 3. We write this in set-builder notation as

$$\{x \mid x \text{ is a real number and } x \neq 3\}$$

Concept Connections

5. Fill in the blank.
   The domain of a rational expression is all real numbers except those that make the _____ zero.

> **PROCEDURE** Finding the Domain of a Rational Expression
>
> **Step 1** Set the denominator equal to zero and solve the resulting equation.
> **Step 2** The domain is the set of real numbers *excluding* the values found in step 1.

> **Example 2** Finding the Domain of Rational Expressions

Find the domain of the expressions.

a. $\dfrac{y-3}{2y+7}$ **b.** $\dfrac{-5}{x}$

Skill Practice

Find the domain.

6. $\dfrac{a+2}{2a-8}$

7. $\dfrac{2}{t}$

Classroom Examples: p. 506, Exercises 14 and 24

Solution:

a. $\dfrac{y-3}{2y+7}$

$2y + 7 = 0$ Set the denominator equal to zero.

$2y = -7$ Solve the equation.

$\dfrac{2y}{2} = \dfrac{-7}{2}$

$y = -\dfrac{7}{2}$ The domain is the set of real numbers except $-\frac{7}{2}$.

Domain: $\{y \mid y \text{ is a real number and } y \neq -\frac{7}{2}\}$

b. $\dfrac{-5}{x}$

$x = 0$ Set the denominator equal to zero.

Domain: $\{x \mid x \text{ is a real number and } x \neq 0\}$ The domain is the set of real numbers except 0.

Instructor Note: Ask the students to find the domain if a rational expression has no variable in the denominator.

For example: $\dfrac{3x+7}{8}$

Answers

5. Denominator
6. $\{a \mid a \text{ is a real number and } a \neq 4\}$
7. $\{t \mid t \text{ is a real number and } t \neq 0\}$

Example 3   Finding the Domain of Rational Expressions

Find the domain of the expressions.

a. $\dfrac{a + 10}{a^2 - 25}$ **b.** $\dfrac{2x^3 + 5}{x^2 + 9}$

Solution:

a. $\dfrac{a + 10}{a^2 - 25}$

$a^2 - 25 = 0$	Set the denominator equal to zero. The equation is quadratic.
$(a - 5)(a + 5) = 0$	Factor.
$a - 5 = 0$ or $a + 5 = 0$	Set each factor equal to zero.
$a = 5$ or $a = -5$	The domain is the set of real numbers except 5 and -5.

Domain: $\{a \mid a \text{ is a real number and } a \neq 5, a \neq -5\}$

b. $\dfrac{2x^3 + 5}{x^2 + 9}$

The quantity x^2 cannot be negative for any real number, x, so the denominator $x^2 + 9$ cannot equal zero. Therefore, no numbers are excluded from the domain.

The domain is the set of all real numbers.

4. Simplifying Rational Expressions to Lowest Terms

In many cases, it is advantageous to simplify or reduce a fraction to lowest terms. The same is true for rational expressions.

The method for simplifying rational expressions mirrors the process for simplifying fractions. In each case, factor the numerator and denominator. Common factors in the numerator and denominator form a ratio of 1 and can be reduced.

Simplifying a fraction: $\dfrac{21}{35} \xrightarrow{\text{Factor}} \dfrac{3 \cdot \overset{1}{\cancel{7}}}{5 \cdot \cancel{7}} = \dfrac{3}{5} \cdot (1) = \dfrac{3}{5}$

Simplifying a rational expression: $\dfrac{2x - 6}{x^2 - 9} \xrightarrow{\text{Factor}} \dfrac{2(\overset{1}{\cancel{x - 3}})}{(x + 3)(\cancel{x - 3})} = \dfrac{2}{(x + 3)}(1) = \dfrac{2}{x + 3}$

Informally, to simplify a rational expression to lowest terms, we simplify the ratio of common factors to 1. Formally, this is accomplished by applying the fundamental principle of rational expressions.

PROPERTY **Fundamental Principle of Rational Expressions**

Let p, q, and r represent polynomials where $q \neq 0$ and $r \neq 0$. Then

$$\frac{pr}{qr} = \frac{p}{q} \cdot \frac{r}{r} = \frac{p}{q} \cdot 1 = \frac{p}{q}$$

Example 4	**Simplifying a Rational Expression to Lowest Terms**

Given the expression $\dfrac{2p - 14}{p^2 - 49}$

a. Factor the numerator and denominator.

b. Determine the domain of the expression.

c. Simplify the expression to lowest terms.

Solution:

a. $\dfrac{2p - 14}{p^2 - 49}$ Factor out the GCF in the numerator.

$= \dfrac{2(p - 7)}{(p + 7)(p - 7)}$ Factor the denominator as a difference of squares.

b. $(p + 7)(p - 7) = 0$ To find the domain restrictions, set the denominator equal to zero. The equation is quadratic.

$p + 7 = 0$ or $p - 7 = 0$ Set each factor equal to 0.

$p = -7$ or $p = 7$ The domain is all real numbers except -7 and 7.

Domain: $\{p \,|\, p \text{ is a real number and } p \neq -7, p \neq 7\}$

c. $\dfrac{2(p \overset{1}{\cancel{- 7}})}{(p + 7)(p \cancel{- 7})}$ Simplify the ratio of common factors to 1.

$= \dfrac{2}{p + 7}$ (provided $p \neq 7$ and $p \neq -7$)

In Example 4, it is important to note that the expressions

$$\frac{2p - 14}{p^2 - 49} \quad \text{and} \quad \frac{2}{p + 7}$$

are equal for all values of p that make each expression a real number. Therefore,

$$\frac{2p - 14}{p^2 - 49} = \frac{2}{p + 7}$$

for all values of p except $p = 7$ and $p = -7$. (At $p = 7$ and $p = -7$, the original expression is undefined.) This is why the domain of an expression is always determined before the expression is simplified.

From this point forward, we will write statements of equality between two rational expressions with the assumption that they are equal for all values of the variable for which each expression is defined.

Skill Practice

Given: $\dfrac{5z + 25}{z^2 + 3z - 10}$

10. Factor the numerator and the denominator.

11. Determine the domain of the expression.

12. Simplify the rational expression to lowest terms.

Classroom Example: p. 506, Exercise 40

Avoiding Mistakes

The domain of a rational expression is always determined *before* simplifying the expression to lowest terms.

Answers

10. $\dfrac{5(z + 5)}{(z + 5)(z - 2)}$

11. $\{z \,|\, z \text{ is a real number and } z \neq -5, z \neq 2\}$

12. $\dfrac{5}{z - 2} \; (z \neq 2, z \neq -5)$

Skill Practice

Simplify to lowest terms.

13. $\dfrac{15q^3}{9q^2}$

Classroom Example: p. 506, Exercise 44

Example 5 Simplifying a Rational Expression to Lowest Terms

Simplify to lowest terms. $\dfrac{18a^4}{9a^5}$

Solution:

$\dfrac{18a^4}{9a^5}$

$= \dfrac{2 \cdot 3 \cdot 3 \cdot a \cdot a \cdot a \cdot a}{3 \cdot 3 \cdot a \cdot a \cdot a \cdot a \cdot a}$ Factor the numerator and denominator.

$= \dfrac{2 \cdot (3 \cdot 3 \cdot a \cdot a \cdot a \cdot a)}{(3 \cdot 3 \cdot a \cdot a \cdot a \cdot a) \cdot a}$ Simplify common factors to lowest terms.

$= \dfrac{2}{a}$

> **TIP:** The expression $\dfrac{18a^4}{9a^5}$ can also be simplified using the properties of exponents.
>
> $$\dfrac{18a^4}{9a^5} = 2a^{4-5} = 2a^{-1} = \dfrac{2}{a}$$

Skill Practice

Simplify to lowest terms.

14. $\dfrac{x^2 - 1}{2x^2 - x - 3}$

Classroom Example: p. 507, Exercise 62

Avoiding Mistakes

Given the expression

$$\dfrac{2c - 8}{10c^2 - 80c + 160}$$

do not be tempted to reduce before factoring. The terms $2c$ and $10c^2$ cannot be "canceled" because they are *terms* not factors.

The numerator and denominator must be in factored form before simplifying.

Example 6 Simplifying a Rational Expression to Lowest Terms

Simplify to lowest terms. $\dfrac{2c - 8}{10c^2 - 80c + 160}$

Solution:

$\dfrac{2c - 8}{10c^2 - 80c + 160}$

$= \dfrac{2(c - 4)}{10(c^2 - 8c + 16)}$ Factor out the GCF.

$= \dfrac{2(c - 4)}{10(c - 4)^2}$ Factor the denominator.

$= \dfrac{2(c - 4)}{2 \cdot 5(c - 4)(c - 4)}$ Simplify the ratio of common factors to 1.

$= \dfrac{1}{5(c - 4)}$

The process to simplify a rational expression to lowest terms is based on the identity property of multiplication. Therefore, this process applies only to factors (remember that factors are multiplied). For example,

Answers

13. $\dfrac{5q}{3}$ **14.** $\dfrac{x - 1}{2x - 3}$

$$\frac{3x}{3y} = \frac{\overset{1}{\cancel{3}} \cdot x}{\cancel{3} \cdot y} = 1 \cdot \frac{x}{y} = \frac{x}{y}$$

↑
Simplify

Terms that are added or subtracted cannot be reduced to lowest terms. For example,

$$\frac{x + 3}{y + 3}$$

↑
Cannot be simplified

The objective of simplifying a rational expression to lowest terms is to create an equivalent expression that is simpler to use. Consider the rational expression from Example 6 in its original form and in its reduced form. If we choose an arbitrary value of c from the domain and substitute that value into each expression, we see that the reduced form is easier to evaluate. For example, substitute $c = 3$:

	Original Expression	**Simplified Expression**
	$\dfrac{2c - 8}{10c^2 - 80c + 160}$	$\dfrac{1}{5(c - 4)}$
Substitute $c = 3$	$= \dfrac{2(3) - 8}{10(3)^2 - 80(3) + 160}$	$= \dfrac{1}{5(3 - 4)}$
	$= \dfrac{6 - 8}{10(9) - 240 + 160}$	$= \dfrac{1}{5(-1)}$
	$= \dfrac{-2}{90 - 240 + 160}$	$= -\dfrac{1}{5}$
	$= \dfrac{-2}{10} \quad \text{or} \quad -\dfrac{1}{5}$	

5. Simplifying a Ratio of −1

When two factors are identical in the numerator and denominator, they form a ratio of 1 and can be reduced. Sometimes we encounter two factors that are opposites and form a ratio of −1. For example,

Simplified Form   **Details/Notes**

$\dfrac{-5}{5} = -1$ The ratio of a number and its opposite is −1.

$\dfrac{100}{-100} = -1$ The ratio of a number and its opposite is −1.

$\dfrac{x + 7}{-x - 7} = -1$ $\dfrac{x + 7}{-x - 7} = \dfrac{x + 7}{-1(x + 7)} = \dfrac{\overset{1}{\cancel{x + 7}}}{-1(\cancel{x + 7})} = \dfrac{1}{-1} = -1$

factor out −1

$\dfrac{2 - x}{x - 2} = -1$ $\dfrac{2 - x}{x - 2} = \dfrac{-1(-2 + x)}{x - 2} = \dfrac{-1(\overset{1}{\cancel{x - 2}})}{\cancel{x - 2}} = \dfrac{-1}{1} = -1$

> **Avoiding Mistakes**
>
> While the expression $2 - x$ and $x - 2$ are opposites, the expressions $2 - x$ and $2 + x$ are *not*.
>
> Therefore $\dfrac{2 - x}{2 + x}$ does not simplify to −1.

Recognizing factors that are opposites is useful when simplifying rational expressions.

── **Skill Practice** ──

Simplify to lowest terms.

15. $\dfrac{2t - 12}{6 - t}$

Classroom Example: p. 507,
Exercise 92

Example 7 Simplifying a Rational Expression to Lowest Terms

Simplify to lowest terms. $\quad \dfrac{3c - 3d}{d - c}$

Solution:

$$\dfrac{3c - 3d}{d - c}$$

$$= \dfrac{3(c - d)}{d - c} \qquad \text{Factor the numerator and denominator.}$$

Notice that $(c - d)$ and $(d - c)$ are opposites and form a ratio of -1.

$$= \dfrac{3(\overset{-1}{\cancel{c - d}})}{\cancel{d - c}} \qquad \underline{\text{Details:}} \quad \dfrac{3(c - d)}{d - c} = \dfrac{3(c - d)}{-1(-d + c)} = \dfrac{3(c - d)}{-1(c - d)}$$

$$= 3(-1) \qquad\qquad\qquad\qquad\qquad\qquad\qquad\qquad = \dfrac{3}{-1} = -3$$

$$= -3$$

TIP: It is important to recognize that a rational expression can be written in several equivalent forms. In particular, two numbers with opposite signs form a negative quotient. Therefore, a number such as $-\frac{3}{4}$ can be written as:

$$-\dfrac{3}{4} \quad \text{or} \quad \dfrac{-3}{4} \quad \text{or} \quad \dfrac{3}{-4}$$

The negative sign can be written in the numerator, in the denominator, or out in front of the fraction. We demonstrate this concept in Example 8.

── **Skill Practice** ──

Simplify to lowest terms.

16. $\dfrac{b - a}{a^2 - b^2}$

Classroom Example: p. 507,
Exercise 98

Example 8 Simplifying a Rational Expression to Lowest Terms

Simplify to lowest terms. $\quad \dfrac{5 - y}{y^2 - 25}$

Solution:

$$\dfrac{5 - y}{y^2 - 25}$$

$$= \dfrac{5 - y}{(y - 5)(y + 5)} \qquad \text{Factor the numerator and denominator.}$$

Notice that $5 - y$ and $y - 5$ are opposites and form a ratio of -1.

$$= \dfrac{\overset{-1}{\cancel{5 - y}}}{(\cancel{y - 5})(y + 5)} \qquad \underline{\text{Details:}} \quad \dfrac{5 - y}{(y - 5)(y + 5)} = \dfrac{-1(-5 + y)}{(y - 5)(y + 5)}$$

$$\qquad\qquad\qquad\qquad\qquad\qquad = \dfrac{-1(y - 5)}{(y - 5)(y + 5)} = \dfrac{-1}{y + 5}$$

Answers

15. -2 **16.** $\dfrac{-1}{a + b}$

$$= \dfrac{-1}{y + 5} \quad \text{or} \quad \dfrac{1}{-(y + 5)} \quad \text{or} \quad -\dfrac{1}{y + 5}$$

Section 7.1 Practice Exercises

Study Skills Exercises

1. Review Section R.2 in this text. Write an example of how to simplify (reduce) a fraction, multiply two fractions, divide two fractions, add two fractions, and subtract two fractions. Then as you learn about rational expressions, compare the operations on rational expressions with those on fractions. This is a great place to use 3×5 cards again. Write an example of an operation with fractions on one side and the same operation with rational expressions on the other side.

2. Define the key terms:

   a. rational expression   **b. domain**

Objective 1: Definition of a Rational Expression

3. **a.** What is a rational number? A number $\dfrac{p}{q}$, where p and q are integers and $q \neq 0$

   b. What is a rational expression? An expression $\dfrac{p}{q}$, where p and q are polynomials and $q \neq 0$

4. **a.** Write an example of a rational number. (Answers will vary.) For example: $\dfrac{2}{3}$

   b. Write an example of a rational expression. (Answers will vary.) For example: $\dfrac{3x^2 + 1}{2x + 5}$

Objective 2: Evaluating Rational Expressions

For Exercises 5–10, substitute the given number into the expression and simplify (if possible). **(See Example 1.)**

5. $\dfrac{1}{x - 6}$; $x = -2$ $-\dfrac{1}{8}$

6. $\dfrac{w - 10}{w + 6}$; $w = 0$ $-\dfrac{5}{3}$

7. $\dfrac{w - 4}{2w + 8}$; $w = 0$ $-\dfrac{1}{2}$

8. $\dfrac{y - 8}{2y^2 + y - 1}$; $y = 8$ 0

9. $\dfrac{(a - 7)(a + 1)}{(a - 2)(a + 5)}$; $a = 2$ Undefined

10. $\dfrac{(a + 4)(a + 1)}{(a - 4)(a - 1)}$; $a = 1$ Undefined

11. A bicyclist rides 24 mi against a wind and returns 24 mi with the same wind. His average speed for the return trip traveling with the wind is 8 mph faster than his speed going out against the wind. If x represents the bicyclist's speed going out against the wind, then the total time, t, required for the round trip is given by

    $$t = \frac{24}{x} + \frac{24}{x + 8}$$ where t is measured in hours.

    a. Find the time required for the round trip if the cyclist rides 12 mph against the wind. $3\frac{1}{5}$ hr or 3.2 hr

    b. Find the time required for the round trip if the cyclist rides 24 mph against the wind. $1\frac{3}{4}$ hr or 1.75 hr

12. The manufacturer of mountain bikes has a fixed cost of $56,000, plus a variable cost of $140 per bike. The average cost per bike, y (in dollars), is given by the equation:

    $$y = \frac{56{,}000 + 140x}{x}$$ where x represents the number of bikes produced.

    a. Find the average cost per bike if the manufacturer produces 1000 bikes. $196

    b. Find the average cost per bike if the manufacturer produces 2000 bikes. $168

    c. Find the average cost per bike if the manufacturer produces 10,000 bikes. $145.60

 Writing   Translating Expression 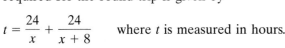 Geometry   Scientific Calculator   Video NS•E

Objective 3: Domain of a Rational Expression

For Exercises 13–24, write the domain. **(See Examples 2–3.)**

13. $\dfrac{5}{k+2}$
$\{k \,|\, k \text{ is a real number and } k \neq -2\}$

14. $\dfrac{-3}{h-4}$
$\{h \,|\, h \text{ is a real number and } h \neq 4\}$

15. $\dfrac{x+5}{(2x-5)(x+8)}$
$\left\{x \,|\, x \text{ is a real number and } x \neq \dfrac{5}{2},\, x \neq -8\right\}$

16. $\dfrac{4y+1}{(3y+7)(y+3)}$
$\left\{y \,|\, y \text{ is a real number and } y \neq -\dfrac{7}{3},\, y \neq -3\right\}$

17. $\dfrac{b+12}{b^2+5b+6}$
$\{b \,|\, b \text{ is a real number and } b \neq -2,\, b \neq -3\}$

18. $\dfrac{c-11}{c^2-5c-6}$
$\{c \,|\, c \text{ is a real number and } c \neq 6,\, c \neq -1\}$

19. $\dfrac{x-4}{x^2+9}$
The set of all real numbers

20. $\dfrac{x+1}{x^2+4}$
The set of all real numbers

21. $\dfrac{y^2-y-12}{12}$
The set of all real numbers

22. $\dfrac{z^2+10z+9}{9}$
The set of all real numbers

23. $\dfrac{t-5}{t}$
$\{t \,|\, t \text{ is a real number and } t \neq 0\}$

24. $\dfrac{2w+7}{w}$
$\{w \,|\, w \text{ is a real number and } w \neq 0\}$

25. Construct a rational expression that is undefined for $x = 2$. (Answers will vary.) For example: $\dfrac{1}{x-2}$

26. Construct a rational expression that is undefined for $x = 5$. (Answers will vary.) For example: $\dfrac{1}{x-5}$

27. Construct a rational expression that is undefined for $x = -3$ and $x = 7$. (Answers will vary.)
For example: $\dfrac{1}{(x+3)(x-7)}$

28. Construct a rational expression that is undefined for $x = -1$ and $x = 4$. (Answers will vary.)
For example: $\dfrac{1}{(x+1)(x-4)}$

29. Evaluate the expressions for $x = -1$.

  a. $\dfrac{3x^2-2x-1}{6x^2-7x-3}$ $\dfrac{2}{5}$ **b.** $\dfrac{x-1}{2x-3}$ $\dfrac{2}{5}$

30. Evaluate the expressions for $x = 4$.

  a. $\dfrac{(x+5)^2}{x^2+6x+5}$ $\dfrac{9}{5}$ **b.** $\dfrac{x+5}{x+1}$ $\dfrac{9}{5}$

31. Evaluate the expressions for $x = 1$.

  a. $\dfrac{5x+5}{x^2-1}$ **b.** $\dfrac{5}{x-1}$
     Undefined   Undefined

32. Evaluate the expressions for $x = 3$.

  a. $\dfrac{2x^2-4x-6}{2x^2-18}$ **b.** $\dfrac{x+1}{x+3}$
     Undefined $\dfrac{2}{3}$

Objective 4: Simplifying Rational Expressions to Lowest Terms

For Exercises 33–42,

  a. Write the domain in set-builder notation.

  b. Simplify the expression to lowest terms. **(See Example 4.)**

33. $\dfrac{3y+6}{6y+12}$ a. $\{y \,|\, y \text{ is a real number and } y \neq -2\}$ b. $\dfrac{1}{2}$

34. $\dfrac{8x-8}{4x-4}$ a. $\{x \,|\, x \text{ is a real number and } x \neq 1\}$ b. 2

35. $\dfrac{t^2-1}{t+1}$ a. $\{t \,|\, t \text{ is a real number and } t \neq -1\}$ b. $t-1$

36. $\dfrac{r^2-4}{r-2}$ a. $\{r \,|\, r \text{ is a real number and } r \neq 2\}$ b. $r+2$

37. $\dfrac{7w}{21w^2-35w}$ a. $\left\{w \,|\, w \text{ is a real number and } w \neq 0,\, w \neq \dfrac{5}{3}\right\}$ b. $\dfrac{1}{3w-5}$

38. $\dfrac{12a^2}{24a^2-18a}$ a. $\left\{a \,|\, a \text{ is a real number and } a \neq 0,\, a \neq \dfrac{3}{4}\right\}$ b. $\dfrac{2a}{4a-3}$

39. $\dfrac{9x^2-4}{6x+4}$ a. $\left\{x \,|\, x \text{ is a real number and } x \neq -\dfrac{2}{3}\right\}$ b. $\dfrac{3x-2}{2}$

40. $\dfrac{8b-20}{4b^2-25}$ a. $\left\{b \,|\, b \text{ is a real number and } b \neq \dfrac{5}{2},\, b \neq -\dfrac{5}{2}\right\}$ b. $\dfrac{4}{2b+5}$

41. $\dfrac{a^2+3a-10}{a^2+a-6}$ a. $\{a \,|\, a \text{ is a real number and } a \neq -3,\, a \neq 2\}$ b. $\dfrac{a+5}{a+3}$

42. $\dfrac{t^2+3t-10}{t^2+t-20}$ a. $\{t \,|\, t \text{ is a real number and } t \neq -5,\, t \neq 4\}$ b. $\dfrac{t-2}{t-4}$

For Exercises 43–84, simplify the expression to lowest terms. **(See Examples 5–6.)**

43. $\dfrac{7b^2}{21b}$ $\dfrac{b}{3}$

44. $\dfrac{15c^3}{3c^5}$ $\dfrac{5}{c^2}$

45. $\dfrac{18st^5}{12st^3}$ $\dfrac{3t^2}{2}$

46. $\dfrac{20a^4b^2}{25ab^2}$ $\dfrac{4a^3}{5}$

47. $\dfrac{-24x^2y^5z}{8xy^4z^3}$ $\quad -\dfrac{3xy}{z^2}$
48. $\dfrac{60rs^4t^2}{-12r^4s^2t^3}$ $\quad -\dfrac{5s^2}{r^3t}$
49. $\dfrac{3(y+2)}{6(y+2)}$ $\quad \dfrac{1}{2}$
50. $\dfrac{8(x-1)}{4(x-1)}$ $\quad 2$

51. $\dfrac{(p-3)(p+5)}{(p+5)(p+4)}$ $\quad \dfrac{p-3}{p+4}$
52. $\dfrac{(c+4)(c-1)}{(c+4)(c+2)}$ $\quad \dfrac{c-1}{c+2}$
53. $\dfrac{(m+11)}{4(m+11)(m-11)}$ $\quad \dfrac{1}{4(m-11)}$
54. $\dfrac{(n-7)}{9(n+2)(n-7)}$ $\quad \dfrac{1}{9(n+2)}$

55. $\dfrac{x(2x+1)^2}{4x^3(2x+1)}$ $\quad \dfrac{2x+1}{4x^2}$
56. $\dfrac{(p+2)(p-3)^4}{(p+2)^2(p-3)^2}$ $\quad \dfrac{(p-3)^2}{p+2}$
57. $\dfrac{5}{20a-25}$ $\quad \dfrac{1}{4a-5}$
58. $\dfrac{7}{14c-21}$ $\quad \dfrac{1}{2c-3}$

59. $\dfrac{4w-8}{w^2-4}$ $\quad \dfrac{4}{w+2}$
60. $\dfrac{3x+15}{x^2-25}$ $\quad \dfrac{3}{x-5}$
61. $\dfrac{3x^2-6x}{9xy+18x}$ $\quad \dfrac{x-2}{3(y+2)}$
62. $\dfrac{6p^2+12p}{2pq-4p}$ $\quad \dfrac{3(p+2)}{q-2}$

63. $\dfrac{2x+4}{x^2-3x-10}$ $\quad \dfrac{2}{x-5}$
64. $\dfrac{5z+15}{z^2-4z-21}$ $\quad \dfrac{5}{z-7}$
65. $\dfrac{a^2-49}{a-7}$ $\quad a+7$
66. $\dfrac{b^2-64}{b-8}$ $\quad b+8$

67. $\dfrac{q^2+25}{q+5}$ Cannot simplify
68. $\dfrac{r^2+36}{r+6}$ Cannot simplify
69. $\dfrac{y^2+6y+9}{2y^2+y-15}$ $\quad \dfrac{y+3}{2y-5}$
70. $\dfrac{h^2+h-6}{h^2+2h-8}$ $\quad \dfrac{h+3}{h+4}$

71. $\dfrac{3x^2+7x-6}{x^2+7x+12}$ $\dfrac{3x-2}{x+4}$
72. $\dfrac{x^2-5x-14}{2x^2-x-10}$ $\dfrac{x-7}{2x-5}$
73. $\dfrac{5q^2+5}{q^4-1}$ $\quad \dfrac{5}{(q+1)(q-1)}$
74. $\dfrac{4t^2+16}{t^4-16}$ $\quad \dfrac{4}{(t-2)(t+2)}$

75. $\dfrac{ac-ad+2bc-2bd}{2ac+ad+4bc+2bd}$ (*Hint:* Factor by grouping.) $\dfrac{c-d}{2c+d}$
76. $\dfrac{3pr-ps-3qr+qs}{3pr-ps+3qr-qs}$ (*Hint:* Factor by grouping.) $\dfrac{p-q}{p+q}$

77. $\dfrac{2t^2-3t}{2t^4-13t^3+15t^2}$ $\dfrac{1}{t(t-5)}$
78. $\dfrac{4m^3+3m^2}{4m^3+7m^2+3m}$ $\dfrac{m}{m+1}$
79. $\dfrac{49p^2-28pq+4q^2}{14p-4q}$ $\dfrac{7p-2q}{2}$
80. $\dfrac{3x-3y}{2x^2-4xy+2y^2}$ $\dfrac{3}{2(x-y)}$

81. $\dfrac{5x^3+4x^2-45x-36}{x^2-9}$ $5x+4$
82. $\dfrac{x^2-1}{ax^3-bx^2-ax+b}$ $\dfrac{1}{ax-b}$
83. $\dfrac{2x^2-xy-3y^2}{2x^2-11xy+12y^2}$ $\dfrac{x+y}{x-4y}$
84. $\dfrac{2c^2+cd-d^2}{5c^2+3cd-2d^2}$ $\dfrac{2c-d}{5c-2d}$

Objective 5: Simplifying a Ratio of -1

85. What is the relationship between $x-2$ and $2-x$? They are opposites.

86. What is the relationship between $w+p$ and $-w-p$? They are opposites.

For Exercises 87–98, simplify to lowest terms. (See Examples 7–8.)

87. $\dfrac{x-5}{5-x}$ $\quad -1$
88. $\dfrac{8-p}{p-8}$ $\quad -1$
89. $\dfrac{-4-y}{4+y}$ $\quad -1$
90. $\dfrac{z+10}{-z-10}$ $\quad -1$

91. $\dfrac{3y-6}{12-6y}$ $\quad -\dfrac{1}{2}$
92. $\dfrac{4q-4}{12-12q}$ $\quad -\dfrac{1}{3}$
93. $\dfrac{k+5}{5-k}$ Cannot simplify
94. $\dfrac{2+n}{2-n}$ Cannot simplify

95. $\dfrac{10x-12}{10x+12}$ $\dfrac{5x-6}{5x+6}$
96. $\dfrac{4t-16}{16+4t}$ $\dfrac{t-4}{4+t}$
97. $\dfrac{x^2-x-12}{16-x^2}$ $-\dfrac{x+3}{4+x}$
98. $\dfrac{49-b^2}{b^2-10b+21}$ $-\dfrac{7+b}{b-3}$

Expanding Your Skills

For Exercises 99–102, factor and simplify to lowest terms.

99. $\dfrac{w^3-8}{w^2+2w+4}$ $\quad w-2$
100. $\dfrac{y^3+27}{y^2-3y+9}$ $\quad y+3$
101. $\dfrac{z^2-16}{z^3-64}$ $\dfrac{z+4}{z^2+4z+16}$
102. $\dfrac{x^2-25}{x^3+125}$ $\dfrac{x-5}{x^2-5x+25}$

Writing   Translating Expression   Geometry   Scientific Calculator   Video   NS&E

Section 7.2 Multiplication and Division of Rational Expressions

Objectives

1. Multiplication of Rational Expressions
2. Division of Rational Expressions

1. Multiplication of Rational Expressions

Recall from Section R.2 that to multiply fractions, we multiply the numerators and multiply the denominators. The same is true for multiplying rational expressions.

> **PROPERTY** Multiplication of Rational Expressions
>
> Let p, q, r, and s represent polynomials, such that $q \neq 0$, $s \neq 0$. Then,
>
> $$\frac{p}{q} \cdot \frac{r}{s} = \frac{pr}{qs}$$

For example:

Multiply the Fractions	Multiply the Rational Expressions
$\dfrac{2}{3} \cdot \dfrac{5}{7} = \dfrac{10}{21}$	$\dfrac{2x}{3y} \cdot \dfrac{5z}{7} = \dfrac{10xz}{21y}$

Sometimes it is possible to simplify a ratio of common factors to 1 *before* multiplying. To do so, we must first factor the numerators and denominators of each fraction.

$$\frac{15}{14} \cdot \frac{21}{10} = \frac{3 \cdot \overset{1}{\cancel{5}}}{2 \cdot \cancel{7}} \cdot \frac{3 \cdot \overset{1}{\cancel{7}}}{2 \cdot \cancel{5}} = \frac{9}{4}$$

The same process is also used to multiply rational expressions.

> **PROCEDURE** Multiplying Rational Expressions
>
> **Step 1** Factor the numerators and denominators of all rational expressions.
> **Step 2** Simplify the ratios of common factors to 1.
> **Step 3** Multiply the remaining factors in the numerator, and multiply the remaining factors in the denominator.

Skill Practice

Multiply.

1. $\dfrac{7a}{3b} \cdot \dfrac{15b}{14a^2}$

Classroom Example: p. 512, Exercise 10

Answer

1. $\dfrac{5}{2a}$

Example 1 Multiplying Rational Expressions

Multiply. $\dfrac{5a^2b}{2} \cdot \dfrac{6a}{10b}$

Solution:

$$\frac{5a^2b}{2} \cdot \frac{6a}{10b}$$

$$= \frac{5 \cdot a \cdot a \cdot b}{2} \cdot \frac{2 \cdot 3 \cdot a}{2 \cdot 5 \cdot b} \qquad \text{Factor into prime factors.}$$

$$= \frac{\overset{1}{\cancel{5}} \cdot a \cdot a \cdot \overset{1}{\cancel{b}}}{2} \cdot \frac{\overset{1}{\cancel{2}} \cdot 3 \cdot a}{\cancel{2} \cdot \cancel{5} \cdot \cancel{b}} \qquad \text{Simplify.}$$

$$= \frac{3a^3}{2} \qquad \text{Multiply remaining factors.}$$

Example 2 Multiplying Rational Expressions

Multiply. $\dfrac{3c - 3d}{6c} \cdot \dfrac{2}{c^2 - d^2}$

Solution:

$\dfrac{3c - 3d}{6c} \cdot \dfrac{2}{c^2 - d^2}$

$= \dfrac{3(c - d)}{2 \cdot 3 \cdot c} \cdot \dfrac{2}{(c - d)(c + d)}$ Factor.

$= \dfrac{\overset{1}{\cancel{3}}(\overset{1}{\cancel{c - d}})}{\cancel{2} \cdot \cancel{3} \cdot c} \cdot \dfrac{\overset{1}{\cancel{2}}}{(\cancel{c - d})(c + d)}$ Simplify.

$= \dfrac{1}{c(c + d)}$ Multiply remaining factors.

Classroom Example: p. 512, Exercise 16

Skill Practice

Multiply.

2. $\dfrac{4x - 8}{x + 6} \cdot \dfrac{x^2 + 6x}{2x}$

Avoiding Mistakes

If all the factors in the numerator reduce to a ratio of 1, a factor of 1 is left in the numerator.

Example 3 Multiplying Rational Expressions

Multiply. $\dfrac{35 - 5x}{5x + 5} \cdot \dfrac{x^2 + 5x + 4}{x^2 - 49}$

Solution:

$\dfrac{35 - 5x}{5x + 5} \cdot \dfrac{x^2 + 5x + 4}{x^2 - 49}$

$= \dfrac{5(7 - x)}{5(x + 1)} \cdot \dfrac{(x + 4)(x + 1)}{(x - 7)(x + 7)}$ Factor the numerators and denominators completely.

$= \dfrac{\overset{1}{\cancel{5}}(\overset{-1}{\cancel{7 - x}})}{\cancel{5}(\cancel{x + 1})} \cdot \dfrac{(x + 4)(\overset{1}{\cancel{x + 1}})}{(\cancel{x - 7})(x + 7)}$ Simplify the ratios of common factors to 1 or −1.

$= \dfrac{-1(x + 4)}{x + 7}$ Multiply remaining factors.

$= \dfrac{-(x + 4)}{x + 7}$ or $\dfrac{x + 4}{-(x + 7)}$ or $-\dfrac{x + 4}{x + 7}$

Classroom Example: p. 512, Exercise 20

Skill Practice

Multiply.

3. $\dfrac{p^2 + 4p + 3}{5p + 10} \cdot \dfrac{p^2 - p - 6}{9 - p^2}$

TIP: The ratio $\dfrac{7 - x}{x - 7} = -1$ because $7 - x$ and $x - 7$ are opposites.

---

2. Division of Rational Expressions

Recall that to divide fractions, multiply the first fraction by the reciprocal of the second.

$$\dfrac{21}{10} \div \dfrac{49}{15} \xrightarrow[\text{of the second fraction}]{\text{multiply by the reciprocal}} \dfrac{21}{10} \cdot \dfrac{15}{49} \xrightarrow{\text{factor}} \dfrac{3 \cdot \overset{1}{\cancel{7}}}{2 \cdot \cancel{5}} \cdot \dfrac{3 \cdot \overset{1}{\cancel{5}}}{\cancel{7} \cdot 7} = \dfrac{9}{14}$$

The same process is used to divide rational expressions.

PROPERTY Division of Rational Expressions

Let p, q, r, and s represent polynomials, such that $q \neq 0$, $r \neq 0$, $s \neq 0$. Then,

$$\dfrac{p}{q} \div \dfrac{r}{s} = \dfrac{p}{q} \cdot \dfrac{s}{r} = \dfrac{ps}{qr}$$

Answers

2. $2(x - 2)$

3. $\dfrac{-(p + 1)}{5}$ or $\dfrac{p + 1}{-5}$ or $-\dfrac{p + 1}{5}$

Skill Practice

Divide.

4. $\dfrac{7y - 14}{y + 1} \div \dfrac{y^2 + 2y - 8}{2y + 2}$

Classroom Example: p. 512, Exercise 32

Avoiding Mistakes

When dividing rational expressions, take the reciprocal of the second fraction and change to multiplication *before* reducing like factors.

Example 4 **Dividing Rational Expressions**

Divide. $\dfrac{5t - 15}{2} \div \dfrac{t^2 - 9}{10}$

Solution:

$\dfrac{5t - 15}{2} \div \dfrac{t^2 - 9}{10}$

$= \dfrac{5t - 15}{2} \cdot \dfrac{10}{t^2 - 9}$ Multiply the first fraction by the reciprocal of the second.

$= \dfrac{5(t - 3)}{2} \cdot \dfrac{2 \cdot 5}{(t - 3)(t + 3)}$ Factor each polynomial.

$= \dfrac{5\overset{1}{\cancel{(t - 3)}}}{\cancel{2}} \cdot \dfrac{\overset{1}{\cancel{2}} \cdot 5}{\cancel{(t - 3)}(t + 3)}$ Simplify the ratio of common factors to 1.

$= \dfrac{25}{t + 3}$

Skill Practice

Divide.

5. $\dfrac{4x^2 - 9}{2x^2 - x - 3} \div \dfrac{20x + 30}{x^2 + 7x + 6}$

Classroom Example: p. 512, Exercise 38

Example 5 **Dividing Rational Expressions**

Divide. $\dfrac{p^2 - 11p + 30}{10p^2 - 250} \div \dfrac{30p - 5p^2}{2p + 4}$

Solution:

$\dfrac{p^2 - 11p + 30}{10p^2 - 250} \div \dfrac{30p - 5p^2}{2p + 4}$

$= \dfrac{p^2 - 11p + 30}{10p^2 - 250} \cdot \dfrac{2p + 4}{30p - 5p^2}$ Multiply the first fraction by the reciprocal of the second.

Factor the trinomial.
$p^2 - 11p + 30 = (p - 5)(p - 6)$

$= \dfrac{(p - 5)(p - 6)}{2 \cdot 5(p - 5)(p + 5)} \cdot \dfrac{2(p + 2)}{5p(6 - p)}$ Factor out the GCF.
$2p + 4 = 2(p + 2)$

Factor out the GCF. Then factor the difference of squares.
$10p^2 - 250 = 10(p^2 - 25)$
$\qquad\qquad = 2 \cdot 5(p - 5)(p + 5)$

Factor out the GCF.
$30p - 5p^2 = 5p(6 - p)$

$= \dfrac{\overset{1}{\cancel{(p - 5)}}\overset{-1}{\cancel{(p - 6)}}}{\cancel{2} \cdot 5\cancel{(p - 5)}(p + 5)} \cdot \dfrac{\overset{1}{\cancel{2}}(p + 2)}{5p\cancel{(6 - p)}}$ Simplify the ratio of common factors to 1 or −1.

$= -\dfrac{(p + 2)}{25p(p + 5)}$

Answers

4. $\dfrac{14}{y + 4}$ **5.** $\dfrac{x + 6}{10}$

Example 6 Dividing Rational Expressions

Divide. $\dfrac{\dfrac{3x}{4y}}{\dfrac{5x}{6y}}$

Skill Practice

Divide.

6. $\dfrac{\dfrac{a^3 b}{9c}}{\dfrac{4ab}{3c^3}}$

Classroom Example: p. 512, Exercise 28

Solution:

$\dfrac{\dfrac{3x}{4y}}{\dfrac{5x}{6y}}$ ⟵ This fraction bar denotes division (÷). This expression is called a complex fraction because it has one or more rational expressions in its numerator or denominator.

$= \dfrac{3x}{4y} \div \dfrac{5x}{6y}$

$= \dfrac{3x}{4y} \cdot \dfrac{6y}{5x}$ Multiply by the reciprocal of the second fraction.

$= \dfrac{3 \cdot \overset{1}{\cancel{x}}}{\cancel{2} \cdot 2 \cdot \cancel{y}} \cdot \dfrac{\overset{1}{\cancel{2}} \cdot 3 \cdot \overset{1}{\cancel{y}}}{5 \cdot \cancel{x}}$ Simplify the ratio of common factors to 1.

$= \dfrac{9}{10}$

Sometimes multiplication and division of rational expressions appear in the same problem. In such a case, apply the order of operations by multiplying or dividing in order from left to right.

Example 7 Multiplying and Dividing Rational Expressions

Perform the indicated operations. $\dfrac{4}{c^2 - 9} \div \dfrac{6}{c - 3} \cdot \dfrac{3c}{8}$

Skill Practice

Perform the indicated operations.

7. $\dfrac{v}{v + 2} \div \dfrac{5v^2}{v^2 - 4} \cdot \dfrac{v}{10}$

Classroom Example: p. 513, Exercise 66

Solution:

In this example, division occurs first, before multiplication. Parentheses may be inserted to reinforce the proper order.

$\left(\dfrac{4}{c^2 - 9} \div \dfrac{6}{c - 3} \right) \cdot \dfrac{3c}{8}$

$= \left(\dfrac{4}{c^2 - 9} \cdot \dfrac{c - 3}{6} \right) \cdot \dfrac{3c}{8}$ Multiply the first fraction by the reciprocal of the second.

$= \left(\dfrac{2 \cdot 2}{(c - 3)(c + 3)} \cdot \dfrac{c - 3}{2 \cdot 3} \right) \cdot \dfrac{3 \cdot c}{2 \cdot 2 \cdot 2}$ Now that each operation is written as multiplication, factor the polynomials and reduce the common factors.

$= \dfrac{\overset{1}{\cancel{2}} \cdot \overset{1}{\cancel{2}}}{\cancel{(c - 3)}(c + 3)} \cdot \dfrac{\overset{1}{\cancel{(c - 3)}}}{2 \cdot \cancel{3}} \cdot \dfrac{\overset{1}{\cancel{3}} \cdot c}{2 \cdot 2 \cdot 2}$

$= \dfrac{c}{4(c + 3)}$ Simplify.

Answers

6. $\dfrac{a^2 c^2}{12}$ 7. $\dfrac{v - 2}{50}$

Section 7.2 Practice Exercises

Boost *your* GRADE at ALEKS.com!

ALEKS® version 3.0

- Practice Problems
- Self-Tests
- NetTutor
- e-Professors
- Videos

For additional exercises, see Classroom Activities 7.2A–7.2B in the *Instructor's Resource Manual* at www.mhhe.com/moh.

Review Exercises

For Exercises 1–8, multiply or divide the fractions.

1. $\dfrac{3}{5} \cdot \dfrac{1}{2}$ $\dfrac{3}{10}$

2. $\dfrac{6}{7} \cdot \dfrac{5}{12}$ $\dfrac{5}{14}$

3. $\dfrac{3}{4} \div \dfrac{3}{8}$ 2

4. $\dfrac{18}{5} \div \dfrac{2}{5}$ 9

5. $6 \cdot \dfrac{5}{12}$ $\dfrac{5}{2}$

6. $\dfrac{7}{25} \cdot 5$ $\dfrac{7}{5}$

7. $\dfrac{\dfrac{21}{4}}{\dfrac{7}{5}}$ $\dfrac{15}{4}$

8. $\dfrac{\dfrac{9}{2}}{\dfrac{3}{4}}$ 6

Objective 1: Multiplication of Rational Expressions

For Exercises 9–24, multiply. **(See Examples 1–3.)**

9. $\dfrac{2xy}{5x^2} \cdot \dfrac{15}{4y}$ $\dfrac{3}{2x}$

10. $\dfrac{7s}{t^2} \cdot \dfrac{t^2}{14s^2}$ $\dfrac{1}{2s}$

11. $\dfrac{6x^3}{9x^6y^2} \cdot \dfrac{18x^4y^7}{4y}$ $3xy^4$

12. $\dfrac{10a^2b}{15b^2} \cdot \dfrac{30b}{2a^3}$ $\dfrac{10}{a}$

13. $\dfrac{4x - 24}{20x} \cdot \dfrac{5x}{8}$ $\dfrac{x - 6}{8}$

14. $\dfrac{5a + 20}{a} \cdot \dfrac{3a}{10}$ $\dfrac{3(a + 4)}{2}$

15. $\dfrac{3y + 18}{y^2} \cdot \dfrac{4y}{6y + 36}$ $\dfrac{2}{y}$

16. $\dfrac{2p - 4}{6p} \cdot \dfrac{4p^2}{8p - 16}$ $\dfrac{p}{6}$

17. $\dfrac{10}{2 - a} \cdot \dfrac{a - 2}{16}$ $-\dfrac{5}{8}$

18. $\dfrac{b - 3}{6} \cdot \dfrac{20}{3 - b}$ $-\dfrac{10}{3}$

19. $\dfrac{b^2 - a^2}{a - b} \cdot \dfrac{a}{a^2 - ab}$ $-\dfrac{b + a}{a - b}$

20. $\dfrac{(x - y)^2}{x^2 + xy} \cdot \dfrac{x}{y - x}$ $-\dfrac{x - y}{x + y}$

21. $\dfrac{y^2 + 2y + 1}{5y - 10} \cdot \dfrac{y^2 - 3y + 2}{y^2 - 1}$ $\dfrac{y + 1}{5}$

22. $\dfrac{6a^2 - 6}{a^2 + 6a + 5} \cdot \dfrac{a^2 + 5a}{12a}$ $\dfrac{a - 1}{2}$

23. $\dfrac{10x}{2x^2 + 3x + 1} \cdot \dfrac{x^2 + 7x + 6}{5x}$ $\dfrac{2(x + 6)}{2x + 1}$

24. $\dfrac{b - 3}{b^2 + b - 12} \cdot \dfrac{4b + 16}{b + 1}$ $\dfrac{4}{b + 1}$

Objective 2: Division of Rational Expressions

For Exercises 25–38, divide. **(See Examples 4–6.)**

25. $\dfrac{4x}{7y} \div \dfrac{2x^2}{21xy}$ 6

26. $\dfrac{6cd}{5d^2} \div \dfrac{8c^3}{10d}$ $\dfrac{3}{2c^2}$

27. $\dfrac{\dfrac{8m^4n^5}{5n^6}}{\dfrac{24mn}{15m^3}}$ $\dfrac{m^6}{n^2}$

28. $\dfrac{\dfrac{10a^3b}{3a}}{\dfrac{5b}{9ab}}$ $6a^3b$

29. $\dfrac{4a + 12}{6a - 18} \div \dfrac{3a + 9}{5a - 15}$ $\dfrac{10}{9}$

30. $\dfrac{8b - 16}{3b + 3} \div \dfrac{5b - 10}{2b + 2}$ $\dfrac{16}{15}$

31. $\dfrac{3x - 21}{6x^2 - 42x} \div \dfrac{7}{12x}$ $\dfrac{6}{7}$

32. $\dfrac{4a^2 - 4a}{9a - 9} \div \dfrac{5}{12a}$ $\dfrac{16a^2}{15}$

33. $\dfrac{m^2 - n^2}{9} \div \dfrac{3n - 3m}{27m}$ $-m(m + n)$

34. $\dfrac{9 - b^2}{15b + 15} \div \dfrac{b - 3}{5b}$ $-\dfrac{b(3 + b)}{3(b + 1)}$

35. $\dfrac{3p + 4q}{p^2 + 4pq + 4q^2} \div \dfrac{4}{p + 2q}$ $\dfrac{3p + 4q}{4(p + 2q)}$

36. $\dfrac{x^2 + 2xy + y^2}{2x - y} \div \dfrac{x + y}{5}$ $\dfrac{5(x + y)}{2x - y}$

37. $\dfrac{p^2 - 2p - 3}{p^2 - p - 6} \div \dfrac{p^2 - 1}{p^2 + 2p}$ $\dfrac{p}{p - 1}$

38. $\dfrac{4t^2 - 1}{t^2 - 5t} \div \dfrac{2t^2 + 5t + 2}{t^2 - 3t - 10}$ $\dfrac{2t - 1}{t}$

 Writing   Translating Expression   Geometry   Scientific Calculator   Video   NS&E

Mixed Exercises

For Exercises 39–64, multiply or divide as indicated.

39. $(w + 3) \cdot \dfrac{w}{2w^2 + 5w - 3}$

$\dfrac{w}{2w - 1}$

40. $\dfrac{5t + 1}{5t^2 - 31t + 6} \cdot (t - 6)$ $\dfrac{5t + 1}{5t - 1}$

41. $\dfrac{\dfrac{5t - 10}{12}}{\dfrac{4t - 8}{8}}$ $\dfrac{5}{6}$

42. $\dfrac{\dfrac{6m + 6}{5}}{\dfrac{3m + 3}{10}}$ 4

43. $\dfrac{q + 1}{5q^2 - 28q - 12} \cdot (5q + 2)$

$\dfrac{q + 1}{q - 6}$

44. $(r - 5) \cdot \dfrac{4r}{2r^2 - 7r - 15}$ $\dfrac{4r}{2r + 3}$

45. $\dfrac{2a^2 + 13a - 24}{8a - 12} \div (a + 8)$ $\dfrac{1}{4}$

46. $\dfrac{3y^2 + 20y - 7}{5y + 35} \div (3y - 1)$ $\dfrac{1}{5}$

47. $\dfrac{y^2 + 5y - 36}{y^2 - 2y - 8} \cdot \dfrac{y + 2}{y - 6}$ $\dfrac{y + 9}{y - 6}$

48. $\dfrac{z^2 - 11z + 28}{z - 1} \cdot \dfrac{z + 1}{z^2 - 6z - 7}$ $\dfrac{z - 4}{z - 1}$

49. $\dfrac{2t^2 + t - 1}{t^2 + 3t + 2} \cdot \dfrac{t + 4}{2t - 1}$ $\dfrac{t + 4}{t + 2}$

50. $\dfrac{3p^2 - 2p - 8}{3p^2 - 5p - 12} \cdot \dfrac{p + 1}{p - 2}$ $\dfrac{p + 1}{p - 3}$

51. $(5t - 1) \div \dfrac{5t^2 + 9t - 2}{3t + 8}$ $\dfrac{3t + 8}{t + 2}$

52. $(2q - 3) \div \dfrac{2q^2 + 5q - 12}{q - 7}$ $\dfrac{q - 7}{q + 4}$

53. $\dfrac{x^2 + 2x - 3}{x^2 - 3x + 2} \cdot \dfrac{x^2 + 2x - 8}{x^2 + 4x + 3}$ $\dfrac{x + 4}{x + 1}$

54. $\dfrac{y^2 + y - 12}{y^2 - y - 20} \cdot \dfrac{y^2 + y - 30}{y^2 - 2y - 3}$

$\dfrac{y + 6}{y + 1}$

55. $\dfrac{\dfrac{w^2 - 6w + 9}{8}}{\dfrac{9 - w^2}{4w + 12}}$ $-\dfrac{w - 3}{2}$

56. $\dfrac{\dfrac{p^2 - 6p + 8}{24}}{\dfrac{16 - p^2}{6p + 6}}$ $-\dfrac{(p - 2)(p + 1)}{4(4 + p)}$

57. $\dfrac{5k^2 + 7k + 2}{k^2 + 5k + 4} \div \dfrac{5k^2 + 17k + 6}{k^2 + 10k + 24}$ $\dfrac{k + 6}{k + 3}$

58. $\dfrac{4h^2 - 5h + 1}{h^2 + h - 2} \div \dfrac{6h^2 - 7h + 2}{2h^2 + 3h - 2}$ $\dfrac{4h - 1}{3h - 2}$

59. $\dfrac{ax + a + bx + b}{2x^2 + 4x + 2} \cdot \dfrac{4x + 4}{a^2 + ab}$ $\dfrac{2}{a}$

60. $\dfrac{3my + 9m + ny + 3n}{9m^2 + 6mn + n^2} \cdot \dfrac{30m + 10n}{5y^2 + 15y}$ $\dfrac{2}{y}$

61. $\dfrac{y^4 - 1}{2y^2 - 3y + 1} \div \dfrac{2y^2 + 2}{8y^2 - 4y}$ $2y(y + 1)$

62. $\dfrac{x^4 - 16}{6x^2 + 24} \div \dfrac{x^2 - 2x}{3x}$ $\dfrac{x + 2}{2}$

63. $\dfrac{x^2 - xy - 2y^2}{x + 2y} \div \dfrac{x^2 - 4xy + 4y^2}{x^2 - 4y^2}$ $x + y$

64. $\dfrac{4m^2 - 4mn - 3n^2}{8m^2 - 18n^2} \div \dfrac{3m + 3n}{6m^2 + 15mn + 9n^2}$ $\dfrac{2m + n}{2}$

For Exercises 65–70, multiply or divide as indicated. **(See Example 7.)**

65. $\dfrac{b^3 - 3b^2 + 4b - 12}{b^4 - 16} \cdot \dfrac{3b^2 + 5b - 2}{3b^2 - 10b + 3} \div \dfrac{3}{6b - 12}$ 2

66. $\dfrac{x^2 - 25}{3x^2 + 3xy} \cdot \dfrac{x^2 + 4x + xy + 4y}{x^2 + 9x + 20} \div \dfrac{x - 5}{x}$ $\dfrac{1}{3}$

67. $\dfrac{a^2 - 5a}{a^2 + 7a + 12} \div \dfrac{a^3 - 7a^2 + 10a}{a^2 + 9a + 18} \div \dfrac{a + 6}{a + 4}$ $\dfrac{1}{a - 2}$

68. $\dfrac{t^2 + t - 2}{t^2 + 5t + 6} \div \dfrac{t - 1}{t} \div \dfrac{5t - 5}{t + 3}$ $\dfrac{t}{5(t - 1)}$

69. $\dfrac{p^3 - q^3}{p - q} \cdot \dfrac{p + q}{2p^2 + 2pq + 2q^2}$ $\dfrac{p + q}{2}$

70. $\dfrac{r^3 + s^3}{r - s} \div \dfrac{r^2 + 2rs + s^2}{r^2 - s^2}$ $r^2 - rs + s^2$

Section 7.3 Least Common Denominator

Objectives

1. Least Common Denominator
2. Writing Rational Expressions with the Least Common Denominator

1. Least Common Denominator

In Sections 7.1 and 7.2, we learned how to simplify, multiply, and divide rational expressions. Our next goal is to add and subtract rational expressions. As with fractions, rational expressions may be added or subtracted only if they have the same denominator.

The **least common denominator (LCD)** of two or more rational expressions is defined as the least common multiple of the denominators. For example, consider the fractions $\frac{1}{20}$ and $\frac{1}{8}$. By inspection, you can probably see that the least common denominator is 40. To understand why, find the prime factorization of both denominators:

$$20 = 2^2 \cdot 5 \qquad \text{and} \qquad 8 = 2^3$$

A common multiple of 20 and 8 must be a multiple of 5, a multiple of 2^2, and a multiple of 2^3. However, any number that is a multiple of $2^3 = 8$ is automatically a multiple of $2^2 = 4$. Therefore, it is sufficient to construct the least common denominator as the product of unique prime factors, in which each factor is raised to its highest power.

$$\text{The LCD of } \frac{1}{20} \text{ and } \frac{1}{8} \text{ is } 2^3 \cdot 5 = 40.$$

> **PROCEDURE** Finding the Least Common Denominator of Two or More Rational Expressions
>
> **Step 1** Factor all denominators completely.
> **Step 2** The LCD is the product of unique prime factors from the denominators, in which each factor is raised to the highest power to which it appears in any denominator.

Skill Practice

Find the LCD for each set of expressions.

1. $\dfrac{3}{8}; \dfrac{7}{10}; \dfrac{1}{15}$

2. $\dfrac{1}{5a^3b^2}; \dfrac{1}{10a^4b}$

Classroom Example: p. 518, Exercise 20

Example 1 Finding the Least Common Denominator of Rational Expressions

Find the LCD of the rational expressions.

a. $\dfrac{5}{14}; \dfrac{3}{49}; \dfrac{1}{8}$ **b.** $\dfrac{5}{3x^2z}; \dfrac{7}{x^5y^3}$

Solution:

a. Factor the denominators, 14, 49, and 8.

	2's	7's
14 =	2	7
49 =		⑦²
8 =	②³	

We circle the factor of 2 raised to its greatest power. We circle the factor of 7 raised to its greatest power. The LCD is their product.

The least common denominator (LCD) is $2^3 \cdot 7^2 = 392$.

b. The denominators are already factored.

	3's	x's	y's	z's
$3x^2z =$	③	x^2		$\textcircled{z}$
$x^5y^3 =$		$\textcircled{x^5}$	$\textcircled{y^3}$	

We circle the factors of $3, x, y$, and z, each raised to its corresponding highest power.

The least common denominator (LCD) is $3^1x^5y^3z^1$ or simply $3x^5y^3z$.

Example 2 — **Finding the Least Common Denominator of Rational Expressions**

Find the LCD for each pair of rational expressions.

a. $\dfrac{a+b}{a^2-25}; \dfrac{1}{2a-10}$

b. $\dfrac{x-5}{x^2-2x}; \dfrac{1}{x^2-4x+4}$

Solution:

a. $\dfrac{a+b}{a^2-25}; \dfrac{1}{2a-10}$

$= \dfrac{a+b}{(a-5)(a+5)}; \dfrac{1}{2(a-5)}$ Factor the denominators.

The LCD is $2(a-5)(a+5)$.   The LCD is the product of unique factors, each raised to its highest power.

b. $\dfrac{x-5}{x^2-2x}; \dfrac{1}{x^2-4x+4}$

$= \dfrac{x-5}{x(x-2)}; \dfrac{1}{(x-2)^2}$ Factor the denominators.

The LCD is $x(x-2)^2$.   The LCD is the product of unique factors, each raised to its highest power.

Skill Practice

Find the LCD.

3. $\dfrac{x}{x^2-16}; \dfrac{2}{3x+12}$

4. $\dfrac{6}{t^2+5t-14};$

$\dfrac{8}{t^2-3t+2}$

Classroom Examples: p. 518, Exercises 26 and 28

Concept Connections

5. Nadia thought that the LCD for the rational expressions $\frac{5}{x}$ and $\frac{1}{x+3}$ should be $x+3$. Explain why this is not the LCD. What is the correct LCD?

2. Writing Rational Expressions with the Least Common Denominator

To add or subtract two rational expressions, the expressions must have the same denominator. Therefore, we must first practice the skill of converting each rational expression into an equivalent expression with the LCD as its denominator. The process is as follows: Identify the LCD for the two expressions. Then, multiply the numerator and denominator of each fraction by the factors from the LCD that are missing from the original denominators.

Answers

3. $3(x-4)(x+4)$
4. $(t+7)(t-2)(t-1)$
5. The unique factors that appear in the denominators are x and $x+3$, both of which are needed in the LCD. The correct LCD is $x(x+3)$.

Example 3 **Converting to the Least Common Denominator**

Find the LCD of each pair of rational expressions. Then convert each expression to an equivalent fraction with the denominator equal to the LCD.

a. $\dfrac{3}{2ab}$; $\dfrac{6}{5a^2}$ **b.** $\dfrac{4}{x+1}$; $\dfrac{7}{x-4}$

Solution:

a. $\dfrac{3}{2ab}$; $\dfrac{6}{5a^2}$ The LCD is $10a^2b$.

$$\frac{3}{2ab} = \frac{3 \cdot 5a}{2ab \cdot 5a} = \frac{15a}{10a^2b}$$

The first expression is missing the factor $5a$ from the denominator.

$$\frac{6}{5a^2} = \frac{6 \cdot 2b}{5a^2 \cdot 2b} = \frac{12b}{10a^2b}$$

The second expression is missing the factor $2b$ from the denominator.

b. $\dfrac{4}{x+1}$; $\dfrac{7}{x-4}$ The LCD is $(x+1)(x-4)$.

$$\frac{4}{x+1} = \frac{4(x-4)}{(x+1)(x-4)} = \frac{4x-16}{(x+1)(x-4)}$$

The first expression is missing the factor $(x-4)$ from the denominator.

$$\frac{7}{x-4} = \frac{7(x+1)}{(x-4)(x+1)} = \frac{7x+7}{(x-4)(x+1)}$$

The second expression is missing the factor $(x+1)$ from the denominator.

Example 4 **Converting to the Least Common Denominator**

Find the LCD of each pair of rational expressions. Then convert each expression to an equivalent fraction with the denominator equal to the LCD.

$$\frac{w+2}{w^2-w-12}; \frac{1}{w^2-9}$$

Solution:

$$\frac{w+2}{w^2-w-12}; \frac{1}{w^2-9}$$

To find the LCD, factor each denominator.

$$\frac{w+2}{(w-4)(w+3)}; \frac{1}{(w-3)(w+3)}$$

The LCD is $(w-4)(w+3)(w-3)$.

$$\frac{w+2}{(w-4)(w+3)} = \frac{(w+2)(w-3)}{(w-4)(w+3)(w-3)}$$

The first expression is missing the factor $(w-3)$ from the denominator.

$$= \frac{w^2-w-6}{(w-4)(w+3)(w-3)}$$

$$\frac{1}{(w-3)(w+3)} = \frac{1(w-4)}{(w-3)(w+3)(w-4)}$$

The second expression is missing the factor $(w-4)$ from the denominator.

$$= \frac{w-4}{(w-3)(w+3)(w-4)}$$

Example 5 Converting to the Least Common Denominator

Find the LCD of the expressions $\dfrac{3}{x-7}$ and $\dfrac{1}{7-x}$.

Solution:

Notice that the expressions $x-7$ and $7-x$ are opposites and differ by a factor of -1. Therefore, we may use either $x-7$ or $7-x$ as a common denominator. Each case is shown below.

Converting to the Denominator $x-7$

$$\dfrac{3}{x-7};\ \dfrac{1}{7-x}$$ Leave the first fraction unchanged because it has the desired LCD.

$$\dfrac{1}{7-x}=\dfrac{(-1)1}{(-1)(7-x)}$$ Multiply the *second* rational expression by the ratio $\frac{-1}{-1}$ to change its denominator to $x-7$.

$$=\dfrac{-1}{-7+x}$$ Apply the distributive property.

$$=\dfrac{-1}{x-7}$$

Converting to the Denominator $7-x$

$$\dfrac{3}{x-7};\ \dfrac{1}{7-x}$$ Leave the second fraction unchanged because it has the desired LCD.

$$\dfrac{3}{x-7}=\dfrac{(-1)3}{(-1)(x-7)};$$ Multiply the *first* rational expression by the ratio $\frac{-1}{-1}$ to change its denominator to $7-x$.

$$=\dfrac{-3}{-x+7}$$ Apply the distributive property.

$$=\dfrac{-3}{7-x}$$

Skill Practice

9. a. Find the LCD of the expressions.

$$\dfrac{9}{w-2};\ \dfrac{11}{2-w}$$

b. Convert each expression to an equivalent fraction with denominator equal to the LCD.

Classroom Example: p. 519, Exercise 48

TIP: In Example 5, the expressions

$$\dfrac{3}{x-7} \quad\text{and}\quad \dfrac{1}{7-x}$$

have opposite factors in the denominators. In such a case, you do not need to include *both* factors in the LCD.

Answers

9. a. The LCD is $(w-2)$ or $(2-w)$.

b. $\dfrac{9}{w-2}=\dfrac{9}{w-2};$

$\dfrac{11}{2-w}=\dfrac{-11}{w-2}$

or

$\dfrac{9}{w-2}=\dfrac{-9}{2-w};$

$\dfrac{11}{2-w}=\dfrac{11}{2-w}$

Section 7.3 Practice Exercises

Boost *your* GRADE at ALEKS.com!   **ALEKS®** version 3.0
- Practice Problems
- Self-Tests
- NetTutor
- e-Professors
- Videos

For additional exercises, see Classroom Activities 7.3A–7.3C in the *Instructor's Resource Manual* at www.mhhe.com/moh.

Study Skills Exercise

1. Define the key term **least common denominator**.

Review Exercises

2. Evaluate the expression for the given values of x. $\dfrac{2x}{x+5}$

    a. $x=1$ **b.** $x=5$ **c.** $x=-5$

    $\dfrac{1}{3}$ 1   Undefined

 Writing   Translating Expression   Geometry   Scientific Calculator   Video   NS&E

For Exercises 3–4, write the domain in set-builder notation. Then reduce the expression to lowest terms.

3. $\dfrac{3x + 3}{5x^2 - 5}$ $\{x \mid x \text{ is a real number and } x \neq 1, x \neq -1\};\ \dfrac{3}{5(x - 1)}$ **4.** $\dfrac{x + 2}{x^2 - 3x - 10}$ $\{x \mid x \text{ is a real number and } x \neq -2, x \neq 5\};\ \dfrac{1}{x - 5}$

For Exercises 5–8, multiply or divide as indicated.

5. $\dfrac{a + 3}{a + 7} \cdot \dfrac{a^2 + 3a - 10}{a^2 + a - 6}$ $\dfrac{a + 5}{a + 7}$ **6.** $\dfrac{6(a + 2b)}{2(a - 3b)} \cdot \dfrac{4(a + 3b)(a - 3b)}{9(a + 2b)(a - 2b)}$ $\dfrac{4(a + 3b)}{3(a - 2b)}$

7. $\dfrac{16y^2}{9y + 36} \div \dfrac{8y^3}{3y + 12}$ $\dfrac{2}{3y}$ **8.** $\dfrac{5b^2 + 6b + 1}{b^2 + 5b + 6} \div (5b + 1)$ $\dfrac{b + 1}{(b + 2)(b + 3)}$

9. Which of the expressions are equivalent to $-\dfrac{5}{x - 3}$? Circle all that apply.

a. $\dfrac{-5}{x - 3}$ **b.** $\dfrac{5}{-x + 3}$ **c.** $\dfrac{5}{3 - x}$ **d.** $\dfrac{5}{-(x - 3)}$ a, b, c, d

10. Which of the expressions are equivalent to $\dfrac{4 - a}{6}$? Circle all that apply.

a. $\dfrac{a - 4}{-6}$ **b.** $\dfrac{a - 4}{6}$ **c.** $\dfrac{-(4 - a)}{-6}$ **d.** $-\dfrac{a - 4}{6}$ a, c, d

Objective 1: Least Common Denominator

11. Explain why the least common denominator of $\frac{1}{x^3}$, $\frac{1}{x^5}$, and $\frac{1}{x^4}$ is x^5.
x^5 is the greatest power of x that appears in any denominator.

12. Explain why the least common denominator of $\frac{2}{y^3}$, $\frac{9}{y^6}$, and $\frac{4}{y^5}$ is y^6.
y^6 is the greatest power of y that appears in any denominator.

For Exercises 13–30, identify the LCD. (See Examples 1–2.)

13. $\dfrac{4}{15}; \dfrac{5}{9}$ 45 **14.** $\dfrac{7}{12}; \dfrac{1}{18}$ 36 **15.** $\dfrac{1}{16}; \dfrac{1}{4}; \dfrac{1}{6}$ 48

16. $\dfrac{1}{2}; \dfrac{11}{12}; \dfrac{3}{8}$ 24 **17.** $\dfrac{1}{7}; \dfrac{2}{9}$ 63 **18.** $\dfrac{2}{3}; \dfrac{5}{8}$ 24

19. $\dfrac{1}{3x^2y}; \dfrac{8}{9xy^3}$ $9x^2y^3$ **20.** $\dfrac{5}{2a^4b^2}; \dfrac{1}{8ab^3}$ $8a^4b^3$ **21.** $\dfrac{6}{w^2}; \dfrac{7}{y}$ w^2y

22. $\dfrac{2}{r}; \dfrac{3}{s^2}$ rs^2 **23.** $\dfrac{p}{(p + 3)(p - 1)}; \dfrac{2}{(p + 3)(p + 2)}$ $(p + 3)(p - 1)(p + 2)$ **24.** $\dfrac{6}{(q + 4)(q - 4)}; \dfrac{q^2}{(q + 1)(q + 4)}$ $(q + 4)(q - 4)(q + 1)$

25. $\dfrac{7}{3t(t + 1)}; \dfrac{10t}{9(t + 1)^2}$ $9t(t + 1)^2$ **26.** $\dfrac{13x}{15(x - 1)^2}; \dfrac{5}{3x(x - 1)}$ $15x(x - 1)^2$ **27.** $\dfrac{y}{y^2 - 4}; \dfrac{3y}{y^2 + 5y + 6}$ $(y - 2)(y + 2)(y + 3)$

28. $\dfrac{4}{w^2 - 3w + 2}; \dfrac{w}{w^2 - 4}$ $(w - 1)(w - 2)(w + 2)$ **29.** $\dfrac{5}{3 - x}; \dfrac{7}{x - 3}$ $3 - x$ or $x - 3$ **30.** $\dfrac{4}{x - 6}; \dfrac{9}{6 - x}$ $x - 6$ or $6 - x$

31. Explain why a common denominator of

$$\frac{b+1}{b-1} \quad \text{and} \quad \frac{b}{1-b}$$

could be either $(b-1)$ or $(1-b)$.

Because $(b-1)$ and $(1-b)$ are opposites; they differ by a factor of -1.

32. Explain why a common denominator of

$$\frac{1}{6-t} \quad \text{and} \quad \frac{t}{t-6}$$

could be either $(6-t)$ or $(t-6)$.

Because $(6-t)$ and $(t-6)$ are opposites; they differ by a factor of -1.

Objective 2: Writing Rational Expressions with the Least Common Denominator

For Exercises 33–56, find the LCD. Then convert each expression to an equivalent expression with the denominator equal to the LCD. **(See Examples 3–5.)**

33. $\dfrac{6}{5x^2}; \dfrac{1}{x}$ $\dfrac{6}{5x^2}, \dfrac{5x}{5x^2}$

34. $\dfrac{3}{y}; \dfrac{7}{9y^2}$ $\dfrac{27y}{9y^2}, \dfrac{7}{9y^2}$

35. $\dfrac{4}{5x^2}; \dfrac{y}{6x^3}$ $\dfrac{24x}{30x^3}, \dfrac{5y}{30x^3}$

36. $\dfrac{3}{15b^2}; \dfrac{c}{3b^2}$ $\dfrac{3}{15b^2}, \dfrac{5c}{15b^2}$

37. $\dfrac{5}{6a^2b}; \dfrac{a}{12b}$ $\dfrac{10}{12a^2b}, \dfrac{a^3}{12a^2b}$

38. $\dfrac{x}{15y^2}; \dfrac{y}{5xy}$ $\dfrac{x^2}{15xy^2}, \dfrac{3y^2}{15xy^2}$

39. $\dfrac{6}{m+4}; \dfrac{3}{m-1}$ $\dfrac{6m-6}{(m+4)(m-1)}, \dfrac{3m+12}{(m+4)(m-1)}$

40. $\dfrac{3}{n-5}; \dfrac{7}{n+2}$ $\dfrac{3n+6}{(n-5)(n+2)}, \dfrac{7n-35}{(n-5)(n+2)}$

41. $\dfrac{6}{2x-5}; \dfrac{1}{x+3}$ $\dfrac{6x+18}{(2x-5)(x+3)}, \dfrac{2x-5}{(2x-5)(x+3)}$

42. $\dfrac{4}{m+3}; \dfrac{-3}{5m+1}$ $\dfrac{20m+4}{(m+3)(5m+1)}, \dfrac{-3m-9}{(m+3)(5m+1)}$

43. $\dfrac{6}{(w+3)(w-8)}; \dfrac{w}{(w-8)(w+1)}$ $\dfrac{6w+6}{(w+3)(w-8)(w+1)}, \dfrac{w^2+3w}{(w+3)(w-8)(w+1)}$

44. $\dfrac{t}{(t+2)(t+12)}; \dfrac{18}{(t-2)(t+2)}$ $\dfrac{t^2-2t}{(t+2)(t-2)(t+12)}, \dfrac{18t+216}{(t+2)(t-2)(t+12)}$

45. $\dfrac{6p}{p^2-4}; \dfrac{3}{p^2+4p+4}$ $\dfrac{6p^2+12p}{(p-2)(p+2)^2}, \dfrac{3p-6}{(p-2)(p+2)^2}$

46. $\dfrac{5}{q^2-6q+9}; \dfrac{q}{q^2-9}$ $\dfrac{5q+15}{(q-3)^2(q+3)}, \dfrac{q^2-3q}{(q-3)^2(q+3)}$

47. $\dfrac{1}{a-4}; \dfrac{a}{4-a}$ $\dfrac{1}{a-4}, \dfrac{-a}{a-4}$ or $\dfrac{-1}{4-a}; \dfrac{a}{4-a}$

48. $\dfrac{3b}{2b-5}; \dfrac{2b}{5-2b}$ $\dfrac{3b}{2b-5}, \dfrac{-2b}{2b-5}$ or $\dfrac{-3b}{5-2b}; \dfrac{2b}{5-2b}$

49. $\dfrac{4}{x-7}; \dfrac{y}{14-2x}$ $\dfrac{8}{2(x-7)}, \dfrac{-y}{2(x-7)}$ or $\dfrac{-8}{2(7-x)}; \dfrac{y}{2(7-x)}$

50. $\dfrac{4}{3x-15}; \dfrac{z}{5-x}$ $\dfrac{4}{3(x-5)}, \dfrac{-3z}{3(x-5)}$ or $\dfrac{-4}{3(5-x)}; \dfrac{3z}{3(5-x)}$

51. $\dfrac{1}{a+b}; \dfrac{6}{-a-b}$ $\dfrac{1}{a+b}, \dfrac{-6}{a+b}$ or $\dfrac{-1}{-a-b}; \dfrac{6}{-a-b}$

52. $\dfrac{p}{-q-8}; \dfrac{1}{q+8}$ $\dfrac{p}{-q-8}, \dfrac{-1}{-q-8}$ or $\dfrac{-p}{q+8}; \dfrac{1}{q+8}$

53. $\dfrac{-3}{24y+8}; \dfrac{5}{18y+6}$ $\dfrac{-9}{24(3y+1)}, \dfrac{20}{24(3y+1)}$

54. $\dfrac{r}{10r+5}; \dfrac{2}{16r+8}$ $\dfrac{8r}{40(2r+1)}, \dfrac{10}{40(2r+1)}$

55. $\dfrac{3}{5z}; \dfrac{1}{z+4}$ $\dfrac{3z+12}{5z(z+4)}, \dfrac{5z}{5z(z+4)}$

56. $\dfrac{-1}{4a-8}; \dfrac{5}{4a}$ $\dfrac{-a}{4a(a-2)}, \dfrac{5a-10}{4a(a-2)}$

Expanding Your Skills

For Exercises 57–60, find the LCD. Then convert each expression to an equivalent expression with the denominator equal to the LCD.

57. $\dfrac{z}{z^2+9z+14}; \dfrac{-3z}{z^2+10z+21}; \dfrac{5}{z^2+5z+6}$ $\dfrac{z^2+3z}{(z+2)(z+7)(z+3)}, \dfrac{-3z^2-6z}{(z+2)(z+7)(z+3)}, \dfrac{5z+35}{(z+2)(z+7)(z+3)}$

58. $\dfrac{6}{w^2-3w-4}; \dfrac{1}{w^2+6w+5}; \dfrac{-9w}{w^2+w-20}$ $\dfrac{6w+30}{(w-4)(w+1)(w+5)}, \dfrac{w-4}{(w-4)(w+1)(w+5)}, \dfrac{-9w^2-9w}{(w-4)(w+1)(w+5)}$

59. $\dfrac{3}{p^3-8}; \dfrac{p}{p^2-4}; \dfrac{5p}{p^2+2p+4}$ $\dfrac{3p+6}{(p-2)(p^2+2p+4)(p+2)}, \dfrac{p^3+2p^2+4p}{(p-2)(p^2+2p+4)(p+2)}, \dfrac{5p^3-20p}{(p-2)(p^2+2p+4)(p+2)}$

60. $\dfrac{7}{q^3+125}; \dfrac{q}{q^2-25}; \dfrac{12}{q^2-5q+25}$ $\dfrac{7q-35}{(q+5)(q^2-5q+25)(q-5)}, \dfrac{q^3-5q^2+25q}{(q+5)(q^2-5q+25)(q-5)}, \dfrac{12q^2-300}{(q+5)(q^2-5q+25)(q-5)}$

 Writing   Translating Expression   Geometry   Scientific Calculator   Video   NSE

Section 7.4 — Addition and Subtraction of Rational Expressions

Objectives

1. Addition and Subtraction of Rational Expressions with the Same Denominator
2. Addition and Subtraction of Rational Expressions with Different Denominators
3. Using Rational Expressions in Translations

1. Addition and Subtraction of Rational Expressions with the Same Denominator

To add or subtract rational expressions, the expressions must have the same denominator. As with fractions, we add or subtract rational expressions with the same denominator by combining the terms in the numerator and then writing the result over the common denominator. Then, if possible, we simplify the expression to lowest terms.

> **PROPERTY Addition and Subtraction of Rational Expressions**
>
> Let p, q, and r represent polynomials where $q \neq 0$. Then,
>
> **1.** $\dfrac{p}{q} + \dfrac{r}{q} = \dfrac{p + r}{q}$ **2.** $\dfrac{p}{q} - \dfrac{r}{q} = \dfrac{p - r}{q}$

Skill Practice

Add or subtract as indicated.

1. $\dfrac{3}{14} + \dfrac{4}{14}$

2. $\dfrac{2}{7d} - \dfrac{9}{7d}$

Classroom Examples: p. 526, Exercises 8 and 12

Example 1 Adding and Subtracting Rational Expressions with a Common Denominator

Add or subtract as indicated.

a. $\dfrac{1}{12} + \dfrac{7}{12}$ **b.** $\dfrac{2}{5p} - \dfrac{7}{5p}$

Solution:

a. $\dfrac{1}{12} + \dfrac{7}{12}$ The fractions have the same denominator.

$= \dfrac{1 + 7}{12}$ Add the terms in the numerators, and write the result over the common denominator.

$= \dfrac{8}{12}$

$= \dfrac{2}{3}$ Simplify to lowest terms.

b. $\dfrac{2}{5p} - \dfrac{7}{5p}$ The rational expressions have the same denominator.

$= \dfrac{2 - 7}{5p}$ Subtract the terms in the numerators, and write the result over the common denominator.

$= \dfrac{-5}{5p}$

$= \dfrac{(-\overset{-1}{\cancel{5}})}{\cancel{5}p}$ Simplify to lowest terms.

$= -\dfrac{1}{p}$

Answers

1. $\dfrac{1}{2}$ **2.** $-\dfrac{1}{d}$

Example 2 **Adding and Subtracting Rational Expressions with a Common Denominator**

Add or subtract as indicated.

a. $\dfrac{2}{3d+5} + \dfrac{7d}{3d+5}$ **b.** $\dfrac{x^2}{x-3} - \dfrac{-5x+24}{x-3}$

Solution:

a. $\dfrac{2}{3d+5} + \dfrac{7d}{3d+5}$ The rational expressions have the same denominator.

$= \dfrac{2+7d}{3d+5}$ Add the terms in the numerators, and write the result over the common denominator.

$= \dfrac{7d+2}{3d+5}$ Because the numerator and denominator share no common factors, the expression is in lowest terms.

b. $\dfrac{x^2}{x-3} - \dfrac{-5x+24}{x-3}$ The rational expressions have the same denominator.

$= \dfrac{x^2 - (-5x+24)}{x-3}$ Subtract the terms in the numerators, and write the result over the common denominator.

$= \dfrac{x^2 + 5x - 24}{x-3}$ Simplify the numerator.

$= \dfrac{(x+8)(x-3)}{(x-3)}$ Factor the numerator and denominator to determine if the rational expression can be simplified.

$= \dfrac{(x+8)\overset{1}{\cancel{(x-3)}}}{\cancel{(x-3)}}$ Simplify to lowest terms.

$= x + 8$

Skill Practice

Add or subtract as indicated.

3. $\dfrac{x^2+2}{x+3} + \dfrac{4x+1}{x+3}$

4. $\dfrac{4t-9}{2t+1} - \dfrac{t-5}{2t+1}$

Classroom Example: p. 526, Exercise 18

Avoiding Mistakes

When subtracting rational expressions, use parentheses to group the terms in the numerator that follow the subtraction sign. This will help you remember to apply the distributive property.

Instructor Note: Have students consider the alternative method for subtraction. Rewrite as addition and change the signs of the numerator in the second fraction.

$\dfrac{x^2}{x-3} - \dfrac{-5x+24}{x-3}$ becomes

$\dfrac{x^2}{x-3} + \dfrac{5x-24}{x-3}$

2. Addition and Subtraction of Rational Expressions with Different Denominators

To add or subtract two rational expressions with unlike denominators, we must convert the expressions to equivalent expressions with the same denominator. For example, consider adding

$$\frac{1}{10} + \frac{12}{5y}$$

The LCD is $10y$. For each expression, identify the factors from the LCD that are missing from the denominator. Then multiply the numerator and denominator of the expression by the missing factor(s).

$$\underbrace{\frac{1}{10}}_{\substack{\text{Missing} \\ y}} + \underbrace{\frac{12}{5y}}_{\substack{\text{Missing} \\ 2}}$$

Answers

3. $x+1$ **4.** $\dfrac{3t-4}{2t+1}$

$$= \frac{1 \cdot y}{10 \cdot y} + \frac{12 \cdot 2}{5y \cdot 2}$$

$$= \frac{y}{10y} + \frac{24}{10y}$$ The rational expressions now have the same denominators.

Avoiding Mistakes

In the expression $\frac{y+24}{10y}$, notice that you cannot reduce the 24 and 10 because 24 is not a factor in the numerator, it is a term. Only factors can be reduced—not terms.

$$= \frac{y + 24}{10y}$$ Add the numerators.

After successfully adding or subtracting two rational expressions, always check to see if the final answer is simplified. If necessary, factor the numerator and denominator, and reduce common factors. The expression

$$\frac{y + 24}{10y}$$

is in lowest terms because the numerator and denominator do not share any common factors.

> **PROCEDURE** **Adding or Subtracting Rational Expressions**
> **Step 1** Factor the denominators of each rational expression.
> **Step 2** Identify the LCD.
> **Step 3** Rewrite each rational expression as an equivalent expression with the LCD as its denominator.
> **Step 4** Add or subtract the numerators, and write the result over the common denominator.
> **Step 5** Simplify to lowest terms.

Skill Practice

Add.

5. $\dfrac{4}{3x} + \dfrac{1}{2x^2}$

Classroom Example: p. 527, Exercise 32

Avoiding Mistakes

Do not reduce after rewriting the fractions with the LCD. You will revert back to the original expression.

Example 3 **Subtracting Rational Expressions with Different Denominators**

Subtract. $\dfrac{4}{7k} - \dfrac{3}{k^2}$

Solution:

$$\frac{4}{7k} - \frac{3}{k^2}$$ **Step 1:** The denominators are already factored.

Step 2: The LCD is $7k^2$.

$$= \frac{4 \cdot k}{7k \cdot k} - \frac{3 \cdot 7}{k^2 \cdot 7}$$ **Step 3:** Write each expression with the LCD.

$$= \frac{4k}{7k^2} - \frac{21}{7k^2}$$

$$= \frac{4k - 21}{7k^2}$$ **Step 4:** Subtract the numerators, and write the result over the LCD.

Answer

5. $\dfrac{8x + 3}{6x^2}$

Step 5: The expression is in lowest terms because the numerator and denominator share no common factors.

Example 4 **Subtracting Rational Expressions with Different Denominators**

Subtract. $\dfrac{2q - 4}{3} - \dfrac{q + 1}{2}$

Solution:

$\dfrac{2q - 4}{3} - \dfrac{q + 1}{2}$ **Step 1:** The denominators are already factored.

                           Step 2: The LCD is 6.

$= \dfrac{2(2q - 4)}{2 \cdot 3} - \dfrac{3(q + 1)}{3 \cdot 2}$ **Step 3:** Write each expression with the LCD.

$= \dfrac{2(2q - 4) - 3(q + 1)}{6}$ **Step 4:** Subtract the numerators, and write the result over the LCD.

$= \dfrac{4q - 8 - 3q - 3}{6}$

$= \dfrac{q - 11}{6}$ **Step 5:** The expression is in lowest terms because the numerator and denominator share no common factors.

Example 5 **Adding Rational Expressions with Different Denominators**

Add. $\dfrac{1}{x - 5} + \dfrac{-10}{x^2 - 25}$

Solution:

$\dfrac{1}{x - 5} + \dfrac{-10}{x^2 - 25}$

$= \dfrac{1}{x - 5} + \dfrac{-10}{(x - 5)(x + 5)}$ **Step 1:** Factor the denominators.

                           Step 2: The LCD is $(x - 5)(x + 5)$.

$= \dfrac{1(x + 5)}{(x - 5)(x + 5)} + \dfrac{-10}{(x - 5)(x + 5)}$ **Step 3:** Write each expression with the LCD.

$= \dfrac{1(x + 5) + (-10)}{(x - 5)(x + 5)}$ **Step 4:** Add the numerators, and write the result over the LCD.

$= \dfrac{x + 5 - 10}{(x - 5)(x + 5)}$

$= \dfrac{\overset{1}{\cancel{x - 5}}}{\cancel{(x - 5)}(x + 5)}$ **Step 5:** Simplify.

$= \dfrac{1}{x + 5}$

Skill Practice

Subtract.

8. $\dfrac{2y}{y-1} - \dfrac{1}{y} - \dfrac{2y+1}{y^2-y}$

Classroom Example: p. 527, Exercise 68

Example 6 **Adding and Subtracting Rational Expressions with Different Denominators**

Add or subtract as indicated. $\quad \dfrac{p+2}{p-1} - \dfrac{2}{p+6} - \dfrac{14}{p^2+5p-6}$

Solution:

$$\dfrac{p+2}{p-1} - \dfrac{2}{p+6} - \dfrac{14}{p^2+5p-6}$$

$$= \dfrac{p+2}{p-1} - \dfrac{2}{p+6} - \dfrac{14}{(p-1)(p+6)} \qquad \textbf{Step 1:} \quad \text{Factor the denominators.}$$

Step 2: The LCD is $(p-1)(p+6)$.

Step 3: Write each expression with the LCD.

$$= \dfrac{(p+2)(p+6)}{(p-1)(p+6)} - \dfrac{2(p-1)}{(p+6)(p-1)} - \dfrac{14}{(p-1)(p+6)}$$

$$= \dfrac{(p+2)(p+6) - 2(p-1) - 14}{(p-1)(p+6)} \qquad \textbf{Step 4:} \quad \text{Combine the numerators, and write the result over the LCD.}$$

$$= \dfrac{p^2 + 6p + 2p + 12 - 2p + 2 - 14}{(p-1)(p+6)} \qquad \textbf{Step 5:} \quad \text{Clear parentheses in the numerator.}$$

$$= \dfrac{p^2 + 6p}{(p-1)(p+6)} \qquad\qquad\qquad \text{Combine } like \text{ terms.}$$

$$= \dfrac{p(p+6)}{(p-1)(p+6)} \qquad\qquad\qquad \text{Factor the numerator to determine if the expression is in lowest terms.}$$

$$= \dfrac{p(p \overset{1}{\cancel{+ 6}})}{(p-1)(p \cancel{+ 6})} \qquad\qquad\qquad \text{Simplify to lowest terms.}$$

$$= \dfrac{p}{p-1}$$

When the denominators of two rational expressions are opposites, we can produce identical denominators by multiplying one of the expressions by the ratio $\frac{-1}{-1}$. This is demonstrated in the next example.

Answer

8. $\dfrac{2y-3}{y-1}$

Example 7   **Adding Rational Expressions with Different Denominators**

Add the rational expressions. $\dfrac{1}{d-7} + \dfrac{5}{7-d}$

Solution:

$\dfrac{1}{d-7} + \dfrac{5}{7-d}$ The expressions $d-7$ and $7-d$ are opposites and differ by a factor of -1. Therefore, multiply the numerator and denominator of *either* expression by -1 to obtain a common denominator.

$= \dfrac{1}{d-7} + \dfrac{(-1)5}{(-1)(7-d)}$ Note that $-1(7-d) = -7 + d$ or $d - 7$.

$= \dfrac{1}{d-7} + \dfrac{-5}{d-7}$ Simplify.

$= \dfrac{1 + (-5)}{d-7}$ Add the terms in the numerators, and write the result over the common denominator.

$= \dfrac{-4}{d-7}$

Classroom Example: p. 527, Exercise 50

Skill Practice

Add or subtract as indicated.

9. $\dfrac{3}{p-8} + \dfrac{1}{8-p}$

---

3. Using Rational Expressions in Translations

Example 8   **Using Rational Expressions in Translations**

Translate the English phrase into a mathematical expression. Then simplify by combining the rational expressions.

The difference of the reciprocal of n and the quotient of n and 3

Solution:

The difference of the reciprocal of n and the quotient of n and 3

The difference of

$$\left(\dfrac{1}{n}\right) - \left(\dfrac{n}{3}\right)$$

The reciprocal of n The quotient of n and 3

$\dfrac{1}{n} - \dfrac{n}{3}$ The LCD is $3n$.

$= \dfrac{3 \cdot 1}{3 \cdot n} - \dfrac{n \cdot n}{3 \cdot n}$ Write each expression over the LCD.

$= \dfrac{3 - n^2}{3n}$ Subtract the numerators.

Skill Practice

Translate the English phrase into a mathematical expression. Then simplify by combining the rational expressions.

10. The sum of 1 and the quotient of 2 and a

Classroom Example: p. 528, Exercise 80

Answers

9. $\dfrac{2}{p-8}$ or $\dfrac{-2}{8-p}$

10. $1 + \dfrac{2}{a}; \dfrac{a+2}{a}$

Section 7.4 Practice Exercises

Boost _your_ GRADE at ALEKS.com!

- Practice Problems
- Self-Tests
- NetTutor

- e-Professors
- Videos

For additional exercises, see Classroom Activities 7.4A–7.4B in the _Instructor's Resource Manual_ at www.mhhe.com/moh.

Review Exercises

1. For the rational expression $\dfrac{x^2 - 4x - 5}{x^2 - 7x + 10}$

a. Find the value of the expression (if possible) when $x = 0, 1, -1, 2,$ and 5. $-\dfrac{1}{2}, -2, 0,$ undefined, undefined

b. Factor the denominator and identify the domain. Write the domain in set-builder notation.
$(x - 5)(x - 2); \{x \,|\, x \text{ is a real number and } x \neq 5, x \neq 2\}$

c. Reduce the expression to lowest terms.
$\dfrac{x + 1}{x - 2}$

2. For the rational expression $\dfrac{a^2 + a - 2}{a^2 - 4a - 12}$

a. Find the value of the expression (if possible) when $a = 0, 1, -2, 2,$ and 6. $\dfrac{1}{6}, 0,$ undefined, $-\dfrac{1}{4},$ undefined

b. Factor the denominator, and identify the domain. Write the domain in set-builder notation.
$(a - 6)(a + 2); \{a \,|\, a \text{ is a real number and } a \neq 6, a \neq -2\}$

c. Reduce the expression to lowest terms.
$\dfrac{a - 1}{a - 6}$

For Exercises 3–4, multiply or divide as indicated.

3. $\dfrac{2b^2 - b - 3}{2b^2 - 3b - 9} \div \dfrac{b^2 - 1}{4b + 6}$ $\dfrac{2(2b - 3)}{(b - 3)(b - 1)}$

4. $\dfrac{6t - 1}{5t - 30} \cdot \dfrac{10t - 25}{2t^2 - 3t - 5}$ $\dfrac{6t - 1}{(t + 1)(t - 6)}$

Objective 1: Addition and Subtraction of Rational Expressions with the Same Denominator

For Exercises 5–26, add or subtract the expressions with like denominators as indicated. **(See Examples 1–2.)**

5. $\dfrac{7}{8} + \dfrac{3}{8}$ $\dfrac{5}{4}$

6. $\dfrac{1}{3} + \dfrac{7}{3}$ $\dfrac{8}{3}$

7. $\dfrac{9}{16} - \dfrac{3}{16}$ $\dfrac{3}{8}$

8. $\dfrac{14}{15} - \dfrac{4}{15}$ $\dfrac{2}{3}$

9. $\dfrac{5a}{a + 2} - \dfrac{3a - 4}{a + 2}$ 2

10. $\dfrac{2b}{b - 3} - \dfrac{b - 9}{b - 3}$ $\dfrac{b + 9}{b - 3}$

11. $\dfrac{5c}{c + 6} + \dfrac{30}{c + 6}$ 5

12. $\dfrac{12}{2 + d} + \dfrac{6d}{2 + d}$ 6

13. $\dfrac{5}{t - 8} - \dfrac{2t + 1}{t - 8}$ $\dfrac{-2(t - 2)}{t - 8}$

14. $\dfrac{7p + 1}{2p + 1} - \dfrac{p - 4}{2p + 1}$ $\dfrac{6p + 5}{2p + 1}$

15. $\dfrac{9x^2}{3x - 7} - \dfrac{49}{3x - 7}$ $3x + 7$

16. $\dfrac{4w^2}{2w - 1} - \dfrac{1}{2w - 1}$ $2w + 1$

17. $\dfrac{m^2}{m + 5} + \dfrac{10m + 25}{m + 5}$ $m + 5$

18. $\dfrac{k^2}{k - 3} - \dfrac{6k - 9}{k - 3}$ $k - 3$

19. $\dfrac{2a}{a + 2} + \dfrac{4}{a + 2}$ 2

20. $\dfrac{5b}{b + 4} + \dfrac{20}{b + 4}$ 5

21. $\dfrac{x^2}{x + 5} - \dfrac{25}{x + 5}$ $x - 5$

22. $\dfrac{y^2}{y - 7} - \dfrac{49}{y - 7}$ $y + 7$

23. $\dfrac{r}{r^2 + 3r + 2} + \dfrac{2}{r^2 + 3r + 2}$ $\dfrac{1}{r + 1}$

24. $\dfrac{x}{x^2 - x - 12} - \dfrac{4}{x^2 - x - 12}$ $\dfrac{1}{x + 3}$

25. $\dfrac{1}{3y^2 + 22y + 7} - \dfrac{-3y}{3y^2 + 22y + 7}$ $\dfrac{1}{y + 7}$

26. $\dfrac{5}{2x^2 + 13x + 20} + \dfrac{2x}{2x^2 + 13x + 20}$ $\dfrac{1}{x + 4}$

Writing   Translating Expression   Geometry   Scientific Calculator   Video   NS E

For Exercises 27–28, find an expression that represents the perimeter of the figure (assume that $x > 0$, $y > 0$, and $t > 0$).

27.
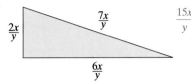

28.
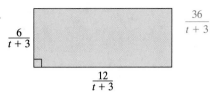

Objective 2: Addition and Subtraction of Rational Expressions with Different Denominators

For Exercises 29–70, add or subtract the expressions with unlike denominators as indicated. **(See Examples 3–7.)**

29. $\dfrac{5}{4} + \dfrac{3}{2a}$ $\dfrac{5a + 6}{4a}$

30. $\dfrac{11}{6p} + \dfrac{-7}{4p}$ $\dfrac{1}{12p}$

31. $\dfrac{4}{5xy^3} + \dfrac{2x}{15y^2}$ $\dfrac{2(6 + x^2y)}{15xy^3}$

32. $\dfrac{5}{3a^2b} + \dfrac{-7}{6b^2}$ $\dfrac{10b - 7a^2}{6a^2b^2}$

33. $\dfrac{2}{s^3t^3} - \dfrac{3}{s^4t}$ $\dfrac{2s - 3t^2}{s^4t^3}$

34. $\dfrac{1}{p^2q} - \dfrac{2}{pq^3}$ $\dfrac{q^2 - 2p}{p^2q^3}$

35. $\dfrac{z}{3z - 9} - \dfrac{z - 2}{z - 3}$ $-\dfrac{2}{3}$

36. $\dfrac{3w - 8}{2w - 4} - \dfrac{w - 3}{w - 2}$ $\dfrac{1}{2}$

37. $\dfrac{5}{a + 1} + \dfrac{4}{3a + 3}$ $\dfrac{19}{3(a + 1)}$

38. $\dfrac{2}{c - 4} + \dfrac{1}{5c - 20}$ $\dfrac{11}{5(c - 4)}$

39. $\dfrac{k}{k^2 - 9} - \dfrac{4}{k - 3}$ $\dfrac{-3(k + 4)}{(k - 3)(k + 3)}$

40. $\dfrac{7}{h + 2} - \dfrac{2h - 3}{h^2 - 4}$ $\dfrac{5h - 11}{(h - 2)(h + 2)}$

41. $\dfrac{3a - 7}{6a + 10} - \dfrac{10}{3a^2 + 5a}$ $\dfrac{a - 4}{2a}$

42. $\dfrac{k + 2}{8k} - \dfrac{3 - k}{12k}$ $\dfrac{5}{24}$

43. $\dfrac{x}{x - 4} + \dfrac{3}{x + 1}$ $\dfrac{(x + 6)(x - 2)}{(x - 4)(x + 1)}$

44. $\dfrac{4}{y - 3} + \dfrac{y}{y - 5}$ $\dfrac{(y + 5)(y - 4)}{(y - 5)(y - 3)}$

45. $\dfrac{6a}{a^2 - b^2} + \dfrac{2a}{a^2 + ab}$ $\dfrac{2(4a - b)}{(a + b)(a - b)}$

46. $\dfrac{7x}{x^2 + 2xy + y^2} + \dfrac{3x}{x^2 + xy}$ $\dfrac{10x + 3y}{(x + y)^2}$

47. $\dfrac{p}{3} - \dfrac{4p - 1}{-3}$ $\dfrac{5p - 1}{3}$ or $\dfrac{-5p + 1}{-3}$

48. $\dfrac{r}{7} - \dfrac{r - 5}{-7}$ $\dfrac{2r - 5}{7}$ or $\dfrac{-2r + 5}{-7}$

49. $\dfrac{4n}{n - 8} - \dfrac{2n - 1}{8 - n}$ $\dfrac{6n - 1}{n - 8}$ or $\dfrac{-6n + 1}{8 - n}$

50. $\dfrac{m}{m - 2} - \dfrac{3m + 1}{2 - m}$ $\dfrac{4m + 1}{m - 2}$ or $\dfrac{-4m - 1}{2 - m}$

51. $\dfrac{5}{x} + \dfrac{3}{x + 2}$ $\dfrac{2(4x + 5)}{x(x + 2)}$

52. $\dfrac{6}{y - 1} + \dfrac{9}{y}$ $\dfrac{3(5y - 3)}{y(y - 1)}$

53. $\dfrac{5}{p - 3} - \dfrac{2}{p - 1}$ $\dfrac{3p + 1}{(p - 3)(p - 1)}$

54. $\dfrac{1}{7x} + \dfrac{5}{2y^2}$ $\dfrac{2y^2 + 35x}{14xy^2}$

55. $\dfrac{y}{4y + 2} + \dfrac{3y}{6y + 3}$ $\dfrac{3y}{2(2y + 1)}$

56. $\dfrac{4}{q^2 - 2q} - \dfrac{5}{3q - 6}$ $\dfrac{12 - 5q}{3q(q - 2)}$

57. $\dfrac{4w}{w^2 + 2w - 3} + \dfrac{2}{1 - w}$ $\dfrac{2(w - 3)}{(w + 3)(w - 1)}$

58. $\dfrac{z - 23}{z^2 - z - 20} - \dfrac{2}{5 - z}$ $\dfrac{3}{z + 4}$

59. $\dfrac{3a - 8}{a^2 - 5a + 6} + \dfrac{a + 2}{a^2 - 6a + 8}$ $\dfrac{4a - 13}{(a - 3)(a - 4)}$

60. $\dfrac{3b + 5}{b^2 + 4b + 3} + \dfrac{-b + 5}{b^2 + 2b - 3}$ $\dfrac{2b}{(b + 1)(b - 1)}$

61. $\dfrac{3x}{x^2 + x - 6} + \dfrac{x}{x^2 + 5x + 6}$ $\dfrac{4x(x + 1)}{(x + 3)(x - 2)(x + 2)}$

62. $\dfrac{x}{x^2 + 5x + 4} - \dfrac{2x}{x^2 - 2x - 3}$ $\dfrac{-x(x + 11)}{(x + 1)(x + 4)(x - 3)}$

63. $\dfrac{3y}{2y^2 - y - 8} - \dfrac{4y}{2y^2 - 7y - 4}$ $\dfrac{-y(y + 8)}{(2y + 1)(y - 1)(y - 4)}$

64. $\dfrac{5}{6y^2 - 7y - 3} + \dfrac{4y}{3y^2 + 4y + 1}$ $\dfrac{8y^2 - 7y + 5}{(3y + 1)(2y - 3)(y + 1)}$

65. $\dfrac{3}{2p - 1} - \dfrac{4p + 4}{4p^2 - 1}$ $\dfrac{1}{2p + 1}$

66. $\dfrac{1}{3q - 2} - \dfrac{6q + 4}{9q^2 - 4}$ $\dfrac{-1}{3q - 2}$

67. $\dfrac{m}{m + n} - \dfrac{m}{m - n} + \dfrac{1}{m^2 - n^2}$ $\dfrac{-2mn + 1}{(m + n)(m - n)}$

68. $\dfrac{x}{x + y} - \dfrac{2xy}{x^2 - y^2} + \dfrac{y}{x - y}$ $\dfrac{x - y}{x + y}$

69. $\dfrac{2}{a + b} + \dfrac{2}{a - b} - \dfrac{4a}{a^2 - b^2}$ 0

70. $\dfrac{-2x}{x^2 - y^2} + \dfrac{1}{x + y} - \dfrac{1}{x - y}$ $\dfrac{-2}{x - y}$

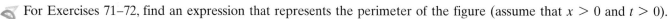

For Exercises 71–72, find an expression that represents the perimeter of the figure (assume that $x > 0$ and $t > 0$).

71.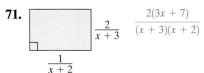
$\dfrac{2}{x+3}$ $\dfrac{2(3x+7)}{(x+3)(x+2)}$

$\dfrac{1}{x+2}$

72.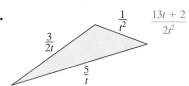
$\dfrac{1}{t^2}$ $\dfrac{13t+2}{2t^2}$

$\dfrac{3}{2t}$ $\dfrac{5}{t}$

Objective 3: Using Rational Expressions in Translations

73. Let a number be represented by n. Write the reciprocal of n. $\dfrac{1}{n}$

74. Write the reciprocal of the sum of a number and 6. $\dfrac{1}{n+6}$

75. Write the quotient of 5 and the sum of a number and 2. $\dfrac{5}{n+2}$

76. Let a number be represented by p. Write the quotient of 12 and p. $\dfrac{12}{p}$

For Exercises 77–80, translate the English phrases into algebraic expressions. Then simplify by combining the rational expressions. **(See Example 8.)**

77. The sum of a number and the quantity seven times the reciprocal of the number. $n+\left(7\cdot\dfrac{1}{n}\right); \dfrac{n^2+7}{n}$

78. The sum of a number and the quantity five times the reciprocal of the number. $n+\left(5\cdot\dfrac{1}{n}\right); \dfrac{n^2+5}{n}$

79. The difference of the reciprocal of n and the quotient of 2 and n. $\dfrac{1}{n}-\dfrac{2}{n}; -\dfrac{1}{n}$

80. The difference of the reciprocal of m and the quotient of $3m$ and 7. $\dfrac{1}{m}-\dfrac{3m}{7}; \dfrac{7-3m^2}{7m}$

Expanding Your Skills

For Exercises 81–86, perform the indicated operations.

81. $\dfrac{-3}{w^3+27}-\dfrac{1}{w^2-9}$ $\dfrac{-w^2}{(w+3)(w-3)(w^2-3w+9)}$

82. $\dfrac{m}{m^3-1}+\dfrac{1}{(m-1)^2}$ $\dfrac{2m^2+1}{(m-1)^2(m^2+m+1)}$

83. $\dfrac{2p}{p^2+5p+6}-\dfrac{p+1}{p^2+2p-3}+\dfrac{3}{p^2+p-2}$ $\dfrac{p^2-2p+7}{(p+2)(p+3)(p-1)}$

84. $\dfrac{3t}{8t^2+2t-1}-\dfrac{5t}{2t^2-9t-5}+\dfrac{2}{4t^2-21t+5}$ $\dfrac{-17t^2-6t+2}{(2t+1)(4t-1)(t-5)}$

85. $\dfrac{3m}{m^2+3m-10}+\dfrac{5}{4-2m}-\dfrac{1}{m+5}$ $\dfrac{-m-21}{2(m+5)(m-2)}$ or $\dfrac{m+21}{2(m+5)(2-m)}$

86. $\dfrac{2n}{3n^2-8n-3}+\dfrac{1}{6-2n}-\dfrac{3}{3n+1}$ $\dfrac{-5n+17}{2(3n+1)(n-3)}$ or $\dfrac{5n-17}{2(3n+1)(3-n)}$

For Exercises 87–90, simplify by applying the order of operations.

87. $\left(\dfrac{2}{k+1}+3\right)\left(\dfrac{k+1}{4k+7}\right)$ $\dfrac{3k+5}{4k+7}$

88. $\left(\dfrac{p+1}{3p+4}\right)\left(\dfrac{1}{p+1}+2\right)$ $\dfrac{2p+3}{3p+4}$

89. $\left(\dfrac{1}{10a}-\dfrac{b}{10a^2}\right)\div\left(\dfrac{1}{10}-\dfrac{b}{10a}\right)$ $\dfrac{1}{a}$

90. $\left(\dfrac{1}{2m}+\dfrac{n}{2m^2}\right)\div\left(\dfrac{1}{4}+\dfrac{n}{4m}\right)$ $\dfrac{2}{m}$

Problem Recognition Exercises

Operations on Rational Expressions

In Sections 7.1–7.4, we learned how to simplify, add, subtract, multiply, and divide rational expressions. The procedure for each operation is different, and it takes considerable practice to determine the correct method to apply for a given problem. The following review exercises give you the opportunity to practice the specific techniques for simplifying rational expressions.

For Exercises 1–20, perform any indicated operations and simplify the expression.

1. $\dfrac{5}{3x+1} - \dfrac{2x-4}{3x+1}$ $\dfrac{-2x+9}{3x+1}$

2. $\dfrac{\dfrac{w+1}{w^2-16}}{\dfrac{w+1}{w+4}}$ $\dfrac{1}{w-4}$

3. $\dfrac{3}{y} \cdot \dfrac{y^2-5y}{6y-9}$ $\dfrac{y-5}{2y-3}$

4. $\dfrac{-1}{x+3} + \dfrac{2}{2x-1}$ $\dfrac{7}{(x+3)(2x-1)}$

5. $\dfrac{x-9}{9x-x^2}$ $-\dfrac{1}{x}$

6. $\dfrac{1}{p} - \dfrac{3}{p^2+3p} + \dfrac{p}{3p+9}$ $\dfrac{1}{3}$

7. $\dfrac{c^2+5c+6}{c^2+c-2} \div \dfrac{c}{c-1}$ $\dfrac{c+3}{c}$

8. $\dfrac{2x^2-5x-3}{x^2-9} \cdot \dfrac{x^2+6x+9}{10x+5}$ $\dfrac{x+3}{5}$

9. $\dfrac{6a^2b^3}{72ab^7c}$ $\dfrac{a}{12b^4c}$

10. $\dfrac{2a}{a+b} - \dfrac{b}{a-b} - \dfrac{-4ab}{a^2-b^2}$ $\dfrac{2a-b}{a-b}$

11. $\dfrac{p^2+10pq+25q^2}{p^2+6pq+5q^2} \div \dfrac{10p+50q}{2p^2-2q^2}$ $\dfrac{p-q}{5}$

12. $\dfrac{3k-8}{k-5} + \dfrac{k-12}{k-5}$ 4

13. $\dfrac{20x^2+10x}{4x^3+4x^2+x}$ $\dfrac{10}{2x+1}$

14. $\dfrac{w^2-81}{w^2+10w+9} \cdot \dfrac{w^2+w+2zw+2z}{w^2-9w+zw-9z}$ $\dfrac{w+2z}{w+z}$

15. $\dfrac{8x^2-18x-5}{4x^2-25} \div \dfrac{4x^2-11x-3}{3x-9}$ $\dfrac{3}{2x+5}$

16. $\dfrac{xy+7x+5y+35}{x^2+ax+5x+5a}$ $\dfrac{y+7}{x+a}$

17. $\dfrac{a}{a^2-9} - \dfrac{3}{6a-18}$ $\dfrac{1}{2(a+3)}$

18. $\dfrac{4}{y^2-36} + \dfrac{2}{y^2-4y-12}$ $\dfrac{2(3y+10)}{(y-6)(y+6)(y+2)}$

19. $(t^2+5t-24)\left(\dfrac{t+8}{t-3}\right)$ $(t+8)^2$

20. $\dfrac{6b^2-7b-10}{b-2}$ $6b+5$

Section 7.5 Complex Fractions

Objectives

1. **Simplifying Complex Fractions (Method I)**
2. **Simplifying Complex Fractions (Method II)**

1. Simplifying Complex Fractions (Method I)

A **complex fraction** is a fraction whose numerator or denominator contains one or more rational expressions. For example,

$$\frac{\dfrac{1}{ab}}{\dfrac{2}{b}} \quad \text{and} \quad \frac{1 + \dfrac{3}{4} - \dfrac{1}{6}}{\dfrac{1}{2} + \dfrac{1}{3}}$$

are complex fractions.

Two methods will be presented to simplify complex fractions. The first method (Method I) follows the order of operations to simplify the numerator and denominator separately before dividing. The process is summarized as follows.

> **PROCEDURE** Simplifying a Complex Fraction (Method I)
>
> **Step 1** Add or subtract expressions in the numerator to form a single fraction. Add or subtract expressions in the denominator to form a single fraction.
> **Step 2** Divide the rational expressions from Step 1 by multiplying the numerator of the complex fraction by the reciprocal of the denominator of the complex fraction.
> **Step 3** Simplify to lowest terms if possible.

Skill Practice

Simplify the expression.

1. $\dfrac{\dfrac{6x}{y}}{\dfrac{9}{2y}}$

Classroom Example: p. 535, Exercise 12

Example 1 Simplifying Complex Fractions (Method I)

Simplify the expression. $\dfrac{\dfrac{1}{ab}}{\dfrac{2}{b}}$

Solution:

Step 1: The numerator and denominator of the complex fraction are already single fractions.

$$\dfrac{\dfrac{1}{ab}}{\dfrac{2}{b}} \longleftarrow \text{This fraction bar denotes division } (\div).$$

$$= \frac{1}{ab} \div \frac{2}{b}$$

$$= \frac{1}{ab} \cdot \frac{b}{2} \qquad \textbf{Step 2:} \quad \text{Multiply the numerator of the complex fraction by the reciprocal of } \frac{2}{b}, \text{ which is } \frac{b}{2}.$$

$$= \frac{1}{a\cancel{b}} \cdot \frac{\cancel{b}^{1}}{2} \qquad \textbf{Step 3:} \quad \text{Reduce common factors and simplify.}$$

$$= \frac{1}{2a}$$

Answer

1. $\dfrac{4x}{3}$

Sometimes it is necessary to simplify the numerator and denominator of a complex fraction before the division can be performed. This is illustrated in Example 2.

Example 2 Simplifying Complex Fractions (Method I)

Simplify the expression.

$$\dfrac{1 + \dfrac{3}{4} - \dfrac{1}{6}}{\dfrac{1}{2} + \dfrac{1}{3}}$$

Solution:

$$\dfrac{1 + \dfrac{3}{4} - \dfrac{1}{6}}{\dfrac{1}{2} + \dfrac{1}{3}}$$

Step 1: Combine fractions in the numerator and denominator separately.

$$= \dfrac{1 \cdot \dfrac{12}{12} + \dfrac{3}{4} \cdot \dfrac{3}{3} - \dfrac{1}{6} \cdot \dfrac{2}{2}}{\dfrac{1}{2} \cdot \dfrac{3}{3} + \dfrac{1}{3} \cdot \dfrac{2}{2}}$$

The LCD in the numerator is 12. The LCD in the denominator is 6.

$$= \dfrac{\dfrac{12}{12} + \dfrac{9}{12} - \dfrac{2}{12}}{\dfrac{3}{6} + \dfrac{2}{6}}$$

$$= \dfrac{\dfrac{19}{12}}{\dfrac{5}{6}}$$

Form a single fraction in the numerator and in the denominator.

$$= \dfrac{19}{\overset{}{\underset{2}{12}}} \cdot \dfrac{\overset{1}{6}}{5}$$

Step 2: Multiply by the reciprocal of $\frac{5}{6}$, which is $\frac{6}{5}$.

$$= \dfrac{19}{10}$$

Step 3: Simplify.

Example 3 Simplifying Complex Fractions (Method I)

Simplify the expression. $\dfrac{\dfrac{1}{x} + \dfrac{1}{y}}{x - \dfrac{y^2}{x}}$

Solution:

$$\dfrac{\dfrac{1}{x} + \dfrac{1}{y}}{x - \dfrac{y^2}{x}}$$

The LCD in the numerator is xy. The LCD in the denominator is x.

$$= \dfrac{\dfrac{1 \cdot y}{x \cdot y} + \dfrac{1 \cdot x}{y \cdot x}}{\dfrac{x \cdot x}{1 \cdot x} - \dfrac{y^2}{x}}$$

Rewrite the expressions using common denominators.

$$= \dfrac{\dfrac{y}{xy} + \dfrac{x}{xy}}{\dfrac{x^2}{x} - \dfrac{y^2}{x}}$$

$$= \dfrac{\dfrac{y + x}{xy}}{\dfrac{x^2 - y^2}{x}}$$

Form single fractions in the numerator and denominator.

$$= \dfrac{y + x}{xy} \cdot \dfrac{x}{x^2 - y^2}$$

Multiply by the reciprocal of the denominator.

$$= \dfrac{\overset{1}{\cancel{y + x}}}{xy} \cdot \dfrac{\overset{1}{\cancel{x}}}{(x + y)(x - y)}$$

Factor and reduce. Note that $(y + x) = (x + y)$.

$$= \dfrac{1}{y(x - y)}$$

Simplify.

2. Simplifying Complex Fractions (Method II)

We will now simplify the expressions from Examples 2 and 3 again using a second method to simplify complex fractions (Method II). Recall that multiplying the numerator and denominator of a rational expression by the same quantity does not change the value of the expression because we are multiplying by a number equivalent to 1. This is the basis for Method II.

PROCEDURE Simplifying a Complex Fraction (Method II)

Step 1 Multiply the numerator and denominator of the complex fraction by the LCD of *all* individual fractions within the expression.

Step 2 Apply the distributive property, and simplify the numerator and denominator.

Step 3 Simplify to lowest terms if possible.

Example 4 Simplifying Complex Fractions (Method II)

Simplify the expression. $\dfrac{1 + \dfrac{3}{4} - \dfrac{1}{6}}{\dfrac{1}{2} + \dfrac{1}{3}}$

Skill Practice

Simplify the expression.

4. $\dfrac{1 - \dfrac{3}{5}}{\dfrac{1}{4} - \dfrac{7}{10} + 1}$

Classroom Example: p. 535,
Exercise 16

Solution:

$\dfrac{1 + \dfrac{3}{4} - \dfrac{1}{6}}{\dfrac{1}{2} + \dfrac{1}{3}}$

The LCD of the expressions $1, \frac{3}{4}, \frac{1}{6}, \frac{1}{2}$, and $\frac{1}{3}$ is 12.

$= \dfrac{12\left(1 + \dfrac{3}{4} - \dfrac{1}{6}\right)}{12\left(\dfrac{1}{2} + \dfrac{1}{3}\right)}$

Step 1: Multiply the numerator and denominator of the complex fraction by 12.

TIP: In step 1, we are multiplying the original expression by $\frac{12}{12}$, which equals 1.

$= \dfrac{12 \cdot 1 + 12 \cdot \dfrac{3}{4} - 12 \cdot \dfrac{1}{6}}{12 \cdot \dfrac{1}{2} + 12 \cdot \dfrac{1}{3}}$

Step 2: Apply the distributive property.

$= \dfrac{12 \cdot 1 + \overset{3}{\cancel{12}} \cdot \dfrac{3}{4} - \overset{2}{\cancel{12}} \cdot \dfrac{1}{6}}{\overset{6}{\cancel{12}} \cdot \dfrac{1}{2} + \overset{4}{\cancel{12}} \cdot \dfrac{1}{3}}$

Simplify each term.

$= \dfrac{12 + 9 - 2}{6 + 4}$

$= \dfrac{19}{10}$ **Step 3:** Simplify.

Example 5 Simplifying a Complex Fraction (Method II)

Simplify the expression. $\dfrac{\dfrac{1}{x} + \dfrac{1}{y}}{x - \dfrac{y^2}{x}}$

Skill Practice

Simplify the expressions.

5. $\dfrac{\dfrac{z}{3} - \dfrac{3}{z}}{1 + \dfrac{3}{z}}$

Classroom Example: p. 535,
Exercise 26

Solution:

$\dfrac{\dfrac{1}{x} + \dfrac{1}{y}}{x - \dfrac{y^2}{x}}$

The LCD of the expressions $\frac{1}{x}, \frac{1}{y}, x$, and $\frac{y^2}{x}$ is xy.

$= \dfrac{xy\left(\dfrac{1}{x} + \dfrac{1}{y}\right)}{xy\left(x - \dfrac{y^2}{x}\right)}$

Step 1: Multiply numerator and denominator of the complex fraction by xy.

Answers

4. $\dfrac{8}{11}$ 5. $\dfrac{z - 3}{3}$

$$= \frac{\cancel{x}y \cdot \dfrac{1}{\cancel{x}} + x\cancel{y} \cdot \dfrac{1}{\cancel{y}}}{xy \cdot x - xy \cdot \dfrac{y^2}{x}}$$

Step 2: Apply the distributive property, and simplify each term.

$$= \frac{y + x}{x^2 y - y^3}$$

$$= \frac{y + x}{y(x^2 - y^2)}$$

Step 3: Factor completely, and reduce common factors.

$$= \frac{\overset{1}{\cancel{y + x}}}{y\cancel{(x + y)}(x - y)}$$

Note that $(y + x) = (x + y)$.

$$= \frac{1}{y(x - y)}$$

Skill Practice

Simplify the expression.

6. $\dfrac{\dfrac{4}{p - 3} + 1}{1 + \dfrac{2}{p - 3}}$

Classroom Example: p. 535,
Exercise 28

Example 6 Simplifying a Complex Fraction (Method II)

Simplify the expression. $\dfrac{\dfrac{1}{k + 1} - 1}{\dfrac{1}{k + 1} + 1}$

Solution:

$$\frac{\dfrac{1}{k + 1} - 1}{\dfrac{1}{k + 1} + 1}$$

The LCD of $\dfrac{1}{k + 1}$ and 1 is $(k + 1)$.

$$= \frac{(k + 1)\left(\dfrac{1}{k + 1} - 1\right)}{(k + 1)\left(\dfrac{1}{k + 1} + 1\right)}$$

Step 1: Multiply numerator and denominator of the complex fraction by $(k + 1)$.

$$= \frac{\overset{1}{\cancel{(k + 1)}} \cdot \dfrac{1}{\cancel{(k + 1)}} - (k + 1) \cdot 1}{\overset{1}{\cancel{(k + 1)}} \cdot \dfrac{1}{\cancel{(k + 1)}} + (k + 1) \cdot 1}$$

Step 2: Apply the distributive property.

$$= \frac{1 - (k + 1)}{1 + (k + 1)}$$

Simplify.

$$= \frac{1 - k - 1}{1 + k + 1}$$

$$= \frac{-k}{k + 2}$$

Step 3: The expression is already in lowest terms.

Answer

6. $\dfrac{p + 1}{p - 1}$

Section 7.5 Practice Exercises

Boost *your* GRADE at
ALEKS.com!

• Practice Problems
• Self-Tests
• NetTutor

• e-Professors
• Videos

For additional exercises, see Classroom Activities
7.5A–7.5B in the *Instructor's Resource Manual*
at www.mhhe.com/moh.

Study Skills Exercise

1. Define the key term **complex fraction**.

Review Exercises

For Exercises 2–3, write the domain in set-builder notation, and simplify the expression.

2. $\dfrac{y(2y+9)}{y^2(2y+9)}$ $\left\{y \mid y \text{ is a real number and } y \neq 0, y \neq -\dfrac{9}{2}\right\}; \dfrac{1}{y}$ 3. $\dfrac{a+5}{2a^2+7a-15}$ $\left\{a \mid a \text{ is a real number and } a \neq \dfrac{3}{2}, a \neq -5\right\};$ $\dfrac{1}{2a-3}$

For Exercises 4–6, perform the indicated operations.

4. $\dfrac{2}{w-2} + \dfrac{3}{w}$ $\dfrac{5w-6}{w(w-2)}$ 5. $\dfrac{6}{5} - \dfrac{3}{5k-10}$ $\dfrac{3(2k-5)}{5(k-2)}$ 6. $\dfrac{x^2-2xy+y^2}{x^4-y^4} \div \dfrac{3x^2y-3xy^2}{x^2+y^2}$ $\dfrac{1}{3xy(x+y)}$

Objectives 1–2: Simplifying Complex Fractions (Methods I and II)

For Exercises 7–34, simplify the complex fractions. **(See Examples 1–6.)**

7. $\dfrac{\frac{7}{18y}}{\frac{2}{9}}$ $\dfrac{7}{4y}$ 8. $\dfrac{\frac{a^2}{2a-3}}{\frac{5a}{8a-12}}$ $\dfrac{4a}{5}$ 9. $\dfrac{\frac{3x+2y}{2y}}{\frac{6x+4y}{2}}$ $\dfrac{1}{2y}$ 10. $\dfrac{\frac{2x-10}{4}}{\frac{x^2-5x}{3x}}$ $\dfrac{3}{2}$

11. $\dfrac{\frac{8a^4b^3}{3c}}{\frac{a^7b^2}{9c}}$ $\dfrac{24b}{a^3}$ 12. $\dfrac{\frac{12x^2}{5y}}{\frac{8x^6}{9y^2}}$ $\dfrac{27y}{10x^4}$ 13. $\dfrac{\frac{4r^3s}{t^5}}{\frac{2s^7}{r^2t^9}}$ $\dfrac{2r^5t^4}{s^6}$ 14. $\dfrac{\frac{5p^4q}{w^4}}{\frac{10p^2}{qw^2}}$ $\dfrac{p^2q^2}{2w^2}$

15. $\dfrac{\frac{1}{8} + \frac{4}{3}}{\frac{1}{2} - \frac{5}{12}}$ $\dfrac{35}{2}$ 16. $\dfrac{\frac{8}{9} - \frac{1}{3}}{\frac{7}{6} + \frac{1}{9}}$ $\dfrac{10}{23}$ 17. $\dfrac{\frac{1}{h} + \frac{1}{k}}{\frac{1}{hk}}$ $k+h$ 18. $\dfrac{\frac{1}{b} + 1}{\frac{1}{b}}$ $1+b$

19. $\dfrac{\frac{n+1}{n^2-9}}{\frac{2}{n+3}}$ $\dfrac{n+1}{2(n-3)}$ 20. $\dfrac{\frac{5}{k-5}}{\frac{k+1}{k^2-25}}$ $\dfrac{5(k+5)}{k+1}$ 21. $\dfrac{2 + \frac{1}{x}}{4 + \frac{1}{x}}$ $\dfrac{2x+1}{4x+1}$ 22. $\dfrac{6 + \frac{6}{k}}{1 + \frac{1}{k}}$ 6

23. $\dfrac{\frac{m}{7} - \frac{7}{m}}{\frac{1}{7} + \frac{1}{m}}$ $m-7$ 24. $\dfrac{\frac{2}{p} + \frac{p}{2}}{\frac{p}{3} - \frac{3}{p}}$ $\dfrac{3(4+p^2)}{2(p-3)(p+3)}$ 25. $\dfrac{\frac{1}{5} - \frac{1}{y}}{\frac{7}{10} + \frac{1}{y^2}}$ $\dfrac{2y(y-5)}{7y^2+10}$ 26. $\dfrac{\frac{1}{m^2} + \frac{2}{3}}{\frac{1}{m} - \frac{5}{6}}$ $\dfrac{2(3+2m^2)}{m(6-5m)}$

27. $\dfrac{\frac{8}{a+4} + 2}{\frac{12}{a+4} - 2}$ $\dfrac{a+8}{a-2} \text{ or } \dfrac{a+8}{2-a}$ 28. $\dfrac{\frac{2}{w+1} + 3}{\frac{3}{w+1} + 4}$ $\dfrac{3w+5}{4w+7}$ 29. $\dfrac{1 - \frac{4}{t^2}}{1 - \frac{2}{t} - \frac{8}{t^2}}$ $\dfrac{t-2}{t-4}$ 30. $\dfrac{1 - \frac{9}{p^2}}{1 - \frac{1}{p} - \frac{6}{p^2}}$ $\dfrac{p+3}{p+2}$

✏ Writing   ↔ Translating Expression   ◆ Geometry   🖩 Scientific Calculator   ⊙ Video   NS&E

31. $\dfrac{t + 4 + \dfrac{3}{t}}{t - 4 - \dfrac{5}{t}}$ $\dfrac{t+3}{t-5}$

32. $\dfrac{\dfrac{9}{4m} + \dfrac{9}{2m^2}}{\dfrac{3}{2} + \dfrac{3}{m}}$ $\dfrac{3}{2m}$

33. $\dfrac{\dfrac{1}{k-6} - 1}{\dfrac{2}{k-6} - 2}$ $\dfrac{1}{2}$

34. $\dfrac{\dfrac{3}{y-3} + 4}{8 + \dfrac{6}{y-3}}$ $\dfrac{1}{2}$

↞↠ For Exercises 35–38, translate the English phrases into algebraic expressions. Then simplify the expressions.

35. The sum of one-half and two-thirds, divided by five. $\dfrac{\dfrac{1}{2} + \dfrac{2}{3}}{5}$; $\dfrac{7}{30}$

36. The quotient of ten and the difference of two-fifths and one-fourth. $\dfrac{10}{\dfrac{2}{5} - \dfrac{1}{4}}$; $\dfrac{200}{3}$

37. The quotient of three and the sum of two-thirds and three-fourths. $\dfrac{3}{\dfrac{2}{3} + \dfrac{3}{4}}$; $\dfrac{36}{17}$

38. The difference of three-fifths and one-half, divided by four. $\dfrac{\dfrac{3}{5} - \dfrac{1}{2}}{4}$; $\dfrac{1}{40}$

39. In electronics, resistors oppose the flow of current. For two resistors in parallel, the total resistance is given by

$$R = \dfrac{1}{\dfrac{1}{R_1} + \dfrac{1}{R_2}}$$

a. Find the total resistance if $R_1 = 2\ \Omega$ (ohms) and $R_2 = 3\ \Omega$. $\dfrac{6}{5}\ \Omega$

b. Find the total resistance if $R_1 = 10\ \Omega$ and $R_2 = 15\ \Omega$. $6\ \Omega$

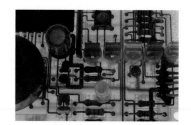

40. Suppose that Joëlle makes a round trip to a location that is d miles away. If the average rate going to the location is r_1 and the average rate on the return trip is given by r_2, the average rate of the entire trip, R, is given by

$$R = \dfrac{2d}{\dfrac{d}{r_1} + \dfrac{d}{r_2}}$$

a. Find the average rate of a trip to a destination 30 mi away when the average rate going there was 60 mph and the average rate returning home was 45 mph. (Round to the nearest tenth of a mile per hour.)
51.4 mph average

b. Find the average rate of a trip to a destination that is 50 mi away if the driver travels at the same rates as in part (a). (Round to the nearest tenth of a mile per hour.)
51.4 mph

c. Compare your answers from parts (a) and (b) and explain the results in the context of the problem.
Because the rates going to and leaving from the destination are the same, the average rate is unchanged. The average rate is not affected by the distance traveled.

Expanding Your Skills

For Exercises 41–48, simplify the complex fractions using either method.

41. $\dfrac{2x^{-1} + 8y^{-1}}{4x^{-1}}$ $\dfrac{y + 4x}{2y}$
$\left(Hint: 2x^{-1} = \dfrac{2}{x} \right)$

42. $\dfrac{6a^{-1} + 4b^{-1}}{8b^{-1}}$ $\dfrac{3b + 2a}{4a}$

43. $\dfrac{(mn)^{-2}}{m^{-2} + n^{-2}}$ $\dfrac{1}{n^2 + m^2}$

44. $\dfrac{(xy)^{-1}}{2x^{-1} + 3y^{-1}}$ $\dfrac{1}{2y + 3x}$

45. $\dfrac{\dfrac{1}{z^2 - 9} + \dfrac{2}{z + 3}}{\dfrac{3}{z - 3}}$ $\dfrac{2z - 5}{3(z + 3)}$

46. $\dfrac{\dfrac{5}{w^2 - 25} - \dfrac{3}{w + 5}}{\dfrac{4}{w - 5}}$ $\dfrac{-3w + 20}{4(w + 5)}$

47. $\dfrac{\dfrac{2}{x - 1} + 2}{\dfrac{2}{x + 1} - 2}$ $-\dfrac{x + 1}{x - 1}$ or $\dfrac{x + 1}{1 - x}$

48. $\dfrac{\dfrac{1}{y - 3} + 1}{\dfrac{2}{y + 3} - 1}$ $-\dfrac{(y + 3)(y - 2)}{(y + 1)(y - 3)}$

For Exercises 49–50, simplify the complex fractions (*Hint:* Use the order of operations and begin with the fraction on the lower right.)

49. $1 + \dfrac{1}{1 + 1}$ $\dfrac{3}{2}$

50. $1 + \dfrac{1}{1 + \dfrac{1}{1 + 1}}$ $\dfrac{5}{3}$

 Writing   Translating Expression   Geometry   Scientific Calculator   Video   NS&E

Rational Equations

1. Introduction to Rational Equations

Thus far we have studied two specific types of equations in one variable: linear equations and quadratic equations. Recall,

$$ax + b = 0, \text{ where } a \neq 0, \text{ is a } \textbf{linear equation}$$

$$ax^2 + bx + c = 0, \text{ where } a \neq 0, \text{ is a } \textbf{quadratic equation}.$$

We will now study another type of equation called a rational equation.

> **DEFINITION** Rational Equation
>
> An equation with one or more rational expressions is called a **rational equation**.

The following equations are rational equations:

$$\frac{y}{2} + \frac{y}{4} = 6 \qquad \frac{1}{x} + \frac{1}{3} = \frac{5}{6} \qquad \frac{6}{t^2 - 7t + 12} + \frac{2t}{t - 3} = \frac{3t}{t - 4}$$

To understand the process of solving a rational equation, first review the process of clearing fractions from Section 2.3. We can clear the fractions in an equation by multiplying both sides of the equation by the LCD of all terms.

Example 1 Solving a Rational Equation

Solve. $\dfrac{y}{2} + \dfrac{y}{4} = 6$

Solution:

$$\frac{y}{2} + \frac{y}{4} = 6 \qquad \text{The LCD of all terms in the equation is } 4.$$

$$4\left(\frac{y}{2} + \frac{y}{4}\right) = 4(6) \qquad \text{Multiply both sides of the equation by 4 to clear fractions.}$$

$$\overset{2}{\cancel{4}} \cdot \frac{y}{2} + \overset{1}{\cancel{4}} \cdot \frac{y}{4} = 4(6) \qquad \text{Apply the distributive property.}$$

$$2y + y = 24 \qquad \text{Clear fractions.}$$

$$3y = 24 \qquad \text{Solve the resulting equation (linear).}$$

$$y = 8$$

$$\underline{\text{Check:}} \quad \frac{y}{2} + \frac{y}{4} = 6$$

$$\frac{(8)}{2} + \frac{(8)}{4} \overset{?}{=} 6$$

$$4 + 2 \overset{?}{=} 6$$

The solution is 8. $\qquad\qquad 6 \overset{?}{=} 6 \checkmark \text{ (True)}$

2. Solving Rational Equations

The same process of clearing fractions is used to solve rational equations when variables are present in the denominator. Variables in the denominator make it necessary to take note of **domain restrictions**. Values for the variable that make any denominator zero are restricted from the domain.

Skill Practice

Solve the equation.

2. $\dfrac{3}{4} + \dfrac{5+a}{a} = \dfrac{1}{2}$

Classroom Example: p. 545, Exercise 26

Example 2 Solving a Rational Equation

Solve the equation. $\dfrac{x+1}{x} + \dfrac{1}{3} = \dfrac{5}{6}$

Solution:

$$\dfrac{x+1}{x} + \dfrac{1}{3} = \dfrac{5}{6}$$

The LCD of all the expressions is $6x$. The domain restriction is $x \neq 0$.

$$6x \cdot \left(\dfrac{x+1}{x} + \dfrac{1}{3}\right) = 6x \cdot \left(\dfrac{5}{6}\right)$$

Multiply by the LCD.

$$\overset{1}{6x} \cdot \left(\dfrac{x+1}{x}\right) + \overset{2}{6x} \cdot \left(\dfrac{1}{3}\right) = \overset{1}{6x} \cdot \left(\dfrac{5}{6}\right)$$

Apply the distributive property.

$$6(x+1) + 2x = 5x$$

Clear fractions.

$$6x + 6 + 2x = 5x$$

Solve the resulting equation.

$$8x + 6 = 5x$$

$$3x = -6$$

$$x = -2$$

-2 is not a restricted value.

Check: $\dfrac{x+1}{x} + \dfrac{1}{3} = \dfrac{5}{6}$

$$\dfrac{(-2)+1}{(-2)} + \dfrac{1}{3} \overset{?}{=} \dfrac{5}{6}$$

$$\dfrac{-1}{-2} + \dfrac{1}{3} \overset{?}{=} \dfrac{5}{6}$$

$$\dfrac{1}{2} + \dfrac{1}{3} \overset{?}{=} \dfrac{5}{6}$$

$$\dfrac{3}{6} + \dfrac{2}{6} \overset{?}{=} \dfrac{5}{6}$$

The solution is -2.

$$\dfrac{5}{6} \overset{?}{=} \dfrac{5}{6} \ \checkmark \ \text{(True)}$$

Answer

2. -4

Example 3 **Solving a Rational Equation**

Solve the equation. $1 + \dfrac{3a}{a-2} = \dfrac{6}{a-2}$

Solution:

$$1 + \frac{3a}{a-2} = \frac{6}{a-2}$$

The LCD of all the expressions is $a - 2$. The domain restriction is $a \neq 2$.

$$(a-2)\left(1 + \frac{3a}{a-2}\right) = (a-2)\left(\frac{6}{a-2}\right)$$

Multiply by the LCD.

$$(a-2)1 + (a\overset{1}{-}2)\left(\frac{3a}{a-2}\right) = (a\overset{1}{-}2)\left(\frac{6}{a-2}\right)$$

Apply the distributive property.

$$a - 2 + 3a = 6$$

Solve the resulting equation (linear).

$$4a - 2 = 6$$

$$4a = 8$$

$$a = 2$$ 2 is a restricted value.

Check: $1 + \dfrac{3a}{a-2} = \dfrac{6}{a-2}$

$$1 + \frac{3(2)}{(2)-2} \overset{?}{=} \frac{6}{(2)-2}$$

$$1 + \frac{6}{0} \overset{?}{=} \frac{6}{0}$$

The denominator is 0 when $a = 2$.

Because the value $a = 2$ makes the denominator zero in one (or more) of the rational expressions within the equation, the equation is undefined for $a = 2$. That is, 2 is not in the domain. No other potential solutions exist for the equation.

The equation $1 + \dfrac{3a}{a-2} = \dfrac{6}{a-2}$ has no solution.

Examples 1–3 show that the steps to solve a rational equation mirror the process of clearing fractions from Section 2.3. However, there is one significant difference. The solutions of a rational equation must not make the denominator equal to zero for any expression within the equation. When $a = 2$ is substituted into the expression

$$\frac{3a}{a-2} \quad \text{or} \quad \frac{6}{a-2}$$

the denominator is zero and the expression is undefined. Therefore, 2 cannot be a solution to the equation

$$1 + \frac{3a}{a-2} = \frac{6}{a-2}$$

Skill Practice

Solve the equations.

3. $\dfrac{x}{x+1} - 2 = \dfrac{-1}{x+1}$

Classroom Example: p. 545, Exercise 32

Instructor Note: Have students explain the difference between

$$\frac{2}{x+3} + \frac{5}{x+2}$$

and

$$\frac{2}{x+3} = \frac{5}{x+2}$$

Answer

3. No solution (the value -1 does not check.)

The steps to solve a rational equation are summarized as follows.

> **PROCEDURE** Solving a Rational Equation
> **Step 1** Factor the denominators of all rational expressions. Determine the domain restrictions.
> **Step 2** Identify the LCD of all expressions in the equation.
> **Step 3** Multiply both sides of the equation by the LCD.
> **Step 4** Solve the resulting equation.
> **Step 5** Check potential solutions in the original equation.

After multiplying by the LCD and then simplifying, the rational equation will be either a linear equation or higher degree equation.

Classroom Example: p. 545, Exercise 24

Skill Practice

Solve the equation.

4. $\dfrac{z}{2} - \dfrac{1}{2z} = \dfrac{12}{z}$

Example 4   **Solving a Rational Equation**

Solve the equation. $1 - \dfrac{4}{p} = -\dfrac{3}{p^2}$

Solution:

$$1 - \dfrac{4}{p} = -\dfrac{3}{p^2}$$

Step 1: The denominators are already factored. The domain restriction is $p \neq 0$.

Step 2: The LCD of all expressions is p^2.

$$p^2\left(1 - \dfrac{4}{p}\right) = p^2\left(-\dfrac{3}{p^2}\right)$$

Step 3: Multiply by the LCD.

$$p^2(1) - p^2\left(\dfrac{4}{p}\right) = p^2\left(-\dfrac{3}{p^2}\right)$$

Apply the distributive property.

$$p^2 - 4p = -3$$

Step 4: Solve the resulting quadratic equation.

$$p^2 - 4p + 3 = 0$$
$$(p - 3)(p - 1) = 0$$

Set the equation equal to zero and factor.

$$p - 3 = 0 \quad \text{or} \quad p - 1 = 0$$

Set each factor equal to zero.

$$p = 3 \quad \text{or} \quad p = 1$$

3 and 1 are not restricted values.

Step 5: Check: $p = 3$

$$1 - \dfrac{4}{p} = -\dfrac{3}{p^2}$$
$$1 - \dfrac{4}{(3)} \stackrel{?}{=} -\dfrac{3}{(3)^2}$$
$$\dfrac{3}{3} - \dfrac{4}{3} \stackrel{?}{=} -\dfrac{3}{9}$$
$$-\dfrac{1}{3} \stackrel{?}{=} -\dfrac{1}{3} \checkmark$$

Check: $p = 1$

$$1 - \dfrac{4}{p} = -\dfrac{3}{p^2}$$
$$1 - \dfrac{4}{(1)} \stackrel{?}{=} -\dfrac{3}{(1)^2}$$
$$1 - 4 \stackrel{?}{=} -3$$
$$-3 \stackrel{?}{=} -3 \checkmark$$

Both solutions 3 and 1 check.

Answer

4. $5, -5$

Example 5 **Solving a Rational Equation**

Solve the equation. $\dfrac{6}{t^2 - 7t + 12} + \dfrac{2t}{t - 3} = \dfrac{3t}{t - 4}$

Solution:

$\dfrac{6}{t^2 - 7t + 12} + \dfrac{2t}{t - 3} = \dfrac{3t}{t - 4}$

$\dfrac{6}{(t - 3)(t - 4)} + \dfrac{2t}{t - 3} = \dfrac{3t}{t - 4}$ **Step 1:** Factor the denominators. The domain restrictions are $t \neq 3$ and $t \neq 4$.

Step 2: The LCD is $(t - 3)(t - 4)$.

Step 3: Multiply by the LCD on both sides.

$(t - 3)(t - 4)\left(\dfrac{6}{(t - 3)(t - 4)} + \dfrac{2t}{t - 3} \right) = (t - 3)(t - 4)\left(\dfrac{3t}{t - 4} \right)$

$(t - 3)(t - 4)\left(\dfrac{6}{(t - 3)(t - 4)} \right) + (t - 3)(t - 4)\left(\dfrac{2t}{t - 3} \right) = (t - 3)(t - 4)\left(\dfrac{3t}{t - 4} \right)$

$6 + 2t(t - 4) = 3t(t - 3)$

$6 + 2t^2 - 8t = 3t^2 - 9t$ **Step 4:** Solve the resulting equation.

$0 = 3t^2 - 2t^2 - 9t + 8t - 6$ Because the resulting equation is quadratic, set the equation equal to zero and factor.

$0 = t^2 - t - 6$

$0 = (t - 3)(t + 2)$

$t - 3 = 0$ or $t + 2 = 0$ Set each factor equal to zero.

$t = 3$ or $t = -2$

3 is a restricted value, but -2 is not restricted. **Step 5:** Check the potential solutions in the original equation.

Check: $t = 3$

3 cannot be a solution to the equation because it will make the denominator zero in the original equation.

$\dfrac{6}{t^2 - 7t + 12} + \dfrac{2t}{t - 3} = \dfrac{3t}{t - 4}$

$\dfrac{6}{(3)^2 - 7(3) + 12} + \dfrac{2(3)}{(3) - 3} \overset{?}{=} \dfrac{3(3)}{(3) - 4}$

$\dfrac{6}{0} + \dfrac{6}{0} \overset{?}{=} \dfrac{9}{-1}$

Zero in the denominator

The only solution is -2.

Check: $t = -2$

$\dfrac{6}{t^2 - 7t + 12} + \dfrac{2t}{t - 3} = \dfrac{3t}{t - 4}$

$\dfrac{6}{(-2)^2 - 7(-2) + 12} + \dfrac{2(-2)}{(-2) - 3} \overset{?}{=} \dfrac{3(-2)}{(-2) - 4}$

$\dfrac{6}{4 + 14 + 12} + \dfrac{-4}{-5} \overset{?}{=} \dfrac{-6}{-6}$

$\dfrac{6}{30} + \dfrac{4}{5} \overset{?}{=} 1$

$\dfrac{1}{5} + \dfrac{4}{5} = 1$ ✔ (True)

$t = -2$ is a solution.

Classroom Example: p. 546, Exercise 42

Skill Practice

Solve the equation.

5. $\dfrac{-8}{x^2 + 6x + 8} + \dfrac{x}{x + 4}$

$= \dfrac{2}{x + 2}$

Concept Connections

6. Explain why it is important to check the solution(s) to a rational equation.

Answers

5. 4; (the value -4 does not check.)

6. It is important to check that the solutions are in the domain of each expression in the equation.

Example 6 **Translating to a Rational Equation**

Ten times the reciprocal of a number is added to four. The result is equal to the quotient of twenty-two and the number. Find the number.

Solution:

Let x represent the number.

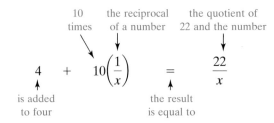

$$4 + \frac{10}{x} = \frac{22}{x}$$ **Step 1:** The denominators are already factored. The domain restriction is $x \neq 0$.

Step 2: The LCD is x.

$$x\left(4 + \frac{10}{x}\right) = x\left(\frac{22}{x}\right)$$ **Step 3:** Multiply both sides by the LCD.

$$4x + 10 = 22$$ Apply the distributive property.

$$4x = 12$$ **Step 4:** Solve the resulting linear equation.

$$x = 3 \text{ is a potential solution.}$$ **Step 5:** 3 is not a restricted value. Substituting $x = 3$ into the original equation verifies that it is a solution.

The number is 3.

3. Solving Formulas Involving Rational Expressions

A rational equation may have more than one variable. To solve for a specific variable within a rational equation, we can still apply principle of clearing fractions.

Example 7 Solving Formulas Involving Rational Equations

Solve for k. $F = \dfrac{ma}{k}$

Solution:

To solve for k, we must clear fractions so that k appears in the numerator.

$$F = \frac{ma}{k} \qquad \text{The LCD is } k.$$

$$k \cdot (F) = k \cdot \left(\frac{ma}{k}\right) \qquad \text{Multiply both sides of the equation by the LCD.}$$

$$kF = ma \qquad \text{Clear fractions.}$$

$$\frac{kF}{F} = \frac{ma}{F} \qquad \text{Divide both sides by } F.$$

$$k = \frac{ma}{F}$$

Skill Practice

8. Solve for t.

$$C = \frac{rt}{d}$$

Classroom Example: p. 546, Exercise 54

Example 8 Solving Formulas Involving Rational Equations

Solve for b. $h = \dfrac{2A}{B + b}$

Solution:

To solve for b, we must clear fractions so that b appears in the numerator.

$$h = \frac{2A}{B + b} \qquad \text{The LCD is } (B + b).$$

$$h(B + b) = \left(\frac{2A}{B + b}\right) \cdot (B + b) \qquad \text{Multiply both sides of the equation by the LCD.}$$

$$hB + hb = 2A \qquad \text{Apply the distributive property.}$$

$$hb = 2A - hB \qquad \text{Subtract } hB \text{ from both sides to isolate the } b \text{ term.}$$

$$\frac{hb}{h} = \frac{2A - hB}{h} \qquad \text{Divide by } h.$$

$$b = \frac{2A - hB}{h}$$

Skill Practice

9. Solve the formula for x.

$$y = \frac{3}{x - 2}$$

Classroom Example: p. 546, Exercise 56

Avoiding Mistakes

Algebra is case-sensitive. The variables B and b represent different values.

Answers

8. $t = \dfrac{Cd}{r}$

9. $x = \dfrac{3 + 2y}{y}$ or $x = \dfrac{3}{y} + 2$

TIP: The solution to Example 8 can be written in several forms. The quantity

$$\frac{2A - hB}{h}$$

can be left as a single rational expression or can be split into two fractions and simplified.

$$b = \frac{2A - hB}{h} = \frac{2A}{h} - \frac{hB}{h} = \frac{2A}{h} - B$$

Skill Practice

10. Solve for h.

$$\frac{b}{x} = \frac{a}{h} + 1$$

Classroom Example: p. 546, Exercise 64

Answer

10. $h = \dfrac{ax}{b - x}$ or $\dfrac{-ax}{x - b}$

Example 9 **Solving Formulas Involving Rational Expressions**

Solve for z. $\quad y = \dfrac{x - z}{x + z}$

Solution:

To solve for z, we must clear fractions so that z appears in the numerator.

$$y = \frac{x - z}{x + z} \qquad \text{LCD is } (x + z).$$

$$y(x + z) = \left(\frac{x - z}{x + z}\right)(x + z) \qquad \text{Multiply both sides of the equation by the LCD.}$$

$$yx + yz = x - z \qquad \text{Apply the distributive property.}$$

$$yz + z = x - yx \qquad \text{Collect } z \text{ terms on one side of the equation and collect terms not containing } z \text{ on the other side.}$$

$$z(y + 1) = x - yx \qquad \text{Factor out } z.$$

$$z = \frac{x - yx}{y + 1} \qquad \text{Divide by } y + 1 \text{ to solve for } z.$$

Section 7.6 Practice Exercises

Boost your GRADE at ALEKS.com!

ALEKS version 3.0

- Practice Problems
- Self-Tests
- NetTutor

- e-Professors
- Videos

For additional exercises, see Classroom Activities 7.6A–7.6B in the *Instructor's Resource Manual* at www.mhhe.com/moh.

Study Skills Exercise

1. Define the key terms:

    a. linear equation   **b. quadratic equation**   **c. rational equation**   **d. domain restrictions**

Review Exercises

For Exercises 2–7, perform the indicated operations.

2. $\dfrac{2}{x - 3} - \dfrac{3}{x^2 - x - 6}$ $\dfrac{2x + 1}{(x - 3)(x + 2)}$

3. $\dfrac{2x - 6}{4x^2 + 7x - 2} \div \dfrac{x^2 - 5x + 6}{x^2 - 4}$ $\dfrac{2}{4x - 1}$

4. $\dfrac{2y}{y - 3} + \dfrac{4}{y^2 - 9}$ $\dfrac{2(y + 2)(y + 1)}{(y - 3)(y + 3)}$

5. $\dfrac{h - \dfrac{1}{h}}{\dfrac{1}{5} - \dfrac{1}{5h}}$ $5(h + 1)$

6. $\dfrac{w - 4}{w^2 - 9} \cdot \dfrac{w - 3}{w^2 - 8w + 16}$ $\dfrac{1}{(w + 3)(w - 4)}$

7. $1 + \dfrac{1}{x} - \dfrac{12}{x^2}$ $\dfrac{(x + 4)(x - 3)}{x^2}$

 Writing   Translating Expression   Geometry   Scientific Calculator   Video   NS&E

Objective 1: Introduction to Rational Equations

For Exercises 8–13, solve the equations by first clearing the fractions. **(See Example 1.)**

8. $\frac{1}{3}z + \frac{2}{3} = -2z + 10$ $\quad 4$

9. $\frac{5}{2} + \frac{1}{2}b = 5 - \frac{1}{3}b$ $\quad 3$

10. $\frac{3}{2}p + \frac{1}{3} = \frac{2p - 3}{4}$ $\quad -\frac{13}{12}$

11. $\frac{5}{3} - \frac{1}{6}k = \frac{3k + 5}{4}$ $\quad \frac{5}{11}$

12. $\frac{2x - 3}{4} + \frac{9}{10} = \frac{x}{5}$ $\quad -\frac{1}{2}$

13. $\frac{4y + 2}{3} - \frac{7}{6} = -\frac{y}{6}$ $\quad \frac{1}{3}$

Objective 2: Solving Rational Equations

14. For the equation

$$\frac{1}{w} - \frac{1}{2} = -\frac{1}{4}$$

    a. Identify the domain.
       $\{w \mid w \text{ is a real number and } w \neq 0\}$

    b. Identify the LCD of the fractions in the equation. $4w$

    c. Solve the equation. 4

15. For the equation

$$\frac{3}{z} - \frac{4}{5} = -\frac{1}{5}$$

    a. Identify the domain.
       $\{z \mid z \text{ is a real number and } z \neq 0\}$

    b. Identify the LCD of the fractions in the equation. $5z$

    c. Solve the equation. 5

16. Identify the LCD of all the denominators in the equation.

$$\frac{x + 1}{x^2 + 2x - 3} = \frac{1}{x + 3} - \frac{1}{x - 1} \quad (x + 3)(x - 1)$$

For Exercises 17–46, solve the equations. **(See Examples 2–5.)**

17. $\frac{1}{8} = \frac{3}{5} + \frac{5}{y}$ $\quad -\frac{200}{19}$

18. $\frac{2}{7} - \frac{1}{x} = \frac{2}{3}$ $\quad -\frac{21}{8}$

19. $\frac{4}{t} = \frac{3}{t} + \frac{1}{8}$ $\quad 8$

20. $\frac{9}{b} - \frac{8}{b} = \frac{1}{4}$ $\quad 4$

21. $\frac{5}{6x} + \frac{7}{x} = 1$ $\quad \frac{47}{6}$

22. $\frac{14}{3x} - \frac{5}{x} = 2$ $\quad -\frac{1}{6}$

23. $1 - \frac{2}{y} = \frac{3}{y^2}$ $\quad 3, -1$

24. $1 - \frac{2}{m} = \frac{8}{m^2}$ $\quad 4, -2$

25. $\frac{a + 1}{a} = 1 + \frac{a - 2}{2a}$ $\quad 4$

26. $\frac{7b - 4}{5b} = \frac{9}{5} - \frac{4}{b}$ $\quad 8$

27. $\frac{w}{5} - \frac{w + 3}{w} = -\frac{3}{w}$
    5; (the value 0 does not check.)

28. $\frac{t}{12} + \frac{t + 3}{3t} = \frac{1}{t}$
    -4; (the value 0 does not check.)

29. $\frac{2}{m + 3} = \frac{5}{4m + 12} - \frac{3}{8}$
    -5

30. $\frac{2}{4n - 4} - \frac{7}{4} = \frac{-3}{n - 1}$
    3

31. $\frac{p}{p - 4} - 5 = \frac{4}{p - 4}$
    No solution; (the value 4 does not check.)

32. $\frac{-5}{q + 5} = \frac{q}{q + 5} + 2$
    No solution; (the value -5 does not check.)

33. $\frac{2t}{t + 2} - 2 = \frac{t - 8}{t + 2}$ $\quad 4$

34. $\frac{4w}{w - 3} - 3 = \frac{3w - 1}{w - 3}$ $\quad 5$

35. $\frac{x^2 - x}{x - 2} = \frac{12}{x - 2}$
    $4, -3$

36. $\frac{x^2 + 9}{x + 4} = \frac{-10x}{x + 4}$
    $-9, -1$

37. $\frac{x^2 + 3x}{x - 1} = \frac{4}{x - 1}$
    -4; (the value 1 does not check.)

38. $\frac{2x^2 - 21}{2x - 3} = \frac{-11x}{2x - 3}$
    -7; (the value $\frac{3}{2}$ does not check.)

39. $\frac{2x}{x + 4} - \frac{8}{x - 4} = \frac{2x^2 + 32}{x^2 - 16}$
    No solution; (the value -4 does not check.)

40. $\frac{4x}{x + 3} - \frac{12}{x - 3} = \frac{4x^2 + 36}{x^2 - 9}$
    No solution; (the value -3 does not check.)

41. $\dfrac{x}{x+6} = \dfrac{72}{x^2-36} + 4$ 4; (the value -6 does not check.)

42. $\dfrac{y}{y+4} = \dfrac{32}{y^2-16} + 3$ 2; (the value -4 does not check.)

43. $\dfrac{5}{3x-3} - \dfrac{2}{x-2} = \dfrac{7}{x^2-3x+2}$ -25

44. $\dfrac{6}{5a+10} - \dfrac{1}{a-5} = \dfrac{4}{a^2-3a-10}$ 60

45. $\dfrac{y-2}{y-3} = \dfrac{11}{y^2-7y+12} + \dfrac{y}{y-4}$ -1

46. $\dfrac{6}{w+1} - \dfrac{3}{w+5} = \dfrac{18}{w^2+6w+5}$ -3

For Exercises 47–50, translate to a rational equation and solve. **(See Example 6.)**

47. The reciprocal of a number is added to three. The result is the quotient of 25 and the number. Find the number.   The number is 8.

48. The difference of three and the reciprocal of a number is equal to the quotient of 20 and the number. Find the number.   The number is 7.

49. If a number added to five is divided by the difference of the number and two, the result is three-fourths. Find the number.   The number is -26.

50. If twice a number added to three is divided by the number plus one, the result is three-halves. Find the number.   The number is -3.

Objective 3: Solving Formulas Involving Rational Expressions

For Exercises 51–68, solve for the indicated variable. **(See Examples 7–9.)**

51. $K = \dfrac{ma}{F}$ for m $m = \dfrac{FK}{a}$

52. $K = \dfrac{ma}{F}$ for a $a = \dfrac{FK}{m}$

53. $K = \dfrac{IR}{E}$ for E $E = \dfrac{IR}{K}$

54. $K = \dfrac{IR}{E}$ for R $R = \dfrac{KE}{I}$

55. $I = \dfrac{E}{R+r}$ for R
$R = \dfrac{E-Ir}{I}$ or $R = \dfrac{E}{I} - r$

56. $I = \dfrac{E}{R+r}$ for r
$r = \dfrac{E-IR}{I}$ or $r = \dfrac{E}{I} - R$

57. $h = \dfrac{2A}{B+b}$ for B
$B = \dfrac{2A-hb}{h}$ or $B = \dfrac{2A}{h} - b$

58. $\dfrac{C}{\pi r} = 2$ for r
$r = \dfrac{C}{2\pi}$

59. $\dfrac{V}{\pi h} = r^2$ for h
$h = \dfrac{V}{r^2\pi}$

60. $\dfrac{V}{lw} = h$ for w
$w = \dfrac{V}{lh}$

61. $x = \dfrac{at+b}{t}$ for t
$t = \dfrac{b}{x-a}$ or $t = \dfrac{-b}{a-x}$

62. $\dfrac{T+mf}{m} = g$ for m
$m = \dfrac{T}{g-f}$ or $m = \dfrac{-T}{f-g}$

63. $\dfrac{x-y}{xy} = z$ for x
$x = \dfrac{y}{1-yz}$ or $x = \dfrac{-y}{yz-1}$

64. $\dfrac{w-n}{wn} = P$ for w
$w = \dfrac{n}{1-Pn}$ or $w = \dfrac{-n}{Pn-1}$

65. $a+b = \dfrac{2A}{h}$ for h $h = \dfrac{2A}{a+b}$

66. $1+rt = \dfrac{A}{P}$ for P
$P = \dfrac{A}{1+rt}$

67. $\dfrac{1}{R} = \dfrac{1}{R_1} + \dfrac{1}{R_2}$ for R
$R = \dfrac{R_1 R_2}{R_2 + R_1}$

68. $\dfrac{b+a}{ab} = \dfrac{1}{f}$ for b
$b = \dfrac{fa}{a-f}$ or $b = \dfrac{-fa}{f-a}$

Problem Recognition Exercises

Comparing Rational Equations and Rational Expressions

Often adding or subtracting rational expressions is confused with solving rational equations. When adding rational expressions, we combine the terms to simplify the expression. When solving an equation, we clear the fractions and find numerical solutions, if possible. Both processes begin with finding the LCD, but the LCD is used differently in each process. Compare these two examples.

Example 1:

Add. $\dfrac{4}{x} + \dfrac{x}{3}$ (The LCD is $3x$.)

$= \dfrac{3}{3} \cdot \left(\dfrac{4}{x}\right) + \left(\dfrac{x}{3}\right) \cdot \dfrac{x}{x}$

$= \dfrac{12}{3x} + \dfrac{x^2}{3x}$

$= \dfrac{12 + x^2}{3x}$ The final answer is a rational expression.

Example 2:

Solve. $\dfrac{4}{x} + \dfrac{x}{3} = -\dfrac{8}{3}$ (The LCD is $3x$.)

$\dfrac{3x}{1}\left(\dfrac{4}{x} + \dfrac{x}{3}\right) = \dfrac{3x}{1}\left(-\dfrac{8}{3}\right)$

$12 + x^2 = -8x$

$x^2 + 8x + 12 = 0$

$(x + 2)(x + 6) = 0$

$x + 2 = 0 \text{ or } x + 6 = 0$

$x = -2 \text{ or } x = -6$ The final answers are numbers. The solutions -2 and -6 both check in the original equation.

For Exercises 1–12, solve the equation or simplify the expression by combining the terms.

1. $\dfrac{y}{2y + 4} - \dfrac{2}{y^2 + 2y}$ $\dfrac{y - 2}{2y}$

2. $\dfrac{1}{x + 2} + 2 = \dfrac{x + 11}{x + 2}$ 6

3. $\dfrac{5t}{2} - \dfrac{t - 2}{3} = 5$ 2

4. $3 - \dfrac{2}{a - 5}$ $\dfrac{3a - 17}{a - 5}$

5. $\dfrac{7}{6p^2} + \dfrac{2}{9p} + \dfrac{1}{3p^2}$ $\dfrac{4p + 27}{18p^2}$

6. $\dfrac{3b}{b + 1} - \dfrac{2b}{b - 1}$ $\dfrac{b(b - 5)}{(b - 1)(b + 1)}$

7. $4 + \dfrac{2}{h - 3} = 5$ 5

8. $\dfrac{2}{w + 1} + \dfrac{3}{(w + 1)^2}$ $\dfrac{2w + 5}{(w + 1)^2}$

9. $\dfrac{1}{x - 6} - \dfrac{3}{x^2 - 6x} = \dfrac{4}{x}$ 7

10. $\dfrac{3}{m} - \dfrac{6}{5} = -\dfrac{3}{m}$ 5

11. $\dfrac{7}{2x + 2} + \dfrac{3x}{4x + 4}$ $\dfrac{3x + 14}{4(x + 1)}$

12. $\dfrac{10}{2t - 1} - 1 = \dfrac{t}{2t - 1}$ $\dfrac{11}{3}$

Section 7.7 Applications of Rational Equations and Proportions

Objectives

1. Solving Proportions
2. Applications of Proportions and Similar Triangles
3. Distance, Rate, and Time Applications
4. Work Applications

1. Solving Proportions

In this section, we look at how rational equations can be used to solve a variety of applications. The first type of rational equation that will be applied is called a proportion.

> **DEFINITION** Proportion
>
> An equation that equates two ratios or rates is called a **proportion**. Thus, for $b \neq 0$ and $d \neq 0$, $\frac{a}{b} = \frac{c}{d}$ is a proportion.

A proportion can be solved by multiplying both sides of the equation by the LCD and clearing fractions.

Skill Practice

Solve the proportion.

1. $\dfrac{10}{b} = \dfrac{2}{33}$

Classroom Example: p. 555, Exercise 10

Example 1 Solving a Proportion

Solve the proportion. $\quad \dfrac{3}{11} = \dfrac{123}{w}$

Solution:

$$\frac{3}{11} = \frac{123}{w} \qquad \text{The LCD is } 11w.$$

$$11w\left(\frac{3}{11}\right) = 11w\left(\frac{123}{w}\right) \qquad \text{Multiply by the LCD and clear fractions.}$$

$$3w = 11 \cdot 123 \qquad \text{Solve the resulting equation (linear).}$$

$$3w = 1353$$

$$\frac{3w}{3} = \frac{1353}{3}$$

$$w = 451$$

Check: $w = 451$

$$\frac{3}{11} = \frac{123}{w}$$

$$\frac{3}{11} \stackrel{?}{=} \frac{123}{(451)}$$

The solution is 451.

$$\frac{3}{11} \stackrel{?}{=} \frac{3}{11} \quad \checkmark \text{ (True)} \qquad \text{Simplify to lowest terms.}$$

Answer

1. 165

TIP: The cross products of any proportion are equal. That is, for $b \neq 0$ and $d \neq 0$, the proportion $\frac{a}{b} = \frac{c}{d}$ is equivalent to $ad = bc$. Some rational equations are proportions and can be solved by equating the cross products. Consider the proportion from Example 1:

$$\frac{3}{11} \diagdown\!\!\!\!\diagup \frac{123}{w}$$

$$3 \cdot w = 11 \cdot 123 \qquad \text{Equate the cross products.}$$

$$3w = 1353 \qquad \text{Solve the resulting equation.}$$

$$\frac{3w}{3} = \frac{1353}{3}$$

$$w = 451 \qquad \text{The solution is 451.}$$

Instructor Note: Remind students that when $\frac{a}{b} = \frac{c}{d}$ is rewritten as $ad = bc$ we still have excluded values $b \neq 0$ and $d \neq 0$. Similarly if we rewrite $\frac{3}{x + 7} = \frac{5}{x + 2}$ as $3(x + 2) = 5(x + 7)$ we still have excluded values -7 and -2 for x.

2. Applications of Proportions and Similar Triangles

Example 2 Using a Proportion in an Application

For a recent year, the population of Alabama was approximately 4.2 million. At that time, Alabama had seven representatives in the U.S. House of Representatives. In the same year, North Carolina had a population of approximately 7.2 million. If representation in the House is based on population in equal proportions for each state, how many representatives did North Carolina have?

Solution:

Let x represent the number of representatives for North Carolina.

Set up a proportion by writing two equivalent ratios.

$$\boxed{\frac{\text{Population of Alabama}}{\text{number of representatives}}} \to \frac{4.2}{7} = \frac{7.2}{x} \leftarrow \boxed{\frac{\text{Population of North Carolina}}{\text{number of representatives}}}$$

$$\frac{4.2}{7} = \frac{7.2}{x}$$

$$7x \cdot \frac{4.2}{7} = 7x \cdot \frac{7.2}{x} \qquad \text{Multiply by the LCD, } 7x.$$

$$4.2x = (7.2)(7) \qquad \text{Solve the resulting linear equation.}$$

$$4.2x = 50.4$$

$$\frac{4.2x}{4.2} = \frac{50.4}{4.2}$$

$$x = 12 \qquad \text{North Carolina had 12 representatives.}$$

Skill Practice

2. A university has a ratio of students to faculty of 105 to 2. If the student population at the university is 15,750, how many faculty members are needed?

Classroom Example: p. 556, Exercise 28

TIP: The equation from Example 2 could have been solved by first equating the cross products:

$$\frac{4.2}{7} \diagdown\!\!\!\!\diagup \frac{7.2}{x}$$

$$4.2x = (7.2)(7)$$

$$4.2x = 50.4$$

$$x = 12$$

Proportions are used in geometry with **similar triangles**. Two triangles are similar if their angles have equal measure. In such a case, the lengths of the corresponding sides are proportional. The triangles in Figure 7-1 are similar. Therefore, the following ratios are equivalent.

$$\frac{a}{x} = \frac{b}{y} = \frac{c}{z}$$

Answer

2. 300

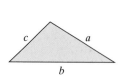
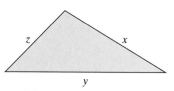

Figure 7-1

Classroom Example: p. 556,
Exercise 34

Skill Practice

3. The two triangles shown are similar triangles. Solve for the lengths of the missing sides.

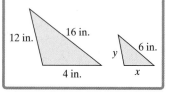

Example 3 **Using Similar Triangles to Find an Unknown Side in a Triangle**

The triangles in Figure 7-2 are similar.

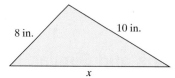

Figure 7-2

a. Solve for x.   **b.** Solve for y.

Solution:

a. The lengths of the upper right sides of the triangles are given. These form a known ratio of $\frac{10}{6}$. Because the triangles are similar, the ratio of the other corresponding sides must be equal to $\frac{10}{6}$. To solve for x, we have:

$$\boxed{\frac{\text{Bottom side from large triangle}}{\text{bottom side from small triangle}}} \rightarrow \frac{x}{9 \text{ in.}} = \frac{10 \text{ in.}}{6 \text{ in.}} \leftarrow \boxed{\frac{\text{Right side from large triangle}}{\text{right side from small triangle}}}$$

$$\frac{x}{9} = \frac{10}{6} \qquad \text{The LCD is 18.}$$

$$\overset{2}{\cancel{18}} \cdot \left(\frac{x}{9}\right) = \overset{3}{\cancel{18}} \cdot \left(\frac{10}{\cancel{6}}\right) \qquad \text{Multiply by the LCD.}$$

$$2x = 30 \qquad \text{Clear fractions.}$$

$$x = 15 \qquad \text{Divide by 2.}$$

The length of side x is 15 in.

b. To solve for y, the ratio of the upper left sides of the triangles must equal $\frac{10}{6}$.

$$\boxed{\frac{\text{Left side from large triangle}}{\text{left side from small triangle}}} \rightarrow \frac{8 \text{ in.}}{y} = \frac{10 \text{ in.}}{6 \text{ in.}} \leftarrow \boxed{\frac{\text{Right side from large triangle}}{\text{right side from small triangle}}}$$

$$\frac{8}{y} = \frac{10}{6} \qquad \text{The LCD is } 6y.$$

$$6\cancel{y} \cdot \left(\frac{8}{\cancel{y}}\right) = \cancel{6}y \cdot \left(\frac{10}{\cancel{6}}\right) \qquad \text{Multiply by the LCD.}$$

$$48 = 10y \qquad \text{Clear fractions.}$$

$$\frac{48}{10} = \frac{10y}{10}$$

$$4.8 = y$$

The length of side y is 4.8 in.

Answer

3. $x = 1.5$ in., and $y = 4.5$ in.

| Example 4 | **Using Similar Triangles in an Application** |

The shadow cast by a yardstick is 2 ft long. The shadow cast by a tree is 11 ft long. Find the height of the tree.

Solution: **Step 1:** Read the problem.

Let x represent the height of the tree.   **Step 2:** Label the variables.

We will assume that the measurements were taken at the same time of day. Therefore, the angle of the Sun is the same on both objects, and we can set up similar triangles (Figure 7-3).

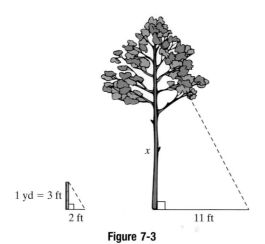

Figure 7-3

┌─ **Skill Practice** ─

4. The Sun casts a 3.2-ft shadow of a 6 ft man. At the same time, the Sun casts a 80-ft shadow of a building. How tall is the building?

Classroom Example: p. 557, Exercise 38

Step 3: Create a verbal model.

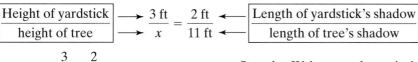

$$\frac{3}{x} = \frac{2}{11}$$

Step 4: Write a mathematical equation.

$$11x\left(\frac{3}{x}\right) = \left(\frac{2}{11}\right)11x$$

Step 5: Multiply by the LCD.

$$33 = 2x$$

Solve the equation.

$$\frac{33}{2} = \frac{2x}{2}$$

$$16.5 = x$$

Step 6: Interpret the results and write the answer in words.

The tree is 16.5 ft high.

3. Distance, Rate, and Time Applications

In Sections 2.7 and 4.4 we presented applications involving the relationship among the variables distance, rate, and time. Recall that $d = rt$.

Answer

4. The building is 150 ft tall.

Skill Practice

5. Alison paddles her kayak in a river where the current of the water is 2 mph. She can paddle 20 mi with the current in the same time that she can paddle 10 mi against the current. Find the speed of the kayak in still water.

Classroom Example: p. 557, Exercise 42

Example 5 **Using a Rational Equation in a Distance, Rate, and Time Application**

A small plane flies 440 mi with the wind from Memphis, TN, to Oklahoma City, OK. In the same amount of time, the plane flies 340 miles against the wind from Oklahoma City to Little Rock, AR (see Figure 7-4). If the wind speed is 30 mph, find the speed of the plane in still air.

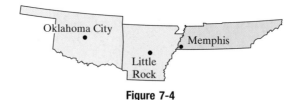

Figure 7-4

Solution:

Let x represent the speed of the plane in still air.

Then $x + 30$ is the speed of the plane with the wind.

$x - 30$ is the speed of the plane against the wind.

Organize the given information in a chart.

	Distance	Rate	Time
With the wind	440	$x + 30$	$\dfrac{440}{x + 30}$
Against the wind	340	$x - 30$	$\dfrac{340}{x - 30}$

Because $d = rt$, then $t = \dfrac{d}{r}$

TIP: The equation

$$\frac{440}{x + 30} = \frac{340}{x - 30}$$

is a proportion. The fractions can also be cleared by equating the cross products.

$$\frac{440}{x + 30} \diagup\!\!\!\!\diagdown \frac{340}{x - 30}$$

$$440(x - 30) = 340(x + 30)$$

The plane travels with the wind for the same amount of time as it travels against the wind, so we can equate the two expressions for time.

$$\left(\begin{array}{c}\text{Time with}\\ \text{the wind}\end{array}\right) = \left(\begin{array}{c}\text{time against}\\ \text{the wind}\end{array}\right)$$

$$\frac{440}{x + 30} = \frac{340}{x - 30}$$

The LCD is $(x + 30)(x - 30)$.

$$(x + 30)(x - 30) \cdot \frac{440}{x + 30} = (x + 30)(x - 30) \cdot \frac{340}{x - 30}$$

$$440(x - 30) = 340(x + 30)$$

$$440x - 13{,}200 = 340x + 10{,}200$$

Solve the resulting linear equation.

$$100x = 23{,}400$$

$$x = 234$$

The plane's speed in still air is 234 mph.

Answer

5. 6 mph

Example 6 Using a Rational Equation in a Distance, Rate, and Time Application

A motorist drives 100 mi between two cities in a bad rainstorm. For the return trip in sunny weather, she averages 10 mph faster and takes $\frac{1}{2}$ hr less time. Find the average speed of the motorist in the rainstorm and in sunny weather.

Solution:

Let x represent the motorist's speed during the rain.

Then $x + 10$ represents the speed in sunny weather.

	Distance	Rate	Time
Trip during rainstorm	100	x	$\dfrac{100}{x}$
Trip during sunny weather	100	$x + 10$	$\dfrac{100}{x + 10}$

Because $d = rt$, then $t = \dfrac{d}{r}$

Because the same distance is traveled in $\frac{1}{2}$ hr less time, the difference between the time of the trip during the rainstorm and the time during sunny weather is $\frac{1}{2}$ hr.

$$\left(\begin{array}{c} \text{Time during} \\ \text{the rainstorm} \end{array} \right) - \left(\begin{array}{c} \text{time during} \\ \text{sunny weather} \end{array} \right) = \left(\frac{1}{2} \text{ hr} \right) \qquad \text{Verbal model}$$

$$\frac{100}{x} - \frac{100}{x + 10} = \frac{1}{2} \qquad \text{Mathematical equation}$$

$$2x(x + 10)\left(\frac{100}{x} - \frac{100}{x + 10} \right) = 2x(x + 10)\left(\frac{1}{2} \right) \qquad \text{Multiply by the LCD.}$$

$$2x(x + 10)\left(\frac{100}{x} \right) - 2x(x + 10)\left(\frac{100}{x + 10} \right) = 2x(x + 10)\left(\frac{1}{2} \right) \qquad \text{Apply the distributive property.}$$

$$200(x + 10) - 200x = x(x + 10) \qquad \text{Clear fractions.}$$

$$200x + 2000 - 200x = x^2 + 10x \qquad \text{Solve the resulting equation (quadratic).}$$

$$2000 = x^2 + 10x$$

$$0 = x^2 + 10x - 2000 \qquad \text{Set the equation equal to zero.}$$

$$0 = (x - 40)(x + 50) \qquad \text{Factor.}$$

$$x = 40 \qquad \text{or} \qquad x = -50$$

Because a rate of speed cannot be negative, reject $x = -50$. Therefore, the speed of the motorist in the rainstorm is 40 mph. Because $x + 10 = 40 + 10 = 50$, the average speed for the return trip in sunny weather is 50 mph.

Skill Practice

6. Harley rode his mountain bike 12 mi to the top of the mountain and the same distance back down. His speed going up was 8 mph slower than coming down. The ride up took 2 hr longer than the ride coming down. Find his speeds.

Classroom Example: p. 558, Exercise 48

Avoiding Mistakes

The equation

$$\frac{100}{x} - \frac{100}{x + 10} = \frac{1}{2}$$

is not a proportion because the left-hand side has more than one fraction. Do not try to multiply the cross products. Instead, multiply by the LCD to clear fractions.

Answer

6. Uphill speed was 4 mph; downhill speed was 12 mph.

4. Work Applications

Example 7 demonstrates how work rates are related to a portion of a job that can be completed in one unit of time.

Skill Practice

7. The computer at a bank can process and prepare the bank statements in 30 hr. A new faster computer can do the job in 20 hr. If the bank uses both computers together, how long will it take to process the statements?

Classroom Example: p. 558, Exercise 54

Example 7   Using a Rational Equation in a Work Problem

A new printing press can print the morning edition in 2 hr, whereas the old printer required 4 hr. How long would it take to print the morning edition if both printers were working together?

Solution:

Let x represent the time required for both printers working together to complete the job.

One method to approach this problem is to determine the portion of the job that each printer can complete in 1 hr and extend that rate to the portion of the job completed in x hr.

- The old printer can perform the job in 4 hr. Therefore, it completes $\frac{1}{4}$ of the job in 1 hr and $\frac{1}{4}x$ jobs in x hr.
- The new printer can perform the job in 2 hr. Therefore, it completes $\frac{1}{2}$ of the job in 1 hr and $\frac{1}{2}x$ jobs in x hr.

	Work Rate	Time	Portion of Job Completed
Old printer	$\dfrac{1 \text{ job}}{4 \text{ hr}}$	x hours	$\dfrac{1}{4}x$
New printer	$\dfrac{1 \text{ job}}{2 \text{ hr}}$	x hours	$\dfrac{1}{2}x$

The sum of the portions of the job completed by each printer must equal one whole job.

$$\begin{pmatrix} \text{Portion of job} \\ \text{completed by} \\ \text{old printer} \end{pmatrix} + \begin{pmatrix} \text{portion of job} \\ \text{completed by} \\ \text{new printer} \end{pmatrix} = \begin{pmatrix} 1 \\ \text{whole} \\ \text{job} \end{pmatrix}$$

$\dfrac{1}{4}x + \dfrac{1}{2}x = 1$ The LCD is 4.

$4\left(\dfrac{1}{4}x + \dfrac{1}{2}x\right) = 4(1)$ Multiply by the LCD.

$\overset{1}{\cancel{4}} \cdot \dfrac{1}{\cancel{4}}x + \overset{2}{\cancel{4}} \cdot \dfrac{1}{\cancel{2}}x = 4 \cdot 1$ Apply the distributive property.

$x + 2x = 4$ Solve the resulting linear equation.

$3x = 4$

$x = \dfrac{4}{3}$ or $x = 1\dfrac{1}{3}$ The time required to print the morning edition using both printers is $1\frac{1}{3}$ hr.

Answer

7. 12 hr

TIP: An alternative approach to solving a "work" problem is to add rates of speed. In Example 7, we could have set up an equation as follows.

$$\left(\begin{array}{c}\text{Rate of speed}\\\text{of old printer}\end{array}\right) + \left(\begin{array}{c}\text{rate of speed}\\\text{of new printer}\end{array}\right) = \left(\begin{array}{c}\text{rate of speed of}\\\text{both working together}\end{array}\right)$$

$$\frac{1\ \text{job}}{4\ \text{hr}} + \frac{1\ \text{job}}{2\ \text{hr}} = \frac{1\ \text{job}}{x\ \text{hr}}$$

$$\frac{1}{4} + \frac{1}{2} = \frac{1}{x}$$

$$4x\left(\frac{1}{4} + \frac{1}{2}\right) = 4x\left(\frac{1}{x}\right) \qquad \text{Multiply by the LCD, } 4x.$$

$$x + 2x = 4$$

$$3x = 4$$

$$x = \frac{4}{3} \qquad \text{The time required for both printers working together is } 1\frac{1}{3} \text{ hr.}$$

Section 7.7 Practice Exercises

Study Skills Exercise

1. Define the key terms:

    a. proportion   **b. similar triangles**

Review Exercises

For Exercises 2–7, determine whether each of the following is an equation or an expression. If it is an equation, solve it. If it is an expression, perform the indicated operation.

2. $\dfrac{b}{5} + 3 = 9$ Equation; 30

3. $\dfrac{m}{m-1} - \dfrac{2}{m+3}$ Expression; $\dfrac{m^2 + m + 2}{(m-1)(m+3)}$

4. $\dfrac{2}{a+5} + \dfrac{5}{a^2 - 25}$ Expression; $\dfrac{2a-5}{(a+5)(a-5)}$

5. $\dfrac{3y+6}{20} \div \dfrac{4y+8}{8}$ Expression; $\dfrac{3}{10}$

6. $\dfrac{z^2 + z}{24} \cdot \dfrac{8}{z+1}$ Expression; $\dfrac{z}{3}$

7. $\dfrac{3}{p+3} = \dfrac{12p+19}{p^2 + 7p + 12} - \dfrac{5}{p+4}$ Equation; 2

8. Determine whether 1 is a solution to the equation. $\dfrac{1}{x-1} + \dfrac{1}{2} = \dfrac{2}{x^2 - 1}$ No

Objective 1: Solving Proportions

For Exercises 9–22, solve the proportions. **(See Example 1.)**

9. $\dfrac{8}{5} = \dfrac{152}{p}$ 95

10. $\dfrac{6}{7} = \dfrac{96}{y}$ 112

11. $\dfrac{19}{76} = \dfrac{z}{4}$ 1

12. $\dfrac{15}{135} = \dfrac{w}{9}$ 1

13. $\dfrac{5}{3} = \dfrac{a}{8}$ $\dfrac{40}{3}$

14. $\dfrac{b}{14} = \dfrac{3}{8}$ $\dfrac{21}{4}$

15. $\dfrac{2}{1.9} = \dfrac{x}{38}$ 40

16. $\dfrac{16}{1.3} = \dfrac{30}{p}$ 2.4375

17. $\dfrac{y+1}{2y} = \dfrac{2}{3}$ 3

18. $\dfrac{w-2}{4w} = \dfrac{1}{6}$ 6

19. $\dfrac{9}{2z-1} = \dfrac{3}{z}$ −1

20. $\dfrac{1}{t} = \dfrac{1}{4-t}$ 2

21. $\dfrac{8}{9a-1} = \dfrac{5}{3a+2}$ 1

22. $\dfrac{4p+1}{3} = \dfrac{2p-5}{6}$ $-\dfrac{7}{6}$

23. Charles' law describes the relationship between the initial and final temperature and volume of a gas held at a constant pressure.

$$\dfrac{V_i}{V_f} = \dfrac{T_i}{T_f}$$

  a. Solve the equation for V_f. $V_f = \dfrac{V_i T_f}{T_i}$

  b. Solve the equation for T_f. $T_f = \dfrac{T_i V_f}{V_i}$

24. The relationship between the area, height, and base of a triangle is given by the proportion

$$\dfrac{A}{b} = \dfrac{h}{2}$$

  a. Solve the equation for A. $A = \dfrac{hb}{2}$

  b. Solve the equation for b. $b = \dfrac{2A}{h}$

Objective 2: Applications of Proportions and Similar Triangles

For Exercises 25–32, solve using proportions.

25. Toni drives her Honda Civic 132 mi on the highway on 4 gal of gas. At this rate how many miles can she drive on 9 gal of gas? **(See Example 2.)**   Toni can drive 297 mi on 9 gal of gas.

26. Tim takes his pulse for 10 sec and counts 12 beats. How many beats per minute is this?
This is 72 beats/min.

27. Suppose a household of 4 people produces 128 lb of garbage in one week. At this rate, how many pounds will 48 people produce in 1 week?   They would produce 1536 lb.

28. Property tax on a $180,000 house is $4000. At this rate, how much tax would be paid on a $216,000 home?
The tax would be $4800.

29. Andrew is on a low-carbohydrate diet. If his diet book tells him that an 8-oz serving of pineapple contains 19.2 g of carbohydrate, how many grams of carbohydrate does a 5-oz serving contain?
5 oz contains 12 g of carbohydrate.

30. Cooking oatmeal requires 1 cup of water for every $\frac{1}{2}$ cup of oats. How many cups of water will be required for $\frac{3}{4}$ cup of oats?   1.5 cups of water would be necessary.

31. A map has a scale of 75 mi/in. If two cities measure 3.5 in. apart, how many miles does this represent?
This represents 262.5 mi.

32. A map has a scale of 50 mi/in. If two cities measure 6.5 in. apart, how many miles does this represent?
This represents 325 mi.

For Exercises 33–36, the figures are similar. Solve for x and y. **(See Example 3.)**

33.

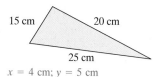

$x = 4$ cm; $y = 5$ cm

34.

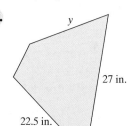

$x = 10$ in.; $y = 18$ in.

35. $x = 3.75$ cm; $y = 4.5$ cm

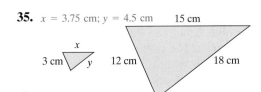

36.

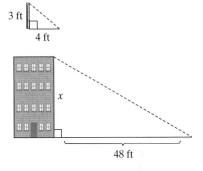

$x = 15$ ft; $y = 21$ ft

37. To estimate the height of a light pole, a mathematics student measures the length of a shadow cast by a meterstick and the length of the shadow cast by the light pole. Find the height of the light pole. **(See Example 4.)** The height of the pole is 7 m.

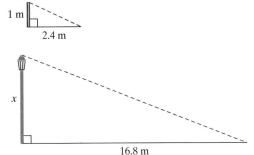

38. To estimate the height of a building, a student measures the length of a shadow cast by a yardstick and the length of the shadow cast by the building. Find the height of the building. The height is 36 ft.

39. A 6-ft-tall man standing 54 ft from a light post casts an 18-ft shadow. What is the height of the light post? The light post is 24 ft high.

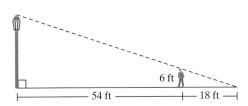

40. For a science project at school, a student must measure the height of a tree. The student measures the length of the shadow of the tree and then measures the length of the shadow cast by a yardstick. Use similar triangles to find the height of the tree. The tree is 36 ft high.

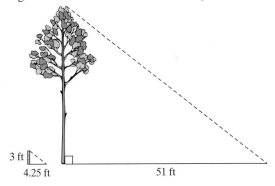

Objective 3: Distance, Rate, and Time Applications

41. A boat travels 54 mi upstream against the current in the same amount of time it takes to travel 66 mi downstream with the current. If the current is 2 mph, what is the speed of the boat in still water? (Use $t = \frac{d}{r}$ to complete the table.) **(See Example 5.)** The speed of the boat is 20 mph.

	Distance	Rate	Time
With the current (downstream)			
Against the current (upstream)			

42. A plane flies 630 mi with the wind in the same time that it takes to fly 455 mi against the wind. If this plane flies at the rate of 217 mph in still air, what is the speed of the wind? (Use $t = \frac{d}{r}$ to complete the table.) The wind speed is 35 mph.

	Distance	Rate	Time
With the wind			
Against the wind			

 Writing   Translating Expression   Geometry   Scientific Calculator   Video   NS&E

43. The jet stream is a fast flowing air current found in the atmosphere at around 36,000 ft above the surface of the Earth. During one summer day, the speed of the jet stream is 35 mph. A plane flying with the jet stream can fly 700 mi in the same amount of time that it would take to fly 500 mi against the jet stream. What is the speed of the plane in still air? The plane flies 210 mph in still air.

44. A fisherman travels 9 mi downstream with the current in the same time that he travels 3 mi upstream against the current. If the speed of the current is 6 mph, what is the speed at which the fisherman travels in still water?   The speed of the boat in still water is 12 mph.

45. An athlete in training rides his bike 20 mi and then immediately follows with a 10-mi run. The total workout takes him 2.5 hr. He also knows that he bikes about twice as fast as he runs. Determine his biking speed and his running speed.   He runs 8 mph and bikes 16 mph.

46. Devon can cross-country ski 5 km/hr faster than his sister Shanelle. Devon skis 45 km in the same time Shanelle skis 30 km. Find their speeds.   Shanelle skis 10 km/hr and Devon skis 15 km/hr.

47. Floyd can walk 2 mph faster than his wife, Rachel. It takes Rachel 3 hr longer than Floyd to hike a 12-mi trail through the park. Find their speeds. **(See Example 6.)**   Floyd walks 4 mph and Rachel walks 2 mph.

48. Janine bikes 3 mph faster than her sister, Jessica. Janine can ride 36 mi in 1 hr less time than Jessica can ride the same distance. Find each of their speeds.   Janine's speed is 12 mph and Jessica's is 9 mph.

49. Sergio rode his bike 4 mi. Then he got a flat tire and had to walk back 4 mi. It took him 1 hr longer to walk than it did to ride. If his rate walking was 9 mph less than his rate riding, find the two rates.   Sergio rode 12 mph and walked 3 mph.

50. Amber jogs 10 km in $\frac{3}{4}$ hr less than she can walk the same distance. If her walking rate is 3 km/hr less than her jogging rate, find her rates jogging and walking (in km/hr).   Amber jogs 8 km/hr and walks 5 km/hr.

Objective 4: Work Applications

51. If it takes a person 2 hr to paint a room, what fraction of the room would be painted in 1 hr? $\frac{1}{2}$ of the room

52. If it takes a copier 3 hr to complete a job, what fraction of the job would be completed in 1 hr? $\frac{1}{3}$ of the job

53. If the cold-water faucet is left on, the sink will fill in 10 min. If the hot-water faucet is left on, the sink will fill in 12 min. How long would it take the sink to fill if both faucets are left on? **(See Example 7.)** $5\frac{5}{11}$ $(5.\overline{45})$ min

54. The CUT-IT-OUT lawn mowing company consists of two people: Tina and Bill. If Tina cuts a lawn by herself, she can do it in 4 hr. If Bill cuts the same lawn himself, it takes him an hour longer than Tina. How long would it take them if they worked together? $2\frac{2}{9}$ $(2.\overline{2})$ hr

55. A manuscript needs to be printed. One printer can do the job in 50 min, and another printer can do the job in 40 min. How long would it take if both printers were used? $22\frac{2}{9}$ $(22.\overline{2})$ min

56. A pump can empty a small pond in 4 hr. Another more efficient pump can do the job in 3 hr. How long would it take to empty the pond if both pumps were used? $1\frac{5}{7}$ (approximately 1.7) hr

57. A pipe can fill a reservoir in 16 hr. A drainage pipe can drain the reservoir in 24 hr. How long would it take to fill the reservoir if the drainage pipe were left open by mistake? (*Hint:* The rate at which water drains should be negative.)   48 hr

58. A hole in the bottom of a child's plastic swimming pool can drain the pool in 60 min. If the pool had no hole, a hose could fill the pool in 40 min. How long would it take the hose to fill the pool with the hole?   120 min

59. Tim and Al are bricklayers. Tim can construct an outdoor grill in 5 days. If Al helps Tim, they can build it in only 2 days. How long would it take Al to build the grill alone? $3\frac{1}{3}$ $(3.\overline{3})$ days

60. Norma is a new and inexperienced secretary. It takes her 3 hr to prepare a mailing. If her boss helps her, the mailing can be completed in 1 hr. How long would it take the boss to do the job by herself? $1\frac{1}{2}$ (1.5) hr

 Writing   Translating Expression   Geometry   Scientific Calculator   Video   NS/E

Expanding Your Skills

For Exercises 61–64, solve using proportions.

61. The ratio of smokers to nonsmokers in a restaurant is 2 to 7. There are 100 more nonsmokers than smokers. How many smokers and nonsmokers are in the restaurant?   There are 40 smokers and 140 nonsmokers.

62. The ratio of fiction to nonfiction books sold in a bookstore is 5 to 3. One week there were 180 more fiction books sold than nonfiction. Find the number of fiction and nonfiction books sold during that week.   There are 450 fiction and 270 nonfiction books sold.

63. There are 440 students attending a biology lecture. The ratio of male to female students at the lecture is 6 to 5. How many men and women are attending the lecture?   There are 240 men and 200 women.

64. The ratio of dogs to cats at the humane society is 5 to 8. The total number of dogs and cats is 650. How many dogs and how many cats are at the humane society?   There are 250 dogs and 400 cats.

Variation

1. Definition of Direct and Inverse Variation

In this section, we introduce the concept of variation. Direct and inverse variation models can show how one quantity varies in proportion to another.

Objectives

1. **Definition of Direct and Inverse Variation**
2. **Translations Involving Variation**
3. **Applications of Variation**

> **DEFINITION** **Direct and Inverse Variation**
>
> Let k be a nonzero constant real number. Then the following statements are equivalent:
>
> 1. y varies **directly** as x.
> y is directly proportional to x. $\Big\}$ $y = kx$
>
> 2. y varies **inversely** as x.
> y is inversely proportional to x. $\Big\}$ $y = \dfrac{k}{x}$
>
> *Note:* The value of k is called the constant of variation.

For a car traveling 30 mph, the equation $d = 30t$ indicates that the distance traveled is *directly proportional* to the time of travel. For positive values of k, when two variables are directly related, as one variable increases, the other variable will also increase. Likewise, if one variable decreases, the other will decrease. In the equation $d = 30t$, the longer the time of the trip, the greater the distance traveled. The shorter the time of the trip, the shorter the distance traveled.

For positive values of k, when two variables are *inversely related*, as one variable increases, the other will decrease, and vice versa. Consider a car traveling between Toronto and Montreal, a distance of 500 km. The time required to make the trip is inversely proportional to the speed of travel: $t = 500/r$. As the rate of speed, r, increases, the quotient $500/r$ will decrease. Thus the time will decrease. Similarly, as the rate of speed decreases, the trip will take longer.

2. Translations Involving Variation

The first step in using a variation model is to translate an English phrase into an equivalent mathematical equation.

 Writing   Translating Expression   Geometry   Scientific Calculator   Video   NS&E

Example 1 Translating to a Variation Model

Translate each expression into an equivalent mathematical model.

a. The circumference of a circle varies directly as the radius.

b. At a constant temperature, the volume of a gas varies inversely as the pressure.

c. The length of time of a meeting is directly proportional to the *square* of the number of people present.

Solution:

a. Let C represent circumference and r represent radius. The variables are directly related, so use the model $C = kr$.

b. Let V represent volume and P represent pressure. Because the variables are inversely related, use the model $V = k/P$.

c. Let t represent time, and let N be the number of people present at a meeting. Because t is directly related to N^2, use the model $t = kN^2$.

Sometimes a variable varies directly as the product of two or more other variables. In this case, we have joint variation.

> **DEFINITION** **Joint Variation**
>
> Let k be a nonzero constant real number. Then the following statements are equivalent:
>
> $\left.\begin{array}{l} y \text{ varies } \textbf{jointly} \text{ as } w \text{ and } z. \\ y \text{ is jointly proportional to } w \text{ and } z. \end{array}\right\} \quad y = kwz$

Example 2 Translating to a Variation Model

Translate each expression into an equivalent mathematical model.

a. y varies jointly as u and the square root of v.

b. The gravitational force of attraction between two planets varies jointly as the product of their masses and inversely as the square of the distance between them.

Solution:

a. $y = ku\sqrt{v}$

b. Let m_1 and m_2 represent the masses of the two planets. Let F represent the gravitational force of attraction and d represent the distance between the planets.

The variation model is $F = \dfrac{km_1m_2}{d^2}$

3. Applications of Variation

Consider the variation models $y = kx$ and $y = k/x$. In either case, if values for x and y are known, we can solve for k. Once k is known, we can use the variation equation to find y if x is known, or to find x if y is known. This concept is the basis for solving many problems involving variation.

> **PROCEDURE** Finding a Variation Model
>
> **Step 1** Write a general variation model that relates the variables given in the problem. Let k represent the constant of variation.
> **Step 2** Solve for k by substituting known values of the variables into the model from step 1.
> **Step 3** Substitute the value of k into the original variation model from step 1.

Example 3 Solving an Application Involving Direct Variation

The variable z varies directly as w. When w is 16, z is 56.

a. Write a variation model for this situation. Use k as the constant of variation.

b. Solve for the constant of variation.

c. Find the value of z when w is 84.

Solution:

a. $z = kw$

b. $z = kw$

$56 = k(16)$ Substitute known values for z and w. Then solve for the unknown value of k.

$\dfrac{56}{16} = \dfrac{k(16)}{16}$ To isolate k, divide both sides by 16.

$\dfrac{7}{2} = k$ Simplify $\dfrac{56}{16}$ to $\dfrac{7}{2}$.

c. With the value of k known, the variation model can now be written as $z = \dfrac{7}{2}w$.

$z = \dfrac{7}{2}(84)$ To find z when $w = 84$, substitute $w = 84$ into the equation.

$z = 294$

Skill Practice

The variable t varies directly as the square of v. When v is 8, t is 32.

6. Write a variation model for this relationship.
7. Solve for the constant of variation.
8. Find t when $v = 10$.

Classroom Examples: p. 564, Exercises 24 and 32

Example 4 Solving an Application Involving Direct Variation

The speed of a racing canoe in still water varies directly as the square root of the length of the canoe.

a. If a 16-ft canoe can travel 6.2 mph in still water, find a variation model that relates the speed of a canoe to its length.

b. Find the speed of a 25-ft canoe.

Solution:

a. Let s represent the speed of the canoe and L represent the length. The general variation model is $s = k\sqrt{L}$. To solve for k, substitute the known values for s and L.

Skill Practice

9. The amount of water needed by a mountain hiker varies directly as the time spent hiking. The hiker needs 2.4 L for a 4-hr hike. How much water will be needed for a 5-hr hike?

Classroom Example: p. 566, Exercise 46

Answers

6. $t = kv^2$ **7.** $\dfrac{1}{2}$
8. 50   **9.** 3 L

$$s = k\sqrt{L}$$

$$6.2 = k\sqrt{16} \qquad \text{Substitute } s = 6.2 \text{ mph and } L = 16 \text{ ft.}$$

$$6.2 = k \cdot 4$$

$$\frac{6.2}{4} = \frac{\cancel{4}k}{\cancel{4}} \qquad \text{Solve for } k.$$

$$k = 1.55$$

$$s = 1.55\sqrt{L} \qquad \text{Substitute } k = 1.55 \text{ into the model } s = k\sqrt{L}.$$

b. $s = 1.55\sqrt{L}$

$$= 1.55\sqrt{25} \qquad \text{Find the speed when } L = 25 \text{ ft.}$$

$$= 7.75 \text{ mph} \qquad \text{The speed is 7.75 mph.}$$

── **Skill Practice** ──

10. The yield on a bond varies inversely as the price. The yield on a particular bond is 5% when the price is $100. Find the yield when the price is $125.

Classroom Example: p. 566, Exercise 48

Example 5 **Solving an Application Involving Inverse Variation**

The loudness of sound measured in decibels varies inversely as the square of the distance between the listener and the source of the sound. If the loudness of sound is 17.92 decibels at a distance of 10 ft from a stereo speaker, what is the decibel level 20 ft from the speaker?

Solution:

Let L represent the loudness of sound in decibels and d represent the distance in feet. The inverse relationship between decibel level and the square of the distance is modeled by

$$L = \frac{k}{d^2}$$

$$17.92 = \frac{k}{(10)^2} \qquad \text{Substitute } L = 17.92 \text{ decibels and } d = 10 \text{ ft.}$$

$$17.92 = \frac{k}{100}$$

$$(17.92)100 = \frac{k}{100} \cdot 100 \qquad \text{Solve for } k \text{ (clear fractions).}$$

$$k = 1792$$

$$L = \frac{1792}{d^2} \qquad \text{Substitute } k = 1792 \text{ into the original model } L = \frac{k}{d^2}.$$

With the value of k known, we can find L for any value of d.

$$L = \frac{1792}{(20)^2} \qquad \text{Find the loudness when } d = 20 \text{ ft.}$$

$$= 4.48 \text{ decibels} \qquad \text{The loudness is 4.48 decibels.}$$

Notice that the loudness of sound is 17.92 decibels at a distance 10 ft from the speaker. When the distance from the speaker is increased to 20 ft, the decibel level decreases to 4.48 decibels. This is consistent with an inverse relationship. For $k > 0$, as one variable is increased, the other is decreased. It also seems reasonable that the further one moves away from the source of a sound, the softer the sound becomes.

Answer

10. 4%

Example 6 Solving an Application Involving Joint Variation

The kinetic energy of an object varies jointly as the weight of the object at sea level and as the square of its velocity. During a hurricane, a 0.5-lb stone traveling at 60 mph has 81 J (joules) of kinetic energy. Suppose the wind speed doubles to 120 mph. Find the kinetic energy.

Solution:

Let E represent the kinetic energy, let w represent the weight, and let v represent the velocity of the stone. The variation model is

$$E = kwv^2$$

$$81 = k(0.5)(60)^2 \qquad \text{Substitute } E = 81 \text{ J}, w = 0.5 \text{ lb, and } v = 60 \text{ mph.}$$

$$81 = k(0.5)(3600) \qquad \text{Simplify exponents.}$$

$$81 = k(1800)$$

$$\frac{81}{1800} = \frac{k(1800)}{1800} \qquad \text{Divide by 1800.}$$

$$0.045 = k \qquad \text{Solve for } k.$$

With the value of k known, the model $E = kwv^2$ can now be written as $E = 0.045wv^2$. We now find the kinetic energy of a 0.5-lb stone traveling at 120 mph.

$$E = 0.045(0.5)(120)^2$$

$$= 324$$

The kinetic energy of a 0.5-lb stone traveling at 120 mph is 324 J.

In Example 6, when the velocity increased by 2 times, the kinetic energy increased by 4 times (note that $324 \text{ J} = 4 \cdot 81 \text{ J}$). This factor of 4 occurs because the kinetic energy is proportional to the *square* of the velocity. When the velocity increased by 2 times, the kinetic energy increased by 2^2 times.

Section 7.8 Practice Exercises

Boost *your* GRADE at ALEKS.com!

ALEKS® version 3.0

- Practice Problems
- Self-Tests
- NetTutor
- e-Professors
- Videos

For additional exercises, see Classroom Activities 7.8A–7.8C in the *Instructor's Resource Manual* at www.mhhe.com/moh.

Study Skills Exercise

1. Define the key terms:

    a. direct variation   **b. inverse variation**   **c. joint variation**

Review Exercises

For Exercises 2–7, perform the indicated operation, or solve the equation.

2. $\dfrac{5p}{p+2} + \dfrac{10}{p+2}$ 5

3. $\dfrac{2y}{3} - \dfrac{3y-1}{5} = 1$ $y = 12$

4. $\dfrac{3}{q-1} \cdot \dfrac{2q^2 + 3q - 5}{6q + 24}$ $\dfrac{2q+5}{2(q+4)}$

Writing   Translating Expression   Geometry   Scientific Calculator   Video   NS⚸E

5. $\dfrac{a}{4} + \dfrac{3}{a} = 2$ *a = 6, a = 2*

6. $\dfrac{3}{b^2 + 5b - 14} - \dfrac{2}{b^2 - 49}$ $\dfrac{b - 17}{(b + 7)(b - 7)(b - 2)}$

7. $\dfrac{a + \dfrac{a}{b}}{\dfrac{a}{b} - a}$ $\dfrac{b + 1}{1 - b}$

Objective 1: Definition of Direct and Inverse Variation

8. In the equation $r = kt$, does r vary directly or inversely with t?   Directly

9. In the equation $w = \dfrac{k}{v}$, does w vary directly or inversely with v?   Inversely

10. In the equation $P = \dfrac{k \cdot c}{v}$, does P vary directly or inversely as v?   Inversely

Objective 2: Translations Involving Variation

For Exercises 11–22, write a variation model. Use k as the constant of variation. **(See Examples 1–2.)**

11. T varies directly as q. $T = kq$

12. W varies directly as z. $W = kz$

13. b varies inversely as c. $b = \dfrac{k}{c}$

14. m varies inversely as t. $m = \dfrac{k}{t}$

15. Q is directly proportional to x and inversely proportional to y. $Q = \dfrac{kx}{y}$

16. d is directly proportional to p and inversely proportional to n. $d = \dfrac{kp}{n}$

17. c varies jointly as s and t. $c = kst$

18. w varies jointly as p and f. $w = kpf$

19. L varies jointly as w and the square root of v. $L = kw\sqrt{v}$

20. q varies jointly as v and the square root of w. $q = kv\sqrt{w}$

21. x varies directly as the square of y and inversely as z. $x = \dfrac{ky^2}{z}$

22. a varies directly as n and inversely as the square of d. $a = \dfrac{kn}{d^2}$

For Exercises 23–28, find the constant of variation, k. **(See Example 3.)**

23. y varies directly as x and when x is 4, y is 18. $k = \dfrac{9}{2}$

24. m varies directly as x and when x is 8, m is 22. $k = \dfrac{11}{4}$

25. p varies inversely as q and when q is 16, p is 32. $k = 512$

26. T varies inversely as x and when x is 40, T is 200. $k = 8000$

27. y varies jointly as w and v. When w is 50 and v is 0.1, y is 8.75. $k = 1.75$

28. N varies jointly as t and p. When t is 1 and p is 7.5, N is 330. $k = 44$

Objective 3: Applications of Variation

Solve Exercises 29–40 using the steps found on page 561. **(See Example 3.)**

29. x varies directly as p. If $x = 50$ when $p = 10$, find x when p is 14. $x = 70$

30. y is directly proportional to z. If $y = 12$ when $z = 36$, find y when z is 21. $y = 7$

31. b is inversely proportional to c. If b is 4 when c is 3, find b when c is 2. $b = 6$

32. q varies inversely as w. If q is 8 when w is 50, find q when w is 125. $q = 3.2$

Writing   Translating Expression   Geometry   Scientific Calculator   Video   NS&E

33. Z varies directly as the square of w. If $Z = 14$ when $w = 4$, find Z when $w = 8$. $\ Z = 56$

34. m varies directly as the square of x. If $m = 200$ when $x = 20$, find m when x is 32. $\ m = 512$

35. Q varies inversely as the square of p. If $Q = 4$ when $p = 3$, find Q when $p = 2$. $\ Q = 9$

36. z is inversely proportional to the square of t. If $z = 15$ when $t = 4$, find z when $t = 10$. $\ z = 2.4$

37. L varies jointly as a and the square root of b. If $L = 72$ when $a = 8$ and $b = 9$, find L when $a = \frac{1}{2}$ and $b = 36$. $\ L = 9$

38. Y varies jointly as the cube of x and the square root of w. $Y = 128$ when $x = 2$ and $w = 16$. Find Y when $x = \frac{1}{2}$ and $w = 64$. $\ Y = 4$

39. B varies directly as m and inversely as n. $B = 20$ when $m = 10$ and $n = 3$. Find B when $m = 15$ and $n = 12$. $\ B = \dfrac{15}{2}$

40. R varies directly as s and inversely as t. $R = 14$ when $s = 2$ and $t = 9$. Find R when $s = 4$ and $t = 3$. $\ R = 84$

For Exercises 41–56, use a variation model to solve for the unknown value. **(See Examples 4–6.)**

41. The amount of medicine that a physician prescribes for a patient varies directly as the weight of the patient. A physician prescribes 3 g of a medicine for a 150-lb person.

a. How many grams should be prescribed for a 180-lb person? $\ 3.6$ g

b. How many grams should be prescribed for a 225-lb person? $\ 4.5$ g

c. How many grams should be prescribed for a 120-lb person? $\ 2.4$ g

42. The number of turkeys needed for a banquet is directly proportional to the number of guests that must be fed. Master Chef Rico knows that he needs to cook 3 turkeys to feed 42 guests.

a. How many turkeys should he cook to feed 70 guests? $\ 5$ turkeys

b. How many turkeys should he cook to feed 140 guests? $\ 10$ turkeys

c. How many turkeys should be cooked to feed 700 guests at an inaugural ball? $\ 50$ turkeys

d. How many turkeys should be cooked for a wedding with 100 guests?
8 turkeys; Note that the answer was rounded up, because it would be better for the chef to have too much food than too little.

43. The unit cost of producing CDs is inversely proportional to the number of CDs produced. If 5000 CDs are produced, the cost per CD is $0.48.

a. What would be the unit cost if 6000 CDs were produced? $\ \$0.40$

b. What would be the unit cost if 8000 CDs were produced? $\ \$0.30$

c. What would be the unit cost if 2400 CDs were produced? $\ \$1.00$

44. An author self-publishes a book and finds that the number of books she can sell per month varies inversely as the price of the book. The author can sell 1500 books per month when the price is set at $8 per book.

a. How many books would she expect to sell if the price were $12? $\ 1000$ books

b. How many books would she expect to sell if the price were $15? $\ 800$ books

c. How many books would she expect to sell if the price were $6? $\ 2000$ books

45. The amount of pollution entering the atmosphere over a given time varies directly as the number of people living in an area. If 80,000 people cause 56,800 tons of pollutants, how many tons enter the atmosphere in a city with a population of 500,000? $\ 355{,}000$ tons

 Writing   Translating Expression   Geometry   Scientific Calculator   Video   NS E

46. The area of a picture projected on a wall varies directly as the square of the distance from the projector to the wall. If a 10-ft distance produces a 16-ft² picture, what is the area of a picture produced when the projection unit is moved to a distance 20 ft from the wall?   64 ft²

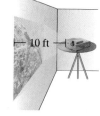

47. The stopping distance of a car varies directly as the square of the speed of the car. If a car traveling at 40 mph has a stopping distance of 109 ft, find the stopping distance of a car that is traveling at 25 mph. (Round your answer to one decimal place.)   42.6 ft

48. The intensity of a light source varies inversely as the square of the distance from the source. If the intensity is 48 lumens at a distance of 5 ft, what is the intensity when the distance is 8 ft?   18.75 lumens

49. The power in an electric circuit varies jointly as the current and the square of the resistance. If the power is 144 W (watts) when the current is 4 A and the resistance is 6 Ω, find the power when the current is 3 A and the resistance is 10 Ω.   300 W

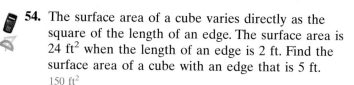

50. The current in a wire varies directly as the voltage and inversely as the resistance. If the current is 9 A (amperes) when the voltage is 90 V (volts) and the resistance is 10 Ω (ohms), find the current when the voltage is 185 V and the resistance is 10 Ω.   18.5 A

51. The resistance of a wire varies directly as its length and inversely as the square of its diameter. A 40-ft wire 0.1 in. in diameter has a resistance of 4 Ω. What is the resistance of a 50-ft wire with a diameter of 0.2 in.?   1.25 Ω

52. Some body-builders claim that, within safe limits, the number of repetitions that a person can complete on a given weight-lifting exercise is inversely proportional to the amount of weight lifted. Roxanne can bench press 45 lb for 15 repetitions.

a. How many repetitions can Roxanne bench with 60 lb of weight?   11 complete repetitions

b. How many repetitions can Roxanne bench with 75 lb of weight?   9 repetitions

c. How many repetitions can Roxanne bench with 100 lb of weight?   6 complete repetitions

53. The weight of a medicine ball varies directly as the cube of its radius. A ball with a radius of 3 in. weighs 4.32 lb. How much would a medicine ball weigh if its radius is 5 in.?   20 lb

54. The surface area of a cube varies directly as the square of the length of an edge. The surface area is 24 ft² when the length of an edge is 2 ft. Find the surface area of a cube with an edge that is 5 ft.
150 ft²

55. The amount of simple interest earned in an account varies jointly as the amount of principal invested and the amount of time the money is invested. If $2500 in principal earns $500 in interest after 4 yr, then how much interest will be earned on $7000 invested for 10 yr? $3500

56. The amount of simple interest earned in an account varies jointly as the amount of principal invested and the amount of time the money is invested. If $6000 in principal earns $840 in interest after 2 yr, then how much interest will be earned on $4500 invested for 8 yr? $2520

Group Activity

Computing Monthly Mortgage Payments

Materials: A calculator

Estimated Time: 15–20 minutes

Group Size: 3

When a person borrows money to buy a house, the bank usually requires a down payment of between 0% and 20% of the cost of the house. The bank then issues a loan for the remaining balance on the house. The loan to buy a house is called a *mortgage*. Monthly payments are made to pay off the mortgage over a period of years.

A formula to calculate the monthly payment, *P*, for a loan is given by the complex fraction:

$$P = \frac{\dfrac{Ar}{12}}{1 - \dfrac{1}{\left(1 + \dfrac{r}{12}\right)^{12t}}}$$

where P is the monthly payment
A is the original amount of the mortgage
r is the annual interest rate
t is the term of the loan in years

Suppose a person wants to buy a $200,000 house. The bank requires a down payment of 20%, and the loan is issued for 30 yr at 7.5% interest for 30 yr.

1. Find the amount of the down payment. _____$40,000_____

2. Find the amount of the mortgage. _____$160,000_____

3. Find the monthly payment (to the nearest cent). _____$1118.74_____

4. Multiply the monthly payment found in question 3 by the total number of months in a 30-yr period. Interpret what this value means in the context of the problem.
$402,746.40; This is the total amount owed to the bank.

5. How much total interest was paid on the loan for the house? _____$242,746.40_____

6. What was the total amount paid to the bank (include the down payment). _____$442,746.40_____

Chapter 7 Summary

Section 7.1 Introduction to Rational Expressions

Key Concepts

A **rational expression** is a ratio of the form $\frac{p}{q}$ where p and q are polynomials and $q \neq 0$.

The **domain** of an algebraic expression is the set of real numbers that when substituted for the variable makes the expression equal a real number. For a rational expression, the domain is all real numbers except those that make the denominator zero.

Simplifying a Rational Expression to Lowest Terms

Factor the numerator and denominator completely, and reduce factors whose ratio is equal to 1 or to -1. A rational expression written in lowest terms will still have the same restrictions on the domain as the original expression.

Examples

Example 1

$$\frac{x + 2}{x^2 - 5x - 14} \quad \text{is a rational expression.}$$

Example 2

To find the domain of $\dfrac{x + 2}{x^2 - 5x - 14}$ factor the denominator: $\dfrac{x + 2}{(x + 2)(x - 7)}$

The domain is $\{x \mid x \text{ is a real number and } x \neq -2, x \neq 7\}$.

Example 3

Simplify to lowest terms. $\dfrac{x + 2}{x^2 - 5x - 14}$

$$\frac{\overset{1}{\cancel{x + 2}}}{\cancel{(x + 2)}(x - 7)} \quad \text{Simplify.}$$

$$= \frac{1}{x - 7} \quad \text{(provided } x \neq 7, x \neq -2\text{).}$$

Section 7.2   Multiplication and Division of Rational Expressions

Key Concepts

Multiplying Rational Expressions

Multiply the numerators and multiply the denominators. That is, if $q \neq 0$ and $s \neq 0$, then

$$\frac{p}{q} \cdot \frac{r}{s} = \frac{pr}{qs}$$

Factor the numerator and denominator completely. Then reduce factors whose ratio is 1 or -1.

Dividing Rational Expressions

Multiply the first expression by the reciprocal of the second expression. That is, for $q \neq 0$, $r \neq 0$, and $s \neq 0$,

$$\frac{p}{q} \div \frac{r}{s} = \frac{p}{q} \cdot \frac{s}{r} = \frac{ps}{qr}$$

Examples

Example 1

Multiply. $\dfrac{b^2 - a^2}{a^2 - 2ab + b^2} \cdot \dfrac{a^2 - 3ab + 2b^2}{2a + 2b}$

$$= \frac{\overset{-1}{\cancel{(b - a)}}(b + a)}{\cancel{(a - b)}(a - b)} \cdot \frac{(a - 2b)\overset{1}{\cancel{(a - b)}}}{2(a + b)}$$

$$= -\frac{a - 2b}{2} \quad \text{or} \quad \frac{2b - a}{2}$$

Example 2

Divide. $\dfrac{x - 2}{15} \div \dfrac{x^2 + 2x - 8}{20x}$

$$= \frac{x - 2}{15} \cdot \frac{20x}{x^2 + 2x - 8}$$

$$= \frac{\overset{1}{\cancel{(x - 2)}}}{\underset{3}{\cancel{15}}} \cdot \frac{\overset{4}{\cancel{20x}}}{\cancel{(x - 2)}(x + 4)}$$

$$= \frac{4}{3(x + 4)}$$

Section 7.3 Least Common Denominator

Key Concepts

Converting a Rational Expression to an Equivalent Expression with a Different Denominator

Multiply numerator and denominator of the rational expression by the missing factors necessary to create the desired denominator.

Finding the Least Common Denominator (LCD) of Two or More Rational Expressions

1. Factor all denominators completely.
2. The LCD is the product of unique factors from the denominators, where each factor is raised to its highest power.

Examples

Example 1

Convert $\dfrac{-3}{x-2}$ to an equivalent expression with the indicated denominator:

$$\frac{-3}{x-2} = \frac{}{5(x-2)(x+2)}$$

Multiply numerator and denominator by the missing factors from the denominator.

$$\frac{-3 \cdot 5(x+2)}{(x-2) \cdot 5(x+2)} = \frac{-15x-30}{5(x-2)(x+2)}$$

Example 2

Identify the LCD. $\dfrac{1}{8x^3y^2z}; \dfrac{5}{6xy^4}$

1. Write the denominators as a product of prime factors:

$$\frac{1}{2^3x^3y^2z}; \frac{5}{2 \cdot 3xy^4}$$

2. The LCD is $2^3 3x^3y^4z$ or $24x^3y^4z$

Section 7.4 Addition and Subtraction of Rational Expressions

Key Concepts

To add or subtract rational expressions, the expressions must have the same denominator.

Steps to Add or Subtract Rational Expressions

1. Factor the denominators of each rational expression.
2. Identify the LCD.
3. Rewrite each rational expression as an equivalent expression with the LCD as its denominator.
4. Add or subtract the numerators, and write the result over the common denominator.
5. Simplify.

Examples

Example 1

Subtract. $\dfrac{c}{c^2-c-12} - \dfrac{1}{2c-8}$

$$= \frac{c}{(c-4)(c+3)} - \frac{1}{2(c-4)}$$

The LCD is $2(c-4)(c+3)$.

$$= \frac{2c}{2(c-4)(c+3)} - \frac{1(c+3)}{2(c-4)(c+3)}$$

$$= \frac{2c-(c+3)}{2(c-4)(c+3)}$$

$$= \frac{2c-c-3}{2(c-4)(c+3)} = \frac{c-3}{2(c-4)(c+3)}$$

Section 7.5 Complex Fractions

Key Concepts

Complex fractions can be simplified by using Method I or Method II.

Method I

1. Add or subtract expressions in the numerator to form a single fraction. Add or subtract expressions in the denominator to form a single fraction.
2. Divide the rational expressions from step 1 by multiplying the numerator of the complex fraction by the reciprocal of the denominator of the complex fraction.
3. Simplify to lowest terms, if possible.

Method II

1. Multiply the numerator and denominator of the complex fraction by the LCD of all individual fractions within the expression.
2. Apply the distributive property, and simplify the result.
3. Simplify to lowest terms, if possible.

Examples

Example 1

Simplify. $\dfrac{1 - \dfrac{4}{w^2}}{1 - \dfrac{1}{w} - \dfrac{6}{w^2}} = \dfrac{\dfrac{w^2}{w^2} - \dfrac{4}{w^2}}{\dfrac{w^2}{w^2} - \dfrac{w}{w^2} - \dfrac{6}{w^2}}$

$= \dfrac{\dfrac{w^2 - 4}{w^2}}{\dfrac{w^2 - w - 6}{w^2}} = \dfrac{w^2 - 4}{w^2} \cdot \dfrac{w^2}{w^2 - w - 6}$

$= \dfrac{(w - 2)(\cancel{w + 2})}{\cancel{w^2}} \cdot \dfrac{\overset{1}{\cancel{w^2}}}{(w - 3)(\cancel{w + 2})}$

$= \dfrac{w - 2}{w - 3}$

Example 2

Simplify. $\dfrac{1 - \dfrac{4}{w^2}}{1 - \dfrac{1}{w} - \dfrac{6}{w^2}} = \dfrac{w^2\left(1 - \dfrac{4}{w^2}\right)}{w^2\left(1 - \dfrac{1}{w} - \dfrac{6}{w^2}\right)}$

$= \dfrac{w^2 - 4}{w^2 - w - 6} = \dfrac{(w - 2)(\cancel{w + 2})}{(w - 3)(\cancel{w + 2})}$

$= \dfrac{w - 2}{w - 3}$

Section 7.6 Rational Equations

Key Concepts

An equation with one or more rational expressions is called a **rational equation**.

Steps to Solve a Rational Equation

1. Factor the denominators of all rational expressions.
2. Identify the LCD of all expressions in the equation.
3. Multiply both sides of the equation by the LCD.
4. Solve the resulting equation.
5. Check each potential solution in the original equation.

Examples

Example 1

Solve. $\dfrac{1}{w} - \dfrac{1}{2w - 1} = \dfrac{-2w}{2w - 1}$

The LCD is $w(2w - 1)$.

$$w(2w - 1)\frac{1}{w} - w(2w - 1)\frac{1}{2w - 1}$$

$$= w(2w - 1)\frac{-2w}{2w - 1}$$

$$(2w - 1)(1) - w(1) = w(-2w)$$

$$2w - 1 - w = -2w^2 \qquad \text{Quadratic equation}$$

$$2w^2 + w - 1 = 0$$

$$(2w - 1)(w + 1) = 0$$

$$w = \tfrac{1}{2} \qquad \text{or} \qquad w = -1$$

$$\text{Does not check.} \qquad\qquad \text{Checks.}$$

Example 2

Solve for I. $q = \dfrac{VQ}{I}$

$$I \cdot q = \dfrac{VQ}{I} \cdot I$$

$$Iq = VQ$$

$$I = \dfrac{VQ}{q}$$

Section 7.7   Applications of Rational Equations and Proportions

Key Concepts

Solving Proportions

An equation that equates two rates or ratios is called a **proportion**:

$$\frac{a}{b} = \frac{c}{d} \quad (b \neq 0, d \neq 0)$$

To solve a proportion, multiply both sides of the equation by the LCD.

Examples 2 and 3 give applications of rational equations.

Examples

Example 1

A 90-g serving of a particular ice cream contains 10 g of fat. How much fat does 400 g of the same ice cream contain?

$$\frac{10 \text{ g fat}}{90 \text{ g ice cream}} = \frac{x \text{ grams fat}}{400 \text{ g ice cream}}$$

$$\frac{10}{90} = \frac{x}{400}$$

$$\overset{40}{\cancel{3600}} \cdot \left(\frac{10}{\cancel{90}}\right) = \left(\frac{x}{\cancel{400}}\right) \cdot \overset{9}{\cancel{3600}}$$

$$400 = 9x$$

$$x = \frac{400}{9} \approx 44.4 \text{ g}$$

Example 2

Two cars travel from Los Angeles to Las Vegas. One car travels an average of 8 mph faster than the other car. If the faster car travels 189 mi in the same time as the slower car travels 165 mi, what is the average speed of each car?

Let r represent the speed of the slower car.
Let $r + 8$ represent the speed of the faster car.

	Distance	Rate	Time
Slower car	165	r	$\dfrac{165}{r}$
Faster car	189	$r + 8$	$\dfrac{189}{r+8}$

$$\frac{165}{r} = \frac{189}{r + 8}$$

$$165(r + 8) = 189r$$

$$165r + 1320 = 189r$$

$$1320 = 24r$$

$$55 = r$$

The slower car travels 55 mph, and the faster car travels $55 + 8 = 63$ mph.

Example 3

Beth and Cecelia have a house cleaning business. Beth can clean a particular house in 5 hr by herself. Cecelia can clean the same house in 4 hr. How long would it take if they cleaned the house together?

Let x be the number of hours it takes for both Beth and Cecelia to clean the house.

Beth can clean $\frac{1}{5}$ of the house in an hour and $\frac{1}{5}x$ of the house in x hr.

Cecelia can clean $\frac{1}{4}$ of the house in an hour and $\frac{1}{4}x$ of the house in x hr.

$$\frac{1}{5}x + \frac{1}{4}x = 1 \quad \text{Together they clean one whole house.}$$

$$20\left(\frac{1}{5}x + \frac{1}{4}x\right) = (1)20$$

$$4x + 5x = 20$$

$$9x = 20$$

$$x = \frac{20}{9}, \text{ or } 2\tfrac{2}{9} \text{ hr working together.}$$

Section 7.8 Variation

Key Concepts

Direct Variation

y varies directly as x.
y is directly proportional to x. $\left.\right\}$ $y = kx$

Inverse Variation

y varies inversely as x.
y is inversely proportional to x. $\left.\right\}$ $y = \dfrac{k}{x}$

Joint Variation

y varies jointly as w and z.
y is jointly proportional to w and z. $\left.\right\}$ $y = kwz$

Steps to Find a Variation Model

1. Write a general variation model that relates the variables given in the problem. Let k represent the constant of variation.
2. Solve for k by substituting known values of the variables into the model from step 1.
3. Substitute the value of k into the original variation model from step 1.

Examples

Example 1

t varies directly as the square root of x.

$t = k\sqrt{x}$

Example 2

W is inversely proportional to the cube of x.

$W = \dfrac{k}{x^3}$

Example 3

y is jointly proportional to x and the square of z.

$y = kxz^2$

Example 4

C varies directly as the square root of d and inversely as t. If $C = 12$ when d is 9 and t is 6, find C if d is 16 and t is 12.

Step 1. $C = \dfrac{k\sqrt{d}}{t}$

Step 2. $12 = \dfrac{k\sqrt{9}}{6} \Rightarrow 12 = \dfrac{k \cdot 3}{6} \Rightarrow k = 24$

Step 3. $C = \dfrac{24\sqrt{d}}{t} \Rightarrow C = \dfrac{24\sqrt{16}}{12} \Rightarrow C = 8$

Chapter 7 Review Exercises

Section 7.1

1. For the rational expression $\dfrac{t - 2}{t + 9}$

   a. Evaluate the expression (if possible) for
   $t = 0, 1, 2, -3, -9$ $\quad -\dfrac{2}{9}, -\dfrac{1}{10}, 0, -\dfrac{5}{6},$ undefined

   b. Write the domain of the expression in set-builder notation. $\quad \{t \mid t$ is a real number and $t \neq -9\}$

2. For the rational expression $\dfrac{k + 1}{k - 5}$

   a. Evaluate the expression for $k = 0, 1, 5,$
   $-1, -2$ $\quad -\dfrac{1}{5}, -\dfrac{1}{2},$ undefined, $0, \dfrac{1}{7}$

   b. Write the domain of the expression in set-builder notation. $\quad \{k \mid k$ is a real number and $k \neq 5\}$

Writing   Translating Expression   Geometry   Scientific Calculator   Video   NS&E

3. Which of the rational expressions are equal to -1? a, c, d

a. $\dfrac{2-x}{x-2}$

b. $\dfrac{x-5}{x+5}$

c. $\dfrac{-x-7}{x+7}$

d. $\dfrac{x^2-4}{4-x^2}$

For Exercises 4–13, write the domain in set-builder notation. Then simplify the expressions to lowest terms.

4. $\dfrac{x-3}{(2x-5)(x-3)}$ $\{x \mid x$ is a real number and $x \neq \frac{5}{2}, x \neq 3\}; \frac{1}{2x-5}$

5. $\dfrac{h+7}{(3h+1)(h+7)}$ $\{h \mid h$ is a real number and $h \neq -\frac{1}{3}, h \neq -7\}; \frac{1}{3h+1}$

6. $\dfrac{4a^2+7a-2}{a^2-4}$ $\{a \mid a$ is a real number and $a \neq 2, a \neq -2\}; \frac{4a-1}{a-2}$

7. $\dfrac{2w^2+11w+12}{w^2-16}$ $\{w \mid w$ is a real number and $w \neq 4, w \neq -4\}; \frac{2w+3}{w-4}$

8. $\dfrac{z^2-4z}{8-2z}$ $\{z \mid z$ is a real number and $z \neq 4\}; -\frac{z}{2}$

9. $\dfrac{15-3k}{2k^2-10k}$ $\{k \mid k$ is a real number and $k \neq 0, k \neq 5\}; -\frac{3}{2k}$

10. $\dfrac{2b^2+4b-6}{4b+12}$ $\{b \mid b$ is a real number and $b \neq -3\}; \frac{b-1}{2}$

11. $\dfrac{3m^2-12m-15}{9m+9}$ $\{m \mid m$ is a real number and $m \neq -1\}; \frac{m-5}{3}$

12. $\dfrac{n+3}{n^2+6n+9}$ $\{n \mid n$ is a real number and $n \neq -3\}; \frac{1}{n+3}$

13. $\dfrac{p+7}{p^2+14p+49}$ $\{p \mid p$ is a real number and $p \neq -7\}; \frac{1}{p+7}$

Section 7.2

For Exercises 14–27, multiply or divide as indicated.

14. $\dfrac{3y^3}{3y-6} \cdot \dfrac{y-2}{y}$ y^2

15. $\dfrac{2u+10}{u} \cdot \dfrac{u^3}{4u+20}$ $\frac{u^2}{2}$

16. $\dfrac{11}{v-2} \cdot \dfrac{2v^2-8}{22}$ $v+2$

17. $\dfrac{8}{x^2-25} \cdot \dfrac{3x+15}{16}$ $\frac{3}{2(x-5)}$

18. $\dfrac{4c^2+4c}{c^2-25} \div \dfrac{8c}{c^2-5c}$ $\frac{c(c+1)}{2(c+5)}$

19. $\dfrac{q^2-5q+6}{2q+4} \div \dfrac{2q-6}{q+2}$ $\frac{q-2}{4}$

20. $\left(\dfrac{-2t}{t+1}\right)(t^2-4t-5)$ $-2t(t-5)$

21. $(s^2-6s+8)\left(\dfrac{4s}{s-2}\right)$ $4s(s-4)$

22. $\dfrac{\dfrac{a^2+5a+1}{7a-7}}{\dfrac{a^2+5a+1}{a-1}}$ $\frac{1}{7}$

23. $\dfrac{\dfrac{n^2+n+1}{n^2-4}}{\dfrac{n^2+n+1}{n+2}}$ $\frac{1}{n-2}$

24. $\dfrac{5h^2-6h+1}{h^2-1} \div \dfrac{16h^2-9}{4h^2+7h+3} \cdot \dfrac{3-4h}{30h-6}$ $-\frac{1}{6}$

25. $\dfrac{3m-3}{6m^2+18m+12} \cdot \dfrac{2m^2-8}{m^2-3m+2} \div \dfrac{m+3}{m+1}$ $\frac{1}{m+3}$

26. $\dfrac{x-2}{x^2-3x-18} \cdot \dfrac{6-x}{x^2-4}$ $\frac{-1}{(x+3)(x+2)}$

27. $\dfrac{4y^2-1}{1+2y} \div \dfrac{y^2-4y-5}{5-y}$ $-\frac{2y-1}{y+1}$

Section 7.3

For Exercises 28–33, identify the LCD. Then write each fraction as an equivalent fraction with the LCD as its denominator.

28. $\dfrac{2}{5a}; \dfrac{3}{10b}$ LCD $= 10ab; \frac{4b}{10ab}, \frac{3a}{10ab}$

29. $\dfrac{7}{4x}; \dfrac{11}{6y}$ LCD $= 12xy; \frac{21y}{12xy}, \frac{22x}{12xy}$

30. $\dfrac{1}{x^2y^4}; \dfrac{3}{xy^5}$ LCD $= x^2y^5; \frac{y}{x^2y^5}, \frac{3x}{x^2y^5}$

31. $\dfrac{5}{ab^3}; \dfrac{3}{ac^2}$ LCD $= ab^3c^2; \frac{5c^2}{ab^3c^2}, \frac{3b^3}{ab^3c^2}$

32. $\dfrac{5}{p+2}; \dfrac{p}{p-4}$ LCD $= (p+2)(p-4); \frac{5p-20}{(p+2)(p-4)}, \frac{p^2+2p}{(p+2)(p-4)}$

33. $\dfrac{6}{q}; \dfrac{1}{q+8}$ LCD $= q(q+8); \frac{6q+48}{q(q+8)}, \frac{q}{q(q+8)}$

34. Determine the LCD.
$$\dfrac{6}{n^2-9}; \dfrac{5}{n^2-n-6}$$
LCD $= (n+3)(n-3)(n+2)$

35. Determine the LCD.
$$\dfrac{8}{m^2-16}; \dfrac{7}{m^2-m-12}$$
LCD $= (m+4)(m-4)(m+3)$

36. State two possible LCDs that could be used to add the fractions. $c-2$ or $2-c$
$$\dfrac{7}{c-2}+\dfrac{4}{2-c}$$

37. State two possible LCDs that could be used to subtract the fractions. $3-x$ or $x-3$
$$\dfrac{10}{3-x}-\dfrac{5}{x-3}$$

Section 7.4

For Exercises 38–49, add or subtract as indicated.

38. $\dfrac{h+3}{h+1}+\dfrac{h-1}{h+1}$ 2

39. $\dfrac{b-6}{b-2}+\dfrac{b+2}{b-2}$ 2

40. $\dfrac{a^2}{a-5}-\dfrac{25}{a-5}$ $a+5$

41. $\dfrac{x^2}{x+7}-\dfrac{49}{x+7}$ $x-7$

Writing ⟷ Translating Expression ⬖ Geometry 🖩 Scientific Calculator 💿 Video NSF

42. $\dfrac{y}{y^2 - 81} + \dfrac{2}{9 - y}$

$\dfrac{-y - 18}{(y - 9)(y + 9)}$ or $\dfrac{y + 18}{(9 - y)(y + 9)}$

43. $\dfrac{3}{4 - t^2} + \dfrac{t}{2 - t}$

$\dfrac{t^2 + 2t + 3}{(2 - t)(2 + t)}$

44. $\dfrac{4}{3m} - \dfrac{1}{m + 2}$

$\dfrac{m + 8}{3m(m + 2)}$

45. $\dfrac{5}{2r + 12} - \dfrac{1}{r}$

$\dfrac{3(r - 4)}{2r(r + 6)}$

46. $\dfrac{4p}{p^2 + 6p + 5} - \dfrac{3p}{p^2 + 5p + 4}$

$\dfrac{p}{(p + 4)(p + 5)}$

47. $\dfrac{3q}{q^2 + 7q + 10} - \dfrac{2q}{q^2 + 6q + 8}$

$\dfrac{q}{(q + 5)(q + 4)}$

48. $\dfrac{1}{h} + \dfrac{h}{2h + 4} - \dfrac{2}{h^2 + 2h}$

$\dfrac{1}{2}$

49. $\dfrac{x}{3x + 9} - \dfrac{3}{x^2 + 3x} + \dfrac{1}{x}$

$\dfrac{1}{3}$

Section 7.5

For Exercises 50–57, simplify the complex fractions.

50. $\dfrac{\dfrac{a - 4}{3}}{\dfrac{a - 2}{3}}$

$\dfrac{a - 4}{a - 2}$

51. $\dfrac{\dfrac{z + 5}{z}}{\dfrac{z - 5}{3}}$

$\dfrac{3(z + 5)}{z(z - 5)}$

52. $\dfrac{\dfrac{2 - 3w}{2}}{\dfrac{2}{w} - 3}$

$\dfrac{w}{2}$

53. $\dfrac{\dfrac{2}{y} + 6}{\dfrac{3y + 1}{4}}$

$\dfrac{8}{y}$

54. $\dfrac{\dfrac{y}{x} - \dfrac{x}{y}}{\dfrac{1}{x} + \dfrac{1}{y}}$

$y - x$

55. $\dfrac{\dfrac{b}{a} - \dfrac{a}{b}}{\dfrac{1}{b} - \dfrac{1}{a}}$

$-(b + a)$

56. $\dfrac{\dfrac{6}{p + 2} + 4}{\dfrac{8}{p + 2} - 4}$

$\dfrac{2p + 7}{2p}$

57. $\dfrac{\dfrac{25}{k + 5} + 5}{\dfrac{5}{k + 5} - 5}$

$-\dfrac{k + 10}{k + 4}$

Section 7.6

For Exercises 58–65, solve the equations.

58. $\dfrac{2}{x} + \dfrac{1}{2} = \dfrac{1}{4}$ -8

59. $\dfrac{1}{y} + \dfrac{3}{4} = \dfrac{1}{4}$ -2

60. $\dfrac{2}{h - 2} + 1 = \dfrac{h}{h + 2}$ 0

61. $\dfrac{w}{w - 1} = \dfrac{3}{w + 1} + 1$ 2

62. $\dfrac{t + 1}{3} - \dfrac{t - 1}{6} = \dfrac{1}{6}$ -2

63. $\dfrac{4p - 4}{p^2 + 5p - 14} + \dfrac{2}{p + 7} = \dfrac{1}{p - 2}$ 3

64. $\dfrac{1}{z + 2} = \dfrac{4}{z^2 - 4} - \dfrac{1}{z - 2}$ No solution; (the value 2 does not check.)

65. $\dfrac{y + 1}{y + 3} = \dfrac{y^2 - 11y}{y^2 + y - 6} - \dfrac{y - 3}{y - 2}$ $-11, 1$

66. Four times a number is added to 5. The sum is then divided by 6. The result is $\frac{7}{2}$. Find the number.   The number is 4.

67. Solve the formula $\dfrac{V}{h} = \dfrac{\pi r^2}{3}$ for h. $h = \dfrac{3V}{\pi r^2}$

68. Solve the formula $\dfrac{A}{b} = \dfrac{h}{2}$ for b. $b = \dfrac{2A}{h}$

Section 7.7

For Exercises 69–70, solve the proportions.

69. $\dfrac{m + 2}{8} = \dfrac{m}{3}$ $\dfrac{6}{5}$

70. $\dfrac{12}{a} = \dfrac{5}{8}$ $\dfrac{96}{5}$

71. A bag of popcorn states that it contains 4 g of fat per serving. If a serving is 2 oz, how many grams of fat are in a 5-oz bag?   10 g

72. Bud goes 10 mph faster on his Harley Davidson motorcycle than Ed goes on his Honda motorcycle. If Bud travels 105 mi in the same time that Ed travels 90 mi, what are the rates of the two bikers?

Ed travels at 60 mph, and Bud travels at 70 mph.

Writing   Translating Expression   Geometry   Scientific Calculator   Video   NS&E

73. There are two pumps set up to fill a small swimming pool. One pump takes 24 min by itself to fill the pool, but the other takes 56 min by itself. How long would it take if both pumps work together?
Together the pumps would fill the pool in 16.8 min.

74 Consider the similar triangles shown here. Find the values of x and y. $x = 11; y = 26$

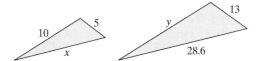

76. Suppose y varies inversely with the cube of x, and $y = 9$ when $x = 2$. Find y when $x = 3$. $y = \dfrac{8}{3}$

77. Suppose y varies jointly with x and the square root of z, and $y = 3$ when $x = 3$ and $z = 4$. Find y when $x = 8$ and $z = 9$. $y = 12$

78. The distance, d, that one can see to the horizon varies directly as the square root of the height above sea level. If a person 25 m above sea level can see 30 km, how far can a person see if she is 64 m above sea level?   48 km

Section 7.8

75. The force applied to a spring varies directly with the distance that the spring is stretched.

  a. Write a variation model using k as the constant of variation. $F = kd$

  b. When 6 lb of force is applied, the spring stretches 2 ft. Find k. $k = 3$

  c. How much force is required to stretch the spring 4.2 ft?   4.5 lb

Chapter 7 Test

For Exercises 1–2,

  a. Write the domain of the rational expressions in set-builder notation.

  b. Reduce the rational expression to lowest terms.

a. $\{a \mid a$ is a real number and $a \neq 0,$

1. $\dfrac{5(x-2)(x+1)}{30(2-x)}$ a. $\{x \mid x$ is a real number and $x \neq 2\}$
b. $-\dfrac{x+1}{6}$

2. $\dfrac{7a^2 - 42a}{a^3 - 4a^2 - 12a}$ $a \neq 6, a \neq -2\}$
b. $\dfrac{7}{a+2}$

3. Identify the rational expressions that are equal to -1.   b, c, d

  a. $\dfrac{x+4}{x-4}$ **b.** $\dfrac{7-2x}{2x-7}$

  c. $\dfrac{9x^2+16}{-9x^2-16}$ **d.** $-\dfrac{x+5}{x+5}$

For Exercises 4–9, perform the indicated operation.

4. $\dfrac{2}{y^2+4y+3} + \dfrac{1}{3y+9}$ $\dfrac{y+7}{3(y+3)(y+1)}$

5. $\dfrac{9-b^2}{5b+15} \div \dfrac{b-3}{b+3}$ $-\dfrac{b+3}{5}$

6. $\dfrac{w^2-4w}{w^2-8w+16} \cdot \dfrac{w-4}{w^2+w}$ $\dfrac{1}{w+1}$

7. $\dfrac{t}{t-2} - \dfrac{8}{t^2-4}$ $\dfrac{t+4}{t+2}$

8. $\dfrac{1}{x+4} + \dfrac{2}{x^2+2x-8} + \dfrac{x}{x-2}$ $\dfrac{x(x+5)}{(x+4)(x-2)}$

9. $\dfrac{1-\dfrac{4}{m}}{m-\dfrac{16}{m}}$ $\dfrac{1}{m+4}$

For Exercises 10–13, solve the equation.

10. $\dfrac{3}{a} + \dfrac{5}{2} = \dfrac{7}{a}$ $\dfrac{8}{5}$

11. $\dfrac{p}{p-1} + \dfrac{1}{p} = \dfrac{p^2+1}{p^2-p}$ 2

12. $\dfrac{3}{c-2} - \dfrac{1}{c+1} = \dfrac{7}{c^2-c-2}$ 1

13. $\dfrac{4x}{x-4} = 3 + \dfrac{16}{x-4}$ No solution. (the value 4 does not check.)

 Writing   Translating Expression   Geometry   Scientific Calculator    Video   NS E

14. Solve the formula $\dfrac{C}{2} = \dfrac{A}{r}$ for r. $r = \dfrac{2A}{C}$

15. If $\frac{3}{2}$ is added to the reciprocal of a number the result is $\frac{2}{5}$ times the reciprocal of that number. Find the number. The number is $-\dfrac{2}{5}$.

16. Solve the proportion.

$$\dfrac{y+7}{-4} = \dfrac{1}{4} \quad -8$$

17. A recipe for vegetable soup calls for $\frac{1}{2}$ cup of carrots for six servings. How many cups of carrots are needed to prepare 15 servings? $1\frac{1}{4}$ (1.25) cups of carrots

18. A motorboat can travel 28 mi downstream in the same amount of time as it can travel 18 mi upstream. Find the speed of the current if the boat can travel 23 mph in still water. The speed of the current is 5 mph.

19. Two printers working together can complete a job in 2 hr. If one printer requires 6 hr to do the job alone, how many hours would the second printer need to complete the job alone? It would take the second printer 3 hr to do the job working alone.

20. Consider the similar triangles shown here. Find the values of a and b. $a = 5.6$; $b = 12$

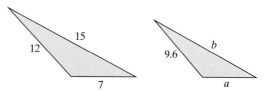

21. Find the LCD of the following pairs of rational expressions.

a. $\dfrac{x}{3(x+3)}, \dfrac{7}{5(x+3)}$ $15(x+3)$ **b.** $\dfrac{-2}{3x^2y}, \dfrac{4}{xy^2}$ $3x^2y^2$

22. The amount of medication prescribed for a patient varies directly as the patient's weight. If a 160-lb person is prescribed 6 mL of a medicine, then how much medicine would be prescribed to a 220-lb person? 8.25 mL

23. The number of drinks sold at a concession stand varies inversely as price. If the price is set at \$1.25 per drink, then 400 drinks are sold. If the price is set at \$2.50 per drink, then how many drinks are sold? 200 drinks are sold.

Chapters 1–7 Cumulative Review Exercises

For Exercises 1–2, simplify completely.

1. $\left(\dfrac{1}{2}\right)^{-4} + 2^4$ 32

2. $|3-5| + |-2+7|$ 7

3. Solve. $\dfrac{1}{2} - \dfrac{3}{4}(y-1) = \dfrac{5}{12}$ $\dfrac{10}{9}$

4. Complete the table.

Set-Builder Notation	Graph	Interval Notation
$\{x \mid x \ge -1\}$		$[-1, \infty)$
$\{x \mid x < 5\}$		$(-\infty, 5)$

5. The perimeter of a rectangular swimming pool is 104 m. The length is 1 m more than twice the width. Find the length and width. The width is 17 m and the length is 35 m.

6. The height of a triangle is 2 in. less than the base. The area is 40 in.² Find the base and height of the triangle. The base is 10 in. and the height is 8 in.

7. Simplify. $\left(\dfrac{4x^{-1}y^{-2}}{z^4}\right)^{-2}(2y^{-1}z^3)^3$ $\dfrac{x^2yz^{17}}{2}$

8. The length and width of a rectangle are given in terms of x.

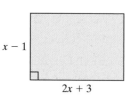

a. Write a polynomial that represents the perimeter of the rectangle. $6x + 4$

b. Write a polynomial that represents the area of the rectangle. $2x^2 + x - 3$

9. Factor completely: $25x^2 - 30x + 9$ $(5x-3)^2$

10. Factor. $10cd + 5d - 6c - 3$ $(2c+1)(5d-3)$

11. Determine the domain of the expression. $\left\{x \mid x \text{ is a real number and } x \ne 5, x \ne -\frac{1}{2}\right\}$

$$\dfrac{x+3}{(x-5)(2x+1)}$$

 Writing   Translating Expression   Geometry   Scientific Calculator   Video   NS E

12. Simplify to lowest terms.

$$\frac{x^2 - 9}{x^2 + 8x + 15} \qquad \frac{x-3}{x+5}$$

13. Divide. $\dfrac{2x - 6}{x^2 - 16} \div \dfrac{10x^2 - 90}{x^2 - x - 12} \qquad \dfrac{1}{5(x+4)}$

14. Simplify.

$$\frac{\dfrac{3}{4} - \dfrac{1}{x}}{\dfrac{1}{3x} - \dfrac{1}{4}} \qquad -3$$

15. Solve.

$$\frac{7}{y^2 - 4} = \frac{3}{y - 2} + \frac{2}{y + 2} \qquad 1$$

16. Solve the proportion.

$$\frac{2b - 5}{6} = \frac{4b}{7} \qquad -\frac{7}{2}$$

17. Determine the x- and y-intercepts.

   a. $-2x + 4y = 8$ x-intercept: $(-4, 0)$; **b.** $y = 5x$
                  y-intercept: $(0, 2)$ x-intercept: $(0, 0)$;
                                     y-intercept: $(0, 0)$

18. Determine the slope

   a. of the line containing the points $(0, -6)$ and $(-5, 1)$ $m = -\dfrac{7}{5}$

   b. of the line $y = -\dfrac{2}{3}x - 6$ $m = -\dfrac{2}{3}$

   c. of a line parallel to a line having a slope of 4. $m = 4$

   d. of a line perpendicular to a line having a slope of 4. $m = -\dfrac{1}{4}$

19. Find an equation of a line passing through the point $(1, 2)$ and having a slope of 5. Write the answer in slope-intercept form. $y = 5x - 3$

20. A group of teenagers buys 2 large popcorns and 6 drinks at the movie theater for $16. A couple buys 1 large popcorn and 2 drinks for $6.50. Find the price for 1 large popcorn and the price for 1 drink. One large popcorn costs $3.50, and one drink costs $1.50.

Writing   Translating Expression   Geometry   Scientific Calculator   Video   NS&E

Radicals

8

CHAPTER OUTLINE

8.1 Introduction to Roots and Radicals   582

8.2 Simplifying Radicals   593

8.3 Addition and Subtraction of Radicals   601

8.4 Multiplication of Radicals   608

8.5 Division of Radicals and Rationalization   614

Problem Recognition Exercises: Operations on Radicals   623

8.6 Radical Equations   624

8.7 Rational Exponents   630

Group Activity: Approximating Square Roots   637

Chapter 8

Chapter 8 is devoted to the study of radicals and their applications. We first present the techniques to add, subtract, multiply, and divide radical expressions. This is followed by radical equations and their applications.

Several of the key terms from the chapter can be found in the word search puzzle. Try to locate the words from the list in the puzzle.

square root   cube   radicand

Pythagorean   index   radical

```
a  c  r  v              d  i  t  l
r  o  e  a           i  x  o  p
d  a  r  n           n  y  o  y
c  l  d  d  n  a  c  i  d  a  r  a
g  x  n  i  t  v  q  u  e  n  e  b
n  j  m  e  c  a  t  e  x  m  r  k
a  p  y  t  h  a  g  o  r  e  a  n
v  j  t  n  o  r  l  e  b  l  u  i
k  t  e  m  g  e  a  u  r  x  q  x
y  h  l  i  t  t  c  o  u  b  s  e
```

Section 8.1   Introduction to Roots and Radicals

Objectives

1. **Definition of a Square Root**
2. **Definition of an *n*th-Root**
3. **Translations Involving *n*th-Roots**
4. **Pythagorean Theorem**

1. Definition of a Square Root

Recall that to square a number means to multiply the number by itself: $b^2 = b \cdot b$. The reverse operation to squaring a number is to find its square roots. For example, finding a square root of 49 is equivalent to asking: "What number when squared equals 49?"

One obvious answer to this question is 7 because $(7)^2 = 49$. But -7 will also work because $(-7)^2 = 49$.

DEFINITION   Square Root

b is a **square root** of a if $b^2 = a$.

Skill Practice

Identify the square roots of the numbers.

1. 64   **2.** -36

3. 36   **4.** $\dfrac{25}{16}$

Classroom Examples: p. 589, Exercises 2 and 4

TIP: All positive real numbers have two real-valued square roots: one positive and one negative. Zero has only one square root, which is 0 itself. Finally, for any negative real number, there are no real-valued square roots.

Example 1   Identifying the Square Roots of a Number

Identify the square roots of

a. 9   **b.** 121   **c.** 0   **d.** -4

Solution:

a. 3 is a square root of 9 because $(3)^2 = 9$.
-3 is a square root of 9 because $(-3)^2 = 9$.

b. 11 is a square root of 121 because $(11)^2 = 121$.
-11 is a square root of 121 because $(-11)^2 = 121$.

c. 0 is a square root of 0 because $(0)^2 = 0$.

d. There are no real numbers that when squared will equal a negative number. Therefore, there are no real-valued square roots of -4.

Recall from Section 1.2 that the positive square root of a real number can be denoted with a radical sign, $\sqrt{}$.

DEFINITION   Notation for Positive and Negative Square Roots

Let a represent a positive real number. Then,

1. $\sqrt{a}$ is the **positive square root** of a. The positive square root is also called the **principal square root**.
2. $-\sqrt{a}$ is the **negative square root** of a.
3. $\sqrt{0} = 0$

Answers

1. $8; -8$
2. There are no real-valued square roots.
3. $6; -6$ **4.** $\dfrac{5}{4}; -\dfrac{5}{4}$

Example 2 Simplifying Square Roots

Simplify the square roots.

a. $\sqrt{36}$ **b.** $\sqrt{225}$ **c.** $\sqrt{1}$ **d.** $\sqrt{\dfrac{9}{4}}$ **e.** $\sqrt{0.49}$

Solution:

a. $\sqrt{36}$ denotes the positive square root of 36. $\sqrt{36} = 6$

b. $\sqrt{225}$ denotes the positive square root of 225. $\sqrt{225} = 15$

c. $\sqrt{1}$ denotes the positive square root of 1. $\sqrt{1} = 1$

d. $\sqrt{\dfrac{9}{4}}$ denotes the positive square root of $\dfrac{9}{4}$. $\sqrt{\dfrac{9}{4}} = \dfrac{3}{2}$

e. $\sqrt{0.49}$ denotes the positive square root. $\sqrt{0.49} = 0.7$

Skill Practice

Simplify the square roots.

5. $\sqrt{81}$ **6.** $\sqrt{144}$

7. $\sqrt{0}$ **8.** $\sqrt{\dfrac{1}{4}}$

9. $\sqrt{0.09}$

Classroom Examples: p. 590, Exercises 18, 20, and 24

The numbers 36, 225, 1, $\frac{9}{4}$, and 0.49 are **perfect squares** because their square roots are rational numbers. Radicals that cannot be simplified to rational numbers are irrational numbers. Recall that an irrational number cannot be written as a terminating or repeating decimal. For example, the symbol $\sqrt{13}$ is used to represent the exact value of the square root of 13. The symbol $\sqrt{42}$ is used to represent the exact value of the square root of 42. These values are irrational numbers but can be approximated by rational numbers by using a calculator.

$$\sqrt{13} \approx 3.605551275 \qquad \sqrt{42} \approx 6.480740698$$

Note: The only way to denote the *exact* values of the square root of 13 and the square root of 42 is $\sqrt{13}$ and $\sqrt{42}$.

A negative number cannot have a real number as a square root because no real number when squared is negative. For example, $\sqrt{-25}$ is *not a real number* because there is no real number, b, for which $(b)^2 = -25$.

TIP: Before using a calculator to evaluate a square root, try estimating the value first.

$\sqrt{13}$ must be a number between 3 and 4 because $\sqrt{9} < \sqrt{13} < \sqrt{16}$.

$\sqrt{42}$ must be a number between 6 and 7 because $\sqrt{36} < \sqrt{42} < \sqrt{49}$.

Example 3 Evaluating Square Roots if Possible

Simplify the square roots if possible.

a. $\sqrt{-100}$ **b.** $-\sqrt{100}$ **c.** $\sqrt{-64}$

Solution:

a. $\sqrt{-100}$ Not a real number

b. $-\sqrt{100}$
$-1 \cdot \sqrt{100}$ The expression $-\sqrt{100}$ is equivalent to $-1 \cdot \sqrt{100}$.
$-1 \cdot 10 = -10$

c. $\sqrt{-64}$ Not a real number

Skill Practice

Simplify the square roots if possible.

10. $\sqrt{-25}$ **11.** $-\sqrt{25}$
12. $\sqrt{-4}$

Classroom Examples: p. 590, Exercises 36 and 38

Answers

5. 9   **6.** 12
7. 0   **8.** $\dfrac{1}{2}$
9. 0.3   **10.** Not a real number
11. −5   **12.** Not a real number

2. Definition of an *n*th-Root

Finding a square root of a number is the reverse process of squaring a number. This concept can be extended to finding a third root (called a cube root), a fourth root, and in general, an *n*th-root.

> **DEFINITION** *n*th-Root
>
> b is an **nth-root** of a if $b^n = a$.

Concept Connections

13. Identify the index and radicand.
$$\sqrt[3]{2xy}$$

The radical sign, $\sqrt{}$, is used to denote the principal square root of a number. The symbol, $\sqrt[n]{}$, is used to denote the principal *n*th-root of a number.

In the expression $\sqrt[n]{a}$, n is called the **index** of the radical, and a is called the **radicand**. For a square root, the index is 2, but it is usually not written ($\sqrt[2]{a}$ is denoted simply as $\sqrt{a}$). A radical with an index of 3 is called a **cube root**, $\sqrt[3]{a}$.

> **DEFINITION** $\sqrt[n]{a}$
>
> **1.** If n is a positive *even* integer and $a > 0$, then $\sqrt[n]{a}$ is the principal (positive) *n*th-root of a.
> **2.** If $n > 1$ is a positive *odd* integer, then $\sqrt[n]{a}$ is the *n*th-root of a.
> **3.** If $n > 1$ is a positive integer, then $\sqrt[n]{0} = 0$.

For the purpose of simplifying radicals, it is helpful to know the following patterns:

Perfect cubes	Perfect fourth powers	Perfect fifth powers
$1^3 = 1$	$1^4 = 1$	$1^5 = 1$
$2^3 = 8$	$2^4 = 16$	$2^5 = 32$
$3^3 = 27$	$3^4 = 81$	$3^5 = 243$
$4^3 = 64$	$4^4 = 256$	$4^5 = 1024$
$5^3 = 125$	$5^4 = 625$	$5^5 = 3125$

Skill Practice

Simplify the expressions if possible.
14. $\sqrt[3]{27}$ **15.** $\sqrt[4]{1}$
16. $\sqrt[3]{216}$ **17.** $\sqrt[5]{-32}$
18. $\sqrt[4]{\dfrac{16}{625}}$ **19.** $\sqrt{0.25}$
20. $\sqrt[4]{-1}$

Classroom Examples: p. 590, Exercises 52 and 66

Example 4   Simplifying *n*th-Roots

Simplify the expressions, if possible.

a. $\sqrt[3]{8}$ **b.** $\sqrt[4]{16}$ **c.** $\sqrt[5]{32}$ **d.** $\sqrt[3]{-64}$

e. $\sqrt[3]{\dfrac{125}{27}}$ **f.** $\sqrt{0.01}$ **g.** $\sqrt[4]{-81}$

Solution:

a. $\sqrt[3]{8} = 2$ Because $(2)^3 = 8$

b. $\sqrt[4]{16} = 2$ Because $(2)^4 = 16$

c. $\sqrt[5]{32} = 2$ Because $(2)^5 = 32$

d. $\sqrt[3]{-64} = -4$ Because $(-4)^3 = -64$

e. $\sqrt[3]{\dfrac{125}{27}} = \dfrac{5}{3}$ Because $\left(\dfrac{5}{3}\right)^3 = \dfrac{125}{27}$

Answers

13. Index: 3; radicand: $2xy$
14. 3   **15.** 1   **16.** 6
17. -2 **18.** $\dfrac{2}{5}$ **19.** 0.5
20. Not a real number.

f. $\sqrt{0.01} = 0.1$ Because $(0.1)^2 = 0.01$

Note: $\sqrt{0.01}$ is equivalent to $\sqrt{\dfrac{1}{100}} = \dfrac{1}{10}$, or 0.1.

g. $\sqrt[4]{-81}$ is not a real number because no real number raised to the fourth power equals -81.

Avoiding Mistakes

When evaluating $\sqrt[n]{a}$, where n is *even*, always choose the principal (positive) root.

$$\sqrt[4]{16} = 2 \quad \text{(not } -2)$$
$$\sqrt{0.01} = 0.1 \quad \text{(not } -0.1)$$

Example 4(g) illustrates that an *n*th-root of a negative number is not a real number if the index is even because no real number raised to an even power is negative.

Finding an *n*th-root of a variable expression is similar to finding an *n*th-root of a numerical expression. However, for roots with an even index, particular care must be taken to obtain a nonnegative solution.

DEFINITION $\sqrt[n]{a^n}$

1. If n is a positive odd integer, then $\sqrt[n]{a^n} = a$
2. If n is a positive even integer, then $\sqrt[n]{a^n} = |a|$

The absolute value bars are necessary for roots with an even index because the variable, a, may represent a positive quantity or a negative quantity. By using absolute value bars, we ensure that $\sqrt[n]{a^n} = |a|$ is nonnegative and represents the principal *n*th-root of a.

Example 5 **Simplifying Expressions of the Form** $\sqrt[n]{a^n}$

Simplify the expressions.

a. $\sqrt{(-3)^2}$ **b.** $\sqrt{x^2}$ **c.** $\sqrt[3]{x^3}$ **d.** $\sqrt[4]{x^4}$ **e.** $\sqrt[5]{x^5}$

Solution:

a. $\sqrt{(-3)^2} = |-3| = 3$ Because the index is *even*, absolute value bars are necessary to make the answer nonnegative.

b. $\sqrt{x^2} = |x|$ Because the index is *even*, absolute value bars are necessary to make the answer nonnegative.

c. $\sqrt[3]{x^3} = x$ Because the index is *odd*, no absolute value bars are necessary.

d. $\sqrt[4]{x^4} = |x|$ Because the index is *even*, absolute value bars are necessary to make the answer nonnegative.

e. $\sqrt[5]{x^5} = x$ Because the index is *odd*, no absolute value bars are necessary.

Skill Practice

Simplify.

21. $\sqrt{(-6)^2}$ **22.** $\sqrt[4]{a^4}$
23. $\sqrt[3]{w^3}$ **24.** $\sqrt[6]{p^6}$
25. $\sqrt[3]{(-2)^3}$

Classroom Examples: p. 591, Exercises 70, 82, and 84

If n is an even integer, then $\sqrt[n]{a^n} = |a|$. However, if the variable a is assumed to be nonnegative, then the absolute value bars may be omitted, that is, $\sqrt[n]{a^n} = a$ provided $a \geq 0$. In many examples and exercises, we will make the assumption that the variables within a radical expression are positive real numbers. In such a case, the absolute value bars are not needed to evaluate $\sqrt[n]{a^n}$.

It is helpful to become familiar with the patterns associated with perfect squares and perfect cubes involving variable expressions.

Answers

21. 6   **22.** $|a|$ **23.** w
24. $|p|$ **25.** -2

The following powers of x are perfect squares:

Perfect squares

$(x^1)^2 = x^2$
$(x^2)^2 = x^4$
$(x^3)^2 = x^6$
$(x^4)^2 = x^8$
$\dots$

> **TIP:** Any expression raised to an even power (multiple of 2) is a perfect square.

The following powers of x are perfect cubes:

Perfect cubes

$(x^1)^3 = x^3$
$(x^2)^3 = x^6$
$(x^3)^3 = x^9$
$(x^4)^3 = x^{12}$
$\dots$

> **TIP:** Any expression raised to a power that is a multiple of 3 is a perfect cube.

Skill Practice

Simplify the expressions. Assume the variables represent positive real numbers.

26. $\sqrt{y^{10}}$ **27.** $\sqrt[3]{x^{12}}$

28. $\sqrt{x^4 y^2}$ **29.** $\sqrt{25c^4}$

Classroom Examples: p. 591, Exercises 92 and 98

Example 6   Simplifying nth-Roots

Simplify the expressions. Assume that the variables are positive real numbers.

a. $\sqrt{c^6}$ **b.** $\sqrt[3]{d^{15}}$ **c.** $\sqrt{a^2 b^2}$ **d.** $\sqrt[3]{64z^6}$

Solution:

a. $\sqrt{c^6}$ The expression c^6 is a perfect square.

    $\sqrt{c^6} = c^3$ This is because $\sqrt{(c^3)^2} = c^3$.

b. $\sqrt[3]{d^{15}}$ The expression d^{15} is a perfect cube.

    $\sqrt[3]{d^{15}} = d^5$ This is because $\sqrt[3]{(d^5)^3} = d^5$.

c. $\sqrt{a^2 b^2} = ab$ This is because $\sqrt{a^2 b^2} = \sqrt{(ab)^2} = ab$.

d. $\sqrt[3]{64z^6} = 4z^2$ This is because $\sqrt[3]{(4z^2)^3} = 4z^2$.

> **TIP:** A perfect nth-root can be simplified by dividing the power of the radicand by the index.
>
> $$\sqrt[3]{d^{15}} = d^{15/3} = d^5$$

---

3. Translations Involving nth-Roots

It is important to understand the vocabulary and language associated with nth-roots. For instance, you must be able to distinguish between the square of a number and the square *root* of a number. The following example offers practice translating between English form and algebraic form.

Answers

26. y^5 **27.** x^4

28. $x^2 y$ **29.** $5c^2$

| Example 7 | **Translating from English Form to Algebraic Form** |

Translate each English phrase into an algebraic expression.

a. The difference of the square of x and the principal square root of 7

b. The quotient of 1 and the cube root of z

Solution:

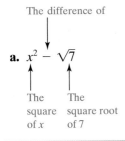

a. $x^2 - \sqrt{7}$

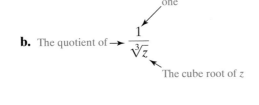

b. The quotient of → $\dfrac{1}{\sqrt[3]{z}}$

Classroom Example: p. 591, Exercise 108

─ **Skill Practice** ─

Translate the English phrases into algebraic expressions.

30. The product of the square of y and the principal square root of x.

31. The sum of 2 and the cube root of y.

4. Pythagorean Theorem

Recall that the **Pythagorean theorem** relates the lengths of the three sides of a right triangle (Figure 8-1).

$$a^2 + b^2 = c^2$$

The principal square root can be used to solve for an unknown side of a right triangle if the lengths of the other two sides are known.

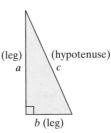

Figure 8-1

| Example 8 | **Applying the Pythagorean Theorem** |

Use the Pythagorean theorem and the definition of the principal square root of a number to find the length of the unknown side.

Solution:

Label the sides of the triangle.

$a^2 + b^2 = c^2$

$a^2 + (8)^2 = (10)^2$ Apply the Pythagorean theorem.

$a^2 + 64 = 100$ Simplify.

$a^2 = 100 - 64$

$a^2 = 36$ This equation is quadratic. One method for solving the equation is to set the equation equal to zero, factor, and apply the zero product rule. However, we can also use the definition of a square root to solve for a.

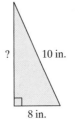

$a = \sqrt{36}$ or $a = -\sqrt{36}$ By definition, a must be one of the square roots of 36 (either 6 or -6). However, because a represents a distance, choose the *positive* (principal) square root of 36.

$a = 6$

The third side is 6 in. long.

─ **Skill Practice** ─

32. Use the Pythagorean theorem to find the length of the unknown side.

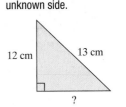

Classroom Example: p. 592, Exercise 114

Answers

30. $y^2\sqrt{x}$ **31.** $2 + \sqrt[3]{y}$

32. 5 cm

Skill Practice

33. A wire is attached to the top of a 20-ft pole. How long is the wire if it reaches a point on the ground 15 ft from the base of the pole?

?   20 ft

15 ft

Classroom Example: p. 592, Exercise 120

Answer

33. The wire is 25 ft long.

Example 9   **Applying the Pythagorean Theorem**

A bridge across a river is 600 yd long. A boat ramp at point R is 200 yd due north of point P on the bridge, such that the line segments $\overline{PQ}$ and $\overline{PR}$ form a right angle (Figure 8-2). How far does a kayak travel if it leaves from the boat ramp and paddles to point Q? Use a calculator and round the answer to the nearest yard.

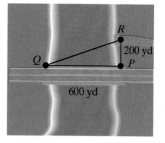

Figure 8-2

Solution:

Label the triangle:

c $a = 200$ yd

$b = 600$ yd

$$a^2 + b^2 = c^2$$

$$(200)^2 + (600)^2 = c^2 \qquad \text{Apply the Pythagorean theorem.}$$

$$40{,}000 + 360{,}000 = c^2 \qquad \text{Simplify.}$$

$$400{,}000 = c^2$$

By definition, c must be one of the square roots of 400,000. Because the value of c is a distance, choose the positive square root of 400,000.

$$c = \sqrt{400{,}000}$$

$$c \approx 632 \qquad$$ Use a calculator to approximate the positive square root of 400,000.

The kayak must travel approximately 632 yd.

Calculator Connections

Topic: Evaluating Square Roots and Higher Order Roots on a Calculator

A calculator can be used to approximate the value of a radical expression. To evaluate a square root, use the $\sqrt{\ }$ key. For example, evaluate: $\sqrt{25}$, $\sqrt{60}$, $\sqrt{\frac{13}{3}}$

Scientific Calculator

Enter:	25 $\sqrt{x}$	Result:	5
Enter:	60 $\sqrt{x}$	Result:	7.745966692
Enter:	13 $\div$ 3 $=$ $\sqrt{x}$	Result:	2.081665999

Graphing Calculator

On the graphing calculator, the radicand is enclosed in parentheses.

```
√(25)
              5
√(60)
      7.745966692
√(13/3)
      2.081665999
```

TIP: The values $\sqrt{60}$ and $\sqrt{\frac{13}{3}}$ are approximated on the calculator to 10 digits. However, $\sqrt{60}$ and $\sqrt{\frac{13}{3}}$ are actually irrational numbers. Their decimal forms are nonterminating and nonrepeating. The only way to represent the exact answers is by writing the radical forms, $\sqrt{60}$ and $\sqrt{\frac{13}{2}}$.

To evaluate cube roots, your calculator may have a $\sqrt[3]{}$ key. Otherwise, for cube roots and roots of higher index (fourth roots, fifth roots, and so on), try using the $\sqrt[x]{y}$ key or $\sqrt[x]{}$ key. For example, evaluate $\sqrt[3]{64}$, $\sqrt[4]{81}$, and $\sqrt[3]{162}$:

Scientific Calculator

Enter: 64 **2ⁿᵈ** $\sqrt[x]{y}$ 3 **=**   **Result:** ⬚ 4

Enter: 81 **2ⁿᵈ** $\sqrt[x]{y}$ 4 **=**   **Result:** ⬚ 3

Enter: 162 **2ⁿᵈ** $\sqrt[x]{y}$ 3 **=**   **Result:** ⬚ 5.451361778

Graphing Calculator

On a graphing calculator, the index is usually entered first.

```
3×√(64)
              4
4×√(81)
              3
3×√(162)
      5.451361778
```

Calculator Exercises

Estimate the value of each radical. Then use a calculator to approximate the radical to three decimal places.
(See the Tip on page 583.)

1. $\sqrt{5}$ 2.236

2. $\sqrt{17}$ 4.123

3. $\sqrt{50}$ 7.071

4. $\sqrt{96}$ 9.798

5. $\sqrt{33}$ 5.745

6. $\sqrt{145}$ 12.042

7. $\sqrt{80}$ 8.944

8. $\sqrt{170}$ 13.038

9. $\sqrt[3]{7}$ 1.913

10. $\sqrt[3]{28}$ 3.037

11. $\sqrt[3]{65}$ 4.021

12. $\sqrt[3]{124}$ 4.987

Section 8.1   Practice Exercises

Boost *your* GRADE at ALEKS.com!

• Practice Problems
• Self-Tests
• NetTutor

• e-Professors
• Videos

For additional exercises, see Classroom Activities 8.1A–8.1D in the *Instructor's Resource Manual* at www.mhhe.com/moh.

Study Skill Exercise

1. Define the key terms:

a. square root

b. positive square root

c. principal square root

d. negative square root

e. perfect square

f. *n*th-root

g. index

h. radicand

i. cube root

j. Pythagorean theorem

Objective 1: Definition of a Square Root

For Exercises 2–9, determine the square roots. **(See Example 1.)**

2. 4 2, −2

3. 144 12, −12

4. −64
There are no real-valued square roots of −64.

5. −49
There are no real-valued square roots of −49.

6. 81 9, −9

7. 0 0

8. $\dfrac{16}{9}$ $\dfrac{4}{3}, -\dfrac{4}{3}$

9. $\dfrac{1}{25}$ $\dfrac{1}{5}, -\dfrac{1}{5}$

 Writing   Translating Expression   Geometry   Scientific Calculator   Video   NS&E

10. a. What is the principal square root of 64?   8

   b. What is the negative square root of 64?   −8

11. a. What is the principal square root of 169?   13

   b. What is the negative square root of 169?   −13

12. Does every number have two square roots? Explain.
No, only positive numbers have two square roots. Zero has only one square root, and negative numbers have no real-valued square roots.

13. Which number has only one square root?   0

14. Which of the following are perfect squares?

   {0, 1, 4, 15, 30, 49, 72, 81, 144, 300, 625, 900}
   0, 1, 4, 49, 81, 144, 625, 900

15. Which of the following are perfect squares?

   {8, 9, 12, 16, 25, 36, 42, 64, 95, 121, 140, 169}
   9, 16, 25, 36, 64, 121, 169

For Exercises 16–31, evaluate the square roots. **(See Example 2.)**

16. $\sqrt{16}$ 4

17. $\sqrt{4}$ 2

18. $\sqrt{81}$ 9

19. $\sqrt{49}$ 7

20. $\sqrt{0.25}$ 0.5

21. $\sqrt{0.16}$ 0.4

22. $\sqrt{0.64}$ 0.8

23. $\sqrt{0.09}$ 0.3

24. $\sqrt{\dfrac{1}{9}}$ $\dfrac{1}{3}$

25. $\sqrt{\dfrac{25}{16}}$ $\dfrac{5}{4}$

26. $\sqrt{\dfrac{49}{121}}$ $\dfrac{7}{11}$

27. $\sqrt{\dfrac{1}{144}}$ $\dfrac{1}{12}$

28. $\sqrt{64 + 36}$ 10

29. $\sqrt{16 + 9}$ 5

30. $\sqrt{169 - 144}$ 5

31. $\sqrt{225 - 144}$ 9

32. Explain the difference between $\sqrt{-16}$ and $-\sqrt{16}$.
$\sqrt{-16}$ is not a real number, and $-\sqrt{16}$ simplifies to −4, which is a real number.

33. Using the definition of a square root, explain why $\sqrt{-16}$ does not have a real-valued square root.
There is no real value of b for which $b^2 = -16$.

34. Evaluate. $-\sqrt{|-25|}$ −5

For Exercises 35–46, evaluate the square roots, if possible. **(See Example 3.)**

35. $-\sqrt{4}$ −2

36. $-\sqrt{1}$ −1

37. $\sqrt{-4}$
Not a real number

38. $\sqrt{-1}$
Not a real number

39. $\sqrt{-\dfrac{4}{49}}$
Not a real number

40. $-\sqrt{-\dfrac{9}{25}}$
Not a real number

41. $-\sqrt{-\dfrac{1}{36}}$
Not a real number

42. $-\sqrt{\dfrac{1}{36}}$ $-\dfrac{1}{6}$

43. $-\sqrt{400}$ −20

44. $-\sqrt{121}$ −11

45. $\sqrt{-900}$
Not a real number

46. $\sqrt{-169}$
Not a real number

Objective 2: Definition of an *n*th-Root

47. Which of the following are perfect cubes?

   {0, 1, 3, 9, 27, 36, 42, 90, 125}   0, 1, 27, 125

48. Which of the following are perfect cubes?

   {6, 8, 16, 20, 30, 64, 111, 150, 216}   8, 64, 216

49. Does $\sqrt[3]{-27}$ have a real-valued cube root?
Yes, −3

50. Does $\sqrt[3]{-8}$ have a real-valued cube root?   Yes, −2

For Exercises 51–66, evaluate the *n*th roots, if possible. **(See Example 4.)**

51. $\sqrt[3]{27}$ 3

52. $\sqrt[3]{-27}$ −3

53. $\sqrt[3]{64}$ 4

54. $\sqrt[3]{-64}$ −4

55. $-\sqrt[4]{16}$ −2

56. $-\sqrt[4]{81}$ −3

57. $\sqrt[4]{-1}$
Not a real number

58. $\sqrt[4]{0}$ 0

59. $\sqrt[4]{-256}$
Not a real number

60. $\sqrt[4]{-625}$
Not a real number

61. $\sqrt[5]{-32}$ −2

62. $-\sqrt[5]{32}$ −2

63. $-\sqrt[6]{1}$ −1

64. $\sqrt[6]{64}$ 2

65. $\sqrt[6]{0}$ 0

66. $\sqrt[6]{-1}$ Not a real number

 Writing   Translating Expression   Geometry   Scientific Calculator   Video   NS&E

For Exercises 67–86, simplify the expressions. **(See Example 5.)**

67. $\sqrt{(4)^2}$ 4

68. $\sqrt{(8)^2}$ 8

69. $\sqrt{(-4)^2}$ 4

70. $\sqrt{(-8)^2}$ 8

71. $\sqrt[3]{(5)^3}$ 5

72. $\sqrt[3]{(7)^3}$ 7

73. $\sqrt[3]{(-5)^3}$ -5

74. $\sqrt[3]{(-7)^3}$ -7

75. $\sqrt[4]{(2)^4}$ 2

76. $\sqrt[4]{(10)^4}$ 10

77. $\sqrt[4]{(-2)^4}$ 2

78. $\sqrt[4]{(-10)^4}$ 10

79. $\sqrt{a^2}$ $|a|$

80. $\sqrt{b^2}$ $|b|$

81. $\sqrt[3]{y^3}$ y

82. $\sqrt[3]{z^3}$ z

83. $\sqrt[4]{w^4}$ $|w|$

84. $\sqrt[4]{p^4}$ $|p|$

85. $\sqrt[5]{x^5}$ x

86. $\sqrt[5]{y^5}$ y

87. Determine which of the expressions are perfect squares. Then state a rule for determining perfect squares based on the exponent of the expression.

$\{x^2, a^3, y^4, z^5, (ab)^6, (pq)^7, w^8x^8, c^9d^9, m^{10}, n^{11}\}$

$x^2, y^4, (ab)^6, w^8x^8, m^{10}$. The expression is a perfect square if the exponent is even.

88. Determine which of the expressions are perfect cubes. Then state a rule for determining perfect cubes based on the exponent of the expression.

$\{a^2, b^3, c^4, d^5, e^6, (xy)^7, (wz)^8, (pq)^9, t^{10}s^{10}, m^{11}n^{11}, u^{12}v^{12}\}$

$b^3, e^6, (pq)^9, u^{12}v^{12}$. The expression is a perfect cube if the exponent is a multiple of 3.

89. Determine which of the expressions are perfect fourth powers. Then state a rule for determining perfect fourth powers based on the exponent of the expression.

$\{m^2, n^3, p^4, q^5, r^6, s^7, t^8, u^9, v^{10}, (ab)^{11}, (cd)^{12}\}$

$p^4, t^8, (cd)^{12}$. The expression is a perfect fourth power if the exponent is a multiple of 4.

90. Determine which of the expressions are perfect fifth powers. Then state a rule for determining perfect fifth powers based on the exponent of the expression.

$\{a^2, b^3, c^4, d^5, e^6, k^7, w^8, x^9, y^{10}, z^{11}\}$

d^5, y^{10}. The expression is a perfect fifth power if the exponent is a multiple of 5.

For Exercises 91–106, simplify the expressions. Assume the variables represent positive real numbers. **(See Example 6.)**

91. $\sqrt{y^{12}}$ y^6

92. $\sqrt{z^{20}}$ z^{10}

93. $\sqrt{a^8b^{30}}$ a^4b^{15}

94. $\sqrt{t^{50}s^{60}}$ $t^{25}s^{30}$

95. $\sqrt[3]{q^{24}}$ q^8

96. $\sqrt[3]{x^{33}}$ x^{11}

97. $\sqrt[3]{8w^6}$ $2w^2$

98. $\sqrt[3]{-27x^{27}}$ $-3x^9$

99. $\sqrt{(5x)^2}$ $5x$

100. $\sqrt{(6w)^2}$ $6w$

101. $-\sqrt{25x^2}$ $-5x$

102. $-\sqrt{36w^2}$ $-6w$

103. $\sqrt[3]{(5p^2)^3}$ $5p^2$

104. $\sqrt[3]{(2k^4)^3}$ $2k^4$

105. $\sqrt[3]{125p^6}$ $5p^2$

106. $\sqrt[3]{8k^{12}}$ $2k^4$

Objective 3: Translations Involving *n*th-Roots

For Exercises 107–110, translate the English phrase to an algebraic expression. **(See Example 7.)**

107. The sum of the principal square root of q and the square of p $\sqrt{q} + p^2$

108. The product of the principal square root of 11 and the cube of x

$\sqrt{11} \cdot x^3$

109. The quotient of 6 and the principal fourth root of x $\dfrac{6}{\sqrt[4]{x}}$

110. The difference of the square of y and 1

$y^2 - 1$

Writing   Translating Expression   Geometry   Scientific Calculator   Video   NS&E

Objective 4: Pythagorean Theorem

For Exercises 111–116, find the length of the third side of each triangle using the Pythagorean theorem. Round the answer to the nearest tenth if necessary. **(See Example 8.)**

111.

9 cm
15 cm
12 cm

112.

10 in.
8 in.
6 in.

113.

5 ft
12 ft   13 ft

114.

3 m
5 m
4 m

115.

6.5 cm   6.9 cm
2.4 cm

116.

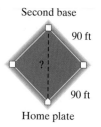

11.6 ft
14.8 ft
9.2 ft

117. Find the length of the diagonal of the square tile shown in the figure. Round the answer to the nearest tenth of an inch.   17.0 in.

12 in.
12 in.

118. A baseball diamond is 90 ft on a side. Find the distance between home plate and second base. Round the answer to the nearest tenth of a foot.   127.3 ft

Second base
90 ft
?
90 ft
Home plate

119. A new plasma television is listed as being 42 in. This distance is the diagonal distance across the screen. If the screen measures 28 inches in height, what is the actual width of the screen? Round to the nearest tenth of an inch.
(See Example 9.)   31.3 in.

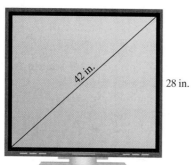

42 in.

28 in.

120. A marine biologist wants to track the migration of a pod of whales. He receives a radio signal from a tagged humpback whale and determines that the whale is 21 mi east and 37 mi north of his laboratory. Find the direct distance between the whale and the laboratory. Round to the nearest tenth of a mile.   42.5 mi

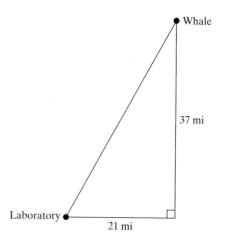

Whale
37 mi
Laboratory
21 mi

121. On a map, the cities Asheville, North Carolina, Roanoke, Virginia, and Greensboro, North Carolina, form a right triangle (see the figure). The distance between Asheville and Roanoke is 300 km. The distance between Roanoke and Greensboro is 134 km. How far is it from Greensboro to Asheville? Round the answer to the nearest kilometer. 268 km

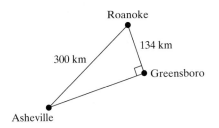

122. Jackson, Mississippi, is west of Meridian, Mississippi, a distance of 141 km. Tupelo, Mississippi, is north of Meridian, a distance of 209 km. How far is it from Jackson to Tupelo? Round the answer to the nearest kilometer. 252 km

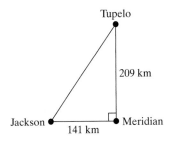

Expanding Your Skills

123. For what values of x will $\sqrt{x}$ be a real number?
$x \geq 0$

124. For what values of x will $\sqrt{-x}$ be a real number? $x \leq 0$

Simplifying Radicals

Section 8.2

1. Multiplication Property of Radicals

You may have already recognized certain properties of radicals involving a product.

Objectives

1. **Multiplication Property of Radicals**
2. **Simplifying Radicals Using the Multiplication Property of Radicals**
3. **Simplifying Radicals Using the Order of Operations**
4. **Simplifying Cube Roots**

> **PROPERTY** Multiplication Property of Radicals
>
> Let a and b represent real numbers such that $\sqrt[n]{a}$ and $\sqrt[n]{b}$ are both real. Then,
>
> $$\sqrt[n]{ab} = \sqrt[n]{a} \cdot \sqrt[n]{b} \quad \textbf{Multiplication property of radicals}$$

The multiplication property of radicals indicates that a product within a radicand can be written as a product of radicals provided the roots are real numbers.

$$\sqrt{100} = \sqrt{25} \cdot \sqrt{4}$$

The reverse process is also true. A product of radicals can be written as a single radical provided the roots are real numbers and they have the same indices.

$$\overset{\text{Same index}}{\sqrt{2} \cdot \sqrt{18}} = \sqrt{36}$$

2. Simplifying Radicals Using the Multiplication Property of Radicals

In algebra, it is customary to simplify radical expressions as much as possible.

DEFINITION Simplified Form of a Radical

Consider any radical expression where the radicand is written as a product of prime factors. The expression is in **simplified form** if all of the following conditions are met:

1. The radicand has no factor raised to a power greater than or equal to the index.
2. There are no radicals in the denominator of a fraction.
3. The radicand does not contain a fraction.

The expression $\sqrt{x^2}$ is not simplified because it fails condition 1. Because x^2 is a perfect square, $\sqrt{x^2}$ is easily simplified.

$$\sqrt{x^2} = x \quad (\text{for } x \geq 0)$$

However, how is an expression such as $\sqrt{x^7}$ simplified? This and many other radical expressions are simplified using the multiplication property of radicals. Examples 1–3 illustrate how nth powers can be removed from the radicands of nth-roots.

Example 1 Using the Multiplication Property to Simplify a Radical Expression

Use the multiplication property of radicals to simplify the expression $\sqrt{x^7}$. Assume $x \geq 0$.

Solution:

The expression $\sqrt{x^7}$ is equivalent to $\sqrt{x^6 \cdot x}$. By applying the multiplication property of radicals, we have

$$\sqrt{x^6 \cdot x} = \sqrt{x^6} \cdot \sqrt{x} \qquad x^6 \text{ is a perfect square because } (x^3)^2 = x^6$$

$$= x^3 \cdot \sqrt{x} \qquad \text{Simplify.}$$

$$= x^3\sqrt{x}$$

In Example 1, the expression x^7 is not a perfect square. Therefore, to simplify $\sqrt{x^7}$, it was necessary to write the expression as the product of the largest perfect square and a remaining, or "leftover," factor: $\sqrt{x^7} = \sqrt{x^6 \cdot x}$.

Example 2 Using the Multiplication Property to Simplify Radicals

Use the multiplication property of radicals to simplify the expressions. Assume the variables represent positive real numbers.

a. $\sqrt{a^{15}}$ b. $\sqrt{x^2 y^5}$ c. $\sqrt{s^9 t^{11}}$

Solution:

The goal is to rewrite each radicand as the product of the largest perfect square and a leftover factor.

a. $\sqrt{a^{15}}$

$= \sqrt{a^{14} \cdot a} \qquad a^{14}$ is the largest perfect square in the radicand.

$= \sqrt{a^{14}} \cdot \sqrt{a} \qquad$ Apply the multiplication property of radicals.

$= a^7 \sqrt{a} \qquad$ Simplify.

b. $\sqrt{x^2 y^5}$

$\qquad = \sqrt{x^2 y^4 \cdot y}$ $\qquad$ $x^2 y^4$ is the largest perfect square in the radicand.

$\qquad = \sqrt{x^2 y^4} \cdot \sqrt{y}$ $\qquad$ Apply the multiplication property of radicals.

$\qquad = xy^2 \sqrt{y}$ $\qquad$ Simplify.

c. $\sqrt{s^9 t^{11}}$

$\qquad = \sqrt{s^8 t^{10} \cdot st}$ $\qquad$ $s^8 t^{10}$ is the largest perfect square in the radical.

$\qquad = \sqrt{s^8 t^{10}} \cdot \sqrt{st}$ $\qquad$ Apply the multiplication property of radicals.

$\qquad = s^4 t^5 \sqrt{st}$ $\qquad$ Simplify.

Each expression in Example 2 involves a radicand that is a product of variable factors. If a numerical factor is present, sometimes it is necessary to factor the coefficient before simplifying the radical.

Example 3 **Using the Multiplication Property to Simplify Radicals**

Use the multiplication property of radicals to simplify the expressions. Assume the variables represent positive real numbers.

a. $\sqrt{50}$ $\qquad$ **b.** $5\sqrt{24a^6}$ $\qquad$ **c.** $\sqrt{81x^4 y^3}$

Solution:

The goal is to rewrite each radicand as the product of the largest perfect square and a leftover factor.

a. Write the radicand as a product of prime factors. From the prime factorization, the largest perfect square is easily identified.

$\sqrt{50} = \sqrt{5^2 \cdot 2}$ $\qquad$ Factor the radicand. $\qquad$ $2\underline{)50}$

$\qquad\qquad$ 5^2 is the largest perfect square. $\qquad$ $5\underline{)25}$

$\qquad\qquad\qquad\qquad\qquad\qquad\qquad\qquad\qquad\qquad\qquad$ 5

$\qquad = \sqrt{5^2} \cdot \sqrt{2}$ $\qquad$ Apply the multiplication property of radicals.

$\qquad = 5\sqrt{2}$ $\qquad$ Simplify.

b. $5\sqrt{24a^6} = 5\sqrt{2^3 \cdot 3 \cdot a^6}$ $\qquad$ Write the radicand as a product of prime factors: $24 = 2^3 \cdot 3$.

$\qquad = 5\sqrt{2^2 a^6 \cdot 2 \cdot 3}$ $\qquad$ $2^2 a^6$ is the largest perfect square in the radicand.

$\qquad = 5\sqrt{2^2 a^6} \cdot \sqrt{2 \cdot 3}$ $\qquad$ Apply the multiplication property of radicals.

$\qquad = 5 \cdot 2a^3 \sqrt{6}$ $\qquad$ Simplify the radical.

$\qquad = 10a^3 \sqrt{6}$ $\qquad$ Simplify the coefficient of the radical.

Skill Practice

Simplify the expressions. Assume the variables represent positive real numbers.

8. $\sqrt{12}$ $\qquad$ **9.** $\sqrt{60x^2}$

10. $7\sqrt{18t^{10}}$

Classroom Examples: pp. 599–600, Exercises 22 and 50

TIP: The expression $\sqrt{50}$ can also be written as:

$\sqrt{25 \cdot 2}$

$= \sqrt{25} \cdot \sqrt{2}$

$= 5\sqrt{2}$

Answers

8. $2\sqrt{3}$ $\qquad$ **9.** $2x\sqrt{15}$

10. $21t^5 \sqrt{2}$

c. $\sqrt{81x^4y^3} = \sqrt{3^4x^4y^3}$ Write the radical as a product of prime factors. *Note:* $81 = 3^4$.

$= \sqrt{3^4x^4y^2 \cdot y}$ $3^4x^4y^2$ is the largest square in the radicand.

$= \sqrt{3^4x^4y^2} \cdot \sqrt{y}$ Apply the multiplication property of radicals.

$= 3^2x^2y \cdot \sqrt{y}$ Simplify the radical.

$= 9x^2y\sqrt{y}$ Simplify the coefficient of the radical.

Avoiding Mistakes

The multiplication property of radicals allows us to simplify a product of factors within a radical. For example,

$$\sqrt{x^2y^2} = \sqrt{x^2} \cdot \sqrt{y^2} = xy \quad \text{(for } x \geq 0 \text{ and } y \geq 0\text{)}$$

However, this rule does not apply to *terms* that are added or subtracted *within* the radical. For example,

$$\sqrt{x^2 + y^2} \quad \text{and} \quad \sqrt{x^2 - y^2}$$

cannot be simplified.

---

3. Simplifying Radicals Using the Order of Operations

Often a radical can be simplified by applying the order of operations. In Example 4, the first step will be to simplify the expression within the radicand.

Example 4   **Simplifying Radicals Using the Order of Operations**

Simplify the expressions. Assume the variables represent positive real numbers.

a. $\sqrt{\dfrac{a^5}{a^3}}$ **b.** $\sqrt{\dfrac{6}{96}}$ **c.** $\sqrt{\dfrac{27x^5}{3x}}$

Solution:

a. $\sqrt{\dfrac{a^5}{a^3}}$ The radical contains a fraction. However, the fraction can be simplified.

$= \sqrt{a^2}$ Reduce the fraction to lowest terms.

$= a$ Simplify the radical.

b. $\sqrt{\dfrac{6}{96}}$ The radical contains a fraction that can be simplified.

$= \sqrt{\dfrac{1}{16}}$ Reduce the fraction to lowest terms.

$= \dfrac{1}{4}$ Simplify.

c. $\sqrt{\dfrac{27x^5}{3x}}$ The fraction within the radicand can be simplified.

$= \sqrt{9x^4}$ Reduce to lowest terms.

$= 3x^2$ Simplify.

Example 5 Simplifying Radical Expressions

Simplify the expressions.

a. $\dfrac{5\sqrt{20}}{2}$ b. $\dfrac{2-\sqrt{36}}{12}$

Solution:

a. $\dfrac{5\sqrt{20}}{2} = \dfrac{5\sqrt{2^2\cdot 5}}{2}$ Following the order of operations, first simplify the radical. 2^2 is the largest perfect square in the radicand.

$= \dfrac{5\sqrt{2^2}\cdot\sqrt{5}}{2}$ Apply the multiplication property of radicals.

$= \dfrac{5\cdot 2\sqrt{5}}{2}$ Simplify the radical.

$= \dfrac{5\cdot 2\sqrt{5}}{2}$ Simplify to lowest terms.

$= 5\sqrt{5}$

b. $\dfrac{2-\sqrt{36}}{12}$

$= \dfrac{2-6}{12}$ Following the order of operations, first simplify the radical.

$= \dfrac{-4}{12}$ Next, simplify the numerator.

$= -\dfrac{1}{3}$ Reduce the fraction to lowest terms.

4. Simplifying Cube Roots

Example 6 Simplifying Cube Roots

Use the multiplication property of radicals to simplify the expressions.

a. $\sqrt[3]{z^5}$ b. $\sqrt[3]{-80}$

Solution:

a. $\sqrt[3]{z^5}$

$= \sqrt[3]{z^3\cdot z^2}$ z^3 is the largest perfect cube in the radicand.

$= \sqrt[3]{z^3}\cdot\sqrt[3]{z^2}$ Apply the multiplication property of radicals.

$= z\sqrt[3]{z^2}$ Simplify.

b. $\sqrt[3]{-80}$

$= \sqrt[3]{-1\cdot 2^4\cdot 5}$ Factor the radicand.

$= \sqrt[3]{-1\cdot 2^3\cdot 2\cdot 5}$ -1 and 2^3 are perfect cubes.

$= \sqrt[3]{-1}\cdot\sqrt[3]{2^3}\cdot\sqrt[3]{2\cdot 5}$ Apply the multiplication property of radicals.

$= -1\cdot 2\cdot\sqrt[3]{10}$ Simplify.

$= -2\sqrt[3]{10}$

$\begin{array}{r}2\,\overline{)80}\\2\,\overline{)40}\\2\,\overline{)20}\\2\,\overline{)10}\\5\end{array}$

Skill Practice

Simplify.

18. $\sqrt[3]{\dfrac{x^{12}}{y^6}}$ **19.** $\sqrt[3]{\dfrac{81}{3}}$

Classroom Example: p. 601, Exercise 82

Example 7   Simplifying Cube Roots

Simplify the expressions.

a. $\sqrt[3]{\dfrac{a^{16}}{a}}$ **b.** $\sqrt[3]{\dfrac{2}{16}}$

Solution:

a. $\sqrt[3]{\dfrac{a^{16}}{a}}$ The radical contains a fraction that can be simplified.

$= \sqrt[3]{a^{15}}$ Reduce to lowest terms.

$= a^5$ Simplify.

b. $\sqrt[3]{\dfrac{2}{16}}$ The radical contains a fraction that can be simplified.

$= \sqrt[3]{\dfrac{1}{8}}$ Reduce to lowest terms.

$= \dfrac{1}{2}$ Simplify.

Answers

18. $\dfrac{x^4}{y^2}$ **19.** 3

Calculator Connections

Topic: Verifying Simplified Radicals

A calculator can support the multiplication property of radicals. For example, use a calculator to evaluate $\sqrt{50}$ and its simplified form $5\sqrt{2}$.

Scientific Calculator

Enter: 50 $\boxed{\sqrt{x}}$ **Result:** 7.071067812

Enter: 2 $\boxed{\sqrt{x}}$ $\boxed{\times}$ 5 $\boxed{=}$ **Result:** 7.071067812

Instructor Note: This is a good opportunity to work on estimation skills. $\sqrt{50}$ is between what two integers? Closer to which one?

Graphing Calculator

```
√(50)
          7.071067812
5*√(2)
          7.071067812
```

TIP: The decimal approximation for $\sqrt{50}$ and $5\sqrt{2}$ agree for the first 10 digits. This in itself does not make $\sqrt{50} = 5\sqrt{2}$. It is the multiplication property of radicals that guarantees that the expressions are equal.

Calculator Exercises

Simplify the radical expressions algebraically. Then use a calculator to approximate the original expression and its simplified form.

1. $\sqrt{125}$ **2.** $\sqrt{18}$ **3.** $\sqrt[3]{54}$ **4.** $\sqrt[3]{108}$

Section 8.2 Practice Exercises

Study Skills Exercise

1. Define the key terms:

   a. simplified form of a radical   **b. multiplication property of radicals**

Review Exercises

2. Which of the following are perfect squares? $\{2, 4, 6, 16, 20, 25, x^2, x^3, x^{15}, x^{20}, x^{25}\}$ $4, 16, 25, x^2, x^{20}$

3. Which of the following are perfect cubes? $\{3, 6, 8, 9, 12, 27, y^3, y^8, y^9, y^{12}, y^{27}\}$ $8, 27, y^3, y^9, y^{12}, y^{27}$

4. Which of the following are perfect fourth powers? $\{4, 16, 20, 25, 81, w^4, w^{16}, w^{20}, w^{25}, w^{81}\}$ $16, 81, w^4, w^{16}, w^{20}$

For Exercises 5–12, simplify the expressions, if possible. Assume the variables represent positive real numbers.

5. $-\sqrt{25}$ -5

6. $\sqrt{-25}$ Not a real number

7. $-\sqrt[3]{27}$ -3

8. $\sqrt[3]{-27}$ -3

9. $\sqrt[4]{a^8}$ a^2

10. $\sqrt[5]{b^{15}}$ b^3

11. $\sqrt{4x^2y^4}$ $2xy^2$

12. $\sqrt{9p^{10}}$ $3p^5$

 13. On a map, Seattle, Washington, is 378 km west of Spokane, Washington. Portland, Oregon, is 236 km south of Seattle. Approximate the distance between Portland and Spokane to the nearest kilometer.   446 km

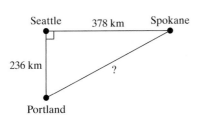

14. A new roof is needed on a shed. How many square feet of tar paper would be needed to cover the top of the roof?   1040 sq. ft

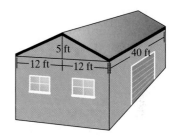

Objective 2: Simplifying Radicals Using the Multiplication Property of Radicals

For Exercises 15–50, use the multiplication property of radicals to simplify the expressions. Assume the variables represent positive real numbers.   **(See Examples 1–3.)**

15. $\sqrt{18}$ $3\sqrt{2}$

16. $\sqrt{75}$ $5\sqrt{3}$

17. $\sqrt{28}$ $2\sqrt{7}$

18. $\sqrt{40}$ $2\sqrt{10}$

19. $6\sqrt{20}$ $12\sqrt{5}$

20. $10\sqrt{27}$ $30\sqrt{3}$

21. $-2\sqrt{50}$ $-10\sqrt{2}$

22. $-11\sqrt{8}$ $-22\sqrt{2}$

23. $\sqrt{a^5}$ $a^2\sqrt{a}$

24. $\sqrt{b^9}$ $b^4\sqrt{b}$

25. $\sqrt{w^{22}}$ w^{11}

26. $\sqrt{p^{18}}$ p^9

27. $\sqrt{m^4n^5}$ $m^2n^2\sqrt{n}$

28. $\sqrt{c^2d^9}$ $cd^4\sqrt{d}$

29. $x\sqrt{x^{13}y^{10}}$ $x^7y^5\sqrt{x}$

30. $v\sqrt{u^{10}v^7}$ $u^5v^4\sqrt{v}$

31. $3\sqrt{t^{10}}$ $3t^5$

32. $-4\sqrt{m^8n^4}$ $-4m^4n^2$

33. $\sqrt{8x^3}$ $2x\sqrt{2x}$

34. $\sqrt{27y^5}$ $3y^2\sqrt{3y}$

35. $\sqrt{16z^3}$ $4z\sqrt{z}$

36. $\sqrt{9y^5}$ $3y^2\sqrt{y}$

37. $-\sqrt{45w^6}$ $-3w^3\sqrt{5}$

38. $-\sqrt{56v^8}$ $-2v^4\sqrt{14}$

39. $\sqrt{z^{25}}$ $\quad z^{12}\sqrt{z}$

40. $\sqrt{25p^{49}}$ $\quad 5p^{24}\sqrt{p}$

41. $-\sqrt{15z^{11}}$ $\quad -z^5\sqrt{15z}$

42. $-\sqrt{6k^{15}}$ $\quad -k^7\sqrt{6k}$

43. $5\sqrt{104a^2b^7}$ $\quad 10ab^3\sqrt{26b}$ **44.** $3\sqrt{88m^4n^{11}}$ $\quad 6m^2n^5\sqrt{22n}$ **45.** $\sqrt{26pq}$ $\quad \sqrt{26pq}$

46. $\sqrt{15a}$ $\quad \sqrt{15a}$

47. $m\sqrt{m^{10}n^{16}}$ $\quad m^6n^8$ **48.** $c^2\sqrt{c^4d^{12}}$ $\quad c^4d^6$ **49.** $\sqrt{48a^3b^5c^4}$ $\quad 4ab^2c^2\sqrt{3ab}$ **50.** $\sqrt{18xy^4z^3}$ $\quad 3y^2z\sqrt{2xz}$

Objective 3: Simplifying Radicals Using the Order of Operations

For Exercises 51–70, use the order of operations, if necessary, to simplify the expressions. Assume the variables represent positive real numbers. **(See Examples 4–5.)**

51. $\sqrt{\dfrac{a^9}{a}}$ $\quad a^4$

52. $\sqrt{\dfrac{x^5}{x}}$ $\quad x^2$

53. $\sqrt{\dfrac{y^{15}}{y^5}}$ $\quad y^5$

54. $\sqrt{\dfrac{c^{31}}{c^{11}}}$ $\quad c^{10}$

55. $\sqrt{\dfrac{5}{20}}$ $\quad \dfrac{1}{2}$

56. $\sqrt{\dfrac{3}{75}}$ $\quad \dfrac{1}{5}$

57. $\sqrt{\dfrac{40}{10}}$ $\quad 2$

58. $\sqrt{\dfrac{80}{5}}$ $\quad 4$

59. $\sqrt{\dfrac{32x^3}{8x}}$ $\quad 2x$

60. $\sqrt{\dfrac{200b^{11}}{2b^5}}$ $\quad 10b^3$

61. $\sqrt{\dfrac{50p^7}{2p}}$ $\quad 5p^3$

62. $\sqrt{\dfrac{45t^9}{5t^5}}$ $\quad 3t^2$

63. $\dfrac{3\sqrt{20}}{2}$ $\quad 3\sqrt{5}$

64. $\dfrac{5\sqrt{18}}{3}$ $\quad 5\sqrt{2}$

65. $\dfrac{5\sqrt{24}}{10}$ $\quad \sqrt{6}$

66. $\dfrac{2\sqrt{27}}{6}$ $\quad \sqrt{3}$

67. $\dfrac{10+\sqrt{4}}{3}$ $\quad 4$

68. $\dfrac{-1+\sqrt{25}}{4}$ $\quad 1$

69. $\dfrac{20-\sqrt{36}}{2}$ $\quad 7$

70. $\dfrac{3-\sqrt{81}}{3}$ $\quad -2$

For Exercises 71–74, find the exact length of the third side of each triangle using the Pythagorean theorem. Write the answer as a simplified radical.

71.

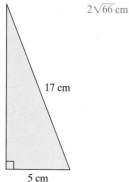

11 ft

11 ft

$11\sqrt{2}$ ft

72.

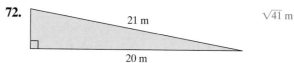

21 m

20 m

$\sqrt{41}$ m

73.

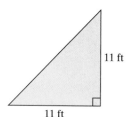

17 cm

5 cm

$2\sqrt{66}$ cm

74.

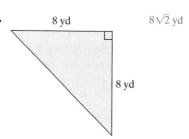

8 yd

8 yd

$8\sqrt{2}$ yd

Objective 4: Simplifying Cube Roots

For Exercises 75–86, simplify the cube roots. **(See Examples 6–7.)**

75. $\sqrt[3]{a^8}$ $a^2\sqrt[3]{a^2}$

76. $\sqrt[3]{8v^3}$ $2v$

77. $7\sqrt[3]{16z^3}$ $14z\sqrt[3]{2}$

78. $5\sqrt[3]{54t^6}$ $15t^2\sqrt[3]{2}$

79. $\sqrt[3]{16a^5b^6}$ $2ab^2\sqrt[3]{2a^2}$

80. $\sqrt[3]{81p^9q^{11}}$ $3p^3q^3\sqrt[3]{3q^2}$

81. $\sqrt[3]{\dfrac{z^4}{z}}$ z

82. $\sqrt[3]{\dfrac{w^8}{w^2}}$ w^2

83. $\sqrt[3]{-\dfrac{32}{4}}$ -2

84. $\sqrt[3]{-\dfrac{128}{2}}$ -4

85. $\sqrt[3]{40}$ $2\sqrt[3]{5}$

86. $\sqrt[3]{54}$ $3\sqrt[3]{2}$

Mixed Exercises

For Exercises 87–110, simplify the expressions. Assume the variables represent positive real numbers.

87. $\sqrt{\dfrac{3}{27}}$ $\dfrac{1}{3}$

88. $\sqrt{\dfrac{5}{125}}$ $\dfrac{1}{5}$

89. $\sqrt{16a^3}$ $4a\sqrt{a}$

90. $\sqrt{125x^6}$ $5x^3\sqrt{5}$

91. $\sqrt{\dfrac{4x^3}{x}}$ $2x$

92. $\sqrt{\dfrac{9z^5}{z}}$ $3z^2$

93. $\sqrt{8p^2q}$ $2p\sqrt{2q}$

94. $\sqrt{6cd^3}$ $d\sqrt{6cd}$

95. $\sqrt{32}$ $4\sqrt{2}$

96. $\sqrt{64}$ 8

97. $\sqrt{52u^4v^7}$ $2u^2v^3\sqrt{13v}$

98. $\sqrt{44p^8q^{10}}$ $2p^4q^5\sqrt{11}$

99. $\sqrt{216}$ $6\sqrt{6}$

100. $\sqrt{250}$ $5\sqrt{10}$

101. $\sqrt[3]{216}$ 6

102. $\sqrt[3]{250}$ $5\sqrt[3]{2}$

103. $\sqrt[3]{16a^3}$ $2a\sqrt[3]{2}$

104. $\sqrt[3]{125x^6}$ $5x^2$

105. $\sqrt[3]{\dfrac{x^5}{x^2}}$ x

106. $\sqrt[3]{\dfrac{y^{11}}{y^2}}$ y^3

107. $\dfrac{-6\sqrt{20}}{12}$ $-\sqrt{5}$

108. $\dfrac{-5\sqrt{32}}{10}$ $-2\sqrt{2}$

109. $\dfrac{-4-\sqrt{25}}{18}$ $-\dfrac{1}{2}$

110. $\dfrac{8-\sqrt{100}}{2}$ -1

Expanding Your Skills

For Exercises 111–114, simplify the expressions. Assume the variables represent positive real numbers.

111. $\sqrt{(-2-5)^2+(-4+3)^2}$ $5\sqrt{2}$

112. $\sqrt{(-1-7)^2+[1-(-1)]^2}$ $2\sqrt{17}$

113. $\sqrt{x^2+10x+25}$ $x+5$

114. $\sqrt{x^2+6x+9}$ $x+3$

Addition and Subtraction of Radicals   Section 8.3

1. Definition of *Like* Radicals

Objectives

1. Definition of *Like* Radicals
2. Addition and Subtraction of Radicals

> **DEFINITION**   *Like* Radicals
>
> Two radical terms are said to be *like* **radicals** if they have the same index and the same radicand.

 Writing   Translating Expression   Geometry   Scientific Calculator   Video   NS E

The following are pairs of *like* radicals:

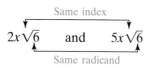

$2x\sqrt{6}$ and $5x\sqrt{6}$ — Indices and radicands are the same.
Both indices are 2. Both radicands are 6.

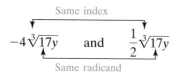

$-4\sqrt[3]{17y}$ and $\frac{1}{2}\sqrt[3]{17y}$ — Indices and radicands are the same.
Both indices are 3. Both radicands are $17y$.

These pairs are *not like* radicals:

$9\sqrt{3}$ and $9\sqrt[3]{3}$ — Radicals have different indices.

$8ab\sqrt{5}$ and $ab\sqrt{10}$ — Radicals have different radicands.

2. Addition and Subtraction of Radicals

To add or subtract *like* radicals, use the distributive property. For example,

$$3\sqrt{7} + 5\sqrt{7} = (3 + 5)\sqrt{7}$$
$$= 8\sqrt{7}$$

$$9\sqrt{2y} - 4\sqrt{2y} = (9 - 4)\sqrt{2y}$$
$$= 5\sqrt{2y}$$

Avoiding Mistakes

The process of adding *like* radicals with the distributive property is similar to adding *like* terms. The numerical coefficients are added and the radical factor is unchanged.

$\sqrt{5} + \sqrt{5}$
$= 1\sqrt{5} + 1\sqrt{5}$
$= 2\sqrt{5}$ Correct

Be careful: $\sqrt{5} + \sqrt{5} \neq \sqrt{10}$
In general,
$$\sqrt{x} + \sqrt{y} \neq \sqrt{x + y}$$

Example 1   **Adding and Subtracting Radicals**

Add or subtract the radicals as indicated. Assume all variables represent positive real numbers.

a. $\sqrt{5} + \sqrt{5}$ **b.** $6\sqrt{15} + 3\sqrt{15} + \sqrt{15}$

c. $\sqrt{xy} - 6\sqrt{xy} + 4\sqrt{xy}$

Solution:

a. $\sqrt{5} + \sqrt{5}$

   $= 1\sqrt{5} + 1\sqrt{5}$ *Note:* $\sqrt{5} = 1\sqrt{5}$.

   $= (1 + 1)\sqrt{5}$ Apply the distributive property.

   $= 2\sqrt{5}$ Simplify.

b. $6\sqrt{15} + 3\sqrt{15} + \sqrt{15}$ The radicals have the same radicand and same index.

   $= 6\sqrt{15} + 3\sqrt{15} + 1\sqrt{15}$ *Note:* $\sqrt{15} = 1\sqrt{15}$.

   $= (6 + 3 + 1)\sqrt{15}$ Apply the distributive property.

   $= 10\sqrt{15}$

c. $\sqrt{xy} - 6\sqrt{xy} + 4\sqrt{xy}$ The radicals have the same radicand and same index.

$= 1\sqrt{xy} - 6\sqrt{xy} + 4\sqrt{xy}$ *Note:* $\sqrt{xy} = 1\sqrt{xy}$.

$= (1 - 6 + 4)\sqrt{xy}$ Apply the distributive property.

$= -1\sqrt{xy}$ Simplify.

$= -\sqrt{xy}$

— **Skill Practice** —

Add or subtract the radicals as indicated. Assume the variables represent positive real numbers.
 1. $3\sqrt{2} + 7\sqrt{2}$
 2. $8\sqrt{x} - \sqrt{x}$
 3. $4\sqrt{ab} - 2\sqrt{ab} - 9\sqrt{ab}$

Classroom Examples: p. 606, Exercises 14 and 18

From Example 1, you can see that the process of adding or subtracting *like* radicals is similar to combining *like* terms.

$$x + x \qquad \text{similarly,} \qquad \sqrt{5} + \sqrt{5}$$
$$= 1x + 1x \qquad\qquad\qquad = 1\sqrt{5} + 1\sqrt{5}$$
$$= 2x \qquad\qquad\qquad\qquad = 2\sqrt{5}$$

The end result is that the numerical coefficients are added or subtracted and the radical factor is unchanged.

Sometimes it is necessary to simplify radicals before adding or subtracting.

Example 2   **Simplifying Radicals before Adding or Subtracting**

Add or subtract the radicals as indicated.

 a. $\sqrt{20} + 7\sqrt{5}$ **b.** $\sqrt{50} - \sqrt{8}$

Solution:

 a. $\sqrt{20} + 7\sqrt{5}$ Because the radicands are different, try simplifying the radicals first.

$= \sqrt{2^2 \cdot 5} + 7\sqrt{5}$ Factor the radicand.

$= 2\sqrt{5} + 7\sqrt{5}$ The terms are *like* radicals.

$= (2 + 7)\sqrt{5}$ Apply the distributive property.

$= 9\sqrt{5}$ Simplify.

 b. $\sqrt{50} - \sqrt{8}$ Because the radicands are different, try simplifying the radicals first.

$= \sqrt{5^2 \cdot 2} - \sqrt{2^2 \cdot 2}$ Factor the radicands.

$= 5\sqrt{2} - 2\sqrt{2}$ The terms are *like* radicals.

$= (5 - 2)\sqrt{2}$ Apply the distributive property.

$= 3\sqrt{2}$ Simplify.

— **Skill Practice** —

Add or subtract the radicals as indicated.
 4. $4\sqrt{18} + \sqrt{8}$
 5. $\sqrt{50} - \sqrt{98}$

Classroom Example: p. 606, Exercise 34

Instructor Note: We can only decide if radicals are *like* if they are simplified. For example, $\sqrt{8}$ and $\sqrt{18}$ are *like* because $\sqrt{8} = 2\sqrt{2}$ and $\sqrt{18} = 3\sqrt{2}$.

Answers

1. $10\sqrt{2}$ **2.** $7\sqrt{x}$ **3.** $-7\sqrt{ab}$
4. $14\sqrt{2}$ **5.** $-2\sqrt{2}$

Example 3 **Simplifying Radicals before Adding or Subtracting**

Add or subtract the radicals as indicated. Assume the variables represent positive real numbers.

a. $-4\sqrt{3x^2} - x\sqrt{27} + 5x\sqrt{3}$

b. $a\sqrt{8a^5} + 6\sqrt{2a^7} + \sqrt{9a}$

Solution:

a. $-4\sqrt{3x^2} - x\sqrt{27} + 5x\sqrt{3}$ Simplify each radical.

$= -4\sqrt{3x^2} - x\sqrt{3^2 \cdot 3} + 5x\sqrt{3}$ Factor the radicands.

$= -4x\sqrt{3} - 3x\sqrt{3} + 5x\sqrt{3}$ The terms are *like* radicals.

$= (-4x - 3x + 5x)\sqrt{3}$ Apply the distributive property.

$= -2x\sqrt{3}$ Simplify.

b. $a\sqrt{8a^5} + 6\sqrt{2a^7} + \sqrt{9a}$ Simplify each radical.

$= a\sqrt{2^3a^5} + 6\sqrt{2a^7} + \sqrt{3^2a}$ Factor the radicals.

$= a\sqrt{2^2a^4 \cdot 2a} + 6\sqrt{a^6 \cdot 2a} + \sqrt{3^2 \cdot a}$

$= a \cdot 2a^2\sqrt{2a} + 6 \cdot a^3\sqrt{2a} + 3\sqrt{a}$ Simplify the radicals.

$= 2a^3\sqrt{2a} + 6a^3\sqrt{2a} + 3\sqrt{a}$ The first two terms are *like* radicals.

$= (2a^3 + 6a^3)\sqrt{2a} + 3\sqrt{a}$ Apply the distributive property.

$= 8a^3\sqrt{2a} + 3\sqrt{a}$

It is important to realize that only *like* radicals can be added or subtracted. The next example provides extra practice for recognizing *unlike* radicals.

Example 4 **Recognizing *Unlike* Radicals**

Explain why the radicals cannot be simplified further by adding or subtracting.

a. $2\sqrt{7} - 5\sqrt{3}$ **b.** $7 + 4\sqrt{5}$

Solution:

a. $2\sqrt{7} - 5\sqrt{3}$ Radicands are not the same.

b. $7 + 4\sqrt{5}$ One term has a radical, and one does not.

Answers

6. $5x\sqrt{3}$

7. $-y\sqrt{7y} + 10\sqrt{7}$

8. One term has a radical and one does not.

9. The radicands are not the same.

Calculator Connections

Title: Verifying Radical Expressions on a Calculator

A calculator can be used to evaluate a radical expression and its simplified form. For example, use a calculator to evaluate the expression on the left and its simplified form on the right: $2\sqrt{5} + 6\sqrt{5} = 8\sqrt{5}$

Scientific Calculator

Enter: 5 $\sqrt{x}$ × 2 = + 5 $\sqrt{x}$ × 6 =   **Result:** 17.88854382

Enter: 5 $\sqrt{x}$ × 8 =   **Result:** 17.88854382

Graphing Calculator

```
2√(5)+6√(5)
          17.88854382
8√(5)
          17.88854382
```

A calculator can help you determine when a rule has been applied *incorrectly*. For example, use a calculator to show that $\sqrt{3} + \sqrt{5} \neq \sqrt{8}$

Scientific Calculator

Enter: 3 $\sqrt{x}$ + 5 $\sqrt{x}$ =   **Result:** 3.968118785 ⎤ Values are
Enter: 8 $\sqrt{x}$ **Result:** 2.828427125 ⎦ not equal.

Graphing Calculator

```
√(3)+√(5)
          3.968118785        Values are
√(8)                         not equal.
          2.828427125
```

Section 8.3   Practice Exercises

Study Skills Exercise

1. Define the key term *like* radicals.

Review Exercises

For Exercises 2–9, simplify the expressions. Assume the variables represent positive real numbers.

2. $\sqrt{25w^2}$ *5w*

3. $\sqrt[3]{8y^3}$ *2y*

4. $\sqrt[3]{4z^4}$ $z\sqrt[3]{4z}$

5. $\sqrt{36x^3}$ $6x\sqrt{x}$

6. $\sqrt{\dfrac{9a^6}{a^2}}$ $3a^2$

7. $\dfrac{\sqrt{25c^6}}{16}$ $\dfrac{5c^3}{4}$

8. $\sqrt{\dfrac{12x^3}{3x}}$ $2x$

9. $\sqrt{-25}$ Not a real number

Writing   Translating Expression   Geometry   Scientific Calculator   Video   NS&E

Objective 1: Definition of *Like* Radicals

10. How do you determine whether two radicals are *like* or *unlike*?
Two radicals are *like* if they have the same radicand and same index.

11. Write two radicals that are considered *unlike*.   For example, $2\sqrt{3}$, $6\sqrt[3]{3}$

12. From the three pairs of radicals, identify the pair of *like* radicals: b

    a. $2\sqrt{x}$ and $8\sqrt[3]{x}$

    b. $\sqrt{5}$ and $-3\sqrt{5}$

    c. $3a\sqrt{3}$ and $3a\sqrt{2}$

13. From the three pairs of radicals, identify the pair of *like* radicals: c

    a. $13\sqrt{5b}$ and $13b\sqrt{5}$

    b. $\sqrt[4]{x^2 y}$ and $\sqrt[3]{x^2 y}$

    c. $-2\sqrt[3]{y^2}$ and $6\sqrt[3]{y^2}$

Objective 2: Addition and Subtraction of Radicals

For Exercises 14–28, add or subtract the expressions, if possible. Assume the variables represent positive real numbers. **(See Example 1.)**

14. $8\sqrt{6} + 2\sqrt{6}$ $10\sqrt{6}$

15. $3\sqrt{2} + 5\sqrt{2}$ $8\sqrt{2}$

16. $4\sqrt{3} - 2\sqrt{3} + 5\sqrt{3}$ $7\sqrt{3}$

17. $5\sqrt{7} - 3\sqrt{7} + 2\sqrt{7}$ $4\sqrt{7}$

18. $\sqrt{11} + \sqrt{11}$ $2\sqrt{11}$

19. $\sqrt{10} + \sqrt{10}$ $2\sqrt{10}$

20. $12\sqrt{x} - 3\sqrt{x}$ $9\sqrt{x}$

21. $15\sqrt{y} - 4\sqrt{y}$ $11\sqrt{y}$

22. $-3\sqrt{a} + 2\sqrt{a} + \sqrt{a}$ 0

23. $5\sqrt{c} - 6\sqrt{c} + \sqrt{c}$ 0

24. $7x\sqrt{11} - 9x\sqrt{11}$ $-2x\sqrt{11}$

25. $8y\sqrt{15} - 3y\sqrt{15}$ $5y\sqrt{15}$

26. $9\sqrt{2} - 9\sqrt{5}$ $9\sqrt{2} - 9\sqrt{5}$

27. $x\sqrt{y} - y\sqrt{x}$ $x\sqrt{y} - y\sqrt{x}$

28. $a\sqrt{b} + b\sqrt{a}$ $a\sqrt{b} + b\sqrt{a}$

For Exercises 29–32, translate the English phrase into an algebraic expression. Then simplify the expression.

29. The sum of three times the cube root of six and eight times the cube root of six.
$3\sqrt[3]{6} + 8\sqrt[3]{6}$; $11\sqrt[3]{6}$

30. The difference of negative two times the cube root of w and five times the cube root of w.
$-2\sqrt[3]{w} - 5\sqrt[3]{w}$; $-7\sqrt[3]{w}$

31. Four times the square root of five, minus six times the square root of five. $4\sqrt{5} - 6\sqrt{5}$; $-2\sqrt{5}$

32. Eight times the square root of two, plus the square root of two. $8\sqrt{2} + \sqrt{2}$; $9\sqrt{2}$

Mixed Exercises

For Exercises 33–62, simplify. Then add or subtract the expressions, if possible. Assume the variables represent positive real numbers. **(See Examples 2 and 3.)**

33. $2\sqrt{12} + \sqrt{48}$ $8\sqrt{3}$

34. $5\sqrt{32} + 2\sqrt{50}$ $30\sqrt{2}$

35. $4\sqrt{45} - 6\sqrt{20}$ 0

36. $8\sqrt{54} - 4\sqrt{24}$ $16\sqrt{6}$

37. $\dfrac{1}{2}\sqrt{8} + \dfrac{1}{3}\sqrt{18}$ $2\sqrt{2}$

38. $\dfrac{1}{4}\sqrt{32} - \dfrac{1}{5}\sqrt{50}$ 0

39. $6p\sqrt{20p^2} + p^2\sqrt{80}$ $16p^2\sqrt{5}$

40. $2q\sqrt{48} + \sqrt{27q^2}$ $11q\sqrt{3}$

41. $-2\sqrt{2k} + 6\sqrt{8k}$ $10\sqrt{2k}$

42. $5\sqrt{27x} - 4\sqrt{12x}$ $7\sqrt{3x}$

43. $11\sqrt{a^4 b} - a^2\sqrt{b} - 9a\sqrt{a^2 b}$
$a^2\sqrt{b}$

44. $-7\sqrt{x^4 y} + 5x^2\sqrt{y} - 6x\sqrt{x^2 y}$
$-8x^2\sqrt{y}$

45. $4\sqrt{5} - \sqrt{5}$ $3\sqrt{5}$

46. $-3\sqrt{10} - \sqrt{10}$ $-4\sqrt{10}$

47. $\dfrac{5}{6}z\sqrt{6} + \dfrac{7}{9}z\sqrt{6}$ $\dfrac{29}{18}z\sqrt{6}$

48. $\dfrac{3}{4}a\sqrt{b} + \dfrac{1}{6}a\sqrt{b}$ $\dfrac{11}{12}a\sqrt{b}$

49. $1.1\sqrt{10} - 5.6\sqrt{10} + 2.8\sqrt{10}$
$-1.7\sqrt{10}$

50. $0.25\sqrt{x} + 1.50\sqrt{x} - 0.75\sqrt{x}$
$\sqrt{x}$

51. $4\sqrt{x^3} - 2x\sqrt{x}$ $2x\sqrt{x}$

52. $8\sqrt{y^9} - 2y^2\sqrt{y^5}$ $6y^4\sqrt{y}$

53. $4\sqrt{7} + \sqrt{63} - 2\sqrt{28}$ $3\sqrt{7}$

54. $8\sqrt{3} - 2\sqrt{27} + \sqrt{75}$ $7\sqrt{3}$

55. $\sqrt{16w} + \sqrt{24w} + \sqrt{40w}$
$4\sqrt{w} + 2\sqrt{6w} + 2\sqrt{10w}$

56. $\sqrt{54y} + \sqrt{81y} - \sqrt{12y}$
$3\sqrt{6y} + 9\sqrt{y} - 2\sqrt{3y}$

57. $\sqrt{x^6 y} + 5x^2\sqrt{x^2 y}$ $6x^3\sqrt{y}$

58. $7\sqrt{a^5 b^2} - a^2\sqrt{ab^2}$ $6a^2 b\sqrt{a}$

59. $4\sqrt{6} + 2\sqrt{3} - 8\sqrt{6}$
$2\sqrt{3} - 4\sqrt{6}$

60. $-7\sqrt{y} - \sqrt{z} + 2\sqrt{z}$
$-7\sqrt{y} + \sqrt{z}$

61. $x\sqrt{8} - 2\sqrt{18x^2} + \sqrt{2x}$
$-4x\sqrt{2} + \sqrt{2x}$

62. $5\sqrt{p^5} - 2p\sqrt{p} + p\sqrt{16p^3}$
$9p^2\sqrt{p} - 2p\sqrt{p}$

For Exercises 63–64, find the exact perimeter of each figure.

63.

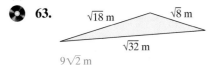

$\sqrt{18}$ m $\sqrt{8}$ m
$\sqrt{32}$ m
$9\sqrt{2}$ m

64.

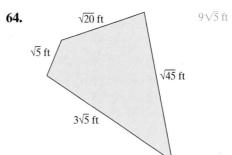

$\sqrt{20}$ ft $9\sqrt{5}$ ft
$\sqrt{5}$ ft
$\sqrt{45}$ ft
$3\sqrt{5}$ ft

65. Find the exact perimeter of a rectangle whose width is $2\sqrt{3}$ in. and whose length is $3\sqrt{12}$ in.
$16\sqrt{3}$ in.

66. Find the exact perimeter of a square whose side length is $5\sqrt{8}$ cm. $40\sqrt{2}$ cm

For Exercises 67–72, determine the reason why the following radical expressions cannot be combined by addition or subtraction. **(See Example 4.)**

67. $\sqrt{5} + 5\sqrt{2}$
Radicands are not the same.

68. $3\sqrt{10} + 10\sqrt{3}$
Radicands are not the same.

69. $3 + 5\sqrt{7}$
One term has a radical. One does not.

70. $-2 + 5\sqrt{11}$
One term has a radical. One does not.

71. $5\sqrt{2} + \sqrt[3]{2}$
The indices are different.

72. $\sqrt[4]{6} - 3\sqrt{6}$
The indices are different.

Expanding Your Skills

73. Find the slope of the line through the points: $(4, 2\sqrt{3})$ and $(1, \sqrt{3})$. $\dfrac{\sqrt{3}}{3}$

74. Find the slope of the line through the points: $(7, 4\sqrt{5})$ and $(2, 3\sqrt{5})$. $\dfrac{\sqrt{5}}{5}$

75. A golfer hits a golf ball at an angle of 30° with an initial velocity of 46.0 meters/second (m/sec). The horizontal position of the ball, x (measured in meters), depends on the number of seconds, t, after the ball is struck according to the equation:

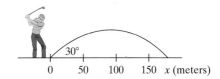

$$x = 23\sqrt{3}t$$

a. What is the horizontal position of the ball after 2 sec? Round the answer to the nearest meter.   80 m

b. What is the horizontal position of the ball after 4 sec? Round the answer to the nearest meter.   159 m

76. A long-jumper leaves the ground at an angle of 30° at a speed of 9 m/sec. The horizontal position of the long jumper, x (measured in meters), depends on the number of seconds, t, after he leaves the ground according to the equation:

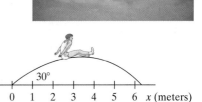

$$x = 4.5\sqrt{3}t$$

a. What is the horizontal position of the long-jumper after 0.5 sec? Round the answer to the nearest hundredth of a meter.   3.90 m

b. What is the horizontal position of the long-jumper after 0.75 sec? Round the answer to the nearest hundredth of a meter.   5.85 m

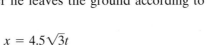

Writing   Translating Expression   Geometry   Scientific Calculator   Video   NS&E

Section 8.4 Multiplication of Radicals

Objectives

1. Multiplication Property of Radicals
2. Expressions of the Form $(\sqrt[n]{a})^n$
3. Special Case Products
4. Multiplying Conjugate Radical Expressions

1. Multiplication Property of Radicals

In this section, we will learn how to multiply radicals that have the same index. Recall from Section 8.2 the multiplication property of radicals.

> **PROPERTY** Multiplication Property of Radicals
>
> Let a and b represent real numbers such that $\sqrt[n]{a}$ and $\sqrt[n]{b}$ are both real. Then,
>
> $$\sqrt[n]{a} \cdot \sqrt[n]{b} = \sqrt[n]{ab}$$

To multiply two radical expressions, use the multiplication property of radicals along with the commutative and associative properties of multiplication.

Skill Practice

Multiply the expressions and simplify the result. Assume the variables represent positive real numbers.

1. $\sqrt{2} \cdot \sqrt{5}$
2. $(-5z\sqrt{6})(4z\sqrt{2})$
3. $(9y\sqrt{x})\left(\frac{1}{3}y\sqrt{xy}\right)$

Classroom Examples: p. 612, Exercises 8, 18, and 22

Example 1 Multiplying Radical Expressions

Multiply the expressions and simplify the result. Assume the variables represent positive real numbers.

a. $\sqrt{3} \cdot \sqrt{2}$ b. $(5\sqrt{3})(2\sqrt{15})$ c. $(6a\sqrt{ab})\left(\frac{1}{3}a\sqrt{a}\right)$

Solution:

a. $\sqrt{3} \cdot \sqrt{2} = \sqrt{6}$ Multiplication property of radicals

b. $(5\sqrt{3})(2\sqrt{15}) = (5 \cdot 2)(\sqrt{3} \cdot \sqrt{15})$ Commutative and associative properties of multiplication

$= 10\sqrt{45}$ Multiplication property of radicals

$= 10\sqrt{3^2 \cdot 5}$ Simplify the radical.

$= 10 \cdot 3\sqrt{5}$

$= 30\sqrt{5}$

c. $(6a\sqrt{ab})\left(\frac{1}{3}a\sqrt{a}\right) = \left(6a \cdot \frac{1}{3}a\right)(\sqrt{ab} \cdot \sqrt{a})$ Commutative and associative properties of multiplication

$= 2a^2\sqrt{a^2b}$ Multiplication property of radicals

$= 2a^2 \cdot a\sqrt{b}$ Simplify the radical.

$= 2a^3\sqrt{b}$

When multiplying radical expressions with more than one term, we use the distributive property.

Answers

1. $\sqrt{10}$ 2. $-40z^2\sqrt{3}$
3. $3xy^2\sqrt{y}$

Example 2 **Multiplying Radical Expressions with Multiple Terms**

Multiply the expressions. Assume the variables represent positive real numbers.

a. $\sqrt{5}(4 + 3\sqrt{5})$ **b.** $(\sqrt{x} - 10)(\sqrt{y} + 4)$ **c.** $(2\sqrt{3} - \sqrt{5})(\sqrt{3} + 6\sqrt{5})$

Solution:

a. $\sqrt{5}(4 + 3\sqrt{5})$

$= \sqrt{5}(4) + \sqrt{5}(3\sqrt{5})$ Apply the distributive property.

$= 4\sqrt{5} + 3\sqrt{5^2}$ Multiplication property of radicals

$= 4\sqrt{5} + 3 \cdot 5$ Simplify the radical.

$= 4\sqrt{5} + 15$

b. $(\sqrt{x} - 10)(\sqrt{y} + 4)$

$= \sqrt{x}(\sqrt{y}) + \sqrt{x}(4) - 10(\sqrt{y}) - 10(4)$ Apply the distributive property.

$= \sqrt{xy} + 4\sqrt{x} - 10\sqrt{y} - 40$ Simplify.

c. $(2\sqrt{3} - \sqrt{5})(\sqrt{3} + 6\sqrt{5})$

$= 2\sqrt{3}(\sqrt{3}) + 2\sqrt{3}(6\sqrt{5}) - \sqrt{5}(\sqrt{3}) - \sqrt{5}(6\sqrt{5})$ Apply the distributive property.

$= 2\sqrt{3^2} + 12\sqrt{15} - \sqrt{15} - 6\sqrt{5^2}$ Multiplication property of radicals

$= 2 \cdot 3 + 11\sqrt{15} - 6 \cdot 5$ Simplify radicals. Combine *like* radicals.

$= 6 + 11\sqrt{15} - 30$

$= -24 + 11\sqrt{15}$ Combine *like* terms.

Skill Practice

Multiply the expressions and simplify the result. Assume the variables represent positive real numbers.

4. $\sqrt{7}(2\sqrt{7} - 4)$
5. $(\sqrt{x} + 2)(\sqrt{x} - 3)$
6. $(2\sqrt{a} + 4\sqrt{6}) \cdot$
$(\sqrt{a} - 3\sqrt{6})$

Classroom Examples: p. 613, Exercises 38 and 42

---

2. Expressions of the Form $(\sqrt[n]{a})^n$

The multiplication property of radicals can be used to simplify an expression of the form $(\sqrt{a})^2$, where $a \geq 0$.

$$(\sqrt{a})^2 = \sqrt{a} \cdot \sqrt{a} = \sqrt{a^2} = a$$

This logic can be applied to *n*th-roots. If $\sqrt[n]{a}$ is a real number, then $(\sqrt[n]{a})^n = a$.

Example 3 **Simplifying Radical Expressions**

Simplify the expressions. Assume $x \geq 0$.

a. $(\sqrt{7})^2$ **b.** $(\sqrt[4]{x})^4$ **c.** $(3\sqrt{2})^2$

Solution:

a. $(\sqrt{7})^2 = 7$ **b.** $(\sqrt[4]{x})^4 = x$ **c.** $(3\sqrt{2})^2 = 3^2 \cdot (\sqrt{2})^2 = 9 \cdot 2 = 18$

Skill Practice

Simplify the expressions.

7. $(\sqrt{13})^2$
8. $(\sqrt[3]{x})^3$
9. $(2\sqrt{11})^2$

Classroom Examples: p. 613, Exercises 46 and 52

Answers
4. $14 - 4\sqrt{7}$
5. $x - \sqrt{x} - 6$
6. $2a - 2\sqrt{6a} - 72$
7. 13 **8.** x **9.** 44

3. Special Case Products

From Example 2, you may have noticed a similarity between multiplying radical expressions and multiplying polynomials.

Recall from Section 5.6 that the square of a binomial results in a perfect square trinomial.

$$(a + b)^2 = a^2 + 2ab + b^2$$

$$(a - b)^2 = a^2 - 2ab + b^2$$

The same patterns occur when squaring a radical expression with two terms.

Skill Practice

Square the radical expression. Assume $p \geq 0$.

10. $(\sqrt{p} + 3)^2$

Classroom Example: p. 613, Exercise 54

Example 4 Squaring a Two-Term Radical Expression

Square the radical expression. Assume the variables represent positive real numbers.

$$(\sqrt{x} + \sqrt{y})^2$$

Solution:

$(\sqrt{x} + \sqrt{y})^2$ This expression is in the form $(a + b)^2$, where $a = \sqrt{x}$ and $b = \sqrt{y}$.

$$\overset{a^2 + 2ab + b^2}{= (\sqrt{x})^2 + 2(\sqrt{x})(\sqrt{y}) + (\sqrt{y})^2}$$ Apply the formula $(a + b)^2 = a^2 + 2ab + b^2$.

$$= x + 2\sqrt{xy} + y$$ Simplify.

TIP: The product $(\sqrt{x} + \sqrt{y})^2$ can also be found using the distributive property.

$$(\sqrt{x} + \sqrt{y})^2 = (\sqrt{x} + \sqrt{y})(\sqrt{x} + \sqrt{y}) = \sqrt{x} \cdot \sqrt{x} + \sqrt{x} \cdot \sqrt{y} + \sqrt{y} \cdot \sqrt{x} + \sqrt{y} \cdot \sqrt{y}$$
$$= \sqrt{x^2} + \sqrt{xy} + \sqrt{xy} + \sqrt{y^2}$$
$$= x + 2\sqrt{xy} + y$$

Skill Practice

Square the radical expression.

11. $(\sqrt{5} - 3\sqrt{2})^2$

Classroom Example: p. 613, Exercise 58

Example 5 Square a Two-Term Radical Expression

Square the radical expression. $(\sqrt{2} - 4\sqrt{3})^2$

Solution:

$(\sqrt{2} - 4\sqrt{3})^2$ This expression is in the form $(a - b)^2$, where $a = \sqrt{2}$ and $b = 4\sqrt{3}$.

$$\overset{a^2 - 2ab + b^2}{(\sqrt{2})^2 - 2(\sqrt{2})(4\sqrt{3}) + (4\sqrt{3})^2}$$ Apply the formula $(a - b)^2 = a^2 - 2ab + b^2$.

$$= 2 - 8\sqrt{6} + 16 \cdot 3$$ Simplify.

$$= 2 - 8\sqrt{6} + 48$$

$$= 50 - 8\sqrt{6}$$

Answers

10. $p + 6\sqrt{p} + 9$
11. $23 - 6\sqrt{10}$

4. Multiplying Conjugate Radical Expressions

Recall from Section 5.6 that the product of two conjugate binomials results in a difference of squares.

$$(a + b)(a - b) = a^2 - b^2$$

The same pattern occurs when multiplying two conjugate radical expressions.

Example 6 **Multiplying Conjugate Radical Expressions**

Multiply the radical expressions. $(\sqrt{5} + 4)(\sqrt{5} - 4)$

Solution:

$(\sqrt{5} + 4)(\sqrt{5} - 4)$ This expression is in the form $(a + b)(a - b)$, where $a = \sqrt{5}$ and $b = 4$.

$$= \overset{a^2 - b^2}{(\sqrt{5})^2 - (4)^2}$$ Apply the formula $(a + b)(a - b) = a^2 - b^2$.

$= 5 - 16$ Simplify.

$= -11$

Skill Practice

Multiply the radical expressions.
12. $(\sqrt{6} - 3)(\sqrt{6} + 3)$

Classroom Example: p. 613, Exercise 62

TIP: The product $(\sqrt{5} + 4)(\sqrt{5} - 4)$ can also be found using the distributive property.

$$(\sqrt{5} + 4)(\sqrt{5} - 4) = \sqrt{5} \cdot (\sqrt{5}) + \sqrt{5} \cdot (-4) + 4 \cdot (\sqrt{5}) + 4 \cdot (-4)$$
$$= 5 - 4\sqrt{5} + 4\sqrt{5} - 16$$
$$= 5 - 16$$
$$= -11$$

Example 7 **Multiply Conjugate Radical Expressions**

Multiply the radical expressions. Assume the variables represent positive real numbers.

$$(2\sqrt{c} - 3\sqrt{d})(2\sqrt{c} + 3\sqrt{d})$$

Solution:

$(2\sqrt{c} - 3\sqrt{d})(2\sqrt{c} + 3\sqrt{d})$ This expression is in the form $(a - b)(a + b)$, where $a = 2\sqrt{c}$ and $b = 3\sqrt{d}$.

$$= \overset{a^2 - b^2}{(2\sqrt{c})^2 - (3\sqrt{d})^2}$$ Apply the formula $(a + b)(a - b) = a^2 - b^2$.

$= 4c - 9d$

Skill Practice

Multiply the radical expressions. Assume the variables represent positive real numbers.
13. $(5\sqrt{a} + \sqrt{b}) \cdot (5\sqrt{a} - \sqrt{b})$

Classroom Example: p. 613, Exercise 70

Answers
12. -3 **13.** $25a - b$

Section 8.4 | Practice Exercises

Study Skills Exercise

1. When writing a radical expression, be sure to note the difference between an exponent on a coefficient and an index to a radical. Write an algebraic expression for each of the following:

   x cubed times the square root of y

   x times the cube root of y

Review Exercises

For Exercises 2–5, perform the indicated operations and simplify. Assume the variables represent positive real numbers.

2. $\sqrt{25} + \sqrt{16} - \sqrt{36}$
   3

3. $\sqrt{100} - \sqrt{4} + \sqrt{9}$
   11

4. $6x\sqrt{18} + 2\sqrt{2x^2}$
   $20x\sqrt{2}$

5. $10\sqrt{zw^4} - w^2\sqrt{49z}$
   $3w^2\sqrt{z}$

6. What three conditions are needed for a radical expression to be in simplified form?   See page 594.

Objective 1: Multiplication Property of Radicals

For Exercises 7–26, multiply the expressions. **(See Example 1.)**

7. $\sqrt{5} \cdot \sqrt{3}$ $\sqrt{15}$

8. $\sqrt{7} \cdot \sqrt{6}$ $\sqrt{42}$

9. $\sqrt{47} \cdot \sqrt{47}$ 47

10. $\sqrt{59} \cdot \sqrt{59}$ 59

11. $\sqrt{b} \cdot \sqrt{b}$ b

12. $\sqrt{t} \cdot \sqrt{t}$ t

13. $(2\sqrt{15})(3\sqrt{p})$ $6\sqrt{15p}$

14. $(4\sqrt{2})(5\sqrt{q})$ $20\sqrt{2q}$

15. $\sqrt{10} \cdot \sqrt{5}$ $5\sqrt{2}$

16. $\sqrt{2} \cdot \sqrt{10}$ $2\sqrt{5}$

17. $(-\sqrt{7})(-2\sqrt{14})$ $14\sqrt{2}$

18. $(-6\sqrt{2})(-\sqrt{22})$ $12\sqrt{11}$

19. $(3x\sqrt{2})(\sqrt{14})$ $6x\sqrt{7}$

20. $(4y\sqrt{3})(\sqrt{6})$ $12y\sqrt{2}$

21. $\left(\frac{1}{6}x\sqrt{xy}\right)(24x\sqrt{x})$

22. $\left(\frac{1}{4}u\sqrt{uv}\right)(8u\sqrt{v})$

23. $(6w\sqrt{5})(w\sqrt{8})$
   $12w^2\sqrt{10}$

24. $(t\sqrt{2})(5\sqrt{6t})$ $10t\sqrt{3t}$

   $4x^3\sqrt{y}$ $2u^2v\sqrt{u}$

25. $(-2\sqrt{3})(4\sqrt{5})$ $-8\sqrt{15}$

26. $(-\sqrt{7})(2\sqrt{3})$ $-2\sqrt{21}$

For Exercises 27–28, find the exact perimeter and exact area of the rectangles.

27.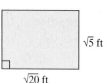
   $\sqrt{5}$ ft

   $\sqrt{20}$ ft

   Perimeter: $6\sqrt{5}$ ft; area: 10 ft^2

28.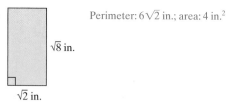
   $\sqrt{8}$ in.

   $\sqrt{2}$ in.

   Perimeter: $6\sqrt{2}$ in.; area: 4 in.^{2}

For Exercises 29–30, find the exact area of the triangles.

29.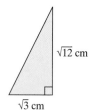
   $\sqrt{12}$ cm

   $\sqrt{3}$ cm

   3 cm^2

30.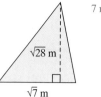
   $\sqrt{28}$ m

   $\sqrt{7}$ m

   7 m^2

For Exercises 31–44, multiply the expressions. Assume the variables represent positive real numbers. **(See Example 2.)**

31. $\sqrt{3w} \cdot \sqrt{3w}$
$3w$

32. $\sqrt{6p} \cdot \sqrt{6p}$
$6p$

33. $(8\sqrt{5y})(-2\sqrt{2})$
$-16\sqrt{10y}$

34. $(4\sqrt{5x})(7\sqrt{3})$
$28\sqrt{15x}$

35. $\sqrt{2}(\sqrt{6} - \sqrt{3})$
$2\sqrt{3} - \sqrt{6}$

36. $\sqrt{5}(\sqrt{10} + \sqrt{7})$
$5\sqrt{2} + \sqrt{35}$

37. $4\sqrt{x}(\sqrt{x} + 5)$
$4x + 20\sqrt{x}$

38. $2\sqrt{y}(3 - \sqrt{y})$
$6\sqrt{y} - 2y$

39. $(\sqrt{3} + 2\sqrt{10})(4\sqrt{3} - \sqrt{10})$
$-8 + 7\sqrt{30}$

40. $(8\sqrt{7} - \sqrt{5})(\sqrt{7} + 3\sqrt{5})$
$41 + 23\sqrt{35}$

41. $(\sqrt{a} - 3b)(9\sqrt{a} - b)$
$9a - 28b\sqrt{a} + 3b^2$

42. $(11\sqrt{m} + 4n)(\sqrt{m} + n)$
$11m + 15n\sqrt{m} + 4n^2$

43. $(p + 2\sqrt{p})(8p + 3\sqrt{p} - 4)$
$8p^2 + 19p\sqrt{p} + 2p - 8\sqrt{p}$

44. $(5s - \sqrt{s})(s + 5\sqrt{s} + 6)$
$5s^2 + 24s\sqrt{s} + 25s - 6\sqrt{s}$

Objective 2: Expressions of the Form $(\sqrt[n]{a})^n$

For Exercises 45–52, simplify the expressions. Assume the variables represent positive real numbers. **(See Example 3.)**

45. $(\sqrt{10})^2$ 10

46. $(\sqrt{23})^2$ 23

47. $(\sqrt[3]{4})^3$ 4

48. $(\sqrt[3]{29})^3$ 29

49. $(\sqrt[4]{t})^4$ t

50. $(\sqrt[4]{xy})^4$ xy

51. $(4\sqrt{c})^2$ $16c$

52. $(10\sqrt{2pq})^2$ $200pq$

Objective 3: Special Case Products

For Exercises 53–60, multiply the radical expressions. Assume the variables represent positive real numbers. **(See Examples 4–5.)**

53. $(\sqrt{13} + 4)^2$
$29 + 8\sqrt{13}$

54. $(6 - \sqrt{11})^2$
$47 - 12\sqrt{11}$

55. $(\sqrt{a} - 2)^2$
$a - 4\sqrt{a} + 4$

56. $(\sqrt{p} + 3)^2$
$p + 6\sqrt{p} + 9$

57. $(2\sqrt{a} - 3)^2$
$4a - 12\sqrt{a} + 9$

58. $(3\sqrt{w} + 4)^2$
$9w + 24\sqrt{w} + 16$

59. $(\sqrt{10} - \sqrt{11})^2$
$21 - 2\sqrt{110}$

60. $(\sqrt{3} - \sqrt{2})^2$
$5 - 2\sqrt{6}$

Objective 4: Multiplying Conjugate Radical Expressions

For Exercises 61–72, multiply the radical expressions. Assume the variables represent positive real numbers. **(See Examples 6–7.)**

61. $(\sqrt{5} + 2)(\sqrt{5} - 2)$ 1

62. $(\sqrt{3} - 4)(\sqrt{3} + 4)$ -13

63. $(\sqrt{x} + \sqrt{y})(\sqrt{x} - \sqrt{y})$ $x - y$

64. $(\sqrt{a} + \sqrt{b})(\sqrt{a} - \sqrt{b})$ $a - b$

65. $(\sqrt{10} - \sqrt{11})(\sqrt{10} + \sqrt{11})$ -1

66. $(\sqrt{3} - \sqrt{2})(\sqrt{3} + \sqrt{2})$ 1

67. $(\sqrt{6} + \sqrt{2})(\sqrt{6} - \sqrt{2})$ 4

68. $(\sqrt{15} - \sqrt{5})(\sqrt{15} + \sqrt{5})$ 10

69. $(8\sqrt{x} - 2\sqrt{y})(8\sqrt{x} + 2\sqrt{y})$
$64x - 4y$

70. $(4\sqrt{s} + 11\sqrt{t})(4\sqrt{s} - 11\sqrt{t})$
$16s - 121t$

71. $(5\sqrt{3} - \sqrt{2})(5\sqrt{3} + \sqrt{2})$ 73

72. $(2\sqrt{7} - 4\sqrt{3})(2\sqrt{7} + 4\sqrt{3})$
-20

Mixed Exercises

For Exercises 73–84, multiply the expressions in parts (a) and (b) and compare the process used. Assume the variables represent positive real numbers.

73. a. $3(x + 2)$ $3x + 6$
   b. $\sqrt{3}(\sqrt{x} + \sqrt{2})$ $\sqrt{3x} + \sqrt{6}$

74. a. $-5(6 + y)$ $-30 - 5y$
   b. $-\sqrt{5}(\sqrt{6} + \sqrt{y})$ $-\sqrt{30} - \sqrt{5y}$

75. a. $(2a + 3)^2$ $4a^2 + 12a + 9$
   b. $(2\sqrt{a} + 3)^2$ $4a + 12\sqrt{a} + 9$

76. a. $(6 - z)^2$ $36 - 12z + z^2$
   b. $(\sqrt{6} - z)^2$ $6 - 2z\sqrt{6} + z^2$

77. a. $(b - 5)(b + 5)$ $b^2 - 25$
   b. $(\sqrt{b} - 5)(\sqrt{b} + 5)$ $b - 25$

78. a. $(3w - 1)(3w + 1)$ $9w^2 - 1$
   b. $(3\sqrt{w} - 1)(3\sqrt{w} + 1)$
     $9w - 1$

79. a. $(x - 2y)^2$ $x^2 - 4xy + 4y^2$
   b. $(\sqrt{x} - 2\sqrt{y})^2$ $x - 4\sqrt{xy} + 4y$

80. a. $(5c + 2d)^2$ $25c^2 + 20cd + 4d^2$
   b. $(5\sqrt{c} + 2\sqrt{d})^2$ $25c + 20\sqrt{cd} + 4d$

81. a. $(p - q)(p + q)$ $p^2 - q^2$
   b. $(\sqrt{p} - \sqrt{q})(\sqrt{p} + \sqrt{q})$ $p - q$

82. a. $(t - 3)(t + 3)$ $t^2 - 9$
   b. $(\sqrt{t} - \sqrt{3})(\sqrt{t} + \sqrt{3})$ $t - 3$

83. a. $(y - 3)^2$ $y^2 - 6y + 9$
   b. $(\sqrt{x - 2} - 3)^2$ $x - 6\sqrt{x - 2} + 7$

84. a. $(p + 4)^2$ $p^2 + 8p + 16$
   b. $(\sqrt{x + 1} + 4)^2$ $x + 8\sqrt{x + 1} + 17$

Writing   Translating Expression   Geometry   Scientific Calculator   Video   NS&E

Section 8.5 Division of Radicals and Rationalization

Objectives

1. Simplified Form of a Radical
2. Division Property of Radicals
3. Rationalizing the Denominator: One Term
4. Rationalizing the Denominator: Two Terms
5. Simplifying Quotients That Contain Radicals

1. Simplified Form of a Radical

Recall the conditions for a radical to be simplified.

DEFINITION Simplified Form of a Radical

Consider any radical expression where the radicand is written as a product of prime factors. The expression is in simplified form if all of the following conditions are met:

1. The radicand has no factor raised to a power greater than or equal to the index.
2. There are no radicals in the denominator of a fraction.
3. The radicand does not contain a fraction.

The basis of the second and third conditions, which restrict radicals from the denominator of an expression, are largely historical. In some cases, removing a radical from the denominator of a fraction will create an expression that is computationally simpler.

The process to remove a radical from the denominator is called **rationalizing the denominator**. In this section, we will show three approaches that can be used to achieve the second and third conditions of a simplified radical.

1. Rationalizing by applying the division property of radicals.
2. Rationalizing when the denominator contains a singe radical term.
3. Rationalizing when the denominator contains two terms involving square roots.

Instructor Note: We call this process rationalizing because the definition of rational is to have a ratio of integers. Rationalization results in an integer denominator.

2. Division Property of Radicals

The multiplication property of radicals enables a product within a radical to be separated and written as a product of radicals. We now state a similar property for radicals involving quotients.

PROPERTY Division Property of Radicals

Let a and b represent real numbers such that $\sqrt[n]{a}$ and $\sqrt[n]{b}$ are both real. Then,

$$\sqrt[n]{\frac{a}{b}} = \frac{\sqrt[n]{a}}{\sqrt[n]{b}} \qquad b \neq 0$$

The division property of radicals indicates that a quotient within a radicand can be written as a quotient of radicals provided the roots are real numbers. For example:

$$\sqrt{\frac{4}{9}} = \frac{\sqrt{4}}{\sqrt{9}}$$

The reverse process is also true. A quotient of radicals can be written as a single radical provided that the roots are real numbers and they have the same indices.

$$\text{same index} \left[\begin{array}{c} \frac{\sqrt[3]{125}}{\sqrt[3]{8}} = \sqrt[3]{\frac{125}{8}} \end{array}\right.$$

In Examples 1 and 2, we will apply the division property of radicals to simplify radical expressions.

Example 1 **Using the Division Property to Simplify Radicals**

Use the division property of radicals to simplify the expressions. Assume the variables represent positive real numbers.

a. $\sqrt{\dfrac{a^{10}}{b^4}}$ **b.** $\sqrt{\dfrac{20x^3}{9}}$

Solution:

a. $\sqrt{\dfrac{a^{10}}{b^4}}$ The radicand contains an irreducible fraction.

$= \dfrac{\sqrt{a^{10}}}{\sqrt{b^4}}$ Apply the division property to rewrite as a quotient of radicals.

$= \dfrac{a^5}{b^2}$ Simplify the radicals.

b. $\sqrt{\dfrac{20x^3}{9}}$ The radicand contains an irreducible fraction.

$= \dfrac{\sqrt{20x^3}}{\sqrt{9}}$ Apply the division property to rewrite as a quotient of radicals.

$= \dfrac{\sqrt{2^2 \cdot 5 \cdot x^2 \cdot x}}{\sqrt{9}}$ Factor the radicand in the numerator to simplify the radical.

$= \dfrac{2x\sqrt{5x}}{3}$ Simplify the radicals in the numerator and denominator. The expression is simplified since it now satisfies all conditions.

Skill Practice

Simplify the expressions.

1. $\sqrt{\dfrac{c^4}{49}}$

2. $\sqrt{\dfrac{12b^5}{25}}$

Classroom Example: p. 621, Exercise 16

Answers

1. $\dfrac{c^2}{7}$ **2.** $\dfrac{2b^2\sqrt{3b}}{5}$

Example 2 Using the Division Property to Simplify Radicals

Use the division property of radicals to simplify the expressions. Assume the variables represent positive real numbers.

a. $\dfrac{\sqrt[3]{9}}{\sqrt[3]{72}}$ **b.** $\dfrac{\sqrt{7y^3}}{\sqrt{y}}$

Solution:

a. $\dfrac{\sqrt[3]{9}}{\sqrt[3]{72}}$ There is a radical in the denominator of the fraction.

$= \sqrt[3]{\dfrac{9}{72}}$ Apply the division property to write the quotient under a single radical.

$= \sqrt[3]{\dfrac{1}{8}}$ Reduce to lowest terms.

$= \dfrac{1}{2}$ Simplify the radical.

b. $\dfrac{\sqrt{7y^3}}{\sqrt{y}}$ There is a radical in the denominator of the fraction.

$= \sqrt{\dfrac{7y^3}{y}}$ Apply the division property to write the quotient under a single radical.

$= \sqrt{7y^2}$ Simplify the fraction.

$= y\sqrt{7}$ Simplify the radical.

Examples 1 and 2 show that radical expressions can sometimes be simplified by using the division property of radicals. However, there are cases where other methods are needed. For example:

$$\dfrac{2}{\sqrt{2}} \text{ and } \dfrac{2}{\sqrt{5} + \sqrt{3}} \text{ are two such cases.}$$

3. Rationalizing the Denominator: One Term

To begin the first case, recall that the nth-root of a perfect nth power is easily simplified. For example:

$$\sqrt{x^2} = x \quad x \geq 0$$

Example 3 Rationalizing the Denominator: One Term

Simplify the expression. $\dfrac{2}{\sqrt{2}}$

Solution:

A square root of a perfect square is needed in the denominator to remove the radical.

$\dfrac{2}{\sqrt{2}} = \dfrac{2}{\sqrt{2}} \cdot \dfrac{\sqrt{2}}{\sqrt{2}}$ Multiply the numerator and denominator by $\sqrt{2}$ because $\sqrt{2} \cdot \sqrt{2} = \sqrt{2^2}$.

$= \dfrac{2\sqrt{2}}{\sqrt{2^2}}$ Multiply the radicals.

$$= \frac{2\sqrt{2}}{2} \quad \text{Simplify.}$$

$$= \frac{2\sqrt{2}}{2} \quad \text{Reduce the fraction to lowest terms.}$$

$$= \sqrt{2}$$

Example 4   **Rationalizing the Denominator: One Term**

Simplify the expression. Assume x represents a positive real number. $\sqrt{\dfrac{x}{5}}$

Solution:

$$\sqrt{\frac{x}{5}} \qquad \text{The radicand contains an irreducible fraction.}$$

$$= \frac{\sqrt{x}}{\sqrt{5}} \qquad \text{Apply the division property of radicals.}$$

$$= \frac{\sqrt{x}}{\sqrt{5}} \cdot \frac{\sqrt{5}}{\sqrt{5}} \qquad \begin{array}{l}\text{Multiply the numerator and denominator by } \sqrt{5} \\ \text{because } \sqrt{5} \cdot \sqrt{5} = \sqrt{5^2}.\end{array}$$

$$= \frac{\sqrt{5x}}{\sqrt{5^2}} \qquad \text{Multiply the radicals.}$$

$$= \frac{\sqrt{5x}}{5} \qquad \text{Simplify the radicals.}$$

> **Skill Practice**
>
> Simplify the expression by rationalizing the denominator.
>
> **6.** $\sqrt{\dfrac{7}{10}}$

Classroom Example: p. 621, Exercise 40

> **Avoiding Mistakes**
>
> In the expression $\dfrac{\sqrt{5x}}{5}$, do not try to "cancel" the factor of $\sqrt{5}$ from the numerator with the factor of 5 in the denominator. $\sqrt{5}$ and 5 are not equal.

Example 5   **Rationalizing the Denominator: One Term**

Simplify the expression. Assume w represents a positive real number.

$$\frac{14\sqrt{w}}{\sqrt{7}}$$

Solution:

$$\frac{14\sqrt{w}}{\sqrt{7}} \qquad \text{Fraction contains a radical in the denominator.}$$

$$= \frac{14\sqrt{w}}{\sqrt{7}} \cdot \frac{\sqrt{7}}{\sqrt{7}} \qquad \begin{array}{l}\text{Multiply the numerator and denominator by } \sqrt{7} \\ \text{because } \sqrt{7} \cdot \sqrt{7} = \sqrt{7^2}.\end{array}$$

$$= \frac{14\sqrt{7w}}{\sqrt{7^2}} \qquad \text{Multiply the radicals.}$$

$$= \frac{14\sqrt{7w}}{7} \qquad \text{Simplify.}$$

$$= \frac{\overset{2}{14}\sqrt{7w}}{\underset{1}{7}} \qquad \text{Reduce to lowest terms.}$$

$$= 2\sqrt{7w}$$

> **Skill Practice**
>
> Simplify the expression by rationalizing the denominator.
>
> **7.** $\dfrac{6y}{\sqrt{3}}$

Classroom Example: p. 621, Exercise 44

> **TIP:** In the expression
>
> $$\frac{14\sqrt{7w}}{7}$$
>
> the factor of 14 and the factor of 7 may be reduced because both factors are outside the radical.
>
> $$\frac{14\sqrt{7w}}{7} = \frac{14}{7} \cdot \sqrt{7w}$$
>
> $$= 2\sqrt{7w}$$

Answers

6. $\dfrac{\sqrt{70}}{10}$ **7.** $2y\sqrt{3}$

Classroom Example: p. 621, Exercise 46

Skill Practice

Simplify the expression by rationalizing the denominator.

8. $\sqrt{\dfrac{z}{18}}$

Example 6 **Rationalizing the Denominator: One Term**

Simplify the expression. Assume w represents a positive real number.

$$\sqrt{\frac{w}{12}}$$

Solution:

$\sqrt{\dfrac{w}{12}}$ The radical contains an irreducible fraction.

$= \dfrac{\sqrt{w}}{\sqrt{12}}$ Apply the division property of radicals.

$= \dfrac{\sqrt{w}}{\sqrt{2^2 \cdot 3}}$ Factor 12 to simplify the radical.

$= \dfrac{\sqrt{w}}{2\sqrt{3}}$ The $\sqrt{3}$ in the denominator needs to be rationalized.

$= \dfrac{\sqrt{w}}{2\sqrt{3}} \cdot \dfrac{\sqrt{3}}{\sqrt{3}}$ Multiply the numerator and denominator by $\sqrt{3}$ because $\sqrt{3} \cdot \sqrt{3} = \sqrt{3^2}$.

$= \dfrac{\sqrt{3w}}{2\sqrt{3^2}}$ Multiply the radicals.

$= \dfrac{\sqrt{3w}}{2 \cdot 3}$ Simplify.

$= \dfrac{\sqrt{3w}}{6}$ This cannot be simplified further because 3 is inside the radical and 6 is not.

4. Rationalizing the Denominator: Two Terms

Recall from the multiplication of polynomials that the product of two conjugates results in a difference of squares.

$$(a + b)(a - b) = a^2 - b^2$$

If either a or b has a square root factor, the expression will simplify without a radical; that is, the expression is rationalized. For example,

$$(\sqrt{5} - \sqrt{3})(\sqrt{5} + \sqrt{3}) = (\sqrt{5})^2 - (\sqrt{3})^2$$
$$= 5 - 3$$
$$= 2$$

Multiplying a binomial by its conjugate is the basis for rationalizing a denominator with two terms involving square roots.

Answer

8. $\dfrac{\sqrt{2z}}{6}$

Example 7 Rationalizing the Denominator: Two Terms

Simplify the expression by rationalizing the denominator.

$$\frac{2}{\sqrt{5} + \sqrt{3}}$$

Solution:

$\frac{2}{\sqrt{5} + \sqrt{3}}$ To rationalize a denominator with two terms, multiply the numerator and denominator by the conjugate of the denominator.

$$= \frac{2}{(\sqrt{5} + \sqrt{3})} \cdot \frac{(\sqrt{5} - \sqrt{3})}{(\sqrt{5} - \sqrt{3})}$$

Conjugates

The denominator is in the form $(a + b)(a - b)$, where $a = \sqrt{5}$ and $b = \sqrt{3}$.

$$= \frac{2(\sqrt{5} - \sqrt{3})}{(\sqrt{5})^2 - (\sqrt{3})^2}$$ In the denominator, apply the formula $(a + b)(a - b) = a^2 - b^2$.

$$= \frac{2(\sqrt{5} - \sqrt{3})}{5 - 3}$$ Simplify.

$$= \frac{2(\sqrt{5} - \sqrt{3})}{2}$$

$$= \frac{2(\sqrt{5} - \sqrt{3})}{2}$$ Reduce to lowest terms.

$$= \sqrt{5} - \sqrt{3}$$

Skill Practice

Simplify the expression by rationalizing the denominator.

9. $\frac{6}{\sqrt{3} - 1}$

Classroom Example: p. 621, Exercise 60

Example 8 Rationalizing the Denominator: Two Terms

Simplify the expression by rationalizing the denominator.

$$\frac{\sqrt{x} + \sqrt{2}}{\sqrt{x} - \sqrt{2}}$$

Solution:

$$\frac{\sqrt{x} + \sqrt{2}}{\sqrt{x} - \sqrt{2}} = \frac{(\sqrt{x} + \sqrt{2})}{(\sqrt{x} - \sqrt{2})} \cdot \frac{(\sqrt{x} + \sqrt{2})}{(\sqrt{x} + \sqrt{2})}$$ Multiply the numerator and denominator by the conjugate of the denominator.

Conjugates

$$= \frac{(\sqrt{x} + \sqrt{2})^2}{(\sqrt{x} - \sqrt{2})(\sqrt{x} + \sqrt{2})}$$

$$= \frac{(\sqrt{x})^2 + 2(\sqrt{x})(\sqrt{2}) + (\sqrt{2})^2}{(\sqrt{x})^2 - (\sqrt{2})^2}$$ Simplify using special case products.

$$= \frac{x + 2\sqrt{2x} + 2}{x - 2}$$ Simplify the radicals.

Skill Practice

Simplify the expression by rationalizing the denominators.

10. $\frac{\sqrt{y} - \sqrt{5}}{\sqrt{y} + \sqrt{5}}$

Classroom Example: p. 622, Exercise 66

Answers

9. $3\sqrt{3} + 3$

10. $\frac{y - 2\sqrt{5y} + 5}{y - 5}$

5. Simplifying Quotients That Contain Radicals

Sometimes a radical expression within a quotient must be reduced to lowest terms. This is demonstrated in the next example.

Skill Practice

Simplify the expression.

11. $\dfrac{6 - \sqrt{24}}{12}$

Classroom Example: p. 622, Exercise 72

Avoiding Mistakes

Remember that it is not correct to reduce *terms* within a rational expression. In the expression

$$\frac{4 - 2\sqrt{5}}{10}$$

do not try to reduce the 4 and the 10. Only common *factors* can be canceled.

Answer

11. $\dfrac{3 - \sqrt{6}}{6}$

Example 9 Simplifying a Radical Quotient to Lowest Terms

Simplify the expression $\dfrac{4 - \sqrt{20}}{10}$.

Solution:

$$\frac{4 - \sqrt{20}}{10}$$
First simplify $\sqrt{20}$ by writing the radicand as a product of prime factors.

$$= \frac{4 - \sqrt{2^2 \cdot 5}}{10}$$

$$= \frac{4 - 2\sqrt{5}}{10}$$
Simplify the radical.

$$= \frac{2(2 - \sqrt{5})}{2 \cdot 5}$$
Factor out the GCF.

$$= \frac{2(2 - \sqrt{5})}{2 \cdot 5}$$
Reduce to lowest terms.

$$= \frac{2 - \sqrt{5}}{5}$$

Section 8.5 Practice Exercises

Study Skills Exercise

1. Define the key term **rationalizing the denominator**.

Review Exercises

For Exercises 2–10, perform the indicated operations. Assume the variables represent positive real numbers.

2. $x\sqrt{45} + 4\sqrt{20x^2}$ $11x\sqrt{5}$

3. $(2\sqrt{y} + 3)(3\sqrt{y} + 7)$
  $6y + 23\sqrt{y} + 21$

4. $(4\sqrt{w} - 2)(2\sqrt{w} - 4)$
  $8w - 20\sqrt{w} + 8$

5. $4\sqrt{3} + \sqrt{5} \cdot \sqrt{15}$ $9\sqrt{3}$

6. $\sqrt{7} \cdot \sqrt{21} + 2\sqrt{27}$ $13\sqrt{3}$

7. $(5 - \sqrt{a})^2$ $25 - 10\sqrt{a} + a$

8. $(\sqrt{z} + 3)^2$ $z + 6\sqrt{z} + 9$

9. $(\sqrt{2} + \sqrt{7})(\sqrt{2} - \sqrt{7})$ -5

10. $(\sqrt{3} + 5)(\sqrt{3} - 5)$ -22

 Writing  Translating Expression   Geometry   Scientific Calculator   Video NS&E

Objective 2: Division Property of Radicals

For Exercises 11–30, use the division property of radicals, if necessary, to simplify the expressions. Assume the variables represent positive real numbers. **(See Examples 1–2.)**

11. $\sqrt{\dfrac{3}{16}}$ $\quad \dfrac{\sqrt{3}}{4}$

12. $\sqrt{\dfrac{7}{25}}$ $\quad \dfrac{\sqrt{7}}{5}$

13. $\sqrt{\dfrac{a^4}{b^4}}$ $\quad \dfrac{a^2}{b^2}$

14. $\sqrt{\dfrac{y^6}{z^2}}$ $\quad \dfrac{y^3}{z}$

15. $\sqrt{\dfrac{c^3}{4}}$ $\quad \dfrac{c\sqrt{c}}{2}$

16. $\sqrt{\dfrac{d^5}{9}}$ $\quad \dfrac{d^2\sqrt{d}}{3}$

17. $\sqrt[3]{\dfrac{x^2}{27}}$ $\quad \dfrac{\sqrt[3]{x^2}}{3}$

18. $\sqrt[3]{\dfrac{c^2}{8}}$ $\quad \dfrac{\sqrt[3]{c^2}}{2}$

19. $\sqrt[3]{\dfrac{y^5}{27y^3}}$ $\quad \dfrac{\sqrt[3]{y^2}}{3}$

20. $\sqrt[3]{\dfrac{7ac}{64c^4}}$ $\quad \dfrac{\sqrt[3]{7a}}{4c}$

21. $\sqrt{\dfrac{200}{81}}$ $\quad \dfrac{10\sqrt{2}}{9}$

22. $\sqrt{\dfrac{80}{49}}$ $\quad \dfrac{4\sqrt{5}}{7}$

23. $\dfrac{\sqrt{8}}{\sqrt{50}}$ $\quad \dfrac{2}{5}$

24. $\dfrac{\sqrt{21}}{\sqrt{12}}$ $\quad \dfrac{\sqrt{7}}{2}$

25. $\dfrac{\sqrt{p}}{\sqrt{4p^3}}$ $\quad \dfrac{1}{2p}$

26. $\dfrac{\sqrt{9t}}{\sqrt{t^5}}$ $\quad \dfrac{3}{t^2}$

27. $\dfrac{\sqrt[3]{z^5}}{\sqrt[3]{z^2}}$ $\quad z$

28. $\dfrac{\sqrt[3]{a^7}}{\sqrt[3]{a}}$ $\quad a^2$

29. $\dfrac{\sqrt[3]{24x^5}}{\sqrt[3]{3x^4}}$ $\quad 2\sqrt[3]{x}$

30. $\dfrac{\sqrt[3]{2y^8}}{\sqrt[3]{54y^7}}$ $\quad \dfrac{\sqrt[3]{y}}{3}$

Objective 3: Rationalizing the Denominator: One Term

For Exercises 31–50, rationalize the denominators. Assume the variable expressions represent positive real numbers. **(See Examples 3–6.)**

31. $\dfrac{1}{\sqrt{6}}$ $\quad \dfrac{\sqrt{6}}{6}$

32. $\dfrac{5}{\sqrt{2}}$ $\quad \dfrac{5\sqrt{2}}{2}$

33. $\dfrac{15}{\sqrt{5}}$ $\quad 3\sqrt{5}$

34. $\dfrac{14}{\sqrt{7}}$ $\quad 2\sqrt{7}$

35. $\dfrac{6}{\sqrt{x+1}}$ $\quad \dfrac{6\sqrt{x+1}}{x+1}$

36. $\dfrac{8}{\sqrt{y-3}}$ $\quad \dfrac{8\sqrt{y-3}}{y-3}$

37. $\sqrt{\dfrac{6}{x}}$ $\quad \dfrac{\sqrt{6x}}{x}$

38. $\sqrt{\dfrac{8}{y}}$ $\quad \dfrac{2\sqrt{2y}}{y}$

39. $\sqrt{\dfrac{3}{7}}$ $\quad \dfrac{\sqrt{21}}{7}$

40. $\sqrt{\dfrac{5}{11}}$ $\quad \dfrac{\sqrt{55}}{11}$

41. $\dfrac{10}{\sqrt{6y}}$ $\quad \dfrac{5\sqrt{6y}}{3y}$

42. $\dfrac{15}{\sqrt{3w}}$ $\quad \dfrac{5\sqrt{3w}}{w}$

43. $\dfrac{9}{2\sqrt{6}}$ $\quad \dfrac{3\sqrt{6}}{4}$

44. $\dfrac{15}{4\sqrt{10}}$ $\quad \dfrac{3\sqrt{10}}{8}$

45. $\sqrt{\dfrac{p}{27}}$ $\quad \dfrac{\sqrt{3p}}{9}$

46. $\sqrt{\dfrac{x}{32}}$ $\quad \dfrac{\sqrt{2x}}{8}$

47. $\dfrac{5}{\sqrt{20}}$ $\quad \dfrac{\sqrt{5}}{2}$

48. $\dfrac{8}{\sqrt{24}}$ $\quad \dfrac{2\sqrt{6}}{3}$

49. $\sqrt{\dfrac{x^2}{y^3}}$ $\quad \dfrac{x\sqrt{y}}{y^2}$

50. $\sqrt{\dfrac{a}{b^5}}$ $\quad \dfrac{\sqrt{ab}}{b^3}$

Objective 4: Rationalizing the Denominator: Two Terms

For Exercises 51–52, multiply the conjugates.

51. $(\sqrt{2}+3)(\sqrt{2}-3)$ $\quad -7$

52. $(\sqrt{3}+\sqrt{7})(\sqrt{3}-\sqrt{7})$ $\quad -4$

53. What is the conjugate of $\sqrt{5}-\sqrt{3}$? Multiply $\sqrt{5}-\sqrt{3}$ by its conjugate. $\quad \sqrt{5}+\sqrt{3};\ 2$

54. What is the conjugate of $\sqrt{7}+\sqrt{2}$? Multiply $\sqrt{7}+\sqrt{2}$ by its conjugate. $\quad \sqrt{7}-\sqrt{2};\ 5$

55. What is the conjugate of $\sqrt{x}+10$? Multiply $\sqrt{x}+10$ by its conjugate. $\quad \sqrt{x}-10;\ x-100$

56. What is the conjugate of $12-\sqrt{y}$? Multiply $12-\sqrt{y}$ by its conjugate. $\quad 12+\sqrt{y};\ 144-y$

For Exercises 57–68, rationalize the denominators. Assume the variable expressions represent positive real numbers. **(See Examples 7–8.)**

57. $\dfrac{4}{\sqrt{2}+3}$ $\quad \dfrac{4\sqrt{2}-12}{-7}$ or $\dfrac{12-4\sqrt{2}}{7}$

58. $\dfrac{6}{4-\sqrt{3}}$ $\quad \dfrac{24+6\sqrt{3}}{13}$

59. $\dfrac{1}{\sqrt{5}-\sqrt{2}}$ $\quad \dfrac{\sqrt{5}+\sqrt{2}}{3}$

60. $\dfrac{2}{\sqrt{3}+\sqrt{7}}$ $\quad \dfrac{\sqrt{3}-\sqrt{7}}{-2}$ or $\dfrac{\sqrt{7}-\sqrt{3}}{2}$

61. $\dfrac{\sqrt{8}}{\sqrt{3}+1}$ $\sqrt{6}-\sqrt{2}$ **62.** $\dfrac{\sqrt{18}}{1-\sqrt{2}}$ $-3\sqrt{2}-6$ **63.** $\dfrac{1}{\sqrt{x}-\sqrt{3}}$ $\dfrac{\sqrt{x}+\sqrt{3}}{x-3}$ **64.** $\dfrac{1}{\sqrt{y}+\sqrt{5}}$ $\dfrac{\sqrt{y}-\sqrt{5}}{y-5}$

65. $\dfrac{2-\sqrt{3}}{2+\sqrt{3}}$ $7-4\sqrt{3}$ **66.** $\dfrac{\sqrt{3}-\sqrt{2}}{\sqrt{3}+\sqrt{2}}$ $5-2\sqrt{6}$ **67.** $\dfrac{\sqrt{5}+4}{2-\sqrt{5}}$ $-13-6\sqrt{5}$ **68.** $\dfrac{3+\sqrt{2}}{\sqrt{2}-5}$

$\dfrac{17+8\sqrt{2}}{-23}$ or $-\dfrac{17+8\sqrt{2}}{23}$

Objective 5: Simplifying Quotients That Contain Radicals

For Exercises 69–76, simplify the expression. **(See Example 9.)**

69. $\dfrac{10-\sqrt{50}}{5}$ $2-\sqrt{2}$ **70.** $\dfrac{4+\sqrt{12}}{2}$ $2+\sqrt{3}$ **71.** $\dfrac{21+\sqrt{98}}{14}$ $\dfrac{3+\sqrt{2}}{2}$ **72.** $\dfrac{3-\sqrt{18}}{6}$ $\dfrac{1-\sqrt{2}}{2}$

73. $\dfrac{2-\sqrt{28}}{2}$ $1-\sqrt{7}$ **74.** $\dfrac{5+\sqrt{75}}{5}$ $1+\sqrt{3}$ **75.** $\dfrac{14+\sqrt{72}}{6}$ $\dfrac{7+3\sqrt{2}}{3}$ **76.** $\dfrac{15-\sqrt{125}}{10}$ $\dfrac{3-\sqrt{5}}{2}$

Recall that a radical is simplified if

1. The radicand has no factor raised to a power greater than or equal to the index.

2. There are no radicals in the denominator of a fraction.

3. The radicand does not contain a fraction.

For Exercises 77–80, state which condition(s) fails. Then simplify the radical.

77. a. $\sqrt{8x^9}$ **b.** $\dfrac{5}{\sqrt{5x}}$ **c.** $\sqrt{\dfrac{1}{3}}$

   a. Condition 1 fails; $2x^4\sqrt{2x}$ b. Condition 2 fails; $\dfrac{\sqrt{5x}}{x}$ c. Condition 3 fails; $\dfrac{\sqrt{3}}{3}$

78. a. $\sqrt{\dfrac{7}{2}}$ **b.** $\sqrt{18y^6}$ **c.** $\dfrac{2}{\sqrt{4x}}$

   a. Condition 3 fails; $\dfrac{\sqrt{14}}{2}$ b. Condition 1 fails; $3y^3\sqrt{2}$ c. Conditions 1 and 2 fail; $\dfrac{\sqrt{x}}{x}$

79. a. $\dfrac{3}{\sqrt{x}+1}$ **b.** $\sqrt{\dfrac{9w^2}{t}}$ **c.** $\sqrt{24a^5b^9}$

   a. Condition 2 fails; $\dfrac{3\sqrt{x}-3}{x-1}$ b. Conditions 1 and 3 fail; $\dfrac{3w\sqrt{t}}{t}$ c. Condition 1 fails; $2a^2b^4\sqrt{6ab}$

80. a. $\sqrt{\dfrac{12}{z^3}}$ **b.** $\dfrac{4}{\sqrt{a}-\sqrt{b}}$ **c.** $\sqrt[3]{27m^3n^7}$

   a. Conditions 1 and 3 fail; $\dfrac{2\sqrt{3z}}{z^2}$ b. Condition 2 fails; $\dfrac{4\sqrt{a}+4\sqrt{b}}{a-b}$ c. Condition 1 fails; $3mn^2\sqrt[3]{n}$

Mixed Exercises

For Exercises 81–96, simplify the radical expressions, if possible. Assume the variables represent positive real numbers.

81. $\sqrt{45}$ $3\sqrt{5}$ **82.** $-\sqrt{108y^4}$ $-6y^2\sqrt{3}$ **83.** $-\sqrt{\dfrac{18w^2}{25}}$ $-\dfrac{3w\sqrt{2}}{5}$ **84.** $\sqrt{\dfrac{8a^2}{7}}$ $\dfrac{2a\sqrt{14}}{7}$

85. $\sqrt{-36}$ Not a real number **86.** $\sqrt{54b^5}$ $3b^2\sqrt{6b}$ **87.** $\sqrt{\dfrac{s^2}{t}}$ $\dfrac{s\sqrt{t}}{t}$ **88.** $\dfrac{x+\sqrt{y}}{x-\sqrt{y}}$ $\dfrac{x^2+2x\sqrt{y}+y}{x^2-y}$

89. $\dfrac{\sqrt{2m^5}}{\sqrt{8m}}$ $\dfrac{m^2}{2}$ **90.** $\dfrac{\sqrt{10w}}{\sqrt{5w^3}}$ $\dfrac{\sqrt{2}}{w}$ **91.** $\sqrt{\dfrac{81}{t^3}}$ $\dfrac{9\sqrt{t}}{t^2}$ **92.** $-\sqrt{a^3bc^6}$ $-ac^3\sqrt{ab}$

93. $\dfrac{3}{\sqrt{11}+\sqrt{5}}$ $\dfrac{\sqrt{11}-\sqrt{5}}{2}$ **94.** $\dfrac{4}{\sqrt{10}+\sqrt{2}}$ $\dfrac{\sqrt{10}-\sqrt{2}}{2}$ **95.** $\dfrac{\sqrt{a}+\sqrt{b}}{\sqrt{a}-\sqrt{b}}$ $\dfrac{a+2\sqrt{ab}+b}{a-b}$ **96.** $\dfrac{\sqrt{x}+1}{\sqrt{x}-1}$ $\dfrac{x+2\sqrt{x}+1}{x-1}$

 Writing   Translating Expression Geometry Scientific Calculator   Video NS&E

Expanding Your Skills

97. Find the slope of the line through the points: $(5\sqrt{2}, 3)$ and $(\sqrt{2}, 6)$. $-\dfrac{3\sqrt{2}}{8}$

98. Find the slope of the line through the points: $(4\sqrt{5}, -1)$ and $(6\sqrt{5}, -5)$. $-\dfrac{2\sqrt{5}}{5}$

99. Find the slope of the line through the points: $(\sqrt{3}, -1)$ and $(4\sqrt{3}, 0)$. $\dfrac{\sqrt{3}}{9}$

100. Find the slope of the line through the points: $(-2\sqrt{7}, -5)$ and $(\sqrt{7}, 2)$. $\dfrac{\sqrt{7}}{3}$

Problem Recognition Exercises

Operations on Radicals

Perform the indicated operations and simplify if possible. Assume that all variables represent positive real numbers.

1. $(\sqrt{3})(\sqrt{6})$ $3\sqrt{2}$

2. $(\sqrt{2})(\sqrt{14})$ $2\sqrt{7}$

3. $\sqrt{3} + \sqrt{6}$
Cannot be simplified further

4. $\sqrt{2} + \sqrt{14}$
Cannot be simplified further

5. $\dfrac{\sqrt{6}}{\sqrt{3}}$ $\sqrt{2}$

6. $\dfrac{\sqrt{14}}{\sqrt{2}}$ $\sqrt{7}$

7. $(3 + \sqrt{z})(3 - \sqrt{z})$ $9 - z$

8. $(4 - \sqrt{y})(4 + \sqrt{y})$ $16 - y$

9. $(2\sqrt{5} + 1)(\sqrt{5} - 2)$ $8 - 3\sqrt{5}$

10. $(4\sqrt{3} - 5)(\sqrt{3} + 4)$ $-8 + 11\sqrt{3}$

11. $2\sqrt{x^2 y} - 3x\sqrt{y}$ $-x\sqrt{y}$

12. $8\sqrt{a^3 b^2} + 3a\sqrt{ab^2}$ $11ab\sqrt{a}$

13. $-3\sqrt{2}\,(4\sqrt{2} + 2\sqrt{3} + 1)$ $-24 - 6\sqrt{6} - 3\sqrt{2}$

14. $-8\sqrt{5}\,(2\sqrt{5} - \sqrt{3} - 2)$ $-80 + 8\sqrt{15} + 16\sqrt{5}$

15. $\dfrac{2}{\sqrt{x} - 7}$ $\dfrac{2\sqrt{x} + 14}{x - 49}$

16. $\dfrac{5}{\sqrt{y} + 4}$ $\dfrac{5\sqrt{y} - 20}{y - 16}$

17. $\dfrac{9}{\sqrt{3}}$ $3\sqrt{3}$

18. $\dfrac{15}{\sqrt{5}}$ $3\sqrt{5}$

19. $\sqrt{\dfrac{7}{x}}$ $\dfrac{\sqrt{7x}}{x}$

20. $\sqrt{\dfrac{11}{y}}$ $\dfrac{\sqrt{11y}}{y}$

21. $\sqrt{y^4 z^{11}}$ $y^2 z^5 \sqrt{z}$

22. $\sqrt{8q^6}$ $2q^3\sqrt{2}$

23. $\sqrt[3]{27p^8}$ $3p^2\sqrt[3]{p^2}$

24. $\sqrt[3]{125u^{11}v^{12}}$ $5u^3 v^4 \sqrt[3]{u^2}$

25. $\dfrac{\sqrt{10x^3}}{\sqrt{x}}$ $x\sqrt{10}$

26. $\dfrac{\sqrt{15y^3}}{\sqrt{5y}}$ $y\sqrt{3}$

27. $6\sqrt{75} - 5\sqrt{12}$ $20\sqrt{3}$

28. $\sqrt{90} - \sqrt{40}$ $\sqrt{10}$

29. $(\sqrt{2} + 7)^2$ $51 + 14\sqrt{2}$

30. $(\sqrt{3} + \sqrt{5})^2$ $8 + 2\sqrt{15}$

31. $\dfrac{x - 5}{\sqrt{x} + \sqrt{5}}$ $\sqrt{x} - \sqrt{5}$

32. $\dfrac{y - 7}{\sqrt{y} + \sqrt{7}}$ $\sqrt{y} - \sqrt{7}$

33. $(4\sqrt{x} + \sqrt{y})(\sqrt{x} - 3\sqrt{y})$ $4x - 11\sqrt{xy} - 3y$

34. $\sqrt[4]{\dfrac{1}{81}}$ $\dfrac{1}{3}$

35. $\sqrt[3]{\dfrac{125}{27}}$ $\dfrac{5}{3}$

36. $(\sqrt{5} - \sqrt{11})^2$ $16 - 2\sqrt{55}$

37. $(\sqrt{x} - 6)^2$ $x - 12\sqrt{x} + 36$

38. $2\sqrt{6} - 5\sqrt{6}$ $-3\sqrt{6}$

39. $5\sqrt{a} + 7\sqrt{a} - \sqrt{a}$ $11\sqrt{a}$

40. $(2\sqrt{3} - 10)(2\sqrt{3} + 10)$ -88

41. $(\sqrt{u} - 3\sqrt{v})(\sqrt{u} + 3\sqrt{v})$ $u - 9v$

42. $x\sqrt{18} + \sqrt{2x^2}$ $4x\sqrt{2}$

43. $4\sqrt{75} - 20\sqrt{3}$ 0

44. $\sqrt{5}(\sqrt{5} + \sqrt{7})$ $5 + \sqrt{35}$

45. $\sqrt{a}(\sqrt{a} + 2)$ $a + 2\sqrt{a}$

46. $(3\sqrt{2} - 4)(5\sqrt{2} + 1)$ $26 - 17\sqrt{2}$

 Writing   Translating Expression   Geometry   Scientific Calculator   Video   NS&E

Section 8.6 Radical Equations

Objectives

1. Solving Radical Equations
2. Translations Involving Radical Equations
3. Applications of Radical Equations

1. Solving Radical Equations

DEFINITION Radical Equation

An equation with one or more radicals containing a variable is called a **radical equation**.

For example, $\sqrt{x} = 5$ is a radical equation. Recall that $(\sqrt[n]{a})^n = a$ provided $\sqrt[n]{a}$ is a real number. The basis to solve a radical equation is to eliminate the radical by raising both sides of the equation to a power equal to the index of the radical.

To solve the equation $\sqrt{x} = 5$, square both sides of the equation.

$$\sqrt{x} = 5$$
$$(\sqrt{x})^2 = (5)^2$$
$$x = 25$$

By raising each side of a radical equation to a power equal to the index of the radical, a new equation is produced. However, it is important to note that the new equation may have **extraneous solutions**; that is, some or all of the solutions to the new equation may *not* be solutions to the original radical equation. For this reason, it is necessary to check *all* potential solutions in the original equation. For example, consider the equation $x = 4$. By squaring both sides we produce a quadratic equation.

Square both sides.
$$x = 4$$
$$(x)^2 = (4)^2 \qquad \text{Squaring both sides produces a quadratic equation.}$$
$$x^2 = 16$$
$$x^2 - 16 = 0$$
$$(x - 4)(x + 4) = 0 \qquad \text{Solving this equation, we find two solutions.}$$
$$x = 4 \quad \text{or} \quad x = -4 \qquad \begin{array}{l}\text{However, } x = -4 \text{ does not check. The value} \\ -4 \text{ is an extraneous solution because it is not a} \\ \text{solution to the original equation, } x = 4.\end{array}$$

PROCEDURE Solving a Radical Equation

Step 1 Isolate the radical. If an equation has more than one radical, choose one of the radicals to isolate.

Step 2 Raise each side of the equation to a power equal to the index of the radical.

Step 3 Solve the resulting equation.

Step 4 Check the potential solutions in the original equation.*

*Extraneous solutions can only arise when both sides of the equation are raised to an *even power*. Therefore, an equation with odd-index roots will not have an extraneous solution. However, it is still recommended that you check *all* potential solutions regardless of the type of root.

Example 1 Solving a Radical Equation

Solve the equation. $\sqrt{2x + 1} + 5 = 8$

Solution:

$$\sqrt{2x + 1} + 5 = 8$$

$$\sqrt{2x + 1} = 8 - 5 \qquad \text{Isolate the radical.}$$

$$\sqrt{2x + 1} = 3$$

$$(\sqrt{2x + 1})^2 = (3)^2 \qquad \text{Raise both sides to a power equal to the index of the radical.}$$

$$2x + 1 = 9 \qquad \text{Simplify both sides.}$$

$$2x = 8 \qquad \text{Solve the resulting equation (the equation is linear).}$$

$$x = 4$$

Check: Check 4 as a potential solution.

$$\sqrt{2x + 1} + 5 = 8$$

$$\sqrt{2(4) + 1} + 5 \overset{?}{=} 8$$

$$\sqrt{8 + 1} + 5 \overset{?}{=} 8$$

$$\sqrt{9} + 5 \overset{?}{=} 8$$

$$3 + 5 \overset{?}{=} 8 \; ✔ \qquad \text{The answer checks.}$$

The solution is 4.

Skill Practice

Solve the equation.

1. $\sqrt{p - 4} - 2 = 4$

Classroom Example: p. 629, Exercise 28

Example 2 Solving a Radical Equation

Solve the equation. $8 + \sqrt{x + 2} = 7$

Solution:

$$8 + \sqrt{x + 2} = 7$$

$$\sqrt{x + 2} = 7 - 8 \qquad \text{Isolate the radical.}$$

$$\sqrt{x + 2} = -1$$

$$(\sqrt{x + 2})^2 = (-1)^2 \qquad \text{Raise both sides to a power equal to the index of the radical.}$$

$$x + 2 = 1 \qquad \text{Simplify.}$$

$$x = -1 \qquad \text{Solve the resulting equation.}$$

Check: Check -1 as a potential solution.

$$8 + \sqrt{x + 2} = 7$$

$$8 + \sqrt{(-1) + 2} \overset{?}{=} 7$$

$$8 + \sqrt{1} \overset{?}{=} 7$$

$$8 + 1 \neq 7 \qquad \text{The value } -1 \text{ does not check. It is an extraneous solution.}$$

There is no solution to the equation.

Skill Practice

Solve the equation.

2. $\sqrt{2y + 5} + 7 = 4$

Classroom Example: p. 629, Exercise 30

TIP: After isolating the radical in Example 2, the equation shows a square root equated to a negative number.

$$\sqrt{x + 2} = -1$$

By definition, a principal square root of any real number must be nonnegative. Therefore, there can be no solution to this equation.

Answers

1. 40
2. No solution (the value 2 does not check.)

- Skill Practice -

Solve the equation.

3. $\sqrt{x + 34} = x + 4$

Classroom Example: p. 629,
Exercise 42

Example 3 **Solving a Radical Equation**

Solve the equation. $p + 4 = \sqrt{p + 6}$

Solution:

$$p + 4 = \sqrt{p + 6}$$ The radical is already isolated.

$$(p + 4)^2 = (\sqrt{p + 6})^2$$ Raise both sides to a power equal to the index.

$$p^2 + 8p + 16 = p + 6$$

$$p^2 + 7p + 10 = 0$$ Solve the resulting equation (the equation is quadratic).

$$(p + 5)(p + 2) = 0$$ Set the equation equal to zero and factor.

$$p + 5 = 0 \quad \text{or} \quad p + 2 = 0$$ Set each factor equal to zero.

$$p = -5 \quad \text{or} \quad p = -2$$ Solve for p.

$\underline{\text{Check: } p = -5}$

$$p + 4 = \sqrt{p + 6}$$

$$(-5) + 4 \stackrel{?}{=} \sqrt{(-5) + 6}$$

$$-1 \stackrel{?}{=} \sqrt{1}$$

$$-1 \neq 1 \quad \text{Does not check.}$$

$\underline{\text{Check: } p = -2}$

$$p + 4 = \sqrt{p + 6}$$

$$(-2) + 4 \stackrel{?}{=} \sqrt{(-2) + 6}$$

$$2 \stackrel{?}{=} \sqrt{4}$$

$$2 \stackrel{?}{=} 2 \; \checkmark \quad \text{The solution checks.}$$

The only solution is -2 (the value -5 does not check).

TIP: Recall that

$(a + b)^2 = a^2 + 2ab + b^2$.

Hence,

$(p + 4)^2$

$\quad = (p)^2 + 2(p)(4) + (4)^2$

$\quad = p^2 + 8p + 16$

- Skill Practice -

Solve the equation.

4. $\sqrt[3]{4p + 1} - \sqrt[3]{p + 16} = 0$

Classroom Example: p. 629,
Exercise 46

Example 4 **Solving a Radical Equation**

Solve the equation. $2\sqrt[3]{2x - 3} - \sqrt[3]{x + 6} = 0$

Solution:

$$2\sqrt[3]{2x - 3} - \sqrt[3]{x + 6} = 0$$

$$2\sqrt[3]{2x - 3} = \sqrt[3]{x + 6}$$ Isolate one of the radicals.

$$\left(2\sqrt[3]{2x - 3}\right)^3 = \left(\sqrt[3]{x + 6}\right)^3$$ Raise both sides to a power equal to the index.

$$(2)^3\left(\sqrt[3]{2x - 3}\right)^3 = \left(\sqrt[3]{x + 6}\right)^3$$ On the left-hand side, be sure to cube both factors, $(2)^3$ and $\left(\sqrt[3]{2x - 3}\right)^3$.

$$8(2x - 3) = x + 6$$ Solve the resulting equation.

$$16x - 24 = x + 6$$

$$15x = 30$$

$$x = 2$$

$\underline{\text{Check:}}$

$$2\sqrt[3]{2x - 3} - \sqrt[3]{x + 6} = 0$$ Check the potential solution, 2.

$$2\sqrt[3]{2(2) - 3} - \sqrt[3]{2 + 6} \stackrel{?}{=} 0$$

$$2\sqrt[3]{4 - 3} - \sqrt[3]{8} \stackrel{?}{=} 0$$

$$2\sqrt[3]{1} - 2 \stackrel{?}{=} 0$$

$$2 - 2 \stackrel{?}{=} 0 \; \checkmark$$ The solution checks.

The solution is 2.

Answers

3. 2 (the value -9 does not check)

4. 5

2. Translations Involving Radical Equations

Example 5 Translating English Form into Algebraic Form

The principal square root of the sum of a number and three is equal to seven. Find the number.

Solution:

Let x represent the number. Label the variable.

$\sqrt{x+3} = 7$ Translate the verbal model into an algebraic equation.

$(\sqrt{x+3})^2 = (7)^2$ The radical is already isolated. Square both sides.

$x + 3 = 49$ The resulting equation is linear.

$x = 46$ Solve for x.

Check: Check 46 as a potential solution.

$\sqrt{x+3} = 7$

$\sqrt{46+3} \stackrel{?}{=} 7$

$\sqrt{49} \stackrel{?}{=} 7$

$7 \stackrel{?}{=} 7$ ✔ The solution checks.

The number is 46.

Skill Practice

5. The principal square root of the sum of a number and 5 is 2. Find the number.

Classroom Example: p. 629, Exercise 48

3. Applications of Radical Equations

Example 6 Using a Radical Equation in an Application

For a small company, the weekly sales, y, of its product are related to the money spent on advertising, x, according to the equation:

$$y = 100\sqrt{x}$$

a. Find the amount in sales if the company spends \$100 on advertising.

b. Find the amount in sales if the company spends \$625 on advertising.

c. Find the amount the company spent on advertising if its sales for 1 week totaled \$2000.

Solution:

a. $y = 100\sqrt{x}$

$= 100\sqrt{100}$ Substitute $x = 100$.

$= 100(10)$

$= 1000$

The amount in sales is \$1000.

Skill Practice

6. If the small company mentioned in Example 6 changes its advertising media, the equation relating money spent on advertising, x, to weekly sales, y, is $y = 100\sqrt{2x}$.
   a. Use the given equation to find the amount in sales if the company spends \$200 on advertising.
   b. Find the amount spent on advertising if the sales for 1 week totaled \$3000.

Classroom Example: p. 629, Exercise 54

Answers

5. The number is -1.
6. a. \$2000 **b.** \$450

b. $y = 100\sqrt{x}$

$= 100\sqrt{625}$ Substitute $x = 625$.

$= 100(25)$

$= 2500$

The amount in sales is \$2500.

c. $y = 100\sqrt{x}$

$2000 = 100\sqrt{x}$ Substitute $y = 2000$.

$\dfrac{2000}{100} = \dfrac{\cancel{100}\sqrt{x}}{\cancel{100}}$ Isolate the radical. Divide both sides by 100.

$20 = \sqrt{x}$ Simplify.

$(20)^2 = (\sqrt{x})^2$ Raise both sides to a power equal to the index.

$400 = x$ Simplify both sides.

<u>Check:</u>   Check 400 as a potential solution.

$y = 100\sqrt{x}$

$2000 \overset{?}{=} 100\sqrt{400}$

$2000 \overset{?}{=} 100(20)$

$2000 \overset{?}{=} 2000$ ✔   The solution checks.

The amount spent on advertising was \$400.

Section 8.6   Practice Exercises

Study Skills Exercise

1. Define the key terms:

   a. radical equation   **b. extraneous solution**

Review Exercises

For Exercises 2–5, rationalize the denominators.

2. $\dfrac{1}{\sqrt{3} - \sqrt{7}}$

$\dfrac{\sqrt{3} + \sqrt{7}}{-4}$ or $-\dfrac{\sqrt{3} + \sqrt{7}}{4}$

3. $\dfrac{1}{\sqrt{2} + \sqrt{10}}$

$\dfrac{\sqrt{2} - \sqrt{10}}{-8}$ or $\dfrac{\sqrt{10} - \sqrt{2}}{8}$

4. $\dfrac{6}{\sqrt{6}}$ $\sqrt{6}$

5. $\dfrac{2\sqrt{2}}{\sqrt{3}}$ $\dfrac{2\sqrt{6}}{3}$

6. Simplify the expression. $\dfrac{10 - \sqrt{75}}{5}$ $2 - \sqrt{3}$

For Exercises 7–10, square the binomials.

7. $(x + 4)^2$ $x^2 + 8x + 16$ **8.** $(3 - y)^2$ $9 - 6y + y^2$ **9.** $(\sqrt{x} + 4)^2$
$x + 8\sqrt{x} + 16$ **10.** $(\sqrt{3} - \sqrt{y})^2$
$3 - 2\sqrt{3y} + y$

For Exercises 11–14, simplify the expressions. Assume the variable expressions represent positive real numbers.

11. $(\sqrt{2x-3})^2$ $2x-3$ **12.** $(\sqrt{m+6})^2$ $m+6$ **13.** $(t+1)^2$
t^2+2t+1

14. $(y-4)^2$
$y^2-8y+16$

Objective 1: Solving Radical Equations

For Exercises 15–47, solve the equations. Be sure to check all of the potential answers. **(See Examples 1–4.)**

15. $\sqrt{t}=6$ 36

16. $\sqrt{p}=5$ 25

17. $\sqrt{x+1}=4$ 15

18. $\sqrt{x-3}=7$ 52

19. $\sqrt{y-4}=-5$ No solution (the value 29 does not check)

20. $\sqrt{p+6}=-1$ No solution (the value −5 does not check)

21. $\sqrt{5-t}=0$ 5

22. $\sqrt{13+m}=0$ −13

23. $\sqrt{2n+10}=3$ $-\frac{1}{2}$

24. $\sqrt{1-q}=15$ −224

25. $\sqrt{6w-8}=-2$ 6

26. $\sqrt{2z-11}=-3$ 32

 27. $\sqrt{5a-4}-2=4$ 8

28. $\sqrt{3b+4}-3=2$ 7

29. $\sqrt{2x-3}+7=3$ No solution (the value $\frac{19}{2}$ does not check)

30. $\sqrt{8y+1}+5=1$ No solution (the value $\frac{15}{8}$ does not check)

31. $5\sqrt{c}=\sqrt{10c+15}$ 1

32. $4\sqrt{x}=\sqrt{10x+6}$ 1

33. $\sqrt{x^2-x}=\sqrt{12}$ 4, −3

34. $\sqrt{x^2+5x}=\sqrt{150}$ −15, 10

35. $\sqrt{9y^2-8y+1}=3y+1$ 0

36. $\sqrt{4x^2+2x+20}=2x$ No solution (the value −10 does not check)

37. $\sqrt{x^2+4x+16}=x$ No solution (the value −4 does not check)

38. $\sqrt{x^2+3x-2}=4$ −6, 3

39. $\sqrt{2k^2-3k-4}=k$ 4, (the value −1 does not check)

40. $\sqrt{6t+7}=t+2$ 3, −1

41. $\sqrt{y+1}=y+1$ 0, −1

 42. $\sqrt{3p+3}+5=p$ 11, (the value 2 does not check)

43. $\sqrt{2m+1}+7=m$ 12, (the value 4 does not check)

44. $\sqrt[3]{3y+7}=\sqrt[3]{2y-1}$ −8

45. $\sqrt[3]{p-5}-\sqrt[3]{2p+1}=0$ −6

46. $\sqrt[3]{2x-8}-\sqrt[3]{-x+1}=0$ 3

47. $\sqrt[3]{a-3}=\sqrt[3]{5a+1}$ −1

Objective 2: Translations Involving Radical Equations

For Exercises 48–53, translate the English sentence to a radical equation and solve the equation. **(See Example 5.)**

48. The square root of the sum of a number and 8 equals 12. Find the number.
$\sqrt{x+8}=12$;   136

49. The square root of the sum of a number and 10 equals 1. Find the number.
$\sqrt{x+10}=1$;   −9

50. The square root of a number is 2 less than the number. Find the number.
$\sqrt{x}=x-2$;   4

51. The square root of twice a number is 4 less than the number. Find the number.
$\sqrt{2x}=x-4$;   8

52. The cube root of the sum of a number and 4 is −5. Find the number.
$\sqrt[3]{x+4}=-5$;   −129

53. The cube root of the sum of a number and 1 is 2. Find the number.
$\sqrt[3]{x+1}=2$;   7

Objective 3: Applications of Radical Equations

54. Ignoring air resistance, the time, t (in seconds), required for an object to fall x feet is given by the equation:

$$t=\frac{\sqrt{x}}{4}$$

a. Find the time required for an object to fall 64 ft.   2 sec

b. Find the distance an object will fall in 4 sec.
256 ft

55. Ignoring air resistance, the velocity, v (in feet per second: ft/sec), of an object in free fall depends on the distance it has fallen, x (in feet), according to the equation:

$$v=8\sqrt{x}$$

a. Find the velocity of an object that has fallen 100 ft.   80 ft/sec

b. Find the distance that an object has fallen if its velocity is 136 ft/sec. **(See Example 6.)**   289 ft

Writing   Translating Expression   Geometry   Scientific Calculator   Video   NS&E

56. The speed of a car, *s* (in miles per hour), before the brakes were applied can be approximated by the length of its skid marks, *x* (in feet), according to the equation:

$$s = 4\sqrt{x}$$

a. Find the speed of a car before the brakes were applied if its skid marks are 324 ft long. 72 mph

b. How long would you expect the skid marks to be if the car had been traveling the speed limit of 60 mph? 225 ft

57. The height of a sunflower plant, *y* (in inches), can be determined by the time, *t* (in weeks), after the seed has germinated according to the equation:

$$y = 8\sqrt{t} \quad 0 \le t \le 40$$

a. Find the height of the plant after 4 weeks. 16 in.

b. In how many weeks will the plant be 40 in. tall? 25 weeks

Expanding Your Skills

For Exercises 58–61, solve the equations. First isolate one of the radical terms. Then square both sides. The resulting equation will still have a radical. Repeat the process by isolating the radical and squaring both sides again.

58. $\sqrt{t + 8} = \sqrt{t} + 2$ 1

59. $\sqrt{5x - 9} = \sqrt{5x} - 3$ $\frac{9}{5}$

60. $\sqrt{z + 1} + \sqrt{2z + 3} = 1$
−1(the value 3 does not check)

61. $\sqrt{2m + 6} = 1 + \sqrt{7 - 2m}$
$\frac{3}{2}$ (the value −1 does not check)

Section 8.7   Rational Exponents

Objectives

1. Definition of $a^{1/n}$
2. Definition of $a^{m/n}$
3. Converting between Rational Exponents and Radical Notation
4. Properties of Rational Exponents
5. Applications of Rational Exponents

1. Definition of $a^{1/n}$

In Sections 5.1–5.3, the properties for simplifying expressions with integer exponents were presented. In this section, the properties are expanded to include expressions with rational exponents. We begin by defining expressions of the form $a^{1/n}$.

> **DEFINITION** $a^{1/n}$
>
> Let *a* be a real number, and let *n* be an integer such that $n > 1$. If $\sqrt[n]{a}$ is a real number, then
>
> $$a^{1/n} = \sqrt[n]{a}$$

Note: $(\sqrt{a})^2 = a$ for $a > 0$ and $(a^{1/2})^2 = a^{2/2} = a$, so $\sqrt{a} = a^{1/2}$.

Writing   Translating Expression   Geometry   Scientific Calculator   Video   NS&E

Example 1 Evaluating Expressions of the Form $a^{1/n}$

Convert the expression to radical notation and simplify, if possible.

a. $9^{1/2}$ **b.** $125^{1/3}$ **c.** $16^{1/4}$ **d.** $-25^{1/2}$ **e.** $(-25)^{1/2}$ **f.** $25^{-1/2}$

Solution:

a. $9^{1/2} = \sqrt{9} = 3$

b. $125^{1/3} = \sqrt[3]{125} = 5$

c. $16^{1/4} = \sqrt[4]{16} = 2$

d. $-25^{1/2}$ is equivalent to $-1 \cdot (25^{1/2})$

  $= -1 \cdot \sqrt{25}$

  $= -5$

e. $(-25)^{1/2}$ is not a real number because $\sqrt{-25}$ is not a real number.

f. $25^{-1/2} = \dfrac{1}{25^{1/2}} = \dfrac{1}{\sqrt{25}} = \dfrac{1}{5}$

Skill Practice

Convert the expression to radical notation and simplify.

1. $36^{1/2}$ 2. $(-27)^{1/3}$
3. $81^{1/4}$ 4. $(-16)^{1/4}$
5. $(16)^{-1/4}$ 6. $-16^{1/4}$

Classroom Examples: pp. 634–635, Exercises 8, 10, 14, and 18

2. Definition of $a^{m/n}$

If $\sqrt[n]{a}$ is a real number, then we can define an expression of the form $a^{m/n}$ in such a way that the multiplication property of exponents holds true. For example:

$$16^{3/4} = \begin{cases} (16^{1/4})^3 = (\sqrt[4]{16})^3 = (2)^3 = 8 \\ (16^3)^{1/4} = \sqrt[4]{16^3} = \sqrt[4]{4096} = 8 \end{cases}$$

TIP: In simplifying the expression $a^{m/n}$ it is usually easier to take the root first. That is, simplify as $(a^{1/n})^m$.

DEFINITION $a^{m/n}$

Let a be a real number, and let m and n be positive integers such that m and n share no common factors and $n > 1$. If $\sqrt[n]{a}$ is a real number, then

$$a^{m/n} = (a^{1/n})^m = (\sqrt[n]{a})^m \quad \text{and} \quad a^{m/n} = (a^m)^{1/n} = \sqrt[n]{a^m}$$

The rational exponent in the expression $a^{m/n}$ is essentially performing two operations. The numerator of the exponent raises the base to the mth-power. The denominator takes the nth-root.

Example 2 Evaluating Expressions of the Form $a^{m/n}$

Convert each expression to radical notation and simplify.

a. $125^{2/3}$ **b.** $100^{-3/2}$ **c.** $(81)^{3/4}$

Solution:

a. $125^{2/3} = (\sqrt[3]{125})^2$ Take the cube root of 125, and square the result.

  $= (5)^2$ Simplify.

  $= 25$

Skill Practice

Convert each expression to radical notation and simplify.

7. $16^{3/4}$ 8. $8^{-2/3}$ 9. $9^{3/2}$

Classroom Examples: p. 36, Exercises 34 and 36

Answers

1. 6   2. -3 3. 3
4. Not a real number.
5. $\dfrac{1}{2}$ 6. -2 7. $(\sqrt[4]{16})^3$; 8
8. $\dfrac{1}{(\sqrt[3]{8})^2}$; $\dfrac{1}{4}$ 9. $(\sqrt{9})^3$; 27

b. $100^{-3/2} = \dfrac{1}{100^{3/2}}$ Take the reciprocal of the base.

$= \dfrac{1}{(\sqrt{100})^3}$ Take the square root of 100, and cube the result.

$= \dfrac{1}{(10)^3}$ Simplify.

$= \dfrac{1}{1000}$

c. $(81)^{3/4} = (\sqrt[4]{81})^3$ Take the fourth root of 81, and cube the result.

$= (3)^3$ Simplify.

$= 27$

3. Converting between Rational Exponents and Radical Notation

Example 3 Using Radical Notation and Rational Exponents

Convert the expressions to radical notation. Assume the variables represent positive real numbers.

a. $x^{3/5}$ **b.** $(2a^2)^{1/3}$ **c.** $5y^{1/4}$ **d.** $p^{-1/2}$

Solution:

a. $x^{3/5} = \sqrt[5]{x^3}$ or $(\sqrt[5]{x})^3$

b. $(2a^2)^{1/3} = \sqrt[3]{2a^2}$

c. $5y^{1/4} = 5\sqrt[4]{y}$ The exponent $\frac{1}{4}$ applies only to y.

d. $p^{-1/2} = \dfrac{1}{\sqrt{p}}$

Example 4 Using Radical Notation and Rational Exponents

Convert each expression to an equivalent expression using rational exponents. Assume that the variables represent positive real numbers.

a. $\sqrt[4]{c^3}$ **b.** $\sqrt{11p}$ **c.** $11\sqrt{p}$

Solution:

a. $\sqrt[4]{c^3} = c^{3/4}$ **b.** $\sqrt{11p} = (11p)^{1/2}$ **c.** $11\sqrt{p} = 11p^{1/2}$

4. Properties of Rational Exponents

In Sections 5.1–5.3, several properties and definitions were introduced to simplify expressions with integer exponents. These properties also apply to rational exponents.

PROPERTY Operations with Exponents

Let a and b be real numbers. Let m and n be rational numbers such that a^m, a^n, and b^n are real numbers. Then,

Description	Property	Example
1. Multiplying like bases	$a^m a^n = a^{m+n}$	$x^{1/3} \cdot x^{4/3} = x^{5/3}$
2. Dividing like bases	$\dfrac{a^m}{a^n} = a^{m-n}$	$\dfrac{x^{3/5}}{x^{1/5}} = x^{2/5}$
3. The power rule	$(a^m)^n = a^{mn}$	$(2^{1/3})^{1/2} = 2^{1/6}$
4. Power of a product	$(ab)^m = a^m b^m$	$(xy)^{1/2} = x^{1/2} y^{1/2}$
5. Power of a quotient	$\left(\dfrac{a}{b}\right)^m = \dfrac{a^m}{b^m} \quad (b \neq 0)$	$\left(\dfrac{4}{25}\right)^{1/2} = \dfrac{4^{1/2}}{25^{1/2}} = \dfrac{2}{5}$

DEFINITION Negative and Zero Exponents

Description	Definition	Example
1. Negative exponents	$a^{-m} = \left(\dfrac{1}{a}\right)^m = \dfrac{1}{a^m} \quad (a \neq 0)$	$(8)^{-1/3} = \left(\dfrac{1}{8}\right)^{1/3} = \dfrac{1}{2}$
2. Zero exponent	$a^0 = 1 \quad (a \neq 0)$	$5^0 = 1$

Example 5 Simplifying Expressions with Rational Exponents

Use the properties of exponents to simplify the expressions. Write the final answers with positive exponents only. Assume the variables represent positive real numbers.

a. $x^{2/3} x^{1/3}$ **b.** $\dfrac{y^{1/10}}{y^{4/5}}$ **c.** $(z^4)^{1/2}$ **d.** $(s^4 t^8)^{1/4}$

Solution:

a. $x^{2/3} x^{1/3} = x^{(2/3)+(1/3)}$ Add exponents.

$\qquad = x^{3/3}$ Simplify.

$\qquad = x$

b. $\dfrac{y^{1/10}}{y^{4/5}} = y^{(1/10)-(4/5)}$ Subtract exponents.

$\qquad = y^{(1/10)-(8/10)}$ The common denominator is 10.

$\qquad = y^{-7/10} = \dfrac{1}{y^{7/10}}$ Simplify and write with a positive exponent.

c. $(z^4)^{1/2} = z^{(4)\cdot(1/2)}$ Multiply exponents.

$\qquad = z^2$ Simplify.

d. $(s^4 t^8)^{1/4} = s^{4/4} t^{8/4}$ Multiply exponents.

$\qquad = st^2$

Skill Practice

Use the properties of exponents to simplify the expressions. Write the answers with positive exponents only. Assume the variables represent positive real numbers.

17. $a^{3/4} \cdot a^{5/4}$ **18.** $\dfrac{t^{2/3}}{t^2}$

19. $(w^{1/3})^{-12}$ **20.** $(y^9 z^{15})^{1/3}$

Classroom Examples: pp. 635–636, Exercises 62, 70, and 74

Answers

17. a^2 **18.** $\dfrac{1}{t^{4/3}}$

19. $\dfrac{1}{w^4}$ **20.** $y^3 z^5$

5. Applications of Rational Exponents

Example 6 Using Rational Exponents in an Application

Suppose P dollars in principal is invested in an account that earns interest annually. If after t years the investment grows to A dollars, then the annual rate of return, r, on the investment is given by

$$r = \left(\frac{A}{P}\right)^{1/t} - 1$$

Find the annual rate of return on \$8000 that grew to \$11,220.41 after 5 years (round to the nearest tenth of a percent).

Solution:

$$r = \left(\frac{A}{P}\right)^{1/t} - 1 \qquad \text{where } A = \$11{,}220.41, \ P = \$8000, \text{ and } t = 5. \text{ Hence,}$$

$$r = \left(\frac{11220.41}{8000}\right)^{1/5} - 1$$

$$= (1.40255125)^{1/5} - 1$$

$$\approx 1.070 - 1$$

$$\approx 0.070 \text{ or } 7.0\%$$

There is a 7.0% annual rate of return.

Answer

21. 2 in.

For the exercises in this set, assume that the variables represent positive real numbers unless otherwise stated.

Review Exercises

1. Given $\sqrt[3]{125}$

    a. Identify the index. 3   **b.** Identify the radicand. 125

2. Given $\sqrt{12}$

    a. Identify the index. 2   **b.** Identify the radicand. 12

For Exercises 3–6, simplify the radicals.

3. $(\sqrt[4]{81})^3$ 27   **4.** $(\sqrt[4]{16})^3$ 8   **5.** $\sqrt[3]{(a+1)^3}$ $a+1$ **6.** $\sqrt[5]{(x+y)^5}$ $x+y$

Objective 1: Definition of $a^{1/n}$

For Exercises 7–18, simplify the expression. **(See Example 1.)**

7. $81^{1/2}$ 9   **8.** $25^{1/2}$ 5   **9.** $125^{1/3}$ 5   **10.** $8^{1/3}$ 2

11. $81^{1/4}$ $\quad 3$ **12.** $16^{1/4}$ $\quad 2$ **13.** $(-8)^{1/3}$ $\quad -2$ **14.** $(-9)^{1/2}$ $\quad$ Not a real number

15. $-8^{1/3}$ $\quad -2$ **16.** $-9^{1/2}$ $\quad -3$ **17.** $36^{-1/2}$ $\quad \dfrac{1}{6}$ **18.** $16^{-1/2}$ $\quad \dfrac{1}{4}$

For Exercises 19–30, write the expressions in radical notation.

19. $x^{1/3}$ $\quad \sqrt[3]{x}$ **20.** $y^{1/4}$ $\quad \sqrt[4]{y}$ **21.** $(4a)^{1/2}$ $\quad \sqrt{4a}$ or $2\sqrt{a}$ **22.** $(36x)^{1/2}$ $\quad \sqrt{36x}$ or $6\sqrt{x}$

23. $(yz)^{1/5}$ $\quad \sqrt[5]{yz}$ **24.** $(cd)^{1/4}$ $\quad \sqrt[4]{cd}$ **25.** $(u^2)^{1/3}$ $\quad \sqrt[3]{u^2}$ **26.** $(v^3)^{1/4}$ $\quad \sqrt[4]{v^3}$

27. $5q^{1/2}$ $\quad 5\sqrt{q}$ **28.** $6p^{1/2}$ $\quad 6\sqrt{p}$ **29.** $\left(\dfrac{x}{9}\right)^{1/2}$ $\quad \sqrt{\dfrac{x}{9}}$ or $\dfrac{\sqrt{x}}{3}$ **30.** $\left(\dfrac{y}{8}\right)^{1/3}$ $\quad \sqrt[3]{\dfrac{y}{8}}$ or $\dfrac{\sqrt[3]{y}}{2}$

Objective 2: Definition of $a^{m/n}$

31. Explain how to interpret the expression $a^{m/n}$ as a radical. $\quad a^{m/n} = \sqrt[n]{a^m}$ or $(\sqrt[n]{a})^m$, provided the roots exist.

32. Explain why $(\sqrt[3]{8})^4$ is easier to evaluate than $\sqrt[3]{8^4}$. $\quad$ It is easier to evaluate a cube root of a smaller number.

For Exercises 33–40, convert the expressions to radical form and simplify. **(See Example 2.)**

33. $16^{3/4}$ $\quad 8$ **34.** $32^{2/5}$ $\quad 4$ **35.** $27^{-2/3}$ $\quad \dfrac{1}{9}$ **36.** $4^{-5/2}$ $\quad \dfrac{1}{32}$

37. $(-8)^{5/3}$ $\quad -32$ **38.** $(-27)^{2/3}$ $\quad 9$ **39.** $\left(\dfrac{1}{4}\right)^{-1/2}$ $\quad 2$ **40.** $\left(\dfrac{1}{9}\right)^{3/2}$ $\quad \dfrac{1}{27}$

Objective 3: Converting between Rational Exponents and Radical Notation

For Exercises 41–48, convert each expression to radical notation. **(See Example 3.)**

41. $y^{9/2}$ $\quad (\sqrt{y})^9$ **42.** $b^{4/9}$ $\quad \sqrt[9]{b^4}$ **43.** $(c^5d)^{1/3}$ $\quad \sqrt[3]{c^5d}$ **44.** $(a^2b)^{1/8}$ $\quad \sqrt[8]{a^2b}$

45. $(qr)^{-1/5}$ $\quad \dfrac{1}{\sqrt[5]{qr}}$ **46.** $(3x)^{-1/4}$ $\quad \dfrac{1}{\sqrt[4]{3x}}$ **47.** $6y^{2/3}$ $\quad 6\sqrt[3]{y^2}$ **48.** $2q^{5/6}$ $\quad 2\sqrt[6]{q^5}$

For Exercises 49–60, write the expressions using rational exponents rather than radical notation. **(See Example 4.)**

49. $\sqrt[3]{x}$ $\quad x^{1/3}$ **50.** $\sqrt[4]{a}$ $\quad a^{1/4}$ **51.** $\sqrt[3]{xy}$ $\quad (xy)^{1/3}$

52. $\sqrt[5]{ab}$ $\quad (ab)^{1/5}$ **53.** $5\sqrt{x}$ $\quad 5x^{1/2}$ **54.** $7\sqrt[3]{z}$ $\quad 7z^{1/3}$

55. $\sqrt[3]{y^2}$ $\quad y^{2/3}$ **56.** $\sqrt[5]{b^2}$ $\quad b^{2/5}$ **57.** $\sqrt[4]{m^3n}$ $\quad (m^3n)^{1/4}$

58. $\sqrt[5]{u^3v^4}$ $\quad (u^3v^4)^{1/5}$ **59.** $4\sqrt[3]{k^3}$ $\quad 4k^{3/3}$ or $4k$ **60.** $6\sqrt[4]{t^4}$ $\quad 6t^{4/4}$ or $6t$

Objective 4: Properties of Rational Exponents

For Exercises 61–80, simplify the expressions using the properties of rational exponents. Write the final answers with positive exponents only. **(See Example 5.)**

61. $x^{1/4}x^{3/4}$ $\quad x$ **62.** $2^{3/5}2^{2/5}$ $\quad 2$ **63.** $(y^{1/5})^{10}$ $\quad y^2$ **64.** $(x^{1/2})^8$ $\quad x^4$

65. $6^{-1/5}6^{6/5}$ $\quad 6$ **66.** $a^{-1/3}a^{2/3}$ $\quad a^{1/3}$ **67.** $(a^{1/3}a^{1/4})^{12}$ $\quad a^7$ **68.** $(x^{2/3}x^{1/2})^6$ $\quad x^7$

69. $\dfrac{y^{5/3}}{y^{1/3}}$ $\quad y^{4/3}$ **70.** $\dfrac{z^2}{z^{1/2}}$ $\quad z^{3/2}$ **71.** $\dfrac{2^{4/3}}{2^{1/3}}$ $\quad 2$ **72.** $\dfrac{5^{6/5}}{5^{1/5}}$ $\quad 5$

73. $(5a^2c^{-1/2}d^{1/2})^2$ $\quad \dfrac{25a^4d}{c}$ **74.** $(2x^{-1/3}y^2z^{5/3})^3$ $\quad \dfrac{8y^6z^5}{x}$ **75.** $\left(\dfrac{x^{-2/3}}{y^{-3/4}}\right)^{12}$ $\quad \dfrac{y^9}{x^8}$ **76.** $\left(\dfrac{m^{-1/4}}{n^{-1/2}}\right)^{-4}$ $\quad \dfrac{m}{n^2}$

77. $\left(\dfrac{16w^{-2}z}{2wz^{-8}}\right)^{1/3}$ $\quad \dfrac{2z^3}{w}$ **78.** $\left(\dfrac{50p^{-1}q}{2pq^{-3}}\right)^{1/2}$ $\quad \dfrac{5q^2}{p}$ **79.** $(25x^2y^4z^3)^{1/2}$ $\quad 5xy^2z^{3/2}$ **80.** $(8a^6b^3c^2)^{2/3}$ $\quad 4a^4b^2c^{4/3}$

Objective 5: Applications of Rational Exponents

81. 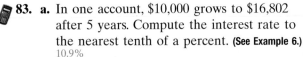 **a.** If the area, A, of a square is known, then the length of its sides, s, can be computed by the formula: $s = A^{1/2}$. Compute the length of the sides of a square having an area of 100 in.^{2}
   10 in.
    b. Compute the length of the sides of a square having an area of 72 in.^{2} Round your answer to the nearest 0.01 in.   8.49 in.

82.  The radius, r, of a sphere of volume, V, is given by
$$r = \left(\dfrac{3V}{4\pi}\right)^{1/3}$$
Find the radius of a spherical ball having a volume of 55 in.^{3} Round your answer to the nearest 0.01 in.   2.36 in.

For Exercises 83–84, use the following information.

If P dollars in principal grows to A dollars after t years with annual interest, then the rate of return is given by

$$r = \left(\dfrac{A}{P}\right)^{1/t} - 1$$

83. **a.** In one account, \$10,000 grows to \$16,802 after 5 years. Compute the interest rate to the nearest tenth of a percent. **(See Example 6.)**
   10.9%
    b. In another account \$10,000 grows to \$18,000 after 7 years. Compute the interest rate to the nearest tenth of a percent.   8.8%

    c. Which account produced a higher average yearly return?   The account in part (a)

84. **a.** In one account, \$5000 grows to \$23,304.79 in 20 years. Compute the interest rate to the nearest whole percent.   8%

    b. In another account, \$6000 grows to \$34,460.95 in 30 years. Compute the interest rate to the nearest whole percent.   6%

    c. Which account produced a higher average yearly return?   The account in part (a)

Expanding Your Skills

85. Is $(a + b)^{1/2}$ the same as $a^{1/2} + b^{1/2}$? Explain why or why not by giving an example.
   No, for example, $(36 + 64)^{1/2} \neq 36^{1/2} + 64^{1/2}$

For Exercises 86–91, simplify the expressions. Write the final answer with positive exponents only.

86. $\left(\dfrac{1}{8}\right)^{2/3} + \left(\dfrac{1}{4}\right)^{1/2}$ $\quad \dfrac{3}{4}$ **87.** $\left(\dfrac{1}{8}\right)^{-2/3} + \left(\dfrac{1}{4}\right)^{-1/2}$ $\quad 6$ **88.** $\left(\dfrac{1}{16}\right)^{-1/4} - \left(\dfrac{1}{49}\right)^{-1/2}$ $\quad -5$

89. $\left(\dfrac{1}{16}\right)^{1/4} - \left(\dfrac{1}{49}\right)^{1/2}$ $\quad \dfrac{5}{14}$ **90.** $\left(\dfrac{x^2y^{-1/3}z^{2/3}}{x^{2/3}y^{1/4}z}\right)^{12}$ $\quad \dfrac{x^{16}}{y^7z^4}$ **91.** $\left(\dfrac{a^2b^{1/2}c^{-2}}{a^{-3/4}b^0c^{1/8}}\right)^8$ $\quad \dfrac{a^{22}b^4}{c^{17}}$

Group Activity

Approximating Square Roots

Materials: A calculator

Estimated Time: 15 minutes

Group Size: 2

Suppose that n represents a positive real number that is *not* a perfect square. To approximate $\sqrt{n}$, we will make repeated use of the formula:

$$\sqrt{n} \approx \frac{1}{2}\left(x + \frac{n}{x}\right) \quad \text{where } x \text{ is an approximation of the square root of } n.$$

We will outline the steps to use this formula and demonstrate by approximating $\sqrt{8}$.

Step 1: Begin by letting x be the nonzero whole number that is closest to the square root of n.

Step 1: To approximate $\sqrt{8}$ we begin with $x = 3$, because 3 is equal to $\sqrt{9}$ which is close to $\sqrt{8}$.

Step 2: Substitute the starting value of x and the value of n into the formula.

Step 2: $\sqrt{n} \approx \frac{1}{2}\left(x + \frac{n}{x}\right)$

$$\sqrt{8} \approx \frac{1}{2}\left(3 + \frac{8}{3}\right)$$

$$\approx 2.83333333333$$

Step 3: Replace x by the answer obtained in step 2. Then apply the formula again, using the new value of x.

Step 3: New value of $x = 2.8\overline{3}$

$$\sqrt{8} \approx \frac{1}{2}\left(2.8\overline{3} + \frac{8}{2.8\overline{3}}\right)$$

$$\approx 2.828431373$$

Step 4: Repeatedly apply step 3, each time using the new value of x from the previous step. You can repeat this process until two consecutive answers differ by less than the desired level of accuracy you want.

Instructor Note: You may want to discuss the use of the *Ans* variable in a calculator.

TIP: You can check to determine if your answer is reasonable by squaring the result.

$$(2.828431373)^2 \approx 8.000024029$$

1. Use the process outlined to approximate $\sqrt{28}$ accurate to 0.0001.   5.2915

2. Use the process outlined to approximate $\sqrt{104}$ accurate to 0.00001.   10.19804

Chapter 8 Summary

Section 8.1 Introduction to Roots and Radicals

Key Concepts

b is a **square root** of a if $b^2 = a$.

The expression $\sqrt{a}$ represents the **principal square root** of a.

b is an nth-root of a if $b^n = a$.

1. If n is a positive *even* integer and $a > 0$, then $\sqrt[n]{a}$ is the principal (positive) nth-root of a.
2. If $n > 1$ is a positive *odd* integer, then $\sqrt[n]{a}$ is the nth-root of a.
3. If $n > 1$ is any positive integer, then $\sqrt[n]{0} = 0$.

$\sqrt[n]{a^n} = |a|$ if n is even.

$\sqrt[n]{a^n} = a$ if n is odd.

$\sqrt[n]{a}$ is not a real number if a is *negative* and n is even.

Pythagorean Theorem

$a^2 + b^2 = c^2$

Examples

Example 1

The square roots of 16 are 4 and -4 because $(4)^2 = 16$ and $(-4)^2 = 16$.

$\sqrt{16} = 4$ \qquad Because $4^2 = 16$

$\sqrt[4]{16} = 2$ \qquad Because $2^4 = 16$

$\sqrt[3]{125} = 5$ \qquad Because $5^3 = 125$

Example 2

$\sqrt{y^2} = |y| \qquad \sqrt[3]{y^3} = y \qquad \sqrt[4]{y^4} = |y|$

Example 3

$\sqrt[4]{-16}$ is not a real number.

Example 4

Find the length of the unknown side.

$a^2 + b^2 = c^2$

$(8)^2 + b^2 = (17)^2$

$64 + b^2 = 289$

$b^2 = 289 - 64$

$b^2 = 225$

$b = \sqrt{225}$ \qquad Because b denotes a length, b must be the positive square root of 225.

$b = 15$

The third side is 15 cm.

Section 8.2   Simplifying Radicals

Key Concepts

Multiplication Property of Radicals

If $\sqrt[n]{a}$ and $\sqrt[n]{b}$ are both real, then

$$\sqrt[n]{ab} = \sqrt[n]{a} \cdot \sqrt[n]{b}$$

Simplifying Radicals

Consider a radical expression whose radicand is written as a product of prime factors. Then the radical is in simplified form if each of the following criteria are met:

1. The radicand has no factor raised to a power greater than or equal to the index.
2. There are no radicals in the denominator of a fraction.
3. The radicand does not contain a fraction.

Examples

Example 1

$$\sqrt{3} \cdot \sqrt{5} = \sqrt{3 \cdot 5} = \sqrt{15}$$

Example 2

$$\sqrt{\frac{b^7}{b^3}} = \sqrt{b^4} = b^2$$

Example 3

$$\sqrt[3]{16x^5y^7} = \sqrt[3]{2^4x^5y^7}$$
$$= \sqrt[3]{2^3x^3y^6 \cdot 2x^2y}$$
$$= \sqrt[3]{2^3x^3y^6} \cdot \sqrt[3]{2x^2y}$$
$$= 2xy^2\sqrt[3]{2x^2y}$$

Section 8.3   Addition and Subtraction of Radicals

Key Concepts

Two radical terms are *like* radicals if they have the same index and the same radicand.

Use the distributive property to add or subtract *like* radicals.

Examples

Example 1

Like radicals. $\sqrt[3]{5z}, \quad 6\sqrt[3]{5z}$

Example 2

$$3\sqrt{7} - 10\sqrt{7} + \sqrt{7}$$
$$= (3 - 10 + 1)\sqrt{7}$$
$$= -6\sqrt{7}$$

Section 8.4   Multiplication of Radicals

Key Concepts

Multiplication Property of Radicals

$\sqrt[n]{a} \cdot \sqrt[n]{b} = \sqrt[n]{ab}$ provided $\sqrt[n]{a}$ and $\sqrt[n]{b}$ are both real.

Special Case Products

$(a + b)(a - b) = a^2 - b^2$

$(a + b)^2 = a^2 + 2ab + b^2$

$(a - b)^2 = a^2 - 2ab + b^2$

Examples

Example 1

$$(6\sqrt{5})(4\sqrt{3}) = 6 \cdot 4\sqrt{5 \cdot 3}$$
$$= 24\sqrt{15}$$

Example 2

$$3\sqrt{2}(\sqrt{2} + 5\sqrt{7} - \sqrt{6}) = 3\sqrt{4} + 15\sqrt{14} - 3\sqrt{12}$$
$$= 3\sqrt{2^2} + 15\sqrt{14} - 3\sqrt{2^2 \cdot 3}$$
$$= 3 \cdot 2 + 15\sqrt{14} - 3 \cdot 2\sqrt{3}$$
$$= 6 + 15\sqrt{14} - 6\sqrt{3}$$

Example 3

$$(4\sqrt{x} + \sqrt{2})(4\sqrt{x} - \sqrt{2}) = (4\sqrt{x})^2 - (\sqrt{2})^2$$
$$= 16x - 2$$

Example 4

$$(\sqrt{x} - \sqrt{5y})^2 = (\sqrt{x})^2 - 2(\sqrt{x})(\sqrt{5y}) + (\sqrt{5y})^2$$
$$= x - 2\sqrt{5xy} + 5y$$

Section 8.5   Division of Radicals and Rationalization

Key Concepts

Division Property of Radicals

If $\sqrt[n]{a}$ and $\sqrt[n]{b}$ are both real, then

$$\sqrt[n]{\frac{a}{b}} = \frac{\sqrt[n]{a}}{\sqrt[n]{b}} \quad b \neq 0$$

Rationalizing the Denominator with One Term

Multiply the numerator and denominator by an appropriate expression to create an nth-root of an nth-power in the denominator.

Rationalizing a Two-Term Denominator Involving Square Roots

Multiply the numerator and denominator by the conjugate of the denominator.

Examples

Example 1

$$\frac{\sqrt{x^{11}}}{\sqrt{x^3}} = \sqrt{\frac{x^{11}}{x^3}} = \sqrt{x^8} = x^4$$

Example 2

$$\frac{10}{\sqrt{5}} = \frac{10}{\sqrt{5}} \cdot \frac{\sqrt{5}}{\sqrt{5}} = \frac{10\sqrt{5}}{\sqrt{5^2}} = \frac{10\sqrt{5}}{5} = 2\sqrt{5}$$

Example 3

$$\frac{\sqrt{2}}{\sqrt{x} - \sqrt{3}} = \frac{\sqrt{2}}{(\sqrt{x} - \sqrt{3})} \cdot \frac{(\sqrt{x} + \sqrt{3})}{(\sqrt{x} + \sqrt{3})}$$
$$= \frac{\sqrt{2x} + \sqrt{6}}{x - 3}$$

Section 8.6   Radical Equations

Key Concepts

An equation with one or more radicals containing a variable is a **radical equation**.

Steps for Solving a Radical Equation

1. Isolate the radical. If an equation has more than one radical, choose one of the radicals to isolate.
2. Raise each side of the equation to a power equal to the index of the radical.
3. Solve the resulting equation.
4. Check the potential solutions in the original equation.

Note: Raising both sides of an equation to an even power may result in extraneous solutions.

Examples

Example 1

Solve. $\sqrt{2x - 4} + 3 = 7$

Step 1: $\sqrt{2x - 4} = 4$

Step 2: $(\sqrt{2x - 4})^2 = (4)^2$

Step 3: $2x - 4 = 16$

$$2x = 20$$

$$x = 10$$

Step 4:

Check:

$$\sqrt{2x - 4} + 3 = 7$$

$$\sqrt{2(10) - 4} + 3 \stackrel{?}{=} 7$$

$$\sqrt{20 - 4} + 3 \stackrel{?}{=} 7$$

$$\sqrt{16} + 3 \stackrel{?}{=} 7$$

$$4 + 3 \stackrel{?}{=} 7 \; ✔ \qquad \text{The solution checks.}$$

The solution is 10.

Section 8.7   Rational Exponents

Key Concepts

If $\sqrt[n]{a}$ is a real number, then

$$a^{1/n} = \sqrt[n]{a}$$

$$a^{m/n} = (\sqrt[n]{a})^m = \sqrt[n]{a^m}$$

Examples

Example 1

$$121^{1/2} = \sqrt{121} = 11$$

Example 2

$$27^{2/3} = (\sqrt[3]{27})^2 = (3)^2 = 9$$

Example 3

$$8^{-1/3} = \frac{1}{8^{1/3}} = \frac{1}{\sqrt[3]{8}} = \frac{1}{2}$$

Chapter 8 Review Exercises

Section 8.1

For Exercises 1–4, state the principal square root and the negative square root.

1. 196
Principal square root: 14;
negative square root: −14

2. 1.44
Principal square root: 1.2;
negative square root: −1.2

3. 0.64
Principal square root: 0.8;
negative square root: −0.8

4. 225
Principal square root: 15;
negative square root: −15

5. Explain why $\sqrt{-64}$ is *not* a real number.
There is no real number b such that $b^2 = -64$.

6. Explain why $\sqrt[3]{-64}$ *is* a real number.
$\sqrt[3]{-64} = -4$ because $(-4)^3 = -64$.

For Exercises 7–18, simplify the expressions, if possible.

7. $-\sqrt{144}$ −12   **8.** $-\sqrt{25}$ −5   **9.** $\sqrt{-144}$
Not a real number

10. $\sqrt{-25}$ **11.** $\sqrt{y^2}$ $|y|$ **12.** $\sqrt[3]{y^3}$ y
Not a real number

13. $\sqrt[4]{y^4}$ $|y|$ **14.** $-\sqrt[3]{125}$ −5   **15.** $-\sqrt[4]{625}$ −5

16. $\sqrt[3]{p^{12}}$ p^4 **17.** $\sqrt[4]{\dfrac{81}{t^8}}$ $\dfrac{3}{t^2}$ **18.** $\sqrt[3]{\dfrac{-27}{w^3}}$ $-\dfrac{3}{w}$

19. The radius, r, of a circle can be found from the area of the circle according to the formula:

$$r = \sqrt{\dfrac{A}{\pi}}$$

a. What is the radius of a circular garden whose area is 160 m²? Round to the nearest tenth of a meter.   7.1 m

b. What is the radius of a circular fountain whose area is 1600 ft²? Round to the nearest tenth of a foot.   22.6 ft

20. Suppose a ball is thrown with an initial velocity of 76 ft/sec at an angle of 30° (see figure). Then the horizontal position of the ball, x (measured in feet), depends on the number of seconds, t, after the ball is thrown according to the equation:

$$x = 38\sqrt{3}t$$

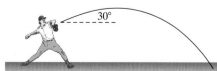

a. What is the horizontal position of the ball after 1 sec? Round your answer to the nearest tenth of a foot.   65.8 ft

b. What is the horizontal position of the ball after 2 sec? Round your answer to the nearest tenth of a foot.   131.6 ft

For Exercises 21–22, translate the English phrases into algebraic expressions.

21. The square of b plus the principal square root of 5. $b^2 + \sqrt{5}$

22. The difference of the cube root of y and the fourth root of x. $\sqrt[3]{y} - \sqrt[4]{x}$

For Exercises 23–24, translate the algebraic expressions into English phrases. (Answers may vary.)

23. $\dfrac{2}{\sqrt{p}}$ The quotient of 2 and the principal square root of p **24.** $8\sqrt{q}$ The product of 8 and the principal square root of q

25. A hedge extends 5 ft from the wall of a house. A 13-ft ladder is placed at the edge of the hedge. How far up the house is the tip of the ladder?
12 ft

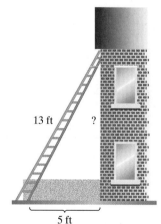

13 ft   ?

5 ft

26. Nashville, TN, is north of Birmingham, Alabama, a distance of 182 miles. Augusta, Georgia, is east of Birmingham, a distance of 277 miles. How far is it from Augusta to Nashville? Round the answer to the nearest mile.
331 mi

Nashville

182 miles

Birmingham   277 miles   Augusta

Section 8.2

For Exercises 27–32, use the multiplication property of radicals to simplify. Assume the variables represent positive real numbers.

27. $\sqrt{x^{17}}$
$x^8\sqrt{x}$

28. $\sqrt[3]{40}$
$2\sqrt[3]{5}$

29. $\sqrt{28}$
$2\sqrt{7}$

30. $5\sqrt{18x^3}$
$15x\sqrt{2x}$

31. $\sqrt[3]{27y^{10}}$
$3y^3\sqrt[3]{y}$

32. $2\sqrt{27y^{10}}$
$6y^5\sqrt{3}$

For Exercises 33–42, use order of operations to simplify. Assume the variables represent positive real numbers.

33. $\sqrt{\dfrac{c^5}{c^3}}$ c

34. $\sqrt{\dfrac{t^9}{t^3}}$ t^3

35. $\sqrt{\dfrac{200y^5}{2y}}$ $10y^2$

36. $\sqrt{\dfrac{18x^3}{2x}}$ $3x$

37. $\sqrt[3]{\dfrac{48x^4}{6x}}$ $2x$

38. $\sqrt[3]{\dfrac{128a^{17}}{2a^2}}$ $4a^5$

39. $\dfrac{5\sqrt{12}}{2}$ $5\sqrt{3}$

40. $\dfrac{2\sqrt{45}}{6}$ $\sqrt{5}$

41. $\dfrac{12-\sqrt{49}}{5}$ 1

42. $\dfrac{20+\sqrt{100}}{5}$ 6

Section 8.3

For Exercises 43–48, add or subtract as indicated. Assume the variables represent positive real numbers.

43. $8\sqrt{6}-\sqrt{6}$ $7\sqrt{6}$

44. $1.6\sqrt{y}-1.4\sqrt{y}+0.6\sqrt{y}$ $0.8\sqrt{y}$

45. $x\sqrt{20}-2\sqrt{45x^2}$ $-4x\sqrt{5}$

46. $y\sqrt{64y}+3\sqrt{y^3}$ $11y\sqrt{y}$

47. $3\sqrt{75}-4\sqrt{28}+\sqrt{7}$ $15\sqrt{3}-7\sqrt{7}$

48. $2\sqrt{50}-4\sqrt{20}-6\sqrt{2}$ $4\sqrt{2}-8\sqrt{5}$

↔ For Exercises 49–50, translate the English phrases into algebraic expressions and simplify.

49. The sum of the principal square root of the fourth power of x and the square of $5x$.
$\sqrt{x^4}+(5x)^2$; $26x^2$

50. The difference of the principal square root of 128 and the square root of 2. $\sqrt{128}-\sqrt{2}$; $7\sqrt{2}$

51. Find the exact perimeter of the triangle. $12\sqrt{2}$ ft

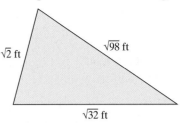

52. Find the exact perimeter of a square whose sides are $3\sqrt{48}$ m. $48\sqrt{3}$ m

Section 8.4

For Exercises 53–62, multiply the expressions. Assume the variables represent positive real numbers.

53. $\sqrt{5}\cdot\sqrt{125}$ 25

54. $\sqrt{10p}\cdot\sqrt{6}$ $2\sqrt{15p}$

55. $(5\sqrt{6})(7\sqrt{2x})$ $70\sqrt{3x}$

56. $(3\sqrt{y})(-2z\sqrt{11y})$
$-6yz\sqrt{11}$

57. $8\sqrt{m}(\sqrt{m}+3)$
$8m+24\sqrt{m}$

58. $\sqrt{2}(\sqrt{7}+8)$
$\sqrt{14}+8\sqrt{2}$

59. $(5\sqrt{2}+\sqrt{13})(-\sqrt{2}-3\sqrt{13})$ $-49-16\sqrt{26}$

60. $(\sqrt{p}+2\sqrt{q})(4\sqrt{p}-\sqrt{q})$ $4p+7\sqrt{pq}-2q$

61. $(8\sqrt{w}-\sqrt{z})(8\sqrt{w}+\sqrt{z})$ $64w-z$

62. $(2x-\sqrt{y})^2$ $4x^2-4x\sqrt{y}+y$

63. Find the exact volume of the box. $10\sqrt{3}$ m^3

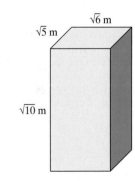

$\sqrt{6}$ m
$\sqrt{5}$ m
$\sqrt{10}$ m

Section 8.5

For Exercises 64–67, use the division property of radicals to write the radicals in simplified form. Assume all variables are positive real numbers.

64. $\dfrac{\sqrt[3]{x^7}}{\sqrt[3]{x^4}}$ x

65. $\dfrac{\sqrt{a^{11}}}{\sqrt{a}}$ a^5

66. $\dfrac{\sqrt{250c}}{\sqrt{10}}$ $5\sqrt{c}$

67. $\dfrac{\sqrt{96y^3}}{\sqrt{6y^2}}$ $4\sqrt{y}$

68. To rationalize the denominator in the expression

$$\dfrac{6}{\sqrt{a}+5}$$

which quantity would you multiply by in the numerator and denominator?   b

a. $\sqrt{a}+5$ **b.** $\sqrt{a}-5$ **c.** $\sqrt{a}$ **d.** -5

69. To rationalize the denominator in the expression

$$\frac{w}{\sqrt{w}-4}$$

which quantity would you multiply by in the numerator and denominator?   b

a. $\sqrt{w}-4$ **b.** $\sqrt{w}+4$

c. $\sqrt{w}$ **d.** 4

For Exercises 70–75, rationalize the denominators. Assume the variables represent positive real numbers.

70. $\dfrac{11}{\sqrt{7}}$ $\frac{11\sqrt{7}}{7}$ **71.** $\sqrt{\dfrac{18}{y}}$ $\frac{3\sqrt{2y}}{y}$ **72.** $\dfrac{\sqrt{24}}{\sqrt{6x^7}}$ $\frac{2\sqrt{x}}{x^4}$

73. $\dfrac{10}{\sqrt{7}-\sqrt{2}}$ **74.** $\dfrac{6}{\sqrt{w}+2}$ **75.** $\dfrac{\sqrt{7}+3}{\sqrt{7}-3}$
$2\sqrt{7}+2\sqrt{2}$ $\frac{6\sqrt{w}-12}{w-4}$ $-8-3\sqrt{7}$

76. The velocity of an object, v (in meters per second: m/sec) depends on the kinetic energy, E (in joules: J), and mass, m (in kilograms: kg), of the object according to the formula:

$$v=\sqrt{\frac{2E}{m}}$$

a. What is the exact velocity of a 3-kg object whose kinetic energy is 100 J? $\frac{10\sqrt{6}}{3}$ m/sec

b. What is the exact velocity of a 5-kg object whose kinetic energy is 162 J? $\frac{18\sqrt{5}}{5}$ m/sec

Section 8.6

For Exercises 77–85, solve the equations. Be sure to check the potential solutions.

77. $\sqrt{p+6}=12$
138

78. $\sqrt{k+1}=-7$
No solution (the value 48 does not check)

79. $\sqrt{3x-17}-10=0$ 39

80. $\sqrt{14n+10}=4\sqrt{n}$ 5

81. $\sqrt{2z+2}=\sqrt{3z-5}$ 7

82. $\sqrt{5y-5}-\sqrt{4y+1}=0$ 6

83. $\sqrt{2m+5}=m+1$ 2 (the value -2 does not check)

84. $\sqrt{3n-8}-n+2=0$ 3, 4

85. $\sqrt[3]{2y+13}=-5$ -69

 86. The length of the sides of a cube is related to the volume of the cube according to the formula: $x=\sqrt[3]{V}$.

a. What is the volume of the cube if the side length is 21 in.?   9261 in.³

b. What is the volume of the cube if the side length is 15 cm?   3375 cm³

Section 8.7

For Exercises 87–92, simplify the expressions.

87. $(-27)^{1/3}$ -3 **88.** $121^{1/2}$ 11   **89.** $-16^{1/4}$ -2

90. $(-16)^{1/4}$
Not a real number

91. $4^{-3/2}$
$\frac{1}{8}$

92. $\left(\dfrac{1}{9}\right)^{-3/2}$
27

For Exercises 93–96, write the expression in radical notation. Assume the variables represent positive real numbers.

93. $z^{1/5}$ $\sqrt[5]{z}$

94. $q^{2/3}$ $\sqrt[3]{q^2}$

95. $(w^3)^{1/4}$ $\sqrt[4]{w^3}$

96. $\left(\dfrac{b}{121}\right)^{1/2}$
$\sqrt{\dfrac{b}{121}}=\dfrac{\sqrt{b}}{11}$

For Exercises 97–100, write the expression using rational exponents rather than radical notation. Assume the variables represent positive real numbers.

97. $\sqrt[5]{a^2}$ $a^{2/5}$

98. $5\sqrt[3]{m^2}$ $5m^{2/3}$

99. $\sqrt[5]{a^2b^4}$ $(a^2b^4)^{1/5}$

100. $\sqrt{6}$ $6^{1/2}$

For Exercises 101–106, simplify using the properties of rational exponents. Write the answer with positive exponents only. Assume the variables represent positive real numbers.

101. $y^{2/3}y^{4/3}$ y^2

102. $a^{1/3}a^{1/2}$ $a^{5/6}$

103. $\dfrac{6^{4/5}}{6^{1/5}}$ $6^{3/5}$

104. $\left(\dfrac{b^4b^0}{b^{1/4}}\right)^4$ b^{15}

105. $(64a^3b^6)^{1/3}$ $4ab^2$

106. $(5^{1/2})^{3/2}$ $5^{3/4}$

 107. The radius, r, of a right circular cylinder can be found if the volume, V, and height, h, are known. The radius is given by

$$r=\left(\frac{V}{\pi h}\right)^{1/2}$$

Find the radius of a right circular cylinder whose volume is 150.8 cm³ and whose height is 12 cm. Round the answer to the nearest tenth of a centimeter.   2.0 cm

Chapter 8   Test

1. State the conditions for a radical expression to be in simplified form.   1. The radicand has no factor raised to a power greater than or equal to the index. 2. There are no radicals in the denominator of a fraction. 3. The radicand does not contain a fraction.

For Exercises 2–7, simplify the radicals, if possible. Assume the variables represent positive real numbers.

2. $\sqrt{242x^2}$ $11x\sqrt{2}$

3. $\sqrt[3]{48y^4}$ $2y\sqrt[3]{6y}$

4. $\sqrt{-64}$ Not a real number

5. $\sqrt{\dfrac{5a^6}{81}}$ $\dfrac{a^3\sqrt{5}}{9}$

6. $\dfrac{9}{\sqrt{6}}$ $\dfrac{3\sqrt{6}}{2}$

7. $\dfrac{2}{\sqrt{5}+6}$ $\dfrac{2\sqrt{5}-12}{-31}$ or $\dfrac{12-2\sqrt{5}}{31}$

 8. Translate the English phrases into algebraic expressions and simplify.

   a. The sum of the principal square root of twenty-five and the cube of five. $\sqrt{25}+5^3$; 130

   b. The difference of the square of four and the principal square root of 16. $4^2-\sqrt{16}$; 12

 9. A baseball player hits the ball at an angle of 30° with an initial velocity of 112 ft/sec. The horizontal position of the ball, x (measured in feet), depends on the number of seconds, t, after the ball is struck according to the equation

$$x = 56\sqrt{3}\,t$$

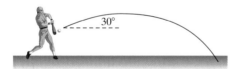

   a. What is the horizontal position of the ball after 1 sec? Round the answer to the nearest foot.   97 ft

   b. What is the horizontal position of the ball after 3.5 sec? Round the answer to the nearest foot.   339 ft

For Exercises 10–19, perform the indicated operations. Assume the variables represent positive real numbers.

10. $6\sqrt{z}-3\sqrt{z}+5\sqrt{z}$ $8\sqrt{z}$

11. $\sqrt{3}(4\sqrt{2}-5\sqrt{3})$ $4\sqrt{6}-15$

12. $\sqrt{50t^2}-t\sqrt{288}$ $-7t\sqrt{2}$

13. $\sqrt{360}+\sqrt{250}-\sqrt{40}$ $9\sqrt{10}$

14. $(3\sqrt{5}-1)^2$ $46-6\sqrt{5}$

15. $(6\sqrt{2}-\sqrt{5})(\sqrt{2}+4\sqrt{5})$ $-8+23\sqrt{10}$

16. $\dfrac{\sqrt{2m^3n}}{\sqrt{72m^5}}$ $\dfrac{\sqrt{n}}{6m}$

17. $(4-3\sqrt{x})(4+3\sqrt{x})$ $16-9x$

18. $\sqrt{\dfrac{2}{11}}$ $\dfrac{\sqrt{22}}{11}$

19. $\dfrac{6}{\sqrt{7}-\sqrt{3}}$ $\dfrac{3\sqrt{7}+3\sqrt{3}}{2}$

 20. A triathlon consists of a swim, followed by a bike ride, followed by a run. The swim begins on a beach at point A. The swimmers must swim 50 yd to a buoy at point B, then 200 yd to a buoy at point C, and then return to point A on the beach. How far is the distance from point C to point A? (Round to the nearest yard.)   206 yd

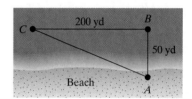

For Exercises 21–23, solve the equations.

21. $\sqrt{2x+7}+6=2$ No solution (the value $\frac{9}{2}$ does not check)

22. $\sqrt{1-7x}=1-x$ 0, −5

23. $\sqrt[3]{x+6}=\sqrt[3]{2x-8}$ 14

24. The height, y (in inches), of a tomato plant can be approximated by the time, t (in weeks), after the seed has germinated according to the equation

$$y = 6\sqrt{t}$$

   a. Use the equation to find the height of the plant after 4 weeks.   12 in.

   b. Use the equation to find the height of the plant after 9 weeks.   18 in.

c. Use the equation to find the time required for the plant to reach a height of 30 in. Verify your answer from the graph. *25 weeks*

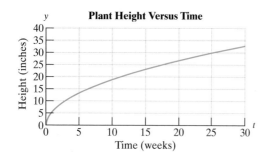

For Exercises 25–26, simplify the expression.

25. $10{,}000^{3/4}$ *1000*

26. $\left(\frac{1}{8}\right)^{-1/3}$ *2*

For Exercises 27–28, write the expressions in radical notation. Assume the variables represent positive real numbers.

27. $x^{3/5}$ $\sqrt[5]{x^3}$ or $(\sqrt[5]{x})^3$

28. $5y^{1/2}$ $5\sqrt{y}$

29. Write the expression using rational exponents: $\sqrt[4]{ab^3}$. (Assume $a \geq 0$ and $b \geq 0$.) $(ab^3)^{1/4}$

For Exercises 30–32, simplify using the properties of rational exponents. Write the final answer with positive exponents only. Assume the variables represent positive real numbers.

30. $p^{1/4} \cdot p^{2/3}$ $p^{11/12}$

31. $\dfrac{5^{4/5}}{5^{1/5}}$ $5^{3/5}$

32. $(9m^2n^4)^{1/2}$ $3mn^2$

Chapters 1–8 Cumulative Review Exercises

1. Simplify. $\dfrac{|-3 - 12 \div 6 + 2|}{\sqrt{5^2 - 4^2}}$ *1*

2. Solve.
$2 - 5(2y + 4) - (-3y - 1) = -(y + 5)$ *−2*

3. Simplify. Write the final answer with positive exponents only.

$$\left(\frac{1}{3}\right)^0 - \left(\frac{1}{4}\right)^{-2}$$ *−15*

4. Perform the indicated operations:
$2(x - 3) - (3x + 4)(3x - 4)$ *$-9x^2 + 2x + 10$*

5. Divide:

$$\frac{14x^3y - 7x^2y^2 + 28xy^2}{7x^2y^2}$$ *$\frac{2x}{y} - 1 + \frac{4}{x}$*

6. Factor completely. $50c^2 + 40c + 8$ *$2(5c + 2)^2$*

7. Solve. $10x^2 = x + 2$ *$-\frac{2}{5}, \frac{1}{2}$*

8. Perform the indicated operations:

$$\frac{5a^2 + 2ab - 3b^2}{10a + 10b} \div \frac{25a^2 - 9b^2}{50a + 30b}$$ *1*

9. Solve for z. $\dfrac{1}{5} + \dfrac{z}{z - 5} = \dfrac{5}{z - 5}$ *No solution (the value 5 does not check)*

10. Simplify:

$$\frac{\dfrac{5}{4} + \dfrac{2}{x}}{\dfrac{4}{x} - \dfrac{4}{x^2}}$$ *$\frac{x(5x + 8)}{16(x - 1)}$*

11. Graph. $3y = 6$

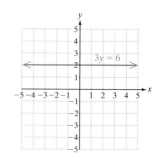

12. The equation $y = 210x + 250$ represents the cost, y (in dollars), of renting office space for x months.

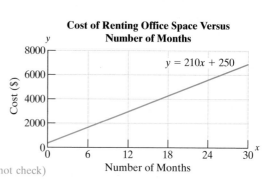

a. Find y when x is 3. Interpret the result in the context of the problem.

$y = 880$; the cost of renting the office space for 3 months is $880.

b. Find x when y is $2770. Interpret the result in the context of the problem.

$x = 12$; the cost of renting office space for 12 months is $2770.

c. What is the slope of the line? Interpret the meaning of the slope in the context of the problem.

$m = 210$; the cost increases at a rate of $210 per month.

d. What is the y-intercept? Interpret the meaning of the y-intercept in the context of the problem.

$(0, 250)$; the down payment of renting the office space is $250.

13. Write an equation of the line passing through the points $(2, -1)$ and $(-3, 4)$. Write the answer in slope-intercept form.   *$y = -x + 1$*

14. Solve the system of equations using the addition method. If the system has no solution or infinitely many solutions, so state:

$$3x - 5y = 23$$
$$2x + 4y = -14 \quad (1, -4)$$

15. Graph the solution to the inequality:
$$-2x - y > 3$$

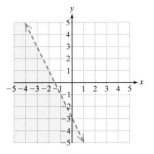

16. How many liters (L) of 20% acid solution must be mixed with a 50% acid solution to obtain 12 L of a 30% acid solution?

8 L of 20% solution should be mixed with 4 L of 50% solution.

17. Simplify. $\sqrt{99}$ *$3\sqrt{11}$*

18. Perform the indicated operation.
$$5x\sqrt{3} + \sqrt{12x^2} \quad 7x\sqrt{3}$$

19. Rationalize the denominator. $\dfrac{\sqrt{x}}{\sqrt{x} - \sqrt{y}}$ *$\dfrac{x + \sqrt{xy}}{x - y}$*

20. Solve. $\sqrt{2y - 1} - 4 = -1$ *5*

Writing   Translating Expression   Geometry   Scientific Calculator   Video   NS&E

More Quadratic Equations

9

CHAPTER OUTLINE

9.1 The Square Root Property 650

9.2 Completing the Square 655

9.3 Quadratic Formula 661

Problem Recognition Exercises: Solving Quadratic Equations 669

9.4 Graphing Quadratic Functions 670

Group Activity: Maximizing Volume 682

Chapter 9

In earlier chapters, we solved quadratic equations by factoring and applying the zero product rule. In Chapter 9, we present additional techniques to solve quadratic equations. These new techniques can be used to solve quadratic equations even if the equations are not factorable.

The clues for the crossword puzzle relate to methods and terms in this chapter. Give it a try!

Across

4. An equation that can be written in the form $ax^2 + bx + c = 0$ is called _____.
6. To apply the quadratic formula, one side of the equation must equal _____.

Down

1. One method of solving a quadratic equation is the _____ root property.
2. The graph of a quadratic function is called a _____.
3. An equation that can be written in the form $ax + b = 0$ is called _____.
5. The first step in solving by completing the square is to make the leading coefficient equal to _____.

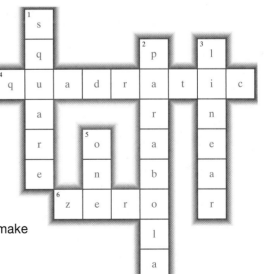

Section 9.1 The Square Root Property

Objectives

1. Review of the Zero Product Rule
2. Solving Quadratic Equations Using the Square Root Property

1. Review of the Zero Product Rule

In Section 6.7, we learned that an equation that can be written in the form $ax^2 + bx + c = 0$, where $a \neq 0$, is a quadratic equation. One method to solve a quadratic equation is to factor the equation and apply the zero product rule. Recall that the zero product rule states that if $a \cdot b = 0$, then $a = 0$ or $b = 0$. This is reviewed in Examples 1–3.

Skill Practice

Solve the quadratic equation by using the zero product rule.

1. $2x^2 + 3x - 20 = 0$

Classroom Example: p. 654, Exercise 6

Example 1 Solving a Quadratic Equation Using the Zero Product Rule

Solve the equation by factoring and applying the zero product rule.

$$2x^2 - 7x - 30 = 0$$

Solution:

$$2x^2 - 7x - 30 = 0 \qquad \text{The equation is in the form } ax^2 + bx + c = 0.$$

$$(2x + 5)(x - 6) = 0 \qquad \text{Factor.}$$

$$2x + 5 = 0 \quad \text{or} \quad x - 6 = 0 \qquad \text{Set each factor equal to zero.}$$

$$2x = -5 \quad \text{or} \quad x = 6 \qquad \text{Solve the resulting equations.}$$

$$x = -\frac{5}{2}$$

The solutions are $-\dfrac{5}{2}$ and 6.

Skill Practice

Solve the quadratic equation by using the zero product rule.

2. $y(y - 1) = 2y + 10$

Classroom Example: p. 654, Exercise 10

Example 2 Solving a Quadratic Equation Using the Zero Product Rule

Solve the equation by factoring and applying the zero product rule.

$$2x(x + 4) = x^2 - 15$$

Solution:

$$2x(x + 4) = x^2 - 15$$

$$2x^2 + 8x = x^2 - 15 \qquad \text{Clear parentheses and combine like terms.}$$

$$x^2 + 8x + 15 = 0 \qquad \text{Set one side of the equation equal to zero. The equation is now in the form } ax^2 + bx + c = 0.$$

$$(x + 5)(x + 3) = 0 \qquad \text{Factor.}$$

Answers

1. $-4, \dfrac{5}{2}$

2. $5, -2$

$$x + 5 = 0 \quad \text{or} \quad x + 3 = 0 \qquad \text{Set each factor equal to zero.}$$

$$x = -5 \quad \text{or} \qquad x = -3 \qquad \text{Solve each equation.}$$

The solutions are -5 and -3.

TIP: The solutions to an equation can be checked in the original equation.

Check: $x = -5$

$$2x(x + 4) = x^2 - 15$$

$$2(-5)(-5 + 4) \overset{?}{=} (-5)^2 - 15$$

$$-10(-1) \overset{?}{=} 25 - 15$$

$$10 \overset{?}{=} 10 \ ✔$$

Check: $x = -3$

$$2x(x + 4) = x^2 - 15$$

$$2(-3)(-3 + 4) \overset{?}{=} (-3)^2 - 15$$

$$-6(1) \overset{?}{=} 9 - 15$$

$$-6 \overset{?}{=} -6 \ ✔$$

Example 3 **Solving a Quadratic Equation Using the Zero Product Rule**

Solve the equation by factoring and applying the zero product rule.

$$x^2 = 25$$

Solution:

$$x^2 = 25$$

$$x^2 - 25 = 0 \qquad \text{Set one side of the equation equal to zero.}$$

$$(x - 5)(x + 5) = 0 \qquad \text{Factor.}$$

$$x - 5 = 0 \quad \text{or} \quad x + 5 = 0 \qquad \text{Set each factor equal to zero.}$$

$$x = 5 \quad \text{or} \qquad x = -5$$

The solutions are 5 and -5.

Skill Practice

Solve the quadratic equation by using the zero product rule.

3. $t^2 = 49$

Classroom Example: p. 654, Exercise 12

2. Solving Quadratic Equations Using the Square Root Property

In Examples 1–3, the quadratic equations were all factorable. In this chapter, we learn techniques to solve *all* quadratic equations, factorable and nonfactorable. The first technique uses the **square root property**.

PROPERTY Square Root Property

For any real number, k, if $x^2 = k$, then $x = \sqrt{k}$ or $x = -\sqrt{k}$.

Note: The solution can also be written as $x = \pm\sqrt{k}$, read as "x equals plus or minus the square root of k."

Answer

3. $7, -7$

Example 4 Solving a Quadratic Equation Using the Square Root Property

Use the square root property to solve the equation.

$$x^2 = 25$$

Solution:

$x^2 = 25$ The equation is in the form $x^2 = k$.

$x = \pm\sqrt{25}$ Apply the square root property.

$x = \pm 5$

The solutions are 5 and -5. Note that this result is the same as in Example 3.

Example 5 Solving a Quadratic Equation Using the Square Root Property

Use the square root property to solve the equation.

$$2x^2 - 10 = 0$$

Solution:

$2x^2 - 10 = 0$ To apply the square root property, the equation must be in the form $x^2 = k$, that is, we must isolate x^2.

$2x^2 = 10$ Add 10 to both sides.

$x^2 = 5$ Divide both sides by 2. The equation is in the form $x^2 = k$.

$x = \pm\sqrt{5}$ Apply the square root property.

Avoiding Mistakes

A common mistake when applying the square root property is to forget the $\pm$ symbol.

Check: $x = \sqrt{5}$

$2x^2 - 10 = 0$

$2(\sqrt{5})^2 - 10 \stackrel{?}{=} 0$

$2(5) - 10 \stackrel{?}{=} 0$

$10 - 10 \stackrel{?}{=} 0$ ✔

Check: $x = -\sqrt{5}$

$2x^2 - 10 = 0$

$2(-\sqrt{5})^2 - 10 \stackrel{?}{=} 0$

$2(5) - 10 \stackrel{?}{=} 0$

$10 - 10 \stackrel{?}{=} 0$ ✔

The solutions are $\sqrt{5}$ and $-\sqrt{5}$.

Example 6 Solving a Quadratic Equation Using the Square Root Property

Use the square root property to solve the equation.

$$(t - 4)^2 = 12$$

Solution:

$(t - 4)^2 = 12$ The equation is in the form $x^2 = k$, where $x = (t - 4)$.

$t - 4 = \pm\sqrt{12}$ Apply the square root property.

Answers

4. ± 8 **5.** $\pm 2\sqrt{3}$

6. $-3 \pm 2\sqrt{2}$

$$t - 4 = \pm\sqrt{2^2 \cdot 3} \qquad \text{Simplify the radical.}$$

$$t - 4 = \pm2\sqrt{3}$$

$$t = 4 \pm 2\sqrt{3} \qquad \text{Solve for } t.$$

Check: $t = 4 + 2\sqrt{3}$ Check: $t = 4 - 2\sqrt{3}$

$$(t - 4)^2 = 12 \qquad\qquad (t - 4)^2 = 12$$

$$(4 + 2\sqrt{3} - 4)^2 \overset{?}{=} 12 \qquad (4 - 2\sqrt{3} - 4)^2 \overset{?}{=} 12$$

$$(2\sqrt{3})^2 \overset{?}{=} 12 \qquad\qquad (-2\sqrt{3})^2 \overset{?}{=} 12$$

$$4 \cdot 3 \overset{?}{=} 12 \qquad\qquad 4 \cdot 3 \overset{?}{=} 12$$

$$12 \overset{?}{=} 12 \;✔ \qquad\qquad 12 \overset{?}{=} 12 \;✔$$

The solutions are: $4 + 2\sqrt{3}$ and $4 - 2\sqrt{3}$.

Example 7   **Solving a Quadratic Equation Using the Square Root Property**

Use the square root property to solve the equation.

$$y^2 = -4$$

Solution:

$$y^2 = -4 \qquad \text{The equation is in the form } y^2 = k.$$

$$y = \pm\sqrt{-4}$$

The expression $\sqrt{-4}$ is not a real number. Thus, the equation, $y^2 = -4$, has no real-valued solutions.

Skill Practice

Use the square root property to solve the equation.

7. $z^2 = -9$

Classroom Example: p. 654, Exercise 42

Answer

7. The equation has no real-valued solutions.

Section 9.1   Practice Exercises

Boost *your* GRADE at ALEKS.com!

 ALEKS® version 3.0

- Practice Problems
- Self-Tests
- NetTutor
- e-Professors
- Videos

For additional exercises, see Classroom Activity 9.1A in the *Instructor's Resource Manual* at www.mhhe.com/moh.

Study Skills Exercise

1. Define the key term **square root property**.

Objective 1: Review of the Zero Product Rule

2. Identify the equations as linear or quadratic.

    a. $2x - 5 = 3(x + 2) - 1$ Linear

    b. $2x(x - 5) = 3(x + 2) - 1$ Quadratic

    c. $ax^2 + bx + c = 0$
       ($a, b,$ and c are real numbers, and $a \neq 0$)
       Quadratic

3. Identify the equations as linear or quadratic.

    a. $ax + b = 0$
       (a and b are real numbers, and $a \neq 0$)
       Linear

    b. $\dfrac{1}{2}p - \dfrac{3}{4}p^2 = 0$ Quadratic

    c. $\dfrac{1}{2}(p - 3) = 5$ Linear

    ✏ Writing   ↔ Translating Expression   �£ Geometry   ▤ Scientific Calculator   ◉ Video   NS&E

For Exercises 4–19, solve using the zero product rule. **(See Examples 1–3.)**

4. $(3z - 2)(4z + 5) = 0$ $\frac{2}{3}, -\frac{5}{4}$

5. $(t + 5)(2t - 1) = 0$ $-5, \frac{1}{2}$

6. $r^2 + 7r + 12 = 0$ $-4, -3$

7. $y^2 - 2y - 35 = 0$ $7, -5$

8. $10x^2 = 13x - 4$ $\frac{1}{2}, \frac{4}{5}$

9. $6p^2 = -13p - 2$ $-2, -\frac{1}{6}$

10. $2m(m - 1) = 3m - 3$ $\frac{3}{2}, 1$

11. $2x^2 + 10x = -7(x + 3)$ $-7, -\frac{3}{2}$

12. $x^2 = 4$ $2, -2$

13. $c^2 = 144$ $12, -12$

14. $(x - 1)^2 = 16$ $-3, 5$

15. $(x - 3)^2 = 25$ $8, -2$

16. $3p^2 + 4p = 15$ $\frac{5}{3}, -3$

17. $4a^2 + 7a = 2$ $\frac{1}{4}, -2$

18. $(x + 2)(x + 3) = 2$ $-4, -1$

19. $(x + 2)(x + 6) = 5$ $-1, -7$

Objective 2: Solving Quadratic Equations Using the Square Root Property

20. The symbol "$\pm$" is read as … Plus or minus

For Exercises 21–44, solve the equations using the square root property. **(See Examples 4–7.)**

21. $x^2 = 49$ ± 7

22. $x^2 = 16$ ± 4

23. $k^2 - 100 = 0$ ± 10

24. $m^2 - 64 = 0$ ± 8

25. $p^2 = -24$ There are no real-valued solutions.

26. $q^2 = -50$ There are no real-valued solutions.

27. $3w^2 - 9 = 0$ $\pm\sqrt{3}$

28. $4v^2 - 24 = 0$ $\pm\sqrt{6}$

29. $(a - 5)^2 = 16$ $9, 1$

30. $(b + 3)^2 = 1$ $-2, -4$

31. $(y - 5)^2 = 36$ $11, -1$

32. $(y + 4)^2 = 4$ $-2, -6$

33. $(x - 11)^2 = 5$ $11 \pm \sqrt{5}$

34. $(z - 2)^2 = 7$ $2 \pm \sqrt{7}$

35. $(a + 1)^2 = 18$ $-1 \pm 3\sqrt{2}$

36. $(b - 1)^2 = 12$ $1 \pm 2\sqrt{3}$

37. $\left(t - \frac{1}{4}\right)^2 = \frac{7}{16}$ $\frac{1}{4} \pm \frac{\sqrt{7}}{4}$

38. $\left(t - \frac{1}{3}\right)^2 = \frac{1}{9}$ $\frac{2}{3}, 0$

39. $\left(x - \frac{1}{2}\right)^2 + 5 = 20$ $\frac{1}{2} \pm \sqrt{15}$

40. $\left(x + \frac{5}{2}\right)^2 - 3 = 18$ $-\frac{5}{2} \pm \sqrt{21}$

41. $(p - 3)^2 = -16$ There are no real-valued solutions.

42. $(t + 4)^2 = -9$ There are no real-valued solutions.

43. $12t^2 = 75$ $\pm\frac{5}{2}$

44. $8p^2 = 18$ $\pm\frac{3}{2}$

45. Check the solution $-3 + \sqrt{5}$ in the equation $(x + 3)^2 = 5$. The solution checks.

46. Check the solution $-5 - \sqrt{7}$ in the equation $(p + 5)^2 = 7$. The solution checks.

For Exercises 47–48, answer true or false. If a statement is false, explain why.

47. The only solution to the equation $x^2 = 64$ is 8. False. -8 is also a solution.

48. There are two real solutions to every quadratic equation of the form $x^2 = k$, where $k \geq 0$ is a real number. False. If $k = 0$, there is only one solution.

49. Ignoring air resistance, the distance, d (in feet), that an object drops in t seconds is given by the equation

$$d = 16t^2$$

a. Find the distance traveled in 2 sec. 64 ft

b. Find the time required for the object to fall 200 ft. Round to the nearest tenth of a second. 3.5 sec

c. Find the time required for an object to fall from the top of the Empire State Building in New York City if the building is 1250 ft high. Round to the nearest tenth of a second. 8.8 sec

50. Ignoring air resistance, the distance, d (in meters), that an object drops in t seconds is given by the equation

$$d = 4.9t^2$$

a. Find the distance traveled in 5 sec. 122.5 m

b. Find the time required for the object to fall 50 m. Round to the nearest tenth of a second. 3.2 sec

c. Find the time required for an object to fall from the top of the Canada Trust Tower in Toronto, Canada, if the building is 261 m high. Round to the nearest tenth of a second. 7.3 sec

 Writing Translating Expression Geometry Scientific Calculator Video NSE

51. A right triangle has legs of equal length. If the hypotenuse is 10 m long, find the length (in meters) of each leg. Round the answer to the nearest tenth of a meter.   7.1 m

 52. The diagonal of a square computer monitor screen is 24 in. long. Find the length of the sides to the nearest tenth of an inch.   17.0 in.

53. The area of a circular wading pool is approximately 200 ft². Find the radius to the nearest tenth of a foot.   8.0 ft

54. According to the International Swimming Federation, the volume of an eight-lane Olympic size pool should be 2500 m³. The length of the pool is twice the width, and the depth is 2 m. Use a calculator to find the length and width of the pool.   The length is 50 m and the width is 25 m.

Completing the Square

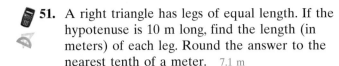

Section 9.2

1. Completing the Square

In Section 9.1, Example 6, we used the square root property to solve an equation in which the square of a binomial was equal to a constant.

$$(t - 4)^2 = 12$$

Square of a binomial → Constant

Objectives

1. **Completing the Square**
2. **Solving Quadratic Equations by Completing the Square**

Furthermore, any equation $ax^2 + bx + c = 0$ $(a \neq 0)$ can be rewritten as the square of a binomial equal to a constant by using a process called **completing the square**.

We begin our discussion of completing the square with some vocabulary. For a trinomial $ax^2 + bx + c$ $(a \neq 0)$, the term ax^2 is called the **quadratic term**. The term bx is called the **linear term**, and the term c is called the **constant term**.

Next, notice that the square of a binomial is the factored form of a perfect square trinomial.

Perfect Square Trinomial   **Factored Form**

$$x^2 + 10x + 25 \longrightarrow (x + 5)^2$$

$$t^2 - 6t + 9 \longrightarrow (t - 3)^2$$

$$p^2 - 14p + 49 \longrightarrow (p - 7)^2$$

✏ Writing   ↔ Translating Expression   ◁ Geometry   📟 Scientific Calculator   ● Video   NS⌕E

Furthermore, for a perfect square trinomial with a leading coefficient of 1, the constant term is the square of half the coefficient of the linear term. For example:

$x^2 + 10x + 25$ ← $\left[\frac{1}{2}(10)\right]^2 = [5]^2 = 25$

$t^2 - 6t + 9$ ← $\left[\frac{1}{2}(-6)\right]^2 = [-3]^2 = 9$

$p^2 - 14p + 49$ ← $\left[\frac{1}{2}(-14)\right]^2 = [-7]^2 = 49$

In general, an expression of the form $x^2 + bx$ will result in a perfect square trinomial if the square of half the linear term coefficient, $(\frac{1}{2}b)^2$, is added to the expression.

─ **Skill Practice** ─

Complete the square for each expression, and then factor the polynomial.

1. $q^2 + 8q$

2. $t^2 - 10t$

3. $v^2 + 3v$

4. $y^2 + \frac{1}{4}y$

Classroom Examples: p. 660, Exercises 6, 14, and 16

Example 1 **Completing the Square** ─

Complete the square for each expression. Then factor the expression as the square of a binomial.

a. $x^2 + 12x$ **b.** $x^2 - 22x$ **c.** $x^2 + 5x$ **d.** $x^2 - \frac{3}{5}x$

Solution:

The expressions are in the form $x^2 + bx$. Add the square of half the linear term coefficient, $(\frac{1}{2}b)^2$.

a. $x^2 + 12x$

$x^2 + 12x + 36$ Add $\frac{1}{2}$ of 12, squared. $[\frac{1}{2}(12)]^2 = (6)^2 = 36$.

$(x + 6)^2$ Factored form

b. $x^2 - 22x$

$x^2 - 22x + 121$ Add $\frac{1}{2}$ of -22, squared. $[\frac{1}{2}(-22)]^2 = (-11)^2 = 121$.

$(x - 11)^2$ Factored form

c. $x^2 + 5x$

$x^2 + 5x + \frac{25}{4}$ Add $\frac{1}{2}$ of 5, squared. $[\frac{1}{2}(5)]^2 = (\frac{5}{2})^2 = \frac{25}{4}$.

$\left(x + \frac{5}{2}\right)^2$ Factored form

d. $x^2 - \frac{3}{5}x$

$x^2 - \frac{3}{5}x + \frac{9}{100}$ Add $\frac{1}{2}$ of $-\frac{3}{5}$, squared.

$$\left[\frac{1}{2}\left(-\frac{3}{5}\right)\right]^2 = \left(-\frac{3}{10}\right)^2 = \frac{9}{100}$$

$\left(x - \frac{3}{10}\right)^2$ Factored form

Answers

1. $q^2 + 8q + 16; (q + 4)^2$

2. $t^2 - 10t + 25; (t - 5)^2$

3. $v^2 + 3v + \frac{9}{4}; \left(v + \frac{3}{2}\right)^2$

4. $y^2 + \frac{1}{4}y + \frac{1}{64}; \left(y + \frac{1}{8}\right)^2$

2. Solving Quadratic Equations by Completing the Square

A quadratic equation can be solved by completing the square and applying the square root property. The following steps outline the procedure.

PROCEDURE Solving a Quadratic Equation in the Form $ax^2 + bx + c = 0$ $(a \neq 0)$ by Completing the Square and Applying the Square Root Property

Step 1 Divide both sides by a to make the leading coefficient 1.
Step 2 Isolate the variable terms on one side of the equation.
Step 3 Complete the square by adding the square of one-half the linear term coefficient to both sides of the equation. Then factor the resulting perfect square trinomial.
Step 4 Apply the square root property and solve for x.

Example 2 Solving a Quadratic Equation by Completing the Square and Applying the Square Root Property

Solve the quadratic equation by completing the square and applying the square root property.

$$x^2 + 6x - 8 = 0$$

Solution:

$x^2 + 6x - 8 = 0$		The equation is in the form $ax^2 + bx + c = 0$.
	Step 1:	The leading coefficient is already 1.
$x^2 + 6x = 8$	**Step 2:**	Isolate the variable terms on one side.
$x^2 + 6x + 9 = 8 + 9$	**Step 3:**	To complete the square, add $\left[\frac{1}{2}(6)\right]^2 = (3)^2 = 9$ to both sides.
$(x + 3)^2 = 17$		Factor the perfect square trinomial.
$x + 3 = \pm\sqrt{17}$	**Step 4:**	Apply the square root property.
$x = -3 \pm \sqrt{17}$		Solve for x.

The solutions are $-3 - \sqrt{17}$ and $-3 + \sqrt{17}$.

Skill Practice

Solve the equation by completing the square and applying the square root property.

5. $t^2 + 6t + 2 = 0$

Classroom Example: p. 660, Exercise 22

Answer

5. $-3 \pm \sqrt{7}$

Classroom Example: p. 660,
Exercise 26

Example 3 Solving a Quadratic Equation by Completing the Square and Applying the Square Root Property

Solve the quadratic equation by completing the square and applying the square root property.

$$2x^2 - 16x - 24 = 0$$

Solution:

$$2x^2 - 16x - 24 = 0$$ — The equation is in the form $ax^2 + bx + c = 0$.

$$\frac{2x^2}{2} - \frac{16x}{2} - \frac{24}{2} = \frac{0}{2}$$ — **Step 1:** Divide both sides by the leading coefficient, 2.

$$x^2 - 8x - 12 = 0$$

$$x^2 - 8x = 12$$ — **Step 2:** Isolate the variable terms on one side.

$$x^2 - 8x + 16 = 12 + 16$$ — **Step 3:** To complete the square, add $[\frac{1}{2}(-8)]^2 = 16$ to both sides of the equation.

$$(x - 4)^2 = 28$$ — Factor the perfect square trinomial.

$$x - 4 = \pm\sqrt{28}$$ — **Step 4:** Apply the square root property.

$$x - 4 = \pm 2\sqrt{7}$$ — Simplify the radical.

$$x = 4 \pm 2\sqrt{7}$$ — Solve for x.

The solutions are $4 + 2\sqrt{7}$ and $4 - 2\sqrt{7}$.

Classroom Example: p. 660,
Exercise 30

Example 4 Solving a Quadratic Equation by Completing the Square and Applying the Square Root Property

Solve the quadratic equation by completing the square and applying the square root property.

$$x(2x - 5) - 3 = 0$$

Solution:

$$x(2x - 5) - 3 = 0$$ — Clear parentheses.

$$2x^2 - 5x - 3 = 0$$ — The equation is in the form $ax^2 + bx + c = 0$.

$$\frac{2x^2}{2} - \frac{5x}{2} - \frac{3}{2} = \frac{0}{2}$$ — **Step 1:** Divide both sides by the leading coefficient, 2.

$$x^2 - \frac{5}{2}x - \frac{3}{2} = 0$$

$$x^2 - \frac{5}{2}x = \frac{3}{2}$$ — **Step 2:** Isolate the variable terms on one side.

Answers

6. $1 \pm 3\sqrt{2}$

7. $\frac{3}{5}, -2$

$$x^2 - \frac{5}{2}x + \frac{25}{16} = \frac{3}{2} + \frac{25}{16}$$

Step 3: Add $\left[\frac{1}{2}\left(-\frac{5}{2}\right)\right]^2 = \left(-\frac{5}{4}\right)^2 = \frac{25}{16}$ to both sides.

$$\left(x - \frac{5}{4}\right)^2 = \frac{24}{16} + \frac{25}{16}$$

Factor the perfect square trinomial. Rewrite the right-hand side with a common denominator and simplify.

$$\left(x - \frac{5}{4}\right)^2 = \frac{49}{16}$$

$$x - \frac{5}{4} = \pm\sqrt{\frac{49}{16}}$$

Step 4: Apply the square root property.

$$x - \frac{5}{4} = \pm\frac{7}{4}$$

Simplify the radical.

$$x = \frac{5}{4} \pm \frac{7}{4}$$

Solve for x.

We have

$$x = \begin{cases} \frac{5}{4} + \frac{7}{4} = \frac{12}{4} = 3 \\ \\ \frac{5}{4} - \frac{7}{4} = -\frac{2}{4} = -\frac{1}{2} \end{cases}$$

The solutions are 3 and $-\frac{1}{2}$.

TIP: Since the solutions to the equation $x(2x - 5) - 3 = 0$ are rational numbers, the equation could have been solved by factoring and using the zero product rule:

$$x(2x - 5) - 3 = 0$$
$$2x^2 - 5x - 3 = 0$$
$$(2x + 1)(x - 3) = 0$$
$$2x + 1 = 0 \quad \text{or} \quad x - 3 = 0$$
$$x = -\frac{1}{2} \quad \text{or} \quad x = 3$$

Section 9.2 Practice Exercises

Study Skills Exercise

1. Define the key terms:

    a. completing the square   **b.** quadratic term

    c. linear term   **d.** constant term

Review Exercises

For Exercises 2–4, solve the quadratic equation using the square root property.

2. $x^2 = 21$ $\pm\sqrt{21}$ **3.** $(x - 5)^2 = 21$ $5 \pm \sqrt{21}$ **4.** $(x - 5)^2 = -21$

There are no real-valued solutions.

Writing   Translating Expression   Geometry   Scientific Calculator   Video   NS&E

Objective 1: Completing the Square

For Exercises 5–16, find the constant that should be added to the expression to make it a perfect square trinomial. Then factor the trinomial. (See Example 1.)

5. $y^2 + 4y$ 4;
$y^2 + 4y + 4 = (y + 2)^2$

6. $w^2 - 6w$ 9;
$w^2 - 6w + 9 = (w - 3)^2$

7. $p^2 - 12p$ 36;
$p^2 - 12p + 36 = (p - 6)^2$

8. $q^2 + 16q$ 64;
$q^2 + 16q + 64 = (q + 8)^2$

9. $x^2 - 9x$ $\frac{81}{4}$;
$x^2 - 9x + \frac{81}{4} = \left(x - \frac{9}{2}\right)^2$

10. $a^2 - 5a$ $\frac{25}{4}$;
$a^2 - 5a + \frac{25}{4} = \left(x - \frac{5}{2}\right)^2$

11. $d^2 + \frac{5}{3}d$ $\frac{25}{36}$;
$d^2 + \frac{5}{3}d + \frac{25}{36} = \left(d + \frac{5}{6}\right)^2$

12. $t^2 + \frac{1}{4}t$ $\frac{1}{64}$;
$t^2 + \frac{1}{4}t + \frac{1}{64} = \left(t + \frac{1}{8}\right)^2$

13. $m^2 - \frac{1}{5}m$ $\frac{1}{100}$;
$m^2 - \frac{1}{5}m + \frac{1}{100} = \left(m - \frac{1}{10}\right)^2$

14. $n^2 - \frac{5}{7}n$ $\frac{25}{196}$;
$n^2 - \frac{5}{7}n + \frac{25}{196} = \left(n - \frac{5}{14}\right)^2$

15. $u^2 + u$ $\frac{1}{4}$;
$u^2 + u + \frac{1}{4} = \left(u + \frac{1}{2}\right)^2$

16. $v^2 - v$ $\frac{1}{4}$;
$v^2 - v + \frac{1}{4} = \left(v - \frac{1}{2}\right)^2$

Objective 2: Solving Quadratic Equations by Completing the Square

For Exercises 17–36, solve the equations by completing the square and applying the square root property.
(See Examples 2–4.)

17. $x^2 + 4x = 12$
2, −6

18. $x^2 - 2x = 8$
4, −2

19. $y^2 + 6y = -5$
−1, −5

20. $t^2 + 10t = 11$
1, −11

21. $x^2 = 2x + 1$
$1 \pm \sqrt{2}$

22. $x^2 = 6x - 2$
$3 \pm \sqrt{7}$

23. $3x^2 - 6x - 15 = 0$
$1 \pm \sqrt{6}$

24. $5x^2 + 10x - 30 = 0$
$-1 \pm \sqrt{7}$

25. $4p^2 + 16p = -4$
$-2 \pm \sqrt{3}$

26. $2t^2 - 12t = 12$
$3 \pm \sqrt{15}$

27. $w^2 + w - 3 = 0$
$-\frac{1}{2} \pm \frac{\sqrt{13}}{2}$

28. $z^2 - 3z - 5 = 0$ $\frac{3}{2} \pm \frac{\sqrt{29}}{2}$

29. $x(x + 2) = 40$
$-1 \pm \sqrt{41}$

30. $y(y - 4) = 10$
$2 \pm \sqrt{14}$

31. $a^2 - 4a - 1 = 0$
$2 \pm \sqrt{5}$

32. $c^2 - 2c - 9 = 0$
$1 \pm \sqrt{10}$

33. $2r^2 + 12r + 16 = 0$
−2, −4

34. $3p^2 + 12p + 9 = 0$
−3, −1

35. $h(h - 11) = -24$
3, 8

36. $k(k - 8) = -7$
1, 7

Mixed Exercises

For Exercises 37–64, solve the quadratic equation by using the zero product rule or the square root property.
(*Hint:* For some exercises, you may have to factor or complete the square first.)

37. $y^2 = 121$
±11

38. $x^2 = 81$
±9

39. $(p + 2)^2 = 2$
$-2 \pm \sqrt{2}$

40. $(q - 6)^2 = 3$
$6 \pm \sqrt{3}$

41. $(k + 13)(k - 5) = 0$
−13, 5

42. $(r - 10)(r + 12) = 0$
10, −12

43. $(x - 13)^2 = 0$
13

44. $(p + 14)^2 = 0$
−14

45. $z^2 - 8z - 20 = 0$
10, −2

46. $b^2 - 14b + 48 = 0$
8, 6

47. $(x - 3)^2 = 16$
7, −1

48. $(x + 2)^2 = 49$
5, −9

49. $a^2 - 8a + 1 = 0$
$4 \pm \sqrt{15}$

50. $x^2 + 12x - 4 = 0$
$-6 \pm 2\sqrt{10}$

51. $2y^2 + 4y = 10$
$-1 \pm \sqrt{6}$

52. $3z^2 - 48z = 6$
$8 \pm \sqrt{66}$

53. $x^2 - 9x - 22 = 0$
11, −2

54. $y^2 + 11y + 18 = 0$
−9, −2

55. $5h(h - 7) = 0$
0, 7

56. $-2w(w + 9) = 0$
0, −9

57. $8t^2 + 2t - 3 = 0$
$\frac{1}{2}, -\frac{3}{4}$

58. $18a^2 - 21a + 5 = 0$
$\frac{5}{6}, \frac{1}{3}$

59. $t^2 = 14$
$\pm \sqrt{14}$

60. $s^2 = 17$
$\pm \sqrt{17}$

61. $c^2 + 9 = 0$
There are no real-valued solutions.

62. $k^2 + 25 = 0$
There are no real-valued solutions.

63. $4x^2 - 8x = -4$
1

64. $3x^2 + 12x = -12$
−2

Expanding Your Skills

For Exercises 65–66, solve by completing the square.

65. To comply with FAA regulations, a piece of luggage must be checked to the luggage compartment of the plane if its combined linear measurement of length, width, and height is over 45 in. Katie's suitcase has a total volume of 4200 in.³ Its length is 30 in., and its width is 4 in. greater than the height. Find the dimensions of the suitcase. Will this suitcase need to be checked?
The suitcase is 10 in. by 14 in. by 30 in. The bag must be checked because 10 in. + 14 in. + 30 in. = 54 in., which is greater than 45 in.

 Writing   Translating Expression   Geometry   Scientific Calculator   Video   NS&E

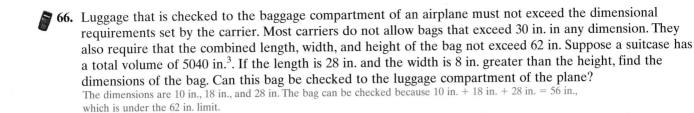

66. Luggage that is checked to the baggage compartment of an airplane must not exceed the dimensional requirements set by the carrier. Most carriers do not allow bags that exceed 30 in. in any dimension. They also require that the combined length, width, and height of the bag not exceed 62 in. Suppose a suitcase has a total volume of 5040 in.^{3}. If the length is 28 in. and the width is 8 in. greater than the height, find the dimensions of the bag. Can this bag be checked to the luggage compartment of the plane?

The dimensions are 10 in., 18 in., and 28 in. The bag can be checked because 10 in. + 18 in. + 28 in. = 56 in., which is under the 62 in. limit.

Quadratic Formula

1. Derivation of the Quadratic Formula

If we solve a general quadratic equation $ax^2 + bx + c = 0$ by completing the square and using the square root property, the result is a formula that gives the solutions for x in terms of a, b, and c.

Objectives

1. Derivation of the Quadratic Formula
2. Solving Quadratic Equations Using the Quadratic Formula
3. Review of the Methods for Solving a Quadratic Equation
4. Applications of Quadratic Equations

$$ax^2 + bx + c = 0$$

Begin with a quadratic equation in standard form.

$$\frac{ax^2}{a} + \frac{b}{a}x + \frac{c}{a} = \frac{0}{a}$$

Divide by the leading coefficient.

$$x^2 + \frac{b}{a}x = -\frac{c}{a}$$

Isolate the terms containing x.

$$x^2 + \frac{b}{a}x + \left(\frac{1}{2} \cdot \frac{b}{a}\right)^2 = \left(\frac{1}{2} \cdot \frac{b}{a}\right)^2 - \frac{c}{a}$$

Add the square of $\frac{1}{2}$ the linear term coefficient to both sides of the equation.

$$\left(x + \frac{b}{2a}\right)^2 = \frac{b^2}{4a^2} - \frac{c}{a}$$

Factor the left side as a perfect square.

$$\left(x + \frac{b}{2a}\right)^2 = \frac{b^2}{4a^2} - \frac{c}{a} \cdot \frac{(4a)}{(4a)}$$

On the right side, write the fractions with the common denominator, $4a^2$.

$$\left(x + \frac{b}{2a}\right)^2 = \frac{b^2 - 4ac}{4a^2}$$

Combine the fractions.

$$x + \frac{b}{2a} = \pm\sqrt{\frac{b^2 - 4ac}{4a^2}}$$

Apply the square root property.

$$x + \frac{b}{2a} = \frac{\pm\sqrt{b^2 - 4ac}}{2a}$$

Simplify the denominator.

$$x = -\frac{b}{2a} \pm \frac{\sqrt{b^2 - 4ac}}{2a}$$

Subtract $\frac{b}{2a}$ from both sides.

$$= \frac{-b \pm \sqrt{b^2 - 4ac}}{2a}$$

Combine fractions.

Concept Connections

Write each quadratic equation in the form $ax^2 + bx + c = 0$. Then identify a, b, and c.

1. $4x^2 - 9x + 1 = 0$
2. $3x^2 = 2x + 8$
3. $x(x + 6) = x + 15$

FORMULA Quadratic Formula

For any quadratic equation of the form $ax^2 + bx + c = 0, (a \neq 0)$ the solutions are

$$x = \frac{-b \pm \sqrt{b^2 - 4ac}}{2a}$$

Answers

1. $4x^2 - 9x + 1 = 0$;
   $a = 4; b = -9; c = 1$
2. $3x^2 - 2x - 8 = 0$;
   $a = 3; b = -2; c = -8$
3. $x^2 + 5x - 15 = 0$;
   $a = 1; b = 5; c = -15$

 Writing  Translating Expression Geometry Scientific Calculator Video NS&E

2. Solving Quadratic Equations Using the Quadratic Formula

Example 1 Solving Quadratic Equations Using the Quadratic Formula

Solve the quadratic equation using the quadratic formula: $3x^2 - 7x = -2$

Solution:

$$3x^2 - 7x = -2$$

$$3x^2 - 7x + 2 = 0 \qquad \text{Write the equation in the form } ax^2 + bx + c = 0.$$

$$a = 3, b = -7, c = 2 \qquad \text{Identify } a, b, \text{ and } c.$$

$$x = \frac{-b \pm \sqrt{b^2 - 4ac}}{2a}$$

$$x = \frac{-(-7) \pm \sqrt{(-7)^2 - 4(3)(2)}}{2(3)} \qquad \text{Apply the quadratic formula.}$$

$$x = \frac{7 \pm \sqrt{49 - 24}}{6} \qquad \text{Simplify.}$$

$$= \frac{7 \pm \sqrt{25}}{6}$$

$$= \frac{7 \pm 5}{6}$$

There are two rational solutions.

$$x = \begin{cases} \dfrac{7 + 5}{6} = \dfrac{12}{6} = 2 \\[2mm] \dfrac{7 - 5}{6} = \dfrac{2}{6} = \dfrac{1}{3} \end{cases}$$

The solutions are 2 and $\frac{1}{3}$.

TIP: Because the solutions to the equation $3x^2 - 7x = -2$ are rational numbers, the equation could have been solved by factoring and using the zero product rule.

$$3x^2 - 7x = -2$$

$$3x^2 - 7x + 2 = 0$$

$$(3x - 1)(x - 2) = 0$$

$$3x - 1 = 0 \quad \text{or} \quad x - 2 = 0$$

$$x = \frac{1}{3} \quad \text{or} \quad x = 2$$

The solutions are $\frac{1}{3}$ and 2.

Example 2 Solving a Quadratic Equation Using the Quadratic Formula

Solve the quadratic equation using the quadratic formula. Then approximate the solutions to three decimal places.

$$2x(x - 1) - 6 = 0$$

Solution:

$2x(x - 1) - 6 = 0$

$2x^2 - 2x - 6 = 0$ Write the equation in the form $ax^2 + bx + c = 0$.

$a = 2, b = -2, c = -6$ Identify a, b, and c.

$x = \dfrac{-b \pm \sqrt{b^2 - 4ac}}{2a}$

$x = \dfrac{-(-2) \pm \sqrt{(-2)^2 - 4(2)(-6)}}{2(2)}$ Apply the quadratic formula.

$= \dfrac{2 \pm \sqrt{4 + 48}}{4}$ Simplify.

$= \dfrac{2 \pm \sqrt{52}}{4}$

$= \dfrac{2 \pm \sqrt{2^2 \cdot 13}}{4}$ Simplify the radical.

$= \dfrac{2 \pm 2\sqrt{13}}{4}$

$= \dfrac{\overset{1}{2}(1 \pm \sqrt{13})}{\underset{2}{4}}$ Simplify the fraction.

$= \dfrac{1 \pm \sqrt{13}}{2}$

$x = \dfrac{1 + \sqrt{13}}{2} \approx 2.303$

$x = \dfrac{1 - \sqrt{13}}{2} \approx -1.303$

The exact solutions are $\dfrac{1 + \sqrt{13}}{2}$ and $\dfrac{1 - \sqrt{13}}{2}$. The approximate solutions are 2.303 and -1.303.

Skill Practice

Solve by using the quadratic formula. Then approximate the solutions to three decimal places.

5. $y(y + 2) = 5$

Classroom Example: p. 668, Exercise 28

Answer

5. $-1 \pm \sqrt{6}$;
$\approx 1.449, \approx -3.449$

Example 3 Solving a Quadratic Equation Using the Quadratic Formula

Solve the quadratic equation using the quadratic formula.

$$\frac{1}{4}w^2 - \frac{1}{2}w - \frac{5}{4} = 0$$

Solution:

$$\frac{1}{4}w^2 - \frac{1}{2}w - \frac{5}{4} = 0$$ To simplify the equation, multiply both sides by 4.

$$4\left(\frac{1}{4}w^2 - \frac{1}{2}w - \frac{5}{4}\right) = 4(0)$$ Clear fractions.

$$w^2 - 2w - 5 = 0$$ The equation is in the form $ax^2 + bx + c = 0$.

$$a = 1, b = -2, c = -5$$ Identify a, b, and c.

$$w = \frac{-b \pm \sqrt{b^2 - 4ac}}{2a}$$

$$w = \frac{-(-2) \pm \sqrt{(-2)^2 - 4(1)(-5)}}{2(1)}$$ Apply the quadratic formula.

$$= \frac{2 \pm \sqrt{4 + 20}}{2}$$ Simplify.

$$= \frac{2 \pm \sqrt{24}}{2}$$

$$= \frac{2 \pm 2\sqrt{6}}{2}$$ The solutions are irrational numbers.

$$= \frac{2(1 \pm \sqrt{6})}{2}$$ Factor and simplify.

$$= 1 \pm \sqrt{6}$$

The solutions are $1 + \sqrt{6}$ and $1 - \sqrt{6}$.

3. Review of the Methods for Solving a Quadratic Equation

Three methods have been presented for solving quadratic equations.

PROCEDURE Methods for Solving a Quadratic Equation
- Factor and use the zero product rule (Section 6.7).
- Use the square root property. Complete the square if necessary (Sections 9.1 and 9.2).
- Use the quadratic formula (Section 9.3).

Answer

6. $\frac{3 \pm \sqrt{57}}{12}$

Using the zero product rule only works if you can factor the equation. The square root property and the quadratic formula can be used to solve any quadratic equation. Before solving a quadratic equation, take a minute to analyze it. Each problem must be evaluated individually before choosing the most efficient method to find its solutions.

Example 4 Solving Quadratic Equations Using Any Method

Solve the quadratic equations using any method.

a. $(x + 1)^2 = 5$ **b.** $t^2 - t - 30 = 0$ **c.** $2x^2 + 5x + 1 = 0$

Solution:

a. $(x + 1)^2 = 5$ — Because the equation is the square of a binomial equal to a constant, the square root property can be applied easily.

$x + 1 = \pm\sqrt{5}$ — Apply the square root property.

$x = -1 \pm \sqrt{5}$ — Isolate x.

The solutions are $-1 + \sqrt{5}$ and $-1 - \sqrt{5}$.

b. $t^2 - t - 30 = 0$ — This equation factors.

$(t - 6)(t + 5) = 0$ — Factor and apply the zero product rule.

$t = 6$ or $t = -5$

The solutions are 6 and -5.

c. $2x^2 + 5x + 1 = 0$ — The equation does not factor. Because it is already in the form $ax^2 + bx + c = 0$, use the quadratic formula.

$a = 2, b = 5, c = 1$ — Identify a, b, and c.

$x = \dfrac{-(5) \pm \sqrt{(5)^2 - 4(2)(1)}}{2(2)}$ — Apply the quadratic formula.

$x = \dfrac{-5 \pm \sqrt{25 - 8}}{4}$ — Simplify.

$x = \dfrac{-5 \pm \sqrt{17}}{4}$

The solutions are $\dfrac{-5 + \sqrt{17}}{4}$ and $\dfrac{-5 - \sqrt{17}}{4}$.

Skill Practice

Solve the equations using any method.

7. $p^2 + 7p + 12 = 0$

8. $5y^2 + 7y + 1 = 0$

9. $(w - 8)^2 = 3$

Classroom Examples: p. 668, Exercises 34, 38, and 44

Answers

7. $-3, -4$

8. $\dfrac{-7 \pm \sqrt{29}}{10}$

9. $8 \pm \sqrt{3}$

4. Applications of Quadratic Equations

Classroom Example: p. 668, Exercise 58

Example 5 Solving a Quadratic Equation in an Application

The length of a box is 2 in. longer than the width. The height of the box is 4 in. and the volume of the box is 200 in.³ Find the exact dimensions of the box. Then use a calculator to approximate the dimensions to the nearest tenth of an inch.

Solution:

Label the box as follows (Figure 9-1):

Width = x

Length = $x + 2$

Height = 4

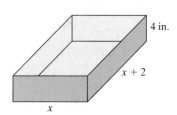

Figure 9-1

The volume of a box is given by the formula: $V = lwh$

$$V = l \cdot w \cdot h$$

$$200 = (x + 2)(x)(4)$$ Substitute $V = 200$, $l = x + 2$, $w = x$, and $h = 4$.

$$200 = (x + 2)4x$$

$$200 = 4x^2 + 8x$$

$$0 = 4x^2 + 8x - 200$$

$$4x^2 + 8x - 200 = 0$$ The equation is in the form $ax^2 + bx + c = 0$.

$$\frac{4x^2}{4} + \frac{8x}{4} - \frac{200}{4} = \frac{0}{4}$$ The coefficients are all divisible by 4. Dividing by 4 will create smaller values of a, b, and c to be used in the quadratic formula.

$$x^2 + 2x - 50 = 0$$ $a = 1$, $b = 2$, $c = -50$

$$x = \frac{-2 \pm \sqrt{(2)^2 - 4(1)(-50)}}{2(1)}$$ Apply the quadratic formula.

$$= \frac{-2 \pm \sqrt{4 + 200}}{2}$$ Simplify.

$$= \frac{-2 \pm \sqrt{204}}{2}$$

$$= \frac{-2 \pm 2\sqrt{51}}{2}$$ Simplify the radical. $\sqrt{204} = \sqrt{2^2 \cdot 51} = 2\sqrt{51}$

$$= \frac{\overset{1}{2}(-1 \pm \sqrt{51})}{\underset{1}{2}}$$ Factor and simplify.

$$= -1 \pm \sqrt{51}$$

We do not use the solution $x = -1 - \sqrt{51}$ because it is a negative number, that is,

$$-1 - \sqrt{51} \approx -8.1$$

The width of an object cannot be negative.

Because the width of the box must be positive, use $x = -1 + \sqrt{51}$.

The width is $(-1 + \sqrt{51})$ in. ≈ 6.1 in.

The length is $x + 2$: $(-1 + \sqrt{51} + 2)$ in. or $(1 + \sqrt{51})$ in. ≈ 8.1 in.

The height is 4 in.

Answer

10. The width is $(-1 + \sqrt{11})$ in. or approximately 2.3 in. The length is $(1 + \sqrt{11})$ in. or approximately 4.3 in.

Calculator Connections

Use the quadratic formula to verify that the solutions to the equation $x(x + 7) + 4 = 0$ are

$$x = \frac{-7 + \sqrt{33}}{2} \quad \text{and} \quad x = \frac{-7 - \sqrt{33}}{2}$$

A calculator can be used to obtain decimal approximations for the irrational solutions of a quadratic equation.

Scientific Calculator

Enter: 7 $+/-$ $+$ 33 $\sqrt{}$ $=$ $\div$ 2 $=$ **Result:** $\boxed{-0.6277186767}$

Enter: 7 $+/-$ $-$ 33 $\sqrt{}$ $=$ $\div$ 2 $=$ **Result:** $\boxed{-6.372281323}$

Graphing Calculator

```
(-7+√(33))/2
          -.6277186767
(-7-√(33))/2
          -6.372281323
```

Calculator Exercises

Use a calculator to obtain a decimal approximation of each expression.

1. $\dfrac{-5 + \sqrt{17}}{4}$ and $\dfrac{-5 - \sqrt{17}}{4}$ **2.** $\dfrac{-40 + \sqrt{1920}}{-32}$ and $\dfrac{-40 - \sqrt{1920}}{-32}$

Section 9.3   Practice Exercises

Boost *your* GRADE at ALEKS.com!

ALEKS® version 3.0

- Practice Problems
- Self-Tests
- NetTutor

- e-Professors
- Videos

For additional exercises, see Classroom Activities 9.3A–9.3B in the *Instructor's Resource Manual* at www.mhhe.com/moh.

Review Exercises

For Exercises 1–4, apply the square root property to solve the equation.

1. $z^2 = 169$ ± 13 **2.** $p^2 = 1$ ± 1 **3.** $(x - 4)^2 = 28$ $4 \pm 2\sqrt{7}$ **4.** $(y + 3)^2 = 7$ $-3 \pm \sqrt{7}$

For Exercises 5–6, solve the equations by completing the square.

5. $3a^2 - 12a - 12 = 0$ $2 \pm 2\sqrt{2}$ **6.** $x^2 - 5x + 1 = 0$ $\dfrac{5}{2} \pm \dfrac{\sqrt{21}}{2}$

Objective 1: Derivation of the Quadratic Formula

7. State the quadratic formula from memory.   For $ax^2 + bx + c = 0$, $x = \dfrac{-b \pm \sqrt{b^2 - 4ac}}{2a}$

8. Can all quadratic equations be solved by using the quadratic formula?   Yes

For Exercises 9–14, write the equations in the form $ax^2 + bx + c = 0$. Then identify the values of a, b, and c.

9. $2x^2 - x = 5$
$2x^2 - x - 5 = 0$; $a = 2$, $b = -1$, $c = -5$

10. $5(x^2 + 2) = -3x$
$5x^2 + 3x + 10 = 0$; $a = 5$, $b = 3$, $c = 10$

11. $-3x(x - 4) = -2x$
$-3x^2 + 14x + 0 = 0$; $a = -3$, $b = 14$, $c = 0$

12. $x(x - 2) = 3(x + 1)$
$x^2 - 5x - 3 = 0$; $a = 1$, $b = -5$, $c = -3$

13. $x^2 - 9 = 0$
$x^2 + 0x - 9 = 0$; $a = 1$, $b = 0$, $c = -9$

14. $x^2 + 25 = 0$
$x^2 + 0x + 25 = 0$; $a = 1$, $b = 0$, $c = 25$

 ✏ Writing   ⟷ Translating Expression   ◿ Geometry   Scientific Calculator   Video   NS E

Objective 2: Solving Quadratic Equations Using the Quadratic Formula

For Exercises 15–32, solve the equations using the quadratic formula. **(See Examples 1–3.)**

15. $t^2 + 16t + 64 = 0$ $\quad -8$

16. $y^2 - 10y + 25 = 0$ $\quad 5$

17. $6k^2 - k - 2 = 0$ $\quad \frac{2}{3}, -\frac{1}{2}$

18. $3n^2 + 5n - 2 = 0$ $\quad -2, \frac{1}{3}$

19. $5t^2 - t = 3$ $\quad \frac{1 \pm \sqrt{61}}{10}$

20. $2a^2 + 5a = 1$ $\quad \frac{-5 \pm \sqrt{33}}{4}$

21. $x(x - 2) = 1$ $\quad 1 \pm \sqrt{2}$

22. $2y(y - 3) = -1$ $\quad \frac{3 \pm \sqrt{7}}{2}$

23. $2p^2 = -10p - 11$ $\quad \frac{-5 \pm \sqrt{3}}{2}$

24. $z^2 = 4z + 1$ $\quad 2 \pm \sqrt{5}$

25. $-4y^2 - y + 1 = 0$ $\quad \frac{1 \pm \sqrt{17}}{-8}$ or $\frac{-1 \pm \sqrt{17}}{8}$

26. $-5z^2 - 3z + 4 = 0$ $\quad \frac{3 \pm \sqrt{89}}{-10}$ or $\frac{-3 \pm \sqrt{89}}{10}$

27. $2x(x + 1) = 3 - x$ $\quad \frac{-3 \pm \sqrt{33}}{4}$

28. $3m(m - 2) = -m + 1$ $\quad \frac{5 \pm \sqrt{37}}{6}$

29. $0.2y^2 = -1.5y - 1$ $\quad \frac{-15 \pm \sqrt{145}}{4}$

30. $0.2t^2 = t + 0.5$ $\quad \frac{5 \pm \sqrt{35}}{2}$

31. $\frac{2}{3}x^2 + \frac{4}{9}x = \frac{1}{3}$ $\quad \frac{-2 \pm \sqrt{22}}{6}$

32. $\frac{1}{2}x^2 + \frac{1}{6}x = 1$ $\quad \frac{-1 \pm \sqrt{73}}{6}$

Objective 3: Review of the Methods for Solving a Quadratic Equation

For Exercises 33–56, choose any method to solve the quadratic equations. **(See Example 4.)**

33. $16x^2 - 9 = 0$ $\quad \frac{3}{4}, -\frac{3}{4}$

34. $\frac{1}{4}x^2 + 5x + 13 = 0$ $\quad -10 \pm 4\sqrt{3}$

35. $(x - 5)^2 = -21$
There are no real-valued solutions.

36. $2x^2 + x + 5 = 0$
There are no real-valued solutions.

37. $\frac{1}{9}x^2 + \frac{8}{3}x + 11 = 0$ $\quad -12 \pm 3\sqrt{5}$

38. $7x^2 = 12x$ $\quad 0, \frac{12}{7}$

39. $2x^2 - 6x - 3 = 0$ $\quad \frac{3 \pm \sqrt{15}}{2}$

40. $4(x + 1)^2 = -15$
There are no real-valued solutions.

41. $9x^2 = 11x$ $\quad 0, \frac{11}{9}$

42. $25x^2 - 4 = 0$ $\quad \frac{2}{5}, -\frac{2}{5}$

43. $(2y - 3)^2 = 5$ $\quad \frac{3 \pm \sqrt{5}}{2}$

44. $(6z + 1)^2 = 7$ $\quad \frac{-1 \pm \sqrt{7}}{6}$

45. $0.4x^2 = 0.2x + 1$ $\quad \frac{1 \pm \sqrt{41}}{4}$

46. $0.6x^2 = 0.1x + 0.8$ $\quad \frac{1 \pm \sqrt{193}}{12}$

47. $9z^2 - z = 0$ $\quad 0, \frac{1}{9}$

48. $16p^2 - p = 0$ $\quad 0, \frac{1}{16}$

49. $r^2 - 52 = 0$ $\quad \pm 2\sqrt{13}$

50. $y^2 - 32 = 0$ $\quad \pm 4\sqrt{2}$

51. $-2.5t(t - 4) = 1.5$ $\quad \frac{-10 \pm \sqrt{85}}{-5}$ or $\frac{10 \pm \sqrt{85}}{5}$

52. $1.6p(p - 2) = 0.8$ $\quad \frac{2 \pm \sqrt{6}}{2}$

53. $(m - 3)(m + 2) = 9$ $\quad \frac{1 \pm \sqrt{61}}{2}$

54. $(h - 6)(h - 1) = 12$ $\quad \frac{7 \pm \sqrt{73}}{2}$

55. $x^2 + x + 3 = 0$
There are no real-valued solutions.

56. $3x^2 - 20x + 12 = 0$ $\quad 6, \frac{2}{3}$

Objective 4: Applications of Quadratic Equations

57. In a rectangle, the length is 1 m less than twice the width and the area is 100 m². Approximate the dimensions to the nearest tenth of a meter. **(See Example 5.)**
The width is 7.3 m. The length is 13.7m

58. In a triangle, the height is 2 cm more than the base. The area is 72 cm². Approximate the base and height to the nearest tenth of a centimeter.
The base is 11.0 cm. The height is 13.0 cm.

59. The volume of a rectangular storage area is 240 ft³. The length is 2 ft more than the width. The height is 6 ft. Approximate the dimensions to the nearest tenth of a foot. The length is 7.4 ft. The width is 5.4 ft. The height is 6 ft.

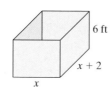

6 ft

x + 2

x

60. In a right triangle, one leg is 2 ft shorter than the other leg. The hypotenuse is 12 ft. Approximate the lengths of the legs to the nearest tenth of a foot.
The legs are 9.4 ft and 7.4 ft.

61. In a rectangle, the length is 4 ft longer than the width. The area is 72 ft^2. Approximate the dimensions to the nearest tenth of a foot.
The width is 6.7 ft. The length is 10.7 ft.

62. In a triangle, the base is 4 cm less than twice the height. The area is 60 cm^2. Approximate the base and height to the nearest tenth of a centimeter.
The height is 8.8 cm. The base is 13.6 cm.

63. In a right triangle, one leg is 3 m longer than the other leg. The hypotenuse is 13 m. Approximate the lengths of the legs to the nearest tenth of a meter.
The legs are 10.6 m and 7.6 m.

Problem Recognition Exercises

Solving Quadratic Equations

For Exercises 1–2, solve the equations using each of the three methods.

a. Factoring and applying the zero product rule

b. Completing the square and applying the square root property

c. Applying the quadratic formula

1. $6x^2 + 7x - 3 = 0$
$\dfrac{1}{3}, -\dfrac{3}{2}$

2. $y^2 + 14y + 49 = 0$
-7

For Exercises 3–22, solve using any method.

3. $x(x - 8) = 6$
$4 \pm \sqrt{22}$

4. $2 - 6y = -y^2$
$3 \pm \sqrt{7}$

5. $(t - 3)^2 - 8 = 0$
$3 \pm 2\sqrt{2}$

6. $(p + 11)^2 - 12 = 0$
$-11 \pm 2\sqrt{3}$

7. $6m^2 = 5m$
$0, \dfrac{5}{6}$

8. $12n^2 = 6n$
$0, \dfrac{1}{2}$

9. $\dfrac{1}{36}w^2 - 1 = 0$ ± 6

10. $\dfrac{25}{49}y^2 - 1 = 0$ $\pm\dfrac{7}{5}$

11. $8x^2 - 22x + 5 = 0$ $\dfrac{5}{2}, \dfrac{1}{4}$

12. $9w^2 - 15w + 4 = 0$ $\dfrac{4}{3}, \dfrac{1}{3}$

13. $p^2 + 20 = 0$ There are no real-valued solutions.

14. $m^2 + 15 = 0$ There are no real-valued solutions.

15. $y(y + 8) = -1$
$-4 \pm \sqrt{15}$

16. $t(t + 6) = -3$
$-3 \pm \sqrt{6}$

17. $\dfrac{1}{2}a^2 - 4a + 8 = 0$
4

18. $\dfrac{1}{3}b^2 + 2b + 3 = 0$
-3

19. $2x^2 - 14x = -20$
$5, 2$

20. $3n^2 - 6n = 9$
$3, -1$

21. $(w + 1)^2 = 100$
$9, -11$

22. $(u - 5)^2 = 64$
$13, -3$

Writing   Translating Expression   Geometry   Scientific Calculator   Video   NS E

Section 9.4 Graphing Quadratic Functions

Objectives

1. **Definition of a Quadratic Function**
2. **Vertex of a Parabola**
3. **Graphing a Parabola**
4. **Applications of Quadratic Functions**

1. Definition of a Quadratic Function

In Chapter 3, we learned how to graph the solutions to linear equations in two variables. Now suppose we want to graph the *nonlinear* equation, $y = x^2$. To begin, we create a table of points representing several solutions to the equation (Table 9-1). These points form the curve shown in Figure 9-2.

Table 9-1

x	y
-3	9
-2	4
-1	1
0	0
1	1
2	4
3	9

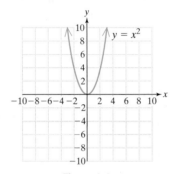

Figure 9-2

The equation $y = x^2$ is a special type of function called a quadratic function, and its graph is in the shape of a **parabola**.

> **DEFINITION Quadratic Function**
>
> Let a, b, and c represent real numbers such that $a \neq 0$. Then a function defined by $y = ax^2 + bx + c$ is called a **quadratic function**.

The graph of a quadratic function is a parabola that opens upward or downward. The leading coefficient, a, determines the direction of the parabola. For the quadratic function defined by $y = ax^2 + bx + c$:

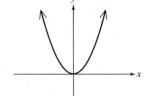

If $a > 0$, the parabola opens *upward*.
For example: $y = x^2$.

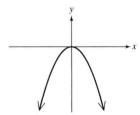

If $a < 0$, the parabola opens *downward*.
For example: $y = -x^2$.

If a parabola opens upward, the **vertex** is the lowest point on the graph. If a parabola opens downward, the **vertex** is the highest point on the graph. For a quadratic function, the **axis of symmetry** is the vertical line that passes through the vertex. Notice that the graph of the parabola is its own mirror image to the left and right of the axis of symmetry.

Concept Connections

1. Identify the function as linear or quadratic.
   a. $y = 2x + 3$
   b. $y = -3x^2 + x + 4$
   c. $y = 5x^2$
   d. $y = 5x$

Answers

1. a. Linear   **b.** Quadratic
   c. Quadratic   **d.** Linear

Here are four quadratic functions and their graphs.

$y = 0.5x^2 + 2x + 3$
$a > 0$
Vertex $(-2, 1)$
Axis of symmetry: $x = -2$

$y = x^2 - 6x + 9$
$a > 0$
Vertex $(3, 0)$
Axis of symmetry: $x = 3$

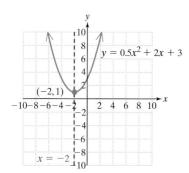

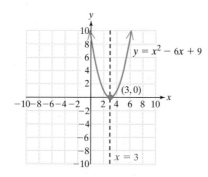

$y = -x^2 + 4x$
$a < 0$
Vertex $(2, 4)$
Axis of symmetry: $x = 2$

$y = -2x^2 - 4$
$a < 0$
Vertex $(0, -4)$
Axis of symmetry: $x = 0$

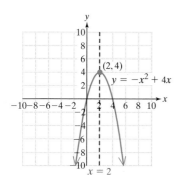

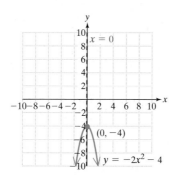

Concept Connections

2. Refer to the graph shown:

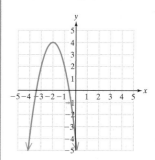

a. Determine the vertex of the parabola.

b. Determine the axis of symmetry of the parabola.

---

2. Vertex of a Parabola

Quadratic functions arise in many applications of mathematics and applied sciences. For example, an object thrown through the air follows a parabolic path. The mirror inside a reflecting telescope is parabolic in shape. In applications, it is often advantageous to analyze the graph of a parabola. In particular, we want to find the location of the x- and y-intercepts and the vertex.

To find the vertex of a parabola defined by $y = ax^2 + bx + c$ ($a \neq 0$), we use the following steps:

PROCEDURE Finding the Vertex of a Parabola

Step 1 The x-coordinate of the vertex of the parabola defined by
$y = ax^2 + bx + c$ ($a \neq 0$) is given by

$$x = \frac{-b}{2a}$$

Step 2 To find the corresponding y-coordinate of the vertex, substitute the value of the x-coordinate found in step 1 and solve for y.

Answers

2. a. $(-2, 4)$ **b.** $x = -2$

Classroom Example: p. 680,
Exercise 46

Skill Practice

3. Given $y = -x^2 - 4x$, perform parts (a)–(g), as in Example 1.

Example 1 Analyzing a Quadratic Function

Given the function defined by $y = -x^2 + 4x - 3$,

a. Determine whether the parabola opens upward or downward.

b. Find the vertex of the parabola.

c. Find the x-intercept(s).

d. Find the y-intercept.

e. Sketch the parabola.

f. Write the domain of the function in interval notation.

g. Write the range of the function in interval notation.

Solution:

a. The function $y = -x^2 + 4x - 3$ is written in the form $y = ax^2 + bx + c$, where $a = -1$, $b = 4$, and $c = -3$. Because the value of a is negative, the parabola opens *downward*.

b. The x-coordinate of the vertex is given by $x = \dfrac{-b}{2a}$.

$$x = \frac{-b}{2a} = \frac{-(4)}{2(-1)} \qquad \text{Substitute } b = 4 \text{ and } a = -1.$$

$$= \frac{-4}{-2} \qquad \text{Simplify.}$$

$$= 2$$

The y-coordinate of the vertex is found by substituting $x = 2$ into the equation and solving for y.

$$y = -x^2 + 4x - 3$$

$$= -(2)^2 + 4(2) - 3 \qquad \text{Substitute } x = 2.$$

$$= -4 + 8 - 3$$

$$= 1$$

The vertex is $(2, 1)$.   Because the parabola opens downward, the vertex is the maximum point on the graph of the parabola.

c. To find the x-intercept(s), substitute $y = 0$ and solve for x.

$$y = -x^2 + 4x - 3$$

$$0 = -x^2 + 4x - 3 \qquad \text{Substitute } y = 0. \text{ The resulting equation is quadratic.}$$

$$0 = -1(x^2 - 4x + 3) \qquad \text{Factor out } -1.$$

$$0 = -1(x - 3)(x - 1) \qquad \text{Factor the trinomial.}$$

Answers

3. a. downward
   b. $(-2, 4)$
   c. $(0, 0)$ and $(-4, 0)$
   d. $(0, 0)$
   e.

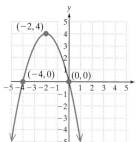

   f. $(-\infty, \infty)$
   g. $(-\infty, 4]$

$x - 3 = 0$ or $x - 1 = 0$ Apply the zero product rule.

$x = 3$ or $x = 1$

The x-intercepts are $(3, 0)$ and $(1, 0)$.

d. To find the y-intercept, substitute $x = 0$ and solve for y.

$y = -x^2 + 4x - 3$

$ = -(0)^2 + 4(0) - 3$ Substitute $x = 0$.

$ = -3$

The y-intercept is $(0, -3)$.

e. Using the results of parts (a)–(d), we have a parabola that opens downward with vertex $(2, 1)$, x-intercepts at $(3, 0)$ and $(1, 0)$, and y-intercept at $(0, -3)$ (Figure 9-3).

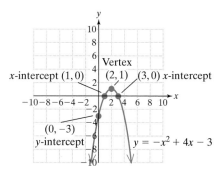

Figure 9-3

f. Because any real number, x, when substituted into the equation $y = -x^2 + 4x - 3$ produces a real number, the domain is all real numbers.

Domain: $(-\infty, \infty)$

g. From the graph, we see that the vertex is the maximum point on the graph. Therefore, the maximum y-value is $y = 1$. The range is restricted to $y \leq 1$.

Range: $(-\infty, 1]$

3. Graphing a Parabola

Example 1 illustrates a process to sketch a quadratic function by finding the location of the defining characteristics of the function. These include the vertex and the x- and y-intercepts. Furthermore, notice that the parabola defining the graph of a quadratic function is symmetric with respect to the axis of symmetry.

To analyze the graph of a parabola, we recommend the following guidelines.

> **PROCEDURE** Graphing a Parabola
>
> Given a quadratic function defined by $y = ax^2 + bx + c$ $(a \neq 0)$, consider the following guidelines to graph the function.
>
> **Step 1** Determine whether the function opens upward or downward.
> - If $a > 0$, the parabola opens upward.
> - If $a < 0$, the parabola opens downward.
>
> **Step 2** Find the vertex.
>
> - The x-coordinate is given by $x = \dfrac{-b}{2a}$
> - To find the y-coordinate, substitute the x-coordinate of the vertex into the equation and solve for y.
>
> **Step 3** Find the x-intercept(s) by substituting $y = 0$ and solving the equation for x.
> - *Note:* If the solutions to the equation in step 3 are not real numbers, then the function has no x-intercepts.
>
> **Step 4** Find the y-intercept by substituting $x = 0$ and solving the equation for y.
>
> **Step 5** Plot the vertex and x- and y-intercepts. If necessary, find and plot additional points near the vertex. Then use the symmetry of the parabola to sketch the curve through the points. (*Note:* The axis of symmetry is the vertical line that passes through the vertex.)

Skill Practice

4. Graph $y = x^2 - 2x + 1$.

Classroom Example: p. 679, Exercise 42

Answer

4.

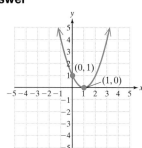

Example 2 Graphing a Parabola

Graph the function defined by $y = x^2 - 6x + 9$.

Solution:

1. The function $y = x^2 - 6x + 9$ is written in the form $y = ax^2 + bx + c$, where $a = 1, b = -6$, and $c = 9$. Because the value of a is positive, the parabola opens upward.

2. The x-coordinate of the vertex is given by

$$x = \frac{-b}{2a} = \frac{-(-6)}{2(1)} = 3$$

Substituting $x = 3$ into the equation, we have

$$y = (3)^2 - 6(3) + 9$$
$$= 9 - 18 + 9$$
$$= 0$$

The vertex is $(3, 0)$.

3. To find the x-intercept(s), substitute $y = 0$ and solve for x.

$$y = x^2 - 6x + 9 \longrightarrow 0 = x^2 - 6x + 9$$
$$0 = (x - 3)^2 \qquad \text{Factor.}$$
$$x = 3 \qquad \text{Apply the zero product rule.}$$

The x-intercept is $(3, 0)$.

4. To find the y-intercept, substitute $x = 0$ and solve for y.

$$y = x^2 - 6x + 9 \longrightarrow y = (0)^2 - 6(0) + 9$$
$$= 9$$

The y-intercept is $(0, 9)$.

5. Sketch the parabola through the x- and y-intercepts and vertex (Figure 9-4).

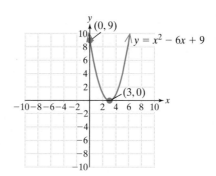

Figure 9-4

> **TIP:** Using the symmetry of the parabola, we know that the points to the right of the vertex must mirror the points to the left of the vertex.

Example 3   **Graphing a Parabola**

Graph the function defined by $y = -x^2 - 4$.

Solution:

1. The function $y = -x^2 - 4$ is written in the form $y = ax^2 + bx + c$, where $a = -1$, $b = 0$, and $c = -4$. Because the value of a is negative, the parabola opens downward.

2. The x-coordinate of the vertex is given by

$$x = \frac{-b}{2a} = \frac{-(0)}{2(-1)} = 0$$

Substituting $x = 0$ into the equation, we have

$$y = -(0)^2 - 4$$
$$= -4$$

The vertex is $(0, -4)$.

3. Because the vertex is below the x-axis and the parabola opens downward, the function cannot have an x-intercept.

> **TIP:** Substituting $y = 0$ into the equation $y = -x^2 - 4$ results in an equation with no real solutions. Therefore, the function $y = -x^2 - 4$ has no x-intercepts.
>
> $$y = -x^2 - 4$$
> $$0 = -x^2 - 4$$
> $$x^2 = -4$$
> $$x = \pm\sqrt{-4} \quad \text{Not a real number}$$

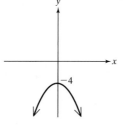

Skill Practice

5. Graph $y = x^2 + 1$.

Classroom Example: p. 680, Exercise 50

Answer

5.

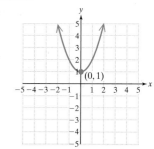

4. The vertex is $(0, -4)$. This is also the *y*-intercept.

5. Sketch the parabola through the *y*-intercept and vertex (Figure 9-5).

To verify the proper shape of the graph, find additional points to the right or left of the vertex and use the symmetry of the parabola to sketch the curve.

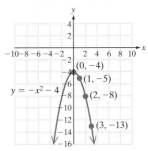

x	*y*
1	-5
2	-8
3	-13

Figure 9-5

4. Applications of Quadratic Functions

Skill Practice

6. A basketball player shoots a basketball at an angle of 45°. The height of the ball *y* (in feet) is given by $y = -16t^2 + 40t + 6$ where *t* is time in seconds. Find the maximum height of the ball and the time required to reach that height.

Classroom Example: p. 681, Exercise 56

Example 4   Using a Quadratic Function in an Application

A golfer hits a ball at an angle of 30°. The height of the ball *y* (in feet) can be represented by

$$y = -16x^2 + 60x$$

where *x* is the time in seconds after the ball was hit (Figure 9-6).

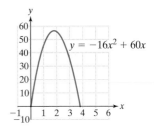

Figure 9-6

Find the maximum height of the ball. In how many seconds will the ball reach its maximum height?

Solution:

The function is written in the form $y = ax^2 + bx + c$, where $a = -16$, $b = 60$, and $c = 0$. Because *a* is negative, the parabola opens downward. Therefore, the maximum height of the ball occurs at the vertex of the parabola.

The *x*-coordinate of the vertex is given by

$$x = \frac{-b}{2a} = \frac{-(60)}{2(-16)} = \frac{-60}{-32} = \frac{15}{8}$$

Substituting $x = \frac{15}{8}$ into the equation, we have

$$y = -16\left(\frac{15}{8}\right)^2 + 60\left(\frac{15}{8}\right)$$

$$= -16\left(\frac{225}{64}\right) + \frac{900}{8}$$

$$= -\frac{225}{4} + \frac{450}{4}$$

$$= \frac{225}{4}$$

The vertex is $\left(\frac{15}{8}, \frac{225}{4}\right)$ or equivalently $(1.875, 56.25)$.

Answer

6. The ball reaches a maximum height of 31 ft in 1.25 sec.

The ball reaches its maximum height of 56.25 ft after 1.875 sec.

Calculator Connections

Some graphing calculators have *Minimum* and *Maximum* features that enable the user to approximate the minimum and maximum values of a function. Otherwise, *Zoom* and *Trace* can be used.

For example, the maximum value of the function from Example 4, $y = -16x^2 + 60x$, can be found using the *Maximum* feature.

The minimum value of the function from Example 2, $y = x^2 - 6x + 9$, can be found using the *Minimum* feature.

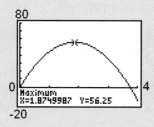 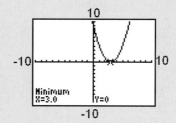

Calculator Exercises

Without using a calculator, find the vertex of each parabola. Then verify the answer using a graphing calculator.

1. $y = x^2 + 4x + 7$ $(-2, 3)$

2. $y = x^2 - 20x + 105$ $(10, 5)$

3. $y = -x^2 - 3x - 4.85$ $(-1.5, -2.6)$

4. $y = -x^2 + 3.5x - 0.5625$ $(1.75, 2.5)$

5. $y = 2x^2 - 10x + \dfrac{25}{2}$ $\left(\dfrac{5}{2}, 0\right)$

6. $y = 3x^2 + 16x + \dfrac{64}{3}$ $\left(-\dfrac{8}{3}, 0\right)$

7. Use a graphing calculator to graph the functions on the same screen. Describe the relationship between the graph of $y = x^2$ and the graphs in parts (b) and (c).   The graph in part (b) is shifted up 4 units. The graph in part (c) is shifted down 3 units.

    a. $y = x^2$ **b.** $y = x^2 + 4$ **c.** $y = x^2 - 3$

8. Use a graphing calculator to graph the functions on the same screen. Describe the relationship between the graph of $y = x^2$ and the graphs in parts (b) and (c).   In part (b), the graph is shifted to the right 3 units. In part (c), the graph is shifted to the left 2 units.

    a. $y = x^2$ **b.** $y = (x - 3)^2$ **c.** $y = (x + 2)^2$

9. Use a graphing calculator to graph the functions on the same screen. Describe the relationship between the graph of $y = x^2$ and the graphs in parts (b) and (c).   In part (b), the graph is stretched vertically by a factor of 2. In part (c), the graph is shrunk vertically by a factor of $\frac{1}{2}$.

    a. $y = x^2$ **b.** $y = 2x^2$ **c.** $y = \frac{1}{2}x^2$

10. Use a graphing calculator to graph the functions on the same screen. Describe the relationship between the graph of $y = x^2$ and the graphs in parts (b) and (c).   In part (b), the graph has been stretched vertically and reflected across the x-axis. In part (c), the graph has been shrunk vertically and reflected across the x-axis.

    a. $y = x^2$ **b.** $y = -2x^2$ **c.** $y = -\frac{1}{2}x^2$

Section 9.4 Practice Exercises

Study Skills Exercise

1. Define the key terms.

    a. quadratic function   **b. parabola**   **c. vertex of a parabola**   **d. axis of symmetry**

Review Exercises

For Exercises 2–8, solve the quadratic equations using any one of the following methods: factoring, the square root property, or the quadratic formula.

2. $3(y^2 + 1) = 10y$ $\frac{1}{3}, 3$ **3.** $3 + a(a + 2) = 18$ $-5, 3$ **4.** $4t^2 - 7 = 0$ $\pm\dfrac{\sqrt{7}}{2}$ **5.** $2z^2 + 4z - 10 = 0$

                                                                                  $1 \pm \sqrt{6}$

6. $(b + 1)^2 = 6$ **7.** $(x - 2)^2 = 8$ **8.** $3p^2 - 12p - 12 = 0$

   $-1 \pm \sqrt{6}$ $2 \pm 2\sqrt{2}$ $2 \pm 2\sqrt{2}$

Objective 1: Definition of a Quadratic Function

For Exercises 9–20, identify the equations as linear, quadratic, or neither.

9. $y = -8x + 3$ Linear   **10.** $y = 5x - 12$ Linear   **11.** $y = 4x^2 - 8x + 22$ **12.** $y = x^2 + 10x - 3$

                                                                      Quadratic   Quadratic

13. $y = -5x^3 - 8x + 14$ **14.** $y = -3x^4 + 7x - 11$ **15.** $y = 15x$ Linear   **16.** $y = -9x$ Linear

  Neither   Neither

17. $y = -21x^2$ Quadratic **18.** $y = 3x^2$ Quadratic   **19.** $y = -x^3 + 1$ Neither   **20.** $y = 7x^4 - 4$ Neither

Objective 2: Vertex of a Parabola

21. How do you determine whether the graph of a function $y = ax^2 + bx + c \ (a \neq 0)$ opens upward or downward?   If $a > 0$ the graph opens upward; if $a < 0$ the graph opens downward.

For Exercises 22–25, identify a and determine if the parabola opens upward or downward. **(See Example 1.)**

22. $y = x^2 - 15$ **23.** $y = 2x^2 + 23$ **24.** $y = -3x^2 + x - 18$ ●**25.** $y = -10x^2 - 6x - 20$

   $a = 1$; upward $a = 2$; upward $a = -3$; downward $a = -10$; downward

26. How do you find the vertex of a parabola?   Find the x-coordinate by $\dfrac{-b}{2a}$. Then substitute the value of x into the equation and solve for y.

For Exercises 27–34, find the vertex of the parabola. **(See Example 1.)**

●**27.** $y = 2x^2 + 4x - 6$ **28.** $y = x^2 - 4x - 4$ **29.** $y = -x^2 + 2x - 5$ **30.** $y = 2x^2 - 4x - 6$

  $(-1, -8)$ $(2, -8)$ $(1, -4)$ $(1, -8)$

31. $y = x^2 - 2x + 3$ **32.** $y = -x^2 + 4x - 2$ **33.** $y = 3x^2 - 4$ **34.** $y = 4x^2 - 1$

  $(1, 2)$ $(2, 2)$ $(0, -4)$ $(0, -1)$

Objective 3: Graphing a Parabola

For Exercises 35–38, find the x- and y-intercepts of the function. Then match each function with a graph. **(See Example 1.)**

35. $y = x^2 - 7$
c; x-intercepts: $(\sqrt{7}, 0)(-\sqrt{7}, 0)$;
y-intercept: $(0, -7)$

a.

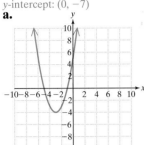

36. $y = x^2 - 9$
b; x-intercepts: $(3, 0)(-3, 0)$;
y-intercept: $(0, -9)$

b.

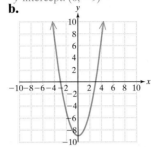

37. $y = (x + 3)^2 - 4$
a; x-intercepts: $(-1, 0)(-5, 0)$;
y-intercept: $(0, 5)$

c.

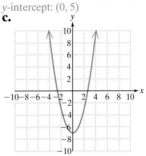

38. $y = (x - 2)^2 - 1$
d; x-intercepts: $(3, 0)(1, 0)$;
y-intercept: $(0, 3)$

d.

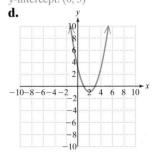

For Exercises 39–50, **(See Examples 1–3.)**

   a. Determine whether the graph of the parabola opens upward or downward.

   b. Find the vertex.

   c. Find the x-intercept(s), if possible.

   d. Find the y-intercept.

   e. Sketch the function.

   f. Identify the domain of the function.

   g. Identify the range of the function.

39. $y = x^2 - 9$ e.
   a. Upward
   b. $(0, -9)$
   c. $(3, 0)(-3, 0)$
   d. $(0, -9)$
   f. $(-\infty, \infty)$
   g. $[-9, \infty)$

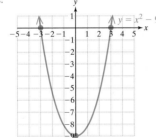

40. $y = x^2 - 4$ e.
   a. Upward
   b. $(0, -4)$
   c. $(2, 0)(-2, 0)$
   d. $(0, -4)$
   f. $(-\infty, \infty)$
   g. $[-4, \infty)$

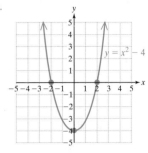

41. $y = x^2 - 2x - 8$ e.
   a. Upward
   b. $(1, -9)$
   c. $(4, 0)(-2, 0)$
   d. $(0, -8)$
   f. $(-\infty, \infty)$
   g. $[-9, \infty)$

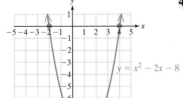

42. $y = x^2 + 2x - 24$ e.
   a. Upward
   b. $(-1, -25)$
   c. $(4, 0)(-6, 0)$
   d. $(0, -24)$
   f. $(-\infty, \infty)$
   g. $[-25, \infty)$

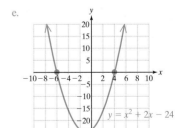

43. $y = -x^2 + 6x - 9$ e.
   a. Downward
   b. (3, 0)
   c. (3, 0)
   d. (0, −9)
   f. (−∞, ∞)
   g. (−∞, 0]

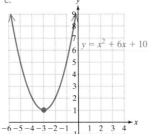

44. $y = -x^2 + 10x - 25$ e.
   a. Downward
   b. (5, 0)
   c. (5, 0)
   d. (0, −25)
   f. (−∞, ∞)
   g. (−∞, 0]

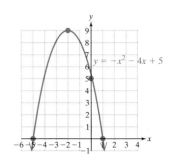

45. $y = -x^2 + 8x - 15$ e.
   a. Downward
   b. (4, 1)
   c. (3, 0)(5, 0)
   d. (0, −15)
   f. (−∞, ∞)
   g. (−∞, 1]

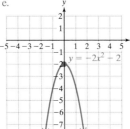

46. $y = -x^2 - 4x + 5$ e.
   a. Downward
   b. (−2, 9)
   c. (−5, 0)(1, 0)
   d. (0, 5)
   f. (−∞, ∞)
   g. (−∞, 9]

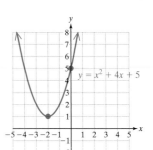

47. $y = x^2 + 6x + 10$ e.
   a. Upward
   b. (−3, 1)
   c. none
   d. (0, 10)
   f. (−∞, ∞)
   g. [1, −∞)

48. $y = x^2 + 4x + 5$ e.
   a. Upward
   b. (−2, 1)
   c. none
   d. (0, 5)
   f. (−∞, ∞)
   g. [1, ∞)

49. $y = -2x^2 - 2$ e.
   a. Downward
   b. (0, −2)
   c. none
   d. (0, −2)
   f. (−∞, ∞)
   g. (−∞, −2]

50. $y = -x^2 - 5$ e.
   a. Downward
   b. (0, −5)
   c. none
   d. (0, −5)
   f. (−∞, ∞)
   g. (−∞, −5]

51. True or False: The function $y = -5x^2$ has a maximum value but no minimum value.   True

52. True or False: The graph of $y = -4x^2 + 9x - 6$ opens upward.   False

53. True or False: The graph of $y = 1.5x^2 - 6x - 3$ opens downward.   False

54. True or False: The function $y = 2x^2 - 5x + 4$ has a maximum value but no minimum value.   False

 Writing   Translating Expression   Geometry   Scientific Calculator   Video   NS E

Objective 4: Applications of Quadratic Functions

55. A child kicks a ball into the air, and the height of the ball, y (in feet), can be approximated by

$y = -16t^2 + 40t + 3$ where t is the number of seconds after the ball was kicked.

   a. Find the maximum height of the ball. **(See Example 4.)**   28 ft

   b. How long will it take the ball to reach its maximum height?   1.25 sec

56. A concession stand at the Arthur Ashe Tennis Center sells a hamburger/drink combination dinner for $5. The profit, y (in dollars), can be approximated by

$y = -0.001x^2 + 3.6x - 400$ where x is the number of dinners prepared.

   a. Find the number of dinners that should be prepared to maximize profit.   1800 dinners

   b. What is the maximum profit? $2840

57. For a fund raising activity, a charitable organization produces calendars to sell in the community. The profit, y (in dollars), can be approximated by

$y = -\dfrac{1}{40}x^2 + 10x - 500$ where x is the number of calendars produced.

   a. Find the number of calendars that should be produced to maximize the profit.   200 calendars

   b. What is the maximum profit? $500

58. The pressure, x, in an automobile tire can affect its wear. Both over-inflated and under-inflated tires can lead to poor performance and poor mileage. For one particular tire, the number of miles that a tire lasts, y (in thousands), is given by

$y = -0.875x^2 + 57.25x - 900$ where x is the tire pressure in pounds per square inch (psi).

   a. Find the tire pressure that will yield the maximum number of miles that a tire will last. Round to the nearest whole unit.   33 psi

   b. Find the maximum number of miles that a tire will last if the proper tire pressure is maintained. Round to the nearest thousand miles.   36 thousand miles or 36,000 miles

Group Activity

Maximizing Volume

Materials: A calculator and a sheet of $8\frac{1}{2}$ by 11 in. paper.

Estimated Time: 25–30 minutes

Group Size: 3

Antonio is going to build a custom gutter system for his house. He plans to use rectangular strips of aluminum that are $8\frac{1}{2}$ in. wide and 72 in. long. Each piece of aluminum will be turned up at a distance of x in. from the sides to form a gutter. See Figure 9-7.

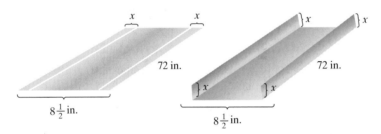

Figure 9-7

Antonio wants to maximize the volume of water that the gutters can hold. To do this, he must determine the distance, x, that should be turned up to form the height of the gutter.

1. To familiarize yourself with this problem, we will simulate the gutter problem using an $8\frac{1}{2}$ by 11 in. piece of paper. For each value of x in the table, turn up the sides of the paper x in. from the edge. Then measure the base, the height, and the length of the paper "gutter," and calculate the volume.

Height, x	Base	Length	Volume
0.5 in.	7.5 in.	11 in.	41.25 in.^{3}
1.0 in.	6.5 in.	11 in.	71.5 in.^{3}
1.5 in.	5.5 in.	11 in.	90.75 in.^{3}
2.0 in.	4.5 in.	11 in.	99 in.^{3}
2.5 in.	3.5 in.	11 in.	96.25 in.^{3}
3.0 in.	2.5 in.	11 in.	82.5 in.^{3}
3.5 in.	1.5 in.	11 in.	57.75 in.^{3}

2. Now follow these steps to find the optimal distance, x, that you should fold the paper to make the greatest volume within the paper gutter.

   a. If the height of the paper gutter is x in., write an expression for the base of the paper gutter. $8.5 - 2x$

   b. Write a function for the volume of the paper gutter. $V = (8.5 - 2x)(x)(11)$ or $V = -22x^2 + 93.5x$

   c. Find the vertex of the parabola defined by the function in part (b).
   $(2.125, 99.34375)$

   d. Interpret the meaning of the vertex from part (c).
   Fold the paper 2.125 in. from the edge. This will produce a maximum volume of 99.34375 in.^{3}

3. Use the concept from the paper gutter to write a function for the volume of Antonio's aluminum gutter that is 72 in. long. $V = (8.5 - 2x)(x)(72)$ or $V = -144x^2 + 612x$

4. Now find the optimal distance, x, that he should fold the aluminum sheet to make the greatest volume within the 72-in.-long aluminum gutter. What is the maximum volume?
Fold the aluminum 2.125 in. from the edge. This will produce a maximum volume of 650.25 in.^{3}

Chapter 9   Summary

Section 9.1   The Square Root Property

Key Concepts

Square Root Property

If $x^2 = k$, then $x = \pm\sqrt{k}$.

The square root property can be used to solve a quadratic equation written as a square of a binomial equal to a constant.

Example

Example 1

$$(x - 5)^2 = 13$$

$x - 5 = \pm\sqrt{13}$ Square root property

$x = 5 \pm \sqrt{13}$ Solve for x.

Section 9.2   Completing the Square

Key Concepts

Solving a Quadratic Equation of the Form $ax^2 + bx + c = 0$ ($a \neq 0$) by Completing the Square and Applying the Square Root Property

1. Divide both sides by a to make the leading coefficient 1.
2. Isolate the variable terms on one side of the equation.
3. Complete the square by adding the square of $\frac{1}{2}$ the linear term coefficient to both sides of the equation. Then factor the resulting perfect square trinomial.
4. Apply the square root property and solve for x.

Example

Example 1

$$2x^2 - 8x - 6 = 0$$

Step 1: $\dfrac{2x^2}{2} - \dfrac{8x}{2} - \dfrac{6}{2} = \dfrac{0}{2}$

$x^2 - 4x - 3 = 0$

Step 2: $x^2 - 4x = 3$

Step 3: $x^2 - 4x + 4 = 3 + 4$ Note that

$(x - 2)^2 = 7$ $[\frac{1}{2}(-4)]^2 = (-2)^2 = 4$

Step 4: $x - 2 = \pm\sqrt{7}$

$x = 2 \pm \sqrt{7}$

Section 9.3 Quadratic Formula

Key Concepts

The solutions to a quadratic equation of the form $ax^2 + bx + c = 0$ $(a \neq 0)$ are given by the **quadratic formula**:

$$x = \frac{-b \pm \sqrt{b^2 - 4ac}}{2a}$$

Three Methods for Solving a Quadratic Equation

1. Factoring
2. Completing the square and applying the square root property
3. Using the quadratic formula

Example

Example 1

$$3x^2 = 2x + 4$$

$$3x^2 - 2x - 4 = 0 \qquad a = 3, b = -2, c = -4$$

$$x = \frac{-(-2) \pm \sqrt{(-2)^2 - 4(3)(-4)}}{2(3)}$$

$$= \frac{2 \pm \sqrt{4 + 48}}{6}$$

$$= \frac{2 \pm \sqrt{52}}{6}$$

$$= \frac{2 \pm 2\sqrt{13}}{6} \qquad \text{Simplify the radical.}$$

$$= \frac{2(1 \pm \sqrt{13})}{6} \qquad \text{Factor.}$$

$$= \frac{1 \pm \sqrt{13}}{3} \qquad \text{Simplify.}$$

Section 9.4   Graphing Quadratic Functions

Key Concepts

Let a, b, and c represent real numbers such that $a \neq 0$. Then a function defined by $y = ax^2 + bx + c$ is called a **quadratic function**.

The graph of a quadratic function is called a **parabola**.

The leading coefficient, a, of a quadratic function, $y = ax^2 + bx + c$, determines if the parabola will open upward or downward. If $a > 0$, then the parabola opens upward. If $a < 0$, then the parabola opens downward.

Examples

Example 1

$y = x^2 - 4x - 3$ is a quadratic function. Its graph is in the shape of a parabola.

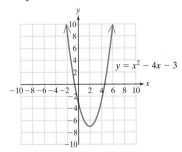

Finding the Vertex of a Parabola

1. The x-coordinate of the vertex of the parabola with equation $y = ax^2 + bx + c$ ($a \neq 0$) is given by

$$x = \frac{-b}{2a}$$

2. To find the corresponding y-coordinate of the vertex, substitute the value of the x-coordinate found in step 1 and solve for y.

If a parabola opens upward, the vertex is the lowest point on the graph. If a parabola opens downward, the vertex is the highest point on the graph.

Example 2

Find the vertex of the parabola defined by $y = 3x^2 + 6x - 1$.

$$x = \frac{-b}{2a} = \frac{-6}{2 \cdot 3} = -1$$

$$y = 3(-1)^2 + 6(-1) - 1 = -4$$

The vertex is $(-1, -4)$.

For the equation $y = 3x^2 + 6x - 1$, $a = 3 > 0$. Therefore, the parabola opens upward. The vertex $(-1, -4)$ represents the lowest point of the graph.

Chapter 9   Review Exercises

Section 9.1

For Exercises 1–4, identify the equations as linear or quadratic.

1. $5x - 10 = 3x - 6$
Linear

2. $(x + 6)^2 = 6$
Quadratic

3. $x(x - 4) = 5x - 2$
Quadratic

4. $3(x + 6) = 18(x - 1)$
Linear

For Exercises 5–12, solve the equations using the square root property.

5. $x^2 = 25$ ± 5

6. $x^2 - 19 = 0$ $\pm\sqrt{19}$

7. $x^2 + 49 = 0$ The equation has no real-valued solutions.

8. $x^2 = -48$ The equation has no real-valued solutions.

9. $(x + 1)^2 = 14$
$-1 \pm \sqrt{14}$

10. $(x - 2)^2 = 60$
$2 \pm 2\sqrt{15}$

11. $\left(x - \dfrac{1}{8}\right)^2 = \dfrac{3}{64}$ $\dfrac{1}{8} \pm \dfrac{\sqrt{3}}{8}$

12. $(2x - 3)^2 = 20$ $\dfrac{3 \pm 2\sqrt{5}}{2}$

Section 9.2

For Exercises 13–16, find the constant that should be added to each expression to make it a perfect square trinomial.

13. $x^2 + 12x$ 36

14. $x^2 - 18x$ 81

15. $x^2 - 5x$ $\dfrac{25}{4}$

16. $x^2 + 7x$ $\dfrac{49}{4}$

For Exercises 17–20, solve the quadratic equations by completing the square and applying the square root property.

17. $x^2 + 8x + 3 = 0$
$-4 \pm \sqrt{13}$

18. $x^2 - 2x - 4 = 0$
$1 \pm \sqrt{5}$

19. $2x^2 - 6x - 6 = 0$
$\dfrac{3}{2} \pm \dfrac{\sqrt{21}}{2}$

20. $3x^2 - 7x - 3 = 0$
$\dfrac{7 \pm \sqrt{85}}{6}$

21. A right triangle has legs of equal length. If the hypotenuse is 15 ft long, find the length of each leg. Round the answer to the nearest tenth of a foot. 10.6 ft

22. A can in the shape of a right circular cylinder holds approximately 362 cm^3 of liquid. If the height of the can is 12.1 cm, find the radius of the can. Round to the nearest tenth of a centimeter. (*Hint:* The volume of a right circular cylinder is given by: $V = \pi r^2 h$) 3.1 cm

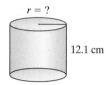

$r = ?$

12.1 cm

Section 9.3

23. Write the quadratic formula from memory.

For $ax^2 + bx + c = 0, x = \dfrac{-b \pm \sqrt{b^2 - 4ac}}{2a}$

For Exercises 24–33, solve the quadratic equations using the quadratic formula.

24. $5x^2 + x - 7 = 0$
$\dfrac{-1 \pm \sqrt{141}}{10}$

25. $x^2 + 4x + 4 = 0$
-2

26. $3x^2 - 2x + 2 = 0$
The equation has no real-valued solutions.

27. $2x^2 - x - 3 = 0$
$\dfrac{3}{2}, -1$

28. $\dfrac{1}{8}x^2 + x = \dfrac{5}{2}$
$-10, 2$

29. $\dfrac{1}{6}x^2 + x + \dfrac{1}{3} = 0$
$-3 \pm \sqrt{7}$

30. $1.2x^2 + 6x = 7.2$ $1, -6$

31. $0.01x^2 - 0.02x - 0.04 = 0$ $1 \pm \sqrt{5}$

32. $(x + 6)(x + 2) = 10$ $-4 \pm \sqrt{14}$

33. $(x - 1)(x - 7) = -18$
The equation has no real-valued solutions.

34. One number is two more than another number. Their product is 11.25. Find the numbers.
The numbers are −2.5 and −4.5, or 2.5 and 4.5.

35. The base of a parallelogram is 1 cm longer than the height, and the area is 24 cm^2. Find the exact values of the base and height of the parallelogram. Then use a calculator to approximate the values to the nearest tenth of a centimeter.

The height is $\dfrac{-1 + \sqrt{97}}{2} \approx 4.4$ cm.   The base is $\dfrac{1 + \sqrt{97}}{2} \approx 5.4$ cm.

36. An astronaut on the moon tosses a rock upward with an initial velocity of 25 ft/sec. The height of the rock, $h(t)$ (in feet), is determined by the number of seconds, t, after the rock is released according to the equation:

$$h(t) = -2.7t^2 + 25t + 5$$

Find the time required for the rock to hit the ground. [*Hint:* At ground level, $h(t) = 0$.] Round to the nearest tenth of a second. 9.5 sec

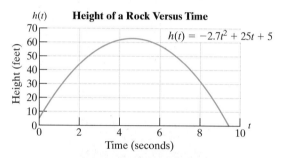

Height of a Rock Versus Time

$h(t) = -2.7t^2 + 25t + 5$

Height (feet)

Time (seconds)

Section 9.4

For Exercises 37–40, identify a and determine if the parabola opens upward or downward.

37. $y = x^2 - 3x + 1$
$a = 1$; upward

38. $y = -x^2 + 8x + 2$
$a = -1$; downward

39. $y = -2x^2 + x - 12$
$a = -2$; downward

40. $y = 5x^2 - 2x - 6$
$a = 5$; upward

For Exercises 41–44, find the vertex for each parabola.

41. $y = 3x^2 + 6x + 4$
Vertex: $(-1, 1)$

42. $y = -x^2 + 8x + 3$
Vertex: $(4, 19)$

43. $y = -2x^2 + 12x - 5$
Vertex: $(3, 13)$

44. $y = 2x^2 + 2x - 1$
Vertex: $\left(-\dfrac{1}{2}, -\dfrac{3}{2}\right)$

For Exercises 45–48,

a. Determine whether the graph of the parabola opens upward or downward.

b. Find the vertex.

c. Find the x-intercept(s) if possible.

d. Find the y-intercept.

e. Sketch the function.

f. Write the domain in interval notation.

g. Write the range in interval notation.

45. $y = 3x^2 + 12x + 9$ e.

a. Upward
b. $(-2, -3)$
c. $(-3, 0)(-1, 0)$
d. $(0, 9)$
f. $(-\infty, \infty)$
g. $[-3, \infty)$

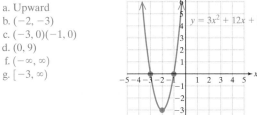

46. $y = -3x^2 + 12x - 10$ e.

a. Downward
b. $(2, 2)$
c. Approximately $(2.82, 0)(1.18, 0)$
d. $(0, -10)$
f. $(-\infty, \infty)$
g. $(-\infty, 2]$

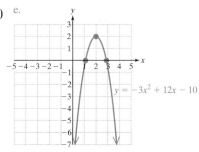

47. $y = -8x^2 - 16x - 12$ e.

a. Downward
b. $(-1, -4)$
c. No x-intercepts
d. $(0, -12)$
f. $(-\infty, \infty)$
g. $(-\infty, -4]$

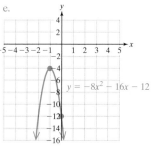

48. $y = 2x^2 + 4x - 1$ e.

a. Upward
b. $(-1, -3)$
c. Approximately $(0.22, 0)(-2.22, 0)$
d. $(0, -1)$
f. $(-\infty, \infty)$
g. $[-3, \infty)$

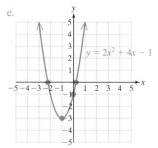

49. An object is launched into the air from ground level with an initial velocity of 256 ft/sec. The height of the object, y (in feet), can be approximated by the function

$$y = -16t^2 + 256t \qquad \text{where } t \text{ is the number of seconds after launch.}$$

a. Find the maximum height of the object.   1024 ft

b. Find the time required for the object to reach its maximum height.   8 sec

Chapter 9   Test

1. Solve the equation by applying the square root property.   The equation has no real-valued solutions.

$$(3x + 1)^2 = -14$$

2. Solve the equation by completing the square and applying the square root property. $4 \pm \sqrt{21}$

$$x^2 - 8x - 5 = 0$$

3. Solve the equation by using the quadratic formula.

$$3x^2 - 5x = -1 \qquad \frac{5 \pm \sqrt{13}}{6}$$

For Exercises 4–10, solve the equations using any method.

4. $5x^2 + x - 2 = 0$ $\frac{-1 \pm \sqrt{41}}{10}$

5. $(c - 12)^2 = 12$ $12 \pm 2\sqrt{3}$

6. $y^2 + 14y - 1 = 0$ $-7 \pm 5\sqrt{2}$

7. $3t^2 = 30$ $\pm\sqrt{10}$

8. $4x(3x + 2) = 15$ $\frac{5}{6}, -\frac{3}{2}$

9. $6p^2 - 11p = 0$ $0, \frac{11}{6}$

10. $\frac{1}{4}x^2 - \frac{3}{2}x = \frac{11}{4}$ $3 \pm 2\sqrt{5}$

11. The surface area, S, of a sphere is given by the formula $S = 4\pi r^2$, where r is the radius of the sphere. Find the radius of a sphere whose surface area is 201 in.^{2} Round to the nearest tenth of an inch.   4.0 in.

$S = 201$ in.^{2}

12. The height of a triangle is 2 m longer than twice the base, and the area is 24 m^2. Find the exact values of the base and height. Then use a calculator to approximate the base and height to the nearest tenth of a meter.

The base is $\frac{-1 + \sqrt{97}}{2} \approx 4.4$ m. The height is $1 + \sqrt{97} \approx 10.8$ m.

13. Explain how to determine if a parabola opens upward or downward.   For $y = ax^2 + bx + c$, if $a > 0$ the parabola opens upward, if $a < 0$ the parabola opens downward.

For Exercises 14–16, find the vertex of the parabola.

14. $y = x^2 - 10x + 25$ **15.** $y = 3x^2 - 6x + 8$
$(5, 0)$ $(1, 5)$

16. $y = -x^2 - 16$ $(0, -16)$

17. Suppose a parabola opens upward and the vertex is located at $(-4, 3)$. How many x-intercepts does the parabola have?
The parabola has no x-intercepts.

18. Given the parabola, $y = x^2 + 6x + 8$

   a. Determine whether the parabola opens upward or downward.   Opens upward

   b. Find the vertex of the parabola.   Vertex: $(-3, -1)$

   c. Find the x-intercepts.
     x-intercepts: $(-2, 0)$ and $(-4, 0)$

   d. Find the y-intercept. y-intercept: $(0, 8)$

   e. Graph the parabola.

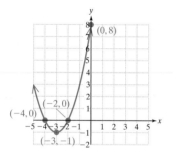

19. Graph the parabola and label the vertex, x-intercepts, and y-intercept.

$$y = -x^2 + 25$$

Vertex: $(0, 25)$; x-intercepts: $(-5, 0)(5, 0)$; y-intercept: $(0, 25)$

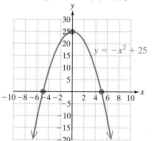

20. The Phelps Arena in Atlanta holds 20,000 seats. If the Atlanta Hawks charge x dollars per ticket for a game, then the total revenue, y (in dollars), can be approximated by

$$y = -400x^2 + 20,000x$$ where x is the price per ticket.

   a. Find the ticket price that will produce the maximum revenue. $25 per ticket

   b. What is the maximum revenue? $250,000

Chapters 1–9   Cumulative Review Exercises

1. Solve. $3x - 5 = 2(x - 2)$ 1

2. Solve for h. $A = \frac{1}{2}bh$ $h = \frac{2A}{b}$

3. Solve. $\frac{1}{2}y - \frac{5}{6} = \frac{1}{4}y + 2$ $\frac{34}{3}$

4. Determine whether 2 is a solution to the inequality. $-3x + 4 < x + 8$
Yes, 2 is a solution.

5. Graph the solution to the inequality: $-3x + 4 < x + 8$. Then write the solution in set-builder notation and in interval notation.

$\{x \mid x > -1\}; (-1, \infty)$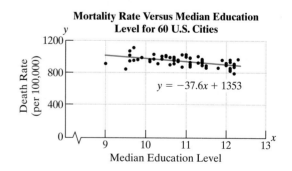

6. The graph depicts the death rate from 60 U.S. cities versus the median education level of the people living in that city. The death rate, y, is measured in number of deaths per 100,000 people. The median education level, x, is a type of "average" and is measured by grade level. (*Source:* U.S. Bureau of the Census)

The death rate can be predicted from the median education level according to the equation:

$$y = -37.6x + 1353$$ where $9 \le x \le 13$.

Mortality Rate Versus Median Education Level for 60 U.S. Cities

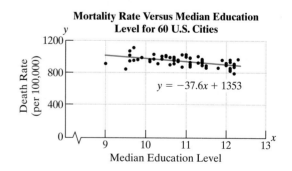

   a. From the graph, does it appear that the death rate increases or decreases as the median education level increases?   Decreases

   b. What is the slope of the line? Interpret the slope in the context of the death rate and education level. $m = -37.6$. For each additional increase in education level, the death rate decreases by approximately 38 deaths per 100,000 people.

c. For a city in the United States with a median education level of 12, what would be the expected death rate?   901.8 per 100,000

d. If the death rate of a certain city is 977 per 100,000 people, what would be the approximate median education level?
10th grade

7. Simplify completely. Write the final answer with positive exponents only.

$$\left(\frac{2a^2b^{-3}}{c}\right)^{-1} \cdot \left(\frac{4a^{-1}}{b^2}\right)^2 \quad \frac{8c}{a^4b}$$

8. Approximately 5.2×10^7 disposable diapers are thrown into the trash each day in the United States and Canada. How many diapers are thrown away each year? 1.898×10^{10} diapers

9. In 1989, the Hipparcos satellite found the distance between Earth and the star, Polaris, to be approximately 2.53×10^{15} mi. If 1 light-year is approximately 5.88×10^{12} miles, how many light-years is Polaris from Earth?
Approximately 430 light-years

10. Perform the indicated operation.
$(2x - 3)^2 - 4(x - 1)$ $4x^2 - 16x + 13$

11. Divide using long division.
$(2y^4 - 4y^3 + y - 5) \div (y - 2)$ $2y^3 + 1 - \dfrac{3}{y - 2}$

12. Factor. $2x^2 - 9x - 35$ $(2x + 5)(x - 7)$

13. Factor completely. $2xy + 8xa - 3by - 12ab$
$(y + 4a)(2x - 3b)$

14. The base of a triangle is 1 m more than the height. If the area is 36 m², find the base and height.
The base is 9 m, and the height is 8 m.

15. Multiply. $\dfrac{x^2 + 10x + 9}{x^2 - 81} \cdot \dfrac{18 - 2x}{x^2 + 2x + 1}$ $-\dfrac{2}{x + 1}$

16. Reduce to lowest terms.
$$\frac{5x + 10}{x^2 - 4} \quad \frac{5}{x - 2}$$

17. Perform the indicated operations.
$$\frac{x^2}{x - 5} - \frac{10x - 25}{x - 5} \quad x - 5$$

18. Simplify completely.
$$\frac{\dfrac{1}{x + 1} - \dfrac{1}{x - 1}}{\dfrac{x}{x^2 - 1}} \quad -\frac{2}{x}$$

19. Solve.
$$1 - \frac{1}{y} = \frac{12}{y^2} \quad 4, -3$$

20. At Lake George, a fishing enthusiast must release any fish that is less than 8 inches in length. If a fish is caught and measured at 20 cm, must the fish be released back to the lake? (*Hint:* 1 in. = 2.54 cm)
20 cm ≈ 7.9 in. The fish must be released.

21. Write an equation of the line passing through the point $(-2, 3)$ and having a slope of $\frac{1}{2}$. Write the final answer in slope-intercept form.
$y = \dfrac{1}{2}x + 4$

For Exercises 22–23,

a. Find the x-intercept (if it exists).

b. Find the y-intercept (if it exists).

c. Find the slope (if it exists).

d. Graph the line.

22. $2x - 4y = 12$
a. $(6, 0)$
b. $(0, -3)$
c. $\frac{1}{2}$

   d.
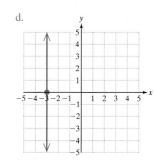

23. $4x + 12 = 0$
a. $(-3, 0)$
b. No y-intercept
c. Slope is undefined.

   d.
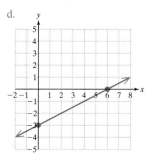

24. Solve the system by using the addition method. If the system has no solution or infinitely many solutions, so state.
$$\frac{1}{2}x - \frac{1}{4}y = \frac{1}{6} \quad \left(1, \frac{4}{3}\right)$$
$$12x - 3y = 8$$

Writing   Translating Expression   Geometry   Scientific Calculator   Video   NSF

25. Solve the system by using the substitution method. If the system has no solution or infinitely many solutions, so state.

$$2x - y = 8 \quad (5, 2)$$

$$4x - 4y = 3x - 3$$

26. In a right triangle, one acute angle is 2° more than three times the other acute angle. Find the measure of each angle. The angles are 22° and 68°.

27. A bank of 27 coins contains only dimes and quarters. The total value of the coins is $4.80. Find the number of dimes and the number of quarters.   There are 13 dimes and 14 quarters.

28. Sketch the inequality, $x - y \le 4$.

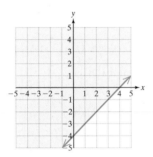

29. Which of the following are irrational numbers?
$\{0, -\frac{2}{3}, \pi, \sqrt{7}, 1.2, \sqrt{25}\}$ $\pi, \sqrt{7}$

For Exercises 30–31, simplify the radicals.

30. $\sqrt{\dfrac{1}{7}}$ $\dfrac{\sqrt{7}}{7}$

31. $\dfrac{\sqrt{16x^4}}{\sqrt{2x}}$ $2x\sqrt{2x}$

32. Perform the indicated operation. $(4\sqrt{3} + \sqrt{x})^2$
$48 + 8\sqrt{3x} + x$

33. Add the radicals. $-3\sqrt{2x} + \sqrt{50x}$ $2\sqrt{2x}$

34. Rationalize the denominator.

$$\frac{4}{2 - \sqrt{a}} \quad \frac{8 + 4\sqrt{a}}{4 - a}$$

35. Solve. $\sqrt{x + 11} = x + 5$
-2, (the value -7 does not check)

36. Factor completely. $8c^3 - y^3$
$(2c - y)(4c^2 + 2cy + y^2)$

37. Which graph defines y as a function of x?   b

a.   **b.**

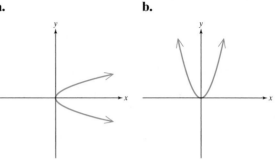

38. Given the functions defined by $f(x) = -\frac{1}{2}x + 4$ and $g(x) = x^2$, find

   a. $f(6)$ **b.** $g(-2)$ **c.** $f(0) + g(3)$
      1   4   13

39. Find the domain and range of the function.
$\{(2, 4), (-1, 3), (9, 2), (-6, 8)\}$
Domain: $\{2, -1, 9, -6\}$; range: $\{4, 3, 2, 8\}$

40. Find the slope of the line passing through the points $(3, -1)$ and $(-4, -6)$. $m = \dfrac{5}{7}$

41. Find the slope of the line defined by
$-4x - 5y = 10$. $m = -\dfrac{4}{5}$

42. What value of k would make the expression a perfect square trinomial?   25

$$x^2 + 10x + k$$

43. Solve the quadratic equation by completing the square and applying the square root property.
$2x^2 + 12x + 6 = 0$. $-3 \pm \sqrt{6}$

44. Solve the quadratic equation by using the quadratic formula. $2x^2 + 12x + 6 = 0$. $-3 \pm \sqrt{6}$

45. Graph the parabola defined by the equation. Label the vertex, x-intercepts, and y-intercept.

$$y = x^2 + 4x + 4$$

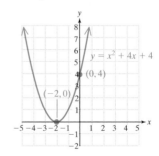

Student Answer Appendix

Chapter R

Chapter Opener Puzzles

$$\underset{1}{\underline{d}}\,\underset{2}{\underline{e}}\,\underset{3}{\underline{c}}\,\underset{4}{\underline{i}}\,\underset{5}{\underline{m}}\,\underset{6}{\underline{a}}\,\underset{7}{\underline{l}}\,\underset{7}{\underline{l}}.$$

Section R.2 Practice Exercises, pp. 17–20

1. Numerator: 7; denominator: 8; proper
3. Numerator: 9; denominator: 5; improper
5. Numerator: 6; denominator: 6; improper
7. Numerator: 12; denominator: 1; improper
9. $\frac{3}{4}$ **11.** $\frac{4}{3}$ **13.** $\frac{1}{6}$ **15.** $\frac{2}{2}$
17. $\frac{5}{2}$ or $2\frac{1}{2}$ **19.** $\frac{6}{2}$ or 3
21. The set of whole numbers includes the number 0 and the set of natural numbers does not.
23. For example: $\frac{2}{4}$ **25.** Prime **27.** Composite
29. Composite **31.** Prime **33.** $2 \times 2 \times 3 \times 3$
35. $2 \times 3 \times 7$ **37.** $2 \times 5 \times 11$ **39.** $3 \times 3 \times 3 \times 5$
41. $\frac{1}{5}$ **43.** $\frac{3}{8}$ **45.** $\frac{7}{8}$ **47.** $\frac{3}{4}$ **49.** $\frac{5}{8}$ **51.** $\frac{3}{4}$
53. False: When adding or subtracting fractions, it is necessary to have a common denominator.
55. $\frac{4}{3}$ **57.** $\frac{2}{3}$ **59.** $\frac{9}{2}$ **61.** $\frac{3}{5}$ **63.** $\frac{5}{3}$
65. $\frac{90}{13}$ **67.** $1050 **69.** 4 graduated with honors.
71. 8 aprons **73.** 8 jars **75.** $\frac{3}{7}$ **77.** $\frac{1}{2}$ **79.** 30
81. 40 **83.** $\frac{7}{8}$ **85.** $\frac{3}{40}$ **87.** $\frac{3}{26}$ **89.** $\frac{29}{36}$
91. $\frac{7}{10}$ **93.** $\frac{35}{48}$ **95.** $\frac{7}{24}$ **97.** $\frac{51}{28}$ or $1\frac{23}{28}$ **99.** 46
101. $\frac{14}{5}$ or $2\frac{4}{5}$ **103.** $\frac{46}{5}$ or $9\frac{1}{5}$ **105.** $\frac{1}{6}$ **107.** $\frac{11}{54}$
109. $\frac{7}{2}$ or $3\frac{1}{2}$ **111.** $\frac{13}{8}$ or $1\frac{5}{8}$ **113.** $\frac{59}{12}$ or $4\frac{11}{12}$ **115.** $\frac{1}{8}$
117. $8\frac{19}{24}$ in. **119.** $1\frac{7}{12}$ hr **121.** $2\frac{1}{4}$ lb **123.** 25 in.

Section R.3 Calculator Connections, p. 26

1. $0.\overline{4}$ **2.** $0.\overline{63}$ **3.** $0.1\overline{36}$ **4.** $0.\overline{384615}$
1.–2.
```
4/9
        .4444444444
7/11
        .6363636364
```
3.–4.
```
3/22
        .1363636364
5/13
        .3846153846
```

Section R.3 Practice Exercises, pp. 27–29

1. Tens **3.** Hundreds **5.** Tenths **7.** Hundredths
9. No, the symbols I, V, X, and so on each represent certain values but the values are not dependent on the position of the symbol within the number. **11.** 0.7 **13.** 0.36
15. $1.\overline{2}$ **17.** $0.\overline{21}$ **19.** 214.1 **21.** 39.268
23. 40,000 **25.** 0.73 **27.** $\frac{9}{20}$ **29.** $\frac{181}{1000}$
31. $\frac{51}{25}$ or $2\frac{1}{25}$ **33.** $\frac{13,007}{1000}$ or $13\frac{7}{1000}$ **35.** $\frac{5}{9}$
37. $\frac{10}{9}$ or $1\frac{1}{9}$ **39.** $0.3, \frac{3}{10}$ **41.** $0.75, \frac{3}{4}$ **43.** $0.0375, \frac{3}{80}$
45. $0.157, \frac{157}{1000}$ **47.** $2.7, \frac{27}{10}$ **49.** Multiply by 100%.
51. 5% **53.** 90% **55.** 120% **57.** 750%
59. 13.5% **61.** 0.3% **63.** 6% **65.** 450%
67. 62.5% **69.** 31.25% **71.** $83.\overline{3}\%$ **73.** $93.\overline{3}\%$
75. $42 **77.** $3375 **79.** 7% **81.** $792
83. $192 **85.** $67,500

Section R.4 Practice Exercises, pp. 37–43

1. b, e, i **3.** 32 m **5.** 17.2 mi **7.** $11\frac{1}{2}$ in.
9. 31.4 ft **11.** a, f, g **13.** 33 cm^2 **15.** 16.81 m^2
17. 84 in.^{2} **19.** 10.12 km^2 **21.** 13.8474 ft^2
23. 66 in.^{2} **25.** 31.5 ft^2 **27.** c, d, h **29.** 307.72 cm^3
31. 39 in.^{3} **33.** 113.04 cm^3 **35.** 1695.6 cm^3
37. 3052.08 in.^{3} **39.** 113.04 cm^3
41. a. $0.25/ft^2 **b.** $104 **43.** Perimeter **45.** 54 ft
47. a. 57,600 ft^2 **b.** 19,200 pieces
49. a. 50.24 in.^{2} **b.** 113.04 in.^{2} **c.** One 12-in. pizza
51. 289.3824 cm^3 **53.** True **55.** True **57.** True
59. Not possible **61.** For example: 100°, 80°
63. a. $\angle 1$ and $\angle 3$, $\angle 2$ and $\angle 4$
b. $\angle 1$ and $\angle 2$, $\angle 2$ and $\angle 3$, $\angle 3$ and $\angle 4$, $\angle 1$ and $\angle 4$
c. $m(\angle 1) = 100°, m(\angle 2) = 80°, m(\angle 3) = 100°$
65. 57° **67.** 78° **69.** 147° **71.** 58°
73. 7 **75.** 1 **77.** 1 **79.** 5
81. $m(\angle a) = 45°, m(\angle b) = 135°, m(\angle c) = 45°,$
$m(\angle d) = 135°, m(\angle e) = 45°, m(\angle f) = 135°, m(\angle g) = 45°$
83. Scalene **85.** Isosceles **87.** True **89.** 45°
91. No, a 90° angle plus an angle greater than 90° would make the sum of the angles greater than 180°. **93.** 40°
95. 37° **97.** $m(\angle a) = 80°, m(\angle b) = 80°, m(\angle c) = 100°,$
$m(\angle d) = 100°, m(\angle e) = 65°, m(\angle f) = 115°, m(\angle g) = 115°,$
$m(\angle h) = 35°, m(\angle i) = 145°, m(\angle j) = 145°$
99. $m(\angle a) = 70°, m(\angle b) = 65°, m(\angle c) = 65°,$
$m(\angle d) = 110°, m(\angle e) = 70°, m(\angle f) = 110°, m(\angle g) = 115°,$
$m(\angle h) = 115°, m(\angle i) = 65°, m(\angle j) = 70°, m(\angle k) = 65°$
101. 82 ft **103.** 36 in.^{2} **105.** 15.2464 cm^2

Chapter 1

Chapter Opener Puzzle

−10	+	2	+	−3	=	−11
+		·		·		−
12	÷	−2	·	3	=	−18
+		+		+		+
6	−	4	−	10	=	−8
=		=		=		=
8	·	0	−	1	=	−1

Section 1.1 Calculator Connections, p. 52

1. ≈ 3.464101615   **2.** ≈ 9.949874371
3. ≈ 12.56637061   **4.** ≈ 1.772453851

Section 1.1 Practice Exercises, pp. 53–55

3.

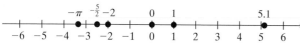

$-\pi$ $-\frac{5}{2}$ -2 0 1 5.1

5. a; rational   **7.** b; rational   **9.** a; rational
11. c; irrational   **13.** a; rational   **15.** a; rational
17. b; rational   **19.** c; irrational
21. For example: $\pi, -\sqrt{2}, \sqrt{3}$
23. For example: $-4, -1, 0$
25. For example: $-\frac{3}{4}, \frac{1}{2}, 0.206$ **27.** $-\frac{3}{2}, -4, 0.\overline{6}, 0, 1$
29. 1   **31.** $-4, 0, 1$ **33. a.** > **b.** > **c.** < **d.** >
35. -18 **37.** 6.1   **39.** $\frac{5}{8}$ **41.** $-\frac{7}{3}$ **43.** 3
45. $-\frac{7}{3}$ **47.** 8   **49.** -72.1 **51.** 2   **53.** 1.5
55. -1.5 **57.** $\frac{3}{2}$ **59.** -10 **61.** $-\frac{1}{2}$
63. False, $|n|$ is never negative.   **65.** True   **67.** False
69. True   **71.** False   **73.** False   **75.** False
77. True   **79.** True   **81.** False   **83.** True
85. True   **87.** True   **89.** For all $a < 0$

Section 1.2 Calculator Connections, p. 62

1. 2   **2.** 91   **3.** 84   **4.** 12   **5.** 49   **6.** 18
1.–3.   **4.–6.**

```
(4+6)/(8-3)
              2
110-5*(2+1)-4
              91
100-2*(5-3)^3
              84
```

```
3+(4-1)²
              12
(12-6+1)²
              49
3*8-√(32+2²)
              18
```

7. 4   **8.** 27   **9.** 0.5   **7.–9.**

```
√(18-2)
              4
(4*3-3*3)^3
              27
(20-3²)/(26-2²)
              .5
```

Section 1.2 Practice Exercises, pp. 63–65

3. $-4, 5.\overline{6}, 0, 4.02, \frac{7}{9}$ **5.** 9.2   **7.** -19 **9.** 15
11. 3   **13.** 9   **15.** $\left(\frac{1}{6}\right)^4$ **17.** a^3b^2 **19.** $(5c)^5$
21. a. x **b.** Yes, 1   **23.** $x \cdot x \cdot x$ **25.** $2b \cdot 2b \cdot 2b$
27. $10 \cdot y \cdot y \cdot y \cdot y$ **29.** $2 \cdot w \cdot z \cdot z$ **31.** 36
33. $\frac{1}{49}$ **35.** 0.008   **37.** 64   **39.** 9   **41.** 2
43. 12   **45.** 4   **47.** $\frac{1}{3}$ **49.** $\frac{5}{9}$ **51.** 20   **53.** 60
55. 8   **57.** 78   **59.** 0   **61.** $\frac{7}{6}$ **63.** 45   **65.** 16
67. 15   **69.** 19   **71.** 3   **73.** 39   **75.** 26   **77.** $\frac{5}{12}$
79. $\frac{5}{2}$ **81.** 57,600 ft²   **83.** 21 ft²   **85.** $3x$
87. $\frac{x}{7}$ or $x \div 7$ **89.** $2 - a$ **91.** $2y + x$
93. $4(x + 12)$ **95.** $3 - Q$ **97.** $2y^3; 16$
99. $|z - 8|; 2$ **101.** $5\sqrt{x}; 10$ **103.** $yz - x; 16$
105. 1   **107.** $\frac{1}{4}$
109. a. $36 \div 4 \cdot 3$ Division must be performed before
    $= 9 \cdot 3$ multiplication.
    $= 27$
 b. $36 - 4 + 3$ Subtraction must be performed
    $= 32 + 3$ before addition.
    $= 35$
111. This is acceptable, provided division and multiplication are performed in order from left to right, and subtraction and addition are performed in order from left to right.

Section 1.3 Practice Exercises, pp. 71–73

3. >   **5.** >   **7.** >
9. -6 **11.** 3   **13.** 3   **15.** -3 **17.** -17
19. 7   **21.** -19 **23.** -23 **25.** -5 **27.** -3
29. 0   **31.** 0   **33.** -5 **35.** -3 **37.** 0
39. -23 **41.** -6 **43.** -3 **45.** 21.3
47. $-\frac{3}{14}$ **49.** $-\frac{1}{6}$ **51.** $-\frac{15}{16}$ **53.** $\frac{1}{20}$
55. -2.4 or $-\frac{12}{5}$ **57.** $\frac{1}{4}$ or 0.25   **59.** 0   **61.** $-\frac{7}{8}$
63. -1 **65.** $\frac{11}{9}$ **67.** -23.08 **69.** -0.002117
71. To add two numbers with different signs, subtract the smaller absolute value from the larger absolute value and apply the sign of the number with the larger absolute value.
73. -1 **75.** 10   **77.** 5   **79.** 1

81. $-6 + (-10); -16$ **83.** $-3 + 8; 5$
85. $-21 + 17; -4$ **87.** $3(-14 + 20); 18$
89. $(-7 + (-2)) + 5; -4$ **91.** $-5 + 13 + (-11); -3°F$
93. $-2 + 6 + (-5); -1$ yd or 1-yd loss
95. a. $52.23 + (-52.95)$ **b.** Yes
97. a. $100 + 200 + (-500) + 300 + 100 + (-200)$ **b.** \$0

Section 1.4 Calculator Connections, pp. 78–79

1. -13 **2.** -2 **3.** 711
1.–3.

```
-8+(-5)
              -13
4+(-5)+(-1)
              -2
627-(-84)
              711
```

4. -0.18 **5.** 11.3   **6.** -990 **7.** -17 **8.** 38
4.–6.

```
-0.06-0.12
              -.18
-3.2+(-14.5)
              -17.7
-472+(-518)
              -990
```

7.–8.

```
-12-9+4
              -17
209-108+(-63)
              38
```

Section 1.4 Practice Exercises, pp. 79–81

3. x^2 **5.** $-b + 2$ **7.** 9   **9.** -3 **11.** -12
13. 4   **15.** -2 **17.** 8   **19.** -8 **21.** 2   **23.** 6
25. 40   **27.** -40 **29.** 0   **31.** -20 **33.** -24
35. 25   **37.** -5 **39.** $-\dfrac{3}{2}$ **41.** $\dfrac{41}{24}$ **43.** $\dfrac{2}{5}$ **45.** $-\dfrac{2}{3}$
47. 9.2   **49.** -5.72 **51.** -10 **53.** -14 **55.** -51
57. -173.188 **59.** 3.243   **61.** $6 - (-7); 13$
63. $3 - 18; -15$ **65.** $-5 - (-11); 6$
67. $-1 - (-13); 12$ **69.** $-32 - 20; -52$
71. $200 + 400 + 600 + 800 - 1000; \$1000$ **73.** 152°F
75. 19,881 m   **77.** 13   **79.** -9 **81.** 5   **83.** -25
85. -2 **87.** $-\dfrac{11}{30}$ **89.** $-\dfrac{29}{9}$ **91.** -2 **93.** -11
95. 2   **97.** -7 **99.** 5   **101.** 5   **103.** 3

Chapter 1 Problem Recognition Exercises, p. 82

1. Add their absolute values and apply a negative sign.
2. Subtract the smaller absolute value from the larger absolute value. Apply the sign of the number with the larger absolute value.
3. 41   **4.** 13   **5.** 31   **6.** 46   **7.** -1.3 **8.** -3.6
9. -16 **10.** -7 **11.** $-\dfrac{1}{12}$ **12.** $\dfrac{7}{24}$ **13.** -36
14. -59 **15.** -12 **16.** -50 **17.** $-\dfrac{19}{6}$ **18.** $-\dfrac{8}{5}$
19. -5 **20.** -32 **21.** 0   **22.** 0   **23.** -7.7
24. -10.5 **25.** -114 **26.** -56 **27.** -32
28. -46 **29.** 0   **30.** 0   **31.** -30 **32.** -400
33. -57 **34.** -26 **35.** -1 **36.** -6 **37.** 4
38. -1 **39.** $\dfrac{1}{36}$ **40.** $\dfrac{1}{64}$

Section 1.5 Calculator Connections, p. 89

1. -30 **2.** -2 **3.** 625   **4.** 625   **5.** -625 **6.** -5.76
1.–3.

```
-6(5)
              -30
-5.2/2.6
              -2
(-5)(-5)(-5)(-5)
              625
```

4.–6.

```
(-5)^4
              625
-5^4
              -625
-2.4²
              -5.76
```

7. 5.76   **8.** -1 **9.** 4   **10.** -36
7.–8.

```
(-2.4)²
              5.76
(-1)(-1)(-1)
              -1
```

9.–10.

```
-8.4/-2.1
              4
90/(-5)(2)
              -36
```

Section 1.5 Practice Exercises, pp. 89–92

3. True   **5.** False   **7.** -56 **9.** -42 **11.** 143
13. 240   **15.** -12.76 **17.** $\dfrac{3}{4}$ **19.** 36   **21.** -36
23. $-\dfrac{27}{125}$ **25.** 0.0016   **27.** -6 **29.** 10   **31.** 2
33. $-\dfrac{1}{5}$ **35.** $(-2)(-7) = 14$ **37.** $-5 \cdot 0 = 0$
39. No number multiplied by zero equals 6.
41. $(-6)(4) = -24$ **43.** 6   **45.** -6 **47.** -6
49. 6   **51.** 8   **53.** -8 **55.** -8 **57.** 8   **59.** 0
61. Undefined   **63.** 0   **65.** 0   **67.** $-\dfrac{3}{2}$ **69.** $\dfrac{3}{10}$
71. -2 **73.** -7.912 **75.** 0.092   **77.** -6 **79.** 2.1
81. 9   **83.** -9 **85.** $-\dfrac{64}{27}$ **87.** -0.008
89. -0.0016 **91.** -29 **93.** 48   **95.** -14.28
97. 340   **99.** $-\dfrac{10}{9}$ **101.** $\dfrac{14}{9}$ **103.** -30 **105.** 96
107. 2   **109.** -1 **111.** $-\dfrac{4}{33}$ **113.** $-\dfrac{4}{7}$ **115.** -24
117. $-\dfrac{1}{20}$ **119.** -23 **121.** 12   **123.** $\dfrac{9}{7}$
125. Undefined   **127.** -48 **129.** -6 **131.** -1
133. 7   **135.** 12   **137.** -40 **139.** $\dfrac{7}{2}$
141. No. The first expression is equivalent to $10 \div (5x)$. The second is $10 \div 5 \cdot x$.
143. $-3.75(0.3); -1.125$ **145.** $\dfrac{16}{5} \div \left(-\dfrac{8}{9}\right); -\dfrac{18}{5}$
147. $-0.4 + 6(-0.42); -2.92$ **149.** $-\dfrac{1}{4} - 6\left(-\dfrac{1}{3}\right); \dfrac{7}{4}$
151. $3(-2) + 3 = -3$; loss of \$3
153. a. -10 **b.** 24   **c.** In part (a), we subtract; in part (b), we multiply.

Section 1.6 Practice Exercises, pp. 102–105

3. 8   **5.** -8 **7.** $-\dfrac{9}{2}$ or -4.5 **9.** 0   **11.** $\dfrac{7}{8}$

13. $-\dfrac{4}{45}$ **15.** $-8 + 5$ **17.** $x + 8$ **19.** $4(5)$

21. $-12x$ **23.** $x + (-3)$; $-3 + x$

25. $4p + (-9)$; $-9 + 4p$ **27.** $x + (4 + 9)$; $x + 13$

29. $(-5 \cdot 3)x$; $-15x$ **31.** $\left(\dfrac{6}{11} \cdot \dfrac{11}{6}\right)x$; x

33. $\left(-4 \cdot -\dfrac{1}{4}\right)t$; t **35.** Reciprocal   **37.** 0

39. $30x + 6$ **41.** $-2a - 16$ **43.** $15c - 3d$

45. $-7y + 14$ **47.** $-\dfrac{2}{3}x + 4$ **49.** $\dfrac{1}{3}m - 1$

51. $-2p - 10$ **53.** $6w + 10z - 16$ **55.** $4x + 8y - 4z$
57. $6w - x + 3y$ **59.** $6 + 2x$ **61.** $24z$ **63.** b
65. i   **67.** g   **69.** d   **71.** h

73.

Term	Coefficient
$2x$	2
$-y$	-1
$18xy$	18
5	5

75.

Term	Coefficient
$-x$	-1
$8y$	8
$-9x^2y$	-9
-3	-3

77. The variable factors are different.
79. The variables are the same and raised to the same power.   **81.** For example: $5y, -2x, 6$ **83.** $2p - 12$

85. $-6y^2 - 7y$ **87.** $7x^3y + xy - 4$ **89.** $3t - \dfrac{7}{5}$

91. $-6x + 22$ **93.** $4w$ **95.** $-3x + 17$
97. $10t - 44$ **99.** -18 **101.** $-2t + 7$

103. $51a - 27$ **105.** -6 **107.** $4q - \dfrac{1}{3}$

109. $6n$ **111.** $2x + 18$ **113.** $32.33z - 30.81$
115. $-2x - 34$ **117.** $9z - 35$ **119.** Equivalent
121. Not equivalent. The terms are not *like* terms and cannot be combined.   **123.** Not equivalent; subtraction is not commutative.   **125.** Equivalent   **127.** $14\frac{2}{7} + (2\frac{1}{3} + \frac{2}{3})$ is easier.   **129. a.** 55   **b.** 210

Chapter 1 Review Exercises, pp. 111–113

1. a. $7, 1$ **b.** $7, -4, 0, 1$ **c.** $7, 0, 1$ **d.** $7, \frac{1}{3}, -4, 0, -0.\overline{2}, 1$
e. $-\sqrt{3}, \pi$ **f.** $7, \frac{1}{3}, -4, 0, -\sqrt{3}, -0.\overline{2}, \pi, 1$
3. 6   **5.** 0   **7.** False   **9.** True   **11.** True

13. True   **15.** $\dfrac{7}{y}$ or $7 \div y$ **17.** $a - 5$ **19.** $13z - 7$

21. 60   **23.** 4   **25.** 225   **27.** $\dfrac{1}{10}$ **29.** $\dfrac{27}{8}$

31. 11   **33.** 10   **35.** 4   **37.** -17 **39.** $-\dfrac{5}{22}$

41. $-\dfrac{27}{10}$ **43.** -4.28 **45.** 8

47. When a and b are both negative or when a and b have different signs and the number with the larger absolute value is negative.

49. -12 **51.** -1 **53.** $-\dfrac{29}{18}$ **55.** -1.2

57. -10.2 **59.** $\dfrac{10}{3}$ **61.** -1 **63.** $-7 - (-18)$; 11

65. $7 - 13$; -6 **67.** $(6 + (-12)) - 21$; -27

69. -170 **71.** -2 **73.** $-\dfrac{1}{6}$ **75.** 0   **77.** 0

79. $-\dfrac{3}{2}$ **81.** -30 **83.** $\dfrac{1}{4}$ **85.** -2 **87.** 17

89. $-\dfrac{7}{120}$ **91.** $-\dfrac{1}{3}$ **93.** -2 **95.** -23 **97.** 70.6

99. False, any nonzero real number raised to an even power is positive.   **101.** True   **103.** True   **105.** For example: $2 + 3 = 3 + 2$ **107.** For example: $5 + (-5) = 0$
109. For example: $5 \cdot 2 = 2 \cdot 5$

111. For example: $3 \cdot \dfrac{1}{3} = 1$

113. $5x - 2y = 5x + (-2y)$, then use the commutative property of addition.
115. $3y, 10x, -12, xy$ **117.** $8a - b - 10$
119. $-8z - 18$ **121.** $p - 2$ **123.** $-14q - 1$
125. $4x + 24$

Chapter 1 Test, pp. 113–114

1. Rational, all repeating decimals are rational numbers.
2.

3. a. False   **b.** True   **c.** True   **d.** True
4. a. $(4x)(4x)(4x)$ **b.** $4 \cdot x \cdot x \cdot x$
5. a. Twice the difference of a and b
b. b subtracted from twice a

6. $\dfrac{\sqrt{c}}{d^2}$ or $\sqrt{c} \div d^2$ **7.** 6   **8.** -19 **9.** -12

10. 28   **11.** $-\dfrac{7}{8}$ **12.** 4.66   **13.** -32 **14.** -12

15. Undefined   **16.** -28 **17.** 0   **18.** 96   **19.** $\dfrac{2}{3}$

20. -8 **21.** 9   **22.** $\dfrac{1}{3}$ **23.** $-\dfrac{3}{5}$

24. a. Commutative property of multiplication
b. Identity property of addition   **c.** Associative property of addition   **d.** Inverse property of multiplication
e. Associative property of multiplication
25. $-12x + 2y - 4$ **26.** $-12m - 24p + 21$

27. $-6k - 8$ **28.** $-4p - 23$ **29.** $4p - \dfrac{4}{3}$

30. 5   **31.** 18   **32.** -6 **33.** -32
34. $12 - (-4)$; 16   **35.** $6 - 8$; -2

Chapter 2

Chapter Opener Puzzle

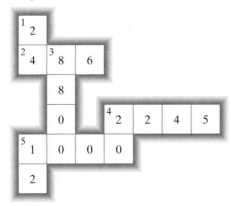

Section 2.1 Practice Exercises, pp. 124–126

3. Expression   **5.** Equation   **7.** Substitute the value into the equation and determine if the right-hand side is equal to the left-hand side.   **9.** No   **11.** Yes
13. Yes   **15.** -1 **17.** 20   **19.** -17 **21.** -16
23. 0   **25.** 1.3   **27.** $\frac{11}{2}$ or $5\frac{1}{2}$ **29.** -2 **31.** -2.13
33. -3.2675 **35.** 9   **37.** -4 **39.** 0   **41.** -15
43. $-\frac{4}{5}$ **45.** -10 **47.** 4   **49.** 41   **51.** -127
53. -2.6 **55.** $-8 + x = 42$; The number is 50.
57. $x - (-6) = 18$; The number is 12.
59. $x \cdot 7 = -63$ or $7x = -63$; The number is -9.
61. $\frac{x}{12} = \frac{1}{3}$; The number is 4.
63. $x + \frac{5}{8} = \frac{13}{8}$; The number is 1.   **65.** 10   **67.** $-\frac{1}{9}$
69. -12 **71.** $\frac{22}{3}$ **73.** -36 **75.** 16   **77.** 2
79. $-\frac{7}{4}$ **81.** 11   **83.** -36 **85.** $\frac{7}{2}$ **87.** 4
89. 3.6   **91.** 0.4084   **93.** Yes   **95.** No   **97.** Yes
99. Yes   **101.** For example: $y + 9 = 15$ **103.** For example: $2p = -8$ **105.** For example: $5a + 5 = 5$
107. -1 **109.** 7

Section 2.2 Practice Exercises, pp. 133–135

3. $-5z + 2$ **5.** $10p - 10$ **7.** To simplify an expression, clear parentheses and combine *like* terms. To solve an equation, use the addition, subtraction, multiplication, and division properties of equality to isolate the variable.   **9.** -3
11. -5 **13.** 2   **15.** 6   **17.** $\frac{5}{2}$ **19.** -42
21. $-\frac{3}{4}$ **23.** 5   **25.** -4 **27.** -26 **29.** 10
31. -8 **33.** $-\frac{7}{3}$ **35.** 0   **37.** -3 **39.** -2
41. $\frac{9}{2}$ **43.** $-\frac{1}{3}$ **45.** 10   **47.** -6 **49.** 0
51. -2 **53.** $-\frac{25}{4}$ **55.** $\frac{10}{3}$ **57.** -0.25

Section 2.3 Practice Exercises, pp. 140–142

3. -2 **5.** -5 **7.** No solution   **9.** 18, 36
11. 100; 1000; 10,000   **13.** 30, 60   **15.** 4   **17.** $-\frac{19}{2}$
19. $-\frac{15}{4}$ **21.** 8   **23.** 3   **25.** 15
27. No solution   **29.** All real numbers   **31.** 5
33. 2   **35.** -15 **37.** 6   **39.** 3
41. All real numbers   **43.** 67   **45.** 90   **47.** 4
49. -3.8 **51.** No solution   **53.** -0.25 **55.** -6
57. $\frac{8}{3}$ or $2\frac{2}{3}$ **59.** -9 **61.** $\frac{1}{10}$ **63.** -2
65. -1 **67.** 2

Answers for the previous section (Section 2.2 continued at top of right column):
59. Contradiction; no solution   **61.** Conditional equation; -15 **63.** Identity; all real numbers
65. One solution   **67.** Infinitely many solutions
69. 7   **71.** $\frac{1}{2}$ **73.** 0   **75.** All real numbers
77. -46 **79.** 2   **81.** $\frac{13}{2}$ **83.** -5
85. No solution   **87.** 2.205   **89.** 10   **91.** -1
93. $a = 15$ **95.** $a = 4$
97. For example: $5x + 2 = 2 + 5x$

Chapter 2 Problem Recognition Exercises, p. 142

1. Expression; $-4b + 18$ **2.** Expression; $20p - 30$
3. Equation; -8 **4.** Equation; -14 **5.** Equation; $\frac{1}{3}$
6. Equation; $-\frac{4}{3}$ **7.** Expression; $6z - 23$
8. Expression; $-x - 9$ **9.** Equation; $\frac{7}{9}$
10. Equation; $-\frac{13}{10}$ **11.** Equation; 20
12. Equation; -3 **13.** Equation; $\frac{1}{2}$ **14.** Equation; -6
15. Expression; $\frac{5}{8}x + \frac{7}{4}$ **16.** Expression; $-26t + 18$
17. Equation; no solution   **18.** Equation; no solution
19. Equation; $\frac{23}{12}$ **20.** Equation; $\frac{5}{8}$
21. Equation; all real numbers
22. Equation; all real numbers
23. Equation; $\frac{1}{2}$ **24.** Equation; 0   **25.** Expression; 0
26. Expression; -1 **27.** Expression; $2a + 13$
28. Expression; $8q + 3$ **29.** Equation; 10
30. Equation; $-\frac{1}{20}$

Section 2.4 Practice Exercises, pp. 149–152

3. $x + 5$ **5.** $3x$ **7.** $3x + 20$ **9.** The number is -4.
11. The number is -3.   **13.** The number is 5.
15. The number is -5.   **17.** The number is 9.
19. a. $x + 1, x + 2$ **b.** $x - 1, x - 2$ **21.** The integers are -34 and -33.   **23.** The integers are 13 and 15.

25. The sides are 14 in., 15 in., 16 in., 17 in., and 18 in.
27. The integers are 42, 44, and 46.   **29.** The integers are 13, 15, and 17.   **31.** The lengths of the pieces are 33 cm and 53 cm.   **33.** Karen's age is 35, and Clarann's age is 23.
35. There were 201 Republicans and 232 Democrats.
37. 4.698 million watch *The Dr. Phil Show.*   **39.** The Congo River is 4370 km long, and the Nile River is 6825 km.
41. The area of Africa is 30,065,000 km². The area of Asia is 44,579,000 km².   **43.** They walked 12.3 mi on the first day and 8.2 mi on the second.   **45.** The pieces are 6 in., 18 in., and 24 in.   **47.** 42, 43, and 44   **49.** Jennifer Lopez made $37 million, and U2 made $69 million.   **51.** The number is 11.
53. The page numbers are 470 and 471.   **55.** The number is 10.   **57.** The deepest point in the Arctic Ocean is 5122 m.
59. The number is $\dfrac{7}{16}$.   **61.** The number is 2.5.

Section 2.5 Practice Exercises, pp. 157–159

3. The numbers are 21 and 22.   **5.** 12.5%   **7.** 85%
9. 0.75   **11.** 1050.8   **13.** 885   **15.** 2200
17. Molly will have to pay $106.99.   **19.** Approximately 231,000 cases   **21.** 2%   **23.** Javon's taxable income was $84,000.   **25.** Pam will earn $420.
27. Bob borrowed $1200.   **29.** The rate is 6%.
31. Penny needs to invest $3302.   **33. a.** $20.40
b. $149.60   **35.** The original price was $470.59.
37. The discount rate is 12%.   **39.** The original cost was $60.   **41.** The tax rate is 5%.   **43.** The ticket price was $67.   **45.** The original price was $210,000.
47. Alina made $4600 that month.   **49.** Diane sold $645 over $200 worth of merchandise.

Section 2.6 Calculator Connections, p. 164

1. 140.056   **2.** 31.831   **3.** 1.273   **4.** 0.455

1–2.
```
880/(2π)
        140.0563499
1600/(π*(4)²)
        31.83098862
```

3–4.
```
20/(5π)
        1.273239545
10/(7π)
        .4547284088
```

Section 2.6 Practice Exercises, pp. 165–168

3. -5 **5.** 0   **7.** -2 **9.** $a = P - b - c$
11. $y = x + z$ **13.** $q = p - 250$ **15.** $b = \dfrac{A}{h}$
17. $t = \dfrac{PV}{nr}$ **19.** $x = 5 + y$ **21.** $y = -3x - 19$
23. $y = \dfrac{-2x + 6}{3}$ or $y = -\dfrac{2}{3}x + 2$
25. $x = \dfrac{y + 9}{-2}$ or $x = -\dfrac{1}{2}y - \dfrac{9}{2}$
27. $y = \dfrac{-4x + 12}{-3}$ or $y = \dfrac{4}{3}x - 4$
29. $y = \dfrac{-ax + c}{b}$ or $y = -\dfrac{a}{b}x + \dfrac{c}{b}$
31. $t = \dfrac{A - P}{Pr}$ or $t = \dfrac{A}{Pr} - \dfrac{1}{r}$

33. $c = \dfrac{a - 2b}{2}$ or $c = \dfrac{a}{2} - b$ **35.** $y = 2Q - x$
37. $a = MS$ **39.** $R = \dfrac{P}{I^2}$
41. The length is 7 ft, and the width is 5 ft.
43. The length is 120 yd and the width is 30 yd.
45. The length is 195 m, and the width is 100 m.
47. The sides are 22 m, 22 m, and 27 m.
49. "Adjacent supplementary angles form a straight angle." The words *Supplementary* and *Straight* both begin with the same letter.   **51.** The angles are 55° and 35°.
53. The angles are 45° and 135°.
55. $x = 20$; the vertical angles measure 37°.
57. The measures of the angles are 30°, 60°, and 90°.
59. The measures of the angles are 42°, 54°, and 84°.
61. $x = 17$; the measures of the angles are 34° and 56°.
63. a. $A = lw$ **b.** $w = \dfrac{A}{l}$ **c.** The width is 29.5 ft.
65. a. $P = 2l + 2w$ **b.** $l = \dfrac{P - 2w}{2}$ **c.** The length is 103 m.
67. a. $C = 2\pi r$ **b.** $r = \dfrac{C}{2\pi}$ **c.** The radius is approximately 140 ft.   **69. a.** 415.48 m² **b.** 10,386.89 m³

Section 2.7 Practice Exercises, pp. 173–177

3. $c = \dfrac{r}{d}$ **5.** 4   **7.** $200 - t$ **9.** $100 - x$
11. $3000 - y$ **13.** 53 tickets were sold at $3 and 28 tickets were sold at $2.   **15.** Josh downloaded 17 songs for $0.90 and 8 songs for $1.50   **17.** Angelina purchased 25 bottles of soda and 20 bottles of flavored water.
19. $x + 7$ **21.** $d + 2000$
23. 20 oz of 50% antifreeze solution
25. The pharmacist needs to use 21 mL of the 1% saline solution.   **27.** The contractor needs to mix 6.75 oz of 50% acid solution.   **29. a.** 300 mi **b.** $5x$ **c.** $5(x + 2)$ or $5x + 60$
31. She walks 4 mph to the lake.   **33.** Bryan hiked 6 mi up the canyon.   **35.** The plane travels 600 mph in still air.
37. The slower car travels 48 mph and the faster car travels 52 mph.   **39.** The speeds of the vehicles are 40 mph and 50 mph.   **41.** The rates of the boats are 20 mph and 40 mph.
43. a. 2 lb **b.** $0.10x$ **c.** $0.10(x + 3) = 0.10x + 0.30$
45. 10 lb of coffee sold at $12 per pound and 40 lb of coffee sold at $8 per pound.   **47.** The boats will meet in $\frac{3}{4}$ hr (45 min).
49. Sam purchased 16 packages of wax and 5 bottles of sunscreen.   **51.** 2.5 quarts of 85% chlorine solution
53. 20 L of water   **55.** The Japanese bullet train travels 300 km/hr and the Acela Express travels 240 km/hr.

Section 2.8 Practice Exercises, pp. 187–192

3. -3

Set-Builder Notation	Graph	Interval Notation
5. $\{x \mid x \geq 6\}$	⊢————→ 6	$[6, \infty)$
7. $\{x \mid x \leq 2.1\}$	←————⊣ 2.1	$(-\infty, 2.1]$
9. $\{x \mid -2 < x \leq 7\}$	⟜————⊣ −2 7	$(-2, 7]$

Set-Builder Notation	Graph	Interval Notation
11. $\left\{x \mid x > \dfrac{3}{4}\right\}$	$\frac{3}{4}$	$\left(\dfrac{3}{4}, \infty\right)$
13. $\{x \mid -1 < x < 8\}$	$-1 \quad 8$	$(-1, 8)$
15. $\{x \mid x \le -14\}$	-14	$(-\infty, -14]$

Set-Builder Notation	Graph	Interval Notation
17. $\{x \mid x \ge 18\}$	18	$[18, \infty)$
19. $\{x \mid x < -0.6\}$	-0.6	$(-\infty, -0.6)$
21. $\{x \mid -3.5 \le x < 7.1\}$	$-3.5 \quad 7.1$	$[-3.5, 7.1)$

23. a. $x = 3$ **b.** $x > 3$ **25. a.** $p = 13$ **b.** $p \le 13$

27. a. $c = -3$ **b.** $c < -3$ **29. a.** $z = -\dfrac{3}{2}$ **b.** $z \ge -\dfrac{3}{2}$

31. $-1 \quad 4$

33. $-3 < x < 5$ $-3 \quad 5$

35. $2 \le x \le 6$ $2 \quad 6$

37. $(-\infty, 1]$ 1

39. $(10, \infty)$ 10

41. $(3, \infty)$ 3

43. $(-\infty, 8]$ 8

45. $(2, \infty)$ 2

47. $(-\infty, -2)$ -2

49. $[14, \infty)$ 14

51. $[-24, \infty)$ -24

53. $[-3, 3)$ $-3 \quad 3$

55. $\left(0, \dfrac{5}{2}\right)$ $0 \quad \frac{5}{2}$

57. $(10, 12)$ $10 \quad 12$

59. $[-1, 4)$ $-1 \quad 4$

61. $[90, \infty)$ 90

63. $(-9, \infty)$ -9

65. $\left[-\dfrac{15}{2}, \infty\right)$ $-\frac{15}{2}$

67. $(-\infty, -3)$ -3

69. $\left[-\dfrac{1}{3}, \infty\right)$ $-\frac{1}{3}$

71. $(-3, \infty)$ -3

73. $(-\infty, 7)$ 7

75. $(-\infty, -5]$ -5

77. $\left(-\infty, \dfrac{15}{4}\right]$ $\frac{15}{4}$

79. $(-\infty, -9)$ -9

81. $(-3, \infty)$ -3

83. $(-\infty, 0]$ 0

85. No   **87.** Yes

89. a. A $[93, 100]$; B+ $[89, 93)$; B $[84, 89)$; C+ $[80, 84)$; C $[75, 80)$; F $[0, 75)$ **b.** B **c.** C   **91.** $L \ge 10$

93. $w > 75$ **95.** $t \le 72$ **97.** $L \ge 8$ **99.** $2 < h < 5$

101. More than 10.2 in. of rain is needed.

103. a. \$1539 **b.** 200 birdhouses cost \$1440. It is cheaper to purchase 200 birdhouses because the discount is greater.

105. a. $2.00x > 75 + 0.17x$ **b.** $x \ge 41$; profit occurs when 41 or more lemonades are sold.

107. $[13, \infty)$ 13

109. $[-4, \infty)$ -4

111. $(14.5, \infty)$ 14.5

Chapter 2 Review Exercises, pp. 199–202

1. a. Equation **b.** Expression **c.** Equation **d.** Equation   **3. a.** Nonlinear **b.** Linear **c.** Nonlinear **d.** Linear   **5.** -8

7. $\dfrac{21}{4}$ **9.** $-\dfrac{21}{5}$ **11.** $-\dfrac{10}{7}$ **13.** The number is 60.

15. The number is -8.   **17.** 1   **19.** $\dfrac{9}{4}$

21. -3 **23.** $\dfrac{3}{4}$ **25.** 0   **27.** $\dfrac{3}{8}$

29. A contradiction has no solution and an identity is true for all real numbers.   **31.** 6   **33.** 13   **35.** -10

37. $\dfrac{5}{3}$ **39.** 2.5   **41.** -4.2 **43.** -312

45. No solution   **47.** All real numbers

49. The number is 30.   **51.** The number is -7.

53. The integers are 66, 68, and 70.

55. The sides are 25 in., 26 in., and 27 in.

57. The minimum salary was \$30,000 in 1980.

59. 23.8   **61.** 12.5%   **63.** 160   **65.** The novel originally cost \$29.50.   **67. a.** \$840 **b.** \$3840

69. $K = C + 273$ **71.** $s = \dfrac{P}{4}$ **73.** $x = \dfrac{y - b}{m}$

75. $y = \dfrac{-2x - 2}{5}$ **77. a.** $h = \dfrac{3V}{\pi r^2}$ **b.** The height is 5.1 in.

79. The angles are 22°, 78°, and 80°.   **81.** The length is 5 ft, and the width is 4 ft.   **83.** The measure of angle y is 53°.   **85.** Gus rides 15 mph and Winston rides 18 mph.

87. They meet in 2.25 hr (2 hr and 15 min).

89. 20 lb of the 40% solder should be used.

91. $\left(-\infty, \frac{1}{2}\right]$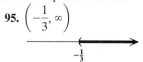

93. a. \$637   **b.** 300 plants cost \$1410, and 295 plants cost \$1416. 300 plants cost less.

95. $\left(-\frac{1}{3}, \infty\right)$

97. $\left(-\infty, -\frac{14}{5}\right]$

99. $(-\infty, 34.5)$

101. $\left(-\infty, \frac{5}{2}\right]$

103. $[-6, 5]$

105. a. $1.50x > 33 + 0.4x$ **b.** $x > 30$; a profit is realized if more than 30 hot dogs are sold.

Chapter 2 Test, pp. 202–203

1. a. b, d   **2. a.** $5x + 7$ **b.** 9   **3.** -16 **4.** 12

5. $-\frac{16}{9}$ **6.** $\frac{7}{3}$ **7.** 15   **8.** $\frac{13}{4}$ **9.** $\frac{20}{21}$

10. No solution   **11.** -3 **12.** -47

13. All real numbers   **14.** $y = -3x - 4$ **15.** $r = \frac{C}{2\pi}$

16. 90   **17.** The numbers are 18 and 13.

18. The sides are 61 in., 62 in., 63 in., 64 in., and 65 in.

19. The cost was \$82.00.   **20.** Each basketball ticket was \$36.32, and each hockey ticket was \$40.64.

21. Clarita originally borrowed \$5000.   **22.** The field is 110 m long and 75 m wide.   **23.** $y = 30$; The measures of the angles are 30°, 39°, and 111°.

24. Paula needs 30 lb of macadamia nuts.

25. One family travels 55 mph and the other travels 50 mph.

26. The measures of the angles are 32° and 58°.

27. a. $(-\infty, 0)$ **b.** $[-2, 5)$

28. $(-2, \infty)$ **29.** $(-\infty, -4]$

30. $\left(-\frac{3}{2}, \infty\right)$ **31.** $[-5, 1]$

32. More than 26.5 in. is required.

Chapters 1–2 Cumulative Review Exercises, pp. 203–204

1. $\frac{1}{2}$ **2.** -7 **3.** $-\frac{5}{12}$ **4.** 16   **5.** 4

6. $\sqrt{5^2 - 9}$; 4   **7.** $-14 + 12$; -2 **8.** $-7x^2y, 4xy, -6$

9. $9x + 13$ **10.** 4   **11.** -7.2

12. All real numbers   **13.** -8 **14.** $-\frac{4}{7}$ **15.** -80

16. The numbers are 77 and 79.   **17.** The cost before tax was \$350.00.   **18.** The height is $\frac{41}{6}$ cm or $6\frac{5}{6}$ cm.

19. $(-2, \infty)$ **20.** $[-1, 9]$

Chapter 3
Chapter Opener Puzzle

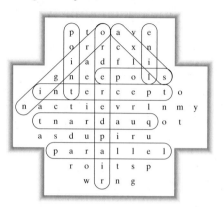

Section 3.1 Practice Exercises, pp. 210–214

3. a. Month 10   **b.** 30   **c.** Between months 3 and 5 and between months 10 and 12   **d.** Months 8 and 9   **e.** Month 3   **f.** 80

5. a. On day 1 the price per share was \$89.25. **b.** \$1.75   **c.** $-\$2.75$

7.

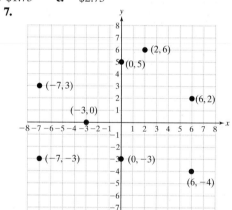

9.

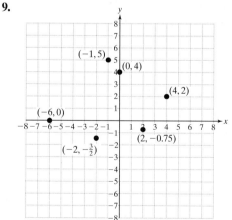

11. IV   **13.** II   **15.** III   **17.** I

19. $(0, -5)$ lies on the y-axis.

21. $\left(\frac{7}{8}, 0\right)$ is located on the x-axis.

23. $A(-4, 2), B(\frac{1}{2}, 4), C(3, -4), D(-3, -4), E(0, -3), F(5, 0)$

25. a. $A(400, 200), B(200, -150), C(-300, -200), D(-300, 250), E(0, 450)$ **b** 450 m

27. a. $(250, 225), (175, 193), (315, 330), (220, 209), (450, 570), (400, 480), (190, 185)$; the ordered pair $(250, 225)$ means that 250 people produce $225 in popcorn sales.

b.

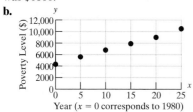

29. a. $(10, 6800)$ means that in 1990, the poverty level was $6800.

b.

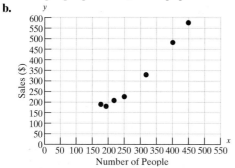

31. a. $(1, -10.2), (2, -9.0), (3, -2.5), (4, 5.7), (5, 13.0),$ $(6, 18.3), (7, 20.9), (8, 19.6), (9, 14.8), (10, 8.7), (11, 2.0),$ $(12, -6.9)$.

b.

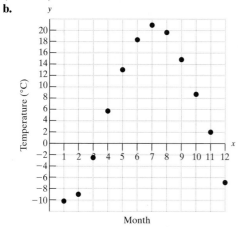

33. a.

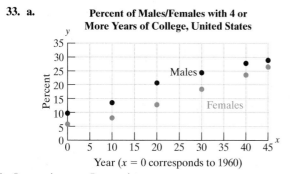

b. Increasing **c.** Increasing

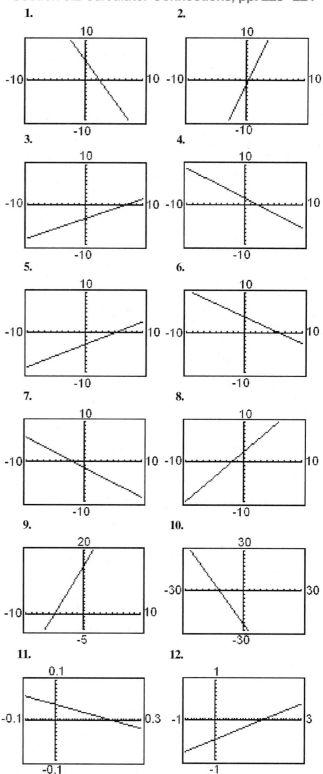

Section 3.2 Practice Exercises, pp. 224–230

3. $(2, 4)$; quadrant I   **5.** $(0, -1)$; y-axis
7. $(3, -4)$; quadrant IV   **9.** Yes   **11.** Yes   **13.** No
15. No   **17.** Yes

19.

x	y
1	-3
-2	0
-3	1
-4	2

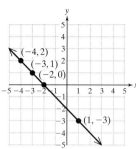

21.

x	y
-2	3
-1	0
-4	9

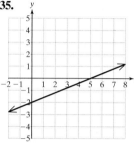

23.

x	y
0	4
2	0
3	-2

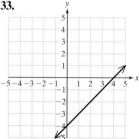

25.

x	y
0	-2
5	-5
10	-8

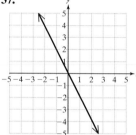

27.

x	y
0	-2
-3	-2
5	-2

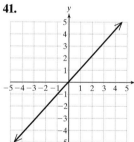

29.

x	y
3/2	-1
3/2	2
3/2	-3

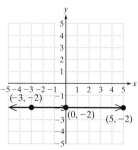

31.

x	y
0	4.6
1	3.4
2	2.2

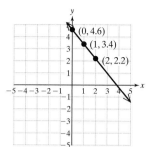

33.

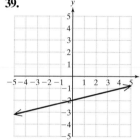

35.

37.

39.

41.

43.

45. y-axis   **47.** x-intercept: $(-1, 0)$; y-intercept: $(0, -3)$

49. x-intercept: $(-4, 0)$; y-intercept: $(0, 1)$

51. x-intercept: $\left(-\frac{9}{4}, 0\right)$;   **53.** x-intercept: $\left(\frac{8}{3}, 0\right)$;

y-intercept: $(0, 3)$ y-intercept: $(0, 2)$

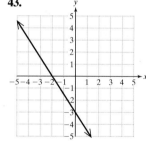

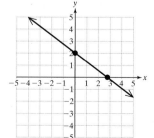

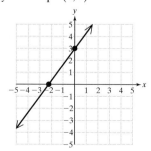

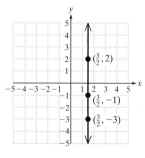

55. x-intercept: $(-4, 0)$;
y-intercept: $(0, 8)$

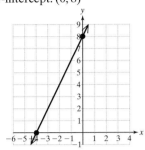

57. x-intercept: $(0, 0)$;
y-intercept: $(0, 0)$

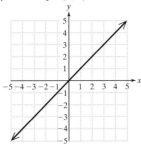

59. x-intercept: $(10, 0)$;
y-intercept: $(0, 5)$

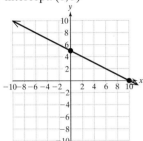

61. x-intercept: $(0, 0)$;
y-intercept: $(0, 0)$

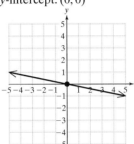

63. True
67. a. Horizontal
c. no x-intercept;
y-intercept: $(0, -1)$

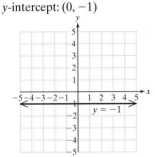

65. True
69. a. Vertical
c. x-intercept: $(4, 0)$;
no y-intercept

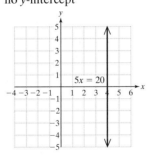

71. a. Horizontal
c. no x-intercept;
y-intercept $(0, -5)$

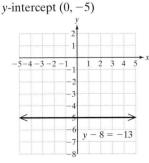

73. a. Vertical
c. All points on the y-axis are
y-intercepts; x-intercept: $(0,0)$

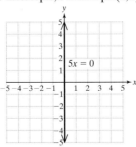

75. A horizontal line may not have an x-intercept.
A vertical line may not have a y-intercept.   **77.** a b, d
79. a. $y = 10,068$ **b.** $x = 3$ **c.** $(1, 10068)$ One year after
purchase the value of the car is $10,068. $(3, 7006)$ Three years
after purchase the value of the car is $7006.

Section 3.3 Practice Exercises, pp. 237–243

3. x-intercept: $(6, 0)$;
y-intercept: $(0, -2)$

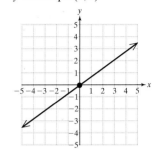

5. x-intercept: $(0, 0)$;
y-intercept: $(0, 0)$

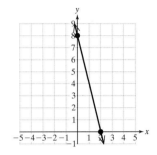

7. x-intercept: $(2, 0)$;
y-intercept: $(0, 8)$

9. $m = \dfrac{1}{3}$ **11.** $m = \dfrac{6}{11}$ **13.** Undefined

15. Positive   **17.** Negative   **19.** Zero
21. Undefined   **23.** Positive   **25.** Negative

27. $m = \dfrac{1}{2}$ **29.** $m = -3$ **31.** $m = 0$

33. The slope is undefined.   **35.** $\dfrac{1}{3}$ **37.** -3

39. $\dfrac{3}{5}$ **41.** Zero   **43.** Undefined   **45.** $\dfrac{28}{5}$ **47.** $-\dfrac{7}{8}$

49. -0.45 or $-\dfrac{9}{20}$ **51.** -0.15 or $-\dfrac{3}{20}$

53. a. -2 **b.** $\dfrac{1}{2}$ **55. a.** 0 **b.** undefined

57. a. $\dfrac{4}{5}$ **b.** $-\dfrac{5}{4}$ **59. a.** undefined **b.** 0

61. Perpendicular   **63.** Parallel   **65.** Neither
67. $l_1: m = 2, l_2: m = 2$; parallel
69. $l_1: m = 5, l_2: m = -\dfrac{1}{5}$; perpendicular

71. $l_1: m = \dfrac{1}{4}, l_2: m = 4$; neither

73. a. $m = 47$ **b.** The number of male prisoners increased
at a rate of 47 thousand per year during this time period.
75. a. $m \approx 0.009$ **b.** The slope $m = 0.009$ means that
postage increased by about $0.009 per year (or equivalently
0.9¢ per year) during this time.

77. $m = \dfrac{3}{4}$ **79.** $m = 0$

81. For example:
$(4, 0)$ and $(-2, -4)$

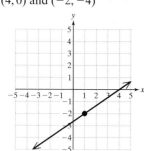

83. For example:
$(3, -1)$ and $(1, 5)$

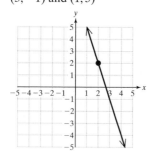

85.

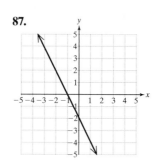

87.

89.

91. $\dfrac{3m - 3n}{2b}$ or $\dfrac{-3m + 3n}{-2b}$ **93.** $\left(\dfrac{c}{a}, 0\right)$

95. For example: $(7, 1)$

Section 3.4 Calculator Connections, p. 247

1. Perpendicular

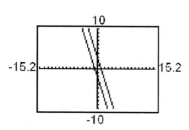

2. Parallel

3. Neither

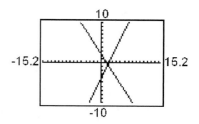

4. The lines may appear parallel; however, they are not parallel because the slopes are different.

5. The lines may appear to coincide on a graph; however, they are not the same line because the y-intercepts are different.

6. The line may appear to be horizontal, but it is not. The slope is 0.001 rather than 0.

Section 3.4 Practice Exercises, pp. 248–251

3. x-intercept: $(10, 0)$; y-intercept: $(0, -2)$

5. x-intercept: none; y-intercept: $(0, -3)$

7. x-intercept: $(0, 0)$; y-intercept: $(0, 0)$

9. x-intercept: $(4, 0)$; ; y-intercept: none

11. $m = -2$; y-intercept: $(0, 3)$ **13.** $m = 1$; y-intercept: $(0, -2)$ **15.** $m = -1$; y-intercept: $(0, 0)$

17. $m = \dfrac{3}{4}$; y-intercept: $(0, -1)$ **19.** $m = \dfrac{2}{5}$; y-intercept: $\left(0, -\dfrac{4}{5}\right)$ **21.** $m = 3$; y-intercept: $(0, -5)$

23. $m = -1$; y-intercept: $(0, 6)$ **25.** Undefined slope; no y-intercept   **27.** $m = 0$; y-intercept: $\left(0, -\dfrac{1}{4}\right)$

29. $m = \dfrac{2}{3}$; y-intercept: $(0, 0)$

31.

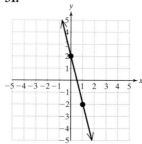

33.

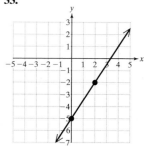

35. b   **37.** e   **39.** c

41. $y = \dfrac{1}{2}x - 3$

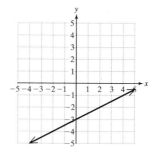

43. $y = -2x + 9$

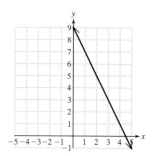

45. $y = -\dfrac{1}{2}x + \dfrac{3}{2}$ **47.** $y = -x$

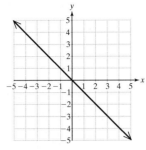

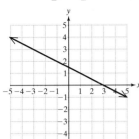

49. $y = \dfrac{4}{5}x$ **51.** $y = -\dfrac{2}{3}$

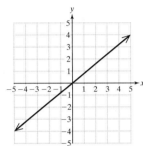

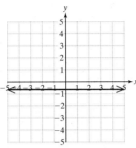

53. Perpendicular   **55.** Neither   **57.** Parallel
59. Perpendicular   **61.** Parallel   **63.** Perpendicular
65. Neither   **67.** Parallel   **69.** $y = -\dfrac{1}{3}x + 2$
71. $y = 10x - 19$ **73.** $y = -11$ **75.** $y = 5x$
77. $y = 6x - 2$ **79. a.** $m = 49.95$. The cost increases
$49.95 per day.   **b.** $(0, 31.95)$. The base fee for renting a car
is $31.95.   **c.** $381.60   **81.** $y = -\dfrac{A}{B}x + \dfrac{C}{B}$; the slope is $-\dfrac{A}{B}$.

83. $m = -\dfrac{6}{7}$ **85.** $m = \dfrac{11}{8}$

Chapter 3 Problem Recognition Exercises, p. 252

1. a, c, d   **2.** b, f, h   **3.** a   **4.** f   **5.** b, f   **6.** c
7. c, d   **8.** f   **9.** e   **10.** g   **11.** b   **12.** h
13. g   **14.** e   **15.** c   **16.** b, h   **17.** e   **18.** e
19. b, f, h   **20.** c, d

Section 3.5 Practice Exercises, pp. 257–260

3.

5.

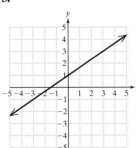

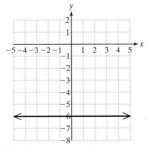

7. 9   **9.** 0   **11.** $y = 3x + 7$ or $3x - y = -7$
13. $y = -4x - 14$ or $4x + y = -14$ **15.** $y = -\dfrac{1}{2}x - \dfrac{1}{2}$
or $x + 2y = -1$ **17.** $y = \dfrac{1}{4}x + 8$ or $x - 4y = -32$

19. $y = 4.5x - 25.6$ or $45x - 10y = 256$ **21.** $y = -2$
23. $y = 2x - 2$ or $2x - y = 2$ **25.** $y = -x - 4$ or
$x + y = -4$ **27.** $y = -0.2x - 2.86$ or
$20x + 100y = -286$ **29.** $y = -2x + 1$
31. $y = 2x + 4$ **33.** $y = \dfrac{1}{2}x - 1$ **35.** $y = 4x + 13$ or
$4x - y = -13$ **37.** $y = -\dfrac{3}{2}x + 6$ or $3x + 2y = 12$
39. $y = -2x - 8$ or $2x + y = -8$ **41.** $y = -\dfrac{1}{5}x - 6$
or $x + 5y = -30$ **43.** $y = 3x - 8$ or $3x - y = 8$
45. iv   **47.** vi   **49.** iii   **51.** $y = 1$ **53.** $x = 2$
55. $y = 2$ **57.** $x = -6$ **59.** $x = -4$

Section 3.6 Calculator Connections, p. 263

1. 13.3

2. −42.3

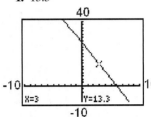

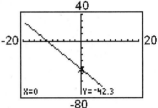

3. 345

4. 95

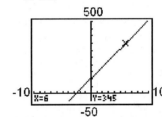

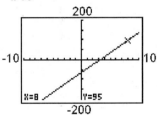

Section 3.6 Practice Exercises, pp. 264–267

3. x-intercept: $(6, 0)$; y-intercept: $(0, 5)$ **5.** x-intercept:
$(-2, 0)$; y-intercept: $(0, -4)$ **7.** x-intercept: none;
y-intercept: $(0, -9)$ **9. a.** $3.00   **b.** $7.20
c. The y-intercept is $(0, 1.6)$. This indicates that the minimum
wage was $1.60 per hour in the year 1970.   **d.** The slope is
0.14. This indicates that the minimum wage has risen
approximately $0.14 per year during this period.
11. a. 30.7°   **b.** 13.4°   **c.** $m = -2.333$. The average
temperature in January decreases at a rate of 2.333° per 1°
of latitude.   **d.** $(53.2, 0)$. At 53.2° latitude, the average
temperature in January is 0°.
13. a. $95   **b.** $190   **c.** $(0, 0)$. For 0 kilowatt-hours used,
the cost is $0.   **d.** $m = 0.095$. The cost increases by $0.095
for each kilowatt-hour used.   **15. a.** $y = -0.08x + 7.6$
b. 5 days   **17. a.** $y = 2.5x + 31$ **b.** 43.5 in.
19. a. $y = 0.25x + 20$ **b.** $84.50   **21. a.** $y = 90x + 105$
b. $1185.00   **23. a.** $y = 0.8x + 100$ **b.** $260.00

Section 3.7 Calculator Connections, p. 275

1.

2.

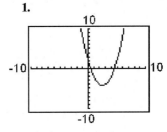

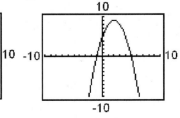

3.

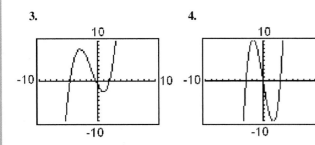

4.

Section 3.7 Practice Exercises, pp. 275–279

3. a. x-intercept: $\left(\frac{5}{3}, 0\right)$; y-intercept: $(0, -5)$ **b.** 3

c.

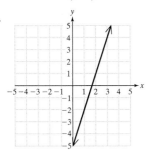

5. Domain: $\{4, 3, 0\}$; range: $\{2, 7, 1, 6\}$
7. Domain: $\{\frac{1}{2}, 0, 1\}$; range: $\{3\}$ **9.** Domain: $\{0, 5, -8, 8\}$;
range: $\{0, 2, 5\}$ **11.** Domain: $\{$Atlanta, Macon, Pittsburgh$\}$;
range: $\{$GA, PA$\}$ **13.** Domain: $\{$New York, California$\}$;
range: $\{$Albany, Los Angeles, Buffalo$\}$ **15.** The relation is
a function if each element in the domain has exactly one
corresponding element in the range.
17. The relations in Exercises 7, 9, and 11 are functions.
19. Yes   **21.** No   **23.** No   **25.** Yes   **27.** Yes
29. $-5, -1, -11$ **31.** $\frac{1}{5}, \frac{1}{4}, \frac{1}{2}$ **33.** 7, 2, 3   **35.** 0, 1, 2
37. b   **39.** c   **41.** The function value at $x = 6$ is 2.
43. The function value at $x = \frac{1}{2}$ is $\frac{1}{4}$.   **45.** $(2, 7)$
47. $(0, 8)$ **49. a.** $s(1) = 32$. The speed of an object
1 sec after being dropped is 32 ft/sec.   **b.** $s(2) = 64$. The
speed of an object 2 sec after being dropped is 64 ft/sec.
c. $s(10) = 320$. The speed of an object 10 sec after being
dropped is 320 ft/sec.   **d.** 294.4 ft/sec
51. a. $h(0) = 3$. The initial height of the ball is 3 ft.
b. $h(1) = 51$. The height of the ball 1 sec after being kicked
is 51 ft.   **c.** $h(2) = 67$. The height of the ball 2 sec after being
kicked is 67 ft.   **d.** $h(4) = 3$. The height of the ball 4 sec after
being kicked is 3 ft.

Chapter 3 Review Exercises, pp. 286–290

1.

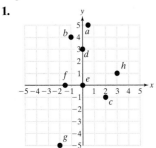

3. III   **5.** IV   **7.** IV   **9.** x-axis
11. a. On day 1, the price was $26.25.   **b.** Day 2   **c.** $2.25
13. No   **15.** Yes

17.

x	y
2	1
3	4
1	-2

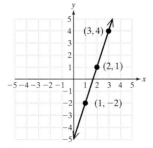

19.

x	y
0	-1
3	1
-6	-5

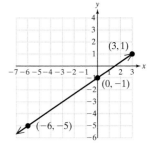

21.

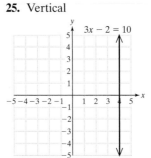

23.

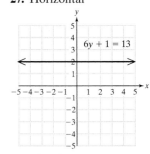

25. Vertical

27. Horizontal

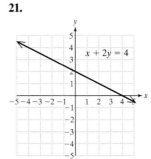

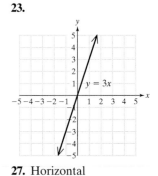

29. x-intercept: $(-3, 0)$; y-intercept: $\left(0, \frac{3}{2}\right)$
31. x-intercept: $(0, 0)$; y-intercept: $(0, 0)$
33. x-intercept: none; y-intercept: $(0, -4)$
35. x-intercept: $\left(-\frac{5}{2}, 0\right)$; y-intercept: none
37. $m = \frac{12}{5}$ **39.** $-\frac{2}{3}$ **41.** Undefined
43. a. -5 **b.** $\frac{1}{5}$ **45.** $m_1 = \frac{2}{3}; m_2 = \frac{2}{3}$; parallel
47. $m_1 = -\frac{5}{12}; m_2 = \frac{12}{5}$; perpendicular   **49.** $y = \frac{5}{2}x - 5$;
$m = \frac{5}{2}$; y-intercept: $(0, -5)$ **51.** $y = \frac{1}{3}x$;
$m = \frac{1}{3}$; y-intercept: $(0, 0)$ **53.** $y = -\frac{5}{2}$; $m = 0$;
y-intercept: $\left(0, -\frac{5}{2}\right)$ **55.** Neither   **57.** Parallel
59. Perpendicular   **61.** $y = 5x$ or $5x - y = 0$
63. For example: $5x + 2y = -4$ **65.** $y - y_1 = m(x - x_1)$

67. For example: $y = -5$ **69.** $y = \frac{2}{3}x + \frac{5}{3}$ or

$2x - 3y = -5$ **71.** $y = -5$

73. $y = 4x + 31$ or $4x - y = -31$

75. a. $m = 137$ **b.** The number of prescriptions increased by 137 million per year during this time period.

c. $y = 137x + 2140$ **d.** 4195 million

77. a. $y = 8x + 700$ **b.** $1340   **79.** Domain: {2}; range: {0, 1, −5, 2}; not a function

81. Domain: $(-\infty, \infty)$; range: $[-2, \infty)$; function

83. Domain: {3, −4, 0, 2}; range: {0, $\frac{1}{2}$, 3, −12}; function

85. a. 0   **b.** 4   **c.** $-\frac{1}{6}$ **d.** $\frac{3}{2}$ **e.** $-\frac{1}{2}$

Chapter 3 Test, pp. 290–292

1. a. II   **b.** IV   **c.** III   **2.** 0   **3.** 0

4. a. (5, 46) At age 5 the boy's height was 46 in. (7, 50) (9, 55) (11, 60)

b.

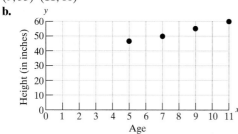

c. 57.5 in.   **d.** No, his height will maximize in his teen years.

5. a. No   **b.** Yes   **c.** Yes   **d.** Yes

6.

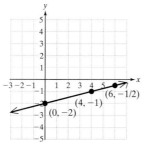

x	y
0	−2
4	−1
6	$-\frac{1}{2}$

7. a. 202 beats per minute   **b.** (20, 200), (30, 190), (40, 180) (50, 170), (60, 160)

8. Horizontal   **9.** Vertical

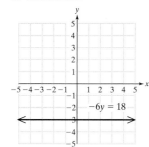

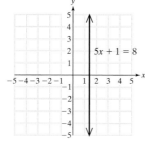

10. x-intercept: $\left(-\frac{3}{2}, 0\right)$; y-intercept: (0, 2)

11. x-intercept: (0, 0); y-intercept: (0, 0)

12. x-intercept: (4, 0); y-intercept: none

13. x-intercept: none; y-intercept: (0, 3)   **14.** $\frac{2}{5}$

15. a. $\frac{1}{3}$ **b.** $\frac{4}{3}$ **16. a.** $-\frac{1}{4}$ **b.** 4

17. a. Undefined   **b.** 0

18.

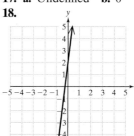

19.

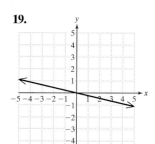

20. Perpendicular   **21.** $y = \frac{1}{4}x + \frac{1}{2}$ or $x - 4y = -2$

22. $y = -\frac{7}{2}x + 15$ or $7x + 2y = 30$ **23.** $y = -6$

24. $y = -\frac{1}{3}x + 1$ or $x + 3y = 3$ **25.** $y = 3x + 8$ or

$3x - y = -8$ **26. a.** $\frac{3}{4}$ **b.** $\frac{1}{2}$ **27. a.** $y = 1.5x + 10$

b. $25   **28. a.** $m = 20$; The slope indicates that there is an increase of 20 thousand medical doctors per year.

b. $y = 20x + 414$ **c.** 1014 thousand or, equivalently, 1,014,000   **29. a.** {0, 2, −15, 4, 9}   **b.** {−1, 3, −8, 4}

c. Function   **30. a.** 6   **b.** 12

Chapters 1–3 Cumulative Review Exercises, p. 293

1. a. Rational   **b.** Rational   **c.** Irrational   **d.** Rational

2. a. $\frac{2}{3}; \frac{2}{3}$ **b.** −5.3; 5.3   **3.** 69   **4.** −13   **5.** 18

6. $\frac{3}{4} \div -\frac{7}{8}; -\frac{6}{7}$ **7.** $(-2.1)(-6)$; 12.6

8. The associative property of addition   **9.** 4   **10.** 5

11. $-\frac{9}{2}$ **12.** −2   **13.** 9241 mi^2 **14.** $a = \frac{c - b}{3}$

15.

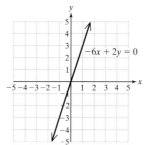

16. x-intercept: (−2, 0); y-intercept: (0, 1)

17. $y = -\frac{3}{2}x - 6$;   slope: $-\frac{3}{2}$; y-intercept: (0, −6)

18. $2x + 3 = 5$ can be written as $x = 1$, which represents a vertical line. A vertical line of the form $x = k$ ($k \neq 0$) has an x-intercept of $(k, 0)$ and no y-intercept.

19. $y = -3x + 1$ or $3x + y = 1$ **20.** $y = \frac{2}{3}x + 6$ or $2x - 3y = -18$

Chapter 4

Chapter Opener Puzzle

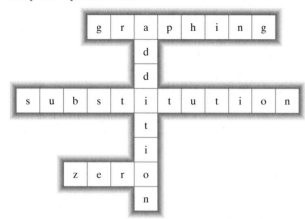

Section 4.1 Calculator Connections, pp. 300–301

1. $(2, 1)$

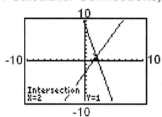

2. $(6, -1)$

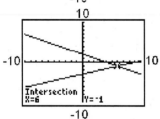

3. $(3, 1)$

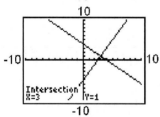

4. $(-2, 0)$

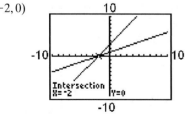

5. No solution

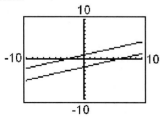

6. Dependent system

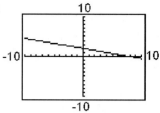

Section 4.1 Practice Exercises, pp. 301–306

3. Yes   **5.** No   **7.** Yes   **9.** No   **11.** b   **13.** d

15. a.   **b.**

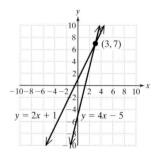

c.

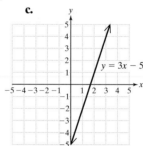

17. c   **19.** a   **21.** a   **23.** b   **25.** c

27.   **29.**

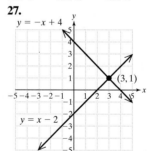

31.   **33.** No solution;
                              inconsistent system

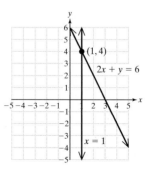

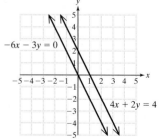

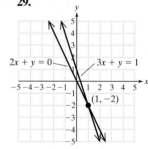

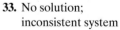

35. Infinitely many solutions; **37.**
$\{(x, y)|-2x + y = 3\}$;
dependent system

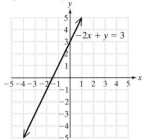

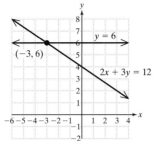

39. Infinitely many solutions; **41.**
$\{(x, y)|y = \frac{5}{3}x - 3\}$;
dependent system

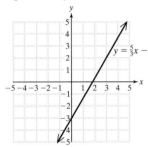

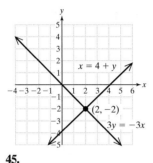

43. No solution; **45.**
inconsistent system

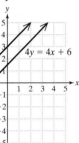

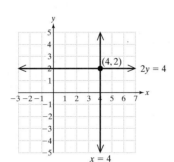

47. No solution; **49.**
inconsistent system

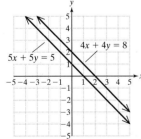

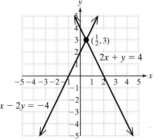

51. Infinitely many solutions; $\{(x, y)|y = 0.5x + 2\}$;
dependent system

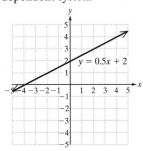

53. 4 lessons will cost $120 for each instructor. **55.** The point of intersection is below the x-axis and cannot have a positive y-coordinate. **57.** For example: $4x + y = 9$; $-2x - y = -5$ **59.** For example: $2x + 2y = 1$

Section 4.2 Practice Exercises, pp. 314–315

1. $y = 2x - 4$; $y = 2x - 4$; coinciding lines
3. $y = -\frac{2}{3}x + 2$; $y = x - 5$; intersecting lines
5. $y = 4x - 4$; $y = 4x - 13$; parallel lines
7. $(3, -6)$ **9.** $(0, 4)$ **11. a.** y in the second equation is easiest to isolate because its coefficient is 1. **b.** $(1, 5)$
13. $(5, 2)$ **15.** $(10, 5)$ **17.** $\left(\frac{1}{2}, 3\right)$ **19.** $(5, 3)$
21. $(1, 0)$ **23.** $(1, 4)$ **25.** No solution; inconsistent system **27.** Infinitely many solutions; $\{(x, y)|2x - 6y = -2\}$; dependent system **29.** $(5, -7)$
31. $\left(-5, \frac{3}{2}\right)$ **33.** $(2, -5)$ **35.** $(-4, 6)$ **37.** $(0, 2)$
39. Infinitely many solutions; $\{(x, y)|y = 0.25x + 1\}$; dependent system **41.** $(1, 1)$ **43.** No solution; inconsistent system **45.** $(-1, 5)$ **47.** $(-6, -4)$
49. The numbers are 48 and 58. **51.** The numbers are 13 and 39. **53.** The angles are 165° and 15°. **55.** The angles are 70° and 20°. **57.** The angles are 42° and 48°.
59. For example: $(0, 3), (1, 5), (-1, 1)$

Section 4.3 Practice Exercises, pp. 322–324

3. No **5.** Yes **7. a.** True **b.** False, multiply the second equation by 5. **9. a.** x would be easier.
b. $(0, -3)$ **11.** $(4, -1)$ **13.** $(4, 3)$ **15.** $(2, 3)$
17. $(1, -4)$ **19.** $(1, -1)$ **21.** $(-4, -6)$
23. $\left(\frac{7}{9}, \frac{5}{9}\right)$ **25.** $\left(\frac{7}{16}, -\frac{7}{8}\right)$
27. There are infinitely many solutions. The lines coincide.
29. The system will have no solution. The lines are parallel.
31. The system will have one solution. The lines intersect at a point whose y-coordinate is 0. **33.** No solution; inconsistent system **35.** Infinitely many solutions; $\{(x, y)|4x - 3y = 6\}$; dependent system **37.** $(2, -2)$
39. $(0, 3)$ **41.** $(5, 2)$ **43.** No solution; inconsistent system **45.** $(-5, 0)$ **47.** $(2.5, -0.5)$
49. $\left(\frac{1}{2}, 0\right)$ **51.** $(-3, 2)$ **53.** $\left(\frac{7}{4}, 3\right)$ **55.** $(0, 1)$
57. No solution; inconsistent system **59.** $(0, -5)$
61. $(4, -2)$ **63.** Infinitely many solutions; $\{(a, b)|a = 5 + 2b\}$; dependent system
65. The numbers are 17 and 19. **67.** The numbers are -1 and 3. **69.** $(1, 3)$ **71.** One line within the system of equations would have to "bend" for the system to have exactly two points of intersection. This is not possible.
73. $A = -5, B = 2$

Chapter 4 Problem Recognition Exercises, p. 325

1. Infinitely many solutions. The equations represent the same line. **2.** No solution. The equations represent parallel lines. **3.** One solution. The equations represent intersecting lines. **4.** One solution. The equations

represent intersecting lines.   **5.** No solution. The equations represent parallel lines.   **6.** Infinitely many solutions. The equations represent the same line.   **7.** $(5, 0)$
8. $(1, -7)$ **9.** $(4, -5)$ **10.** $(2, 3)$ **11.** $(2, 0)$
12. $(8, 10)$ **13.** $\left(2, -\dfrac{5}{7}\right)$ **14.** $\left(-\dfrac{14}{3}, -4\right)$
15. No solution; inconsistent system   **16.** No solution; inconsistent system   **17.** $(-1, 0)$ **18.** $(5, 0)$
19. Infinitely many solutions; $\{(x, y) \mid y = 2x - 14\}$; dependent system   **20.** Infinitely many solutions; $\{(x, y) \mid x = 5y - 9\}$; dependent system
21. $(2200, 1000)$ **22.** $(3300, 1200)$ **23.** $(5, -7)$
24. $(2, -1)$ **25.** $\left(\dfrac{2}{3}, \dfrac{1}{2}\right)$ **26.** $\left(\dfrac{1}{4}, -\dfrac{3}{2}\right)$

Section 4.4 Practice Exercises, pp. 331–335

1. $(-1, 4)$ **3.** $\left(\dfrac{5}{2}, 1\right)$ **5.** The numbers are 4 and 16.
7. The angles are 80° and 10°.   **9.** DVDs are $10.50 each, and CDs are $15.50 each.   **11.** Technology stock costs $16 per share, and the mutual fund costs $11 per share.
13. Patricia bought forty 42¢ stamps and ten 59¢ stamps.
15. Shanelle invested $3500 in the 10% account and $6500 in the 7% account.   **17.** $9000 was borrowed at 6%, and $3000 was borrowed at 9%.   **19.** Invest $12,000 in the bond fund and $18,000 in the stock fund.   **21.** 15 gal of the 50% mixture should be mixed with 10 gal of the 40% mixture.   **23.** 12 gal of the 45% disinfectant solution should be mixed with 8 gal of the 30% disinfectant solution.
25. She should mix 20 mL of the 13% solution with 30 mL of the 18% solution.   **27.** The speed of the boat in still water is 6 mph, and the speed of the current is 2 mph.   **29.** The speed of the plane in still air is 300 mph, and the wind is 20 mph.   **31.** The speed of the plane in still air is 525 mph and the speed of the wind is 75 mph.   **33.** There are 17 dimes and 22 nickels.   **35. a.** 835 free throws and 1597 field goals   **b.** 4029 points   **c.** Approximately 50 points per game   **37.** The speed of the plane in still air is 160 mph, and the wind is 40 mph.   **39.** 12 lb of candy should be mixed with 8 lb of nuts.   **41.** $15,000 was invested in the 5.5% account, and $45,000 was invested in the 6.5% account.
43. Dallas scored 30 points, and Buffalo scored 13 points.   **45.** There were 300 women and 200 men in the survey.

Section 4.5 Practice Exercises, pp. 341–345

3.

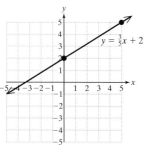

5. When the inequality symbol is ≤ or ≥
7. All of the points in the shaded region are solutions to the inequality.   **9.** a

11. For example:
$(0, 5)\ (2, 7)\ (-1, 8)$

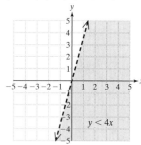

13. For example:
$(1, -1)\ (3, 0)\ (-2, -9)$

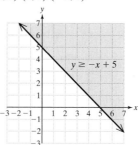

15. For example:
$(0, 0)\ (0, 2)\ (-1, -3)$

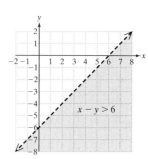

17.

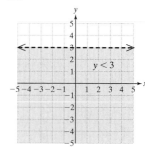

19.

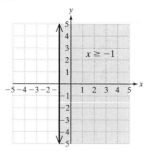

21.

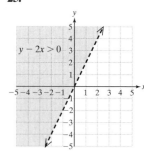

23.

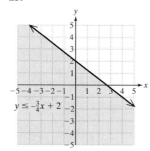

25.

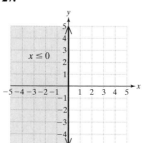

27.

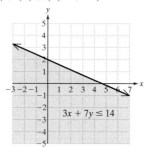

29.

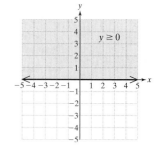

31.

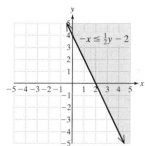

33.

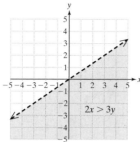

53.

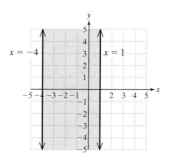

35. **a.** The set of ordered pairs above the line $x + y = 4$, for example, $(6, 3)(-2, 8)(0, 5)$ **b.** The set of ordered pairs on the line $x + y = 4$, for example, $(0, 4)(4, 0)(2, 2)$ **c.** The set of ordered pairs below the line $x + y = 4$, for example, $(0, 0)(-2, 1)(3, 0)$

Chapter 4 Review Exercises, pp. 352–355

1. Yes   **3.** No   **5.** Intersecting lines (the lines have different slopes)   **7.** Parallel lines (the lines have the same slope but different y-intercepts)   **9.** Coinciding lines (the lines have the same slope and same y-intercept)
11. $(0, -2)$

37.

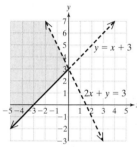

39.

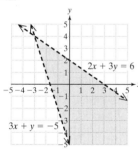

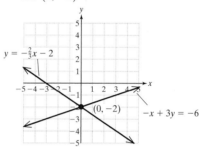

13. Infinitely many solutions; $\{(x, y) \mid 2x + y = 5\}$; dependent system

41.

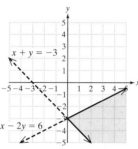

43.

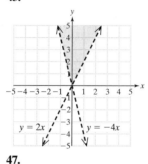

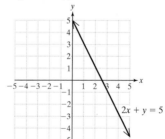

45.

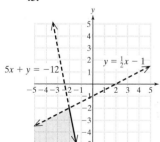

47.

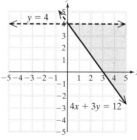

15. $(1, -1)$ **17.** No solution; inconsistent system

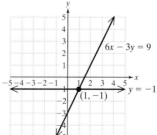

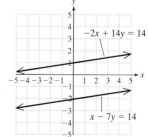

49.

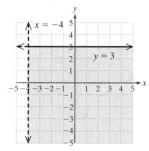

51.

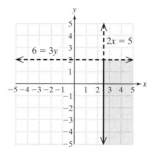

19. 200 mi   **21.** $(-4, 1)$ **23.** Infinitely many solutions; $\{(x, y) \mid y = -2x + 2\}$; dependent system   **25. a.** y in the second equation is easiest to isolate because its coefficient is 1.   **b.** $\left(\frac{9}{2}, 3 \right)$ **27.** $(0, 4)$ **29.** No solution; inconsistent system   **31.** The angles are $42°$ and $48°$.
33. See page 317.   **35. b.** $(2, 2)$ **37.** $(-6, 2)$

39. $\left(\dfrac{1}{4}, -\dfrac{2}{5}\right)$ **41.** No solution; inconsistent system

43. $(1, 0)$ **45. b.** $(-2, -1)$ **47.** He should invest $75,000 at 12% and $525,000 at 4%.   **49.** The speed of the boat is 18 mph, and that of the current is 2 mph.

51. A hot dog costs $4.50 and a drink costs $3.50

53. The score was 72 on the first round and 82 on the second round.

55. For example: **57.** For example:
$(5, 5)(4, 0)(0, 7)$ $(0, 0)(0, -5)(-1, 1)$

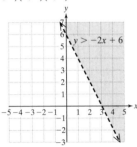

59.   **61.**

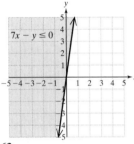   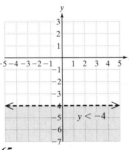

63.   **65.**

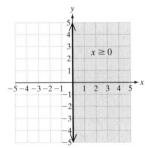   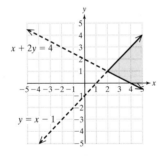

67.

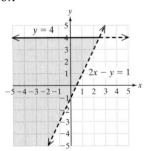

Chapter 4 Test, pp. 356–357

1. $y = -\dfrac{5}{2}x - 3; y = -\dfrac{5}{2}x + 3$; Parallel lines

2.

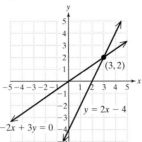

3. Infinitely many solutions; $\{(x, y)\,|\,2x + 4y = 6\}$; dependent system

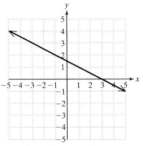

4. $(-2, 0)$ **5.** Swoopes scored 614 points and Jackson scored 597.   **6.** $\left(2, -\dfrac{1}{3}\right)$ **7.** 12 mL of the 50% acid solution should be mixed with 24 mL of the 20% solution.

8. a. No solution   **b.** Infinitely many solutions   **c.** One solution   **9.** $(-5, 4)$ **10.** No solution   **11.** $(3, -5)$ **12.** $(-1, 2)$ **13.** Infinitely many solutions; $\{(x, y)\,|\,10x + 2y = -8\}$ **14.** $(1, -2)$ **15.** CDs cost $8 each and DVDs cost $11 each.   **16. a.** $18 was required.   **b.** They used 24 quarters and 12 $1 bills.   **17.** $1200 was borrowed at 10%, and $3800 was borrowed at 8%.

18. He scored 155 receiving touchdowns and 10 touchdowns rushing.   **19.** The plane travels 500 mph in still air, and the wind speed is 45 mph.   **20.** The cake has 340 calories, and the ice cream has 120 calories.

21. 60 mL of 10% solution and 40 mL of 25% solution.

22.   **23.**

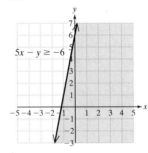   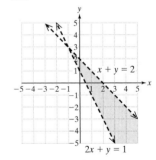

Chapters 1–4 Cumulative Review Exercises, pp. 357–358

1. $\dfrac{11}{6}$ **2.** $-\dfrac{21}{2}$ **3.** No solution   **4.** $y = \dfrac{3}{2}x - 3$

5. $\left[\dfrac{3}{11}, \infty\right)$ ⊢────────▶
                        $\dfrac{3}{11}$

6. The angles are 37°, 33°, and 110°.   **7.** The rates of the hikers are 2 mph and 4 mph.   **8.** Jesse Ventura received approximately 762,200 votes.   **9.** 36% of the goal has been achieved.   **10.** The angles are 36.5° and 53.5°.

11. $x = 5z + m$ **12.** $y = \frac{2}{3}x - 2$ **13. a.** $-\frac{2}{3}$ **b.** $\frac{3}{2}$

14. $y = -3x + 3$

15. c. $(0, 2)$ **16.** $(0, 2)$

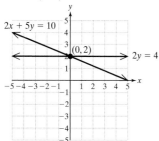

17. a.

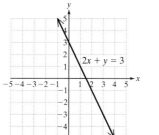

b.

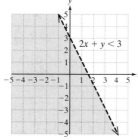

c. Part (a) represents the solutions to an equation. Part (b) represents the solutions to a strict inequality.   **18.** 20 gal of the 15% solution should be mixed with 40 gal of the 60% solution.   **19.** x is 27°; y is 63°   **20. a.** 1.4 **b.** Between 1920 and 1990, the winning speed in the Indianapolis 500 increased on average by 1.4 mph per year.

Chapter 5

Chapter Opener Puzzle

1	5	6	2	4	^{A}3
2	4	3	5	1	6
4	3	1	6	^{B}5	2
5	6	^{C}2	4	3	1
3	2	4	1	6	5
^{D}6	1	5	3	2	^{E}4

Section 5.1 Calculator Connection, p. 366

1.–3.

```
(1.06)^5
        1.338225578
(1.02)^40
        2.208039664
5000(1.06)^5
        6691.127888
```

4.–6.

```
2000(1.02)^40
        4416.079327
3000(1+.06)^2
        3370.8
1000(1+.05)^3
        1157.625
```

Section 5.1 Practice Exercises, pp. 367–369

3. Base: x; exponent: 4   **5.** Base: 3; exponent: 5
7. Base: -1; exponent: 4   **9.** Base: q; exponent: 1
11. v **13.** 1   **15.** $(-6b)^2$ **17.** $-6b^2$ **19.** $(y + 2)^4$
21. $\frac{-2}{t^3}$ **23.** No; $-5^2 = -25$ and $(-5)^2 = 25$

25. Yes; $-2^5 = -32$ and $(-2)^5 = -32$ **27.** Yes; $\left(\frac{1}{2}\right)^3 = \frac{1}{8}$
and $\frac{1}{2^3} = \frac{1}{8}$ **29.** Yes; $\left(\frac{3}{10}\right)^2 = \frac{9}{100}$ and $(0.3)^2 = 0.09$

31. 16   **33.** -1 **35.** $\frac{1}{9}$ **37.** $-\frac{4}{25}$ **39.** 48
41. 4   **43.** 9   **45.** 50   **47.** -100 **49.** 400
51. 1   **53.** 1   **55.** 1000   **57.** -800
59. a. $(x \cdot x \cdot x \cdot x)(x \cdot x \cdot x) = x^7$
b. $(5 \cdot 5 \cdot 5 \cdot 5)(5 \cdot 5 \cdot 5) = 5^7$ **61.** z^8 **63.** a^9
65. 4^{14} **67.** $\left(\frac{2}{3}\right)^4$ **69.** c^{14} **71.** x^{18}

73. a. $\dfrac{p \cdot p \cdot p \cdot p \cdot p \cdot p \cdot p \cdot p}{p \cdot p \cdot p} = p^5$
b. $\dfrac{8 \cdot 8 \cdot 8 \cdot 8 \cdot 8 \cdot 8 \cdot 8 \cdot 8}{8 \cdot 8 \cdot 8} = 8^5$
75. x^2 **77.** a^9 **79.** 7^7 **81.** 5^7 **83.** y **85.** h^4
87. 7^7 **89.** 10^9 **91.** $6x^7$ **93.** $40a^5b^5$ **95.** $13r^8s^5$
97. s^9t^{16} **99.** $-30v^8$ **101.** $16m^{20}n^{10}$ **103.** $2cd^4$
105. z^4 **107.** $\dfrac{25hjk^4}{12}$ **109.** $-8p^7q^9r^6$ **111.** $-3stu$
113. \$5724.50   **115.** \$4764.06   **117.** 201 in.^{2}
119. 268 in.^{3} **121.** x^{2n+1} **123.** p^{2m+3} **125.** z
127. r^3

Section 5.2 Practice Exercises, pp. 373–374

1. 4^9 **3.** a^{20} **5.** d^9 **7.** 7^6 **9.** When multiplying expressions with the same base, add the exponents. When dividing expressions with the same base, subtract the exponents.   **11.** 5^{12} **13.** 12^6 **15.** y^{14} **17.** w^{25}
19. a^{36} **21.** y^{14} **23.** They are both equal to 2^6.
25. $2^{(2^4)} = 2^{16}$ $(2^2)^4 = 2^8$; the expression $2^{(2^4)}$ is greater than $(2^2)^4$.   **27.** $25w^2$ **29.** $s^4r^4t^4$
31. $\dfrac{16}{r^4}$ **33.** $\dfrac{x^5}{y^5}$ **35.** $81a^4$ **37.** $-27a^3b^3c^3$
39. $-\dfrac{64}{x^3}$ **41.** $\dfrac{a^2}{b^2}$ **43.** $6^3u^6v^{12}$ or $216\,u^6v^{12}$
45. $5x^8y^4$ **47.** $-h^{28}$ **49.** m^{12} **51.** $\dfrac{4^5}{r^5s^{20}}$ or $\dfrac{1024}{r^5s^{20}}$
53. $\dfrac{3^5p^5}{q^{15}}$ or $\dfrac{243p^5}{q^{15}}$ **55.** y^{14} **57.** x^{31} **59.** $200a^{14}b^9$
61. $16p^8q^{16}$ **63.** $-m^{35}n^{15}$ **65.** $25a^{18}b^6$ **67.** $\dfrac{4c^2d^6}{9}$

69. $\dfrac{c^{27}d^{31}}{2}$ **71.** $-\dfrac{27a^9b^3}{c^6}$ **73.** $16b^{26}$ **75.** x^{2m}

77. $125a^{6n}$ **79.** $\dfrac{m^{2b}}{n^{3b}}$ **81.** $\dfrac{3^n a^{3n}}{5^n b^{4n}}$

Section 5.3 Practice Exercises, pp. 381–383

3. c^9 **5.** y **7.** 3^6 or 729 **9.** $7^4 w^{28} z^8$ or $2401 w^{28} z^8$
11. a. 1 **b.** 1 **13.** 1 **15.** 1 **17.** -1 **19.** 1
21. 1 **23.** -7 **25. a.** $\dfrac{1}{t^5}$ **b.** $\dfrac{1}{t^5}$ **27.** $\dfrac{343}{8}$ **29.** 25
31. $\dfrac{1}{a^3}$ **33.** $\dfrac{1}{12}$ **35.** $\dfrac{1}{16b^2}$ **37.** $\dfrac{6}{x^2}$ **39.** $\dfrac{1}{64}$
41. $-\dfrac{3}{y^4}$ **43.** $-\dfrac{1}{t^3}$ **45.** a^5 **47.** $\dfrac{x^4}{x^{-6}} = x^{4-(-6)} = x^{10}$
49. $2a^{-3} = 2 \cdot \dfrac{1}{a^3} = \dfrac{2}{a^3}$ **51.** $\dfrac{1}{x^4}$ **53.** 1 **55.** y^4
57. $\dfrac{n^{27}}{m^{18}}$ **59.** $\dfrac{81k^{24}}{j^{20}}$ **61.** $\dfrac{1}{p^6}$ **63.** $\dfrac{1}{r^3}$ **65.** a^8
67. $\dfrac{1}{y^8}$ **69.** $\dfrac{1}{7^7}$ **71.** 1 **73.** $\dfrac{1}{a^4 b^6}$ **75.** $\dfrac{1}{w^{21}}$
77. $\dfrac{1}{27}$ **79.** 1 **81.** $\dfrac{1}{64x^6}$ **83.** $-\dfrac{16y^4}{z^2}$ **85.** $-\dfrac{a^{12}}{6}$
87. $80c^{21}d^{24}$ **89.** $\dfrac{p^{27}}{8}$ **91.** $\dfrac{4d^{16}}{c^2}$ **93.** $\dfrac{9y^2}{2x}$

95. $\dfrac{9}{20}$ **97.** $\dfrac{9}{10}$ **99.** $\dfrac{5}{4}$ **101.** $\dfrac{10}{3}$

Section 5.4 Calculator Connections, p. 386

1.–2.
```
(5.2E6)*(4.6E-3)
            23920
(2.19E-8)*(7.84E
-4)
     1.71696E-11
```

3.–4.
```
(4.76E-5)/(2.38E
9)
            2E-14
(8.5E4)/(4.0E-1)
           212500
```

5.
```
((9.6E7)*(4.0E-3
))/(2.0E-2)
         19200000
```

6.
```
((5.0E-12)*(6.4E
-5))/((1.6E-8)*(
4.0E2))
           5E-11
```

Section 5.4 Practice Exercises, pp. 387–389

3. b^{13} **5.** 10^{13} **7.** $\dfrac{1}{y^5}$ **9.** $\dfrac{x^{20}}{y^{12}}$ **11.** w^4 **13.** 10^4

15. Move the decimal point between the 2 and 3 and multiply by 10^{-10}; 2.3×10^{-10}.   **17.** 5×10^4
19. 2.08×10^5 **21.** 6.01×10^6 **23.** 8×10^{-6}
25. 1.25×10^{-4} **27.** 6.708×10^{-3} **29.** 1.7×10^{-24} g
31. $\$2.7 \times 10^{10}$ **33.** 6.8×10^7 gal; 1.0×10^2 miles
35. Move the decimal point nine places to the left; 0.000 000 0031.   **37.** 0.00005   **39.** 2800   **41.** 0.000603
43. 2,400,000   **45.** 0.019   **47.** 7032
49. 0.000 000 000 001 g   **51.** 1600 calories and 2800 calories

53. 5.0×10^4 **55.** 3.6×10^{11} **57.** 2.2×10^4
59. 2.25×10^{-13} **61.** 3.2×10^{14} **63.** 2.432×10^{-10}
65. 3.0×10^{13} **67.** 6.0×10^5 **69.** 1.38×10^1
71. 5.0×10^{-14} **73.** 3.75 in.   **75.** $\$2.97 \times 10^{10}$
77. a. 6.5×10^7 **b.** 2.3725×10^{10} days
c. 5.694×10^{11} hr   **d.** 2.04984×10^{15} sec

Chapter 5 Problem Recognition Exercises, p. 389

1. t^8 **2.** 2^8 or 256 **3.** y^5 **4.** p^6 **5.** $r^4 s^8$
6. $a^3 b^9 c^6$ **7.** w^6 **8.** $\dfrac{1}{m^{16}}$ **9.** $\dfrac{x^4 z^3}{y^7}$ **10.** $\dfrac{a^3 c^8}{b^6}$
11. 1.25×10^3 **12.** 1.24×10^5 **13.** 8.0×10^8
14. 6.0×10^{-9} **15.** p^{15} **16.** p^{15} **17.** $\dfrac{1}{v^2}$
18. $c^{50} d^{40}$ **19.** 3 **20.** -4 **21.** $\dfrac{b^9}{2^{15}}$ **22.** $\dfrac{81}{y^6}$
23. $\dfrac{16y^4}{81x^4}$ **24.** $\dfrac{25d^6}{36c^2}$ **25.** $3a^7 b^5$ **26.** $64x^7 y^{11}$
27. $\dfrac{y^4}{x^8}$ **28.** $\dfrac{1}{a^{10} b^{10}}$ **29.** $\dfrac{1}{t^2}$ **30.** $\dfrac{1}{p^7}$ **31.** $\dfrac{8w^6 x^9}{27}$
32. $\dfrac{25b^8}{16c^6}$ **33.** $\dfrac{q^3 s}{r^2 t^5}$ **34.** $\dfrac{m^2 p^3 q}{n^3}$ **35.** $\dfrac{1}{y^{13}}$ **36.** w^{10}
37. $-\dfrac{1}{8a^{18} b^6}$ **38.** $\dfrac{4x^{18}}{9y^{10}}$ **39.** $\dfrac{k^8}{5h^6}$ **40.** $\dfrac{6n^{10}}{m^{12}}$

Section 5.5 Practice Exercises, pp. 395–398

3. $\dfrac{45}{x^2}$ **5.** $\dfrac{2}{t^4}$ **7.** $\dfrac{1}{3^{12}}$

9. 4.0×10^{-2} is in scientific notation in which 10 is raised to the -2 power. 4^{-2} is not in scientific notation and 4 is being raised to the -2 power.   **11.** $-7x^4 + 7x^2 + 9x + 6$
13. Binomial; $10; 2$ **15.** Monomial; $6; 2$
17. Binomial; $-1; 4$ **19.** Trinomial; $12; 4$
21. Monomial; $23; 0$ **23.** Monomial; $-32; 3$
25. The exponents on the x-factors are different.
27. $35x^2 y$ **29.** $10y$ **31.** $8b^5 d^2 - 9d$

33. $4y^2 + y - 9$ **35.** $4a - 8c$ **37.** $a - \dfrac{1}{2}b - 2$

39. $\dfrac{4}{3}z^2 - \dfrac{5}{3}$ **41.** $7.9t^3 - 3.4t^2 + 6t - 4.2$ **43.** $-4h + 5$
45. $2m^2 - 3m + 15$ **47.** $-3v^3 - 5v^2 - 10v - 22$
49. $-8a^3 b^2$ **51.** $-53x^3$ **53.** $-5a - 3$ **55.** $16k + 9$
57. $2s - 14$ **59.** $3t^2 - 4t - 3$ **61.** $-2r - 3s + 3t$
63. $\dfrac{3}{4}x + \dfrac{1}{3}y - \dfrac{3}{10}$ **65.** $-\dfrac{2}{3}h^2 + \dfrac{3}{5}h - \dfrac{5}{2}$
67. $2.4x^4 - 3.1x^2 - 4.4x - 6.7$
69. $4b^3 + 12b^2 - 5b - 12$ **71.** $-3x^3 - 2x^2 + 11x - 31$
73. $4y^3 + 2y^2 + 2$ **75.** $3a^2 - 3a + 5$
77. $9ab^2 - 3ab + 16a^2 b$ **79.** $4z^5 + z^4 + 9z^3 - 3z - 2$
81. $2x^4 + 11x^3 - 3x^2 + 8x - 4$ **83.** $-2w^2 - 7w + 18$
85. $-p^2 q - 4pq^2 + 3pq$ **87.** 0 **89.** $-5ab + 6ab^2$
91. $11y^2 - 10y - 4$ **93.** For example, $x^3 + 6$
95. For example, $8x^5$ **97.** For example, $-6x^2 + 2x + 5$

Section 5.6 Practice Exercises, pp. 404–407

3. $-2y^2$　**5.** $-8y^4$　**7.** $8uvw^2$　**9.** $7u^2v^2w^4$

11. $-12y$　**13.** $21p$　**15.** $12a^{14}b^8$　**17.** $-2c^{10}d^{12}$

19. $16p^2q^2 - 24p^2q + 40pq^2$　**21.** $-4k^3 + 52k^2 + 24k$

23. $-45p^3q - 15p^4q^3 + 30pq^2$　**25.** $y^2 - y - 90$

27. $m^2 - 14m + 24$　**29.** $12p^2 - 5p - 2$

31. $12w^2 - 32w + 16$　**33.** $p^2 - 14pw + 33w^2$

35. $12x^2 + 28x - 5$　**37.** $8a^2 - 22a + 9$

39. $9t^2 - 18t - 7$　**41.** $3m^2 + 28mn + 32n^2$

43. $5s^3 + 8s^2 - 7s - 6$　**45.** $27w^3 - 8$

47. $p^4 + 5p^3 - 2p^2 - 21p + 5$

49. $6a^3 - 23a^2 + 38a - 45$

51. $8x^3 - 36x^2y + 54xy^2 - 27y^3$　**53.** $9a^2 - 16b^2$

55. $81k^2 - 36$　**57.** $\dfrac{1}{4} - t^2$　**59.** $u^6 - 25v^2$

61. $4 - 9a^2$　**63.** $\dfrac{4}{9} - p^2$　**65.** $a^2 + 10a + 25$

67. $x^2 - 2xy + y^2$　**69.** $4c^2 + 20c + 25$

71. $9t^4 - 24st^2 + 16s^2$　**73.** $t^2 - 14t + 49$

75. $16q^2 + 24q + 9$　**77. a.** 36　**b.** 20

c. $(a + b)^2 \neq a^2 + b^2$ in general.　**79. a.** $9x^2 + 6xy + y^2$

b. $9x^2y^2$　**81.** $36 - y^2$　**83.** $49q^2 - 42q + 9$

85. $r^3 - 15r^2 + 63r - 49$　**87.** $3t^3 - 12t$

89. $81w^2 - 16z^2$　**91.** $25s^2 - 30st + 9t^2$

93. $10a^2 - 13ab + 4b^2$　**95.** $s^2 + \dfrac{147}{5}s - 18$

97. $4k^3 - 4k^2 - 5k - 25$　**99.** $u^3 + u^2 - 10u + 8$

101. $w^3 + w^2 + 4w + 12$　**103.** $\dfrac{4}{25}p^2 - \dfrac{4}{5}pq + q^2$

105. $4v^4 + 48v^3$　**107.** $4h^2 - 7.29$　**109.** $5k^6 - 19k^3 + 18$

111. $6.25y^2 + 5.5y + 1.21$　**113.** $h^3 + 9h^2 + 27h + 27$

115. $8a^3 - 48a^2 + 96a - 64$　**117.** $6w^4 + w^3 - 15w^2 - 11w - 5$

119. $30x^3 + 55x^2 - 10x$　**121.** $2y^3 - y^2 - 15y + 18$

123. $x + 6$　**125.** $k = 6$ or -6

Section 5.7 Practice Exercises, pp. 412–414

1. $6z^5 - 10z^4 - 4z^3 - z^2 - 6$　**3.** $10x^2 - 29xy - 3y^2$

5. $11x - 2y$　**7.** $y^2 - \dfrac{3}{4}y + \dfrac{1}{2}$　**9.** $a^3 + 27$

11. Use long division when the divisor is a polynomial with two or more terms.　**13.** $5t^2 + 6t$　**15.** $3a^2 + 2a - 7$

17. $x^2 + 4x - 1$　**19.** $3p^2 - p$　**21.** $1 + \dfrac{2}{m}$

23. $-2y^2 + y - 3$　**25.** $x^2 - 6x - \dfrac{1}{4} + \dfrac{2}{x}$

27. $a - 1 + \dfrac{b}{a}$　**29.** $3t - 1 + \dfrac{3}{2t} - \dfrac{1}{2t^2} + \dfrac{2}{t^3}$

31. a. $z + 2 + \dfrac{1}{z + 5}$　**33.** $t + 3 + \dfrac{2}{t + 1}$

35. $7b + 4$　**37.** $k - 6$　**39.** $2p^2 + 3p - 4$

41. $k - 2 + \dfrac{-4}{k + 1}$　**43.** $2x^2 - x + 6 + \dfrac{2}{2x - 3}$

45. $y^2 + 2y + 1 + \dfrac{2}{3y - 1}$　**47.** $a - 3 + \dfrac{18}{a + 3}$

49. $4x^2 + 8x + 13$　**51.** $w^2 + 5w - 2 + \dfrac{1}{w^2 - 3}$

53. $n^2 + n - 6$　**55.** $x - 1 + \dfrac{-8}{5x^2 + 5x + 1}$

57. Multiply $(x - 2)(x^2 + 4) = x^3 - 2x^2 + 4x - 8$, which does not equal $x^3 - 8$.　**59.** Monomial division; $3a^2 + 4a$

61. Long division; $p + 2$　**63.** Long division;

$t^3 - 2t^2 + 5t - 10 + \dfrac{4}{t + 2}$　**65.** Long division;

$w^2 + 3 + \dfrac{1}{w^2 - 2}$　**67.** Long division; $n^2 + 4n + 16$

69. Monomial division; $-3r + 4 - \dfrac{3}{r^2}$　**71.** $x + 1$

73. $x^3 + x^2 + x + 1$　**75.** $x + 1 + \dfrac{1}{x - 1}$

77. $x^3 + x^2 + x + 1 + \dfrac{1}{x - 1}$

Chapter 5 Problem Recognition Exercises, p. 415

1. $2x^3 - 8x^2 + 14x - 12$　**2.** $-3y^4 - 20y^2 - 32$

3. $x^2 - 1$　**4.** $4y^2 + 12$　**5.** $36y^2 - 84y + 49$

6. $9z^2 + 12z + 4$　**7.** $36y^2 - 49$　**8.** $9z^2 - 4$

9. $16x^2 + 8xy + y^2$　**10.** $4a^2 + 4ab + b^2$　**11.** $16x^2y^2$

12. $4a^2b^2$　**13.** $-x^2 - 3x + 4$　**14.** $5m^2 - 4m + 1$

15. $-7m^2 - 16m$　**16.** $-4n^5 + n^4 + 6n^2 - 7n + 2$

17. $8x^2 + 16x + 34 + \dfrac{74}{x - 2}$

18. $-4x^2 - 10x - 30 + \dfrac{-95}{x - 3}$

19. $6x^3 + 5x^2y - 6xy^2 + y^3$　**20.** $6a^3 - a^2b + 5ab^2 + 2b^3$

21. $x^3 + y^6$　**22.** $m^6 + 1$　**23.** $4b$　**24.** $-12z$

25. $a^4 - 4b^2$　**26.** $y^6 - 36z^2$　**27.** $64u^2 + 48uv + 9v^2$

28. $4p^2 - 4pt + t^2$　**29.** $4p + 4 + \dfrac{-2}{2p - 1}$

30. $2v - 7 + \dfrac{29}{2v + 3}$　**31.** $4x^2y^2$　**32.** $-9pq$

33. $10a^2 - 57a + 54$　**34.** $28a^2 - 17a - 3$

35. $\dfrac{9}{49}x^2 - \dfrac{1}{4}$　**36.** $\dfrac{4}{25}y^2 - \dfrac{16}{9}$

37. $-\dfrac{11}{9}x^3 + \dfrac{5}{9}x^2 - \dfrac{1}{2}x - 4$　**38.** $-\dfrac{13}{10}y^2 - \dfrac{9}{10}y + \dfrac{4}{15}$

39. $1.3x^2 - 0.3x - 0.5$　**40.** $5w^3 - 4.1w^2 + 2.8w - 1.2$

Chapter 5 Review Exercises, pp. 420–423

1. Base: 5; exponent: 3　**3.** Base: -2; exponent: 0

5. a. 36　**b.** 36　**c.** -36　**7.** 5^{13}　**9.** x^9　**11.** 10^3

13. b^8　**15.** k　**17.** 2^8　**19.** Exponents are added only when multiplying factors with the same base. In such a case, the base does not change.　**21.** $7146.10

23. 7^{12}　**25.** p^{18}　**27.** $\dfrac{a^2}{b^2}$　**29.** $\dfrac{5^2}{c^4d^{10}}$　**31.** $2^4a^4b^8$

33. $-\dfrac{3^3x^9}{5^3y^6z^3}$　**35.** a^{11}　**37.** $4h^{14}$　**39.** $\dfrac{x^6y^2}{4}$　**41.** 1

43. -1　**45.** 2　**47.** $\dfrac{1}{z^5}$　**49.** $\dfrac{1}{36a^2}$　**51.** $\dfrac{17}{16}$

53. $\dfrac{1}{t^8}$ **55.** $\dfrac{2y^7}{x^6}$ **57.** $\dfrac{n^{16}}{16m^8}$ **59.** $\dfrac{k^{21}}{5}$ **61.** $\dfrac{1}{2}$

63. a. 9.74×10^7 **b.** 4.2×10^{-3} in.   **65.** 9.43×10^5
67. 2.5×10^8 **69.** $\approx 9.5367 \times 10^{13}$. This number is too big to fit on most calculator displays.
71. a. $\approx 5.84 \times 10^8$ mi   **b.** $\approx 6.67 \times 10^4$ mph
73. a. Trinomial   **b.** 4   **c.** 7   **75.** $7x - 3$
77. $14a^2 - 2a - 6$ **79.** $10w^4 + 2w^3 - 7w + 4$
81. $-2x^2 - 9x - 6$ **83.** For example, $-5x^2 + 2x - 4$
85. $6w + 6$ **87.** $18a^8b^4$ **89.** $-2x^3 - 10x^2 + 6x$
91. $20t^2 + 3t - 2$ **93.** $2a^2 + 4a - 30$
95. $b^2 - 8b + 16$ **97.** $-2w^3 - 5w^2 - 5w + 4$
99. $12a^3 + 11a^2 - 13a - 10$ **101.** $\dfrac{1}{9}r^8 - s^4$
103. $2h^5 + h^4 - h^3 + h^2 - h + 3$ **105.** $4y^2 - 2y$
107. $-3x^2 + 2x - 1$ **109.** $x + 2$
111. $p - 3 + \dfrac{5}{2p + 7}$ **113.** $b^2 + 5b + 25$
115. $y^2 - 4y + 2 + \dfrac{9y - 4}{y^2 + 3}$ **117.** $w^2 + w - 1$

Chapter 5 Test, pp. 423–424

1. $\dfrac{(3 \cdot 3 \cdot 3 \cdot 3) \cdot (3 \cdot 3 \cdot 3)}{3 \cdot 3 \cdot 3 \cdot 3 \cdot 3 \cdot 3} = 3$ **2.** 9^6 **3.** q^8
4. $27a^6b^3$ **5.** $\dfrac{16x^4}{y^{12}}$ **6.** 1   **7.** $\dfrac{1}{c^3}$ **8.** 14
9. $49s^{18}t$ **10.** $\dfrac{4}{b^{12}}$ **11.** $\dfrac{16a^{12}}{9b^6}$
12. a. 4.3×10^{10} **b.** $0.000\,0056$ **13. a.** $2.4192 \times 10^8 \, \text{m}^3$
b. $8.83008 \times 10^{10} \, \text{m}^3$ **14.** $5x^3 - 7x^2 + 4x + 11$ **a.** 3
b. 5   **15.** $24w^2 - 3w - 4$ **16.** $15x^3 - 7x^2 - 2x + 1$
17. $-10x^5 - 2x^4 + 30x^3$ **18.** $8a^2 - 10a + 3$
19. $4y^3 - 25y^2 + 37y - 15$ **20.** $4 - 9b^2$
21. $25z^2 - 60z + 36$ **22.** $15x^2 - x - 6$
23. $y^3 - 11y^2 + 32y - 12$ **24.** Perimeter: $12x - 2$;
area: $5x^2 - 13x - 6$ **25.** $-3x^6 + \dfrac{x^4}{4} - 2x$ **26.** $2y - 7$
27. $w^2 - 4w + 5 + \dfrac{-10}{2w + 3}$ **28.** $3x^2 + x - 12 + \dfrac{15}{x^2 + 4}$

Chapters 1–5 Cumulative Review Exercises, pp. 424–425

1. $-\dfrac{35}{2}$ **2.** 4   **3.** $5^2 - \sqrt{4}; 23$ **4.** $\dfrac{28}{3}$
5. No solution   **6.** Quadrant III   **7.** y-axis
8. The measures are $31°, 54°, 95°$.   **9. a.** 12 in.
   b. 19.5 in.   **c.** 5.5 hr   **10.** $(-3, 4)$
11. $[-5, \infty)$ ———[———→
          -5
12. $5x^2 - 9x - 15$ **13.** $-2y^2 - 13yz - 15z^2$
14. $16t^2 - 24t + 9$ **15.** $\dfrac{4}{25}a^2 - \dfrac{1}{9}$
16. $-4a^3b^2 + 2ab - 1$ **17.** $4m^2 + 8m + 11 + \dfrac{24}{m - 2}$
18. $\dfrac{c^2}{16d^4}$ **19.** $\dfrac{2b^3}{a^2}$ **20.** 4.1×10^3

Chapter 6

Chapter Opener Puzzle

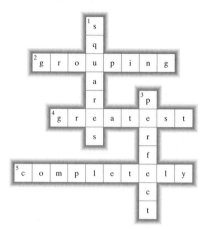

Section 6.1 Practice Exercises, pp. 435–437

3. 7   **5.** 6   **7.** y **9.** $4w^2z$ **11.** $2xy^4z^2$
13. $(x - y)$ **15. a.** $3x - 6y$ **b.** $3(x - 2y)$
17. $4(p + 3)$ **19.** $5(c^2 - 2c + 3)$ **21.** $x^3(x^2 + 1)$
23. $t(t^3 - 4 + 8t)$ **25.** $2ab(1 + 2a^2)$ **27.** $19x^2y(2 - y^3)$
29. $6xy^5(x^2 - 3y^4z)$ **31.** The expression is prime because it is not factorable.   **33.** $7pq^2(6p^2 + 2 - p^3q^2)$
35. $t^2(t^3 + 2rt - 3t^2 + 4r^2)$ **37. a.** $-2x(x^2 + 2x - 4)$
b. $2x(-x^2 - 2x + 4)$ **39.** $-1(8t^2 + 9t + 2)$
41. $-15p^2(p + 2)$ **43.** $-q(q^3 - 2q + 9)$
45. $-1(7x + 6y + 2z)$ **47.** $(a + 6)(13 - 4b)$
49. $(w^2 - 2)(8v + 1)$ **51.** $7x(x + 3)^2$
53. $(2a - b)(4a + 3c)$ **55.** $(q + p)(3 + r)$
57. $(2x + 1)(3x + 2)$ **59.** $(t + 3)(2t - 5)$
61. $(3y - 1)(2y - 3)$ **63.** $(b + 1)(b^3 - 4)$
65. $(j^2 + 5)(3k + 1)$ **67.** $(2x^6 + 1)(7w^6 - 1)$
69. $(y + x)(a + b)$ **71.** $(vw + 1)(w - 3)$
73. $5x(x^2 + y^2)(3x + 2y)$ **75.** $4b(a - b)(x - 1)$
77. $6t(t - 3)(s - t^2)$ **79.** $P = 2(l + w)$
81. $S = 2\pi r(r + h)$ **83.** $\dfrac{1}{7}(x^2 + 3x - 5)$
85. $\dfrac{1}{4}(5w^2 + 3w + 9)$ **87.** For example, $6x^2 + 9x$
89. For example, $16p^4q^2 + 8p^3q - 4p^2q$

Section 6.2 Practice Exercises, pp. 441–443

3. $4xy^5(x^2y^2 - 3x^3 + 2y^3)$ **5.** $(a + 2b)(x - 5)$
7. $(x + 8)(x + 2)$ **9.** $(z - 9)(z - 2)$
11. $(z - 6)(z + 3)$ **13.** $(p - 8)(p + 5)$
15. $(t + 10)(t - 4)$ **17.** Prime   **19.** $(n + 4)^2$
21. a   **23.** c   **25.** They are both correct because multiplication of polynomials is a commutative operation.
27. The expressions are equal and both are correct.
29. Descending order   **31.** $(x - 15)(x + 2)$
33. $(w - 13)(w - 5)$ **35.** $(t + 18)(t + 4)$
37. $3(x - 12)(x + 2)$ **39.** $8p(p - 1)(p - 4)$
41. $y^2z^2(y - 6)(y - 6)$ or $y^2z^2(y - 6)^2$
43. $-(x - 4)(x - 6)$ **45.** $-(m + 2)(m - 3)$
47. $-2(c + 2)(c + 1)$ **49.** $xy^3(x - 4)(x - 15)$
51. $12(p - 7)(p - 1)$ **53.** $-2(m - 10)(m - 1)$
55. $(c + 5d)(c + d)$ **57.** $(a - 2b)(a - 7b)$ **59.** Prime

61. $(q - 7)(q + 9)$ **63.** $(x + 10)^2$ **65.** $(t + 20)(t - 2)$

67. The student forgot to factor out the GCF before factoring the trinomial further. The polynomial is not factored completely, because $(2x - 4)$ has a common factor of 2.

69. $x^2 + 9x - 52$ **71.** $(x^2 + 1)(x^2 + 9)$

73. $(w^2 + 5)(w^2 - 3)$ **75.** $7, 5, -7, -5$

77. For example: $c = -16$

Section 6.3 Practice Exercises, pp. 450–452

3. $3ab(7ab + 4b - 5a)$ **5.** $(n - 1)(m - 2)$

7. $6(a - 7)(a + 2)$ **9.** a   **11.** b

13. $(2y + 1)(y - 2)$ **15.** $(3n + 1)(n + 4)$

17. $(5x + 1)(x - 3)$ **19.** $(4c + 1)(3c - 2)$

21. $(10w - 3)(w + 4)$ **23.** $(3q + 2)(2q - 3)$

25. Prime   **27.** $(5m + 2)(5m - 4)$

29. $(6y - 5x)(y + 4x)$ **31.** $2(m + 4)(m - 10)$

33. $y^3(2y + 1)(y + 6)$ **35.** $-(a + 17)(a - 2)$

37. $10(4m + p)(2m - 3p)$ **39.** $(x^2 + 1)(x^2 + 9)$

41. $(w^2 + 5)(w^2 - 3)$ **43.** $(2x^2 + 3)(x^2 - 5)$

45. $-2(z - 9)(z - 1)$ **47.** $(q - 7)(q - 6)$

49. $(2t + 3)(3t - 1)$ **51.** $(2m - 5)^2$ **53.** Prime

55. $(2x - 5y)(3x - 2y)$ **57.** $(4m + 5n)(3m - n)$

59. $5(3r + 2)(2r - 1)$ **61.** Prime   **63.** $(2t - 5)(5t + 1)$

65. $(7w - 4)(2w + 3)$ **67.** $(x + 9)(x - 2)$

69. $(a - 12)(a + 2)$ **71.** $(r + 8)(r - 3)$

73. $(x + 5y)(x + 4y)$ **75.** Prime

77. $(a + 20b)(a + b)$ **79.** $(t - 7)(t - 3)$

81. $d(5d^2 + 3d - 10)$ **83.** $4b(b - 5)(b + 4)$

85. $y^2(x - 3)(x - 10)$ **87.** $-2u(2u + 5)(3u - 2)$

89. $(2x^2 + 3)(4x^2 + 1)$ **91.** $(5z^2 - 3)(2z^2 + 3)$

93. a. $(x - 12)(x + 2)$ **b.** $(x - 6)(x - 4)$

95. a. $(x - 6)(x + 1)$ **b.** $(x - 2)(x - 3)$

Section 6.4 Practice Exercises, pp. 456–457

3. $(y + 5)(8 + 9y)$ **5.** $12, 1$ **7.** $-8, -1$ **9.** $5, -4$

11. $9, -2$ **13.** $(x + 4)(3x + 1)$ **15.** $(w - 2)(4w - 1)$

17. $(m + 3)(2m - 1)$ **19.** $(4k + 3)(2k - 3)$

21. $(2k - 5)^2$ **23.** Prime   **25.** $(3z - 5)(3z - 2)$

27. $2(7y + 4)(y + 3)$ **29.** $(6y - 5z)(2y + 3z)$

31. $-(3w - 5)(5w + 1)$ **33.** $-4(x - y)(3x - 2y)$

35. $6y(y + 1)(3y + 7)$ **37.** $(a^2 + 2)(a^2 + 3)$

39. $(3x^2 - 5)(2x^2 + 3)$ **41.** $(8p^2 - 3)(p^2 + 5)$

43. $(5p - 1)(4p - 3)$ **45.** $(3u - 2v)(2u - 5v)$

47. $(4a + 5b)(3a - b)$ **49.** $(h + 7k)(3h - 2k)$

51. Prime   **53.** $(8w + 1)(2w + 1)$ **55.** $(q - 1)^2$

57. $-2(a - 1)(a - 9)$ **59.** $(m + 3)(m - 2)$

61. $2(10y + 1)(y - 4)$ **63.** $w^3(w - 4)(w - 7)$

65. $r(4r - 5)(r + 2)$ **67.** $4q(q - 5)(q + 4)$

69. $b^2(a + 10)(a + 3)$ **71.** $-1(m - 2)(m + 17)$

73. $-2(h - 9)(h - 5)$ **75.** $(m^2 + 3)(m^2 + 7)$

77. No. $(5x - 10)$ contains a common factor of 5.

Section 6.5 Practice Exercises, pp. 463–464

3. $(3x - 1)(2x - 5)$ **5.** $5xy^5(3x - 2y)$

7. $(x + b)(a - 6)$ **9.** $(y + 10)(y - 4)$

11. $x^2 - 25$ **13.** $4p^2 - 9q^2$ **15.** $(x - 6)(x + 6)$

17. $3(w + 10)(w - 10)$ **19.** $(2a - 11b)(2a + 11b)$

21. $(7m - 4n)(7m + 4n)$ **23.** Prime

25. $(y + 2z)(y - 2z)$ **27.** $(a - b^2)(a + b^2)$

29. $(5pq - 1)(5pq + 1)$ **31.** $(c^3 - 5)(c^3 + 5)$

33. $2(5 - 4t)(5 + 4t)$ **35.** $(z + 2)(z - 2)(z^2 + 4)$

37. $\left(a + \dfrac{1}{3}\right)\left(a - \dfrac{1}{3}\right)\left(a^2 + \dfrac{1}{9}\right)$ **39.** $(x + 3)(x - 3)(x + 5)$

41. $(c + 5)(c - 5)(c - 1)$ **43.** $(2 + y)(x + 3)(x - 3)$

45. $(x + 2)(x - 2)(y + 3)(y - 3)$ **47.** $9x^2 + 30x + 25$

49. a. $x^2 + 4x + 4$ is a perfect square trinomial.

b. $x^2 + 4x + 4 = (x + 2)^2; x^2 + 5x + 4 = (x + 1)(x + 4)$

51. $(x + 9)^2$ **53.** $(5z - 2)^2$ **55.** $(7a + 3b)^2$

57. $(y - 1)^2$ **59.** $5(4z + 3w)^2$ **61.** $(3y + 25)(3y + 1)$

63. $2(a - 5)^2$ **65.** Prime   **67.** $(2x + y)^2$

69. $y(y - 6)$ **71.** $(2p - 5)(2p + 7)$

73. $(-t + 2)(t + 6)$ or $-(t - 2)(t + 6)$

75. $(-2b + 15)(2b + 5)$ or $-(2b - 15)(2b + 5)$

77. a. $a^2 - b^2$ **b.** $(a - b)(a + b)$

Section 6.6 Practice Exercises, pp. 469–470

3. $5(2 - t)(2 + t)$ **5.** $(t + u)(2 + s)$

7. $(3v - 4)(v + 3)$ **9.** $-(c + 5)^2$

11. $x^3, 8, y^6, 27q^3, w^{12}, r^3s^6$ **13.** $(a + b)(a^2 - ab + b^2)$

15. $(y - 2)(y^2 + 2y + 4)$ **17.** $(1 - p)(1 + p + p^2)$

19. $(w + 4)(w^2 - 4w + 16)$

21. $(x - 10)(x^2 + 10x + 100)$ **23.** $(4t + 1)(16t^2 - 4t + 1)$

25. $(10a + 3)(100a^2 - 30a + 9)$

27. $\left(n - \dfrac{1}{2}\right)\left(n^2 + \dfrac{1}{2}n + \dfrac{1}{4}\right)$

29. $(5m + 2)(25m^2 - 10m + 4)$ **31.** $(x^2 - 2)(x^2 + 2)$

33. Prime   **35.** $(t + 4)(t^2 - 4t + 16)$

37. Prime   **39.** $4(b + 3)(b^2 - 3b + 9)$

41. $5(p - 5)(p + 5)$ **43.** $(\frac{1}{4} - 2h)(\frac{1}{16} + \frac{1}{2}h + 4h^2)$

45. $(x - 2)(x + 2)(x^2 + 4)$ **47.** $(q - 2)(q^2 + 2q + 4)$

$(q + 2)(q^2 - 2q + 4)$ **49.** $\left(\dfrac{2x}{3} - w\right)\left(\dfrac{2x}{3} + w\right)$

51. $(x^3 + 4y)(x^6 - 4x^3y + 16y^2)$

53. $(2x + 3)(x - 1)(x + 1)$

55. $(2x - y)(2x + y)(4x^2 + y^2)$

57. $(3y - 2)(3y + 2)(9y^2 + 4)$

59. $(a + b^2)(a^2 - ab^2 + b^4)$ **61.** $(x^2 + y^2)(x - y)(x + y)$

63. $(k + 4)(k - 3)(k + 3)$ **65.** $2(t - 5)(t - 1)(t + 1)$

67. $\left(\dfrac{4}{5}p - \dfrac{1}{2}q\right)\left(\dfrac{16}{25}p^2 + \dfrac{2}{5}pq + \dfrac{1}{4}q^2\right)$

69. $(a^4 + b^4)(a^8 - a^4b^4 + b^8)$ **71. a.** The quotient is $x^2 + 2x + 4$.   **b.** $(x - 2)(x^2 + 2x + 4)$

73. $x^2 + 4x + 16$ **75.** $2x + 1$

Chapter 6 Problem Recognition Exercises, pp. 470–472

1. A prime factor cannot be factored further.   **2.** Factor out the GCF.   **3.** Look for a difference of squares: $a^2 - b^2$, a difference of cubes: $a^3 - b^3$, or a sum of cubes: $a^3 + b^3$.

4. Grouping   **5. a.** Difference of squares

b. $2(a - 9)(a + 9)$ **6. a.** Nonperfect square trinomial

b. $(y + 3)(y + 1)$ **7. a.** None of these   **b.** $6w(w - 1)$

8. a. Difference of squares   **b.** $(2z + 3)(2z - 3)(4z^2 + 9)$

9. a. Nonperfect square trinomial   **b.** $(3t + 1)(t + 4)$

10. a. Sum of cubes   **b.** $5(r + 1)(r^2 - r + 1)$

11. a. Four terms-grouping   **b.** $(3c + d)(a - b)$

12. a. Difference of cubes   **b.** $(x - 5)(x^2 + 5x + 25)$

13. a. Sum of cubes   **b.** $(y + 2)(y^2 - 2y + 4)$

14. a. Nonperfect square trinomial   **b.** $(7p - 1)(p - 4)$
15. a. Nonperfect square trinomial   **b.** $3(q - 4)(q + 1)$
16. a. Perfect square trinomial   **b.** $-2(x - 2)^2$
17. a. None of these   **b.** $6a(3a + 2)$
18. a. Difference of cubes   **b.** $2(3 - y)(9 + 3y + y^2)$
19. a. Difference of squares   **b.** $4(t - 5)(t + 5)$
20. a. Nonperfect square trinomial   **b.** $(4t + 1)(t - 8)$
21. a. Nonperfect square trinomial   **b.** $10(c^2 + c + 1)$
22. a. Four terms-grouping   **b.** $(w - 5)(2x + 3y)$
23. a. Sum of cubes   **b.** $(x + 0.1)(x^2 - 0.1x + 0.01)$
24. a. Difference of squares   **b.** $(2q - 3)(2q + 3)$
25. a. Perfect square trinomial   **b.** $(8 + k)^2$
26. a. Four terms-grouping   **b.** $(t + 6)(s^2 + 5)$
27. a. Four terms-grouping   **b.** $(x + 1)(2x - y)$
28. a. Sum of cubes   **b.** $(w + y)(w^2 - wy + y^2)$
29. a. Difference of cubes   **b.** $(a - c)(a^2 + ac + c^2)$
30. a. Nonperfect square trinomial   **b.** Prime
31. a. Nonperfect square trinomial   **b.** Prime
32. a. Perfect square trinomial   **b.** $(a + 1)^2$
33. a. Perfect square trinomial   **b.** $(b + 5)^2$
34. a. Nonperfect square trinomial   **b.** $-1(t + 8)(t - 4)$
35. a. Nonperfect square trinomial   **b.** $-p(p + 4)(p + 1)$
36. a. Difference of squares   **b.** $(xy - 7)(xy + 7)$
37. a. Nonperfect square trinomial   **b.** $3(2x + 3)(x - 5)$
38. a. Nonperfect square trinomial   **b.** $2(5y - 1)(2y - 1)$
39. a. None of these   **b.** $abc^2(5ac - 7)$
40. a. Difference of squares   **b.** $2(2a - 5)(2a + 5)$
41. a. Nonperfect square trinomial   **b.** $(t + 9)(t - 7)$
42. a. Nonperfect square trinomial   **b.** $(b + 10)(b - 8)$
43. a. Four terms-grouping   **b.** $(b + y)(a - b)$
44. a. None of these   **b.** $3x^2y^4(2x + y)$
45. a. Nonperfect square trinomial   **b.** $(7u - 2v)(2u - v)$
46. a. Nonperfect square trinomial   **b.** Prime
47. a. Nonperfect square trinomial   **b.** $2(2q^2 - 4q - 3)$
48. a. Nonperfect square trinomial   **b.** $3(3w^2 + w - 5)$
49. a. Sum of squares   **b.** Prime
50. a. Perfect square trinomial   **b.** $5(b - 3)^2$
51. a. Nonperfect square trinomial   **b.** $(3r + 1)(2r + 3)$
52. a. Nonperfect square trinomial   **b.** $(2s - 3)(2s + 5)$
53. a. Difference of squares   **b.** $(2a - 1)(2a + 1)(4a^2 + 1)$
54. a. Four terms-grouping   **b.** $(p + c)(p - 3)(p + 3)$
55. a. Perfect square trinomial   **b.** $(9u - 5v)^2$
56. a. Sum of squares   **b.** $4(x^2 + 4)$
57. a. Nonperfect square trinomial   **b.** $(x - 6)(x + 1)$
58. a. Nonperfect square trinomial   **b.** Prime
59. a. a. Four terms-grouping   **b.** $2(x - 3y)(a + 2b)$
60. a. Nonperfect square trinomial
b. $m(4m + 1)(2m - 3)$ **61. a.** Nonperfect square trinomial   **b.** $x^2y(3x + 5)(7x + 2)$
62. a. Difference of squares   **b.** $2(m^2 - 8)(m^2 + 8)$
63. a. Four terms-grouping   **b.** $(4v - 3)(2u + 3)$
64. a. Four terms-grouping   **b.** $(t - 5)(4t + s)$
65. a. Perfect square trinomial   **b.** $3(2x - 1)^2$
66. a. Perfect square trinomial   **b.** $(p + q)^2$
67. a. Nonperfect square trinomial   **b.** $n(2n - 1)(3n + 4)$
68. a. Nonperfect square trinomial   **b.** $k(2k - 1)(2k + 3)$
69. a. Difference of squares   **b.** $(8 - y)(8 + y)$
70. a. Difference of squares   **b.** $b(6 - b)(6 + b)$
71. a. Nonperfect square trinomial   **b.** Prime
72. a. Nonperfect square trinomial   **b.** $(y + 4)(y + 2)$
73. a. Nonperfect square trinomial   **b.** $(c^2 - 10)(c^2 - 2)$

Section 6.7 Practice Exercises, pp. 476–478

3. $4(b - 5)(b - 6)$ **5.** $(3x - 2)(x + 4)$ **7.** $4(x^2 + 4y^2)$
9. Neither   **11.** Quadratic   **13.** Linear   **15.** $-3, 1$
17. $\dfrac{7}{2}, -\dfrac{7}{2}$ **19.** -5 **21.** $0, \dfrac{1}{5}$ **23.** The polynomial must be factored completely.   **25.** $5, -3$ **27.** $-12, 2$
29. $4, -\dfrac{1}{2}$ **31.** $\dfrac{2}{3}, -\dfrac{2}{3}$ **33.** $6, 8$ **35.** $0, -\dfrac{3}{2}, 4$
37. $-\dfrac{1}{3}, 3, -6$ **39.** $0, 4, -\dfrac{3}{2}$ **41.** $0, -\dfrac{9}{2}, 11$
43. $0, 4, -4$ **45.** $-6, 0$ **47.** $\dfrac{3}{4}, -\dfrac{3}{4}$ **49.** $0, -5, -2$
51. $-\dfrac{14}{3}$ **53.** $5, 3$ **55.** $0, -2$ **57.** -3 **59.** $-3, 1$
61. $\dfrac{3}{2}$ **63.** $0, \dfrac{1}{3}$ **65.** $0, 2$ **67.** $3, -2, 2$ **69.** $-5, 4$
71. $-5, -1$

Chapter 6 Problem Recognition Exercises, p. 478

1. a. $(x + 7)(x - 1)$ **b.** $-7, 1$ **2. a.** $(c + 6)(c + 2)$
b. $-6, -2$ **3. a.** $(2y + 1)(y + 3)$ **b.** $-\dfrac{1}{2}, -3$
4. a. $(3x - 5)(x - 1)$ **b.** $\dfrac{5}{3}, 1$ **5. a.** $\dfrac{4}{5}, -1$
b. $(5q - 4)(q + 1)$ **6. a.** $-\dfrac{1}{3}, \dfrac{3}{2}$ **b.** $(3a + 1)(2a - 3)$
7. a. $-8, 8$ **b.** $(a + 8)(a - 8)$ **8. a.** $-10, 10$
b. $(v + 10)(v - 10)$ **9. a.** $(2b + 9)(2b - 9)$ **b.** $-\dfrac{9}{2}, \dfrac{9}{2}$
10. a. $(6t + 7)(6t - 7)$ **b.** $-\dfrac{7}{6}, \dfrac{7}{6}$
11. a. $-\dfrac{3}{2}, -\dfrac{1}{2}$ **b.** $2(2x + 3)(2x + 1)$
12. a. $-\dfrac{4}{3}, -2$ **b.** $4(3y + 4)(y + 2)$
13. a. $x(x - 10)(x + 2)$ **b.** $0, 10, -2$
14. a. $k(k + 7)(k - 2)$ **b.** $0, -7, 2$
15. a. $-1, 3, -3$ **b.** $(b + 1)(b - 3)(b + 3)$
16. a. $8, -2, 2$ **b.** $(x - 8)(x + 2)(x - 2)$
17. $(s - 3)(2s + r)$ **18.** $(2t + 1)(3t + 5u)$
19. $-\dfrac{1}{2}, 0, \dfrac{1}{2}$ **20.** $-5, 0, 5$ **21.** $0, 1$ **22.** $-3, 0$
23. $\dfrac{1}{3}$ **24.** $-\dfrac{7}{12}$ **25.** $(2w + 3)(4w^2 - 6w + 9)$
26. $(10q - 1)(100q^2 + 10q + 1)$ **27.** $-\dfrac{7}{5}, 1$
28. $-6, -\dfrac{1}{4}$ **29.** $-\dfrac{2}{3}, -5$ **30.** $-1, -\dfrac{1}{2}$ **31.** 3
32. 0 **33.** $-4, 4$ **34.** $-\dfrac{2}{3}, \dfrac{2}{3}$ **35.** $1, 6$ **36.** $-2, -12$

Section 6.8 Practice Exercises, pp. 483–486

3. $0, -\dfrac{2}{3}$ **5.** $6, -1$ **7.** $-\dfrac{5}{6}, 2$
9. The numbers are 7 and -7.   **11.** The numbers are 10 and -4.   **13.** The numbers are -9 and -7, or 7 and 9.

15. The numbers are 5 and 6, or -1 and 0.   **17.** The room is 15 yd by 20 yd.   **19. a.** The slab is 7 m by 4 m.
b. 22 m   **21.** The base is 7 ft and the height is 4 ft.
23. 3 sec   **25.** 0 sec and 1.5 sec
27.

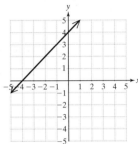

29. $c = 25$ cm   **31.** $a = 15$ in.

33. The brace is 20 in. long.   **35.** The kite is 19 yd high.
37. The bottom of the ladder is 8 ft from the house. The distance from the top of the ladder to the ground is 15 ft.
39. The hypotenuse is 10 m.

Chapter 6 Review Exercises, pp. 493–495

1. $3a^2b$ **3.** $2c(3c - 5)$ **5.** $2x(3x + x^3 - 4)$
7. $t(-t + 5)$ or $-t(t - 5)$ **9.** $(b + 2)(3b - 7)$
11. $(w + 2)(7w + b)$ **13.** $3(4y - 3)(5y - 1)$
15. $(x - 3)(x - 7)$ **17.** $(z - 12)(z + 6)$
19. $3w(p + 10)(p + 2)$ **21.** $-(t - 8)(t - 2)$
23. $(a + b)(a + 11b)$ **25.** Different   **27.** Both positive   **29.** $(2y + 3)(y - 4)$ **31.** $(2z + 5)(5z + 2)$
33. Prime   **35.** $10(w - 9)(w + 3)$ **37.** $(3c - 5d)^2$
39. $(v^2 + 1)(v^2 - 3)$ **41.** The trinomials in Exercises 37 and 38.   **43.** $-3, -5$ **45.** $(y + 3)(4y + 1)$
47. $x(x + 5)(4x - 3)$ **49.** $(p - 3q)(p - 5q)$
51. $10(4s - 5)(s + 2)$ **53.** $2z^4(z + 7)(z - 3)$
55. $(3p + 2)(2p + 5)$ **57.** $(3w - z)^2$
59. $(a - b)(a + b)$ **61.** $(a - 7)(a + 7)$
63. $(10 - 9t)(10 + 9t)$ **65.** Prime   **67.** $(y + 6)^2$
69. $(3a - 2)^2$ **71.** $-3(v + 2)^2$ **73.** $2(c^2 - 3)(c^2 + 3)$
75. $(p + 3)(p - 4)(p + 4)$ **77.** $(a + b)(a^2 - ab + b^2)$
79. $(4 + a)(16 - 4a + a^2)$ **81.** $(p^2 + 2)(p^4 - 2p^2 + 4)$
83. $6(x - 2)(x^2 + 2x + 4)$ **85.** $x(x - 6)(x + 6)$
87. $4(2h^2 + 5)$ **89.** $(x + 4)(x + 1)(x - 1)$
91. $n(2 + n)(4 - 2n + n^2)$ **93.** $(x - 3)(2x + 1) = 0$ can be solved directly by the zero product rule because it is a product of factors set equal to zero.
95. $9, \dfrac{1}{2}$ **97.** $0, 7, \dfrac{9}{4}$ **99.** $-\dfrac{1}{4}, 6$ **101.** $5, -5$
103. $4, 2$ **105.** $-\dfrac{2}{3}$ **107.** $\dfrac{11}{2}, -12$ **109.** $0, 2, -2$
111. The ball is at ground level at 0 and 1 sec.
113. The legs are 6 ft and 8 ft; the hypotenuse is 10 ft.
115. The numbers are 29 and 30, or -2 and -1.

Chapter 6 Test, p. 495

1. $3x(5x^3 - 1 + 2x^2)$ **2.** $(a - 5)(7 - a)$
3. $(6w - 1)(w - 7)$ **4.** $(13 - p)(13 + p)$
5. $(q - 8)^2$ **6.** $(2 + t)(4 - 2t + t^2)$
7. $(a + 4)(a + 8)$ **8.** $(x + 7)(x - 6)$
9. $(2y - 1)(y - 8)$ **10.** $(2z + 1)(3z + 8)$
11. $(3t - 10)(3t + 10)$ **12.** $(v + 9)(v - 9)$
13. $3(a + 6b)(a + 3b)$ **14.** $(c - 1)(c + 1)(c^2 + 1)$
15. $(y - 7)(x + 3)$ **16.** Prime   **17.** $-10(u - 2)(u - 1)$
18. $3(2t - 5)(2t + 5)$ **19.** $5(y - 5)^2$ **20.** $7q(3q + 2)$
21. $(2x + 1)(x - 2)(x + 2)$ **22.** $(y - 5)(y^2 + 5y + 25)$
23. $(mn - 9)(mn + 9)$ **24.** $16(a - 2b)(a + 2b)$
25. $(4x - 3y^2)(16x^2 + 12xy^2 + 9y^4)$ **26.** $3y(x - 4)(x + 2)$
27. $\dfrac{3}{2}, -5$ **28.** $0, 7$ **29.** $8, -2$ **30.** $\dfrac{1}{5}, -1$

31. $3, -3, -10$ **32.** The tennis court is 12 yd by 26 yd.
33. The two integers are 5 and 7, or -5 and -7.
34. The base is 12 in., and the height is 7 in.
35. The shorter leg is 5 ft.

Chapters 1–6 Cumulative Review Exercises, p. 496

1. $\dfrac{7}{5}$ **2.** -3 **3.** $y = \dfrac{3}{2}x - 4$
4. There are 10 quarters, 12 nickels, and 7 dimes.
5. $[-4, \infty)$ **6. a.** Yes   **b.** 1
c. $(0, 4)$ **d.** $(-4, 0)$ **e.**

7. a. Vertical line   **b.** Undefined   **c.** $(5, 0)$ **d.** Does not exist   **8.** $y = 3x + 14$ **9.** $(5, 2)$
10. $-\dfrac{7}{2}y^2 - 5y - 14$ **11.** $8p^3 - 22p^2 + 13p + 3$
12. $4w^2 - 28w + 49$ **13.** $r^3 + 5r^2 + 15r + 40 + \dfrac{121}{r - 3}$
14. c^4 **15.** 1.6×10^3 **16.** $(w - 2)(w + 2)(w^2 + 4)$
17. $(a + 5b)(2x - 3y)$ **18.** $(2x - 5)(2x + 1)$
19. $(y - 3)(y^2 + 3y + 9)$ **20.** $0, \dfrac{1}{2}, -5$

Chapter 7

Chapter Opener Puzzle

$$\underset{4}{\mathrm{p}}\ \underset{1}{\mathrm{r}}\ \underset{2}{\mathrm{o}}\ \underset{}{\mathrm{c}}\ \underset{1}{\mathrm{r}}\ \underset{5}{\mathrm{a}}\ \underset{3}{\mathrm{s}}\ \underset{}{\mathrm{t}}\ \underset{5}{\mathrm{i}}\ \underset{3}{\mathrm{n}}\ \underset{2}{\mathrm{a}}\ \underset{}{\mathrm{t}}\ \underset{}{\mathrm{i}}\ \underset{}{\mathrm{o}}\ \underset{}{\mathrm{n}}$$

Section 7.1 Practice Exercises, pp. 505–507

3. a. A number $\dfrac{p}{q}$, where p and q are integers and $q \neq 0$
b. An expression $\dfrac{p}{q}$, where p and q are polynomials and $q \neq 0$ **5.** $-\dfrac{1}{8}$ **7.** $-\dfrac{1}{2}$ **9.** Undefined
11. a. $3\dfrac{1}{5}$ hr or 3.2 hr   **b.** $1\dfrac{3}{4}$ hr or 1.75 hr
13. $\{k \mid k$ is a real number and $k \neq -2\}$
15. $\left\{x \mid x \text{ is a real number and } x \neq \dfrac{5}{2}, x \neq -8\right\}$
17. $\{b \mid b$ is a real number and $b \neq -2, b \neq -3\}$
19. The set of all real numbers   **21.** The set of all real numbers   **23.** $\{t \mid t$ is a real number and $t \neq 0\}$
25. For example: $\dfrac{1}{x - 2}$ **27.** For example: $\dfrac{1}{(x + 3)(x - 7)}$
29. a. $\dfrac{2}{5}$ **b.** $\dfrac{2}{5}$ **31. a.** Undefined   **b.** Undefined
33. a. $\{y \mid y$ is a real number and $y \neq -2\}$ **b.** $\dfrac{1}{2}$

35. a. $\{t \mid t \text{ is a real number and } t \neq -1\}$ **b.** $t - 1$

37. a. $\{w \mid w \text{ is a real number and } w \neq 0, \ w \neq \frac{5}{3}\}$

b. $\dfrac{1}{3w - 5}$ **39. a.** $\{x \mid x \text{ is a real number and } x \neq -\frac{2}{3}\}$

b. $\dfrac{3x - 2}{2}$ **41. a.** $\{a \mid a \text{ is a real number and } a \neq -3, \ a \neq 2\}$

b. $\dfrac{a + 5}{a + 3}$ **43.** $\dfrac{b}{3}$ **45.** $\dfrac{3t^2}{2}$ **47.** $-\dfrac{3xy}{z^2}$

49. $\dfrac{1}{2}$ **51.** $\dfrac{p - 3}{p + 4}$ **53.** $\dfrac{1}{4(m - 11)}$ **55.** $\dfrac{2x + 1}{4x^2}$

57. $\dfrac{1}{4a - 5}$ **59.** $\dfrac{4}{w + 2}$ **61.** $\dfrac{x - 2}{3(y + 2)}$ **63.** $\dfrac{2}{x - 5}$

65. $a + 7$ **67.** Cannot simplify   **69.** $\dfrac{y + 3}{2y - 5}$

71. $\dfrac{3x - 2}{x + 4}$ **73.** $\dfrac{5}{(q + 1)(q - 1)}$ **75.** $\dfrac{c - d}{2c + d}$

77. $\dfrac{1}{t(t - 5)}$ **79.** $\dfrac{7p - 2q}{2}$ **81.** $5x + 4$ **83.** $\dfrac{x + y}{x - 4y}$

85. They are opposites.   **87.** -1 **89.** -1 **91.** $-\dfrac{1}{2}$

93. Cannot simplify   **95.** $\dfrac{5x - 6}{5x + 6}$ **97.** $-\dfrac{x + 3}{4 + x}$

99. $w - 2$ **101.** $\dfrac{z + 4}{z^2 + 4z + 16}$

Section 7.2 Practice Exercises, pp. 512–513

1. $\dfrac{3}{10}$ **3.** 2 **5.** $\dfrac{5}{2}$ **7.** $\dfrac{15}{4}$ **9.** $\dfrac{3}{2x}$ **11.** $3xy^4$

13. $\dfrac{x - 6}{8}$ **15.** $\dfrac{2}{y}$ **17.** $-\dfrac{5}{8}$ **19.** $-\dfrac{b + a}{a - b}$

21. $\dfrac{y + 1}{5}$ **23.** $\dfrac{2(x + 6)}{2x + 1}$ **25.** 6 **27.** $\dfrac{m^6}{n^2}$ **29.** $\dfrac{10}{9}$

31. $\dfrac{6}{7}$ **33.** $-m(m + n)$ **35.** $\dfrac{3p + 4q}{4(p + 2q)}$ **37.** $\dfrac{p}{p - 1}$

39. $\dfrac{w}{2w - 1}$ **41.** $\dfrac{5}{6}$ **43.** $\dfrac{q + 1}{q - 6}$ **45.** $\dfrac{1}{4}$

47. $\dfrac{y + 9}{y - 6}$ **49.** $\dfrac{t + 4}{t + 2}$ **51.** $\dfrac{3t + 8}{t + 2}$ **53.** $\dfrac{x + 4}{x + 1}$

55. $-\dfrac{w - 3}{2}$ **57.** $\dfrac{k + 6}{k + 3}$ **59.** $\dfrac{2}{a}$ **61.** $2y(y + 1)$

63. $x + y$ **65.** 2 **67.** $\dfrac{1}{a - 2}$ **69.** $\dfrac{p + q}{2}$

Section 7.3 Practice Exercises, pp. 517–519

3. $\{x \mid x \text{ is a real number and } x \neq 1, x \neq -1\}; \ \dfrac{3}{5(x - 1)}$

5. $\dfrac{a + 5}{a + 7}$ **7.** $\dfrac{2}{3y}$ **9.** a, b, c, d

11. x^5 is the greatest power of x that appears in any denominator.   **13.** 45 **15.** 48 **17.** 63 **19.** $9x^2y^3$

21. w^2y **23.** $(p + 3)(p - 1)(p + 2)$ **25.** $9t(t + 1)^2$

27. $(y - 2)(y + 2)(y + 3)$ **29.** $3 - x$ or $x - 3$

31. Because $(b - 1)$ and $(1 - b)$ are opposites; they differ by a factor of -1.

33. $\dfrac{6}{5x^2}; \dfrac{5x}{5x^2}$ **35.** $\dfrac{24x}{30x^3}; \dfrac{5y}{30x^3}$ **37.** $\dfrac{10}{12a^2b}; \dfrac{a^3}{12a^2b}$

39. $\dfrac{6m - 6}{(m + 4)(m - 1)}; \dfrac{3m + 12}{(m + 4)(m - 1)}$

41. $\dfrac{6x + 18}{(2x - 5)(x + 3)}; \dfrac{2x - 5}{(2x - 5)(x + 3)}$

43. $\dfrac{6w + 6}{(w + 3)(w - 8)(w + 1)}; \dfrac{w^2 + 3w}{(w + 3)(w - 8)(w + 1)}$

45. $\dfrac{6p^2 + 12p}{(p - 2)(p + 2)^2}; \dfrac{3p - 6}{(p - 2)(p + 2)^2}$

47. $\dfrac{1}{a - 4}; \dfrac{-a}{a - 4}$ or $\dfrac{-1}{4 - a}; \dfrac{a}{4 - a}$

49. $\dfrac{8}{2(x - 7)}; \dfrac{-y}{2(x - 7)}$ or $\dfrac{-8}{2(7 - x)}; \dfrac{y}{2(7 - x)}$

51. $\dfrac{1}{a + b}; \dfrac{-6}{a + b}$ or $\dfrac{-1}{-a - b}; \dfrac{6}{-a - b}$

53. $\dfrac{-9}{24(3y + 1)}; \dfrac{20}{24(3y + 1)}$ **55.** $\dfrac{3z + 12}{5z(z + 4)}; \dfrac{5z}{5z(z + 4)}$

57. $\dfrac{z^2 + 3z}{(z + 2)(z + 7)(z + 3)}; \dfrac{-3z^2 - 6z}{(z + 2)(z + 7)(z + 3)};$

$\dfrac{5z + 35}{(z + 2)(z + 7)(z + 3)}$

59. $\dfrac{3p + 6}{(p - 2)(p^2 + 2p + 4)(p + 2)}; \dfrac{p^3 + 2p^2 + 4p}{(p - 2)(p^2 + 2p + 4)(p + 2)};$

$\dfrac{5p^3 - 20p}{(p - 2)(p^2 + 2p + 4)(p + 2)}$

Section 7.4 Practice Exercises, pp. 526–528

1. a. $-\dfrac{1}{2}, -2, 0,$ undefined, undefined

b. $(x - 5)(x - 2); \{x \mid x \text{ is a real number and } x \neq 5, x \neq 2\}$

c. $\dfrac{x + 1}{x - 2}$ **3.** $\dfrac{2(2b - 3)}{(b - 3)(b - 1)}$ **5.** $\dfrac{5}{4}$ **7.** $\dfrac{3}{8}$

9. 2 **11.** 5 **13.** $\dfrac{-2(t - 2)}{t - 8}$ **15.** $3x + 7$

17. $m + 5$ **19.** 2 **21.** $x - 5$ **23.** $\dfrac{1}{r + 1}$

25. $\dfrac{1}{y + 7}$ **27.** $\dfrac{15x}{y}$ **29.** $\dfrac{5a + 6}{4a}$ **31.** $\dfrac{2(6 + x^2y)}{15xy^3}$

33. $\dfrac{2s - 3t^2}{s^4t^3}$ **35.** $-\dfrac{2}{3}$ **37.** $\dfrac{19}{3(a + 1)}$

39. $\dfrac{-3(k + 4)}{(k - 3)(k + 3)}$ **41.** $\dfrac{a - 4}{2a}$ **43.** $\dfrac{(x + 6)(x - 2)}{(x - 4)(x + 1)}$

45. $\dfrac{2(4a - b)}{(a + b)(a - b)}$ **47.** $\dfrac{5p - 1}{3}$ or $\dfrac{-5p + 1}{-3}$

49. $\dfrac{6n - 1}{n - 8}$ or $\dfrac{-6n + 1}{8 - n}$ **51.** $\dfrac{2(4x + 5)}{x(x + 2)}$

53. $\dfrac{3p + 1}{(p - 3)(p - 1)}$ **55.** $\dfrac{3y}{2(2y + 1)}$ **57.** $\dfrac{2(w - 3)}{(w + 3)(w - 1)}$

59. $\dfrac{4a - 13}{(a - 3)(a - 4)}$ **61.** $\dfrac{4x(x + 1)}{(x + 3)(x - 2)(x + 2)}$

63. $\dfrac{-y(y + 8)}{(2y + 1)(y - 1)(y - 4)}$ **65.** $\dfrac{1}{2p + 1}$

67. $\dfrac{-2mn + 1}{(m + n)(m - n)}$ **69.** 0 **71.** $\dfrac{2(3x + 7)}{(x + 3)(x + 2)}$

73. $\dfrac{1}{n}$ **75.** $\dfrac{5}{n + 2}$ **77.** $n + \left(7 \cdot \dfrac{1}{n}\right); \dfrac{n^2 + 7}{n}$

79. $\dfrac{1}{n} - \dfrac{2}{n}; -\dfrac{1}{n}$ **81.** $\dfrac{-w^2}{(w + 3)(w - 3)(w^2 - 3w + 9)}$

83. $\dfrac{p^2 - 2p + 7}{(p + 2)(p + 3)(p - 1)}$

85. $\dfrac{-m - 21}{2(m + 5)(m - 2)}$ or $\dfrac{m + 21}{2(m + 5)(2 - m)}$ **87.** $\dfrac{3k + 5}{4k + 7}$

89. $\dfrac{1}{a}$

Chapter 7 Problem Recognition Exercises, p. 529

1. $\dfrac{-2x + 9}{3x + 1}$ **2.** $\dfrac{1}{w - 4}$ **3.** $\dfrac{y - 5}{2y - 3}$

4. $\dfrac{7}{(x + 3)(2x - 1)}$ **5.** $-\dfrac{1}{x}$ **6.** $\dfrac{1}{3}$ **7.** $\dfrac{c + 3}{c}$

8. $\dfrac{x + 3}{5}$ **9.** $\dfrac{a}{12b^4c}$ **10.** $\dfrac{2a - b}{a - b}$ **11.** $\dfrac{p - q}{5}$

12. 4 **13.** $\dfrac{10}{2x + 1}$ **14.** $\dfrac{w + 2z}{w + z}$ **15.** $\dfrac{3}{2x + 5}$

16. $\dfrac{y + 7}{x + a}$ **17.** $\dfrac{1}{2(a + 3)}$ **18.** $\dfrac{2(3y + 10)}{(y - 6)(y + 6)(y + 2)}$

19. $(t + 8)^2$ **20.** $6b + 5$

Section 7.5 Practice Exercises, pp. 535–536

3. $\left\{a \mid a \text{ is a real number and } a \neq \dfrac{3}{2}, a \neq -5\right\}; \dfrac{1}{2a - 3}$

5. $\dfrac{3(2k - 5)}{5(k - 2)}$ **7.** $\dfrac{7}{4y}$ **9.** $\dfrac{1}{2y}$ **11.** $\dfrac{24b}{a^3}$ **13.** $\dfrac{2r^5 t^4}{s^6}$

15. $\dfrac{35}{2}$ **17.** $k + h$ **19.** $\dfrac{n + 1}{2(n - 3)}$ **21.** $\dfrac{2x + 1}{4x + 1}$

23. $m - 7$ **25.** $\dfrac{2y(y - 5)}{7y^2 + 10}$ **27.** $-\dfrac{a + 8}{a - 2}$ or $\dfrac{a + 8}{2 - a}$

29. $\dfrac{t - 2}{t - 4}$ **31.** $\dfrac{t + 3}{t - 5}$ **33.** $\dfrac{1}{2}$ **35.** $\dfrac{\frac{1}{2} + \frac{2}{3}}{5}; \dfrac{7}{30}$

37. $\dfrac{3}{\frac{2}{3} + \frac{3}{4}}; \dfrac{36}{17}$ **39. a.** $\dfrac{6}{5}\,\Omega$ **b.** $6\,\Omega$ **41.** $\dfrac{y + 4x}{2y}$

43. $\dfrac{1}{n^2 + m^2}$ **45.** $\dfrac{2z - 5}{3(z + 3)}$ **47.** $-\dfrac{x + 1}{x - 1}$ or $\dfrac{x + 1}{1 - x}$

49. $\dfrac{3}{2}$

Section 7.6 Practice Exercises, pp. 544–546

3. $\dfrac{2}{4x - 1}$ **5.** $5(h + 1)$ **7.** $\dfrac{(x + 4)(x - 3)}{x^2}$

9. 3 **11.** $\dfrac{5}{11}$ **13.** $\dfrac{1}{3}$

15. a. $\{z \mid z \text{ is a real number and } z \neq 0\}$ **b.** $5z$ **c.** 5

17. $-\dfrac{200}{19}$ **19.** 8 **21.** $\dfrac{47}{6}$ **23.** $3, -1$ **25.** 4

27. 5; (the value 0 does not check.)   **29.** -5

31. No solution; (the value 4 does not check.)   **33.** 4

35. $4, -3$ **37.** -4; (the value 1 does not check.)

39. No solution; (the value -4 does not check.)

41. 4; (the value -6 does not check.)   **43.** -25

45. -1 **47.** The number is 8.   **49.** The number is -26.

51. $m = \dfrac{FK}{a}$ **53.** $E = \dfrac{IR}{K}$

55. $R = \dfrac{E - Ir}{I}$ or $R = \dfrac{E}{I} - r$

57. $B = \dfrac{2A - hb}{h}$ or $B = \dfrac{2A}{h} - b$ **59.** $h = \dfrac{V}{r^2\pi}$

61. $t = \dfrac{b}{x - a}$ or $t = \dfrac{-b}{a - x}$ **63.** $x = \dfrac{y}{1 - yz}$ or $x = \dfrac{-y}{yz - 1}$ **65.** $h = \dfrac{2A}{a + b}$ **67.** $R = \dfrac{R_1 R_2}{R_2 + R_1}$

Chapter 7 Problem Recognition Exercises, p. 547

1. $\dfrac{y - 2}{2y}$ **2.** 6 **3.** 2 **4.** $\dfrac{3a - 17}{a - 5}$ **5.** $\dfrac{4p + 27}{18p^2}$

6. $\dfrac{b(b - 5)}{(b - 1)(b + 1)}$ **7.** 5 **8.** $\dfrac{2w + 5}{(w + 1)^2}$ **9.** 7

10. 5 **11.** $\dfrac{3x + 14}{4(x + 1)}$ **12.** $\dfrac{11}{3}$

Section 7.7 Practice Exercises, pp. 555–559

3. Expression; $\dfrac{m^2 + m + 2}{(m - 1)(m + 3)}$ **5.** Expression; $\dfrac{3}{10}$

7. Equation; 2 **9.** 95 **11.** 1 **13.** $\dfrac{40}{3}$ **15.** 40

17. 3 **19.** -1 **21.** 1 **23. a.** $V_f = \dfrac{V_i T_f}{T_i}$

b. $T_f = \dfrac{T_i V_f}{V_i}$ **25.** Toni can drive 297 mi on 9 gal of gas.

27. They would produce 1536 lb.   **29.** 5 oz contains 12 g of carbohydrate.   **31.** This represents 262.5 mi.

33. $x = 4$ cm; $y = 5$ cm   **35.** $x = 3.75$ cm; $y = 4.5$ cm

37. The height of the pole is 7 m.   **39.** The light post is 24 ft high.   **41.** The speed of the boat is 20 mph.

43. The plane flies 210 mph in still air.   **45.** He runs 8 mph and bikes 16 mph.   **47.** Floyd walks 4 mph and Rachel walks 2 mph.   **49.** Sergio rode 12 mph and walked 3 mph.   **51.** $\dfrac{1}{2}$ of the room   **53.** $5\frac{5}{11}\,(5.\overline{45})$ min

55. $22\frac{2}{9}\,(22.\overline{2})$ min   **57.** 48 hr   **59.** $3\frac{1}{3}\,(3.\overline{3})$ days

61. There are 40 smokers and 140 nonsmokers.

63. There are 240 men and 200 women.

Section 7.8 Practice Exercises, pp. 563–566

3. $y = 12$ **5.** $a = 6; a = 2$ **7.** $\dfrac{b + 1}{1 - b}$ **9.** Inversely

11. $T = kq$ **13.** $b = \dfrac{k}{c}$ **15.** $Q = \dfrac{kx}{y}$ **17.** $c = kst$

19. $L = kw\sqrt{v}$ **21.** $x = \dfrac{ky^2}{z}$ **23.** $k = \dfrac{9}{2}$

25. $k = 512$ **27.** $k = 1.75$ **29.** $x = 70$ **31.** $b = 6$

33. $Z = 56$ **35.** $Q = 9$ **37.** $L = 9$ **39.** $B = \dfrac{15}{2}$

41. a. 3.6 g   **b.** 4.5 g   **c.** 2.4 g   **43. a.** \$0.40   **b.** \$0.30
c. \$1.00   **45.** 355,000 tons   **47.** 42.6 ft
49. 300 W   **51.** 1.25 Ω   **53.** 20 lb   **55.** \$3500

Chapter 7 Review Exercises pp. 574–577

1. a. $-\dfrac{2}{9}, -\dfrac{1}{10}, 0, -\dfrac{5}{6}$, undefined

b. $\{t\,|\,t \text{ is a real number and } t \neq -9\}$ **3.** a, c, d
5. $\{h\,|\,h \text{ is a real number and } h \neq -\frac{1}{3}, h \neq -7\}; \frac{1}{3h+1}$
7. $\{w\,|\,w \text{ is a real number and } w \neq 4, w \neq -4\}; \frac{2w+3}{w-4}$
9. $\{k\,|\,k \text{ is a real number and } k \neq 0, k \neq 5\}; -\frac{3}{2k}$
11. $\{m\,|\,m \text{ is a real number and } m \neq -1\}; \frac{m-5}{3}$

13. $\{p\,|\,p \text{ is a real number and } p \neq -7\}; \frac{1}{p+7}$ **15.** $\dfrac{u^2}{2}$

17. $\dfrac{3}{2(x-5)}$ **19.** $\dfrac{q-2}{4}$ **21.** $4s(s-4)$

23. $\dfrac{1}{n-2}$ **25.** $\dfrac{1}{m+3}$ **27.** $-\dfrac{2y-1}{y+1}$

29. LCD $= 12xy; \dfrac{21y}{12xy}, \dfrac{22x}{12xy}$

31. LCD $= ab^3c^2; \dfrac{5c^2}{ab^3c^2}, \dfrac{3b^3}{ab^3c^2}$

33. LCD $= q(q+8); \dfrac{6q+48}{q(q+8)}, \dfrac{q}{q(q+8)}$

35. LCD $= (m+4)(m-4)(m+3)$
37. $3 - x$ or $x - 3$ **39.** 2   **41.** $x - 7$

43. $\dfrac{t^2+2t+3}{(2-t)(2+t)}$ **45.** $\dfrac{3(r-4)}{2r(r+6)}$ **47.** $\dfrac{q}{(q+5)(q+4)}$

49. $\dfrac{1}{3}$ **51.** $\dfrac{3(z+5)}{z(z-5)}$ **53.** $\dfrac{8}{y}$ **55.** $-(b+a)$

57. $-\dfrac{k+10}{k+4}$ **59.** -2 **61.** 2   **63.** 3   **65.** $-11, 1$

67. $h = \dfrac{3V}{\pi r^2}$ **69.** $\dfrac{6}{5}$ **71.** 10 g   **73.** Together the
pumps would fill the pool in 16.8 min.   **75. a.** $F = kd$
b. $k = 3$ **c.** 4.5 lb   **77.** $y = 12$

Chapter 7 Test pp. 577–578

1. a. $\{x\,|\,x \text{ is a real number and } x \neq 2\}$ **b.** $-\dfrac{x+1}{6}$

2. a. $\{a\,|\,a \text{ is a real number and } a \neq 0, a \neq 6, a \neq -2\}$

b. $\dfrac{7}{a+2}$ **3.** b, c, d   **4.** $\dfrac{y+7}{3(y+3)(y+1)}$

5. $-\dfrac{b+3}{5}$ **6.** $\dfrac{1}{w+1}$ **7.** $\dfrac{t+4}{t+2}$

8. $\dfrac{x(x+5)}{(x+4)(x-2)}$ **9.** $\dfrac{1}{m+4}$ **10.** $\dfrac{8}{5}$ **11.** 2

12. 1   **13.** No solution. (the value 4 does not check.)

14. $r = \dfrac{2A}{C}$ **15.** The number is $-\dfrac{2}{5}$.   **16.** -8

17. $1\frac{1}{4}$ (1.25) cups of carrots   **18.** The speed of the
current is 5 mph.   **19.** It would take the second printer
3 hr to do the job working alone.   **20.** $a = 5.6; b = 12$
21. a. $15(x+3)$ **b.** $3x^2y^2$ **22.** 8.25 mL
23. 200 drinks are sold.

Chapters 1–7 Cumulative Review Exercises pp. 578–579

1. 32   **2.** 7   **3.** $\dfrac{10}{9}$

4.

Set-Builder Notation	Graph	Interval Notation
$\{x\,|\,x \geq -1\}$	(ray right from -1)	$[-1, \infty)$
$\{x\,|\,x < 5\}$	(ray left from 5)	$(-\infty, 5)$

5. The width is 17 m and the length is 35 m.

6. The base is 10 in. and the height is 8 in.   **7.** $\dfrac{x^2yz^{17}}{2}$

8. a. $6x + 4$ **b.** $2x^2 + x - 3$ **9.** $(5x - 3)^2$
10. $(2c+1)(5d-3)$
11. $\left\{x\,|\,x \text{ is a real number and } x \neq 5, x \neq -\dfrac{1}{2}\right\}$

12. $\dfrac{x-3}{x+5}$ **13.** $\dfrac{1}{5(x+4)}$ **14.** -3 **15.** 1   **16.** $-\dfrac{7}{2}$

17. a. x-intercept: $(-4, 0)$; y-intercept: $(0, 2)$
b. x-intercept: $(0, 0)$; y-intercept: $(0, 0)$

18. a. $m = -\dfrac{7}{5}$ **b.** $m = -\dfrac{2}{3}$ **c.** $m = 4$ **d.** $m = -\dfrac{1}{4}$

19. $y = 5x - 3$ **20.** One large popcorn costs \$3.50, and
one drink costs \$1.50.

Chapter 8

Chapter Opener Puzzle

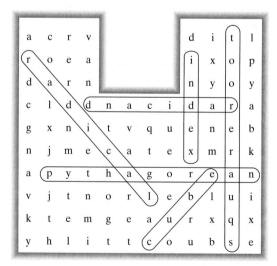

Section 8.1 Calculator Connections, p. 589

1. 2.236   **3.** 7.071   **5.** 5.745   **7.** 8.944
9. 1.913   **11.** 4.021

Section 8.1 Practice Exercises, pp. 589–593

3. $12, -12$ **5.** There are no real-valued square roots of
-49.   **7.** 0   **9.** $\dfrac{1}{5}, -\dfrac{1}{5}$ **11. a.** 13   **b.** -13 **13.** 0

15. 9, 16, 25, 36, 64, 121, 169   **17.** 2   **19.** 7   **21.** 0.4

23. 0.3 **25.** $\dfrac{5}{4}$ **27.** $\dfrac{1}{12}$ **29.** 5 **31.** 9

33. There is no real value of b for which $b^2 = -16$.
35. -2 **37.** Not a real number **39.** Not a real number
41. Not a real number **43.** -20 **45.** Not a real number
47. 0, 1, 27, 125 **49.** Yes, -3 **51.** 3 **53.** 4
55. -2 **57.** Not a real number **59.** Not a real number
61. -2 **63.** -1 **65.** 0 **67.** 4 **69.** 4 **71.** 5
73. -5 **75.** 2 **77.** 2 **79.** $|a|$ **81.** y **83.** $|w|$
85. x **87.** x^2, y^4, $(ab)^6$, w^8x^8, m^{10}. The expression is a perfect square if the exponent is even. **89.** p^4, t^8, $(cd)^{12}$. The expression is a perfect fourth power if the exponent is a multiple of 4. **91.** y^6 **93.** a^4b^{15} **95.** q^8 **97.** $2w^2$
99. $5x$ **101.** $-5x$ **103.** $5p^2$ **105.** $5p^2$
107. $\sqrt{q} + p^2$ **109.** $\dfrac{6}{\sqrt[4]{x}}$ **111.** 9 cm **113.** 5 ft

115. 6.9 cm **117.** 17.0 in. **119.** 31.3 in.
121. 268 km **123.** $x \geq 0$

Section 8.2 Calculator Connections, p. 598

1.
```
√(125)
        11.18033989
5*√(5)
        11.18033989
```
2.
```
√(18)
        4.242640687
3*√(2)
        4.242640687
```
3.
```
³√(54)
        3.77976315
3*³√(2)
        3.77976315
```
4.
```
³√(108)
        4.762203156
3*³√(4)
        4.762203156
```

Section 8.2 Practice Exercises, pp. 599–601

3. 8, 27, y^3, y^9, y^{12}, y^{27} **5.** -5 **7.** -3 **9.** a^2
11. $2xy^2$ **13.** 446 km **15.** $3\sqrt{2}$ **17.** $2\sqrt{7}$
19. $12\sqrt{5}$ **21.** $-10\sqrt{2}$ **23.** $a^2\sqrt{a}$ **25.** w^{11}
27. $m^2n^2\sqrt{n}$ **29.** $x^7y^5\sqrt{x}$ **31.** $3t^5$ **33.** $2x\sqrt{2x}$
35. $4z\sqrt{z}$ **37.** $-3w^3\sqrt{5}$ **39.** $z^{12}\sqrt{z}$ **41.** $-z^5\sqrt{15z}$
43. $10ab^3\sqrt{26b}$ **45.** $\sqrt{26pq}$ **47.** m^6n^8
49. $4ab^2c^2\sqrt{3ab}$ **51.** a^4 **53.** y^5 **55.** $\dfrac{1}{2}$
57. 2 **59.** $2x$ **61.** $5p^3$ **63.** $3\sqrt{5}$ **65.** $\sqrt{6}$
67. 4 **69.** 7 **71.** $11\sqrt{2}$ ft **73.** $2\sqrt{66}$ cm
75. $a^2\sqrt[3]{a^2}$ **77.** $14z\sqrt[3]{2}$ **79.** $2ab^2\sqrt[3]{2a^2}$ **81.** z
83. -2 **85.** $2\sqrt[3]{5}$ **87.** $\dfrac{1}{3}$ **89.** $4a\sqrt{a}$
91. $2x$ **93.** $2p\sqrt{2q}$ **95.** $4\sqrt{2}$ **97.** $2u^2v^3\sqrt{13v}$
99. $6\sqrt{6}$ **101.** 6 **103.** $2a\sqrt[3]{2}$ **105.** x
107. $-\sqrt{5}$ **109.** $-\dfrac{1}{2}$ **111.** $5\sqrt{2}$ **113.** $x + 5$

Section 8.3 Practice Exercises, pp. 605–607

3. $2y$ **5.** $6x\sqrt{x}$ **7.** $\dfrac{5c^3}{4}$ **9.** Not a real number
11. For example, $2\sqrt{3}$, $6\sqrt[3]{3}$ **13.** c **15.** $8\sqrt{2}$
17. $4\sqrt{7}$ **19.** $2\sqrt{10}$ **21.** $11\sqrt{y}$ **23.** 0
25. $5y\sqrt{15}$ **27.** $x\sqrt{y} - y\sqrt{x}$ **29.** $3\sqrt[3]{6} + 8\sqrt[3]{6}$; $11\sqrt[3]{6}$

31. $4\sqrt{5} - 6\sqrt{5}$; $-2\sqrt{5}$ **33.** $8\sqrt{3}$ **35.** 0 **37.** $2\sqrt{2}$
39. $16p^2\sqrt{5}$ **41.** $10\sqrt{2k}$ **43.** $a^2\sqrt{b}$ **45.** $3\sqrt{5}$
47. $\dfrac{29}{18}z\sqrt{6}$ **49.** $-1.7\sqrt{10}$ **51.** $2x\sqrt{x}$ **53.** $3\sqrt{7}$
55. $4\sqrt{w} + 2\sqrt{6w} + 2\sqrt{10w}$ **57.** $6x^3\sqrt{y}$
59. $2\sqrt{3} - 4\sqrt{6}$ **61.** $-4x\sqrt{2} + \sqrt{2x}$ **63.** $9\sqrt{2}$ m
65. $16\sqrt{3}$ in. **67.** Radicands are not the same.
69. One term has a radical. One does not.
71. The indices are different. **73.** $\dfrac{\sqrt{3}}{3}$ **75. a.** 80 m
b. 159 m

Section 8.4 Practice Exercises, pp. 612–613

3. 11 **5.** $3w^2\sqrt{z}$ **7.** $\sqrt{15}$ **9.** 47 **11.** b
13. $6\sqrt{15p}$ **15.** $5\sqrt{2}$ **17.** $14\sqrt{2}$ **19.** $6x\sqrt{7}$
21. $4x^3\sqrt{y}$ **23.** $12w^2\sqrt{10}$ **25.** $-8\sqrt{15}$
27. Perimeter: $6\sqrt{5}$ ft; area: 10 ft^2 **29.** 3 cm^2 **31.** $3w$
33. $-16\sqrt{10y}$ **35.** $2\sqrt{3} - \sqrt{6}$ **37.** $4x + 20\sqrt{x}$
39. $-8 + 7\sqrt{30}$ **41.** $9a - 28b\sqrt{a} + 3b^2$
43. $8p^2 + 19p\sqrt{p} + 2p - 8\sqrt{p}$ **45.** 10 **47.** 4
49. t **51.** $16c$ **53.** $29 + 8\sqrt{13}$ **55.** $a - 4\sqrt{a} + 4$
57. $4a - 12\sqrt{a} + 9$ **59.** $21 - 2\sqrt{110}$ **61.** 1
63. $x - y$ **65.** -1 **67.** 4 **69.** $64x - 4y$ **71.** 73
73. a. $3x + 6$ **b.** $\sqrt{3x} + \sqrt{6}$ **75. a.** $4a^2 + 12a + 9$
b. $4a + 12\sqrt{a} + 9$ **77. a.** $b^2 - 25$ **b.** $b - 25$
79. a. $x^2 - 4xy + 4y^2$ **b.** $x - 4\sqrt{xy} + 4y$
81. a. $p^2 - q^2$ **b.** $p - q$ **83. a.** $y^2 - 6y + 9$
b. $x - 6\sqrt{x - 2} + 7$

Section 8.5 Practice Exercises, pp. 620–623

3. $6y + 23\sqrt{y} + 21$ **5.** $9\sqrt{3}$ **7.** $25 - 10\sqrt{a} + a$
9. -5 **11.** $\dfrac{\sqrt{3}}{4}$ **13.** $\dfrac{a^2}{b^2}$ **15.** $\dfrac{c\sqrt{c}}{2}$ **17.** $\dfrac{\sqrt[3]{x^2}}{3}$
19. $\dfrac{\sqrt[3]{y^2}}{3}$ **21.** $\dfrac{10\sqrt{2}}{9}$ **23.** $\dfrac{2}{5}$ **25.** $\dfrac{1}{2p}$ **27.** z
29. $2\sqrt[3]{x}$ **31.** $\dfrac{\sqrt{6}}{6}$ **33.** $3\sqrt{5}$ **35.** $\dfrac{6\sqrt{x+1}}{x+1}$
37. $\dfrac{\sqrt{6x}}{x}$ **39.** $\dfrac{\sqrt{21}}{7}$ **41.** $\dfrac{5\sqrt{6y}}{3y}$ **43.** $\dfrac{3\sqrt{6}}{4}$
45. $\dfrac{\sqrt{3p}}{9}$ **47.** $\dfrac{\sqrt{5}}{2}$ **49.** $\dfrac{x\sqrt{y}}{y^2}$ **51.** -7
53. $\sqrt{5} + \sqrt{3}$; 2 **55.** $\sqrt{x} - 10$; $x - 100$
57. $\dfrac{4\sqrt{2} - 12}{-7}$ or $\dfrac{12 - 4\sqrt{2}}{7}$ **59.** $\dfrac{\sqrt{5} + \sqrt{2}}{3}$
61. $\sqrt{6} - \sqrt{2}$ **63.** $\dfrac{\sqrt{x} + \sqrt{3}}{x - 3}$ **65.** $7 - 4\sqrt{3}$
67. $-13 - 6\sqrt{5}$ **69.** $2 - \sqrt{2}$ **71.** $\dfrac{3 + \sqrt{2}}{2}$
73. $1 - \sqrt{7}$ **75.** $\dfrac{7 + 3\sqrt{2}}{3}$ **77. a.** Condition 1 fails;
$2x^4\sqrt{2x}$ **b.** Condition 2 fails; $\dfrac{\sqrt{5x}}{x}$ **c.** Condition 3 fails;
$\dfrac{\sqrt{3}}{3}$ **79. a.** Condition 2 fails; $\dfrac{3\sqrt{x} - 3}{x - 1}$ **b.** Conditions
1 and 3 fail; $\dfrac{3w\sqrt{t}}{t}$ **c.** Condition 1 fails; $2a^2b^4\sqrt{6ab}$
81. $3\sqrt{5}$ **83.** $-\dfrac{3w\sqrt{2}}{5}$ **85.** Not a real number

87. $\dfrac{s\sqrt{t}}{t}$ **89.** $\dfrac{m^2}{2}$ **91.** $\dfrac{9\sqrt{t}}{t^2}$ **93.** $\dfrac{\sqrt{11}-\sqrt{5}}{2}$

95. $\dfrac{a+2\sqrt{ab}+b}{a-b}$ **97.** $-\dfrac{3\sqrt{2}}{8}$ **99.** $\dfrac{\sqrt{3}}{9}$

Chapter 8 Problem Recognition Exercises, p. 623

1. $3\sqrt{2}$ **2.** $2\sqrt{7}$ **3.** Cannot be simplified further
4. Cannot be simplified further   **5.** $\sqrt{2}$ **6.** $\sqrt{7}$
7. $9-z$ **8.** $16-y$ **9.** $8-3\sqrt{5}$
10. $-8+11\sqrt{3}$ **11.** $-x\sqrt{y}$ **12.** $11ab\sqrt{a}$
13. $-24-6\sqrt{6}-3\sqrt{2}$ **14.** $-80+8\sqrt{15}+16\sqrt{5}$
15. $\dfrac{2\sqrt{x}+14}{x-49}$ **16.** $\dfrac{5\sqrt{y}-20}{y-16}$ **17.** $3\sqrt{3}$
18. $3\sqrt{5}$ **19.** $\dfrac{\sqrt{7x}}{x}$ **20.** $\dfrac{\sqrt{11y}}{y}$ **21.** $y^2z^5\sqrt{z}$
22. $2q^3\sqrt{2}$ **23.** $3p^2\sqrt[3]{p^2}$ **24.** $5u^3v^4\sqrt[3]{u^2}$ **25.** $x\sqrt{10}$
26. $y\sqrt{3}$ **27.** $20\sqrt{3}$ **28.** $\sqrt{10}$ **29.** $51+14\sqrt{2}$
30. $8+2\sqrt{15}$ **31.** $\sqrt{x}-\sqrt{5}$ **32.** $\sqrt{y}-\sqrt{7}$
33. $4x-11\sqrt{xy}-3y$ **34.** $\dfrac{1}{3}$ **35.** $\dfrac{5}{3}$
36. $16-2\sqrt{55}$ **37.** $x-12\sqrt{x}+36$ **38.** $-3\sqrt{6}$
39. $11\sqrt{a}$ **40.** -88 **41.** $u-9v$ **42.** $4x\sqrt{2}$
43. 0 **44.** $5+\sqrt{35}$ **45.** $a+2\sqrt{a}$
46. $26-17\sqrt{2}$

Section 8.6 Practice Exercises, pp. 628–630

3. $\dfrac{\sqrt{2}-\sqrt{10}}{-8}$ or $\dfrac{\sqrt{10}-\sqrt{2}}{8}$ **5.** $\dfrac{2\sqrt{6}}{3}$
7. $x^2+8x+16$ **9.** $x+8\sqrt{x}+16$ **11.** $2x-3$
13. t^2+2t+1 **15.** 36 **17.** 15 **19.** No solution
(the value 29 does not check)   **21.** 5 **23.** $-\dfrac{1}{2}$
25. 6 **27.** 8 **29.** No solution (the value $\frac{19}{2}$ does not check)   **31.** 1 **33.** $4,-3$ **35.** 0
37. No solution (the value -4 does not check)
39. 4, (the value -1 does not check)   **41.** $0,-1$
43. 12, (the value 4 does not check)   **45.** -6 **47.** -1
49. $\sqrt{x+10}=1;\ -9$ **51.** $\sqrt{2x}=x-4;\ 8$
53. $\sqrt[3]{x+1}=2;\ 7$ **55. a.** 80 ft/sec   **b.** 289 ft
57. a. 16 in.   **b.** 25 weeks   **59.** $\dfrac{9}{5}$
61. $\dfrac{3}{2}$ (the value -1 does not check)

Section 8.7 Practice Exercises, pp. 634–636

1. a. 3   **b.** 125   **3.** 27   **5.** $a+1$ **7.** 9   **9.** 5
11. 3   **13.** -2 **15.** -2 **17.** $\dfrac{1}{6}$ **19.** $\sqrt[3]{x}$
21. $\sqrt{4a}$ or $2\sqrt{a}$ **23.** $\sqrt[5]{yz}$ **25.** $\sqrt[3]{u^2}$ **27.** $5\sqrt{q}$
29. $\sqrt{\dfrac{x}{9}}$ or $\dfrac{\sqrt{x}}{3}$ **31.** $a^{m/n}=\sqrt[n]{a^m}$ or $(\sqrt[n]{a})^m$, provided the
roots exist.   **33.** 8   **35.** $\dfrac{1}{9}$ **37.** -32 **39.** 2
41. $(\sqrt{y})^9$ **43.** $\sqrt[3]{c^5d}$ **45.** $\dfrac{1}{\sqrt[5]{qr}}$

47. $6\sqrt[3]{y^2}$ **49.** $x^{1/3}$ **51.** $(xy)^{1/3}$ **53.** $5x^{1/2}$
55. $y^{2/3}$ **57.** $(m^3n)^{1/4}$ **59.** $4k^{3/3}$ or $4k$ **61.** x
63. y^2 **65.** 6   **67.** a^7 **69.** $y^{4/3}$ **71.** 2
73. $\dfrac{25a^4d}{c}$ **75.** $\dfrac{y^9}{x^8}$ **77.** $\dfrac{2z^3}{w}$ **79.** $5xy^2z^{3/2}$
81. a. 10 in.   **b.** 8.49 in.   **83. a.** 10.9%   **b.** 8.8%
c. The account in part (a)   **85.** No, for example,
$(36+64)^{1/2} \neq 36^{1/2}+64^{1/2}$ **87.** 6   **89.** $\dfrac{5}{14}$ **91.** $\dfrac{a^{22}b^4}{c^{17}}$

Chapter 8 Review Exercises, pp. 642–644

1. Principal square root: 14; negative square root: -14
3. Principal square root: 0.8; negative square root: -0.8
5. There is no real number b such that $b^2=-64$.   **7.** -12
9. Not a real number   **11.** $|y|$ **13.** $|y|$ **15.** -5
17. $\dfrac{3}{t^2}$ **19. a.** 7.1 m   **b.** 22.6 ft   **21.** $b^2+\sqrt{5}$
23. The quotient of 2 and the principal square root of p
25. 12 ft   **27.** $x^8\sqrt{x}$ **29.** $2\sqrt{7}$ **31.** $3y^3\sqrt[3]{y}$
33. c   **35.** $10y^2$ **37.** $2x$ **39.** $5\sqrt{3}$ **41.** 1
43. $7\sqrt{6}$ **45.** $-4x\sqrt{5}$ **47.** $15\sqrt{3}-7\sqrt{7}$
49. $\sqrt{x^4+(5x)^2};\ 26x^2$ **51.** $12\sqrt{2}$ ft   **53.** 25
55. $70\sqrt{3x}$ **57.** $8m+24\sqrt{m}$ **59.** $-49-16\sqrt{26}$
61. $64w-z$ **63.** $10\sqrt{3}$ m^3 **65.** a^5 **67.** $4\sqrt{y}$
69. b   **71.** $\dfrac{3\sqrt{2y}}{y}$ **73.** $2\sqrt{7}+2\sqrt{2}$
75. $-8-3\sqrt{7}$ **77.** 138   **79.** 39   **81.** 7
83. 2, (the value -2 does not check)
85. -69 **87.** -3 **89.** -2 **91.** $\dfrac{1}{8}$
93. $\sqrt[5]{z}$ **95.** $\sqrt[4]{w^3}$ **97.** $a^{2/5}$ **99.** $(a^2b^4)^{1/5}$
101. y^2 **103.** $6^{3/5}$ **105.** $4ab^2$ **107.** 2.0 cm

Chapter 8 Test, pp. 645–646

1. 1. The radicand has no factor raised to a power greater
than or equal to the index. 2. There are no radicals in the
denominator of a fraction. 3. The radicand does not contain a
fraction.   **2.** $11x\sqrt{2}$ **3.** $2y\sqrt[3]{6y}$ **4.** Not a real number
5. $\dfrac{a^3\sqrt{5}}{9}$ **6.** $\dfrac{3\sqrt{6}}{2}$ **7.** $\dfrac{2\sqrt{5}-12}{-31}$ or $\dfrac{12-2\sqrt{5}}{31}$
8. a. $\sqrt{25}+5^3;\ 130$ **b.** $4^2-\sqrt{16};\ 12$ **9. a.** 97 ft
b. 339 ft   **10.** $8\sqrt{z}$ **11.** $4\sqrt{6}-15$ **12.** $-7t\sqrt{2}$
13. $9\sqrt{10}$ **14.** $46-6\sqrt{5}$ **15.** $-8+23\sqrt{10}$
16. $\dfrac{\sqrt{n}}{6m}$ **17.** $16-9x$ **18.** $\dfrac{\sqrt{22}}{11}$ **19.** $\dfrac{3\sqrt{7}+3\sqrt{3}}{2}$
20. 206 yd   **21.** No solution (the value $\frac{9}{2}$ does not check)
22. $0,-5$ **23.** 14   **24. a.** 12 in.   **b.** 18 in.   **c.** 25 weeks
25. 1000   **26.** 2   **27.** $\sqrt[5]{x^3}$ or $(\sqrt[5]{x})^3$ **28.** $5\sqrt{y}$
29. $(ab^3)^{1/4}$ **30.** $p^{11/12}$ **31.** $5^{3/5}$ **32.** $3mn^2$

Chapters 1–8 Cumulative Review Exercises, pp. 646–647

1. 1   **2.** -2 **3.** -15 **4.** $-9x^2+2x+10$
5. $\dfrac{2x}{y}-1+\dfrac{4}{x}$ **6.** $2(5c+2)^2$ **7.** $-\dfrac{2}{5},\dfrac{1}{2}$ **8.** 1
9. No solution (the value 5 does not check)

10. $\dfrac{x(5x+8)}{16(x-1)}$ **11.**

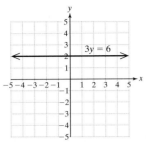

12. a. $y = 880$; the cost of renting the office space for 3 months is $880. **b.** $x = 12$; the cost of renting office space for 12 months is $2770. **c.** $m = 210$; the cost increases at a rate of $210 per month. **d.** $(0, 250)$; the down payment of renting the office space is $250. **13.** $y = -x + 1$

14. $(1, -4)$ **15.**

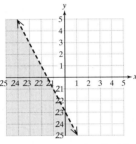

16. 8 L of 20% solution should be mixed with 4 L of 50% solution. **17.** $3\sqrt{11}$ **18.** $7x\sqrt{3}$

19. $\dfrac{x + \sqrt{xy}}{x - y}$ **20.** 5

Chapter 9

Chapter Opener Puzzle

Section 9.1 Practice Exercises, pp. 653–655

3. a. Linear **b.** Quadratic **c.** Linear **5.** $-5, \dfrac{1}{2}$

7. $7, -5$ **9.** $-2, -\dfrac{1}{6}$ **11.** $-7, -\dfrac{3}{2}$ **13.** $12, -12$

15. $8, -2$ **17.** $\dfrac{1}{4}, -2$ **19.** $-1, -7$ **21.** ± 7

23. ± 10 **25.** There are no real-valued solutions.
27. $\pm\sqrt{3}$ **29.** $9, 1$ **31.** $11, -1$ **33.** $11 \pm \sqrt{5}$

35. $-1 \pm 3\sqrt{2}$ **37.** $\dfrac{1}{4} \pm \dfrac{\sqrt{7}}{4}$ **39.** $\dfrac{1}{2} \pm \sqrt{15}$

41. There are no real-valued solutions. **43.** $\pm\dfrac{5}{2}$

45. The solution checks. **47.** False. -8 is also a solution.
49. a. 64 ft **b.** 3.5 sec **c.** 8.8 sec **51.** 7.1 m
53. 8.0 ft

Section 9.2 Practice Exercises, pp. 659–661

3. $5 \pm \sqrt{21}$ **5.** $4; y^2 + 4y + 4 = (y + 2)^2$
7. $36; p^2 - 12p + 36 = (p - 6)^2$

9. $\dfrac{81}{4}; x^2 - 9x + \dfrac{81}{4} = \left(x - \dfrac{9}{2}\right)^2$

11. $\dfrac{25}{36}; d^2 + \dfrac{5}{3}d + \dfrac{25}{36} = \left(d + \dfrac{5}{6}\right)^2$

13. $\dfrac{1}{100}; m^2 - \dfrac{1}{5}m + \dfrac{1}{100} = \left(m - \dfrac{1}{10}\right)^2$

15. $\dfrac{1}{4}; u^2 + u + \dfrac{1}{4} = \left(u + \dfrac{1}{2}\right)^2$ **17.** $x = 2, x = -6$

19. $y = -1, y = -5$ **21.** $x = 1 \pm \sqrt{2}$

23. $x = 1 \pm \sqrt{6}$ **25.** $-2 \pm \sqrt{3}$ **27.** $-\dfrac{1}{2} \pm \dfrac{\sqrt{13}}{2}$

29. $-1 \pm \sqrt{41}$ **31.** $2 \pm \sqrt{5}$ **33.** $-2, -4$
35. $3, 8$ **37.** ± 11 **39.** $-2 \pm \sqrt{2}$ **41.** $-13, 5$
43. 13 **45.** $10, -2$ **47.** $7, -1$ **49.** $4 \pm \sqrt{15}$

51. $-1 \pm \sqrt{6}$ **53.** $11, -2$ **55.** $0, 7$ **57.** $\dfrac{1}{2}, -\dfrac{3}{4}$

59. $\pm\sqrt{14}$ **61.** There are no real-valued solutions.
63. 1 **65.** The suitcase is 10 in. by 14 in. by 30 in. The bag must be checked because 10 in. + 14 in. + 30 in. = 54 in., which is greater than 45 in.

Section 9.3 Calculator Connections, p. 667

1.
```
(-5+√(17))/4
         -.2192235936
(-5-√(17))/4
         -2.280776406
```

2.
```
(-40+√(1920))/-3
2
         -.1193063938
(-40-√(1920))/-3
2
         2.619306394
```

Section 9.3 Practice Exercises, pp. 667–669

1. ± 13 **3.** $4 \pm 2\sqrt{7}$ **5.** $2 \pm 2\sqrt{2}$

7. For $ax^2 + bx + c = 0$, $x = \dfrac{-b \pm \sqrt{b^2 - 4ac}}{2a}$

9. $2x^2 - x - 5 = 0; a = 2, b = -1, c = -5$
11. $-3x^2 + 14x + 0 = 0; a = -3, b = 14, c = 0$
13. $x^2 + 0x - 9 = 0; a = 1, b = 0, c = -9$

15. -8 **17.** $\dfrac{2}{3}, -\dfrac{1}{2}$ **19.** $\dfrac{1 \pm \sqrt{61}}{10}$ **21.** $1 \pm \sqrt{2}$

23. $\dfrac{-5 \pm \sqrt{3}}{2}$ **25.** $\dfrac{1 \pm \sqrt{17}}{-8}$ or $\dfrac{-1 \pm \sqrt{17}}{8}$

27. $\dfrac{-3 \pm \sqrt{33}}{4}$ **29.** $\dfrac{-15 \pm \sqrt{145}}{4}$

31. $\dfrac{-2 \pm \sqrt{22}}{6}$ **33.** $\dfrac{3}{4}, -\dfrac{3}{4}$

35. There are no real-valued solutions.   **37.** $-12 \pm 3\sqrt{5}$

39. $\dfrac{3 \pm \sqrt{15}}{2}$ **41.** $0, \dfrac{11}{9}$ **43.** $\dfrac{3 \pm \sqrt{5}}{2}$

45. $\dfrac{1 \pm \sqrt{41}}{4}$ **47.** $0, \dfrac{1}{9}$ **49.** $\pm 2\sqrt{13}$

51. $\dfrac{-10 \pm \sqrt{85}}{-5}$ or $\dfrac{10 \pm \sqrt{85}}{5}$ **53.** $\dfrac{1 \pm \sqrt{61}}{2}$

55. There are no real-valued solutions.   **57.** The width is 7.3 m. The length is 13.7 m   **59.** The length is 7.4 ft. The width is 5.4 ft. The height is 6 ft.   **61.** The width is 6.7 ft. The length is 10.7 ft.   **63.** The legs are 10.6 m and 7.6 m.

Chapter 9 Problem Recognition Exercises, p. 669

1. $\dfrac{1}{3}, -\dfrac{3}{2}$ **2.** -7 **3.** $4 \pm \sqrt{22}$ **4.** $3 \pm \sqrt{7}$

5. $3 \pm 2\sqrt{2}$ **6.** $-11 \pm 2\sqrt{3}$ **7.** $0, \dfrac{5}{6}$ **8.** $0, \dfrac{1}{2}$

9. ± 6 **10.** $\pm \dfrac{7}{5}$ **11.** $\dfrac{5}{2}, \dfrac{1}{4}$ **12.** $\dfrac{4}{3}, \dfrac{1}{3}$

13. There are no real-valued solutions.   **14.** There are no real-valued solutions.   **15.** $-4 \pm \sqrt{15}$ **16.** $-3 \pm \sqrt{6}$

17. 4   **18.** -3 **19.** $5, 2$ **20.** $3, -1$ **21.** $9, -11$

22. $13, -3$

Section 9.4 Calculator Connections, p. 677

1.

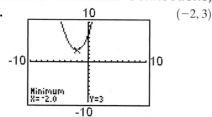

$(-2, 3)$

2.

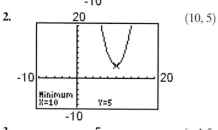

$(10, 5)$

3.

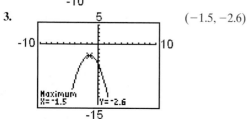

$(-1.5, -2.6)$

4.

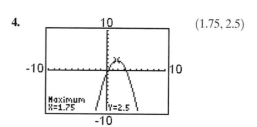

$(1.75, 2.5)$

5.

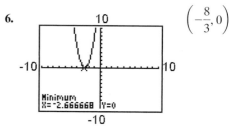

$1 \pm \sqrt{6}$

6.

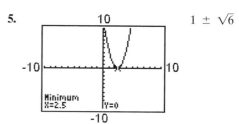

$\left(-\dfrac{8}{3}, 0\right)$

7. The graph in part (b) is shifted up 4 units. The graph in part (c) is shifted down 3 units.

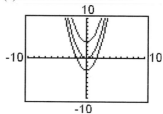

8. In part (b), the graph is shifted to the right 3 units. In part (c), the graph is shifted to the left 2 units.

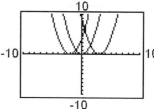

9. In part (b), the graph is stretched vertically by a factor of 2. In part (c), the graph is shrunk vertically by a factor of $\frac{1}{2}$.

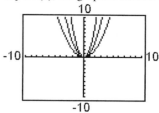

10. In part (b), the graph has been stretched vertically and reflected across the *x*-axis. In part (c), the graph has been shrunk vertically and reflected across the *x*-axis.

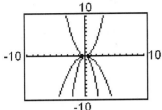

Section 9.4 Practice Exercises, pp. 678–681

3. $-5, 3$ **5.** $-1 \pm \sqrt{6}$ **7.** $2 \pm 2\sqrt{2}$
9. Linear   **11.** Quadratic   **13.** Neither
15. Linear   **17.** Quadratic   **19.** Neither
21. If $a > 0$ the graph opens upward; if $a < 0$ the graph opens downward.   **23.** $a = 2$; upward   **25.** $a = -10$; downward   **27.** $(-1, -8)$ **29.** $(1, -4)$ **31.** $(1, 2)$
33. $(0, -4)$ **35.** c; *x*-intercepts: $(\sqrt{7}, 0)(-\sqrt{7}, 0)$;
y-intercept: $(0, -7)$ **37.** a; *x*-intercepts: $(-1, 0)(-5, 0)$;
y-intercept: $(0, 5)$
39. **a.** Upward   **b.** $(0, -9)$ **c.** $(3, 0)(-3, 0)$ **d.** $(0, -9)$
e.   **f.** $(-\infty, \infty)$ **g.** $[-9, \infty)$

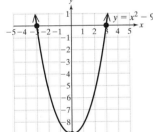

41. **a.** Upward   **b.** $(1, -9)$ **c.** $(4, 0)(-2, 0)$ **d.** $(0, -8)$
e.   **f.** $(-\infty, \infty)$ **g.** $[-9, \infty)$

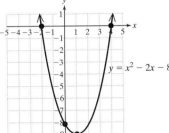

43. **a.** Downward   **b.** $(3, 0)$ **c.** $(3, 0)$ **d.** $(0, -9)$
e.   **f.** $(-\infty, \infty)$ **g.** $(-\infty, 0]$

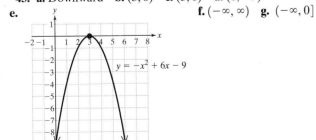

45. **a.** Downward   **b.** $(4, 1)$ **c.** $(3, 0)(5, 0)$ **d.** $(0, -15)$
e.   **f.** $(-\infty, \infty)$ **g.** $(-\infty, 1]$

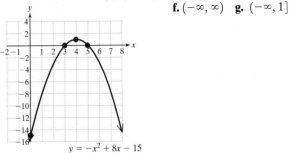

47. **a.** Upward   **b.** $(-3, 1)$ **c.** none   **d.** $(0, 10)$
e.   **f.** $(-\infty, \infty)$ **g.** $[1, -\infty)$

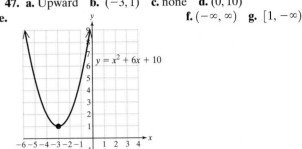

49. **a.** Downward   **b.** $(0, -2)$ **c.** none   **d.** $(0, -2)$
e.   **f.** $(-\infty, \infty)$ **g.** $(-\infty, -2]$

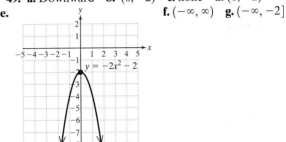

51. True   **53.** False   **55.** **a.** 28 ft   **b.** 1.25 sec
57. **a.** 200 calendars   **b.** $500

Chapter 9 Review Exercises, pp. 685–687

1. Linear   **3.** Quadratic   **5.** ± 5 **7.** The equation has no real-valued solutions.   **9.** $-1 \pm \sqrt{14}$
11. $\frac{1}{8} \pm \frac{\sqrt{3}}{8}$ **13.** 36   **15.** $\frac{25}{4}$ **17.** $-4 \pm \sqrt{13}$
19. $\frac{3}{2} \pm \frac{\sqrt{21}}{2}$ **21.** 10.6 ft   **23.** For $ax^2 + bx + c = 0$,
$x = \frac{-b \pm \sqrt{b^2 - 4ac}}{2a}$ **25.** -2 **27.** $\frac{3}{2}, -1$
29. $-3 \pm \sqrt{7}$ **31.** $1 \pm \sqrt{5}$
33. The equation has no real-valued solutions.
35. The height is $\frac{-1 + \sqrt{97}}{2} \approx 4.4$ cm.   The base is
$\frac{1 + \sqrt{97}}{2} \approx 5.4$ cm.   **37.** $a = 1$; upward   **39.** $a = -2$; downward   **41.** Vertex: $(-1, 1)$ **43.** Vertex: $(3, 13)$
45. **a.** Upward   **b.** $(-2, -3)$ **c.** $(-3, 0)(-1, 0)$ **d.** $(0, 9)$

e.

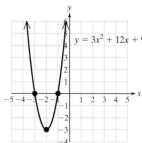

f. $(-\infty, \infty)$ **g.** $[-3, \infty)$

47. a. Downward   **b.** $(-1, -4)$ **c.** No x-intercepts
d. $(0, -12)$

e.

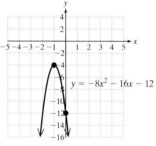

f. $(-\infty, \infty)$ **g.** $(-\infty, -4]$

49. a. 1024 ft   **b.** 8 sec

Chapter 9 Test, pp. 687–688

1. The equation has no real-valued solutions.
2. $4 \pm \sqrt{21}$ **3.** $\dfrac{5 \pm \sqrt{13}}{6}$ **4.** $\dfrac{-1 \pm \sqrt{41}}{10}$
5. $12 \pm 2\sqrt{3}$ **6.** $-7 \pm 5\sqrt{2}$ **7.** $\pm\sqrt{10}$
8. $\dfrac{5}{6}, -\dfrac{3}{2}$ **9.** $0, \dfrac{11}{6}$ **10.** $3 \pm 2\sqrt{5}$ **11.** 4.0 in.
12. The base is $\dfrac{-1 + \sqrt{97}}{2} \approx 4.4$ m. The height is
$1 + \sqrt{97} \approx 10.8$ m.   **13.** For $y = ax^2 + bx + c$, if $a > 0$
the parabola opens upward, if $a < 0$ the parabola opens
downward.   **14.** $(5, 0)$ **15.** $(1, 5)$ **16.** $(0, -16)$
17. The parabola has no x-intercepts.
18. a. Opens upward   **b.** Vertex: $(-3, -1)$
c. x-intercepts: $(-2, 0)$ and $(-4, 0)$ **d.** y-intercept: $(0, 8)$
e.

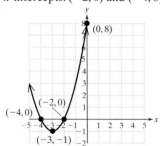

19. Vertex: $(0, 25)$; x-intercepts: $(-5, 0)(5, 0)$; y-intercept:
$(0, 25)$

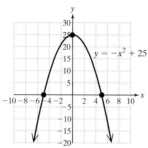

20. a. \$25 per ticket   **b.** \$250,000

Chapters 1–9 Cumulative Review Exercises, pp. 688–689

1. 1   **2.** $h = \dfrac{2A}{b}$ **3.** $\dfrac{34}{3}$ **4.** Yes, 2 is a solution.
5. $\{x \mid x > -1\}$; $(-1, \infty)$ ⟶ (number line at -1)

6. a. Decreases   **b.** $m = -37.6$. For each additional
increase in education level, the death rate decreases by
approximately 38 deaths per 100,000 people.
c. 901.8 per 100,000   **d.** 10th grade
7. $\dfrac{8c}{a^4 b}$ **8.** 1.898×10^{10} diapers
9. Approximately 430 light-years   **10.** $4x^2 - 16x + 13$
11. $2y^3 + 1 - \dfrac{3}{y - 2}$ **12.** $(2x + 5)(x - 7)$
13. $(y + 4a)(2x - 3b)$ **14.** The base is 9 m, and the
height is 8 m.   **15.** $-\dfrac{2}{x + 1}$ **16.** $\dfrac{5}{x - 2}$ **17.** $x - 5$
18. $-\dfrac{2}{x}$ **19.** $4, -3$ **20.** 20 cm ≈ 7.9 in. The fish
must be released.   **21.** $y = \dfrac{1}{2}x + 4$
22. a. $(6, 0)$ **b.** $(0, -3)$ **c.** $\frac{1}{2}$
d.

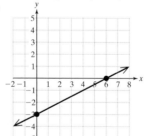

23. a. $(-3, 0)$ **b.** No y-intercept   **c.** Slope is undefined.
d.

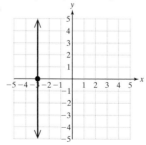

24. $\left(1, \dfrac{4}{3}\right)$ **25.** $(5, 2)$ **26.** The angles are 22° and 68°.

27. There are 13 dimes and 14 quarters.

28.

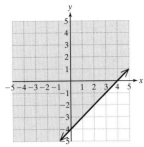

29. $\pi, \sqrt{7}$ **30.** $\dfrac{\sqrt{7}}{7}$ **31.** $2x\sqrt{2x}$

32. $48 + 8\sqrt{3x} + x$ **33.** $2\sqrt{2x}$ **34.** $\dfrac{8 + 4\sqrt{a}}{4 - a}$

35. -2, (the value -7 does not check)

36. $(2c - y)(4c^2 + 2cy + y^2)$ **37.** b

38. a. 1   **b.** 4   **c.** 13

39. Domain: $\{2, -1, 9, -6\}$; range: $\{4, 3, 2, 8\}$

40. $m = \dfrac{5}{7}$ **41.** $m = -\dfrac{4}{5}$ **42.** 25

43. $-3 \pm \sqrt{6}$ **44.** $-3 \pm \sqrt{6}$

45.

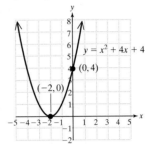

Credits

Application Index

Biology/Health/Life Sciences

Age vs. height, 266, 290
Age vs. systolic blood pressures, 213
Average length of hospital stay, 266
Blood pumped daily, 387
Calories in cake vs. ice cream, 357
Cancer diagnoses, 278
Cancer rates, 157
Concentration of chemicals, 170–171
Fat content of popcorn, 576
Fat in ice cream, 573
Fish length, 190
Grams of carbohydrates, 556
Height of plant, 630, 645
Height vs. age, 290
Height vs. arm length, 266
Hospice care patients per month, 210
Length of fish, 689
Maximum recommended heart rate vs. age, 279, 291
Milk mixture, 354
Number of medical doctors, 292
Oatmeal production, 556
Pulse rate, 556
Ratio of dogs to cats at shelter, 559
Roses planted in garden, 28
Time required for plant growth, 646
Weight loss, 20

Business

Amount of money borrowed, 334
Base salary, 159
Checkbook balancing, 70, 73
Commission earned, 159
Consumer goods price before discount, 158
Consumer purchase with sales tax, 157
Cost of bikes, 505
Cost of concession stand vs. product, 262, 267
Cost of consumer items before discount, 156
Cost of membership, 159
Cost of office space rental, 646
Cost of posters, 263
Discount on books, 174
Discount on geraniums, 201–202
Discount on T-shirts, 191
Discount on wooden birdhouses, 191
Electric company bill calculation, 266
Fundraising event profits, 421
Housing permits issued per year, 210
Internet sales, 151
Laptop computer tax rate, 159
Loading platform dimensions, 482–483
Milk production, 333
Minimum wage earnings vs. year, 264
M&Ms sold, 421
Monthly costs of business, 267
Movie releases, 207
Movie ticket sales, 383–385
Movie ticket versus popcorn sales, 212–213
Number of book sales, 565
Number of hybrid cars sold, 262
Original price of novel, 200
Percentage of funds raised, 358
Profit and loss, 73
Profit on concessions, 681
Profit on fundraisers, 681
Profit on hot dogs, 202
Profit on lemonade, 191
Profit on movie tickets, 148–149
Profit on sports ticket sales, 688
Property tax, 556
Quantity of food purchased, 174
Ratio of fiction to nonfiction books sold, 559
Rental car costs, 191
Salary, 186
Sales commissions, 159
Sales tax, 26, 28, 153, 200, 203
Server profit, 152
Sign dimensions, 479–480
Social security, 28
Star earnings, 152
Taxable income, 158
Tea mixture, 334
Television commercial mixtures, 335
Ticket sales, 174
Tickets sold, 169–170
Time required for building grill, 558
Time required for copying, 558
Time required for housecleaning, 573
Time required for lawn mowing, 558
Time required for mailing, 558
Time required for print job, 554–555, 558, 578
Total paycheck, 154–155
TV show episodes produced, 152
Unit cost of CD production, 565
Wealth of Bill Gates, 387

Chemistry

Mixing acid solutions, 356, 647
Mixing alcohol solutions, 328–329
Mixing antifreeze solutions, 333, 358
Mixing disinfectant solutions, 333
Mixing salt solutions, 33

Construction

Area of pool, 369
Building completion, 28
Circumference of fountain, 168
Cost of carpet, 39
Diagonal of tile, 592
Dimensions of concrete slab, 484
Dimensions of fence, 40
Dimensions of ladder, 486, 642
Dimensions of parking area, 166
Dimensions of workbench, 166
Hammer strikes, 19
Height of building, 642
Length of rope, 151
Lengths of board, 19, 147–148, 151
Lengths of pipe, 152
Lengths of wire, 152
Perimeter of lot, 166
Perimeter of yard, 40
Radius of fountain, 642
Radius of garden, 642
Sprinkler system layout, 212
Tar paper required for roof coverage, 599
Time required to paint room, 558

Consumer Applications

Additional cashews needed, 174
Amount of food items purchased, 177
Amount of goods created, 19
Amount of money borrowed, 356
Amount of mortgage paid, 29
Amount of shrimp, 20
Amount of turkey per sandwich, 20
Area of pizza, 40, 369
Calories in cake vs. ice cream, 357
Candy/nut mixture, 334
Carrots required for soup, 578
Completion time of task, 28
Concentration of beef, 201
Concentration of chemicals, 175, 177, 201
Cookies remaining to bake, 174

Cost of CDs, 332
Cost of CDs vs. DVDs, 356
Cost of cellular phone plan, 263
Cost of food, 326
Cost of food items, 176
Cost of home, 28
Cost of hot dog vs. soft drink, 354
Cost of movie tickets, 332
Cost of paint, 39
Cost of popcorn vs. drink, 326, 579
Cost of rental car, 251, 267, 353
Cost of state fair, 292
Cost of storage space rental, 267
Cost of suit, 204
Cost of trolley ride, 356
Cost of video games vs. DVD
    rentals, 332
Cost of zoo admission, 354
Diagonal of a TV, 592
Diagonal of computer monitor, 655
Dimensions of painting, 484
Dimensions of pool, 655
Dimensions of room, 484
Dimensions of storage area, 668
Dimensions of suitcase, 660, 661
Discounted cost of suit, 28
Fat content of popcorn, 576
Fat in ice cream, 573
Holiday aprons, 19
Home sales, 209
Jeopardy scores, 73, 80
Lottery winnings, 389
Lottery winnings vs. price, 92
Mixture of nuts, 203
Moving cots, 261
MP3 song purchase, 174
Number of cars in parking lot, 186
Number of coins/bills by denomination,
    33–34, 356, 496
Number of drink sales, 578
Number of necessary turkeys, 565
Phone bill calculation, 267
Radius of can, 681
Taxes, 28
Ticket costs, 266
Tip left, 28
US. Postal Service rates, 241
Volume of balloon, 369
Volume of coffee mug, 40
Volume of gravel, 39
Volume of soup can, 40
Weight of staple, 388
Width of futon, 20
Wind energy, 267

Distance/Speed/Time

Area of tower, 40
Average rate of trip, 536
Average speed of cars, 573
Average speed of trip, 553

Dimensions of loading platform,
    482–483
Dimensions of photo, 166
Distance between Earth and
    Polaris, 689
Distance between Earth and Sun,
    389, 422
Distance between Mercury and
    Sun, 422
Distance covered by falling object,
    629, 654
Distance covered by properly-inflated
    tires, 681
Distance covered by triathlon, 645
Distance hiked, 176
Distance of boats, 486
Distance of runway vs. speed of
    plane, 290
Distance traveled by airplane, 177
Distance traveled by boat, 177
Distance traveled by car, 203
Distance traveled by car vs. gallons of
    gas, 556
Distance traveled by kayak, 588
Distance traveled for picnic, 172–173
Distance traveled in moving
    vehicles, 175
Distance walked, 20
Height and view of horizon, 577
Height of building, 557
Height of dropped object vs. time,
    480–481
Height of football vs. time, 278
Height of light pole, 557
Height of rock on moon vs. time, 686
Height of thrown object, 687
Height of tree, 551, 557
Height required for amusement park
    ride, 186
Length of hike, 190
Map scales, 556
Miles driven in car, 486
Perimeter of garden, 166
Position of ball, 642, 645
Speed of airplane, 172, 176, 330–331,
    333, 334, 357, 558
Speed of airplane discounting wind,
    552, 557, 558
Speed of bicycle, 171–172
Speed of biker, 201
Speed of boat, 176, 333, 354
Speed of boat discounting current,
    557, 558
Speed of canoe, 176, 330, 561–562
Speed of car, 358
Speed of car vs. length of skid
    marks, 630
Speed of current, 330, 333, 354, 578
Speed of football vs. time, 278
Speed of hikers, 357
Speed of motorcycles, 576

Speed of object in free-fall, 278
Speed of train, 171, 177
Speed of truck, 201
Speed of vehicle, 176
Speed of walking, 176
Speed of wind, 330–331, 333, 334, 357
Stopping distance of car, 566
Thickness of paper, 388, 421
Time needed to travel set distance, 201
Time required for bicycle trip, 505
Time required for dropped object to
    hit, 480–481, 484
Time required for object shot straight
    up to hit, 485
Time required for object to fall, 629, 654
Time required for rocket to
    hit ground, 485
Time required for sink filling, 558
Time required to empty pond, 558
Time required to empty pool, 577
Time required to empty reservoir, 558
Time required to fill pool, 558
Time spent hiking, 201
Velocity of falling object, 629

Environment

Areas of land, 152
Average temperature, 49–50
Average temperatures, 111, 265
Depths of ocean, 81, 151, 152
Depths of water, 151
Dimensions of pyramid, 168
Distance between cities, 593, 599, 642
Distance between lightning strike and
    observer, 241
Distance between whale and
    laboratory, 592
Elevations of cities, 54
Garbage production, 556
Height of cave, 190
Height of mountains, 8, 151
Height of pyramids, 157
Height of tree, 551, 557
Hurricane power vs. wind speed, 563
Kinetic energy, 563
Lengths of rivers, 151
Location of fire observation towers, 209
Locations of park visitors, 212
Number of tigers, 261
Pollution levels vs. population, 565
Rain amounts, 20, 191, 202
Rain amounts vs. time, 292
Slope of ski run, 230–231
Snowfall, 191, 203, 424
Speed of current, 333, 354, 578
Speed of wind, 190, 333, 334, 357, 563
Temperature, 190
Temperature and volume of a gas, 556
Temperature averages, 214
Temperature changes, 66

Temperature differences, 73, 81, 111
Temperature lows, 209
Volume of Niagara Falls, 423
Volume of oil spill, 387
Water temperature, 190

Investment

Compound interest, annual
    compounding, 365–366, 369, 421
Division of investments, 354
Family budget, 29
Interest earned, 29, 566, 636
Interest paid, 28
Money in the bank, 174, 175
Original purchase price, 159
Salary saved, 28
Salary saved monthly, 18
Simple interest
    calculation of rate, 158
    determining amount borrowed, 155,
        158, 203, 332
    determining initial deposit, 327–328,
        332, 334
    interest due on loan, 158
    interest earned, 158, 200
Stock price per share, 332
Stock prices over time, 210, 211
Value of Bill Gates' stock, 389

Politics

Gender representation in
    US Senate, 151
Number of men vs. women voting, 354
Party representation differences, 151
Poverty threshold, 213
Representation in U.S. House by
    state, 549
Salaries of politicians, 387
Votes for Jesse Ventura, 357
Voting age, 186

School

Applicants accepted, 26
Athletes with honors, 18
Grade requirements, 187
Pages of report printed, 18
Ratio of male to female students at
    lecture, 559

Test questions, 26, 154
Test Scores, 190
Test scores, 186
Test scores as function of study
    time, 274

Science

Amount of medicine prescribed,
    565, 578
Area of projected picture, 566
Current in wire, 566
Decibel levels, 562
Determining tire pressure, 681
Dimensions of box, 666
Dinosaur extinction, 389
Distance between Earth and Sun, 422
Distance between Mercury and
    Sun, 422
Forces on springs, 577
Height of rock on moon vs. time, 686
Intensity of light source, 566
Kinetic energy vs. speed, 644
pH scale, 421
Power in electric circuit, 566
Resistance in wire, 566

Sports

Baseball salary, 200
Baskets by Kareem Abdul-Jabbar, 334
Cost of bikes, 505
Cost of tennis instruction, 306
Dimensions of football field, 40
Dimensions of soccer field, 40
Distance between home plate and
    second base, 592
Distance covered by kayak, 588
Distance covered by triathlon, 645
Golf score, 54, 354
Height of ball, 681
Height of golf ball, 676
Hike length, 176
Meeting point of hikers, 357
Points scored by Cynthia Cooper vs.
    Sheryl Swoopes, 356
Points scored by player, 268
Points scored per minute by
    Shaquille O'Neal and
    Allen Iverson, 265

Points scored per team, 335
Pool depth, 190
Pool dimensions, 578
Pool perimeter, 166
Position of baseball, 645
Position of golf ball, 607
Position of long jumper, 607
Repetitions of weightlifting, 566
Revenue on ticket sales, 688
Soccer field perimeter, 168, 203
Speed of canoe, 561–562
Speed of cyclist, 558
Speed of runner, 558
Speed of skiers, 558
Speed of walkers, 558
Sports preferences, 154
Tennis court dimensions, 495
Ticket prices, 203
Time required for bicycle trip, 505
Touchdown receiving vs.
    rushing, 357
Weight of medicine ball, 566
Yards gained in football, 70, 73

Statistics/Demographics

Age differences, 151
College completion, 214–215
Distance hiked, 152
Hours worked per week, 158
Median income for males in U.S.,
    236–237
Morality rate vs. education level,
    688–689
Number of drug arrests, 284
Number of drug rehab admits,
    206–207
Number of female federal and state
    prisoners, 241
Number of jail inmates, 264
Number of male federal and state
    prisoners, 240
Participants in survey, 335
Population of U.S. colonies, 213
Ratio of smokers to nonsmokers, 559
Relation of women with
    children, 269
State populations, 200, 236
TV audiences, 151
World populations, 200

Subject Index

A

$a^{1/n}$, 630–631
Absolute value, 50–52, 107
AC-method to factor trinomials, 452–456, 489
Acute angles, 34
Acute triangles, 36
Addition
　associative property of, 94, 95
　commutative property of, 93, 94
　distributive property of multiplication over, 96–98, 110
　of fractions, 12–15
　identity and inverse properties of, 95–96
　of polynomials, 391–392, 419
　of radicals, 602–604, 639
　of rational expressions, 520–525, 570
　of real numbers, 66–70, 108
　symbol for, 56
Addition method
　explanation of, 316
　to solve systems of linear equations, 316–321, 349
Addition property
　of equality, 117–119, 193
　of inequality, 182–183
Additive identity, 95
Algebraic expressions
　evaluation of, 56–57, 88
　explanation of, 56
　simplification of, 69–70, 98–102
　translating English expressions to, 60–61, 69–70
Alternate exterior angles, 35
Alternate interior angles, 35
$a^{m/n}$, 631–632
Angles
　complementary, 34, 162–163
　explanation of, 34
　method to find unknown, 36
　properties of, 35
　in triangles, 36, 37
　types of, 34–35
Applications
　addition of real numbers in, 70
　costs in, 169–170, 326
　discount and markup in, 156
　distance, rate, and time in, 329–331, 551–553
　functions in, 274
　geometry in, 159–164, 197, 403

　linear equations in, 143–149, 196, 261–263, 284
　mixtures in, 170–171, 198, 328–329
　percents in, 25–26, 153–155, 197
　plotting points in, 209
　polynomials in, 395
　principal and interest in, 327–328
　quadratic equations in, 479–481, 492, 666–667
　quadratic functions in, 676
　radical equations in, 627–628
　rational equations in, 548–555, 573
　rational exponents in, 634
　slope in, 231, 236–237
　substitution method in, 312–313
　subtraction of real numbers in, 76
　systems of linear equations in, 326–331, 350
　uniform motion in, 171–173, 198
　variation in, 560–563
　volume in, 33–34
　work, 554–555
Approximation
　of irrational numbers on calculators, 52
　of repeating decimals on calculators, 26
　of square roots, 637
Area
　of circle, 32
　explanation of, 30
　formulas for, 30
　method to find, 31–32
Associative properties of real numbers, 94–95, 110
Axis of symmetry, 670, 671

B

b^0, 375, 378, 418
Bases
　explanation of, 57, 108, 360
　multiplication and division of common, 362–364, 417
Binomials. *See also* Polynomials
　explanation of, 390
　factoring, 429, 432, 459, 466–468
　finding greatest common, 429
　product of two, 401–402
　square of, 402–403
b^{-n}, 376, 378, 418
Body mass index (BMI), 192
Braces, 59
Brackets, 59

C

Calculators
　exponential expressions on, 62, 89, 366
　irrational numbers on, 52
　linear equations on, 223–224
　minimum and maximum values on, 677
　operations with signed numbers on, 78–79
　pi (π) key on, 164
　radical expressions on, 605
　repeating decimals on, 26
　scientific notation on, 386
　simplified radicals on, 598
　square roots and higher-order roots on, 588–589
　systems of linear equations in two variables on, 300–301
　using *evaluate* feature on, 264
　ZSquare option on, 247
Circles, 30, 32
Circumference
　explanation of, 29
　method to find, 30
　solving geometry application involving, 164
Clearing decimals, 139–140, 195
Clearing fractions, 136–137, 195, 253
Coefficients
　explanation of, 98, 110, 390
　leading, 390
Commutative property, of real numbers, 93–94, 110
Complementary angles
　explanation of, 34
　solving geometry application involving, 162–163
Completing the square
　explanation of, 655–657, 683
　solving quadratic equations by, 657–659
Complex fractions, 530–534, 571
Composite numbers, 7
Compound inequalities
　explanation of, 179–180
　method to solve, 185–186
Compound interest, 365–366
Conditional equations, 131, 132, 194
Congruent angles, 34
Conjugate radical expressions, 611
Conjugates, 401, 402
Consecutive even integers, 145

Consecutive integers, 145–147
Consecutive odd integers, 145, 146
Consistent systems of linear equations, 297, 347
Constant
    explanation of, 56, 108
    of variation, 559
Constant terms, 98
Contradictions, 131, 132, 194
Corresponding angles, 35
Cost applications, 169–170, 326
Cube roots
    explanation of, 584
    simplification of, 597–598
Cubes
    factoring sum or difference of, 464–466, 490
    perfect, 465, 584, 586
    volume of, 32, 33

D

Data, 206, 280
Decimals
    clearing, 139–140, 195
    converted to fractions, 22–23
    converted to percents, 24–25
    converting fractions to, 21–22
    converting percents to, 23–24
    place value system and, 20–21
    repeating, 21
    rounding, 22
    solving linear equations that contain, 139–140
    terminating, 21–23
Degree of polynomials, 390
Degree of the term, 390
Denominators. *See also* Least common denominator (LCD)
    adding and subtracting fractions with, 12, 13
    explanation of, 6
    of rational expressions, 520–525
    rationalizing the, 614, 616–619
Dependent systems of linear equations
    explanation of, 297, 347
    graphs of, 300
    substitution to solve, 311
Difference
    of cubes, 464–466, 490
    explanation of, 56
    of squares, 401–402, 458–460, 464, 490
Direct variation
    applications of, 561–562
    explanation of, 559, 574
Discount, 156
Distance applications, 329–331, 551–553
Distributive property
    to add and subtract like terms, 99, 100

of multiplication over addition, 96–98, 110
    polynomials and, 399–400
Division
    of common bases, 362–364
    of fractions, 11–12
    of polynomials, 407–412, 420
    of rational expressions, 509–511, 569
    of real numbers, 85–87, 109
    in scientific notation, 385
    symbol for, 56
Division property
    of equality, 119–123, 193
    of inequality, 183–184
    of radicals, 614–616, 640
Domain
    of function, 273
    of rational expressions, 499–500, 568
    of relation, 268, 269, 285
    restrictions from, 538

E

Elimination method. *See* Addition method
Equal angles, 34
Equality
    addition and subtraction properties of, 117–119, 193
    multiplication and division properties of, 119–123, 136, 193
Equations. *See also* Linear equations; Linear equations in one variable; Linear equations in two variables
    conditional, 131, 132, 194
    factoring to solve, 474–476 (*See also* Factors/factoring)
    of lines, 253–257, 283
    literal, 159–161, 197
    quadratic, 472, 474–475, 479–481, 537, 650–653, 657–659, 662–667
    radical, 624–628, 641
    rational, 537–544, 548–555, 572, 573
    solutions to, 116
Equilateral triangles, 36
Equivalent fractions, 14–15
Exponential expressions
    on calculators, 62, 89
    evaluation of, 57–58, 62, 84–85, 360–362
    explanation of, 57, 84
    simplification of, 364–365, 376–378
Exponential notation, 360
Exponents
    applications of, 365–366
    explanation of, 57, 108, 360, 417
    negative, 376–377, 633
    power rule for, 370, 371, 417
    properties of, 370–371, 378–381, 417
    rational, 630–634, 641

simplifying expressions with, 371–373, 379–381
    zero, 375, 633
Expressions
    algebraic, 56–57, 60–61, 69–70, 98–102
    explanation of, 56, 108
    exponential, 57–58, 84–85, 360–362
    radical, 605, 608–610
    rational, 498–504, 508–511, 514–525, 568–574
Extraneous solutions, 624

F

Factors/factoring
    binomial, 429, 432, 459, 466–468
    difference of squares, 458–460, 490
    explanation of, 7, 428
    greatest common, 9, 428–431, 488
    by grouping, 432–434, 488
    negative, 431–432
    perfect square trinomials, 460–462, 490
    polynomials, 428–434
    to solve equations, 474–476
    steps in, 470
    trinomials, 437–441, 443–449, 488
First-degree equations. *See* Linear equations
Fractions
    addition of, 12–15
    basic definitions of, 6–7
    clearing, 136–138, 195
    complex, 530–534
    converted to decimals, 21–22
    converted to percents, 24–25
    converting decimals to, 22–23
    converting percents to, 23–24
    division of, 11–12
    fundamental principle of, 8
    improper, 7, 15, 16
    linear equations that contain, 135–138
    in lowest terms, 8–9
    multiplication of, 10–11, 508
    operations on mixed numbers, 15–16
    prime factorization, 7–8
    proper, 7
    subtraction of, 12–15
    writing equivalent, 14–15
Function notation, 271, 285
Functions
    applications of, 274
    on calculators, 275
    determining if relation is, 269–270
    domain and range of, 273
    evaluation of, 271
    explanation of, 269–270, 285
    minimum and maximum value of, 677

G

Geometry
   angles and, 34–36, 162–163
   area and, 30–32
   formulas and applications of,
      159–164, 197
   perimeter and, 29–30, 161–162
   polynomials and, 395, 403
   quadratic equations and, 479–481
   substitution method in, 313
   triangles and, 36–37, 163
   volume and, 32–34
Graphing calculators. *See* Calculators
Graphs
   of functions, 275
   of horizontal lines, 222
   interpretation of, 206–207, 209
   of linear equations in two variables,
      216–224
   of linear inequalities, 178–180,
      335–339
   of lines using slope-intercept form,
      244–245
   of parabolas, 673–676
   of quadratic functions, 673–676, 685
   of systems of linear equations,
      297–301, 347
   of systems of linear inequalities,
      339–340
   of vertical lines, 222–223
Greatest common factor (GCF)
   explanation of, 9, 428, 488
   factoring out, 430–431
   identification of, 428–429
Grouping
   factoring by, 432–434, 488
   symbols for, 59

H

Higher-degree polynomials, 449, 476
Horizontal lines
   explanation of, 221, 281
   finding slope of, 234
   graphs of, 222
   linear equations and, 256

I

Identities, 131–132, 194
Identity property of real numbers,
   95–96, 110
Improper fractions, 7, 15, 16
Inconsistent systems of linear equations
   explanation of, 297, 347
   graphs of, 299
   substitution to solve, 310
Independent systems of linear
   equations, 297, 347
Index, of radical, 584
Inequalities. *See also* Linear inequalities

addition and subtraction properties
   of, 182–183
compound, 179–180, 185–186
explanation of, 49–50
graphs of linear, 178–180
multiplication and division properties
   of, 183–184
translating expressions involving, 186
Inequality signs, 49
Infinity symbol, 180
Integers
   consecutive, 145–147
   explanation of, 107
   set of, 46
   subtraction of, 74
Interest
   compound, 365–366
   simple, 155, 197, 327–328, 365
Interval notation, 180–181, 198
Inverse property of real numbers,
   96, 110
Inverse variation
   applications, 562
   explanation of, 559, 574
Irrational numbers
   on calculators, 52
   explanation of, 47–48, 107
Isosceles triangles, 36

J

Joint variation
   applications of, 563
   explanation of, 474, 560

L

Leading coefficient of 1, to factor
   trinomials, 437–441, 488
Least common denominator (LCD)
   explanation of, 14, 136
   of rational expressions, 514–515,
      521–525, 570
   writing rational expressions with,
      515–517
Least common multiple (LCM), 13–14
Like radicals
   addition and subtraction of, 602–604
   explanation of, 601–602
Like terms
   combining, 100, 101, 102
   distributive property add and
      subtract, 99, 100
   explanation of, 99, 110, 391
Linear equations. *See also* Systems of
   linear equations
   applications of, 143–149, 196,
      261–263, 284
   on calculators, 223–224
   clearing decimals to solve,
      139–140, 195

clearing fractions to solve,
   136–138, 195
with decimals, 139–140
explanation of, 537
forms of, 256
with fractions, 135–138
modeling, 279–280, 284
that involve multiple steps, 127–128
translating to, 123–124, 144–145
written from observed data
   points, 262
Linear equations in one variable
   explanation of, 116–119, 193
   method to solve, 129–131, 137, 194
Linear equations in two variables
   explanation of, 215–216, 281
   on graphing calculators, 223–224
   horizontal and vertical lines in graphs
      of, 221–223
   interpretation of, 261
   plotting points in graphs of, 216–219
   x-intercepts and y-intercepts in
      graphs of, 219–221
Linear inequalities. *See also* Systems of
   linear inequalities
   addition and subtraction properties
      of, 182–183
   applications of, 186–187
   graphs of, 178–180, 335–339
   in one variable, 178–180, 198
   solving applications with, 187
   test point method to solve,
      335–337, 351
   in two variables, 335–339, 351
Linear models, 345–346
Lines
   equations of, 253–257, 283
   finding x- and y-intercepts of,
      220–221
   horizontal, 221, 222, 234, 281
   parallel, 35, 235, 236, 245–246,
      255, 282
   perpendicular, 235, 236, 245–246,
      255–256, 282
   slope-intercept form of, 243–247, 283
   slope of, 230–237, 282
   vertical, 221–223, 234, 281
Literal equations, 159–161, 197
Long division, to divide polynomials,
   408–412
Lowest terms
   simplifying fractions to, 8–9, 531–533
   simplifying rational expressions to,
      500–504

M

Markup, 156
Mixed numbers
   explanation of, 7
   operations on, 15–16

Mixture applications, 170–171, 198, 328–329
Modeling, linear equations, 279–280, 284
Monomials
    division of polynomials by, 407–408
    explanation of, 390
    multiplication of polynomials by, 399
Mortgage payments, 567
Multiplication
    associative property of, 94, 95
    of common bases, 362–364
    commutative property of, 93, 94
    distributive property over addition, 96–98, 110
    of fractions, 10–11, 508
    identity and inverse properties of, 95–96
    of polynomials, 398–401, 419
    of radicals, 608–611, 640
    of rational expressions, 508–509, 511, 569
    of real numbers, 82–84, 109
    in scientific notation, 385
    symbol for, 56
Multiplication property
    of equality, 119–123, 136, 193
    of inequality, 183–184
    of radicals, 593–598, 608–609, 639
Multiplicative identity, 95

N

Natural numbers
    explanation of, 6, 107
    set of, 46
Negative exponents, 376–377, 633
Negative factors, factoring out, 431–432
Negative square roots, 582
Notation. *See* Symbols and notation
*n*th-roots
    explanation of, 584
    simplification of, 584–586, 616
    translations involving, 586–587
Number line
    of inequalities, 178, 179
    real, 46, 66, 74
Numbers. *See also* Real numbers
    composite, 7
    irrational, 47–48, 107
    mixed, 7, 15–16
    natural, 6, 46, 107
    prime, 7
    rational, 47, 107
    sets of, 46
    whole, 6, 46, 107
Numerators, 6
Numerical coefficient, 98

O

Obtuse angles, 34
Obtuse triangles, 36
Opposites, of real numbers, 50, 107
Ordered pairs
    explanation of, 207, 208, 280
    as solution to system of linear equations, 296
Order of operations
    application of, 59–60, 77–78, 87–88
    explanation of, 31, 59, 108
    to simplify radicals, 596–597
Origin, 207, 280

P

Parabolas
    explanation of, 670
    graphs of, 673–676
    vertex of, 671–673, 685
Parallel lines
    explanation of, 35, 235, 236, 282
    point-slope formula and, 255
    slope-intercept form and, 245–246
Parallelograms, area of, 30, 31
Parentheses
    clearing, 100, 101, 102, 138, 392
    explanation of, 59, 76
Percents
    applications of, 25–26, 153–155, 197
    converted to decimals and fractions, 23–24
    converting decimals and fractions to, 24–25
    explanation of, 23, 153
Perfect cubes, 465, 584, 586
Perfect fifth powers, 584
Perfect fourth powers, 584
Perfect squares, 458, 583, 584, 586
Perfect square trinomials
    explanation of, 401–402
    factoring, 460–462, 490
    identification of, 462
Perimeter
    explanation of, 29–30
    method to find, 30
    solving geometry application involving, 161–162
Perpendicular lines
    explanation of, 235, 236, 282
    point-slope formula and, 255–256
    slope-intercept form and, 245–246
Pi (π), on calculators, 164
Place value, 20–21
Plotting points
    applications of, 209
    graphing linear equations in two variables by, 216–219
    in rectangular coordinate system, 208–209

Point-slope formula
    equations of line using, 253–256
    explanation of, 253, 256, 283
Polynomials. *See also* Binomials; Factors/factoring; Trinomials
    addition of, 391–392, 419
    applications of, 395, 403
    degree of, 390
    division of, 407–412, 420
    explanation of, 390
    finding opposite of, 393
    geometry and, 395, 403
    higher-degree, 449, 476
    identifying parts of, 390–391
    multiplication of, 398–401, 419
    prime, 431, 440, 444
    special case products of, 401–403
    subtraction of, 393–395, 419
Positive square roots, 582
Power, 57
Power rule for exponents, 370, 371, 378, 417
Prime factorization, 7–8, 428
Prime numbers, 7
Prime polynomials, 431, 440, 444
Principal square roots, 582, 587, 638
Problem-solving strategies, 196
Products
    explanation of, 56
    power of, 417
    special case, 401–403, 419, 610
Proper fractions, 7
Proportions
    applications of, 549–551
    explanation of, 548, 573
    methods to solve, 548–549
Pythagoras, 481
Pythagorean theorem
    applications using, 587–588
    explanation of, 416, 481–483, 587

Q

Quadrants, 207, 280
Quadratic equations
    applications of, 479–483, 492, 666–667
    explanation of, 472, 537
    methods to solve, 474–475, 650–653, 657–659, 664–665
    quadratic formula to solve, 662–664
    translations of, 479
Quadratic formula
    derivation of, 661
    explanation of, 684
    to solve quadratic equations, 662–664, 684
Quadratic functions
    applications of, 676
    explanation of, 670–671
    graphs of, 673–676, 685
    vertex of parabola and, 671–673

Quotients
    explanation of, 56, 85, 86
    of nonzero real numbers, 121
    power of, 378, 417
    of radicals, 615, 620

R

Radical equations
    applications of, 627–628
    explanation of, 624, 641
    methods to solve, 624–626
    translations involving, 627
Radical expressions
    on calculators, 605
    conjugate, 611
    multiplication of, 608–609
    simplification of, 609
    squaring two-term, 610
Radical notation, 632
Radical quotients, 615, 620
Radicals
    addition and subtraction of,
        602–604, 639
    division property of, 614–616, 640
    like, 601–604
    multiplication of, 608–611, 640
    multiplication property of, 593–598,
        608–609, 639
    simplification of, 593–598, 603–604,
        616, 639
    simplified form of, 594, 614
    unlike, 604
Radical sign, 58, 584
Radicand, 584
Range
    of function, 273
    of relation, 268, 269, 285
Rate applications, 329–331, 551–553
Rate of change
    applications of, 236–237
    explanation of, 263
    writing linear equations given, 263
Rational equations
    applications of, 548–555, 573
    explanation of, 537, 572
    methods to solve, 537–542, 572
    solving formulas involving, 542–544
    translation to, 542
Rational exponents
    applications of, 634
    explanation of, 630–632, 641
    properties of, 632–633
    radical notation and, 632
Rational expressions
    addition and subtraction of,
        520–525, 570
    in complex fractions, 530–534, 571
    division of, 509–511, 569
    domain of, 499–500
    evaluation of, 498

explanation of, 498, 568
    least common denominator of,
        514–517, 521–525, 570
    multiplication of, 508–509, 511, 569
    simplification of, 500–504
    in translations, 525
Rationalizing the denominator
    explanation of, 614, 640
    one term, 616–618, 640
    two term, 618–619, 640
Rational numbers, 47, 107
Ratios, 503–504. See also Proportions
Real number line
    explanation of, 46, 66, 74, 107
    plotting points on, 46
Real numbers
    absolute value of, 50–52
    addition of, 66–70
    associative properties of, 94–95
    commutative properties of, 93–94,
        110
    distributive property of
        multiplication over addition and,
        96–98, 110
    division of, 85–88
    explanation of, 46, 107
    identity and inverse properties of,
        95–96
    inequalities and, 49–50
    multiplication of, 82–85
    opposite of, 50
    order of operations and, 59–60
    properties of, 93–98, 110
    quotient of nonzero, 121
    reciprocal of, 85
    set of, 46–49, 107
    simplifying algebraic expressions
        and, 98–102
    subtraction of, 74–78
    symbols for, 49
Reciprocals, 11, 85
Rectangles, 30
Rectangular coordinate system
    explanation of, 207, 280
    plotting points in, 208–209
Rectangular solids, 32
Relation
    explanation of, 268, 285
    finding domain and range of,
        268, 269
    as function, 269–270
Repeating decimals, 21
Right angles, 34
Right circular cones, 32, 33
Right circular cylinders, 32–34
Right triangles, 36, 416, 481–483
Roots
    on calculators, 588–589
    cube, 597–598, 584
    nth, 584–587
    square, 58, 108, 582–583, 587–589, 638

S

Sales tax, 153, 197
Scalene triangles, 36
Scientific notation
    on calculators, 386
    explanation of, 383, 418
    multiplying and dividing numbers
        in, 385
    writing numbers in, 383–384
Set-builder notation, 180–181, 198
Sets
    explanation of, 46
    of integers, 46
    of natural numbers, 46
    of real numbers, 46–49, 107
    of whole numbers, 46
Sign rules, for factoring trinomials,
    439, 446
Similar triangles
    applications of, 549–551
    explanation of, 549
Simple interest
    application involving, 327–328, 365
    explanation of, 155, 197
Simplification
    of algebraic expressions, 69–70,
        98–102
    of complex fractions, 530–534, 571
    of cube roots, 597–598
    of exponential expressions, 364–365,
        376–378
    of expressions with exponents,
        371–373, 379–381
    of expressions with rational
        exponents, 633
    of fractions to lowest terms, 8–9
    of nth-roots, 584–586, 616
    of radicals, 593–598, 603–604, 616, 939
    of rational expressions, 500–504, 568
    of square roots, 58, 583
Slope
    applications of, 231, 236–237
    explanation of, 230–231, 282
    formula for, 231–232, 253
    method to find, 232–234
    of parallel, 235–236
    of perpendicular lines, 235–236
Slope-intercept form
    explanation of, 243–244, 256, 257, 283
    graphs of, 244–245
    parallel and perpendicular lines and,
        245–246
    writing equation of line using,
        246–247, 299, 300
Solutions
    to equations, 116
    extraneous, 624, 625
    to system of linear equations, 296, 347
        (See also Systems of linear
        equations)

Solution set
   explanation of, 119, 178
   of inequalities, 178, 179
Special case products, 401–403, 419, 610
Spheres, 32
Square root property
   explanation of, 651, 683
   quadratic equations solved using,
      651–653, 657–659
Square roots
   approximation of, 637
   explanation of, 58, 108, 582, 638
   negative, 582
   positive, 582
   principal, 582, 587, 638
   Pythagorean theorem and, 587–588
   simplification of, 58, 583
Squares
   area of, 30
   of binomials, 402–403
   completing, 655–659, 683
   difference of, 401–402, 464, 490
   perfect, 458, 583, 584, 586
   of radical expressions, 610
Standard form, 243, 256, 257, 384–385
Straight angles, 34
Subsets, 46
Substitution method
   applications of, 312–313
   explanation of, 307
   to solve systems of linear equations,
      307–313, 348
Subtraction
   of fractions, 12–15
   of polynomials, 393–394, 419
   of radicals, 602–604, 639
   of rational expressions, 520–525, 570
   of real numbers, 74–79, 109
   symbol for, 56
Subtraction property
   of equality, 117–119, 193
   of inequality, 182–183
Sum
   of angles in triangles, 37
   of cubes, 464–465, 490
   explanation of, 56
   of squares, 458–460
Supplementary angles, 34
Symbols and notation
   basic operations, 56
   function, 271
   grouping, 59
   infinity, 180
   interval, 180–181, 198
   nth root, 584
   radical notation, 632
   radical sign, 58, 584
   real numbers, 49
   scientific notation, 383–386, 418
   set-builder, 180–181, 198
   square bracket, 178
   square root, 582

Systems of linear equations
   addition method to solve,
      316–321, 349
   applications of, 326–331, 350
   consistent, 297, 347
   dependent, 297, 300, 311, 347
   explanation of, 296
   graphing method to solve,
      296–301, 347
   inconsistent, 297, 299, 310, 347
   independent, 297, 347
   interpreting solutions to, 310
   substitution method to solve,
      307–313, 348
   in two variables, 300–301, 326–331
Systems of linear inequalities,
   339–340, 351

T

Terminating decimals
   converted to fractions, 22–23
   explanation of, 21
Terms. *See also* Lowest terms
   degree of, 390
   explanation of, 110
   like, 99–102, 391
   unlike, 99, 391
Test point method, 335–337, 351
Test points, 183, 336
Time applications, 329–331, 551–553
Translations
   addition of real numbers, 69–70
   English form to algebraic form,
      60–61, 587
   linear equations, 123–124, 143–145
   linear inequalities, 186
   nth-roots, 586–587
   radical equations, 627
   rational equations, 542
   rational expressions, 525
   subtraction of real numbers, 75–76
   variation, 559–560
Trapezoids, area of, 30, 31
Trial-and-error method, to factor
     trinomials, 443–449, 489
Triangles
   area of, 30
   explanation of, 36–37
   right, 36, 416, 481–483
   similar, 549–551
   sum of angles in, 37
Trinomials. *See also* Polynomials
   AC-method to factor, 452–456, 489
   explanation of, 390
   higher-degree, 449, 456
   leading coefficient of 1 to factor,
     437–441, 488
   perfect square, 401–402, 460–462, 490
   trial-and-error method to factor,
     443–449, 489

U

Uniform motion, 171–172, 198
Unlike radicals, 604
Unlike terms, 99, 391

V

Variables
   explanation of, 56, 108
   of polynomials, 391
   solving for indicated, 160, 161
Variable terms, 98
Variation
   applications of, 560–563
   constant of, 559
   direct, 559, 574
   inverse, 559, 574
   joint, 560, 563, 574
   translations involving, 559–560
Vertex, of parabolas, 670, 671, 685
Vertical angles, 35
Vertical lines
   explanation of, 221–222, 281
   finding slope of, 234
   graphs of, 222–223
   linear equations and, 256
Vertical line test
   explanation of, 270, 285
   use of, 270, 271
Volume
   explanation of, 32
   maximizing, 682–683
   method to find, 33–34

W

Whole numbers
   explanation of, 6, 107
   set of, 46
Work applications, 554–555

X

x-axis, 207, 219, 280
x-coordinates, 207
x-intercepts, 219–221, 230, 281

Y

y-axis, 207, 219, 280
y-coordinates, 207
y-intercepts, 219–221, 230, 246,
   247, 281

Z

Zero exponents, 375, 633
Zero product rule
   explanation of, 472
   to solve equations, 375, 473, 491,
     650–651, 665

Perimeter and Circumference

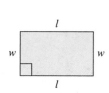

Rectangle

$P = 2l + 2w$

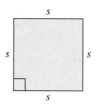

Square

$P = 4s$

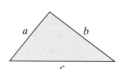

Triangle

$P = a + b + c$

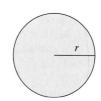

Circle

Circumference: $C = 2\pi r$

Area

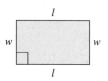

Rectangle

$A = lw$

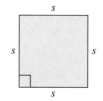

Square

$A = s^2$

Parallelogram

$A = bh$

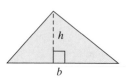

Triangle

$A = \frac{1}{2}bh$

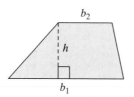

Trapezoid

$A = \frac{1}{2}(b_1 + b_2)h$

Circle

$A = \pi r^2$

Volume

Rectangular Solid

$V = lwh$

Cube

$V = s^3$

Right Circular Cylinder

$V = \pi r^2 h$

Right Circular Cone

$V = \frac{1}{3}\pi r^2 h$

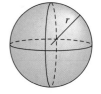

Sphere

$V = \frac{4}{3}\pi r^3$

Angles

- Two angles are **complementary** if the sum of their measures is 90°.

- Two angles are **supplementary** if the sum of their measures is 180°.

- $\angle a$ and $\angle c$ are vertical angles, and $\angle b$ and $\angle d$ are vertical angles. The measures of vertical angles are equal.

- The sum of the measures of the angles of a triangle is 180°.

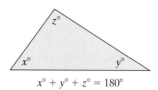

$x° + y° + z° = 180°$